TELETRAFFIC ISSUES
in an Advanced Information Society

ITC-11

STUDIES IN TELECOMMUNICATION

VOLUME 5

International Teletraffic Congress
International Advisory Council
The Development and Application of Teletraffic Theory

The first International Teletraffic Congress, entitled 'On the Application of the Theory of Probability in Telephone Engineering and Administration', took place in 1955 in Copenhagen. Every 3 years since then, specialists from tele-administrations, industry, and universities have gathered to present new methodologies and applications of the theory of teletraffic and teleplanning. Their audience is composed of experts and users, and the main focus is on issues of telecommunication traffic, as they affect customer service, and efficient telecommunication equipment loading, with special emphasis on probabilistic and other mathematical handling of traffic problems.

The International Advisory Council – with representatives from each past Congress country and the two future Congress countries – is responsible for the Congresses. Up till now, the following have served on the Council:

Denmark: Arne Jensen (Chairman, since 1955).
The Netherlands: L. Kosten, J.W. Cohen.
France: R. Fortet, P. Le Gall.
United Kingdom: E.P.G. Wright, J. Povery, A.C. Cole.
United States: R. Wilkinson, Walt Hayward, S. Katz.
Germany: Konrad Rohde, P. Kühn (Vice-Chairman since 1985).
Sweden: Chr. Jacobäus, Bengt Wallström.
Australia: Clem Pratt.
Spain: Eduardo Villar.
Canada: P. O'Shaughnessy.
USSR: V. Neiman.
Japan: H. Inose, M. Akiyama.
Italy: P. de Ferra.

CCITT: E.P.G. Wright (U.K.); Clem Pratt (Australia); Ingvar Tånge (Sweden);
G. Gosztony (Hungary); A. Lewis (Canada).

NORTH-HOLLAND – AMSTERDAM • NEW YORK • OXFORD • TOKYO

TELETRAFFIC ISSUES

in an Advanced Information Society

ITC-11

Proceedings of the Eleventh International Teletraffic Congress
Kyoto, Japan, September 4-11, 1985

Edited by:

Minoru AKIYAMA

Department of Electrical Engineering
The University of Tokyo
Tokyo, Japan

Part 1

N·H
P·C

1985

NORTH-HOLLAND – AMSTERDAM • NEW YORK • OXFORD • TOKYO

ISBN Part 1: 0 444 87917 X
ISBN Part 2: 0 444 87918 8
ISBN Set: 0 444 87919 6

Published by:
ELSEVIER SCIENCE PUBLISHERS B.V.
P.O. BOX 1991
1000 BZ AMSTERDAM
THE NETHERLANDS

Sole distributors for the U.S.A. and Canada:
ELSEVIER SCIENCE PUBLISHING COMPANY, INC.
52 VANDERBILT AVENUE
NEW YORK, N.Y. 10017
U.S.A.

Legal Notice:
All opinions expressed in these proceedings are those of the authors and are not binding on the International Advisory Council of the International Teletraffic Congress.

PRINTED IN THE NETHERLANDS

INTRODUCTION

HISTORICAL BACKGROUND

The first International Teletraffic Congress was organized by Prof. Dr. Arne Jensen and held in Copenhagen in 1955. This first meeting was a great success and the continuance of the Congress was assured. Since that time it has taken place every three years with subsequent Congresses being held in The Hague, Paris, London, New York, Munich, Stockholm, Melbourne, Torremolinos and Montreal.

THE ELEVENTH INTERNATIONAL TELETRAFFIC CONGRESS

Meeting place and time
The Congress was held in Kyoto, from Wednesday, September 4, 1985 to Wednesday, September 11, 1985.

Scope of the Congress
As in previous Congresses, the Kyoto Congress dealt with the development and application of teletraffic theory. Within this general framework, the field of teletraffic theory and its applications to engineering, operations and planning related to the general fields of telecommunications and computer systems was covered.

The proposed major theme, around which discussions were focused was *Teletraffic Issues in an Advanced Information Society.*

NEXT CONGRESS

The Twelfth International Teletraffic Congress will be held in Torino, Italy, in June, 1988.

SUMMARY OF ITC 11

For ITC 11, a total of 172 papers in the technical sessions, 35 papers in the special interest group meetings and 3 special lectures, from 30 countries, were submitted. More than 400 delegates and 100 accompanying persons participated in the congress.

The technical sessions of ITC 11 covered mainly ten areas, ISDN, LAN, network planning, routing, switching systems, queueing systems, data communications, traffic administrations, grade of services, and fundamental teletraffic theories. ITC 11 has contributed greatly to the progress of teletraffic engineering, and also to the realization of the coming advanced information network society. The session structure of ITC 11 is summarized in the contents list of the proceedings.

The proceedings of ITC 11 contain a tremendous amount of information. Indeed, it would be pretentious to summarize them. The following comments are my impressions:

1. *New communication services and network design*
Generally speaking, one of the features of ITC 11 is that discussion around the topics of new communication services and network design (such as ISDN, LAN, and related traffic problems) has increased so much.

In the ISDN sessions, 10 papers were presented together with a number of important research results which will directly contribute to the design policy of ISDN. Value added services in ISDN, ISDN protocol, logical network structure as a basis for traffic design, the concept of grade of services in an integrated network, and integrated communications with heterogeneous traffic were discussed in depth. Several new traffic problems, such as an advanced channel reservation and multipoint connections for the application of video conferencing were also pointed out.

In the session, Voice and Data Systems, 5 papers were presented, and several novel voice-data integrated traffic handling systems were proposed. Voice and data integration by packetized switching will raise a new traffic problem which will be worthy of further investigation.

In the session on Local Area Networks, 10 papers were presented. LAN is one of the rapidly expanding areas connected to that of rapidly expanding business communications. In this area, integrated traffic handling mutual connections and modular structure to increase flexibility are important topics. Modeling and statistics based on traffic demand are also important discussion points. Besides computer communication type networks, such as bus or ring, digital PBX will play an essential role, especially in its application to office automation. Circuit and packet combined switching will become an attractive theme to be investigated.

In the session entitled Satellite and Radio Systems, 9 papers were presented. Major topics in this session were the traffic problems on cellular mobile radio systems and satellite communication network design.

2. *Data communication networks*
In the session entitled Data Communication Networks, 13 papers were presented. Besides fundamental research on stochastic traffic models, a number of papers covered large scale data networks, taking traffic handling performance, reliability and flexibility into account as evaluation factors. There were a number of interesting papers

which investigated such topics as an algorithm to decide network topology, hierarchical data network design, point-to-multipoint communications using satellites, wide area data and voice services, computer aided planning tools, and so on. There were also some interesting papers which covered network behavior based on actual measured parameters, and on subscriber behavior in the choice of digital data services.

3. *Network planning and routing*

In the session entitled Network Planning and Design, 10 papers were presented from 8 countries. Most of the papers dealt with digital communication network design problems by taking real factors into account in an attempt to realize the ISDN of each country. Since the problem became very complicated due to the uncertainty of many of its factors, the use of mathematical programming, case studies or heuristic approaches has become indispensable. It is important to establish a new design method in addition to the conventional stochastic approach in this field.

In the session Alternate Routing and Dynamic Routing, 14 papers were presented. Since ITC 10 was held in Montreal, dynamic routing has become a major topic in this field. Various types of novel routing algorithms including real time dynamic non-hierarchical routing, centralized or decentralized routing policies, multi-hour dimensioning, learning automata based routing and an application to the international communication system were presented.

4. *Traffic administration and measurement*

In the sessions under the title Traffic Administration and Traffic Measurement 10 papers were presented from 6 countries. They discussed the present state of traffic control and measured data in each country. As for the peak traffic, for example, it was reported that data measured by time consistent busy hour and moving daily peak hour are some what different. Detailed probability distributions of calls based on field data were also reported.

In the coming advanced information society, networks must handle various types of new services. Subscribers' behavior will change. Networks will still be expanding. A lot of uncertain factors will remain. In such circumstances, traffic forecasting will become more and more important. In the session entitled Traffic Forecasting, several new forecasting procedures, such as a filtering method, were proposed.

The problem of traffic demand forecasting was also discussed in the session entitled CCITT Related Problems. One of the topics worth mentioning here is the research which is taking place on the relation between tariff and traffic profiles. It was analyzed by using actual measured traffic data. The design algorithm of cluster engineering for alternate routing was also reported in this session.

5. *Switching systems*

17 papers were presented in the area of switching systems, in the sessions, Switching Systems and Modeling, and Overload Control. The topics can be classified into four areas. These are modeling and analysis of SPC, scheduling and overload control, simulation systems, and integrated traffic handling. One interesting trend in this field is that research work concerned with control system design has increased in comparison to that of switching network analyses.

In the session entitled Switching Networks various types of switching networks were proposed. The problem of whether wide and narrow band traffic should be integrated or separated still remained. A broadcast switching network was analyzed. It will become an attractive theme to be further studied – especially for its application to future video communication networks.

6. *Fundamental theories*

Fundamental traffic problems were discussed in several sessions. Problems such as numerical calculation methods, approximations, derivative functions of fundamental formulae, and mixed traffic problems were discussed in the sessions, Traffic Theory, and Traffic Models and Methodologies.

The session title, Overflow Traffic, is a traditional one. However, topics of interest seem to be changing. Most of the papers in this session discussed reaching new targets, digital networks and non-hierarchical communication networks.

In the session, Grade of Services, some interesting subjects were discussed. Problems concerned with counter-measures to reduce lost calls due to the called subscriber and subscriber retrieval behavior, the definition and calculation of end-to-end GOS parameters and service quality in failure conditions, and so on, were raised.

In the session entitled Queueing Systems, 13 papers were presented. Fundamental traffic problems concerned with approximations, numerical calculations, priority queues, traffic modeling for new communications, such as packet switching, and facsimile communication were discussed. In the session, Queueing Networks, approximate analyses on large scale queueing networks, and queueing networks with priority services or population size constraints were investigated.

7. *Special sessions*

On the last day of the technical program, we had three special technical sessions, inviting leading experts in each field. The session titles were Telecommunications Services and Traffic Engineering, Telecommunications Policies and Traffic Engineering, and International Cooperation in Traffic Engineering.

In order to increase communication between congress members, special interest group meetings were held in parallel with the normal oral presentation of paper sessions. These overlapped with coffee breaks. It was the first trial of the ITC. In these meetings a total 26 papers were presented – there were many earnest and deep discussions in an informal atmosphere.

The last session was that of panel discussions entitled, Mutual Influence of Information Technology and Teletraffic Engineering. ITC 11 was summarized and future trends in technology were discussed.

Besides the technical sessions, we had three special lectures in the opening and closing sessions. The titles of the special lectures were, Tradition and Development, Advanced Information Society and Telecommunications, and Civilization in an Advanced Information Society.

8. *Conclusion*

The modern information technology we have now at hand is more than an extension of that of the past, and will grow even more powerfully in the years to come. It will integrate not only telecommunication services but also various socio-economic activities and even cultures. The circumstances surrounding telecommunication networks will change substantially in the coming advanced information society.

The distribution of traffic will change as telecommunication services diversify. Subscribers' behavior will also change. Telecommunication networks are growing widely and deeply. The necessity for traffic engineers, in a competitive environment, to involve marketing and tariff making activities has increased. The forecasting of traffic demand, especially for new services, has become more difficult. Traffic engineering should adapt to the change of circumstances. We are now at the turning point. The role of traffic engineers will become more and more important.

Finally, I would like to thank all of the members of the International Advisory Committee and the National Committees, and all authors and participants for their support and cooperation to ITC 11.

Minoru Akiyama
Chairman
Technical Committee

SPONSORS

Nippon Telegraph and Telephone Corporation (NTT)
Kokusai Denshin Denwa Co., Ltd. (KDD)

in conjunction with

The Institute of Electronics and Communication Engineers of Japan
Operations Research Society of Japan

supported by

The Ministry of Posts and Telecommunications
Fujitsu Limited
Hitachi, Ltd.
NEC Corporation
Oki Electric Industry Co., Ltd.

ITC 11 ORGANIZATIONS

INTERNATIONAL ADVISORY COUNCIL

Chairman:	Prof. Dr. A. Jensen	Denmark
Members:	Prof. Dr. J.W. Cohen	The Netherlands
	Mr. A.C. Cole	United Kingdom
	Ing. P. de Ferra	Italy
	Dr. G. Gosztony	Hungary
	Mr. W.S. Hayward	U.S.A.
	Prof. Dr. H. Inose	Japan
	Dr. C. Jacobaeus	Sweden – Honorary Member
	Prof. Dr. L. Kosten	The Netherlands – Honorary Member
	Prof. Dr. P. Kühn	Federal Republic of Germany
	Dr. P. Le Gall	France
	Mr. A. Lewis	Canada
	Dr. V.I. Neiman	U.S.S.R.
	Mr. J.J. O'Shaughnessy	Canada
	Dr. C.W. Pratt	Australia
	Mr. J.E. Villar	Spain
	Prof. Dr. B. Wallström	Sweden

NATIONAL COMMITTEE

Chairman:	Hiroshi Inose	The University of Tokyo
Vice Chairmen:	Haruo Yamaguchi	NTT
	Yoh Matsumoto	KDD
Advisors:	Haruo Akimaru	Toyohashi University of Technology
	Masaya Fujiki	Nippon University
	Eiichi Gambe	The University of Electro-Communications
	Yusai Okuyama	The Ministry of Posts and Telecommunications
Members:	Minoru Akiyama	The University of Tokyo
	Goro Emori	Hitachi, Ltd.
	Jun Jinguji	Oki Electric Industry Co., Ltd.
	Hiroshi Kaji	KDD
	Toshiro Kunihiro	NEC Corporation
	Moriji Kuwabara	NTT

	Hidenori Morimura	Tokyo Institute of Technology
	Tadashi Nishimoto	KDD
	Bun-ichi Oguchi	Fujitsu Limited
	Shoji Yoshida	NTT
Secretaries:	Hiromasa Ikeda	NTT
	Kinji Ono	KDD

EXECUTIVE COMMITTEE

Chairman:	Shoji Yoshida	NTT
Vice Chairmen:	Minoru Akiyama	The University of Tokyo
	Masato Chiba	NTT
	Hiromasa Ikeda	NTT
	Kunishi Nosaka	KDD
Members:	Toshiharu Aoki	NTT
	Koji Hirose	Oki Electric Industry Co., Ltd.
	Masaki Kawakami	NTT
	Yasuhiko Kawasumi	KDD
	Zenya Koono	Hitachi, Ltd.
	Yuji Matsuo	NTT
	Hiroshi Matsuura	Fujitsu Limited
	Kinji Ono	KDD
	Toshio Terashima	NEC Corporation
Secretaries:	Masataka Akagawa	KDD
	Kunio Kodaira	NTT
	Hiromichi Mori	KDD
	Ichiro Sakakibara	NTT

TECHNICAL COMMITTEE

Chairman:	Minoru Akiyama	The University of Tokyo
Vice Chairmen:	On Hashida	NTT
	Hiromichi Mori	KDD
Members:	Hiroyasu Itoh	NEC Corporation
	Konosuke Kawashima	NTT
	Masao Mori	Tokyo Institute of Technology
	Shuzo Morita	Fujitsu Laboratories Limited
	Yasushi Okita	Oki Electric Industry Co., Ltd.
	Hidehiko Sanada	Osaka University
	Hideyoshi Tominaga	Waseda University
	Koichi Yamamoto	Hitachi, Ltd.

Secretaries:	Tadao Saito	The University of Tokyo
	Jun Matsuda	NTT
	Jun Matsumoto	KDD

ORGANIZING COMMITTEE

Chairman:	Masato Chiba	NTT
Vice Chairman:	Yasuhiko Kawasumi	KDD
Members:	Masataka Akagawa	KDD
	Sumitoshi Ando	KDD
	Hisabumi Honda	Hitachi, Ltd.
	Takashi Kobayashi	Fujitsu Limited
	Kunio Kodaira	NTT
	Fumio Moriguchi	Oki Electric Industry Co., Ltd.
	Ichiro Sakakibara	NTT
	Tohru Ueda	NTT
	Kazuo Watanabe	NEC Corporation
Secretaries:	Toshitaka Fujii	NTT
	Tohru Kizuka	KDD

SECRETARIAT

Secretary General:	Hiromasa Ikeda	NTT
Deputy Secretary General:	Toshiharu Aoki	NTT
Members:	Toshitaka Fujii	NTT
	Shun-ichi Iisaku	KDD
	Tetsuji Isogai	Oki Electric Industry Co., Ltd.
	Kunio Kodaira	NTT
	Sen Nakabayashi	NEC Corporation
	Kunitake Ohara	NTT
	Keizo Takagi	Fujitsu Limited
	Tohru Ueda	NTT
	Yasushi Wakahara	KDD
	Katsuhiko Wakisaka	Hitachi, Ltd.
	Makoto Yoshida	NTT

CONTENTS

SESSION 1.4: TRAFFIC MODELS AND METHODOLOGIES

Chairperson: B. Wallstroem (Sweden)
Vice-Chairperson: J. Matsuda (Japan)

SESSION 2.1A: ISDN AND NEW SERVICES II

Chairperson: H. Terada (Japan)
Vice-Chairperson: D.R. Mansfield (Canada)

SESSION 2.1B: OVERLOAD CONTROL

Chairperson: N.W. MacFadyen (United Kingdom)
Vice-Chairperson: T. Hoshi (Japan)

SESSION 2.3A: NETWORK PLANNING AND DESIGN I
Chairperson: S. Katz (U.S.A.)
Vice-Chairperson: T. Takemura (Japan)

SESSION 2.3B: SATELLITE AND RADIO SYSTEMS I
Chairperson: A. Myskja (Norway)
Vice-Chairperson: H. Okada (Japan)

SESSION 2.4A: NETWORK PLANNING AND DESIGN II
Chairperson: V.B. Iversen (Denmark)
Vice-Chairperson: T. Yamaguchi (Japan)

SESSION 3.1B: TRAFFIC ADMINISTRATION

Chairperson: M. Bonatti (Italy)
Vice-Chairperson: T. Takenaka (Japan)

SESSION 3.2A: QUEUEING SYSTEMS II

Chairperson: R.G. Schehrer (F.R.G.)
Vice-Chairperson: Y. Takahashi (Japan)

SESSION 3.2B: TRAFFIC MEASUREMENT

Chairperson: K. Bø (Norway)
Vice-Chairperson: S. Iisaku (Japan)

SESSION 3.4B: TRAFFIC FORECASTING II
Chairperson: F. Schreiber (F.R.G.)
Vice-Chairperson: K. Murakami (Japan)

MEETING S2.1K: TRAFFIC FORECASTING AND ADMINISTRATION

MEETING S2.2K: SWITCHING NETWORKS

PART 2

SESSION 4.1A: LOCAL AREA NETWORKS I
Chairperson: A. Descloux (U.S.A.)
Vice-Chairperson: N. Shiratori (Japan)

SESSION 4.1B: CCITT RELATED PROBLEMS
Chairperson: G. Gosztony (Hungary)
Vice-Chairperson: H. Murata (Japan)

SESSION 4.2A: LOCAL AREA NETWORKS II
Chairperson: G. Pujolle (France)
Vice-Chairperson: S. Morita (Japan)

SESSION 4.2B: GRADE OF SERVICE
Chairperson: A. Lewis (Canada)
Vice-Chairperson: F. Machihara (Japan)

SESSION 4.3A: DYNAMIC ROUTING I
Chairperson: D.G. Haenschke (U.S.A.)
Vice-Chairperson: I. Akiyoshi (Japan)

SESSION 4.4B: SWITCHING SYSTEMS AND MODELING II

Chairperson: G. Miranda (Italy)
Vice-Chairperson: Y. Takahashi (Japan)

SESSION 5.1A: ALTERNATE ROUTING

Chairperson: H. Akimaru (Japan)
Vice-Chairperson: R.J. Harris (Australia)

SESSION 5.1B: SWITCHING SYSTEMS AND MODELING III

Chairperson: H. Hofstetter (F.R.G.)
Vice-Chairperson: Y. Sakita (Japan)

TRADITION AND DEVELOPMENT

Michio NAGAI

Senior Adviser to the Rector
United Nations University

1. A CONFUCIAN INDUSTRIAL SOCIETY

Japan's traditional culture embraces Buddhism, Confucianism and Shinto, of which Shinto can be regarded as indigenous to Japan. In this sense, Japanese culture can be more accurately described as being a heterogeneous culture rather than a homogenous one. In addition, it can also be said that the ideas of democratic social systems, the underpinnings of which are science, technology and the ideas of liberty, equality and fraternity, have also become a part of Japan's culture in little more than one century of contact with Western civilization.

America is often described as a melting pot of different races. By comparison, Japan can be described as a melting pot of different cultures. In other words, the culture of contemporary Japan is made up of a mixture of Eastern and Western elements.

Before describing the nature of this cultural heterogeneity and its inherent problems, I would like to begin first by setting out briefly the relationship between Confucianism, as one element of this cultural melting pot, and the modern industrial society.

Many people from other parts of the globe who visit East Asia today are interested in the connection between tradition and development. One reason for this interest is the significant economic development of the region, not only Japan, especially when compared with other parts of the world.

At present, Japan is the only nation in the non-western world which is counted among the advanced industrialized nations. According to the OECD, however, South Korea, Taiwan, Singapore and Hong Kong are among the so-called "newly industralized countries (regional districts)", i.e., NICs. They are undoubtedly areas which deserve our attention. The economic success of these East Asian countries stands out clearly when compared with the more limited economic progress of the NICs of Latin America, Mexico, Brazil, and the NICs of Europe, Spain, Portugal, Greece and Yugoslavia, whose economic performance has suffered from considerable difficulties since the early 1980s.

Two young American scholars, Roy Calder and Kent E. Hofheinz, have taken note of this difference in their book entitled, East Asian Edge, published by Basic Books of New York. They add North Korea and the People's Republic of China to the five East Asian countries I have just mentioned and refer to all seven of them collectively as "Confucian countries and areas".

To be sure, these countries enjoy high rates of real GNP growth, investment and savings. And they make wise use of advanced technologies. The moral foundation of these countries is Confucianism. In the Confucian ethic, a stable family is the starting point of life, and husband, wife and children all pitch in and work together. As to learning, Chinese characters are difficult to master, but potentially 1.3 billion people in these countries have studied or are studying them along with mathematics, science and other subjects. At school and at work, the people of these Confucian countries and areas are known to be diligent, and have contributed greatly to their countries' rapid economic growth through co-operation and solidarity. Moreover, the governments of these countries have made considerable efforts to provide effective administration under the precept that leaders must first assure the welfare of the people before pursuing their personal pleasure. This is at least the ideal, if not the actuality. One distinctive feature of these countries in comparison with others in different regions is that their governments are relatively stable and their authority is respected by their people.

If a passion for learning, the virtues of diligence and co-operation, respect for authority and a preference for working together on the basis of a national consensus are the salient characteristics of Confucianism, then it is true that a Confucian sphere exists in East Asia, one which merits attention on a global level.

In this Confucian sphere, on does not find the same view of man or ideas that are found in a Christian context. But one does find a mutual respect for human dignity based on the ideas of solidarity and hard work that prevail in societies and schools and the high value placed on the education of children in every household.

Clark Kerr in his recent book, The Future of Industrial Societies, cites this sphere as an example of region outside the West where modernization and industrialization have succeeded to a large degree. He also sees this region as reaching a stage where the contradictory Western ideologies of free enterprise and socialism may come to convergence. In his view, the Scandinavian countries represent another area in the West that may reach a stage of convergence similar to that of the Confucian sphere.

Max Weber once wrote in his Protestant Ethic and the Spirit of Capitalism, that mammonism did not give rise to capitalism, rather the new and free capitalist society was brought about by ascetic diligence and rationalism.

I think it is safe to say that we are now seeing the emergence of a new industial society that is supported by Confucian ethics, family management and statesmanship.

2. METAMORPHOSIS OF TRADITION-DIALOG BETWEEN THE PAST AND THE FUTURE

The linkage between Confucianism, a philosophy dating back to ancient times, and industrialization, a relatively new social process, was by no means easy to effect.

The relationship between tradition and development has certainly not been a simple one in Western history either. Development involves building towards the future, yet it does not mean discarding everything related to the past. Instead, one must select and try to make the most of those elements from the past that are compatible with the ideas, hopes and desires for the future. The handing down of tradition does not mean merely preserving the past in an unchanged state. In some cases, we must have the courage to leave behind certain things from the past. In other cases, certain elements are to be utilized to promote future development. This is the real meaning of the transmission of tradition.

In other words, in the midst of our present day activities, a dialogue is taking place between the cultural heritage from the past and the ideas, hopes and desires for the future. To simplify this relationship, the present acts as an intermediary between the past and the future. Such is the relationship between tradition and development. In the West, one can also find excellent examples of this relationship, the foremost being the Renaissance and the Reformation.

The relationship between tradition and development in the non-western world has been much more complicated than viewers on the outside can imagine. The principal reason for this lies in the fact that until recently social changes in the non-western world were quickened by stimuli and threats applied form the outside. As a simple factual illustration of this, I would like to draw your attention to the following example. Before the nations of East Asia and Southeast Asia and the insular countries of the Pacific achieved their independence following World War II, they had been dominated by Western and Japanese imperialism. Thailand and Japan are,of course, two exceptions, since both were able to retain their independence. Yet in their case as well, the path of social development was laden with hardships.

In 1868, Japan embarked on a new start as a modern sovereign nation. What is known as the Meiji Restoration was in fact a type of national revolution undertaken at a time when India and China, two countries long admired by Japan, had already become a colony and a semicolony respectively. In such a situation, when a concerted effort was being made to promote future development, there was an extremely powerful force that operated to negate the culture heritage that had been built up in the society. When the Meiji Government proposed building the country's first national university in Tokyo, some people advocated that the school should be established around a core philosophy of either Confucianism, Buddhism or Shinto. After long deliberation, however, the Government took the decisive step to establish the university on the foundations of Western science, technology, social science and the humanities.

In other words, instead of scholarship and education designed to perpetuate tradition, Japan entered into an era of learning and education centered on Westernization which marked a break with tradition. The revolution that occurred in Meiji Japan was not from inside and below, but from outside and above.

This should not be construed, however, as indicating that Confucianism, Buddhism and Shinto were completely discarded. On the contrary, they continued to thrive among the people and have been sustained down to the present day. In very simple terms, Westernization was like a brilliant spectacle that took place on Japan's mainstreets, and despite the pressure it brought to bear, tradition continued to live tenaciously on the country's side streets. The actual situation was, of course, much more complex.

When we look at Japan's modern history, we see that wealth and power have at times assumed an independent existence divorced from the country's tradition and also from the cultural tradition of the West. And at such times, their pursuit has been advocated in utterly secular terms. Moreover, tradition has sometimes been used in ways convenient to wealth and power. For instance, the Confucian idea of a vertical social order was used to strengthen the power of authority. Also, the authority of the Japanese Emperor was made absolutely unquestionable by giving it the backing of Shinto. This became what was known as "state Shinto", and the result was the militarism and expansionism which Japan pursued throughout East Asia and the Pacific before and during World War II. In this case, the culture did not move the system; the system took advantage of certain aspects of the culture.

This historical example that I have just given you attests to the difficulties inherent in the relationship between tradition and development in the non-western world - difficulties that were not so clearly present in the West, but can still be found in Japan and many other areas of the non-Western world. Those difficulties involve the complicated issue of how to deal with the continuity and discontinuity of culture in terms of ideology and social systems.

It is very intersting to note that the close scrutiny and examination of traditional culture that took place under the impact of social change was not confined only to Japan. In China as well, where Confucianism originated, the question of what to do with Confucianism has been one of the more perplexing issues during the turbulent social changes of this century. In the early part of this century, the rejection of Confucianism was the dominant current of new thought espoused by the some movements that arose to establish a new culture. At the time of the Great Cultural Revolution during the 1960s and 1970s, Chairman Mao repeatedly attacked and criticized Confucious. However, since the beginning of the 1980s, shrines dedicated to Confucious have been re-opened in a number of places and Chinese are studying seriously their own cultural history including Confucianism.

This, of course, does not mean that the Chinese people today are going to revert to classical Confucianism. While I said earlier that the countries of East Asia were Confucian in character, the meaning and content of Confucianism has been modified during the processes of social change. Just as in the West, where the new ideas of the Reformation breathed new life into the Christian tradition, Confucianism, too, has undergone changes in keeping with the times.

In a similar vein, "state Shinto" was discredited following Japan's defeat in World War II and the Emperor himself issued a statement denying his divinity. Yet the ideas of reverence for nature and of harmony between nature and man, which are a part of the Shinto tradition, are still very much alive in Japan. In fact, the attachment of the Japanese people to nature deepened appreciably following the considerable destruction of the natural and cultural environment caused by the rapid industrialization of the 1960s.

Various changes have also been seen in the world of Buddhism. One indication of this is the strong activity of the many new religions related to Buddhism. A religious reformation in the true sense of the term, however, is still an issue for the future.

3. A MELTING POT OF HETEROGENEOUS CULTURES

What is traditional Japanese culture? When this question is put forward, people from other countries as well as the Japanese themselves are likely to respond that it is the culture excluding those elements adopted from the West. Consequently, they are apt to regard Chinese characters, Confucianism and Buddhism as representing Japanese culture. However, Confucianism and Chinese characters originally came from China, and Buddhism originated in India and was introduced to Japan. These elements were originally a part of a foreign culture, i.e., not originally Japanese. In this sense, Japan did not import elements of its culture solely from Western civilization. Rather, cultural elements from other parts of Asia and from the West were imported to Japan, the only difference being the time when these elements were transmitted. Cultural importation form other parts of Asia took place mainly from the sixth century onward, while that from the West occurred from the sixteenth century onward.

If we look for differences between Japan and other Asian countries, we can point to the fact that Japan established a modern sovereign nation at a relatively early stage of world history and that Japan made a serious effort to create a melting pot of heterogeneous cultures that included Asian and Western culture as well. In this context, we should not overlook how the Japanese approached Christianity and the military technology of firearms, which the Spanish and Portuguese transmitted to Japan in the sixteenth century, and how they approached science such as medicine and astronomy, which they learned from the Dutch in the latter half of the eighteenth century.

Japan's political leaders in the sixteenth century were apprehensive about the propagation of Christianity and took steps to prohibit it. Subsequently, Japan entered a long period of national isolation. However, those same political leaders were deeply interested in firearms and military technology. That interest was later connected to the intellectual curiosity the Japanese showed towards science in the latter half of the eighteenth century. As a result, Japanese intellectuals came to possess an understanding of Western science and technology from a relatively early time as compared with their counterparts in other Asian countries.

In 1853, when Commodore Perry brought his squadron of four warships, two of which were steam-powered, to Japan to pressure the Japanese to open the country's ports, the government at that time was intimidated by the enormous power of this foreign force. At the same time , it did not take the Japanese long to realize that behind those four warships lay the European and American industrial revolution and that this significant historical change was related to the progress of science and technology. Not only that, but those Japanese with more farsighted wisdom realized the expansion of the industrial revolution was also related to political democracy, efficient governmental administration, an effective educational system and free enterprise, and other important social institutions of Western countries.

Following the success of the Meiji Restoration, however, Japan made the mistake of pursuing course of militaristic expansionism during this century. After the country's defeat in World War II, a program of reform was carried out under the direction of the Occupation Forces, which instituted a political system based on the theory of popular sovereignty and infused democratic principles into the economic and educational systems. As a result, Japan reached a level of social development that went far beyond that which was achieved at the time of the Meiji Restoration.

When we look at the past four centuries in this manner, it is clear that the Japanese have regarded Western culture, including its ideologies and systems, as an important model for the development of their own society and have continually studied Western culture with great eagerness and enthusiasm.

In this historical process, however, serious discord ensued between the culture transmitted from other parts of Asia, Japan's indigenous culture and the culture transmitted from the West. As an illustration of the intensity of this discord, a leading educational figure at the beginning of the Meiji Period even went so for as to propose that Japan abandon the Japanese language and adopt English as the national language which would be used in the public educational system. On the opposite side, there were not a few Japanese intellectuals who were ardently attached to the indigenous culture handed down from ancient times.

The historical result of that discord can be seen in the fact that today the Japanese live in a heterogeneous culture which can be viewed as a melting pot of diverse cultures.

This does not mean, however, that the path of Japan's cultural history has reached its culminating point. As I see it, there is still a long journey that lies ahead.

Japanese society has already passed through the industrialization stage characterized by rapid economic growth and is on the verge of entering the post-industrial phase. The age of mass production and mass consumption has given way to consumption that stresses personal, individualized preferences and this trend will lead to a situation where culture may be prized even more. At that point, I think we may see a revival of the Japanese tradition of aesthetic creativity that flourished prior to the Meiji period. During the three hundred years of peace that prevailed during the Tokugawa Period preceding Meiji Japan, there was a flowering of a brilliant aesthetic culture. Drama, literature, poetry (including haiku), architecture, the Japanese Kimono, and painting - all of these creations and many more that still today give pleasure to the eye or to the ear or to one's mind resulted from the growth that took place during the Tokugawa Period. The Japanese love of nature and the tradition of seeking unity and harmony between man and nature also prevailed during that period. The samurai, who were the military men of the day, lived their lives as administrators and educators and tried to contribute to peace instead of pursuing their original profession as warriors. These historical facts are of great interest today at a time when people everywhere desire world peace.

Until recently, the Japanese have lived within the limited framework of Japan as their native land. In contrast to this, from this point forward the Japanese must at the same time transcend this national framework and live as citizens of a World confronted with many difficult and complex problems. The critical point is whether or not they can do this.

Around the periphery of Japan today we see the newly industrializing countries of East Asia along with the Southeast Asian nations and the island countries of the Pacific. China is proceeding along a path of modernization and on the Korean Peninsula there are signs pointing towards a reconciliation between North and South. On the far shores of the Pacific as well as to the South lie countries of the Western Civilization. Naturally, the traditions of these regions are not limited to Confucianism and Buddhism. Moreover, Confucianism and Buddhism are also undergoing change. Besides these religions we also find Islam, Christianity and Hinduism. And in many areas, there are folk beliefs that date back to ancient times.

The discord between tradition and development can be found in all of these regions, and the people caught up in this discord are seeking a way out of their anguish.

At this conference, various scientific and technical problems in the field of communications will be examined. I am sure that remarkable progress will be made, but that alone will not open up new avenues for the future of mankind.

It is my hope that the Japanese cultural melting pot will become even more open to the outside and that it will contribute to the progress and happiness of people the world over.

By way of conclusion, I would also like to express my hope that the dialog between Eastern and Western cultures, which is being repeated every minute of every hour of every day within the Japanese culture, will provide a new prospect of the future for us all.

Thank you for your kind attention.

ADVANCED INFORMATION SOCIETY AND TELECOMMUNICATIONS

Yasusada KITAHARA

Nippon Telegraph and Telephone Corporation
Tokyo, Japan

ABSTRACT

As the information-oriented society advances in the various fields of industry, society, and the home, telecommunications services must be improved to meet increasing and more diversified customer demands.

This paper introduces the recent direction of telecommunications, the approach to the INS (Information Network System), and problems in the telecommunications fields.

1. INTRODUCTION

Ever increasing progress is being made toward the "advanced information society" on a worldwide scale. This trend began gaining importance about two decades ago when on-line real time data communications made its debut through the integration of computers and telecommunications. Prior to this, telecommunications mainly took the form of information transmission, with computers simply processing information. With the advent of data communications, however, it became possible to both process and transmit information to any place, instantaneously. Moreover, this integration of computers and telecommunications has had a tremendous impact on all related areas, thereby resulting in the rapid development of both information communications and processing.

One of the main factors supporting this trend has been the remarkable progress made in the field of electronics, especially with regard to LSI technology. Moreover, the costs associated with information processing and transmission have dropped significantly, while reliability has improved greatly. In data processing, for example, costs have been reduced to one-two hundredth (1/200) while processing speed has increased 400 times over the last thirty years (IBM figures).

The ever-increasing demands for information and communications are largely due to the growing realization that efficient utilization of information will play an extremely important role in promoting both individual and national prosperity. In addition, a lack of information may create serious problems due to lost opportunities.

With this as background, it is clear that customer demands for information and communications processing services will continue to increase, and that the computer and telecommunications technology of the future will be far more advanced than it is today. Therefore, it is believed that the trend toward a more information-oriented society will intensify even further in the future.

2. FROM MONOPOLY TO COMPETITION

Although society is moving more and more toward the integration of information processing and communications, the telecommunications and computer fields have historically developed along different lines. Telecommunications services, centered primarily around the telephone service, have been provided under public or monopolistic management in almost every country except the United States. On the other hand, the information processing field based on computers has been an arena for keen competiton among private enterprises. This competition has resulted in rapid progress in services.

Telephone system are naturally characterized by monopolistic tendencies due to the extremely large initial investment required, the need to provide homogeneous service on a wide scale, and the insuring of connectivity. Consequently, most countries have operated their telephone systems as monopolistic enterprises. With the diversification of customer needs and the advent of a variety of non-telephone services, however, the traditional monopolistic system cannot meet rapidly expanding demands. For this reason, Japan liberalized its telecommunications market on April 1, 1985 after more than thirty years, and privatized the Nippon Telegraph and Telephone Public Corporation to form the new private Nippon Telegraph and Telephone Corporation (NTT).

The following two points are believed important in order to promote the sound development of information and communications processing services as well as to provide truly useful services to customers. First, it is important to recognize that free competition will result in both price reductions and the diversification of services, as excessive regulation tends to hamper the interest of users and the nation as a whole. This is especially true in the business field. The United States, United Kingdom and Japan have chosen free competition as the medium for providing information and communication services. It is expected that these services will become even more advanced and that customers will be able to enjoy less-expensive, more versatile services. For this reason, it is considered important to make competition a priority in the business field. Second, looking at information and communications services as a whole, some services are believed to be naturally less suitable when it comes to opening up the market to competition. Telecommunications forms one of the most important infrastructures of our society, and its benefits must be shared equally by all people. However, looking at present telecommunications services, it cannot be denied that there still exists profitable areas centering around large urban areas and unprofitable areas centering around rural areas. Therefore, if telecommunications is to be an infrastructure equally benefiting everyone, it is necessary to insure that there are no regional differences with regard to the basic telecommunications service provided, at least in the case of basic telephone service.

This aim is clearly set forth in our laws. The Telecommunications Business Law governing telecommunications business activities specifies that "the purpose of the business is to promote the public welfare". Moreover, the Nippon Telegraph and Telephone Public Corporation Law upon which NTT exists today clearly states NTT's responsibility as the "provision of telephone services essential to the national wellbeing under fair and proper conditions".

Although free competition is a basic pre-requisite in the various fields of telecommunications, it is necessary to take measures to enforce regulations in those fields that compose a distinct infrastructure. This is the reason for the stipulation which requires that the laws be reviewed three years after their enforcement.

Before discussing the merits and demerits of regulation and deregulation, however, the key question which must be asked of the telecommunications business is· "How can the telecommunications industry most benefit the nation?".

3. APPROACH TO THE INS

NTT has set the establishment of the INS as one of its major management objectives for realizing the above concept. The INS concept is aimed at providing "more diverse services that can be used economically anytime and anywhere irrespective of distance". In addition, the formation of this system has nothing to do with the management structure of NTT.

The process for realizing the INS concept will take the form as shown below. Rapidly advancing technology such as digital, optical fiber, and satellite communications technologies will be used, and an even greater number of economical digital facilities will be installed, gradually replacing existing analog facilities. Digitization will be increasingly introduced into existing individual networks to facilitate smooth integration and to promote the formation of the INS foundation.

In concrete terms, commercial INS service will be introduced in the Tokyo, Osaka, and Nagoya areas in 1985. This reflects the results of the operations on the INS model system which began in September of 1984 and the INS system trials at the International Science and Technology Exposition (EXPO '85) which opened in March of 1985. Furthermore, INS services are scheduled to be provided on a nationwide basis to customers in 1988. This will be initiated by furnishing digital switching functions to prefectural capital-level cities and by interconnecting these cities by means of a digital network.

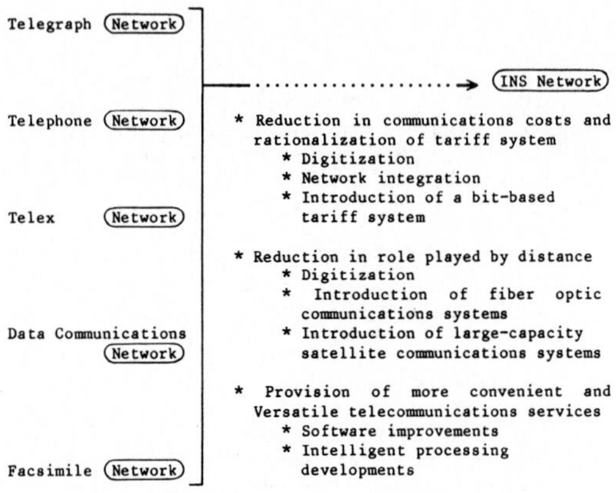

Fig. The Process of Achieving the INS

4. PROBLEMS IN THE TELECOMMUNICATIONS FIELD

4.1 From a Telephone Society to a Video Society

Telephone services, the main component of existing information and communications services, will continue to play an important public role. Furthermore, the technology for supporting these telephone services and standards have almost reached their full capabilities. It can reasonably be said that the telephone system based on current telecommunications technology has reached an advanced stage of development.

On the other hand, from 60 to 80% of the information received is visual, and the present telephone syatem network cannot meet the new demands for visual information services. As a result, it will be necessary to develop public broadband services utilizing video technology. For this purpose, visual services must be provided at a much lower cost.

A band 1 000 times broader than that for voice signals is required for the transmission of video singals. However, no customer would use the video service if the rates for it were 1 000 times higher than those for the telephone service. To reduce video service costs, much effort is even now being put into the development of less-expensive broadband transmission technology, coding technology, and video (broadband) technology. However, the current activities are not adequate in reducing costs to a feasible level. To achieve a video-based society, further technological development efforts are required. In addition, psychological and human engineering studies should be conducted.

4.2 Insuring Telecommunications Connectivity

The most important function of a telecommunications system is to insure that everyone can make use of it at any time.

Needless to say, as the rangee widens for interconnection of equipment, the utility of the network also increases. The function corresponding to this communications range is called connectivity, and is ranked as the most important function in any telecommunications system. The connectivity of a network is high if a customer can communicate freely with anyone, anywhere, and with any type of Customer Premises Equipment (CPE).

From this point of view, the connectivity of the existing analog telephone network is rated highly because it fulfills the requirements for voice communications around the world.

The connectivity problem has been eliminated for all practical purposes. There are many instances, however, where communications between different kinds of computers or CPEs is impossible. Even for facsimile communications using telephone networks, communications is not possible between all types of facsimile equipment. With regard to voice communications, problems will arise when telephone networks are digitized and plural coding systems are introduced, although all telephones worldwide could be interconnected when there existed only analog networks. Therefore, great expectations exist for the standardization efforts now being made at CCITT, ISO, and IEC to solve these problems.

There is, however, one problem which we must be borne in mind. It concerns the relation between connectivity and the development of customer premises equipment such as personal computers and word processors, for performing information processing.

This equipment has been rapidly developed and made different by companies in order to promote competition. That is to say, the goal is to make equipment more sophisticated and less-expensive than equipment manufactured by other companies. This contrasts sharply with the concept of connectivity. Thus, the development of CPE technology may be hampered if too much emphasis is placed on connectivity.

On the other hand, it is also true that new and substantial added value is achieved when mature customer premises equipment has connectivity and communications functions. Therefore, an important point of study and analysis will be how to rectify these two contradictory factors and at what point in time total connectivity will be assured.

4.3 New Technology and System Life-time

The driving force behind the present advances in the information and communications processing industry is the wide variety of new technology supported by the revolutionary progress in LSIs. This technology has resulted in lower cost, move compact, lighter, less energy consuming equipment. Moreover, this equipment is more sophisticated and has higher reliability, thus contributing to the spread of information and communications systems.

Moreover, in order to better meet the needs of the forthcoming advanced information society, further efforts must be made to develop more advanced technology and to actively introduce it. Particular emphasis must be placed on the development of high-density and high-speed LSIs, low-cost optical fiber, large-capacity satellite communications systems, software, and intelligent processing technology.

However, as the pace of technological development accelerates, equipment will become obsolete at an even faster rate. Therefore, those involved in telecommunications management will be required to better estimate the life-time of each piece of equipment in a system when constructing networks. The life-time of each piece of hardware will depend on technological innovations, which will lead to a drastic cost reduction and less emphasis on general life-time.

It must be particularly emphasized that research and development of new technology should be conducted considering long-term objectives. It must always be kept in mind that no matter how sophistcated the technology is, it is worthless if not used by customers.

5. CONCLUSION

Present trends toward the realization of the advanced information society are increasing on a worldwide scale. The world is undergoing dramatic changes, but I am confident that it is moving toward a brighter future. These changes will first appear in the business field, and eventually will extend to all people.

This trend can be observed not only in industrialized nations but also in developing countries. As a result, equal access to information has been in strong demand during information order discussions between the North and South held under the direction of UNESCO since 1976 (i.e. the discussions of the MacBright Committee). This was also stressed in the report, the "Lost Link", recently published by the Independent Commission for Worldwide Telecommunications Development of the ITU.

Looking toward telecommunications in the 21st century, it will be necessary for private enterprises, telecommunications organizations, and countries to exchange ideas on worldwide competition and cooperation.

Similarly, the information-oriented society of the future will be more complex than today's society, and many problems are certain to arise. Examples range from invasion of privacy, information control, and excesses of information to various types of inherent problems, including alienation in society.

No one can deny the fact that progress in science and technology has greatly contributed to improvements in life styles. However, it is also true that although science and technology has provided us with knowledge unimaginable in the past, it can also be a danger depending on how it is used. Therefore, to insure that the advanced information society will actually benefit mankind, it will be imperative to consider not only the technological aspect, but also to fully discuss the issues in the fields of natural, social, and cultural sciences in a coordinated, effective manner. Only in this way, true harmony can be achieved between science and humanity in which people can cooperate with each other to create a better world.

This concept, integrating the progress in science and technology with the welfare of mankind, may very well be philosophy which will lead to a truly advanced information society in which the INS is a communication infrastructure.

REFERENCE

[1] Y. Kitahara, " Information network system -- Telecommunications in the twenty-first century -- ", WILLIAM HEINEMANN LTD., London, 1983.

CIVILIZATION IN ADVANCED INFORMATION SOCIETY

Hiroshi INOSE

The Faculty of Engineering
University of Tokyo
Tokyo, Japan

ABSTRACT

Among a great many aspects relative to civilization in advanced information society, some of them which seem to the author as conspicuous and significant are brought forward as the subjects for discussion. These are, rise of communities of interest, selectivity and authenticity of information, changing work pattern, learning through life and creation of culture. Several proposals are made for the purpose of alleviating problems and taking full benefit of the capability of emerging information technologies.

Ever since the mankind formed a society, its civilization has always been dependent upon information. The use of languages spoken and written, made communications between individuals possible, and invention of paper and printing helped recording and dissemination of written languages. Telegraph and telephone permitted communication to take place trascending geographical distances. Phonograph made possible the recording of spoken languages and music, while photograph and motion picture enabled the recording of still and moving images.

The forthcoming society which is often referred to an advanced information society may not be considered as being totally different from the society of the past, in the sense that it continuingly depends upon information. However, the modern information technologies we have at hand are more than the extensions of the information technologies of the past, and will grow even more powerful in the years to come. Information technologies are rapidly merging into one common digital art and in so doing, penetrating into all aspects of socio-economic activities. The impact of this art is so great that the changes it will bring about in our civilization may characterize the forthcoming society to be as distinct from the past one.

Information technologies are integrating diversified socio-economic activities which in the past were often functionally or geographically separate. Information technologies are changing the industrial structure by bringing about new industrial sectors that produce highly value-added products while improving the productivity of traditional industirial sectors. Information technologies promise to improve societal infrastructure such as medical care and transportation, making it possible to provide them efficiently and economically. Information technologies can integrate various information services that have been separate in the past, and thereby provide users with diversified information more effectively and at less cost. Information technologies provide powerful tools for understanding and preserving the human cultural heritage and for enhancing cultural creativity of mankind.

Although civilization in an advanced information society has a great many aspects, the present paper addresses at some of these, namely, rise of communities of interest, selectivity and authenticity of information, changing work pattern, learning through life and creation of culture.

1. RISE OF COMMUNITIES OF INTEREST

In the ancient era, each isolated hamlet was a small world. In such a territorial community, each member knew the rest of the members perfectly well and therefore everyone was able to take part in all events. The people or events in other regions made their appearances in the life of that community only in legends or in rumors. It was, so to speak, a homeostatic society. Today, all such small worlds have been closely woven together to form a complex and huge unified body as a result of the developements made in communication and transportation. From now on, people increasingly communicate, overcoming terrestrial distance, with those with whom they share interests or concern, while rejecting our neighbors, using privacy and the right of anonymity as a shield. Thus, the community of advanced information society will be formed by intellectual commonness rather than by territorial intimacy.

The fact that the people of the world belong to several of many communities of interest and that such communities are growing each year is, we may say, providing mankind with new opportunities. This is because such a trend means that a large number of routes of communication, which are of value to civilization, are being formulated and intensified on a global scale. In contrast to the fact that the nations which are formal communities having their roots in territorial relationship, are gradually losing their problem-solving capability, the communities of interest, which are informal communities that are bound by common interest, are now capable of solving problems better on the basis of mutual understanding and trust. It may be said that the advanced information society is characterized by the lowering of national barriers and the rising of communities of interest.

On the other hand, the collapse of communities bound by territorial intimacy may be taken as a threat to civilization. The reason is that this would lead to the loss of the elements constituting the cultural indentity, such as language, dialect, custom and festival, and of functions of towns and villages which are the people's places to encounter.

In an advanced information society, it will be easier to humans in learning about cultures of different territorial communities. However, exposure to different cultures brings both advantages and disadvantages. It helps one territorial community to understand the cultures of others, and by selectively borrowing from these, a territorial community will be able to enhance its own culture. If on the other hand, one territorial community is exposed to a powerful culture and absorbs it without discrimination, the cultural identity of that territorial commnity will be lost. This is particularly true when a remote rural community is suddenly exposed to an advanced civilization. Because of the conveniences it provides, an advanced civilization may quickly overwhelm the community, and the traditional culture of the community may be lost.

In the past, we have seen many instances in which an advanced civilization was successfully adapted by a number of territorial communities and flourished through the endogenous efforts of such communities. However, we have also seen other instances in which the introduction of advanced civilization merely caused chaos or complete loss of identity. Thus, an effort should be made to strengthen the endogenous culture of territorial communities and to maximize the profit and minimize the loss due to such interaction.

2. SELECTIVITY AND AUTHENTICITY OF INFORMATION

One of the aspects of advanced information society is information explosion. The volume of information being generated is far greater than can possibly be consumed. Many television and radio programs are scarcely watched or heard, and a large fraction of newspapers, journals, commercial catalogs and books are thrown away unread. Indeed, a quantitative study on communications flows carried out in the United States and Japan revealed that more than 95% of information supplied is through radio and television of which less than 1% is being consumed, and that information supply grows three times as fast as information consumption. People get frustrated to feel that they do not have means for selecting the information they need and for acquiring accurate information they can trust from among the flood of information.

For the purpose of providing selective access to information, databases play a major role. Databases however, require an enormous efforts in collecting, abstracting,indexing, entering and updating of information over a long period of time. Very few institutions can afford to do this by themselves, especially when the information has little commercial value. Therefore, building such databases requires public support, at least until the database becomes commercially valuable. In view of the fact that governments collect enormous amount of invaluable data for their administrative use, they should be urged to provide their data to the public as much as possible, to the extent that no infringement on privacy and national security occurs. Information is an invisible commodity. Like visible commodity, it requires an appropriate distribution mechanism in order to reach end users. The governments therefore, should assist information providers and vendors in developing their distribution channels.

For the purpose of providing accurate information, database services must collect, collate, compile and process information with utmost care to assure its authenticity and validity. All data should be provided with source quotations, audit trails and the identities of persons who control the files. Integrity of information is one of the most important issues in an advanced information society where all socio-economic activities are heavily dependent upon information. When information concerns the background, knowledge or activities of individuals or corporations, integrity implies something else. In addition to authenticity and validity of information, it implies privacy and confidentiality. Insofar as possible, such information should be gathered only for very compelling reasons, should be kept only as long as necessary and should be guarded diligently against ill use.

3. CHANGING WORK PATTERN

Another aspect of advanced information society is the drastic change in industrial structure which has been and will be brought about by the extensive use of information technologies. Whenever structural changes in industry have taken place, workers in traditional industrial sectors have lost their jobs while newly emerging industrial sectors have suffered from a shortage of workers. A change to an information-oriented industrial structure can be no exception.

Skilled workers, such as those doing routine machining jobs in industrial plants, may lose their jobs because of the increasing use of industrial robots. General office workers without special talent may also lose their jobs because of the rapid penetration of word processors and other office automation equipment. On the other hand, enormous job opportunities will be created in the area of software production, since much new software is needed to improve productivity in traditional industrial sectors and to support knowledge-intensive products and services. No matter the extent of computer support in this effort, software is ultimately written by human beings, and the need is ever increasing.

What is clear at this moment is that extensive education and training should be provided for the younger generation to undertake software production and that sufficient opportunities of continuing education and retraining should be given to existing engineers and technicians so that a smooth shift of labor force from decaying to emerging areas can be made.

Production of the enormous amount of software that is envisaged will be possible only through an international division of labor. In particular, application programs should be produced locally to meet end users' needs. Developing countries, especially those that have a well-developed higher-education system, should be encouraged to participate in software production. The unusually large proportion of Americans of Chinese ancestry in the field of software technology in the United States seems to suggest that Chinese people are in some way especially well suited to software production. Whatever the reason, China itself and Hong Kong, Taiwan and Singapore, as well as countries where overseas Chinese play an important economic role such as Thailand, Malaysia, Indonesia and the Philippines, have a tremendous potential for software production. It is conceivable that the East Asian region may take the lead in the industry in the next century.

It should also be noted that software can be produced away from an office and on part-time basis, and that computer games and other entertainment by personalized computers are increasingly attracting interest of all ages and thereby familiarlizing computer programming which in the past was considered to be difficult and boring.

4. LEARNING THROUGH LIFE

As has been metioned earlier, continuing education and retraining are indispensable factors for the smooth shift of labor force from declining to emerging industrial sectors. In an advanced information society in which the pace of technological innovation will be intensified, knowledges gained will very quickly become obsolete. People in an affluent society increasingly place importance on spiritual rather than materialistic satisfaction and thereby wish to know and think more than they did in the past.

Formal education which has been in existence over centuries may need a drastic reorganization in an advanced information society. The present higher education is something like a cathode of an electron tube that gives initial velocity to charged particles. To reach their target, the emitted particles need an appropriate mechanism which provides energy and guides them along the right path. Appropriate means of refleshing the knowledge of the graduates, especially in science and technology, should be provided periodically over their life time, so that they can keep up with the rapid progress. The continuing education and retraining for displaced labor force should be made, for fairness, by public rather than private cost, and should be integrated to the formal education system. Means for learning should be provided to the general public including house wives, senior citizens and handicapped so as to fulfill their desire to be exposed to the changing intellectual world which has been unknown to them.

Advanced information society when motivated to do so, has technologies at hand to help such reorganization. A broadcasting satellite, through its wideband trasmission capability and broad coverage, can enhance a university-on-the-air to provide nationwide, educational programs not only in liberal arts but also in high technology topics and thereby provide engineers and technicians to keep abreast of rapid technological innovations. A digital network that provide nationwide service can attain a sufficient economy of scale by permitting shared use of highly sophisticated and expesive software for computer-aided-instruction in diversified disciplines. Main frames and supercomputers shared by way of a digital network, or personalized microcomputers can provide a great variety of opportunities to the general public of all ages to be exposed to a highly structured thought through their screens and key boards, and thereby acquire computer literacy. Huge databases of various disciplines can provide information selectively through digital network, not only for specialists and scholars but also for the general public. Above all, modern information technologies that support diversified handling of information in various forms are providing educators and students with a great deal of opportunities to improve their ways of teaching and learning.

5. CREATION OF CULTURE

An advanced information society could be truly epoch-making in the history of civilization, if it can contribute significantly to the enrichment of human culture. Culture is a highly complex compound of thought, appreciation, aspiration and entertainment, covering arts and letters. As the invention of printing made literacy widespread and stimulated talented authors to write great books, and as the progress of musical instrument provided powerful tools for creative composers to produce their masterpieces, emerging information technologies which are far more powerful than that of the past, will provide a great deal of opportunities to the artists in the future.

However, no matter how powerful, information technologies are, like other technologies, merely tools for the mankind. If an artist without inspiration have dabbled in information technology, the results have often been something like a child's banging on the piano. Contribution of technology for the enrichment of culture generally takes a slower process than that for industrial buildup, because it requires serious attention of creative artists.

It should be noted however, that some musical artists and composers, are exploring very seriously the potentialities of information technologies, ranging from the use of electronic synthesizers and computer composition of music to computer production of musical scores. Computers enable architects to explore extensively the appearance, structure and cost of their design. Computer graphics can be used to create still and moving images of objects that exist only in the mind of the artist. Numerically controled machine tools can be used to fashion individual sculpture and ornament through interactive conversation with the artist. Interactive computer games are now providing opportunities for younger generations to enter a world of structured thought and creation while enjoying the games. Sophisticated simulators as typically represented by interactive flight trainers may open up possibilities of creating a totally artificial environment that provide people with highly artistic experience, something like wandering through a well layed-out garden.

All the above and other developments in an advanced information society may bring about enormous challenges and opportunities for teletraffic engineering.

REFERENCES

[1] S. MacBride, "Many Voices, One World," UNESCO, Paris, 1980.

[2] H. Inose, "An Introduction to Digital Integrated Communications Systems," Univ. of Tokyo Press, Tokyo, 1980.

[3] H. Inose and J. R. Pierce, "Information Technology and Civilization," W. H. Freeman, New York, 1984.

[4] I. de Sola Pool et al, "Communications Flows," Univ. of Tokyo Press, Tokyo, 1984.

TELETRAFFIC ISSUES in an Advanced Information Society
ITC-11
Minoru Akiyama (Editor)
Elsevier Science Publishers B.V. (North-Holland)

ISDN AND VALUE - ADDED SERVICES IN PUBLIC AND PRIVATE NETWORKS

Lester A. Gimpelson

ITT Europe , Inc.
Brussels , Belgium

Abstract

Integration of information processing with the communications of ISDN represents the next step in the union of these two disciplines. Future traffic work by the communications and processing sectors must precede the probable scenario for progress beyond ISDN transport: starting with integration of voice and data services and then integrating a broad range of value-added services based upon information processing. Unknown traffic levels and characteristics of these services lead to urgent requirements for advanced exchange and network designs and for new management schemes. It's recommended that the ITC take the lead in directing efforts assuring that traffic advances supply timely support to initial and augmented ISDN implementations. This paper also describes the development and implementation of ISDN and ISDN augmented with value-added services, shows how the same functional services can be offered by public and private networks (both separately and jointly) and suggests traffic advances that will be needed to assure successful ISDN introduction.

Introduction

The substantial, but generally independent progress that has been made in the technologies of communication and information processing is finally being applied to systems taking advantage of advances in both areas. ("Information processing" is used here in a very broad sense, from batch computing to translation processes between dissimilar systems.) Indeed the boundary points between communication and information processing are no longer clearly definable. Tasks or functions or services that are essentially similar from the viewpoint of users are now provided by quite different implementations due to their realization from either communications or computing sources; an example is the service offering of a LAN and a ISDN-capable PBX. Whether implementations are distinctly different, or as is happening, are converging as the two underlying technologies themselves advance and converge, users, facility providers and operators are faced with new questions -- and problems -- of performance, design to grade-of-service and reliability and survivability standards, implementation and operations economy, provision and forecasting of new services and their characteristics, etc. These questions are a direct consequence of the substantial expansion of services being offered to users; many of these new services were not even considered some years ago for implementation on separate, dedicated data facilities; and now they will be routinely contending for the same, limited resources. Indeed even the analytic technologies developed by the communications and computing disciplines, starting from quite different bases, have progressed independently until their recent convergence.

The projected offering of an enormous breadth of new services that constitute what is now termed "the information society" imposes an opportunity and challenge on designers to build capable, robust and economical combined communications and computing systems; and they have the clear obligation to understand and protect the operation of these combined systems for a society that will become increasingly dependent upon these new services, as society is today dependent upon the availability of electrical power and basic telephone service.

When communications migrated from progressive control to central or common control, attendant understanding of these new exchanges and analytic means for designing the new networks lagged behind the physical implementations, as seen by the severe network and exchange overloads that occurred; in a sense, intuition had failed network designers and communications operators. A parallel overload situation occurred when the first national packet switching network was initiated (in the U.S.) -- by the "computer fraternity" -- even though the telephone network's experiences, problems and remedies were well known by that time. (And considering that there are a number of currently "troubled" packet networks, all the solutions in that control area do not seem to have been developed or applied as yet. In fact there has been a recent "crash" of the nationwide packet network in Europe following an overload situation.)

An impressive body of design and analytic technology has been developed for the communications sector (exchanges, networks) and for data networks. But at present there is not a similar wealth of either experience or technology for the ISDN era, the combination of the traditional voice communications with a broad variety of data services.

ISDN is usually defined as the transport of data and voice services; this is too restrictive, since it does not include the offering and access of the so-called value-added services directly via public and private networks and it does not recognize the potential for offering private-network features by public exchanges ("centrex" with ISDN and value-added services). Since neither the traffic volumes nor characteristics of these services can be forecasted at present, the equipment supplier, the designer and the network operator have the obligation to provide services and network structures that are robust in the sense of meeting adequate performance criteria regardless (well, almost) of the eventual traffic development. This situation is distinctly different from the previous orderly development of voice telephony services and the somewhat more volatile emergence of data services.

The obligation now is to develop network designs

and tariff structures that will encourage the growth of new transport and value-added services. Initial indications from several countries are that ISDN subscribers will profit from the digitalization of public networks via usage-sensitive charges for 64 kb/s data calls similar to those for voice calls, and with termination charges about double those for voice-only subscribers (which is logical considering the potential for initiating three simultaneous calls as each line). Should this trend be general throughout operating administrations, the new industry -- users, terminals, carriers, -- will develop rapidly. (It was shown in a previous paper, [1], that the cost for carrying a 64 kb/s ISDN data call is essentially the same as for an analog or digital voice call. This sharp reduction in cost over current digital data network costs is one of the two primary advantages of ISDN; the other is the offering of a multiplicity of simultaneous services, two 64 kb/s channels and one 16 kb/s channel.)

It would be easy to emphasize the potential problems attendant with ISDN introduction by focusing on the uncertainties that such a change brings to the basic telecommunications system. Indeed, as will be pointed out, there are many questions that will need to be addressed if the transition to ISDN is to be economic and operationally convenient for both the service vendor and the user. Later sections of this paper will cover these issues. However it will be useful to precede those discussions with a description (albeit, forecast) of the services expansion that will mark this transition; the next section describes the transition to ISDN, and the section that follows expands ISDN in the sense of both digital transport and value-added services and then indicates how these services will be extended to private networks and derived private networks (that is, derived from public facilities).

Transition to ISDN

Initially the digitalization of the telephone network (and it was just a telephone network, with some acoustically coupled or modemed data traffic) was motivated by improved transmission quality and reduced overall network cost (largely thru the integration of transmission and switching functions facilitating the elimination of numerous concentration and expansion points). It was assumed that there would first be a transition to an integrated digital network, IDN, and then a transition to a network that integrated voice and a variety of data services, ISDN, integrated services digital network. (The word "integrated" thus has quite different meanings in IDN and ISDN.)

The motivation for ISDN was economy of scale and operational convenience [2]. The economy aspect can be approached via the question: Is the marginal cost for carrying data services on the huge network base formed on the digital telephone network less than the cost incurred by providing separate network facilities for data services? The construction in Reference 1 answers this question in the affirmative by demonstrating the economy of the integration of the majority of data services onto the digital telephone base, forming a general digital transport network, DTN, for carrying any digital signal. (That construction does place an important constraint on the switching points with regard to overhead expenses incurred for non-ISDN traffic; but since at least one switching system has successfully demonstrated a zero-overhead implementation, the proof is valid [3].)

Since the ISDN will transport digital voice and data streams equally, the usage cost to the service provider is essentially the same. So in contrast to currently separate and expensive 64 kb/s data services, an ISDN implementation has the potential for offering data services at the same price as voice services. (One needs to differentiate the cost to the communications provider from the price or tariff to the user.) Further it has been demonstrated that virtually all existing twisted-pair loops (both within buildings from PBXs and thru streets from public exchanges) are capable of carrying the standard ISDN subscriber package of 2B + D channels or 2 x 64 kb/s + 16 kb/s = 144 kb/s [4]. As a result the in-place wiring can be used to offer three simultaneous services to the user (where previously 9.6 kb/s or so was considered the loop limit). This simultaneity attribute is a further element in the economy of ISDN over separate voice and data facilities, as the latter would require a separate entry means to the user.

ISDN service is actually defined as n x 64 kb/s, and one ISDN-capable digital exchange has demonstrated its ability to carry up to 2 Mb/s, the equivalent to n = 30 channels [5]. This advance allows the much heralded opportunity for the user to specify the "bandwidth" needed for each data call by specifying n from 1 thru 30. (Traffic problems arising from contentious mixtures of traffic with large differences in their bandwidth requirements are treated in [6]; integration of voice and data is dealt with in [7].) And, finally, ISDN exchanges (like the one referenced above) can be fully capable packet switching points (as signaling is a packet transfer operation), allowing the integration of circuit and packet switching into the same node [3].

The concept and realization of ISDN has produced a potential for a remarkable reduction in the cost of supplying data services and a substantial improvement in operational convenience, since a very wide variety of data and voice services are available thru a single access point. Provisioning (dimensioning and inventory) of the switching points has also been markedly simplified since a single exchange type carries all services (circuit, asynchronous, packet) and the user (or line) differentiation becomes a matter of installing the relevant line module: voice only, ISDN with n = 2, ISDN with packet capabilities, ISDN with n = 30; and, of course, these assignments are readily changed by substituting line modules as user demand changes. The transition scenario has thus become: analog to generic digital transport, rather than using an intermediate step (analog to IDN to ISDN).

The ISDN accomplishment in terms of economy and convenience is indeed impressive. But, first, looking ahead, it's clear that for many years there will be no further major cost reduction for data and voice services. Second, ISDN as described thus far, is a transport mechanism only: deliver bits to a distant point, unaltered and in proper sequence -- the lower layers of the ISO communications model. In effect, the independence of the communications and computing sectors remains, with the former providing the latter with a geographically pervasive and very economical transport network. It remains for the two technologies to combine, extending the capabilities of the ISDN. (Before that expansion, however, it should be noted that adequate techniques are still not available for the analysis, design and operation of the "classical ISDN" of 2B + D transport offering, as will be discussed later.)

ISDN and value-added services

The usual reference to an ISDN implies an integration of services in the sense of the transport of digitalized voice and data, the former with telephonic signaling and the latter either gaining access via telephonic signaling and then employing data signaling

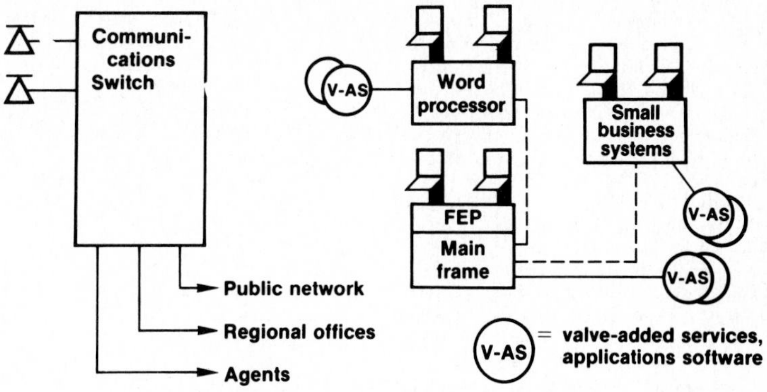

Fig 1 : Typical office configuration, at present

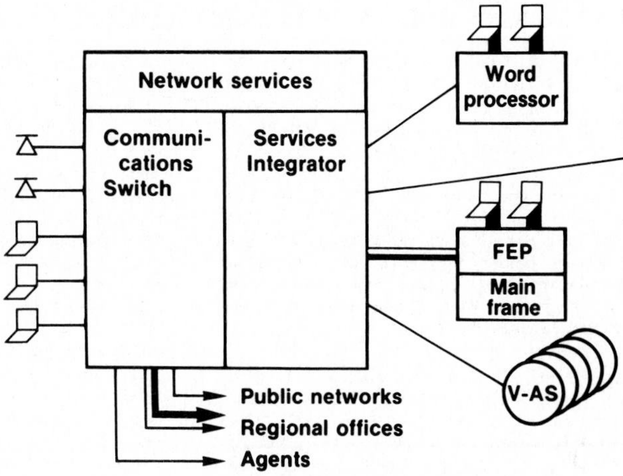

Fig 2 : Integrated office configuration

or via data signaling only. Circuit and message, synchronous and asynchronous, packet -- all traffic types are to be accommodated. Packet switching is considered by some as a value-added service, but by others (including this author) as a transport mechanism, which can be combined within a communications node with circuit traffic.

The potential for combining telecommunications and computer technologies becomes apparent when the delivered signal is not just a replica of the input signal, but has been processed in a useful manner. It can be debated on operational, economical, technical and political (regulatory) grounds whether such processing should be included within the public and private network itself, treated and offered as a network service, or whether these functions should be external to the network, provided at its periphery. Certainly both forms of implementation will exist, and techniques for analysis and design are needed for both. The case of networks themselves providing value-added services represents the new challenge for the communications carrier and the network designer [8 and 9].

To prove the usefulness of this ISDN enhancement, it's convenient to present an example that illustrates the potential for integrating value-added functions in the network itself. For this example a private or corporate network is used as it avoids, for the moment, certain regulatory questions, but the example will then be extended for application to the public network.

The current situation at most network locations, Figure 1, consists of a PBX (or a cluster of PBXs for large sites with campus-like configurations) or PBXs at remote locations forming a PBX network. Additionally there are several separate networks for computing (large mainframes), for word processing, for specialized processing (like process control, research groups, clusters of PCs with application processors), for information resources (like management information systems), and telex-related services. Several of the separate data networks may be linked via a mainframe proprietary system (like SNA), but generally such linkages are not readily used, for example, by either word-processor operators or executives with PC's on their desks. Indeed offices (and desks) are becoming increasingly littered with a multitude of different terminals, each connected to its own network and each incompatible with the others. This situation results from the lack of a single international or even national standard, plus the multiplicity of manufacturers' standards and the multiple, multi-option standards produced by standards organizations [10]. (Even considering the many signaling systems in use, dial arrangements and different eras of switching and transmission equipment operating around the world, telephony presents a picture of simplicity and discipline compared with the data world today.) Substantial and sincere efforts are being made to resolve these problems in the data-related area, but that it will be some years before the problem is greatly reduced and even then it seems as though there will be two remaining standards, rather than one.

The integration of different data networks, their integration with the digital telephony network and the integration of value-added services need not wait until the single or dual-standard era arrives, since the conversion between systems and networks is one of the important value-added services that networks can provide [11].

Rather than maintaining PBX networks separate from a member of independent or coupled-with-difficulty data networks, future configurations will consist of a general communications hub to which both traditional telephones and terminals will be attached, Figure 2. In effect terminals will migrate from their current assignment on individual networks to the "line" side of the communications switch, which will give them "switched" access to different networks and services. The terminals may be "dumb" or intelligent (PCs) and will eventually have integrated voice communications: a single equipment on the desk offering access to voice communications and a wide variety of data services and value-added services. The communications hub will provide both the access to different networks and systems and the necessary conversions to give the user the same convenience currently available with the voice network. Services provided by the hub can be divided into several groupings (where these groupings are not necessarily separate implementations):

1) Communications switch: physical connection for telephones and terminals (data, text, image, etc.) to the services integrator (see below), to other links in the same network, to other private networks, to the public network, to directly attached processors, etc. These are the functions that are now considered as traditional ISDN-PBX communications functions.

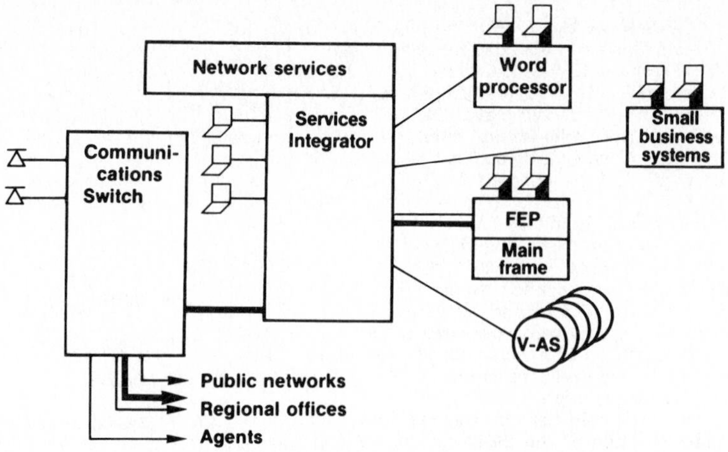

Fig 3 : Integrated office configuration - separated PBX

2) Services integrator: an interface between the communications switch and networks and processors, providing access to different data networks and to large mainframes, specialized processors, data bases, word processing systems, etc. This access facility will be an intelligent one including conversions needed to allow the unsophisticated computer user to utilize applications on these systems, transfer data between them and send messages, documents or images to other users. In effect this is a class of services that the user should not know about (as, at present, the telephone user doesn't select a carrier or signaling system).

3) Value-added services modules: specialized data bases, data-management systems, electronic and voice mail, messaging systems in general (for voice, text and images), applications software, specialized processors, etc. These are services that users will request by name or function, but again with the services integrator providing a standard user interface and the convenience of interchange between services. (The actual path of access will be from the user's terminal, thru the communications switch and services integrator to the value-added services, which are installed on directly coupled or remote processors.) It's important to note that users should access services on data bases by their functions or contexts and without knowledge of location or format, rather then by specifying software or processors or networks or data bases; users should then be able to move that information into any application or document, with the services integrator performing necessary conversions.

4) Management module: network management functions for the network administrator, including maintenance and grade-of-service testing (end-to-end!), traffic statistics, traffic controls, restoration and reconfiguration, accounting (charging to departments), etc. Management now takes on a new significance for data networks. (It's parallel to the situation some years ago when telephone administrations "gave control" of their long-distant networks to subscribers who could dial their own national and international calls.) Since terminals will no longer be affixed to particular processors or networks, but will access different network resources via the communications hub, network administrators will have an increasingly difficult task in controlling and diagnosing their networks and in preparing growth forecasts. These difficulties are compounded by the diverse traffic characteristics of the different message types.

An alternative to the approach for integration of voice and data services, Figure 3, integrates most data services (again providing the functions attributed to a services integrator for users' convenience), but voice services remain separate and associated with a "traditional" PBX (which will nevertheless be digital and offer a wide variety of voice-related services). The reasons for this implementation vary from the justifiable, like technical difficulties with existing PBXs (digital, but not ISDN capable), to financial (remaining depreciable lifetime of current PBXs), to organizational (existing corporate structures placing communications and data processing in different fiefdoms), to reasons which appear to be simply technological dogma.

In both Figures 2 and 3 the heavy lines between units represent the 2 Mb/s capability noted earlier for n x 32 kb/s, with n controllable from 1 thru 30. When this service is offered by a PBX, it permits these high bit rates to be brought to certain terminals (for example, some graphics devices) and also to be switched on a temporary basis between corporate locations for mainframe-to-mainframe interchange and similar applications.

The configuration just presented of an integrated communications and processing system represents a major step in the combining of the communications and computing technologies. As time progresses networks accessed by such a hub will tend to integrate, resulting in fewer networks and more processing capability directly associated with communications functions. Indeed it will become increasingly difficult to identify the boundaries between communications and computing.

For convenience of presentation, the construction thus far has used ISDN-capable PBX's in private networks to demonstrate ISDN's potential beyond transport mechanisms for directly integrating data and value-added services. It also illustrated the new network management and related traffic issues attendant with the combination of diverse traffic types and the dissociation of terminals from specific networks. These concepts need to be extended to the public network, which is readily done.

First, it should be clear that access to data networks and value-added services can be provided via ISDN-capable public exchanges in national networks in the same manner as via ISDN-capable PBXs in corporate networks. In both cases the value-added services are the two types: those addressed by the user requesting specific application, and those of which the user has no knowledge, like protocol or format conversions. The ISDN-equipped subscriber in the public network should have the potential to access services like electronic mail, word processing, data-base management, applications software, etc.; and he would then be charged for his rental or usage of those resources as he is now charged for usage of long-distance circuits. The implementation of these resources could be via integrating them directly on public exchanges (as with PBXs) or accessed by subscribers via public exchanges to services and resources provided at administrations' or private sites. Of course, not all resources need to be implemented at each exchange, with some locally provided and some placed at central locations; this is essentially a dimensioning exercise.

The final step in this evolution will be the implementation of the ISDN-capable PBX as a centrex feature of a public exchange, again with a full spectrum of value-added services. This offering of augmented-ISDN services via centrex would allow even small business locations the complete service potential of large PBXs. Or one could envision corporate networks employing augmented-ISDN PBX's at their large locations, while equivalent ISDN-PBX services are obtained via centrex for their smaller sites. The economy is clear for the corporate network and for the administration providing augmented-ISDN for individual ISDN lines and for centrex arrangements. However there is now a further consideration for the network management and network design tasks, because, in addition to the mixtures of traffic types noted earlier, public and private network traffic and services requests would contend for the same resources.

These design possibilities may be viewed as a hierarchy or nesting of functions associated with LAN's, PBXs and exchanges, Figure 4:

- o LAN functions - these are actually the same functions as are available from an ISDN-capable PBX (with full data features, like packet switching and up to 2 Mb/s data rates to terminals, as was discussed earlier); LAN functions can thus be nested in an ISDN PBX;
- o ISDN-PBX functions - can be nested in an ISDN public exchange via a centrex arrangement.
- o Public exchange functions - can thus include LAN and ISDN-PBX functions as well as ISDN and augmented-ISDN functions to individual subscribers.

The designer of public and private networks will choose from these different implementations to build future networks, their service offerings and their control mechanisms.

The teletraffic challenge

The preceding sections have described a probable course of development for communications, information processing and their integration. While the eventual scenario may be somewhat different, there will surely be an increasing intermingling of quite disparate traffic types, some of which are not even available at present for study and measurement; this diversity will be coupled with the increased (switched) access from terminals to networks, processors, value-added services, data bases, applications software, etc. How will these future systems be designed, measured, dimensioned, controlled, tariffed? Even considering the transport-layer ISDN, that is, without the network-level value-added services integration described above, the present lack of traffic level and traffic characteristic data makes design, forecasting and control problems very difficult indeed. The only approaches, and they are not wholly independent, have two aspects; robust exchange and network designs that are insensitive to traffic characteristics and traffic mixtures, and truly modular design of network communication and processing components allowing rapid and convenient rearrangement and reprovisioning to maintain communications and processing performance standards during periods of growth.

Taking the telecommunications viewpoint now, both administrations and suppliers are approaching the major changes associated with an introduction of basic ISDN features to meet quite legitimate demands of subscribers, initially mostly from the business community. It will be necessary to reconsider the mass of accumulated experience and intuition developed for the traffic engineering and operation of voice telephony: basic assumptions, models, formulae, algorithms, design and analytic tools. While initially exchanges and PBXs will be the primary focus, network aspects will eventually require the most reappraisal and network management attention.

This is not the first such reappraised for the traffic community:

- o introduction of conditional selection replaced step principles, first in exchanges and subsequently in networks;
- o repeated attempts phenomena and the whole field of exchange and network overload diagnosis and control;
- o development of electronic control replacing electromechanical control, and associated emphasis on queueing theory rather than loss theory;
- o common channel signaling techniques;
- o time division switching for voice and data replacing space division that had been used

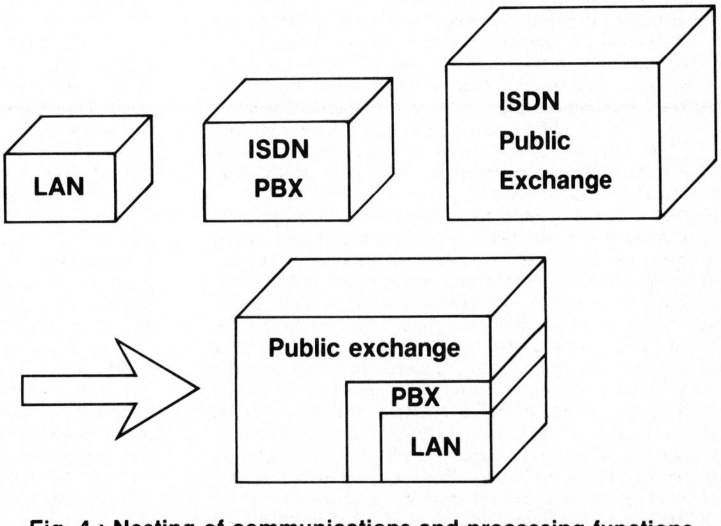

Fig. 4 : Nesting of communications and processing functions

almost exclusively for voice switching;

- and now distributed and progressive control are returning to displace central control (see first item above).

None of these changes individually caused a replacement of the previous wealth of traffic knowledge; each was an occasion to increase the store of traffic knowledge, techniques and experience. But not always was the traffic progress sufficiently timely to adequately support the technology of implementations. Overload control is probably the best (or worst) example of the traffic field's delay in understanding a phenomenon and providing a solution. The advance from progressive to common control was a major step forward in providing subscribers with improved and economical service, particularly in the area of subscriber controlled long-distance dialing. At that time traffic work was directed toward the design of common control exchanges and the economics of metropolitan and national network design; dimensioning techniques for both exchanges and networks were largely sought to reduce overall cost, for example by the introduction of multi-alternate routing to provide heavily loaded initial choice routes and then to take advantage of the efficiencies obtained by combining a number of overflow parcels from these first-choice routes. The resulting network efficiencies were indeed high -- and then aimed even higher via routing techniques that took advantage of traffic parcels' non-coincident busy hours. Of course, each technique that increased efficiency in individual exchanges or networks also reduced the exchanges' and networks' capacities to accommodate overload situations "gracefully" (an interesting euphemism applied at the time network problems were recognized). As a member then of the traffic fraternity which marched bravely into these unexpected problems, this author is not criticizing his fellows working at that time. The point is that a non-forecasted phenomenon developed, it was not understood, it took time to analyze. When the problem was diagnosed there was a delay before control means could be invented and routinely installed; these are complete network management systems that maintain exchange and network efficiencies at very high levels during normal conditions and also at times marked by overloads, skewed traffic distributions and network component failures. It might be asked: should, or even could, the overload problem have been foreseen when common control equipment and then subscriber long-distance dialing were introduced? Probably not. (This author is sure that he will now receive a reference to a paper written contemporaneously with these developments and warning of the impending difficulties!)

We appear to be approaching a similar situation with the introduction of ISDN, first in its transport level implementation and then when it is expanded to integrate information processing. As with the introduction of common control, the present introduction of ISDN raises a disturbing problem: What will be the traffic-related and operational problems that will develop as these services are introduced? Indeed we are not even certain as to the problems themselves. Certainly the need for an enhanced network management system has been identified; also techniques and tools are needed for dimensioning multi-service networks; adequate parameter sets need to be defined for traffic measurements to allow real-time control and long-term forecasting; robust, non-traffic-characteristics-sensitive network structures and design techniques are needed; and this list continues. One worrying aspect is whether some important factor is being overlooked; and another is whether sufficient attention is being devoted to ISDN and augmented-ISDN network management. For the second question, it seems now that current efforts may not be adequate to yield definitive progress in time for the initiation of public and private integrated services networks. However, together with this worry, it's important to add that, as someone who has worked with the development of ISDN from its early stages (for example, as a participant in the CCITT's 1974 Special D meetings (precursor of Study Group XVIII) in the same conference hall as the eleventh ITC in Kyoto in 1985), and who has supported ISDN introduction as a most beneficial direction for subscribers, this author welcomes these changes and has great confidence in this direction. This is accompanied with confidence that the traffic community, now increasingly comprising both communications and information processing specialists will advance our traffic knowledge, techniques, tools and experience to assure subscribers of quality ISDN services.

Directions for ISDN traffic technology

It is a difficult task to forecast the next directions for the traffic field because of the uncertainties noted in the previous sections. (It's also a somewhat dangerous exercise in advance of the publication of the Eleventh Teletraffic Congress's proceedings, and the specialized meeting on ISDN-related traffic issues scheduled for Belgium in 1986.) However due to the very large scope of ISDN (in comparison with the analog voice telephony network) an overall structure is needed to both define all aspects of this work and to identify areas which may not be receiving adequate attention. It's not clear where the responsibility for drawing up and administering such a structure rests: with the ITC (forming an appropriate experts committee) or with bodies like the CCITT or ISO (which already have the organizational means to direct the effort)? This author believes that a committee of ITC experts should at least provide the leadership in carrying this task thru the structure definition stage. A structure might be developed to parallel this hierarchy:

- services provided (basic voice and data delivery, ISO application layers; narrow-, wide- and broad-band; single and multiple resource; assigned and contended resources; ...)
- users (single and multiple terminal; cluster; by application or resource usage; human or device originators; fixed or mobile; ...)
- communications hubs (switching, integration and management function sets: the manufacturers' responsibilities)
- networks (terminal, distribution or local, transit; PBX, LAN, computer; information services; signaling; specialized, integrated; circuit, packet, mixed; cellular, etc.: the carriers' responsibilities)

For each of these categories and the several items within each category there is needed: traffic characterization; traffic measurement sets; forecasting methods; grade-of-service definitions (link-by-link and end-to-end); user, terminal and line traffic profiles, characterizations and models; busy hour concept for mixtures of users, terminals and applications; control theory and practice.

Network aspects will need special attention for a number of reasons:

- initially ISDN will link several existing data networks (either with the digital voice network integration or not) and traffic will now flow between networks, rather than just between terminals;
- as integration of data services proceeds, mixtures

of quite different traffic types will contend for network resources;

- o there will always be a multiplicity of networks since the characteristics of their traffic and services or bandwidth disparities or reliability requirements will make integration unsuitable; however traffic interchanges will be needed;
- o integration of processing within the networks themselves leads to new situations of resource provisioning; also there will be choices of central versus distributed placement of processing resources;
- o network architectures themselves need revaluation (and possibly innovation) to determine optimum topologies, routing, resource placement, measurements and network management functions;
- o theory, tools, simulation and analytic techniques are need for network and resources planning [12];
- o basically new approaches would be appropriate to deal with the integration of information delivery communications functions and information processing functions.

Most of the points above have counterparts in current traffic theory and practice, except most notably the new dimension represented by the last point wherein communication and processing are no longer distinct entities. (Recent similar situations required new traffic developments when the separate space and time switching stages of exchanges' digital matrices were replaced by a single, integrated matrix without identifiable space and time stages [13]; or when overload controls were applied to distributed systems [14].) But even with the similarities (or counterparts above), the problems are considerably more complicated and will require new techniques which are more sophisticated than those used for telephony; for example, multiple parameters are now required where formerly a single one was sufficient in the characterization of ISDN traffic, as is proposed in [15]. The control and guarantee of end-to-end performance needs new network disciplines for data traffic traversing different networks from point of origin to destination with intermediate processing, say for protocol or format conversion, or for combining data from a remote resource with a document message passing between two points other than the data location.

While the sophisticated operations described in this paper and the examples noted in this section for the higher layer ISO functions will need study, techniques and engineering methods, the first step is efforts toward providing adequate traffic expertise for the initial, basic ISDN services. However this work should be done in an open-ended manner to facilitate later advanced studies that will yield results in advance of industry's and administrations' progress to the augmented ISDN defined here.

Recommendations

The structure for ISDN traffic work, as proposed in the previous section, is needed as a guide for immediate and long-range traffic research and traffic application progress. Immediate attention should be directed toward the network aspects associated with the first ISDN stage of integrating digital voice and various data transport mechanisms, which will include access arrangements to still separate networks. Work already in progress on characterization of data traffic (including field measurements) should lead to the definition of ISDN traffic source models that include the simultaneity features of ISDN sources. Also urgently needed is network control and management procedures for mixed services networks and for the period of uncertainty ahead with largely unknown traffic levels, mixtures and characteristics. It's extremely important for the early success of ISDN operation that network management techniques guarantee a high performance level at all times to the new ISDN subscribers; less than excellent initial performance experiences will seriously delay the migration of subscribers and businesses from separate networks to an integrated network. Initial success resulting from excellent performance and attractive tariffs will have the positive effect of accelerating progress toward the introduction and integration of the higher-layer services (with the substantial additional traffic work that these advances imply). These steps require the joint efforts of administrations and manufacturers, and of the communications and information processing sectors. The four groups come together at these Teletraffic Congresses and it is thru the ITC that the technological progress can be directed and advanced, as well as reported.

References:

1] L. A. Gimpelson, "An ISDN approach to integrated corporate networks", "Data communications in the ISDN era", March 1985, Tel Aviv.

2] G. Robin, S. R. Treves, "An introduction to integrated services digital networks", Electrical Communication, Vol. 56, No. 1, 1981. Note: entire issue is on subject of ISDN.

3] A. Chalet, R. Drignath, "Data module architecture including packet operations (for System 12)", Electrical Communication, Vol. 59, No.1/2, 1985. Note: entire issue on System 12.

4] L. Gasser, H. W. Renz, "Transmission at 144 kb/s on digital subscriber loops", Electrical Communication, Vol. 59, No.1/2, 1985.

5] S. R. Treves, D.C. Up, "Technique for wideband ISDN applications (for System 12)", Electrical Communication, Vol. 59, No.1/2, 1985.

6] L. A. Gimpelson, "Analysis of mixtures of wide- and narrow-band traffic", IEEE Transactions on Communications Technology, Vol. 13, No.3, 1965.

7] Session 20 at ICC '78, with papers by W. Hsieh, H. Rudin, H. Miyahare, etc.; and also papers at the 10th ITC.

8] L. A. Gimpelson, S. R. Treves, "Telecommunications networks beyond ISDN transport", Electrical Communication, Vol. 59, No.1/2, 1985.

9] S. S. Katz, "Introduction of new network services", ITC-10.

10] D. Becker, H.-J. Bergs, M. Scham, "Implementation of open communications services in the ISDN", "Data communications in the ISDN era", March 1985, Tel Aviv.

11] L. A. Gimpelson, S. R. Kimbleton, P. S. -C. Wang, "Networking for public and private network applications", ICC '84, Amsterdam.

12] O. Gonzalez Soto, T. Borja; "A step to ISDN planning: from user traffic to service cost comparison", ITC-11.

13] J. R. de los Mozos, R. Buchheister, "Blocking calculations methods for digital switching networks with step-by-step hunting and retrials", ITC-10.

14] G. Morales Andres, M. Villen Altamirano, "Traffic overload control (for System 12)", Electrical Communication, Vol. 59, No.1/2, 1985.

15] O. Gonzalez Soto, "On traffic modeling and characterization of ISDN users", "Data communications in the ISDN era", March 1985, Tel Aviv.

TELETRAFFIC ISSUES in an Advanced Information Society
ITC-11
Minoru Akiyama (Editor)
Elsevier Science Publishers B.V. (North-Holland)

TRAFFIC MODELS FOR TELECOMMUNICATION SERVICES WITH ADVANCE CAPACITY RESERVATION

James ROBERTS and LIAO Keqiang

Centre National d'Etudes des Télécommunications,
CNET/PAA/ATR, 38-40 rue du Général Leclerc,
92131 ISSY LES MOULINEAUX, FRANCE

ABSTRACT
Certain new telecommunications services such as videoconferencing are based on the advance reservation of transmission channels. The corresponding traffic process is quite different to that of traditional telecommunications services and usual teletraffic models seem inadequate for system performance evaluation and network dimensioning. We discuss implied teletraffic issues and attempt to describe the reservation process by mathematical models. In particular we derive formulae useful for dimensioning a videoconference network to a low blocking probability grade of service standard.

1. INTRODUCTION

In analysing the performance of the Telecom 1 satellite integrated services network [1], we encountered a new class of teletraffic problems relating to telecommunications services provided on an advanced reservation basis. For such services, users ensure the availability of the necessary transmission channels for a planned communication in a preliminary dialogue with a system reservation centre. The interval between reservation and start of communication can conceivably range from a matter of minutes to several weeks.

Such reservation services are not of course confined to Telecom 1. AT&T has provided since July 1982 a 3Mbit/s High Speed Switched Digital Service on a reservation basis [2]. In CCITT I-Series Recommendations, circuit switched bearer services at 64, 384, 1536 and 1920 Kbit/s are doted with the reservation establishment attribute. Proposed videoconference networks [3,4] are naturally based on the use of advance reservation.

While the development of reservation based services seems then to be gathering momentum, little is to be found in the literature on the implied teletraffic issues. In [5] the author evaluates one aspect of a particular reservation service implementation; in [4], grade of service standards for reserved videoconference calls are proposed but no indication is given of how the network should be dimensioned to ensure that these are satisfied. In this paper we hope at least to show how the use of reservation introduces important and challenging performance evaluation problems and to incite further studies by the teletraffic community.

We begin with a brief discussion of what is implied by the reservation set up option before presenting mathematical models for analysing the performance of certain implementations. The videoconference service is an important case for which we have obtained some original results; derivations are given in an appendix. Lastly, we draw some conclusions from the studied models and suggest areas calling for further research.

2. THE RESERVATION SERVICE

In this section we discuss the reasons for creating an advanced reservation service, describe the way in which such a service can be implemented and indicate some characteristics of reservation traffic.

2.1 Need for reservation

We identify two main reasons for implementing telecommunications services with advanced reservation:
- users require guaranteed service (i.e. no blocking) for scheduled calls;
- it is impossible to offer certain high bit rate services on a demand basis in the present digital network.

The former would apply, for example, to a video conference service where users plan a meeting in advance and must often book to use conference studios. It would clearly be intolerable that a conference be refused at the appointed time for want of available transmission channels.

The second reason no doubt explains why, in the CCITT I-Series Recommendations, high bit rate bearer services do not receive the demand establishment attribute. When demand is known in advance the necessary trunks and paths through switching networks can be pre-assigned (and busied) in periods of light traffic.

2.2 Implementation of reservation services

To implement a reservation service it is obviously necessary to have a network reservation centre (or centres) with which users communicate in the preliminary reservation phase. This centre records details of the reservation request and is responsible for setting up and clearing down the transmission link at the appropriate moments.

In dedicated networks with reservation recently described in the literature [2,3,4] the reservation centre keeps up to date a record of available network resources over a period (whose length corresponds to the maximum allowed notice) and checks for access conflicts before accepting a new reservation request. In case of congestion the request might simply be refused (loss systems) or a new start time proposed, after (delay systems) or before (negative delay systems?) the original start time. User behaviour will be manifested by "repeat attempts" for different start times in case of blocking.

Note in passing that, since resources are first allocated to calls only in the reservation centre logic, the use of reservation gives total flexibility in the choice of call routing strategy (dynamic routing, rearrangement,...).

In a large integrated network, it may not be feasible to keep a record of all available network resources over a sufficiently long period, especially if reserved calls only account for a small proportion of carried traffic. It is conceivable that the reservation centre records only the individual call details and attempts to establish the desired connection before the appointed start time. Set up is "guaranteed" by starting the attempt a sufficient time in advance and waiting for channels to clear [5] or by making a sufficient number of repeat attempts. We must assume here that reservation traffic is indeed so low compared to available capacity that congestion never occurs in practice.

Existing digital switches were not designed to handle multi-channel calls. To set up such calls, in the early stages of the ISDN it can be necessary to assign trunks and paths through the switching network when the system is empty (the blocking probability would be too high in normal traffic conditions). In practice, reserved links would be set up well before the appointed start time in a period of light traffic and maintained until the end of the call.

2.3 Reservation traffic characteristics

If the purpose of teletraffic studies may be summarised as to find the functional relationship between system capacity, traffic offered and grade of service, to evaluate reservation services we must first know how to quantify this type of traffic and decide what grade of service standards should be applied.

Telecommunications traffic is traditionally described in terms of an arrival process and a holding time distribution. To characterise a reservation traffic process it is also necessary to specify the distribution of the interval between reservation request and desired start time. Indeed, in many practical applications (e.g. videoconference), start time and holding time may be deterministic (users reserve a given "time slot") and the random nature of the process is wholly contained in the notice interval distribution.

Two view points may be considered appropriate for fixing grade of service standards for reservation services. For high speed digital link services, reservation may be seen as a way of avoiding the worst effects of congestion in an under-provided network in the first stages of digitalisation. In this case a high blocking probability for short notice reservations is perhaps acceptable. On the other hand, a purpose built videoconference network should arguably be dimensioned to have a very low blocking probability. Users who have made large investments in expensive studio equipment and codecs will not gladly accept that a conference between two available studios be refused for want of a free transmission link.

We believe that the teletraffic issues raised by the use of reservation are not trivial and, what is more, are not immediately amenable to analysis by usual methods.

3. RESERVATION AS A GENERALISED QUEUEING SYSTEM

Here we consider reservation traffic systems as continuous time processes, like classical queueing systems, with an infinite succession of arrivals and departures. In this sense classical queues are particular reservation systems where customers give no notice. We would like to be able to characterise the behaviour of such systems under appropriate simplifying assumptions by expressing performance as a function of traffic and capacity parameters. In fact it proves extremely difficult to say the least intelligent thing about the simplest non trivial system.

3.1 A process with two parameters

If the occupancy state of a system may be represented by a vector with a sufficient number of components $\underline{K}$, the associated reservation stochastic process consists in indexing the random value of this variable on <u>two</u> parameters. We define $\underline{K}(t,u)$ as the system state at time u corresponding to reservations received before time t (t u). The essential difficulty in analysing such systems stems from the need to account for this double time dependence.

Even in the stationary limit when $t \to \infty$ we still have to deal with a random <u>function</u>

$$\underline{K}(w) = \lim_{t\to\infty} \underline{K}(t,t+w)$$

Further, to determine blocking probabilities it is not enough to know the distribution of $\underline{K}(w)$ for any given w; we must characterise the evolution of $\underline{K}(w)$ over an <u>interval</u>.

To see the difficulty in treating such systems it is instructive to consider a simple example.

3.2 A single server loss system with reservation

Requests for a single server arrive according to some random process. A request at time t demands a service of duration D starting at time T. D and the notice interval $T-t$ are independent random variables with distribution functions:

$$F(d) = Pr(D<d) \quad \text{and} \quad G(x) = Pr(T-t<x).$$

System state takes one of two values:

$$K(t,u) = \begin{cases} 0, & \text{if at } t \text{ the server is not reserved for } u, \\ 1, & \text{otherwise.} \end{cases}$$

If a request cannot be satisfied (i.e. $K(t,u)=1$ for some $u \in (T,T+D)$), then it is cleared from the system.

Ideally, we would like to calculate the probability of loss as a function of D and the interval $T-t$. In fact, for a non-deterministic notice distribution G(x), we have been unable even to determine the probability that the server is busy for whatever (simplifying) choice of request arrival process or call service time distribution.

We might attempt to consider the system as a preemptive priority queue where the relative priority of two customers is determined by their request arrival epochs. This approach fails for at least two reasons:

- with respect to a test customer, arrivals (i.e. start times) of higher priority customers constitute a non-homogeneous process, even if the request arrival process is homogeneous;

- the reservation process in reality allows "retrospective acceptance" of lower priority customers if a pre-empting customer is subsequently pre-empted itself.

3.3 Reservation in a congested system
Consider now a general reservation system where the notice interval has the same constant value S for all customers and where we exclude the possibility of bringing forward the starting time in case of blocking. It is clear that the system will behave exactly like the corresponding real time (loss, delay, repeat attempt, ...) system whatever the value of S since relative customer priorities are now determined, as usual, by their start times.

It also seems intuitively obvious that, again excluding the possibility of advancing blocked customers, system behaviour with a notice distribution G(x) will be the same as that for a distibution G(x-c), for any positive constant c. This observation has the following consequence when we consider the use of reservation in a congested system.

We have implicitly assumed above that the distribution G(x) is independent of system behaviour. In reality we can expect customers to react to a congested system by lengthening their notice interval. Since all customers will manifest the same tendancy, we will see a shift in the distribution G. If this shift were unaccompanied by any distorsion i.e. if G(x) were replaced by G(x-c), the system would fall into exactly the same congested state as before. Customers have only to further increase their notice...

Overall system performance perceived by customers is progressively and inexorably degraded, although underlying traffic volume remains the same. We believe the same general effect will be observed even accounting for distortion to the notice interval distribution.

4. RESERVATION AS A BIRTH PROCESS

In [1] we identified a class of "discrete time" reservation systems where the period requested for a given communication is assimilated to a dimension of the service system resources. We characterised the arrival process of requests for a given period as a state dependent non-homogeneous Poisson process. This model applies to traffic in a videoconference network where conferences are reserved for one or more time slots of say, 1 hour each. We can also apply a birth process model in an approximate analysis of the effects of using a common trunk group for reservation and demand services in an ISDN. In this formulation the notice interval distribution is accounted for by the time varying arrival rate of requests for a considered time slot.

4.1 Service integration
Circuit switched bearer services in the ISDN may be provided on demand (i.e. with lost call operation) or with the reservation establishment attribute (CCITT Rec I 211). The latter is presently the only set up mode recommended for multi-channel calls (6, 24 or 30 x 64 Kbit/s unrestricted service). In this section we evaluate some of the effects of grouping the two types of traffic on a common trunk group.

We assume that all reservation traffic is for communications with the same bit rate of d x 64 Kbit/s and that reservations are made exclusively for periods encompassing the demand busy hour. By this assumption the reserved channels constitute a constant background load from the point of view of the on demand calls.

We consider a trunk group of M 30-channel PCM systems. The arrival process of requests for d channels, concerning a given busy hour, is assumed to be Poisson with time dependent rate $\lambda(t)$. Demand calls are for single channels and arrive in a homogeneous Poisson process.

The expected number of reservation requests is:

$$(4.1) \qquad R = \int_{-\infty}^{T} \lambda(t)\, dt$$

where T is the busy hour starting time. The distribution of the number of requests N_r is:

$$(4.2) \qquad P_n = Pr\{N_r = n\} = \frac{R^n}{n!} e^{-R}$$

We assume R is small compared to available capacity so that reservation blocking probability is negligible. The number of channels remaining for the demand service is $30M-N_r d$. We assume statistical equilibrium conditions in the busy hour so that the expected proportion of blocked calls when $N_r=n$ is given by the Erlang loss formula:

$$(4.3) \qquad b(n) = \frac{A^{(30N-nd)}}{(30N-nd)!} \Big/ \Big\{ \sum_{i=0}^{30N-nd} \frac{A^i}{i!} \Big\}$$

where A is the demand traffic offered. Expected blocking probability is then

$$(4.4) \qquad B = \sum_n P_n\, b(n)$$

Now, if we know R and A it is possible to dimension the trunk group so that B is less than some grade of service standard. Suppose that a 6-PCM system trunk group is offered 30-channel reservation traffic and single channel demand traffic. We assume a loss probability of 0.01 for demand calls.

In figure 1 we show the traffic capacity (expressed as mean channel occupancy: (A+30R)/30M) as a function of the percentage of reservation traffic.

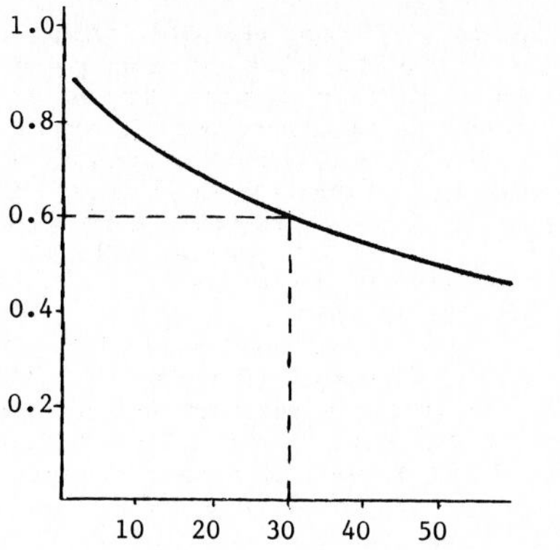

Figure 1. Traffic capacity against percentage reservation traffic.

Capacity decreases from 90% to 60% as the reservation percentage increases from 0% to 30%. The 30% figure corresponds to a mean reservation traffic of R = 1.1 (i.e. just over one reserved 30-channel call per busy hour).

Demand grade of service is not the same qualitatively as in a dedicated capacity system since the expected blocking probability varies randomly from day to day. In the table below we give the P_n and B(n) values corresponding to the case of 30% reservation traffic.

n	0	1	2	3	4	5	6
P_n	.332	.366	.201	.074	.020	.004	.001
b(n)	10^{-22}	10^{-12}	10^{-6}	.016	.254	.618	1.00

We see that in 90% of busy hours expected blocking is less than 10^{-6}. However, in about 3% of busy hours, blocking will exceed 0.25 and, once or twice every year, demand service is effectively interrupted when all, or all but 1, PCM systems are pre-reserved.

The assumptions regarding the traffic process are of course open to question. In particular, the stability of reservation traffic is by no means established and day to day variations may well be greater than those known for telephone traffic. For a 100% overload expected blocking increases from 1% to 8.5%. If a 100% overload were sustained then realised blocking would exceed 60% once every 13 busy hours.

This simple analysis at least points to certain dangers inherent in the unrestricted mixing of reservation and demand traffics, especially when the former concerns multi-channel calls.

4.2 Traffic in a videoconference network

We consider here a dedicated network for setting up transmission links (probably at 2 Mbit/s) between essentially private videoconference facilities. We assume links are reserved in advance for an integral number of time slots of say, one hour and consider the traffic offered for the busiest hour of the day (or week).

The total population of videoconference studios is N and these studios are distributed among M "zones" with N_i studios located in zone i, $1 \leq i \leq M$. Our aim is to characterise the traffic between all the different zones with the ultimate aim of deriving appropriate network design guidelines and dimensioning procedures. The model presented below takes account of the finite population of studios (traffic sources) and their geographical distribution. We assume all conferences involve only two participating studios. Outline derivations of the main results are given in the appendix.

Let $K_{ij}(t)$ $(= K_{ji}(t))$ be the random number of conferences reserved at time t between zones i and j (or within zone i if i=j) and let

(4.5) $\sigma_i = 2\,K_{ii}(t) + \sum_{i \neq j} K_{ij}(t)$

be the number of busy studios in zone i. We assume that reservation requests for conferences between zones i and j arrive in a Poisson process with instantaneous rate:

(4.6) $$\lambda_{ij}(t) = \begin{cases} \alpha(t)\,(N_i-\sigma_i)\,(N_j-\sigma_j), & i=j \\ \alpha(t)\,\binom{N_i-\sigma_i}{2}, & i=j \end{cases}$$

where $\alpha(t)$ is an arrival rate per free pair of studios.

This choice of "symmetric" traffic (i.e. no dependence on studio locations) is a compromise between an accurate representation of the (unknown) request arrival process and mathematical tractability.

Let X(t) be the mean number of conferences reserved for the busy hour in question at prior time t: $X(t) = \Sigma\sigma_i/2$. Let $\rho(t)$ be the mean studio "occupancy":

(4.7) $\rho(t) = 2\,E\{X(t)\}\,/\,N$

Since, in practice, we will more easily estimate studio utilisation than the underlying arrival process $\alpha(t)$, we use $\rho(t)$ as an independent variable.

In the appendix we show that the variance of the overall traffic X(t) is very closely approximated by:

(4.8) $$v(t) = \frac{N}{6}(1-\rho)(3\rho - 3\rho^2 + \rho^3)$$

By virtue of the symmetric traffic assumption the moments of the individual inter- and intra-zone demands can be expressed simply in terms of the moments of X(t).

Using the standard notation:

$(m)_r = m\,(m-1)\ldots(m-r+1)$,

the factorial moments of the traffic K_{ij} may be expressed:

(4.9) $$E\{(K_{ij})_r\} = \frac{2^r\,(N_i)_r\,(N_j)_r}{(N)_{2r}}\,E\{(X)_r\}$$

for $i \neq j$ with

(4.10) $$E\{(K_{ii})_r\} = \frac{(N_i)_{2r}}{(N)_{2r}}\,E\{(X)_r\}.$$

All the demands K_{ij} are negatively correlated and to calculate the variance of combinations (e.g. $K_{ij}+K_{rs}$) we also need the following relations:

$$(4.11)\quad E\{K_{ij}K_{rs}\} = \frac{4\, N_i N_j N_r N_s}{(N)_4} E\{(X)_2\}$$

and

$$(4.12)\quad E\{K_{ij}K_{is}\} = \frac{4\,(N_i)_r\, N_j N_s}{(N)_4} E\{(X)_2\}.$$

Approximation (4.8) and expressions (4.9) to (4.12) thus allow us to calculate the mean and variance of combinations of offered traffic streams as functions of expected studio utilisation.

To approximate the demand distribution we have successfully fitted a discretised normal distribution. This approximation allows us, in particular, to determine the capacity required to carry all the offered traffic in say, 99% of cases. Direct use of the model to dimension a network to more severe grade of service standards (e.g. as in [4]) is probably not possible in view of the disruptive effects of congestion on the request arrival process.

5. CONCLUSIONS

The possibility of reserving transmission channels in advance for a planned communication is a new facility offered by existing or planned specialised networks (notably for videoconferencing) and included in the CCITT specifications for the ISDN. Traditional methods for studying traffic in telecommunication networks are not sufficient when reservation is used, not least because of the need to take into account the distribution of the notice interval between reservation request time and desired start time. It is however becoming increasingly urgent to evaluate the performance of networks with reservation traffic to ensure that the right implementation decisions are made and to provide the basis of network dimensioning procedures.

In the present paper we have considered two approaches to modelling the reservation traffic process: as a generalised queueing system and as a birth process. The first approach proves extremely difficult and we have been unable to obtain significant analytical results for even the simplest system. We are led however, by qualitative arguments, to question the wisdom of operating a reservation system with a high level of congestion: users will tend collectively to increase indefinitely the notice necessary for a successful request (by each individually lengthening his own notice) with no change to the congested state of the network (cf. section 3.3).

The second approach, by means of a birth process, is more promising. We have considered the integration of reservation and demand traffic in an ISDN and shown, under certain assumptions, that large fluctuations in demand traffic grade of service can occur due to the pre-allocation of channels to reservation requests (cf. section 4.1).

Finally, we have modelled the traffic in a videoconference network as a multidimensional birth process. We have derived simple closed expressions allowing the calculation of the first two moments of offered traffic streams as functions of the number and distribution of studios and their expected utilisation rate. Fitting a normal distribution with the same mean and variance allows us to estimate the number of channels required to carry the offered traffic with a low blocking probability (cf. section 4.2).

Further work is necessary before we can answer all the important questions raised by the introduction of the reservation facility. The choice of the reservation attribute for the establishment of high bit rate calls in the ISDN is largely due to technical limitations in existing switching systems; is the cost of implementation, including the cost of necessary trunk group extensions, in fact, less than the cost of providing a demand service (with necessary modifications to exchange logic and equipment)? In the case of a videoconference network, what grade of service standards are appropriate? how can we dimension for a high blocking probability? what is the optimal network architecture taking into account the flexible routing strategies (e.g. use of rearrangement) made possible by the use of reservation?

We hope this paper will incite others to examine these questions and to reflect on the teletraffic models which will enable their resolution.

REFERENCES

[1] ROBERTS J. "Teletraffic models for the Telecom 1 integrated services network". ITC10, Montreal, June 1983.

[2] LONDON H.S, GUIFFRIDA T.S. "High Speed Switched Digital Service". IEEE Comm Mag. Vol 21, Part 2, March 1983.

[3] MORRISON D.G. "A switching system for the UK videoconferencing network". ISS'84, Florence 1984.

[4] TANAKA T. "Video conference system for private use". Japan Telecom Review. July 1983.

[5] LOTITO N. "Performance analysis of a PCM switching system under combined voice/video loads". CSELT Rapporti Tecnici. Vol XII, N°1, February 1984.

APPENDIX: OUTLINE DERIVATION OF VIDEOCONFERENCE TRAFFIC FORMULAE.

(Notation is as defined in section 4.2)

A.1 Symmetric traffic

The choice of symmetric request arrivals given by the intensities (4.6) leads to the following two simplifications:

- the total number of reserved conferences X(t) is a one dimensional birth process with arrival rate, when X(t)=x, of

$$\text{(A.1)}\qquad \lambda_x(t) = \binom{N-2x}{2}\,\alpha(t)$$

- all microscopic states (i.e. where we identify individual studios) with a given total number of reserved conferences are equiprobable.

That the overall arrival rate is independent of the individual $K_{ij}(t)$ and is given by (A.1) may be verified on summing the intensities (4.6) for i=1...M and j=i...M and setting $\Sigma\sigma_i=2x$.

The second simplification allows the distributions of the K_{ij} to be studied by combinatorial methods (cf. section A.5).

A.2 Distribution of the number of reservations

Let $P_x(t)$ be the distribution of X(t). $P_x(t)$ satisfies the birth equations:

$$\text{(A.2)}\qquad \frac{dP_x}{dt} = \lambda_{x-1}(t)P_{x-1} - \lambda_x(t)P_x$$

The substitution $u=\int\alpha(t)dt$ yealds the homogeneous equations:

$$\text{(A.3)}\qquad \frac{dP_x}{du} = \binom{N-2x-2}{2}P_{x-1} - \binom{N-2x}{2}P_x$$

with initial condition:

$$P_x(0) = \begin{cases} 1, & x=0 \\ 0, & x\geq 1. \end{cases}$$

The direct solution of (A.3) (cf. Syski, p.153) leads to a formula for $P_x(u)$ which is unstable numerically for a value of N greater than 40. An alternative formulation leads to the following calculation method which is stable for all N.

Let $R_m(u)$ be the distribution of the number of events in a Poisson process of rate $\binom{N}{2}$:

$$\text{(A.4)}\qquad R_m(u) = \left[\binom{N}{2}u\right]^m/m!\ \exp\left\{-\binom{N}{2}u\right\}$$

and let $p_x(m)$ be the probabilities defined by the recurrence relations:

$$p_0(0) = 1;\qquad p_1(0) = 0;$$

$$\text{(A.5)}\qquad p_x(m) = \frac{\binom{N-2x-2}{2}}{\binom{N}{2}}\,p_{x-1}(m-1)+\left[1-\frac{\binom{N-2x}{2}}{\binom{N}{2}}\right]p_x(m-1)$$

for x≥1 and m≥1.

Then $P_x(u)$ is given by

$$\text{(A.6)}\qquad P_x(u) = \sum_m p_x(m)\,R_m(u).$$

It is quite easy to verify by differentiating the r.h.s. of (A.6) that $P_x(u)$ so defined is a solution of equations (A.3).

A.3 Variance of the number of conferences

Let m(u), v(u), r(u) be the first three central moments of X(u):

$$\text{(A.7)}\qquad \begin{aligned} m(u) &= E\{X(u)\} \\ v(u) &= E\{[X(u)-m(u)]^2\} \\ r(u) &= E\{[X(u)-m(u)]^3\} \end{aligned}$$

From (A.3) we derive the following two equations:

$$\text{(A.8)}\qquad \frac{dm}{du} = \binom{N-2m}{2} + 2v$$

$$\text{(A.9)}\qquad \frac{dv}{du} = \binom{N-2m}{2} - 4v(N-2m-1) + 4r$$

Dividing (A.9) by (A.8) we have:

$$\text{(A.10)}\qquad \frac{dv}{dm} = \frac{1 - \dfrac{8v}{(N-2m)} + \dfrac{8r}{(N-2m)(N-2m-1)}}{1 + \dfrac{4v}{(N-2m)(N-2m-1)}}\ .$$

For large N the last terms in numerator and denominator of the r.h.s. of (A.10) are both very much smaller than the other terms. This suggests the approximation $v = v_0$ where v_0 satisfies:

$$\text{(A.11)}\qquad \frac{dv_0}{dm} = 1 - \frac{8v_0}{(N-2m)}\ .$$

The solution satisfying initial condition $v_0(0)=0$ is:

$$v_0 = \frac{N}{6}\left(1-\frac{2m}{N}\right)\left(1-\left(1-\frac{2m}{N}\right)^3\right).$$

In terms of the mean studio utilisation we have then, the approximation:

$$\text{(A.12)}\qquad v \simeq \frac{N}{6}(1-\rho)(3\rho - 3\rho^2 + \rho^3).$$

This approximation is in fact extremely good. The table below gives a sample of results where exact values are determined directly from the distribution $P_x(u)$ given by (A.6).

N = 100			N = 300		
ρ	v		ρ	v	
	exact	approx		exact	approx
.24	7.10	7.09	.23	20.94	20.93
.50	7.32	7.31	.49	22.21	22.21
.76	3.97	3.98	.75	12.33	12.33
.91	1.54	1.54	.90	4.79	4.79

A.4 Distribution of traffic streams

Let $Q_x(\underline{k})$ be the probability that the network is in state $\underline{k}$ just after the xth reservation request. The $Q_x(\underline{k})$ are thus state probabilities of an imbedded Markov chain and satisfy the transition equations:

$$\text{(A.13)}\qquad Q_x(\underline{k}) = \sum_{i\leq j} \frac{\lambda_{ij}[\underline{k}-\underline{e}_{ij}]}{\Sigma\,\lambda_{rs}[\underline{k}-\underline{e}_{rs}]}\,Q_{x-1}(\underline{k}-\underline{e}_{ij})$$

where $\underline{k}-\underline{e}_{ij}$ is the vector $\{k_{11}\ldots k_{ij}-1\ldots k_{MM}\}$ and the λ_{ij} are given by (4.6).

These equations can be used recursively (starting with $Q_0(0)=1$) to evaluate all the probabilities $Q_x(\underline{k})$. Since we know the distribution of X (section A.2), we can thus calculate exactly the joint distribution of the K_{ij}, at least for small networks. For large networks the dimensions of $Q_x(\underline{k})$ are too great for practical computation. In the next paragraph we derive explicit expressions for the moments of the K_{ij}.

A.5 Moments of the K_{ij}.

The $Q_x(\underline{k})$ may be interpreted as conditional probabilities:

$$Q_x(\underline{k}) = \Pr\{\underline{K}=\underline{k} \ / \ X=x\}.$$

By the second remark in section A.1, these probabilities may be expressed directly as:

$$\text{(A.14)} \quad Q_x(\underline{k}) = \frac{\text{number of distinguishable ways of choosing x studio pairs so that the occupancy state is } \underline{k}.}{\text{number of distinguishable ways of choosing x studio pairs.}}$$

It may be verified by substitution that this definition is consistent with equations (A.13).

The number in the denominator of (A.14) is:

$$\binom{N}{2x} \frac{(2x)!}{x! \ 2^x} .$$

There are $\binom{N}{2x}$ ways of choosing the 2x studios. For a given permutation of these studios $\{s_1,s_2 \ldots s_{2x}\}$ we can form x pairs by linking adjacent studios together: $\{(s_1,s_2)\ldots(s_{2x-1},s_{2x})\}$. Each of the (2x)! permutations does not give a distinguishable set of pairs however since for x! ways of arranging the pairs and 2 ways of arranging the members of each pair we have the same occupancy state. The total number of distinguishable ways of linking the 2x studios is therefore $(2x)!/x! \ 2^x$.

By a similar argument we derive the following expression for the numerator:

$$\prod_{i=1}^{M} \left\{ \binom{N_i}{2k_{ii} \ k_{i1} \ldots k_{iM}} \frac{2k_{ii}!}{k_{ii}! \ 2^{k_{ii}}} \prod_{i<j} k_{ij}! \right\}$$

where

$$\binom{N}{m_1 m_2 \ldots m_r} = \frac{N!}{m_1! \ m_2! \ldots m_r! \ (N-\Sigma \ m_i)!}$$

Substituting in (A.14) we find:

$$\text{(A.15)} \quad Q_x(\underline{k}) = \frac{N_1! \ldots N_M! x! (N-2x)!}{N!} \ \frac{2^{\left(\sum_{i<j} k_{ij}\right)}}{\prod_{i<j} k_{ij}! \ \prod_i (N_i-\sigma_i)!}$$

Since, for given x, these probabilities must add up to 1, we deduce the identity:

$$\text{(A.16)} \quad \sum_{\underline{k}\epsilon\Phi_x\{N_i\}} \frac{2^{\left(\sum_{i<j} k_{ij}\right)}}{\prod_{i<j} k_{ij}! \ \prod_i (N_i-\sigma_i)!} = \frac{N!}{\prod_i N_i! \ x! \ (N-2x)!}$$

where the summation domain $\Phi_x\{N_i\}$ is such that

$$\underline{k} \ \epsilon \ \Phi_x\{N_i\} \iff \sum_i \sigma_i = 2x \text{ and } 0 \leqslant \sigma_i \leqslant N_i \text{ for } i \leqslant M.$$

The rth factorial moment of the traffic K_{nm}, given a total of x conferences, is:

$$E\{(K_{nm})_r/X=x\} = \sum_{\underline{k}\epsilon\Phi_x\{N_i\}} k_{nm}(k_{nm}-1)\ldots(k_{nm}-r+1)Q_x(\underline{k})$$

Insertion of (A.15) gives:

$$\text{(A.17)} \quad E\{(K_{nm})_r \ /X=x\} = \frac{x!(N-2x)! \ \Pi \ N_i!}{N!} \sum_{\underline{k}} \frac{2^{\left(\sum_{i<j} k_{ij}\right)}}{\prod_{\substack{i<j \\ (i,j)=(n,m)}} k_{ij}!(k_{nm}-r)! \ \prod_i (N_i-\sigma_i)!}$$

Writing:

$$\begin{aligned} k_{nm}' &= k_{nm}-r, \\ k_{ij}' &= k_{ij}, \quad \text{for } (i,j)\neq(n,m), \\ N_n' &= N_n-r, \\ N_m' &= N_m-r, \\ N_i' &= N_i, \quad \text{for } i\neq n \text{ and } i\neq m \\ \sigma_i' &= 2k_{ii}' + \sum_{i=j} k_{ij}', \end{aligned}$$

the sum in (A.17) may be expressed:

$$2^r \sum_{\underline{k}\epsilon\Phi_{x-r}\{N_i'\}} \frac{2^{\left(\sum_{i<j} k_{ij}'\right)}}{\prod_{i<j} k_{ij}'! \ \prod_i (N_i'-\sigma_i')!}$$

Now, by identity (A.16), this expression is equal to:

$$2^r \frac{(N-2r)!}{\Pi \ N_i'! \ (x-r)! \ (N-2x)!}$$

We deduce:

$$E\{(K_{nm})_r/X=x\} = \frac{2^r \ (N_n)_r \ (N_m)_r}{(N)_{2r}} (x)_r$$

Finally then, we see then that the corresponding unconditional moment:

$$E\{(K_{nm})_r\} = \sum_x E\{(K_{nm})_r/X=x\} \ P_x$$

is given by expression (4.9). The other moment expressions (4.10) to (4.12) may be derived in a similar way.

TELETRAFFIC ISSUES in an Advanced Information Society
ITC-11
Minoru Akiyama (Editor)
Elsevier Science Publishers B.V. (North-Holland)

TRAFFIC MODELS FOR LARGE ISDN-PABX'S

Helga HOFSTETTER Dietmar WEBER

SIEMENS AG

Federal Republic of Germany

ABSTRACT

A traffic model is a collection of data characterizing the traffic to be handled by an exchange. This paper deals with a traffic model for large-scale ISDN-PABX's. It concerns the telephone, teletex, facsimile, interactive videotex and data dialog services. Based on measurement results or - if none were available - on general reflections and assumptions, these ISDN services are described by their call mixes and by the mean holding times of the terminals involved. The volume of ISDN traffic anticipated for the future in the Federal Republic of Germany is taken as an example so as to extend the traffic model shown to an overall traffic model for complete ISDN-PABX's. In this way the model helps to formulate traffic design objectives for new ISDN-PABX's in Germany.

1 INTRODUCTION

Modern telephone PABX's work on a digital basis, i.e. the voice is transmitted through the exchange via PCM-coded signals (one speech highway with 64 kbps). Moreover, if such a PABX is designed as recommended by CCITT [1], each station terminal can be given so-called **basic access** with a capacity of 144 kbps. This basic access contains two user-channels (B-channels) with 64 kbps each and one 16 kbps-channel (D-channel) mainly used for signalling.

This design allows far more services - in addition to voice - to be included into the area of application of one and the same PABX. Such PABX's are called service-integrating PABX's or ISDN-PABX's [2,3].

The integrated services of prime importance are [4,5]:

- voice,
- teletex,
- facsimile,
- interactive videotex and
- data dialog.

In contrast to public exchanges, PABX's are particulary well suited to the introduction of ISDN services because of the following facts:

- the required terminals are already present today to a great extent (of course not yet integrated!)
- access to different public and private networks already exists
- office communication offers a great variety of applications.

For the moment, large ISDN-PABX's will be used as outlined in the configuration shown in figure 1:

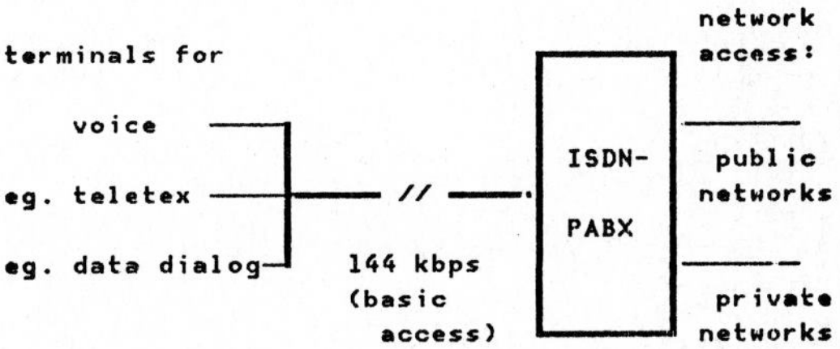

Figure 1: ISDN integration in the PABX itself

Here service-specific terminals are used, many already implemented today. The service integration takes place in the PABX itself. In comparison with present implementation (separate exchanges for the different services such as voice, teletex, etc.), this integration leads to a better usage of the exchange and its resources as well as to improved operating (just <u>one</u> exchange!). But, apart from a few new features, there is no noticeable change for the users. For them a change will only be apparent with the integration of the services in one so-called multifunctional terminal (figure 2).

With the introduction of multifunctional terminals, it is possible to use several different services from one and the same terminal device; within the framework of the ISDN basic access even simultaneously.

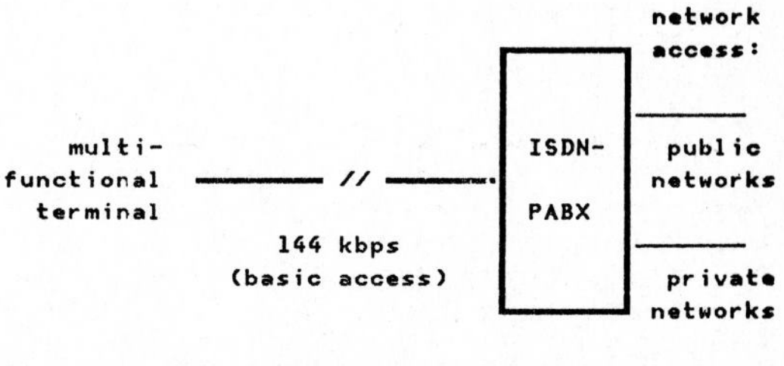

Figure 2: ISDN integration in the terminal

As a basis for the development of optimum-cost ISDN-PABX's, it is necessary to know the size and the properties of the different traffic types that are switched by the ISDN-PABX. These values determine among other things

- the required capability of the system control (size of the control areas for the different microcomputers involved; structure of the system control)
- the number of resources necessary to handle the traffic (number of trunk modules or code receivers and so on).

Size and structure of the traffic to be handled by an ISDN-PABX are determined by the behaviour of the station users connected to the exchange. For the case characterized in figure 1 (ISDN-integration in the PABX itself), user behaviour will be essentially the same as in today's service-specific exchanges. However, for the case characterized in figure 2, changes in user behaviour have to be expected, because then all users will be equipped with completely new and more powerful terminals.

The basic data on the traffic behaviour of users are summarized in a so-called traffic model (cf.[6]). In this paper a traffic model for large ISDN-PABX's is presented. It was derived to support the development of new communication systems in our Company. It refers firstly to the configuration in figure 1, where the ISDN integration is implemented in the exchange itself, consisting of separate models for each service. When, in future, experiences have been gained from the field trials with multifunctional terminals, the traffic model will have to be extended to configurations in accordance with figure 2.

The following traffic model refers to services handled by the B-channels of a basic access and therefore also implies assumptions about the traffic on the D-channel. Services that are handled exclusively by the D-channel (such as telemetry) are not considered in this paper.

2 PARAMETERS OF TRAFFIC MODELS FOR ISDN-PABX'S

A traffic model describes the characteristic properties of the call attempts to be handled by an ISDN-PABX. A call attempt refers always to a terminal of the PABX requesting a service A trunk leading a request from outside to the PABX will also be considered in this context as a terminal. Examples of call attempts are:

- request for connection with a certain station,
- request for connection with a certain device (such as videotex data base, main frame, outgoing trunk),
- request for automatic callback.

Due to the new features that are possible in modern PABX's, call attempts become much more complicated than in conventional exchanges. More and more compound call attempts, i.e. call attempts consisting of several call attempts in the conventional sense, occur. The call forwarding feature may serve as an example of this.

In view of the application of the traffic model for dimensioning control capacity, it is useful to break down compound call attempts into their components, called **basic call attempts**. These have to be considered in the model as independent call attempts. From now on, all call attempts are to be understood in this sense.

The most important parameters of a traffic model describing the basic call attempts for a service in an ISDN-PABX are:

- General parameters independent to a large extent of the particular application of the PABX:
  - The call mix, i.e. the classification of all call attempts of a terminal into possible call types, such as successful calls and the different unsuccessful calls.
  - The mean holding time of the call attempts of a terminal in a service. The holding time constitutes the whole time for which the considered terminal is not accessible for other calls. For example for outgoing telephone calls, this holding time corresponds to the time from the "off hook"-event to the "on hook"-event. For incoming calls it corresponds to the time from starting the ringing tone to the "off hook"-event, or - if the called party gives no answer, to the moment the calling party hangs up.

  Of course, the call mix depends on the actual traffic load of the exchange (cf.[7]). We shall consider in the following sections only the normal load case.

- Parameters depending on the particular application of the PABX:
  - The time of the service-dependent busy hours.
  - The mean number of busy hour call attempts per terminal in a specific service, or the mean traffic load per terminal.
  - The splitting of the holding times of a terminal into the different traffic types in any service (eg. internal traffic, incoming/outgoing traffic).

Section 3 (following) considers the above-mentioned general parameters. A traffic model results characterizing the different ISDN-services. Section 4 shows how to dimension system components of an ISDN-PABX with this fundamental data. In section 5 the parameters depending on the particular application of the PABX are considered. From this, traffic design objectives for large ISDN-PABX's can be developed.

3 TRAFFIC MODEL FOR THE INDIVIDUAL SERVICES OF AN ISDN-PABX

3.1 Telephone

Table 1 shows the general parameter values of our telephone traffic model. Three main traffic types are considered: Internal traffic, outgoing external traffic and incoming external traffic. For each of them the call mix of the busy hour call attempts and the mean holding times are given in table 1.

The traffic parameters of table 1 are based on three groups of sources:

- Measurements of the telephone traffic in telephone PABX's [8,9,10,11]
- Measurements of the telephone traffic in public exchanges [7]
- Modification of the results of these measurements with regard to modified user behaviour in PABX's offering new features.

Traffic type / Call type	Call mix		Mean holding time sec	
	no.1	no.2	A-prty	B-prty
internal traffic				
successful call	53 %	18 %	10+T	6+T
B-party no answer	10 %	8 %	24	20
enter voice mail	-	2 %	30	-
retrieve voice mail	-	2 %	-	33
B-party busy	30 %	23 %	8	-
activate callback	-	21 %	2	-
callback convers.	-	21 %	12+T	6+T
incomplete dialling	7 %	5 %	3	-
outg.external traffic				
successful call	65 %	52 %	20+T	-
B-party no answer	6 %	5 %	34	-
B-party busy	17 %	16 %	18	-
blocked call	5 %	5 %	12	-
activate SNR	-	8 %	2	-
SNR conversation	-	8 %	20+T	-
incomplete dialling	7 %	6 %	4	-
inc.external traffic				
successful call	58 %	58 %	-	6+T
B-party no answer	10 %	10 %	-	20
B-party busy	30 %	30 %	-	-
incomplete dialling	2 %	2 %	-	-

T: Mean conversation time
Call mixes:
no.1 - PABX without modern features
no.2 - PABX with modern features (example)

Table 1: Traffic model for telephone

Following the measurement results of [8,11] we assume the following mean conversation times T:

Internal traffic: 80 sec,
External traffic: 160 sec.

The measured values here have been reduced by about 20 % in order to take into account that the new features of modern PABX's lead to shorter conversation times. By way of example we assume that about 10 % of all terminals in a modern PABX are executive/secretary terminals.

The values of the third column in table 1 refer to a model PABX. Its call mix is based on the following assumptions:

- In 30 % of the "B-party no answer" cases on internal call attempts the voice mail feature is used.
- In 90 % of the "B-party busy" cases on internal call attempts the automatic callback feature is used.
- In 30 % of all unsuccessful completed call attempts in outgoing traffic the saved number redial service (SNR) is used.

3.2 Teletex

The teletex service has the same traffic types as the telephone service, namely internal traffic, outgoing external traffic and incoming external traffic.

Modern teletex terminals contain a memory where incoming messages can be stored (cf.[12]). Therefore "B-party no answer" cannot occur. "B-party busy" can only occur if a teletex terminal is just sending or receiving a message when a call arrives. The holding time of a teletex terminal comprises only that time during which an originating connection is set up and information is sent or received. The time when the terminal is used in the so-called local mode is not included in the holding time because then the terminal is able to accept calls. As the holding times of the teletex service are rather short, the "B-party busy" cases are neglected in the call mix.

In order to determine the teletex traffic model parameters, measurements were carried out in the teletex PABX of our Company [13]. About 260 teletex terminals are connected to this PABX. The data transmission rate is 2400 bps. The observed traffic types were internal traffic, outgoing and incoming external traffic. The measurement results lead to the teletex traffic model shown in table 2.

Traffic type / Call type	Call mix	Mean holding time sec		Number of ISO A4 pages
		Apty	Bpty	
internal traffic				
successful call	100 %	T	T	approx 1.5
outg.external traffic				
successful call	100 %	2+T	-	approx 0.8
inc.external traffic				
successful call	100 %	-	T	approx 0.8

T: Mean transmission time
internal traffic - 15 sec
external traffic - 8 sec

Table 2: Traffic model for teletex

3.3 Facsimile

At the present time, facsimile terminals of groups 2 and 3 (fa2- and fa3-terminals) are in use. The information is transmitted after being coded into analog signals. To transmit one ISO A4-page, fa2- and fa3-terminals need 180 sec and 60 sec respectively. Future facsimile terminals (group 4) will work on a digital basis. They will need only 10 sec to transmit one ISO A4-page (see[4]).

Fa2- and fa3-terminals are downward compatible,i.e. communicating with a fa2-terminal, a fa3-terminal behaves like a fa2-terminal. For communication between a fa2- or fa3-terminal and a fa4-terminal, a code and transmission speed conversion is needed. For this, an additional connection has to be established (additional basic call attempt !).

Like the telephone service, the facsimile service also includes internal traffic as well as outgoing and incoming external traffic. The call mix of the facsimile traffic model is equivalent to that of telephone traffic, but without the typical telephone features. The call type "B-party no answer" will not occur, as facsimile terminals are able to work without an operator. "B-party busy" will no longer occur for call attempts using a converter as mentioned above because this device is able to store the information.

Table 3 below summarizes our assumptions on the facsimile traffic model.

No measurements concerning the mean durations T of facsimile connections are known to us. Based on our experiences within our Company and in agreement with [14] we assume the mean length of a transmitted document to amount to 1.5 ISO A4-pages. This means the parameters T of table 3 will have the following values:

$$T = \begin{cases} 240 \text{ sec} & \text{for fa2-terminals} \\ 240 \text{ sec} & \text{for fa3-terminals connected to a fa2-terminal} \\ 90 \text{ sec} & \text{for fa3-terminals otherwise} \\ 15 \text{ sec} & \text{for fa4-terminals} \end{cases}$$

3.4 Interactive videotex

Certainly interactive videotex is the service which up to now has been the least in use in PABX's. For this reason there are no measurements available concerning videotex user behaviour. The statements on the videotex traffic model are only based on our own assumptions.

Interactive videotex is handled in so-called sessions. An established videotex connection will remain through-connected throughout the complete session.

Interactive videotex calls are always originating calls. There are two different operation modes possible. They use transmission rates of 2400 bps and 64 kbps respectively (cf.[6,15]).

The interactive videotex service with 2400 bps is offered mainly by external videotex centres, whereas interactive videotex with 64 kbps will be offered firstly by private systems (in-house). The call mix of the interactive videotex service is therefore based on the call mix of outgoing telephone traffic and internal telephone traffic respectively, as given in table 4.

Traffic type/ Call type	Call mix				holding time in sec			
	A-party fa2,3		A-party fa4		A-party		B-party	
	*1	*2	*1	*2	Fa 2,3	Fa 4	Fa 2,3	Fa 4
internal traffic								
successf. call	65 %	48 %	70 %	50 %	4+T	T	T	T
B-party busy	30 %	–	30 %	–	8	4	–	–
incompl.dialling	5 %	4 %	–	–	3	–	–	–
outg.ext.traffic								
successf.call	73 %	–	78 %	–	14+T	T	–	–
B-party busy	17 %	–	17 %	–	18	4	–	–
blocked call	5 %	–	5 %	–	12	4	–	–
incompl.dialling	5 %	–	–	–	4	–	–	–
inc.ext.traffic								
successful call	68 %	49 %	70 %	50 %	–	–	T	T
B-party busy	30 %	–	30 %	–	–	–	–	–
incompl.dialling	2 %	2 %	–	–	–	–	–	–

T: Mean transmission time
*1: without conversion
*2: with conversion

Table 3: Traffic model for facsimile

Traffic type/ Call type	Call mix	Mean holding time sec	
		2.4 kbps	64 kbps
ext. videotex centre			
successful call	96 %	14+T	T
blocked call	6 %	12	4
private videotex centre			
successful call	100 %	4+T	T

T: Mean duration of a session

Table 4: Traffic model for interactive videotex

The mean duration T of an interactive videotex session is assumed to be 300 sec.

3.5 Data dialog

Data dialog occurs in many different areas of application, which means that data dialog traffic can have many different structures. In our ISDN traffic model, all the possible kinds of data dialog traffic are represented in two classes:

Class I: Dialog sessions of long duration, occuring for example with use of in-house computers connected to the company's PABX.

Class II: Dialog sessions with only very few dialog steps (eg. information retrieval systems: question - answer).

The holding time of a class I data dialog terminal is the time between LOGON (start of the process in the computer) and LOGOFF (end of the process in the computer).

Measurements have been carried out in our Company to analyse class I data dialog traffic: Five different main frames were examined processing very different tasks (system software development, data base, commercial applications) [16]. When the measurements were performed about 500 data terminals were connected to the main frames. Each of them could communicate with any of the five computers. The measurements were performed throughout one month. It was observed that the mean holding time of the dialogs amounted to 36 minutes (compare the results of [17]: mean holding time equals 45 - 50 minutes).

For class II data dialog traffic, extensive measurement results were already available [18,19,20]. We assume the holding time of a class II data dialog terminal - with respect to the ISDN-PABX - to be the time for completing one task, i.e. duration of one transaction (cf.[19]). All data dialog connections through the PABX are successful calls. The parameter values of the data dialog traffic model can be seen in table 5.

Traffic type/ Call type	Call mix	Mean holding time sec	
		class I	class II
Successful calls	100 %	2160	60

Table 5: Traffic model for data dialog

4 PRACTICAL IMPORTANCE OF THE GENERAL PARAMETERS OF THE TRAFFIC MODEL

The tables in section 3 form the basis for a traffic analysis of a PABX, which is given by the kind and number of its terminals and their traffic values.

With the aid of tables 1-5 in the previous section, in principle it is possible to obtain the amounts of all essential traffic loads that need to be known for the dimensioning of the different system components. Examples of such system components are:

- trunk groups leading to the different public and private networks (eg. public telephone network, IDN),
- device pools (eg. code receivers),
- memory (eg. for automatic callback, voice mail memory),
- control components in the system,
- signalling channels in the system (cf.[21] for telephone traffic only).

To obtain the traffic load of such a system component one has to proceed as follows:

For each service realized in the PABX under consideration, the tables 1 to 5 of the previous section are to extended by one column. There, the mean holding times for this component for each kind of traffic and each call type have to be listed up. These values have to be determined either by system-dependent measurements or by analysis.

Given these completed tables,the terminal configuration and the terminal usage (traffic values or busy hour call attempts), the load of the system component in question can be calculated. The evaluation is performed effectively with the aid of a computer program. For this, we use a dialog-oriented PASCAL program running on a main frame.

5 TRAFFIC DESIGN OBJECTIVES FOR LARGE ISDN-PABX'S IN GERMANY

In this section the traffic requirements of future ISDN-PABX's are estimated. For this purpose, those traffic model parameters depending strongly on the use of the PABX (see section 2) are also considered. Thus, a traffic model results which characterizes not only the individual ISDN services but also the overall traffic of an ISDN-PABX.

The values of the traffic model parameters are chosen in such a way that the resulting overall traffic model is a typical model for the traffic in large ISDN-PABX's used in the Federal Republic of Germany. In this way, the overall traffic model provides traffic design objectives for new ISDN-PABX's in this country. In Germany, the traffic design objectives for PABX's have usually been given as maximum admissible loads. Similarly the following parameter values of the overall traffic model are understood as maximum admissible values.

In particular, the overall traffic model assumes that the busy hours of all ISDN services existing in an ISDN PABX occur at exactly the same time. The measurement results given in [13,17,22] show that this assumption is realistic.

5.1 Model parameters for the terminal load

In Germany, the maximum tolerated station traffic loads for the telephone service in PABX's are given in the Telephone Regulations of the German Postal Authorities [23]. The load rules relate to all those services which are carried on the public telephone network. Depending on the size of the PABX (i.e. the number of stations connected) the maximum traffic loads amount to between 0.12 and 0.17 Erl per station. These values include the loads for internal traffic with a maximum of 0.05 Erl, 0.075 Erl, or 0.10 Erl depending on the application of the PABX.

The existing telephone traffic design objectives will be valid in ISDN-PABX's and are therefore accepted in the overall traffic model. For the model the same call mix is assumed as is given as an example for an ISDN-PABX with new features in table 1.

The existing telephone design traffic loads will also apply to the facsimile services in an ISDN-PABX. We assume, however, that in large PABX's a substantial number of facsimile terminals (about 80 %) will be used as "no-trunk-access"-terminals and will only be qualified to communicate the considerable amount of internal facsimile traffic. For those terminals a higher load can be admitted. We accept in our traffic model a maximum traffic load of 0.4 Erl per facsimile terminal (cf.[24]). The traffic model assumes the following mix of terminals in large ISDN-PABX's:

fa2:fa3:fa4 = 10:60:30.

For the interactive videotex service we presume 0.3 Erl traffic load per videotex terminal.

For the teletex service the overall traffic model assumes a mean terminal traffic load of 0.05 Erl at a data transmission rate of 2400 bps. A ratio of external traffic : internal traffic = 50 : 50 is fixed in the traffic model.

The class I data dialog traffic has the measured values of [16] as traffic data in the overall traffic model, giving a mean terminal load of 0.5 Erl. In consideration of [19] this value is also applied to the class II dialogs in the traffic model. It is further assumed that in large PABX's 90 % of all data terminals are operated as pure class I terminals.

On the basis of the above parameter values the overall traffic model for ISDN-PABX's in Germany is characterized in the following table 6.

Service	Mean traffic load per terminal Erl	Mean number of BHCA per terminal
Telephone	0.12...0.17	4.8...7.2
Teletex	0.05	13.0
Facsimile		
trunk access 20%	0.12...0.17	4.1...8.0
no trk acs 80%	0.40	12.1
Mean value	0.35	10.5...11.3
Videotex		
ext.Vdx centre	0.30	3.7
priv.Vdx centre	0.30	3.6
Mean value	0.30	3.7
Data Dialog		
class I (90 %)	0.50	0.8
class II (10 %)	0.50	30.0
Mean value	0.50	3.7

Table 6: Traffic load and number of BHCA per terminal

5.2 Model parameters for the load of the basic accesses

In ISDN PABX's [3,25] the station users may be connected to a basic access with two B-channels (see section 1). Generally each ISDN user is able to connect several terminals for different services to his basic access. The basic access of a user is characterized by a single call number.

In our overall traffic model we assume the following equipment of a large ISDN-PABX with service-specific terminals (table 7):

Service	BA's with terminals for the service	Mean number of BHCA per basicaccess	Mean traffic load per BA Erl
Telephone	100 %	4.8...7.2	0.12...0.17
Teletex	10 %	1.3	0.005
Facsimile	10 %	1.1	0.035
Videotex	30 %	1.1	0.100
Data Dialog	35 %	1.3	0.175
Total		9.6...12.0	0.44...0.49

BA: Basic Access

Table 7: Traffic load and number of BHCA per Basic Access

6 CONCLUSION

The traffic parameters postulated in the preceding paragraph show a mean traffic load of about 0.46 Erl per basic access as design objective for large ISDN-PABX's in the Federal Republic of Germany. This value results in 0.23 Erl for each B-channel. The number of calls offered to the system control amounts to between 9.6 and 12.0 BHCA depending on the size of the exchange. This number is nearly twice as high as the number of BHCA which are at the present adopted for pure telephone service.

The traffic model given in this paper forms a basis for the development of new PABX's. The work refining and correcting of this model is being continued. Control measurements with the existing ISDN services will have to be performed continuously. We are prepared to accept that the user behaviour may change completely when multifunctional terminals are introduced on a broad basis. In this case our traffic model will also have to be fundamentally changed to adapt itself to the new conditions.

REFERENCES

[1] CCITT Recommendations of the I-series, Red Book, 1985, to be published

[2] J. Swoboda, "Digitale Nebenstellenanlagen im dienstintegrierten digitalen Netz ISDN", Informatik Spektrum, Vol.7, pp.138-153, 1984

[3] System HICOM, Produktschrift der SIEMENS AG, 1984, Bestellnr. A19100-K3161-G430-02

[4] G. Arndt, H.-J. Rothamel, "Services in the ISDN", telcom report, Vol.6, No.5, pp.197-201, 1983

[5] G. Robin, S.R. Treves, "An introduction to integrated services digital networks", Electrical Communications, Vol.56, No.1, pp.4-16, 1981

[6] NTG-Empfehlung 0903, Nachrichtenverkehrstheorie, Begriffe, 1984

[7] G. Dietrich, "Telephone Traffic Model for Common Control Investigations", 7th International Teletraffic Congress, Stockholm 1973, paper no.331

[8] A. Myskja, O. Walman, "An Investigation of Telephone-User Habits by Means of Computer Techniques", IEEE Transactions on Commun., Vol.21, pp.663-671, 1973

[9] R. Evers, "Das Verhalten der Teilnehmer einer Nebenstellenanlage - Teil 1: Die Häufigkeit von Wiederholungen nach erfolglosen Anrufversuchen", Technischer Bericht Nr.143, Heinrich-Hertz-Institut Berlin, 1971

[10] R. Evers, "Das Verhalten der Teilnehmer einer Nebenstellenanlage - Teil 2: Zeitintervalle innerhalb der Belegungen", Technischer Bericht Nr.146, Heinrich-Hertz-Institut Berlin, 1971

[11] R. Evers, "Das Verhalten der Teilnehmer einer Nebenstellenanlage - Teil 3: Belegungsdauer, Belegungsabstände und Wiederholungsabstände", Technischer Bericht Nr.158, Heinrich-Hertz-Institut Berlin, 1972

[12] R. Rüggeberg, "The Development of the Teletex Service", IEEE International Conference on Communication, Boston, USA, 1983, pp.781-785

[13] G. Niestegge, "Auswertung von Verkehrsmessungen von Teletex-Verkehr", Interner Bericht, Siemens AG, 1984

[14] R. Rüggeberg, "Künftige Textkommunikationsdienste", VDE-Fachbericht, Vol.31, VDE-Kongreß, Oktober 1980, Berlin

[15] H.J. Rothamel, "Videotex supported by the integrated services digital network", telcom report, Vol.7, No.6, pp.270-273, 1984

[16] G. Niestegge, "Auswertung von Verkehrsmessungen von Datendialog-Verkehr", Interner Bericht, Siemens AG, 1984

[17] G. Bryan, "JOSS: 20,000 hours at a console - a statistical summary", Proc. AFIPS Conf. 1967

[18] H.-D. Südhofen, "Benutzerverhalten in Fernverarbeitungssystemen: Dokumentation wichtiger Meßergebnisse 1974-1978", DFV-Bericht, RWTH Aachen, 1981

[19] H.-D. Südhofen, "Benutzerverhalten in Fernverarbeitungssystemen: Meßergebnisse zur tageszeitabhängigkeit charakteristischer Größen von Einzelbenutzern und Benutzerbündeln", Interner Bericht, RWTH Aachen, Dezember 1984

[20] P. Pawlita, "Traffic Measurements in Data Networks, Recent Measurement Results, and some Implications", IEEE Transactions on Communications, Vol.29, No.4, pp.525-535, 1981

[21] J. Beckmann, "Belastung von Vermittlungssteuerungen in Telephon-Nebenstellenanlagen", Nachrichtentechnische Zeitschrift, Vol.29, No.8, pp.582-584, 1976

[22] R. Evers, "A survey of subscriber behaviour including repeated call attempts - results of measurements in two PABX's", Sixth International Symposium on Human Factors in Telecommunication, Stockholm, June 1972

[23] Fernmeldeordnung der Deutschen Bundespost, Rahmenregelung für mittlere und große W-Anlagen, 123D5, Stand 8.84

[24] I. Richer, M. Steiner, M. Sengoku, "Office Communications and the Digital PBX", North Holland Publishing Company, Computer Networks, Vol.5, pp.411-422, 1981

[25] K. Raab, "Office Communication Systems Based on the Capability of Today's PABX Technology", IRECON, Sydney, 1983

TELETRAFFIC ISSUES in an Advanced Information Society
ITC-11
Minoru Akiyama (Editor)
Elsevier Science Publishers B.V. (North-Holland)

ISDN PROTOCOL AND ARCHITECTURE MODELS

R. M. Potter

Bell Communications Research
Red Bank, New Jersey, U.S.A.

ABSTRACT

This paper provides an overview of current concepts and models, being communicated by the community of CCITT SG XVIII readers and participants which will significantly influence the teletraffic studies and models for the next generation. The concepts described here are offered as possible building blocks and a source of terminology for future quality of service and performance traffic studies in the ISDN era.

1. INTRODUCTION

The generic pieces of any model for teletraffic studies can be thought of as comprising:

i. the underlying services for applications requested by users which produce the input traffic demand distribution to the relevant network(s) system(s),

ii. the signalling control provided by the network (and/or the user) for

   a. *connections* needed for the transport of the demand traffic, the routing, availability of resources as well as guaranteeing a requested or expected grade of service,

   b. *calls* possible since ISDN provides user controlled functionality, flexibility of possible configurations and versatility of access to the network(s) or remote data bases currently being designed into both ISDN access protocols and internetwork protocols,

iii. the implicit level of performance of the signalling network(s), protocols, network switches, relay points, databases, interworking units and resource availability, and

iv. the cost to provide a particular level of service (interactions between realistic cost and service and demand characteristics).

A dominant theme of this paper is the recognition in ISDN of the importance of both a "user perspective" of all services, protocols, performance and cost as well as a "control perspective" of providing the flexible and dynamic connection and call processing possibilities, which allow the many integrated services to be available to an ISDN user. Both perspectives, their interactions and influence on modelling scenarios are discussed.

In Section 2., the CCITT I-Series classification of ISDN services, from both a "user" and "control" perspective, is considered fundamental. Necessary terminology and concepts are introduced. This highlights the conceptual differences between an *integrated services* digital network and today's more service specific networks. The generic structure adopted in CCITT for discussing ISDN services is introduced in terms of applications, layering, attributes and capabilities. The role of quality of service attributes within telecommunication service definitions, with possible relevant quantifiable parameters and their dependence on "user" and/or "control" perspective is considered. The effect of many services simultaneously to be available in an ISDN, their interaction and different existing service level expectations are studied.

In Section 3. modelling concepts contained in the CCITT Recommendation I.320 "ISDN Protocol Reference Model" are introduced and their importance in providing a foundation to ISDN traffic and performance studies are considered. This construct is found to be a useful tool in understanding the role of user, control and management when providing connections and calls for the many applications in ISDN. Several "communication contexts" are introduced. Protocol layers, planes and blocks, key features of the ISDN protocol model are described. Their role is considered within an appropriate "communication context" in the determination of protocol (signalling and user information) performance, an intrinsic part of overall expected network performance.

Understanding the interactions between different protocol elements, layers, switching points and communication contexts for possible ISDN information flows is necessary to provide estimates for appropriate performance attribute values. Examples illustrating these concepts for some ISDN information flows are given.

In Section 4. ISDN relevant concepts of calls and connections are discussed. Configurations of control systems currently being developed for providing calls and connections in ISDN(s) and their associated impact on possible network performance are highlighted. Issues associated with performance classification for the many possible control systems and their interworking are raised. Concepts from the CCITT Recommendation I.310 on "Connection Types" are introduced for classification and estimation of parameters for network performance in conjunction with application of Hypothetical Reference Configurations (HRXs) to designate values for quality of service parameters expected for all calls and associated connection elements.

The final section raises issues which become relevant when considering teletraffic provisioning for the ISDN era.

2. SERVICES

An ISDN provides a set of network capabilities which are defined by standardized protocols and functions which enable *Telecommunication services* to be offered to users.(Service classification and description are independent of the different possible arrangements for ownership and provisioning to the customer of the means required to support a service.)

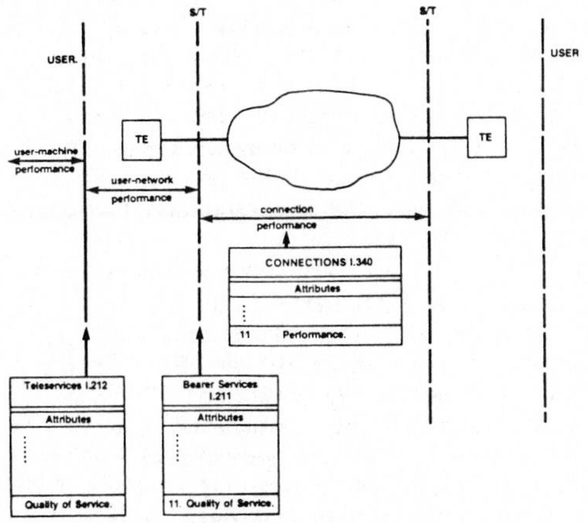

Figure 1: Reference Points Associated With Telecommunications Services

From a **user's perspective** Telecommunications services are described by *attributes* of service characteristics as they apply *at a given reference point* where the customer accesses the service. Figure 1 illustrates the appropriate reference points for the classification of Telecommunication services into *Bearer Services* [1] and *Teleservices* [2] . Each classification can further be considered as being either basic or basic + supplementary where a supplementary service modifies the basic service and thus can not be offered as a stand alone service. Tables 1 and 2 illustrate values of attributes characterizing Bearer and Teleservices respectively.

From a **network perspective** Telecommunication services are supported by *network capabilities*: Low Layer Capabilities (LLC) which relate to Bearer Services and High Layer Capabilities (HLC) which together with LLC relate to Teleservices. The Low Layer Capabilities to support user access are to be defined by a set of low layer functions (LLFs) relating to the lower layers of the access protocol which provide the capability for the transport of user information over the access. In turn, these LLFs are broken down as basic low layer functions (BLLFs) which are to support the necessary requirement to establish a connection and convey user information over the connection; and additional low layer functions (ALLFs) which are necessary to support the low layer requirements of supplementary services. High Layer Capabilities are defined as a set of High Layer Functions (HLFs) generally associated with those additional layers appropriate for transfer, storage and processing of user messages with respect to other users, retrieval centres or network service centres.

Table 1

Values of Bearer Service Attributes

	ATTRIBUTES.*	POSSIBLE VALUES OF ATTRIBUTES.							
INFORMATION TRANSFER ATTRIBUTES	INFORMATION TRANSFER MODE 1.	CIRCUIT					PACKET		
	INFORMATION TRANSFER RATE 2.	BIT RATE KBIT/S					THROUGHPUT		
		64	384	536	1920	Other values for further study	OPTIONS FOR FURTHER STUDY		
	INFORMATION TRANSFER CAPABILITY 3.	unrestricted digital information	speech	3.1kHz audio	7kHz audio	15kHz audio	video	others for further study	
	STRUCTURE 4.	8kHz integrity		Service data unit integrity		unstructured			
	ESTABLISHMENT OF COMMUNICATION 5.	DEMAND		RESERVED		PERMANENT			
	COMMUNICATION CONFIGURATION 6.	point-to-point		multipoint		broadcast			
	SYMMETRY 7.	undirectional		bi-directional symmetric		bi-directional asymmetric			
ACCESS ATTRIBUTES	ACCESS CHANNEL & RATE 8.	D(16)	D(64)	E	B	H0	H11	H12	others for further study
	SIGNAL ACCESS PROTOCOL 9.1	I.440	I.451	SS#7	I.462	others for further study			
	INFORMATION ACCESS PROTOCOL 9.2	G.711	G.721	I.460	I.461	X.25	others for further study		
GENERAL ATTRIBUTES	Supplementary Services provided 10.	under study							
	Quality of service 11.								
	Interworking scenarios 12.								
	Operational and commercial 13.								

Table 2

Values of Teleservice Attributes

ATTRIBUTES*	POSSIBLE VALUES OF ATTRIBUTES							
INFORMATION TRANSFER ATTRIBUTES & ACCESS ATTRIBUTES	Refer to Recommendation I.211							
TYPE OF USER INFORMATION	SPEECH	SOUND	TEXT	fac-simile	text-facsimile	video text	video	others
Layer 4 Protocol	X.224			T.70			others	
Layer 5 Protocol	X.225			T.62			others	
Layer 6 Protocol	T.73		T.61		T.6	T.100	others	
resolution	200		240		300	400	others	
graphic mode	alphamosaic		geometric		photographic		others	
Layer 7 Protocol	T.60		T.5		T.72		others	
general attributes	under study							

Quality of service (QoS) attributes from a "users perspective" can be a subjective expectation of certain key service parameters (e.g. network response time for a particular Telecommunications transaction, time to access the system) independent of knowledge about other classes of user, topology, technology, protocol(s), information encoding and other network design characteristics. Whereas quality of service from a "control" perspective can be thought of as a hierarchical combination of technical network performance parameters appropriate for the layer of protocol(s), signalling systems, technology and connection configuration etc. to provide the requested user service. Therefore a model to relate the two perspectives is needed. One approach is discussed in [3].

3. PROTOCOL MODELLING

In ISDN, many information flows (user information and signalling information[1]) within a particular communication context need to be investigated before traffic models with appropriate performance characterization can be developed. Information flows between users have been modelled [4] by representative users and control functions. See Figure 2.

Information flows for circuit switched connections under the control of common channel signalling; packet switched communication over B ,D and H channels[2]; signalling between users and network based facilities; end to end signalling between users (e.g. to change the mode of communication over an already established connection) and combinations of all the previously listed information flows (whereby several simultaneous modes of communications can take place under common signalling control), necessitate elaboration of existing models and development of new modelling concepts. Understanding and development of a reference model is necessary for providing structure to relationships between signalling flows for control of particular ISDN connections and configurations as well as for user information flows.

The possibility to control diverse connections (e.g. multimedia calls; renegotiation of connection characteristics of calls; network management and multiple related protocols) by associated and common channel signalling, influences all performance parameters associated with set up time estimates, delay and blocking, of a communication path between users. The performance complexity of interworking between the many inband and out-of-band signalling systems (as enumerated in section 4.) and their associated concatenated connection elements influence the customer perceived quality of service.

Key modelling concepts introduced in [6] to reflect the diversity and complexity of information flows related to various Telecommunications services in ISDN are those of *communication contexts, protocol planes, layers and blocks* .

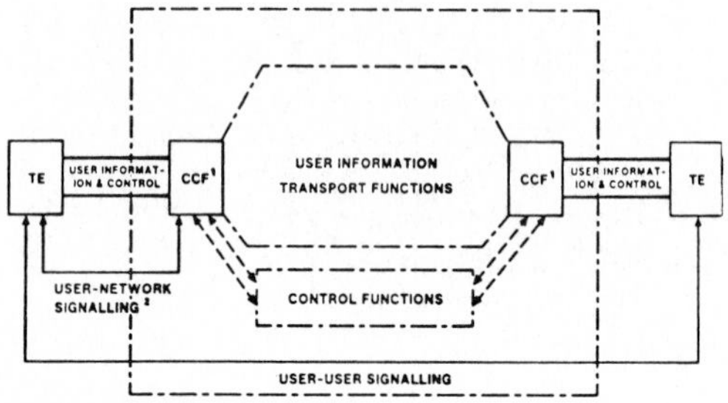

NOTES:
1. CCF1 Connection Control Function [4]
2. INCLUDES Connection Control and Communication with e.g. Operations Centre

Figure 2: Information Flows

3.1 Communication Contexts

It has been apparent that traditional layered modelling [7] of information flows across networks has an implicit communication context in which the network is considered "transparent" to point-to-point user information applications other than providing appropriate routing and addressing functions. Figure 3 illustrates other communication contexts key to developing performance models for ISDN traffic and information flows across connections involving circuit switching, packed switching, out-of-band signalling, access to remote data bases and/or other networks where

- Context A: realm of user-to-user signalling over a connection;
- Context B: user-network signalling between calling TE[3] and the ISDN;
- Context C: intra-network signalling;
- Context D: user-network signalling between called TE and ISDN.

The choice of communication context A for user-to-user signalling being carried across the signalling network has implicitly the excellent performance characteristics of a signalling network rather than those of a public network. Elaboration of appropriate communication contexts for circuit switched calls is given in section 3.2 .

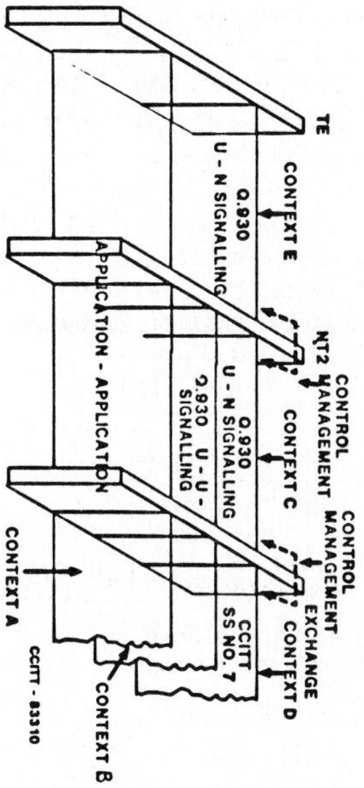

Figure 3: Examples of Communication Contexts

Communication contexts may be chained together simultaneously or sequentially in *time* with information passing from one context to another through an application process or a management entity.

Within a particular communication context, e.g. user-to-user; user-to-originating switch; user-to-data base within network; control entity to control entity, the generic communication context of layered protocol communication between end protocol blocks is a useful concept. With its associated performance parameters it can be considered the basic building block for quantifying relationships between different protocol layers and primitives. Time sequential or simultaneous models can then be built up to represent flows of information across multiple communication contexts with appropriate performance estimates.

3.2 Protocol Planes, Layers and Blocks

Once a communication context appropriate for modelling the representative information flow has been established, it may be considered from an *end system* perspective is either being

in context i.e. belonging to the *user plane* U, or **out of context** i.e., belonging to the *control plane* C. The dialogue in each plane is independently described in terms of separate layer structures depending at this time on the signalling system(s) chosen for providing the appropriate connection element(s) for the specified context. Within a particular context, current protocol models [6], [7] consider 7 layered end system protocol blocks with 3 layered intermediate system protocol blocks. Currently there is an awareness that the present development of signalling systems for ISDN access[4] and intra-network[5] communication are structured in layers which however are not consistent with existing OSI layers. A generic protocol block is illustrated in figure 4. Hence to obtain a realistic approach to estimate values of performance attributes one currently needs to understand the relationships between the associated layers of other protocols as well as their "peer" layers. This interworking between layers (rather than purely a single peer layer) of associated concatenated contexts' protocols will influence performance studies when evaluating switching or relay influences on end to end performance.

However, a layered model allowing for communication between appropriate peer layers for the two planes U and C within the appropriate communication context, and providing for possible concatenated and simultaneously occurring contexts is a useful tool for obtaining a first estimate of expected network performance values relating to call setup, clearing and reconfiguration.

Examples of applications of this modelling approach are illustrated in Figures 5, 6 and 7 for providing packet switched connections via a B channel, packet

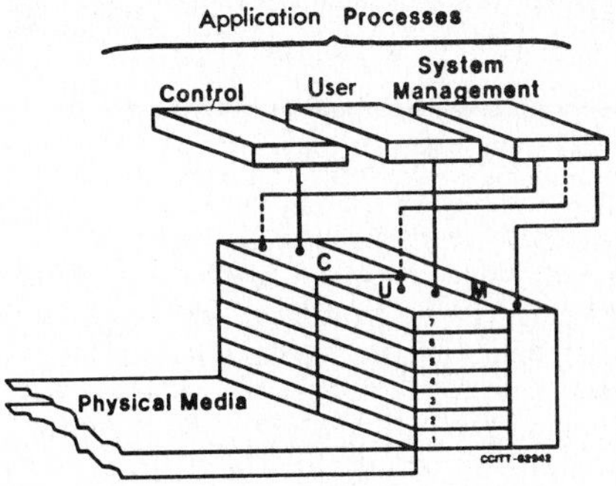

Peer-to-Peer Protocols Associated with U and C are not shown

Figure 4: A Generic Protocol Block

switched connections via a D channel, and circuit switched connections via the B channel respectively. In particular, estimates of performance values for each plane's layers within a chosen communication context can now be obtained through detailed knowledge of the chosen protocol procedures in that context (possible protocols are considered for providing particular connection types in Section 4.) in conjunction with suitable functions describing interactions between the U and C planes for that context. Each communication context has associated with it an implicit connection configuration.

4. CONNECTIONS

Users specify only appropriate service requests whereas the network is to allocate resources to set up a connection of an appropriate type to support the information flows associated with the requested service. Some of these information flows may or may not be associated with a specific ISDN call (e.g. flows associated with obtaining access to a remote data base for authorization, registration to supplementary services). *Means* (transmission and switching), *control* and *protocols* (signalling, flow/congestion control, routing), and *operations* and *maintenance* are functions associated with a connection.

In order to allocate network performance parameters it is necessary to specify appropriate connection elements capable of providing the correct connection type to form the designated connections, e.g. a 64 kb/s circuit switched connection within a national ISDN can be thought of as comprising two access connection elements and a transit network element with each connection element being of the "64 kb/s circuit switched" type. Each connection element can be thought of as representing a particular communication context as described in the previous section. However, [12] describes many ISDN connections, connection elements and connection types with associated attributes. Each connection element of a given connection type has associated signalling configurations. For example, Table 3 illustrates for access network and intra network context signalling systems being developed to provide various connection types in ISDN. The complexity and number of possible connection needed to support multiple services for a given connection is simplified by designation of appropriate Hypothetical Reference Connections (HRXs). Possible HRXs [8] for the international section of an ISDN connection are illustrated in Figure 9. However the network performance attributes for appropriate communication context(s) and associated network elements need to incorporate at least the signalling performance (as described in section 3.) parameter values for providing the required connection type.

In the ISDN environment there is recognised the need to distinguish between *connections* as described previously and *calls*. Calls can exist prior to the establishment of a connection (implicitly understood to be an end to end connection. The performance consequences from a network perspective of providing for such capabilities within associated and common control signalling procedures needs to be addressed.

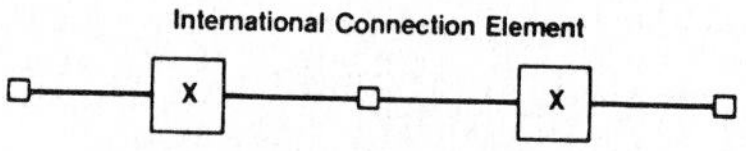

Figure 9(a): International Connection Element Model Associated with a Moderate Length HRX

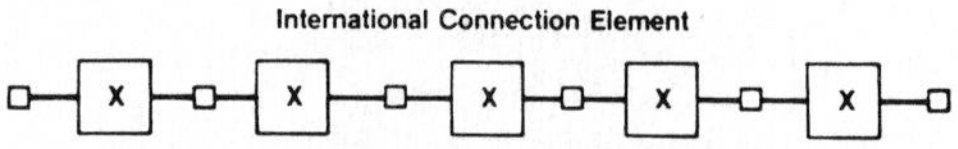

Figure 9(b): International Connection Element Model Associated with the Longest Length HRX

KEY: [X] = Digital Exchange (ISC)
□ = Digital Link

5. ISSUES

This paper contains a high level perspective of possible influences and characteristics that the varied services and complex control structures for providing appropriate network capabilities could place on traditional modelling techniques for ISDN. Traffic resulting from an information flow(s) within a designated communication context perspective needs to be considered. Appropriate classification, frequency of occurrence, control traffic interaction, priority classes, mixtures of traffic relevant to the different protocol model communication contexts for simultaneous applications all need to be addressed. Performance consequences interactions between information flows between concatenated and parallel networks providing additional capabilities to today's networks need quantifying.

---

1. User information: e.g. digitized voice, data and information transmitted between user. Control information: information which is acted upon e.g. in controlling a network connection, controlling the use of an already established network connection.
2. Channels are defined in [5] as: B channels - 64 kb/s; D channels - 16 kb/s for basic voice interface or 64 kb/s for primary voice interface ; H channel - 384 kb/s for H0 channels, 1536 kb/s for H1 channels, 2048 kb/s for H12 channels.
3. TE : terminal equipment as defined in [1].
4. Q.921 (I.441) and Q.931 (I.451) defined in [9].
5. Signalling System No. 7 : Q.711 through to Q.764 as defined in [10], [11].

The traditional understanding of measures such as "busy" and "congestion" due to blocking and delay needs expansion to include individual customer definition of "busy" and its associated engineering implications. The expected increase in customer flexibility increases forecast uncertainty in many areas, viz., demand characteristics, service mixtures and interactions, busy hour estimation and provisioning of functional entities needed to provide the expected control functionality.

The challenge to ISDN traffic and performance models is to reflect the ability of individual customers to optimize, within their applications, expected quality of service parameters, costs and requested flexibility while reflecting the appropriate network(s) mechanism to provide for user action, uncertainty in demand, controlling users' diverse connections and calls in a cost effective manner with an understanding of the network(s)' performance.

ACKNOWLEDGEMENTS

I acknowledge the help and assistance of Leonard. J. Forys, the support and inspiration provided by my colleagues in SG XVIII, XI and in particular, the encouragement of Warren S. Gifford.

REFERENCES

[1] SG XVIII Report R32 "Final Report to the VIIIth CCITT Plenary Assembly (Part V)", Recommendation I.211, pp. 40-69, 1984.

[2] SG XVIII Report R32 ibid, Recommendation I.212, pp. 69-72.

[3] COM XVIII - No. D.57, "An Approach to Quality of Service and Network Performance in ISDN", Federal Republic of Germany, January, 1985.

[4] SG XVIII Report R32, ibid, Recommendation I.310 pp. 72-82.

[5] SG XVIII, Report R32, ibid, Recommendation I.412, pp. 144-151.

[6] SG XVIII, Report R32, ibid, Recommendation I.320, pp. 83-102.

[7] SG VII, Report R34, "Final Report on the Work of SG VII During the Study Period 1981-1984 (Part III.6)" Recommendation X.200, pp. 2-78.

[8] COM XVIII - No. D.31, "Proposals for Reference Configurations for ISDN Connection Types", British Telecom, January 1985.

[9] SG XI, Report R58, "Final Report of Study Group XI to the VIIth CCITT Plenary Assembly ", Part III.10, Digital Access Signalling System, pp. 6-254, 1984.

[10] SG XI, Report R56, ibid, Part III.8, Signalling System No. 7, SCCP, pp. 4-120, 1984.

[11] SG XI, report R57, ibid, Part III.9, Signalling System No. 7, ISDN, pp 4-125, 1984.

[12] SG XII, Report R32, ibid, Recommendation I.340, pp. 117-131.

Connection Configuration	Access Protocol D-ch protocol/X.25	Network Signalling S.S. No. 7/X.25	Switching Mode
TE — B-ch 64kbps — LS — (Network 64 kbps	D-ch protocol	ISDN-UP	Circuit-Switched
TE — Sub-rate 8/16/32 — LS — (64 kbps	D-ch protocol (channel identification)	ISDN-UP	
TE — Sub-rate 8/16/32 — LS — (Sub-rate 8/16/32kbps	D-ch protocol (channel identification)	ISDN-UP	
TE — Wide-band H./H 64 x n — LS — (Wide-band H./H 64 x n kbps	D-ch protocol (channel identification)	ISDN-UP	
TE — B-ch Packet — LS (64K) LS — B Packet — TE	B-ch set-up: D-ch X.25 on B-ch	ISDN-UP X75 (X.25) on circuit	
TE — B-ch Packet — LS (64kbps) IP (	B-ch set-up: D-ch X.25 on B-ch (I.462)	ISDN-UP X.25 on circuit	
TE — D-ch Packet — [LS — PH] (PSDN/ISDN Packet	LAPD + X.25 (I.462)	X.25/X.75+ I.462 (via Packet Handler)	Packet-Switched
TE — B-ch Packet — [LS — PH] (PSDN/ISDN Packet	B-ch set-up: D-ch X.25 on B-ch (I.462)	X.25/X.75 + I.462 (via Packet Handler)	
TE — B-ch Packet — LS (Signalling Network e.g. No.7	LAPD + X.25 (I.462)	SCCP	
TE — D-ch Signalling — LS (Signalling Network e.g. No.7	D-ch protocol Signalling connection	SCCP	Signalling
TE — Asymmetric Circuit — LS — Asymmetric Circuit	D-ch protocol (channel identification)	ISDN-UP	Circuit-Switched
TE — B-ch / D-ch Packet — LS (Circuit / Packet	D-ch protocol	ISDN-UP X.25 etc.	Mixed
TE — B-ch / Signalling — LS { Circuit / Signalling	D-ch protocol	ISDN-UP SCCP etc.	
TE — Multiple Connections — LS (	D-ch protocol	ISDN-UP etc.	Various Types
TE — Multi-cast — LS (Destn 1, Destn 2, Destn 3	D-ch protocol	ISDN-UP etc.	
Reserved connections			

Table 3: Connection Configurations with Supporting Signalling Capabilities

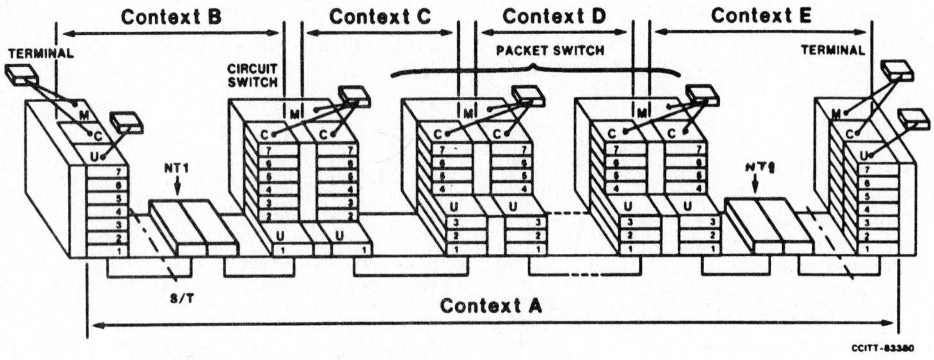

Figure 5: Packet Switched Connections
Via a B Channel

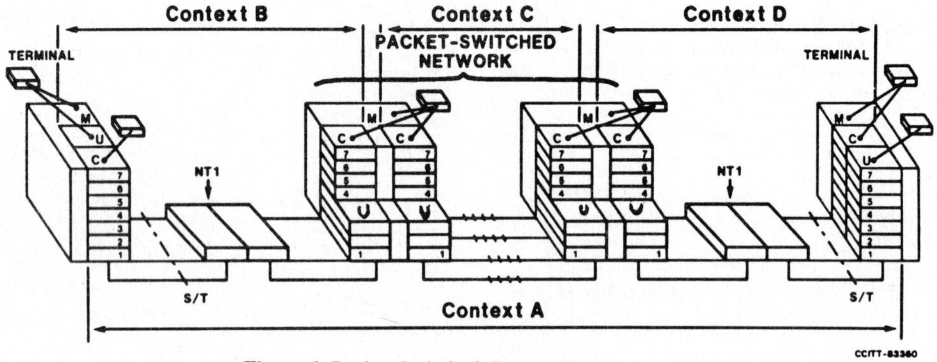

Figure 6: Packet Switched Connections
Via a D Channel

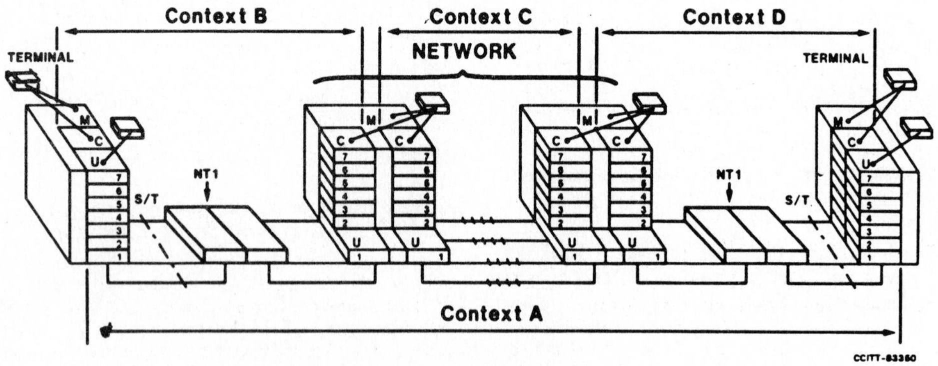

Figure 7: Circuit Switched Connections
Via a B Channel

TELETRAFFIC ISSUES in an Advanced Information Society
ITC-11
Minoru Akiyama (Editor)
Elsevier Science Publishers B.V. (North-Holland)

A STEP TO ISDN PLANNING: FROM USER TRAFFIC TO SERVICE COST COMPARISON

Oscar GONZALEZ SOTO and Teodoro BORJA

Centro de Investigación de Standard Eléctrica, S.A.
Madrid, Spain

ABSTRACT

A traffic characterization of multiservice users in ISDN is employed as a step in the planning process for selecting network strategies. A user-to-user communication path modeling is given for dimensioning of the resources being used by our observed end-to-end traffic plus the interference traffic of other users at the common network levels.

A cost assignment made as a function of the resource utilization for each traffic class allows the comparison of network implementations such as circuit vs. packet switching modes or dedicated versus partially or fully integrated networks.

1. INTRODUCTION

The high potential of ISDN relies on the resource and cost savings for the provision of multiservice communication when compared to the proliferation of many specialized networks. The type of integration considered is the one being standardized by CCITT /1/ for the public networks with either low or high levels of integration.

A unified characterization developed for multiservice environment /2/ is taken as a starting point for traffic definition at the user and network interfaces with applicability to packet and circuit switching modes. The following five essential traffic demand parameters: call rate, erlang traffic, information volume, packet rate and transmission erlang traffic (erlangbits), are selected to describe the traffic mixes at the different network interfaces. At those interfaces where the communication or protocol procedures change, such as gateways and internal transfer mechanisms within a node, an application of the protocol layering efficiencies is made to derive the gross traffic associated with each resource and to permit the corresponding dimensioning.

Parallel modeling of the capacity and cost is addressed for the purpose of cost assignment to each service and for the comparison of integration alternatives. To handle the complexity of the complete network analysis the problem is partitioned in four related hierarchical levels in a similar way to the network performance by hierarchical modeling in /3/ but also introducing the cost evaluation associated to each level. Figure 1 represents the capacity/cost modeling partition and results associated with the hierarchical modeling.

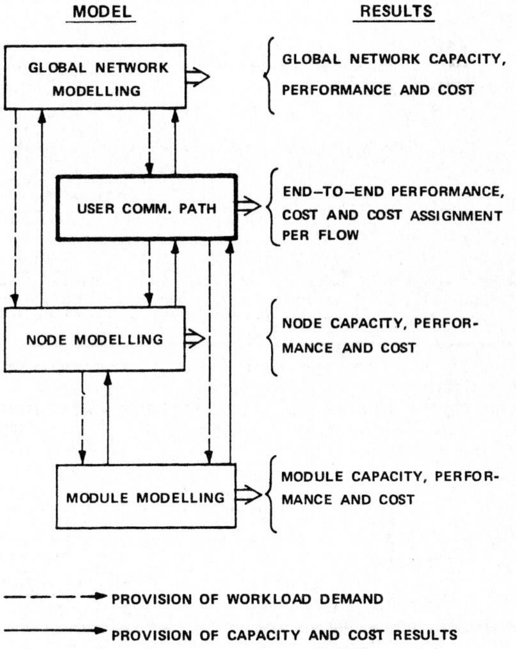

FIG. 1.– CAPACITY / COST HIERARCHICAL MODELLING

The ISDN traffic and cost modeling associated with the communication path or third hierarchical level is the object of the present contribution which identifies the traffic and cost parameters associated with the path. The simultaneous consideration of capacity and cost allows the cost assignment for all the services which handle integrated traffic according to the call composition of a given service and to the resources being utilized. Reported work in dedicated networks compare packet and circuit switching delays /4/ /5/, costs and delays /6/ and tariffs /7/. Integrated switching structures are explored in cost at /8/.

2. TRAFFIC PARAMETERS FOR ISDN PLANNING

The planning and engineering of Integrated Services Digital Networks requires consideration of the corresponding multiservice flows and traffic characterization of its users with very different levels of activity, speeds, information volumes and holding times. Shared use of the network for many services implies important economic advantages from an engineering point of view always that the characterization may provide the workload demand for network resources.

As many of the services may be carried by either packet or circuit switched mode, a common characterization developed in /2/ is followed to allow resource dimensioning and further comparison of switching modes without "a priori" loosing of parameters. The characterization associated with the call content in the information flow is generalized for all the services to obtain an integrated flow modeling. A summary of the characterization follows.

2.1 Per service characterization. Call components

In order to be able to represent the specific origin-destination flows in the network and derive the resource dimensioning associated with each service (i) a "per service characterization" is done. Traffic variables and values relate to each of the services defined by CCITT by their corresponding attributes such as information transfer mode, configuration, protocol, quality of service, etc. The potential combinations of attributes to calls is very large, causing many variants of call types. With the objective of a tractable number of cases, all the variants may be grouped by their dominant traffic characteristics: bandwith, throughput class, traffic level, etc giving origin to the traffic class (j) or set of traffics having similar dominant traffic characteristics through the network.

Among the traffic units used at different communication procedures: call, transaction, message, packet, etc the unit call is selected as the basic reference for the services with aggregation of the associated components (packets, blocks, bits, etc) as required. A unified definition of parameters for circuit mode and packet mode calls is used although the ulterior impact on the dimensioning procedures will be different.

A call of service i class j is characterized at a given network interface by the following traffic parameters referred to the engineering period.

λ_{ij} = call rate or virtual call rate associated with circuit or packet mode switching respectively.

h_{ij} = call or virtual call holding time for circuit and packet mode calls.

ρ_{ij} = total call traffic given in erlangs for circuit mode traffic through the network and virtual erlangs for packet mode traffic.

I_{ij} = call information volume expressed in bits per flow direction (Forward or Backward).

τ_{ij} = transmission traffic or part of the traffic in which the user is effectively transmitting information (Erlangbits) per flow direction.

ν_{ij} = component of the virtual call traffic in which the call is active but without transmitting information per flow direction.

B_{ij} = block rate of information units (message/s, packet/s, etc) traveling through the network per flow direction.

S_{ij} = average size of block B_{ij} per flow direction.

Additional traffic parameters which influence the call components are given in /2/.

2.2 Relation between information and traffic

It is assumed that the basic need of a user on the communication is for transmission of I_{ij} units of information at a given minimum rate. If this is considered as an "invariant" of the user relative to the alternatives in the communication procedures, the network traffic parameters associated with each call are obtained by the following process of Fig.2 between information and traffic.

The interrelation among I τ and ρ becomes a bridge between symbol information as defined in information theory and traffic with the corresponding relationship between the two fields.

Each of the two processes, coding and protocol layering have important effect on the final gross information flowing through the network and consequently on the actual traffic demand at the user and network interfaces.

- The derivation of the net source information from the original user symbols rate may be obtained by the classical Information Theory approaches due to Shannon /9/. On the basis of Markovian behaviour of the sources and a stationary process we have: a message of M independent symbols with probability of occurrence P_i and N symbols/message produces on average I(M) bits given by:

$$I(M) = \sum_{1}^{M} N\, P_i \log_2 (1/P_i)$$

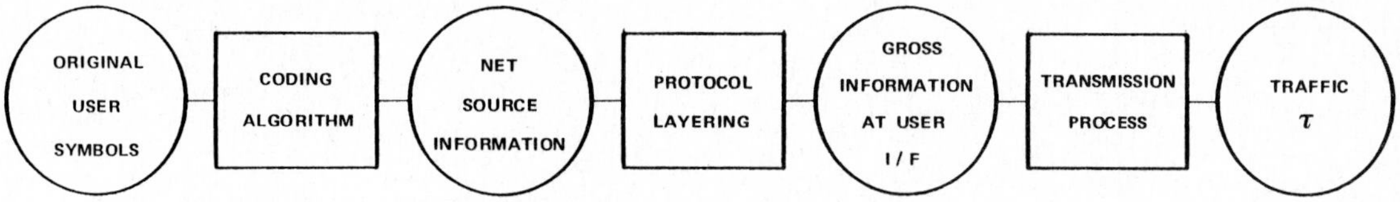

FIG. 2.– PROCESSES BETWEEN INFORMATION AND TRAFFIC

The message entropy or average information content per symbol is given by:

$$H(M) = I(M)/N = \sum_{1}^{M} P_i \log_2 (1/P_i) ;$$

Being r_s the symbol rate at the source, the average information rate results in:

$$R = r_s H \text{ information bits/s}$$

The output of the coding algorithm will be a function of the coding efficiency "e" at the encoder which is defined as the relation between the source entropy H and the actual average output bit rate of the encoder H_M for a given message length and coding technique /10/.

$$e = H/H_M$$

Practical coding algorithms and compression will define the values of e_{ij} for each service according to the variety of symbols, size of messages, coding technique, etc.

The source information volume and rate to the network will be:

$$I_{ij}(\text{source}) = I_{ij}\ (\text{user})/e_{ij}$$
$$R_{ij}(\text{source}) = R_{ij}\ (\text{user})/e_{ij}$$

° The protocol layering process models the evolution of information and idle periods within a call through an Incremental Information and Time Flow due to the contribution of each protocol layer L. Defining:

I_{ij}^L, τ_{ij}^L, ν_{ij}^L, B_{ij}^L as the information volume, transmission traffic, idle traffic and block rate observed at protocol layer L for the traffic class j of the service i. With L=1,2, ...7 for OSI layers, L=u at the user side and L=s at the source to the network, and:

IO_{ij}^L, τO_{ij}^L, ρO_{ij}^L as the protocol information overhead, marginal transmission and marginal total traffics respectively.

The following interrelations are obtained among traffic, information and protocol:

Accumulated information volume and marginal transmission traffic per layer L:

$$I_{ij}^L = I_{ij}^s + \sum_{k=7}^{L} IO_{ij}^k \quad ; \quad \tau O_{ij}^L = IO_{ij}^L / C_{ij}$$

Layering efficiency:

$$T_{ij}^L = \tau_{ij}^s / (\sum_{k=7}^{L} \tau O_{ij}^k + \tau_{ij}^s)$$

Traffic efficiency:

$$E_{ij}^L = \tau_{ij}^s / (\sum_{k=7}^{L} \rho O_{ij}^k + \tau_{ij}^s)$$

T_{ij}^L give a measure of gross to net information traffic and consequently of the adequacy of the protocol procedures for each service while E_{ij}^L indicate the intensity of time utilization within a call and capability of the call to share network resources. Both are function of the user, protocol and network behaviour and under a set of assumptions may be obtained following an extension of the procedures in /11/.

The protocol layering process applies to packet and circuit switching modes with very similar components in the high layers and in the call establishment and disconnection phases. An exception is made for stable call phase where marginal transmission traffic at layers 2 and 3 are zero for circuit switching mode.

The joint coding and protocol layering processes originate the following relationships between traffic and information demands:

$$\rho_{ij}^L = \lambda_{ij} I_{ij}^u (1/e_{ij})(1/E_{ij}^L)(1/C_{ij})$$

erlangs

$$\tau^{L}_{ij}=\lambda_{ij} I^{u}_{ij}(1/e_{ij})(1/T^{L}_{ij})(1/C_{ij})$$

erlangbits

which become interesting equations for traffic demand derivation when communication procedures or media change and ensure consistency among the different parameters involved in a service.

An illustration of a projection made for traffic services and classes in a business area in 1990 is summarized in Fig. 3 and 4 which represent average values.

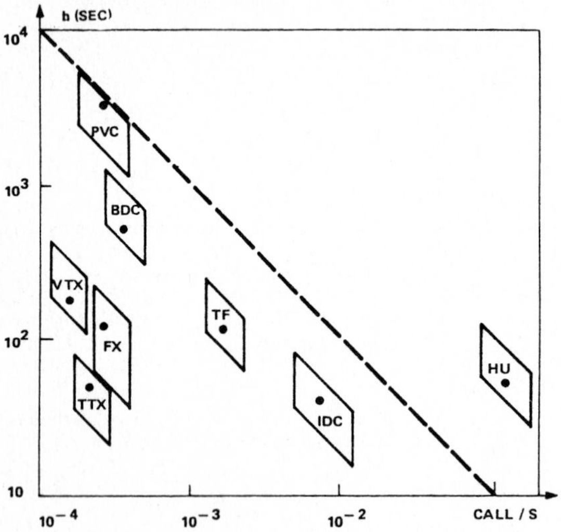

FIG. 3.– HOLDING TIME VERSUS CALL RATE

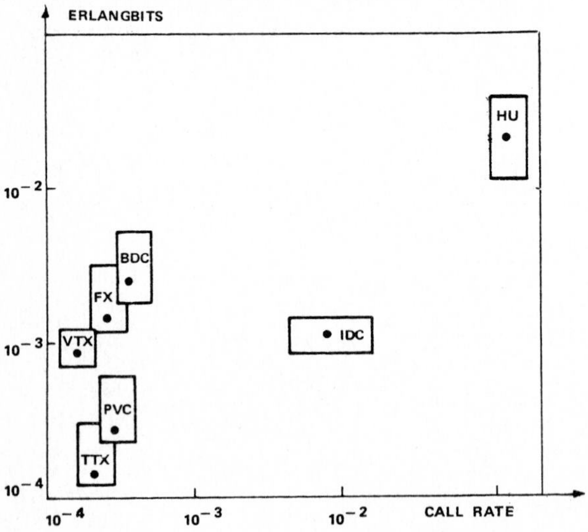

FIG. 4.– ERLANGBITS PER SERVICE AND CLASS

where:

TP = Voice telephone service
VTX = Videotex service
TTX = Teletex service
FX = Facsimile service
IDC = Interactive Data class
BDC = Batch Data class
PVC = Permanent Virtual Call class
HU = Heavy User class for access ports to host computers

2.3 Integrated Call Attempt

For those network interfaces or resources which handle a mix of services like the Basic Access, Primary Access, switching nodes and in general the high hierarchy levels of the network, all the services and traffic classes are grouped to simplify the treatment. An equivalent Integrated Call Attempt is defined as the weighted average of each traffic class demand for the convergent flows with their corresponding penetrations δ_{ij}. The ICAs are used for node and network dimensioning and are obtained by:

$$\lambda = \sum_i \sum_j \lambda_{ij}\, \delta_{ij} \;;\quad h = \sum_i \sum_j \frac{\lambda_{ij}}{\lambda} h_{ij}\, \delta_{ij}$$

$$\rho = \sum_i \sum_j \rho_{ij}\, \delta_{ij} \;;\quad \tau = \sum_i \sum_j \tau_{ij} \delta_{ij}$$

$$I = \sum_i \sum_j I_{ij} \frac{\lambda_{ij}}{\lambda} \delta_{ij} \;; B = \sum_i \sum_j B_{ij}\, \delta_{ij}$$

3. USER-TO-USER REFERENCE PATH CONFIGURATION

For the purpose of capacity/cost modeling a communication path is defined as the set of user and network resources directly or indirectly assigned to the user-user traffic flows and which impact path capacity and cost. The communication path subnetwork is composed of the source and destination equipments as well as all the nodes and links belonging to the K networks which provide their communication.

Fig. 5 represents a general configuration of user-to-user path which is supported by K different networks with L(k) nodes being used within each network k. The complete path comprises

$$\sum_{k=1}^{K} L(k) \text{ nodes and } \sum_{k=1}^{K} L(k)-1 \text{ links.}$$

This configuration aims to represent several network scenarios such as a single dedicated network, mixed networks with a medium level of integration or fully integrated networks in the long term in which all the nodes will have multiservice handling capacity.

The degree of modeling at the path level covers the variety of networks and nodes encountered in packet, circuit and integrated networks to allow the comparison of alternatives without entering in detailed analysis specific to each type treated at other hierarchical levels in

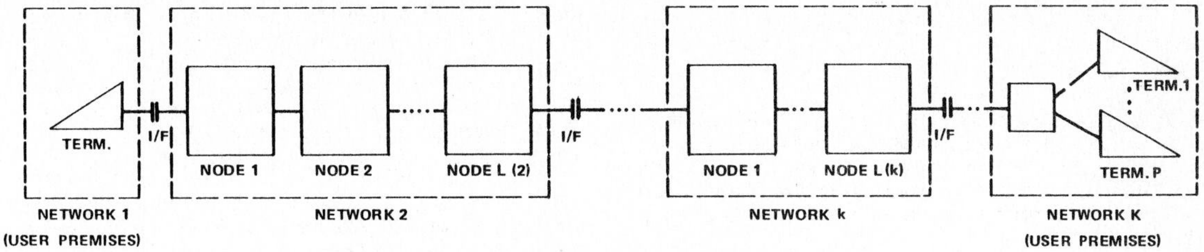

FIG. 5.– END TO END COMMUNICATION PATH

the modeling. Fig. 6 shows the block partition in the modeling of a hop. A node is represented by the collection of incoming, outgoing and common resources assigned to the path flows through the node. The transmission resources in the forward direction complete the composition of a hop. The block of common resources includes both those being also used directly by other flows and modules in the node and the overheads like operation, maintenance, etc which contribute to the costs but are not directly related to a basic call.

Cost and capacity values are assigned to the modules and flows throughout the path as explained in following paragraphs 4 and 5 to relate path cost to total and marginal flows.

4. TRAFFIC DEMAND AT NETWORK INTERFACES AND PATH DIMENSIONING

The different nodes in the path are connected by means of interfaces and different protocols may be used in each hop. In order to obtain the traffic demand at the interfaces where the protocol changes an iterative application of the incremental information process is build over the net user information I^{u}_{ij}.

Inside a node a frequent case for the packet mode flow is to have transparent mechanisms which send information from incoming to outgoing modules by adding internal information overheads to the external layer 3 information, when this is the case, in order to obtain the traffic demand at the node the efficiency procedure is applied assuming that the layer 3 information I^{3}_{ij} is conserved.

Performance models to derive the behaviour of each module are solved at the first hierarchical level of Fig. 1 for each design and the results are inputs to the path model. For the degree of detail required in the comparison of alternatives it is sufficient to evaluate the capacities associated with a given end-to-end loss probability in the circuit mode flow and to a given delay in the packet mode flow. This simplified modeling is based on the following assumptions:

- ° All the flows in the path have statistical independence among them.
- ° Calls arrive in a Poissonian law to all the modules in the path.
- ° Traffic assignment to the available resources in the path is considered homogeneous.

The simplified models for resources dimensioned under loss or delay criteria are summarized:

4.1 Simplified Switching Model

The offered traffic to a generic module klx for x=m,q,n,t is obtained from the flow traffic demand per service and traffic class that are served by each

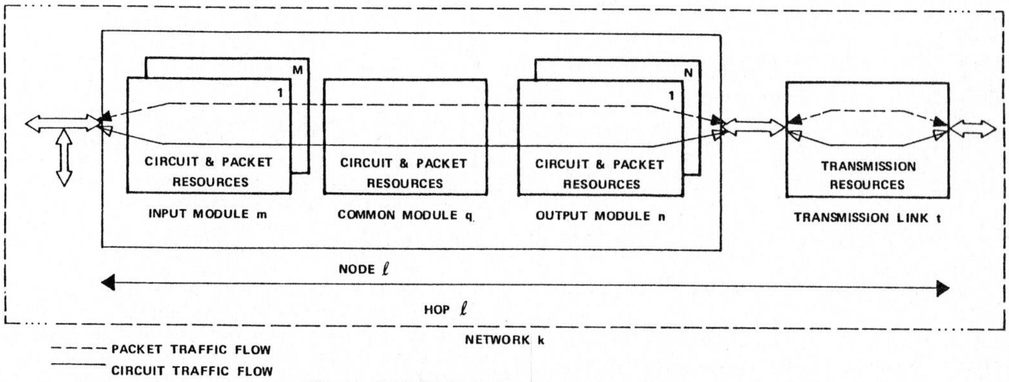

FIG. 6.– DEGREE OF DETAIL IN A HOP MODELLING

module as follows:

$$A = \Lambda(c)\ .\ G(c) + \tau^1(c) + \tau^1(p)$$

where:

$\Lambda(c)$ = Total call rate for the circuit mode calls at each module.

$G(c)$ = Weighted average holding time of the circuit mode calls.

$\tau^1(c) = \sum_i \sum_j \tau^1(c)_{ij}\ \delta(c)_{ij}$ = weighted average of transmission traffic for the packet mode traffic (erlangbits) generated by the circuit mode call signalling at protocol layer 1.

$\tau^1(p) = \sum_i \sum_j \tau^1(p)_{ij}\ \delta(p)_{ij}$ = weighted average of transmission traffic (erlangbits) for the packet mode traffic generated by the packet mode calls.

$\delta(c)$, $\delta(p)$ penetration of the services by circuit mode and packet mode calls.

Assuming that the objective to be fulfilled for the user is to reach an end-to-end loss probability lower than a specified value P_e, the following condition must hold for the K networks of the path:

$$1-\prod_{k=1}^{K}(1-P_k) \leq P_e$$

and, for the modules of the network k

$$1-\prod_{l=1}^{l(k)}(1-P_{klm})\ (1-P_{klq})\ (1-P_{kln}) \leq P_k$$

being P_{klm}, P_{klq}, P_{kln} the loss probabilities associated to the incoming, common and outgoing modules at the node l of the k network. The application to each module of the classical dimensioning procedures (i.e. Erlang and Engset formulae) relate the module sizes (VS_{klx}) and the carried traffics $A_{klx}(1-P_{klx})$. for x=m,n,q module types.

The partial traffic demand associated to each module klx due to our end-to-end user flow is:

$$\rho(c,U) = \sum_{\forall i,j \in Path} [\lambda(c)_{i,j} \times h(c)_{ij} + \tau^1(c)_{i,j}] \quad ; \quad \text{and}$$

$$\tau(p,U) = \sum_{\forall i,j \in Path} \tau^1(p)_{i,j}$$

for circuit mode and packet mode calls respectively.

Offered traffic to a module klx subject to dimensioning by loss criteria is obtained from the user demand and the corresponding loss probabilities of the preceding modules in the path. This allows determination of the ratio of user-to-user path traffic to the total carried traffic used in the cost assignment process.

4.2 Simplified Processing Model

The traffic demand for processing resources at each module klx for x=m,q,n is obtained from the demand per service and traffic class as:

$$D = \Lambda(p).s(p) + \Lambda(c).s(c) + B(p) + B(p)/w$$

where:

$\Lambda(c)$, $\Lambda(p)$ = Average call rates for circuit and packet mode switching.

$s(c)$, $s(p)$ = Mean number of control packets produced by a circuit or packet mode call.

By application to each module of the queueing modeling with classes associated to the control, data and acknowledgment packets with their respective processing times /6/, /12/, the relation among volume of resources per module VP_{klx}, delays and carried traffics are derived subject to nominal module capacity limits. Additional end-to-end performance modeling /3/, /13/ is done for the path flow to verify that end-to-end delays for control and data packets through all the nodes of the path is lower than the grade of service specification.

The traffic volume at the processing resources is given by:

$$F = \Lambda(p).s(p).ptc + \Lambda(c).s(c).pts + B(p).ptd + (B(p)/w).pta$$

where ptc, pts, ptd, pta are the processing times for control, signalling, data and ack packets respectively.

Partial traffic demands associated with the processing resources per module due to our observed end-to-end flow are:

$$f(c,U) = \sum_{\forall i,j \in Path} s(c)_{ij}\ \lambda(c)_{ij}\ .\ pts$$

$$f(p,U) = \sum_{\forall i,j \in Path} [s(p)_{ij}\ \lambda(p)_{ij}\ .\ ptc + B^3(p)_{ij}ptd + (B^3(p)_{ij}/w^3)pta]$$

The offered traffic due to the module resources subject to dimensioning by module and path delay, the corresponding occupancies and the sizes VP_{klx} are provided as inputs to the cost assignment process of the next section.

5. COST MODELING AND ASSIGNMENT

The purpose of cost modeling in the path is to find the total cost of resources associated with the end-to-end path and the partial cost to be assigned to a service or a group of services.

Criteria to assign the cost are directly associated with the volume of resources of switching, processing and transmission type through the path.

- Total path configuration costs are directly obtained from the module and resource dimensioning in the previous section and the function of the module cost versus module size.

$$\text{PATH COST}=\sum_k \sum_l [\sum_x (\text{S.COST}_{klx}(\text{VS}_{klx}) + \text{P.COST}_{klx}(\text{VP}_{klx}))+\text{T.COST}_{klt}(\text{VT}_{klt})]$$

where x = module m,q,n and S.COST, P.COST, T.COST are the switching, processing and transmission costs respectively.

- End-to-end partial costs for a specific service in the path associated with a user are allocated in proportion to the relative use of path resources and consequently to the ratio of service traffics ρ_{ij}, τ_{ij} to total module capacity for all the modules in the path.
The following assignment is obtained for a circuit mode service:

$$\text{COST}(c)_{ij}=\sum_k \sum_l [\sum_x (\text{S.COST}_{klx} \cdot \rho(c)_{ijklx}/A_{klx} + \text{P.COST}_{klx} \cdot f(c)_{ijklx}/F_{klx})] + \text{T.COST}_{klt}\, \rho(c)_{ijklt}/A_{klt}$$

For a set of end-to-end services sharing the same path the cost is:

$$\text{COST}(U)= \sum_{\forall i,j \in \text{Path}} [\text{COST}(c)_{ij}+\text{COST}(p)_{ij}]$$

By comparison of total and partial costs for a specific performance in each designed configuration the most economic may be obtained according to the call composition. Dedicated versus different levels of integration may be checked as well as the communication procedures within an integrated network such as circuit or packet mode, throughput class alternatives, etc.

For a general comparison including leased lines and semipermanent connections a common monthly charge period is taken as a reference. In this case costs comparisons provide a guide to tariff policy determination for the multiservice flows.

6. CONCLUSIONS

A unified characterization of multiservice traffic flows is used by relating the traffic demands to the original information volume demands, this establishes a relationship between information and traffic fields.

This characterization which includes the call content is the base for the capacity-cost comparison in the end-to-end configurations associated with the user path in different levels of integration. The finding of the most adequate alternatives for each traffic service and class form a basic step in ISDN network planning. Partial costs derivation also provide help in the establishment of tariff policies for the multiservice environment.

REFERENCES

/1/ CCITT SG. XVIII Rec. I.200, I.300 series.

/2/ O. González Soto, "On traffic modeling and characterization of ISDN users", Data Communication in the ISDN Era, Y. Perry, North-Holland, IFIP 1985.

/3/ O. González Soto, D. Gutiérrez, L. Martínez Miguez, "Analysis of a user communication path in a Store & Forward Network", X ITC, Montreal, 1983.

/4/ H. Miyahara, T. Hasegawa, Y. Teshigawara, "A comparative evaluation of switching methods in computer communication networks", ICC, 1975.

/5/ P. Kermani, L. Kleinrock, "A traffic study of switching systems", IEEE 1979.

/6/ K. Kummerle, H. Rudin, "Packet and Circuit Switching: Cost/Performance boundaries", Computer Networks, 1978.

/7/ H.C. Ratz, J.A. Field, "Economic comparisons of Data Communication services", Computer Networks, 1980.

/8/ I. Gitman, H. Frank, "Economic analysis of integrated voice and data networks. A case study", Proceedings of the IEEE. Nov. 1978.

/9/ C.E. Shannon, "Mathematical Theory of Communication", The BSTJ, Vol.27, 1948.

/10/ N. Abramson, "Information theory and coding", McGraw Hill, N.Y., 1963.

/11/ O. González Soto, L.M. Miguez, "Parameters and communication efficiencies in the modeling of a packet switched network", IX ITC, Torremolinos, 1979.

/12/ L. Kleinrock, "Queueing Systems", Vol.2, Willey, N.Y., 1975.

/13/ M. Schwartz, "Performance Analysis of the SNA Virtual Route Pacing Control", IEEE Trans. on Comm., Vol. 30, nr.1, 1982.

TELETRAFFIC ISSUES in an Advanced Information Society
ITC-11
Minoru Akiyama (Editor)
Elsevier Science Publishers B.V. (North-Holland)

WHO INTRODUCED THE BIRTH- AND DEATH TECHNIQUE?

J.W. COHEN

MATHEMATICAL INSTITUTE, UNIVERSITY OF UTRECHT
UTRECHT, THE NETHERLANDS

ABSTRACT

The technique of Birth- and Death equations plays a vital role in the theoretical investigations concerning the performance of models encounterd in teletraffic and computer systems. The Danish mathematician Erlang is usually credited for being the originator of this analytic technique. His ideas have been of paramount importance in the development of the analysis of the basic teletraffic models. However, he seems never to have used explicitly the Birth- and Death technique, and it may be questioned whether he applied this type of analysis.

In Erlang's papers no explicit indications are available that he has applied the Birth- and Death techniqeu. So the question arises who developed this type of analytical approach. It seems difficult to give a definite answer. In the early twenties Fry and Molina undoubtedly had fully mastered this technique, but also in the researches of the biological statisticians Mckandrick and Yule basic principles of this analytic technique can be found.

1. INTRODUCTION

The technique of Birth- and Death equations plays a vital role in the theoretical investigations concerning the performance of models encountered in teletraffic and computer systems. The Danish mathematician A.K. Erlang is usually credited for being the originator of this analytic technique. His ideas have been of paramount importance in the development of the analysis of the basic teletraffic models. However, he seems never to have used explicitly the Birth- and Death technique and it may be questioned whether he has applied this type of analysis.

In the present study we intend to throw some light un this question. Herefore we shall start with a derivation of the stationary distribution of the number of busy servers in an M/G/∞ model, a problem already considered by Erlang. Our derivation, which is based on geometrical considerations and which uses an idea stemming from D. van Dantzig, is extremely simple and provides a sharp insight in several phenomena such as insensitivity, reversibility, and the use of local and global properties of sample functions in the deduction of mathematical relations, cf. [2]. Comparison of the present derivation with the arguments as used by Erlang does strongly conjecture that Erlang mainly uses the global property of the sample functions, viz. "upcrossing intensity is equal to downcrossing intensity".

If as it seems very likely Erlang did no knew the Birth- and Death technique, so the question arises who developed this type of analytical approach. A definite answer seems to be difficult to give. In the early twenties Fry and Molina undoubtedly had fully mastered this technique, but also in the researches of the biological statisticians McKendrick and Yule basic principles of this analytical technique can be found. It may be conjectured that Fry and Molina on the one side, McKendrick and Yule on the other side independently developed the principals of the Birth- and Death technique; they all used the technique of formulating equations between probabilities by means of recursive relations, a type of argumentation already applied by P.S. Laplace in his famous book: Théorie Analytique des Probabilités.

2. ON A GEOMETRIC MODEL OF M/G/∞.

We shall start with a Poisson point process with intensity λ on the horizontal line. Its points will be marked as "1" or "2" points, independently of each other with probabilities p_1 and p_2, $p_1 + p_2 = 1$. To a "1" point a line segment of length τ_1 is assigned, similarly to a "2" point one with length τ_2; these linesegments are plotted vertically, see figure 1, and their end points constitute a point process in the plane.

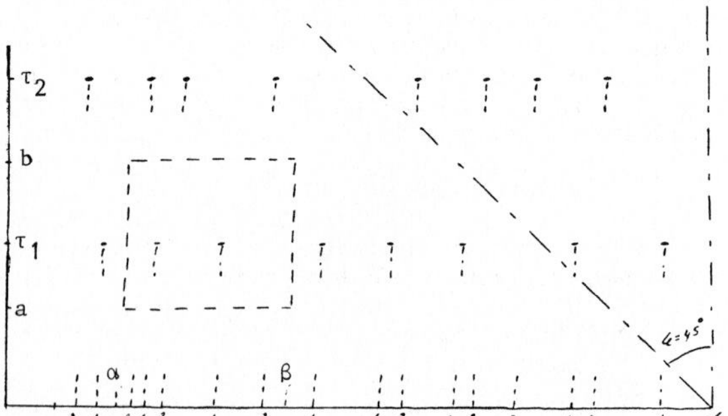

fig. 1

It is readily verified that the number of points contained in any rectangle $(\alpha,\beta) \times (a,b)$ has a Poisson distribution with intensity depending on $(\beta-\alpha)$ and on the position of a and b with respect to τ_1 and τ_2. Further it is evident that the number of points in disjoint rectangles are independent. By applying Carathéodory's extension theorem (cf. [1]) it follows that the distribution of the number of points in any measurable set in the plane is Poisson, i.e. the point process in the plane is a nonhomogeneous Poisson process. In particular it follows that the nubmer of points in the infinite wedge of angle $\phi = 45°$ to the vertical axis and with its top at the point t has a Poisson distribution with intensity $\lambda(p_1\tau_1+p_2\tau_2)$, as a simple calculation shows.

Let us next rotate every point of the point process in the plane clockwise by 90° around its projection on the horizontal line. The point process on the line so obtained is the sum of two independent Poisson processes with intensities λp_1 and λp_2, and as such is again a Poisson process with intensity λ. Obviously, the latter process may be regarded as the departure process of a service facility with an infinite number of servers and Poisson arrivals. The line segments with endpoints in the wedge fall on its top t after rotation and represent the service times of the customers being served at time t. Hence their number has a Poisson distribution. The service time distribution is obviously given by

$$p_1U(\tau-\tau_1) + p_2U(\tau-\tau_2),$$

here $U(\tau-\sigma)$, $\tau \in (-\infty,\infty)$ being the at σ-degenerated probability distributoin; i.e. $U(\tau-\sigma) = 0$ for $\tau < \sigma$, $= 1$ for $\tau > \sigma$.

The arguments used apply equally to a service time distribution

$$\sum_{i=1}^{m} p_iU(\tau-\tau_i),$$

with $o < \tau_1 < \ldots < \tau_m$; $\sum_{i=1}^{m} p_i = 1$. Their extension to a general service time distribution is rather obvious.

Hence we reach the conclusion that in an M/G/∞ system the stationary distribution of the number of busy servers is Poisson, and the departure process is a Poisson process; these properties being independent of the type of service time distribution. This latter phenomenon is the so called *insensitivity property* of the M/G/∞ model.

Another evident consequence of the argument above is the reversibility of the process, i.e. the stochastic structure of the process formed by the number of busy servers at time t is identical for t increasing or decreasing (if the process is stationary).

3. ERLANG'S PRINCIPLE OF STATISTICAL EQUILIBRIUM.

Next consider two wedges (as in figure 2) a distance Δt apart, again with two levels τ_1 and τ_2.

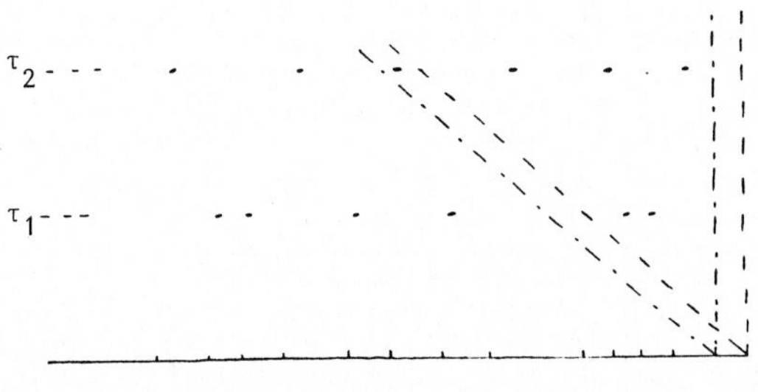

Obviously, the probability that the vertical strip formed by the vertical boundaries of the wedges contains just one point is $\lambda\Delta t + o(\Delta t)$. To calculate the probability that the sloping strip between the sides of the wedges contains one point, note htat the probability of having K points in the wedge at t with n points at level τ_2 is equal to

$$\frac{(\lambda p_2\tau_2)^n}{n!}\frac{(\lambda p_1\tau_1)^{k-n}}{(k-n)!}e^{-\lambda(p_1\tau_1+p_2\tau_2)}.$$

Given that at level τ_2 there are n points, then they are all uniformly distributed on $[0,\tau_2]$, and similarly for level τ_1. Hence the probability of having K points in the wedge at t and no points in the sloping strip is equal to

$$\sum_{n=0}^{k}\{\frac{\tau_2-\Delta t}{\tau_2}\}^n\{\frac{\tau_1-\Delta t}{\tau_1}\}^{k-n}\frac{(\lambda p_2\tau_2)^n}{n!}\frac{(\lambda p_1\tau_1)^{k-n}}{(k-n)!}\cdot$$

$$\cdot\, e^{-\lambda(p_1\tau_1+p_2\tau_2)} =$$

$$\{1-\frac{\Delta t}{p_1\tau_1+p_2\tau_2}\}^k\frac{\{\lambda(p_1\tau_1+p_2\tau_2)\}^k}{k!}\cdot$$

$$\cdot\, e^{-\lambda(p_1\tau_1+p_2\tau_2)} =$$

$$\{1-\frac{k\Delta t}{p_1\tau_1+p_2\tau_2}+o(\Delta t)\}\Pr\{\underline{x}_t=k\} \qquad \text{for } \Delta t\downarrow 0,$$

where $\underline{x}_t$ denotes the number of points in the wedge at time t. It follows that for $\Delta t \downarrow 0$.

$$\frac{k}{p_1\tau_1+p_2\tau_2}\Delta t + o(\Delta t)$$

is the conditional probability that the sloping strip contains just one point when the wedge at time t contains K; that of containing two points is of order $(\Delta t)^2$.

Consider two time points u and w with $w > u$ and such that $\underline{x}_u = 0$, $\underline{x}_w = 0$, and $\underline{x}_v > 0$ for some $v \in (u,w)$. Then the number of upward jumps from $\underline{x}_t = x$ to $\underline{x}_t = k+1$ for t lying between u and v is equal to that of the downward jumps from $k+1$ to k, with probability one; note that jumps larger than one have probabilities of $o(\Delta t)$. This is actually the principle of *statistical equilibrium* as applied by Erlang. It holds if points u and w exist with probability one. Application of this principle shows that the intensity of upward jumps from k to $k+1$ is equal to that of downward jumps from $k+1$ to k.

Using the probabilities derived above, the equality of these intensities is expressed by

$$(1)\quad \lambda \Pr\{\underline{x}_t=k\} = \frac{k+1}{p_1\tau_1+p_2\tau_2}\Pr\{\underline{x}_t=k+1\},\ k=0,1,2,\ldots.$$

From this relation it is readily seen that $\underline{x}_1$ has a Poisson distribution with parameter $\lambda(p_1\tau_1+p_2\tau_2)$, the same result as derived in the preceding section.

The principle of statistical equilibrium, which in present day terminology is called "partial balance" obviously applies to the sample functions of the number of busy servers in the M/G/N loss model, which includes the M/D/N and the M/M/N case. Here it is readily seen that the upward jump intensity at $\underline{x}_t = k$ is equal to $\lambda \Pr\{\underline{x}_t=k\}$, but it is rather hard to prove that the downward jump intesity at $\underline{x}_t=k+1$ is equal to $\{(k+1)/(p_1\tau_1+p_2\tau_2)\}\Pr\{\underline{x}_t=k+1\}$. The finiteness of N, the number of servers, complicates the stochastic structure of the process, and the argument used above for $N=\infty$ does not apply. Only if the service times are negative exponentially distributed (the case M/M/N) is a simple deduction possible, because of the memory less properties of this distribution. Erlang's principle of statistical equilibrium is not questioned, but its application fails according to the difficulty of transforming it into a mathematical equation. It took nearly sixty years before the basic idea of transforming global sample function propoerties into quantitative relations between probabilities was applied again in queueing analysis, cf. [2], [3], [10].

4. ON THE BIRTH- AND DEATH EQUATION

A stochastic process $\underline{x}_t$, $t \in (0,\infty)$ with a denumerable state space is a Birth- and Death process if the process is Markovian and if in every time interval $t \div t + \Delta t$ only transitions to a neighbouring state are possible (i.e. with probability $1-o(\Delta t)$). It is this stochastic structure which motivates the formulation of the so called Birth- and Death eqauations for the transition probabilities $p_{ij}(t,s) = \Pr\{\underline{x}_s=j\,|\,\underline{x}_t=i\}$, $s > t$. Obviously, these Birth- and Death equations are based on a local property of the sample functions of the process $\underline{x}_t$. For the stationary distribution $\Pr\{\underline{x}_t = x\}$, $x = 0,1,2,\ldots$, of the M/M/∞ queueing model with arrival rate λ and average service time β the Birth- and Death equations read

$$(2)\quad -\lambda \Pr\{\underline{x}_t=0\} + \frac{1}{\beta}\Pr\{\underline{x}_t=1\} = 0,$$
$$-(\lambda+\frac{k}{\beta})\Pr\{\underline{x}_t=k\} + \lambda\Pr\{\underline{x}_t=k-1\} +$$
$$+\frac{k+1}{\beta}\Pr\{\underline{x}_t=k+1\} = 0, \quad k = 1,2,\ldots .$$

It is readily seen that

$$(3)\quad \lambda\Pr\{\underline{x}_t=k\} = \frac{k+1}{\beta}\Pr\{\underline{x}_t=k+1\}, \qquad k=0,1,2,\ldots,$$

is a first "integral" of the system (2) of second order "recurrence" equations.

Actually, the relation (3) is the same as (1), see the preceding section, which has been derived from the global property: "the rate of upcrossings is equal to the rate of downcrossings".

In the present day literature Erlang is usually regarded as the first researcher who introduced the technique of Birth- and Death equations in the analysis of stochastic processes. The correctness of this viewpoint is questionable. A careful reading of Erlang's work (cf. [4]). shows that Erlang has never formulated equations which have the typical structure of the Birth- and Death equations. Erlang is rather sparing in providing explicit mathematical arguments. His most explicit argumenst occurs on page 141 of [4], and here his arguments are much closer to the use of the global property of up- and downcrossings than to the local property on which the birth- and death equations are based.

The question arises: where do we find in literature the first explicit introduction of the Birth- and Death technique in the analysis of stochastic processes.

In Fry's important book, cf. [5], published in 1928 the Birth- and Death technique is the standard method in the analysis of telephone traffic models. A similar explicit use of this technique is to be found in Molina's paper [6] of 1927. Fry's discussion of kinetic gas theory in his book points to the conjecture that he was strongly influenced in his analysis of stochastic processes by the ideas of kinetic gastheory. Explicit references concerning the origin of the Birth- and Death technique, however, do not occur in his book. There is no doubt, however, that both Molina and Fry completely understood this type of analysis.

The name Birth- and Death process presumably has been introduced by Feller. It raises the conjecture that such processes have been first encountered in medical and biological statistics. For an interesting historical account on this subject see the study [7]. Here a study of McKendrick [8] is mentioned in which the author is concerned with the analysis of a probability distribution related to the phenomenon of phagocytosis. In this study we encounter arguments and equations which have a structure that resembles that of the Birth- and Death analysis. Like Fry's McKendrick's reasoning seems to be influenced by the Methods of kinetic gastheory. Next to McKendrick's work that of Yule needs to be mentioned. The influence of their studies (around 1914-1930) on the development of the theory of stochastic processes in biology and medicine is discussed in [9], and it turns out that both McKendrick and Yule apply a type of analysis which comes very close to that what at present is called the Birth- and Death technique. Whether Molina and Fry, on the one hand, and McKendrick and Yule on the other hand knew of each others researches is not clear. My personal feeling is that Fry and Molina should be credited for the first shaping of the Birth- and Death type of analysis.

REFERENCES

[1] Loève, M. Probability Theory, D. van Nostrand Company, Inc., New York, 1960, second edition.

[2] Cohen, J.W. On Regenerative Processes in Queueing Theory, Lect. Notes, Econ. Math. Springer, Berlin, 1976.

[3] Cohen, J.W. On up- and downcrossings, J. Appl. Prob. 14 (1977) 405-410.

[4] Brockmeyer, E., Halstrøm, H.L., Jensen, A., The Life and Works of A.K. Erlang, Trans. Acad. Techn. Sciences, 1948, no. 2, Copenhagen, 1948, see also A.K. Erlang, Solutions of some problems in the theory of probability of significance in automatic telephone exchanges, Elektrotechnikeren 13 (1917) 5.

[5] Fry, Th. C. Probability and its Engineering Uses , D. van Nostrand, New York, 1928.

[6] Molina, E.C. Application of the theory of probability to telephone trunking problems, Bell Syst. Tech. J. 6 (1927) 461-494.

[7] Irwin, J.D. The place of mathematics in medical and biological statistics, J. Roy. Stat. Soc. A. 126 (1963) 1-44.

[8] McKendrick, A.G. Studies on the theory of continuous probabilisties with sperial reference to its bearing on natural phenomena of a progressive nature, Proc. London Math. Soc. 13 (1914) 401-406.

[9] Irwin, J.O. The contributions of G.U. Yule and A.G. McKendrick to stochastic process methods in medicine and Biology in: Stochastic Models in Medicine and Biology, ed. J. Gurland. Proc. Symp. Math. Res. Center, Univ. Wisconsin, 1963, Madison, Univ. of Wisconsin Press, 1964, p. 147-165.

[10] Stidham, S. A last word on $L = \lambda W$, Op. Res. 22 (1974) 417-421.

TELETRAFFIC ISSUES in an Advanced Information Society
ITC-11
Minoru Akiyama (Editor)
Elsevier Science Publishers B.V. (North-Holland)

AN ANALYSIS OF THE COOPERATIVE SERVICE SYSTEM USING NEARLY COMPLETELY DECOMPOSABILITY

Hidenori MORIMURA
Tokyo Institute of Technology
Tokyo, Japan

Toshihiko TSUNAKAWA
Fuji Xerox Co.,Ltd
Tokyo, Japan

ABSTRACT

A Markovian model of cooperative service system for voice messages and data stream is analysed using the nearly completely decomposability of the transition matrix. Numerical examples show that this method is applicable to a large number of channels case with sufficient accuracy. It has also a flexibility in the sense that the slot size for voice message may be defferent from the one for data message. These advantages make the method fit for practical use.

1 INTRODUCTION

Cooperative service systems for voice messages and data stream are often considered in literatures. Various approaches are treated for the modeling and analysis of them. Within the limits of the movable boundary schema which is our subject, several authors proposed analytical methods for the models, e. g., M.J.Fischer and T.C.Harris[2], H.Yasuda et al.[8], D.P. Gaver and J.P.Lehoczky[5] and A.Leon-Garcial et al.[7].

Here, we shall describe more precisely the system and model of the movable boundary schema of a cooperative service system. In it two types of call will be transmitted. We shall call them CS(Circuit Switching) calls and PS(Packet Switching) calls. Sometimes we shall call them voice messages and data, respectively. The communication channel is sliced temporally by time intervals with common length which are named as frames. Each frame is further sliced temporally by common length time intervals named as slots and is shared by k CS calls and n PS calls ($k \cdot I_c + n \cdot I_p \leqq N$), where N is the total number of slots in a frame and I_c and I_p are some positive integers. Each CS(PS) call will be assigned to $I_c(I_p)$ slots.

A certain portion of the frame capacity i.e., s slots, is allocated to CS calls. This class of traffic is treated as a loss system and is loaded into the frame on FCFS basis at the beginning of the frame period. The virtual time slot to which it is assigned is retained indefinitely during succeeding frames until the termination of the connection. Then, PS calls are assigned to the remaining slots. These calls may occupy the slots during a frame period.

Although the maximum number of slots for CS calls is constant s for all frames, the maximum number of PS calls may exceed $N-s$ occasionally. Thus we shall call the system as the movable boundary schema.

The relation between frame and slot and the allocation schema of these slots are illustrated in Fig.1.

An example given by Fischer-Harris[2] has the following characteristics.

frame period: 10 ms
frame length: 15440 bits
1 CS call occupies 80 bits/frame
1 PS call occupies 400 bits/frame
size of voice region(s): 50 channels = 4000 bits
size of data region($N-s$): 11200 bits
common channel inter-office signaling: 240 bits

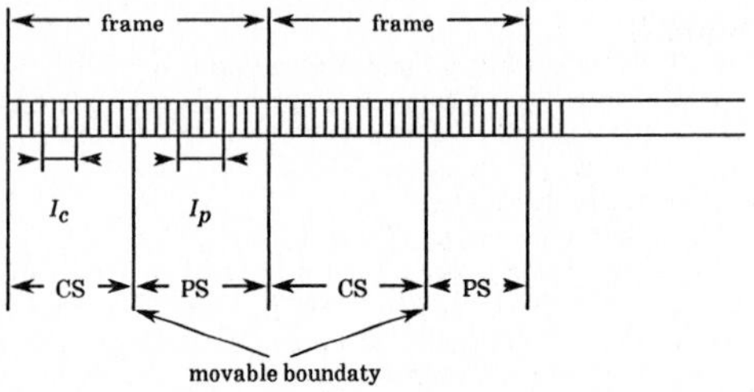

Fig.1 Time Sharing Schema

In the present paper we shall consider this example as an example for practical use and give some numerical examples whose charasteristics are close to it.

In Section 2, the performance model of the multiplex structure is described. It was analysed in some special cases by several researchers. We shall give a brief survey of them and point out their defects in numerical computation for practical use. Next section (Section 3) is devoted to present our Markovian model. An analysis of it using nearly completely decomposability is done in Section 4. And through some numerical examples in Section 5, the accuracy of our computational method is checked. In the last section, some considerations are added.

2 PERFORMANCE MODEL

The cooperative service system can be formulated as the following discrete time queueing system with two type arrivals, CS call(voice message) and PS call(data), and N channels. As illustrated in Fig.2, each CS call occupies I_c servers if there are idle servers more than I_c at the arrival epoch, and is lost otherwise. For CS calls s servers are available. It is not allowable to assign servers for CS calls beyond s.

Each PS call needs I_p servers instead of I_c and may be assigned to some servers beyond $N-s$ when there are idle servers for CS calls.

Let $\tau > 0$ be a real constant which represents the length of a frame period. All calls arrive randomly but at discrete time $k\tau(k=0,1,2,...)$. This means that the arrivals of CS calls and PS calls make binomial streams with mean arrival rate a_c and a_p, respectively.

In practical situation, arrivals may occur at any time. Thus we allow the situation to the outside of this queueing system, but support that all arrivals can occur only at the start of frame because the practical system may have some buffer memories.

A CS call arrived at the epoch when I_c servers for it are free will be served for tX, where X is a geometrically distributed random variable with $E[X]=1/d_c$. It is assumed by the practical situation that $1/d_c$ may be sufficiently larger than τ.

The other hand, each PS call requests I_p servers for the duration τ and waits if neccessary and feasible. However, if all servers except less than I_p and waiting room are occupied

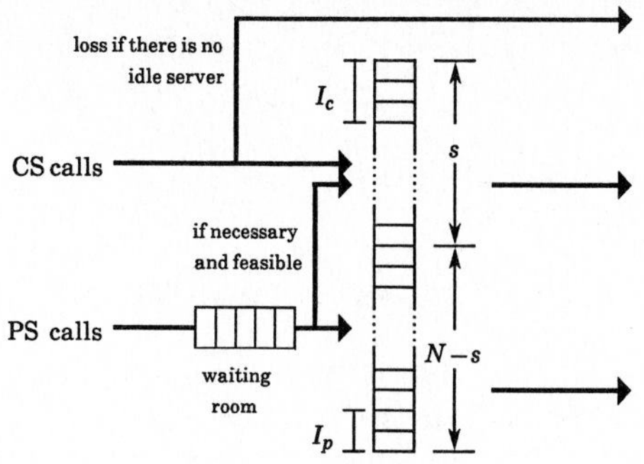

Fig.2 Queueing Model

completely by other calls, the arrival PS call will be lost. It is assumed here that the total number of PS calls in the system is limited by M.

As illustrated in Fig.2, the size of assignment of servers for CS call may be different from the one for PS call. We say the situation as flexibility. But, almost all former studies were under more serious conditions. In the other words, they are slightly inferior in flexibility.

Most remarkable feature of cooperative service system dealt with here is movable boundary by which the number of available servers for PS calls may change depend on the number of CS calls in the system. Thus we must consider the numbers of CS calls and PS calls jointly. This causes an exact analysis of the system to be difficult. In the past, several approximation methods have been proposed with numerical comparison to be estimated values got by simulation. We can cite Kummerle[6], Fischer-Harris[8], Fisher[3], Weistein-Malpass[9], Lehoczky-Gaver[5], Yasuda-Okada-Nakanishi[8], etc. as representative examples of the approximated analysis. But unfortunately they are not sufficient in accuracy, computational effort and/or flexibility.

It is meaningful to see the assumptions and computational techniques in above studies. We shall quote here [2] and [8] since we can proceed numerical computation by them.

The assumption in [2] are that (i) $M=\infty$, i.e., there is an infinite waiting room and (ii) $I_p=nI_c$, where n is a positive integer. The later asumption seems to be flexible, but is not so. Because, in a stage of computation, we have to find the zeros of

$$z^{[\frac{s}{n}]+N-s}\, e^{a_p\tau(1-z)} - \sum_{k=0}^{[\frac{s}{n}]} \eta_k z^k \tag{1}$$

in the unit circle and it is difficult to cmputate it effectively. In the formula(1), we put

$$\eta_k = \sum_{i=s-k(n+1)+1}^{s-kn} v_c(i), \tag{2}$$

where $\{v_c(i)\}$ is the equilibrium distribution of the number of CS calls in the system. If we put as $n=1$, the formula (1) become

$$z^N e^{a_p\tau(1-z)} - \sum_{k=0}^{s} v_c(k) z^k. \tag{3}$$

And since if we neglect the effect of PS calls, the behavior of CS calls may be analysed through M/M/s/s queueing system, we can regard the equilibrium distribution in the queueing system as $\{v_c(k)\}$. Hence, the second term in (3) is approximated by

$$\sum_{k=0}^{s} v_c(k) z^k \simeq \frac{a\sum_{k=0}^{s}\frac{(\rho_c z)^k}{k!}}{b\sum_{k=0}^{s}\frac{\rho_c^{\,k}}{k!}} \simeq e^{-\rho_c(1-z)} \tag{4}$$

When we accept the approximation, we can reduce (3) to the formula

$$z^N e^{a_p\tau(1-z)} - e^{-\rho_c(1-z)} \tag{5}$$

It is well known(e.g.[4]) how to compute the $z+N-s$ zeros of (5) in the unit circle. Additinal assumption of independence between CS calls and PS calls is requested there. This is not a good assumption since the effect of PS calls is essentially important. In fact, Yasuda et al.[2] shows that the numerical values got by the method are far from the corresponding ones got by simulation.

Yasuda et al.[8] attempted to treat the system exactly as far as possible and to approach via a Markov chain. They assumed there that (i) any plural CS calls cannot arrive in a frame, (ii) $n=1$. They also assumed that CS calls behave as a bulk and are independent from PS calls. Based on the assumptions an approximation procedure is proposed in which $2N-s$ zeros of

$$(1-a'\tau)(1-d'\tau)-(1-a'\tau)z^c e^{a_p\tau(1-z)} - (a'd'\tau^2+1-d'\tau)z^N e^{a_p\tau(1-z)} + z^N e^{2a_p\tau(1-z)} \tag{6}$$

are requested, where a' and d' are the effective arrival rate and effective service rate of a bulk of CS calls respectively.

As stated above, these two studies put assumptions which are convenient for simplification but not so suitable for the original model, hence the accuracies of approximations may be not so good in some cases. Further they need to compute zeros of some functions. Note taht computations of zeros are difficult in both sense of accuracy and effort for high power(i.e., large s). It may be serious for practical use.

3 A MORKOVIAN MODEL

Watching only the starts of frames(SOF), we can find a temporary homogeneous finite Markov chain for the pair of the number of CS calls and PS calls in the system. In that case, we support that there is a suitable size buffer memory for a little wait of arrived calls and open the "gate" at each epoch of SOF to introduce these calls in th system. All calls finished their service depart from the system just before SOF, and arrived calls in the previous frame may start to be served or to wait in the waiting room just after it except to be lost. At that time(i.e., just after SOF), let (k,j) be the state that there are k CS calls and j PS calls in the system and S be the state space. That is

$$S = \{(k,j);\, 0\leq k\leq s,\, 0\leq j\leq M\} \tag{7}$$

The pair of the number of PS and CS calls makes a Markov chain with the following transition probabilities $g\{(k,i),(l,j)\}$:

$$g\{(k,i),(l,j)\} = p(k,l)a_p^f(j,M) \qquad ;i\leq U(k),\, j\leq M \tag{8}$$

$$= p(k,l)a_p^f(j-(i-U(k)),M-(i-U(k)))$$

$$;\, U(k)<i\leq M,\, i-U(k)\leq j\leq M \tag{9}$$

$$= 0 \qquad ;\, U(k)<i\leq M, j<i-U(k), \tag{10}$$

where

$$U(k) = N-s+\left\lfloor \frac{I_c(s-k)}{I_p} \right\rfloor \tag{11}$$

$$p(k,l) = \sum_{x=0}^{min(k,l)} \gamma(k,x) a_c^f(l-x, s-x) \tag{12}$$

$$\gamma(k,x) = \frac{k!}{(k-x)!x!}(1-e^{-d_c\tau})^{k-x} e^{-xd_c\tau} \quad ; x \leqq k, x \leqq s$$

$$= 0 \quad ; otherwise \tag{13}$$

$$a_c^f(k,K) = e^{-a_c\tau}\frac{(a_c\tau)^k}{k!} \quad ; k<K$$

$$= 1 - \sum_{y=0}^{K-1} e^{-a_c\tau}\frac{(a_c\tau)^y}{y!} \quad ; k=K$$

$$= 0 \quad ; k>K \tag{14}$$

$$a_p^f(k,K) = e^{-a_p\tau}\frac{(a_p\tau)^k}{k!} \quad ; k<K$$

$$= 1 - \sum_{y=0}^{K-1} e^{-a_p\tau}\frac{(a_p\tau)^y}{y!} \quad ; k=K$$

$$= 0. \quad ; k>K \tag{15}$$

In below, we shall explain how to get these formulae. The condition $i \leqq U(k)$ means that there is no waiting PS call in the previous frame. Thus, the transition from (k,i) to (l,j) may occur with probability which is product of $p(k,l)$ and $a_p^f(k,M)$, where $p(k,l)$ given by (12) is the transition probability from k CS calls to l and $a_p^f(k,M)$ given by (15) is the probability of k PS calls arrived in the previous frame. This is (8). When $U(k) < i \leqq M$, since there are $i - U(k)$ PS calls waiting in the previous frame, these calls may be served in the next frame. Then, arrivals of $j-(i-U(k))$ PS calls result in existing of j PS calls in the next frame. Hence, (9) gives the transition probability in that case. It is also evident that the case such as (10) cannot occur.

Furthermore, note that if x of k CS calls existing in the previous frame retain their slots in the next frame and new $l-x$ CS calls are added, then there exist l CS calls in the next frame. Thus, we have $p(k,l)$ summing up these probabilities respect to x. It is easy to see that $\gamma(k,x)$ is given by (13).

If we put

$$u = kM+j \quad ; 0 \leqq k \leqq s, 0 \leqq j \leqq M \tag{16}$$

we can reduce the pair $\{(k,j)\}$ to the one-dimensional state $\{u\}$ in a lexicographical order. Let Q be the transition matrix arranged by the order and ν be the equilibrium distribution of the chain. It is well known that ν can be found solving the equation

$$\nu Q = \nu \tag{17}$$

under the normalization condition. However, it is impossible to solve it numerically in the sense of computational time, memory capacity and/or accuracy when the size of S is large. For example, if $s=3$, $N=10$, $M=100$ total number of states is 404, hence the direct method of solving has no any value in practical use. This is the reason that we need some approximation method to solve (17).

4 NEARLY COMPLETELY DECOMPOSABILITY

Let P be the transition matrix of the number of CS calls. i.e.,

$$P = [p(k,l)] \quad ; k,l = 0,1,\ldots,s \tag{18}$$

and Q^* be a matrix

$$Q^* = \begin{bmatrix} Q_0^* & 0 & \cdots & \cdots & 0 \\ 0 & Q_1^* & 0 & \cdots & 0 \\ \vdots & & \ddots & & \vdots \\ 0 & \cdots & \cdots & 0 & Q_s^* \end{bmatrix}, \tag{19}$$

where,

$$Q_k^* = \begin{bmatrix} a_k(0,M) & a_k(1,M) & \ldots & a_k(M,M) \\ \vdots & \vdots & & \vdots \\ a_k(0,M) & a_k(1,M) & \ldots & a_k(M,M) \\ 0 & a_k(0,M) & \ldots & a_k(M\text{-}1,M\text{-}1) \\ & & & \\ 0 & 0\ldots0 & a_k(M\text{-}U(k),U(k)) \ldots & a_k(U(k),U(k)) \end{bmatrix} \tag{20}$$

$U(k)$ rows

Then, we can represent Q as

$$Q = \begin{bmatrix} Q_{00} & Q_{01} & \cdots & \cdots & Q_{0M} \\ Q_{10} & Q_{11} & & \cdots & \vdots \\ \vdots & & \ddots & & \vdots \\ Q_{1M} & \cdots & \cdots & \cdots & Q_{MM} \end{bmatrix} \tag{21}$$

$$Q_{kl} = p(k,l) Q_k^* \tag{22}$$

Further, we shall introduce the following notations.

$$\varepsilon_k = \sum_{l \neq k} p(k,l) = 1 - p(k,k) \tag{23}$$

$$\varepsilon = max(\varepsilon_0, \varepsilon_1, \ldots, \varepsilon_s) \tag{24}$$

$$c_{kk} = -\frac{\varepsilon_k}{\varepsilon} Q_k^* \tag{25}$$

$$c_{kl} = -\frac{p(k,l)}{\varepsilon} Q_k^* \quad ; k \neq l \tag{26}$$

$$C = [c_{kl}]. \tag{27}$$

Then, we can rewrite as

$$Q = Q^* + \varepsilon C. \tag{28}$$

And ε may be evaluated as follows.

$$\varepsilon = 1 - min\{p(k,k); k\}$$

$$\leqq 1 - min\{\sum_{x=0}^{k} \gamma(k,x) a_c(k-x, s-x); k\}$$

$$\leqq 1 - min\{min\{e^{-d_c\tau k - a_c\tau}; 0 \leqq k \leqq s-1\}, e^{-d_c\tau s}\}$$

$$\leqq max\{1 - e^{-d_c\tau(s-1) - a_c\tau}, 1 - e^{-d_c\tau s}\} \tag{29}$$

(29) shows that ε is sufficiently small provided a_c and d_c are sufficiently small and then Q can be approximated by Q^* using the concept of nearly completely decomposability.

In [1], P.J.Croutois shows that the equilibrium distribution $\{\pi_{iI}\}$ may be approximated by $\{x_{iI}\}$ given by the product

$$x_{iI} = X(I) \cdot \nu_I(i) \quad ; i = 0,\ldots,M, I = 0,\ldots,s \tag{30}$$

where $X(I)$ is the equilibrium distribution of the Markov chain of the aggregate states whose transition probability is P and $\{\nu_I(i)\}$ is the equilibrium distribution in each aggregate state I. The order of the difference between $\{\pi_{iI}\}$ and $\{x_{iI}\}$ is ε^2.

In comparison with the direct computation, the computational efforts, memory capacity and computational time are drastic decreasing for the large size of transition matrix. For example, these quantities are evaluated in the case of $s=3$, $N=10$, $M=100$, $I_c=I_p$ as the following table. In the table and heareafter we denote the direct method and the method using nearly completely decomposability as DM and NCD, respectively. Above discrepancy may rapidly increase as N, s

method	memory	computational effort
DM	about 160k	one 400 dimensional linear equation
NCD	about 10k	four 100 and one 4 dimensinal linear equations

Table 1 Comparison of Computational Effort

and M increase and it may be impossible soonly to find the equilibrium distribution directly. But our approximation method may be hopeful in that cases.

Now, we shall find some measures of effectiveness of performance. Hereafter, we shall denote the equilibrium distribution v such as

$$v = \{ v(i,j) \; ; 0 \leqq i \leqq s, \; 0 \leqq j \leqq M \}.$$

In other words, $v(i,j)$ means the steady state probability that there are i CS calls and j PS calls in the system. First, let us find the loss probability of CS calls. If there are j CS calls in the system at a frame and $j-i$ calls finish their service, then $s-i$ servers for CS calls will be free at SOF of the next frame. Thus, the mean possible number of CS calls to be started their service just after SOF of the next frame is given by

$$\sum_{j=0}^{s} \sum_{i=0}^{j} v(j) \gamma(j,i) \sum_{k=0}^{s-i} k a_c^f(k, s-i)$$

from which we can find the loss probability l_c, i.e.,

$$l_c = 1 - \tau^{-1} \sum_{j=0}^{s} v(j) \sum_{i=0}^{j} \gamma(j,i) \sum_{k=0}^{s-i} k a_c^f(k, s-i). \qquad (31)$$

Similarlly the loss probability l_p for PS calls is given by

$$l_p = 1 - \tau^{-1} \sum_{j=0}^{s} \{ \sum_{i=0}^{U(j)} v(j,i) \sum_{k=0}^{M} k a_p^f(k, M)$$

$$+ \sum_{i=U(j)+1}^{M} v(j,i) \sum_{k=0}^{M-(i-U(j))} k a_p^f(k, M-(i-U(j))) \} \qquad (32)$$

Further, we can find the mean number of CS calls in system(L_c), the mean system time of PS calls(W_p) and so on, and may use these as measures of effectiveness of the system performance. For example

$$W_p = \sum_{j=0}^{s} \sum_{i=0}^{M} i v(j,i) / (1 - l_p) \qquad (33)$$

In either case, only finding the equilibrium distribution $v(i,j)$ may be requested.

5 COMPUTATIONAL EXAMPLES

In order to know how effect the approximation method using nearly completely decomposability does, we shall show some numerical examples and compare with the results using direct method and the approximation methods proposed by Yasuda et al.[2]. The later method is used here as a superior one in some approximation techniques got in the past. We shall call it here as the approxiamtion method VCA(Virtual Call Approximation) which is slightly modified to a convenient form in the comparison with our method. More precisely, in VCA the number of PS calls is limited by $M < \infty$ and the computational capability is extended to the case $n \neq 1$. Let us introduce the following notations.

L_p: mean number of PS calls in the system

Example	1	2	3	4	5	6
s	2	3	3	3	8	16
c	2	7	7	7	8	4
I_c	1	1000	1000	1000	1	1
I_p	1	1000	1000	1000	2	2
M	30	50	50	50	100	100
$a_c\tau$	.001	.0035	.007	.015	.0004 .0008	.0004 .0008
$1/d_c\tau$	500	500	500	500	5000	5000

Table 2 Parameters of Examples

example	1	2,3,4	5	6
frame length(sec.)	.1	.1	.01	.01
frame length(bits)	10000	10000	15440	15440
speed(bps)	100k	100k	1544k	1544k
number of voice slots	2	3	8	16
number of packet slots	2	7	8	4
voice slot length(bits)	2500	1000	640	640
packet slot length(bits)	2500	1000	1280	1280
limit number of voice in system	2	3	8	16
limit number of packet in system	30	50	100	100
arrival rate(/sec.)	.01	.035 .07 .15	.04 .08	.04 .08
average service time of voice(sec.)	50	50	50	50
average service time of packet(sec.)	.1	.1	.01	.01

Tabel 3 Interpretation of Table 2

a_h: arrival rate of virtual CS calls
$1/d_h$: mean service time for virtual CS calls
ρ_h: utilization factor of virtual CS calls

Suppose that the following relations are valid in a M/M/1 queueing system.

$$d_h \rho_h = d_c L_c \qquad (34)$$

$$\rho_h = L_c / s \qquad (35)$$

$$\rho_h = a_h / (a_h + d_h - a_h d_h) \qquad (36)$$

From (34), (35) and (36) we get

$$a_h = s d_c L_c / (s - L_c) + s d_c L_c \qquad (37)$$

$$d_h = s d_c \qquad (38)$$

Putting a transition probability matrix of order $2(M+1)$ as

$$Q^* = \begin{bmatrix} (1-a_h\tau)Q_0^* & a_h\tau Q_0^* \\ (1-a_h\tau)d_h\tau Q_s^* & (1-d_h\tau + a_h d_h \tau^2) Q_s^* \end{bmatrix} \qquad (39)$$

the equilibrium distributions of virtual CS calls are calculated.

DM computes the equilibrium distribution by solving the equation(17) under the condition of normality:

$$\sum_{j=0}^{s} \sum_{i=0}^{M} v(j,i) = 1 \tag{40}$$

using a well-known technique. In this case, the size of the transition matrix Q is $(s+1)(M+1) \times (s+1)(M+1)$. We utilize the result calculated by the DM as the exact values.

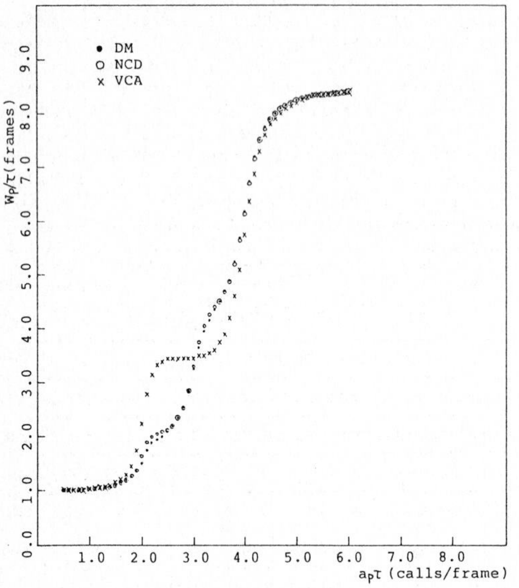

Fig.3 Example 1

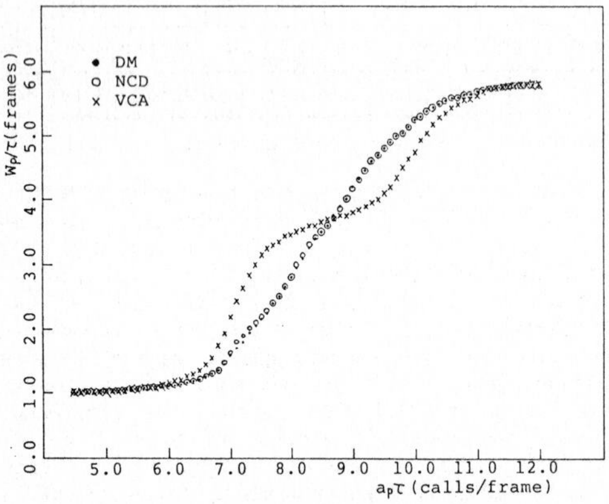

Fig.4 Example 2

Six examples are shown here whose parameters given in Table 2. Since Examples 5 and 6 have too many states to compute the equilibrium distributions by DM, we find them in these cases via simulation. The programs of the above computations are written using Fortran(double precision) and Pascal with a pseudo-randodm number generated by multiplicative congruential method.

In addition, the parameters in Table 2 may be interpreted to these in Table 3 as practical data in a multiplex switching system.

In Fig.3 to 8, these numerical results computed by three method are illustrated. We can see that NCD gives sufficiently good approximation through all examples, but VCA does not for some values of a_p. In addition, as mentioned above, the

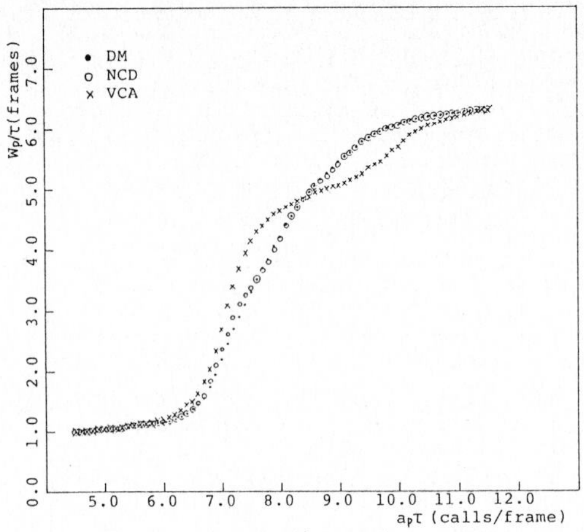

Fig.5 Example 3

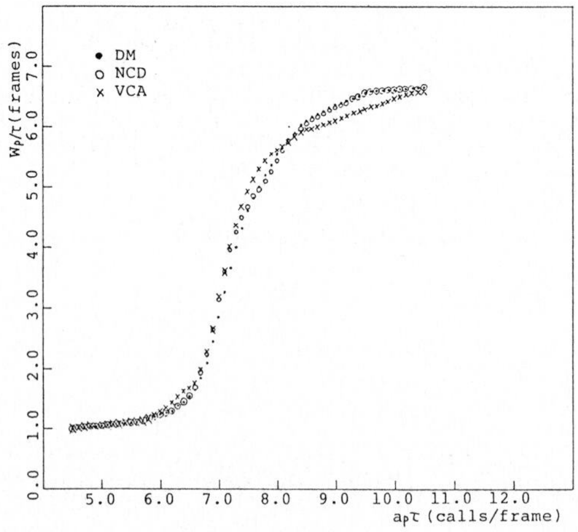

Fig.6 Example 4

computational effort for NCD is considerably decreasing compared with the one for VCA. To see this, in Table 4, the sizes of linear equations which provide an essential part of the computational efforts are summarized. The maximum degree of coupling ε given in (24), the mean number of CS calls in the system L_c and the loss probability of CS calls l_c given in (31) are also added in Table 4.

Ex	size of linear equations			ε	L_c	l_c
	DM	VCA	NCD			
1	93	62	31×3	.00399	.46200	.07692
2	204	102	51×4	.00075	1.44806	.17262
3	204	102	51×4	.00110	2.09282	.40211
4	204	102	51×4	.00190	2.56893	.65751
5	simulation		101 ×9	.00180	1.99848	.00086
6	simulation		101 ×17	.00220	3.87871	.03042

Table 4 Size of Linear Equations

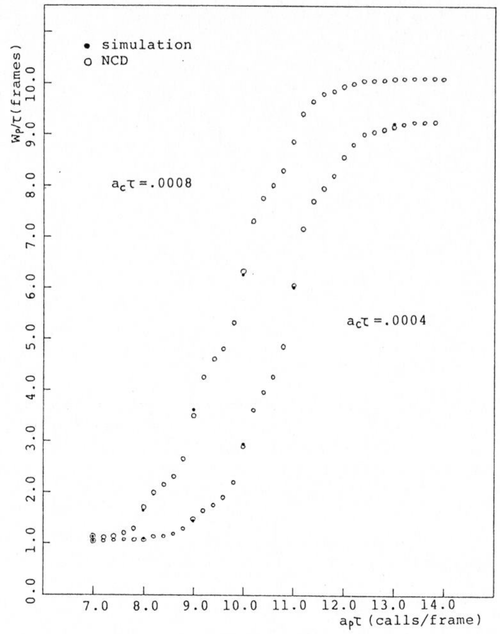

Fig.7 Example 5

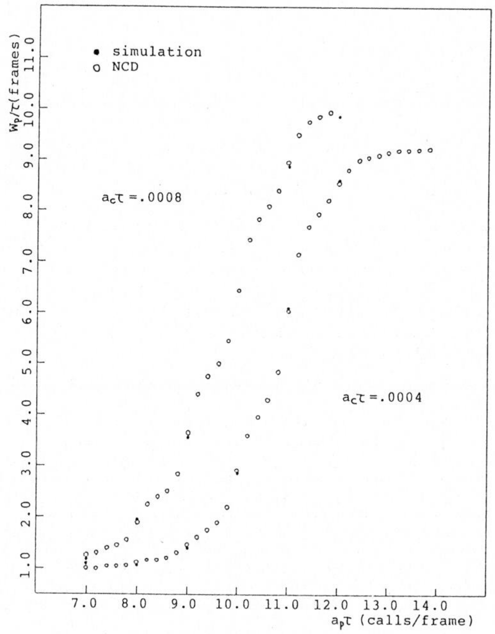

Fig.8 Example 6

6 SOME CONSIDERATIONS

As seen in the previous section, even through the size of the transition probability matrix is too large to compute by other methods, numerical computation by NCD may be possible with considerably high accuracy. The case of Example 5 and 6 are practical examples of multiplex switching system in the communication system PCS-24. We use a middle size computer for these examples. Thus, the possible range of computation may extend from now on.

However, since it is an approximation method, it may be requested to check whether the degree of accuracy changes with a_p or not. Fig.3 to 6 shows that the discrepancies between values by NCD and ones by DM are slightly large at near of the saturation point of L_p in each aggregare state, e.g., $a_p=2$ and $a_p=3$ in Example 1. This may be caused by the fact that mean number of PS calls in the system in each aggregate state are close to M at near of these values of a_p, hence this strongly effects to the shape of global L_p. Fig.9 illustrates modally the circumstance.

From above considerations, the approximation errors may be dependent of M. Therefore, for some values of M and a_p, we calculate the error and summarize them in Table 5. As expected, the errors become to be large for large M. But we know where they are large and that the biases are positive, then a modification method of these may be developed.

a_p	$M=15$			$M=30$			$M=60$		
	DM	NCD	error rate	DM	NCD	error rate	DM	NCD	error rate
1.2	1.0415	1.0416	.0001	1.0415	1.0416	.0001	1.0415	1.0416	.0001
1.6	1.1141	1.1153	.0011	1.1151	1.1165	.0012	1.1151	1.1165	.0012
2.0	1.3354	1.3483	.0097	1.5318	1.6369	.0687	1.6871	2.2148	.3128
2.4	1.5819	1.5897	.0049	2.0383	2.0798	.2040	2.8667	3.0558	.0660
2.8	1.8894	1.8911	.0009	2.5472	2.5180	.0114	3.6952	3.4225	-.0738
3.2	2.3529	2.3599	.0030	3.9577	4.0573	.0252	7.0932	7.7823	.0972
3.6	2.7683	2.7700	.0006	4.6701	4.6592	-.0023	8.2927	8.1809	-.0135
4.0	3.2372	3.2350	.0007	6.2247	6.1732	-.0083	12.5510	12.0461	-.0402
4.4	3.6679	3.6665	.0004	7.7312	7.7249	-.0008	16.1902	16.1823	-.0005
4.8	3.9238	3.9233	.0001	8.1460	8.1452	-.0001	16.6252	16.6245	.0000
5.2	4.0534	4.0533	.0000	8.2912	8.2910	.0000	16.7706	16.7704	.0000

Table 5 Accuracy of NCD

It is known[1] that the approximation errors of NCD are dependent of ε and are the order of ε^2. But, since the knowledge of the order is not useful for practical use, we find out the values of the relative errors when ε changes and the utilization factor of CS calls to be fixed. These values are summalized in Table 6. A tendancy of goodness of approximation by ε can be easily seen. In a usual multiplex switching system, the accuracy of the aproximation may be sufficient, since ε is far smaller than 0.01. Note that the examples in Table 5 and 6 are based on Example 1. Beside the features of NCD mensioned above, this method is superior in its flexibility. Because, since its object is a Markov chain, the applicable region of objects is extensive. Further an extension to the case in which a PS call requests several frames may be proceeded.

In the present paper, we assume taht a "gate" opens at SOF and all arrived calls can wait until the epoch. This assumption is introduced for the sake of simplification of our model, however, it may be slightly unnatural to represent any hardware in practice. Since we predict that such simplification causes only a little bias for their results, we accept the assumption. If one wants to use a more suitable model to a practical system, he can proceed his analysis in an analogous way to the present study.

In addition, we shall note that we can simplify our computation procedure if we use some contrivances.

a_p	a_c=.001 d_c=.002. ε=.0004			a_c=.01 d_c=.02 ε=.004			a_c=.1 d_c=.2 ε=.0388		
	DM	NCD	error rate	DM	NCD	error rate	DM	NCD	error rate
1.2	1.0416	1.0416	.0000	1.0415	1.0416	.0001	1.0412	1.0421	.0008
1.6	1.1162	1.1163	.0001	1.1151	1.1165	.0012	1.1077	1.1178	.0091
2.0	1.6223	1.6360	.0084	1.5318	1.6369	.0687	1.3117	1.6468	.2555
2.4	2.0740	2.0782	.0021	2.0383	2.0798	.2040	1.7878	2.0959	.1723
2.8	2.5202	2.5164	-.0015	2.5472	2.5180	-.0114	2.5025	2.5348	.0129
3.2	4.0438	4.0550	.0028	3.9577	4.0573	.0252	3.5917	4.0802	.1360
3.6	4.6580	4.6569	-.0002	4.6701	4.6592	-.0023	4.7828	4.6819	-.0211
4.0	6.1769	6.1715	-.0009	6.2247	6.1732	-.0083	6.5445	6.1903	-.0541
4.4	7.7244	7.7237	-.0001	7.7312	7.7249	-.0008	7.7881	7.7370	-.0066
4.8	8.1442	8.1442	.0000	8.1460	8.1452	-.0001	8.1632	8.1560	-.0009
5.2	8.2900	8.2900	.0000	8.2912	8.2910	.0000	8.3033	8.3015	-.0002

Table 6 Accuracy of NCD

ACKNOLEDGEMENT

The authors indebted to Professor H. Okada of Kobe University for his help to know an example of practical systems.

REFERENCES

[1] P.J.Crutois, "Decomposability." Academic Press, 1977.
[2] M.J.Fisher and T.C.Harris, "A model for evaluating the performance of an integrated cirduit- and packet-switched multiplex structure", IEEE Trans. on Com., COM-24, no.2, 1976.
[3] M.J.Fischer, "Data performance in a system where data packets are trasmitted during voice silent periods-single channel case," IEEE Trans. on Com., COM-27, no.9, 1979.
[4] M.Fujiki and E.Ganbe, "Theory of communication traffic", (in Japanese) Maruzen, 1980.
[5] D.P.Gaver and J.P.Lehoczky, "Diffusion approximations for the cooperative service of voice and data messages," J. of Appl. Prob., vol.18, 1981.
[6] K.Kummerle, "Multiplexer performance for intergrated line and packet switched traffic," ICCC, 1974.
[7] A.Leon-Garcia, R.H.Kwong and G.F.Williams, "Perfromance evaluation methods for an integrated voice/data link," IEEE Trans. on Com., COM-30, no.8, 1982.
[8] H.Yasuda, H.Okada and Y.Nakanishi, "A queueing model of a time sliced cooperative switching system",(in Japanese) Memoir, Inst. Math. Sci., Kyoto Univ., 1980.
[9] C.J.Weinstein, MlL.Malpass and M.J.Fischer, "Data traffic performance of an integrated circuit- and packet-switched multiplex structure," IEEE Trans. on Com., COM-28, no.6, 1980.

TELETRAFFIC ISSUES in an Advanced Information Society
ITC-11
Minoru Akiyama (Editor)
Elsevier Science Publishers B.V. (North-Holland)

DELAYS IN QUEUES, PROPERTIES AND APPROXIMATIONS

Shlomo HALFIN

Bell Communications Research
Morristown, New Jersey 07960, USA

ABSTRACT

We consider approximations for the moments of the delay in a $GI/G/1$ queue which are obtained by assuming that the idle time distribution is the same as the equilibrium excess distribution of interarrival time. The resulting approximation for the expected delay depends on the first three moments of the interarrival time and the first two moments of the service time. The approximation for the second moment of the delay uses one additional moment of each distribution. As a partial justification for such approximations we prove theorems about the behavior of the idle time distribution in heavy traffic, and about the monotonicity of the expected delay for the $K_2/G/1$ family with respect to the third moment of the interarrival time. Next we show how to improve the approximations. Finally, we present numerical results which show that such approximations work well for various queues with medium to heavy loads.

1. INTRODUCTION

The problem of finding bounds and approximations for the delay in the $GI/G/1$ queue has been extensively studied. Results which were obtained in the 1960s are summarized in Kingman [6]. Those include the Kingman [5] and Marshall [10] bounds on the expected waiting time, and exponential bounds for the distribution itself. A current comprehensive survey of this area can be found in Daley et al. [4].

In recent years the need for algorithms for performance evaluation of networks of queues led to renewed interest in approximations. Non-Markorian networks cannot generally be treated by exact analysis at the current state of the art. Kuehn [8] and Whitt [11], [12] developed methods to analyze non-Markovian queues by considering few (usually two) moments of the external and internal flows. Thus it is important to find out to what extent these moments determine the congestion and delays at the nodes (queues), and develop approximations which depend on few moments. Whitt [13], [14] addressed those issues. He found that the expected number in queue (and therefore also the expected delay) depends mainly on the first two moments of the service time, and on the first three moments of the interarrival time. The dependence on the third moment is negligible in heavy traffic, but quite pronounced for moderate loads. Approximations for the expected delay which do not depend on the third moment of the interarrival time, like that of Kraemer and Langenbach-Belz [7], work well for certain regions of the parameters, but are off target in other regions.

In the current work we pursue the idea that for long busy periods the subsequent idle periods tend to be distributed as the equilibrium excess distribution of the interarrival time. Since idle times are intimately related to delays, this idea yields approximations for the moments of the delay. In particular, the approximation for the expected delay depends on the first three moments of the interarrival time and on the first two moments of the service time. Notation is introduced in Section 2. The approximations are presented and discussed in Section 3. In Section 4 we show that the idle times in the $GI/M/1$ queue indeed converge to the equilibrium excess distribution of the interarrival time in heavy traffic. In Section 5 we show that for the family $K_2/G/1$ the expected delay is a monotone decreasing function of the third moment of the interarrival time, when the first two moments are kept fixed. Numerical results for other families are presented in Section 6. A useful lower bound for the delay in $H_2/G/1$ queues is obtained in Section 7. We introduce an "improved" approximation for the expected delay in Section 8. Finally, the results are summarized and discussed in Section 9.

2. NOTATION

We denote by A,B and I variables possessing the distributions of the interarrival time, service time, and idle time, respectively. Let $a_i = E(A^i), b_i = E(B^i), i = 0,1,...$ and let $\rho = b_1/a_1$, be the load coefficient. We always assume that $\rho < 1$, and that enough of the a_i's and b_i's exist. Let D denote a variable possessing the stationary delay distribution, and let Y be a variable whose distribution is that of the difference between an interarrival time and a service time. Finally, let $a(s)$, $b(s)$, *and* $i(s)$ be the moment generating functions of A,B and I, respectively.

3. APPROXIMATIONS FOR THE MOMENTS OF THE DELAY.

Lemoine presented in [9] relations between the moments of the delay and the idle time. They can be rewritten as follows

$$\sum_{i=0}^{r} (-1)^j \binom{r+1}{j} E(D^j) E(Y^{r+1-j}) = E(Y) \frac{E(I^{r+1})}{E(I)}. \quad (3.1)$$

In particular for $r = 1$ and $r = 2$, one gets

$$E(D) = \frac{E(Y^2)}{2E(Y)} - \frac{E(I^2)}{2E(I)} \quad (3.2)$$

and

$$E(D^2) = \frac{E(I^3)}{3E(I)} + E(D) \frac{E(Y^2)}{E(Y)} - \frac{E(Y^3)}{3E(Y)}. \quad (3.3)$$

Formula (3.2), which appears in [10], gives rise to Kingman's bound $E(D) \leqslant \frac{E(Y^2)}{2E(Y)}$.

We examine the approximations which are produced by assuming that I is distributed as the equilibrium excess distribution of A. Specifically, we replace $E(I^r)$ by $\frac{a_{r+1}}{(r+1)a_1}$ in (3.1). In particular, we define by substituting in (3.2) and (3.3)

$$d_1^* = \frac{a_2 - 2a_1b_1 + b_2}{2(a_1 - b_1)} - \frac{a_3}{3a_2} \quad (3.4)$$

$$d_2^* = \frac{a_4}{6a_2} + d_1^* \frac{a_2 - 2a_1b_1 + b_2}{2(a_1 - b_1)} - \frac{a_3 - 3a_2b_1 + 3a_1b_2 - b_3}{2(a_1 - b_1)} \quad (3.5)$$

and proceed to check the behavior of the approximations

$$E(D) \sim d_1^* \quad (3.6)$$

and

$$E(D^2) \sim d_2^* . \quad (3.7)$$

Clearly, (3.6) and (3.7) are exact whenever the interarrival time is exponential, or exponential with an atom at 0, i.e. for the $M/G/1$ queue, and the $M/G/1$ queue with batch arrivals, where the batches have a geometric distribution. On the other hand, there are situations where these approximations are completely off. For instance d_1^* can become negative if a_3 is large.

Intuitively, one would expect that the approximations would work well if the busy periods are mostly long, i.e. the heavier the traffic, the better the approximation (except in the singular case of the $D/D/1$ queue).

4. THE $GI/M/1$ CASE

We prove the following theorem

Theorem 4.1 For a family of $GI/M/1$ queues with a fixed interarrival time distribution, the idle time distribution converges to the equilibrium excess distribution of the interarrival time when $\rho \to 1$.

Proof: Following [3], formula (5.194), we have

$$i(s) = \frac{b_1^{-1}(1-a(s)) - \sigma^*}{s - \sigma^*},$$

where σ^* is the unique solution in the interval (0,1) of the functional equation

$$a(\sigma^*) = 1 - b_1\sigma^* . \quad (4.1)$$

It is well known (see also the proof of Theorem 4.2) that when $\rho \to 1$, (*i.e.* $b_1 \to a_1$,) then $\sigma^* \to 0$. Thus

$$\lim_{\rho \to 1} i(s) = \frac{1 - a(s)}{a_1 s}$$

But the right hand side is the equilibrium excess distribution of A. This completes the proof.

For the mean delay we can prove the following

Theorem 4.2 Under the conditions of Theorem 4.1 we have

$$E(D) = d_1^* + O(1-\rho)$$

Proof: Using the expansion

$$a(s) = 1 - a_1 s + \frac{a_2}{2}s^2 + \frac{a_3}{6}s^3 + \cdots$$

equation (4.1) can be rewritten as

$$h(\sigma^*) = a_1(1-\rho) \quad (4.2)$$

where

$$h(s) = \frac{a_2}{2} - \frac{a_3}{6}s^2 + \cdots . \quad (4.3)$$

Following [2], we have that if $y = h(s)$, then the inverse function $s = h^{-1}(y)$ has the following expansion

$$s = \frac{2}{a_2}y + \frac{4}{3}\frac{a_3}{a_2^3}y^2 + O(y^3) . \quad (4.4)$$

Putting $y = a_1(1-\rho)$ in (4.4), and using (4.2) we get

$$\sigma^* = \frac{2a_1}{a_2}(1-\rho) + \frac{4}{3}\frac{a_1^2 a_3}{a_2^3}(1-\rho)^2 + O((1-\rho)^3) \quad (4.5)$$

or

$$\sigma^* = \frac{2a_1}{a_2}(1-\rho)\left[1 + \frac{2}{3}\frac{a_1a_3}{a_2^2}(1-\rho) + O((1-\rho)^2)\right] .$$

Thus

$$(\sigma^*)^{-1} = \frac{a_2}{2a_1(1-\rho)}\left[1 - \frac{2}{3}\frac{a_1a_3}{a_2^2}(1-\rho) + O((1-\rho)^2\right] . \quad (4.6)$$

But it well known that

$E(D) = (\sigma^*)^{-1} - b_1$, so by substituting (4.6) we get

$$E(D) = \frac{a_2 - 2a_1b_1(1-\rho)}{2a_1(1-\rho)} - \frac{a_3}{2a_2} + O(1-\rho) .$$

Finally, since $a_1(1-\rho) = a_1 - b_1$, and $b_2 = 2b_1^2$ for the exponential distribution, we get by comparing to (3.4)

$E(D) = d_1^* + O(1-\rho)$.

This completes the proof.

5. THE $K_2/G/1$ CASE

Following [3], we denote by K_2 the family of distributions whose moment generating functions are rational, with the degree of the denominator being at most 2. Thus for this family $a(s) = \frac{\alpha_1(s)}{\alpha_2(s)}$, where $\alpha_1(s) = us + v_2$ and $\alpha_2(s) = s^2 + v_1 s + v_2$. This three parameter family includes as subfamilies H_2, and E_2, the exponential, as well as other distributions.

If we consider subfamilies of K_2 for which the first two moments are fixed, then we get for all cases, except for the exponential distribution (as we will see later), a one parameter family, where the parameter can be taken as the third moment. We prove the following

Theorem 5.1 For the family $K_2/G/1$, if the first two moments of the interarrival time and the service distribution are kept fixed, then $E(D)$ is a monotone decreasing function of the third moment of the interarrival time.

Proof: Let $q_i = \frac{a_i}{i!}$, $i=1,2,3$. It is somewhat more convenient to use the q_is instead of the a_is. Differentiating the identity

$$(s^2 + v_1 s + v_2)\, a(s) = us + v_2$$

three times, and each time evaluating at $s = 0$, we get three equations, from which we solve v_1, v_2, and q_3 in terms of u, q_1, *and* q_2

$$v_1 = \frac{uq_2 - q_1}{q_2 - q_1^2};\ v_2 = \frac{uq_1 - 1}{q_2 - q_1^2};\ q_3 = \frac{uq_2^2 - 2q_1q_2 + q_1^3}{uq_1 - 1}. \quad (5.1)$$

It is easy to verify that $uq_1 - 1 = 0$ if and only if $q_2 - q_1^2 = 0$, and that happens if and only if the interarrival distribution is exponential. But in that case q_3 and u are constants, and there is nothing to prove. So from now on we assume that $q_2 - q_1^2 \neq 0$ and $uq_1 - 1 \neq 0$.

By differentiating (5.1) we get

$$\frac{dq_3}{du} = -\left[\frac{q_2 - q_1^2}{uq_1 - 1}\right]^2 < 0\,.$$

Thus q_3 is strictly decreasing as a function of u. We complete the proof by showing that $E(D)$ is strictly increasing as a function of u.

By [3], formula (5.205) we have

$$E(D) = \frac{\rho}{b_1(1-\rho)}\left[\frac{a_2 + b_2}{2} + b_1\frac{\alpha_1'(0)}{\alpha_1(0)} - a_1\frac{\alpha_2'(0)}{\alpha_2(0)}\right] + \delta^{-1}, \quad (5.2)$$

where δ is the (unique) positive root of

$a(-x)\,b(x) = 1$.

Substituting (5.1) into (5.2) and collecting terms we get

$$E(D) = c - u\frac{q_2 - q_1^2}{uq_1 - 1} + \delta^{-1} \quad (5.3)$$

where $c = \frac{\rho}{b_1(1-\rho)}\frac{a_2 + b_2}{2} - \frac{q_1}{1-\rho}$, is a constant

not dependent on u. Thus

$$\frac{dE(D)}{du} = \frac{q_2 - q_1^2}{(uq_1 - 1)^2} - \delta^{-2}\frac{d\delta}{du} \quad (5.4)$$

The proof will be completed if we show that

$$\frac{d\delta}{du} < \delta^2\frac{q_2 - q_1^2}{(uq_1 - 1)^2}\,. \quad (5.5)$$

Solving $b(\delta) = \frac{\alpha_2(-\delta)}{\alpha_1(-\delta)}$ for a non-zero root is equivalent to solving

$$b^*(\delta) = \alpha^*(u,\delta) \quad (5.6)$$

where

$b^*(\delta) = \frac{1 - b(\delta)}{\delta}$, and

$$\alpha^*(u,\delta) = \frac{\alpha_1(-\delta) - \alpha_2(-\delta)}{\delta\,\alpha_1(-\delta)} = \frac{-\delta(q_2 - q_1^2) + (uq_1 - 1)q_1}{-u\delta(q_2 - q_1^2) + uq_1 - 1}\,.$$

Note that $b^*(s) = b_1\tilde{b}(s)$, where $\tilde{b}$ is a moment generating function. Thus $b^*(s)$ is strictly decreasing, and $b^*(0) = b_1$.

By the implicit function theorem we get from (5.6)

$$\frac{d\delta}{du} = \frac{\partial\alpha^*}{\partial u}\left[\frac{db^*}{d\delta} - \frac{\partial\alpha^*}{\partial\delta}\right]^{-1}. \quad (5.7)$$

Calculating the derivatives we get

$$\frac{\partial\alpha^*}{\partial u} = \frac{-\delta^2(q_2 - q_1^2)^2}{(-u\delta(q_2 - q_1^2) + uq_1 - 1)^2}$$

and

$$\frac{\partial\alpha^*}{\partial\delta} = \frac{(q_2 - q_1^2)\,(uq_1 - 1)^2}{(-u\delta(q_2 - q_1^2) + uq_1 - 1)^2}\,.$$

Consider the inequality

$$\frac{db^*}{d\delta} - \frac{\partial\alpha^*}{\partial\delta} < -\frac{\partial\alpha^*}{\partial\delta}\,. \quad (5.8)$$

Detailed analysis of how the functions b^* and α^* intersect, which is left to the reader, shows that either both sides of (5.8) are positive (when $q_2 - q_1^2 < 0$), or both sides are negative (when $q_2 - q_1^2 > 0$). Thus in both cases

$$\left[\frac{db^*}{d\delta} - \frac{\partial\alpha^*}{\partial\delta}\right]^{-1} > -\left[\frac{\partial\alpha^*}{\partial\delta}\right]^{-1}$$

Substituting in (5.7), and noticing that $\frac{\partial\alpha^*}{\partial u}$ is negative we get

$$\frac{d\delta}{du} < -\frac{\partial\alpha^*}{\partial u} \Big/ \frac{\partial\alpha^*}{\partial\delta} = \delta^2\frac{q_2 - q_1^2}{(uq_1 - 1)^2}\,.$$

So (5.5) is satisfied, which completes the proof.

6. NUMERICAL RESULTS FOR OTHER QUEUES

We compared simulation results of Albin [1] for the first two moments of the delay of various queues, to those obtained by using approximations (3.6) and (3.7). Tables 1 and 2 summarize the comparisons for $\rho = 0.7$ and $\rho = 0.9$, respectively.

Four service time distributions are used. The first two, a hyperexponential and a lognormal, have equal first and second moments, with the common squared coefficient of variation being $c_s^2 = 5$. Their third moments, however, are quite different. The results of the simulations show that indeed for both these distributions the first moments of the delay are close, while the second moments are quite distinct. The other two service time distributions, an exponential shifted by a constant, and an Erlang of order 2, also have

Table 1 Comparison of simulation and approximation results for $\rho = 0.7$.

A	S	$H_2 - 5$		LNOR-5		M+D - 0.5		E_2 - 0.5	
		$E(D)$	$E(D^2)$	$E(D)$	$E(D^2)$	$E(D)$	$E(D^2)$	$E(D)$	$E(D^2)$
$H_2 - 3.1$	Simulation	6.9	140	6.8	176	2.7	17	2.8	19
	Approximation	6.3	132	6.3	180	2.6	17	2.6	17
M+D - 0.58	Simulation	4.5	73	4.4	101	0.75	1.9	0.75	1.8
	Approximation	4.4	71	4.4	119	0.75	1.9	0.75	1.8

Table 2 Comparison of simulation and approximation results for $\rho = 0.9$.

A	S	$H_2 - 5$		LNOR-5		M+D - 0.5		E_2 - 0.5	
		$E(D)$	$E(D^2)$	E(D)	$E(D^2)$	E(D)	$E(D^2)$	E(D)	$E(D^2)$
$H_2 - 4.6$	Simulation	41	3800	39	3700	20	790	21	930
	Approximation	39	3400	39	3700	20	860	20	860
M+D - 0.52	Simulation	23	1200	22	1600	3.9	34	4.0	35
	Approximation	22	1200	22	1500	3.9	36	3.9	35

equal first and second moments, with $c_s^2 = 0.5$. However, unlike for the first pair, their third moments are pretty close. Here, as expected, the simulated results for both first and second moments of the delay differ very little between the two distributions. Two interarrival time distributions are used for each ρ, one being a hyperexponential, and the other a shifted exponential. Thus we have a variety of combinations of interarrival and service distributions, which hopefully cover a sizable portion of "real life" situations. The load range of 0.7 to 0.9 corresponds to medium to high traffic, a range where calculating delays is important.

Observing the numbers in Table 1 and 2, we conclude that (3.6) and (3.7) work well for these queues.

7. A LOWER BOUND FOR $E(D)$ FOR THE $H_2/G/1$ QUEUE.

Let

$$a(s) = p\,\frac{1}{1+m_1 s} + (1-p)\,\frac{1}{1+m_2 s} \tag{7.1}$$

with $0 < p < 1$, and $m_1 > m_2$.

Clearly we have

$$i(s) = p^{*}\,\frac{1}{1+m_1 s} + (1-p^{*})\,\frac{1}{1+m_2 s}, \tag{7.2}$$

where p^{*} is the probability that the customer which ends a busy period is followed by a "long" (exponential with mean m_1) interarrival time.

Thus

$p \geqslant P$ (customer is followed by a "long" interarrival time & customer starts a busy period) =

P (customer is followed by a "long" interarrival time, given that the customer starts a busy period). P (customer starts a busy period).

But

$$P\text{ (customer starts a busy period)} = \frac{E(Y)}{E(I)}$$

(see [10]).

Thus

$$p \geqslant p^{*}\,\frac{E(Y)}{E(I)} = p^{*}\,\frac{a_1(1-\rho)}{p^{*}(m_1-m_2)+m_2}. \tag{7.3}$$

Define

$$\bar{p} = \frac{m_2 p}{(1-\rho)m_2-\rho(m_2-m_1)p}. \tag{7.4}$$

From (7.3) it follows that if the denominator of the right hand side of (7.4) is positive, then

$$p^{*} \leqslant \bar{p}. \tag{7.5}$$

But

$$\frac{E(I^2)}{2E(I)} = \frac{m_2^2 + p^{*}(m_1^2-m_2^2)}{m_2 + p^{*}(m_2-m_1)}, \tag{7.6}$$

and the right hand side of (7.6) is monotone increasing in p^{*}. Thus if (7.5) is valid, we get

$$\frac{E(I^2)}{2E(I)} \leqslant \frac{m_2^2 + \bar{p}(m_1^2-m_2^2)}{m_2 + \bar{p}(m_2-m_1)}. \tag{7.7}$$

This proves the following theorem

Theorem 7.1 If $(1-\rho)m_2 - \rho(m_2-m_1)p > 0$, then

$$E(D) \geqslant \frac{E(Y^2)}{2E(Y)} - \frac{m_2^2 + \bar{p}(m_1^2 - m_2^2)}{m_2 + \bar{p}(m_2 - m_1)}, \qquad (7.8)$$

where $\bar{p}$ is given by (7.4).

The importance of Theorem (7.1) is that it enables us to use an alternative approximation for $E(D)$ when (3.6) fails due to a_3 being very large. If $a_2 > 2a_1^2$, we can fit a H_2 distribution to the first three moments of A, and use the corresponding bound (7.8), in case where it is larger than value which is produced by (3.6).

8. AN IMPROVED APPROXIMATION

The idea here is to take into account that some busy periods are short and the distribution of idle periods that follows may differ substantially from the equilibrium excess distribution of A.

Specifically, the probability of having a busy period of length 1 is

$$p_1 = P(Y > 0) = \int F_S(y)\, dF_A(y) \qquad (8.1)$$

where F_A and F_S are the distribution functions of A and S, respectively.

The distribution of an idle period I_1, which follows a busy period of length 1 is

$$P(I_1 \leqslant z) = P(Y \leqslant z | Y > 0)\,. \qquad (8.2)$$

The "improved approximation" follows by making the assumption that idle periods either follow a busy period of length 1, in which case they are distributed as (8.2), or else they are distributed as the equilibrium excess distribution of A.

The problem with this approach is the resulting approximation for the expected delay depends on more than just a few moments of A and S. However, in many special cases it is possible to obtain the corresponding approximation for the expected delay. For example, for $GI/M/1$, the expression for the delay depends on the usual moments of A and S, and also on $a(b_1)$.

Finally, Tables 3 and 4 present numerical results for the original and the improved approximations for $H_2/M/1$ distributions. Table 3 has $a_1 = 1$ $a_2 = 13$ and a_3 varying, with $\rho = 0.7$. Table 4 has $a_1 = 1$, $a_2 = 3$, $a_3 = 16.2$, and ρ is varying.

Table 3 - Relative errors for approximations of $E(D)$ for $H_2/M/1$ queues, with $a_1 = 1, a_2 = 13, \rho = 0.7$.

a_3	Original approximation	Improved approximation	Kraemer, Langenbach-Belz
254	0	0	0.40
281	0	0	0.37
313	0	0	0.33
352	0.01	0.01	0.28
401	0.02	0.01	0.21
468	0.05	0	0.09
565	0.15	0.04	0.13
1032	0.68*	0.68*	1.39
1946	0.40*	0.40*	2.67
18287	0.05*	0.05*	4.11

* These values were obtained by using the lower bound (7.8).

Table 4 - Relative errors for approximations of $E(D)$ for $H_2/M/1$ queues with $a_1 = 1$, $a_2 = 3$, $a_3 = 16.2$.

ρ	Original approximation	Improved approximation	Kraemer Langenbach-Belz
0.4	0.30	0.05	0.15
0.5	0.12	0.01	0.13
0.6	0.05	0	0.11
0.7	0.02	0.01	0.08
0.8	0.01	0.01	0.06
0.9	0	0	0.03

9. SUMMARY AND CONCLUSIONS

We presented various approximations for the first two moments of the delay, and obtained theoretical results for special cases. At the penalty of using an additional moment of the interarrival time we get approximations which seem to work well in those ranges of load and peakedness which are of interest to practitioners of traffic theory.

References:

[1] S. L. Albin, "Approximating Queues with Superposition Arrival Processes", Doctoral dissertation, Dept. of O.R. and Ind. Eng., Columbia U., 1981.

[2] C. Carathéodory, "Theory of Functions", Second English Edition, Vol. I, page 231, 1958.

[3] J. W. Cohen, "The Single Server Queue", Revised Edition, North-Holland, 1982.

[4] D. J. Daley, A. Ya. Kreinin and C. D. Trengrove, "Inequalities concerning the waiting time in single server queues; a survey," Statistics Department, Australian National University, 1983.

[5] J. F. C. Kingman, "Some inequalities for the queue GI/G/1," Biometrica 49, 315-324, 1962.

[6] ______, "Inequalities in the theory of queues," J. Roy, Stat. Soc. Ser. B32, 102-110, 1970.

[7] W. Kraemer and M. Langenbach-Belz, "Approximate formulae for the delay in the queueing system GI/G/1," Eighth Int. Teletraffic Cong. Melbourne, 235-1-8, 1976.

[8] P. J. Kuehn, "Approximate analysis of general queueing networks by decomposition," IEEE Trans. Commun., COM-27, No. 1 113-126, 1979.

[9] A. J. Lemoine, "On random walks and stable GI/G/1 queues," Math. of OR 1, 159-164. 1976.

[10] K. T. Marshall, "Some inequalities in queueing," Oper. Res. 16 651-665, 1968.

[11] W. Whitt, "The queueing network analyzer," B. S. T. J. 62 2779-2815, 1983.

[12] ______, "Approximations for networks of queues," Tenth Int. Teletraffic Cong. Montreal. 4.2.2, 1983.

[13] ______, "On approximations for queues, I: Extremal Distributions," AT&T Bell Lab. Tech. J. 63 115-138, 1984.

[14] ______, "On approximations for Queues, III: Mixture of Exponential Distributions," AT&T Bell Labs. Tech. J. 63 163-176, 1984.

TELETRAFFIC ISSUES in an Advanced Information Society
ITC-11
Minoru Akiyama (Editor)
Elsevier Science Publishers B.V. (North-Holland)

APPROXIMATIONS TO THE WAITING TIME PERCENTILES IN THE M/G/c QUEUE

L.P. SEELEN and H.C. TIJMS

Dept. of Actuarial Sciences and Econometrics, Vrije Universiteit
Amsterdam, The Netherlands

1. INTRODUCTION AND MAIN RESULT

In the last decade several useful approximations have been obtained for the average waiting time in the M/G/c queue, see [9] and [10] for discussion and further references. An appealing two-moment approximation to $E(W_q)$ is given by

$$E_{app}(W_q)=(1-c_S^2)E_{det}(W_q)+c_S^2E_{exp}(W_q), \qquad (1)$$

where c_S denotes the coefficient of variation (=ratio of standard deviation and mean) of the service time and $E_{det}(W_q)$ and $E_{exp}(W_q)$ represent the average waiting time for the respective cases of deterministic service and exponential service with the same means $E(S)$. This approximation agrees with the Pollaczek-Khintchine formula for the special case of c=1 server. In [2] it was already noticed that the Pollaczek-Khintchine formula can be written in the form (1) involving a linear interpolation on the squared coefficient of variation of the service time and it was pointed out that such a linear interpolation might yield useful approximations to more complex queueing models. The approximation (1) for the M/G/c queue was investigated in [1] and [5], where the reference [1] also presents a simple but accurate approximation to $E_{det}(W_q)$ (see also [9] or [10] for a correction to a misprint in the latter result). The two-moment approximation (1) shows an excellent performance provided c_S^2 is not too large (say, $0\leq c_S^2\leq 2$); for larger values of c_S^2 it is no longer true that measures of system performance are fairly insensitive to more than the first two moments of service time.

A natural question is whether an approximation of the form (1) also applies to the waiting time probabilities in the M/G/c queue. The answer is in the negative, but it appears that an approximation like (1) can be used when considering the waiting time *percentiles* rather than the waiting time probabilities. Letting W_q be the waiting time of a customer (excluding service time) when the system is in statistical equilibrium, the p-th waiting time percentile $\xi(p)$ is defined by

$$P\{W_q\leq\xi(p)\} = p.$$

Note that $\xi(p)$ is only defined for $\Pi_W\leq p<1$ with Π_W denoting the delay probability $P\{W_q>0\}$. Also, let $\xi_{det}(p)$ and $\xi_{exp}(p)$ denote the waiting time percentile $\xi(p)$ for the particular cases of the M/D/c queue (deterministic service) and the M/M/c queue (exponential service) with the same average service times $E(S)$. Then, the two-moment approximation

$$\xi_{app}(p)=(1-c_S^2)\xi_{det}(p)+c_S^2\xi_{exp}(p) \qquad (2)$$

performs quite well for all values of p provided c_S^2 is not too large (say, $0\leq c_S^2\leq 2$). A similar statement applies to the conditional waiting time percentiles $\eta(p)$ defined by

$$P\{W_q\leq\eta(p)|W_q>0\} = p, \quad 0<p<1.$$

Note that, by the approximation for the waiting time percentiles, we can indirectly approximate the waiting time probabilities.

The two-moment approximation (2) is of practical value only when it is easy to compute the particular waiting time percentiles $\xi_{det}(p)$ and $\xi_{exp}(p)$. The percentiles $\xi_{exp}(p)$ (and $\eta_{exp}(p)$) are trivial to calculate, since for the M/M/c queue with arrival rate λ and service rate $\mu=1/E(S)$ we have the well-known explicit result

$$P\{W_q>x\} = \Pi_W(\exp)e^{-(c\mu-\lambda)x}, \quad x\geq 0,$$

where $\Pi_W(\exp)$ is Erlang's delay probability for which a simple explicit expression is available. In the next section we show that the waiting time percentiles $\xi_{det}(p)$ and $\eta_{det}(p)$ can rather easily be calculated by using a tailor-made numerical method for the M/D/c queue.

2. SPECIAL-PURPOSE ALGORITHM FOR THE M/D/c QUEUE

In this section we discuss a special-purpose algorithm for calculating the state probabilities and the waiting time probabilities in the M/D/c queue with Poisson arrivals at rate λ and a deterministic service time D such that the server utilization $\rho=\lambda D/c$ is smaller than 1. The algorithm is extremely efficient and easy to program, and requires very small computing times even for a very large number of servers and high traffic. The method uses a number of fundamental results that go essentially back to Crommelin's paper [3]. The algorithm calculates first the steady-state probabilities and calculates next the waiting time probabilities by using the values of the state probabilities. A special iterative method is applied to calculate the state probabilities from the equilibrium equations derived in [3]. This iterative method will be described later. Suppose for the moment that the state probabilities have been computed. In the original paper [3] of Crommelin two closed-form expressions are given for the waiting time probability $P\{W_q \leq x\}$. The first of these two expressions is a finite sum involving terms that alternate in sign and the second one is an infinite sum involving positive terms only. Unfortunately, as already realized by Crommelin, both representations offer numerical difficulties when the traffic is non-light. Incidentally, for multi-server queues the server utilization ρ is in general not a suitable measure for the traffic load on the system and for that purpose one should use the delay probability Π_W rather than ρ. For non-light traffic, the numerical evaluation of the above mentioned sum with terms alternating in sign will be hampered by roundoff errors due to loss of significance, while for the other closed-form representation the numerical problem is the slow convergence of the infinite sum where the calculations may be halted by the occurrences of underflow and overflow before convergence is achieved. Fortunately, we can provide a practically useful alternative for the calculation of the waiting time probabilities (percentiles). A computationally more useful representation of the waiting time probabilities is actually contained in Crommelin's paper [3], but was apparently overlooked in the aim at a closed-form solution. By an ingenious probabilistic argument, he obtained the following result. Letting p_j be the steady-state probability of having j customers in the system and using the representation

$$x = mD + u \quad \text{for some integer } m \geq 0 \text{ and } 0 \leq u < D,$$

the waiting time probability $P\{W_q \leq x\}$ may be expressed as

$$P\{W_q \leq x\} = b_{mc+c-1}(u), \tag{3}$$

where the $b_\ell(u)$'s satisfy the equation

$$\sum_{i=0}^{j} p_j = \sum_{k=0}^{j} b_{j-k}(u) e^{-\lambda u} \frac{(\lambda u)^k}{k!}, \quad j=0,1,\ldots, \tag{4}$$

allowing for a recursive computation of the $b_\ell(u)$'s starting with $b_0(u) = e^{\lambda u} p_0$. Although this recursion scheme also involves the taking of differences, it offers considerably less numerical difficulties than the closed-form expressions discussed earlier. Actually, a computationally better form of the recursion relation (4) is obtained by rewriting it as

$$b_j(u) = e^{\lambda u} p_j - \sum_{k=0}^{j-1} b_k(u) \frac{(\lambda u)^{j-1-k}}{(j-1-k)!} \left(\frac{\lambda u}{j-k} - 1\right)$$

$$\text{for } j=0,1,\ldots . \tag{5}$$

Thus the desired waiting time probability $P\{W_q \leq x\}$ may be calculated by applying the recursion scheme (5) until $b_{mc+c-1}(u)$ is obtained. This recursion scheme however should not be applied blindly, since it will also ultimately be hampered by roundoff errors for large values of λu when the traffic is non-light. The recursion scheme (5) should be used in an appropriate combination with the asymptotic expansion

$$P\{W_q > x\} \approx \alpha e^{-\delta x} \quad \text{for } x \text{ large.} \tag{6}$$

Extensive numerical experiments show that in practical applications the asymptotic expansion (6) is already very accurate long before the recursion scheme (5) offers numerical difficulties. Our empirical finding is that for practical purposes the asymptotic expansion (6) may be used for

$$x \geq D/\sqrt{c}$$

provided the delay probability Π_W is not too small (say, $\Pi_W \geq 0.2$). Also, we found that for any $p \geq 0.9$ the p-th conditional waiting time percentile $\eta(p)$ may be calculated by using (6) provided the traffic is not too light. The coefficients α and δ of the asymptotic expansion (6) are easily computed. It is well-known that the constant δ is the unique positive solution to the equation (cf. [3] and [8])

$$\lambda(e^{\delta D/c} - 1) = \delta. \tag{7}$$

Letting

$$\tau = 1 + \delta/\lambda, \tag{8}$$

it follows from results in [8] and [9] that

$$\alpha = \frac{\eta\delta}{\lambda(\tau-1)^2 \tau^{c-1}}, \tag{9}$$

where η, being defined as $\lim_{j\to\infty} \tau^j p_j$, is given by

$$\eta = \left[\frac{c}{\tau} - \lambda D\right]^{-1} \sum_{i=0}^{c-1} p_i (\tau^{i-1} - \tau^{c-1}). \tag{10}$$

Notice that the amplitude factor α needs only the first c state probabilities.

To the end of this section, we discuss a special-purpose iterative method for the calculation of the state probabilities p_j, $j \geq 0$ from the equilibrium equations

$$p_j = e^{-\lambda D}\frac{(\lambda D)^j}{j!}\sum_{k=0}^{c} p_k + \sum_{k=c+1}^{c+j} e^{-\lambda D}\frac{(\lambda D)^{j-k+c}}{(j-k+c)!}\, p_k \qquad \text{for } j=0,1,\ldots, \tag{11}$$

together with the normalizing equation

$$\sum_{j=0}^{\infty} p_j = 1. \tag{12}$$

A common approach for solving this infinite system of linear equations is to truncate first the system by a sufficiently large chosen integer L such that $\Sigma_{j=L}^{\infty} p_j \leq 10^{-8}$ (say), where L is found by using explicit results for the state probabilities in the M/M/c queue, and to solve next the truncated system of linear equations by the standard successive overrelaxation method, cf. [4]. This computational approach may be considerably improved in two respects. Firstly, a successive overrelaxation method with a dynamically adjusted relaxation factor may be used in order to avoid the difficulty of not knowing on beforehand the optimal value of the relaxation factor. Secondly, in view of the theoretical result

$$\frac{p_{j-1}}{p_j} \approx \tau \qquad \text{for all j sufficiently large} \tag{13}$$

with τ given by (8), the infinite system of linear equations (11)-(12) may be reduced to a finite system of linear equations by using the asymptotic estimate

$$p_j \approx \tau^{N-j} p_N \qquad \text{for } j \geq N \tag{14}$$

when N is chosen sufficiently large. The asymptotic expansion (13) applies usually already for relatively small values of j, in particular when the traffic load on the system increases. Thus, for non-light traffic, an integer N such that (14) is sufficiently accurate will typically be much smaller than the truncation integer L discussed earlier. On the contrary for very light traffic situations the asymptotic expansion (14) may apply not before the state probabilities are negligibly small, but the algorithm below is designed in such a way that in those situations it operates automatically as if a truncation integer L as above would be used. A good choice of N is usually not known on beforehand and an initial guess with N very large would be inefficient. In the algorithm the problem of determining an appropriate value of N is solved by using an adaptive scheme starting with a "low" estimate of N and increasing this estimate when necessary. By a specially designed successive overrelaxation method with a variable relaxation factor, a sequence of finite systems of linear equations with increasing sizes is solved. Here the solution of the system associated with some value of N is used as starting point for the system associated with the next value of N. The efficiency of the algorithm is further improved by choosing adaptively the accuracy number of the stopping criterion of the iterative method; this accuracy number ε is chosen smaller as N increases. We next describe the details of the algorithm sketched above.

Supposing an estimate for the integer N such that (14) holds and replacing the probabilities p_j by $p_N \tau^{N-j}$ for $j \geq N$, we obtain from (11) and (12) after some manipulations the following system of linear equations

$$p_j = \sum_{\substack{k=0 \\ k \neq j}}^{N} a_{jk} p_k, \qquad j=0,\ldots,N \tag{15}$$

$$\sum_{j=0}^{N} p_j + \frac{\tau}{\tau-1} p_N = 1 \tag{16}$$

where

$$a_{jk} = \frac{a[\min(j,j-k+c)]}{1-a[\min(j,c)]}, \quad 0 \leq j \leq N-1,\ 0 \leq k \leq \min(c+j,N-1),$$

$$a_{jk} = \frac{\sum_{k=N}^{c+j} \tau^{N-k} a[j-k+c]}{1-a[\min(j,c)]}, \quad N-c \leq j \leq N-1,\ k=N,$$

$$a_{jk} = \frac{a[\min(j,j-k+c)]}{1-\sum_{k=N}^{c+j} \tau^{N-k} a[j-k+c]}, \quad j=N,\ 0 \leq k \leq N-1,$$

$$a_{jk} = 0 \qquad , \text{otherwise.}$$

Here $a[\ell]=e^{-\lambda D}(\lambda D)^{\ell}/\ell!$, $\ell \geq 0$. Following the usual notation for the successive overrelaxation method, the operator B_ω associated with a relaxation factor ω transforms each vector $x=(x_0,\ldots,x_N)$ into the vector $B_\omega x$ whose components $(B_\omega x)_i$ are recursively defined by

$$(B_\omega x)_i = (1-\omega)x_i + \omega\{\sum_{j=0}^{i-1} a_{ij}(B_\omega x)_j + \sum_{j=i+1}^{N} a_{ij} x_j\}$$

$$\text{for } i=0,1,\ldots,N.$$

Assuming that the integer N is sufficiently large so that (15) has a solution, then this solution is an eigenvector of B_ω with associated eigenvalue 1. Letting $\lambda_1(\omega)$ be the eigenvalue having the largest absolute value among the eigenvalues of B_ω unequal to 1, the standard successive overrelaxation method with a fixed relaxation factor ω converges only if $|\lambda_1(\omega)|<1$. Moreover, the standard overrelaxation method has the best convergence rate for that value of ω for which $|\lambda_1(\omega)|$ is smallest. It should be noted that the optimal value of ω may be rather sensitive to the parameters of the specific problem considered and in some cases will be close to 1; in the algorithm below we keep ω always between 1 and 2. In case $\lambda_1(\omega)$ is real, it is possible to estimate $\lambda_1(\omega)$ after some iterations of the overrelaxation method (this is done by the parameter r^h in the algorithm below). This estimate provides a method to formulate a successive overrelaxation algorithm in which the relaxation factor is dynamically adjusted in order to search for that value of ω for which $|\lambda_1(\omega)|$ is smallest. In [6] such an approach was proposed to solve the balance equations arising in the continuous-time Markov chain analysis of multi-server queueing systems; the resulting algorithm was instrumental in compiling the tablebook [7]. For the linear equations (15) and (16) this approach needs a modification in order to avoid divergence problems when the estimate of N is too low.

We now give the steps of the algorithm for the calculation of the state probabilities in the M/D/c queue.

Special-purpose overrelaxation method for the M/D/c queue.

Step 0. Choose $N>c$ and $x^0 \geq 0$ with $\Sigma_{i=0}^{N} x_i^0 + \tau(\tau-1)^{-1} x_N^0 = 1$. Also, $h:=0$ and $\omega:=1.20$.

Step 1. $\omega^{old}:=0$, $\lambda(\omega^{old}):=1$, $f^h:=r^h:=\infty$.

Step 2. $h:=h+1$. Compute the vectors

$$\tilde{x}^h := B_\omega x^{h-1}, \quad x_i^h = \left[\sum_{i=0}^{N} |\tilde{x}_i^h| + \frac{\tau}{\tau-1}|\tilde{x}_N^h|\right]^{-1} \tilde{x}^h$$

and the scalar

$$f^h := \frac{1}{N}\sum_{i=0}^{N}\left|\frac{\tilde{x}_i^h - \tilde{x}_i^{h-1}}{\tilde{x}_i^h}\right|.$$

If $f^h < \varepsilon_N$ with ε_N a prespecified accuracy number, then go to step 4. Otherwise

$$r^h := \frac{f^h}{f^{h-1}}.$$

If $r^h \geq 1$ or $h \geq 10$, then ω is likely too large and decrease ω as $\omega:=1+\frac{1}{2}(\omega-1)$, put $x^0:=x^h$ and $h:=0$, and go to step 1. If $r^h<1$ and r^h has sufficiently converged according to $|(r^h-r^{h-1})/r^h|<0.025$, then go to step 3; otherwise return to step 2.

Step 3. $\lambda(\omega):=r^h$. Test for one of the following four possibilities: (a) $\omega>\omega^{old}$ and $\lambda(\omega)>\lambda(\omega^{old})$; (b) $\omega>\omega^{old}$ and $\lambda(\omega)\leq\lambda(\omega^{old})$; (c) $\omega<\omega^{old}$ and $\lambda(\omega)>\lambda(\omega^{old})$; (d) $\omega<\omega^{old}$ and $\lambda(\omega)\leq\lambda(\omega^{old})$. For the cases (a) and (d),

$$\omega^{old}:=\omega, \ \lambda(\omega^{old}):=\lambda(\omega), \ \omega:=1+0.85(\omega-1),$$

whereas for the cases (b) and (c),

$$\omega^{old}:=\omega, \ \lambda(\omega^{old}):=\lambda(\omega), \ \omega:=1+1.25(\omega-1).$$

Next, $x^0:=x^h$, $h:=0$, $f^h:=r^h:=\infty$, and go to step 2.

Step 4. If

$$\left|\sum_{i=0}^{N} \tilde{x}_i^h + \frac{\tau}{\tau-1}\tilde{x}_N^h - 1\right| < 10^{-5}$$

and

$$\left|\sum_{i=0}^{c-1} i x_i^h + c\left(1-\sum_{i=0}^{c-1} x_i^h\right) - c\rho\right| < 10^{-5},$$

then the algorithm is stopped and the state probabilities p_i are obtained from $p_i = x_i^h$ for $0\leq i\leq N$ and $p_i = \tau^{N-i} x_N^h$ for $i>N$ (the above stopping criteria use that the probabilities sum to 1 and that the average number of busy servers equals $c\rho$). Otherwise

$$x_i^0 := x_i^h \text{ for } 0\leq i\leq N, \quad x_i^0 := \frac{1}{\tau} x_{i-1} \text{ for } N<i\leq N+10$$

$$N:=N+10, \ h:=0,$$

and go to step 1.

3. NUMERICAL RESULTS

In this section we present some numerical results showing that the two-moment approximation (2) performs quite well for practical purposes provided c_S^2 is not too large (say, $0\leq c_S^2\leq 2$). For Erlang-2 service ($c_S^2=0.5$) and H_2 service with balanced means and $c_S^2=2$, table 1 gives the approximate and (nearly) exact values of the conditional waiting time percentiles $\eta(p)$ with p=0.5, 0.9, 0.95 and 0.99 for several values of the number c of servers. The number c of servers is varied as 2, 5, 10, 25 and 50. In all examples we assume a server utilization $\rho=0.8$ and the normalization $E(S)=1$. The (nearly) exact values of the conditional waiting time percentiles for Erlang-2 and H_2 services were computed from an extremely accurate approximation of the waiting time distribution function by a sum of a "sufficiently" number of exponential functions whose coefficients were obtained from the exactly computed state probabilities, see [6] and [7]; the accurateness of this approximation has been tested by using the exact relation $E[L_q(L_q-1)\dots(L_q-k+1)]=\lambda^k E[W_q^k]$, $k\geq 1$ for the M/G/1 queue. Also, we include for convenience in table 1 the exact values of the delay probability Π_W.

Table 1. Numerical results.

	p	Erlang-2 service (c_S^2=0.5) 0.5	0.9	0.95	0.99	Π_W
c=2	exa	1.341	4.286	5.554	8.498	0.7087
	app	1.332	4.282	5.551	8.497	
c=5	exa	0.554	1.735	2.242	3.420	0.5484
	app	0.556	1.732	2.239	3.417	
c=10	exa	0.285	0.881	1.135	1.724	0.4021
	app	0.286	0.877	1.131	1.721	
c=25	exa	0.118	0.364	0.467	0.703	0.2033
	app	0.119	0.365	0.466	0.701	
c=50	exa	0.061	0.189	0.241	0.361	0.0840
	app	0.061	0.191	0.243	0.362	
		H_2-service with c_S^2=2				
	p	0.5	0.9	0.95	0.99	Π_W
c=2	exa	2.361	8.818	11.62	18.11	0.7146
	app	2.534	8.705	11.37	17.55	
c=5	exa	0.890	3.409	4.526	7.123	0.5617
	app	0.967	3.445	4.510	6.981	
c=10	exa	0.426	1.634	2.186	3.479	0.4182
	app	0.468	1.699	2.231	3.466	
c=25	exa	0.161	0.603	0.811	1.314	0.2159
	app	0.179	0.651	0.865	1.362	
c=50	exa	0.077	0.280	0.376	0.612	0.0902
	app	0.085	0.309	0.412	0.658	

REFERENCES

[1] G.P. Cosmetatos, "Some approximate equilibrium results for the multiserver queue M/G/1", Oper. Res. Quart., vol. 27, pp. 615-620, 1976.

[2] D.R. Cox, "The statistical analysis of congestion, J. Roy. Stat. Soc., Series A, vol. 118, pp. 324-335, 1955.

[3] C.D. Crommelin, "Delay probability formulae when the holding times are constant", P.O. Elect. Engr. J., vol. 25, pp. 41-50, 1932.

[4] P.J. Kühn, "Tables on Delay Systems", Institute of Switching and Data Technics, University of Stuttgart, 1976.

[5] E. Page, "Queueing Theory in O.R.", Butterworths London, 1972.

[6] L.P. Seelen, "An algorithm for Ph/Ph/c queues", Research Report no. 131, Dept. of Actuarial Sciences and Econometrics, Vrije Universiteit, Amsterdam, 1984.

[7] L.P. Seelen, H.C. Tijms and M.H. Van Hoorn, "Tables for Multi-Server Queues", North-Holland, Amsterdam, 1985.

[8] Y. Takahashi, "Asymptotic exponentiality of the tail of the waiting time distribution in a Ph/Ph/c queue", Adv. Appl. Prob., vol. 13, pp. 619-630, 1981.

[9] H.C. Tijms, "Stochastic Operations Research: A Computational Approach", Wiley, Chichester, 1986.

[10] M.H. Van Hoorn, "Algorithms and Approximations for Queueing Systems", CWI Tract no. 8, CWI, Amsterdam, 1984.

TELETRAFFIC ISSUES in an Advanced Information Society
ITC-11
Minoru Akiyama (Editor)
Elsevier Science Publishers B.V. (North-Holland)

A GENERALIZATION OF THE CLASSICAL TELETRAFFIC THEORY

Villy Baek IVERSEN

Technical University of Denmark
Lyngby, Denmark

ABSTRACT

A loss system with a state-dependent Poisson arrival process, state-dependent probability of blocking, and homogeneous servers is shown to be insensitive to the holding time distribution. This model comprises all the classical loss systems. The model is generalized to cover multi-dimensional models of Erlang's loss system, Engset's loss system, the Negative Binomial case, the machine interference model, Palm-Jacobæus' formula, Erlang's interconnection formula, the generalized Engset formula, the A-formula, etc. The mathematical derivations are elementary and easy to comprehend for switching engineers. They are based on results for Cox distributions and the well-known Erlang phase method.

1. A GENERALIZED LOSS SYSTEM /1/

We consider a loss system with n homogeneous servers, and define the state of the system to be the number of busy servers. The arrival process is a Poisson process the intensity of which may depend on the state of the system. The probability of blocking may also depend on the state of the system.

When the system is in state y, the arrival intensity is denoted by λ_y'. Thus the net arrival intensity of accepted calls in state y is:

$$\lambda_y = \lambda_y' \cdot u(y) \ , \quad 0 \le y \le n \tag{1}$$

where $u(y) = 1 - w(y)$. (2)

u(y) is called the passage probability.

1.1 Macro State probabilities

If the state probabilities of the system are independent of the holding time distribution, we may obtain the state probabilities by assuming exponential distributed service times (intensity μ for every server). The state transition diagram of this system is shown in fig. 1.

Under the assumption of statistical equilibrium we have the following state probabilities

$$P(y) = \frac{\lambda_0 \cdot \lambda_1 \cdot \, .. \, \cdot \lambda_{y-1}}{\mu \cdot 2\mu \cdot \, ... \, y\mu} \cdot P(0) \ , \quad 0 \le y \le n \tag{3}$$

where P(0) is obtained from the normalization condition

$$\sum_{y=0}^{n} P(y) = 1 \tag{4}$$

P(y) is equal to the proportion of time, the system spends in state y. These probabilities are in general different from the probabilities experienced by arriving calls.

The time congestion may be defined as:

$$E = \sum_{y=0}^{n} P(y) \cdot w(y) \tag{5}$$

The state probabilities at call arrival epochs for all call attempts are obtained as follows:

$$Q(y) = \frac{\lambda_y' \cdot P(y)}{\sum_{i=0}^{n} \lambda_i' \cdot P(i)} \ , \quad 0 \le y \le n \tag{6}$$

The call congestion becomes:

$$B = \sum_{y=0}^{n} Q(y) \cdot w(y) \tag{7}$$

Other performance characteristics are also obtained from the state probabilities. Wallström /2/ has derived the distribution of the overflow traffic and its moments for an infinite overflow group (Kosten system).

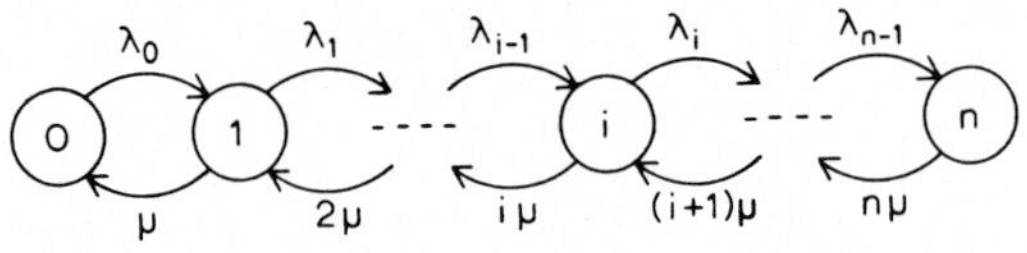

Fig. 1: State transition diagram of a generalized loss system.

1.2 Cox distributions

To prove that the above-mentioned model is insensitive to the holding time distribution we introduce Cox distributions (branching Erlang distributions). An example is shown in fig. 2. A Cox distribution has a rational Laplace transform and can be approximated arbitrarily close to any distribution function.

In the sense of weak convergence /3/ a Cox distribution is equivalent to a general distribution. If we are able to prove that a system is insensitive to Cox distributions, then it is insensitive to any distribution. We are able to deal with Cox distributions by means of birth and death processes. This approach is simple to understand and is in accordance with the classical tools of teletraffic theory based on Erlang's phase method and statistical equilibrium.

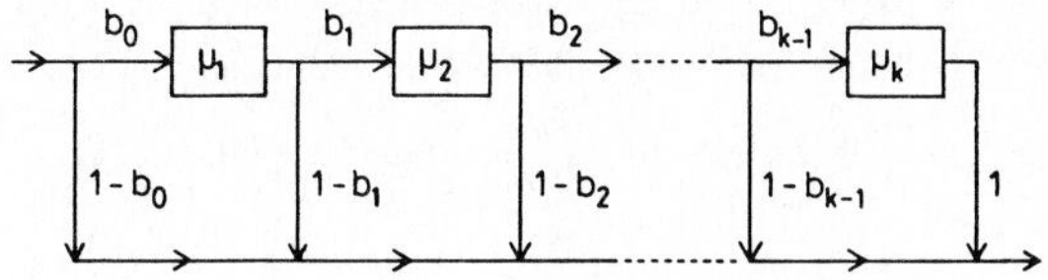

Fig. 2: A phase diagram representation of a Cox-k distribution.

The mean value of the distribution shown in fig. 2 becomes:

$$m = \sum_{i=1}^{k} \frac{q_i}{\mu_i} = \sum_{i=1}^{k} m_i \tag{8}$$

where

$$q_i = \prod_{\nu=0}^{i-1} b_\nu \,, \quad 1 \le i \le k \tag{9}$$

is the probability of reaching phase i.
$m_i = q_i/\mu_i$ is the contribution of phase i to the mean value. The relative contribution from phase i to the mean value is m_i/m.

The second moment can be written as

$$m_2 = 2 \cdot \sum_{i=1}^{k} \{ (\sum_{j=1}^{i} \frac{1}{\lambda_j}) \cdot \frac{q_i}{\lambda_i} \} \tag{10}$$

and Palm's form factor is defined as

$$\varepsilon = \frac{m_2}{m^2} \,, \quad 1 < \varepsilon < \infty \tag{11}$$

Cox distributions are extremely useful in teletraffic theory /4/. They include the negative exponential distribution, Erlang distributions, hyper-exponential distributions etc. Fig. 3 illustrates a decomposition of an exponential phase into a Cox-2 distribution /4/. By this decomposition a k-phase hyper-exponential distribution is easily transformed to a Cox-k distribution. An interrupted Poisson process (IPP) is equivalent to a two-phase hyper-exponential distributed inter-arrival time /5/, which is transformed to a Cox-2 distribution. As the hyper-exponential distribution has $\varepsilon \ge 2$, whereas the Cox-2 distribution has $\varepsilon \ge 1.5$, we notice that a Cox-2 distribution is more general than an IPP arrival process.

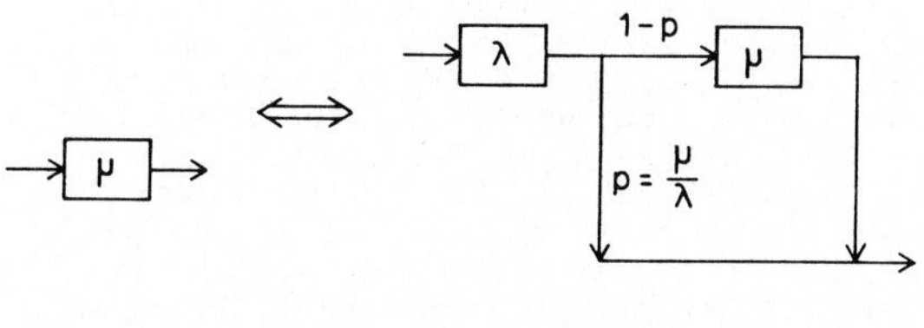

Fig. 3: Decomposition of an exponential distribution into a Cox-2 distribution. To obtain a true branching probability we must require that $\lambda \ge \mu$.

1.3 Micro states

We want to show that the above-mentioned generalized loss system is insensitive to the holding time distribution. By insensitive we mean that the state probabilities (3) depend only upon the mean holding time, but is independent of the actual distribution.

We assume that the holding time distribution is a Cox distribution. To describe the state of the system we must not only know the number of busy servers y (macro states), but also the number of servers in phase i, y_i (micro states). For a system with n servers and Cox-k distributed holding times we get:

$$0 \le y_i \le n \,, \qquad y = \sum_{i=1}^{k} y_i \tag{12}$$

If the system is insensitive, then every portion of the holding time has equal importance for the state probabilities. For a given holding time, the probability of observing the server in phase i is proportional to the relative contribution of this phase to the mean holding time (m_i/m (8)). Thus we have a polynomial experiment with k outcomes.

If the system is in state y, then this macro state is split up in a number of micro states given by the polynomial distribution:

$$P\{y|y_1,y_2, \ldots ,y_k\} = \binom{y}{y_1\, y_2 \ldots y_k} \cdot \left(\frac{m_1}{m}\right)^{y_1} \cdot \left(\frac{m_2}{m}\right)^{y_2} \cdot \ldots \cdot \left(\frac{m_k}{m}\right)^{y_k} \cdot P(y) \tag{13}$$

where

$$\binom{y}{y_1\, y_2 \ldots y_k} = \frac{y!}{y_1! \cdot y_2! \cdot \ldots \cdot y_k!} \tag{14}$$

is the polynomial coefficient, and P(y) is the macro state probability given by (3).

In appendix A we prove that these state probabilities fulfil the equilibrium equations.

1.4 Examples

In this section we mention some classical models, which are particular cases of the above model and thus are insensitive to the holding time distribution.

A. Full availability: $w(y) = 0 \,, \quad 0 \le y \le n$

a.1: Erlang's loss system:

$$\lambda_y' = \lambda$$

a.2: Engset's loss system:

$$\lambda_y' = (N - y) \cdot \gamma$$

a.3: Negative Binomial model:

$$\lambda_y' = (N + y) \cdot \gamma$$

B. Limited availability k:

b.1: Erlang's interconnection formula:

$$\lambda_y' = \lambda$$

$$w(y) = \binom{y}{k}/\binom{n}{k} \,, \quad k \le y \le n$$
$$w(y) = 0 \,, \quad 0 \le y < k \tag{15}$$

b.2: Gradings and link systems.

Many methods of calculation only takes account of the total number of busy servers by using e.g. a passage factor. This goes for e.g. Jacobæus' method for link systems. In section 1.5 we consider a generalization of Erlang' interconnection formula.

C. The machine interference model /6/,/7/

This is an example, where we interprete the model in a different way by interchanging the arrival process and the holding time process. In this way we get a waiting time system with a limited number of sources. We consider a model of n machines and m (< n) repair-men. We assume the operating time of a machine is exponential distributed (intensity μ) and that every repair-man services a machine at rate γ. If we define the state of the system to be the number of operating machines, then we get:

$$\lambda'_y = m \cdot \lambda \quad , 0 \le y \le n-m$$
$$\lambda'_y = (n - y)\cdot\lambda \ , n-m \le y \le n$$

Thus the machine interference model is insensitive to the operating time distribution of the machines. The distribution of the waiting time for an idle repair-man is also independent of the operating time distribution.

1.5 Generalization of Erlang's Ideal Grading (EIG) to several traffic streams with individual hunting capacity /8/

The assumptions of Erlang's ideal grading (also called Erlang's Interconnection Formula) (EIG) are as follows. The offered traffic is pure chance traffic type 1 (PCT-1). The system is in statistical equilibrium, and lost calls are cleared. The number of trunks is n, and the availability is k. A call is allowed to hunt k different trunks, which are chosen at random.

If the system is in state y when a call arrives, then the probability that the call is accepted is given by the passage probability u(y) (2), where w(y) is given by (15). In statistical equilibrium the state probabilities are given by

$$P(y) = \frac{Q_y \cdot \frac{A^y}{y!}}{\sum_{j=0}^{n} Q_j \cdot \frac{A^j}{j!}} \quad , \ y = 0,1, \ldots ,n \qquad (16)$$

where

$$Q_y = \prod_{j=0}^{y-1} \{1 - w(j)\} \ , \ y = 1, 2, \ldots , n \qquad (17)$$
$$Q_0 = 1$$

The arrival process is a Poisson process and therefore the time congestion E is equal to the call congestion B:

$$E = B = \sum_{j=0}^{n} w(j) \cdot P(j) \qquad (18)$$

We can generalize the above system to several traffic streams. We consider the following system:

n = total number of (homogeneous) trunks
g = number of traffic streams
$A_\nu = \lambda_\nu/\mu$ = offered traffic (PCT-1) from stream number ν , $\nu = 1, 2, \ldots , g$
k_ν = availability of stream number ν
E_ν = time congestion for stream number ν

Under the assumption of statistical equilibrium we get the following equations:

$$(i+1)\mu\cdot P(i+1) = \{ \sum_{j=1}^{g} \lambda_j (1-w(i,j)) \}\cdot P(i) \qquad (19)$$

where $i = 0, 1, \ldots , n-1$, and w(i,j) is the blocking probability for stream number j when the system is in state i:

$$w(i,j) = 0 \quad , 0 \le i < k_j$$
$$w(i,j) = \binom{i}{k_j} / \binom{n}{k_j} , k_j \le i \le n \qquad (20)$$

P(0) is obtained from the normalization condition (4).

For traffic stream ν the congestion becomes:

$$E_\nu = B_\nu = \sum_{j=0}^{n} P(j)\cdot w(j,\nu) \qquad (21)$$

For g = 1 or $k_\nu = k$ we obtain the ordinary EIG. We notice that the generalized EIG is insensitive to the holding time distribution. EIG gives the lower limit of the blocking probability for gradings with random hunting. The model is eg. applicable to radio communication systems, where the customers have individual availability.

The above system can easily be generalized to a system, where every traffic stream is a state-dependent Poisson process, i.e. to Engset traffic etc.

In part 2 of the paper EIG is generalized in a different way.

1.6 The single source model

If we consider the case n = 1 server (fig.1), then this may be interpreted as a single traffic source having access to one server. Let the offered traffic from this source when it is idle be

$$\gamma = \lambda/\mu \qquad (22)$$

where $1/\lambda$ is the mean inter-arrival time and $1/\mu$ is the mean holding time. It is easy to see that this model is insensitive for both the inter-arrival time distribution and the holding time distribution. We shall return to this example in part 2.

1.7 Other works

Many authors have lately dealt with the property of insensitivity in queueing theory. We shall only comment on a few works with a background in teletraffic theory. In general, the approaches are highly mathematical and often based on generalized semi-Markov schemes. This holds good for the works of König et al /9/.

An applicable approach was published by Basharin & Kokotushkin /10/. On the basis of a theorem of Kowalenko, they give necessary and sufficient conditions (Basharin's theorem) for insensitivity, a property which they call strong statistical equilibrium. Theses conditions are fulfilled for systems with state-dependent Poisson arrival processes which are 'quasi full available'. The conditions are similar to the conditions for V.E. Benes' thermodynamic model of telephone traffic. In /10/ it is also shown that the overflow traffic from insensitive systems is sensitive with respect to the holding time distribution.

In /11/ Hordijk et al. showed that the semi-Markov process corresponding to fig. 1 has state probabilities, which depend on the holding time distribution only through their mean values.

Henderson /12/ has also dealt with the subject and given practical examples of insensitive systems.

Cohen /13/ has given a survey of the subject with a historical background.

2. MULTI-DIMENSIONAL LOSS SYSTEMS

We consider a loss system with n homogeneous servers (trunks). This system is offered N independent traffic streams. Every traffic stream is a generalized loss system as described in part 1. The state of the system is described by the vector

$$(y_1, y_2, \dots, y_N)$$

where y_i is the number of servers occupied by the i'th traffic stream. Traffic stream number i is allowed to occupy at most n_i servers:

$$0 \le n_i \le n \tag{23}$$

If

$$\sum_{i=1}^{N} n_i \le n \tag{24}$$

then we have N independent loss systems, and the state space is described by a multi-dimensional reversible Markov process /14/, which is insensitive to the holding time distribution.

If we truncate the state space by constraints as

$$\sum_{i=1}^{N} c_i \cdot y_i \le n \tag{25}$$

$$\sum_{i \in X} c_i \cdot y_i \le n_x \le n \tag{26}$$

where X is a subset of the N traffic streams, then the relative values of the remaining states are unchanged, and the system described by the restricted state space is still insensitive to the holding time distribution.

Besides the passage probabilities of the individual traffic streams we may introduce a common passage factor for a set of traffic streams, which is a function of the total number of servers occupied by this set of traffic streams.

The above model is useful for combining the traffic from several trunk groups onto one common group, for introducing virtual circuit protection, for investigating systems with re-arrangement of calls /15/, for analysing systems with traffic streams with differing capacity requirements etc.

Multi-dimensional models have previously been dealt with by a.o. Erlang & Arne Jensen /16/, Gimpelson /17/, Lam /18/, Aein /19/, Roberts /20/.

2.1 Erlang's multi-dimensional loss system /16/

As an example we may consider a system with N PCT-1 traffic processes (λ_1,μ_1), (λ_2,μ_2), ... , (λ_N,μ_N).

We denote the offered traffic by $A_i = \lambda_i/\mu_i$.

Under the assumption of statistical equilibrium we get the state probabilities:

$$P(y_1,y_2, \dots ,y_N) = C \cdot \frac{A_1^{y_1}}{y_1!} \cdot \frac{A_2^{y_2}}{y_2!} \cdot \dots \cdot \frac{A_N^{y_N}}{y_N!} \tag{27}$$

where $0 \le y_i \le n_i$

and (25) is fulfilled. C is a normalization constant.

This is Erlang's multi-dimensional loss system /16/. The probability of loss for a traffic stream is obtained by adding together appropriate state probabilities. We may restrict the state space further by restrictions like (26). X may e.g. be the traffic streams which have access to occupy a trunk group (link) in a connection between two exchanges.

An example of this is given in /21/.

2.2 The A-Formula /22/

A simple N-dimensional case is obtained by using the single source model of section 1.6. Let us introduce the passage probabilities from Erlang's ideal grading (formula (2) & (15)) for the total number of calls. Then the system is described by the following parameters:

Structure: N = number of sources (dimension),
n = number of servers (homogeneous),
k = availability.

Traffic: $\gamma(i) = \lambda(i)/\mu(i) = m(i)/\ell(i)$
= offered traffic from source i when this source is idle,
$\ell(i) = 1/\lambda(i)$ = mean inter-arrival time for source number i,
$m(i) = 1/\mu(i)$ = mean holding time for source number i.

In general we have $k \le n \le N$.
This system has the following state probabilities:

$$P(y) = \frac{Q_y \cdot \sum\limits_{1,2,\dots,y}^{N} \left(\prod\limits_{i=1}^{y} \gamma(i)\right)}{\sum\limits_{y=0}^{n} Q_y \cdot \sum\limits_{1,2,\dots,y}^{N} \left(\prod\limits_{i=1}^{y} \gamma(i)\right)} \tag{28}$$

The left side denotes the probability of finding a total of y sources busy. The inner summation $\sum\limits_{1,2,\dots,y}^{N}$ is to be extended over all $\binom{N}{y}$ combinations of y sources. Q_y is given by (17) & (15). The time congestion is given by (18). The call congestion for source number i becomes equal to the time congestion when this source is removed from the system.

For k = n we obtain the Generalized Engset Formula of Cohen /23/. Many more details on micro states etc. are given in /22/.

This model is insensitive to the inter-arrival time distribution and the holding time distribution, <u>if</u> the sources are independent. This is, however, only fulfilled for exponential distributed inter-arrival times!

The above-mentioned system can be further generalized to several groups of inhomogeneous sources with individual availability for the groups (cf. section 1.5).

2.3 Applications to simulation and traffic measurements

In the above-mentioned systems we are free to choose the holding time distribution. When simulating these systems we should choose a constant holding time as this results in minimal variance of the observed statistics /11/. Similar considerations are valid for traffic measurements.

All other systems (e.g. O'Dell gradings, overflow systems) are sensitive to the holding time distribution. When the arrival process is a (state-dependent) Poisson process, then the worst case (maximum call congestion) is constant holding times. This is confirmed by the exact calculation of small gradings and extensive numerical simulations.

If the Poisson arrival process is deformed, then the systems mentioned in this paper also become sensitive to the holding time distribution.

If the arrival process is more smooth than the Poisson process, then constant holding times yield minimum call congestion.

If the arrival process is more peaked than the Poisson process (overflow traffic), then constant holding times yield maximum call congestion. These results are confirmed in /24/, where exact numerical solutions are obtained for the queueing system Cox/Cox/n,m. For m = 0 waiting positions we get a general full available loss system.

For real loss systems (e.g. telephone networks) a worst case is constant holding times. however, the sensitivity is in general very small.

The time-true simulation of loss systems using constant holding times has several advantages and no drawbacks /25/:

- Worst case, i.e. maximum call congestion,
- Minimum variance
- Elimination of event scheduling, both for arrivals and departures, and, of course,
- fast and exact generation of holding times.
- time-true simulation has automatic disconnection of calls.

For these reasons 'constant holding time'-simulation becomes more effective than the Markov Chain (roulette) simulation method. The above-mentioned principles have been implemented in a software package SIMSOR for simulating large hierarchical telephone networks /26/.

2.4 Final remarks

We have seen that all the classical teletraffic models for loss systems are insensitive to the holding time distribution. As the calls are generated by a large number of independent subscribers, the arrival process is a Poisson process (Palm-Khintchine theorem). This explains why the teletraffic theory has been very successful in applications.

3. REFERENCES

/1/ V. B. Iversen, "The Classical Teletraffic Theory Revisited", Proc. Third International Seminar on Teletraffic Theory ("Fundamentals of Teletraffic Theory"), Moscow, pp. 169-180, 1984.

/2/ B. Wallström, "Congestion Studies in Telephone Systems with Overflow Facilities", Ericsson Technics, Vol. 22, no. 3, pp. 187-351, 1966.

/3/ A. Hordijk and R. Schassberger, "Weak Convergence of Generalized Semi-Markov Processes", Stochastic Processes and their Applications, Vol. 12, pp. 271-291, 1982

/4/ V. B. Iversen and B. Friis Nielsen, "Some Properties of Coxian Distributions with Applications", To appear in Proc. International Conf. on Modelling Techniques and Tools for Performance Analysis, Paris, 1985.

/5/ A. Kuczura, "The Interrupted Poisson Process as an Overflow Process", The Bell System Technical Journal, Vol. 52, no. 3, pp. 437-448, 1973.

/6/ V. B. Iversen, "Palm's Machine Interference Model alias Erlang's Loss System", Proc. Performance of Data Communication Systems and Their Applications, Paris, pp. 251-260, 1981.

/7/ V. B. Iversen, "Data og Teletrafik Teori" (In Danish), Den Private Ingeniørfond, Copenhagen 1985.

/8/ V. B. Iversen, "Erlang's ideelle gradering med asymmetrisk belastning og individuel søgekapacitet" (In Danish), IMSOR Research Report no. 13, 14 pp. 1982.

/9/ D. König, K. Matthes and K. Nawrotzki,"Unempfindlichkeitseigenschaften von Bedienungsprozessen", pp. 357-445 in "B. W. Gnedenko & I. N. Kowalenko, Einführung in die Bedienungstheorie", R. Oldenburg Verlag, München Wien, 1971.

/10/ G. P. Basharin and V. A. Kokotushkin, "Conditions for Strong Statistical Equilibrium of Complex Mass Servicing Systems", Problems of Information Transmission, Vol. 7, no. 3, 242-248, 1971.

/11/ A. Hordijk, D. L. Iglehart and R. Schassberger, "Discrete Time Methods for Simulating Continuous Time Markov Chains", Adv. Appl. Prob., Vol. 8, pp. 772-788, 1976.

/12/ W. Henderson, "Insensitivity and Reversed Markov Processes", Adv. Appl. Prob., Vol. 15, pp. 752-768, 1983.

/13/ J. W. Cohen, "Sensitivity and Insensitivity", Delft Progress Report, Vol. 5, no.3, pp. 159-173, 1980.

/14/ F. P. Kelly, "Reversibility and Stochastic Networks", J. Wiley & Sons, 1979.

/15/ V. B. Iversen and M. G. Nielsen, "Intelligent Hunting", IMSOR Research Report no. 9, 13 pp. 1980.

/16/ Arne Jensen, "An Elucidation of A. K. Erlang's Statistical Works through the Theory of Stochastic Processes", pp. 23-100 in E. Brockmeyer, H. L. Halstrøm and Arne Jensen, The Life and Works of A. K. Erlang", Copenhagen 1948.

/17/ L. A. Gimpelson, " Analysis of Mixtures of Wide- and Narrow-Band Traffic", IEEE Trans. on Communication Technology, Vol. 13, no. 3, pp. 258-266. 1965.

/18/ S. S. Lam, "Queueing Networks with Population Size Constraints", IBM J. Res. Develop., Vol. 21, pp. 370-378, 1977.

/19/ A. M. Aein, "A Multi-User Class, Blocked-Calls-Cleared, Demand Access Model", IEEE Trans. on Communications, Vol. 26, no. 3, pp. 378-385, 1978.

/20/ J. W. Roberts, "A Service System with Heterogeneous User Requirements - Application to Multi-Services Telecommunications Systems", Proc. Performance of Data Communication

Systems and Their Applications, Paris, pp. 423-431, 1981.

/21/ D. Y. Burman, J. P. Lehoczky and Y. Lim, "Insensitivity of Blocking Probabilities in a Circuit-Switching Network", J. Appl. Prob., Vol. 21, pp. 850-859, 1984.

/22/ V. B. Iversen, "The A-Formula", Teleteknik, English ed., 1980, no. 2, pp. 64-79. 1980.

/23/ J. W. Cohen, "The generalized Engset formulae", Philips Telecommunication Review, Vol. 18, no. 4, pp. 158-170, 1957.

/24/ S. Hansen, "Phase-Type Distributions in Queueing Theory", IMSOR, Ph.D. thesis, Technical University of Denmark, 209 pp. 1983.

/25/ V. B. Iversen, "Simulating Telephone Networks Using Constant Holding Times, IMSOR, Researh Report no. 13, 7 pp., Technical Univ. of Denmark, 1983.

/26/ M. G. Nielsen, "Simulationsprogrammet SIMSOR" (In Danish), IMSOR, Research Rep. no. 12, 107 pp., Technical University of Denmark, 1982.

APPENDIX A

In this appendix we show that under the assumption of statistical equilibrium the state probabilities $P\{y|y_1,y_2, \dots ,y_k\}$ (13) fulfil the equilibrium equations. The general course of the proof is illustrated by the following example.

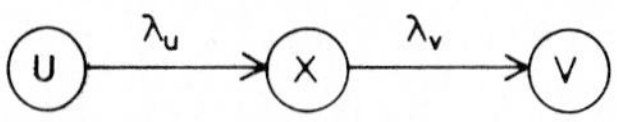

We consider the state probability $P(x)$ under the assumption of statistical equilibrium. After normalization of the equilibrium equation by $P(x)$ we get:

$$\text{Flow in} = \frac{P(u)}{P(x)} \cdot \lambda_u$$

$$= \text{"ratio"} \cdot \text{"rate"}$$

$$\text{Flow out} = \frac{P(x)}{P(x)} \cdot \lambda_v = \lambda_v$$

$$= \text{"rate"}$$

The course of action is similar to the one used in /22/. In the following we find the total flow in and the total flow out of the state $P\{y|y_1,y_2, \dots ,y_k\}$, and show that these flows are of equal size.

Three different types of events give rise to state transitions:

(1) acceptance of a call attempt,
(2) phase shift in a service time, and
(3) departure of a call.

Every event results in either a flow in or a flow out.

A.1 An arriving call results in a positive holding time

Lost calls have no influence upon the state of the system. Therefore we only consider accepted calls (net arrival intensity λ_y). We might also include the probability b_o of the Cox distribution. However, for different reasons this is not done.

a. Flow in: F_{1a}

State transition:

$$P\{y-1|y_1-1,y_2,\dots,y_k\} \rightarrow$$
$$P\{y|y_1,y_2,\dots,y_k\} \quad , \quad 0 < y_1 \le y$$

$$\text{Ratio} = \frac{y_1}{y} \cdot \frac{m}{m_1} \cdot \frac{P(y-1)}{P(y)}$$

$$\text{Rate} = \lambda_{y-1} \cdot b_o$$

$$F_{1a} = \frac{y_1}{y} \cdot \frac{m}{m_1} \cdot \frac{y\mu}{\lambda_{y-1}} \cdot \lambda_{y-1} \cdot b_o$$

The mean holding time is $m = \frac{1}{\mu}$, and we have (8):

$$m_1 = \frac{q_1}{\mu_1} = \frac{b_o}{\mu_1}$$

By inserting this we get:

$$F_{1a} = y_1 \cdot \mu_1 \qquad (a.1)$$

b. Flow out: F_{1b}

State transition:

$$P\{y|y_1,y_2,\dots,y_k\} \rightarrow$$
$$P\{y+1|y_1+1,y_2,\dots y_k\} \quad , \quad 0 \le y_1 < y$$

$$F_{1b} = \lambda_y \cdot b_o \qquad (a.2)$$

A.2 Phase shift in the departure process

The service time of a busy server shifts from phase j to phase (j + 1) ($1 \le j < k$). The service continues in the succeeding phase.

a. Flow in: F_{2a}

State transition:

$$P\{y|y_1,\dots,y_j+1,y_{j+1}-1,\dots,y_k\}$$
$$P\{y|y_1,\dots y_j,y_{j+1},\dots y_k\} \quad , \quad 1 \le j < k$$

$$\text{Ratio} = \frac{y_{j+1}}{y_j+1} \cdot \frac{m_j}{m_{j+1}}$$

$$= \frac{y_{j+1}}{y_j+1} \cdot \frac{\mu_{j+1}}{\mu_j} \cdot \frac{1}{b_j}$$

$$\text{Rate} = (y_j + 1) \cdot \mu_j \cdot b_1$$

By adding contributions from all existing calls under service we get:

$$F_{2a} = \sum_{j=1}^{k-1} \left(\frac{y_{j+1}}{y_j+1} \cdot \frac{\mu_{j+1}}{\mu_j} \cdot \frac{1}{b_j}\right) \cdot (y_j + 1) \cdot \mu_j \cdot b_j$$

$$= \sum_{j=1}^{k-1} y_{j+1} \cdot \mu_{j+1}$$

$$F_{2a} = \sum_{j=2}^{k} y_j \cdot \mu_j \qquad (a.3)$$

b. Flow out: F_{2b}

State transition:

$P\{y|y_1,y_2, \ldots ,y_k\} \rightarrow$
$P\{y|y_1, \ldots ,y_j-1,y_{j+1}+1, \ldots ,y_k\}, \; 1 \le j < k$

$$F_{2b} = \sum_{j=1}^{k} y_j \cdot \mu_j \cdot b_j \qquad (a.4)$$

(Notice that $b_k = 0$).

A.3 Departure process: a service time terminates

In this case a service time terminates, and the state of the system is reduced by one.

a. Flow in: F_{3a}

State transition:

$P\{y+1|y_1, \ldots ,y_j+j, \ldots ,y_k\}$
$P\{y|y_1, \ldots ,y_j, \ldots ,y_k\} \quad , \qquad 1 \le j \le k$

$$\text{Ratio} = \frac{P(y+1)}{P(y)} \cdot \frac{y+1}{y_j+1} \cdot \frac{m_j}{m}$$

$$= \frac{\lambda_y}{(y+1)\cdot\mu} \cdot \frac{y+1}{y_j+1} \cdot \frac{m_j}{m}$$

$$\text{Rate} = (y_j+1) \cdot \mu_j \cdot (1-b_j)$$

$$F_{3a} = \sum_{j=1}^{k} \lambda_y \cdot m_j \cdot \mu_j \cdot (1-b_j)$$

$$= \lambda_y \cdot \sum_{j=1}^{k} q_j \cdot (1-b_j)$$

$$F_{3a} = \lambda_y \cdot b_o \qquad (a.5)$$

b. Flow out: F_{3b}

State transition:

$P\{y|y_1,y_2, \ldots ,y_k\} \rightarrow$
$P\{y-1|y_1, \ldots ,y_j-1, \ldots ,y_k\} \quad , \qquad 1 \le j \le k$

$$F_{3b} = \sum_{j=1}^{k} y_j \cdot \mu_j \cdot (1-b_j) \qquad (a.6)$$

A.4 Balance equations

We want to show that the total flow in:

$$F_a = F_{1a} + F_{2a} + F_{3a}$$

equals the total flow out:

$$F_b = F_{1b} + F_{2b} + F_{3b}$$

We get:

$$F_a = y_1 \cdot \mu_1 + \sum_{j=2}^{k} y_j \cdot \mu_j + \lambda_y \cdot b_o$$

$$F_a = \sum_{j=1}^{k} y_j \cdot \mu_j + \lambda_y \cdot b_o$$

$$F_b = \lambda_y \cdot b_o + \sum_{j=1}^{k} y_j \cdot \mu_j \cdot b_j$$

$$+ \sum_{j=1}^{k} y_j \cdot \mu_j \cdot (1-b_j)$$

$$F_b = \sum_{j=1}^{k} y_j \cdot \mu_j + \lambda_y \cdot b_o$$

Thus we get: $\qquad F_a = F_b \qquad$ q.e.d.

TELETRAFFIC ISSUES in an Advanced Information Society
ITC-11
Minoru Akiyama (Editor)
Elsevier Science Publishers B.V. (North-Holland)

ON MODELING AND PERFORMANCE OF SYSTEMS WITH INTERDEPENDENT TRAFFIC SOURCES AND DYNAMIC ONE-WAY CHANNEL ALLOCATION

Holger DAHMS

Institute for Electronic Systems and Switchings, University of Dortmund

Dortmund, FRG

ABSTRACT

This paper deals with systems with interdependent traffic sources which are suitable to increase the channel efficiency and the admissible offered traffic by dynamic one-way channel allocation. The number of one-way channels is assumed to be equal or less than the number of subscribers. A limited number of storage places may be provided for reducing the loss of speech samples. Such systems may, e. g., be suitable for overload protection in PABX-Systems. The state probabilities and the characteristic traffic values are calculated exactly by means of closed form solutions and iterative methods. Results for these models are presented in examples and diagrams.

1. INTRODUCTION

In order to increase the channel efficiency in communication systems, some modern multichannel TDM-systems admit a number of sources to share a smaller number of channels by voice-activated switching. Publications about the analysis of such systems assume that there doesn't exist any dependence between two subsribers. It is, however, well known that certain interdependencies between the two traffic sources of a connection can be recognized concerning their talking and listening periods (active and passive periods).
In this paper models and analyzing methods for two systems are presented which take into account this particular interdependence of a pair of traffic sources. After the description of the considered system configuration in section 2, a model for the talking activities of a pair of sources is derived in section 3. The equations of steady state and some characteristic values for model No. 1 - systems without any waiting places - are given in section 4 and those of model No. 2 - systems with a limited number of waiting places in section 5. Finally, both models are extended to the more general case of a variable number of conversations being in progress as a function of time in section 6. A conclusion follows in section 7.

2. SYSTEM CONFIGURATIONS

In the sequel, TDM-switching systems are considered which admit dynamic one-way channel allocation. During each conversation three different microstates can be defined taking into account the number of active sources. The first model deals with a given number of conversations being in progress, each conversation concerning the connection of a pair of sources. An extension of this model leads to systems in which more than two sources may take part in a connection. The number of one-way channels is assumed to be equal or less than the number of subscribers. In the latter case, it is possible that speech samples are lost.
The second model deals with a system which enables the storage of speech samples for the fact that the number of channels is not sufficient for immediate transmission of all speech samples. If the number of waiting places is, however, too small a loss of speech samples is possible again. The whole system comprising channels as well as waiting places is operating as a delay-loss system in the usual sense with FIFO service discipline.

3. MODELS FOR THE TALKING ACTIVITIES

Many investigations of systems with dynamic channel allocation are based upon the description of speech and pause intervals of one speaker only given by a binomial distribution /1/-/5/. During a telephone connection between two subscribers four different states are possible at any time considering the number of speakers just being active. In the early sixties P.T. Brady has carried through measurements of 16 telephone conversations during 137.4 minutes using special speech detectors with variable thresholds between signals representing speech and pause intervals. His analyses /6/ lead to the definition of transition probabilities between the states mentioned above which are shown in Table 1.

Table 1 Transition probabilities of changing state in a 5-msec Period for the -40 dBm Threshold Condition

From \ To	Neither	A	B	Both
Neither	0.98940	0.00529	0.00530	0.00001
A	0.00387	0.99486	0.00001	0.00126
B	0.00367	0.0	0.99510	0.00123
Both	0.00005	0.00885	0.01015	0.98095

The values of Table 1 admit a certain modification of these transition probabilities. In this paper only three states are considered, in which no, one or two speech sources are active. The corresponding transition probabilities are denoted by $q(i,j)$ with $0 \le i \le 2$, $0 \le j \le 2$ and $q(0,2)=q(2,0)=0$. Considering connections between $n>2$ speech sources transition probabilities can be formally defined by $q(i,j)$ with $0 \le i \le n$, $0 \le j \le n$ and $q(i,j)=0$ if $j>i+1$ or $j<i-1$, respectively.

4. SYSTEMS WITHOUT WAITING PLACES

The models of these systems are based on the transition probabilities according to section 3. Transition rates with respect to the input process and the terminating process (referring to new calls and terminating conversations) are negligible in the sequel because these transition rates are much smaller than the transition probabilities caused by speech activities. Therefore at first a model for a given number of conversations being in progress is investigated taking into account this particular dependence of each pair of traffic sources.

4.1 Steady states and their probabilities

In the case of x connections all possible steady states can be described by a vector with three parameters a_i, (i=0,1,2) indicating the number of conversations with i active sources, respectively. A state in which both sources are inactive at a_0 pairs, one source is active at a_1 pairs and both sources are active at a_2 pairs be denoted by $\{a_0,a_1,a_2\}$ and the probability of this state by $p(a_0,a_1,a_2)$. The parameters a_i are defined such that

$$\sum_{i=0}^{2} a_i = x \qquad (1a) \qquad \text{and} \qquad y = \sum_{i=0}^{2} i a_i \qquad (1b)$$

where y is the number of occupied timeslots. A change of this number can occur by random events at discrete times defined by the TDM frames. Consequently this stochastic process can be described by a homogenous Markov chain.

4.1.1 One pair of sources

Fig. 1 shows the steady state transition diagram of the Markov chain for an example of one pair of traffic sources.

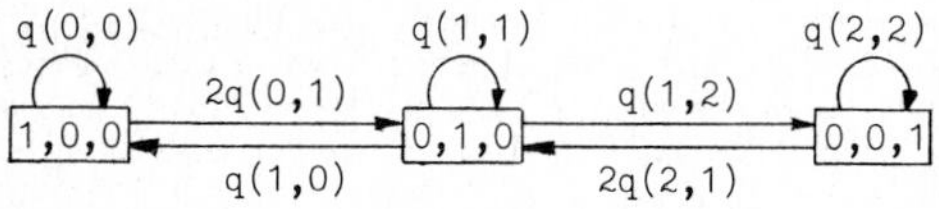

Fig. 1 Markov chain for a pair of sources

For this case the following equations of state are obtained:

$$2q(2,1)p(0,0,1) = q(1,2)p(0,1,0) \qquad (2a)$$

$$(q(1,2)+q(1,0))p(0,1,0) = 2q(0,1)p(1,0,0) + 2q(2,1)p(0,0,1) \qquad (2b)$$

$$q(1,0)p(0,1,0) = 2q(0,1)p(1,0,0) \qquad (2c)$$

with the normalizing condition

$$p(1,0,0) + p(0,1,0) + p(0,0,1) = 1 \qquad (2d)$$

The solution of this set of equations is

$$p(1,0,0) = \frac{1}{1 + \frac{2q(0,1)}{q(1,0)} + \frac{q(0,1)q(1,2)}{q(1,0)q(2,1)}} \qquad (3a)$$

$$p(0,1,0) = \frac{2q(0,1)}{q(1,0)}\, p(1,0,0) \qquad (3b)$$

$$p(0,0,1) = \frac{q(0,1)q(1,2)}{q(1,0)q(2,1)}\, p(1,0,0) \qquad (3c)$$

4.1.2 Several pairs of sources

If x pairs of traffic sources have established a connection, the set of eqs. (3a)-(3c) are valid for each of the x pairs. The combination of possible active subscribers belonging to these x pairs leads to the states shown in the state transition diagram in Fig. 2 later on. Their probabilities, however, can be calculated exactly even now in the following way.

Under the condition that the pairs of sources do not depend upon each other the probabilities that the microstates {1,0,0}, {0,1,0}, {0,0,1} occur at the same time at these x pairs are given by

$$p(x,0,0) = p^x(1,0,0) \qquad (4a)$$

$$p(0,x,0) = p^x(0,1,0) \qquad (4b)$$

$$p(0,0,x) = p^x(0,0,1) \qquad (4c)$$

Taking into account all possible combinations which may form the state $\{a_0,a_1,a_2\}$ leads to the formula

$$p(a_0,a_1,a_2) = \frac{\left[\frac{2q(0,1)}{q(1,0)}\right]^{a_1} \left[\frac{q(0,1)q(1,2)}{q(1,0)q(2,1)}\right]^{a_2}}{\left[1 + \frac{2q(0,1)}{q(1,0)} + \frac{q(0,1)q(1,2)}{q(1,0)q(2,1)}\right]^x} \cdot bc$$

$$\text{with} \quad bc = \binom{x}{a_0}\binom{x-a_0}{a_1}\binom{x-a_0-a_1}{a_2} \qquad (5)$$

By inserting eqs. (3a)-(3c) and eq. (1) in eq. (5) it can easily be shown that eq. (5) represents a multinomial distribution

$$p(a_0,a_1,a_2) = \frac{x!}{a_0!a_1!a_2!}\, p(1,0,0)^{a_0}\, p(0,1,0)^{a_1}\, p(0,0,1)^{a_2} \qquad (6)$$

4.1.3 Steady state transition diagram

The state diagram in Fig. 2 shows the microstates for an example with x=3 connections.

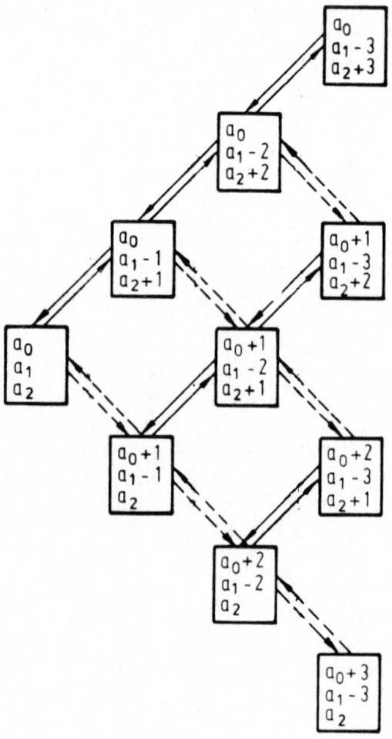

Fig. 2 Steady state transition diagram

The number of microstates X(x) at x connections can be calculated by a sum of natural numbers.

$$X(x) = \sum_{i=1}^{x+1} i = \frac{(x+1)(x+2)}{2} = \binom{x+2}{2} \qquad (8)$$

Summing over x from x=0 to x=n yields the total number X of microstates which are possible at up to n activ pairs of sources:

$$X = \sum_{x=0}^{n} X(x) = \sum_{x=0}^{n} \binom{x+2}{2} = \binom{n+3}{3} \qquad (9)$$

The numbers calculated in eqs. (8) and (9) are special cases of the so called figured numbers, known as "triangle numbers" and "tetrahedron numbers" /7/.

For the microstates the following state equation can be obtained with the aid of Fig. 2.

$$\begin{aligned} p(a_0,a_1,a_2)2a_0q(0,1)&+a_1(q(1,2)+q(1,0))+2a_2q(2,1)= \\ &p(a_0\ ,a_1+1,a_2-1)\ (a_1+1)\ q(1,2)\ + \\ &p(a_0+1,a_1-1,a_2\ )2(a_0+1)\ q(0,1)\ + \\ &p(a_0-1,a_1+1,a_2\ )\ (a_1+1)\ q(1,0)\ + \\ &p(a_0\ ,a_1-1,a_2+2)2(a_2+1)\ q(2,1) \end{aligned} \qquad (10)$$

Inserting eq. (5) in eq. (10) it can easily be shown that equation (5) is the solution of this general state equation. The steady state probabilities of the considered system are therefore given by the multinomial distribution of eq. (5).

4.1.4 Modified transition probabilities

In horizontal groups, Fig. 2 shows several parallel microstates with the same number of y occupied one-way channels but without any direct transitions. The microstates of a horizontal group can be combined to a macrostate {y} comprising all microstates corresponding to the same value of y. The probability p(y) for y timeslots is given by

$$p(y) = \sum_{i=1}^{z(x,y)} p_i(y), \quad z(x,y) = \begin{cases} [y/2]+1 & , y \le x \\ [(2x-y)/2]+1, & y \ge x \end{cases} \qquad (11)$$

z(x,y) is the greatest possible number of parallel microstates in the case of x pairs of sources and y occupied timeslots. [a] be the greatest integer number less or equal a. $p_i(y)$ is equal to the probability of the above mentioned states given by eq. (5). The parameters a_0, a_1 and a_2 can be expressed as a function of x and y in the following manner with $1 \le i \le z(x,y)$:

$$\left.\begin{aligned} a_0&=x-y+i \\ a_1&=y-2i \\ a_2&=i \end{aligned}\right\} y \le x \quad (12a) \qquad \left.\begin{aligned} a_0&=i \\ a_1&=2(x-i)-y \\ a_2&=y-x+i \end{aligned}\right\} y \ge x \quad (12b)$$

The symmetry of the arrangement of the microstates of Fig. 2 can be seen in eqs. (12a) and (12b), too, if y is replaced in eq. (12b) by Y=2x-y for y ≥ x. Eqs. (5), (11), (12) and the abbriviations

$$A1 = 2q(0,1)/q(1,0)\ , \quad A2 = q(1,2)/2q(2,1) \qquad (13)$$

yield

$$p(y) = \frac{x!}{(1+A1+A1A2)^x} \sum_{i=0}^{[Y/2]} \frac{A1^{Y-2i}}{(Y-2i)!} \frac{(A1A2)^i}{i!} \frac{1}{(x-Y+i)!} \qquad (14)$$

$$\text{with } Y = \begin{cases} y & , y \le x \\ 2x-y & , y \ge x \end{cases}$$

With the aid of eq. (14) a simplified steady state transition diagram can be developed which in contrast to Fig. 2 is one-dimensional as shown in Fig. 3:

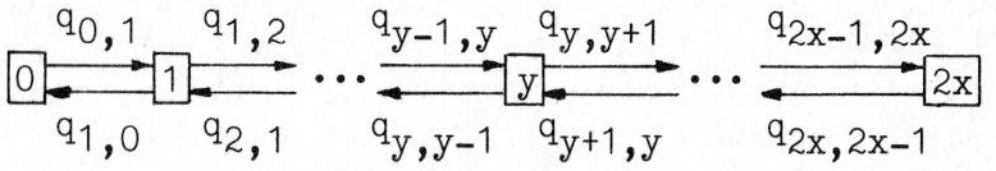

Fig. 3 Transition diagram of macrostates

The new transition probabilities $q_{y,y+1}$ and $q_{y+1,y}$ can be calculated by the probabilities $p_i(y)$, $q_i(y,y+1)$ and $q_i(y+1,y)$.

$$q_{y,y+1} = \frac{\sum_{i=0}^{z(x,y)-1} p_i(y)q_i(y,y+1)}{\sum_{i=0}^{z(x,y)-1} p_i(y)} \qquad (15a)$$

$$q_{y+1,y} = \frac{\sum_{i=0}^{z(x,y+1)-1} p_i(y+1)q_i(y+1,y)}{\sum_{i=0}^{z(x,y+1)-1} p_i(y)} \qquad (15b)$$

Eqs. (15a) and (15b) yield the probabilities for these transitions between the states with y and y+1 active sources as the number of occupied timeslots is equivalent to the number of active talkers. These equations are therefore the fundamental formulae for the second model, too, which is described in section 5.

4.1.5 Influence of the transition probabilities concerning one pair of talkers

In section 4.1.4 a one-dimensional process has been obtained with transitions only to neighboring states characterizing the number of talkers being active. This model is based on the transition probabilities of Table 1 taking into account the interdependence between the two talkers of each pair of sources.

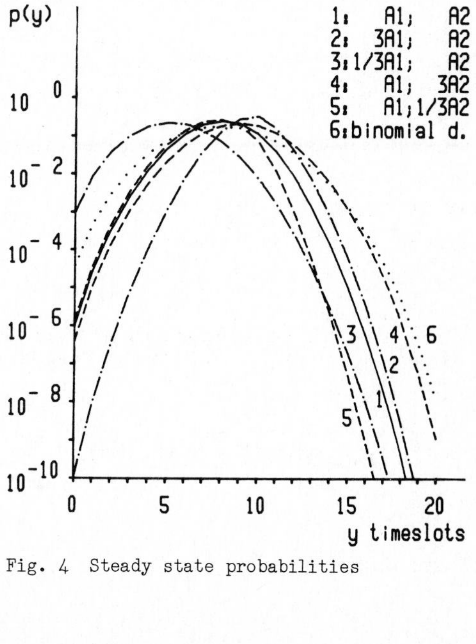

Fig. 4 Steady state probabilities

As the values of Table 1 represent measurements of talkers from the English-speaking world, the influence of a variation of A1 and A2 given by eq. (13) upon the steady state probabilities p(y) is shown in Fig. 4 (dashed and dashed dotted lines). In addition the curve for a binomial distribution is plotted as a dotted line (curve No. 6) with the same mean as the multinomial distribution shown in curve No. 1 as a solid line.

4.2 Cutout fraction

The preceeding models are based on the fact that there are always enough timeslots being allocatable by activ talkers. If x pairs of sources have established a connection maximally 2x timeslots are required to prevent a loss of speech samples. In the sequel, only $n < 2x$ channels be available, so that a loss of some samples may arise. The probability for k talkers being active is given by eq. (14). The cutout fraction - denoted by cof - can be defined by the ratio of the number of samples getting lost and the total number of samples arriving:

$$\mathrm{cof} = \frac{\sum_{k=n+1}^{2x} (k-n)p(k)}{\sum_{i=0}^{2x} ip(i)} \tag{16}$$

4.3 Transmission advantage

In conventional TDM-systems the ratio of the number 2x of busy sources to the number n of timeslots provided for these busy sources is $2x/n=1$. Systems with voice-activated switching and one-way channel allocation yield a so called transmission advantage tra, which is given by

$$\mathrm{tra} = 2x/n^*(\mathrm{cof}) - 1 \tag{17}$$

where $n^*(\mathrm{cof})$ means the dynamic number of timeslots required to maintain the cutout fraction below a desired threshold.

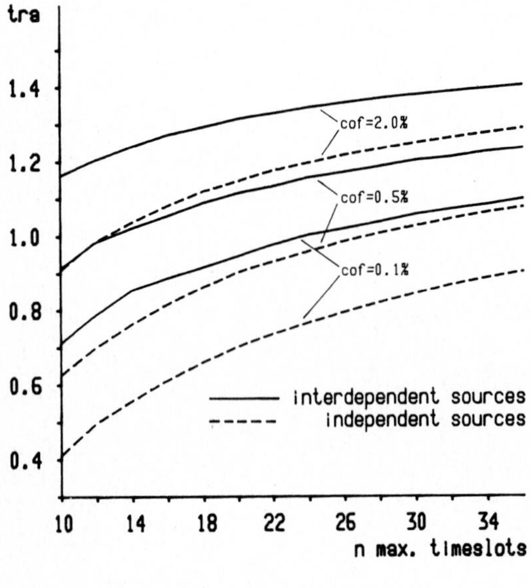

Fig. 5 Transmission advantage

Listening tests /1/ have shown that cutout fractions of 0.5 percent up to 2 percent are perceptible. Fig. 5 shows the values of tra for a maximal cutout fraction of 0.1, 0.5 and 2 percent (solid lines) in comparison with results based on a binomial distribution with the same mean known from literature /3/ (dashed lines).
Taking into account the interdependence between two talkers the transmission advantage exceeds the values based on a binomial distribution about 10 to 30 percent.

4.4 Several n-tupel of sources

An extension of eq. (5) yields formulae which enable the calculation of the steady state probabilities for m connections each between an n-tupel of sources /9/. These results can, however, not be presented here for lack of space.

5. SYSTEMS WITH WAITING PLACES

For systems with waiting places for speech samples, the service discipline applied can be described as follows:

1. As long as not all timeslots are occupied (and all waiting places are idle), incoming samples can be directly allocated to the timeslots of a frame.
2. If there are more samples than available one-way channels in a frame, samples of all talkers are transferred into a buffer and the n timeslots of the frame are allocated to the first n samples of the buffer.
3. Henceforth only samples are transmitted which have been taken from waiting places until the buffer is empty again.
4. As long as the buffer is not empty again the special case may arise that traffic sources will occupy even more than one timeslot within a frame.
5. The total system comprising channels as well as waiting places is operating as a delay-loss system in the usual sense with FIFO-service discipline.

5.1 Steady state transition diagram

States of this model be denoted by three parameters

1. the number of active sources k ($0 \leq k \leq 2x$),
2. the number of allocated channels y ($0 \leq y \leq n$),
3. the number of occupied waiting places j, ($0 \leq j \leq s$).

Here x means the number of pairs of sources, n and s the maximal numbers of timeslots and waiting places, respectively. Based on the model of a Markov chain according Fig. 3 a steady state diagram can be derived which is shown schematically in Fig. 6. From Fig. 6 it can be seen that three areas can be distinguished in this state diagram which are denoted as area I, area II and area III. Fig. 7a to 7c show states of each area with their possible transitions in more detail.
In the state diagram, described in Figs. 6 and 7a to 7c, transitions between non-neighboring states can occur (in contrary to the state diagram shown in Fig. 1 in which only transitions to neighboring states are possible).

5.2 The equations of state

As can be seen in Fig. 6 the states of area I and III are characterized by the fact that no waiting places are occupied (j=0), those of area II by the fact that all timeslots are occupied (y=n).

Therefore the state probabilities of area I and III be denoted by po(k,y) and those of area II by pn(k,j). The values k, y, j be again the number of active sources, occupied timeslots and occupied waiting places, respectively. With the aid of Figs. 7a - 7c and eqs. (15a) and (15b) the following equations of state are obtained for the system considered:

Area I:

$$po(0,0)q_{0,1}=q_{1,0}\sum_{i=1}^{n} po(1,i)+q_{0,0}\sum_{i=1}^{n} po(0,i) \qquad (21a)$$

$$po(k,k)(q_{k,k+1}+q_{k+1,k})=q_{k+1,k}\sum_{i=k+1}^{n} po(k+1,i)+ \qquad (21b)$$
$$q_{k,k}\sum_{i=k+1}^{n} po(k,i) + q_{k,k-1}\sum_{i=k-1}^{n} po(k-1,i)\ , \quad 0<k<n$$

$$po(n,n)(q_{n,n+1}+q_{n+1,n})=q_{n-1,n}\sum_{i=n-1}^{n} po(n-1,i) \qquad (21c)$$

Area III:

$$po(0,y)(q_{0,1}+q_{0,0})=q_{1,0}pn(1,y)+q_{0,0}pn(0,y) \qquad (21d)$$

$$po(k,y) = q_{k+1,k}pn(k+1,y-k) + q_{k,k}pn(k,y-k) + q_{k-1,k}pn(k-1,y-k)\ , \quad 0<k<n,\ k<y\le n \qquad (21e)$$

Area II:

$$pn(0,j)(q_{0,1}+q_{0,0}) = q_{1,0}pn(1,j+n)+q_{0,0}pn(0,j+n) \qquad (21f)$$

$$pn(k,j) = q_{k+1,k}pn(k+1,j+n-k) + q_{k,k}pn(k,j+n-k) +$$
$$q_{k-1,k}pn(k-1,j+n-k)\ , \begin{cases} 0<k<n+1\ , & 1\le j\le s-\sum_{i=1}^{n-k} i \\ k=n+1\ , & 1<j<s \\ n+1<k<2x, & \sum_{i=1}^{k-n} i\le j<s \end{cases} \qquad (21g)$$

$$pn(k,j)=0 \begin{cases} k<n\ , & j>s-\sum_{i=1}^{n-k} i \\ k>n\ , & j<\sum_{i=1}^{k-n} i \\ k>2x, & 1\le j\le s \end{cases} \qquad (21h)$$

$$pn(n+1,1) = q_{n,n+1}po(n,n) \qquad (21i)$$

$$pn(2x,j)(q_{2x,2x}+q_{2x,2x-1}) = q_{2x,2x}p(2x,j+n-2x) +$$
$$q_{2x-1,2x}pn(2x,j+n-2x)\ , \quad \sum_{i=1}^{2x-n} i\le j<s \qquad (21j)$$

$$pn(k,s)=q_{k+1,k}\sum_{i=s+n-k}^{s} pn(k+1,i)+q_{k,k}\cdot\sum_{i=s+n-k}^{s} pn(k,i)$$
$$+q_{k-1,k}\sum_{i=s+n-k}^{s} pn(k-1,i)\ , \quad n<k\le 2x \qquad (21k)$$

and the normalizing condition

$$\sum_{k=0}^{n}\sum_{y=k}^{n} po(k,y) + \sum_{k=0}^{2x}\sum_{j=1}^{n} pn(k,j)=1 \qquad (21l)$$

The set of eqs. (21a)-(21l) can, e. g., be solved iteratively according to the successive overrelaxation (SOR) method /8/.

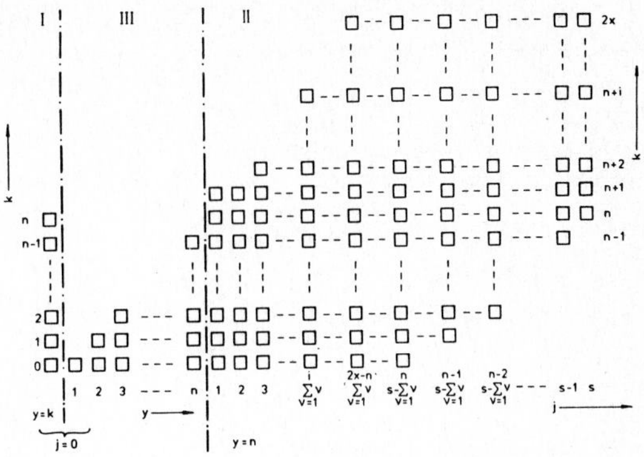

Fig. 6 Steady state diagram

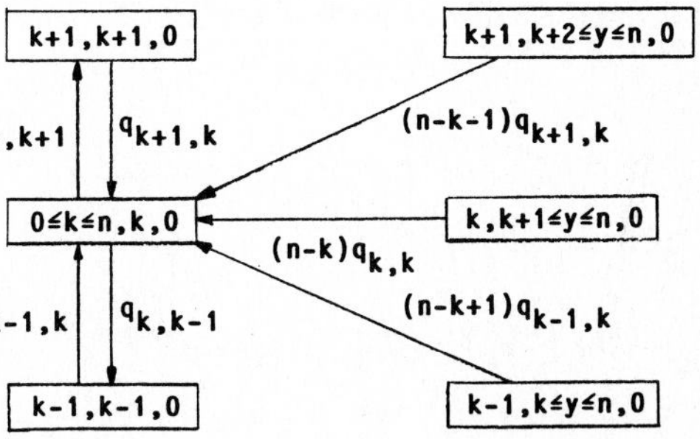

Fig. 7a State transitions in area I

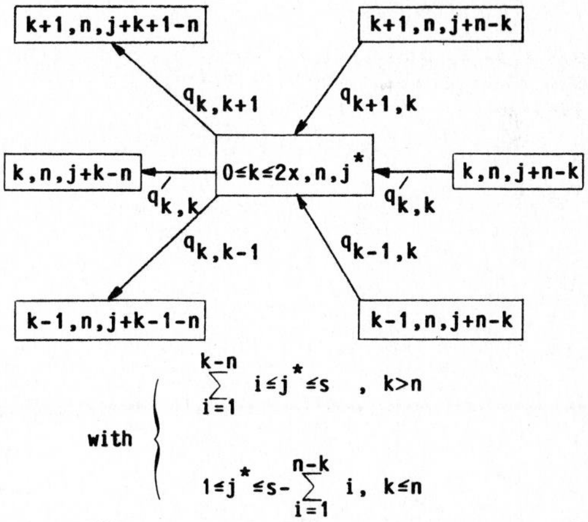

Fig. 7b State transitions in area II

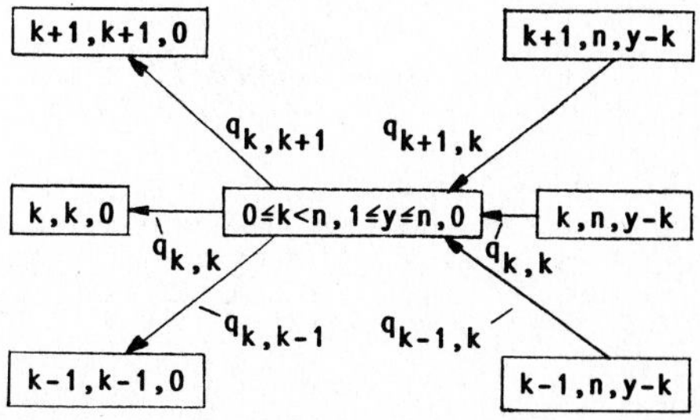

Fig. 7c State transitions in area III

5.3 Characteristic values

5.3.1 Overflow probability and cutout fraction

A limitation of the number of waiting places may cause an overflow and a loss of several speech samples. The probability for an overflow of i samples denoted by pso(i) is obtained as

$$pso(i) = \sum_{k=n+i}^{2x} \{ pn(k+1,s-k+n+1)q_{k+1,k} + pn(k,s-k+n+1)q_{k,k} + pn(k-1,s-k+n+1)q_{k-1,k} \} \quad (25)$$

with $0 < i \leq 2x-n$

The cutout fraction denoted by cof_W is the ratio of the means of the number of lost samples and the total number of samples arriving. It holds

$$cof_W = \frac{\sum_{i=1}^{2x-n} i\, pso(i)}{\sum_{k=0}^{n} k \sum_{y=k}^{n} po(k,y) + \sum_{k=0}^{2x} k \sum_{j=1}^{s} pn(k,j)} \quad (26)$$

5.3.2 Delay time

For calculating the delay time of single speech samples the system is regarded at the time t. The state $\{k,n,j\}$ with $0 \leq k \leq 2x$ and $0 < j \leq s$ is characterized by the following attributes:

1. k speech samples arrive from k active talkers
2. n samples out of the buffer are transmitted
3. j waiting places are occupied

A stored burst of k samples will be delayed by the time td(k,j) which is just the mean of the delay-time of one speech sample.

$$td(k,j) = \frac{[j-k/2]}{n} + 1 \quad (27)$$

The mean delay-time tdm of all speech samples arriving in a frame is then given by

$$tdm = \sum_{k=1}^{2x} pn(k,j)\cdot k \cdot td(k,j) \quad (28)$$

The mean delay time with respect to waiting samples be denoted by tdm_W, and the mean delay time with respect to arbitrary speech samples arriving at the system by tdm_{sy}. It holds

$$tdm_W = \frac{\sum_{j=1}^{s} \sum_{k=1}^{2x} pn(k,j)\cdot k \cdot td(k,j)}{\sum_{j=1}^{s} \sum_{k=1}^{2x} pn(k,j)\cdot k} \quad (29)$$

$$tdm_{sy} = \frac{\sum_{j=1}^{s} \sum_{k=1}^{2x} pn(k,j)\cdot k \cdot td(k,j)}{\sum_{k=0}^{2x} k \sum_{j=1}^{s} pn(k,j) + \sum_{k=0}^{n} k \sum_{y=k}^{n} po(k,y)} \quad (30)$$

5.3.3 Mean probability of a delay

The mean probability W that an arriving sample has to wait until it will be transmitted can be calculated in the following way where w(j,k) is the probability that an arbitrary sample out of a block of k samples has to wait. For

$j \geq k$: all new samples have to wait, w(j,k)=1,

$j < k$: a) all k samples are transmitted, w(j,k)=0
b) a part of the k samples are transmitted and j of k samples are stored, w(j,k)=j/k.

$$W = \frac{\sum_{k=0}^{2x} k \sum_{j=1}^{s} pn(k,j)w(j,k)}{\sum_{k=0}^{2x} k \sum_{j=1}^{s} pn(k,j) + \sum_{k=0}^{n} k \sum_{y=k}^{n} po(k,y)} \quad (31)$$

Inserting the values for w(j,k) yields

$$W = \frac{\sum_{k=0}^{2x} \sum_{j=1}^{k-1} pn(k,j)j + \sum_{k=1}^{2x} \sum_{j=k}^{s} pn(k,j)k}{\sum_{k=0}^{2x} k \sum_{j=1}^{s} pn(k,j) + \sum_{k=0}^{n} k \sum_{y=k}^{n} po(k,y)} \quad (32)$$

5.3.4 The mean queue length of waiting samples and the mean number of occupied timeslots

The term for the mean queue length may be denoted by Q and for the mean number of occupied timeslots by E[y]. It holds

$$Q = \sum_{k=0}^{2x} \sum_{j=1}^{s} pn(k,j)j \quad (33)$$

$$E[y] = \sum_{k=0}^{2x} n \sum_{j=1}^{s} pn(k,j) + \sum_{k=0}^{n} \sum_{y=k}^{n} po(k,y)y \quad (34)$$

6. VARIABLE NUMBER OF CONVERSATIONS IN PROGRESS

The models of section 4 and 5 deal with the case of a fixed number of connections. The transition rates concerning the input process and the terminating process, however, are not taken into account up to now. In the sequel the steady state probabilities calculated by the formulae in section 4 and 5 will be weighted with the probabilities q(x) that x connections are established. q(x) may be derived, e. g., from Erlang's formula /10/. Usually it is reasonable to admit a maximal number of connections (channels) xmax for a given offered traffic A and a maximal loss probability B. In systems with dynamic one-way channel allocation and without waiting places the probability q_y that y timeslots are reserved can be calculated by

$$q_y = \sum_{x=0}^{xmax} p(y)q(x) \quad (35)$$

The cutout fraction of the total system coft=f(x) can be calculated by the following formulae which is similar to eq. (16).

$$coft = \frac{\sum_{x>[n/2]}^{xmax} \sum_{k=n+1}^{2x} (k-n)p(k)q(x)}{\sum_{x=0}^{xmax} \sum_{k=1}^{2x} kp(k)q(x)} \quad (36)$$

For systems with waiting places the cutout fraction $coft_W$ is derived from eqs. (25) and (26) by multiplying the overflow probabilities pso(i) and the state probabilities with q(x):

$$coft_W = \frac{\sum_{x>[n/2]}^{xmax} \sum_{i=1}^{2x-n} i\, pso(i)\, q(x)}{\sum_{x=0}^{xmax} q(x) \left\{ \sum_{k=0}^{n} k \sum_{y=k}^{n} po(k,y) + \sum_{k=0}^{2x} k \sum_{j=0}^{s} pn(k,j) \right\}} \quad (37)$$

The expected advantage concerning an increased admissible offered traffic for systems with dynamic one-way channel allocation compared to common systems is investigated by calculating the following values:

1. the maximal offered traffic A=f(B,xmax) according to Erlang's loss formula,
2. the number n<2xmax of one-way channels leading to a cutout fraction less than a given threshold with the same offered traffic A,
3. the maximal offered traffic A_m=f(B,m) with m=n/2.

The enhancement of the permissible offered traffic can therefore be described by means of the ratio A/A_m which is shown in Fig. 8 for the systems with waiting places (dashed lines) and without waiting places (solid lines) as a function of the number of channels, the loss probability B and the cutout fraction coft or $coft_w$.

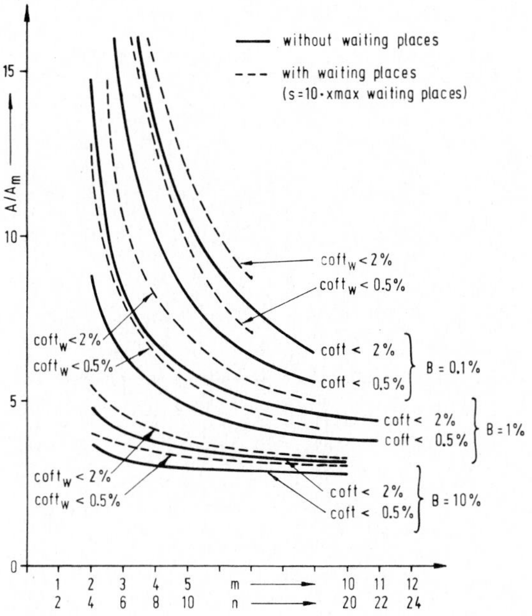

Fig. 8 Enhancement of the offered traffic

As to be seen in Fig. 8 the admissible offered traffic may be about 5 times greater than in usual systems. The corresponding mean delay time can be calculated by eqs. (29) and (30) by multiplying the steady state probabilities in these formulae with q(x), again. The values for the mean delay times are less than 20 TDM frames for this special example with s=10xmax waiting places.

7. CONCLUSION

The first model considered in this paper leads to closed form solutions not only for pairs but also for n-tupels of interdependent sources if the transition probabilities are known, e. g., from measurements. Based on these solutions the reduction of the state transition diagram yields the second model for systems with waiting places which can be calculated exactly by means of iterative methods. The investigations have shown that dynamic one-way channel allocation leads to higher channel efficiency which is even greater than obtained in a model assuming independence of all traffic sources.

These results lead to the idea that one-way channel allocation may be a suitable way for overload protection, too. In PABX-systems, e. g., the external traffic can be switched in the usual way, whereas the internal traffic (where a slightly decreased speech quality may be acceptable) could be switched by one-way channel allocation.

Such service disciplines seem to be suitable for applications to systems with possible overload situations.

REFERENCES

/1/ Bullington, K., Fraser, J. "Engineering Aspects of TASI", Bell Syst. Techn. J., vol. 38, pp. 353-364, March 1959

/2/ Weinstein, C. J., "Fractional Speech Loss and Talker Activity Model for TASI and for Packet-Switched Speech", IEEE Trans. Commun., vol. COM-26, pp. 1253-1257, August 1978

/3/ Harrington, E. A., "Voice/Data Integration Using Circuit Switched Networks", IEEE Trans. Commun., vol. COM-28, pp. 781-793, June 1980

/4/ Nakhla, M. S., Black, D. H. A., "Analysis of a TASI System Employing Speech Storage", IEEE, Trans. Commun., vol. COM-30, pp. 780-785, April 1982

/5/ Wessels, G., Digitale Sprachübertragung in paketvermittelnden Netzen, Minerva-Publikation, München, 1982, (Minerva-Fachserie Technik, ISBN 3-597-10143-7)

/6/ Brady, P. T., "A Statistical Analysis of On-Off Patterns in 16 Conversations", Bell Syst. Techn. J., vol. 47, pp. 73-91, Jan. 1968

/7/ Flachsmeyer, J., Kombinatorik, VEB Deutscher Verlag der Wissenschaften, Berlin 1969

/8/ Törnig, W., Numerische Mathematik für Ingenieure und Physiker, Band 1, Springer Verlag, Berlin-Heidelberg-New York, 1979

/9/ Dahms, H., Vermittlungssysteme mit dynamischer Kanalzuordnung, Ph. D. Thesis, University of Dortmund, 1985

/10/ Brockmeyer, E., Halstrøm, H., Jensen, A., The Life and Works of A. K. Erlang, Acta Polytechnica Scandinavia, 1960, Nr. 287

TELETRAFFIC ISSUES in an Advanced Information Society
ITC-11
Minoru Akiyama (Editor)
Elsevier Science Publishers B.V. (North-Holland)

PERFORMANCE ANALYSIS OF INTEGRATED COMMUNICATION SYSTEM WITH HETEROGENEOUS TRAFFIC

Shun-ichi IISAKU and Yoshiyori URANO

Research and Development Laboratories, KDD
Tokyo, Japan

ABSTRACT

This paper describes several capacity assignment strategies for integrated communication systems with heterogeneous traffic, and proposes the approximate methods to evaluate analytically the performance in terms of blocking probability of demand call for lost-call-cleared mode and that of reservation request for reservation mode. In the case of reservation system , this paper newly introduces the multiqueue model and the multiserver queueing model , in addition to the modified Erlang B model and the arrival process model.

1. INTRODUCTION

According to the increasing demands of business user for integrated services such as voice , data , and image information , multi-service satellite communication systems which offer integrated digital services among small earth stations located on or near customer premises have been introduced [1]-[3].

To meet user demands , we have developed an Integrated Circuit / Packet Switching System based on Demand Assignment Control (ISSDA) offering integrated services via the TDMA satellite link[3]. The main feature of the ISSDA is that the modified No.7 signalling system with the ISDN user part is adopted as a demand assignment control , which assigns satellite communication channel to a user only when needed , to utilize high-speed satellite link effectively and to cope with advanced future communication requirements. The adoption of No. 7 signalling system will facilitate the integration of multi-service satellite communication system into ISDN.

According to communication purposes and its applications , this system can offer three kinds of switching services , i.e. circuit switching service , packet switching service and broadband service. In order to provide these three switching services , it becomes very important to use the satellite link efficiently.

The capacity assignment strategies for above-mentioned multi-service satellite communication systems are an important factor on traffic design to integrate a variety of services having various capacity requests and different call set up procedures such as real-time base and reservation base. The research of this problem has just been started [1]. Performance and quality of service of this system are evaluated in terms of blocking probability of demand call for lost-call-cleared mode and that of reservation request for reservation mode.

The main purpose of this paper is to evaluate analytically the blocking probability for each traffic group which has a pool of dedicated capacity units.

In particular, we discuss the reservation system like the teleconference service, which requests starting time of service, service holding time, required transmission capacity. With regard to performance evaluation, in contrast to demand base, it is necessary to consider the time duration between time instant of reservation request and starting time of service, and the procedure to treat a blocked request which has been refused by the system at the time of reservation. It is very difficult, however, to analyze the general reservation system exactly, because each call has different distribution in regard to the above mentioned time duration.

In order to simplify the analysis, we assume that reservation service is performed by FCFS (first come first served) discipline in request arrival order, all the reservation calls request the same capacity and the blocked reservation request is cleared. In accordance with the above assumption, we present the modified Erlang B model and the arrival process model. In addition , in the case that the blocked reservation requests arrive with some probability again , we present the approximate methods including the multiqueue model and the multiserver queueing model.

2. CAPACITY ASSIGNMENT STRATEGY

First , users are divided in two user classes, according to traffic characteristics , that is , lost-call-cleared traffic and reservation traffic. Moreover , each class of users is divided in several user groups according to capacity demand. And the time slots have two parts , one part is dedicated to lost-call-cleared traffic and the other part is dedicated to reservation traffic. The boundary between the two user classes may be fixed.

The capacity assignment problem can be defined as follows : The capacity , N , measured in units of time slots is to be dynamically shared by a user class divided in several groups, according to traffic characteristics, for example lost-call-cleared-mode, reservation mode, and capacity demand, for example variable band-width request. Each user group in two user classes , M , has equal average arrival rate , average call duration and equal number of capacity units requested for a single call.

On the other hand, several capacity assignment strategies have been proposed and studied in the papers [4]-[8].

(1) Complete sharing
Every user class and/or user group has free access to the whole capacity on a First-Come-First-Served basis.

(2) Complete partitioning
Every user class and/or user group has a pool of dedicated capacity units.

(3) Sharing with maximum allocation
User classes and/or user groups share the common pool of resources, but there is a limit on the number of users of a group and/or a class that are allowed in the system.

(4) Sharing with minimum allocation
A given number of dedicated capacity units is assigned to each user class and/or user group and the rest being shared by all on a First-Come-First-Served basis.

(5)Sharing with priorities
Priorities are assigned to user classes and/or user groups. Users may preempt other users of lower priority.

We define a system state as the number of calls in progress for each users group. That is, a system state is given by the vector $n=(n_1, n_2, \cdots, n_M)$, where n_i is the number of the ith group calls in progress. The ith group, $i=1, \cdots, M$, is characterized by calling rate from an infinite population, the traffic statistics are assumed Poisson distribution with mean λ_i, average call duration $1/\mu_i$, number of capacity units requested for a single ith group user call K_i, maximum number of ith group user calls allowed N_i, number of user calls with dedicated time slots to the ith group L_i. Access disciplines are characterized by assigning to every group a pair of numbers (N_i, L_i), that is, the maximum number of user calls and their dedicated time slots. This is an important subset of possible access disciplines. Resource size imposes a physical constraint for each user class:

$$\Sigma_{i=1}^{M} K_i \cdot n_i \leqq N. \qquad (1)$$

where n_i is the number of ith group user calls. Fig.1 provides a scheme of the system model. Several strategies are as follows.

In the case of complete sharing, every user group has free access to the whole capacity on a First-Come-First-Served basis.

$$L_i = 0 \ , \ N_i = [N/K_i]^+ \ ,$$

$$i=1, \cdots, M. \qquad (2)$$

where $[A]^+$ is the integer part of A. As for complete partitioning, every user group has a pool of dedicated capacity units.

$$N_i = L_i \ ,$$

$$\Sigma_{i=1}^{M} L_i K_i = N \ , \ i=1, \cdots, M. \qquad (3)$$

For sharing with maximum allocation,

$$L_i = 0 \ , \ N_i \leqq P_i \ ,$$

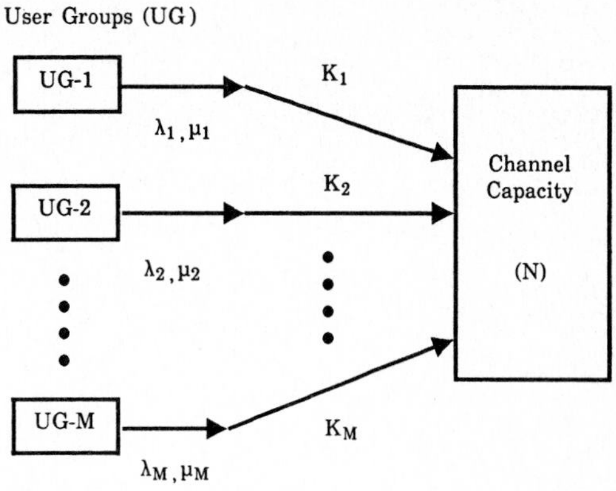

Fig.1 System model

$$i=1, \cdots, M. \qquad (4)$$

where P_i is the maximum number of time slots from the common pool allowed for each user call. For sharing with minimum allocation,

$$L_i \geqq 0 \ , \ N_i \leqq [(N - \Sigma_{i=1}^{M} L_i K_i)/K_i]^+ \ ,$$

$$i=1, \cdots, M. \qquad (5)$$

This is a generalization of the above two disciplines (both complete sharing and complete partitioning).

3. LOST-CALL-CLEARED MODE

In this section, we discuss only lost-call-cleared traffic when the boundary between the lost-call-cleared mode and reservation mode is fixed.

3.1 CALL HAVING THE SAME CAPACITY

Assume that lost-call traffic originates according to a Poisson distribution with mean rate λ, the holding time is negative exponential with mean $1/\mu$, the frame time duration is b seconds and the number of lost-call traffic slots per frame is N. In the following, we show that, under the condition that $\lambda b \ll 1$ and $N\mu b \ll 1$, the blocking probability for the lost-call traffic can be closely approximated by the well-known Erlang B equation:

$$B = a^N / N! / \Sigma_{i=0}^{N} a^i / i! \qquad (6)$$

where $a=\lambda/\mu$ is the offered lost-call traffic load in Erlang.

3.2 CALL HAVING THE VARIOUS CAPACITY

In the case of lost-call having various transmission capacity requests, we assume that lost-call traffic originates according to a Poisson distribution with mean rate λ_i ($i=1, \cdots, M$), the holding time is negative exponential with mean $1/\mu_i$ ($i=1, \cdots, M$), number of capacity units requested for a single ith user group call K_i, calls whose requirements can not be satisfied are blocked and depart without affecting the system further. If the constraint is given by eq. (2), we call such a strategy as complete sharing. In this section, we consider in detail complete sharing strategy. Define the overall occupancy distribution

$$Q(n)=Pr(\Sigma_{i=1}^{M} K_i n_i = n) \qquad (7)$$
$$0 \leqq n \leqq N$$

so that, let the time congestion of ith group calls be E_i,

$$E_i = \sum_{n \geqq N-Ki} Q(n). \qquad (8)$$
$$i=1, \cdots, M.$$

Now, for Poisson arrivals for all i, we have the simple recurrence relation which provides the distribution of the number of the time slots occupied for the complete sharing strategy

$$n\,Q(n)=\Sigma_{i=1}^{M} \lambda_i K_i Q(n-K_i)/\mu_i$$

$$=\Sigma_{i=1}^{M} a_i K_i Q(n-K_i),$$
$$0 \leqq n \leqq N \qquad (9)$$

where $a_i=\lambda_i/\mu_i$, $Q(n)=0$ for $n < 0$ and

$$\Sigma_{n=0}^{N} Q(n)=1. \qquad (10)$$

The blocking probability PB_i for each user group can be written

$$PB_i= \Sigma_{i=0}^{Ki-1} Q(n-i) \qquad (11)$$
$$i=1, \cdots, M.$$

In Fig. 2, we present a sample of the results obtained for this model. Suppose that the number of total time slots are 6, (K1, K2) = (1, 2) and $a_1=0.25$. This figure gives blocking probability as a function of offered load (a_2). Table 1 shows the state distribution (Q(n), n=0, ⋯, 6) of the number of time slots occupied for the complete sharing system of 6 time slots in the case of offering two kinds of traffic.

In heterogeneous traffic with various bit rates, the blocking probability (PB) for high bit rate calls is grater than that for low bit rate calls, when time slots are shared between them. As for the evaluation of the blocking probability for each user group, several criteria may be proposed for the optimization problem.

(1) Average blocking probability is minimized

$$\text{Min } PB=\Sigma_{i=1}^{M} \lambda_i PB_i / \Sigma_{i=1}^{M} \lambda_i \qquad (12)$$

(2)Minimization of the maximum PB_i

$$\text{Min Max}(PB_1, \cdots, PB_M) \qquad (13)$$

On the other hand, there is the capacity allocation problem in reverse way. In this problem, the blocking requirements for each user group is given, and the best strategy which achieves minimum capacity size is obtained.

One method has been envisaged in order to put the blocking probability of high bit rate calls close to that of low bit rate calls, the so-called trunk reservation, restricting the number of time slots for the low bit rate call unless sufficient capacity is available for the high bit rate call.

4. RESERVATION MODE

In this chapter, we discuss only reservation traffic on condition that the boundary between the lost-call-cleared mode and reservation mode is fixed. We consider the reservation system which requests starting time of communication and predetermined service time at the time of reservation such as the teleconference service and the reserved circuit switched service with X.21. In general, this system is characterized by three elements :

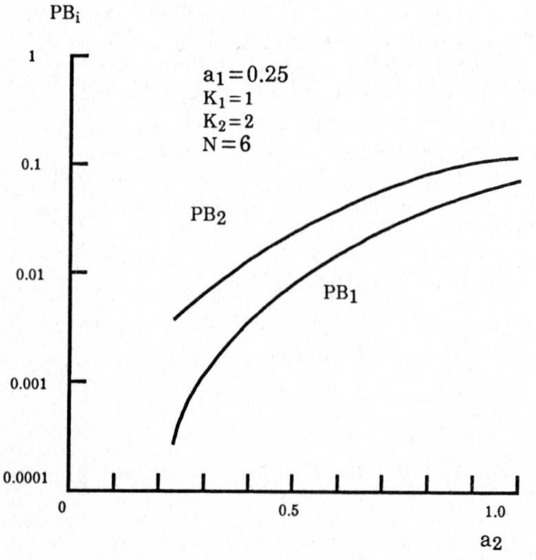

Fig.2 Blocking probability

	(K1, K2)=(1, 2)	(K1, K2)=(1, 3)
Q(0)	0.476	0.476
Q(1)	0.119	0.119
Q(2)	0.252	0.03
Q(3)	0.061	0.241
Q(4)	0.067	0.06
Q(5)	0.015	0.012
Q(6)	0.012	0.061

(a1, a2)=(0.25, 0.5)
N=6

Table 1 State distribution of the number of time slots

starting time of service, service holding time, required transmission capacity. With regard to performance evaluation, in contrast to demand base, it is necessary to consider the notice time (the time duration between time instant of reservation request and starting time of service) and the procedure to treat a blocked request which would be refused by the system at the time of reservation.

4.1 RESERVATION SYSTEM

The reservation system which we discuss is characterized by the following control information elements such as required capacity C , service time H and communication starting time T.

Actual reservation procedures are as follows. User makes reservation by means of speech conversation or data communication between user's reservation terminal and network reservation center , and requests a circuit by indicating communication starting time , the service time , the required capacity and so on. When the reservation is succeeded , that is , capacity is available at T , the system informs the user of the results of the reservation. When a user's reservation request is refused by the system , that is , capacity is not available at T , user may abandon the reservation request or inquire about available time of the system.

Blocking probability that a reservation request would be refused is a function of required capacity , service time and the notice time.

4.2 CALL HAVING THE SAME CAPACITY

We consider reservation system having relatively long interval between time instant of reservation request and starting time of actual communication. Mathematical model is as follows. The time interval $0 \leqq t \leqq \infty$ is divided into contiguous fixed intervals (e.g. one day) : $(j-1)D \leqq t \leqq jD$, $1 \leqq j \leqq \infty$. Moreover , each fixed interval is divided into fixed time segments (e.g. one hour) d_i ($i=1, \cdots, R$) , each of which accommodates S reservation calls. S is the number of time slots per one time segment and d is the length of the time segment. Reservation is made for a time slot in the time segment. It is very difficult, however, to analyze the reservation system exactly. In order to simplify reservation system , we assume reservation service is performed by FCFS (first come first served) discipline in request arrival order and all the reservation calls request the same capacity. And we examine a few approximate methods as described below.

We deal with the case that all blocked reservation requests are cleared.

First , we consider the M / G / S model as shown in Fig.3. Supposing that the reservation request arrives at random by a Poisson process with the rate λ per D and reservation request per one time segment arrives according to the Poisson distribution with mean rate λ / R. The holding time is general distribution with mean g and the number of servers is S which is equivalent to that of time slots. In this model , mean holding time is approximated by the sum of constant service time d and mean notice time f.

$$g=d+f \tag{14}$$

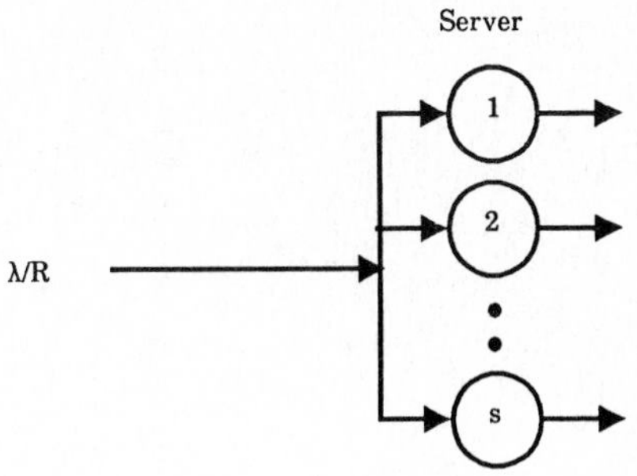

Fig.3 M / G / S model

Therefore , the blocking probability (Br) for the reservation request can be approximated by the Erlang B formula :

$$Br = a^S / S! / \sum_{i=0}^{S} a^i / i! \tag{15}$$

where $a=\lambda(d+f)/R$ is the offered traffic load in Erlang.

This approximate method is a simplified model. Because of the mutual dependence between arrival process and holding time distribution , this model may not be valid in the strict sense.

Second , we discuss the arrival process model of reservation requests for one time segment during the interval I (e.g. one month) in which user can make reservation. Let the Poisson process λ / R denote the mean reservation request arrival rate for one time segment during I. Then , the probability of n reservation request arrivals for a time segment during I is

$$P_n(I) = e^{-\lambda I/R} \cdot (\lambda I/R)^n / n! \quad n = 0, 1, 2, \cdots \tag{16}$$

Since S is the number of the time slots per time segment , blocking state occurs in case that more than S reservation requests arrive. Therefore , the blocking probability of reservation request (Br) is

$$Br = \sum_{i=1}^{\infty} P_{S+i}(I) \cdot i/(S+i)$$
$$= \sum_{i=1}^{\infty} e^{-\lambda I/R} \cdot i(\lambda I/R)^{S+i} / (S+i)(S+i)! \tag{17}$$

where $i/(S+i)$ is conditional blocking probability in the case that S+i reservation requests arrive during I.

Third , we consider the model of multiserver queueing system with finite waiting room. This model is shown in Fig. 4. This model consists of a first stage queue of unlimited capacity , a second stage queue having S finite capacity and S servers. The server attaches to the second stage queue. The arrivals at first stage queue occur in accordance with Poisson process with mean λ / R and the service time has constant distribution with mean d. The server's operation is assumed to start at the beginning of the time segment. Operation is as follows. The instant a switch is closed , reservation calls in the first stage queue

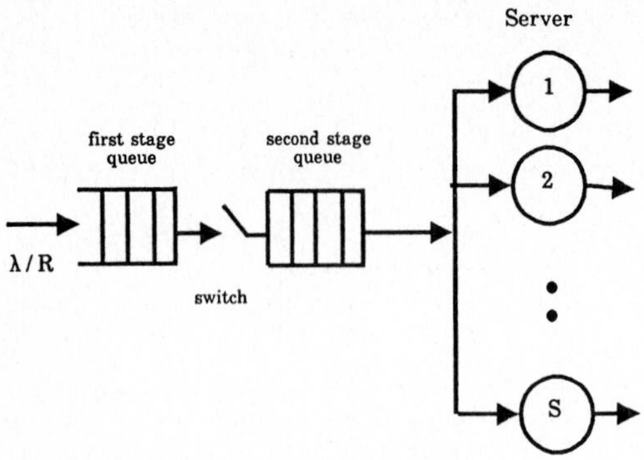

Fig. 4 Multiserver queueing model with finite capacity

enter the second stage queue and the servers begin the reservation service. However , as the maximum number of capacity of second stage queue are S , reservation requests more than S are blocked and the blocked calls are cleared. Switching period is 1 / d. This model is similar to the above mentioned model which considers the arrival process of reservation request.

Next , we deal with the case that blocked reservation requests are scheduled for other time segments.

First , we consider the blocking probability of reservation request in (17). Now , we assume that if a new reservation request is blocked, the blocked reservation request arrives repeatedly up to R times according to the Poisson process. From the above assumption , total arrival rate (Λ) for one time segment during I is given by

$$\Lambda = \lambda (1 - B^{R+1}) / R (1 - B) \qquad (18)$$

Applying the total arrival rate in (18) to (17) , the blocking probability of reservation request including blocked requests is obtained. In this model , as blocked call arrival process is treated by Poisson process , total arrival rate is smaller than actual arrival rate.

Second , we consider the model of multiqueue model with finite waiting room. This system is similar to the polling system consisting of a number of buffered input terminals connected to a computer by transmission line. Multiqueue model of reservation system is shown in Fig. 5. This system consists of R first stage queues having S finite capacity , R second stage queues having S finite capacity and S servers. The arrivals at each first stage queue occur in accordance with independent Poisson process (mean arrival rate to first stage queue i is denoted by λ_i , i = 1 , ··· , R) and the service time of each queue has constant distribution (mean service time is d). The service operation is assumed to start at the beginning of the time segment d_i (i = 1 , ··· , R). When the servers arrive at second stage queue , switch is closed in an instant , all reservation calls in the first stage queue enter the second stage queue and switch is opened. Switching period is 1/D. Each second stage queue is served by servers cyclically.

In this model , in case a reservation request is refused in the first stage queue , the modified

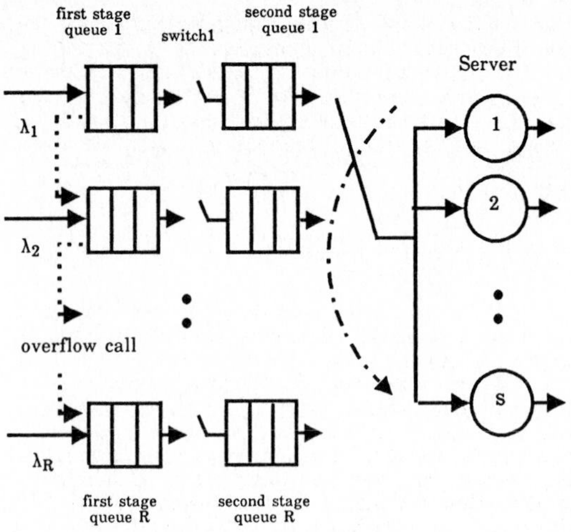

Fig. 5 Multiqueue model with overflow call

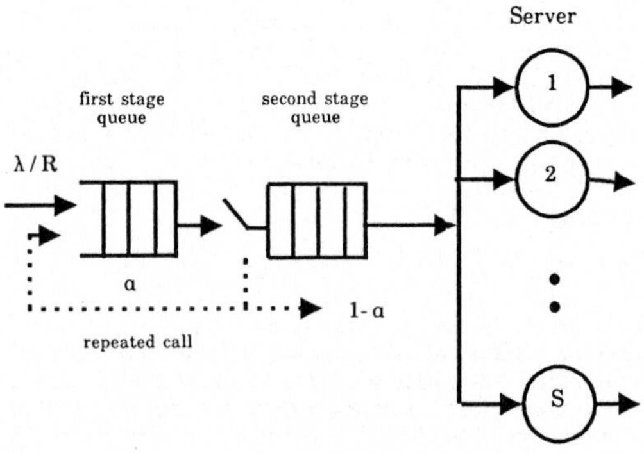

Fig. 6 Multiserver queueing model with repeated call

request arrives at another first stage queue immediately with some probability again. In accordance with the above assumption and operation , it is shown that reservation system can be modeled as a multiqueue model with the overflow call phenomenon. Although arrival process of the overflow call from the first stage queue i to queue j loses the nature of random process , the variance of the overflow call in the queueing system is smaller than that of loss system. Therefore , in this model , the arrival process of the overflow call is approximated by the Poisson process. Most actual reservation system in which blocked reservation requests are scheduled for other time segments is well reflected in this approximate method. As for the analysis of the blocking probability , this model is independent of servers. Therefore , it is sufficient to consider only the arrival process , the number of capacity and the finite queue.

Third , from the view point of the multiserver queueing model with finite capacity as shown in Fig. 4 , we examine an approximate method. We assume that the modified request arrives with some probability (α) again in case a request is refused. In accordance with the above assumption, reservation system can be modeled as a multiserver queueing model with repeated call phenomenon which relates to the loss system , as shown in Fig.6. This model may be analyzed by the approximate method in reference[13].

5.SHARING BETWEEN LOST-CALL AND RESERVATION CALL

In this section , we discuss the capacity assignment strategy sharing between lost-call and reservation call.

If a reservation service shares a transmission facility with lost-call services (complete sharing policy) , it seems that the quality of service of the latter may be inferior to the former and vary widely as a function of random fluctuations of carried reservation traffic. However , it is expected to improve resource utilization by sharing resources between lost-call and reservation call. Therefore, it is necessary to consider the traffic model that priority is assigned to lost-call (the sharing with priorities), with some schemes, for example, trunk reservation. On the other hand , all the lost calls are blocked as long as all shared time slots are reserved by reservation traffic. As for this point , it is necessary to consider the strategy which allocates a given number of dedicated time slots exclusively to lost call and shares the other time slots. This strategy is the sharing with minimum allocation.

6. CONCLUSION

In this paper , we have described several capacity assignment strategies for integrated communication system with heterogeneous traffic and discussed the approximate method giving the blocking probability of demand call for lost-call-cleared mode and that of reservation request for reservation mode.

In particular, we have investigated the reservation system in which all reservation calls request the same capacity. And we have presented the simple approximate methods such as the modified Erlang B model , the arrival process model , the multiqueue model with overflow phenomenon and the multiserver queueing model with repeated call phenomenon.

The results of this study will be applied to a multi-service digital satellite network based on DA / TDMA and a terrestrial-based ISDN.

ACKNOWLEDGEMENTS

The authors express their gratitude to Dr. H. Kaji , director of KDD R and D Laboratories , Dr. K. Nosaka , vice director and Dr. K. Ono , deputy director , for their continuous guidance.

REFERENCES

[1] J.W. Robert, "Teletraffic model for the Telecom 1 integrated services network", ITC 10, Montreal 1983.

[2] J.S. Lee, "Symbiosis between a terrestrial-based integrated services digital network and a digital satellite network", IEEE J. select. areas commun. , vol. SAC-1, pp.103-109, Jan. 1983.

[3] S. Iisaku , K. Matsuo, and K. Ono, "An integrated circuit / packet communication system based on TDMA demand assignment control", I.E.C.E. , Tech. Report , SE-67 , Oct. 1984.

[4] J.M. Aein and G.S. Kosevych, "Satellite capacity allocation", Proc. IEEE , vol. 65, pp. 332-342, March 1977.

[5] G. Barreris and R. Brignolo, "Capacity allocation in a DAMA satellite system", IEEE Trans. Commun. , vol. COM-30, pp.1750-1757, July 1982.

[6] L. Green, "A queueing system in which customers require a random number of servers", Opns. Res. , vol. 28, No 6, pp. 1334-1346, 1980.

[7] J.S. Kaufman, "Blocking in a shared resource environment", IEEE Trans. Commun. , vol. COM-29, pp. 1474-1481, Oct. 1981.

[8] J.M. Aein, "A multi user-class , blocked-calls-cleared demand access model", IEEE Trans. Commun. , vol. COM-26, pp.378-385, March 1978.

[9] L.E.N. Delbrouck, "A unified approximate evaluation of congestion functions for smooth and peaky traffics", IEEE Trans. Commun. , vol. COM-29, pp.85-91, Feb. 1981.

[10] L.E.N. Delbrouck, "On the steady-state distribution in a service facility carrying mixtures of traffic with different peakedness factors and capacity requirements", IEEE Trans. Commun. , vol. COM-31, pp.1209-1211, Nov. 1983.

[11] E.A. Vandoorn, "A note on Delbrouck's approximate solution to the heterogeneous blocking problem", IEEE Trans. Commun. , vol. COM-32, pp.1210-1211, Nov. 1984.

[12] G.J. Foschini and B. Gopinath, "Sharing memory optimally", IEEE Trans. Commun. , vol. COM-31, pp. 352-360, Mar. 1983.

[13] O. Hashida and K. Kawashima, "Buffer behavior with repeated calls", I.E.C.E. , vol. J62-B, no.3 pp.222-228, Mar. 1979.

TELETRAFFIC ISSUES in an Advanced Information Society
ITC-11
Minoru Akiyama (Editor)
Elsevier Science Publishers B.V. (North-Holland)

FLEXIBLE TIME SLOT ASSIGNMENT
A PERFORMANCE STUDY FOR THE INTEGRATED SERVICES DIGITAL NETWORK

V. RAMASWAMI and K. ASWATH RAO

Bell Communications Res. Inc. & AT&T Bell Laboratories

New Jersey, USA

ABSTRACT: There are 24 time slots on the DS1 Signal, and a user who is assigned x time slots can transmit at the rate of $64x$ kilo bits per second (bps). The Signal is to be shared between two types of users who transmit at the rate of 384K bps and 64K bps respectively; these need 6 time slots each and 1 time slot each respectively. We compare the CCITT-standards-based Fixed Scheme, which requires that the time slots assigned to a 384K bps call should form one of four pre-defined 'channels' with six time slots each, to the Flexible Scheme which imposes no restrictions. Based on this study, the CCITT Study Group XVIII adopted a resolution to include the Flexible Scheme in the CCITT standards. The queueing model for the Fixed Scheme may be of wider interest to computer performance analysis because it deals with a problem similar to that arising in memory allocation schemes using fixed partition files.

1. INTRODUCTION

The DS1 Signal is a transmission system used for digital communication. What is transmitted on the DS1 Signal are 8000 frames per second where each frame consists of 193 bits. Leaving the one bit per frame that is used for frame identification and assuming that signalling is done out of band, 192 bits per frame are available to the users. These 192 bits are divided into 24 "time slots" of 8 bits each. Thus a caller who is allocated x time slots on the DS1 Signal transmits at the rate of $8000 \times 8 \times x$ or $64x$ kilo bits per second (bps).

One of the channel allocation schemes under the ISDN (Integrated Services Digital Network) standards permits the sharing of the DS1 Signal by two sets of users - one transmitting at the rate of 64K bps (B-channel) and one transmitting at the rate of 384K bps (H_0 channel). These two types of calls respectively need 1 and 6 time slots respectively, and when they share a DS1 Signal, interesting questions arise as to how the time slots should be assigned to these calls. These were some of the topics of discussion for the Study Group XVIII of the International Telegraph and Telephone Consultative Committee (CCITT) in its meeting in Brasilia in February 1984.

Two schemes of time slot assignment were under consideration. One of these is a "flexible scheme" which places no restrictions on the way time slots are assigned, and the other a "fixed scheme" described as follows. The 24 time slots are divided into four pre-defined groups, say, 1-6, 7-12, 13-18 and 19-24. When a H_0 call arrives, the system determines if any of the four groups is idle and if so assigns one of them to that call. In addition, a "call packing" is performed for the B-channel calls - that is, when a B-channel call arrives and there is a choice of more than one group to assign it to, the system would assign the call to the group with the most number of busy time slots among those with at least one idle time slot.

While it is clear that the flexible scheme would result in a lower blocking probability for the H_0 channel, it is not clear how significant the difference between the two schemes would be. This study was undertaken to develop models for both schemes of time slot allocation and to make a comparison. Based on the results obtained here which show that the flexible scheme is significantly superior, a draft resolution was adopted by the Study Group recommending that the flexible scheme be made part of the CCITT standard.

2. MODELLING ASSUMPTIONS

To obtain tractable models of both schemes we assumed the following:

a) Calls of both types arrive as independent Poisson processes. λ_1 and λ_2 denote the rates per unit time of arrivals of H_0 and B-channel calls respectively.

b) The holding times of the two types of calls are exponentially distributed with rates μ_1 for H_0 and μ_2 for B-channel calls respectively.

c) In addition to the above, it was also assumed that calls which cannot be carried will be blocked.

It is possible to relax these assumptions, and we will discuss this later in Section 6.

Under the above assumptions, one obtains for each of the schemes a Markovian queueing model of 24 "servers" serving two types of "customers". These models and their analyses are presented in Sections 4 and 5. A numerical example is presented in the next section.

3. A NUMERICAL EXAMPLE

Using the Markovian queueing models and the algorithms presented in Sections 4 and 5, we developed an APL program which, for a given input of values of the parameters $\lambda_1, \lambda_2, \mu_1$, and μ_2, computes a number of performance measures for both the flexible and the fixed schemes. The computed performance measures included the following steady state quantities:

- The joint and marginal distributions of the number of calls of both types in the system
- The expected number and variance of the number of calls of either type
- The blocking probabilities for each type of call
- The conditional blocking probabilities for each type of call given the number of slots busy, given the number of type 1 calls in progress, and given the number of type 2 calls in progress.

For brevity we present our results for only one set of sample runs that correspond to the choice of parameter values $\mu_1=\mu_2=1$ and $\lambda_1=0.05\lambda_2$ for various values of λ_2. The superiority of the flexible scheme exhibited by the data below was, however, seen to hold for all choices of parameter values that we tried.

In Table 1 below we present the following performance measures only:

- Blocking probability for 384K bps calls (B_1)
- Blocking probability for 64K bps calls (B_2)
- Conditional blocking probability for a 384K bps call given that at least one such call is in progress (B_1^*)
- Conditional blocking probability for a 64K bps call given that at least one 384K bps call is in progress (B_2^*)
- Probability that at least one 384 K bps call is in progress (P^*)

The blocking probabilities (both conditional and unconditional) given in Table 1 are plotted in Figure 1 and Figure 2 for an easy comparison. These results show the superiority of the flexible scheme in a number of ways. Consider for example the case $\lambda_2=9$. The blocking probability B_1 for the 384K bps calls under the fixed scheme is .1528, and this can be reduced to .0831 by adopting the flexible scheme. This improvement is accomplished at a negligible bad effect on the blocking probability B_2 for the 64K bps calls which goes up from .0038 to .0065 when one changes from the fixed to the flexible scheme. The improvement that results by changing over to the flexible scheme is more pronounced in the conditional blocking probability B_1^* for a 384K bps call given that at least one such call is in progress. Thus for example, when $\lambda_2=9$, about 35% of the time the system will have at least one 384K bps call in progress under both schemes (33.7% and 35.1% to be precise). Under the fixed scheme 384K bps calls that arrive during such congested times have a 42% chance of being blocked ($B_1^*=.422$), whereas under the flexible scheme, this chance is only 23% ($B_1^*=.232$). In practical terms what this shows is that the fixed scheme results in a system that can behave very badly during transient periods of congestion. The flexible scheme on the other hand results in a system wherein the deterioration of service that occurs under congestion is not so drastic.

FIGURE 1: BLOCKING PROBABILITIES FOR 384 Kbps TRAFFIC

— UNCONDITIONAL BLOCKING PROBABILITY B_1

----- CONDITIONAL BLOCKING PROBABILITY B_1^*

$\mu_1=\mu_2=1, \ \lambda_1=0.05\lambda_2$

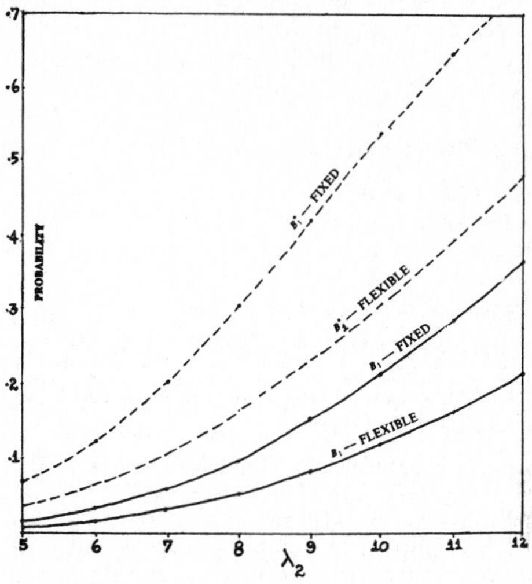

FIGURE 2: BLOCKING PROBABILITIES FOR 64 Kbps TRAFFIC

— UNCONDITIONAL BLOCKING PROBABILITY B_2

----- CONDITIONAL BLOCKING PROBABILITY B_2^*

$\mu_1=\mu_2=1, \ \lambda_1=0.05\lambda_2$

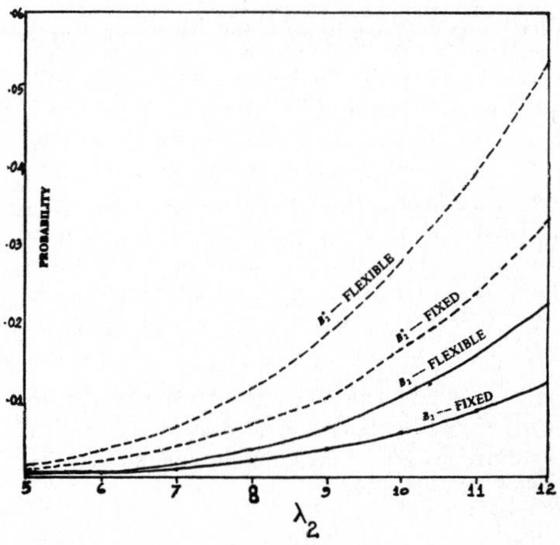

As mentioned earlier, the improvement that results from adopting the flexible scheme shown in these data was seen to occur for all parameter values. Based on our intuitive insights backed up by theoretical considerations, we also believe that if the call blocking assumption is weakened, then the difference in performance between the two schemes will become larger. These were the reasons that led to our recommendation that the flexible scheme should be made part of the CCITT standards.

4. MODEL FOR THE FLEXIBLE SCHEME

The queueing model for the flexible scheme has 24 servers serving two types of customers. The two types of arrivals form independent Poisson processes with rates λ_1 and λ_2, and have exponentially distributed service times with parameters μ_1 and μ_2 respectively. Customers of Type 1 need 6 servers each while those of Type 2 need 1 server each. There are no restrictions in allocating servers to customers. Customers who cannot find enough free servers are lost.

The Markov process describing this queue is the process (N_t, J_t) where N_t is the number of type 1 customers and J_t is the number of type 2 customers in the system at time t. The state space of this Markov process

$$\Xi = \{ (n,j) : 0 \leq n \leq 4,\ 0 \leq j \leq 24,\ 6n+j \leq 24 \}$$

consists of 65 states, as one may easily verify. It is also easy to see that the infinitesimal generator P of this Markov process is such that

$$P[(n,j)\ ;\ (m,k)] = \begin{cases} \lambda_1 & \text{if } m=n+1,\ j=k \\ \lambda_2 & \text{if } k=j+1,\ m=n \\ n\mu_1 & \text{if } m=n-1,\ j=k \\ j\mu_2 & \text{if } k=j-1,\ m=n \end{cases}$$

The off-diagonal elements of P not specified above are all zero, and the diagonal elements of P are such that all row sums of P are zero.

The unique solution π of the linear equations

$$\pi P = 0, \quad \pi e = 1,$$

where e is a column vector of $1's$, is the vector of steady state probabilities of this Markov process. The element $\pi(n,j)$ of π may be interpreted both as the long run proportion of time spent by the system in the state (n,j) as well as [6] the stationary probability that an arrival of either type sees the system in state (n,j). From these, it is now easy to obtain many performance measures such as the blocking probabilities for either type of arrival, the conditional blocking probabilities given the number of either type of customers in the system etc. We do not present the relevant formulas which are easy to derive.

It suffices to say that the analysis of the flexible scheme is routine, leads to the consideration of a Markov process with a small number of states, and poses no special difficulties.

5. MODEL FOR THE FIXED SCHEME

The queueing model for the fixed scheme may be described as follows. A set of 24 servers is divided into 4 groups of 6 servers each, say, 1-6, 7-12, 13-18, and 19-24. (The general case with NK servers divided into N groups of K servers each can be handled along the lines discussed below; for ease of exposition we do not consider the problem at this generality). To this set of servers are offered two streams of customers who arrive according to independent Poisson streams with rates λ_1 and λ_2 respectively. The service times of these two types are distributed as exponential random variables with parameters μ_1 and μ_2 respectively. While the customers of the first type need an entire group of servers whom they release simultaneously upon service completion, each customer of the second type needs only one server. Customers who cannot be carried are blocked and lost for ever. When a type 2 customer arrives if more than one of the four groups has a free server, then that customer is assigned to a server in one of the groups among those which have the largest number of busy servers. This sort of "packing" clearly reduces blocking for type 1 customers.

The server assignment rules described above are fairly complicated. Queueing models of this type naturally arise when considering memory allocation schemes under the restriction that the information contained in a record must be placed in contiguous locations. These models are typically quite complicated. However, the present problem, as will be shown below, has interesting simplifications making an exact analysis feasible.

We describe the state of the system by the vector

$$S(t) = (N(t), N_0(t), \cdots, N_6(t))$$

where $N(t)$ is the number of server groups serving type 1 customers at time t, $N_0(t)$ is the number of groups that have all servers idle at time t, and for $1 \leq i \leq 6$, $N_i(t)$ is the number of groups at time t that have exactly i busy servers serving type 2 customers. It is clear that $\{ S(t) : t \geq 0 \}$ is a Markov process with state space

$$\Omega = \{ (n, n_0, \cdots, n_6) : n \geq 0,\ n_i \geq 0,\ n + \sum_{i=0}^{6} n_i = 4 \}.$$

The number of states in Ω is the same as the number of ways of throwing 4 indistinguishable balls into 8 urns. Recalling [1], p38 that the number of ways of throwing r indistinguishable balls into c urns is given by the combinatorial formula ${}_{c+r-1}C_{c-1}$, we have that Ω contains ${}_{11}C_7 = 330$ states. As done in the model for the flexible scheme we can in principle compute the steady state probability vector of this Markov process and from that obtain the relevant performance measures. However, unlike in that case, the computational problem here is quite involved due to the large dimensionality of the state space and the complicated nature of the state description. We devise below a scheme for computing the infinitesimal generator Q of this Markov process recursively without listing all the states and computing its entries one at a time.

Structure of Q

We assume that the states in Ω are listed in the lexicographic order - that is, in the order $(0,...,0,4)$, $(0,...,1,3)$, $(0,...,2,2)$, $\cdots$,$(4,0,...,0)$. Denote by $\underline{k}$ the subset of Ω given by

$$\underline{k} = \{ (n,n_0,...,n_6) \in \Omega : \ n=k \ \}, \quad 0 \leqslant k \leqslant 4.$$

The subset $\underline{k}$ consists of all states that correspond to situations when k type 1 customers are in service, and will be called level $\underline{k}$. The matrix to be discussed below is the matrix A whose off-diagonal elements are the same as those of Q and whose diagonal elements are all zero. Having computed A, one gets Q by a trivial modification of the diagonal elements of A by noting that $Qe = \mathbf{0}$, where e is a column vector of $1's$.

Partitioning the matrix A according to the levels $\underline{k}$ of the state space, it is easily seen that A has the block tri-diagonal form

$$A = \begin{bmatrix} A_{00} & A_{01} & 0 & 0 & 0 \\ A_{10} & A_{11} & A_{12} & 0 & 0 \\ 0 & A_{21} & A_{22} & A_{23} & 0 \\ 0 & 0 & A_{32} & A_{33} & A_{34} \\ 0 & 0 & 0 & A_{43} & A_{44} \end{bmatrix} \qquad (1)$$

The sub-matrix A_{ij} in (1) contains the transition rates from states in level $\underline{i}$ to states in level $\underline{j}$. Note that the number of states in level $\underline{i}$ is given by the combinatorial formula $_{10-i}C_6$ which is also the number of ways of throwing $4-i$ indistinguishable balls into 7 urns; denote this number by m_i. We have

$$m_0 = 210, \quad m_1 = 84, \quad m_2 = 28, \quad m_3 = 7, \quad m_4 = 1. \ (2)$$

In terms of these, the dimensions of the various blocks in (1) are obtained by observing that the dimension of A_{ii} is $m_i \times m_i$.

The matrices $A_{i,i+1}$ will be called the *upper blocks*, A_{ii} the *diagonal blocks*, and $A_{i,i-1}$, the *lower blocks*. We discuss below the computation of these blocks.

The Upper blocks

Denote by level $\underline{ij}$ the set of states in Ω given by

$$\underline{ij} = \{ (n,n_0,...,n_6) \in \Omega : \ n=i, \ n_0=j \ \},$$

and observe that level $\underline{ij}$ consists of those states corresponding to situations when exactly i type 1 customers are in service and exactly j groups are idle. The number of states in level $\underline{ij}$ is given by the combinatorial formula $_{9-i-j}C_5$ which is also the number of ways of throwing $4-i-j$ indistinguishable balls into 6 urns. Let this number be denoted by m_{ij}.

An upward transition in levels occurs when a type 1 customer arrives to the system while there is at least one idle group. Thus, a transition from $\underline{i}$ to $\underline{i+1}$ is a transition from $\underline{ij}$ to $\underline{i+1,j-1}$ for some $j \geqslant 1$. A transition from $\underline{i}$ to $\underline{i+1}$ is not possible from states in $\underline{i0}$, for, these states correspond to having no idle groups. Where possible, the rate of transition from $\underline{i}$ to $\underline{i+1}$ is λ_1, and such a transition leaves the remaining coordinates $n_1,...,n_6$ of the state unaltered. From these considerations and by partitioning the matrix $A_{i,i+1}$ according to sub-levels $\underline{ij}$ of $\underline{i}$ and to sub-levels $\underline{i+1,j}$ of $\underline{i+1}$ it is easily seen that the matrix $A_{i,i+1}$ is as given below, where we denote by I_k the identity matrix of order $k \times k$ and by $\mathbf{0}_{a \times b}$ the zero matrix of order $a \times b$.

$$A_{i,i+1} = \begin{bmatrix} \mathbf{0}_{m_{i0} \times m_{i+1}} \\ \\ \lambda_1 I_{m_{i+1}} \end{bmatrix} \qquad i = 0,1,2,3, \qquad (3)$$

where m_i are given by (2), and m_{i0} are given by

$$m_{00} = 126, \quad m_{10} = 56, \quad m_{20} = 21, \quad m_{30} = 6. \qquad (4)$$

The Lower Blocks

The block $A_{i,i-1}$ consists of transition rates from states in level $\underline{i}$ to level $\underline{i-1}$. A transition from $\underline{i}$ to $\underline{i-1}$ is indeed a transition from a sub-level $\underline{ij}$ to a sub-level $\underline{i-1,j+1}$. Such a transition occurs when one of the i type 1 customers in service leaves; the rate of this is clearly $i\mu_1$. Also, such a transition leaves the other components $n_1,...,n_6$ of the state vector unaltered. These considerations and appropriate partitionings of $\underline{i}$ and $\underline{i-1}$ lead to the formula

$$A_{i,i-1} = \begin{bmatrix} \mathbf{0}_{m_i \times m_{i-1,0}} & i\mu_1 I_{m_i} \end{bmatrix} \qquad i = 1,2,3,4. \qquad (5)$$

The Diagonal Blocks

The diagonal blocks A_{ii} are much more complicated. In fact, partioning the levels $\underline{i}$ into sub-levels $\underline{ij}$ is not particularly helpful in the determination of these matrices. We shall show that under a different partitioning of the levels $\underline{i}$, these matrices may be constructed recursively starting from A_{33}. Note that A_{44}, which is a scalar, is a diagonal element of A and therefore equal to zero.

There are only 7 states in level $\underline{3}$. These, listed in the lexicographic order, are as follows: (300 000 01), (300 000 10), (300 001 00), (300 010 00), (300 100 00), (301 000 00), (310 000 00). The 7×7 matrix A_{33} is easily seen to be given by

$$A_{33} = \begin{bmatrix} 0 & 6\mu_2 & 0 & 0 & 0 & 0 & 0 \\ \lambda_2 & 0 & 5\mu_2 & 0 & 0 & 0 & 0 \\ 0 & \lambda_2 & 0 & 4\mu_2 & 0 & 0 & 0 \\ 0 & 0 & \lambda_2 & 0 & 3\mu_2 & 0 & 0 \\ 0 & 0 & 0 & \lambda_2 & 0 & 2\mu_2 & 0 \\ 0 & 0 & 0 & 0 & \lambda_2 & 0 & \mu_2 \\ 0 & 0 & 0 & 0 & 0 & \lambda_2 & 0 \end{bmatrix} \qquad (6)$$

We now turn to the computation of A_{ii}, $i=0,1,2$. To this end, for $i=0,1,2$, partition level $\underline{i}$ into sub-levels $\underline{ik^*}$, $0 \leqslant k \leqslant 6$, where

$$\underline{ik^*} = \{ (n,n_0,...,n_6) \in \Omega : n=i, \ n_j=0 \text{ for } 0 \leqslant j \leqslant k-1, \ n_k > 0 \ \}.$$

The sub-level $\underline{ik^*}$ consists of those states which correspond to the situations when i type 1 customers are in service $(n=i)$ and, in addition, the remaining $4-i$ groups are such that each of them has at least k busy servers $(n_0= \cdots =n_{k-1}=0)$ serving type 2 customers and that the number of groups among them with exactly k type 2 customers is positive $(n_k > 0)$. The lexicographic ordering of Ω entails that these sub-levels will be ordered in the decreasing order of the index k. We now partition the matrix A_{ii} according to sub-levels

$\underline{ik^*}$, $k=6,5,...,0$ and note that we get a block tri-diagonal form given by

$$A_{ii} = \begin{bmatrix} D_{i6} & U_{i6} & 0 & 0 & 0 & 0 & 0 \\ L_{i5} & D_{i5} & U_{i5} & 0 & 0 & 0 & 0 \\ 0 & L_{i4} & D_{i4} & U_{i4} & 0 & 0 & 0 \\ 0 & 0 & L_{i3} & D_{i3} & U_{i3} & 0 & 0 \\ 0 & 0 & 0 & L_{i2} & D_{i2} & U_{i2} & 0 \\ 0 & 0 & 0 & 0 & L_{i1} & D_{i1} & U_{i1} \\ 0 & 0 & 0 & 0 & 0 & L_{i0} & D_{i0} \end{bmatrix} \quad (7)$$

In (7) the matrix D_{ij} corresponds to transitions from sub-level $\underline{ij^*}$ to itself, L_{ij} to transitions from sub-level $\underline{ij^*}$ to sub-level $\underline{i(j+1)^*}$, and U_{ij} to transitions from sub-level $\underline{ij^*}$ to sub-level $\underline{i(j-1)^*}$. Note that (7) does take into account the lexicographic ordering.

Let r_{ij} denote the number of states in sub-level $\underline{ij^*}$. It is easily verified that $r_{ij} = {}_{10-i-j}C_{6-j} - {}_{9-i-j}C_{5-j}$. The following table of r_{ij}, $0 \leq i \leq 2$, $0 \leq j \leq 6$, is used for the determination of the dimensions of the blocks in (7).

Table of r_{ij}

i j:	0	1	2	3	4	5	6
0	84	56	35	20	10	4	1
1	28	21	15	10	6	3	1
2	7	6	5	4	3	2	1

The matrix L_{ij}

We have noted that L_{ij} governs transitions from states in $\underline{ij^*}$ to $\underline{i(j+1)^*}$. The dimension of this matrix is $r_{ij} \times r_{i,j+1}$. Consider now any state in $\underline{ij^*}$. It is of the form $(i,0,...,0,n_j,...,n_6)$ with $n_j \neq 0$. Clearly, a transition from this state to a state in $\underline{i(j+1)^*}$ can happen only if $n_j=1$, for, no co-ordinate can change by more than one unit in one transition. Also, such a transition is the result of a type 2 arrival which is assigned a server from the group which prior to that arrival had j busy servers serving type 2 customers. The packing rule described earlier implies that a type 2 customer will be assigned to a group with j busy servers only if $n_{j+1}=...=n_5=0$. In other words, a transition from $\underline{ij^*}$ to $\underline{i(j+1)^*}$ must be a transition from the state for which $n=i$, $n_j=1$, $n_6=3-i$ to the state for which $n=i$, $n_{j+1}=1$, and $n_6=3-i$. From these considerations, it follows that the only non-zero element of L_{ij} is the element located in its first row and first column. That element is clearly λ_2 since λ_2 is the arrival rate for type 2 customers. This characterizes the matrix L_{ij} completely.

The matrix U_{ij}

The matrix U_{ij} which governs the transitions from $\underline{ij^*}$ to $\underline{i(j-1)^*}$ is of order $r_{ij} \times r_{i,j-1}$. Any transition of this type occurs due to the departure of a type 2 customer from one of the n_j groups each of which is currently serving j type 2 customers. It therefore is a transition from a state of the form $(i,0,...,0,n_j,n_{j+1},...,n_6)$, $n_j \neq 0$, to the state $(i,0,...,0,1,n_j-1,n_{j+1},...,n_6)$, and its rate is clearly $jn_j\mu_2$.

The set $\underline{i6^*}$ has only one state namely $(i,0,0,0,0,0,0,4-i)$. From this state if a transition occurs to the set $\underline{i5^*}$ then it must be to the state $(i,0,0,0,0,0,1,3-i)$; further, the rate of this is $6(4-i)\mu_2$. Thus the block U_{i6} is given by

$$U_{i6} = \begin{bmatrix} 6(4-i)\mu_2 & \mathbf{0} \end{bmatrix} \quad i=0,1,2, \quad (8)$$

where $\mathbf{0}$ is a row vector of zeroes of appropriate dimension; the zeroes in the right side of (8) indicate the impossibility of entry into states in $\underline{i5^*}$ for which $n_5 \geq 2$.

To determine the matrices U_{ij}, $1 \leq j \leq 5$, it is convenient to partition the states in $\underline{ij^*}$ according to the possible values of the co-ordinate n_j. Let the number of states in $\underline{ij^*}$ with $n_j=k$ be denoted be $r_{ij}(k)$, $1 \leq k \leq 4-i$. Clearly, $r_{ij}(k)$ is the same as the number of ways of throwing $4-i-k$ indistinguishable balls into $8-(j+2)$ urns and therefore given by ${}_{9-i-j-k}C_{5-j}$. It is easily verified that $\sum_{k=1}^{4-i} r_{ij}(k) = r_{ij}$. One also notes that $r_{ij}(k) \equiv r_{0j}(k+i)$. The following table of $r_{0j}(\cdot)$ is used in the determination of the numbers of $r_{ij}(\cdot)$.

Table of $r_{0j}(k)$

k j:	1	2	3	4	5
1	35	20	10	4	1
2	15	10	6	3	1
3	5	4	3	2	1
4	1	1	1	1	1

Consider now a state in $\underline{ij^*}$. When $n_j=k$ there are k groups each with exactly j type 2 customers, and the departure of any one of these kj type 2 customers results in a transition to $\underline{i(j-1)^*}$. The rate of this is clearly $kj\mu_2$. Also, the state so entered in $\underline{i(j-1)^*}$ must be such that $n_{j-1}=1$. These considerations lead to the following formula for U_{ij}, $0 \leq i \leq 2$, $1 \leq j \leq 5$.

$$U_{ij} = \begin{bmatrix} j\mu_2 I_{r_{ij}(1)} & 0 & \cdot & \cdot & 0 & \mathbf{0} \\ 0 & 2j\mu_2 I_{r_{ij}(2)} & \cdot & \cdot & 0 & \mathbf{0} \\ \cdot & & \cdot & & \cdot & \cdot \\ \cdot & & & \cdot & \cdot & \cdot \\ 0 & 0 & \cdot & \cdot & (4-i)j\mu_2 I_{r_{ij}(4-i)} & \mathbf{0} \end{bmatrix} \quad (9)$$

The last column in (9) is a block of zero columns which indicate the impossibility of entering a state with $n_{j-1} \geq 2$. The number of such zero columns is clearly $r_{i,j-1} - r_{ij}$.

The matrix D_{ij}

We now turn our attention to the matrix D_{ij} which governs transitions from $\underline{ij^*}$ to itself. The dimension of this matrix is $r_{ij} \times r_{ij}$. The main result we prove is the following theorem which shows that the matrix D_{ij} is simply the principal sub-matrix of $A_{i+1,i+1}$ - that is, it is comprised of the elements in the first r_{ij} rows and r_{ij} columns of the latter. This theorem along with the results in the previous two sub-sections permits us to compute A_{22} from A_{33}, A_{11} from A_{22}, and finally A_{00} from A_{11}.

Theorem: For any matrix B, let $B\langle r\rangle$ denote the principal sub-matrix of order r of B - that is, $B\langle r\rangle$ is the $r \times r$ matrix formed by the elements in the first r rows and r columns of B. For $i=0,1,2$, we have

$$D_{ij} = A_{i+1,i+1}\langle r_{ij}\rangle \,. \quad (10)$$

Proof: The diagonal elements of both sides of (10) are zero, for, these are in turn diagonal elements of A. A typical off-diagonal element of D_{ij} is the rate of transition from a state $(i,0,...,0,n_j,n_{j+1},...,n_6)$ to a state $(i,0,...,0,m_j,m_{j+1},...,m_6)$ where $n_j, m_j \geq 1$. It is elementary to see that this rate must be the same as the rate of transition from the state $(i+1,0,...,0,n_j-1,n_{j+1},...,n_6)$ to the state $(i+1,0,...,0,m_j-1,m_{j+1},...,m_6)$. The latter is an off-diagonal element of $A_{i+1,j+1}$. Equation (10) now follows from the lexicographic ordering assumed by us.

The above results completely determine the matrix A and thereby the infinitesimal generator Q of the Markov process $\{S(t) : t \geq 0\}$. The implementation of the resulting algorithm is quite straightforward. Its particular appeal lies in the fact that it can easily be generalized to the case of NK servers who are divided into N groups of K servers each.

The Steady State Probabilities

Let θ denote the unique solution of the system of linear equations

$$\theta Q = 0, \quad \theta e = 1, \tag{11}$$

where e is a column vector of 1's. The elements of the vector θ yield the long run fraction of time spent by the system in the various states; these are [6] also the probabilities that an arrival of either type finds the system in various states.

For the problem at hand, we computed θ using an algorithm due to Wachter [5], which converts (11) to a non-singular system of linear equations. For the general problem with NK servers, the resulting system of equations may be large; however, one may exploit the block tri-diagonal structure and sparsity of Q and devise efficient iterative techniques of the block Gauss-Seidel type. These are well-known and will not be discussed here.

Performance Measures

Having computed the steady state probability vector θ one may obtain many interesting performance measures. These require the summation of certain selected elements of θ, and we illustrate this with some examples.

Assume that the elements of θ are numbered from 0 to 329 - that is

$$\theta = (\theta_0, \cdots, \theta_{329})$$

Let B_i denote the blocking probability of a type i customer, $i=1,2$. It is easy to verify that these blocking probabilities are given as follows:

$$B_1 = \sum_{i=0}^{125}\theta_i + \sum_{i=0}^{55}\theta_{210+i} + \sum_{i=0}^{20}\theta_{294+i} + \sum_{i=0}^{5}\theta_{322+i} + \theta_{329}$$

$$B_2 = \theta_0 + \theta_{210} + \theta_{294} + \theta_{322} + \theta_{329}$$

Some other performance measures which may be derived are: the probability distributions and moments of the number of customers of either type in service, the conditional blocking probability of either type of customer given the number of customers of either or both types in service at a point of arrival, etc. The formulas for these in terms of θ are fairly easy to derive and will not be presented.

6. RELAXATION OF MODEL ASSUMPTIONS

We assumed that both types of customers will be blocked when they cannot find the needed number of servers. In some cases, such as those where the transmission is of data, one may wish to consider models where waiting is permitted. Further, the model with waiting throws some light on the effect of re-attempts which is usually significant. The analysis of the resulting models is possible by resorting to some recent matrix-analytic methods.

Consider, for example, the case where one of the streams can be buffered. Here, besides the configuration of the servers/server groups, one needs to keep track of the number of customers waiting as well. This leads to a model of the Quasi-birth-and-death type which can be analysed by the matrix-geometric techniques, c.f., Neuts [2], Chapter 3. The results in [4] provide efficient techniques to determine the waiting time distribution for the stream that can be buffered. In such a case, the algorithm presented here can be used to generate the block matrices in the infinitesimal generator. We refer the reader to [2] and [4] for a discussion of the general methodology and computational algorithms.

If both streams of customers can be buffered, even then we can apply the matrix-geometric techniques as long as one of the streams has a finite waiting room. Such models usually require the specification of some priority rules for deciding what type of customer will enter service when servers become available. These could be used as approximations for cases when both streams have infinite waiting rooms.

The relaxation of the assumption of exponentiality for service times is more difficult. If one of the two streams has a phase type distribution [2] (these include well-known special cases such as mixtures of exponentials and generalized Erlangs), an analysis is still possible. The explosion in dimensionality that results due to the need to keep track of the service phases requires substantial care as in the multiserver queue with phase type servers. We refer the reader to [3] for relevant techniques.

References

[1] Feller,W: *An introduction to probability theory and its applications,* John Wiley & sons, New York, 1969.

[2] Neuts, M.F: *Matrix-geometric solutions in stochastic models,* Johns Hopkins University Press, Baltimore, MD, 1981.

[3] Ramaswami,V & Lucantoni,D.M: Algorithms for the multi-server queue with phase type service - *to appear*

[4] Ramaswami,V & Lucantoni,D.M: Stationary waiting time distribution in queues with phase type service and in quasi-birth-and-death processes - *to appear in* Stochastic Models, 1985.

[5] Wachter,P.G: Solving certain systems of homogeneous equations with special reference to Markov chains, M.Sc. Thesis, Department of Mathematics, McGill University, 1973..

[6] Wolff,R.W: Poisson arrivals see time averages, Operations Research, 30, 223-231, 1982.

TABLE 1: PERFORMANCE MEASURES

$\mu_1 = \mu_2 = 1, \ \lambda_1 = 0.05\lambda_2$

(Values in parantheses correspond to the flexible scheme)

λ_2	B_1	B_2	B_1^*	B_2^*	P^*
5	.0149	.0002	.0675	.0010	.2200
	(.0077)	(.0004)	(.0348)	(.0017)	(.2207)
6	.0317	.0005	.1232	.0021	.2562
	(.0168)	(.0009)	(.0653)	(.0038)	(.2578)
7	.0592	.0012	.2028	.0041	.2885
	(.0317)	(.0020)	(.1082)	(.0067)	(.2922)
8	.0997	.0022	.3054	.0070	.3159
	(.0535)	(.0037)	(.1641)	(.0116)	(.3234)
9	.1528	.0038	.4224	.0111	.3371
	(.0831)	(.0065)	(.2324)	(.0186)	(.3510)
10	.2167	.0059	.5407	.0166	.3512
	(.1209)	(.0105)	(.3110)	(.0279)	(.3743)
11	.2887	.0087	.6486	.0238	.3566
	(.1661)	(.0158)	(.3958)	(.0397)	(.3928)
12	.3666	.0124	.7390	.0335	.3524
	(.2174)	(.0224)	(.4819)	(.0539)	(.4061)

λ_1= *Arrival rate of* 384*K bps calls* (*Type* 1)
λ_2= *Arrival rate of* 64*K bps calls* (*Type* 2)
B_i= *Blocking probability* for *Type i calls*, $i=1,2$
B_i^*= *Conditional blocking probability* for *Type i calls given that at least* 1 *Type* 1 *call is in progress*, $i=1,2$
P^*= *Probability that at least* 1 *Type* 1 *call is in progress*

TELETRAFFIC ISSUES in an Advanced Information Society
ITC-11
Minoru Akiyama (Editor)
Elsevier Science Publishers B.V. (North-Holland)

TRAFFIC CHARACTERISTICS OF CONTROL SIGNALS IN AN ISDN SWITCHING SYSTEM

Hatsuho MURATA

NEC Corporation, Tokyo, Japan

ABSTRUCT

In an ISDN switching system, traffic characteristics of signalling information messages via the D-channel is investigated according to the network layer protocol described in the CCITT I/Q series recommendations. A simple traffic model for the signalling information messages is proposed and analyzed. Using this model, signalling information traffic via the D-channel can be estimated accurately enough to design a D-channel handler for an ISDN switching system.

1. INTRODUCTION

The Integrated Services Digital Network (ISDN) attracts attention as the communication network technology of the 21st century. In 1984, the CCITT introduced the CCITT I series recommendations for standardized ISDN USER-NETWORK INTERFACES. An ISDN switching system with the above-mentioned standardized interface will be marketed in a few years.

One of the differences between an ISDN switching system and a traditional switching system is the adoption of a frame structured signalling protocol via the D-channel. The D-channel is one of the channels which are designated for user-network interface in the CCITT I/Q series recommendations. Between user terminals and the switching system, the D-channel conveys signalling information for circuit switched connection control, packet switched connection control, user packets, and so forth.

The protocols for signalling information via D-channel are defined as follows, based on the OSI reference model. Data link layer (layer 2) is described in Q.920 (I. 440) and Q. 921 (I. 441), and network layer (layer 3) is described in Q.930 (I. 450) and Q. 931 (I.451).

In this paper, signalling information traffic between user terminals and the switching system is investigated according to the network layer protocol described in the CCITT recommendations.

2. SIGNALLING VIA THE D-CHANNEL

Fig.1 shows an example of the ISDN basic access. On a customer's premises, user terminals are installed and connected to the Network Termination (NT) through connectors.
In a subscriber loop between the NT and a switching system, two B-channels (64Kbps) for data/voice flow and one D-channel (16Kbps) for signalling flow are utilized. Every terminal will be able to access these channels, depending on the customer's requirements.

In the switching system, the D-channel signal is picked up from the subscriber loop, and sent into a D-channel handler which processes signalling information in order to control user-network connections.

An example of logical links in the D-channel is shown in Fig.2. An endpoint in a user terminal and an endpoint in the switching system are connected by a logical link. A user terminal or the switching system may have multiple endpoints. In addition, every terminal has a special endpoint which is connected with

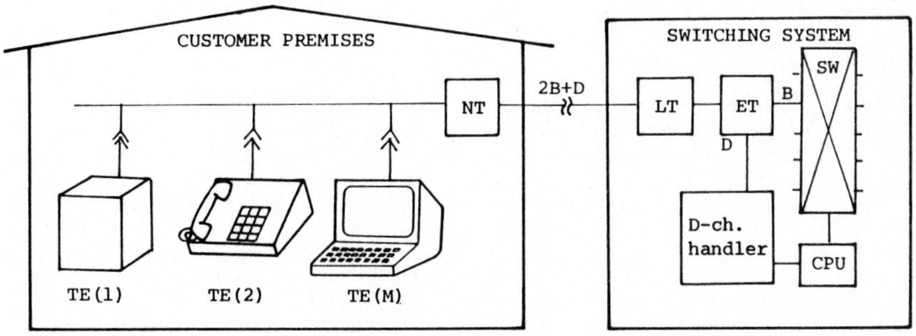

Fig. 1 An example of the ISDN basic access

the same endpoint in the switching system by a broadcast data link.

Fig.3 indicates a typical frame format of signalling information in the D-channel. One frame starts a flag sequence "01111110" and ends another flag sequence "01111110". The layer 2 field in the frame contains a data link connection identifier which indicates the logical link number in the D-channel.
The layer 3 field consists of a protocol discriminator, a call reference identifying a call on the user-network interface, a message type, and information elements.

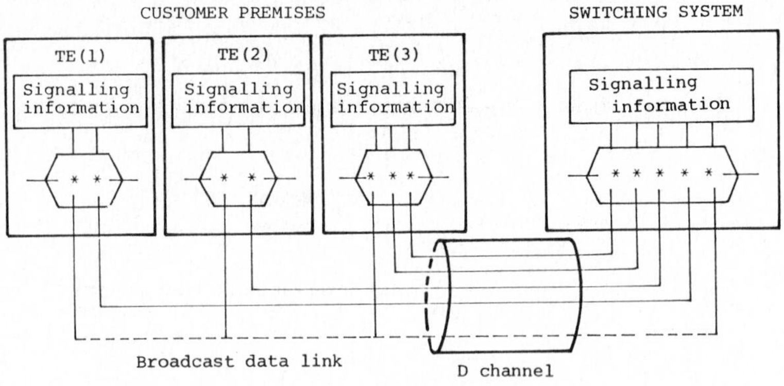

Fig. 2 An example of logical links in the D-channel

These information elements carry signalling information in the D-channel between terminals and the switching system. Here, we simply define the frame including these information elements as a "message."

Traffic characteristics of "messages" can be estimated more precisely than that of ordinary packet communications, because every transfer sequence of the messages is described in CCITT Recommendation Q.931.

(1) In many cases, when a endpoint sends a message via a logical link, the endpoint will not send another message to the other endpoint of the link untill it receives a response message from the other endpoint. Thus, the ability to send messages depends on the reception of messages from the other endpoint of the logical link.

(2) Massages for signalling in the D-channel are classified into two types. Type A is a class of messages which will be automatically sent after the reception of a preceding message. Type B is a class of messages which will be sent when a trigger input is given directly by a user, by a CPU in the user terminal, or by a message from another switching system.

In Fig. 4, examples of type A messages are RELEASE, RELEASE COMPLETE, and so forth. Examples of type B messages are SETUP, CONNECT, DISCONNECT, etc.
Therefore, once a message begins to be sent, the messages in the link continue to be sent and received for a while. Once the message terminates, no messages will be sent for a while.

(3) Because of the "multi-link" capability of the D-channel, it is necessary to consider a worst case senario where many terminals communicate with a D-channel handler at the same time. In addition, the number of terminals in a subscriber loop is ordinarily limited to 8 in the basic access, which is not enough to bring about the large number

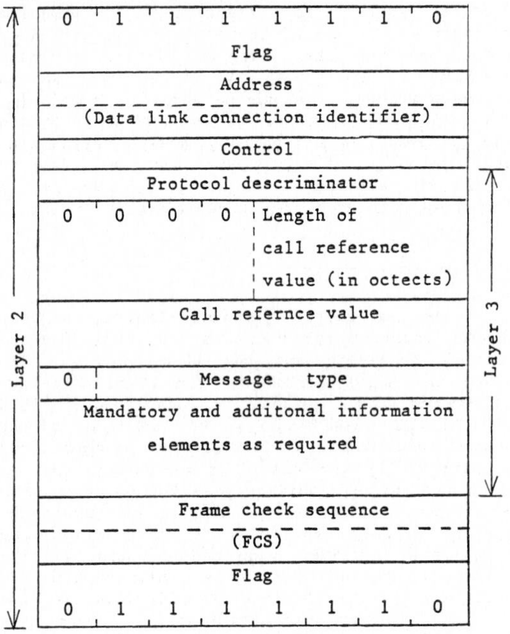

Fig. 3 A typical frame format

effect for D-channel traffic.

(4) The message traffic in a link congests at a specific sequence in the D-channel protocol. An example is a sequence concerning called terminals: Many terminals return the CALL PROCEEDING and ALERT messages as a response to SETUP message, and after the reception of CONNECT from a terminal, the switching system sends a RELEASE message to all the other terminals. These terminals send RELEASE COMPLETE messages in return.

Signalling information messages in the D-channel are classified into the above mentioned types as shown in Table 1.

The message traffic in the D-channel is described with the continuous message bursts which start with a type B message and continue with a type A message. Table 2 provides a list of these continuous message bursts and shows that the average length of the burst is about 2-3 messages in consideration of the frequency of each burst.

Although the average time spent to receive, process, and send a message using D-channel signalling format is not clearly described in the CCITT recommendations, the CCITT Q. 514 (Performance and availability design objectives for a digital local and combined local/transit exchange) suggests design objective values of about 250 msec - 400 msec as the average time for dial tone sending delay, exchange call set up delay, exchange call release delay, and so forth.

Therefore, the average spent time of a message in the system is estimated as about 100 msec - 200 msec.

On the other hand, a signalling message with a minimum length of 11 octets and an average length of about 20 octets, takes about 10 msec to be sent or received via a 16Kbps D-channel. Even if 50 msec is assigned to the processing in a terminal/switching system, the total spent time in the system (70 msec) is not of sufficient length to compare with the design objectives.

Accordingly, an investigation concerning the message processing capabilities of terminals and the switching system, number of connectable terminals in a subscriber loop, and the average spent time in a system, becomes necessary in consideration of the traffic characteristics of signalling information messages in the D-channnel.

3. TRAFFIC MODEL

In regard to the afore-mentioned traffic characteristics, the traffic model illustrated in Fig. 5 is proposed here.

Terminals: The service time in a terminal has an exponential distribution with service rate μ_1. Although each terminal makes its service independenly, one queue with m servers is presented rather than m queues with one server, to simplify the traffic model. This simplification is justified by the fact that the messages arriving at the queue do not need to wait, because the number of messages in the traffic model does not exceed the number of servers in the queue, as will be discussed later.

The switching system: A service time in a switching system is also proposed as having an exponential distribution, with service rate μ_2. Messages between terminals and the switching system are transfered in a physical link and processed by a D-channel handler. Therefore, a queue with a server represents the switching system. The number of messages waiting in the queue has no limit.

An idle state: In order to illustrate "massage bursts," an idle state queue is presented. When a message remains in the idle state queue, the continuous exchanges of

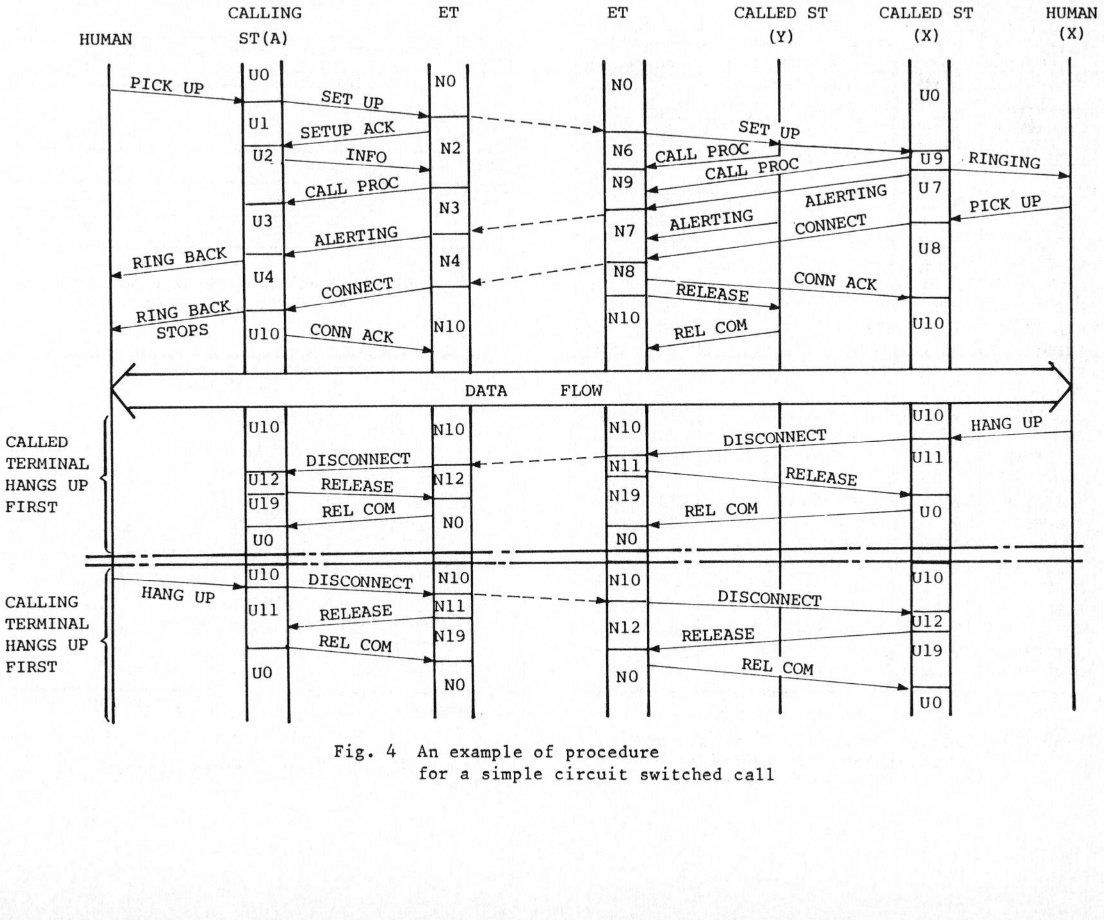

Fig. 4 An example of procedure for a simple circuit switched call

messages in the link is stopped for a while.

Afterwards, the next message burst begins when the message departs from this queue.

A service in this queue has an exponential distribution with service rate μ_3.

Total number of messages in the model: To reflect the stuation that a finite number of logical links are connected with a D-channel handler, and that an endpoint which does send/receive/process a continuous message burst has less probability of makeing/receiving a new message burst, the total number of messages in the model is proposed as being equal to the number of active logical links. Accordingly, the state of each logical link in the D-channel is indicated by the name of the queue where the message remains. The number of active logical links may be simply regarded as the number of terminals.

message	message type A	message type B	circuit mode connection	temporary user-to-user signalling connections	permanent user-to-user signalling connections	support of X.25 packet-mode connections via the D-channel	support of X.25 packet-mode connections via the B-channel	preceding message
ALERT	x		x	x			x	SETUP
CALL PROC	x		x	x			x	SETUP/INFO
CANC		x	x					
CANC ACK	x		x					CANC
CANC REJ	x		x					CANC
CONG CONT		x	x	x	x			
CONN		x	x				x	
CONN ACK	x		x	x			x	CONN
DET		x	x				x	DISC
DET ACK	x		x					DET
DISC		x	x				x	
FAC		x	x					
FAC ACK	x		x					FAC
FAC REJ	x		x					FAC
INFO		x	x					
REG		x	x					
REG ACK	x		x					REG
REG REJ	x		x					REG
REL	x		x	x		x	x	DISC
REL COMP	x		x	x		x	x	REL
RES		x	x					
RES ACK	x		x					RES
RES REJ	x		x					RES
SETUP		x	x	x		x		
SETUP ACK	x		x			x		SETUP
STAT		x	x	x	x	x		
SUSP		x	x					
SUSP ACK	x		x					SUSP
SUSP REJ	x		x					SUSP
USER INFO		x	x	x	x			

Table 1 Message types of the signalling information in the D-channel.

Distribution rate: The distribution rate r_{ij} represents the probability that a message leaving queue i will arrive at queue j. The following equation is satisfied in the model.

$$\sum_{j=1}^{3} r_{ij} = 1 \qquad i=1,2,3 \qquad (1)$$

Type B message	Type A messages following type B	number of messages/burst
CANC →	CANC ACK ←	2
	CANC REJ ←	2
CONG CONT →	none	1
CONN →	CONN ACK ←	2
DISC →	REL ← REL COMP →	3
	DET ← DET ACK →	3
FAC →	FAC ACK ←	2
	FAC REJ ←	2
INFO →	none	1
REG →	REG ACk ←	2
	REJ REJ ←	2
RES →	RES ACK ←	2
	RES REJ ←	2
SETUP →	SETUP ACK ← INFO x n → CALL PROC ← ALERT ←	4+n
	CALL PROC ← ALERT ←	3
STAT →	none	1
SUSP →	SUSP ACK ←	2
	SUSP REJ ←	2

Table 2 Continuous message bursts

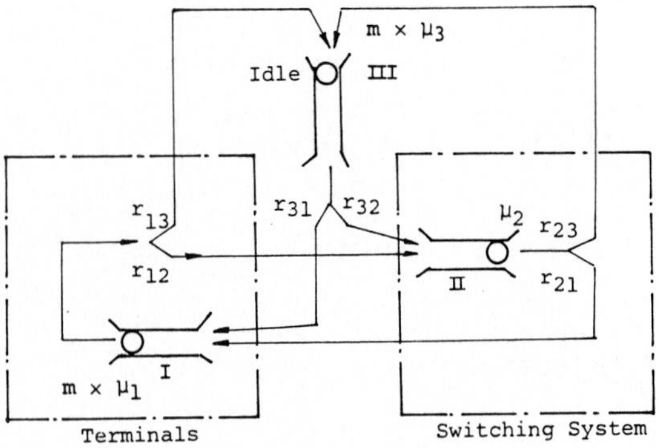

Fig. 5 Traffic model proposed

Fig. 5 also shows that r_{11}, r_{22} and r_{33} are equal to zero in the model.

An example of the message exchange process: a logical link remains in the idle state initially.

(1) A message is sent from a terminal to the switching system: the message leaves queue III and arrives at queue II according to the distribution rate r_{32}.
(2) The message is processed in the switching system: the message waits and gets service in queue II.
(3) A response message is sent from the switching system to the terminal: the message leaves queue II and arrives at queue I according to the distribution rate r_{21}.
(4) The message is processed in the terminal: the message is served with service rate μ_1 at the queue I.
(5) Another message is sent from the terminal to the switching system: the message leaves queue I and arrives at queue II.
(5')The continuous message burst is stopped: the message leaves queue I and arrives at queue III.

Messages are exchanged between the terminal and the switching system in certain times depending on the distribution rate r_{ij}.

4. MODEL ANALYSIS

Initially, distribution rates are calculated according to the average occurences of message transfer.

Since a continuous message burst in the model indicates a round trip upon leaving the idle state queue until returning to the idle state queue, the average occurences of message transfer is calculated as follows:

If the first message of the burst is sent from a terminal, the probability P_n that this burst will stop after n occurences of message transfer is obtained by the following formula:

$$P_n = \begin{cases} r_{21}^{k}\, r_{12}^{k}\,(1-r_{21}) & n=2k+1,\ k=0,1,2,\ldots \quad (2) \\ r_{21}^{k}\, r_{12}^{k-1}\,(1-r_{12}) & n=2k,\ k=1,2,3,\ldots \quad (3) \end{cases}$$

Average occurences of message transfer m_t is,

$$m_t = \sum_{n=1}^{\infty} n\, P_n = \frac{1 + r_{21}}{1 - r_{12}\, r_{21}} \quad (4)$$

In the same manner, if the first message of the burst is sent from the switching system, the probatility P_n is

$$P_n = \begin{cases} r_{12}^{k}\, r_{21}^{k}\,(1-r_{12}) & n=2k+1,\ k=0,1,2,\ldots \quad (5) \\ r_{12}^{k}\, r_{21}^{k-1}\,(1-r_{21}) & n=2k,\ k=1,2,3,\ldots \quad (6) \end{cases}$$

and average occurences of message transfer m_s is

$$m_s = \sum_{n=1}^{\infty} n\, P_n = \frac{1 + r_{12}}{1 - r_{12}\, r_{21}} \quad (7)$$

When m_s and m_t are given, the distribution rate r_{12}, r_{21} is calculated by solving the equations (4), (7).

$$\begin{cases} r_{12} = (m_s - 1) / m_t & (8) \\ r_{21} = (m_t - 1) / m_s & (9) \end{cases}$$

State equations: the following equations describe the behavior of the equilibrium distribution of messages in the model.

$$p(k_1,k_2,k_3) \sum_{i=1}^{3} \delta_{k_i-1}\, \alpha_i(k_i)\, \mu_i = \sum_{i=1}^{3} \sum_{j=1}^{3} \delta_{k_j-1}\, \alpha_i(k_i+1)\, \mu_i\, r_{ij}\, p(\ ,k_j-1,k_i+1,) \quad (10)$$

$$\delta_k = \begin{cases} 1 & k=0,1,2,3,\ldots \quad (11) \\ 0 & k<0 \quad (12) \end{cases}$$

$$\alpha_i(k_i) = \begin{cases} k_i & k_i \leq m_i \quad (13) \\ m_i & k_i \geq m_i \quad (14) \end{cases}$$

$$m_1 = m_3 = m\,, \quad m_2 = 1$$

δ_k is introduced to indicate the fact that the service rate must be zero when a given queue is empty and $\alpha_i(k_i)$ merely gives the number of messages in the ith queue when there are k_i messages at that queue.

Here, consider a set of numbers $\{x_i\}$, which are solutions to the following set of linear equations:

$$\mu_i\, x_i = \sum_{j=1}^{3} \mu_j\, x_j\, r_{ji} \qquad i=1,2,3 \quad (15)$$

r_{11}, r_{22} and r_{33} are all zero in this model. The solutions of these linear equations are given as follows.

$$x_2 = \frac{\mu_1\ \ r_{32} + r_{12}\, r_{31}}{\mu_2\ \ r_{31} + r_{21}\, r_{32}}\, x_1 \quad (16)$$

$$x_3 = \frac{\mu_1\ \ r_{23} + r_{13}\, r_{21}}{\mu_3\ \ r_{21} + r_{31}\, r_{23}}\, x_1 \quad (17)$$

Since there are only 2 independent equations, x_1 remains at the right side of the solutions. According to the closed Markovian network investigation by Gordon and Newell(3),(4), the solution of equation (10) can be shown to equal

$$p(k_1,k_2,k_3) = \frac{1}{G(K)} \prod_{i=1}^{3} \frac{x_i^{k_i}}{\beta_i(k_i)} \quad (18)$$

$$\beta_i(k_i) = \begin{cases} k_i! & k_i \leq m_i \quad (19) \\ m_i!\, m_i^{k_i-m_i} & k_i \geq m_i \quad (20) \end{cases}$$

and the normalization constant is computed by

$$G(K) = \left(\sum_{K \in A} \prod_{r=1}^{3} \frac{x_i^{k_i}}{\beta_i(k_i)} \right)^{-1} \quad (21)$$

Here,

$$A = \{ K=(k_1,k_2,k_3) \mid \sum_{i=1}^{3} k_i = m \ , \ k_i \geq 0 \} \quad (22)$$

The average message length in the queue is defined as:

$$\bar{k}_i = \sum_{K \in A} k_i \ p(k_1,k_2,k_3) \qquad i=1,2,3 \quad (23)$$

and P_i is defined as the probability that at least one message will be served in queue.

$$P_i = \sum_{k_i \neq 0} p(k_1,k_2,k_3) \qquad i=1,2,3 \quad (24)$$

Therefore the arrival rate at queue j is defined as:

$$\lambda_j = \sum_{i=1}^{3} \mu_i \ P_i \ r_{ij} \qquad j=1,2,3 \quad (25)$$

and the average time spent in the queue i is computed as follows:

$$T_i = \bar{k}_i \ / \ \lambda_i \qquad i=1,2,3 \quad (26)$$

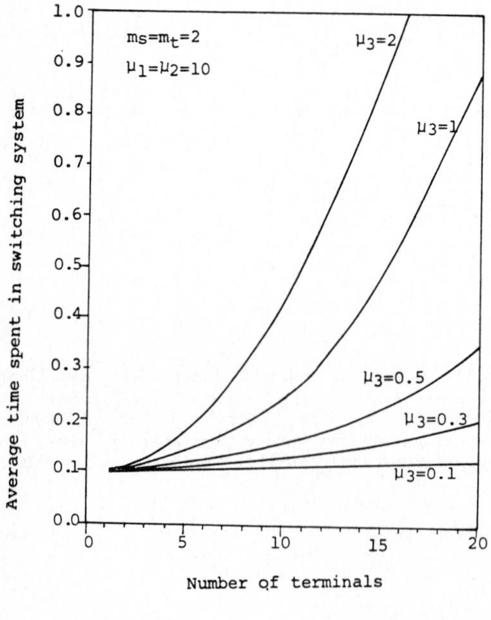

Fig. 6 An example of average time spent in the switching system

By calculating these equations, the following figures are obtained:

Fig. 6 shows an example of the average time spent in the switching system (T_2) where the average occurences of message transfer m_s and m_t equal 2, service rates of a terminal and the switching system (μ_1, μ_2) equal 10 messages/sec or 20 messages/sec.

If the design objective for the average time spent in the switching system is less than 150msec/message, the number of active logical links in the D-channel should be less than 9 in a case where the service rate of a continous message burst μ_3 equals 0.5.

Fig. 6 also shows that the active logical links in the D-channel should be less than 14 in the case of μ_3=0.3.

Fig. 7 shows a calculation example of the average time spent in the switching system vs. the services rate of a teminal μ_1. When the service rate of a terminal μ_1 increases, the average time spent in the switching system increse slightly.

The sum of the average time spent in the terminal and the average time spent in the switching system is shown in Fig. 8. This figure shows that the message processing capability in the D-channel is increased according to the increase of service rate of the terminal. But the effect is decreased when many messages are to be processed in the switching system.

The traffic model is compared with other traffic models. One of them is the M/M/1 queueing model: The service rate of the switching system is also μ_2 , but the arrival rate λ is approximated as follows:

$$\lambda = \frac{r_{31} \ m_t + r_{32} \ m_s}{2} \ \mu_3 \quad (27)$$

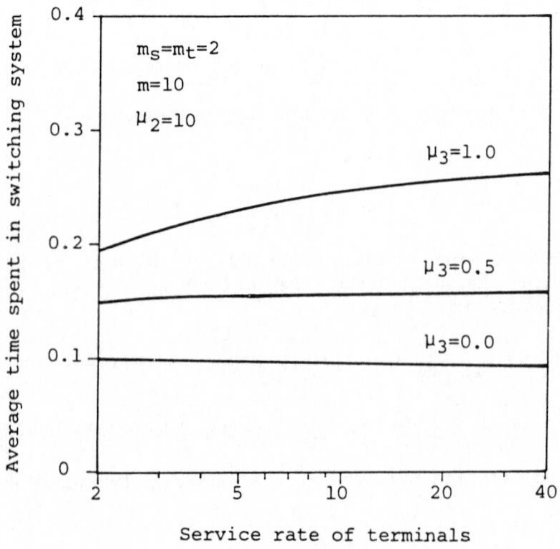

Fig. 7 Average time vs. service rate

$$0 \leqq \rho = \frac{m\lambda}{\mu_2} < 1 \tag{28}$$

Thus, the average time spent in the switching system T is:

$$T = 1 / (\mu_2 - m \lambda) . \tag{29}$$

The other model is the M/M/1//m queueing model[3]: Using same arrival rate λ , the probability P(k) that k messages are in the model is determined as follows:

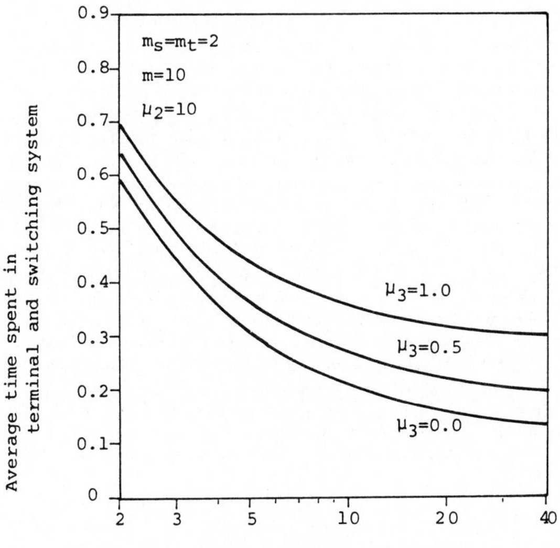

Fig. 8 Average time spent in a terminal and the switching system

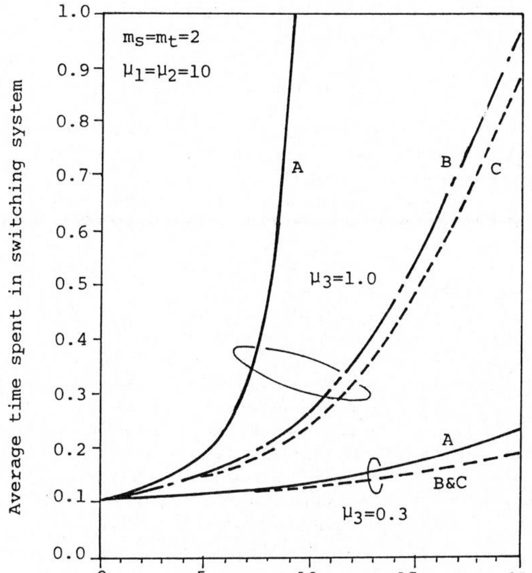

Fig. 9 Comparison of various models

$$p(k) = p(0) \left(\frac{\lambda}{\mu_2}\right)^k \frac{m\,!}{(m-k)\,!} \qquad k=0,1,\ldots,m \tag{30}$$

$$p(0) = \left[\sum_{k=0}^{m} \left(\frac{\lambda}{\mu_2}\right)^k \frac{m\,!}{(m-k)\,!} \right]^{-1} \tag{31}$$

Thus, the average time spent in the switching system T is

$$T = \frac{\bar{k}}{m - \bar{k}} \; \frac{1}{\lambda} . \tag{32}$$

$$\text{where} \quad \bar{k} = \sum_{k=0}^{m} k\, p(k) \tag{33}$$

According to Fig. 9, M/M/1//m queueing model (B) gives sufficient approximation except for a heavy traffic situation.

On the other hand, M/M/1 queueing model (A) does not give an accurate approximation for the traffic model (C) proposed in this paper.

5. CONCLUSION

The traffic characteristics of signalling information messages via the D-channel is investigated according to the network layer protocol described in the CCITT I/Q series recommendations.
A simple traffic model for the signalling information messages is proposed and analyzed

(1) Approximate calculation of the average number of the messages and the average spent time in the switching system is shown.
(2) Except for the heavy message traffic cases, the M/M/1//m queueing model provides accurate approximation to calculate the above-mentioned values.

Accordingly, signalling information traffic via the D-channel can be estimated accurately enough to design a D-channel handler for an ISDN switching system. However, this approximation method should be confirmed using the field test data from an ISDN switching system in the future.

REFERENCES

(1) CCITT Recommendations; I.120, I.410 - I.431, red book, 1984.
(2) CCITT Recommendations; Q.920, Q.921, Q.930, Q.931, red book, 1984.
(3) L. Kleinrock, "Queueing Systems, Volume 1: Theory," John Wiley & Sons, New York, 1975.
(4) W. J. Gordon and G. F. Newell, "Closed Queueing Systems with Exponential Servers," Operations Research, Vol.15, pp.256-265, 1967.

TELETRAFFIC ISSUES in an Advanced Information Society
ITC-11
Minoru Akiyama (Editor)
Elsevier Science Publishers B.V. (North-Holland)

CIRCUIT-SWITCHED MULTIPOINT SERVICE PERFORMANCE MODELS

L.G. MASON *Y. DESERRES* *C. MEUBUS*

INRS-Telecommunications, Montreal, Canada

Abstract

Network requirements for several video teleconferencing services are described. A circuit-switched ISDN architecture for a mix of point-to-point and multipoint traffic is proposed. For the multipoint connections, two levels of switching are involved, namely call set up and call rearrangement. Approximate and exact conference set up delay models are considered and compared for the multichannel transmission links where blocked point-to-point traffic is lost and multipoint traffic is delayed. A non-preemptive priority structure is assumed where the background point-to-point traffic has a higher priority. End-to-end connection delay models are then given in terms of the constituent link delay distributions for both progressive and concurrent call set up protocols. A state space formulation for call rearrangement is described along with state complexity results. Two example designs illustrate the methodology. Open problems are discussed and on-going work is outlined.

1. INTRODUCTION

The impact of multimedia, multilocation teleconferencing on the design and operating characteristics of an IDSN based on circuit-switching has until recently not received much attention because of the low penetration of such services. The volume of network traffic attributable to such services may be expected to grow rapidly in the near future, however, as a result of technology improvements, competitive incentive for business to enhance productivity, changing travel economics, and other socio-economic factors. The present paper can be viewed as a contribution to the design and analysis of a variety of such emerging services where the distinguishing feature is the requirement for multipoint connections.

In this paper investigations into the design and performance evaluation of circuit-switched networks carrying a heterogeneous mix of point-to-point and multipoint services are described. Associated teletraffic problems are addressed, with emphasis on statistical and combinatorial aspects. Issues related to transmission performance, numbering plan, signaling scheme, protocol specification and validation are not considered.

In section 2. results are summarized for the optimal topologies, and functional requirements for a variety of teleconferencing services reported in [1]. A possible network architecture for the integrated handling of multipoint and point-to-point circuit-switched traffic is then outlined. In section 3. several approximate and exact performance models for the multichannel transmission links carrying a mixture of narrow band and wideband traffic are described. The models differ in respect to priority structure for the two traffic classes, release mechanism for multiple channel calls, and channel holding time distributions. The performance results include set up delay and probability of wait for the wideband traffic, and loss probability for narrow band traffic. The approximate and exact models are compared for the case with a non-preemptive priority structure, where narrow band traffic has a higher priority and where blocked narrow band traffic is lost and blocked wideband traffic is delayed. In section 4. set up delay models for multipoint connections are given in terms of the delay models of the constituent links for both progressive and concurrent modes of call set up. Section 5. discusses a state space model and complexity results for the rearrangement phase of the teleconference. The above models and results serve as elements in an overall design procedure which is illustrated by way of specific examples in section 6. Section 7. concludes with a discussion of on-going work.

2. TELECONFERENCING SERVICES AND NETWORK ARCHITECTURE

In Reference [1] a gamut of multimedia multipoint services was described along with optimal topological designs for service networks which are permanently dedicated to a fixed set of conference sites. Table 1. displays information on the optimal service network topologies, number of channels per link as a function of the number of conference sites and network functional requirements to implement a range of multipoint services in a circuit-switched ISDN. These design results were presented in a different form in Reference [1] for the case of FDX and SX links. The HDX case has been included for completeness. For the present study, the FDX case is most relevant, although designs based on SX links would appear attractive for some services if and when SX links are available in the public switched network. The HDX case is of less practical interest, as long haul connections would require switchable repeaters. There may be some potential for HDX transmission for local area teleconferencing; however, this requires more study. We note that for local area networks, the savings in transmission line costs resulting from HDX transmission are less significant than in long haul applications.

A global view of the considered network architecture is shown in Figure 1.

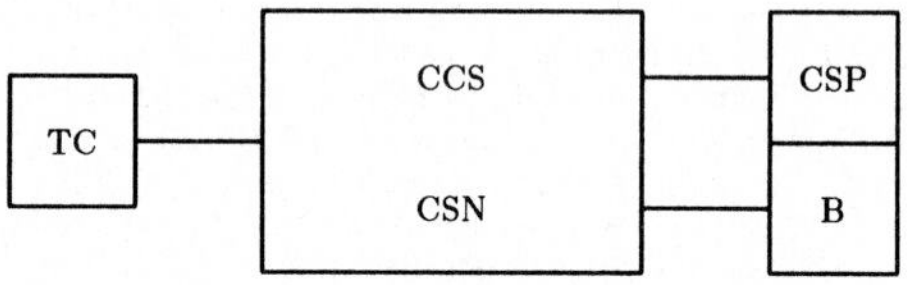

Fig. 1 Global Network Architecture

It is assumed that video and audio teleconferencing transmission will be handled over a circuit-switched network, while the necessary signaling for call-set up and clearing will be conveyed over either the CCS network or some alternative packet switching network. For several of the teleconferencing services, two levels of switching are required. In the set up phase the various non-blocking service networks are reserved where these reservation requests are in contention with background point-to-point traffic. Following set up of the appropriate service network, a rearrangement phase is entered where the interconnection pattern of the transmission links reserved in the set up phase is varied to satisfy the instantaneous signal flow requirements without blocking.

The rationale for separating the set up and rearrangement functions is twofold. First this separation admits greater flexibility as some services do not require rearrangement. Alternatively, in dedicated networks the set-up function is not required. Secondly, in cases where both set up and rearrangement are required, the switching specifications for these levels are very different, leading to different implementations.

The set up phase is most naturally implemented via the CCS network. Since the set up delay specifications for conferencing services will likely be no more stringent than those for point-to-point services, the CCS network or an alternative packet switched network should be adequate.

The rearrangement phase, on the other hand, has a small address space , (the number of active conference sites). This suggests an abbreviated addressing scheme which is dynamically assigned to sites at the time of call set-up. This will reduce rearrangement delays which are very stringent, as the reconfiguration delay should ideally be imperceptable to the conferencees. The possibility for fast rearrangement of terrestrial networks gives them a potential advantage over satellite implementations where rearragement delays on the order of 0.27 seconds are unavoidable.

In addition to the availability of an existing circuit-switched network, CSN, a packet switching signaling network, CCS, and the standard terminal equipment such as microphones, speakers, cameras , monitors and so on , three new modules will be required. We have called these units terminal controllers (TC), teleconferencing bridges (B), and call set up processors (CSP). The terminal controller performs two functions. In conjunction with the CSP(s) and the CCS network it provides for multipoint call set up and clearing. The second function is to provide signaling for call rearrangement. This is done in conjunction with the bridge controllers via the circuit-switched network. The conference bridges terminate transmission lines and vary the interconnection pattern among these lines upon request from the terminal controllers for call rearrangement. For some services, multiple signal copies are also required. In Table 1. we refer to this function as splitting. The functions of the CSP(s) include determining an appropriate routing and channel requirement for the teleconference on the basis of information forwarded from the terminal controllers as well as to set up and clear the multipoint connection in the circuit-switched network via signaling over the CCS network. For reasons of flexibility it would appear desireable to implement the various functions in separate modules as not all services require all functions.

3. LINK PERFORMANCE MODELS

Transmission links for the integrated handling of point-to-point and multipoint calls are modeled as multiserver queues carrying two traffic classes. We shall designate the traffic which requires a single channel or server per link as narrow band while calls requiring more than one server per link is wideband traffic. The requirement for multiple channel seizures can arise from the fact that a conference video signal may require 6, 12, 24, or more voice channels depending on the specifications on image quality. For some multipoint service networks, several channels are required on each link to provide for nonblocking, even though a single voice channel can support the audio or video signal. This is due to the fact that more than one signal must be transmitted simultaneously on each link. Examples are the dynamic allocation and the switched complete broadcast 56 kb/s video services.

Several models have been considered to evaluate the link set-up delay performance for wideband traffic and the blocking performance of the background narrow band traffic [2,3,4]. They differ in respect to 1) release mechanism for multiserver requests: simultaneous, and independent channel release, 2) priority structure: FIFO, non-preemptive priority to narrow band traffic, 3) disposition of narrow band traffic: blocked calls lost, blocked calls delayed, wideband traffic which is blocked is delayed in all cases considered, 4) service time distributions: exponential with equal means for both traffic classes, exponential with class dependent means.

3.1 Equal Service Rates, Independent Channel Release

Cases with equal service rates and independent channel release admit a simplified analysis using the approach first suggested by Green [5]. By identifying the first position in queue with a server in an embedded M/G/1 queue, the results of Welch [6] for the M/G/1 with exceptional service for the first customer can be applied. The delay distribution for the embedded system corresponds to the waiting time distribution in the original queue. In [2] Green's model has been applied to develop expressions for the mean weighting time, and probability of wait, for a Poisson arrival process to a FIFO M-server queue where each arrival requests a random number of servers. In [3] this analysis has been extended to include a non-preemptive priority structure where narrow band traffic has priority and may be either blocked or delayed. The analysis approach follows that of Green [5], with the exception that a cycle of the renewal process consists of a non-queue and a queuing period of the low priority queue only. The analysis of the queuing and the non-queue period is based on the transient Markov chain for the number of busy servers plus the number of high priority customers queued. Below we summarize the results for the case of most interest here, namely, where narrow band blocked calls are lost.

The mean delay for wideband traffic is given by the general expression first given in Green [5].

$$\bar{W}_2 = (1-p_q)p_d d_1 + p_q\left(b_1 + \frac{d_2}{2d_1} + \frac{\lambda_2^+ b_2}{2(1-\lambda_2^+ b_1)}\right) \quad (1)$$

where λ_i is the arrival rate of calls requesting i channels and λ_2^+ is the arrival rate of wideband traffic. p_q is the steady state probability that the queue of low priority customers is not empty. p_d is the probability that a low priority customer arriving during a non- queue period initiates a queuing period. d_n and b_n are the moments of the time spent in the first position in queue for the call which initiated the queue and subsequent wideband arrivals during the queuing period, respectively. By noting that cycles formed by the epochs at which the low priority queue empties constitute a renewal process we obtain

$$p_q = \frac{E(Q)}{E(Q)+E(\bar{Q})} \quad (2)$$

$$p_d = \frac{1}{\lambda_2^+ b_1} \quad (3)$$

where the expected duration of the queuing period is given by

$$E(Q) = \frac{d_1}{1-\lambda_2^+ b_1} \quad (4)$$

The expected duration of the non-queue period is obtained through analysis of the transient Markov chain as

$$E(\bar{Q}) = \sum_{j=0}^{M} \frac{V_{m,j}}{\lambda + j\mu} \quad (5)$$

where $V_{m,j}$ is the mean number of visits to the state with j servers busy during a non-queue period which begins with m servers busy. $V = (I-P)^{-1}$, where P is the transition rate matrix of the embedded Markov chain for the non-queue period.

The moments d_1, d_2, b_1, b_2 are obtained from an analysis of first passage times of the embedded Markov chain for the number of busy servers during a queuing period.

$$d_1 = \sum_{n=1}^{m-1} q_n \sum_{k=m-n+1}^{n} \frac{\alpha_k}{\alpha_{m-n+1}^+} T_n^{m-k} + q_m \sum_{k=2}^{m} \frac{\alpha_k}{\alpha_2^+} T_m^{m-k} \quad (6)$$

$$d_2 = \sum_{n=1}^{m-1} q_n \sum_{k=m-n+1}^{m} \frac{\alpha_k}{\alpha_{m-n-1}^+} U_n^{m-k} + q_m \sum_{k=2}^{m} \frac{\alpha_k}{\alpha_2^+} U_m^{m-k} \quad (7)$$

$$b_1 = \sum_{k=2}^{m} \frac{\alpha_k}{\alpha_2^+} T_m^{m-k} \quad (8)$$

$$b_2 = \sum_{k=2}^{m} \frac{\alpha_k}{\alpha_2^+} U_m^{m-k} \quad (9)$$

where

$$T_n^k = 1/\mu \sum_{i=k+1}^{n} \sum_{j=0}^{m-i} \frac{(i-1)!}{(i+j)!}\left(\frac{\lambda_1}{\mu}\right)^j \quad (10)$$

and

$$U_n^k = 2/\mu \sum_{i=k+1}^{n} \sum_{j=0}^{m-i} \frac{(i-1)!}{(i+j)!}\left(\frac{\lambda_1}{\mu}\right)^j T_{i+j}^k \quad (11)$$

α_k is the probability that an arrival requests k channels and α_k^+ is the probability an arrival requests at least k channels. q_n is the probability of there being n channels busy given that a queuing period has just commenced, where

$$q_n = \frac{\lambda^+_{m-n+1} V_{m,n}}{\lambda + n\mu} \qquad 1 \leq n \leq m-1$$

$$q_n = \frac{\lambda_2^+ V_{m,n}}{\lambda_2^+ + m\mu} \qquad n = m \tag{12}$$

The probability of waiting for wideband calls is given by

$$P_w = p_q + \frac{E(\bar{Q})}{E(Q) + E(\bar{Q})} p_d \tag{13}$$

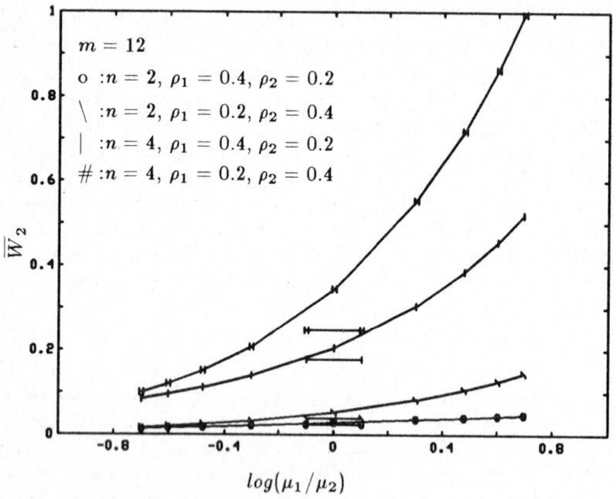

Fig. 2 Mean waiting time for wideband traffic

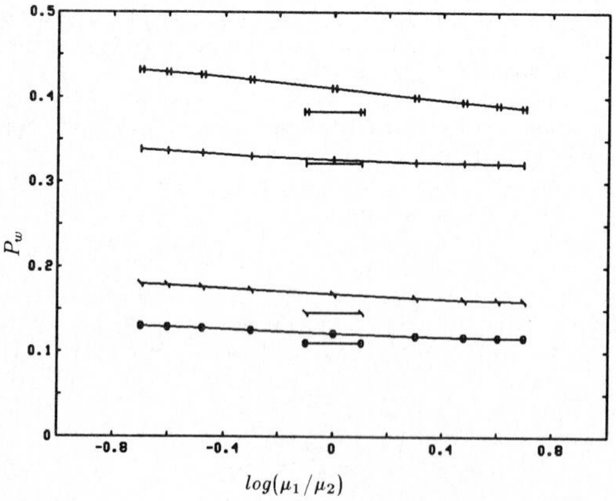

Fig. 3 Probability of waiting for wideband traffic

The probability of blocking P_b for narrow band traffic has also been obtained in [3] by an analysis of the transient Markov chain. Due to space limitations we do not give the analytic derivation here.

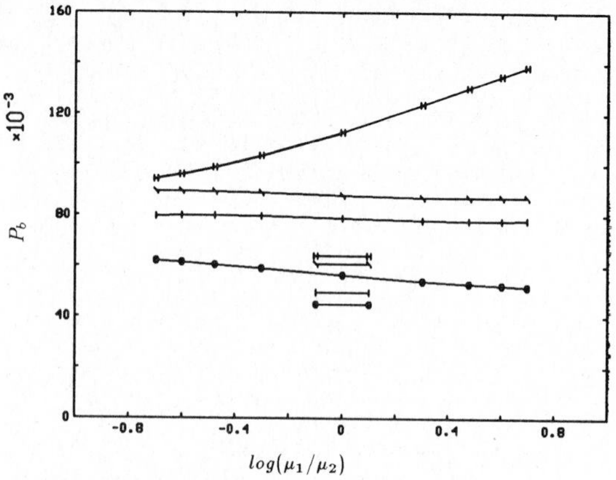

Fig. 4 Probability of blocking for narrow-band traffic

3.2 Non-Equal Service Rates and Simultaneous Release

The approach given above is not applicable when the channel service times are class-dependent. As this case is of practical importance, since one would expect conferences to last longer on average than background point-to-point calls, a state-space model has been developed and implemented in software [4]. Moreover, as the simultaneous release mechanism is easily incorporated in the state space formulation, this model can be employed to evaluate the accuracy of the independent release approximation in the case where channel holding times are not class-dependent. The balance equations were solved by standard techniques.

3.3 Comparison of Results

Performance results for the priority scheme are depicted in Figures 2-4, for the approximate and exact models as a function of the ratio of mean channel holding times for the two traffic classes for various traffic mixes. The unit of time is the holding time of the narrow band traffic $1/\mu_1$.

To use the approximate model in cases where the service rates of the two traffic classes were not equal, we have replaced the true holding time for wideband traffic with that for the narrow band traffic while at the same time using an adjusted arrival rate for wideband traffic which leaves the offered traffic in Erlangs constant with respect to changes in the ratio of service times. Accordingly the performance measures for this approximation are independent of the service time ratio, while the exact performance results do depend upon the ratio of service times. In particular, the mean waiting time increases when the holding time of the wideband traffic increases. On the other hand, the probabilities of waiting and blocking are less sensitive to the holding time ratio for a fixed level of offered traffic. The difference in performance predicted by the two models, where the service times are equal, reflects the error introduced by assuming an independent channel release mechanism if the channels are actually released together. Differences in the curves for service rates not equal also reflect errors introduced by modifying the true arrival rate and service rates of the wideband traffic as described above. These results indicate that the approximate model introduces significant error in the set-up delay when the holding times for the two traffic classes are not equal. The blocking probabilities show less sensitivity to holding time ratio. The approximate model gives a reasonable prediction of the probability of waiting for the wideband calls; however, there is a significant error in the loss probability of narrow band traffic which arises from the independent channel release assumption.

4. END-TO-END CONNECTION DELAY

Given the link delay distributions, the set-up time for a multipoint connection is obtained by composing the involved link distributions according to the setup protocol employed. We have considered two cases, namely progressive setup where links are reserved sequentially, one at a time, and concurrent setup where link reservation requests are made simultaneously for all links.

4.1 Progressive Set-up

Assuming link independence, the end-to-end connection delay distribution is given by the convolution of the delay distributions of the constituent links. For the general case where the delay distributions are link-dependent, the convolution operation is performed numerically. For the uniform complete network, the convolution can be expressed in a simpler closed form. For example, if we are considering an n-fold tandem connection where the n link delay distributions are all equal and given by

$$Pr\{W < x | x > 0\} \tag{14}$$

$$Pr\{W = 0\} = P_w \tag{15}$$

the end-to-end connection delay distribution is given by

$$Pr\{T = 0\} = (1 - P_w)^n \tag{16}$$

$$Pr\{0 < T < x\} = \sum_{i=1}^{n} \binom{n}{i} P_w^i (1 - P_w)^{n-i} Pr\{iW < x | W > 0\} \tag{17}$$

While the delay distribution for a link is, in general, a complex function it can be approximated by low order moment matching. Empirical results have shown that for the priority scheme with independent release, the link delay distribution is well approximated by the exponential function

$$Pr\{W < x | W > 0\} = 1 - e^{\frac{-P_w}{W} x} \tag{18}$$

which implies that

$$Pr\{iW < x | W > 0\} = 1 - e^{\frac{-P_w}{W} x} \sum_{r=0}^{i-1} \frac{1}{r!} \left(\frac{P_w x}{W}\right)^r \tag{19}$$

For Green's model with FIFO it can be shown that the coefficient of variation of waiting time is large for light traffic and approaches unity asymptotically as the traffic increases. To model such delay distributions the hyperexponential function has been used. For a coefficient of variation less than unity we have employed the Gamma distribution and have matched the first two moments. Space limitations prevent us from including the explicit formulae for these cases.

4.2 Concurrent Set-up

The end-to-end connection delay distribution is given as the maximum of the individual link distributions. If there are ν links in the multipoint connection then we have

$$T = \max_{i=1,..\nu} \{W_i\} \tag{20}$$

and

$$t_n = E(T^n) = \int_0^\infty x^n d\left(\prod_{i=1}^{\nu} Pr\{W_i \leq x\}\right) \tag{21}$$

where link independence has been assumed in evaluating the moments t_n.

In general this integral must be evaluated numerically. In the special case where all link delays have the same exponential distribution given by

$$Pr\{W_i \leq x\} = 1 - \alpha e^{-\beta x} \tag{22}$$

one obtains a closed form analytic expression for the moments of connection delay.

$$t_n = \sum_{i=1}^{\nu} \binom{\nu}{i} (-1)^{i+1} \frac{n! \alpha^i}{(\beta i)^n} \tag{23}$$

5. A STATE SPACE MODEL FOR REARRANGEMENT

Beneš [7] has advanced a state space model for the operation of circuit- switched networks carrying point-to-point traffic as well as a group theoretic approach to reduce the number of states through generation of the equivalence classes [8]. This state reduction technique is important for generating the set of states to determine if indeed a network is non-blocking and, if it is not, to determine the state probabilities necessary for computing exact performance measures for blocking networks. For networks of reasonable size there are an enormous number of states and exact calculations are usually not possible even with state equivalencing. The rearrangement phase of a multipoint teleconference appears to be an exception as typically there are relatively few sites involved in a conference, and the multipoint topologies associated with some conference service types possess a high degree of symmetry. In [9] a state space model has been formulated for circuit-switched networks which carry multipoint traffic. In this model the admissable states consist of connections which form disjoint spanning trees in the network's dual graph. The dual graph of a network is formed by replacing crosspoints by edges with a value 0 when the crosspoint is open and 1 when it is closed. The terminals and links of the given network on the other hand form the vertices of the dual graph.

Table 2. lists the cardinality of the original assigment and state set along with the bounds on the number of equivalence classes for the complete network as a function of the number of conference sites. Each assignment or looking pattern maps into a single network state because the optimal routing is deterministic.

The number of looking patterns in an n-node network is given by

$$|States| = (n-1)^n \tag{24}$$

The symmetry group G_η associated with the various network topologies η of the teleconferencing service networks is used to generate the equivalence classes under the assumption of symmetry among conference sites. It is also employed in the derivation of upper and lower bounds on the number of equivalence classes for various conference services and supporting network topologies. For complete networks, G_{mesh} is the complete permutation group S_n containing $n!$ elements. For star topologies G_{star} is S_{n-1} containing $(n-1)!$ elements. For ring topologies containing n nodes the symmetry group is the dihedral group D_n containing $2n$ elements. In general the symmetry group associated with the minimum spanning tree may only contain the identity permutation, and no state reduction is possible.

A closed form expression for an upper bound on the number of equivalence classes for the complete network has been derived [10] to be

$$UB_{mesh} = \left\lfloor \sum_{\alpha=1}^{\lfloor n/2 \rfloor} (n-\alpha-1)!(n-2\alpha+1)/(\alpha-1)! \right\rfloor \tag{25}$$

A lower bound on the number of equivalence classes for the complete network is given by

$$LB_{mesh} = \lceil (n-1)^n / n! \rceil \tag{26}$$

The upper and lower bounds on the number of equivalence classes for other network topologies η are easily obtained from corresponding bounds for the complete network by the expressions

$$UB_\eta = \left\lfloor UB_{mesh} \frac{|G_{mesh}|}{|G_\eta|} \right\rfloor \tag{27}$$

$$LB_\eta = \left\lceil LB_{mesh} \frac{|G_{mesh}|}{|G_\eta|} \right\rceil \tag{28}$$

where $|G_\eta|$ denotes the number of elements in the symmetry group for network topology η.

6. EXAMPLE DESIGNS

We will now describe the use of the previous models for two different teleconferencing services, namely the dynamic allocation service, and the switched complete broadcast service. In these examples we are assuming a low bit rate video signal of 56 kb/s.

6.1 Dynamic Allocation

As we are assuming the public switched network employs FDX transmission, reference to Table 1. indicates that the optimal service network topology for dynamic allocation is the ring interconnecting the n teleconference sites. The optimal multipoint circuit routing is provided by the solution of the traveling salesman problem for these n sites. While this is an NP-complete problem, typically the number of conference sites is small which enables an exact solution. For conferences involving a large number of sites, a heuristic solution to the TSP can be used.

The call set-up procedure which is most natural for this case is decentralized progressive control, where links are reserved sequentially around the ring. It is also natural to employ a progressive control procedure to effect the rearrangement, with some form of token passing to remove conflicts in rearrangement requests.

For this example let us assume that the supporting network is completely connected with each link containing $m = 12$ channels. Let us assume that the number of conference sites involved is $n = 5$. Reference to Table 1. reveals that for this service network, $(5-1)/2 = 2$ channels must be reserved per link in the ring to provide a nonblocking service. For arrival rates $\lambda_1 = 4.8$, $\lambda_2 = .24$ and service rate parameters of $\mu_1 = 1$ and $\mu_2 = .2$ the total offered traffic per link in Erlangs is $12(\rho_1 + \rho_2) = 7.2$ where $\rho_1 = \lambda_1/12\mu_1 = 0.4$, and $\rho_2 = \lambda_2/12\mu_2 = 0.2$. The mean waiting time per link is $\bar{W}_2 = 0.0465$, the probability of wait is $P_w = 0.116$ and the loss probability is $P_b = .0519$. Application of Equations (16) and (17) yields the distribution of waiting time to set up the ring. The mean waiting time is $T = n\bar{W}_2 = .2325$, while the probability of wait is $1 - (1 - P_w)^n = .460$.

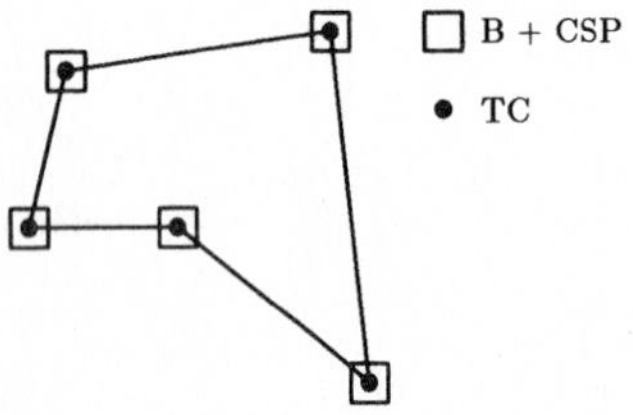

Fig. 5 DAN Network Layout

For this example the number of connection states in the rearrangement phase is $(5-1)^5 = 1024$, while the upper and lower bounds on the number of equivalence classes are 336 and 102 respectively.

To evaluate the benefit of service integration we have computed the performance for a segregated system subjected to the same traffic demands where the 12 channels per link were partitioned with 4 channels dedicated to multipoint traffic. The link performance figures for this system were computed by the Erlang formulae to be $\bar{W}_2 = 2.8125$, $P_w = 0.450$, and $P_b = 0.0609$. It is evident from this example that the integrated system provides a substantially better performance than dedicated systems for the same transmission capacity.

6.2 Switched Complete Broadcast

The optimal service network topology is a star with the bridge located at the vertex median of the conference sites. The most natural call set-up mode for this case is the centralized concurrent mode, where the caller signals the CSP which is collocated with the bridge. All n links including the source node are then set up concurrently. Table 1, reveals that for FDX links, $n - 1 = 4$ channels must be reserved for each branch of the star to set up the nonblocking service network. The holding times are as in the previous case, while the arrival rates were chosen as $\lambda_1 = 4.8$, $\lambda_2 = 0.12$. For the integrated system the performance figures for each of the $n - 1$ links, are $\bar{W}_2 = 0.520$, $P_w = 0.321$ and $P_b = 0.0780$. For the segregated system with 4 channels per link dedicated to conference traffic the performance figures are $\bar{W}_2 = 7.5$,$P_w = 0.6$, and $P_b = 0.0609$. Again the advantage of service integration is evident. The mean time to set up the star connection is given by Equation (23) as 1.924 time units. For progressive reservation the mean set-up time is 2.6 time units.

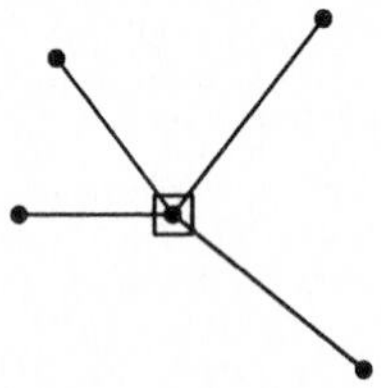

Fig. 6 CSB Network Layout

As far as the rearrangenent phase is concerned the centralized bridge modifies the instantaneous interconnection pattern upon request from the conference sites. As there is only a single bridge no interbridge coordination is necessary and conflicts in rearrangement requests do not occur if the rearrangements are made on a FIFO basis.

Here once again the state space for the number of connection patterns contains 1024 elements. Equivalencing is more effective here than in the ring network because of the greater symmetry of the star. The upper and lower bounds on the number of equivalence classes are 150 and 40 respectively.

7. CONCLUSION

In this paper a possible network architecture for the integrated handling of point-to-point and multipoint traffic is described. Performance models have been developed which have demonstrated the requirement for models which take into account the multichannel nature and long holding times of teleconferences. These preliminary results have also shown that service integration can lead to cost reduction and/or performance improvement relative to providing dedicated networks for the two traffic classes.

For the simple example designs considered in this paper the results on optimal service network layout and capacities of Reference [1] were applicable as a single conference group for a particular conference type was considered and the supporting network topology and traffic was assumed to be uniform. When more than one conference group and/or more than one conference type is involved or where the supporting network is not uniform, the optimal design problem is substantially more complex and one would expect that optimal service network designs will include both access and core network segments. For nonuniform networks the marginal link capacity as well as the link length must be taken into account when determining the optimal routing for the multipoint connection. Moreover, the optimal deployment of CSPs may differ from that of the conference bridges. Ongoing work addresses these questions along with procedures for dimensioning the supporting ISDN for the nonuniform case.

8. REFERENCES

[1] M.J. Ferguson, L.G. Mason "Network Design for a Large Class of Teleconferencing Services", *IEEE Trans. on Comm.*, VOL. COM-32, No. 7, July, 1984.

[2] C. Meubus, "Performance Models for Multi-Media Multi-location Teleconferencing Call Set-up on a Circuit-Switched Network", *INRS- Telecommunications Tech. Rep.*, No. 83-20, Sept. 1983.

[3] Y. DeSerres, "A Non-Preemptive Priority Queueing System in which Customers Require a Random Number of Servers",

INRS-Telecommunications, Tech. Rep. No. 84-32, August 1984.

[4] Y. DeSerres, "A Multi-server Non-preemptive Cutoff Priority Queue in which some Customers require more than one server", *INRS-Telecommunications* Tech. Rep. No. 85-09, March, 1985.

[5] L. Green, "A Queueing System in which Customers Require a Random Number of Servers", *Operations Research*, Vol-28, No. 6, pp 1335-1346, Nov-Dec 1980.

[6] P.D. Welch, "On a Generalized M/G/1 Queueing Process in which the First Customer of Each Busy Period Receives Exceptional Service", *Operations Research* 12, pp 736-752, 1964.

[7] V.E. Beneš, *The Theory of Connecting Networks and Telephone Traffic* ,Academic Press, New York, London, 1965.

[8] V.E. Beneš, "Reduction of Network States under Symmetries", *Bell System Technical Journal*, Vol. 57, No. 1, Jan. 1978.

[9] L.G. Mason, "State Space Models and Complexity Results for Teleconferencing Networks", *INRS-Telecommunications* Tech. Rep. No. 82-14 ,Oct. 1982.

SERVICE	OPTIMAL SERVICE NETWORK FDX	SX	HDX	SPLITTING	REARRANGE
Fully Connected	Complete 1	Complete 1	Complete 1	N/A	N/A
Dynamic Allocation	Ring $(n-1)/2$, odd n $n/2, n/2-1$, even n	MST n	Mesh $1, 2$	N/A	Yes
Complete Broadcast	Ring $(n-1)/2$, odd n $n/2, n/2-1$, even n	MST n	MST n	Yes	N/A
Switched Complete Broadcast	Median Star $n-1$	Median Star $2(n-1)$	Median Star $2(n-1)$	Yes	Yes
Switched Broadcast	MST 1	Ring 1	MST 1	Yes	Yes

Table 1 Optimal Service Networks

State Reduction for complete networks			
n	States	UB_{mesh}	LB_{mesh}
2	1	1	1
3	8	2	2
4	81	7	4
5	1024	28	9
6	15625	139	21
7	279936	822	56

Table 2 Complexity vs Number of Sites

TELETRAFFIC ISSUES in an Advanced Information Society
ITC-11
Minoru Akiyama (Editor)
Elsevier Science Publishers B.V. (North-Holland)

ANALYSIS OF OVERLOAD PERFORMANCE FOR A CLASS OF M/D/1 PROCESSOR QUEUEING DISCIPLINES

by

H. Heffes

AT&T Bell Laboratories
Holmdel, NJ

ABSTRACT

We consider the overload performance of a class of queueing and service disciplines aimed at controlling the load offered to a processor in such a way as to keep delays, for served customers, small. In particular, we analyze one such control scheme, the M/D/1 LIFO queueing discipline, where the customer rejection mechanism corresponds to arrivals to a full buffer pushing out the oldest customer in the queue. Comparisons with the M/D/1 FIFO finite buffer scheme and related M/M/1 results show the significant effect of the queueing discipline, and the variability of unloading the input buffer, on system performance. We present results for the delay distribution of served customers and the throughput-delay tradeoffs. We also present an approximation for the M/G/1 LIFO-pushout scheme.

In some overloaded call processing systems, long delays can result in customers abandoning or turning "bad" (e.g., dialing before receiving dial-tone). For the situation where customers in queue turn "bad" at a random time after their arrival we present results for the throughput of good customers. Here, the results show a strong dependence on the mechanism for customers turning bad.

1. INTRODUCTION

When a call processing system is overloaded, long delays can result in either poor service given to the customer or can result in customers, abandoning or turning "bad" [1]. An example of this occurs when the service provided is the giving of dial-tone. If the customer starts dialing before receiving dial tone, due to long dial-tone delays, the system will not receive all the digits and an unsuccessful call results. This can lead to the system expending real time on unsuccessful calls and therefore reduces the effective throughput. Thus there is a need for control schemes which reduce the load offered to the processor by selectively refusing service to some customers in such a way as to keep delays, for served customers, small. This fact has been recognized for some time and has resulted in improved strategies for local switches [2,3].

In this paper we analyze one such control scheme, the M/D/1 LIFO queueing discipline where the customer rejection mechanism corresponds to arrivals to a full buffer pushing out the oldest customer in the queue. The LIFO-pushout sheme has been shown to have desirable performance, in overload, for the M/M/1 queue when compared to other queueing disciplines [4,5]. However, for some control mechanisms, e.g., a rate based control scheme [6], where the rate of accepting new customers into the system is limited (e.g., one customer admitted every T seconds), a deterministic "service time" assumption is more appropriate.

For the M/D/1 LIFO-pushout scheme, we present results for the delay distribution of served customers and for the throughput-delay tradeoff operating characteristic, which is a useful basis for comparisons; improved schemes resulting in smaller delays for a given throughput. In an environment where a customer in queue, unknown to the system, can turn "bad" at a random time after its arrival, we present results for the throughput of successful services ("goodput" [1]). For the results in this paper we have used the same service time for good and bad customers and consider two distribution functions for the time a customer in queue remains good. The results exhibit a strong dependence on the mechanism for customers turning bad and thus indicate the importance in characterizing the mechanism in such an environment.

Comparisons are made with the FIFO; finite buffer, discipline which we show to be "equivalent" to an infinite buffer FIFO scheme where customers time out after spending an appropriately defined time in queue. These disciplines are defined as follows:

(i) LIFO-Pushout (LIFO-PO)

Last-in-first-out service, a finite buffer of size $N-1$; a customer arriving to see a full buffer pushes out the oldest customer in the buffer. This is a component of the overload strategy discussed in [3].

(ii) FIFO-Blocking (FIFO-BL)

First-in-first-out (FIFO) service; a finite buffer of size $N-1$; a customer arriving to see a full buffer leaves immediately. This is the M/D/1/N queue.

(iii) FIFO-Timeout (FIFO-TO)

FIFO discipline, infinite buffer; every arriving customer joins the queue but will leave at a time T_0 after arrival if it is still in the buffer at that time.

In addition to comparing the above schemes we make comparisons with results [5] for exponentially distributed service times to determine the sensitivity of performance to the service distribution. Based on the observed sensitivity an approximation for the M/G/1 case is suggested. These results can be used to estimate the effect of overload control strategies with varying degrees of regularity of unloading the input buffer (e.g., window based strategies [6]).

The M/D/1 LIFO-pushout discipline is analyzed by considering the partial differential-difference equations for the probability a tagged customer in a given queue position gets served with remaining delay not exceeding a given value, conditioned on the elapsed service time of the customer in service. The solution is in terms of boundary coefficients obtained by solving a system of linear equations. The throughput of good customers for the FIFO case, which we use for comparison, is obtained by recognizing it's equivalence to an appropriately defined timeout problem, and using level crossing ideas [7].

We note that although the results are developed for a single server system they can be used to approximately analyze overload control schemes which control access to distributed systems (e.g., analysis of dial-tone delays for a pre-dial control of a distributed switching machine [6]). They can also be used to study the class of SPC overload controls treated in [13], under the LIFO-pushout discipline.

2. ANALYSIS OF CONTROL SCHEME

The positions in the buffer are numbered 2, 3, ... N with the processor numbered 1. An arriving customer goes into service (position 1) if the system is empty, otherwise it goes into position 2. If an arrival occurs while a customer is waiting in position 2, the waiting customer moves into position 3 and the new arrival moves into position 2. Of course a service completion brings the positions of all waiting customers down by 1. If a customer is in position N and an arrival occurs then the waiting customer gets pushed out and the arrival joins position 2.

2.1 Delay Distribution of Served Customers Let

P_s = P{an arrival gets served},

p_o = P{an arrival sees an empty system},

$f_s(t)$ = density function for the waiting time of customers who get served,

$$G(j,x,t) = P\left\{\begin{matrix}\text{a customer in position j gets served}\\ \text{and its remaining waiting time} \le t\end{matrix} \middle| z = x\right\},$$

where z denotes the elapsed service time of the customer in service, and

$$g(j,x,t) = \frac{\partial}{\partial t} G(j,x,t)\,.$$

Denoting λ as the arrival rate of the Poisson arrival process and T as the deterministic service time, we have

$$f_s(t) = \frac{1-p_o}{P_s} \int_0^T g(2,x,t)\,\frac{dx}{T}\,, \quad t > 0.$$

We denote

$$\bar{g}(2,t) = \int_0^T g(2,x,t)\,\frac{dx}{T}$$

which has the interpretation as the marginal density of delay *and* being served for those customers arriving to a busy system. Using Little's Law we have

$$\lambda\, T\, P_s = 1 - p_o.$$

and obtain

$$f_s(t) = \lambda\, T\, \bar{g}(2,t)\,, \quad t > 0, \qquad \text{(2-1a)}$$

with an atom at zero

$$F_s(0) = \frac{p_o}{P_s}. \qquad \text{(2-1b)}$$

The delay distribution determination reduces to the determination of g(2,x,t) since p_o can be obtained from the M/D/1 finite buffer problem [8] or as a byproduct of the analysis presented here. To analyze the system we have

$$G(j,x,t) = G(j,x+h,t-h)\,(1-\lambda h) + G(j+1,x+h,t-h)\,\lambda h + o(h),$$

$$(1<j\le N,\ 0<x<T)$$

where we identify $G(N+1, x+h, t-h) = 0$, from which we obtain the partial differential-difference equations,

$$\frac{\partial}{\partial t} G(j,x,t) - \frac{\partial}{\partial x} G(j,x,t) + \lambda\, G(j,x,t) = \lambda\, G(j+1,x,t),$$

$$(1<j\le N,\ 0<x<T) \qquad \text{(2-2)}$$

where $G(N+1, x, t) = 0$.

From (2-2) we obtain

$$\frac{\partial}{\partial x} g^*(j,x,s) = (\lambda+s)\, g^*(j,x,s) - \lambda\, g^*(j+1,x,s),$$

$$(1<j\le N,\ 0<x<T)$$

with $g^*(N+1, x, s) = 0$, and where the Laplace-Stieltjes transform

$$g^*(j,x,s) = \int_0^\infty e^{-st}\, dG(j,x,t).$$

This set of differential - difference equations can be solved backwards (with respect to j), in terms of the boundary conditions, $g^*(j,o,s)$, to yield

$$g^*(j,x,s) = e^{(\lambda+s)x} \sum_{i=j}^{N} \frac{(-\lambda x)^{i-j}}{(i-j)!}\, g^*(i,0,s). \qquad \text{(2-3)}$$

Using the boundary conditions

$$g^*(j-1,0,s) = g^*(j,T,s), \quad 2 \le j \le N$$

with

$$g^*(1,0,s) = 1,$$

we obtain the set of linear equations for the desired boundary conditions

$$g^*(j-1,0,s) = e^{(\lambda+s)T} \sum_{i=j}^{N} \frac{(-\lambda T)^{i-j}}{(i-j)!}\, g^*(i,0,s)$$

$$2 < j \le N \qquad \text{(2-4a)}$$

and

$$1 = e^{(\lambda+s)T} \sum_{i=2}^{N} \frac{(-\lambda T)^{i-2}}{(i-2)!}\, g^*(i,0,s). \qquad \text{(2-4b)}$$

For a given s, these equations (2-4a) can be recursively solved backwards, using (2-4b) for normalizing the solution. Finally, we denote the Laplace transform of $\bar{g}(2,t)$

$$\bar{g}^*(2,s) = \int_0^T g^*(2,x,s)\,\frac{dx}{T} = \ell[\bar{g}(2,t)]$$

and use (2-3) to obtain

$$\bar{g}^*(2,s) = \frac{1}{T} \sum_{i=2}^{N} \left[\left(\sum_{k=2}^{i} \left(\frac{\lambda}{\lambda+s}\right)^{k-2} \frac{(-\lambda T)^{i-k}}{(i-k)!}\, \frac{e^{(\lambda+s)T}}{(\lambda+s)}\right) - \left(\frac{\lambda}{\lambda+s}\right)^{i-2} \frac{1}{(\lambda+s)}\right] g^*(i,0,s)\,. \qquad \text{(2-5)}$$

Inversion of (2-5) yields $\bar{g}(2,t)$ which, together with (2-1), gives the desired density function for the waiting time of served customers. To obtain the atom at zero we can use the normalization condition

$$1 = \frac{p_o}{P_s} + \int_{0^+}^{\infty} f_s(t)dt = \frac{p_o}{P_s} + \lambda T\, \bar{g}^*(2,0)$$

which gives

$$F_s(0) = 1 - \lambda T\, \bar{g}^*(2,0). \qquad \text{(2-6)}$$

The throughput is then given by

$$\lambda P_s = \frac{\lambda}{1+\lambda T(1-\bar{g}^*(2,0))}. \qquad (2\text{-}7)$$

2.2 Mean Delay of Served Customers To obtain the mean delay of served customers,

$$M = \lambda T \int_0^\infty t\bar{g}(2,t)dt = -\lambda T \frac{d}{ds}\bar{g}^*(2,s)\Big|_{s=0},$$

we differentiate (2-5) which gives

$$\bar{g}^{*'}(j,0) = \frac{1}{T}\sum_{i=j}^{N}\left[\left(\sum_{k=j}^{i}\frac{(-\lambda T)^{i-k}}{(i-k)!}\frac{e^{\lambda T}}{\lambda}\right)-\frac{1}{\lambda}\right]g^{*'}(i,0,0) +$$

$$\frac{1}{T}\sum_{i=j}^{N}\left[\left(\sum_{k=j}^{i}\frac{(-\lambda T)^{i-k}}{(i-k)!}\frac{e^{\lambda T}}{\lambda}\left\{T-\frac{(k-j+1)}{\lambda}\right\}\right)\right.$$

$$\left. + \frac{(i-j+1)}{\lambda^2}\right] g^*(i,0,0) \Bigg] g^*(i,0,0) \qquad (2\text{-}8)$$

The quantities $g^*(i,0,0)$ are obtained by solving the linear equations (2-4) at $s = 0$ and the quantities $g^{*'}(i,0,0)$ are obtained from the following linear equations obtained by differentiating (2-4):

$$-T e^{-\lambda T} g^*(j-1,0,0) = -e^{-\lambda T} g^{*'}(j-1,0,0)$$

$$+ \sum_{i=j}^{N}\frac{(-\lambda T)^{i-j}}{(i-j)!} g^{*'}(i,0,0); \qquad (2\text{-}9a)$$

for $2 < j \le N$ and

$$-T e^{-\lambda T} = \sum_{i=2}^{N}\frac{(-\lambda T)^{i-2}}{(i-2)!} g^{*'}(i,0,0). \qquad (2\text{-}9b)$$

3. "GOODPUT" RESULTS

The next set of results correspond to the situation where a customer in queue, unknown to the system, turns "bad" at a random time after its arrival [1]. Thus serving a customer with delay in excess of this random time results in an unsuccessful service. For the results in this paper we have used the same service time for good and bad customers. Clearly the delay distribution of served customers and the distribution of the time at which a customer turns bad determine the rate at which the system serves good customers (goodput). Defining

$$P(t) = \Pr\left[\text{customer in queue for t seconds is good}\right]$$

we consider two cases:

$$\text{Case I: } P(t) = e^{-\alpha t}$$

and

$$\text{Case II: } P(t) = \begin{cases} 1 & t \le \tau \\ 0 & t > \tau \end{cases}.$$

The goodput, V, is given by

$$V = \lambda P_s \int_{0^-}^{\infty} P(t)\, dF_s(t)$$

which results in

$$V_I = \lambda p_o + \lambda(1-p_o)\bar{g}^*(2,\alpha).$$

From (2-7) and, $p_o = 1 - \lambda T P_s$, we obtain

$$V_I = \frac{\lambda}{1+\lambda T(1-\bar{g}^*(2,0))}\left[1 + \lambda T(\bar{g}^*(2,\alpha) - \bar{g}^*(2,0))\right]$$

(3-1)

For Case II we clearly have

$$V_{II} = \frac{\lambda}{1+\lambda T(1-\bar{g}^*(2,0))} F_s(\tau). \qquad (3\text{-}2)$$

4. NUMERICAL RESULTS

In this section we present system performance measures for the queuing discipline studied, in overload, and compare the results with the M/D/1 FIFO-BL scheme. We also make comparisons with the M/M/1 LIFO-PO queue to see the sensitivity of the performance measures to the service distribution. Specifically we present numerical results for the tails of the delay distribution, throughput-delay tradeoffs and the effect of customers turning "bad".

Figure 1 shows the conditional mean delay results as a function of the offered load ($\rho=\lambda T$), ranging into overload conditions. The unit of time in all the numerical results is the service time. The LIFO-PO scheme as well as the FIFO-BL scheme (obtained from tabulated results in [9]) are shown for three values of N. We observe the significant range of results with the LIFO-PO scheme being more than an order of magnitude smaller than the FIFO-BL scheme for $N = 11$ and $\rho = 2$. We also note the decreasing behavior as a function of ρ, for the pushout scheme, as ρ becomes large. While the FIFO results clearly approach $(N-1)T$, the limiting pushout behavior is explained as follows. As ρ becomes large, an arriving customer enters the first waiting position and either goes into service, if the next arrival occurs after a service completion, or gets pushed back to the second waiting position, where his chances of being served become vanishingly small. As ρ gets large it is the last arrival during a service time that gets hold of the server and the mean remaining service is the mean delay experienced by this last arrival. Thus

$$E[W|\text{served}] \underset{\text{LIFO-PO}}{\rightarrow} \frac{1}{\lambda}.$$

This limiting behavior has been observed numerically, requiring up to $\rho = 5$, for some cases. We thus have

$$\frac{E[W|\text{served}]_{\text{LIFO-PO}}}{E[W|\text{served}]_{\text{FIFO-BL}}} \rightarrow \frac{1}{(N-1)\rho}$$

as ρ becomes large.

In Figure 2 we look at the results for a given value of $\rho = 1.6$ and plot the throughput as a function of the conditional mean delay. These throughput-delay tradeoff curves, generated by varying N, can be used to compare the performance of the schemes for a given maximum allowable throughput or processor utilization. Conversely for a given maximum mean delay e.g., $E[W|\text{served}] = 1$, the throughput of the LIFO-PO scheme is more than 5 percent greater than that of the FIFO-BL scheme. Figure 3 shows the throughput-delay tradeoffs for $\rho = 1.0$. If the maximum allowable processor utilization is e.g., 0.95 then the corresponding mean delays are 2.6 for the LIFO-PO scheme as compared to 4.5 for the FIFO-BL scheme. When comparing Figures 2 and 3, it should be kept in mind that to achieve a given throughput one requires a larger N for the smaller ρ.

The sensitivity of the throughput-mean delay tradeoff curves to the service distribution is shown in Figures 4 and 5 which show the results for the LIFO-PO schemes with deterministic service and exponential service [5] for $\rho = 1.0$ and 1.5 respectively. We note from Figure 4 that, for a given throughput, say 0.95, the mean delays even for the LIFO-PO scheme are quite sensitive to the service distribution with E[W|served] = 2.9 for deterministic service and E[W|served] = 6.0 for exponentially distributed service. It is interesting to note that *for a given throughput* the mean delays for the deterministic service times are approximately $\frac{1}{2}$ the mean delays for exponential service times. The factor $\frac{1}{2}$ is the ratio of the second moments of the service distribution. The approximate tradeoff curves for LIFO-PO (M/D/1), generated in this manner, are also shown in Figures 4 and 5 and suggest obtaining approximate M/G/1 LIFO-PO tradeoffs by scaling the M/M/1 mean delay results (for a given throughput) with the ratio of the second moments of the service distribution. We note further that scaling the M/M/1 FIFO-BL tradeoffs in the same manner results in a fairly good approximation to the M/D/1 FIFO-BL tradeoffs. This is shown in Figure 6 and suggests using the same scaling to approximate the tradeoffs for the other queuing disciplines studied in [5] for more general service distributions. For a given buffer size, approximate mean delays can then be obtained by evaluating the throughput and reading off the mean delay for the approximate tradeoff. We note that the throughput can generally be simply evaluated since for a given service distribution and buffer size the throughput is often independent of the service discipline e.g., the throughputs for FIFO-BL, LIFO-PO, FIFO-PO can be evaluated using the simpler FIFO-BL discipline. To test out the approximation we draw the exact tradeoff curve for Erlangian service time with two stages, i.e., the M/E2/1 FIFO-BL system, using results in [9] and compare it with the approximate tradeoff obtained by scaling the mean delays for the M/M/1 FIFO-BL system at each throughput by the ratio of the second moments of the service distributions 0.75. Figure 7 shows the excellent degree of approximation.

The results have shown the significant advantages of using the LIFO-PO discipline in addition to the significant advantage of more regular unloading of the input queue, a fact that was observed when comparing average dial-tone delay in a distributed switching machine for a rate based and a window based overload control. [6].

In Figures 8 and 9 we show the delay distribution tails for the overload case $\rho = 1.6$, and for N = 11 and N = 2 respectively. We note that the FIFO-BL scheme serves all accepted customers within (N−1) service times and this scheme is exactly *equivalent*, with respect to the distribution of the waiting time of served customers, *to a FIFO-Timeout scheme with the timeout interval* $T_0 = (N-1)T$. We note that the shape of the distribution tail in Figure 8 is very similar to the FIFO-Timeout results in [5]. For the LIFO-PO scheme it is possible for a customer's position to oscillate and therefore we have the small tail for $t > 10$. As expected we see that most of the customers get served very quickly for the LIFO-PO scheme (approximately 80 percent get served within a service time), whereas less than 1 percent of the customers wait less than 5 service times for the FIFO-BL scheme. The corresponding mean delays are 9.0 and 0.97 for the FIFO-BL and LIFO-PO schemes respectively. The complementary delay distributions, $\bar{F}_s(t)$, for N = 2 can be shown to satisfy

$$\bar{F}_s(t)\Big|_{LIFO-PO} \Big/ \bar{F}_s(t)\Big|_{FIFO-BL} = e^{-\lambda t}, \; 0 \le t \le T.$$

In the previous examples we compared the delay and throughput performance. Here we look at the goodput performance for the LIFO-PO and FIFO-BL schemes, where we choose the control parameters (buffer size) to maximize the goodput, as a function of the severity of the overload. Figure 10 shows the results for Case I, for both deterministic and exponential service times, where the average time a customer remains good is $\alpha^{-1} = 10$ service times. We see the superiority of the LIFO-PO scheme over FIFO-BL scheme as well as the sensitivity to the service time distribution. The goodput results for the M/D/1 FIFO-BL scheme are obtained using the previously recognized equivalence with the FIFO-Timeout case, and level crossing arguments [7] to analyze the Timeout case. Details are in the Appendix.

In Figure 11 we show the corresponding maximum goodput results for Case II which corresponds to the case where a customer in queue turns bad exactly τ time units after its arrival. We choose $\tau = 10$, which gives the same average time for a customer in queue remaining good as the Case I examples. The maximum goodput results for the M/D/1 FIFO-BL scheme are obtained by choosing N = 11, which ensures an accepted customer initiates service within 10 service times. We note however (see Figure 1) that the average time spent in queue by a customer is large (9.0 service times at $\rho=1.6$). The M/D/1 LIFO-PO scheme, which has excellent delay performance, has more limited goodput for this case. In heavy overload ($\rho=1.6$) the small (0.85 percent) goodput reduction can be traded of by a 9:1 reduction in the average time a customer (good or bad) spends in queue for the LIFO-PO scheme. At a smaller overload ($\rho=1.1$) the 7.1 percent goodput reduction is traded off by only a 2:1 reduction in the average time a customer (good or bad) spends in queue. We again note the sensitivity of the results to the service distribution, particularly for the FIFO-BL scheme. The M/M/1 results are explainable by observing that crossing of the complimentary conditional waiting time distributions, for LIFO-PO and FIFO-BL, occurs further out on the tail, as ρ increases. For deterministic service times, the crossing of the maximum goodput curves does not occur due to the finite support of the FIFO-BL distribution and the selection of the maximizing buffer size.

The numerical results have demonstrated some potential advantages of the overload control queueing discipline studied. We note that in order to reduce this scheme to practice, and realize its significant potential, one needs to consider implementation questions related to the application. Examples of such issues are the effect of retrials, methods for rewarding "patient customers", the possibility that good and bad customers have different service times and the use of appropriate distributions for when customers turn bad. These issues are discussed in [3], for a given application, and similar approaches may be applicable to address some of these issues here. We note, however, that whether one is faced with effectively different service times for good and bad customers depends on the application. If this scheme is used to model a rate based overload control mechanism[6] where the "service time" corresponds to the reciprocal of the controlled rate over a time interval, say 10 seconds, then lower real time requirements for unsuccessful calls would serve to slightly lower the "service times" for *all customers, good and bad*, over the next 10 second time interval.

5. ACKNOWLEDGMENT

The author wishes to thank B. T. Doshi for his helpful discussions.

REFERENCES

[1] B. T. Doshi, E. H. Lipper, "Comparison of Service Disciplines with Delay Dependent Behavior," *Applied Probability - Computer Science The Interface Volume II*, ed. R. L. Disney, T. J. Ott, Birkhauser, 1982, pp. 269-301.

[2] J. Borchering, L. J. Forys, A. A. Fredericks, E. Hejny, "Coping with Overload," Bell Labs Record, July-August, 1981.

[3] L. J. Forys, "Performance Analysis of a New Overload Strategy," *Proceedings of the Tenth International Teletraffic Congress*, Montreal, Canada, June 1983.

[4] P. J. Kühn, "On a Combined Delay and Loss System with Different Queue Disciplines," *Transactions of the Sixth Prague Conference on Information Theory, Statistical Decision Functions, Random Processes*. Czechoslovak Academy of Sciences, Prague (1973), pp 501-528.

[5] B. T. Doshi, H. Heffes, "Overload Performance Comparisons for Several Processor Queueing Disciplines for the M/M/1 Queue," to appear.

[6] B. T. Doshi, H. Heffes, "Analysis of Overload Control Schemes for a Class of Distributed Switching Machines," *Proceedings of the Tenth International Teletraffic Congress*, Montreal, Canada, 1983.

[7] P. H. Brill, M. J. M. Posner, "Level Crossings in Point Processes Applied to Queues: Single Server Case," Operations Research, Vol. 25, No. 4, July-August, 1977, pp. 662-674.

[8] H. Kobayashi, "*Modeling and Analysis: An Introduction to System Performance Evaluation Methodology*, Addison-Wesley, Philippines, 1978.

[9] P. J. Kühn, *Tables on Delay Systems*, Institute of Switching and Data Technics, University of Stuttgart, 1976.

[10] B. T. Doshi, "An M/G/1 Queue With a Hybrid Discipline," BSTJ, Vol. 62, No. 5, (May-June, 1983), pp. 1251-1271

[11] S. Karlin, H. Taylor, *A First Course in Stochastic Processes*, Academic Press, New York, 1975.

[12] D. L. Jagerman, "An Inversion Technique for the Laplace Transform with Application to Approximation," BSTJ, Vol. 57, No. 3, March, 1978, pp. 669-710.

[13] B. Wallström, "On a Simple Overload Control for SPC Computers," *Proceedings of the Third International Seminar on Teletraffic Theory*, Moscow, USSR, June 1984, pp. 455-463.

Appendix

We evaluate the Case I goodput results for the M/D/1 FIFO-BL scheme. For the probability of getting served, the throughput and the waiting time for customers who get served the FIFO-BL scheme, with buffer size $N-1$, is equivalent to the FIFO-TO scheme with timeout interval $T_0 = (N-1)T$. This in turn is equivalent to the following scheme:

Let x_t denote the work in the system (processor and buffer) at time t. If an arrival occurs at time t and $x_t < T_0$, then it joins the buffer. It will be served and its waiting time will be x_t. If, on the other hand, an arrival at time t sees $x_t > T_0$ then it will leave.

Let J denote the distribution of work in the system. Then

$$P_s = P\{x < T_0\},$$

$$F_s(o) = \frac{J(o)}{P_s},$$

and

$$f_s(t) = \frac{J'(t)}{P_s} = \frac{j(t)}{P_s}, \quad 0 < t < T_0.$$

We note that P_s and $F_s(o)$ (and therefore $J(o)$) are identical with the corresponding quantities for the LIFO-PO scheme and are thus known. It suffices to obtain $j(t)$. By level crossing arguments [7] we obtain

$$j(t) = \lambda \int_{0^+}^{\min(t,T_0)} j(y)\,\overline{G}(t-y)dy + \lambda\, J(0)\,\overline{G}(t) \quad \text{(A-1a)}$$

with normalization condition

$$J(0) + \int_{0+}^{T_0+T} j(y)dy = 1, \quad \text{(A-1b)}$$

where $\overline{G}$ is the complementary service time distribution,

$$\overline{G}(x) = \begin{cases} 1 & x < T \\ 0 & x > T. \end{cases}$$

In general one can solve the integral equation (A-1) by converting it to a first order differential equation which can then be solved over successive intervals of length T using the solution from the previous interval. Our concern here is to evaluate the Case I goodput, given by

$$V_I = \lambda\, P_s \int_{0^-}^{T_0} e^{-\alpha t}\, dF_s(t).$$

Using arguments similar to those in [10] we define

$$h(x) = \lambda\, \overline{G}(x)$$

and let $m(x)$ be the renewal density for $h(x)$. Then $m(x)$ satisfies [11]

$$m(x) = h(x) + \int_0^x h(x-y)\, m(y)dy. \quad \text{(A-2)}$$

Equation (A-1) can now be written as, for $0 < t < T_0$,

$$j(t) = J(o)\, h(t) + \int_{0^+}^{t} j(y)\, h(t-y)\, dy. \quad \text{(A-3)}$$

This is a renewal equation whose solution is

$$j(t) = J(o)\, h(t) + J(o) \int_0^t h(t-x)\, m(x)dx, \quad 0 < t < T_0$$

which simplifies, upon using (A-2), to

$$j(t) = J(o)\, m(t), \quad o < t < T_0. \quad \text{(A-4)}$$

We note that

$$\int_0^\infty h(x)dx = \lambda\, T = \rho$$

and the function $m(t)$ is well defined for any finite x irrespective of the value of $o < \rho < \infty$. Using (A-4) in the goodput expression gives

$$V_I = \lambda\, J(0) + \lambda\, J(0) \int_0^{T_0} e^{-\alpha t}\, m(t)dt. \quad \text{(A-5)}$$

Since $J(o)$ is known we only need to evaluate the integral in (A-5). To evaluate this we define

$$y(t) = \int_0^t e^{-\alpha x}\, m(x)dx$$

with Laplace transform

$$\hat{y}(s) = \ell\,[y(t)] = \frac{1}{s}\,\hat{m}\,(s+\alpha) \quad \text{(A-6)}$$

where $\hat{m}(s)$, the Laplace transform of $m(t)$, is given by

$$\hat{m}(s) = \frac{\hat{h}(s)}{1-\hat{h}(s)} = \frac{\lambda(1-e^{-sT})}{s-\lambda+\lambda e^{-sT}}.$$

We evaluate the integral by numerically inverting the transform [12] in (A-6) at the point $t = T_0$.

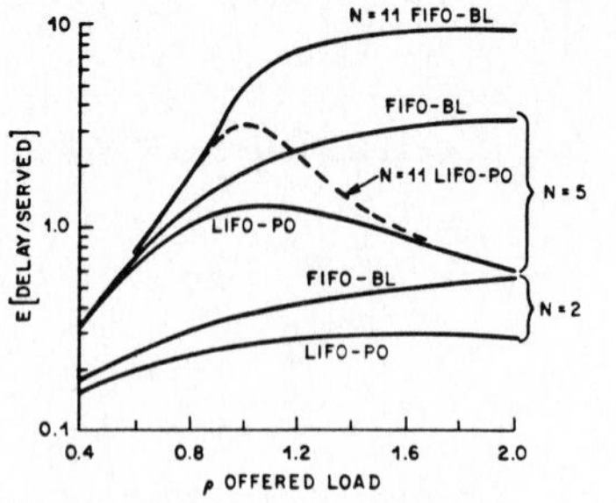

FIGURE 1 MEAN DELAY CHARACTERISTICS (Q DISCIPLINE COMPARISONS)

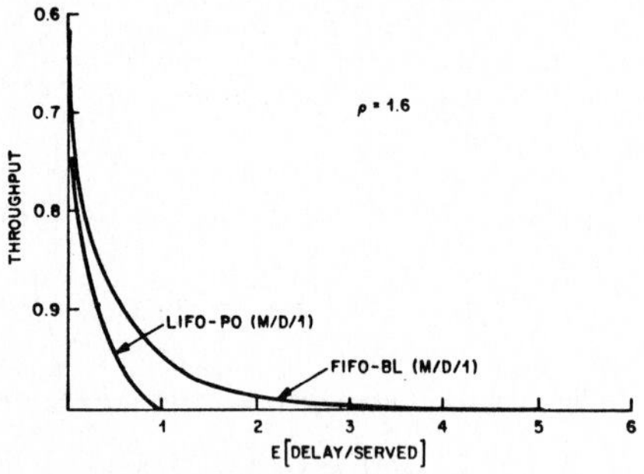

FIGURE 2 THROUGHPUT-MEAN DELAY TRADEOFFS (ρ = 1.6) (Q DISCIPLINE COMPARISONS)

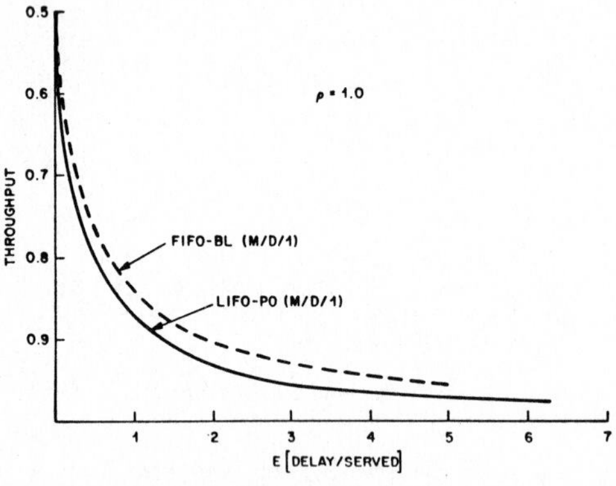

FIGURE 3 THROUGHPUT-MEAN DELAY TRADEOFFS (ρ = 1.0) (Q DISCIPLINE COMPARISONS)

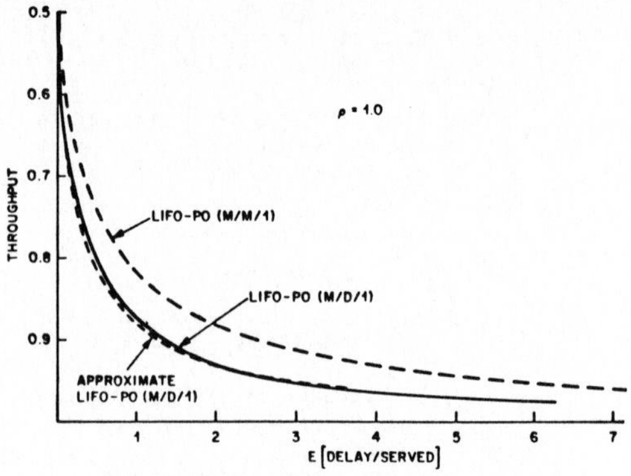

FIGURE 4 THROUGHPUT-MEAN DELAY TRADEOFFS (ρ = 1.0) (SERVICE DISTRIBUTION COMPARISONS)

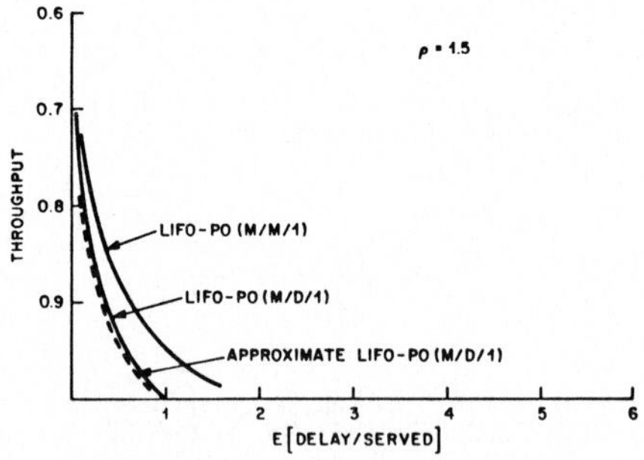

FIGURE 5 THROUGHPUT - MEAN DELAY TRADEOFFS (ρ = 1.5) (SERVICE DISTRIBUTION COMPARISONS)

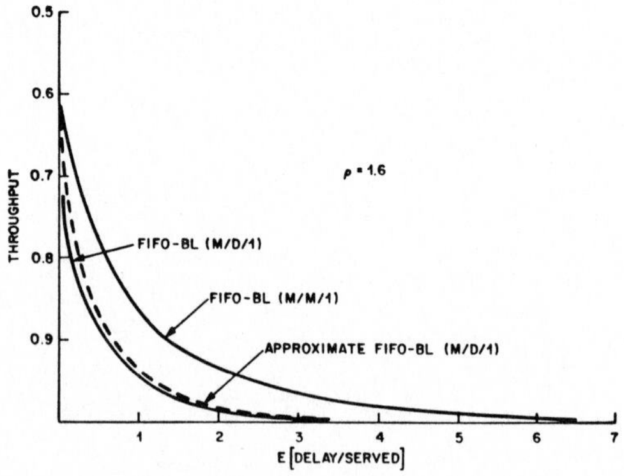

FIGURE 6 THROUGHPUT-MEAN DELAY TRADEOFFS (ρ = 1.6) (SERVICE DISTRIBUTION COMPARISONS)

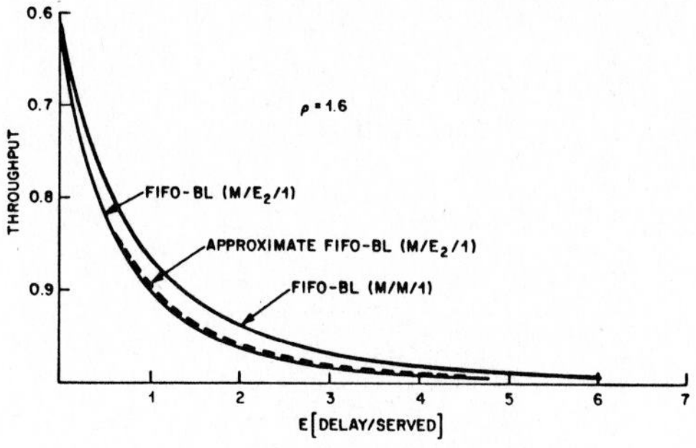

FIGURE 7 THROUGHPUT-MEAN DELAY TRADEOFF ($\rho = 1.6$) (SERVICE DISTRIBUTION COMPARISONS)

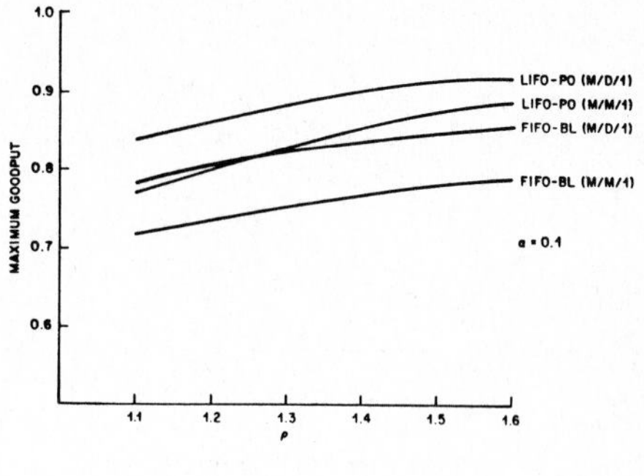

FIGURE 10 MAXIMUM GOODPUT COMPARISONS - CASE I

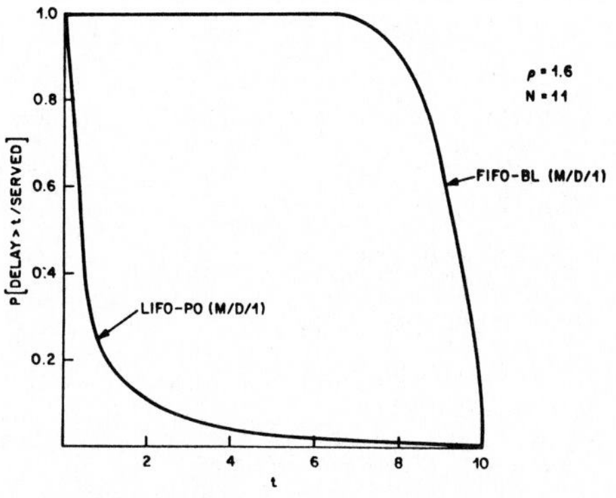

FIGURE 8 DELAY DISTRIBUTION COMPARISONS ($\rho = 1.6$)

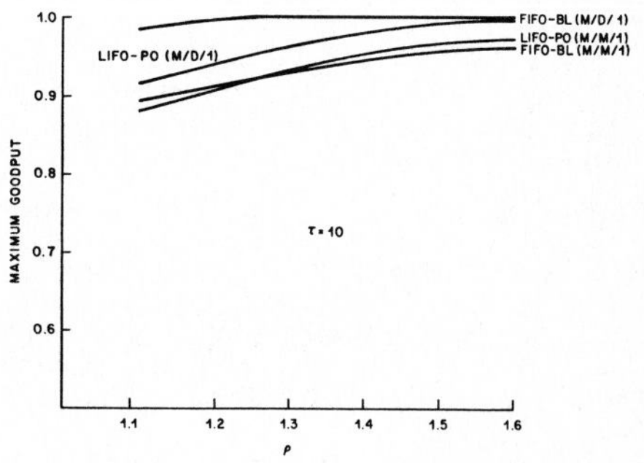

FIGURE 11 MAXIMUM GOODPUT COMPARISONS - CASE II

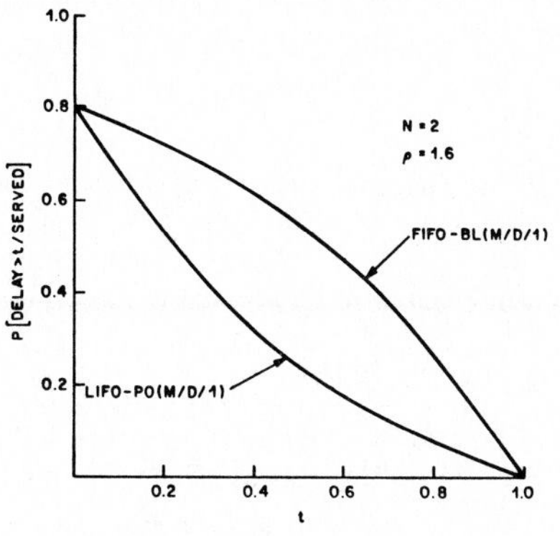

FIGURE 9 DELAY DISTRIBUTION COMPARISONS ($\rho = 1.6$)

TELETRAFFIC ISSUES in an Advanced Information Society
ITC-11
Minoru Akiyama (Editor)
Elsevier Science Publishers B.V. (North-Holland)

ON APPLICATION AND PERFORMANCE OF CUT-OFF PRIORITY QUEUES IN SWITCHING SYSTEMS WITH OVERLOAD PROTECTION

Hans - Dieter IDE and Rudolf G. SCHEHRER

Institute for Electronic Systems and Switching,
University of Dortmund, Dortmund, FRG

ABSTRACT

In this paper two types of delay-loss systems with cut-off priorities for overload protection are investigated.

The first part of the paper deals with systems in which waiting calls may be abandoned due to the impatience of customers.

In the second part of the paper, a new model of dynamic cut-off priorities is presented which enables an improvement of system performance.

In both parts, exact methods for the calculation of characteristic traffic values and waiting time distributions are presented. Results are given in several diagrams.

1. INTRODUCTION

This paper deals with delay-loss systems with cut-off priorities. In the systems considered there are n servers, s waiting places and two types of offered traffic A_1 and A_2. For calls of A_1 the systems act as loss systems with full access to the n trunks, and for calls of A_2 as delay-loss systems with FIFO discipline. Calls of A_2 can, however, only be switched if less than m ($m<n$) servers are busy (m can be interpreted as a restriction parameter).

It is well known that systems of this type can be applied in switching systems for protecting the traffic A_1 against overload of the traffic A_2 in paticular local overload situations [1,5,6,7, 8,9,10] as well as for other purposes [2].

The traffic A_1 and the traffic A_2 are supposed to have the same mean holding time h. The holding times and the interarrival times are assumed to be negative exponentially distributed with the terminating rate and the arrival rates λ_1 and λ_2. The total arrival rate be denoted by λ and the total offered traffic by A. This leads to the equations

$$A_1 = \lambda_1 h, \quad (1)$$

$$A_2 = \lambda_2 h, \quad (2)$$

$$A = A_1 + A_2, \quad (3)$$

$$\lambda = \lambda_1 + \lambda_2, \quad (4)$$

$$\varepsilon = 1/h. \quad (5)$$

A state in which x servers and z waiting places are occupied is referred to as a state (x,z), with the probability p(x,z). Waiting places can only be occupied if at least m servers are busy, i. e., states with $x<m$ and $z>0$ do not exist.

Section 2 deals with a system in which calls may be abandoned due to the impatience of customers. Then, in section 3, a new model with dynamic cut-off priorities is presented and investegated. A conclusion follows in section 4.

2. SYSTEMS WITH IMPATIENT CUSTOMERS

2.1 System configuration

In delay - loss systems which are, e. g., used in telephone switching systems, waiting calls may be abandoned after a time T_{ab} due to the impatience of customers. Therefore, in this section a system with cut - off priorities and impatient customers is investigated. It is assumed that the times T_{ab} are distributed negative exponentially with the mean h'. Thus, h' can be interpreted as the mean waiting time of calls before being abandoned (provided that the call is not served before). The quotient h/h' be denoted by η:

$$\frac{h}{h'} = \eta \quad (6)$$

For the terminating rate ε' with respect to a waiting call due to abandonment holds

$$\varepsilon' = \frac{1}{h'} \quad (7)$$

2.2 Equations of state

For the systems discribed, the following equations of state hold true in statistical equilibrium

$$A\, p(x,0) = (x+1)\, p(x+1,0), \quad 0 \le x < m, \quad (8)$$

$$(d_1A_1 + d_2A_2 + x + \eta z)\, p(x,z)$$
$$= d_1(x+1)\, p(x+1,z) + d_3A_1\, p(x-1,z)$$
$$+ d_4A\, p(m-1,0) + d_5A_2\, p(x,z-1)$$
$$+ d_2\eta\,(z+1)\, p(x,z+1) + d_6m\, p(m,z+1), \quad (9)$$
$$m \le x \le n,\ 0 \le z \le s,$$

where

$d_1=1$, if $x \neq n$, $d_1=0$ otherwise, (10)

$d_2=1$, if $z \neq s$, $d_2=0$ otherwise, (11)

$d_3=1$, if $x \neq m$, $d_3=0$ otherwise, (12)

$d_4=1$, if $x=m$ and $z=0$, $d_4=0$ otherwise, (13)

$d_5=1$, if $z \neq 0$, $d_5=0$ otherwise, (14)

$d_6=1$, if $x=m$ and $z \neq s$, $d_6=0$ otherwise, (15)

with the normalizing condition

$$\sum_{x=0}^{m-1} p(x,0) + \sum_{x=m}^{n} \sum_{z=0}^{s} p(x,z) = 1. \qquad (16)$$

These equations can be solved with the aid of iterative methods, e. g., the successive over-relaxation (SOR) method [4].

2.3. Characteristic traffic values

From the state probabilities calculated as shown in section 2.2, the following characteristic traffic values can be determined.

The total traffic Y carried by the systems is

$$Y = \sum_{x=0}^{m-1} xp(x,0) + \sum_{x=m}^{n} \sum_{z=0}^{s} xp(x,z) . \qquad (17)$$

For the calls of the traffic A_1, the loss probability B_1 and the carried traffic Y_1 are determined as follows

$$B_1 = \sum_{z=0}^{s} p(n,z), \qquad (18)$$

$$Y_1 = A_1(1-B_1). \qquad (19)$$

For calls of A_2 the waiting probability p_W, the mean queue length Q, the loss probability B_2, the carried traffic Y_2 and the probability p_{ab} of abandonment the following formulae hold true

$$p_W = \sum_{x=m}^{n} \sum_{z=0}^{s-1} p(x,z), \qquad (20)$$

$$Q = \sum_{x=m}^{n} \sum_{z=1}^{s} z\, p(x,z), \qquad (21)$$

$$B_2 = \sum_{x=m}^{n} p(x,s), \qquad (22)$$

$$Y_2 = Y - Y_1 , \qquad (23)$$

$$p_{ab} = (A_2 - Y_2 - A_2 B_2)/A_2 . \qquad (24)$$

The loss probability B_2 refers to such calls of A_2 which are rejected because all waiting places are occupied, whereas the probability of abandonment p_{ab} refers to calls which at first start waiting and then are abandoned.

The mean waiting times t_W with respect to arbitrary calls and t_W^* with respect to waiting calls are

$$t_W = \frac{Q}{A_2} h , \qquad (25)$$

$$t_W^* = \frac{Q}{A_2\, p_W} h. \qquad (26)$$

Example No. 1

As an example, a system with n=30 servers, s=11 waiting places and the traffic values $A_1=A_2=11,35$ Erlangs is considered. In fig. 1 the loss probabilities B_1 and B_2, the probability of abandonment p_{ab}, the waiting probability p_W and the waiting times t_W^*/h (for waiting calls) are shown by solid lines as a function of the parameter η . (This example has been chosen such that the values $B_1 \approx 0.5\%$ and $B_2 \approx 1\%$ are obtained approximately for $\eta = 0$, i. e. if calls are not abandoned). Furthermore, these values are shown by dashed lines for the case of overload of A_2 ($A_1=11.35$ Erlangs, $A_2=2A_1=22.7$Erlangs) in fig. 1.

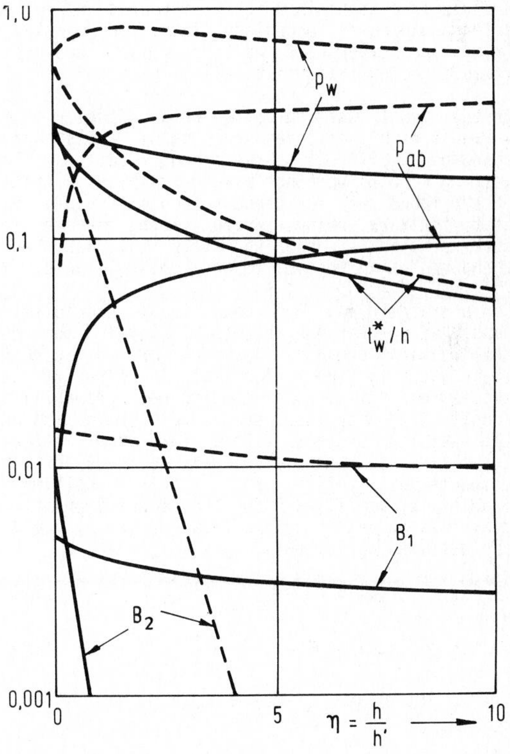

Fig.1: Characteristic traffic values as a function of
——— normal load ($A_2=A_1$)
- - - - - overload ($A_2=2A_1$)

In case of overload and $\eta = 0$ (corresponding to a system in which calls are not abandoned) the loss probability B_2 of the overloaded traffic A_2 increases by a factor of about 32, whereas the loss probability B_1 (corresponding to the traffic

A_1 which is not concerned by overload) is increased only by a factor of about 3. Thus, the traffic A_1 can be protected against overload of the traffic A_2.

For increasing values of $\eta = h/h'$, fig. 1 shows a slight decrease of the loss probabilities B_1 and a remarkable decrease of the loss probability B_2. On the other hand, the probability of abandonment increases.

The decrease of the curves representing the mean waiting times t_w^* and the solid curve for the waiting probability p_w are obvious. The maximum of the upper curve for the loss probability p_w can be explained as follows: for smaller values of η, p_w decreases because more calls are lost, and for larger values of η because of the increasing abandonment of calls.

2.4 Waiting time distributions

2.4.1 Waiting time distribution for arbitrary calls

For the calculation of the waiting time distribution, a test call is considered which arrives in a state in which $x \geq m$ trunks and $u < s$ waiting places are occupied. Thus, in a state (x,u), the number u of waiting calls does not include the arriving test call.

In the sequel, waiting calls which finally are served are refered to as successful calls, and abandoned calls as unsuccessful calls. The waiting process terminates when it enters a state of the taboo set. The taboo set consists of the state "call is being served" with respect to successful calls and of the state "call has been abandoned" with respect to unsuccessful calls.

For a state (x,u), the conditional probability density $\varepsilon_{H,s}$ for the successful termination of the waiting process (i.e., for a direct transition to the taboo set in case of a successful call), the conditional probability density $\varepsilon_{H,u}$ for the unsuccessful termination of the waiting process (i.e., for the direct transition to the taboo set in case of an unsuccessful call) and the conditional probability density ε_H for the direct transition to the taboo set in case of an arbitrary call, are obtained as follows

$$\varepsilon_{H,s}(x,u) = d_7\, m\varepsilon\,, \tag{27}$$

$$\varepsilon_{H,u}(x,u) = \varepsilon'\,, \tag{28}$$

$$\varepsilon_H(x,u) = \varepsilon_{H,s}(x,u) + \varepsilon_{H,u}(x,u) \tag{29}$$

with

$$d_7 = 1 \text{ if } x=m \text{ and } u=0,\; d_7=0 \text{ otherwise.} \tag{30}$$

If the system is in a state (x,u), transitions avoiding the taboo set are only possible to the states $(x+1,u)$, $(x-1,u)$ and $(x,u-1)$. For the conditional probability densities $r_{x+1}(x,u)$, $r_{x-1}(x,u)$ and $r_{u-1}(x,u)$, respectively, which correspond to these transitions, the following equations hold true

$$r_{x+1}(x,u) = d_1\lambda_1\,, \tag{31}$$

$$r_{x-1}(x,u) = d_3 x\varepsilon\,, \tag{32}$$

$$r_{u-1}(x,u) = u\varepsilon' + d_8 m \tag{33}$$

with

$$d_8 = 1 \text{ if } x=m \text{ and } u>0,\; d_8=0 \text{ otherwise.} \tag{34}$$

The constants d_1 and d_3 are given by the equations (10) and (12).

The conditional probability density $r(x,u)$ that a state (x,u) is left (by a transition to any other state which may or may not belong to the taboo set) is

$$r(x,u) = r_{r+1}(x,u) + r_{r-1}(x,u) + r_{u-1}(x,u) + \varepsilon_H(x,u). \tag{35}$$

The probability that an arbitrary call (which may be successful or unsuccessful) waits longer than a time t under the condition, that this call arrives in a state (x,u) be denoted by $w(t|x,u)$. For this conditional (complementary) distribution function $w(t|x,u)$, the following set of differential equations holds true which is well known as the so-called Kolmogorov backward equation

$$\frac{d}{dt}w(t|x,u) = r(x,u)w(t|x,u) + r_{x+1}(x,u)w(t|x+1,u) + r_{x-1}(x,u)w(t|x-1,u) + r_{u-1}(x,u)w(t|x,u-1), \tag{36}$$
$$m \leq x \leq n,\quad 0 \leq u \leq s-1.$$

The definition of a test call, implying that this call starts waiting, leads to the initial conditions

$$w(0|x,u) = 1\,. \tag{37}$$

With these initial conditions the set of differential equations (36) can be solved, e.g., with the aid of the Runge-Kutta method [5].

Now the absolute (complementary) distribution function $w^*(t)$ of the waiting time of waiting calls is obtained as

$$w^*(t) = \frac{1}{p_w}\sum_{x=m}^{n}\sum_{u=0}^{s-1} p(x,u)w(t|x,u)\,. \tag{38}$$

2.4.2 Waiting time distribution of successful calls and unsuccessful calls.

The conditional probability that a call is successful and waits longer than a time t be denoted by $w_s(t|x,u)$. These conditional (complementary) distribution functions fulfil the same differential equation (36) as the function $w(t|x,u)$. The initial conditions $w_s(0|x,u)$ for successful calls are, however, different from the initial conditions $w(0|x,u)$ for arbitrary calls.

It can be shown [7] that the following equation holds true

$$\lim_{t\to 0} \frac{d}{dt} w_s(t|x,u) = -\varepsilon_{H,s}(x,u) \tag{39}$$

which in connection with equation (36) for for t=0 leads to the following set of equations

$$-r(x,u)w_s(0|x,u)+r_{x+1}(x,u)w(0|x+1,u)$$

$$+r_{x-1}(x,u)w_s(0|x-1,u)+r_{u-1}(x,u)w(0|x,u-1)$$

$$=-\varepsilon_{H,s}(x,u) \,, \tag{40}$$

$m \le x \le n, \quad 0 \le u \le s-1.$

From these equations the initial conditions $w_S(t|x,u)$ can, e. g., be calculated iteratively by means of the successive overrelaxtion (SOR) method which has shown to converge rather fast in this case. A recursive solution for the equations (40) has also been derived [7] which can not be presented here for lack of space.

With the initial conditions calculated in this way, the set of differential equations (36) for the functions $w_S(t|x,u)$ can be solved, e. g., by means of the Runge - Kutta method. This leads to the conditional waiting time distributions $w_S(t|x,u)$ of successful calls.

The conditional probability that a call is unsuccessful and waits longer than a time t be denoted by $w_u(t|x,u)$. The waiting time distributions $w_u(t|x,u)$ of unsuccessful calls can be calculated by means of the equation

$$w_u(t|x,u) = w\,(t|x,u) - w_s(t|x,u). \tag{41}$$

The absolute complementary distribution function $w_S^*(t)$ for successful calls (and $w_u^*(t)$ for unsuccessful calls, respectivly) can be calculated in analogy to equation (38).

Example No. 2

For a system as considered in example No. 1 with n=30, s=11, A_1=A_2=11.35 Erlangs and η =1, the waiting time distributions $w^*(t)$, $w_S^*(t)$ and $w_u^*(t)$ for waiting calls, for successful and unsuccessful calls, respectively, are shown by solid lines in fig. 2. Furthermore, these values are shown by dashed lines in fig. 2 for the case of overload of A_2 (A_1=11.35 Erlangs, A_2=2A_1=22.7 Erlangs).

3. CUT-OFF PRIORITIES WITH STATE DEPENDENT SERVER RESERVATION

In this section a new strategy for overload protection with cut-off priorities and dynamic (state dependent) server reservation is presented. This new strategy concernes mainly the service discipline of calls of the traffic A_2 which are denoted as 2-calls in the sequel.

In principle, the normal service discipline as described in section 1 enables an efficient overload control in systems as considered here. If, however, many waiting places are occupied and there are still several servers (n-m) idle it seems reasonable to admit further 2-calls to be served in order to achieve a better server utilization. Therefore, a modified service discipline as described in section 3.1 has been developed which enables a better server utilization and shorter waiting times.

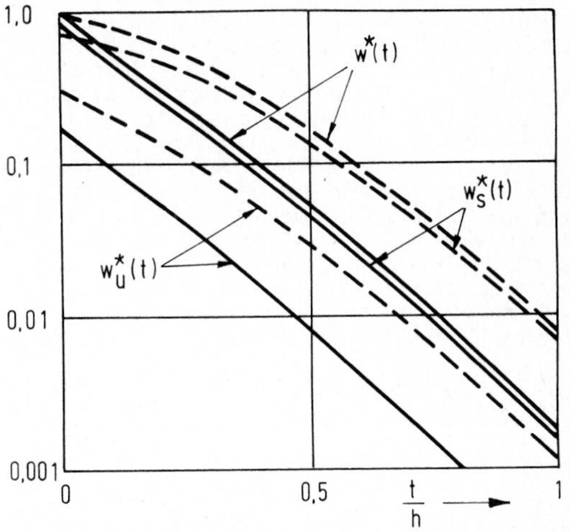

Fig.2: Waiting time distributions for waiting calls, successful calls and unsuccessful calls.
——— normal load (A_2=A_1)
- - - - overload (A_2=2A_1)

3.1 System description

In the systems considered in section 3, at most s waiting places are available for the queue of waiting 2-calls from which (if exactly m servers are busy) k waiting places ($1 \le k \le s$) are allowed to be occupied.

The waiting places are allocated dynamically, depending on the state of the system. If x servers ($x \ge m$) are busy a certain number c(x) of the total s waiting places can be occupied by waiting 2-calls. The function c(x) can be chosen e. g. according to the following formula in which a(x) represents an arbitrary integer function

$$c(x) = \min\,(\,k+a(x)\,,\,s\,), \tag{42}$$

with $x \ge m$, $a(x) \in N$.

If the system is in a state in which x servers are busy ($x \ge m$) and c(x) waiting places are occupied (c(x)< s), and a new 2-call arrives, then a certain number j(x) of the waiting 2-calls is served according to the FIFO discipline and the new 2-call starts waiting at the end of the queue. The function j(x) can be chosen (like a(x)) as an arbitrary integer function.

In this way the service of 2-calls is controlled by arriving new 2-calls, which "push waiting 2-calls into the system" to be served. This mechanism therefore is called PUSH-IN and the modified FIFO service discipline is called FIFO-PI in the sequel.

If the system is in a state in which x servers are busy ($x \ge m$) and more than c(x-1) waiting places are occupied (c(x-1)< s) and if in this state an arbitrary call ends then a waiting

2-call is served according to the FIFO service discipline.

If c(x) equals s and all s waiting places are occupied and a new 2-call arrives in this state, the call is lost.

For the practical application of cut-off priorities for overload control a synthesis algorithm for system dimensioning has been developed [8]. It can be shown that the cut-off parameter m should be chosen relatively near to the total number n of servers for an efficient overload control (see fig. 3).

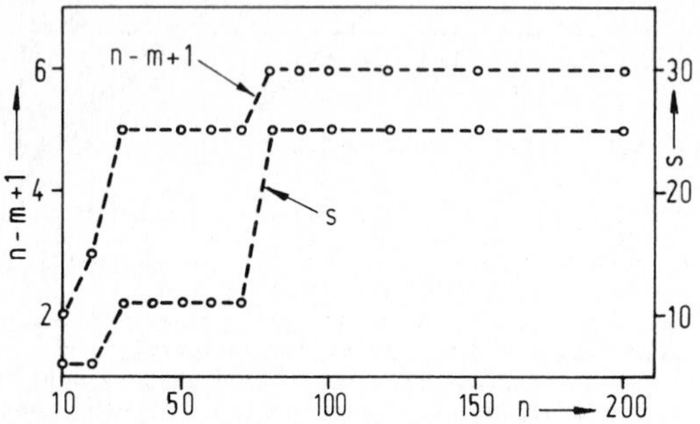

Fig.3: Number of exclusively reserved servers (n-m+1) and waiting places (s) as functions of the total number of servers (n) with $B_1 \approx 0.5\%$ and $B_2 \approx 1\%$

Thus, for the investegation of the general behavior of the described overload protection strategie, it seems reasonable to choose the following functions a(x) and j(x)

$$a(x) = x-m, \tag{43}$$

$$j(x) = 1. \tag{44}$$

3.2 Equations of state

For the states of the system described above the following differential equations hold true

$$\frac{d}{dt} p(x,0,t) = a_1(\lambda_1+\lambda_2)\; p(x-1,0,t) +(x+1)\varepsilon\; p(x+1,0,t)-(\lambda_1+\lambda_2+x\varepsilon)p(x,0,t), \tag{45}$$

$z = 0$ and $0 \le x \le m-1$,

$$\frac{d}{dt} p(m,z,t) = a_2\lambda_2\; p(m,z-1,t) +(m+1)\varepsilon\; p(m+1,z,t) + m\varepsilon\; p(m,z+1,t) +a_3\lambda\, p(m-1,z,t)-(\lambda_1+\lambda_2+m\varepsilon)p(m,z,t), \tag{46}$$

$x = m$ and $0 \le z \le k-1$,

$$\frac{d}{dt} p(x,z,t) = \lambda_2\; p(x,z-1,t) +(x+1)\varepsilon p(x+1,z,t)-(\lambda_1+a_4\lambda_2+x\varepsilon)p(x,z,t), \tag{47}$$

$m \le x \le m+s-k$ and $z = x-m+k$,

$$\frac{d}{dt} p(x,z,t) = a_2\lambda_2\; p(x,z-1,t) +a_5(x+1)\varepsilon\, p(x+1,z,t) +\lambda_1\; p(x-1,z,t) -(a_5\lambda_1+a_4\lambda_2+x\varepsilon)p(x,z,t), \tag{48}$$

$m+1 \le x \le n$ and $0 \le z \le x-m+k-1 \le s$,

where
$$a_1 = 0 \text{ if } x = 0,\; a_1 = 1 \text{ otherwise}, \tag{49}$$
$$a_2 = 0 \text{ if } z = 0,\; a_2 = 1 \text{ otherwise}, \tag{50}$$
$$a_3 = 1 - a_2, \tag{51}$$
$$a_4 = 0 \text{ if } z = s,\; a_4 = 1 \text{ otherwise}, \tag{52}$$
$$a_5 = 0 \text{ if } x = n,\; a_5 = 1 \text{ otherwise}, \tag{53}$$

with the normalizing condition

$$\sum_{x=0}^{n} p(x,0,t) + \sum_{x=m}^{m+s-k}\; \sum_{z=1}^{k+x-m} p(x,z,t) + \sum_{x=m+s-k+1}^{n}\; \sum_{z=1}^{s} p(x,z,t)=1. \tag{54}$$

3.2.1 Solution of the state equations under steady state conditions

In this case the state probabilities are not dependent from the time t and the differential quotient becomes zero. These equations can be solved iteratively, e. g., according to the successive overrelaxation (SOR) method [4].

3.2.2 Recursive solution

In addition to the iterative method, a recursive solution for the state probabilities has also been derived [8,9] which can, however, not be described in this paper for lack of space.

This recursive solution seems to be well suited for practical application. It is up to 100 times faster than the iterative solution and has shown to be fairly insensitive against rounding errors in all examples calculated up to now.

3.2.3 Characteristic traffic values

The loss probability B_1 of the offered traffic A_1 and the loss probability B_2 of the offered traffic A_2, respectively, can be calculated according to the following formulae

$$B_1 = \sum_{z=0}^{s} p(n,z), \tag{55}$$

$$B_2 = \sum_{x=m+s-k}^{n} p(x,s). \tag{56}$$

The total traffic carried by the system is

$$Y = A - A_1 B_1 - A_2 B_2. \tag{57}$$

Furthermore, the waiting probability p_W and the mean queue length Q can be calculated as follows

$$p_W = \sum_{z=0}^{k} \sum_{x=m}^{n} p(x,z) + \sum_{z=k+1}^{s-1} \sum_{x=m+z-k}^{n} p(x,z), \qquad (58)$$

$$Q = \sum_{z=0}^{k} \sum_{x=m}^{n} z p(x,z) + \sum_{z=k+1}^{s-1} \sum_{x=m+z-k}^{n} z\, p(x,z). \qquad (59)$$

The mean waiting time t_W^* of waiting 2-calls and the mean waiting time t_W of arbitrary 2-calls can be calculated according to the equations (25) and (26).

Example No. 3

In this example the same system as in example No. 1 and No. 2 is considered. This system has loss probabilities $B_1 \approx 0.5\%$ and $B_2 \approx 1\%$ if a static cut-off priority (scop) (k=11) is applied. In the following diagramm the total server utilization Y/n and the mean waiting time t_W^* of waiting calls are shown as functions of the overload factor $o = A_2/A_{2n}$ with $A_1 = A_{2n} = 11.35$ Erlangs for a system with a static cut-off priority (scop) (k=11) and systems with state dependent server reservation (sdsr) (k=8,9,10).

As the diagram shows, the server utilization is better and the mean waiting time of waiting calls is lower for systems with state dependent server reservation.

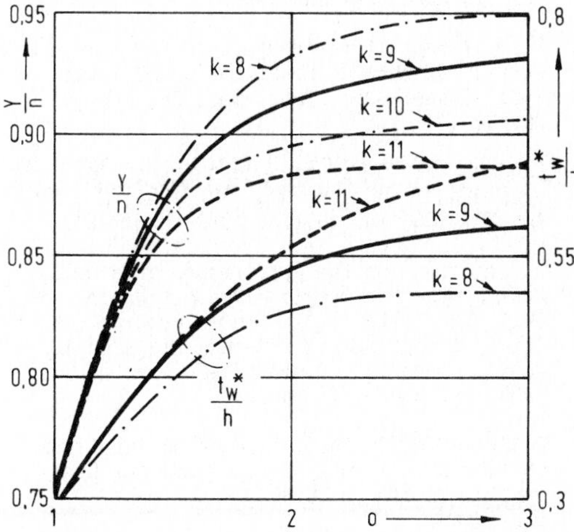

Fig.4: Server utilization and mean waiting time as functions of the overload factor o

3.2.5 Solution of the differential equations for the state probabilities

In this section the solution of the differential equations for the state probabilities is presented. With this solution the transient behavior of the system and its reaction on overload pulses and jumps can be investegated. These differential equations (eq. (45) - eq. (53)) have been solved numerically with regard to eq. (54) by means of the Runge-Kutta method of 4th order [3].

In fig. 5 the server utilization for systems as considered in example No.3 is shown for an instationary calling rate λ_2 of the traffic A_2.

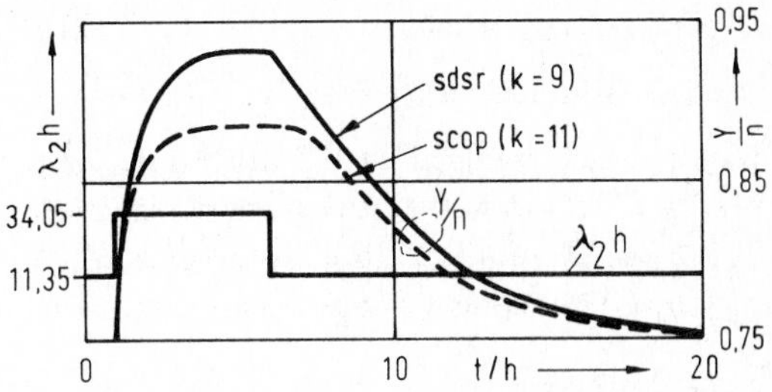

Fig.5: Server utilization Y/n in reaction on an overload pulse of the traffic A_2

3.3 Waiting time distribution

In this section the waiting time distribution of the described overload protection method with state dependent server reservation is calculated. Therefore a test call which arrives in a state (x,u) in which $x \geq m$ servers are busy and $u < s$ waiting places are occupied is considered.

Due to the push-in mechanism calls arriving after the considered test call can influence the waiting time of the test call. Therefore, considering the calls which arrive after the testcall, a third dimension v is introduced to describe the state-space for the random walk of the waiting process [8,9].

The taboo set consists of all states in which a call is served immediatelly, i. e. all states which can be reached from a state with u=0 in one single step.

The (complementary) waiting time distribution function for each state be denoted by $w(t|x,u,v)$. For these distribution functions the following set of differential equations holds true

$$\frac{d}{dt} w(t|x,u,v) = -(b_1\lambda_1 + b_2\lambda_2 + x\varepsilon) w(t|x,u,v) + x\varepsilon w(t|x-1,u,v) + b_1\lambda_1 w(t|x+1,u,v) + b_2\lambda_2 w(t|x,u,v+1), \qquad (60)$$

$$m+s-k+1 \leq x \leq n\,, \quad 0 \leq u \leq s-1\,, \quad 0 \leq v \leq s-u-1,$$

$$\frac{d}{dt} w(t|x,u,v) = -(\lambda_1 + \lambda_2 + x\varepsilon)\, W(t|x,u,v) + \lambda_1\, w(t|x+1,u,v) + \lambda_2\, w(t|x,u,v+1) + x\varepsilon\, w(t|x-1,u,v), \qquad (61)$$

$$m+1 \leq x \leq m+s-k,\; 0 \leq u \leq k+x-m-1,\; 0 \leq v \leq k+x-m-u-2,$$

$$\frac{d}{dt} w(t|m,u,v) = -(\lambda_1 + \lambda_2 + m\varepsilon)\, w(t|m,u,v) + \lambda_1\, w(t|m+1,u,v) + \lambda_2\, w(t|m,u,v+1) + b_3 m\varepsilon\, w(t|m,u-1,v), \qquad (62)$$

$$x = m, \quad 0 \leq u \leq k-1\,, \quad 0 \leq v \leq k-u-2,$$

$$\frac{d}{dt} w(t|x,u,v) = -(\lambda_1+b_4\lambda_2+x\varepsilon)w(t|x,u,v)$$
$$+\lambda_1 w(t|x+1,u,v) + b_3 b_4 \lambda_2 w(t|x+1,u-1,v+1)$$
$$+ b_3 x\varepsilon w(t|x,u-1,v), \qquad (63)$$

$$m \le x \le m+s-k, \quad 0 \le u \le k+x-m-1, \quad v = k+x-m-u-1,$$

with
$$b_1=0 \text{ if } x=n, \quad b_1=1 \text{ otherwise}, \qquad (64)$$
$$b_2=0 \text{ if } v=s-u-1, \quad b_2=1 \text{ otherwise}, \qquad (65)$$
$$b_3=0 \text{ if } u=0, \quad b_3=1 \text{ otherwise}, \qquad (66)$$
$$b_4=0 \text{ if } x=m+s-k, \quad b_4=1 \text{ otherwise}. \qquad (67)$$

In the considered system each waiting call is successful. Thus, the initial conditions $w(0|x,u,v)$ are

$$w(0|x,u,v) = 1. \qquad (68)$$

The absolute (complementary) waiting time distribution function of waiting calls be denoted by $w^*(t)$ and is obtained as

$$w^*(t) = \left[\sum_{x=m}^{m+s-k} \sum_{u=0}^{k+x-m-1} w(t|x,u,0)\, p(x,u) \right.$$
$$+ \sum_{x=m+s-k+1}^{n} \sum_{u=0}^{s-1} w(t|x,u,0)\, p(x,u)$$
$$\left. + \sum_{u=k-1}^{s-2} w(t|m+u-k+2,u,0)\, p(m+u-k+1,u+1) \right] / p_W . \qquad (69)$$

Example No.4

In this example the total waiting time distribution functions of waiting calls are shown for systems as considered in the example No. 3.

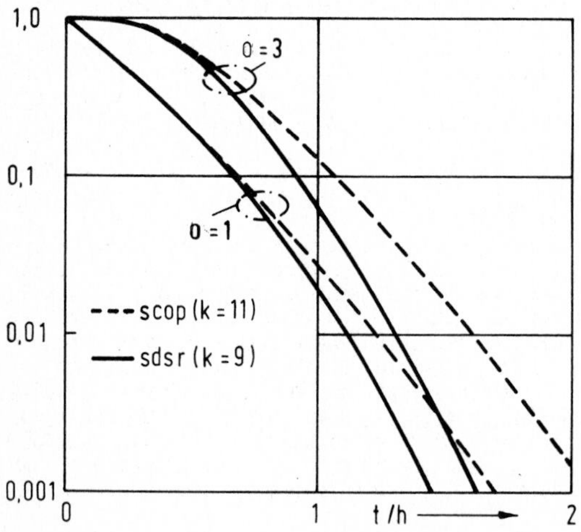

Fig.6: Waiting time distributions of a system with a static cut-off priority (scop) and a system with state dependent server reservation (sdsr) for o=1 and o=3

The diagramm shows that the waiting time decreases when applying state dependent server reservation.

4. CONCLUSION

In this paper two different systems with cut-off priorities are investegated which are suitable for overload protection of switching systems in particular local overload situations.

For the calculation of the state probabilities and characteristic traffic values, exact solutions are presented. Furthermore, the waiting time distributions are calculated with the aid of the Runge-Kutta method. Results are presented in diagrams.

5. REFERENCES

[1] K. Kawashima: Efficient Numerical Solutions for a Unified Trunk Reservation System with Two Classes, Rev. of the Electr. Com. Lab., Vol. 31, 1983, No.3, pp. 419-429.

[2] I. D. S. Taylor, J. G. C. Templeton: Waiting Time In a Multi-Server Cutoff-Priority Queue, and Its Application to an Urban Ambulance Service, Oper. Research, Vol. 38, 1980, pp. 1168-1188.

[3] A. Ralston, H. S. Wilf: Mathematische Methoden für Digitalrechner, Oldenbourg Verlag, München-Wien 1969.

[4] D. M. Young: Iterative Methods for Solving Partial Difference Equations of Elliptic Type, Trans. Amer. Math. Soc. 76, 1954, pp. 92-111.

[5] R. Schehrer: On a Delay-Loss System for Overload Protection, AEÜ, 38, 1984, No. 3, pp. 201-206, and 10.ITC, 1983, Montreal.

[6] R. Schehrer: On the Performance of a Delay-Loss System for Overload Protection in Switching Systems with Stored Program Control, International Seminar "Fundamentals of Teletraffic Theorie", Moscow, 1984, and AEÜ 39, 1985, No. 3 or 4 (in print).

[7] R. Schehrer: On the Influence of Impatient Customers in a Delay-Loss System for Overload Protection, AEÜ 39, 1985, to be published.

[8] H.-D. Ide: Performance Analysis of a Delay Loss System with Cut-Off Priority and State Dependent Server Reservation, AEÜ 39, 1985 to be published.

[9] H.-D. Ide: Über ein Warte-Verlust System mit dynamischer Zuteilung der Bedienungseinheiten zur Abwehr lokaler Überlastsituationen, Ph. D. Thesis, University of Dortmund, 1985.

[10] G. Brune: On Delay and Loss in a Switching System for Voice and Data with Internal Overflow, ITC 11, 1985, Kyoto.

TELETRAFFIC ISSUES in an Advanced Information Society
ITC-11
Minoru Akiyama (Editor)
Elsevier Science Publishers B.V. (North-Holland)

PERFORMANCE ANALYSIS OF CONGESTION AND FLOW CONTROL PROCEDURES FOR SIGNALLING NETWORKS

Patrick BROWN, Prosper CHEMOUIL, Brigitte DELOSME

Centre National d'Etudes des Télécommunications
Issy-les-Moulineaux, France

ABSTRACT

This paper deals with the problem of designing control procedures which are needed when introducing new services in signalling networks. Performance evaluation methods are required in order to analyze congestion control effects on the network, and to adapt flow control procedures to the network operating conditions. The exact solution of a congestion control procedure model enables us to develop a fast simulation method. Simulations are carried out on a representative testbed network with different traffic hypotheses. Analysis of results shows the adequacy of the congestion control procedure, and highlights how flow control is involved.

1. INTRODUCTION

The evolution of telecommunication networks towards an Integrated Services Digital Network (ISDN) for circuit and packet switched networks, implies the use of a signalling network able to connect the switching exchanges, and also data bases, Operation and Maintenance (O&M) centres, gateway nodes and other specialized nodes. Signalling networks must provide all the basic call or connection control procedures, but may also allow the transfer of information between different nodes, and even may be used to fully support some low-bandwidth consuming new services. Then a wide range of communications should share the same network, while different grades of service are required.
This paper deals with the problem of designing control procedures for such networks, which must cater with different service performance requirements, and where the traffic characteristics evolve with the addition of new functions and services.

Signalling networks were mainly designed to carry signalling and control information for telephone networks, and then to provide very high performances (short transfer delay, no loss of messages, high availability). Therefore the topology and the dimensioning of the network as well as a full set of failure recovery procedures (changeover, rerouting, ...) have been defined in order to ensure the security which meets the grade of service requirements in a large number of link or node failure cases [1]. This results in a hierarchical network with several possible routes for each origin-destination flow, and with a very low utilization rate of the links (digital 64 kbits/s channels) for basic signalling traffic. But as every failure and overload configuration cannot be taken into account when dimensioning the network, congestion control procedures are needed to alleviate possible overloads. On the other hand, preventive network access flow control mechanisms do not seem indispensable for the signalling traffic as it is naturally limited by the capacity of the telephone network to enter and carry calls.

As new traffic is added to the network, we must consider the statistical nature of the offered flows and the service requirements. For example, file transfer type applications, which utilize a great part of the transmission resources during a long time period, imply a great variation of the load on network links. Then the potential peak traffic is much higher than the traffic rate one can sensibly use when planning the network. In this case, flow and access control mechanisms are obviously needed to prevent few users from saturating the network. However, even if a very short packet transfer delay is not needed, flowing a file transfer within a fixed maximum delay may be required. Then traffic to be throttled should be selected according to the grade of service constraints and to the congestion state of the network.
Control procedures at different levels must cooperate in a way which ensures the required performance for each type of traffic, and achieves a good efficiency of the network.

In a previous paper [2], a congestion control procedure has been proposed and its consistency was shown.
In this paper we focus on the congestion information propagation towards the network access nodes. Different network access control mechanisms are then discussed according to the information available at these nodes. As the behaviour of control procedures is strongly dependent on the network topology and the offered traffic configurations, it is necessary to simulate these procedures on a representative network. Therefore we propose a fast simulation method which allows to deal with large networks.

2. COMMON CHANNEL SIGNALLING NETWORKS

A common channel signalling network consists in a transit network, composed of signalling transfer points (STP) fully interconnected, and in source and destination nodes, the signalling points (SP) which are connected to two STPs. The set of SPs connected to the same pair of STPs is called a cluster. See Fig.1.

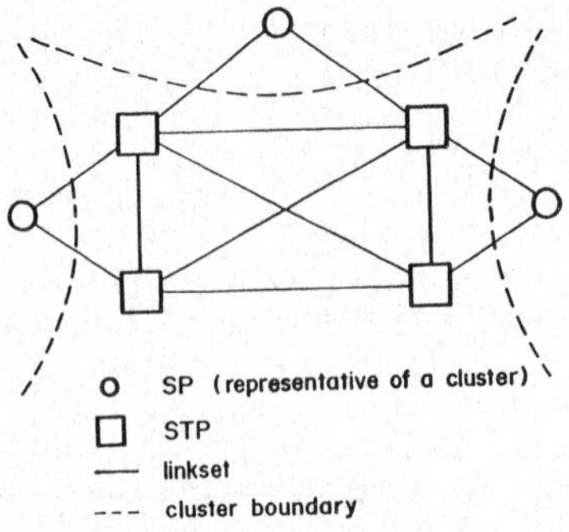

Figure 1 - Signalling Network Structure -

Signalling networks are structured in a Message-Transfer Part (MTP) which provides the network layer functions, and User-Parts (UP) which are defined for each application. The upper level functions may be layered according to the OSI model, but it is not always required. The Telephone User Part (TUP) needs not be layered as the MTP has been conceived for the signalling application. For other applications a Signalling Connection Control Part (SCCP) has been added to provide standard end-to-end services. This layer can be used for flow control, as it sets up the new transactions. See Fig.2.

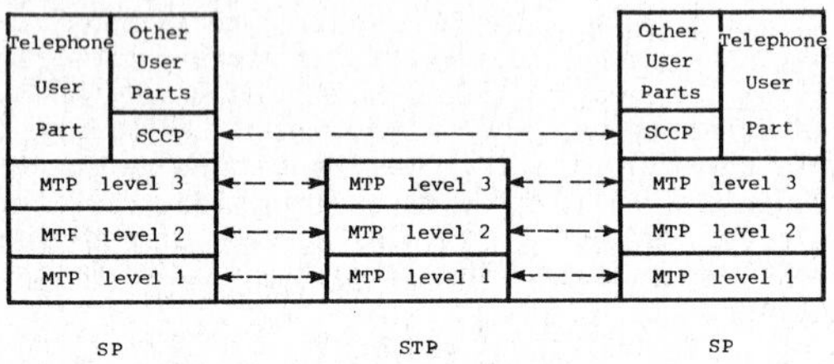

Figure 2 - Signalling Network Architecture -

Message routing is performed by the MTP according to the routing label included in each message header. This label contains the origin and destination nodes and a field called SLS, used to share the load on the links. The messages with the same destination and SLS are routed on the same path. The MTP should then ensure message sequencing with a high probability. The SLS is associated with the telephone circuit number in the basic signalling application. The SCCP provides a mean to assign at each call a connection number or a SLS, for every application, and handles the routing of virtual circuits. Moreover the SCCP manages several classes of service, ranging from the datagram service to more or less complete virtual circuit (VC) services (using the sequencing capability of the MTP). The basic VC's service is complemented with segmenting and reassembling capabilities, message numbering, missequencing control, window flow control and error recovery mechanisms.

3. CONGESTION CONTROL

Congestion control procedures are based on the observation of the load over the linksets and nodes of the network. In the following, congestion problems are restricted to link congestion which are presumed to be much more frequent. Control mechanisms then detect overload and end of overload conditions according to the length of outgoing queues. As signalling networks are designed to carry in the same manner all the messages, priorities do not exist inside the network, and thus selective reject of messages is not allowed. A statistic on the messages queueing before a congested linkset would allow to realize a fair control by the detection of traffic flows involved in the congestion. But each linkset may carry thousands of flows, implying from over ten to a thousand sources and it was thought difficult to manage.
Different control methods have then been investigated :

- The first ones are based on the regulation of flows routed over congested links. A method was proposed which utilizes the random selection and the analysis of one message among n. It consists then in sending a traffic restriction demand to the source node of each selected message, the receipt of this demand triggering an access restriction mechanism.
The main drawback of this method is to transfer the solution of a network congestion problem to the network access level, and then to lead to traffic over-restrictions.

- The second ones are based on the rerouting of flows over spare capacities available in the network. The sets of flows to be diverted (or eventually restricted) are predetermined according to traffic forecasts. The major drawback of this approach is its risks of unfair actions, as it acts on given flows whatever their part in congestion is.
It is a method of this type that we have developed, as it allows to solve efficiently a lot of congestion cases in the network, without making use of access control mechanisms.

The main features of the method that we propose are now briefly described; details may be found in [2].
When a congestion occurs the first actions aim at solving the problem locally. Diversions of selected traffic flows are attempted over alternative routes. When not possible, a second action consists in sending a control message to adjacent nodes in order to divert these traffic flows upstream. Last, origin nodes may initiate traffic restriction.
This procedure acts like the one used in case of failure, but only on some selected flows. Congested routes are decreed "restricted" for the chosen destinations and the traffic towards these destination points is rerouted over alternative routes, as it would be when a route is "prohibited" due to a failure. If the alternative route is not available, a control message called Transfer-Restricted (TR) message, similar to the Transfer-Prohibited (TP) message, is sent upstream but only to some selected nodes. This message indicates the signalling point which sends the message and the destination for which a

traffic diversion is asked. When recovering from congestion, or failure, Transfer-Allowed messages (TA) are sent to the alerted points. The selection of the points which are either the destination for which the routes are restricted, or the upstream nodes which receive a traffic diversion demand, allows to adapt to the congestion development the traffic to be diverted (or restricted at a source node if no route is available). The congestion control mechanism performs step by step, increasing or decreasing the set of flows to act on.

As the network is overdimensioned, traffic reroutings should increase the utilization rate of the network and, acting quickly, they decrease message loss. However they entail risks of message missequencing and thus may affect the grade of service of the diverted traffic flows. Hence reroutings must be scarce.

The critical point of such a procedure is the partitioning of flows which must allow a gradual action, so that the link load may fit in a chosen range of traffic.
The flows must be aggregated in a limited number of sets with even nominal traffics. Then the diversion of a set of flows which is not involved in the congestioning will not create congestions on alternate routes, and the number of action steps is limited so that any overloaded set will be reached quickly.

The mechanisms to detect congestion are designed to absorb random traffic peaks and to react fast enough to avoid message loss. They are based on threshold crossings and the observation of queue lengths after time periods.

It was shown in [2] that parameters could be set to obtain a desired traffic range where risks of actions are negligible.

4. NETWORK ACCESS CONTROL

Flow control may associate mechanisms at different layers, but common network access procedures must be designed to ensure the global functioning of the network and the grade of service required for each class of service offered to the user.

Access control procedures are twofold :
- they must make use of the receipt of congestion information to adapt the traffic to the network state,
- they should prevent network overload by controlling the offered flows.

Congestion Resorption Methods

When rerouting attempts of originating flows have failed, a restriction demand is received in the access node. The control message only carries information on the destination difficult to reach due to network congestion, and the source must undertake the access restriction without further knowledge. As each traffic is carried in the same way throughout the network, and as the grade of service requirements are different, traffic should be controlled at the source according to the application characteristics.

Therefore traffic is to be divided into several priority classes. Two main classes can be considered :

- the higher priority class includes signalling messages and all urgent message-oriented traffics as alarms and control messages.
- the other class contains all the other Operation and Maintenance, and data traffics.

Inside these two main classes of priorities, it may be envisaged to divide each class into subclasses. On one hand it could be needed to reduce the amount of signalling and urgent messages traffic. The most logical way to restrict this traffic is to limit the access of new calls, i.e. the Telephone User Part must reject Initial Address Messages (IAMs). On the other hand concerning less urgent O&M and data services, different priorities can be given according to the urgence of message delivery and the bandwidth of the application. Therefore the lowest priority will concern applications without short delay requirements, such as O&M file transfers, and some data transfers.
A class of priority will then be assigned to each application type of traffic. The priorities should also be associated with the classes of service requested by the applications; the class of service information is included in the first message sent to set up the path for a connection (connection request). Specific requirements may involve the upper levels of the application.

One problem of flow control is how to re-establish the traffic when congestion has vanished. Generally the traffic which was reduced needs to be transmitted on the network and will be offered anew at a different time. It must be avoided to produce again the same traffic matrix offered to the same network. If not, the same process of congestion - restriction - restoring - congestion would occur. In order to avoid these oscillations several choices can be made :

- One consists in waiting different delays (random or fixed per node) after the receipt of a Transfer-Allowed message before re-establishing traffic.

- A similar method may be based on the time delay between receipts of Transfer-Restricted and Transfer-Allowed messages (small delays between TR and TA receipts can suggest an oscillation cycle has started). The drawback is that a mechanism is needed at each origin signalling point, which activates timers and stores the last control message per destination.

- Another solution is that the SCCP handles the re-establishing of application flows. Then, new transactions set-up and restricted calls restoring may be associated.

Preventive Methods

Some causes of congestion may be eliminated by using different techniques of end-to-end flow control which limit the traffic offered to the network.

- The throughput between two terminating nodes may be regulated by the use of window mechanisms. It avoids saturating the receiver when negotiating the window size. It is also possible to limit temporarily the window size allowed for a source when a restriction request has been asked towards a destination node. This mechanism should be managed by the SCCP.

• The SCCP allows the consulting of network management data bases before establishing a virtual circuit. If the number of possible large bandwidth flows is small, a table containing each existing VC may be stored, at one central node or at several nodes (these tables being updated in a distributed manner). Then a VC connection is accepted if its routing is theoretically possible according to the previous routing of the existing VCs. The request is rejected if the calculated load of a link is greater than a fixed maximum value.

• A last mechanism would be to control the traffic between sets of nodes (defined in accordance with the network structure) using a credit mechanism. When a VC connection is requested, the SCCP consults the local data base of its set. A token towards the destination node's set is given by the database if there exists some left between the terminating sets. Tokens are given back at the end of connections. This method does not require network load evaluation, and as tokens are locally assigned, global information is not needed.

The last methods control virtual circuit connections according to an estimation of the load on network elements.
If there are no serious failure combinations in the network, it is expected that rerouting procedures should be efficient enough to recover from congestion.

Congestion information should be used by the flow control mechanisms. For instance, the receipt of a restriction demand may interact with the window mechanism provided by a class of service. The origin node will decrease its window size towards the concerned destination, and hence reduce its throughput. On the other hand the management of tokens towards the congested destination's set allows a fast traffic restriction after a TR receipt, and a gradual restoration upon receipt of a transfer-allowed message.

When designing the flow and access control procedures, the main elements which must be examined are the number of steps before a total restriction of traffic towards a destination, and the duration of unavailability for an application, depending on the choice of its priority and class of service. The first element characterizes the rapidity to quench the congestion. The second one allows to link the application requirements with the network grade of service specifications.

5. SIMULATION METHOD

As the routing may depend on the state of the network, analytical models using product form are not relevant. Because we are concerned with the behaviour of the congestion control on large networks, simulations on a message event basis are not realistic. The analysis method we have then used, is based upon simulations in which the events are the changes in the status of linksets or counters, and where the transitions times and probabilities have been calculated by solving exact queueing models [2], [3].
The occurence of a congestion detection on a link depends on the elementary variations of its queue length. The modelling of the link behaviour provides a mean to reduce this phenomenon to the quantities of interest: the distribution of detection occurence times, the mean message loss. A congestion detection will result in a change in route statuses. As the routing of traffic depends only on the route statuses, the simulation consists in modifying the loads offered to the links and in generating the overloads and underloads which result.

Mean and variance of detection times, and mean message loss on a link are obtained with a Poisson arrival and exponentially distributed emission times hypotheses. Load fluctuations may create an overload or an underload. To determine the nature of the next detection both delays are drawn. Because of the detection mechanism adjustment, a given load will fall in a range leading quickly to overload or to underload, or in a stable range where both eventualities are highly unprobable. Either both delays are very long or they differ greatly in magnitude. Then the choice of the event with the shortest delay seems reasonable.
Modification of route statuses can be achieved after a fixed or random delay, which may take into account congestion information transfer.

This simulation method could mask some phenomena which are due to instantaneous queue length values as the correlation of lengths of queues on adjacent links, or the transient behaviour on a queue when changing the offered load. Their consequences are limited because there are many flows on a link and routing is diversified, and because the congestion control procedure reacts to average traffic.

6. SIMULATION RESULTS AND ANALYSIS

Simulations of control procedures are carried out on a testbed network which is a forecast of the French North-East signalling network. This network is composed of 10 STPs and about 250 SPs which constitute 14 clusters (see Fig.3). This structure has been determined by the use of a planning software optimizing network availability [4]. The network is representative of most configurations which can be found when the constitution of clusters depends on traffic affinities.

The traffic offered to signalling networks is divided in two parts:

- a message-oriented traffic, where the messages are isolated or are not strongly time-correlated when they are related to the same connection. This traffic represents the signalling traffic (which is the major part of the overall traffic), and the traffic of some new services and O&M functions.
- a packet-oriented traffic, where the information is segmented in packets which are sent at

correlated instants. It consists in large bandwidth consuming flows of data for O&M use, as collect of statistics, remote loading and file transfers. But this traffic should be scarce.

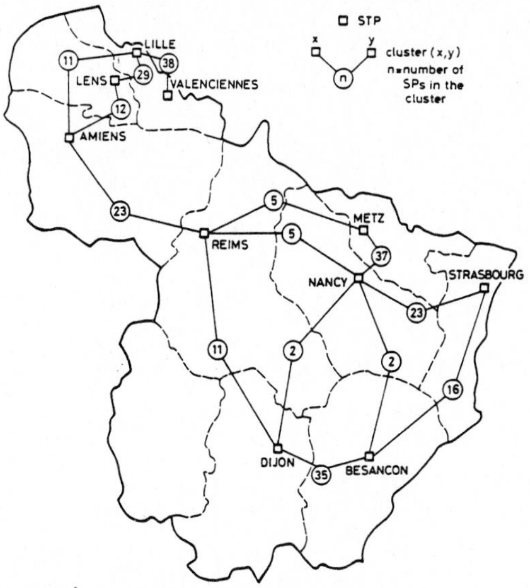

Figure 3 - French North-East Signalling Network Topology -

The network is assumed to be dimensioned for the basic signalling application, which is a realistic hypothesis for many years. Then, if only nominal traffic forecasts are considered, even in multiple failure configurations, the control procedures have a very low probability to act. Consequently very critical traffic assumptions are made in order to test the behaviour of the control mechanisms.

The traffic scenarios we consider are the following :

• the traffic flows are distributed as the signalling traffic is (according to a gravitary model). Each SP generates a load per link which is randomly drawn, according to a given mean load entering the network. The traffic may be ill-balanced on the links outgoing a SP. The average load ρ on a link outgoing a SP is uniformly increased in order to create congestions and defines the scenarios A.1-A.3, conducting to a general overload.

• Failures on network elements are added to the traffic conditions A.1. Some worst cases of complete linkset failures and multiple failures define the scenarios B.1-B.3.

• Some large bandwidth flows are added to the basic scenario A.1. They converge towards a given destination in scenario C. They are calculated in order to create inter-STPs congestion in scenarios D.

The network dimensioning uses a maximum load per link of 0.2, but results in a network with an average load ρ = 0.01 for signalling traffic.

The basic scenario is :
- A.1, where ρ =0.3 with even load sharing on the links.
The other scenarios differ as follows :
- A.2 : uneven load sharing (10%-90%);
- A.3 : ρ =0.4;
- B.1 : failures on the linksets Lille-Nancy and Dijon to a SP in the (Dijon,Nancy) cluster;
- B.2 : failures on the linksets Lille-Nancy, Lille-Dijon and Lille-Strasbourg;
- B.3 : failure of the node Reims;
In B.1-B.3 scenarios failures occur at the beginning of the simulation. In B.1 and B.2, repair is assumed after 2 hours, in B.3 after 4 hours.
- C : addition of 5 data flows of 0.15 E, towards a SP in cluster (Strasbourg,Besançon);
- D : addition of 20 data flows of 0.1E in average, from the North part to the East part of the network overloading 8 inter-STP linksets and the linksets towards a SP.

The main characteristics of the simulation are :
- duration = 4 hours;
- stable load range from ρ_{min}=0.45 to ρ_{max}=0.6;
- linkset buffer size = 32 messages;
- partitioning of flows according to the clusters, such as a set contains at most 10 SPs;
- first restriction of a flow concerns 50% of its traffic.

The general grade of service parameters, along with the results of the different scenario configurations are listed in Table 1.

The global efficiency stays very high even in the worst configurations. It justifies the use of rerouting procedures based on the network overdimensioning and secure topology.

• The rerouting actions are not very frequent (their number being measured during 4 hours), they are distributed and do not imply excessive route status modifications in a node. Apart the case of B.3, which is explained below, the maximum number of reroutings for a flow stays under 4. This proves that missequencing risks due to the congestion control procedure are very low.
Comparing the actions in A.2 with the ones in A.1, it can be seen that an uneven load sharing, which causes greater overloads on some linksets, is quickly solved by the rerouting procedure.
The values given in the configurations with failures must be compared with the ones which depend only on the failure management procedure in Table 2. The rerouting procedure copes with the B.1 and B.2 failures, although they are severe. Regarding the B.3 values, they are due to an oscillation cycle which happens on the Amiens-Dijon linkset, as the congestion controls acts on a set of flows which exceeds the load gap between ρ_{min} and ρ_{max}. Then, acting on SPs which share the failed STP Reims, restriction of these flows underloads the congested linkset, and permits a traffic restoration. This causes an oscillation detection, which spaces out the restriction and restoration actions, but must be solved better at the access control level. A relevant control should inhibit the traffic restoration as long as one STP of the SP's cluster is out of order.

To summarize, the rerouting procedures do not require heavy management as actions are spaced

Table 1- Global Performances of the Network -

QOS parameters		Configurations A.1	A.2	A.3	B.1	B.2	B.3	C	D	D'
efficiency (%)		100	100	99.84	100	100	98.10	99.86	99.74	99.76
restricted flows	mean efficiency (%)	-	-	60.17	-	-	56.41	75.83	62.50	59.03
	total number	0	0	736	0	0	4414	133	407	8
	max nb. per node	0	0	18	0	0	65	1	14	1
rerouting actions	% of destinations	25.70	24.50	67.07	44.58	61.85	91.57	26.10	52.61	56.27
	% of acting nodes	28.57	29.34	86.49	58.69	58.69	91.51	88.42	75.68	75.68
	mean nb. per node	11	11	32	42	78	122	5	30	28
	max nb. per node	50	21	81	159	238	507	50	188	178
	max nb. for a flow	2	1	3	3	4	47	3	4	4
number of messages lost		0	10.98	4.41	0.82	2.05	107.9	13.14	14.47	20.36
loss probability (per 10^9 mess.)		0.00	4.18	1.16	0.30	0.66	56.4	5.73	4.70	7.6
overhead (per 10^6 mess.)		0.24	0.28	3.13	3.81	5.26	23.75	0.49	2.13	2.18

(Results for B-configurations take into account the failure management procedure)

Table 2 - Rerouting Actions Due to Failures -

	B.1	B.2	B.3
% of destinations	26.51	42.97	91.57
% of acting nodes	0.77	30.50	20.46
mean nb. per node	66	52	172
max nb. per node	130	187	212
max nb. for a flow	1	1	1

and do not involve complex algorithms. The overhead due to control messages is very small as most of the actions are undertaken locally. The message loss is negligible, even with a maximum queue length of 32 (the standard value being 128), and moreover it respects the grade of service requirements of 10^{-7}.

Next, congestion evolution on the inter-STP network is displayed on Fig.4.
Two configurations were chosen, one with a strong generalized overload (A.3, see Fig.4-a), and the other with severe failures (B.2, see Fig.4-b).
Although we are considering extreme cases, the bigger proportion of the inter-STP linksets are in the least loaded part, and only a few linksets are highly loaded (ρ>0.6).
We can see that the number of overloaded linksets decreases along with time, and stabilizes (it means that ρ is very close to ρ_{max} for these linksets, and then the congestion detection mean time is about 6 hours, and would occur after the end of the simulation).
In B.2 configuration, the failures happen at the beginning of the simulation,and last two hours; at this time only 2 linksets exceed ρ_{max}. Just after the recovery, the overloads vanish. In A.3 configuration, only 4 linksets are loaded a little above 0.6.
Decreases of the number of highly loaded linksets cause simultaneous decreases of the least loaded portion of the network. Traffic on the least loaded linksets has increased, and subsequently, some of these linksets enter the medium loaded part. Overloads are then reduced and oscillations are avoided since congested linksets enter the medium loaded part. Their traffic is in the stable load range, and thus a diverted traffic will not be changed back.
The rerouting control procedure balances the traffic over the linksets and improves the network utilization.

• The restriction procedure which is simulated is the simplest mechanism making use of the congestion control message receipt. A TR receipt cuts off the lowest priority traffic which is restored upon a TA receipt. Restrictions act on 50% of the traffic representing the packet-oriented traffic.
This basic procedure allows a fast restriction since the efficiency is close to 50% for most of the restricted flows. A small proportion of flows are restricted, and the global efficiency is still high. We may conclude that the congestion control procedure selects flows involved in the congestion.
In scenario A.3, representing a general overload, the spreading of the congestion causes many reroutings, but congestion is quenched by only restricting the flows between two clusters.
In scenario B.3 reroutings and restrictions allow to carry about 60% of the traffic which was handled by the failed STP.
In scenario C, where overloads focus towards a given destination, restrictions concern only flows towards this destination, and less than 30% of reroutings are added to the basic scenario A.1.
In scenario D, several flows towards the overloaded destination, including the added data flows, are restricted. But, as the rest of the overload is distributed, restriction spreads over a large number of flows, representing the set of flows between two clusters.
The choice of a uniform restriction rate of 50% shows how the congestion control procedure behaves when a flow restriction depends only on its size, and not on class priorities. In fact, most of the traffic outgoing a SP is the basic, high-priority signalling traffic since only some SPs can receive or generate large bandwidth

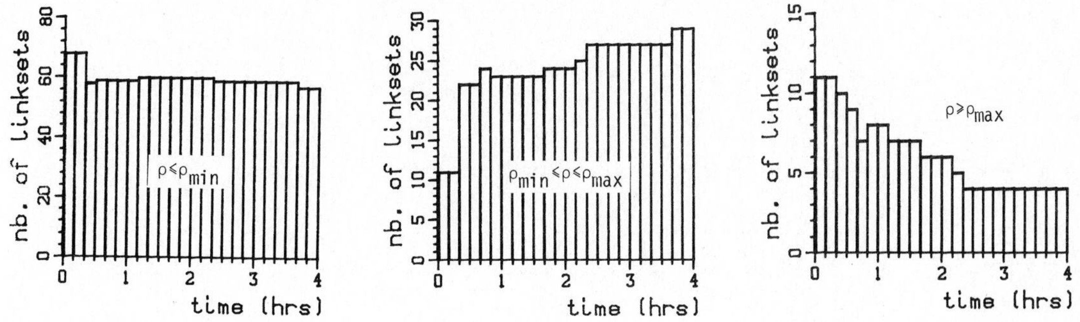

a. Scenario A.3

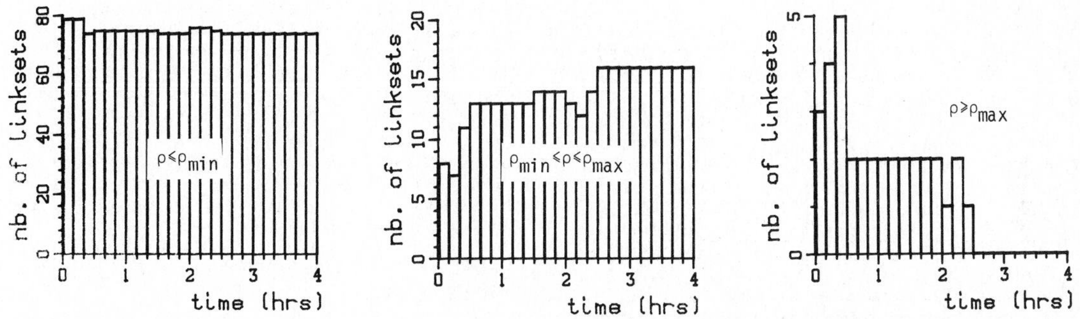

b. Scenario B.2

Figure 4 - Evolution of the Load Distribution on the Inter-STP Network.

traffic, and only few are assumed working at the same time. Therefore a variant D' of scenario D was considered, where only the additional data flows can be restricted, the restriction rate still being 50%. It can be seen that only the flows towards the overloaded SP are restricted and that it is sufficient to alleviate the congestion. But as more time is needed to reach efficiently the sources, more messages are lost. On the other hand, the number of reroutings is very similar and then congestion does not spread. It can then be expected that the use of class of priorities should decrease the number of restricted flows without major drawbacks. The assignment of a lowest priority to the bigger and unpredictable flows seems the best solution, they will be restricted first when a congestion occurs.
Partitioning the traffic in many classes of priorities may not be necessary as long as applications involving very high and peaked traffic, and requiring different grades of service are not introduced.

In conclusion, the congestion control procedure solves most of the congestion problems by using only its rerouting capabilities. It provides sufficient means to restrict efficiently the traffic when it is needed. However priority classes and preventive flow and access control procedures should ensure grade of service requirements of future applications. They depend on the administration decisions concerning the use of signalling networks.

7. CONCLUSION

The presented study proves the efficiency of the proposed congestion and network access control methods for signalling networks. The analyses of the effects of the control mechanisms on the network quality of service are helpful to design and to enhance the control procedures at the different steps of the network growth and of the integration of additional services. It allows to link the service requirements with the network performances and to define the upper level control according to each application needs. A general interest of this paper is to provide a method of performance analysis of control procedures which may be very useful to the designers of large packet-switched networks.

REFERENCES

[1] CCITT recommendations Q701-Q741, "Specifications of Signalling Systems N°7",Red Book, 1984,ITU, Geneva.

[2] P. Brown, P. Chemouil, B. Delosme, "A Congestion Control Policy for Signalling Networks", ICCC'84, Nov. 1984, Sydney.

[3] P. Brown, J. Labetoulle, "First Passsage Times in Markovian Queues", 3rd Int. Seminar on Teletraffic Theory, ITC, June 1984, Moscow.

[4] B. Delosme, A. Vidal-Madjar, "Designing a Target Signalling Network", Networks'83, March 1983, Brighton.

TELETRAFFIC ISSUES in an Advanced Information Society
ITC-11
Minoru Akiyama (Editor)
Elsevier Science Publishers B.V. (North-Holland)

STATOR - STATISTICAL OVERLOAD REGULATION - AND TAIL - TIME ACCOUNT INPUT LIMITATION - TWO CONCEPTS FOR OVERLOAD REGULATION IN SPC SYSTEMS

Georg Daisenberger, Jörg Oehlerich, Gerhard Wegmann

SIEMENS AG
Munich, Federal Republic of Germany

ABSTRACT

The phenomenon of overload in SPC systems cannot generally be counteracted by over dimensioning, at least not in an economical manner. As a result, a new discipline has arisen in recent years, devoted to investigating effective methods of overload protection that are easy to implement. Depending on system architecture, solutions often differ greatly.

This paper describes two methods developed for the EWSD system: STATOR, located in the coordination processor CP, controls any necessary peripheral restrictions on the basis of current call rate and instantaneous CPU load, without itself being stressed by rejecting calls; TAIL makes the decision to accept or reject a call in the event of overload in a peripheral group processor GP on the basis of a predetermined CPU work load.

1. INTRODUCTION

1.1 The Nature of the Problem

The load placed on switching processors by current signaling procedures stems from the processing of calls that have been accepted a few seconds previously. This "time shift" effect is explained by the fact that the greater part of the work load ("secondary tasks" such as digit reception, routing etc.) does not occur until several seconds after the call has been accepted whereas the processing associated with acceptance ("initial task") contributes very little to the work load (Fig. 1).

The CPU load, and secondary effects resulting from it such as queues and task waiting times, are therefore less suitable for use as indicators for a quick-acting regulation mechanism for the prevention of overload. As known from common feedback systems, the time shift between cause and response leads to unacceptable oscillation effects which can only be avoided by means of sufficiently high attenuation, i.e. slow counteraction.

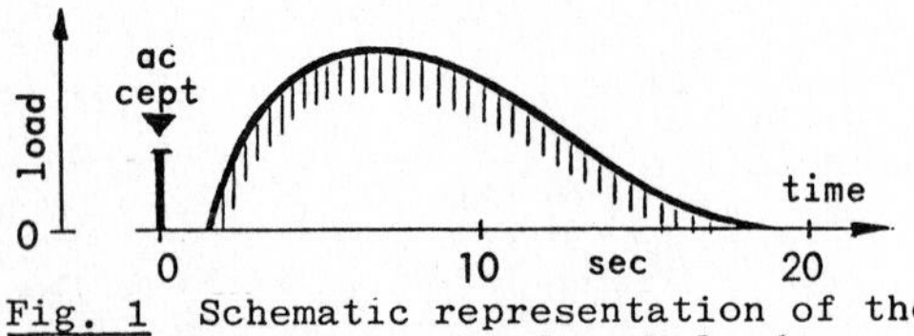

Fig. 1 Schematic representation of the dispersion (in time) of work load associated with the processing of a call ("work load cluster")

Modern overload protection concepts, e.g. [1]...[7] are therefore usually directly linked to the call arrival rate so as to ensure the rapid reponse of regulation procedures.

Unfortunately the arrival rate only gives a rough idea of the expected actual load, because the work load caused by a call varies according to the type of call and subscriber behavior etc. The actual work load per call is therefore also dependent on the instantaneous traffic and call mix which can vary from exchange to exchange and from one time to another. The reference value for a call rate indicator therefore has to be continually updated according to the observed load.

Since however, changes in the traffic and call mix generally only occur slowly, a slow-acting loop can be used for the updating, one that is sufficiently slow to avoid the oscillation effect mentioned earlier.

1.2 Brief Characteristics of the Concepts STATOR *) and TAIL *)

STATOR is a regulation system for protection against overload in centralized processors with relatively high call arrival rates (in EWSD, the coordination processor CP, Fig. 2). The function it performs derives from the concept that the centrally located regulation mechanism is not involved in actual prevention (throttling, rejection) but shares in the task of controlling peripheral restrictors that reduce the input to a given percentage in the event of overload. Advantage: both acknowlegement and management of non-accepted calls are dispensed with, i.e the CP is kept free of blind load.

---

*) International patents applied for

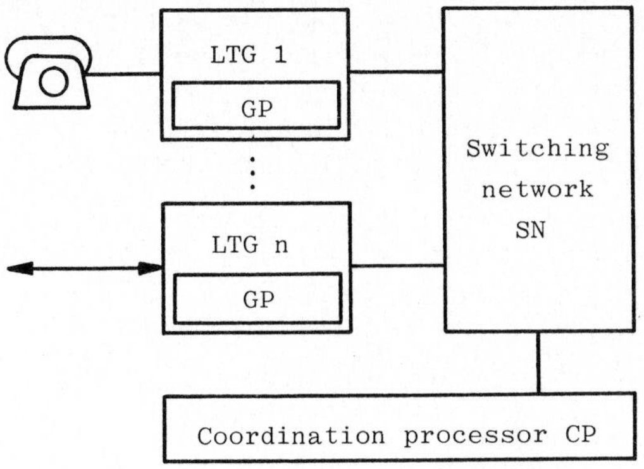

LTG: line trunk group
GP: group processor

Fig. 2 Simplified schematic of EWSD

The regulation system is based on two cooperating control loops. A fast "control loop" controls the call restrictions in the periphery on the basis of the call arrival rate; a slow acting "adjustment loop" updates the reference value (indicator) of the control loop on the basis of the observed CPU load.

In contrast to the relatively complex STATOR, TAIL is an elementary algorithm for limiting input. It is a particular advantage of the TAIL algorithm that, in contrast to the usual arrival count methods, it can also be used in cases where the arrival rate is low, and is therefore particularly suited to peripheral processors such as the GP in EWSD (Fig. 2).
Moreover, TAIL is also of theoretical interest, because it enables fundamental characteristics of overload protection in SPC systems to be made transparent and calculatable.

2. THE STATOR CONCEPT

2.1 Introductory Remarks

To simplify the description, the operation of STATOR will be explained for a mono-processor system; STATOR is however with minor modification also suitable for multiprocessor systems.

The regulation mechanism (RM) developed for STATOR is only active during an over load condition and there are special criteria to determine the "start" and "end" of this condition these will be explained later. The description of the operation therefore always assumes the presence of a peripheral overload.

The RM's function is to control peripheral restrictors which work on a "quota system". The restrictors reduce external input to an acceptable quota (e.g. 75%) determined by the RM. The CP does not "thin out" the traffic, i.e. all calls arriving at the CP are fully processed and there is therefore no blind load.

The acceptance quotas (AQ) can be verified, for instance by means of selection algorithms (which will not be explained further here) or by excluding parts of the periphery from service There are two possible types of AQ, each requiring different regulation strategies a) continuous or b) graduated. STATOR operates with graduated quotas, e g. 100%, 75%, 50%, 25% (In the real system there are more than 4 levels). These AQs, starting with zero (≙ 100%), are assigned to "restriction levels L", thus in our example L = 0,1,2,3.

Graduated restriction gives rise to a fundamental task, namely also to generate intermediate AQs (e g. 62%) in order to always make full use of the CPU's capacity. STATOR solves this problem with the fast oscillating control loop (2.2), which effects rapid switching between different restriction levels

proportioned in such a way that the statistical average of the levels passed through equals the AQ required for full usage of the CPU, and

- so rapidly that the fluctuations of the AQ only produce an insignificant fluctuation in the CPU load due to the dispersion effect (1.1).

2.2 Control Loop

The control loop is responsible for controlling the restriction levels of the peripheral devices. It makes use of a reference variable n* that is continually updated by the adjustment loop (2.3) and defines the acceptable number of calls per control interval (CI). CI is a time interval of a selected, suitable constant length, e.g. 1 sec.

The basic principle of the regulation system is very simple each call accepted by the periphery is reported to the CP by means of a message ("initial task"). Let n be the number of initial tasks counted during one CI while the overload process is active, and L the instantaneous restriction level. At the end of each CI, L is redetermined using the following rule:

if n > n* then L = min(m,L+1)
else L = max(0,L-1)

(in the example m = 3)

"n>n*" results in the restriction level L being incremented by 1, "n<=n*" leads to a corresponding decrement, thereby lowering and raising the AQ. Given that CI is small enough and the response of the restrictors (in the GPs) is fast enough (2.1), this produces oscillations that on the one hand verify, in the statistical mean, the necessary intermediate values of AQ, but on the other hand have little influence on the CPU load, due to the dispersion effect (see 2.7 for example).

It should be noted, however, that the above indication algorithm should be regarded as only the simplest of several possible methods of restriction control. This "easy solution" leads to "distortion effects" (bias) at the extremes of the AQ range, which can be avoided by using more sophisticated (integrating) algorithms. However, simulations have demonstrated that the distortion is sufficiently compensated for by the automatic action of the adjustment loop (2.3) and so a more sophisticated solution does not appear to be essential.

2.3 Adjustment Loop

The function of the adjustment loop is to supply the reference variable n* to the control loop. In contrast to the control loop, it is possible to use known methods here:

The current total load y of the CP is determined per "adjustment interval" (AI). It is practical to select a value for AI such that it is an integer multiple of CI large enough to be able to produce a sufficiently reliable measurement. For statistical and practical reasons, values of 1 sec for CI and 4 sec for AI have been found suitable for arrival rates greater than 20 calls/sec.

Assuming that the CP load y is to be regulated to a desired value (high load)

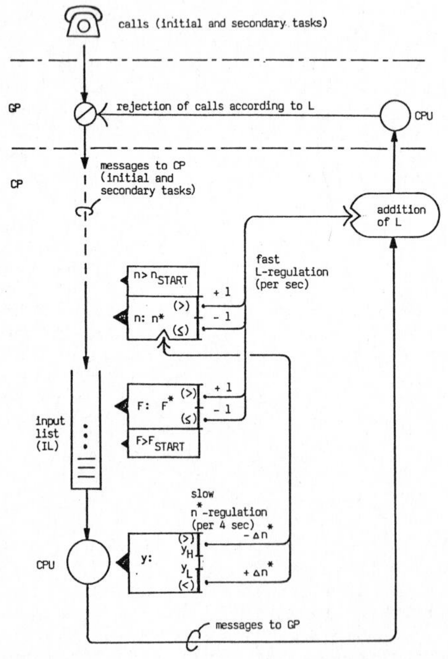

L : restriction level
n : actual number of initial tasks
n^*: reference variable of the indicator
n_{START}: start level of the indicator
Δn^*: adjustment step
y : actual CP-occupancy
F : actual length of the IL
F^*: control level of the IL-indicator
F_{START}: start level of the IL-indicator
y_H, y_L: high and low value of the adjustment interval

Fig. 3 Operating principle of STATOR

of e.g. y_{HL} = 0.95 Erl. n* is redefined at the end of every AI in accordance with the value found for y:

if $y > y_H$ then $n^* = n^* - \Delta n^*$

else if $y < y_L$ then $n^* = n^* + \Delta n^*$

Thus n* is not changed in the range $y_L \leq y \leq y_H$. Suitable values for y_L and y_H are values at a distance of $\pm$ 0.01 ... 0.02 Erl from y_{HL}. Δn^* is a constant "step", representing a compromise between two opposing objectives: on the one hand, Δn^* should be large enough to allow the fastest possible correction of n*; on the other hand, it should not be so high that it causes the uncontrolled oscillations mentioned in 1.1.

A suitable value has proved to be one where n*, expressed as a general formula, does not vary by more than approx. 1% ... 1.5% per second. If AI is 4 sec, then Δn^* is approx. 4% ... 6% of the number of calls per sec at high load n_{HL}.

2.4 Activation and Deactivation

As already mentioned, STATOR is only active when overload actually exists. There are two reasons for this:

- When the load falls below y_L at the end of an overload phase, the adjustment loop would adjust the reference variable n* to unrealistically high values if the protection system were permanently active. n* would then be unusable at the start of the next overload phase.

- Assuming that n* corresponds to the correct high load value n_{HL}, randomly occurring peaks would cause an unacceptably frequent occurrence of the condition "n>n*", thus initiating restriction unnecessarily, i.e. too many calls would be rejected.

For these reasons, activation and deactivation are controlled by special criteria:

Activation: If $n > n_{START}$, STATOR is activated and n* is set to the initial value n_I, which is, due to the time shift between initial and secondary tasks, 10%... 20% higher than n_{HL} (the exact value depends on the required quality of service). The start indicator n_{START} is set high enough to ensure a very low probability that unpermitted action occurs, e.g. 0.05%, assuming a Poisson distribution of n.

Deactivation: STATOR is switched off when restriction level L has continually equalled zero for a given guard period (approx. 6 sec). After deactivation, n* is re-initialized to the initial value n_I.

In addition, when STATOR is not active, the initialization parameters n_{START} and

n_I are adapted to changing traffic and call mixes by means of load measurement and arrival rate counting, but at less frequent intervals, e.g. approx. 1 min ("idle adjustment loop").

2.5 Transmission of Restriction Levels

To avoid placing an additional burden on the message paths during overload conditions, the restriction levels L are embedded in the messages that are to be sent to the periphery in any case ("piggy-back" method). Since the peripheral controls (GPs) that place the greatest proportion of load on the CP are also those that receive the most messages, this means that the "load focal points" are informed first. This "partial" informing also has the effect of spreading the restriction levels over the periphery in a more or less random fashion, which in turn creates additional smoothing of the control loop oscillation (by means of spatial dispersion), and the mean AQ moves towards the desired value.

Nevertheless, the "piggy-back" method reaches its limit in the event of sudden overload occurring in a low-load situation because in this case not enough messages are being generated for the pe-

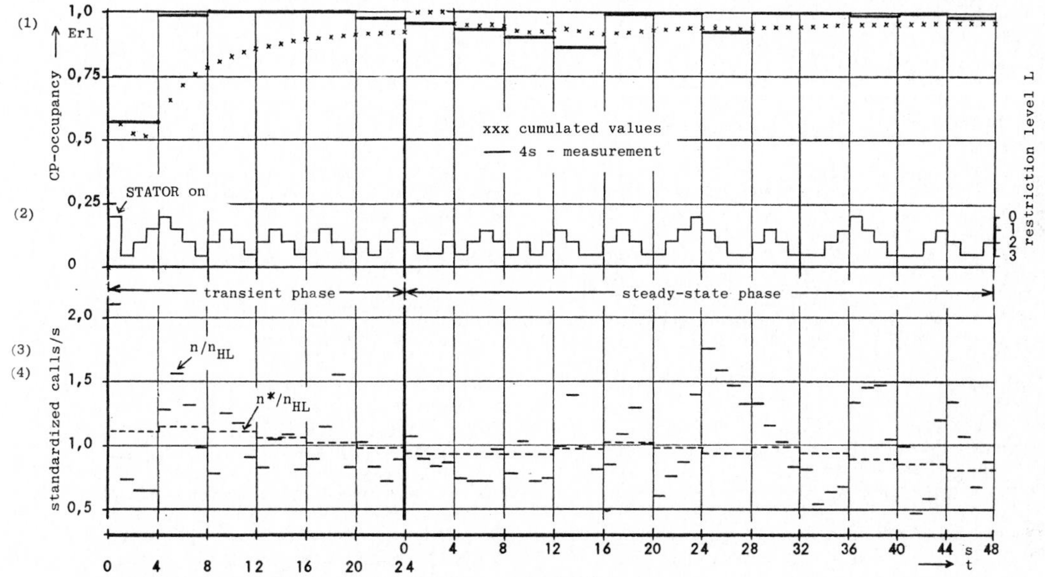

a) Call offering of 210%(110% overload) offered to a completely idle system

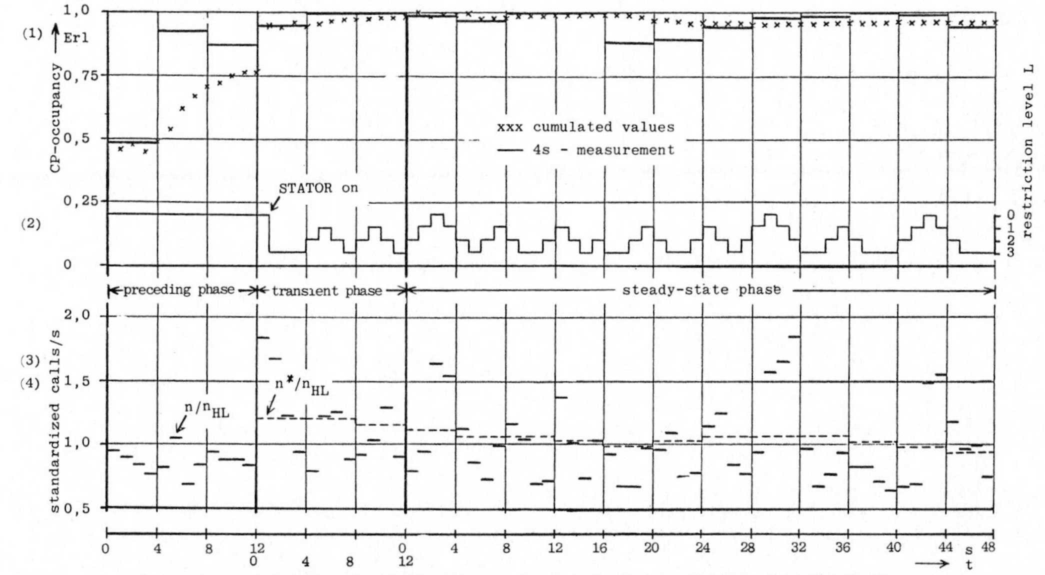

b) Call offering of 210%(110% overload) offered to a system that already has a stationary load of 0,90 Erl

Fig.4 Two examples of Simulation results of STATOR

riphery. For this reason, the current value of L is sent in a broadcast message when (and only when) STATOR is activated. The variation with time of the mean AQ across the entire periphery Q(t), assuming that the message flow behaves as independent Poisson streams, is defined by

$$dQ(t)/dt = \beta \cdot (Q_o(t) - Q(t))$$

where $Q_o(t)$ is the AQ ordered by the regulation mechanism at time t (by specifying L) and β is the message rate per GP (messages per time unit that transfer the current L to a considered GP).

2.6 Internal, Very Slowly Rising Overload or Very Light Overload

Internal overload (by O&M or safeguarding), very slowly rising ("creeping") overload and very light overload are not detected by the arrival rate indicator n_{START}, or are not detected soon enough.

For this reason, a supplementary occupation level monitor for the CP input list (IL) has been implemented (Fig. 3): here too a start value (F_{START}) and a supervision value (F*). F_{START} and F* are fixed values, F_{START} being approx. 2/3 of F*. This is chosen so that, in the event of creeping or very light overload, protection is activated via F_{START} but regulation is then linked to the arrival rate indicator n*. Only if n* regulation has no effect (e.g. in the case of internal overload) and the IL continues to fill up, are the restriction levels controlled via F* in the same manner as via n*:

```
if (n>n* or F>F*) then L = min (m, L + 1)
                  else L = max (0, L - 1)
```

2.7 Simulations

The limited space only allows us to show two examples out of the large number of simulations. They are based on a coordination processor CP that can handle about 240 000 BHCA ($n_{HL} \approx 67$ call/sec ≙ 100%) at high load (y_{HL} = 0.95 Erl).

Both examples deal with a sudden call offering of 210% (≙ 110% overload)

- in one case offered to a completely idle system (Fig. 4a)
- and in the other case offered to a system that already has a stationary load of 0.90 Erl (Fig. 4b),

The figures show the preceding and the overload phase, the latter divided into transient and steady-state phases.

The diagram (1) illustrates the CP load: the cumulated values are represented by crosses, the 4s-measurement values by horizontal lines. The load, controlled by the adjustment loop, balances out to the desired load of 0.95 Erl. Fluctuations in the 4s measured values mainly reflect the random character of the call arrival rate. STATOR only has a minor influence on such fluctuations, as has been confirmed by control simulations at high load without STATOR.

In all simulations, the overload led neither to an unacceptably high start peak (too high occupancy level of CP input list) or to a "load gap" (due to excessive restriction). It was proven that, even at very high surplus input e.g. 110% (input level 210%) with the CP load regulated to 0.95 Erl, the CCITT requirements for response times for reference load B (Rec. Q.504/514) were complied with for the accepted calls.

Diagram (2) shows the variation of restriction level L. Since 52% of the calls must be rejected at 210% offered calls, regulation operates almost entirely at levels L = 1 ...3.

The last two diagrams (3) and (4) show the number n of calls accepted by the system (CP) per second and the reference variable n* respectively, in both cases standardized to high load n_{HL}.

3. THE TAIL ALGORITHM

3.1 Basic Philosophy

A switching processor accepts a call "on credit", in other words on the assumption that sufficient capacity will be available when the main part (tail) of the processing work load arises some time later. The basic philosophy of TAIL is to create a "credit account" (W) in which to enter all "undone work", thus providing a reliable basis for making an immediate decision to accept or reject an arriving call. The following basic algorithm, which can be modified in a variety of ways or extended, is suitable for managing and analyzing the account:

a) on the arrival of each call:

   if $W > \Omega$: reject

   - else ($W \leq \Omega$): accept and increment W by D

b) permanently:

   - while W > 0: decrement W at a constant rate η

D may be set equal to the probable total CPU time (T) required to completely process the call, η equal to the processing capacity (y_{lim}) available for call processing tasks in Erl and Ω a suitable threshold.

In real applications, D can be for example a value determined by known program runtimes. A more sophisticated approach would be to have D continuously updated by automatic adjustment action.

"Account withdrawals" are assumed to be continual and can be approximately verified by means of timed routines. Another possibility is to calculate the value using the time elapsed since the previous call arrival.

It is immediately obvious that the effect does not change if the set of parameters (D,η,Ω) is multiplied by a positive constant. Later on we will make use of this possibility to set $\eta=1$.

3.2 Fundamental Aspects

The basic algorithm given in 3.1 can be enhanced with a variety of additional facilities, for example:

- giving priority to urgent call types - by assigning different values of Ω.
- a waiting facility for calls that cannot immediately be accepted - by combining with a limited queue.

The considerations in the following, however, refer to the basic version only.

As shown in 3.3, the TAIL algorithm guarantees that the load limit η cannot be exceeded under statistical equilibrium. The threshold value Ω is of no significance here. Its importance is that it determines the "selectivity" of the overload protection (OLP). This means the ability to distinguish between "overload" and "non-overload", and thus avoid undesirable rejections in the case of "non-overload" (see 3.4). Unnecessary protection action results from fundamental laws applying to every OLP, but this action tends to remain hidden in more complex regulation mechanisms, whereas it is revealed transparently in the case of TAIL. This could be one of the reasons why degradation of service due to unnecessary action by the OLP has hitherto not received enough attention in studies on grade of service.

3.3 Limiting Guarantee

Fig. 5 shows a section of the time sequence of the credit account (credit process) showing how it may vary according to the call arrival process. Generalizing, the values recorded for D are assumed to be random variables, dependent neither on each other nor on the "balance" W(t). The same applies to the interarrival times Z (which have a negative exponential distribution in the case of Poisson arrival).

The limiting guarantee can be explained as follows with the aid of Fig. 5:
Let Z^* be the interarrival time and R^* the reduction in W(t) caused by "withdrawals" between two successive arrivals. Then it is obvious that $R^* \leq Z^*$. This reduction is counteracted by an increase D. Since the account is limited, reductions and increases must be in balance over the long-term average. Thus for the average values (marked with a superimposed line) $\bar{D} = \bar{R}^* \leq \bar{Z}^*$. Consequently for the acceptance rate (λ^*) and the load (y) resulting from accepted calls,

$$\lambda^* = 1/\bar{Z}^* \leq \eta/\bar{D} \qquad (1)$$

$$y = \bar{T}/\bar{Z}^* = \bar{D}/\bar{Z}^* \leq \eta = y_{lim} \qquad (2)$$

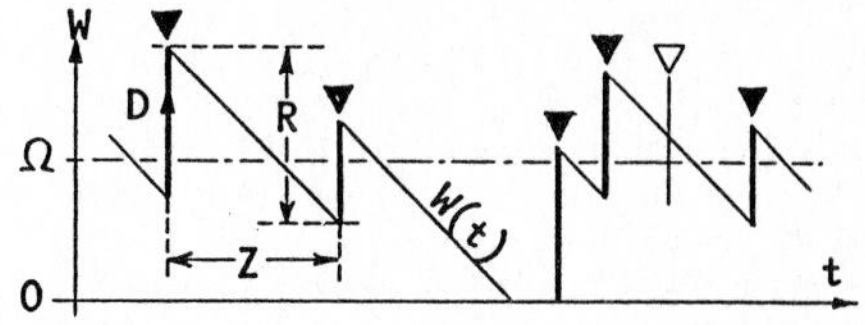

Fig. 5 Credit process: ▼ accepted, ▽ rejected call; Z interarrival time; D increment; R decrement from call to call

3.4 Stationary Behavior

The behavior of TAIL in connection with a stationary arrival process can be calculated using formulae that are relatively easy to handle (3.6) - a programmable pocket calculator suffices. Fig. 6 shows in schematic form the variation of the acceptance rate λ^* with respect to the given arrival rate (λ) and how it is affected by parameters $\bar{D}$ and Ω (when $\eta=1$). The acceptance limit (λ_{lim}) is not defined by a threshold (Ω) - which may surprise the newcomer - but by the mean increment $\bar{D}$.

In practice, the significance of the stationary analysis is that it provides information on the frequency of unnecessary rejections when there is no overload ("selectivity"). In this context, Fig. 6 demonstrates that there is always a certain region around λ_{lim} in which the OLP - inevitably - causes more rejections than would be expected under ideal circumstances. The deviation from ideal behavior can be influenced by means of the threshold Ω. An example to illustrate this: if D is constant and the arrival rate $\lambda=0.8\lambda_{lim}$ and $\Omega/\bar{D} = 5$ or 10, then the loss due to unnecessary rejections is 2.2% or 0.34%.

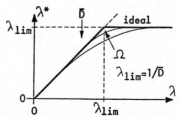

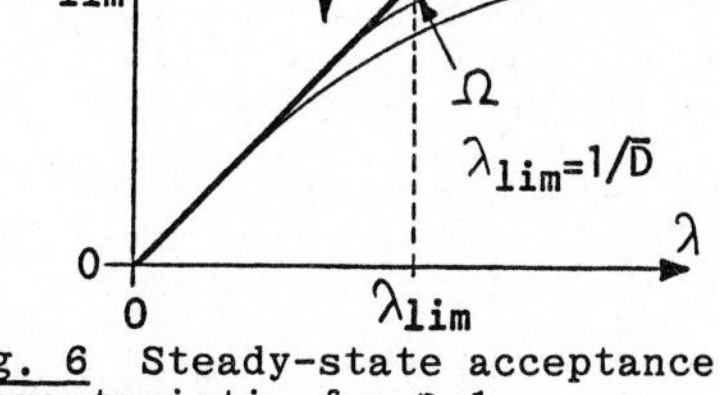

Fig. 6 Steady-state acceptance characteristic for $\eta=1$

3.5 Start Peaks

The obvious solution would appear to

be simply to set Ω as high as possible. But this leads to another undesirable effect - the greater Ω is, the longer it takes for the credit limit to be reached in the event of overload and for the input limit to start to operate. Until this time, the excess calls offered are not prevented from entering the system being protected. As a result, the secondary tasks cause a correspondingly delayed "start peak" which increases in proportion to Ω.

This correlation clearly implies that the OLP requires a certain "orientation" phase in order to be sufficiently sure that its action is genuinely called for. The less time it is allowed, the more frequent the occurrence of unnecessary action. Fig. 7 shows an example of how the start peak is created.

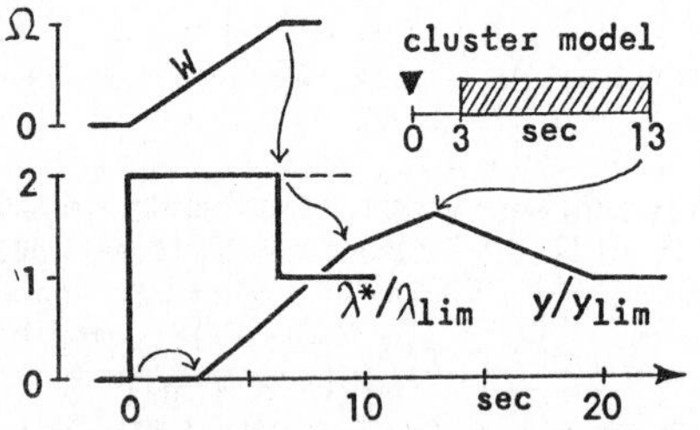

Fig. 7 Start peak, illustrated by the way of a fluid model with uniformly dispersed work load clusters (initial tasks ignored). The max. possible height of the peak is independent of the amount of overload and depends only on Ω

3.6 Excerpt of the Stochastic Analysis

Given that the paramters of TAIL have been chosen such that $\eta=1$, we observe W(t) at each arrival epoch before incrementation takes place. The values of W(t) then form an embedded Markov chain W_ν ($\nu=1,2,\ldots$). Assuming Poisson arrivals, appropriate convolutions lead to the equilibrium equation

$$F'(w) = \lambda \int_{x=0}^{\min(w,\Omega)} dF(x)G(w-x) \qquad (w>0) \qquad (3)$$

where F(w) and G(t) are the distribution functions of W_ν and D. For the limit $\Omega\to\infty$, F(w) is identical to a solution applying to an "analogous M/G/1" (having D as as holding time). The probability of rejection follows from this as

$$B = P(W_\nu > \Omega) = \bar{F}_o(\Omega)\,\frac{1-A}{1-A\bar{F}_o(\Omega)} \qquad (4)$$

where $A=\lambda\bar{D}$ and $\bar{F}_o=1-F_o$, F_o being a solution applying to the analogous M/G/1. Let g(s) represent the Laplace Stieltjes transform of $G(t/\bar{D})$. $\bar{F}_o(w)$ can then be represented by

$$\bar{F}_o(x/\bar{D}) = \phi(x)\, e^{\alpha x} \qquad (x\geq 0) \qquad (5)$$

where α is defined by

$$(1-g(\alpha))/\alpha = 1/A \qquad (6)$$

and $\phi(x)$ is a function within boundaries

$$\phi(0)=A \quad \text{and} \quad \phi(\infty)=(A-1)/(Ag'(\alpha)+1)$$

(5) may be found e.g. by means of a residuum analysis of Pollaczek's equation. α is negative for $A<1$. Analytical continuation applies for $A\geq 1$.

$\phi(\infty)$ is monotonous in A and equal to 1 for A=1. If we simply set A for $\phi(x)$

$$B \doteq e^{\alpha\omega}(1-A)/(1-A^2 e^{\alpha\omega}) \;,\; \omega=\Omega/\bar{D} \qquad (7)$$

is obtained, which is exact for exponential D, or otherwise a pretty accurate approximation within the range of practical interest. In the (worst) case of D being constant, $\phi(\infty)/A=1.32$ for $A=0.5$, for example.

REFERENCES

[1] E. Abdou, J. Yan, "Overload Modelling of Switching Systems and the Evaluation of Overload Controls" IEEE-Proc. B4.4.1-4.4.5, NTC 1981

[2] L.J. Forys, "Performance Analysis of a New Overload Strategy", Proc. 10th ITC, Montreal, Ses. 5.2, Paper 4 , 1983

[3] F.C. Schoute, "Adaptive Overload Control for an SPC Exchange", Proc. 10th ITC, Montreal, Ses. 5.2, Paper 3, 1983

[4] P. Somoza, A. Guerrero, "Dynamic Processor Overload Control and its Implemention in Certain Single-Processor and Multiprocessor SPC Systems", Proc. 9th ITC, Torremolinos, Paper 532, 1979

[5] A. Toda, H. Kawashima, M. Asama, "Automatic Overload Control of the D-10 Local Electronic Switching System", Int. Switch. Symp., Kyoto, Paper 433-1, pp. 1-8, 1976

[6] P. Tran-Gia, "Subcall-Oriented Modelling of Overload Control in SPC Switching Systems", Proc. 10th ITC, Montreal, Session 5.2, Paper 1, 1983

[7] K. Wildling, T. Karlstedt, "Call Handling and Control of Processor Load in a SPC - A Simulation Study", Proc. 9th ITC, Torremolinos, Paper 537.1979

TELETRAFFIC ISSUES in an Advanced Information Society
ITC-11
Minoru Akiyama (Editor)
Elsevier Science Publishers B.V. (North-Holland)

NEW OVERLOAD ISSUES IN A DIVESTED ENVIRONMENT

L. J. Forys

Bell Communications Research
Red Bank, New Jersey, U.S.A.

ABSTRACT

The new economic environment resulting from divestiture has put continued emphasis on the Operating Telephone Companies (OTCs) to reduce capital expenditure and efficiently utilize existing equipment. Of particular importance is real time capacity. We examine the capacity of local exchanges which have incorporated the new overload strategy described by the author at the 10th ITC using an approximate analytic model, simulation and field data. Particular attention is paid to the effects of nonstationary traffic inputs. An improvement of about 5% is reported. Other issues examined by the paper deal with the impact that service problems experienced by one Interexchange Carrier can affect other traffic. An analytic model is developed to quantify the impact in special cases and the need for selective overload strategies is indicated.

1. INTRODUCTION

The divestiture of the "Bell System" on January 1, 1984 has of course had enormous consequences in a number of areas. The new economic environment, for example, has resulted in increased stress on capital expenditure and efficient use of resources by the Operating Telephone Companies (OTCs). This has encouraged continued examination of the capacities of a number of traffic sensitive components, in particular, the real time capacities of local exchanges. At the 10th ITC we presented results (see ref. [1]) for a newly implemented overload control for local exchanges. At that time we alluded to a possible increase in capacity for No. 1 ESS^{TM} local exchanges. In this paper we will examine this issue more closely by presenting a model which accounts for nonstationary traffic variations. This model incorporates the thinking presented at the 9th ITC (see ref. [2]) and is applicable not only to the analysis of No. 1 ESS^{TM} exchanges, but to a wide class of local switching exchanges. Our analysis shows that a substantial capacity increase is possible by incorporating the new overload strategy. It indicates that the new strategy is much more robust to traffic variations than previous strategies.

The model results are verified both by simulation and recently analyzed field data. The agreement between theory and practice is shown to be excellent. Although designed originally for real time overload protection, the new strategy will also be invaluable for handling hardware overloads such as digit receivers, where the same considerations prevail.

Another important impact of divestiture on overload controls has been the introduction of "equal access" into SPC exchanges. Because of equal access, a number of competing Interexchange Carriers (ICs) will be using common resources at both local exchanges and access tandems. Although the recently introduced overload strategy handles excessive real time loads (as well as receiver overloads) in an efficient manner, it may not provide sufficiently selective response to causes. In this paper we will discuss a number of overload scenarios caused by service difficulties experienced by a given IC and demonstrate how these service difficulties can impact the service of other ICs. This motivates the need for more selective overload controls which identify the cause of the service impairments and takes appropriate actions to control the cause without impacting other traffic.

The paper is organized as follows. In Section 2 we begin by reviewing the improved overload strategy and how it differs from previous controls. In Section 3 the issue of nonstationary traffic and its impact on real time capacity is examined, and previous results summarized. An approximate analytic model is described in Section 4 which determines "worst case" traffic patterns for the improved strategy. The results of the model are verified and extended using simulations. This is presented in Section 5. Finally, in Section 6 recent field results are analyzed which confirm our predictions and indicate substantial (about 5%) real time capacity increases can be realized.

The need for selective overload controls to mitigate the impact of IC specific service problems is discussed in Section 7 and the results of possible scenarios quantified in 8. Finally, in section 9 we discuss how network management type controls which are present in many equal access exchanges can be used to mitigate the problem.

2. REVIEW OF OVERLOAD STRATEGIES

Until fairly recently, new originations for service were handled in a first-in-first-out (FIFO) manner or random order of service (sometimes a combination of both) by nearly all local exchanges. Unfortunately, these methods of servicing do not properly account for customer behavior patterns and can lead to significant reduction in the system's ability to properly complete calls. At the 10th ITC, Burkard et al [3] presented the results of a field study which investigated the behavior of originating customers when subjected to dial tone delays. The results show that as many as 40% of the customers started dialing and an additional 45% abandoned or flashed before dial tone when dial tone was randomly delayed by 3 seconds. These would impact the switching system in adverse ways since the system would interpret the "early dialers" as partial dials and the flashes as false seizures. Sizable real time would be "wasted" processing these ineffective attempts who would then often reattempt their bids, thus further inflating the load.

The overload strategy described by the author in [1] consciously incorporates customer behavior phenomena in its design. New originations are served in a last-in-first-out (LIFO) manner unless they are still in the off hook state after a period of 20 to 30 seconds. At this point, the "patient" customer is treated as a new attempt and moved to the start of the queue. Customer activity such as flashing or abandoning is disregarded when in queue. This strategy has the effect that most customers experience no dial tone delays under overloads and hence will most likely be properly processed. The 20 to 30 second "time out" is introduced to insure that the early dialers (or "flashers") have completed dialing and have abandoned their requests. The automatic reattempt after 20 to 30 seconds insures that patient customers will be given priority. This strategy was shown to give vastly superior throughput performance under overloads by both analytic models, simulations and field studies. It was conjectured in [1] that the strategy would also give improved real time capacity. We examine this point in what follows. It is important to note that not only does the above LIFO strategy give superior performance for real time overload control, but that it has important application in controlling hardware overloads as well. For example, a version of LIFO has been shown to give superior performance in controlling digit receiver overloads as well since the same customer behavior phenomenon are pertinent. Access to many switching networks may also benefit from these considerations.

3. EFFECTS OF NONSTATIONARY TRAFFIC PATTERNS

At the 9th ITC, we discussed the results of the affects of nonstationary traffic patterns (see reference [2]) . This paper pointed out the presence of significant nonstationary Poisson traffic patterns present in many exchanges. It also addressed the question of how to simply characterize these traffic patterns. The characterization would have to be simple enough to allow measurement of only a few parameters of the process on an ongoing basis. The characterization chosen was to use a mean and variance of call arrivals in small (10-30 second) time intervals for a fixed time interval, 15 minutes. The process is modelled as being piecewise stationary Poisson process. Thus, if λ_i, $i = 1, ..., n$ is the mean intensity of a piecewise stationary process with peakedness factor z, then the expectation of the measured sample variance over an interval of length T is given by (see [5]):

$$(2z - 1)\lambda T + \frac{T^2}{n-1}\sum_{1=1}^{n}(\lambda_i - \lambda)^2 \quad (1)$$

where

$$\lambda = \frac{1}{n}\sum_{i=1}^{n}\lambda_i \quad (2)$$

Unfortunately, the peakedness and mean do not uniquely characterize a nonstationary Poisson process. There are a multitude of such processes which produce the same mean and peakedness. Each of these processes would presumably influence a traffic sensitive system in different ways.

The method used to characterize the traffic was to select the process which would produce the "worst case" results for the system of interest. In our case, the system investigated was a single server FIFO queueing system which served to model many local exchanges with the "pre-LIFO" overload strategy. The traffic model which resulted from this analysis was simply approximated by a two level Poisson process where the values of the two levels and their duration were determined by the peakedness and mean. It was noted that the "optimum" solution was robust in that small deviations from the optimal produced similar results. It was hypothesized that it would take an unusual traffic pattern in order that the effects of traffic variation be minimal for queueing systems. It was also stated that blocking systems would presumably be less sensitive. We will pursue this point in the next section.

Based on our results, a trial measurement was introduced into versions of a No. 1 ESS^{TM}. Instead of measuring a sample variance, a transformation was used termed the "variability ratio" defined below:

$$VR = \frac{V(M) + 1}{2} \quad (3)$$

where $V(M)$ is given by

$$V(M) = \frac{n}{n-1}v/m + 1 \quad (4)$$

Here, v is the measured biased sample variance (see eq. (9)) and m is the sample mean of the traffic process. Our experience with the trial measurement indicated that the variability ratio was almost always less than 2.

4. DETERMINATION OF "WORST CASE TRAFFIC PATTERNS"

As in [2] we will use an approximate analysis to guide us in determining a worst case traffic pattern with which to analyze the capacity of systems with the LIFO/Time-Out strategy. The behavior of this overload strategy under heavy loads causes a new origination request to either be served almost immediately, or after a sufficiently long delay so that the "early dialer" has abandoned. Thus the system appears quite similar to a "blocked calls cleared" system with possible reattempts.

To simplify the analysis, we will consider a discrete time approximation to the system. We consider a finite time interval T which is subdivided into n equal subintervals. The time varying underlying call process is modeled as a vector $\underline{N}$, whose k^{th} component is N_k. The term N_k represents the number of calls arriving in the kth subinterval. We assume that the system can process at most M calls in any subinterval. If more than M calls arrive in any subinterval, the excess will be "blocked" and a fraction, p, of these will reattempt immediately in the next subinterval. In reality, calls which are "blocked" will of course reattempt at some random time in the future. However, by choosing the subinterval length equal to a mean abandonment time (say 30 seconds), we hope to capture the essentials of the phenomenon.

With these assumptions, the total number of calls, including reattempts, arriving in the subinterval k is given by C_k :

$$C_k = N_k + p(C_{k-1} - M)^+ \tag{5}$$

The real time capacity of local exchanges is governed by a constraint on the percentage of calls which have delays exceeding a certain threshold, for example 3 seconds. In our model we will equate the number of calls exceeding 3 seconds by those calls which are "blocked". $N_B(T)$ will denote the number of calls blocked in [0,T] and $N_S(T)$ will denote those served. Then, the fraction of calls blocked, $P_B(T)$ is given by

$$P_B(T) = \frac{N_B(T)}{N_B(T) + N_S(T)} \tag{6}$$

Since $P_B(T)$ is a strictly monotonically increasing function of $N_B(T)$, to determine a "worst case" traffic pattern it suffices to maximize the total number blocked in [0,T]. In accordance with our assumptions, this is simply

$$N_B(T) = \sum_{j=1}^{n} (C_j - M)^+ \tag{7}$$

Let us denote by m and v the mean and variance of the underlying calls process (we use the "biased" sample variance for consistency with our work on the FIFO control strategy). Thus,

$$m = \frac{1}{n}\sum_{j=1}^{n} N_j \tag{8}$$

$$v = \frac{1}{n}\sum_{j=1}^{n} (N_j - m)^2 \tag{9}$$

We want to therefore maximize the expression in equation (7) subject to the constraints of equations (8) and (9).

We solve the problem in two steps. For a fixed integer r, $1 \neq r \neq n$, we first find the underlying traffic pattern which produces the most total blocking in r subintervals, and then maximize with respect to r. Without loss of generality, the blocking can be assumed to occur in a contiguous set of subintervals. Also, the first subinterval in such a contiguous set can be assumed to be first subinterval of [0,T]. Thus, we seek to maximize N_B^r defined by:

$$N_B^r(T) = \sum_{j=1}^{r} (C_j - M) \tag{10}$$

Using equation (5) recursively, we can determine that:

$$N_B^r(T) = \sum_{j=1}^{r} A_j N_j - M \sum_{j=0}^{r-1} (r-j)p^j \tag{11}$$

where,

$$A_j = \frac{1 - p^{j+1}}{1 - p} \tag{12}$$

for $j = 0, 1, ..., r-1$ and $p \neq 1$

As a final step to setting up the appropriate optimization problem, we will "relax" the variance constraint in equation (9) by imposing an inequality. We will solve this "relaxed" problem and show that the solution obtained satisfies the equality condition as well. Thus our optimization problem is:

maximize

$$N_B^r(T) \tag{13}$$

subject to

$$M(N_1, \cdots, N_n) = mn \tag{14}$$

$$V(N_1, ..., N_n) \leq vn \tag{15}$$

where,

$$M(N_1, ..., N_n) = \sum_{j=1}^{n} N_j \tag{16}$$

$$V(N_1, ..., N_n) = \sum_{j=1}^{n} (N_j - m)^2 \tag{17}$$

We can solve this problem by using classical Calculus of Variation methods. Thus, the "Lagrangian" is given by:

$$L = N_B^r(T) + \lambda_1\{M(N_1, ..., N_n) - mn\} + \lambda_2\{V(N_1, ..., N_n) - vn\} \tag{18}$$

Equating the partial derivatives with respect to $N_1, ..., N_n$ equal to zero one obtains:

$$\lambda_2(N_j - m) = -\frac{(A_{r-j} + \lambda_1)}{2} \qquad j = 1, ..., r \tag{19}$$

$$\lambda_2(N_j - m) = -\frac{\lambda_1}{2} \qquad j = r+1, ..., n \tag{20}$$

Summing the above equations, and using equation (8) we obtain:

$$\lambda_1 = -\frac{\sum_{j=0}^{n-1} A_j}{n} \tag{21}$$

Squaring both sides of equations (19) and (20), summing and using (21) yields:

$$4\lambda_2^2 \sum_{j=1}^{n} (N_j - m)^2 = \sum_{j=0}^{n-1} A_j^2 - \left(\sum_{j=0}^{n-1} A_j\right)^2 / n \tag{22}$$

With a little algebra one can show that for $p \neq 0$, the right hand side of equation (22) is positive, and so λ_2 cannot be zero. Therefore, the condition defined by (21) is sufficient for the mean constraint to be satisfied. Moreover, if,

$$\lambda_2^2 \geq \frac{\sum_{j=0}^{n-1} A_j^2 - \left(\sum_{j=0}^{n-1} A_j\right)^2 / n}{4nv} \tag{23}$$

then the variance inequality constraint is satisfied as well. Clearly, if equality holds in (23), then equality will hold for the constraint.

Let λ_2^- denote the negative value of λ_2 corresponding to equality in (23), and let $H(L)$ denote the Hessian matrix of the Lagrangian. Performing the appropriate differentiations results in :

$$H(L) = 2\lambda_2^- I \tag{24}$$

Here, I is the identity matrix. Thus $H(L)$ is positive definite and the usual sufficiency conditions (Kuhn-Tucker) for a local maximum are fulfilled. Because of the convexity of $N_B^r(T)$ and the constraint set, the solution is also a global maximum. The maximum point must also lie on the boundary of the constraint set and so the solution provides for equality of the variance constraint.

By solving the respective equations we observe that:

$$N_j - N_{j+1} = -\frac{p^{r-j}}{2\lambda_2^-} \qquad j = 1, \cdots, r \tag{25}$$

$$N_j - N_{j+1} = 0 \qquad j = r+1, ..., n \tag{26}$$

From (25) we see that the "worst case" call stream is strictly monotonically decreasing ($\lambda_2^- \leq 0$) for the first r intervals and remains constant for the remainder of the time interval [0,T]. Also, the rate of decrease increases with r.

The solution obtained has the following physical interpretation. Since we assume that blocked calls reattempt in the next interval, calls arriving earliest in the interval are provided with the largest number of subintervals in which to reattempt. Thus from the point of view of maximizing the total number of blocked calls, the largest possible number should arrive in the first subinterval, the second largest in the second subinterval, etc. .

The optimization problem is completed by computing numerically the value of the maximum blocking for each r, and simply selecting the r which produces the most blocking.

Finally, we note that for "flat" traffic, i.e. $v = 0$, we obtain:

$$P_B(T) = \frac{(m - C)^+}{1 - p}\left\{n - \frac{p(1 - p^n)}{1 - p}\right\} \tag{27}$$

$$\text{for } p \neq 1$$

This value was used as a benchmark to determine the relative degradation of capacity with increasing variances. Note that, while the case $p = 1$ is unrealistic, a simple closed form solution exists to our problem in this event:

$$P_B(T) = \frac{n(n+1)}{2}(m - C)^+ \qquad (28)$$

5. SIMULATION RESULTS

The "worst case" deterministic traffic was used as the mean arrival rate for a piecewise stationary Poisson process. This was then used as an input to a detailed simulation of a No. 1 ESS^{TM} and the results compared to both our analytic approximation and to the results obtained by using the worst case traffic model in reference [2] for the FIFO control strategy.

The values of T, N and p were chosed to be 900 seconds 30 and .8 respectively. Thus the subinterval sizes were 30 seconds long. Using M = 27,500 calls per quarter hour and a variability ratio of 2, the results of the analytic model and simulation were compared. With m = 27,500, 26,750 and 26,250 calls per quarter hour, the analytic model predicted blockings of 23.5%, 18% and 13% respectively. The simulation values obtained were 26.5%, 15% and 7.5%. The disparity between the results can be considered mild in view of the simplicity of the analytic model. Note also that there is close agreement for higher loads since it is in this region that the system more closely resembles a blocking system.

We should emphasize that we are interested in comparing *relative* changes in performance, and that the analytic results are to be used as guides to selecting stochastic inputs to a simulation.

In figure 1 we plot the load-service relationships obtained from our simulation for a range of variability ratios.

The high day criterion used to determine real time capacity in local SPC exchanges is 20% dial tone delay in excess of 3 seconds. With this criterion we see that for a variability ratio of 2, the reduction of capacity is 2% compared with stationary Poisson traffic. As we indicated earlier, variability ratios in excess of 2 are uncommon.

These results are dramatic when the results of the FIFO strategy are compared. The results from [2] were used to determine a suitable "worst case" traffic model. It is a two level time varying Poisson process. This was used as input to the same simulation, but with the FIFO strategy. In this case, a capacity reduction of 10% for the same parameters was determined. The differences in relative performance are not so surprising. Blocking type systems generally are less sensitive to traffic mean changes than are queueing type systems.

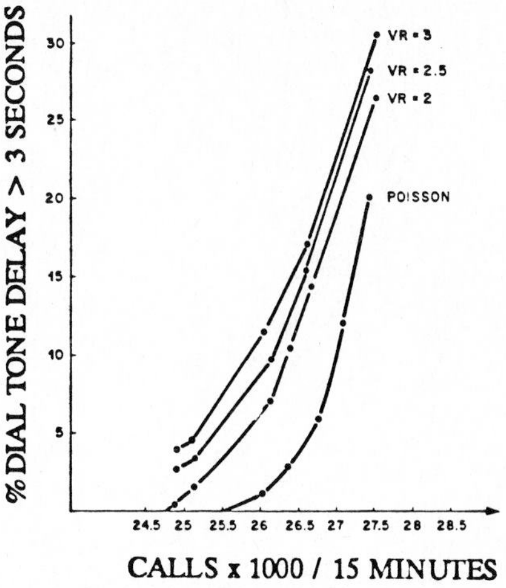

FIGURE 1: DIAL TONE DELAY VERSUS LOAD (SIMULATION)

6. FIELD DATA

Dial Tone Delay data were gathered for all No. 1 ESS^{TM} local Bell Operating Company exchanges for the period 1979 through 1984 inclusive. The exchanges were divided into categories according to whether the systems had the LIFO strategy or the pre-LIFO (or FIFO) strategy. Offices having hardware shortages and not just real time shortages were eliminated from the study. The measured % of Dial Tone Delay was plotted versus the measured number of E-E cycles for various types of exchanges. An E-E cycle is the number of times the central processor cycles through its base level activities (see [6]). It was found that insufficient information existed to analyze offices not equipped with an auxiliary Signal Processor. For Signal Processor equipped exchanges however, the field study results compared favorably with the simulation results. Figure 2 indicates the simulation results with various variability ratios. (Recall that a variability ratio of more than 2 is unusual.)

Figure 3 depicts the results of the field investigation. A statistical curve fit to the data is indicated, with the LIFO (-1) indicating the curve fit when one data point was deleted. Note that the scales of figures 2 and 3 differ by a factor of 4. The simulation curves for a variability ratio of 2 are an excellent approximation (a slight upper bound, as expected) to the field data!

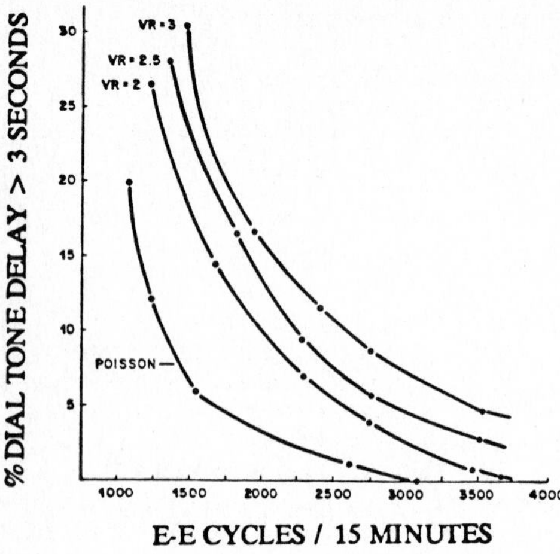

FIGURE 2: DIAL TONE DELAY VS E-E CYCLES (SIMULATION)

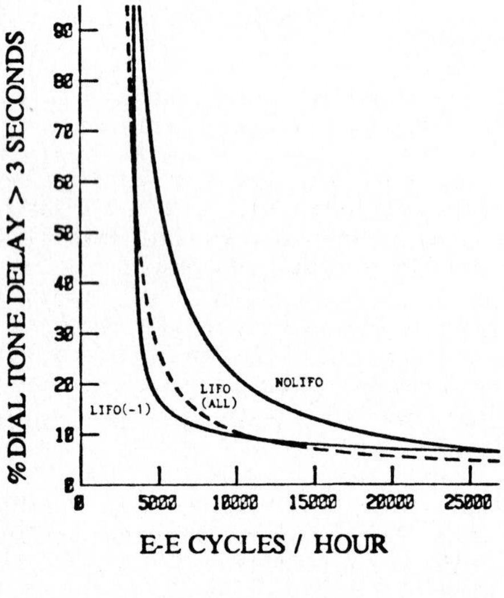

FIGURE 3: LIFO AND FIFO FIELD DATA

It was determined that a LIFO switch must slow down at least an additional 4600 E-E cycles per hour before it reaches 20% dial tone delay as compared with a NOLIFO (or FIFO) switch.

In figure 4 we plot simulation results indicating the relationship between E-E cycles and load for a range of variability ratios for the LIFO case.

Note that there is nearly a linear relationship between E-E cycles and load except at the high values of load. This is the region where overload control actions limit the rate at which originations can be served by the system. Using the linear relationship for various values of parameters, it was determined that the LIFO strategy yields a 3-7% capacity improvement over the pre LIFO strategy. The precise improvement is dependent on the amount of overhead in the office.

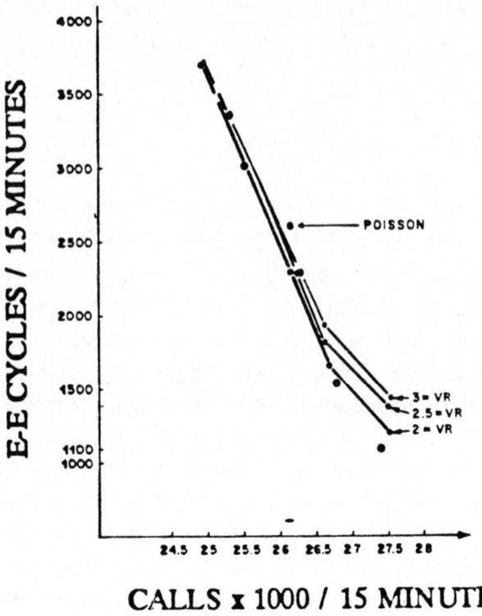

FIGURE 4: E-E CYCLES VERSUS LOAD (SIMULATION)

7. SELECTIVE OVERLOAD CONTROL ISSUES

In a divested environment one is faced with the situation that several Interexchange Carriers (ICs) will be sharing common resources both in local exchanges as well as access tandems. Although the LIFO strategy does an excellent job of protecting switch performance for general overloads, it does so by controlling all traffic. For cases where the overloads are caused by specific circumstances, it does not take selective action to mitigate the specific problem.

There are many scenarios whereby service difficulties experienced by one IC can possibly affect the service of other ICs unless specific corrective actions are taken. These scenarios are produced by examining the resources the ICs share in common and focusing on specific types of service difficulties. Among resources shared in common are real time, digit receivers, switching network, transmitters (outpulsers), memory and in some cases even trunks.

Among the possible service difficulties that might impact the equal access switch are the failure of the trunk group to the IC, the failure of an IC switch or other problems internal to the IC's network .

We will examine a subset of the possibilities and quantify the magnitude of the impact and the new overload control issues it presents. Specifically we will focus on what happens when the IC switch which is accessed by a trunk group from the equal access switch experiences a total failure. We will assume that the equal access switch has no knowledge of the failure.

As we indicated before, one item which can be shared by many ICs (as well as other local interoffice calls) is the common group of transmitters. If the transmitters are Multifrequency transmitters, their holding time is typically 2.5 seconds. In the event of the IC switch failure, the equal access switch will seize a trunk to the IC, and wait for a "wink" signal which is used to denote that outpulsing of digits may begin. Since the IC switch has failed, no wink will be forthcoming and the transmitter will be held until a time-out threshold is exceeded. A typical time-out mechanism is that if more than 4 to 8 seconds have elapsed before detection of wink, that transmitter may be preempted (reused) by another call requiring that type of transmitter. If no call arrives to preempt the transmitter, then the transmitter will time-out in 16 seconds and the timed-out call will be sent to reorder. The reason for the range of 4 to 8 seconds is that the determination of which transmitters are preemptable is done only every 4 seconds by switches such as a No. 1 ESS^{TM}. Other time-out mechanisms will produce similar results and so we will concentrate on this mechanism. There is normally no queueing for transmitters. Thus, calls destined to the failed IC will on average, hold the MF transmitters for 6 seconds or more.

Not only is the holding time thus increased for these calls, but after preemption or time-out, the calls will often reattempt. In addition, calls not destined for the failed switch will find that there are no free or preemptable transmitters available and they will be sent to reorder, and will often reattempt.

8. ANALYSIS OF TRANSMITTER BLOCKING

We will assume that the reattempt probability for all calls is a constant, p. Without loss of generality we can assume that calls which are timed-out after 16 seconds in effect have the same holding time as a preempted call. This will not affect the number of calls blocked or preempted since a transmitter can only time-out if there are no calls arriving to preempt it. Again, without loss of generality, we assume that all calls to the failed IC will receive a uniform holding time between 4 and 8 seconds (the minimum preemption interval) which of courses yields a mean holding time of $1/\mu_F = 6$ seconds. All calls not destined to the failed IC will receive a normal holding time with mean $1/\mu_O = 2.5$ seconds.

We assume that reattempts occur sufficiently far in the future so that they can be modelled as an independent Poisson process. Thus, the entire stream is Poisson. In a blocked calls cleared system offered Poisson traffic, the blocking is independent of holding time distribution , depends only on its mean, and is given by the classical Erlang B formula. Thus we can derive a nonlinear equation to be solved which involves one unknown, the total offered load. This can be solved numerically using the method of successive approximations.

This analysis indicates that the impact can be dramatic. If the fraction of MF calls to a failed IC is only 10%, the overall blocking will be 20% (at ABSBH loads) if the reattempt rate were .8.

9. POSSIBLE CONTROL STRATEGIES

Two general approaches to mitigating these effects are to provide additional equipment or to require that failed IC switches send a failure indication. A similar alternative is to monitor completion ratios to each IC network. If the completion ratio were zero, these calls could be blocked from attempting to seize a transmitter. If the ratio were nonzero, caution must be exercised. "Call gapping" controls could then be used to limit the rate at which calls would be sent to the IC in question. It is more difficult to quantify the effects of IC failures on queueing systems. Because of the effects of abandonments, different techniques must be employed. An approach to this kind of problem has recently been proposed by Sze in [4]. In [4], the problem of non exponential holding times is also considered.

ACKNOWLEDGEMENTS

We wish to thank J. M. Bennett of Bell Communications Research for his analysis of the field data and E. H. Lipper of AT&T Bell Laboratories for his work on the worst case traffic model.

REFERENCES

[1] L. J. Forys "Performance Analysis Of A New Overload Strategy", Proceedings of 10th ITC, Montreal, Canada, 1983.

[2] L. J. Forys "A Characterization Of Traffic Variability For SPC Systems", Proceedings of the 9th ITC, Torremolinos, Spain, 1979.

[3] L. Burkard, J. J. Phelan, M. D. Weekly, "Customer Behavior And Unexpected Dial Tone Delay", Proceedings of th 10th ITC, Montreal, Canada, 1983.

[4] D. Y. Sze "A Queueing Model For Overload Analysis", submitted for presentation at the International Seminar On Computer Networking and Performance Evaluation", Tokyo, Japan, 1985.

[5] J. M. Holtzman, D. L. Jagerman, "Estimating Peakedness From Arrival Counts", Proceedings of the 9th ITC, Torremolinos, Spain, 1979.

[6] J. A. Harr, Mrs. E. S. Hoover, R. B. Smith, " Organization Of The No. 1 ESS Stored Program", The Bell System Technical Journal, pp1923-1960, September 1964.

TELETRAFFIC ISSUES in an Advanced Information Society
ITC-11
Minoru Akiyama (Editor)
Elsevier Science Publishers B.V. (North-Holland)

MODELS FOR SWITCHING NETWORKS WITH INTEGRATED VOICE AND DATA TRAFFIC

A. DESCLOUX

Bell Communications Research
Morristown, New Jersey, U.S.A.

ABSTRACT

Digital networks designed to handle voice and data traffic simultaneously must satisfy several requirements. In particular voice traffic should suffer only very short and infrequent delays to avoid clipping and data traffic must be practically error-free when it reaches its destination. In this paper we investigate models for burst switching, a form of circuit switching in which each speech burst is assigned to a virtual (transparent) channel for its whole duration. We assume throughout that voice is given priority over data. Under this and other conditions stated in the text, we present formulas for the computation of the delay distributions for both voice and data traffic.

1. INTRODUCTION

Digitized voice and data transmissions through switching networks can be achieved in many different ways. There are two extreme modes of operation : datagram-switching and circuit-switching. In the first instance, the digitized information is split into small blocks (single packets or datagrams involving a fixed number of bits) which are sent one by one through the network. In the case of circuit switching exclusive use of a channel is provided for the whole duration of a transaction between two end-points.

Packet switching involves some overhead which includes the time needed for packetization and headers specifying the packet destinations. In addition, since the packets of a given message may follow different routes through the network, the packets may not arrive in chronological order at the terminating nodes and a fixed delay of sufficient length must be introduced at these end-points to allow for accurate message reconstruction. In datagram networks the reconstruction delay may be the principal component of the overall end-to-end delay and some form of virtual circuit-switching may be used to decrease or eliminate its impact.

Circuit switching is best suited for relatively long messages and is still the main approach used today in voice transmission. With the advent of new technologies, however, the assignment of a channel for the whole duration of a telephone conversation is wasteful. The speech activity (i.e. the proportion of time during which active talkers generate sounds) usually falls well below 50% and so-called burst or fast circuit-switching systems are now being investigated [2] in which a channel (circuit) is dedicated to a burst only for the duration of the burst. (This technique, also known as Time Assignment Speech Interpolation or TASI for short, was first used to increase the capacity of the transatlantic undersea telephone cables.) In voice transmission the reconstruction delays must be avoided. As a result the channels connecting any two nodes must have exactly the same transmission speed and, in multinode networks, all the bursts of a given conversation must pass through the same nodes. When compared to datagram switching, this fixed assignment of a conversation to a fixed end-to-end route , however, makes for a somewhat less efficient use of the network capacity. But there is a compensating factor, namely that the bursts are generally much longer than the datagrams and therefore more header-efficient.

It is generally recognized that single-packet transactions are ideal for datagrams and that relatively long messages (speech-bursts fall in that category) are best handled on a virtual-circuit basis. (For intermediate modes of operation see Ref. [5].)

In burst-switching systems, the talkers are not aware of the clipping and/or of the loss of their respective talk-spurts. Hence, from the talkers' point of view, the burst-streams are unaffected by the momentary states of the system and do not interfere with each other and the system appears to be of infinite capacity. In the telephone traffic literature, this situation is referred to as the "blocked call held assumption". From the listener's point of view, the delayed bursts that are longer than their respective waiting-times are clipped and those that are shorter than the time needed for a circuit to become available are lost (not heard at all). The quality of voice transmission is, therefore, directly related to the talk-spurt delay-distribution. In the sequel the terms delay and waiting-time always refer to the time intervals during which bursts are denied access to a channel and may be either mutilated or even lost. We stress that the portion of a burst that originates when all the channels are busy is not transmitted unless queueing is allowed.

In this paper we first consider the situation where the talk spurts that arrive when all the channels are busy are either clipped or even totally lost. Alterations of this sort must be relatively infrequent for speech intelligibility to remain of acceptable quality. This can be achieved by making sure that the talk-spurt access-delays are sufficiently short and, if necessary, by giving priority to voice traffic and/or recoding (such as dropping the least significant bit). Allowing a talker's bursts to queue up for a short period of time and/or recoding at lower bit rates may help improve speech integrity by reducing clipping. These are other possible options which we shall also investigate. By contrast data communication is not as adversely affected by short delays but all the bits must be delivered as accurately as possible. Here we shall consider the case where data are transmitted only if some channels are not needed for voice transmission.

The main purpose of this paper is to derive formulas for the delay distributions for voice and data. More specifically we shall obtain the voice access delay-distributions first when there is no provision for queueing and then examine the case where buffers provide limited waiting capability. For the reasons given earlier, we shall always suppose that voice is given priority over data. Information pertaining to the data-delays will be given in terms of the conditional voice busy-period distributions.

We shall assume throughout that the voice bursts and data segments arrive according to independent Poisson processes. In all cases we first assume that the burst lengths are exponentially distributed and then indicate whether this condition can be relaxed or not. In particular we shall show that the standard burst-switching model can be completely analyzed for arbitrary burst-length distributions.

Space limitations makes it impossible to give complete derivations of the formulas. Full details and expanded presentation of numerical results will be published elsewhere.

2. BURST-SWITCHING WITHOUT STORAGE BUFFERS

In this section, we first describe the simplest traffic model for burst switching. The underlying assumptions are as follows:

1. bursts (talk-spurts) arrive according to a Poisson process of intensity α,
2. the burst lengths are exponentially distributed with mean set equal to 1,
3. bursts generated when all the circuits are busy are either partially served or not at all: they are partially served (i.e., some of their leading bits are lost) if a circuit becomes available before they terminate and are totally lost otherwise.
4. access to the circuits is on a first-come first-served basis,
5. the maximum number of unserved bursts at any one time cannot exceed m, a fixed but arbitrary limit.

Let c be the number of circuits and write p_n, $n=0,1,...,c+m$, for the equilibrium state-probabilities. Then[3]

$$p_n = q \cdot \alpha^n / n!, \; n=0,1,\ldots,c+m,$$

where

$$q^{-1} = \sum_{n=0}^{c+m} p_n .$$

Next, let $W_n(t)$ be the probability that a burst which on arrival finds $c+n$, $n \geqslant 0$, demands in the system must wait at least t before it either has access to the system or leaves it without being served at all. This conditional probability is usually expressed as follows (see Ref.[3]) :

$$W_n = e^{-(c+1)t} \frac{c(c+1)\ldots(c+n)}{n!} \sum_{k=0}^{n} (-1)^k \binom{n}{k} e^{-kt} (c+k)^{-1}.$$

This formula is not well suited for computation as it involves terms that are alternately positive and negative. It can however be shown that it is equivalent to the following expression:

$$W_n(t) = c\binom{c+n}{n} e^{-(c+1)t} \sum_{i=0}^{n} \binom{n}{i} (1-e^{-t})^i (c+1)^{-1} \binom{c+n}{n-i}^{-1},$$

and the unconditional probability that a burst must wait at least t is given by the formula

$$W(t) = \sum_{n=0}^{m-1} p_{c+n} W_n(t)$$

$$= cp_c e^{-(c+1)t} \sum_{n=0}^{m-1} \frac{\alpha^n}{n!} \sum_{i=0}^{n} \binom{n}{i} (1-e^{-t})^i (c+i)^{-1} \binom{c+n}{n-i}^{-1},$$

which is the delay-distribution implemented in the subjective-test simulator. (In the preceding expression and in the sequel bursts arriving when the system is in state $c+m$ are ignored and the p's normalized accordingly. The possible states at arrival instants are, therefore, 0,1,...,c+m-1).

Two other delay probabilities are also of interest: the probability, $W_S(t)$, that a burst is eventually served but clipped for a time interval of length t at least and the probability, $W_L(t)$, that a burst waits at least t but defects before a circuit becomes available. Since

$$W_S(t) = \int_t^\infty e^{-u} \, dW(u)$$

and

$$W_L(t) = \int_t^\infty e^{-u} \, W(u) \cdot du ,$$

we have :

$$W_S(t) = cp_c e^{-(c+1)t} \sum_{n=0}^{m-1} \frac{\alpha^n}{n!} \sum_{i=0}^{n} \binom{n}{i} (1-e^{-t})^i (c+i+1)^{-1} \binom{c+i+1}{n-i}^{-1}$$

and

$$W_L(t) = cp_c e^{-(c+1)t} \sum_{n=0}^{m-1} \frac{\alpha^n}{n!} \sum_{i=0}^{n} \binom{n}{i} (1-e^{-t})^i \left[(c+i)^{-1} \binom{c+n}{n-i} - (c+i+1) \binom{c+n+1}{n-i}^{-1} \right].$$

The probability, P_S, that a burst receives partial service and the probability, P_L, that a burst which enters the system is completely lost are given, respectively, by the formulas

$$P_S = W_S(0) = \frac{c}{c+1} p_c \sum_{n=0}^{m-1} \frac{\alpha^n}{n!} \binom{c+n+1}{c+1}^{-1}$$

and

$$P_L = W_L(0) = cp_c \sum_{n=0}^{m-1} \frac{\alpha^n}{n!} \left[\frac{1}{c} \binom{c+n}{c}^{-1} - \frac{1}{c+1} \binom{c+n+1}{c+1}^{-1} \right].$$

From these two relations we find that

$$P_S - P_L = p_c \sum_{n=0}^{m-1} \frac{\alpha^n}{n!} \binom{c+n}{c}^{-1} \cdot \frac{c-(n+1)}{c+n+1}$$

$$= \sum_{n=0}^{m-1} p_{c+n} \cdot \frac{c-(n+1)}{c+n+1}.$$

Hence, as one would expect, P_S is larger than P_L if and only if, during periods of congestion, circuit releases are more likely to occur than defections from waiting positions. A sufficient condition for P_S to be larger than P_L is that the number of circuits be at least as large as the number of waiting positions.

For fixed values of α and c, the ratio

$$P_L/P_S = \frac{c+1}{c} \sum_0^{m-1} \frac{\alpha^n}{(c+n)!} / \sum_0^{m-1} \frac{\alpha_n}{(c+n+1)!} - 1$$

increases as m increases and, therefore, cannot exceed

$$\lim_{m \infty} \frac{c+1}{c} \sum_0^\infty \frac{\alpha^n}{(c+n)!} / \sum_0^\infty \frac{\alpha^n}{(c+n+1)!} - 1$$

$$= \left(\frac{c+1}{c}\right) \alpha \cdot \frac{1-P(c-1,\alpha)}{1-P(c,\alpha)} - 1$$

$$= (c+1)\alpha \frac{\int_0^\alpha e^{-u} u^{c-1} du}{\int_0^\alpha e^{-u} u^c du} - 1,$$

where $P(n,\alpha)$ is the Poisson summation:

$$P(n,\alpha) = \sum_0^n \frac{\alpha^u}{n!} e^{-\alpha}.$$

The computation of $W(t)$ and of its associated frequency function, $w(t) = -dW(t)/dt$, was carried out over a wide parameter range by John Cavanaugh and his results were used as an input to subjective tests of burst switching. [1] Some of these frequency functions are reproduced in Figures 1 and 2. As can be seen from these graphs, the shape of these curves show strong dependence on the parameter choices. Clearly the assumption often made that $W(t)$ is exponential (or at least approximately so) is not tenable.

The behavior depicted in Figure 2 is particularly interesting and is sufficiently odd to require an explanation. To this end we note that $W(t)$ is a mixture of two probability distributions, namely W_S and W_L with weights P_S and P_L, respectively. The density functions w_S and w_L corresponding to W_S and W_L behave quite "normally". The following properties can be proved:

1. w_L is always monotonically decreasing as t increases,
2. w_S is a monotonically decreasing function of t whenever $\alpha \leqslant c+1$,
3. for $\alpha > c+1$, and as t increases, w_S first increases, reaches a maximum value for some $t>0$ and then decreases monotonically.

As noted earlier, the probability P_L increases as m increases and we therefore expect the steady decrease of w_L over $[0,t]$ to have a more pronounced effect (even create the observed dips) on the overall behavior of $w(t)$ as m increases. This fact is indeed corroborated by the examples presented in Figure 2.

Next we extend the previous formulas to arbitrary burst-length distributions except that their means are assumed to be finite and set equal to 1. Leaving all the other features of the model unchanged, it can be shown that the equilibrium probabilities are the same as before (see Ref. [6]). Furthermore let

$$P(n,x_1,x_2,\ldots,x_n)$$

be the probability that, at a random instant, there are n bursts in the system and that their residual stays are $x_1,x_2,\cdots,x_n$. Let H be the cumulative burst-length distribution. Then we have [6]

$$P(n,x_1,x_2,\ldots,x_n) = p_n \prod_1^n [1-H^*(x_i)] \ ,$$

where

$$H^*(x) = \int_0^x [1-H(u)]\cdot du \ .$$

We now consider an hypothetical test burst of unlimited duration and let V be the probability that it waits at least t. Then making use of the identity

$$\binom{n}{k}\binom{k}{i} = \binom{n}{i}\binom{n-i}{k-i}$$

V is given by the following expression:

$$V(t) = cp_c\,[1-H^*(t)]^c \sum_0^{m-1} \frac{\alpha^n}{n!} \cdot \sum_{k=0}^{n} \binom{n}{k} \binom{c+n}{n-k}^{-1} (c+k)^{-1}[H^*(t)]^k \ .$$

and the equilibrium probability that a burst must wait at least t is given by the following formula:

$$W(t) = [1-H(t)]\cdot V(t)$$

with V(t) as above.

The probabilities W_S and W_L can be expressed, as before, in terms of $W(t)$:

$$W_S(t) = \int_t^\infty [1-H(u)]\, dW(u)$$

and

$$W_L(t) = \int_t^\infty W(u)\, dH(u)$$

3. BURST-SWITCHING WITH FLOW-THROUGH BUFFERS

In this section we consider a group of c circuits and assume that the bursts arrive according to a Poisson process of intensity λ, that they are served in order of arrival, and that their lengths are exponentially distributed. Throughout the mean burst-duration is taken as unit of time.

Input buffers are provided to allow the bursts to enter the system even if all the channels are occupied and we make the assumption (easily removed) that the number of buffers is sufficiently large so that losses due to all-buffer busy conditions are negligible.

So long as channels are available, the bit streams of the talk-spurts are not delayed at all and flow unimpeded through their assigned buffers. The individual bits of a delayed talk- spurt are placed sequentially in a buffer and the bits of a talk-spurt that arrive when their assigned buffer is full displace the oldest waiting bits to accommodate the newer bits. Thus a delayed burst is served in its entirety if and only if its leading bit does not have to wait longer than θ, the capacity of the buffer expressed in time units (individual bits cannot wait longer than θ).

Although the talk spurts are digitized and, perhaps transmitted on a time-division basis with fixed time slot-sizes, we shall treat the burst flow as if it were a time-continuous process. As far as burst-switching is concerned this simplification has no appreciable effect on the results presented below.

As noted above, the bursts fall into three categories depending on whether they are fully served, clipped and partially served or not served at all. Statistics for the latter can be investigated separately since we have assumed that losses due to buffer shortages are negligible. All the other bursts will then have service times which are exponentially distributed with mean λ. Hence, as far as delays are concerned, all we have to do is to remove from the traffic flow the bursts that are not served and, as we shall see, the problem is then reduced to a study of a delay-system with state-dependent arrival rates which we now proceed to derive.

Consider a random instant at which q bursts that will eventually be served are waiting for transmission. The probability $W_q(t)$ that an hypothetical burst arriving at that instant would have to wait at most t is given by the formulas [3] :

$$W_q(t) = 1 - e^{-ct} \sum_{k=0}^{q} \frac{1}{k!}(ct)^k = \frac{1}{q!} \int_0^c t\, e^{-u}\, \mu^q du. \qquad (1)$$

This would be the delay distribution of a bursts allowed to wait as long as necessary to get served. Our assumptions preclude this to happen and the probability γ_q that the burst has access to a free circuit is equal to :

$$\gamma_q = \int_0^\theta dW_q(t) + \int_\theta^\infty pr[S \geqslant t-\theta] dW_q(t), \qquad (2)$$

where S stands for the service time. γ_q is the conditional probability that a burst arriving when q other bursts are waiting for service will have access to a circuit.

This last formula is easily proved. The first term in (2) is the probability that a delayed burst is served without clipping. For this to happen the waiting time must not exceed θ, the capacity of the buffer. The integrand of the second term is equal to the (infinitesimal) probability that the delay falls between t and t and $t+dt$ and that, at time t, bits are still in the buffer waiting to be served. The second integral is thus equal to the probability that a burst is clipped but eventually reaches a circuit for partial service.

By means of (1) we obtain the following expression for γ_q:

$$\gamma_q = \frac{1}{q!}\int_0^{c\theta} e^{-u}\,\mu^q du + \frac{e^\theta}{q!}\left[\frac{c}{c+1}\right]^{q+1}\int_{\theta(c+1)}^{\infty} e^{-u}\,\mu^q\,du \quad (3)$$

$$= 1 - \left[1 + c\theta + (c\theta)^2 + \ldots \frac{(c\theta)^q}{q!}\right] e^{-c\theta}$$

$$+ \left[\frac{c}{c}+1\right]^{q+1}\left[1+(c+1)\theta + \frac{[(c+1)\theta]^2}{2!} + \ldots + [(c+1\theta]^2 + \frac{[(c+1)\theta]^q}{q!}\right] e^{-c\theta}.$$

Let N be the number of bursts either waiting or being served at a random instant. We stress that the bursts that do not get any service are instantly dismissed and are not counted. Let

$$P_n \equiv Pr\left[N = n\right]$$

The equilibrium state probabilities, p_n, n = 0,1, . . ., are then readily found:

$$P_n = p_0\frac{\lambda^n}{n!}, \quad n = 0,1,\ldots,c, \quad (4)$$

$$P_{c+q} = \dot{p}_0\frac{\lambda^{c+q}}{c!}\,\frac{\gamma_0\,\gamma_1\,\gamma_q}{c^q}, \quad q = 0,1,\ldots,$$

where p_0 is such that $\sum_{n=0} p_n$ ~=~ 1.

The preceding formulas allow us to compute the following quantities:

(i) the probability, P_L, that a burst is not served at all (i.e. entirely lost):

$$P_L = \sum_{q=0} p_{c+q}\left[1-\gamma_q\right], \quad (5)$$

(ii) the probability, P_S, that a burst is served without clipping

$$P_S = \sum_{n=0}^{c-1} p_n + \sum_{q=0} p_{c+q}\left[1 - \sum_{m=o}^{q}\frac{(c\theta)^m}{m!}.e^{-c\theta}\right\}, \quad (6)$$

(iii) the probability, P_{CS}, that a burst is served but clipped

$$P_{CS} = \sum_{q=o} p_{c+q}\left[\frac{c}{c+1}\right]^{q+1}\left[1 - \sum_{m=0}^{q}\frac{[(c+1)\theta]^m}{m!}\,e^{-c\theta}\right], \quad (7)$$

(iv) the probability, $W(t)$, that a burst is delayed and eventually served after waiting at most t:

$$W(t) = \sum_{q=0} p_{c+q}\,W_q\,(t) \quad (8a)$$

$$= \sum_{q=0} p_{c+q}\left[1 - \sum_{n=0}^{q}\frac{(ct)^n}{n!}\,e^{-ct}\right], \quad \text{if } 0 \leqslant t \leqslant \theta$$

and

$$W(t) = W(\theta) + \sum_{q=0} p_{c+q}\int_\theta^t e^{-(u-\theta)}dW(u) \quad (8b)$$

$$= W(\theta) + \sum_{q=0} p_{c+q}\left[\frac{c}{c+1}\right]^q \left\{\sum_{n=0}^{q}\frac{[(c+1)\theta]^n}{n!}\,e^{-c\theta} - \sum_{n=0}^{q}\frac{[(c+1)t]^n}{n!}\,e^{-ct}\right\}, \quad t \geqslant \theta.$$

4. BURST-SWITCHING WITH FLUSHING BUFFERS

The models of this and of the preceding sections differ only as far as buffer clearing is concerned. There the buffers are no longer emptied bit-by-bit as the need to accommodate new bits arises: we shall now assume that the buffers are instantaneously freed of their entire content as soon as their capacity is exceeded. The analysis that follows is similar to that of Section 2 and we shall retain the notation used there, adding however stars to quantities that may differ numerically.

The probability, γ^*_q, that a burst arriving when q other bursts are waiting is eventually served is now given by the formula:

$$\gamma^*_q = \int_0^\theta dW_q\,(t) + \int_\theta^\infty Pr\left[S \geqslant \left[t/\theta\right].\theta\right]\,dW_q\,(t) \quad (9)$$

where $[t/\theta]$ is equal to the largest integer that does not exceed t/θ

Let

$$\Pi_{qn} = \frac{1}{q!}\int_{n\theta/c}^{(n+1)\theta/c} e^{-u}\,u^q\,du.$$

Then

$$\gamma^*_q = \sum_{n=0}^{\infty}\Pi_{qn}\,e^{-n\theta} \quad (10)$$

and the equilibrium state probabilities are now given by the formulas

$$P^*_n = p^*_0\frac{\lambda^n}{n!}, n = 0,1,\ldots, c,$$

$$p^*_{c+q} = p^*_0\frac{\lambda^n}{c!}\,\frac{\gamma^*_0\,\gamma^*_1\cdots\gamma^*_q}{c^q}, \quad q = 0,1,\cdots$$

where p_0 is such that $\sum_{n=0} p^*_n = 1$.

The probabilities P^*_L, P^*_S are given by formulas (5), (6) and (7), respectively, with the p's replaced by the p^*'s of this section.

Finally we note that

$$W^*(t) = \sum_{q=0} p^*_{c+q}\,W_q\,(t) \quad \text{if } 0 = t \leqslant \theta$$

$$W^*(t) = W^*(\theta) + \sum_{q=0} p_{c+q}\int_\theta^t e^{-[u/\theta].\theta}\,dW_q\,(u)$$

$$= W^*(\theta) + \sum_{q=0} p^*_{c+q}\left\{\sum_{n=1}^{[t/\theta]-1} e^{-n\theta}\int_{n\theta}^{(n+1)\theta} dW_q\,(u)\right.$$

$$+ e^{-(n+1)\theta} \int_{(n+1)\theta}^{t} dW_q(u) \Big\}.$$

5. ON THE IMPACT OF ROUTING

In this section we show how the delay distribution can be computed when the end-to-end connections can be established via different sets of nodes. The following considerations are restricted to the case where two routes (referred to as route 1 and 2, respectively) are available to set up a new call, each new call being assigned to the route with the least amount of traffic.

Let p(n,m) be the equilibrium probability that there are n calls in progress on route 1 and m calls in progress on route 2. Assuming the input to be random with intensity 2λ and the burst-lengths to be exponentially distributed, we then have the following set of equilibrium equations :

$$(2\lambda+n+m)p(n,m) = (n+1)p(n+1,m) + (m+1)p(n,m+1) + 2\lambda p(n-1,m) \quad , n-1 < m,$$

$$(2\lambda+n+m)p(n,m) = (n+1)p(n+1,m) + (m+1)p(n,m+1) + 2\lambda p(n,m-1) \quad , m-1 < n,$$

$$(2\lambda+n+m)p(n,m) = (n+1)p(n+1,m) + (m+1)p(n,m+1) + \lambda p(n-1,m) \quad , n-1 = m,$$

$$(2\lambda+n+m)p(n,m) = (n+1)p(n+1,m) + (m+1)p(n,m+1) + \lambda p(n,m-1) \quad , n = m-1.$$

By themselves these equations cannot be solve recursively but we can proceed as follows. Let

$$G(x,y) \equiv \sum_{n,m} p(n,m)x^n y^m .$$

Under the present assumptions, G is a symmetric function of x and y and it is then readily proved that it satisfies the following partial differential equation of the first order :

$$\lambda(2-x-y)G(x,y) + (x-1)\partial G(x,y)/\partial x + (y-1)\partial G(x,y)/\partial y = 0$$

whose general solution is of the form

$$G(x,y) = e^{-\lambda(2-x-y)} \cdot f(u), \quad u = (1-x)/(1-y),$$

and where f is an arbitrary function of u.

By Faa di Bruno formula (see Ref. [5]) we have

$$\left[\frac{\partial^n G(x,y)}{\partial x^n}\right]_{x=y=0} = \sum_{k=0}^{n} \binom{n}{k} \lambda^{n-k} \left[\frac{\partial^k f(u)}{\partial x^k}\right]_{u=1}$$

$$= \sum \binom{n}{k}^{n-k} \lambda^{n-k} (-1)^n f_k .$$

and

$$\left[\frac{\partial^n G(x,y)}{\partial y^n}\right]_{x=y=0} = \sum \binom{n}{k} \lambda^{(n-k)} \sum \frac{n! f_k}{k_1! \cdots k_n!} ,$$

where $f_k \equiv [\, d^k f(u)/du^k \,]_{u=1}$, $k = k_1+k_2+ \cdots k_n$ and the sum over all solutions in non-negative integers of $k_1+2k_2+ \cdots +nk_n = n$.

Since G is a symmetric function of its arguments we must have

$$\frac{\partial^n G(x,y)}{\partial x^n} = \frac{\partial^n G(x,y)}{\partial y^n}$$

so that

$$\sum \binom{n}{k} \lambda^{n-k} \left[(-1)^k f_k - \sum \frac{n!\, f_k}{k_1!+k_2! \cdots + k_n!} \right], \qquad (11)$$

$$n=1,2,3,\cdots .$$

Let $L_{n,k}$ be the coefficient of f_k in the summation of the preceding equation. It can be shown that

$$L_{n+1,k} = (n+k)\, L_{n,k} + L_{n,k-1} \,, n \geqslant k \,.$$

Since $L_{1,1} = 1$ and $L_{1,j} = 0$ for $j > 1$, the f_k are Lah's numbers taken positively (see Ref. [4])

Relations (11) , together with the equilibrium state equations, can be used to compute the state probabilities step-by-step and the delay distribution can be determined as in Section 2. The details are omitted.

Clearly the probabilities

$$P_k \equiv \sum_{n+m=k} p(n,m)$$

are independent of the burst-length distribution but the p(n,m)'s do depend on that distribution. The results of this section should, however, provide useful information on the impact of routing. They could also be used iteratively to obtain bounds for the cases where there are more than two routes.

6. DATA-HANDLING CAPABILITIES

Since we have assumed that voice is always given priority over data, several parameters and distributions pertaining to the latter are readily obtained :

1. The equilibrium distribution of the maximum amount of data-traffic that can be carried at any instant.
2. The distribution of the time intervals when at least one channel is not occupied by voice traffic and during which data may be transmitted.
3. The distribution of the maximum amount of traffic that can be transmitted during such intervals.

Insight into the delays encountered by the data-traffic may be gained by considering the voice busy-period distribution. Let $W_D(t)$ be the probability that a data-packet or segment has to wait at least t and let $B(t)$ be the complementary distribution of the voice busy period. Then we always have :

$$W_D \geqslant B(t) \,, \text{ for } t \geqslant 0 \,.$$

The distribution B(t) is given by the formula

$$B(t) = \exp \left[-(c+\lambda) + \lambda(1-e^{-x}) \right]$$

The exact cumulative waiting-time distribution of the leading bit of a data segment is related to B(t). If we designate it by $\tilde{W}(t)$ we have:

$$\tilde{W}(t) = \sum_{i=0}^{\infty} p_{c+i} \int_{0}^{t} F_{c,i}(t-y)\, d(1-B_c(y))$$

where $F_{c,i}$ stands for the distribution of the first return to state c from state $c+i$. Either of the two preceding distributions can be used to decide whether the delays suffered by the data traffic are of acceptable magnitude or not.

7. SUMMARY

In this paper we have described the main features and options of burst switching. We have also presented several formulas that can be used - and are being used - to analyze the performance of burst switching systems from a traffic point of view. Results of this investigation will appear in a more comprehensive document.

REFERENCES

[1] J. R. Cavanaugh and J. R. Rosenberger, Results of study of the speech transmission performance of burst and packet mode switching - phase 1 - burst mode, to be published.

[2] E. F. Haselton, A PCM frame switching concept leading to burst switching network architecture, IEEE Communication Magazine, Vol. 21, No. 6, pp. 13-19, September 1983.

[3] J. Riordan, Stochastic Service Systems, Wiley, New York, 1962.

[4] J. Riordan, An Introduction to Combinatorial Analysis, Wiley, New York, 1958.

[5] R. D. Rosner, Packet Switching, Lifetime Learning Publications, Belmont, CA, 1982.

[6] L. Takács, On Erlang's formula, Annals of Math. Statist., Vol. 40, No. 1, pp. 71-78, 1969.

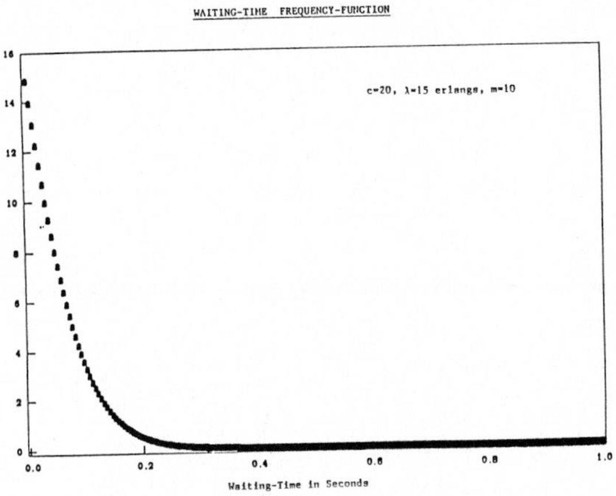

Fig. 1

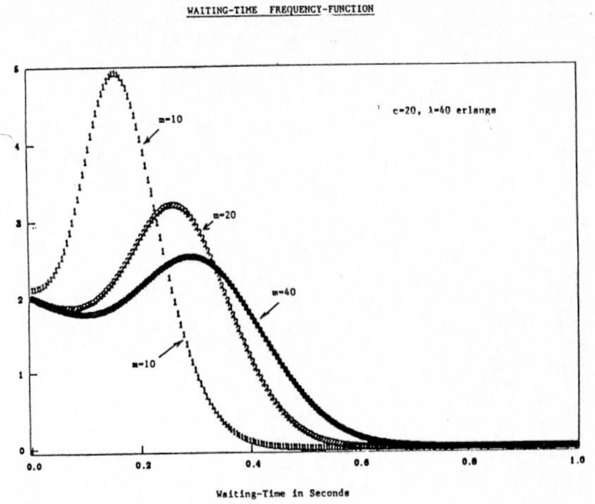

Fig. 2

TELETRAFFIC ISSUES in an Advanced Information Society
ITC-11
Minoru Akiyama (Editor)
Elsevier Science Publishers B.V. (North-Holland)

ON DELAY AND LOSS IN A SWITCHING SYSTEM FOR VOICE AND DATA WITH INTERNAL OVERFLOW

Georg BRUNE

Institute for Electronic Systems and Switching,
University of Dortmund, Dortmund, FRG

ABSTRACT

This paper deals with two types of combined delay loss systems which are suitable for overload protection in switching systems with two different types of offered traffic. Besides the protection of priority calls against overload of ordinary calls by means of cut-off priorities, an additional protection of ordinary calls against overload of priority calls is achieved by means of internal overflow. An important application of such systems is, e. g., possible in integrated switching systems for voice and data.

Therefore, besides a performance analysis of systems with the same mean holding time for all calls, a method for systems with different mean holding times of ordinary and priority calls is also presented.

Exact values of state probabilities and characteristic traffic values as well as waiting time distributions are calculated. Results are presented in diagrams.

1. INTRODUCTION

In many switching systems with two types of offered traffic, A_1 and A_2, only the traffic A_2 is usually concerned by overload. It is well known that in this case so called cut-off priorities are suitable for the protection of traffic A_1 against overload of the traffic A_2 [1, 2, 5, 6, 7, 8, 9, 10]. If, however, an overload of the traffic A_1 can also occur, the delay and loss probabilities for the traffic A_2 can increase to a very high level in such systems.

Therefore, this paper deals with systems which additionaly are suitable for a certain protection of the traffic A_2 in situations in which A_1 is overloaded. The models considered operate as combinations of a loss system and a delay-loss system. The protection of traffic A_1 against overload of A_2 is achieved by means of a certain kind of trunk reservation with cut-off priorities, and the protection of traffic A_2 against overload of A_1 with the aid of internal overflow.

After a detailed description of the modeling of the considered systems in section 2, the characteristic traffic values are calculated exactly in section 3, and the distribution of waiting times in section 4, for the case of equal holding times of the traffic parts A_1 and A_2. Because of the importance for the application to combined systems for voice and data, the characteristic traffic values for systems with different mean holding times are investigated in section 5. A method for the exact calculation of the corresponding waiting time distribution is presented in section 6.

Results are shown in several diagrams.

2. MODEL DESCRIPTION

In both systems considered here, two independent types of traffic A_1 and A_2 arrive at a group of n trunks of a switching system. From the n trunks, k are reserved for calls of the traffic A_2. For the traffic A_1, the remaining n-k trunks act as a full access group, operating as a loss system. For the traffic A_2, both models operate as a delay-loss system with s waiting places. Waiting calls are served according to the discipline "first in first out" (FIFO). Calls of the traffic A_2 can, however, only be switched via one of these n-k trunks if less than m ($m < n-k$) of these n-k trunks are busy. The two models presented here differ with respect to the service of arriving calls of the traffic A_2.

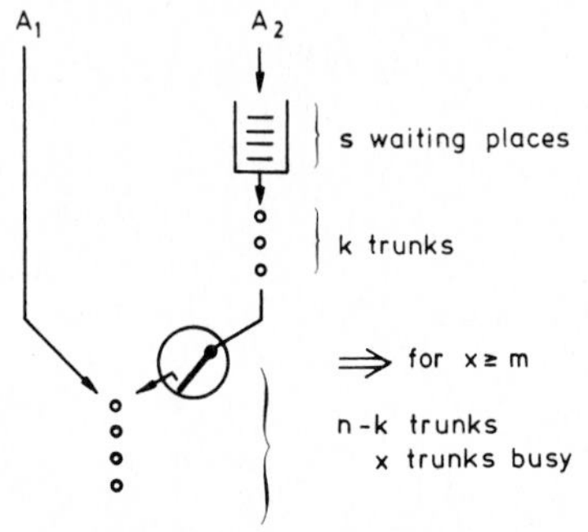

Fig. 1 Model No. 1

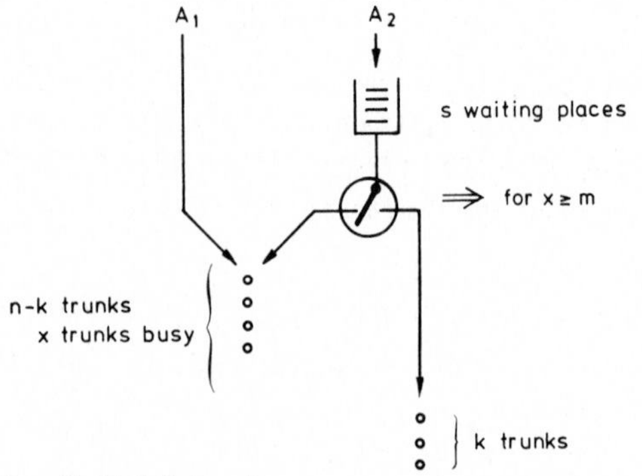

Fig. 2 Model No. 2

In model No. 1, as shown schematically in fig. 1,

arriving calls of the traffic A_2 are first offered to the group of k trunks which is hunted in a full acces mode. If all k trunks are busy, arriving calls of A_2 are overflowing to the group of n-k trunks, provided that less than m trunks are busy in this group. Calls arriving in times when m or more trunks are busy in this group and all k trunks are busy, start waiting if less than s waiting places are occupied. If all k trunks are busy and all s waiting places are occupied, overflowing calls of the traffic A_2 are lost.

In model No. 2, as shown schematically in fig. 2, calls of A_2 are switched via one of the n-k trunks if less than m trunks are busy in this group. If m or more trunks are busy, arriving calls of A_2 are overflowing to the group of k trunks which is hunted in a full access mode. Calls overflowing, when all k trunks are busy, start waiting if less than s waiting places are occupied. If all k trunks are busy and all s waiting places are occupied, overflowing calls of the traffic A_2 are lost in analogy to model No. 1.

In both models, waiting calls of the traffic A_2 can be served if one of the k trunks becomes idle or if one of the n-k trunks becomes idle in a state in which m of these n-k trunks are busy.

3. CHARACTERISTIC TRAFFIC VALUES FOR SYSTEMS WITH UNIFORM MEAN HOLDING TIMES

3.1. Preconditions

It is assumed that the traffic parts A_1 and A_2 are of Poisson type with the same mean holding time h and with the mean arrival rates λ_1 and λ_2. The total offered traffic be denoted by A and the mean terminating rate by ε, so that the following equations holds true:

$$A_1 = \lambda_1 h \qquad (1)$$

$$A_2 = \lambda_2 h \qquad (2)$$

$$A = A_1 + A_2 \qquad (3)$$

$$\varepsilon = \frac{1}{h} \qquad (4)$$

The state of system can be described by the values (x,z), where x is the number of busy trunks in the full access group of n-k trunks, whereas z denotes the sum of the number of busy trunks of the k exclusive trunks for traffic A_2 and the number (z-k) of occupied waiting places in the queue. The stationary probability that the state (x,z) exists be denoted by p(x,z).

3.2. State Equations of Model No. 1

The state diagram for model No. 1 is shown in fig. 3. According to this diagram the following equations are obtained

$$Ap(0,0) = p(1,0)+p(0,1) \qquad (5a)$$

$$(A+x)p(x,0) = A_1p(x-1,0)+(x+1)p(x+1,0) +(z+1)p(x,z+1), \quad 0<x<n-k \qquad (5b)$$

$$(A_2+n-k)p(n-k,0) = A_1p(n-k-1,0) +(z+1)p(n-k,z+1) \qquad (5c)$$

$$(A+z)p(0,z) = p(1,z)+(z+1)p(0,z+1) +A_2p(0,z-1), \quad 0<z<k \qquad (5d)$$

$$(A+k)p(0,k) = p(1,k)+A_2p(0,k-1) \qquad (5e)$$

$$(A+x+z)p(x,z) = A_1p(x-1,z) +(x+1)p(x+1,z)+(z+1)p(x,z+1) +A_2p(x,z-1), \quad 0<x<n-k, \ 0<z<k \qquad (5f)$$

$$(A_2+n-k+z)p(n-k,z) = A_1p(n-k-1,z) +(z+1)p(n-k,z+1)+A_2p(n-k,z-1), \quad 0<z<k \qquad (5g)$$

$$(A+x+k)p(x,k) = Ap(x-1,k)+(x+1)p(x+1,k) +A_2p(x,k-1), \quad 0<x<m \qquad (5h)$$

$$(A+m+k)p(m,k) = Ap(m-1,k)+(m+1)p(m+1,k) +(k+m)p(m,k+1)+A_2p(m,k-1) \qquad (5i)$$

$$(A+m+k)p(m,z) = (m+1)p(m+1,z) +(k+m)p(m,k+1)+A_2p(m,k-1), \quad k<z<k+s \qquad (5k)$$

$$(A+x+k)p(x,z) = A_1p(x-1,z)+(x+1)p(x+1,z) +kp(x,z+1)+A_2p(x,z-1), \quad m<x<n-k, \ k\le z<k+s \qquad (5l)$$

$$(A_2+n)p(n-k,z) = A_1p(n-k-1,z) +kp(n-k,z+1)+A_2p(n-k,z-1), \quad k\le z<k+s \qquad (5m)$$

$$(A_1+m+k)p(m,k+s) = (m+1)p(m+1,k+s) +A_2p(m,k+s-1) \qquad (5n)$$

$$(A_1+x+k)p(x,k+s) = A_1p(x-1,k+s) +(x+1)p(x+1,k+s)+A_2p(x,k+s-1), \quad m<x<n-k \qquad (5o)$$

$$np(n-k,k+s) = A_1p(n-k-1,k+s) +A_2p(n-k,k+s-1) \qquad (5p)$$

including the normalizing condition

$$\sum_{x=0}^{m-1}\sum_{z=0}^{k} p(x,z)+\sum_{x=m}^{n-k}\sum_{z=0}^{k+s} p(x,z)=1. \qquad (5q)$$

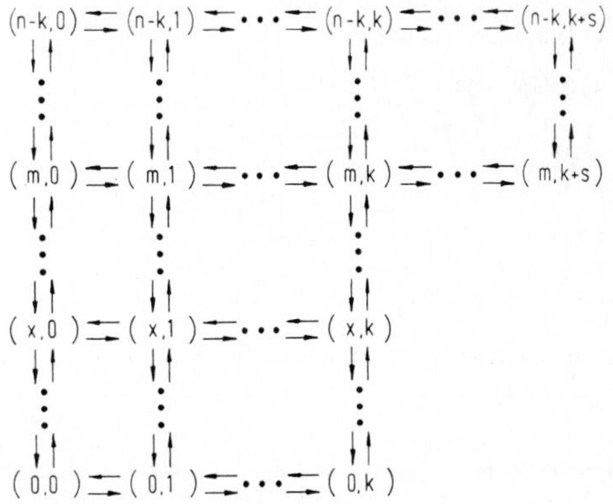

Fig. 3 State diagram of model No. 1

3.3. State equations of model No. 2

The second model differs from the first model by the service discipline as described above. Therefore the following set of state equations is obtained according to the schematic state diagram in fig 4.

$$Ap(0,0) = p(1,0)+p(0,1) \qquad (6a)$$

$$(A+x)p(x,0) = Ap(x-1,0)+(x+1)p(x+1,0) +p(x,1), \quad 0<x<m \qquad (6b)$$

$$(A+x)p(x,0) = A_1p(x-1,0)+(x+1)p(x+1,0) +p(x,1), \quad m<x<n-k \qquad (6c)$$

$$(A_2+n-k)p(n-k,0) = A_1p(n-k-1,0)+p(n-k,1) \qquad (6d)$$

$$(A+z)p(0,z) = p(1,z)+(z+1)p(0,z+1),\quad 0<z<k \tag{6e}$$

$$(A+k)p(0,k) = p(1,k) \tag{6f}$$

$$(A+x+z)p(x,z) = Ap(x-1,z)+(x+1)p(x+1,z) +(z+1)p(x,z+1),\quad 0<x<m,\ 0<z<k \tag{6g}$$

$$(A+m+z)p(m,z) = Ap(m-1,z)+(m+1)p(m+1,z) +(z+1)p(m,z+1)+A_2p(m,z-1),\quad 0<z<k \tag{6h}$$

$$(A+x+z)p(x,z) = A_1p(x-1,z)+(x+1)p(x+1,z) +(z+1)p(x,z+1)+A_2p(x,z-1),\quad m<x<n-k,\ 0<z<k \tag{6i}$$

$$(A_2+n-k+z)p(n-k,z) = A_1(p(n-k-1,z) +(z+1)p(n-k,z+1)+A_2p(n-k,z-1),\quad 0<z<k \tag{6k}$$

$$(A+x+k)p(x,k) = Ap(x-1,k)+(x+1)p(x+1,k),\quad 0<x<m \tag{6l}$$

$$(A+m+k)p(m,k) = Ap(m-1,k)+(m+1)p(m+1,k) +(k+m)p(m,k+1)+A_2p(m,k-1) \tag{6m}$$

$$(A+m+k)p(m,z) = (m+1)p(m+1,z) +(k+m)p(m,z+1)+A_2p(m,z-1),\quad k<z<k+s \tag{6n}$$

$$(A_1+m+k)p(m,k+s) = (m+1)p(m+1,k+s) +A_2p(m,k+s-1) \tag{6o}$$

$$(A+x+k)p(x,z) = A_1p(x-1,z)+(x+1)p(x+1,z) +kp(x,z+1)+A_2p(x,z-1),\quad m<x<n-k,\ k\le z<k+s \tag{6p}$$

$$(A_2+n)p(n-k,z) = A_1p(n-k-1,z)+kp(n-k,z+1) +A_2p(n-k,z-1),\quad k\le z<k+s \tag{6q}$$

$$(A_1+x+k)p(x,k+s) = A_1p(x-1,k+s) +(x+1)p(x+1,k+s)+A_2p(x,k+s-1),\quad m<x<n-k \tag{6r}$$

$$np(n-k,k+s) = A_1p(n-k-1,k+s) +A_2p(n-k,k+s-1) \tag{6s}$$

with the normalizing condition

$$\sum_{x=0}^{m-1}\sum_{z=0}^{k}p(x,z)+\sum_{x=m}^{n-k}\sum_{z=0}^{k+s}p(x,z) = 1 \tag{6t}$$

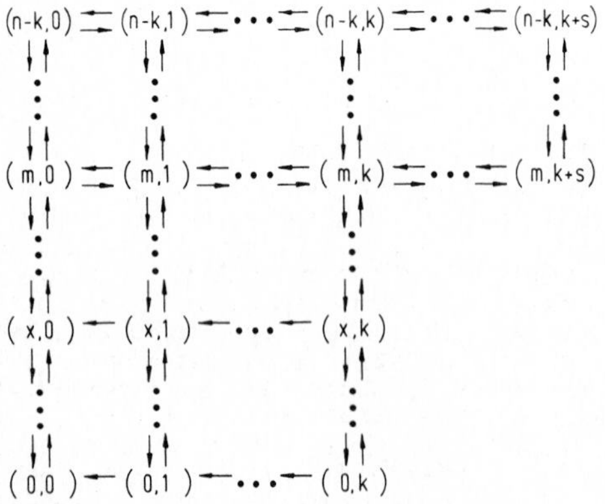

Fig. 4 State diagram for model No. 2

3.4. Calculation of state probabilities

Exact values of the state probabilities for both systems have been calculated iteratively by means of the so-called successive overrelaxation (SOR) method [3].

3.5. Characteristic traffic values

The loss probabilities B_1, B_2 corresponding to the both types of traffic A_1 and A_2 can be determined according to the equations

$$B_1 = \sum_{z=0}^{k+s} p(n-k,z) \tag{7}$$

$$B_2 = \sum_{x=m}^{n-k} p(x,k+s)\ . \tag{8}$$

The total loss probabilty B_t results in

$$B_t = \frac{A_1B_1+A_2B_2}{A}\ . \tag{9}$$

For the waiting probability p_W of calls of the traffic A_2 the following formula holds true

$$p_W = \sum_{x=m}^{n-k}\sum_{z=k}^{k+s-1} p(x,z)\ . \tag{10}$$

The mean queue length Q results in

$$Q = \sum_{z=k+1}^{k+s}\sum_{x=m}^{n-k} (z-k)p(x,z)\ . \tag{11}$$

According to Little's theorem the following formulae are obtained for the mean waiting time t_W of all incoming calls of the traffic A_2 and for the mean waiting time t^*_W of all waiting calls

$$t_W = \frac{Q}{\lambda_2} = \frac{Q}{A_2}h \tag{12}$$

$$t^*_W = \frac{Q}{\lambda_2p_W} = \frac{Q}{A_2p_W}\ h. \tag{13}$$

3.6. Example No. 1

As an example, a system according to model No. 1 (with n=30 trunks, restriction parameter m=13 and s=12 waiting places) and a system according to model No. 2 (with n=30, m=12 and s=15) be considered. The basic offered traffic values (without overload) are $A_{1n} = A_{2n} = 11$ Erlangs in both cases. The diagram in fig. 5 shows the behavior of the loss probabilities B_1 and B_2 and the total loss probability B_t,the waiting probability p_W and the mean waiting time of waiting calls t^*_W/h as a function of the overload factor A_2/A_{2n}, if k = 10 trunks are used exclusively by calls of A_2. The values for model No. 1 are shown by solid lines, and the values of model No. 2 by dashed lines. From fig. 5 it can be seen that, in case of both models, there is an effective protection of calls of A_1 against overload of the offered traffic A_2 (for an overload factor $A_2/A_{2n} = 2$, B_1 is slightly enlarged by a factor of about 1.57 whereas B_2 increases by a factor of about 36.9 in case of model No. 1), as it is well known from systems with cut-off priorities [1, 2, 5, 6, 7, 8, 9, 10].

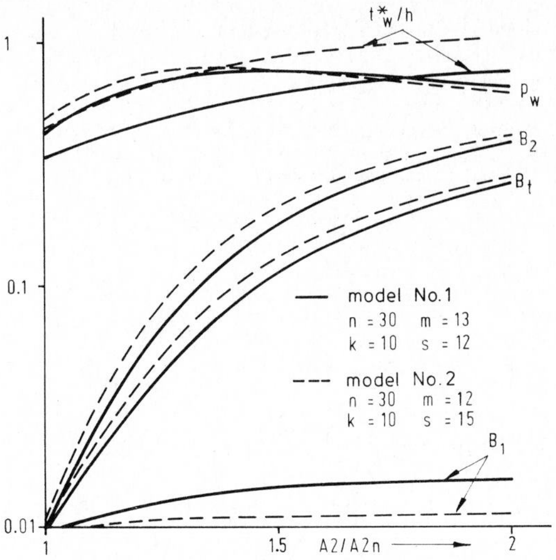

Fig. 5 B_1, B_2, B_t, p_w and t^*_w/h versus overload factor A_2/A_{2n}

For the case of overload of the traffic A_1 in the system considered here, fig. 6. shows the effect of the protection (due to the internal overflow) of the traffic A_2 which, in this case, is not concerned by overload. If A_1 is increased in model No. 1 to a value $A_1 = 2A_{1n}$, the loss probabily B_1 increases by a factor of 22.6 whereas the loss probability of A_2 calls increases only by a factor of about 9.1. By means of the particular dimensioning of the system considered in this example, a lesser intensity of this overload protection by internal overflow has intentionaly been chosen as compared with the overload protection of priority calls by means of cut-off priorities. Of course, the intensities of service protection for A_1 and A_2, respectively, can be adjusted by suitable choice of the values k, m and s. Values of p_w and t^*_w are also shown in fig. 6.

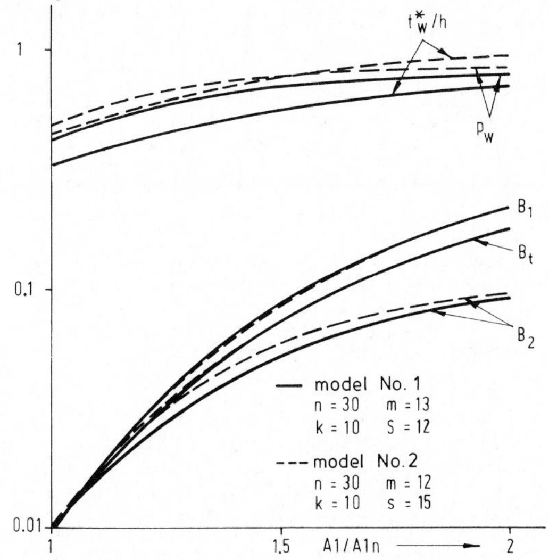

Fig. 6 B_1, B_2, B_t, p_w and t^*_w/h versus overload factor A_1/A_{1n}

From the diagrams shown in fig. 5 and fig. 6 it can be seen that model No. 1 as well as model No. 2 are suitable for the realizing of the desired overload protection. In the example considered above, however, model No. 1 may be given preference as it enables a slightly larger carried total traffic (24.75 Erlangs in model No. 1 versus 24.09 Erlangs in model No. 2) in case of overload of A_1.

4. WAITING TIME DISTRIBUTION FOR SYSTEMS WITH UNIFORM MEAN HOLDING TIME

Calls of the traffic A_2 have to wait if m or more of the n-k trunks and all of the k trunks (which are used exclusively for calls of A_2) are busy and less than s waiting places are occupied, in both models. It is supposed that waiting calls do not give up before being served. The state spaces of the waiting processes, which are identical for the service disciplines in both models, are shown in fig. 7.

4.1. The waiting process

It is well known that the waiting process can be considered as a random walk of waiting calls. The waiting time starts with the arrival of a call of A_2 in a state (x,z) on waiting place (x,z+1) and is terminated by a transition into the taboo set, which is equivalent to the states (x,k) indicated at the left boarder of the state space of the waiting process shown in fig. 7. During this process x and z may vary in the ranges

$$m \le x \le n-k \quad \text{and} \quad k+1 \le z \le k+s.$$

It is obvious that this process is not concerned by further waiting calls, arriving after a considered test call, because the calls are served according to "FIFO" discipline.

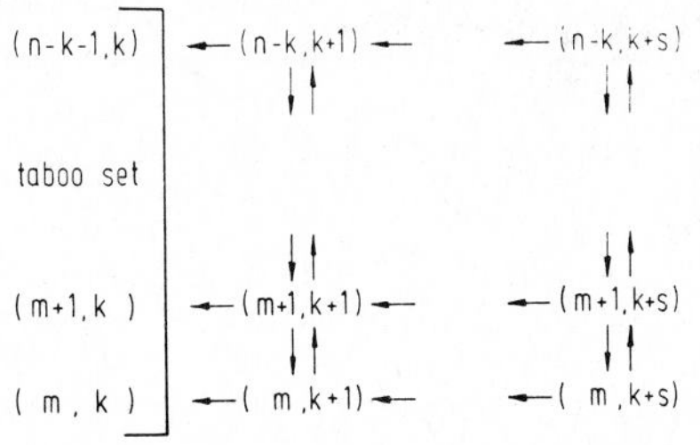

Fig. 7 State diagram of the waiting process

4.2 The waiting time distribution

In the sequel the probability that a waiting call lasts longer than the time t under the condition that the call starts waiting in a state (x,z) be denoted by w(t|x,z). The conditional probability for a transition from a state (x,z) to a state (x',z') not belonging to the taboo set (and avoiding the taboo set) be denoted by q(x,z,x',z') and the conditional probability that a state (x,z) is left by any transition to another state (which may also belong to the taboo set) by q(x,z). According to the state diagram in fig. 7 the following values for these transition coefficients are obtained:

$$q(x,z,x+1,z) = \lambda_1 \;, \quad m \le x < n-k,\, k+1 \le z \le k+s \tag{14a}$$

$$q(x,z,x-1,z) = x\varepsilon \;, \quad m < x \le n-k,\, k+1 \le z \le k+s \tag{14b}$$

$q(x,z,x,z-1) = k\varepsilon$,
$m<x\le n-k, k+1<z\le k+s$ (14c)

$q(m,z,m,z-1) = (m+k)\varepsilon$, $k+1<z\le k+s$ (14d)

$q(x,z) = (x+k)\varepsilon + \lambda_1$,
$m\le x<n-k, k+1\le z\le k+s$ (14e)

$q(n-k,z) = n\varepsilon$, $k+1\le z\le k+s$. (14f)

All other values $q(x,z,x',z')$ are equal to zero. For the conditional (complementary) distribution funktion $w(t|x,z)$, the following set of differential equations holds true which is well known as the so-called Kolmogorov backward equation:

$$\frac{d}{dt} w(t|x,z) = -q(x,z)w(t|x,z) + q(x,z,x+1,z)w(t|x+1,z) + q(x,z,x-1,z)w(t|x-1,z) + q(x,z,x,z-1)w(t|x,z-1) \quad . \quad (15)$$

The fact that the waiting time of waiting calls is greater than 0 leads to the following formula for the initial conditions $w(0|x,z)$

$$w(0|x,z) = 1, \; m\le x\le n-k, k+1\le z\le k+s. \quad (16)$$

With the set of differential equations (15) and these initial conditions (16) the conditional waiting times have been calculated with the aid of the Runge-Kutta method [3]. Then the absolute (complementary) distribution funktion $w(t)$ of the waiting time of arbitrary calls can be determined by the following equation:

$$w(t) = \sum_{x=m}^{n-k} \sum_{z=k+1}^{k+s} p(x,z-1)\; w(t|x,z) \quad . \quad (17)$$

The absolute (complementary) distribution funktion $w^*(t)$ of the total waiting time with respect to waiting calls is

$$w^*(t) = \frac{1}{p_W}\, w(t) \quad . \quad (18)$$

4.3. Example No. 2

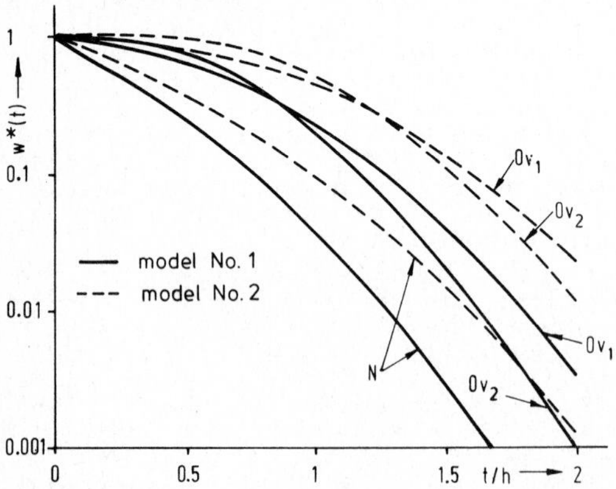

Fig. 8 Waiting time distribution

For the systems considered in example No. 1, the waiting time distributions $w^*(t)$ for waiting calls according to model No. 1 (by solid lines) and model No. 2 (by dashed lines) are shown for a situation without overload N ($A_1 = A_{1n}$, $A_2 = A_{2n}$) and for both overflow situations Ov_1 ($A_1 = 2A_{1n}$) and Ov_2 ($A_2 = 2A_{2n}$) in fig. 8.

5. SYSTEMS WITH UNEQUAL MEAN HOLDING TIMES

This section deals with systems in which the offered traffic A_1 and the offered traffic A_2 have different holding times h_1 and h_2 and, consequently, different terminating rates $\varepsilon_1=1/h_1$ and $\varepsilon_2=1/h_2$, respectively. For this case of different holding times, both models (as discribed in section 2) have been investigated [10]. In this paper, however, only the calculation methods for model No. 1 can be presented for lack of space.

In this case a new discription of states is necessary. Now a triple of parameters (x,u,z) is required to describe a state of system, where x denotes the number of calls of A_1 being served by the group of n-k trunks, u is the number of calls of A_2 being served in this group and z the total number of calls which are being served by one of the k trunks or waiting. The corresponding state diagram is shown in fig. 9.

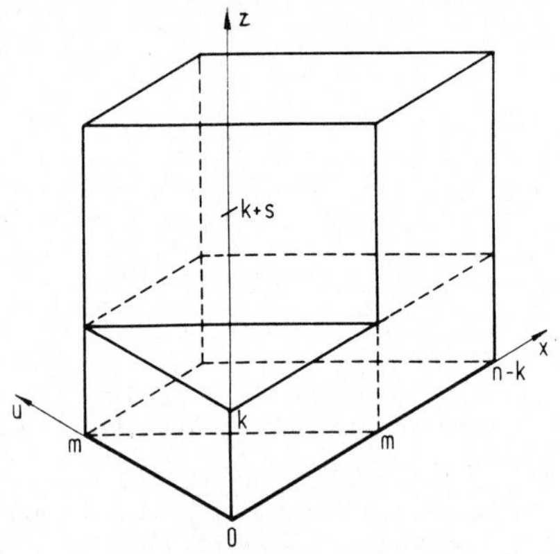

Fig. 9 State diagram for different mean holding times

5.1. The equations of state

Let $p(x,u,z)$ be the probability that the system is in a state (x,u,z). λ_1 and λ_2 are the arrival rates and ε_1 and ε_2 are the terminating rates of the two kinds of offered traffic A_1 ($A_1=\lambda_1/\varepsilon_1$) and A_2 ($A_2=\lambda_2/\varepsilon_2$) respectively.

In the sequel a boolean expression $d_{i,j}$ is used with the meaning

$$d_{i,j} = 0 \text{ if } i = j \quad \text{and} \quad (19)$$

$$d_{i,j} = 1 \text{ if } i \neq j \; . \quad (20)$$

$\overline{d_{i,j}}$ is the boolean complement of $d_{i,j}$.

For the considered system according to state diagram shown in fig. 9 the following equations of state are obtained

$$(\lambda_1+\lambda_2+x\varepsilon_1+(u+z)\varepsilon_2)p(x,u,z) = d_{z,0}\,\lambda_2 p(x,u,z-1) +(z+1)\varepsilon_2 p(x,u,z+1)+(u+1)\varepsilon_2 p(x,u+1,z) +d_{x,0}\,\lambda_1 p(x-1,u,z)+(x+1)\varepsilon_1 p(x+1,u,z),$$
$$0\le x<m,\; 0\le u<m,\; x+u<m,\; 0\le z<k \quad (21a)$$

$$(\lambda_1+\lambda_2+x\varepsilon_1+(u+k)\varepsilon_2)p(x,u,k) = \lambda_2 p(x,u,k-1)+d_{u,0}\,\lambda_2 p(x,u-1,k) +(u+1)\varepsilon_2 p(x,u+1,k)+d_{x,0}\,\lambda_1 p(x-1,u,k) +(x+1)\varepsilon_1 p(x+1,u,k),\quad 0\le x<m,\; 0\le u<m,$$
$$x+u<m \quad (21b)$$

$$(d_{x,n-k}\lambda_1+\lambda_2+x\varepsilon_1+(u+z)\varepsilon_2)p(x,u,z) =$$
$$d_{z,0}\lambda_2 p(x,u,z-1)+(z+1)\varepsilon_2 p(x,u,z+1)$$
$$+d_{x+u,n-k}d_{u,m}(u+1)\varepsilon_2 p(x,u+1,z)$$
$$+d_{x,0}\lambda_1 p(x-1,u,z)$$
$$+d_{x+u,n-k}(x+1)\varepsilon_1 p(x+1,u,z),$$
$$0\le x\le n-k,\ 0\le u\le m,\ m\le x+u\le n-k,$$
$$0\le z<k \qquad (21c)$$

$$(d_{x,n-k}\lambda_1+d_{z,k+s}\lambda_2+x\varepsilon_1+(u+k)\varepsilon_2)p(x,u,z) =$$
$$\lambda_2 p(x,u,z-1)$$
$$+d_{z,k+s}(\bar{d}_{x+u,m}u+k)\varepsilon_2 p(x,u,z+1)$$
$$+d_{x+u,n-k}d_{u,m}(u+1)\varepsilon_2 p(x,u+1,z)$$
$$+d_{x,0}(d_{x+u,m}\lambda_1+\bar{d}_{z,k}\lambda_1)p(x-1,u,z)$$
$$+d_{x+u,n-k}(x+1)\varepsilon_1 p(x+1,u,z)$$
$$+\bar{d}_{x+u,m}d_{u,0}d_{z,k+s}(x+1)\varepsilon_1 p(x+1,u-1,z+1),$$
$$0\le x\le n-k,\ 0\le u\le m,\ m\le x+u\le n-k, \qquad (21d)$$
$$k\le z\le k+s$$

with the normalizing condition

$$\sum_{z=0}^{k}\sum_{u=0}^{m-1}\sum_{x=0}^{m-1-u} p(x,u,z)$$
$$+\sum_{z=0}^{k+s}\sum_{u=0}^{m}\sum_{x=m-u}^{n-k-u} p(x,u,z) = 1 \qquad (21e)$$

These equations have been solved with the aid of the successive overrelaxation (SOR) method [3].

5.2. Characteristic traffic values

From the steady state probabilities as explained above, the following characteristic traffic values can be calculated.

The loss probabilities B_1 and B_2 can be determined as follows:

$$B_1 = \sum_{u=0}^{m}\sum_{z=0}^{k+s} p(n-k-u,u,z), \qquad (22)$$

$$B_2 = \sum_{u=0}^{m}\sum_{x=m-u}^{n-k-u} p(x,u,k+s). \qquad (23)$$

The total loss probability results in

$$B_t = \frac{A_1B_1+A_2B_2}{A_1+A_2}\ . \qquad (24)$$

Furthermore, the following formulae are obtained for the waiting probability p_W

$$p_W = \sum_{z=k}^{k+s-1}\sum_{u=0}^{m}\sum_{x=m-u}^{n-k-u} p(x,u,z), \qquad (25)$$

the mean queue length Q

$$Q = \sum_{z=k+1}^{k+s}\sum_{u=0}^{m}\sum_{x=m-u}^{n-k-u} p(x,u,z), \qquad (26)$$

the mean waiting time t_W with respect to arbitrary calls

$$t_W = \frac{Q}{\lambda_2} = \frac{Q\varepsilon_2}{\lambda_2}h_2 \qquad (27)$$

and the mean waiting time with respect to waiting calls

$$t^*_W = \frac{t_W}{p_W} \qquad (28)$$

5.3 Example No. 3

For model No. 1 with the parameters of example No. 1 and the mean holding time ratio $h_1/h_2=10$ the following loss probabilities are obtained: $B_1=0.0046$ and $B_2=0.0352$ without overload (N), $B_1=0.2091$ and $B_2=0.1086$ when A_1 is overloaded (Ov_1) and $B_1=0.0047$ and $B_2=0.4175$ when A_2 is overloaded (Ov_2). In the other case of $h_1/h_2=0.1$ there are the following results: $B_1=0.0252$ and $B_2=0.0004$ without overload (N), $B_1=0.2519$ and $B_2=0.0295$ when A_1 is overloaded (Ov_1) and $B_1=0.0915$ and $B_2=0.2528$ when A_2 is overloaded (Ov_2). This example shows that there is a strong dependence of the behavior of system from the mean holding time ratio as can be seen in comparison with example No. 1. The decrease of the loss probability of the traffic part with the larger holding time and the simultaneous increase of the other loss probability is caused by the effect that in the common group of trunks, more often trunks becomes idle by the termination of calls with lesser holding time which then, can be occupied by a call with a larger holding time.
For the application to systems with voice (larger holding time) and data (smaller holding time) the first case is the more important because voice-calls must not wait and data-calls usually can wait without loss of service quality.

6.- WAITING TIME DISTRIBUTION

6.1. Waiting process

For systems with different holding times of A_1 and A_2, a triple of parameters (x,u,z) is used to describe the random walk of waiting calls. In this model a call arriving in a state (x,u,z) of system in which $x+u \ge m$ trunks of the common group and $k\le z<k+s$ of the k exclusive trunks and s waiting places are occupied, is assumed to start the waiting process in a state (x,u,z+1) of the random walk. The waiting process terminates when it enters a state of the taboo set (x,u,k), which is equivalent to the states in which the considered call is served.
In the sequal the probability that an arbitrary call, which starts waiting in state (x,u,z), lasts longer than t be denoted by $w(t|x,u,z)$. The conditional probability density, that a state (x,u,z) is left by a transition to any other state, which may or may not belong to the taboo set, be denoted by $q(x,u,z)$ and the conditional probability density for a transition from the state (x,u,z) to a state (x',u',z') not belonging to the taboo set by $q(x,u,z,x',u',z')$. These conditional probability densities have the following values:

$$q(x,u,z) = d_{x+u,n-k}\lambda_1+x\varepsilon_1+(u+k)\varepsilon_2 \qquad (29a)$$

$$q(x,u,z,x+1,u,z) = d_{x,n-k}\lambda_1 \qquad (29b)$$

$$q(x,u,z,x+1,u,z) = d_{x+u,n-k}\varepsilon_1 \qquad (29c)$$

$$q(x,u,z,x,u-1,z) = d_{x+u,m} u \varepsilon_2 \qquad (29d)$$

$$q(x,u,z,x,u,z-1) = d_{z,k+1}(\overline{d}_{x+u,m} u \varepsilon_2 + k \varepsilon_2) \qquad (29e)$$

$$q(x,u,z,x-1,u+1,z-1) = d_{z,k+1}\overline{d}_{x+u,m} x \varepsilon_1. \qquad (29f)$$

The probability densities for all other transitions are equal to zero. (For $d_{i,j}$ see equation (19) and (20)).
For the conditional (complementary) distribution functions $w(t|x,u,z)$, the following set of differential equations holds true, according to the Kolmogorov backward equation:

$$\frac{d}{dt} w(t|x,u,z) = -q(x,u,z)w(t|x,u,z)$$
$$+q(x,u,z,x+1,u,z)w(t|x+1,u,z)$$
$$+q(x,u,z,x-1,u,z)w(t|x-1,u,z)$$
$$+q(x,u,z,x,u-1,z)w(t|x,u-1,z)$$
$$+q(x,u,z,x,u,z-1)w(t|x,u,z-1)$$
$$+q(x,u,z,x-1,x+1,z-1)w(t|x-1,u+1,z-1), \qquad (30)$$

with the initial conditions

$$w(0|x,u,z) = 1,\ 0 \le u \le m,\ m-u \le x \le n-k-u,$$
$$k+1 \le z \le k+s. \qquad (31)$$

With this set of differential equations and the corresponding initial conditions the conditional distribution functions can be calculated, e.g., with the aid of the Runge-Kutta method [3].
Finally the absolute (complementary) distribution function $w(t)$ of the waiting time of arbitrary calls is obtained as

$$w(t) = \sum_{u=0}^{m} \sum_{x=m-u}^{n-k-u} \sum_{z=k+1}^{k+s} p(x,u,z-1)w(t|x,u,z). \qquad (32)$$

The absolut (complementary) distribution function $w^*(t)$ of the total waiting time with respect to waiting calls is

$$w^*(t) = \frac{1}{p_W} w(t). \qquad (33)$$

6.2. Example No. 4

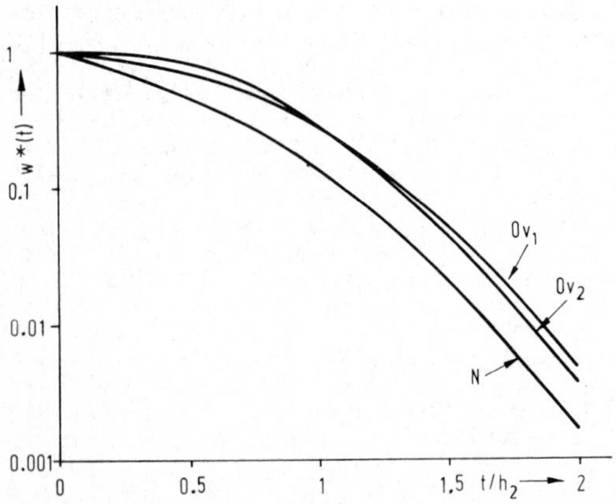

Fig 10. Waiting time distribution

For the system considered in example No. 3, with the holding time ratio $h_1/h_2=10$, the waiting time distributions $w^*(t)$ for waiting calls according to model No. 1 are shown for a situation without overload N and for both overflow situations Ov_1 and Ov_2 in fig. 10.

7. CONCLUSION

In this paper, two different models of switching systems are presented, which are not only able to protect calls of the traffic A_1 against overload of traffic A_2, but also to prevent extremly high impairment of calls of the traffic A_2 if A_1 is overloaded.
Such systems may, e. g., be applied at the integration of voice and data, therefore, besides a performance analysis of both models with equal holding times a different holding time model is investigated.
For all models, the state probabilities and the resulting characteristic traffic values as well as the waiting time distributions are calculated. Results are illustrated in examples and diagrams.

8. REFERENCES

[1] K. Kawashima,"Efficient Numerical Solutions for a Unified Trunk Reservation System with Two Classes," Rev. of the Electr. Com. Lab, Vol. 31, 1983, No.3, pp. 419-429.

[2] I. D. S. Taylor, J. G. C. Templeton, "Waiting Time In a Multi-Server Cutoff-Priority Queue, and Its Application to an Urban Ambulance Service," Oper. Research, Vol. 38, 1980, pp. 1168-1188.

[3] W. Törnig, "Numerische Mathematik für Ingenieure und Physiker," Band I/II. Springer, Berlin.

[4] A. Ralston, H. S. Wilf, "Mathematische Methoden für Digitalrechner," Oldenburg Verlag, München-Wien 1969.

[5] R. Schehrer,"On a Delay-Loss System for Overload Protection," AEÜ, 38 (1984), 3, pp. 201-206, and 10.ITC, 1983, Montreal.

[6] R. Schehrer,"On the Performance of a Delay-Loss System for Overload Protection in Switching Systems with Stored Program Control," International Seminar "Fundamentals of Teletraffic Theorie", Moscow, Moscow, 1984, and AEÜ 39, 1985, No. 3 or 4 (in print).

[7] R. Schehrer, "On the Influence of Impatient Customers in a Delay-Loss System for Overload Protection," AEÜ 39, 1985, to be published.

[8] H.-D. Ide," Performance Analysis of a Delay Loss System with Cut-Off Priority and State Dependent Server Reservation," AEÜ 39, 1985 to be published.

[9] R. Schehrer, H.-D. Ide, "On Application and Performance of Cut-off Priority Queues in Switching Systems with Overload Protection" 11th ITC, Kyoto, 1985

[10] G. Brune, "Über ein System zur Überlastabwehr in Vermittlungssystemen mit zwei Prioritätsklassen und unterschiedlicher Leitungsreservierung" Ph. D. Thesis, University of Dortmund, 1985.

TELETRAFFIC ISSUES in an Advanced Information Society
ITC-11
Minoru Akiyama (Editor)
Elsevier Science Publishers B.V. (North-Holland)

AN OPTIMAL CONTROL OF AN INTEGRATED CIRCUIT-AND PACKET-SWITCHING SYSTEM

*Tadayoshi SHIOYAMA, **Katsuhisa OHNO and ***Hisashi MINE

* Osaka Prefectural Industrial Research Institute, Osaka
** Konan University, Kobe *** Kyoto University, Kyoto, Japan

ABSTRACT

We consider an optimal control of the integrated circuit-and packet-switched multiplexing system. The system is dynamically controlled by rejecting or accepting a new arrival voice call according to its current state. The control problem is formulated as the Markov decision process with the cost structure consisting of the holding and rejection costs. The properties of an optimal control policy are analyzed.

1. INTRODUCTION

Integrated switching, which includes circuit-switching for voice and packet-switching for data, is motivated by the desire to share transmission and switching facilities efficiently and to provide service to a variety of subscribers. The slotted enveloped network (SENET) [1] is an important early development. In SENET, a fixed number of time slots in a synchronous time division multiplexer (STDM) frame is reserved for voice while the rest of the frame is allocated to data. The traffic performance of the SENET-concept integrated voice and data system has been analyzed by many researchers [2]-[4].

Janakiraman, Pagurek and Neilson [5] have considered a variable-frame strategy and compared its performance with a SENET-like fixed-frame scheme.

Gitman,Hsieh and Occhiogrosso [6], Avellaneda, Hayes and Nassehi [7] and Woodside [8] have solved a capacity-allocation problem in which TDM slots in many links of a network are optimally allocated between voice and data to minimize a blocking probability for voice subject to a specified maximum mean delay to data. Konheim and Pickholtz [9] have proposed and analyzed a model of a moving-boundary, fixed frame length, integrated multiplexer. In their model, the assignment of slots within a frame to the voice and data sources is established anew at the beginning of each frame using an allocation function depending jointly on the number of voice and data packets stored at the start of the frame.

Weinstein, Malpass and Fisher [10] have investigated a scheme in which a voice flow rate is controlled depending on voice channel utilization to reduce data packet delays. However, they have not discussed an optimization of their scheme.

In this paper, we consider an optimal control of the integrated circuit-and packet-switched multiplexing system. The system is dynamically controlled by rejecting or accepting a new arrival voice call according to its current state. This control problem is formulated as a Markov decision process, and the properties of an optimal control policy are analyzed.

2. MODEL

The multiplexing technique under investigation is a time division scheme in which each fixed duration frame is partitioned into M time slots. For notational simplicity, the frame length is taken as a unit time. A certain number of time slots in the frame are allocated to the transmission of digtized vois. The voice is circuit-switched by assigning synchronous slots and is under the Lost-Calls-Cleared assumption. The voice traffic is modeled by a Poisson call arrival process with average arrival rate λ (calls/frame) and exponentially distributed call holding times with average service rate μ (calls/frame). It is assumed that $\lambda<<1$ and $M\mu<<1$. This assumption is satisfied in practical systems [6].

The remaining time slots in the frame are dedicated to the transmission of data packets. Each data packet has a fixed size of one time slot and arrives in a Poisson process with average arrival rate λ_p (packets/frame). If no time slots are available for data packets, they wait for transmission in a buffer of size L.

The system is observed at the beginning of each frame. According to the observed state, the system is controlled by rejecting or accepting a new arrival voice call so as to minimize an expected cost. The following costs are taken into consideration. A rejection cost α is incurred by one rejected voice call and a loss cost α_p by one packet lost due to the fully occupied buffer. A holding cost of one data packet during one frame is taken as one unit cost. Then, an expected immediate cost in one frame is determined by three costs mentioned above. An optimal control to be analyzed in this paper is one that minimizes the total expected discounted cost over an infinite horizon.

3. FORMULATION

The optimal control problem stated in the preceeding section is formulated as a Markov decision process. The state of the system is described by $\mathbf{i}=(i_1,i_2)$, where i_1 represents the number of voice calls in service and i_2 the number of data packets in service or in queue. The state space S is defined as a set $\{\mathbf{i}=(i_1,i_2); i_1=0,1,\ldots,M, i_2=0,1,\ldots,L+M-i_1\}$. When the system is observed in state $\mathbf{i}$ at the beginning of frame, a control action $a(\mathbf{i})$ is chosen from "0" or "1". Here, the action "0" or "1" means to reject or accept a new arrival voice call if it occurs during the frame. A policy is defined as any rule $\{a(\mathbf{i});\mathbf{i}\in S\}$ depending on observed states.

Let $P(\mathbf{i},\mathbf{j};a(\mathbf{i}))$ denote the conditional probability that the system is in state $\mathbf{j}=(j_1,j_2)$ at the beginning of the next frame given that an action $a(\mathbf{i})$ is taken in state $\mathbf{i}$ at the beginning of the current frame. For notational simplicity, define p_ℓ by

$$p_\ell = \lambda_p^\ell \exp(-\lambda_p)/\ell!$$

$$\text{for } \ell=0,1,\ldots,L+M-j_1-(i_2-M+i_1)^+-1$$

and

$$p_{L+M-j_1-(i_2-M+i_1)^+} = \sum_{\ell=L+M-j_1-(i_2-M+i_1)^+}^{\infty} \lambda_p^\ell \exp(-\lambda_p)/\ell!.$$

The assumptions in the preceeding section imply that for $\mathbf{i}\in\{\mathbf{i}=(i_1,i_2);i_1=1,2,\ldots,M,\ i_2=0,1,\ldots,L+M-i_1\}$ and $j_2=(i_2-M+i_1)^+,\ldots,L+M-(i_1-1)$

$$P(\mathbf{i},(i_1-1,j_2);a(\mathbf{i}))$$

$$=\{\lambda(1-a(\mathbf{i}))+(1-\lambda)\}i_1\mu p_{j_2-(i_2-M+i_1)^+} \tag{1}$$

and that for $\mathbf{i}\in S$

$$P(\mathbf{i},(i_1,j_2);a(\mathbf{i}))$$

$$=[\lambda a(\mathbf{i})i_1\mu+\{\lambda(1-a(\mathbf{i}))+(1-\lambda)\}(1-i_1\mu)]$$

$$\times p_{j_2-(i_2-M+i_1)^+} \tag{2}$$

$$\text{for } j_2=(i_2-M+i_1)^+,\ldots,L+M-i_1$$

and

$$P(\mathbf{i},(i_1+1,j_2);a(\mathbf{i}))$$

$$=\lambda a(\mathbf{i})(1-i_1\mu)p_{j_2-(i_2-M+i_1)^+} \tag{3}$$

$$\text{for } j_2=(i_2-M+i_1)^+,\ldots,L+M-(i_1+1),$$

where $(x)^+=\max(0,x)$.

When the system is in state $\mathbf{i}$ and an action $a(\mathbf{i})$ is taken, a holding cost $(i_2-M+i_1)^+$ is incurred by data packets during the frame and an expected rejection cost $\alpha\lambda(1-a(\mathbf{i}))$ by a rejected voice call. When the buffer is full, i.e., $i_1+i_2=L+M$, an expected loss cost α_p is incurred by lost data packets. Consequently, an expected immediate cost during the frame, denoted by $C(\mathbf{i};a(\mathbf{i}))$, is given by

$$C(\mathbf{i};a(\mathbf{i}))=(i_2-M+i_1)^+ +\alpha\lambda(1-a(\mathbf{i}))+\alpha_p\lambda_p\delta(\mathbf{i}), \tag{4}$$

where $\delta(\mathbf{i})=0$ for $i_1+i_2<L+M$, $=1$ for $i_1+i_2=L+M$ and $a(\mathbf{i})=0$ at $i_1=M$.

Let $v_\mathbf{i}(n)$ be the minimal expected discounted cost over n frames given that the initial state is $\mathbf{i}$. For simplicity, $v_\mathbf{i}(0)$ is defined as 0 for $\mathbf{i}\in S$. Then, $v_\mathbf{i}(n)$ satisfies the following recursive relation [11]: for $n=0,1,\ldots$ and $\mathbf{i}\in S$,

$$v_\mathbf{i}(n+1)=\min[v_\mathbf{i}(n+1;0),v_\mathbf{i}(n+1;1)], \tag{5}$$

where for $a=0$ or 1 and discounted factor β,

$$v_\mathbf{i}(n+1;a)=C(\mathbf{i};a)+\beta\sum_\mathbf{j} P(\mathbf{i},\mathbf{j};a)v_\mathbf{j}(n). \tag{6}$$

Moreover, it holds that for the minimal total expected discounted cost $v_\mathbf{i}^*$,

$$v_\mathbf{i}^*=\lim_{n\to\infty} v_\mathbf{i}(n)$$

and that

$$v_\mathbf{i}^*=\min_a[C(\mathbf{i};a)+\beta\sum_\mathbf{j} P(\mathbf{i},\mathbf{j};a)v_\mathbf{j}^*]. \tag{7}$$

Thus, the problem is to find a policy minimizing the right hand side of the above optimality equation. An optimal control policy can be determined by a modified policy iteration algorithm with a suboptimality test [12].

4. CONTROL LIMIT POLICY

Let a partial ordering $\leqq$ on the state space S be defined by

$$\mathbf{i}=(i_1,i_2)\leqq\mathbf{j}=(j_1,j_2) \text{ if and only if } i_1\leqq j_1 \text{ and } i_2\leqq j_2.$$

A policy $\{a(\mathbf{i});\mathbf{i}\in S\}$ satisfying for $\hat{\mathbf{i}}\in S$

$$a(\mathbf{i})=1 \quad \text{for } \mathbf{i}\leqq\hat{\mathbf{i}}$$
$$=0 \quad \text{for } \mathbf{i}>\hat{\mathbf{i}},$$

is called a control limit policy, and the state $\hat{\mathbf{i}}$ is called its control limit [13]. Moreover, define $\mathbf{F}$ as the set all non-negative increasing functions with respect to the partial order $\leqq$ on S.

Lemma [14]

(a) Let $\mathbf{K}$ be the set of all subsets K of S such that if $\mathbf{i}\in K$ and $\mathbf{i}\leqq\mathbf{i}'$, then $\mathbf{i}'\in K$. Then, for each $f(\mathbf{i})\in\mathbf{F}$ there exists a non-negative sequence $\{\alpha_K\}$ such that $f(\mathbf{i})=\sum_{K\in\mathbf{K}}\alpha_K\mathbf{I}_K(\mathbf{i})$, where $\mathbf{I}_K(\mathbf{i})=1$ for $\mathbf{i}\in K$, $=0$ otherwise.

(b) Suppose that $\sum_{\mathbf{j}\in S} P(\mathbf{i},\mathbf{j})\mathbf{I}_K(\mathbf{j})\in\mathbf{F}$ for $K\in\mathbf{K}$ and transition probabilities $P(\mathbf{i},\mathbf{j})$. Then, for $f(\mathbf{i})\in\mathbf{F}$, $\sum_{\mathbf{j}\in S} P(\mathbf{i},\mathbf{j})f(\mathbf{j})\in\mathbf{F}$.

Let S_0 denote the set $\{\mathbf{i};(i_2-M+i_1)^+=0\}$ and S_+ the set $\{\mathbf{i};(i_2-M+i_1)^+>0\}$. It follows from (4) and (5) that

$$v_\mathbf{i}(1)=C(\mathbf{i};a(\mathbf{i}))\in\mathbf{F}. \tag{8}$$

Moreover, for each $\ell=(i_2-M+i_1)^+,\ldots,L+M-j_1$ and $j_1=i_1-1,i_1,i_1+1$

$$\sum_{j_2=\ell}^{L+M-j_1} p_{j_2-(i_2-M+i_1)^+}$$

$$=\sum_{k=\ell-(i_2-M+i_1)^+}^{\infty} \lambda_p^k\exp(-\lambda_p)/k!\in\mathbf{F}. \tag{9}$$

In the following it is assumed that

$$\min\{\exp(-\lambda_p),\lambda_p^{L+M}\exp(-\lambda_p)/(L+M)!\}$$

$$>M\mu/(1-M\mu). \tag{10}$$

Then, it follows that for each $\ell=(i_2-M+i_1+1)^+,\ldots,L+M-j_1$ and $\mathbf{i}\in S_+$

$$\{1-(i_1+1)\mu\}\sum_{j_2=\ell}^{L+M-j_1} p_{j_2-(i_2-M+i_1+1)^+}$$

$$-(1-i_1\mu)\sum_{j_2=\ell}^{L+M-j_1} p_{j_2-(i_2-M+i_1)^+}$$

$$=\{1-(i_1+1)\mu\}\sum_{k=\ell-(i_2-M+i_1+1)^+}^{\infty} \lambda_p^k\exp(-\lambda_p)/k!$$

$$-(1-i_1\mu)\sum_{k=\ell-(i_2-M+i_1)^+}^{\infty}\lambda_p^k\exp(-\lambda_p)/k!$$

$$=\{1-(i_1+1)\mu\}p_{\ell-(i_2-M+i_1+1)^+}-\mu\sum_{k=\ell-(i_2-M+i_1)^+}^{\infty}\lambda_p^k\exp(-\lambda_p)/k!$$

$$=\{1-(i_1+1)\mu\}p_{\ell-(i_2-M+i_1+1)^+}$$

$$-\mu\{1-\sum_{k=0}^{\ell-(i_2-M+i_1)^+-1}\lambda_p^k\exp(-\lambda_p)/k!\}$$

$$=\{(1-i_1\mu)p_{\ell-(i_2-M+i_1+1)^+}-\mu\}$$

$$+\mu\sum_{k=0}^{\ell-(i_2-M+i_1)^+-2}\lambda_p^k\exp(-\lambda_p)/k!$$

$$>0. \tag{11}$$

That is, for each $\ell=(i_2-M+i_1+1)^+,\ldots,L+M-j_1$ and $\mathbf{i}\in S_+$,

$$\sum_{j_2=\ell}^{L+M-j_1}(1-i_1\mu)p_{j_2-(i_2-M+i_1)^+}\in\mathbf{F}. \tag{12}$$

On the other hand, for $\mathbf{i}\in S_+$ and $\bar{\mathbf{i}}\in S_0$,

$$(1-i_1\mu)\sum_{j_2=\ell}^{L+M-j_1}p_{j_2-(i_2-M+i_1)^+}$$

$$-(1-\bar{i}_1\mu)\sum_{j_2=\ell}^{L+M-j_1}p_{j_2}$$

$$\geqq(1-M\mu)\sum_{j_2=\ell}^{L+M-j_1}p_{j_2-1}-\sum_{j_2=\ell}^{L+M-j_1}p_{j_2}$$

$$=(1-M\mu)\{p_{\ell-1}+M\mu/(1-M\mu)\sum_{j_2=0}^{\ell-1}p_{j_2}-M\mu/(1-M\mu)\}$$

$$\geqq(1-M\mu)\{p_{\ell-1}-M\mu/(1-M\mu)\}$$

$$>0. \tag{13}$$

In the following, states in S_0 are regarded as an aggregate state. Then, (1)-(3),(9) and (11)-(13) imply that for $K\in\mathbf{K}$,

$$\sum_{\mathbf{j}\in S}P(\mathbf{i},\mathbf{j};a(\mathbf{i}))I_k(\mathbf{j})\in\mathbf{F}. \tag{14}$$

By mathematical induction, lemma,(5),(6),(8) and (14) lead to the following relation:

$$v_{\mathbf{i}}(n)\in\mathbf{F}\quad\text{for any } n. \tag{15}$$

Define $\tilde{P}(\mathbf{i},\mathbf{j})$ as

$$\tilde{P}(\mathbf{i},\mathbf{j})=P(\mathbf{i},\mathbf{j};1)-P(\mathbf{i},\mathbf{j};0) \tag{16}$$

From (1)-(3) and (14), it follows that

$$\tilde{P}(\mathbf{i},\mathbf{j})=-\lambda i_1\mu p_{j_2-(i_2-M+i_1)^+}\quad\text{for } j_1=i_1-1; \tag{17}$$

$$=-\lambda(1-2i_1\mu)p_{j_2-(i_2-M+i_1)^+}\quad\text{for } j_1=i_1; \tag{18}$$

$$=\lambda(1-i_1\mu)p_{j_2-(i_2-M+i_1)^+}\quad\text{for } j_1=i_1+1. \tag{19}$$

Therefore, it holds that for each $\ell=(i_2-M+i_1)^+,\ldots$ $\ldots,L+M-j_1$

$$\sum_{j_1=i_1-1}^{i_1+1}\sum_{j_2=\ell}^{L+M-j_1}\tilde{P}(\mathbf{i},\mathbf{j})=0, \tag{20}$$

$$\sum_{j_1=i_1}^{i_1+1}\sum_{j_2=\ell}^{L+M-j_1}\tilde{P}(\mathbf{i},\mathbf{j})=\lambda i_1\mu\sum_{j_2=\ell}^{L+M-j_1}p_{j_2-(i_2-M+i_1)^+} \tag{21}$$

and

$$\sum_{j_1=i_1+1}^{i_1+1}\sum_{j_2=\ell}^{L+M-j_1}\tilde{P}(\mathbf{i},\mathbf{j})$$

$$=\lambda(1-i_1\mu)\sum_{j_2=\ell}^{L+M-j_1}p_{j_2-(i_2-M+i_1)^+}. \tag{22}$$

Equations (9),(12),(13) and (20)-(22) yield for $K\in\mathbf{K}$,

$$\sum_{\mathbf{j}\in S}\tilde{P}(\mathbf{i},\mathbf{j})I_K(\mathbf{j})\in\mathbf{F}. \tag{23}$$

Since (6) leads to

$$v_{\mathbf{i}}(n+1;1)-v_{\mathbf{i}}(n+1;0)=-\alpha\lambda+\beta\sum_{\mathbf{j}}\tilde{P}(\mathbf{i},\mathbf{j})v_{\mathbf{j}}(n), \tag{24}$$

(15),(23) and lemma imply that $v_{\mathbf{i}}(n+1;1)-v_{\mathbf{i}}(n+1;0)$ increases with respect to the partial order $\leqq$. Thus, there exists a control limit.

5. NUMERICAL RESULTS

The optimal control policy was computed by the modified policy iteration algorithm [11] on a FACOM M-382 computer at the Data Processing Center, Kyoto University. In the computation, the values of parameters μ,L,M,α and α_p were set as

$$\mu=10^{-3}, L=M=10 \text{ and } \alpha=100\alpha_p.$$

Figure 1-(a) shows control limit lines of the optimal policies in cases where $\beta=0.9,\rho(=\lambda/\mu)=10,\lambda_p=3$ and $\alpha=1,10,100$. In the figure, action 'reject' is optimal in states above and on the control limit lines. The figure shows that the control limit lines go up as the rejection cost α increases. Figure 1-(b) shows a change of the control limit lines as the value of λ_p changes 1 to 3 in the case of $\beta=0.9,\rho=10$ and $\alpha=100$. The control limit lines go down with increasing λ_p. Figure 1-(c) describes the dependence of the control limit line on the value of β. The larger β leads to the lower control limit.

6. CONCLUSIONS

We considered an optimal control of the integrated circuit-and packet-switched multiplexing system. The control problem is formulated as the Markov decision process with the cost structure consisting of the holding and rejection costs. Under condition (10), it is proved that a control limit policy is optimal. The control limit is affected by a rejection cost, an offered load of data packet and a discounted factor.

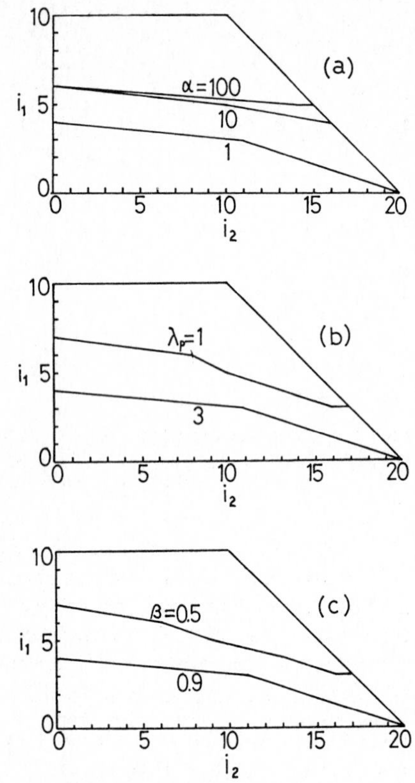

Fig. 1 Control limit line. (a)Influence by α where β=0.9,ρ=10 and λ_p=3; (b)Influence by λ_p where β=0.9,ρ=10 and α=1; (c)Influence by β where ρ=10,λ_p=3 and α=1.

ACKNOWLEDGEMENT

We would like to thank Professor T.Hasegawa and Dr.Y.Takahashi in Kyoto University for their invaluable suggestions.

REFERENCES

[1] G.J.Coviello and P.A.Vena,"Integration of circuit/packet switching by a SENET (Slotted Envelope NEtwork) concept," in Nat. Telecommun. Conf. Rec., New Orleans, LA, vol.2, pp.42.12-42.17, 1975.

[2] M.J.Fischer and T.C.Harris,"A model for evaluating the performance of an integrated circuit-and packet-switched multiplex structure," IEEE Trans. on Commun., vol.COM-24, no.2, pp.195-202, 1976.

[3] D.P.Gaver and J.P.Lehoczky,"Channels that cooperatively service a data stream and voice messages," IEEE Trans. on Commun., vol.COM-30, no.5, pp.1153-1161, 1982.

[4] A.L.Garcia, R.H.Kwong and G.F.Williams,"Performance evaluation methods for an integrated voice/data link," IEEE Trans. on Commun., vol. COM-30, no.8, pp.1848-1858, 1982.

[5] N.Janakiraman, B.Pagurek and J.E.Neilson,"Performance analysis of an integrated switch with fixed or variable frame rate and movable voice/data boundary," IEEE Trans. on Commun., vol.COM-32, no.1, pp.34-39, 1984.

[6] I.Gitman, W.N.Hsieh and B.J.Occhiogrosso, "Analysis and design of hybrid switching networks," IEEE Trans. on Commun. vol.COM-29, no.9, pp.1290-1300, 1981.

[7] O.A.Avellaneda, J.F.Hayes and M.M.Nassehi, "A capacity allocation problem in voice-data networks," IEEE Trans. on Commun. vol. COM-30, no.7, pp.1767-1773, 1982.

[8] C.M.Woodside,"An incremental capacity-allocation algorithm for voice/data networks," IEEE Trans. on Commun. vol.COM-31, no.9, pp.1113-1114, 1983.

[9] A.G.Konheim and R.L.Pickhotz,"Analysis of integrated voice/data multiplexing," IEEE Trans. on Commun. vol.COM-32, no.2, pp.140-147, 1984.

[10] C.J.Weinstein, M.L.Malpass and M.J.Fisher, "Data traffic performance of an integrated circuit-and packet-switched multiplex structure," IEEE Trans. on Commun. vol.COM-28, no.6, pp.873-878, 1980.

[11] C.Derman,"Finite State Markovian Decision Processes," Academic Press, New York and London, 1970.

[12] K.Ohno,"A unified approach to algorithms with a suboptimality test in discounted semi-Markov decision process," J.Operations Research Society of Japan, vol.24, no.4, pp. 296-323, 1981.

[13] S.S.Lam and L.Kleinrock,"Packet switching in a multiaccess broadcast channel: dynamic control procedures," IEEE Trans. on Commun. vol.COM-23, no.9, pp.891-904, 1975.

[14] C.C.White,"Bounds on optimal cost for a replacement problem with partial observations," Naval Research Logistics Quarterly, vol.26, pp.415-422, 1979.

TELETRAFFIC ISSUES in an Advanced Information Society
ITC-11
Minoru Akiyama (Editor)
Elsevier Science Publishers B.V. (North-Holland)

The Interaction between Queueing and Voice Quality in Variable Bit Rate Packet Voice Systems

J. M. Holtzman

AT&T Bell Laboratories
Holmdel, New Jersey 07733, USA

ABSTRACT

In packetized voice systems, a possible overload scheme is to decrease the bit rate when the traffic load increases past a threshold. This paper continues the research into such systems with particular emphasis on the interaction between the traffic performance and voice quality. The adaptive bit rate strategy can be an effective overload control but at the possible expense of voice quality. The object of the paper is to see the effect of traffic-sensitive bit rates upon voice quality. The analysis uses a spectral approach.

1. INTRODUCTION AND OVERVIEW

In packetized voice systems, a possible overload scheme is to decrease the bit rate when the traffic load increases past a threshold. Such schemes are considered, e.g., in [1] and [2] and the references therein. This paper continues the research into such systems with particular emphasis on the interaction between the traffic performance and voice quality. The adaptive bit rate strategy can be an effective overload control but at the possible expense of voice quality. The object of the paper is to see the effect of traffic-sensitive bit rates upon voice quality.

The problem considered here is the interaction between the queueing system/overload control strategy and the coding algorithm and speech perception. This paper ties these disparate phenomena together in an overall model. The analysis uses a spectral approach to analyze the effect of the queueing system/overload control upon voice quality.

The rationale for choice of voice quality measure is given in Section 2. Most importantly, it must quantitatively reflect subjective quality tests. A discussion of the relation between subjective quality tests and objective quality measures is given. Secondarily, but crucial to the present analysis, is that the chosen measure be analyzable in terms of the queueing model.

The methodology for analyzing the effect of reduced bit rates is discussed in Section 3. We combine a queueing analysis with consideration of a voice quality measure to characterize the voice quality degradation. A state dependent queueing model reflects the traffic sensitive bit rates. Then the queueing model is combined with the voice quality measure by using their spectral characteristics as a common denominator. In this way both the traffic performance and voice quality are jointly analyzed.

Multiple bit dropping levels are considered in Section 4. Concluding remarks are given in Section 5.

2. VOICE QUALITY

As mentioned, a modelling approach is being presented to relate voice quality to traffic load. The modelling is intended to be consistent with subjective quality test results and be able to extrapolate beyond the specific parameters of those tests.

We combine a queueing analysis with consideration of a distortion measure to characterize the voice quality degradation. We use a scaled L_2 norm of the difference of the log spectra as the distance between two speech segments S_1 and S_2:

$$d(S_1,S_2) = k\left\|\log(F_1)-\log(F_2)\right\|_2$$

$$= k\left\{\frac{1}{W}\int_0^W[\log F_1(\omega)-\log F_2(\omega)]^2 d\omega\right\}^{1/2} \quad (2.1)$$

where F_i is the spectrum of S_i and W is a maximum frequency ω of interest. The parameter k is a normalizing factor to be discussed. There are a number of other distortion measures (some better than others for specific applications; see [3], [4], [5]). The approach used is not strictly the use of an objective measure, as opposed to a subjective measure, so that there is a degree of robustness in the choice of distortion measure.

A good discussion of some of the characteristics of objective and subjective measures is given in [6]. The dilemma of using subjective testing for system design, briefly stated, is as follows:

1. Subjective quality testing is necessary for determining the perceptual effects of new modes of voice degradation.
2. Subjective quality testing is time consuming and does not lead naturally to the functional relationships needed for system design and optimization.
3. Objective measures, appropriately chosen, are well suited to system design but cannot be generally expected to track subjective measures well over a wide variety of conditions.

The solution proposed here is to use an objective measure but one which is anchored to subjective results. An objective measure is chosen which available evidence indicates has a reasonable relationship to subjective results. The connection with subjective results is strengthened via incorporation of those results into the objective measure. Specifically, the distortion measure for variable bit rates is normalized using the parameter k by results for *steady* bit rates so that it replicates those results when the rate is constant. That is, the norm is scaled so that it equals the difference in mean opinion scores* corresponding to steady bit rates r_2 and r_1. Thus, the model is anchored to known results.

We distinguish this incorporation of subjective results from two other procedures:

1. Tuning up of a given objective measure to improve correlation with subjective results. This may be appropriate in some cases but our approach uses subjective results as a more integral ingredient of the objective measure.
2. Taking all the existing subjective results and forcing an objective measure to replicate those results. This is little more than curve fitting and would not yield much predictive power. We use the minimum amount of subjective results so that the resulting objective measure can predict the rest (and thus be validated as a predictor for new conditions).

* The mean opinion score is probably the simplest and most commonly used subjective voice quality measure. See Appendices E and F of [7] for more discussion and further references.

3. EFFECT OF BIT DROPPING ON THE SPECTRA VIA A QUEUEING MODEL

To illustrate the approach, consider the following state dependent model. When the queue size q (not including the packet in service) reaches the threshold Q, the bit rate drops from r_1 to r_2. When q drops below Q, the r_1 rate is resumed. Note that in this first illustrative model, only one lower bit rate is used (as opposed to multiple thresholds) and no hysteresis is used on the thresholding. We refer to continual time periods when $q \geq Q$ and $q < Q$ as brisk and slack periods, respectively. The analysis assumes that the brisk and slack periods are not too long, in particular, not so long that the effect of switching between them is possibly worse than a steady r_2 rate.

During the brisk periods there is a lowered bit rate while during the slack periods there is considered to be essentially no degradation. That is, the r_1 rate is used as a benchmark. A simple way to see the effect of the bit dropping on the noise is to consider it as a random switch. During the brisk period, additional quantization noise is let through a switch. Thus the noise is multiplied by a random function which equals zero for a random time with mean t_s and which is one for a random time with mean t_b. t_s and t_b are the mean slack and brisk periods, respectively. Then, since this 0-1 random function is (to a good approximation) independent of the noise,* the effect is to multiply the noise autocorrelation function by its autocorrelation function.† The composite power spectrum is then determined by Fourier transform. Then the spectrum of interest is

$$L(\omega) = \Gamma[R(\tau)N(\tau)]\,|H(\omega)|^2 \tag{3.1}$$

where Γ means Fourier transform, $R(\tau)$ is the autocorrelation of the random switch, $N(\tau)$ is the autocorrelation function of the lowered bit rate noise, and $H(\omega)$ is an auditory transfer function. The distortion measure becomes

$$d = k\left\{\frac{1}{W}\int_0^W \left[\log\{[(S(\omega)+\Gamma(R(\tau)N(\tau))]|H(\omega)|^2\} - \log[S(\omega)|H(\omega)|^2]\right]^2 d\omega\right\}^{1/2} \tag{3.2}$$

* We are investigating the effect of multi-users on a single user.

† With $n(t)$ and $s(t)$ the noise and random switch time functions, resp., and using independence,

$$E[s(t)n(t)s(t+\tau)n(t+\tau)] = E[s(t)s(t+\tau)]E[n(t)n(t+\tau)]$$
$$= R(\tau)N(\tau).$$

$$\approx k\left\{\frac{1}{W}\int_0^W\left[\frac{\Gamma(R(\tau)N(\tau))}{S(\omega)}\right]^2 d\omega\right\}^{1/2} \quad (3.3)$$

where $S(\omega)$ is an averaged or long term speech spectrum. The last approximation, using $ln(1+x) \approx x$ (for relatively low noise) and renormalizing, shows most clearly how the bit dropping effect enters. This approximation can get crude for large bit rate reductions.

3.1 Brisk and Slack Periods

Reference 8 discusses the calculation of t_s and t_b for state dependent service and arrival rates. Such a model can be used with a higher service rate when in the brisk period. This assumes that bit dropping is done on packets *served* during the brisk period. The case of dropping bits on packets *arriving* during the brisk period can be modelled (approximately) with a state dependent arrival rate. The latter case is not as quick to react in overloads. See [9] and the references therein for queueing models of superpositions of packet voice streams (Section II of [9] describes several models).

3.2 Autocorrelation Function

To gain some insight, consider modelling the brisk and slack periods as independent exponential random variables with means determined from the queueing analysis. (The exponentiality is an approximation). The autocorrelation function of the 0-1 function is

$$R(\tau) = \frac{t_b}{(t_b + t_s)^2}\left[t_b + t_s e^{-(t_s^{-1}+t_b^{-1})\tau}\right] \quad (3.4)$$

$$= f^2 + f(1-f)e^{-\tau/t_s f} \quad (3.5)$$

where

$$f = \frac{t_b}{(t_b+t_s)} \quad (3.6)$$

is the fraction of time at the reduced bit rate.

We then get

$$\Gamma[R(\tau)N(\tau)] = f^2\eta(\omega) + f(1-f)\eta(\omega+(t_s f)^{-1}) \quad (3.7)$$

for insertion into (3.2), where η is the spectrum of the quantization noise N.

Equation (3.5) shows how the variable bit rate effect is affected by the time scale of the switching. In particular, for fixed f, $R(\tau)$ monotonically increases with t_s and

$$\lim_{t_s \to 0} R(\tau) = f^2 . \quad (3.8)$$

The result for very small t_s is physically intuitive. For very small t_s and t_b (we are keeping f fixed in this discussion), the sampling of the noise is so fast that it is like multiplying the noise amplitude by a factor of f or the spectrum by f^2.

4. ANALYSIS OF MULTIPLE LOWER BIT RATES

We shall extend the model to dropping the bit rate to two lower levels, from r_1 to r_2 and r_3. It will be clear how to generalize the model to any number of levels. While we are doing this, we will also generalize the queueing model to any model for which we can conveniently obtain transition probabilities between states.

Let x be the queue size (now including the packet in service) and define the following:

$$\textit{slack period} = \textit{time continually in states } (0, X_1-1) \quad (4.1)$$

$$\textit{brisk period} = \textit{time continually in states } (X_1, X_2-1) \quad (4.2)$$

$$\textit{very brisk periods} = \textit{time continually in states } (X_2, \infty) \quad (4.3)$$

Thus, X_1 and X_2 are the thresholds for dropping the bit rate to r_2 and r_3, respectively.

Define a continuous time Markov process with states 0, 1, and 2 corresponding to the slack, brisk, and very brisk periods, respectively. Let p_i be the stationary probability* that $x(t)=i$ and let p_{ij} be the transition probability

$$p_{ij}(\tau) = P[x(t+\tau)=i \mid x(t)=j] \quad (4.4)$$

Then, the autocorrelation function needed for the analysis is

$$R(\tau) = \sum_{i=X_1}^{X_2-1} p_i \sum_{j=X_1}^{X_2-1} p_{ji}(\tau) \quad (4.5)$$

$$+\alpha\left[\sum_{i=X_1}^{X_2-1} p_i(\tau)\sum_{j=X_2}^{\infty} p_{ji}(\tau) + \sum_{i=X_2}^{\infty} p_i \sum_{j=X_1}^{X_2-1} p_{ji}(\tau)\right] + \alpha^2 \sum_{i=X_2}^{\infty} p_i \sum_{j=X_2}^{\infty} p_{ji}(\tau)$$

---

* Satisfaction of conditions for existence of these probabilities is implicitly assumed.

where $\alpha \geq 2$ represents the nonlinear effect of reducing bit rates (the mean opinion score decreases faster than linearly as a function of decreased bit rate; see [10]).

For any queueing system for which the above stationary and transition probabilities are available (calculable), the relevant autocorrelation function can be calculated.

Remark. This assumes that the noise spectra have the same shape for the r_2 and r_2 bit rates. This is an approximation since it is known that as the number of bits increases, the noise becomes more white ([11], p. 185). A simple approximation for multiple bit dropping levels, which can use more than one noise spectrum, is to superimpose the effect of single levels.

5. CONCLUDING REMARKS

Preliminary results using this approach to predict subjective quality tests of mean opinion scores showed reasonably good results. The general approach allows use of different distortion measures and queueing models. For example, some applications might use frequency weighting with the L_2 measure. The general approach may also be applicable to some digital speech interpolation (DSI) systems not using packet format. Section 6.4.4 of [7] gives a brief discussion of DSI with references, as well as additional references on varying the bit rate.

ACKNOWLEDGEMENT

I thank M. R. Aaron and N. S. Jayant for their comments.

REFERENCES

[1] T. Bially, B. Gold, and S. Seneff, "A Technique for Adaptive Voice Flow Control in Integrated Packet Networks", IEEE Trans. on Communications, vol. COM-28, no. 3, pp. 325-333, March 1980.

[2] M. Listanti and F. Villani, "Voice Communication Handling in X.25 Packet Switching Networks", Globecom '83, San Diego, Calif., Nov. 28 - Dec. 1, 1983, Paper 2.4.

[3] B. McDermott, C. Scagliola, D. Goodman, "Perceptual and Objective Evaluation of Speech Processed by Adaptive Differential PCM," BSTJ, vol. 57,, No. 5, pp. 1597-1618, May-June 1978.

[4] R. M. Gray et al, "Distortion Measures for Speech Processing", IEEE Transactions on Acoustics, Speech, and Signal Processing, vol. ASSP-28, no. 4, pp. 367-376, August, 1980.

[5] K. Itoh, N. Kitawaki, K. Kakeni, "Objective Quality Measures for Speech Waveform Coding Systems", Review of the Electrical Communication Laboratories, vol. 32, No. 2, pp. 220-228, 1984.

[6] T. P. Barnwell, III, "On the Standardization of Objective Measures for Speech Quality Testing," Workshop on Standardization for Speech Technology, Natl. Bureau of Standards, Gaithersburg, Md., March 18-19, 1982.

[7] N. S. Jayant and P. Noll, "Digital Coding of Waveforms," Prentice-Hall, 1984.

[8] Y. K. Tham and J. N. P. Hume, "Analysis of Voice and Low-priority Data Traffic by Means of Brisk Periods and Slack Periods," Computer Communications, vol. 6, no. 1, pp. 14-33, February 1983.

[9] J. N. Daigle and J. D. Langford, "Queueing Analysis of a Packet Voice Communication System," IEEE INFOCOM '85, Washington, D.C., March 25-27, 1985.

[10] W. R. Daumer, "Subjective Evaluation of Several Efficient Coders", IEEE Trans. on Comm., vol. COM-30, no. 4, pp. 655-662, April 1982.

[11] L. R. Rabiner, "Digital Processing of Speech Signals", Prentice-Hall, 1978.

TELETRAFFIC ISSUES in an Advanced Information Society
ITC-11
Minoru Akiyama (Editor)
Elsevier Science Publishers B.V. (North-Holland)

PERFORMANCE EVALUATION OF BURST-SWITCHED INTEGRATED VOICE-DATA NETWORKS

J. D. Morse

GTE Laboratories Incorporated
40 Sylvan Road, Waltham, Massachusetts 02254, USA

ABSTRACT

The traffic characteristics of two burst-switched network topologies are evaluated. These ring and tree configurations have potential application in the local-access-area of the public switched telephone network. Performance aspects are addressed primarily through simulation. A brief overview of burst switching is given and network models are described. Methodology and results from a performance evaluation of the two integrated networks are presented and discussed. Sensitivity of voice freezeout, data delay, and data queue sizes to traffic loads are investigated for the two network topologies. Finally, a promising control scheme for the tree topology is evaluated.

1. INTRODUCTION

Efficient integration of voice and data in a digital network is a goal for network designers. A new network architecture based on the burst switching concept [1,2,3] shows promise as a basis for such an integrated network. Presented in this paper are performance evaluation analyses and results for the burst-switched local-access-area derived from mathematical models and computer simulations. Mathematical modeling is limited due to the advanced features of burst switching and the complexity of voice and data interactions. An overview of burst switching is given below followed by a presentation of network topological design considerations. Two topologies are investigated herein. Simulation models for a *ring network* and a *tree network* are discussed in Section 2. Voice and data traffic models used in the simulations are described in Section 3.

Ring and tree traffic performance results and a comparison of the two topologies will be given in Section 4. In all cases in this paper, the results assume zero community of interest (COI), i.e., all calls terminate outside the local-access-area. This limitation is considered appropriate since an assumed zero COI produces worst case performance. Also, various estimates of voice traffic COI for the group size we are considering range from only 1% to 10%, depending on population density and numerous other factors.

Another parameter included in the simulation models but not investigated in this paper is intentional voice delay. This parameter was investigated previously in the burst switched ray configuration [3] and it was shown that a small degree of improvement in voice freezeout performance could be obtained with intentional voice delay in a range around 10 ms. These voice-only results are similar for the ring and tree configurations.

Computer simulation is a powerful tool which makes possible the study of numerous network conditions and combinations of parameters. Simulation results and exact mathematical formulae should be compared where possible to test modeling assumptions. Mathematical results applicable to the burst-switched local-access-area are available for only idealized situations. We use these results where possible (Section 4) but have chosen not to limit use of the simulation to those cases. In fact, the simulations allow many possible areas of study beyond the scope of this paper and these are listed in Section 5. Also in Section 5 is a summary of the conclusions of this study.

1.1 Review of Burst Switching

Burst switching has been a significant area of research at GTE Laboratories for over four years. It employs a highly distributed network architecture which places link switches in the outside plant. In a burst switched network, information is transmitted in variable-length packets called bursts which can contain either voice or data. Transmission is channelized and channels are occupied only when speech energy or data are present. Speech/silence detectors are placed at each voice port in the switching node. Voice has priority over data but will not preempt a data burst. If there are more active sources than available channels, speech will be lost (freezeout) or data will be queued only for as long as congestion exists. Otherwise, bursts are sent through the switching nodes (link switches) with negligible delay and storage. That is, burst switching differs from conventional packet switching in that the information is not routinely stored in the switching nodes. Channels are dynamically allocated as bursts pass through intermediate link switches. End-to-end delay is small due to a burst header of only four bytes for voice and six bytes for data.

1.2 Network Topological Design Considerations

Selection of a network topology must take into account many factors, such as:

- **Node Complexity** — Includes both the processing capability required and the number of processors as determined by the degree of the node.
- **Path Length and Delay** — Usually a function of the number of hops in the path (however, an example is shown later of a short path which does not reduce delay).
- **Network fault tolerance** — How well a network is able to withstand component failures.
- **Outside Plant Engineering** — The physical connectivity of switches as opposed to the logical connectivity. Two objectives in the digitalization of the public switched telephone network are: (1) use existing outside plant, and (2) minimize customer loop length.

- **Modularity** — How easily new links and switches can be added to an existing network.
- **Traffic** — The throughput, delay, and availability characteristics of a network.

No topology has a clear advantage for all applications. Two basic topologies, the tree and ring, are selected for investigation in this paper. Relative to the ring topology, the tree generally offers delay and modularity advantages but disadvantages in node complexity and network fault tolerance. Outside plant considerations are heavily dependent on the application. The many factors which play a role in network design suggest that there is a need for topological alternatives and evaluation of these alternatives. Certain traffic considerations for two of these alternative configurations are examined in this paper.

2. LOCAL-ACCESS-AREA NETWORK MODELS

The first configuration considered consists of a variable number of link switches (LS) connected in a ring by T1-rate (1.544 Mb/s) transmission links (Figure 1). The bidirectional ring is connected to the outside world by two adjacent interface link switches (ILS) which do not support traffic sources. This entire configuration is referred to as a link group.

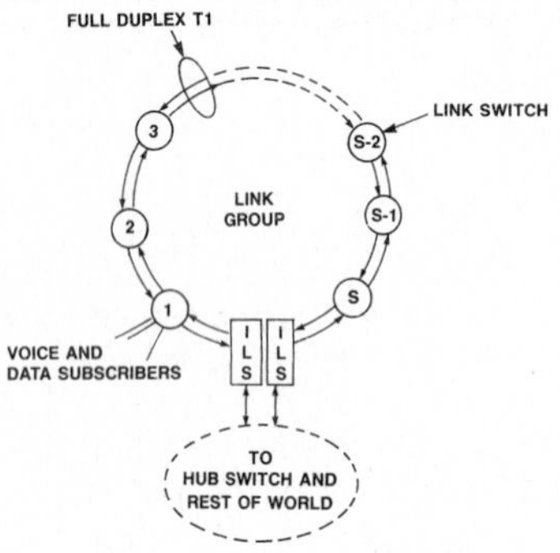

Figure 1. Burst switching local access area ring configuration

The second configuration considered consists of two connected binary trees each with link switches again connected by T1 links. Only binary trees are considered in order to minimize the degree of the nodes. The tree link group is shown in Figure 2. Two ILSs are also present in the tree configuration. The path to an ILS in the tree generally has a small number of hops. This fact and the fact that the two trees are connected makes possible alternate routing of congested speech bursts (as opposed to conventional circuit-switched telephone networks which perform alternate routing of calls).

Traffic on both the ring and tree may be classified as either internal or external. Internal traffic both originates and terminates on the link group and is routed to avoid the ILSs. External traffic is routed via the shortest path, i.e., through the nearest ILS. External voice traffic sources are assumed to be in pairwise conversation, so there is exactly one source outside the link group associated with each external voice source on the link group. Only external traffic is considered in this paper.

Both network models are represented by discrete event simulations developed in the GPSSH language. The simulations can be used to examine behavior of the network under various conditions, e.g., number of link switches, transmission channels, and number of voice and data sources.

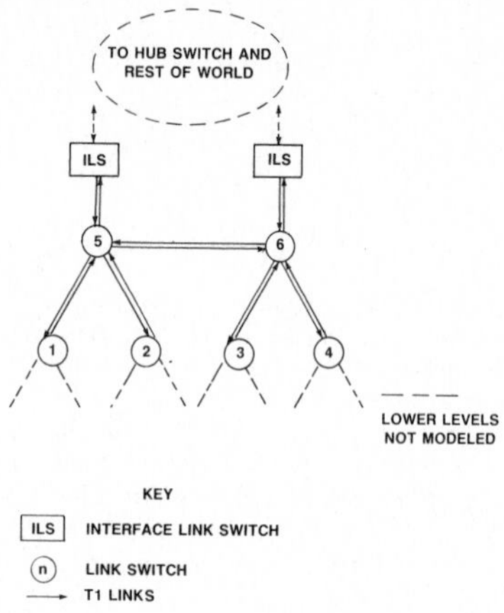

Figure 2. Tree configuration with six link switches

In the ring simulation, eight link switches are connected by links with 24 T1 channels. The number of switches simulated is not varied because the ring results are identical for any number of switches greater than six when there is zero COI and other parameters are held constant. The reason is that blockage occurs in no more than three switches in tandem, even in the congested cases which will be considered. The addition of more than six switches (three on each half of the ring) will not affect the distribution of congestion on the links. Another constant condition in the ring simulation is the location of the host computers. They are assumed to be connected to link switches 4 and 5, so that the longest possible path is taken by bursts going off the link group from the hosts and also by external bursts coming on the link group to the hosts.

The tree simulation models six switches. Only two levels of the binary trees are modeled because congestion does not occur in the lower levels of the tree for the cases considered in this paper.

In both models each half acts independently of the other when neither COI nor rerouting for congestion is present. In both the ring and tree simulations error-free channels are assumed. Also, traffic sources are uniformly distributed among the switches in the ring. The tree and ring have the same number of sources at the nodes adjacent to the Interface Link Switches in order to get a direct comparison of the two configurations. The remaining sources in the tree are uniformly distributed.

3. VOICE AND INTERACTIVE DATA MODELS

The voice traffic model is based on empirical results published in [4]. This model represents measured English conversational speech through use of talkspurt and pause length probability distributions. Voice burst mean and standard deviation are 284 ms and 241 ms, respectively. Silent interval mean and standard deviation are 480 ms and 1379 ms, respectively. The resultant speech activity level is 37.1%. It is important to use a realistic voice model. O'Reilly [5] shows that data performance in an integrated voice-data environment is greatly affected by the assumed voice talkspurt and pause distributions.

The interactive data traffic model is derived from measurements of actual computer-terminal traffic in a scientific environment found in [6]. In order to incorporate this data model in a communication network simulation, it is decomposed into: (i) a model of traffic generated by the user at a terminal, and (ii) a model of traffic generated by the computer in response to users.

Interarrival times are the sum of computer reaction time, user time, and transmission periods in both directions. The transmission periods are 0.2% of the total interval and are ignored. Therefore, the single user interarrival time distribution is formed by convolving the distributions for the two remaining times.

The single user model is used to form an aggregate model for multiple users. The random arrival of bursts from a single terminal can be modeled as a stationary renewal point process. Therefore the aggregate model is formed by a composition of renewal processes as described in [7]. Users are assumed to be independent. The aggregate model represents traffic from a number of interactive terminals on a common controller. We model *32 terminals* on a common controller which has *a single data port* at the link switch. Mean values for the model parameters, assuming 64 kb/s channels, are as follows:

mean burst length (terminal to host) = 7 ms (448 bits)
mean burst length (host to terminal) = 36 ms (2304 bits)

mean interarrival time (single terminal, both directions) = 27323 ms
mean interarrival time (32 terminals, both directions) = 854 ms

Note that the total channel utilization is only 0.008 and 0.042 in the two directions for the 32 interactive sessions. The simulation allows any mix of terminal and host traffic to enter and leave the link group; however, all simulation results in this paper have the same number of host and terminal sources both entering and leaving the link group.

The procedure for forming an aggregate model can be applied to any number of terminals. However, as the number of terminals increases the interarrival distribution approaches the exponential distribution. The general distribution from the actual measurements is used in this paper.

4. RING AND TREE PERFORMANCE RESULTS

Link switches support both voice and data traffic sources. Performance of integrated voice and data is evaluated using an interactive data model, following a brief discussion below of voice-only analyses. Voice performance is measured by cutout fraction which is the fraction of speech lost due to freezeout. Freezeout occurs as front-end clipping of the speech burst. Data performance is expressed primarily in terms of end-to-end queueing delay in the link group and also maximum queue length at a link switch.

In general, the ring and tree models yield similar voice-only freezeout results. Results are also very similar to those from the ray configuration discussed in [3]. Weinstein [8] shows that voice freezeout in the single node case is dependent only on the number of active sources, the mean activity level, and the number of channels. When one considers the multi-node case, *the total mean freezeout is independent of the number and topology of link switches* (see [9]) as long as the traffic is entirely external, the number of channels does not vary between nodes, and the particular local access area (LAA) considered has one link to the rest of the world. To obtain voice-only freezeout of sources at each link switch separately, the method of [9] can be used.

4.1 DATA DELAY AND QUEUE LENGTH RESULTS

Link group delays are investigated in order to determine regions of operation where the burst-switched local-access-area will not contribute significantly to response times for interactive data users. Also, we will look at the relationship between delay and freezeout. Queue lengths provide indications of memory requirements in link switches for various levels of supported traffic.

Delay sensitivity to a variation in data traffic volume will be investigated first. Delay is defined to be total queueing delay through all link switches in the burst path within the local-access-area. Processing delays are assumed negligible due to the cut-through nature of burst switching and low burst overhead. The 95th percentile of the delay distribution is used as a figure-of-merit.

The number of off-hook voice sources is held constant at 98 while the number of data sources is varied. The 95% data queueing delays resulting from simulation are shown in Figure 3.* Delay increases steadily up to about 4800 data terminals as shown in Figure 3. With heavier data traffic, delay increases more rapidly. The number of data terminals was varied for both the ring and tree in the same increments to get a direct comparison between the two configurations. As seen in Figure 3, delays for the two configurations are very similar over the entire range of data terminals. This result occurs even though the bursts in the tree are traversing fewer links than in the ring. Speech freezeout is increasing, of course, while data sources are being added. This result is shown in Figure 4 for the same cases as in Figure 3. Freezeout level in Figure 4 is nearly a linear function of the number of data sources increasing at a rate of only 0.00011 for each additional 100 data terminals. Freezeout for both the ring and tree reaches 0.0108 ± 0.0010 with 98 voice sources and 6400 terminals. The system is stressed beyond the CCITT objective of 0.005 [10] because this objective is considered to be conservative.**

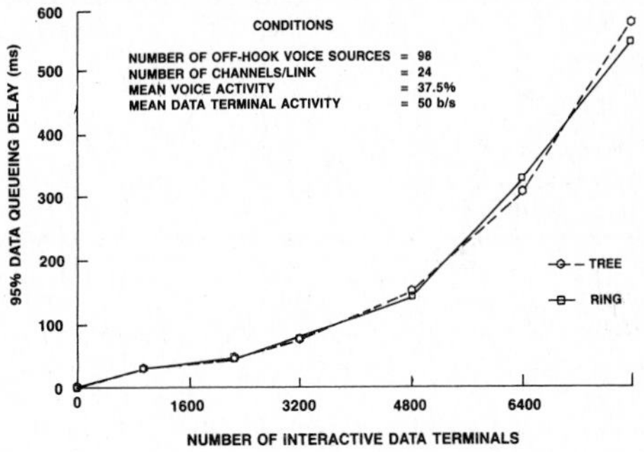

Figure 3. 95th percentile data queueing delay vs number of data terminals

*The 90% confidence interval half-lengths for the six cases shown in Figure 3 range from 7 ms to 30 ms. The same random number sequences were used to generate the voice traffic in all cases.

**As argued in [3], the 0.005 objective may be conservative because current speech detectors [4] produce bursts of shorter average duration than assumed when the CCITT freezeout objective was established. Therefore more bursts are clipped but most are clipped a negligible amount (less than 20 ms) at the same 0.005 level. Since the resultant impairment in speech quality is less, the 0.005 objective is considered conservative.

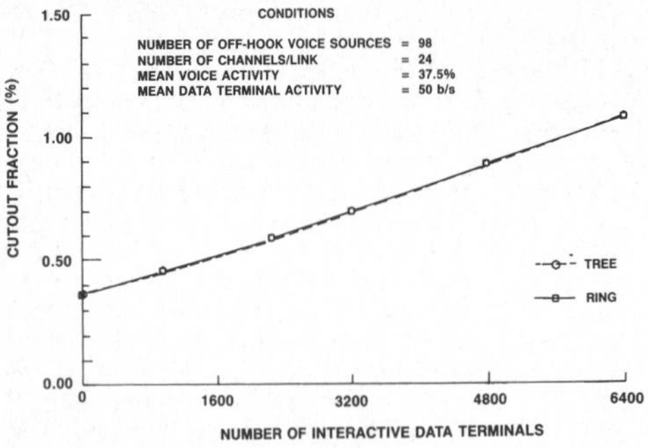

Figure 4. Variation of freezeout with number of data Terminals

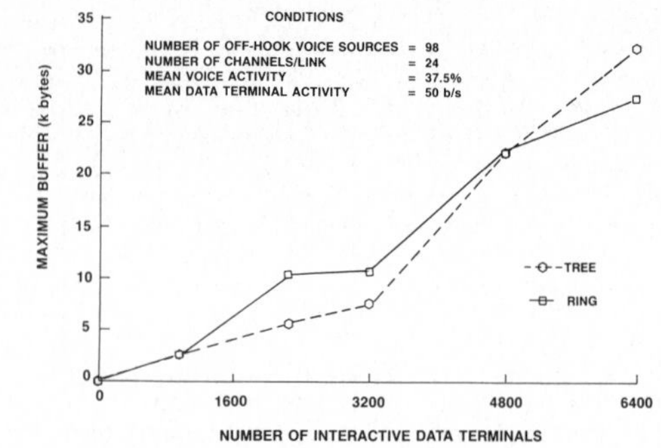

Figure 5. Maximum buffer length vs number of data terminals

Also of interest are the lengths of the queues which build up in the link switches. These results are useful for buffer sizing. There are no restrictions on queue size in the simulation in order to determine the maximum queue buildup. The lengths of the largest queues in the same simulation cases as in Figure 3 are shown in Figure 5. Maximum buffers in both the ring and tree increase steadily with increasing data load in Figure 5. The trend toward a nonlinear increase seen with delay at heavy loads in Figure 3 is not seen in Figure 5. Note that maximum buffer is a single sample in a run and therefore is not a meaningful statistic but rather an observation. Its confidence interval can not be computed. However, each of the samples in Figure 5 comes from simulations of about 200,000 voice and data bursts; the maximum buffer was observed to be in steady state in these cases.

While 95% data queueing delay is a convenient figure of merit, the total cumulative distribution of delays provides a more detailed performance picture. The cumulative distributions of queueing delays on the link group for the ring and tree baseline cases of 98 voice sources and 960 data terminals are shown in Figure 6. These two cases were chosen as baseline because they just meet the 0.005 freezeout objective. Note that over 90% of the data bursts in the baseline simulation runs experience zero queueing delay. Mean data delays for the ring and tree are 5.1 ms and 4.7 ms, respectively.

The comparison above was done in order to see the effect of spreading the data queues in the ring as opposed to the tree. Because traffic leaving the tree link group merges at LS 5 and LS 6 (Figure 2), it is not surprising that congestion is concentrated at these two switches. In the ring (Figure 1) one would expect congestion concentrated in switches 1 and S but also perhaps spread out along switches leading into these two switches. Some blocking of bursts was observed in the ring at switches 2 and S-1. However, for the cases considered in this paper, this spreading out of the blocking in the ring was not enough to cause any statistically significant differences in performance between the tree and ring.

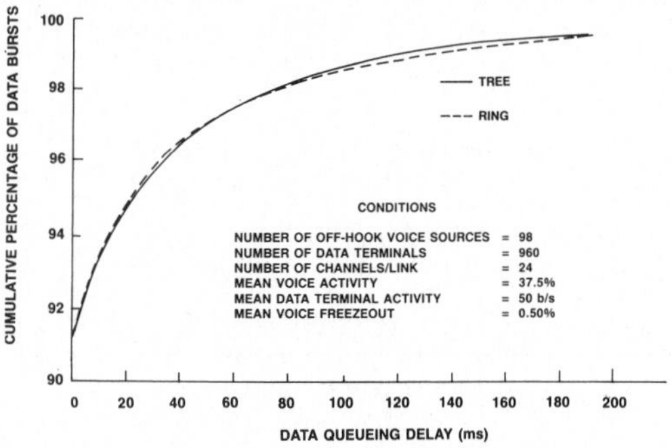

Figure 6. Cumulative frequency distribution of data delays for ring and tree baseline cases

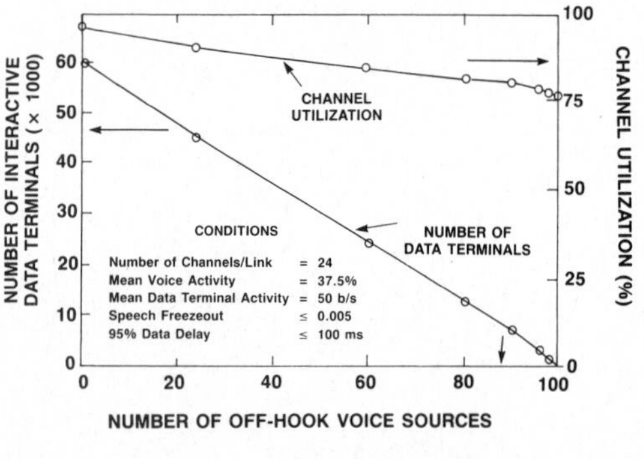

Figure 7. Number of voice and interactive data sources

4.2 Maximum Number of Data and Voice Sources on a Link Group

Consider now what happens when we incrementally add data traffic to the T1 links in a manner that maintains acceptable speech quality and also stays within acceptable data delay limits. We use here the 0.005 CCITT freezeout objective and an arbitrarily chosen 95% link group data queueing delay limit of 100 ms. Results of varying the number of voice and data sources are shown in Figure 7 and Table 1 and are discussed below. The relation between the number of active data terminals and off-hook voice sources is virtually the same for both the ring and tree. For this reason, one may refer to Figure 7 for both topologies. This curve extends to 100 voice sources with no data sources present.

Table 1 Ring simulation voice and data performance

No. of Off-Hook Voice	No. of Data Ports	No. of Data Terminals	Channel Utilization	Mean Data Delay (ms)	95% Data Delay (ms)	Speech Freezeout (%)
100	0	0	0.77	0	0	0.45
98	30	960	0.77	5	32	0.46
98	42	1344	0.78	6	37	0.49
96	94	3008	0.79	8	44	0.49
90	220	7040	0.81	13	74	0.48
80	400	12800	0.82	18	93	0.36
60	768	24576	0.85	13	78	0.24
24	1410	45120	0.90	15	84	0.18
0	1880	60160	0.95	21	94	0.0

From the voice and data model parameters it is known that 15 data ports (480 terminals) produce approximately the same traffic volume (number of bits) as a single voice source. So, as expected, speech freezeout remains fairly constant when the number of voice sources is decreased by 2 (from 100 to 98) and 30 data ports are added to the ring. However, when the number of voice sources is decreased by 10, 220 data ports can be supported at the same level of voice performance; the ratio of data ports to replaced voice sources has increased from 15:1 to 22:1. Note in Table 1 and Figure 7 that utilization of the channels into and out of the interface link switches (ILSs) has increased from 0.77 to 0.81 and also that data delay has increased but is still within our stated limit. This result satisfies our intuition that *data bursts fill in the gaps between voice bursts and increase utilization.* Voice performance is not greatly impacted because the data bursts are an order of magnitude shorter than the voice bursts. Therefore voice bursts, which have nonpreemptive priority, have a relatively short wait for any channels previously seized by data bursts.

As the number of data ports is increased even further while maintaining freezeout <0.005, channel utilization continues to increase but at the expense of increasing data delay. Therefore, as the number of voice sources decreases to 80 and below, *data delay replaces speech freezeout as the limiting constraint.* Channel utilization continues to increase until it reaches 95% in an all data environment. At this point the link group supports 1880 data ports, equivalent to 60160 data terminals, and the 95% data queueing delay is 94 ms. We recognize that such a scenario is somewhat unrealistic. That is, it is unlikely one would find such a large number of terminals in an area which would be reasonably served by a single link group.

4.3 Rerouting for Congestion

A capability of the tree simulation is Rerouting for Congestion (RC). RC is applied to external voice traffic on a burst by burst basis as opposed to a call by call basis. Each talkspurt issued by a talker is free to take an independent path to its destination regardless of the paths of previous bursts. All bursts will attempt to follow their primary route first, but if that route is blocked the burst will follow its alternate route through the other ILS. Only one change of direction is allowed. Internal data traffic is not rerouted to avoid passing through the hub switch and also to avoid the (usually) most congested set of channels leading into the hub switch. Data traffic of any type is not rerouted because potential sequencing problems could result if a burst is queued.

RC was not implemented in the ring simulation. Most congestion is found at the last link switch and a ring could have a maximum of 16 LSs or 32 LSs. Therefore, most rerouted bursts in the ring would end up with a lengthened path taking them through nearly all the link switches.

Results from an all voice, zero COI case in Figure 8 show a sizable decrease in freezeout when RC is used in the tree simulation. With 100 off-hook talkers the reduction is about 83%, from 0.0046 to 0.0008. Most of this blocking occurs at the four channel sets leading to or from the ILSs. This is due to the fact that all sources on the LAA are competing for those last sets of channels leading to the rest of the world. The marked reduction in freezeout as a result of the RC option makes possible an increase in the number of off-hook subscribers that an LAA could support at the same freezeout level. It is seen in Figure 8 that this number could be increased from 100 to 107 and still meet the CCITT objective of 0.005 freezeout.

To implement RC in an actual burst-switched system would be feasible, albeit nontrivial, since at least one additional bit for routing control would need to be assigned and processed. A potential problem in using RC is that of voice bursts arriving out of sequence at the destination. As long as the minimum pause length between bursts is greater than the additional processing delay caused by passing through a larger number of LS's, the bursts will always be in sequence entering the hub switch. A minimum pause can be ensured by the speech detector in order to avoid a sequencing problem on the link group. It is assumed that additional processing delays will be small so the integrity of the silence intervals will be preserved.

5. CONCLUSIONS

In summary, the performance of two burst switching network topologies are evaluated in terms of their traffic carrying capabilities. This paper gives insight into the effective design of an efficient burst switched network. An integrated voice and data environment was investigated via simulation. Mixing voice and interactive data in the ring and tree simulation models produces similar speech freezeout, data delay, and maximum data queue results over a wide range of traffic loads.

It was shown that channel utilization increases significantly with an increasing percentage of interactive data traffic relative to voice traffic. Keeping speech freezeout and data queueing delay below stated thresholds, channel utilization on the ring increased from 0.77 in an all voice environment to 0.95 in an all data environment. In the region of interest around the baseline case (98 off-hook voice sources and 960 interactive data terminals), the 95th percentile of the data queueing delay increases only about 40 ms with the addition of 2000 interactive data terminals. The maximum link switch buffer size used to queue data in the baseline case simulation runs was less than 5 kbytes.

Rerouting of blocked voice bursts was implemented in the tree simulation. Speech freezeout was greatly reduced as a result (Figure 8). In an all-voice situation while maintaining a 0.005 freezeout level, the number of off-hook voice sources in the simulated tree network could be increased from 100 to 107 with rerouting for congestion.

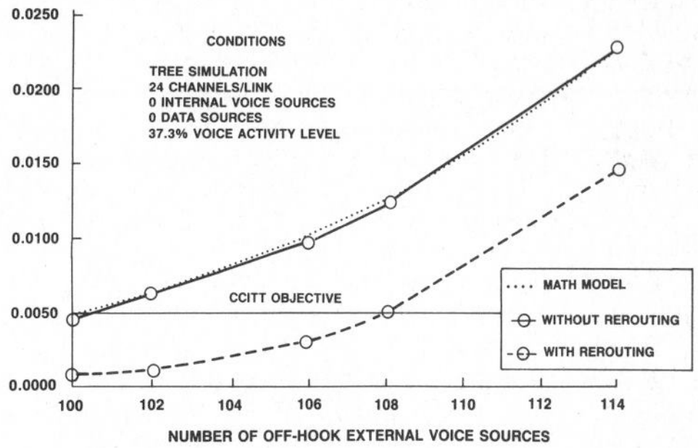

Figure 8. Rerouting for congestion improvement

The scope of the studies described in this paper has been performance aspects of voice and data in the local access area. Performance modeling activities need to be extended into the exchange area, which includes a single hub switch, and eventually into network-wide multi-hub studies. Possible extensions to past performance evaluation studies are to consider:

1. Multirate voice and data,
2. Traffic which both originates and terminates on the link group,
3. Other forms of congestion control,
4. Bulk data traffic on dedicated channels using a variable bandwidth concept, and
5. Variable host computer locations.

6. ACKNOWLEDGMENTS

The author wishes to thank S. Kopec for simulation development efforts and C. Jack for generating simulation performance results.

7. REFERENCES

[1] E.F. Haselton, "A PCM Frame Switching Concept Leading to Burst Switching Architecture," Int. Conf.on Communicatons, Paper E6.7, June 19 - 22, 1983.

[2] S. Amstutz, "Burst Switching—A Method for Dispersed and Integrated Voice and Data Switching," Int. Conf. on Communications, Paper A7.8, June 19 - 22, 1983.

[3] J. Morse and S. Kopec, "Performance Evaluation of a Distributed Burst-Switched Communications System," Phoenix Conf. on Computers and Communications, March 1983.

[4] Y. Yatsuzuka, "Highly Sensitive Speech Detector and High-Speed Voiceband Discrimination in DSI-ADPCM System," *IEEE Trans. on Communications COM-30* [4] (April 1982).

[5] P. O'Reilly, "A Fluid-Flow Approach to Performance Analysis of Integrated Voice-Data Systems with DSI," submitted to *IEEE Trans. on Communications*.

[6] P. Pawlita, "Traffic Measurements in Data Networks, Recent Measurement Results and Some Implications," *IEEE Trans. on Communications COM-29* [4] (April 1981).

[7] P. Kuehn, "Approximate Analysis of General Queueing Networks by Decomposition," *IEEE Trans. on Communications COM-27* [1] (January 1979).

[8] C. Weinstein, "Fractional Speech Loss and Talker Activity Model for TASI and for Packet-Switched Speech," *IEEE Trans. on Communications COM-26* [8] (August 1978).

[9] Y. Lim, "End-to-End Packet Loss Fractions in a Tandem Packet Voice Network," Proc. of INFOCOMM 85, March 25 - 28, 1985.

[10] Minutes of the CCITT Working Party XVIII/2 (Speech Processing) meeting, June 24, 1982.

TELETRAFFIC ISSUES in an Advanced Information Society
ITC-11
Minoru Akiyama (Editor)
Elsevier Science Publishers B.V. (North-Holland)

OVERFLOW TRAFFIC COMBINATION AND CLUSTER ENGINEERING

Pierre LE GALL

CNET-DICET, French PTT,
PARIS, FRANCE

ABSTRACT

We derive practical numerical calculation procedures for determining the individual losses experienced by the different traffic streams in large hierarchical or non-hierarchical telephone networks in conditions of statistical equilibrium. We show that each partial traffic may be characterized by its congestion factor : the ratio of call congestion to time congestion for the stream considered in isolation and for an equivalent capacity to be determined.

In the case of "peaky" traffics, this factor is more representative than the usual "peakedness factor". We therefore amend the Hayward loss formula and present a relatively simple calculation method applicable to telephone, data and multi-channel (ISDN) traffics.

1. INTRODUCTION

At the last ITC [7], we presented a number of very theoretical considerations on the combination of overflow traffics in the case of the lost call model. In this paper we present practical numerical calculation methods for evaluating the individual loss experienced by each traffic stream. We consider hierarchical and non-hierarchical networks in conditions of statistical equilibrium. Call holding times have the negative exponential distribution. We consider calls of different types (telephone, data), including the case of multi-channel calls in the ISDN.

In the above mentioned paper we showed that the usual approach consisting in defining the overall overflow traffic offered as a one dimensional point process of overflow arrivals, composed of the sum of x partial overflow processes, is not rigorously correct. We showed that it is, in fact, necessary to define an offered process of x dimensions in which we take account of the simultaneous evolution of all primary trunk groups, rather than simply considering each partial overflow traffic and its own primary trunk group separately. An approximate arrival process cannot be defined uniquely by the instants of served arrivals ; it should also be defined by the instants of non served arrivals which (as opposed to the case of a Poisson process), depend on the blocking states of all (primary and overflow) trunk groups.

To come back to the traditional point of view, that is to the possibility of partitioning the state equations, it appeared that the equivalent distributions for each traffic stream must be modified in certain blocking configurations. Calculations are then very complex but provide an important theoretical result. The possibility of having approximately a "product form" solution for the combination of partial states means that each traffic stream can, in practice, conserve its own time scale. The only slight influence of disparities in the mean holding times of different call types (telephone, data) occurs through the normalizing factors. We assume therefore, without introducing a significant supplementary error, that mean call holding times are the same for all types.

To derive a practical calculation method, we now come back to the usual problem concerning the superposition of x one-dimensional regenerative arrival processes to approximately represent the overall overflow arrival process. We have already presented this study at the 7th ITC [5] with the following major conclusion : in a blocking situation, partial traffics seem to be generated by a modified offered process, thus explaining the "softening" of the traffic.

Usual methods of fitting and combining the first moments of partial overflow traffics do not take account of the above property and assume, on the contrary, an infinite capacity overflow trunk group. Our study is therefore incompatible with these methods which, moreover, lead to complicated calculations if 3rd and 4th moments are used.

In the presentation of the results which we deduce below from the above mentioned study, it will be seen that each partial overflow traffic can be characterized by just two quantities : its intensity and its "congestion factor". The latter is defined as the ratio of call congestion to time congestion for this traffic offered in isolation to a trunk group of equivalent capacity, determined by means of a relatively simple system of equations.

The method of R.I. Wilkinson [11] uses the overall variance or the "overall peakedness factor" to evaluate mean call congestion. To evaluate individual losses, we propose to replace this peakedness factor by the above mentioned congestion factor. This substitution can also be made in "Hayward's loss formula" which then has excellent accuracy, even in the case of low blocking.

The proposed method is particularly well adapted to "very peaky" overflow traffics and takes account of the way the traffic peaks are affected on contact with other overflow traffics.

It also allows us to present a simple rule for cluster engineering in hierarchical networks.

We begin by recalling the above mentioned study.

2. THE TRUNK GROUP

2.1. Assumptions and notations

a) Lost calls are cleared, set up times are negligible and we assume stationarity.

b) x renewal processes are offered to the same group of n trunks. The inter-arrival distribution of the i th process is F_i (t) with mean value $(1/\lambda_i)$.

c) All call holding times are independent of the arrival processes and of each other and have the same negative exponential distribution whose mean value is equal to the unit of time. Note, moreover, that at the end of [6], we proved the non-influence of more general distributions on the overflow occupation distribution at arbitrary and overflow instants, in the stationary case and for an infinite overflow group.

d) Calls of the i th offered process occupy d_i channels. Except in section 5, we assume $d_i = 1$ (i = 1,...,x) corresponding to ordinary calls.

2.2. Blocking states

2.2.1. Notation

Offered traffic intensity for the i th process is :

$$A_i = \lambda_i \qquad (1)$$

Similarly, we write for the characteristic function of inter-arrival times :

$$\varphi_i(z) = \int_0^\infty e^{-zt}\, dF_i(t)\ , \quad (R(z) \geq 0) \qquad (2)$$

A further important function is :

$$\gamma_i(z) = \frac{\varphi_i(z)}{1-\varphi_i(z)} = \int_0^\infty e^{-zt}\left[dF_i(t) + dF_i(t) \circledast dF_i(t) + \dots\right] \qquad (3)$$

The quantity $\gamma_i(\nu)$ gives the mean number of (partial) calls arriving during the simultaneous occupation of ν given busy trunks.

Denote by $P(k_1 \dots k_x)$ the probability that, at an arbitrary instant, there are simultaneously k_i calls of type i in progress (i = 1,...,x) and let :

$$S(\nu_1 \dots \nu_x) = \sum_{k_1 \dots k_x} \binom{k_1}{\nu_1} \cdots \binom{k_x}{\nu_x} P(k_1 \dots k_x) \qquad (4)$$

If we divide this expression by $\dfrac{n!}{\nu_1!\, \nu_2! \dots \nu_x!}$

we find (see [5]) the probability that (at an arbitrary instant) ν_i given trunks are busy in the i th traffic stream (i = 1,...,x). For the blocking state, where $k_1 + \dots + k_x = n$, we have :

$$P(k_1 \dots k_x) = S(k_1 \dots k_x) \qquad (5)$$

At a call arrival instant in the i th traffic stream, denote the quantity equivalent to $S(\nu_1 \dots \nu_x)$ by $B_i(\nu_1 \dots \nu_x)$

2.2.2. Case of an isolated trunk group

In the case of an infinite capacity trunk group and a single traffic stream (x = 1), the equivalents of B and S are $b_\infty(\nu)$ and $s_\infty(\nu)$ given by :

$$\begin{cases} b_\infty(0) = 1\ , & b_\infty(\nu) = \prod_{j=0}^{\nu} \gamma(j)\ , \\ s_\infty(0) = 1\ , & s_\infty(\nu) = \dfrac{A}{\nu} \cdot b_\infty(\nu - 1)\ . \end{cases} \qquad (6)$$

If v is the variance of the (carried) traffic, the "peakedness factor" has the expression :

$$z = \frac{v}{A} = \frac{2\, s_\infty(2)}{A} + \frac{s_\infty(1)}{A} - \frac{[s_\infty(1)]^2}{A} = \gamma(1) + 1 - A \qquad (7)$$

If the trunk group has only n trunks, the equivalents of B and S are written $b_n(\nu)$ and $s_n(\nu)$ and we recall the relation :

$$b_n(\nu) = f(\nu) \cdot s_n(\nu)\ , \qquad (8)$$

where :

$$\begin{cases} f(0) = 1\ , \\ \boxed{f(\nu) = \dfrac{\nu \cdot \gamma(\nu)}{A}} \end{cases} \qquad (9)$$

We call f(n) the congestion factor of the considered traffic : it is equal to the ratio of the call congestion $b_n(n)$ to the time congestion $s_n(n)$ for a group of n trunks.

For Poisson traffic (z = 1), we have f(n) = 1. For "peaky" traffic (z > 1), we have f(n) > 1. If Z is large enough (very peaky traffic), we have f(n) > Z if n is large enough.

2.2.3. Case of x traffic streams

For the i th traffic stream, and recalling expression (7), let

$$\begin{cases} D_i(1) = 1 + \gamma_i(1) - A_i = z_i\ , \\ D_i(\nu) = 1 + \nu \cdot \gamma_i(\nu) - (\nu - 1) \cdot \gamma_i(\nu - 1)\ . \end{cases} \qquad (10)$$

Taking account of (9), this can be written

$$\boxed{D_i(\nu) = 1 + A_i\left[f_i(\nu) - f_i(\nu - 1)\right]} \qquad (11)$$

For Poisson traffic we have $\nu \cdot \gamma(\nu) = A$ and therefore : $f(\nu) \equiv D(\nu) \equiv 1$.

In [5], chapter (IV.3), we made use of a remark by Morris and Wolman [8] on the properties of statistical equilibrium. It follows that, in a finite capacity trunk group, and in a congestion situation, the frequency of the ends of calls in progress is equal to the frequency of the starts of congestion states for a given partial traffic stream, independently of the occupancy distribution in the other congested traffic streams. This results from the mutual independence of arrival processes and call holding times. In the above mentioned paper, we deduced expressions (122) and (123) which, taking account of (9) and (11) above, can be written in the following form, for $\nu_1 + \dots + \nu_x = n$:

$$\begin{cases} S(\nu_1 \dots \nu_x) = K \cdot \prod_{i=1}^{x} \left\{ \dfrac{A_i^{\nu_i}}{\nu_i!} \cdot \prod_{\lambda=1}^{\nu_i - 1} \dfrac{f_i(\lambda)}{D_i(\lambda + 1)} \right\} \\ B_i(\nu_1 \dots \nu_x) = f_i(\nu_i) \cdot S(\nu_1 \dots \nu_x) \end{cases} \qquad (12)$$

where $f_i(\nu_i)$ is expression (9) relative to the i th partial traffic stream. This is its "congestion factor" for a capacity ν_i. In (12) we set $\prod_{\lambda=1}^{\nu} \left(\frac{f_i}{D_i}\right) \equiv 1$ for $\nu = 0$ and $\nu = 1$.

K is a constant factor whose exact expression is not important for the time being.

Expressions (12) are remarkable in that they do not explicitly contain the trunk group capacity n. Further, for the i th traffic stream, let :

$$\begin{cases} \Phi_i(z) = \dfrac{z\, \gamma_i(z)}{1 + (z+1)\, \gamma_i(z+1)}\ , \\ \Gamma_i(z) = \dfrac{\Phi_i(z)}{1 - \Phi_i(z)} = \dfrac{\gamma_i(z)}{D_i(z+1)}\ , \end{cases} \qquad (13)$$

$$\text{and : } \lim_{z \to 0} z\, \Gamma_i(z) = \frac{A_i}{D_i(1)} = \frac{A_i}{z_i} \qquad (14)$$

The first expression (12) can then be written :

$$\begin{cases} S(\nu_1 \dots \nu_x) = K \cdot \prod_{i=1}^{x} S_{\infty,i}(\nu_i)\ , \\ \text{with :} \\ S_{\infty,i}(\nu_i) = \dfrac{1}{\nu_i}\left(\dfrac{A_i}{z_i}\right) \cdot \prod_{\lambda=1}^{\nu_i - 1} \Gamma_i(\lambda) \end{cases} \qquad (15)$$

In a congested situation, the system behaves as if the trunk group were of infinite capacity, on condition that we make the substitution $\gamma_i(z) \to \Gamma_i(z)$. In other words, congestion states appear to be relative to mutually independent partial traffic streams generated by renewal arrival processes defined by $\Phi_i(z)$ and not by $\varphi_i(z)$.

If we have $D_i(z) \equiv 1$ (case of Poisson arrivals), we again have the case of infinite n. $D_i(z)$ expresses the correlation between traffic streams but does not depend on the trunk group capacity n.

2.2.4. Individual loss equations

Taking account of (12), the time congestion is :

$$\pi = \sum_{\nu_1 \dots \nu_x} S(\nu_1 \dots \nu_x) \quad , \quad \left(\sum_{i=1}^{x} \nu_i = n\right) \qquad (16)$$

Similarly, the call congestion of the i th partial traffic stream is given by

$$B_i = \sum_{\nu_1 \dots \nu_x} f_i(\nu_i) \cdot S(\nu_1 \dots \nu_x) , \quad \left(\sum_{i=1}^{x} \nu_i = n\right) \qquad (17)$$

Can we relate B_i to π in a simple manner ? If so, we would be able to deduce the individual loss distribution. Consider the term of (16) in the neighbourhood of the maximal point $(\nu_{1,0} \dots \nu_{x,0})$. This point is such that

$$\frac{S(\nu_{1,0} \dots \nu_{i,0} \dots \nu_{j,0} \dots \nu_{x,0})}{S(\nu_{1,0} \dots \nu_{i,0}-1 \dots \nu_{j,0}+1 \dots \nu_{x,0})} \simeq 1$$

Expression (12) then gives, for $\sum_{i=1}^{x} \nu_{i,0} = n$:

$$\frac{A_i}{\nu_{i,0}} \cdot \frac{f_i(\nu_{i,0})}{D_i(\nu_{i,0}+1)} = C^t \qquad (i = 1, \dots, x)$$

This gives the following equivalent capacity equations (for i = 1, ..., x) :

$$\boxed{\frac{\nu_{i,0}}{A_i \cdot \dfrac{f_i(\nu_{i,0})}{D_i(\nu_{i,0}+1)}} = \frac{n}{\sum_{k=1}^{x} A_k \cdot \dfrac{f_k(\nu_{k,0})}{D_k(\nu_{k,0}+1)}}} \qquad (18)$$

or, taking account of (9) and (13) :

$$\boxed{\frac{\nu_{i,0}}{\nu_{i,0} \cdot \Gamma_i(\nu_{i,0})} = \frac{n}{\sum_{k=1}^{x} \nu_{k,0} \cdot \Gamma_k(\nu_{k,0})}} \qquad (18a)$$

The summations (16) and (17) depend essentially on the leading terms in the neighbourhood of the maximal point. Moreover, in general, the functions $f(\nu)$, $D(\nu)$ and $\Gamma(\nu)$ vary more slowly than the function $\{A^\nu/\nu!\}$, as ν varies about the maximum. From (17) then, we deduce the approximation

$$\boxed{B_i \simeq \pi \cdot f_i(\nu_{i,0})} \qquad (19)$$

Equations (18) may be solved by successive iterations and, for each partial traffic stream, give the equivalent capacity $\nu_{i,0}$ and the individual congestion factor $f_i(\nu_{i,0})$

2.2.5. Time congestion

It remains to find π in order to be able to use (19). Let B be the mean call congestion. Total rejected traffic C, given by the sum of partial rejected traffics, is :

$$C = A \cdot B = \sum_{i=1}^{x} A_i \cdot B_i = \pi \cdot \sum_{i=1}^{x} A_i \cdot f_i(\nu_{i,0}) \; ,$$

where $A = \sum_{i=1}^{x} A_i$. We deduce the following relation for the overall traffic :

$$B = \pi \cdot Z_0 \qquad (20)$$

where we let :

$$\boxed{Z_0 = \sum_{i=1}^{x} \frac{A_i}{A} \cdot f_i(\nu_{i,0})} \qquad (21)$$

Z_0 is the mean congestion factor

The overall traffic carried, in a congestion situation, can be defined by its first moments (in a congestion situation). The system behaves as if this traffic were the result of $x = Z_0$ (provisionally assumed to be a positive integer) identical partial traffics having the same call congestion $B_i = B$. Equations (18) for the partial equivalent capacity give $\nu_{i,0} = \left(\frac{n}{Z_0}\right)$ without changing the trunk load.

We could have chosen a value of x other than Z_0 but expression (20) would not then have been satisfied. To see this simply, consider the analogy of a single traffic stream for which calls need Z_0 channels. We again have relation (20) and the reduced equivalent capacity (n/Z_0) for the equivalent (single channel) calls undergoing the same call congestion.

Finally, the mean call congestion is a function of the form :

$$B = B\left(\frac{n}{Z_0} \, , \, \frac{A}{Z_0}\right) \qquad (22)$$

where successive moments must be introduced appropriately. From (20) we deduce the time congestion :

$$\pi \simeq \frac{1}{Z_0} \cdot B\left(\frac{n}{Z_0} \, , \, \frac{A}{Z_0}\right) \qquad (23)$$

In the next section we apply these results to the case of overflow traffic.

3. HIERARCHICAL NETWORKS

3.1. The circuit cluster

The basic element of any hierarchical network is the "cluster" : x primary trunk groups overflow on to a (second choice) overflow group of n trunks ; primary group i is offered Poisson traffic of intensity a_i and consists of m_i circuits (i = 1, ..., x). The overflow group cannot itself overflow on to a primary group because of the supposed hierarchy.

From [5], section IV.4, we know that expression (1) for the intensity of the i th overflow (2nd choice) traffic is :

$$A_i = b_i = a_i \cdot E_{m_i}(a_i) \qquad (24)$$

where $E_m(a)$ is the Erlang loss formula. Similarly, we have :

$$\boxed{f_i(\nu) = \frac{1}{E_{m_i}(a_i)} \cdot \frac{\sigma_\nu(m_i; a_i)}{\sigma_{\nu+1}(m_i; a_i)}} \qquad (25)$$

where $\sigma_\nu(m; a)$ has the generating function

$$g_\nu(u) = \sum_{m=0}^{\infty} \sigma_\nu(m; a) \cdot u^m = e^{au} \cdot (1-u)^{-\nu} \qquad (26)$$

This function was originally introduced by L. Kosten [4] and then by E. Brockmeyer [1].

In [11] we gave the following recurrence relations which can be used to determine this term :

$$\begin{cases} \sigma_0(m, a) = \dfrac{a^m}{m!} \; , \quad \sigma_1(m; a) = \dfrac{\sigma_0(m; a)}{E_m(a)} \; , \\ \nu \cdot \sigma_{\nu+1}(m; a) = (m + \nu - a)\,\sigma_\nu(m; a) + a\,\sigma_{\nu-1}(m; a) \end{cases} \qquad (27)$$

In the case of (first choice) Poisson traffic directly offered to the overflow group we have m = o and therefore b = a, $f(\nu) \equiv D(\nu) \equiv 1$.

3.2. State probabilities method

We can directly apply expressions (12), (16) and (17). To evaluate K we can use the relation due to N.I. Bech which we gave as formula (52) in [7] :

$$\sum_{i=1}^{x} \frac{a_i}{a} \sum_{j_1=0}^{m_1} \dots \sum_{j_x=0}^{m_x} \frac{p(j_1 \dots m_i \dots j_x ; n)}{E_{j_1 + \dots + m_i + \dots + j_x + n}(a)} = 1 \, . \qquad (28)$$

Ex	m_1	a_1	d_1	b_1	z_1	m_2	a_2	d_2	b_2	z_2	b	Z
1	54	51.50	1	3.93	4.00	7	14.30	1	7.98	1.52	11.91	2.34
2	54	51.50	1	3.93	4.00	0	4.00	1	4.00	1	7.93	2.49
3	54	25.75	2	2.96	2.97	0	4.00	1	4.00	1	6.96	1.84
4	54	8.58	6	1.74	1.97	0	4.00	1	4.00	1	5.74	1.29

Table 1

Considered examples (data)

Case of two primary trunk groups

where $p(j_1 \dots j_x ; n)$ is the probability of having j_i $(i = 1, \dots, x)$ calls in progress in the i th trunk group and congestion in the overflow group. In the Annex of [11] we derived the following relation relative to a single unlimited overflow traffic :

$$\Delta_\infty(j,\nu) = \Delta_\infty(\nu) \cdot \frac{\sigma_\nu(j;a)}{\sigma_{\nu+1}(m;a)} \tag{29}$$

Expression (12), for $\nu_1 + \dots + \nu_x = n$, now becomes :

$$P(j_1,\nu_1;\dots;j_x,\nu_x) = S(\nu_1 \dots \nu_x) \prod_{i=1}^{x} \frac{\sigma_{\nu_i}(j_i;a_i)}{\sigma_{\nu_i+1}(m_i;a_i)} \tag{30}$$

Using expressions (12) and (30) and taking account of (28), we can calculate the constant K and the expressions (16) and (17) giving the values of the time congestion π and of the individual overall loss probabilities :

$$P_i = E_{m_i}(a_i) \cdot B_i \tag{31}$$

This still rather complex calculation method has allowed us to verify the accuracy of the results obtained for known examples, e.g. those quoted in [10] and even the cases of small capacities originally given by C. Palm [9].

In table 1 we note two examples (n° 1 and 2) taken from [10] since the first traffic (i = 1) here is especially "peaky" ($z_1 = 4$). The results given by the above method ("method 1") are compared against the exact solutions of the Chapman-Kolmogorov equations given in [10] : see table 2. The accuracy is excellent except for the "less peaky" traffic in the case of high blocking.

The close accuracy obtained for more typical congestion values justifies the approximations made and the use of the special "product form" in (12) which was originally given in [5]. We can thus, in practice, retain the time scale appropriate for each traffic stream . In other words, the results justify the approximating assumption of the same mean call holding time in all traffic streams. Lastly, using expressions (12) we can now consider the "equivalent capacities" approximation method.

3.3. The equivalent capacities method

The works of R.I. Wilkinson [11] showed that the overall traffic, and consequently its mean blocking probability B, were practically defined by its first two moments, i.e. by only two constraints. It follows that expression (22) necessarily reduces to Erlang's loss formula :

$$B \simeq E_{\frac{n}{z_0}}\left(\frac{b}{z_0}\right) \tag{32}$$

We recognize "Hayward's loss formula" [3] where the overall peakedness factor Z is replaced by the mean congestion factor z_0 defined by (21).

From (23), the time congestion π of the overflow group is approximately :

$$\pi \simeq \frac{1}{z_0} \cdot E_{\frac{n}{z_0}}\left(\frac{b}{z_0}\right) \tag{33}$$

Let

$$f_i(\nu) = \frac{\sigma_\nu(m_i;a_i)}{\sigma_{\nu+1}(m_i;a_i)} \tag{34}$$

By (24) and (25) we can write :

$$A_i\, f_i(\nu) = a_i \cdot f_i(\nu) \tag{35}$$

Expression (11) is now

$$D_i(\nu) = 1 + a_i[f_i(\nu) - f_i(\nu-1)] \tag{36}$$

In (18), let the equivalent capacity of the i th partial overflow traffic be $n_i = \nu_{i,0}$.

Taking account of (34) and (36), the equivalent capacity equations (18), in the overflow group, are now :

$$\frac{n_i}{n} = \frac{a_i f_i(n_i)/D_i(n_i+1)}{\sum_{k=1}^{x} a_k f_k(n_k)/D_k(n_k+1)} \tag{37}$$

For the successive iterations (two iterations already give a good approximation), one set of values (n_i) is used to calculate the right hand sides of (37), thus estimating the left hand sides giving the next set (n_i). As starting values we could set $n_{i,0} = n \cdot (a_i / A)$, writing

$$A = \sum_{k=1}^{x} a_k \,, \quad b = \sum_{i=1}^{x} b_i = \sum_{i=1}^{x} a_i \cdot E_{m_i}(a_i) \tag{38}$$

Expressions (21) and (35) then give the mean congestion factor to be used in (33) :

$$z_0 = \sum_{i=1}^{x} \frac{a_i}{b} \cdot f_i(n_i) \tag{39}$$

where the (n_i) are a solution of (37). The expression of (33) then gives π, on taking account of (38). The individual loss probability of the i th overflow traffic is then given by (19) and (25) :

$$B_i = \pi \cdot f_i(n_i) \tag{40}$$

The results obtained from these expressions are quite

						Method exact			Method (1)		Method (2)			
Ex	n	Z	$E_{\frac{n}{z}}(\frac{b}{z})$	Z_o	B_o	B	B_1	B_2	B_1	B_2	B_1	B_2	$f_1(n_1)$	$f_2(n_2)$
1	27	2.340	0.006	2.687	0.009	0.009	0.019	0.005	0.019	0.005	0.019	0.005	5.37	1.36
	18	"	0.088	2.323	0.087	0.087	0.163	0.059	0.161	0.051	0.16	0.050	4.36	1.32
	13	"	0.238	2.049	0.223	0.215	0.357	0.171	0.359	0.136	0.39	0.14	3.60	1.29
2	22	2.487	0.005	3.245	0.011	0.011	0.019	0.004	0.019	0.004	0.019	0.004	5.53	1.00
	15	"	0.061	2.808	0.074	0.079	0.131	0.030	0.13	0.029	0.12	0.026	4.65	1.00
	10	"	0.224	2.375	0.218	0.225	0.352	0.107	0.35	0.098	0.35	0.092	3.77	1.00
3	31	1.841		2.377	0.010				0.027	0.004	0.028	0.004	4.24	1.00
	20	"		2.088	0.081				0.18	0.041	0.17	0.039	3.56	1.00
	13	"		1.824	0.228				0.40	0.14	0.40	0.12	2.94	1.00
4	48	1.295		1.553	0.011				0.043	0.008	0.043	0.007	2.82	1.00
	31	"		1.439	0.081				0.23	0.065	0.22	0.056	2.44	1.00
	20	"		1.332	0.222				0.45	0.21	0.42	0.17	2.09	1.00

Table 2

Comparative results concerning the examples of table 1

(Case of two primary trunk groups overflowing on to a group of n trunks)

B : mean call congestion on the overflow group
B_o : approximate value of B : $B_o = E_{\frac{n_1+n_2}{z_o}}\left(\frac{b}{z_o}\right)$
B_i : call congestion of the i th traffic stream (i = 1, 2)
$f_i(n_i)$: congestion factor of the i th traffic stream (i = 1, 2), formula (25)

Exact method : Kolmogorov state equations method
Method (1) : state probabilities method, formulae (12), (28), (30) and (59)
Method (2) : Equivalent capacities method, formulae (33), (37) or (53); (39), (40) or(60).

accurate as shown by examples 1 and 2 in table 2. Note that the Hayward loss formula

$$B \simeq E_{\frac{n}{z}}\left(\frac{b}{z}\right) \tag{41}$$

gives far too low an estimation for low congestion while expression (32) is a good approximation in all cases : "Hayward's loss formula" should indeed be used with Z_o and not Z. It is then convenient and can be used generally to calculate π in (33).

Note that for low congestion, very peaky traffic (i = 1) corresponds to a congestion factor $f_1(n_1) = 5{,}4$ greater than the peakedness factor z_1 (= 4) which decreases with n : the "peaks" of this very peaky traffic are smoothed on contact with the other traffic stream.

3.4. Third choice traffic

In case of a second overflow, the intensity of the i th 3rd choice traffic is

$$c_i = \left(a_i\, E_{m_i}(a_i)\right) \cdot B_i = \alpha_i\, P_i \tag{42}$$

Denote by $f_i'(\nu)$ its congestion factor. For its evaluation, let us consider the overall traffic of the overflowing group. Taking account of (25) and (32), its congestion factor is

$$Z_o \cdot \frac{1}{B} \cdot \frac{\sigma_\nu\left(\frac{n}{Z_o}\,;\frac{b}{Z_o}\right)}{\sigma_{\nu+1}\left(\frac{n}{Z_o}\,;\frac{b}{Z_o}\right)}$$

For each congestion duration in the overflow group (of n circuits), it overflows an average of calls multiplied by Z_o if we compare with the case of a Poisson traffic. In the case of the i th traffic considered, an average of calls multiplied by $f_i(n_i)$ instead of Z_o is overflowing. The congestion factor of this i th 3rd choice traffic is

$$f_i'(\nu) = \frac{f_i(n_i)}{E_{\frac{n}{Z_o}}\left(\frac{b}{Z_o}\right)} \cdot \frac{\sigma_\nu\left(\frac{n}{Z_o}\,;\frac{b}{Z_o}\right)}{\sigma_{\nu+1}\left(\frac{n}{Z_o}\,;\frac{b}{Z_o}\right)} \tag{43}$$

This expression depends on n_i and above all on (n/Z_o). It results in a higher congestion factor. Equations (18) can then be used again for this 2nd overflow, and so on.

3.5. Cluster engineering

Until now we simply set a standard $B = B_o$ for the average call congestion on the overflow group and applied Wilkinson's method [11] to evaluate just B and not the individual B_i.

The possibility of simultaneously observing an entire circuit cluster, due to the extensive introduction of electronic switching, now allows us to envisage a more economic network management by imposing standards for individual overall loss probabilities, $P_i = P_o$, and for direct offered traffic, $\pi = \pi_o$. We have $\pi_o > P_o$ since the direct traffic generally has a supplementary "service protection". Expressions (31), (33), (40) and (34) then lead to the fundamental relation :

$$f_i(n_i) = \frac{P_o}{\pi_o} = C \tag{44}$$

The function $f(\nu)$ may easily be evaluated with the following recurrence relation, deduced from (27) :

$$\begin{cases} f(0) = E_m(a) \quad , \\ \dfrac{\nu}{f(\nu)} = (m+\nu-a) + a\, f(\nu-1) \quad . \end{cases} \tag{45}$$

For example, take $P_0 = 0.01$ and $\pi_0 = 0.03$ giving $C = 1/3$. Expression (39), taking account of (38), may now be written : $Z_0 = C \cdot \dfrac{A}{b}$.

Expression (33) and the standard $\pi = \pi_0$ then allow us to calculate the number n of overflow trunks. The "equivalent capacity" equations (37) become :

$$\boxed{\frac{n_i}{n} = \frac{a_i / D_i(n_i+1)}{\sum_{k=1}^{x} a_k / D_k(n_k+1)}} \tag{46}$$

It is then convenient to draw up a table taking account of (44) and giving n_i and $D_i\ (n_i + 1)$ for given C, a_i and m_i . Table 3 is a reduced example for C = 1/3. We can proceed in the following manner. Initially assume $D_i\ (n_i + 1) = 1$ and therefore $n_i = n.(a_i / A)$. By interpolation on n_i , the table gives the corresponding values of m_i and $D_i\ (n_i + 1)$ which are inserted in (46) to obtain a new set of values (n_i), and so on. In practice, two iterations are sufficient and even the first frequently gives an acceptable value for m_i . We thus solve simply the fundamental cluster engineering problem concerning the addition of a new traffic stream.

To construct an entirely new network we must use a new iteration : to each value of n there corresponds a set $\{m_i\}$ and therefore values of b and π . The iteration should satisfy the condition $\pi = \pi_0$. As to the value of C, its choice depends on economic considerations.

a=31.00					
	! m !	23	24	25	26
	! e !	0.316	0.290	0.264	0.240
	! b !	9.792	8.983	8.194	7.429
	! n !	0.381	0.952	1.512	2.065
	! D !	2.194	2.115	2.044	1.980

Table 3 Table for Cluster engineering

- Equation (44) : f (n) = C for C = 0.333

- e = $E_m(a)$, b = a $E_m(a)$, D = D (n + 1)

4. NON HIERARCHICAL NETWORKS

4.1. Mutual overflow

The basic element consists of the mutual overflow between two groups of N_i trunks respectively (i = 1, 2) for Poisson traffics of intensity a_i(i = 1,2). These two trunk groups also carry other hierarchical overflow traffics. For the first step, we suppose that the mutual overflow traffics are Poisson. The "equivalent capacities" method gives a first set of capacities for the various traffic streams. Then we use the expression (43) and, by successive iterations, we deduce the capacities at first choice and at second choice for the traffic a_i (i = 1, 2) and the partial losses.

If the influence of the hierarchical overflow traffics is weak, we have to take account of the very peaky nature of mutual overflow traffics, in order to evaluate their second choice blocking probabilities. In a simple manner, we may suppose (provisionally) a general mutual overflow between the two trunk groups (and for any traffic). It results in a supplementary (non-existent) overflow traffic T for the original hierarchical traffics. For the traffics a_i (i = 1, 2) the system approximately behaves as if an overall traffic (A - T), with a mean congestion factor Z_0, would be offered to the group of $(N_1 + N_2)$ trunks, A being the total traffic really offered (excluding internal overflow traffics). Finally, the (first + second) choice blocking probability for the traffics a_i (i = 1, 2) is approximately by excess :

$$P \simeq \frac{1}{Z_0} \cdot E_{\frac{N_1+N_2}{Z_0}}(A-T) \tag{47}$$

taking account of (33).

4.2. Third choice traffic

We suppose now that, after mutual overflow on the first two trunk groups, the traffic overflows (hierarchically) on to another trunk group. The "equivalent capacities" method may again be applied. The traffic a_i gives rise to a third choice traffic of intensity :

$$c_i = a_i P \tag{48}$$

Taking account of (43) and (47), its "congestion factor" is approximately by default :

$$f''(\nu) = \frac{1}{E_{\frac{N_1+N_2}{Z_0}}(A-T)} \cdot \frac{\sigma_\nu\left(\frac{N_1+N_2}{Z_0} ; \frac{A-T}{Z_0}\right)}{\sigma_{\nu+1}\left(\frac{N_1+N_2}{Z_0} ; \frac{A-T}{Z_0}\right)} \tag{49}$$

In other words, the system behaves as if this traffic were overflowing from a primary group of

$$m = \frac{N_1 + N_2}{Z_0} \tag{50}$$

trunks. Thus, while mutual overflow on the first two choices improves the accessibility of the traffic and more evenly shares overloads between the first two trunk groups, there results a very peaky third choice traffic for which it is preferable to come back to a hierarchical structure : it is necessary to find the best trade - off between the flexible routing provided by non-hierarchical networks and the resistance to local overloads of hierarchical networks.

5. MULTI-CHANNEL CALLS

5.1. The equivalent average call

Suppose now that $d_i \geqslant 1$ (i = 1, 2, ..., x). We therefore have the case of the ISDN where a call of the ith stream requires d_i channels operated in circuit switching mode. We come back to the hierarchical network and the circuit cluster of section 3.

Expressions (12), where n does not appear, are still valid for the calls themselves on condition that we satisfy the new condition :

$$\sum_{i=1}^{x} \nu_i d_i = n \tag{51}$$

For a call of the ith stream, the system behaves as if the capacity of the primary trunk group were only (m_i / d_i). Expressions (25) and (34) become :

$$f_i(\nu) = \frac{\rho_i(\nu)}{\rho_i(0)} \; , \; \rho_i(\nu) = \frac{\sigma_\nu\left(\frac{m_i}{d_i} ; a_i\right)}{\sigma_{\nu+1}\left(\frac{m_i}{d_i} ; a_i\right)} \tag{52}$$

Expression (36) for $D_i\ (\nu)$ is unchanged. Equations (18) are still valid for the left hand side. However, in order to take account of the new condition (51), it is necessary to multiply numerator and denominator of the ith ratio by d_i before adding the numerators together to equal n and then adding the denominators. Lastly, instead of (37), the equivalent capacity equations expressed in numbers of calls (and not in number of trunks) may be written, on taking account of (52) and (36) :

$$\boxed{\frac{n_i}{n} = \frac{a_i \cdot f_i(n_i)/D_i(n_i+1)}{\sum_{k=1}^{x} a_k \cdot d_k \cdot f_k(n_k)/D_k(n_k+1)}} \qquad (53)$$

The intensity (in numbers of calls) of the overall overflow traffic may be written :

$$b = \sum_{i=1}^{x} b_i = \sum_{i=1}^{x} a_i \, E_{\frac{m_i}{d_i}}(a_i) \qquad (54)$$

The mean congestion factor Z_0 is still defined by (39) taking account of (52) and (54). Total equivalent capacity (in numbers of calls) of the overall overflow traffic is :

$$\sum_{i=1}^{x} n_i = n_0 (< n) \qquad (55)$$

where the n_i are the solution of (53). Expression (32), taking account of (54) and (55), provides an evaluation of the mean blocking probability B of the average equivalent call in the overflow group :

$$B \simeq E_{\frac{n_0}{Z_0}}\left(\frac{b}{Z_0}\right) \qquad (56)$$

Expression (33) then gives the time congestion of the overflow group :

$$\boxed{\Pi \simeq \frac{1}{Z_0} \cdot E_{\frac{n_0}{Z_0}}\left(\frac{b}{Z_0}\right)} \qquad (57)$$

5.2. Individual loss probabilities

Assume that a given call of the ith traffic stream requires only one channel. Expression (40) then gives the overall call congestion :

$$P_i = \left\{\frac{1}{d_i} \cdot E_{\frac{m_i}{d_i}}(a_i)\right\} \cdot \left\{\Pi \cdot f_i(n_i)\right\} = E_{\frac{m_i}{d_i}}(a_i) \cdot Q_i$$

where the factor $(1/d_i)$ comes from the fact that the loss is now evaluated for a single primary trunk and not d_i . Let $M = \sum_{i=1}^{x} b_i d_i$, $D = \frac{1}{M} \cdot \sum_{i=1}^{x} b_i d_i^2$,

$$\left\{ K = \left(\frac{n}{M}\right)^{1/D} , \quad H_i = \frac{1}{d_i} \cdot \frac{K^{d_i} - 1}{K - 1} \right. . \qquad (58)$$

In [2], formula (19), where only Poisson partial traffics are considered, the following approximate expression for the overall call congestion when the call requires d_i channels is given :

$$P_i \cdot \frac{K^{d_i} - 1}{K - 1} = E_{\frac{m_i}{d_i}}(a_i) \cdot Q_i \cdot H_i \qquad (59)$$

In other words, the individual call congestion of the ith traffic stream in the overflow group is approximately :

$$\boxed{B_i \simeq \Pi \cdot f_i(n_i) \cdot H_i} \qquad (60)$$

where we have taken into account (52), (57) and (58). For $d_i = 1$ we have $H_i = 1$.

In short, we must reason on the number of calls and not on the number of busy trunks, so that all the considerations in the previous sections still apply, whether the network be hierarchical or not, on condition that we make the substitution $m_i \rightarrow m_i/d_i$, for the capacity (in calls) of the primary trunk groups. In particular, expressions (44) and (48) are still valid on taking account of (52). Only the denominator of the right hand side of (53) contains the term d_k explicitly.

Tables 1 and 2 give examples 3 and 4 extending example 2. The first traffic stream corresponds successively to $d_1 = 2$ (example 3) and $d_1 = 6$ (example 4) so that the load offered to the primary trunk group remains the same : $a_1 . d_1 = 51.50$. We give the results provided by the state equations method (section 3.2) where, in addition to substitution (61), we have replaced n by the integer part of n_0 defined by (55). We compare these results with the values obtained by the present equivalent capacity (in numbers of calls) method. The results of the two methods are practically equivalent.

6. CONCLUSION

The "equivalent capacities" method converges rapidly to give good approximations. It can be used to analyze large hierarchical or non-hierarchical networks and covers the case of multi-channel calls.

It is based on the distribution of blocking states, thus bringing us to replace the traffic's "peakedness factor" by its "congestion factor", justifying a more extended use of "Hayward's loss formula".

By (3) and (9), this factor is directly related to the nature of the arrival process and not simply to the arrival intensity. This shows that it is not sufficient to try to approximate the process by modifying the arrival intensity as a function of the occupancy state of the overflow group. Furthermore, the traffic combination law has shown the necessity to consider an auxiliary function D (ν) which comes from the modification to the nature of the process of refused arrivals. Any approximate method must take account of this modification instead of uniquely considering the process of served calls.

REFERENCES

[1] E. Brockmeyer, "Det simple overflowproblem i telefontraficteorien", Teleteknik, 5, n° 4, pp 361-374, 1954.

[2] G. Fiche, P. Le Gall, S. Ricupero, "Study of blocking for multislot connections in digital link systems", I.T.C. 11, Kyoto, 1985.

[3] A.A. Fredericks, "Congestion in blocking systems - A simple approximation technique", Bell Syst. Tech. J., 59, n° 6, pp 805-827, 1980.

[4] L. Kosten, "Sur la probabilité d'encombrement dans les multiplages gradués", Annales des PTT, Paris, pp 1002-1019, 1937.

[5] P. Le Gall, "Random processes applied to traffic theory and engineering", ITC 7, Stockholm, 1973, and : Commut. Electron., Paris, n° 43, pp 5-51, 1973.

[6] P. Le Gall, "Further studies on lost call telephone traffic", ITC 9, Torremolinos, 1979 ; and : Ann. Telecommun., Fr., 35, n° 3-4, pp 103-112, 1980.

[7] P. Le Gall, "End to end dimensioning of hierarchical telephone networks and traffic combination", ITC 10, Montreal, 1983 ; and : Ann. Telecommun., Fr., 39, n° 3-4, pp 129-141, 1984.

[8] Morris, E. Wolman, "A note on statistical equilibrium", Oper. Res., 9, pp 751-753, 1961.

[9] C. Palm, "Calcul exact de la perte dans les groupes de circuits échelonnés", Ericsson Technics, 4, p 41, 1936.

[10] B. Wallström, L. Reneby, "On Individual losses in overflow systems", ITC 9, Torremolinos, 1979.

[11] R.I. Wilkinson, "Theory of toll traffic engineering in the USA", Bell Syst. Techn. J., 35, n° 2, pp 421-514, 1956.

TELETRAFFIC ISSUES in an Advanced Information Society
ITC-11
Minoru Akiyama (Editor)
Elsevier Science Publishers B.V. (North-Holland)

ON NON-RENEWAL TELEPHONE TRAFFIC AND ITS PEAKEDNESS

C.E.M. PEARCE

Applied Mathematics Department, Adelaide
University, Adelaide, Australia

ABSTRACT

Some results for loss systems with renewal inputs are extended to the case of a general ergodic input traffic. The formulae obtained provide a theoretical framework which can be used for some practical calculations. It is assumed throughout that trunk holding times are negative exponential. For more general holding time models see Descloux [1], Le Gall [2-5] and Pollaczek [6,7].

1. INTRODUCTION

Dimensioning problems in teletraffic are frequently attacked via Wilkinson's equivalent random method [8] or its extension to smooth traffic (Mina [9], Bretschneider [10]). See, for example, references [11-17]. These presuppose that Poisson traffic is offered to the trunk group from which the traffic under consideration is imagined to have overflowed. The Poisson assumption is not always entirely satisfactory and alternatives have been put forward (see Rahko [18-20] for the use of a normal distribution, Rahko [18,20] for a Weibull distribution and Schehrer [21], Harris and Rubas [22,23] for binomial offered traffic).

A difficulty with assuming a specific structural form of offered traffic is that the traffic at hand may not necessarily be able to be produced by the overflow of an input of the type considered from a finite trunk group. To cope with this problem Potter [24] has produced an *equivalent non-random method* wherein an unspecified renewal offered traffic is posited.

The theoretical basis for these ideas lies in known formulae due to Palm [25], Takács [26, 27] and Cohen [28] for the distribution of carried traffic on primary and secondary trunk groups for renewal offered traffic. See references [29-34].

The implementation of the equivalent random method is in practice often effected by means of approximate formulae found by Rapp [35] from numerical calculations. The present author has shown [36] how Rapp's formulae may be derived analytically as heavy traffic approximations of exact formulae of Cohen and Takács. In fact the arguments of [36] enable Rapp's formulae to be extended to general renewal offered traffic.

In this paper we consider traffic which constitutes a general strictly stationary and metrically transitive stream. For brevity we shall term such traffic *ergodic*. We deal with some of the preliminaries required for handling such traffic analogously to ordinary renewal traffic. This level of generality provides certain theoretical advantages. To begin, we recall the result of Palm [25] and Takács [27] that the overflow traffic resulting from offered renewal traffic to a negative exponential trunk group is itself renewal. It is a severe limitation to the applicability of queueing models to teletraffic networks that the pooling of renewal streams does not, in general, result in a renewal stream. However it is readily shown that the result of any sequence of pooling or overflowing of ergodic streams is still an ergodic stream. This is indeed even the case when correlations exist amongst streams being pooled. The presence of correlations has normally precluded analyses even in the restricted context of Poisson streams (though see the work of Neal [37]). For a treatment of pooled renewal streams see Bech [38], Le Gall [39], Padgett and Tsokos [40], Kuczura [41] and the author [42].

The case of an ergodic stream has received very limited attention in the literature. Franken [43] and Neuts and Chen [44] consider a semi-Markov process. For the general case see Fortet [45-47] and Le Gall [4,5,48-50] who employ a stochastic integral approach and Finch [51] and the author [42,52] who proceed combinatorially. The cost of the generality involved is often that it is difficult to recover even known formulae for simple cases. Thus Le Gall notes that some of the general formulae remain extremely complicated even when the arrival process is renewal. However in [50] some simple expressions are derived for the case of stationary Poisson offered traffic. In this respect the present combinatorial approach seems to be fairly satisfactory.

In the next section the ergodic stream overflowing from a finite negative exponential trunk group is considered for ergodic offered traffic. The overflow overflow sream is characterised in terms of the offered stream. The trunk occupancy distribution arising when ergodic traffic is offered to a hypothetical infinite trunk group is determined in section three. In section four some initial formulae are presented and comments made on the practically significant question of peakedness. Attention is given to methods of deriving formulae suitable for numerical calculation from empirical data.

2. ERGODIC OVERFLOW TRAFFIC

Arrivals to a group of N negative exponential trunks occur at instants $\tau(0)<\tau(1)<\ \ldots$ of an ergodic stream. We define $t(j,k)=\tau(j)-\tau(j-k)$, $j\geq k>0$, and denote by $\theta(n)$ a random variable with the unconditional distribution of $t(j,n)$. It is supposed that

$$P[0<t(j,1)<\infty]=1.$$

The arrival stream may then be characterized by the functionals

$$\psi_0(s_1,..,s_j)=E[\exp(-\sum_{1\leq m\leq j} s_m t(m,1))], j\geq 1, \Re(s_m)\geq 0.$$

For clarity, first suppose that N=1. The arrival at time $\tau(n+1)$ will result in an overflow if the call occupying the trunk at time $\tau(n)+0$ is still present, which for a given sequence of arrival instants will occur with probability $\exp[-\mu t(n+1,1)]$, where $1/\mu$ is the mean holding time. Hence conditional on an overflow event at time $\tau(n_0)$, the next j overflows occur at the instants

$$\tau(n_1)<\tau(n_2)<..<\tau(n_j)$$

with probability

$$V(n_0,n_1,..,n_j)\equiv$$

$$\sum_{1\leq r\leq j} e^{-\mu t(n_r,1)} \prod_{k\in(n_0,n_j)\setminus\{n_i\}} [1-e^{-\mu t(k,1)}] ,$$

where (n_0,n_j) denotes the set of integers n satisfying $n_0<n<n_j$.

If the overflow stream is characterized by the functionals $\psi_1(s_1,..,s_j)$, $j\geq 1$, then

$$\psi_1(s_1,..,s_j)$$

$$= \sum_{n_1=n_0+1}^{\infty} \sum_{n_2=n_1+1}^{\infty} .. \sum_{n_j=n_{j-1}+1}^{\infty} E[V(n_0,..,n_j)\exp(-\sum_{1\leq r\leq j} s_r t(n_r,n_r-n_{r-1}))]$$

$$\equiv\sum_{n_1} ..\sum_{n_j} E[\exp\{-\sum_{r=1}^{j} [s_r t(n_r,n_r-n_{r-1})+\mu t(n_r,1)]\}\times$$

$$\sum_{\ell=0}^{n_j-n_0-j} (-1)^{\ell} \sum_{\substack{k_1<..<k_\ell \\ k_i\in(n_0,n_j)\setminus\{n_r\}}} \exp[-\mu\sum_{k_i} t(k_i,1)]]$$

$$= \sum_{m_1=1}^{\infty} .. \sum_{m_j=m_{j-1}+1}^{\infty} \sum_{\ell=0}^{m_j-j} (-1)^{\ell}.$$

$$\sum_{\substack{h_1<..<h_\ell \\ h_i\in(0,m_j)\setminus\{m_r\}}} \psi_0(u_1,..,u_{m_j}), \qquad (2.1)$$

where

$$u_a=\begin{cases} s_i \text{ if } a\in(m_{i-1},m_i)\setminus\{m_r|1\leq r\leq\ell\} \\ s_i+\mu \text{ if } a\in\{m_i\}\cup((m_{i-1},m_i)\cap\{h_r\}) \end{cases}$$

and we take $m_0=0$.

In the special case of a renewal iput

$$\psi_0(u_1,..,u_p) = \prod_{1\leq\ell\leq p} \psi_0(u_\ell)$$

and (2.1) becomes

$$\psi_1(s_1,..,s_j)= \prod_{r=1}^{j} \{ \sum_{p_r=1}^{\infty} \sum_{\ell=0}^{p_r-1} \binom{p_r-1}{\ell}(-1)^{\ell} \times$$

$$[\psi_0(s_r)]^{p_r-\ell-1} [\psi_0(s_r+\mu)]^{\ell+1}\}$$

$$\prod_{r=1}^{j} \frac{\psi_0(s_r+\mu)}{1-\psi_0(s_r)+\psi_0(s_r+\mu)} .$$

This product form agrees with the known renewal character of the overflow as shown by Palm and the formula

$$\psi_1(s) = \frac{\psi_0(s+\mu)}{1-\psi_0(s)+\psi_0(s+\mu)}$$

given by Takács [27] for the overflow from a single trunk.

By induction or by direct combinatorial argument, we may derive the following generalization of (2.1) to the overflow from a group of N trunks, with corresponding functionals $\psi_N(s_1,..,s_j)$, $j\geq 1$.

Suppose j to be a fixed positive integer. Let $n^{(N)}$ be a set of integers

$$0=n_0^{(N)}<n_1^{(N)}<..<n_j^{(N)}$$

and suppose $n^{(p)}=\{n_0^{(p)}, n_1^{(p)}, ..\}$ for $0<p<N$ are similar sets, subject to the constraints $n^{(p)}\subset n^{(p-1)}$ for $0<p\leq N$, with

$$n^{(0)}=\{0,1,..,n_j^{(N)}\} .$$

(If we imagine the trunks to be ordered 1,2,..,N, with the overflow from trunk i being offered to trunk i+1, then for p>0 the elements of $n^{(p)}$ are just the indexes n of the time points $\{\tau(n)\}$ at which an overflow occurs from trunk p.) We represent a typical collection $\{n^{(p)};0\leq p\leq N\}$ by g and denote by G the class of all such g. For each g we define H=H(g) to be the class of all families $h=h(g)\equiv\{m^{(p)};1\leq p\leq N\}$ with $m^{(p)}\subset n^{(p-1)}\setminus n^{(p)}$. The set $m^{(p)}$ may be empty even when $n^{(p-1)}\setminus n^{(p)}$ is not. Finally we take

$$\ell(h)= \sum_{1\leq p\leq N} |m^{(p)}|$$

and write the elements of $(m^{(p)}\cup n^{(p-1)})\setminus \{0\}$ as

$$n_{a(1)}^{(p-1)} < n_{a(2)}^{(p-1)} < \ldots .$$

Then

$$\psi_N(s_1,..,s_j)= \sum_{n_1^{(N)}=1}^{\infty} .. \sum_{n_j^{(N)}=n_{j-1}^{(N)}+1}^{\infty} \sum_{g\in G} \sum_{h\in H}$$

$$(-1)^{\ell(h)}\psi_0(u(1),u(2),..,u(n_j^{(N)})), \qquad (2.2)$$

where, for a particular g and h,

$$u(r)= \sum_{1\leq i\leq j} s_i I_r((n_{i-1}^{(N)},n_i^{(N)})]+$$

$$\mu \sum_{1\leq p\leq N} \sum_{i} I_r((n_{a(i)-1}^{(p-1)},n_{a(i)}^{(p-1)}]),$$

I_r representing the indicator function

$$I_r(A) = \begin{cases} 1 \text{ if } r\in A \\ 0 \text{ otherwise} \end{cases}$$

3. TRAFFIC CARRIED BY AN INFINITE GROUP

For convenience, we suppose that the call arriving at $\tau(0)$ finds the trunk group empty. Let

$$\phi_r = \sum_{1\leq i_1<..<i_r} E[\exp(-\mu \sum_{1\leq m\leq r} \theta(i_m))],\ r\geq 1,$$

$$\phi_0 = 1\ ,$$

$$S_{r,n}(\tau) = \sum_{0\leq i_1<..<i_r\leq n} \prod_{\ell=1}^{r} \exp[-\mu(\tau-\tau(i_\ell))],$$

$$S_{0,n}(\tau) = 1,$$

where

$$\beta(\tau) \equiv \sup\{k:\tau(k)\leq \tau\} = n.$$

Suppose $Q_{r,n}$ is the unconditional mean of $S_{r,n}(\tau)$ over $(\tau(n),\tau(n+1))$, this interval being weighted in proportion both to its length and to its frequency in a population of intervals. We shall show that $Q_{r,n}$ has a well-defined limit as $n\to\infty$, which may be interpreted as the expected value of $S_{r,n}$ at a randomly selected time point under steady-state conditions. If $E(t(n,1)) = T$, we have by stationarity that for $r\geq 1$

$$Q_{r,n} = E(\int_{\tau(n)}^{\tau(n+1)} S_{r,n}(\tau)d\tau/(r\mu T)$$

$$= E[\sum_{0\leq i_1<..<i_r\leq n} \{\prod_{1\leq \ell\leq r} e^{-\mu(\tau(n)-\tau(i_\ell))} - \prod_{1\leq \ell\leq r} e^{-\mu(\tau(n+1)-\tau(i_\ell))}\}]/r\mu T$$

$$= E[\sum_{0\leq i_1<..<i_r=n} \prod_{\ell=1}^{r} e^{-\mu(\tau(n)-\tau(i_\ell))} - \sum_{0=i_1<..<i_r\leq n} \prod_{\ell=1}^{r} e^{-\mu(\tau(n+1)-\tau(i_\ell))}]/(r\mu T)$$

$$\equiv (R_1-R_2)/(r\mu T)\ \text{, say.}$$

Now

$$R_1 \to E[\sum_{1\leq k_2<..<k_r} \exp(-\mu \sum_{m=2}^{r} \theta(k_m))] = \phi_{r-1} \quad \text{as } n\to\infty .$$

Also, by the Schwarz inequality,

$$R_2 \leq \sum_{0<i_2<..<i_r\leq n} \{E(e^{-2\mu(r-1)\theta(1)})\}^{\frac{1}{2}(n+1-i_r)} \times \{E[\exp(-2\mu\ell\theta(1))]\}^{\frac{1}{2}} \times \prod_{\ell=2}^{r} \{E[e^{-2\mu(\ell-2)\theta(1)}]\}^{\frac{1}{2}(i_\ell-i_{\ell-1})}$$

$$\leq \{E[e^{-2\mu\theta(n+1)}]\}^{\frac{1}{2}} \prod_{\ell=1}^{r-1} \{E[e^{-2\mu\ell\theta(1)}]\}^{\frac{1}{2}} \div [1-\{E[e^{-2\mu\ell\theta(1)}]\}^{\frac{1}{2}}]$$

$$\to 0 \text{ as } n\to\infty,$$

since from the ergodicity assumption

$$\theta(n+1)/(n+1)\to T \text{ a.s. as } n\to\infty .$$

Here we interpret the product over ℓ as unity when $r=1$.

Hence

$$Q_{r,n} \to Q_r = \phi_{r-1}/(r\mu T) \text{ as } n\to\infty \text{ for } r\geq 1. \tag{3.1}$$

Also

$$Q_{0,n} \equiv 1.$$

The probability generating function $P(z,\tau)$ for the number of occupied trunks at time τ is readily seen to be

$$P(z,\tau) = \prod_{\tau(i)\leq\tau} [1+(z-1)e^{-\mu(\tau-\tau(i))}],\ |z|\leq 1,$$

so that

$$(d^k/dz^k)P(z,\tau)|_{z=1} = k!\ S_{k,n}(\tau). \tag{3.2}$$

Suppose that the probability that j trunks are occupied at time τ is $P_j(\tau)$, so that

$$P(z,\tau) = \sum_{0\leq j\leq\infty} P_j(\tau)z^j, \qquad |z|\leq 1 .$$

Relations (3.2) may be inverted to provide

$$P_j(\tau) = \sum_{j\leq r\leq\infty} (-1)^{r-j}\binom{r}{j} S_{r,n}(\tau),\ j\geq 0.$$

(*Cf.* section three of [33].)

An asymptotic version of this result may be obtained via (3.1). Suppose the distribution of the number of occupied trunks at a randomly selected time point in the steady-state is $\{P_j; 0\leq j\leq\infty\}$. Then we have

$$P_j = \sum_{j=r}^{\infty} (-1)^{r-j}\binom{r}{j}\phi_{r-1}/(r\mu T),\ j\geq 1. \tag{3.3}$$

The normalization condition $\Sigma_0^\infty P_j = 1$ provides

$$P_0 = 1 + \sum_{r=1}^{\infty} (-1)^r \phi_{r-1}/(r\mu T). \tag{3.4}$$

Equations (3.3),(3.4) generalize the corresponding formulae known in the renewal case (*cf.* [33]).

The steady-state version of (3.2) incidentally provides useful information about moments. If the distribution $\{P_j\}$ has probability generating function $P(z)$ and M,V,Z denote respectively its mean, variance and peakedness, then

$$M = P'(1) = \phi_0/(\mu T) = 1/(\mu T)\quad , \tag{3.5}$$

$$V = P''(1)+P'(1)-P'(1)^2 = \phi_1/(\mu T)+M-M^2\quad , \tag{3.6}$$

so that

$$Z = \phi_1+1-M \tag{3.7}$$

(*cf.* Theorem 1 of van Doorn [53]).

4. PEAKEDNESS PROPERTIES

By definition, we have

$$\phi_1 = \sum_{n=1}^{\infty} E[\exp(-\mu\theta(n))]\quad . \tag{4.1}$$

From Jensen's inequality (see, for example [54],

page 152), $E[\exp(-\mu\theta(n))]$ is, for a given mean value $n\mu T = n/M$, minimised when the distribution of $\mu\theta(n)$ is concentrated at that mean, so that

$$\phi_1 \geq \sum_{n\geq 1} \exp(-n/M) = 1/[\exp(1/M)-1] .$$

Hence for offered traffic with a given mean M, we have from (3.7) that

$$Z \geq [1-\exp(-1/M)]^{-1} - M$$

(*cf.* [53]). It is evident that equality obtains only in the case of deterministic traffic. This result had earlier been established in the case of renewal traffic (see refs [41,54]). For some further work on peakedness in the context of renewal traffic see [55-62].

A basic property noted in [54] is that in the heavy traffic limit $\mu \to 0$ the peakedness $Z(\mu)$ converges to

$$Z(0) = \frac{\psi''(0)}{2[\psi'(0)]^2} , \qquad (4.2)$$

where the distribution function F(.) of inter-arrival times is characterized by the functional

$$\psi(s) = \int_0^\infty \exp(-sx)\, dF(x), \ \mathcal{R}(s) \geq 0.$$

Here we assume the first two moments of F exist. The result may be written

$$Z(0) = \tfrac{1}{2}(1+c) ,$$

where c is the coefficient of variation of the distribution of inter-arrival times. This property is, of course, inherited by the overflow traffic from a finite trunk group, as is readily seen.

We note that this result needs some modification in the case of non-renewal traffic. A simple example of a non-renewal traffic is prescribed by the finite order moving average

$$t(n,1) = \sum_{\nu=0}^{p} b_\nu x_{n-\nu} , \qquad (4.3)$$

where p is a fixed positive integer, the b's are constants summing to unity and (x_n) is an independent and identically distributed sequence of random variables. The b's need not all be non-negative if the support of the distribution function of the x's is bounded away from zero. Let this distribution function be characterized by the functional $\zeta(s)$. Relations (3.7),(4.1) give

$$Z(\mu) = 1 + \sum_{n\geq 1} E[\exp(-\mu\theta(n))] - 1/(\mu T) . \qquad (4.4)$$

We may substitute for $\theta(n)$ using (4.3) and imitate the development of [36] to show that

$$Z(\mu) \to \frac{\zeta''(0)}{2[\zeta'(0)]^2} = \tfrac{1}{2}(1+c') \text{ as } \mu \to 0 \qquad (4.5)$$

(with the usual moment assumption).

This result generalizes (4.2), but the coefficient of variation c' now refers to the common distribution of the x's rather than that of the inter-arrival times for calls. These are connected through

$$c = c'\Sigma b_\nu^2 . \qquad (4.6)$$

Now suppose an ergodic stream with mean, variance and peakedness M,V,Z respectively is offered to a single trunk to produce an overflow with corresponding quantities M',V',Z'. We have

$$M' = ME[\exp(-\mu\theta(1))] , \qquad (4.7)$$

$$Z-1 = E[\sum_{n\geq 1} \exp(-\mu\theta(n))] - M . \qquad (4.8)$$

The probability that the n-th overflow occurs at $\tau(k)$ is

$$W(k) \equiv \sum_{0<h_1<..<h_n=k} \Big\{ \prod_{\substack{j\in(0,k) \\ j\notin\{k_i\}}} [1-e^{-\mu t(j,1)}] \Big\} \times \prod_{i=1}^{n} e^{-\mu t(h_i,1)} .$$

Hence if $\theta_1(n)$ is a random variable distributed as the sum of n consecutive inter-arrival times in the overflow stream in the steady-state, then

$$E[\sum_{n\geq 1} e^{-\mu\theta_1(n)}] = E[\sum_{n\geq 1} \sum_{k\geq n} W(k) e^{-\mu\theta(k)}] .$$

We may interchange summations and perform the inner summation to derive

$$E[\sum_{n\geq 1} \exp(-\mu\theta_1(n))] = E[\sum_{k\geq 1} \exp\{-\mu(\theta(1)+\theta(k))\}],$$

where $\theta(n),\theta(1)$ share their most recent offered traffic inter-event time. Hence

$$Z'-1 = E[\sum_{n\geq 1} \exp\{-\mu(\theta(1)+\theta(n))\}] - M' . \qquad (4.9)$$

In the event that

$$E[\sum_{n\geq 1} e^{-\mu\{\theta(1)+\theta(n)\}}] \geq E[e^{-\mu\theta(1)}] E[\sum_{n\geq 1} e^{-\mu\theta(n)}], \qquad (4.10)$$

a comparison of the useful formulae (4.7),(4.9) provides

$$\frac{Z'-1}{M'} \geq \frac{Z-1}{M} \qquad (4.11)$$

(compare Theorem 4 of [53]). Relation (4.10) holds in the case of a renewal input as a trivial consequence of the Schwarz inequality, but we cannot expect it to obtain in general.

If (4.11) holds for overflows from a single trunk then it is immediately inherited for overflows from finite trunk groups. Since M'<M, it also implies the result

$$Z'-1 > \min(Z-1,0)$$

noted in [54] for the renewal case. The inequality M'<M in conjunction with (4.11) also implies

$$Z'/M' > Z/M ,$$

that is, the coefficient of variation of the trunk occupancy distribution for the overflow exceeds that of the offered traffic, a result which again is noted in [54] for the renewal case.

4.1 Approximate Methods

In the context of renewal traffic, the equivalent random method is a convenient computational device to exploit "nice" properties of the Poisson stream. It is natural to seek a comparable device for non-renewal traffic. Most crudely, one can simply ignore correlation effects and employ the equivalent random method regardless. If it is desired to incorporate correlation effects, there is need for some class of non-renewal traffic which is relatively straightforward to manipulate algebraically and can be offered to a finite trunk group in place of the Poisson traffic of the equivalent random method. Such a class is provided by the moving average streams, which are only marginally harder than renewal streams to handle analytically (see [42]). Most simply, the sequence (x_n) of (4.3) can be taken to have a common negative exponential distribution. The analysis of [36] can be paralleled to derive Rapp-type approximate formulae. An alternative and more direct approach is not to regard the non-renewal traffic of interest as the overflow from a fictitious trunk group but to approximate it directly by a moving average stream. If it is desired to allow for correlations between intervals t(j,1) and t(j-i,1) for $1 \leq i \leq q$ say, but not for i>q, then a moving average with p=q can be utilized.

Suppose we take the simple case p=1 and consider the heavy traffic approximation

$$\zeta(\mu)= 1-\mu T+(\mu T)^2\alpha_2-(\mu T)^3\alpha_3+O(\mu^4), \quad \mu T=A^{-1} \text{ small},$$

so that

$$\alpha_i= \frac{\zeta^{(i)}(0)}{i!\, T^i} \quad , i=2,3.$$

A substitution from (4.3) into (4.4) yields

$$Z(\mu)= \alpha_2+A^{-1}(b_1^2\alpha_2+b_0b_1+b_0^2\alpha_2^2+\alpha_2^2-\alpha_3-\alpha_2)+O(A^{-2}). \tag{4.12}$$

If the distribution of (x_n) is taken to be negative exponential then $\alpha_2 = \alpha_3 = 1$, and (4.12) reduces to

$$Z = 1 - A^{-1}b_0b_1 + O(A^{-2}) \; . \tag{4.13}$$

In the spirit of the equivalent random method we take A as the measured mean traffic and use a measured peakedness Z to calculate b_0,b_1 from the approximation

$$Z = 1 - A^{-1}b_0b_1 \tag{4.14}$$

with $b_0+b_1=1$.

Since the support of the negative exponential distribution is not bounded away from zero we should assume $b_0,b_1 \geq 0$. Relation (4.14) thus provides physical values for the coupling coefficients b_0,b_1 for smooth traffic with peakedness in the range

$$Z \in [1-\tfrac{1}{4}A^{-1},1] \tag{4.15}$$

The above development suggests that some smooth traffics may actually be Poisson traffics with positive correlations between the lengths of successive inter-event times. It may bear fruit to examine this possibility empirically. In any case the above mathematical device may be used to handle some smooth traffics without departing from the use of the negative exponential distribution or working with the negative trunks of the extended equivalent random method.

If (4.15) is not satisfied, then a more general distribution for the (x_n) may be employed for which the constants α_2 and α_3 lead to a different range for Z from that given by 4.15).

REFERENCES

[1] A. Descloux, "On Markovian servers with recurrent input," Proc. 6th Int. Teletraffic Congress, Munich, pp. 331/1-6, 1970.

[2] P. Le Gall,"Méthodes de calcul de l'encombrement dans les systèmes téléphoniques automatiques a marquage," Ann. Télécom., vol. 12, no. 11, pp. 374-386, 1957.

[3] P. Le Gall,"Les calculs d'organes dans les centraux téléphoniques modernes," Collection Tech. et Scient. du C.N.E.T., édit. de la revue d'optique, pp. 1-76, 1959.

[4] P. Le Gall,"Random processes applied to traffic theory and engineering," Proc. 7th Int. Teletraffic Congress, Stockholm, pp. 221/1-26, 1973.

[5] P. Le Gall," General telecommunications traffic without delay," Proc. 8th Int. Teletraffic Congress, Melbourne, pp. 125/1-8, supp. 1-7, 1976.

[6] F. Pollaczek," Problèmes de calcul des probabilités relatifs à des systèmes téléphoniques sans possibilité d'attente," Ann. de l'Inst. H. Poincaré, vol. 12, no. 2, pp. 57-96, 1951.

[7] F. Pollaczek,"Généralisation de la théorie probabiliste des systèmes téléphoniques sans dispositif d'attente," C. R. Acad. Sci. Paris, vol. 236, no. 15, pp. 1469-1470, 1953.

[8] R. I. Wilkinson,"Theory of toll traffic engineering in the U.S.A.," Bell System. Tech. J., vol. 35, no. 3, pp. 421-514, 1956.

[9] R. R. Mina,"A solution to the problem of smooth traffic," Proc. 4th Int. Teletraffic Congress, London, pp. 76/1-16 and addendum 1-4, 1964.

[10] G. Bretschneider,"Extension of the E.R. method to smooth traffics," Proc. 7th Int. Teletraffic Congress, Stockholm, pp. 411/1-9, 1973.

[11] E. G. Wormald,"Consecutive overflow telephone traffic graphs," I.E. Aust., E.E. Trans., pp. 168-172, Sept. 1968.

[12] K. Rahko,"Dimensioning tables for smooth, Poisson and peaked traffic based on EERT," Tech. Res. Cent. of Finland, Telecomm. Lab., tiedonanto, 1976.

[13] R. Schehrer,"On higher order moments of overflow traffic behind groups of full access,"Proc. 8th Int. Teletraffic Congress, Melbourne, pp. 422/1-8, 1976.

[14] D. T. Nightingale,"Computations with smooth traffics and the Wormald chart," Proc. 8th Int. Teletraffic Congress, Melbourne, pp. 145/1-7, 1976.

[15] B. Sagerholm,"A method for the calculation of traffic offered to an alternative routing arrangement,"Proc. 8th Int. Teletraffic Congress, Melbourne, pp. 142/1-6, 1976.

[16] B. Sanders, W. H. Haemers and R. Wilcke, "Simple approximation techniques for congestion functions for smooth and peaked traffic,"Proc. 10th Int. Teletraffic Congress, Montréal, pp. 4.4b.1/1-7, 1983.

[17] A. A. Friedricks,"Congestion in blocking systems- a simple approximation technique," Bell Syst. Tech. J., vol. 59, pp. 805-827, 1980.

[18] K. Rahko,"Dimensioning of traffic routes according to the EERT-method and corresponding methods,"Proc. 8th Int. Teletraffic Congress, Melbourne, pp. 143/1-9, 1976.

[19] K. Rahko,"The dimensioning of local telephone traffic routes based on the distribution of the traffic carried," Acta Polytech. Scand., pp. El.14,95, 1967.

[20] K. Rahko,"Tables of congestion based on normal distribution," Tech. Res. Cent. of Finland, Telecomm. Lab., tiedonanto 6,86, 1974.

[21] R. Schehrer,"On the calculation of overflow systems with full availability groups and finite source traffic," 10th Rep. on Studies in Congestion Theory, Univ. of Stuttgart, 1973.

[22] R. J. Harris and J. Rubas,"Calculation of overflow traffic moments for full and limited availability systems with binomially distributed offered traffic," Aust. Tele. Res., vol. 8, no. 2, pp. 42-45, 1974.

[23] J. Rubas,"Dimensioning of alternative routing networks offered smooth traffic," Proc. 8th Int. Teletraffic Congress, Melbourne, pp. 144/1-6, 1976.

[24] R. M. Potter,"The equivalent non-random method and restrictions imposed on renewal overflow systems by the specification of a finite number of overflow traffic moments," Proc. 9th Int. Teletraffic Congress, Torremolinos, pp. 247/1-6, 1979.

[25] C. Palm,"Intensitätsschwangkungen in Fernsprechvehrkehr," Ericsson Tech., vol. 44, pp. 1-189, 1943.

[26] L. Takács,"On the generalisation of Erlang's formula," Acta Math. Acad. Sci. Hung., vol. 7, pp. 419-433, 1956.

[27] L. Takács," On the limiting distribution of the number of coincidences concerning telephone traffic," Ann. Math. Statist., vol. 30, pp. 131-141, 1959.

[28] J. W. Cohen,"The full availability group of trunks with an arbitrary distribution of the inter-arrival times and the negative exponential holding time distribution," Simon Stevin Wis-en Natuurkundig Tijdschrift, vol. 26, pp. 169-181, 1957.

[29] L. Kosten,"Über Sperrungswahrscheinlichkeiten bei Staffelschaltungen," Elek. Nach. Tech., vol. 14, pp. 5-12, 1937.

[30] E. Brockmeyer,"Det simple Overflowproblem i Telefontrafikteorien," Teleteknik, vol. 5, pp. 361-374, 1954. (Engl. transl. vol 8, pp. 92-105, 1957.)

[31] L. Takács,"Introduction to the theory of queues," (chapters 3,4), Oxford University Press, Oxford, 1962.

[32] B. Wallström,"Congestion studies in telephone systems with overflow facilities," Ericsson Tech., vol. 22, no. 3, pp. 187-351, 1966.

[33] C. E. M. Pearce and R. M. Potter,"Some formulae old and new for overflow traffic in telephony," Aust. Telecomm. Res., vol. 11, no. 1, pp. 92-97, 1977.

[34] R. M. Potter,"Explicit formulae for all overflow traffic moments to the Kosten and Brockmeyer systems with renewal input," Aust. Telecomm. Res., vol. 13, no. 2, pp. 39-49, 1980.

[35] Y. Rapp,"Planning of junction network in a multi-exchange area," Ericsson. Tech., vol. 20, pp. 77-130, 1964.

[36] C. E. M. Pearce,"A generalisation of Rapp's formula," J. Austral. Math. Soc., Ser. B, vol. 23, pp. 291-296, 1982.

[37] S. R. Neal,"Combining correlated streams of non-random traffic," Bell Syst. Tech. J., vol. 50, no. 6, pp. 2015-2037, 1971.

[38] N. I. Bech,"Metode til beregning af spaerring i alternativ trunking-og gradingsystemes," Teleteknik, vol. 5, no. 5, 1954.

[39] P. Le Gall,"Le trafic de débordement," Proc. 3rd Int. Teletraffic Congress, Paris, pp. 26/1-23 and three appendices, 1961.

[40] W. J. Padgett and C. P. Tsokos,"On a stochastic integral equation of the Volterra type in telephone traffic theory," J. Appl. Prob., vol. 8, pp. 269-275, 1971.

[41] A. Kuczura,"Loss systems with mixed renewal and Poisson inputs," Proc. 7th Int. Teletraffic Congress, Stockholm, pp. 412/1-5, 1973.

[42] C. E. M. Pearce,"Invariance and secondary stochastic processes," Proc. 14th Conference of Oper. Res. Soc. of New Zealand, vol. 1, pp. 19-29, 1978.

[43] P. Franken,"Erlangsche Formeln für Semimarkowschen Eingang," Elek. Informat. Kyber., vol. 4, pp. 197-204, 1968.

[44] M. F. Neuts and S. -Z. Chen,"The infinite server queue with semi-Markovian arrivala and negative exponential services," J. Appl. Prob., vol. 9, pp. 178-184, 1972.

[45] R. Fortet,"Random distributions with an application to telephone engineering," Proc. 3rd Berkeley Symp. on Math. Stat. and Prob., vol. 2, pp. 81-88, 1956.

[46] R. Fortet and B. Canceill,"Probabilités de perte en selection conjugée," Proc. 1st Int. Teletraffic Congress, Copenhagen, Teleteknik, vol. 1, no. 1, pp. 41-55, 1957.

[47] R. Fortet,"Applications of characteristic functionals in traffic theory," Proc. 5th Int. Teletraffic Congress, New York, p. 287, 1967.

[48] P. Le Gall,"Processus stochastiques appliqués à la theorie et à l'exploitation du trafic," Commut. et Electron., no. 43, pp.5-51, 1973.

[49] P. Le Gall,"Stochastic integral equations applied to telecommunications traffic without delay," Stoch. Proc. and Applic., vol. 2, pp. 261-280, 1974.

[50] P. Le Gall,"Le trafic téléphonique à appels perdus," Commut. et Electron., no. 54, pp. 5-30, 1976.

[51] P. D. Finch,"The co-incidence problem in telephone traffic with non-recurrent arrival process," J. Austral. Math. Soc., vol. 3, pp. 237-240, 1963.

[52] C. E. M. Pearce,"A queueing system with non-recurrent input and batch servicing," J. Appl. Prob., vol. 2, pp. 442-448, 1965.

[53] E. A. van Doorn,"Some analytical aspects of the peakedness concept," Proc. 10th Int. Teletraffic Congress, Montréal, pp. 4.4b 5/1-8, 1983.

[54] W. Feller,"An introduction to probability theory and its applications, vol. 2," Wiley and Sons, New York, 1966.

[55] J. M. Holtzman,"The accuracy of the equiv--alent random method with renewal input," Proc. 7th Int. Teletraffic Congress, Stockholm, pp. 414/1-6, 1973.

[56] H. Heffes and J. M. Holtzman,"Peakedness with traffic carried by a finite trunk group with renewal input," Bell System Tech. J., vol. 52, pp. 1617-1642, 1973.

[57] H. Heffes and J. M. Holtzman,"Peakedness in switching machines: its effect and estimation," Proc. 8th Int. Teletraffic Congress, Melbourne, pp. 343/1-8, 1976.

[58] A. E. Eckberg,"generalised peakedness theory," Proc. 10th Int. Teletraffic Congress, Montréal, pp. 4.4b 5/1-7, 1983.

[59] W. Whitt,"Heavy-traffic approximations for service systems with blocking," AT & T Bell Labs Tech. J., vol. 63, pp. 689-708, 1984.

[60] D. L. Jagerman,"Methods in traffic calculations," AT & T Bell Labs Tech. J., vol. 63, pp. 1283-1310, 1984.

[61] J. M. Holtzman and D. L. Jagerman,"Estimating peakedness from arrival counts," Proc. 9th Int. Teletraffic Congress, Torremolinos, pp. 134/1-4, 1979.

[62] P. J. Burke,"The limit of the blocking as offered load decreases with fixed peakedness," Bell System Tech. J., vol. 61, pp. 2911-2916, 1982.

TELETRAFFIC ISSUES in an Advanced Information Society
ITC-11
Minoru Akiyama (Editor)
Elsevier Science Publishers B.V. (North-Holland)

BLOCKING OF OVERFLOW TRAFFIC COMPONENTS

J. de Boer

AT&T en Philips Telecommunicatie Bedrijven B.V.
Hilversum, the Netherlands

ABSTRACT

Two methods for approximating the blocking probabilities of overflow traffic components are compared for the cases of two and more than two components. A conjecture about the peakednesses of the blocked traffics is stated. A new approximation of the equivalent traffic of Wilkinson's Equivalent Random Theory is given.

1. INTRODUCTION

A number of independent overflow traffic streams is offered to a common overflow route. The determination of the blocking (loss) probabilities of the traffic components on this overflow route is an important and difficult problem. When the blocked traffics are offered afterwards to an other route their peakednesses are also important.
Several approximations for the blocking probabilities of the traffic components are given in the literature. Here we are concerned with a comparison of two methods, namely the Method of Manfield and Downs (MMD) and the Method of Akimaru and Takahashi (MAT). Both use an Interrupted Poisson Process (IPP) as a model for overflow processes. In MMD an IPP is used for each traffic stream whereas in MAT the total overflow traffic is represented by an IPP.
In Section 2 the problem of the blocking probabilities is defined and some notation is given. In Sections 3 and 4 MMD and MAT are treated respectively and in Section 5 they are compared. Section 6 states a conjecture about the peakednesses of the blocked traffics and Section 7 contains the conclusions. Some mathematical details are given in the Appendix.

2. PROBLEM AND NOTATION

We restrict ourselves in the sequel almost always to two traffic streams. Then the loss system drawn in Fig. 1 is considered.
For shortness we often use an index i in the sequel where it is tacitly understood that i can be 1 as well as 2.
Two mutually independent traffics A_1 and A_2 are considered. A_i is a Poisson traffic with negative exponential distribution of holding times which is offered to a primary full availability group of L_i lines (L_i may be zero). The blocked traffic with mean M_i and peakedness (=variance/mean) Z_i is offered to an overflow full availability group of N lines. It is not relevant here whether A_i and L_i are explicitly given as characteristics of a real network or whether they are calculated from M_i and Z_i by Wilkinson's Equivalent Random Theory (ERT) as described in [1]

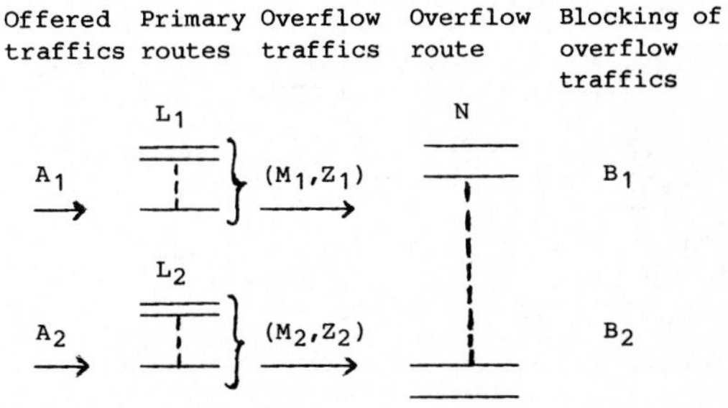

Fig. 1 Loss system with two overflow traffics offered to a common overflow route

The total overflowing traffic has mean M and peakedness Z given by

$$M=M_1+M_2$$
$$Z=(M_1Z_1+M_2Z_2)/M$$

The combined traffic M has a loss probability B on the N lines, while M_i has a loss probability B_i. Thus

$$BM=B_1M_1+B_2M_2 \quad (1)$$

It is well known that B can be determined with sufficient accuracy by ERT or - when M/Z is not too small - by Hayward's approximation ([2] and [3]). In Section 3 an other approximation is mentioned. The problem here, however, is the approximate calculation of B_1 and B_2. In the literature several - mostly heuristic - procedures are known. In e.g. [4] some are mentioned. It is not surprising that an exact treatment is virtually impossible because in [5] it is shown that the interarrival time distribution of the traffic (M_i,Z_i) is a mixture of L_i+1 negative exponential distributions.
Here we are concerned with two related methods, namely that of Manfield and Downs [6] and that of Akimaru and Takahashi [7] respectively, which we treat below.

3. METHOD OF MANFIELD AND DOWNS (MMD)

The overflow process (M_i,Z_i) is modeled as an Interrupted Poisson Process (IPP) described in [8]. Three parameters are chosen as functions of M_i,Z_i and A_i such that the first three moments of both processes are equal. Formulas for the parameters are given in Section A1 of the Appendix. One of the parameters is λ_i which is the product of the intensity of the IPP when it is on, and the mean holding time (this implies $\lambda_i \geq M_i$). As shown in Section A2 the formulas of MMD lead to

$$\frac{B_1}{B_2} = \frac{M_1(Z_1-1)f_1+M}{M_2(Z_2-1)f_2+M} \qquad (2)$$

where

$$f_i \overset{df}{=} \frac{N(\lambda_i-M_i)}{M_i\{(N-1)(Z_i-1)+\lambda_i-M_i\}} \qquad (3)$$

In order to use (2) λ_1 and λ_2 have to be determined. When A_1 and A_2 are given they can be found with formula (12) of Section A1. If not, A_1 and A_2 must be calculated or estimated. This can be done by ERT or e.g. by Rapp's formula

$$A_i=M_iZ_i+3Z_i(Z_i-1)$$

or by solving a cubic equation derived in Section A3 which will be discussed in Section 4. B is calculated by modeling the superposition of the two overflow processes as an IPP. The formula for B is given in [6] and [7] and can be written as

$$B^{-1}=\sum_{j=0}^{N}\left[\binom{N}{j}\Big/\prod_{r=1}^{j}\left\{\frac{M}{r}+\frac{(Z-1)(\lambda-M)}{(r-1)(Z-1)+\lambda-M}\right\}\right] \qquad (4)$$

where λ is calculated in the same way as λ_i above. In cases like those in [6] where A_1 and A_2 are known and hence the first three moments of the total overflow traffic can be calculated, λ can be found by equating these moments to the corresponding moments of the IPP (In general λ is not equal to $\lambda_1+\lambda_2$). Then B_1 and B_2 follow from (1)-(4).
For any two streams i and j out of k (>2) streams (2) is generalized to

$$\frac{B_i}{B_j} = \frac{M_i(Z_i-1)f_i+M}{M_j(Z_j-1)f_j+M}$$

For N=1 and any value of k we get

$$\frac{B_i}{B_j} = \frac{M+Z_i-1}{M+Z_j-1}$$

Some results of the method for cases mentioned in [6] are discussed in Section 5.

4. METHOD OF AKIMARU AND TAKAHASHI (MAT)

Here B is calculated for the total overflow process by (4) where especially

$$\lambda=MZ+3Z(Z-1) \qquad (5)$$

is taken, i.e. a two moment fit is made. This particular choice of is not essential for the method. In order to determine B_1 and B_2 two steps are taken:

- The special case of an overflow traffic (M_1,Z_1) and a Poisson traffic $(M_2,1)$ is considered. Because B_2 is then equal to the time congestion on the N lines, B_2 and hence also B_1, can be found in this special case.
- By a simple extension B_1 and B_2 are found for the case of one overflow traffic (M_1,Z_1) together with an other overflow traffic (M_2,Z_2) which is the addition of two independent traffics, namely (M_1,Z_1) and $(M_2-M_1,1)$.

The formula for B_1/B_2 in the second case is then generally used. It is as follows

$$\frac{B_1}{B_2} = \frac{1+(Z_1-1)f}{1+(Z_2-1)f} \qquad (6)$$

where

$$f \overset{df}{=} \frac{N(\lambda-M)}{M\{(N-1)(Z-1)+\lambda-M\}} \qquad (7)$$

The similarity with the formulas (2) and (3) is evident, but there is only equality in degenerated cases. For any two streams i and j out of k(>2) streams B_i/B_j is found from (6) with 1 replaced by i and 2 by j.
This method is easier to apply than MMD because no λ_i are needed beside λ. If (5) is used the calculation is very simple and only B requires some more work.
However, two points of criticism can be raised:

- First the decomposition of an overflow traffic according to the description above is in general not possible and it is not easy to see how a similar procedure valid for any two overflow traffics could be found. However, on heuristic grounds (6) and (7) can also be used for the general situation of two or more overflow traffics, and in some - or perhaps even many - cases the results are (very) good. E.g. in Table 1 where MAT is compared with exact results for N=1. For this value of N (6) becomes

$$\frac{B_1}{B_2} = \frac{M+Z_1-1}{M+Z_2-1}$$

Here and in the sequel exact results are either taken from literature or obtained by solving the system of state equtions. The exact results for L_2=0 in Table 1 are given in [9].
Further simulations with 3 and 5 traffic streams (see Section 5) suggest that for more streams the accuracy of MAT may improve.

A_1	A_2	L_2	B_1		B_2	
			MAT	Exact	MAT	Exact
0.05	0.05	0	.0034	.0034	.0497	.0497
0.1	0.1	0	.0124	.0122	.0980	.0981
0.5	0.5	0	.1569	.1515	.3922	.3939
2.5	2.5	0	.5932	.5877	.8008	.8047
7.5	7.5	0	.8270	.8256	.9312	.9324
0.09	0.01	1	.0449	.0449	.0119	.0120
0.45	0.05	1	.2121	.2119	.1304	.1312
1.8	0.2	1	.5768	.5766	.5372	.5305
9	1	1	.8936	.8968	.9023	.8939

Table 1. Comparison of blocking values from MAT with exact values for L_1=N=1

- The second point of criticism concerns the choice of λ according to (5) which is more suitable for A than for λ. It can therefore be argued that A be calculated or estimated like A_1 and A_2 in MMD. Here this is done in the following way.
  It is shown in Section A3 that for N=1 or 2

  $$B=E_{L+N}(A)/E_L(A)$$

  where $E_L(A)$ is the Erlang loss formula for A erlangs on L lines and L and A are the fictitious number of primary lines and the fictitious traffic respectively of ERT. This means that for N=1 and N=2 B agrees with ERT. The equality is proved for N=1 and N=2 by showing that both sides of the equation are the same function of M,Z and A. For N=3 there is only equality if A is a root of a cubic equation (18) which lies between Z(M+Z-1) and $+\infty$.

It appears numerically in Section A4 that this root lies closer to the equivalent random traffic of ERT than MZ+3Z(Z-1). It is in fact a very good approximation of the ERT traffic value and it is easier to obtain than by manipulation of the Erlang formula with non-integer numbers of lines.
Substitution of the root in (12) gives the corresponding value of λ. Table 2 gives a comparison between exact values of B_1 and B_2, calculations with λ according to (5), and λ according to (18) and (12), in some cases mentioned in [7] (the exact values in Table 2 of [7] are most likely misquoted). The second kind of calculation is called MMAT (Modified Method of Akimaru and Takahashi).
It appears that MMAT is better than MAT if $L_2 \neq 0$, but there is hardly any difference between the two methods if $L_2=0$. The performance of both methods is not very good. Errors of 5% in B_1 or B_2 are common.
One can think of an other modification of MAT, namely computing λ from (12) with A=MZ+3Z(Z-1). This modification has not been considered here, however, because a numerical investigation showed that MMAT performs better. It appeared also that λ, and therefore also λ-M, is highly sensitive to small changes in A which in turn has a great influence on B_1/B_2 but much less on B.

A_1	A_2	L_1	L_2	N	Method	B_1	B_2	B
5	10	5	0	20	1	.0113	.0076	.0081
					2	.0113	.0076	.0081
					3	.0132	.0079	.0085
10	20	10	0	30	1	.0396	.0240	.0255
					2	.0392	.0240	.0254
					3	.0412	.0240	.0257
15	15	10	0	30	1	.0292	.0191	.0220
					2	.0283	.0191	.0218
					3	.0294	.0193	.0222
5	15	5	15	10	1	.0461	.0693	.0613
					2	.0427	.0610	.0547
					3	.0398	.0669	.0575
10	15	10	10	10	1	.2296	.2078	.2134
					2	.2237	.2048	.2097
					3	.2155	.2028	.2061
15	10	10	15	10	1	.1311	.1581	.1326
					2	.1269	.1500	.1282
					3	.1264	.1547	.1280

Table 2. Comparison of blocking values from MAT (1), MMAT (2) and exact calculation (3)

5. COMPARISON OF MMD AND MMAT

For some cases mentioned in [6] and [7] MMD and MMAT are applied and compared with each other and with an exact calculation. In all cases λ is calculated via (18) and (12). This leads for MMD to blocking values which sometimes deviate slightly from those in [6]. If necessary, λ is found with (12).
For both methods it should be noted that it may sometimes be easier - and for not too small M/Z equally accurate - to calculate B by Hayward's approximation

$$B= E_{N/Z}(M/Z)$$

than by (4).
The results of the comparison are given in Table 3.

The following conclusions can be drawn from this table.
- The approximation of B by MMD and MMAT is good. It seems to be an exception when the error is 5%.
- When there is one Poisson traffic ($L_2=0$) MMAT is better than MMD which may show errors of 30% in B_1 or B_2.
- In cases with $L_2 \neq 0$ the errors of MMD in B_1 or B_2 are mostly in the order of some percents and exceptionally 10%. The errors of MMAT in B_1 or B_2 are generally larger and often in the range of 5-10%.
- In cases with $L_2 \neq 0$ where the ratio between M_1/Z_1 and M_2/Z_2 differs considerably from 1, MMD is better than MMAT (e.g. where $A_1=22$).
- When the ratio between M_1/Z_1 and M_2/Z_2 is in the order of 1, MMD and MMAT perform equally well. This happens e.g. for $A_1=8$, $A_2=12$ and $A_1=14$, $A_2=20$.

Some simulations with three and five traffic streams suggest that the above mentioned trends may still be present but less pronounced and that the relative errors in B_1 and B_2 are smaller than for two streams. If this is true, MMAT may be preferable for more streams because of its greater simplicity.

6. CONJECTURE

So far we were only concerned with the blocking probabilities of the k different traffic streams. The mean M_{oi} of the traffic of stream i that flows over from the N lines can be found from

$$M_{oi}=B_i M_i \qquad (8)$$

A_1	A_2	L_1	L_2	N	Method	B_1	B_2	B
5	10	5	0	20	1	.0095	.0079	.0081
					2	.0113	.0076	.0081
					3	.0132	.0079	.0085
10	20	10	0	30	1	.0300	.0250	.0254
					2	.0392	.0240	.0254
					3	.0412	.0240	.0257
15	15	10	0	30	1	.0252	.0204	.0218
					2	.0283	.0191	.0218
					3	.0294	.0193	.0222
5	15	5	15	10	1	.0420	.0613	.0547
					2	.0427	.0610	.0547
					3	.0398	.0669	.0575
10	15	10	10	10	1	.2096	.2098	.2097
					2	.2237	.2048	.2097
					3	.2155	.2028	.2061
15	10	10	15	10	1	.1280	.1321	.1282
					2	.1269	.1500	.1282
					3	.1264	.1547	.1280
8	12	10	15	6	1	.1017	.1219	.1121
					2	.1022	.1214	.1121
					3	.0961	.1268	.1119
14	12	10	15	6	1	.3658	.4009	.3715
					2	.3601	.4301	.3715
					3	.3600	.3996	.3664
22	12	10	15	6	1	.6184	.6761	.6227
					2	.6127	.7458	.6227
					3	.6190	.6451	.6209
8	16	10	15	6	1	.2480	.2901	.2807
					2	.2582	.2872	.2807
					3	.2433	.2855	.2761
14	16	10	15	6	1	.4702	.5168	.4885
					2	.4624	.5290	.4885
					3	.4638	.5054	.4801
22	16	10	15	6	1	.6589	.7218	.6722
					2	.6476	.7640	.6722
					3	.6624	.6925	.6687
8	20	10	15	6	1	.4098	.4500	.4448
					2	.4383	.4458	.4448
					3	.4067	.4461	.4410
14	20	10	15	6	1	.5699	.6061	.5900
					2	.5699	.6061	.5900
					3	.5658	.5976	.5835
22	20	10	15	6	1	.7030	.7534	.7203
					2	.6933	.7719	.7203
					3	.7084	.7336	.7170

Table 3. Comparison of blocking values from MMD (1), MMAT (2) and exact calculation (3) for two traffic streams

The conjecture is that the peakedness Z_{oi} of this traffic can be approximated by using the formula

$$B_i \cdot \frac{M_{oi}+Z_{oi}-1}{M_i+Z_i-1} = B \cdot \frac{M_o+Z_o-1}{M+Z-1} \quad (i=1,2,\ldots,k) \qquad (9)$$

where M_o and Z_o are the mean and peakedness of the total overflow from the N lines. Now

$$M_o = BM \qquad (10)$$

and Z_o can be found from Hayward's approximation. By assuming that the traffic M/Z offered to N/Z lines is Poissonian, the peakedness of the overflow can be calculated by

$$1-\frac{BM}{Z}+\frac{M/Z}{\frac{N}{Z}+1+\frac{BM}{Z}-\frac{M}{Z}}$$

and Z_o can be found as Z times this peakedness. Hence

$$Z_o = Z - M_o + \frac{MZ}{N+Z+M_o-M} \qquad (11)$$

Thus by applying (10), (11), (8) and (9) Z_{oi} can be calculated. Simulations as well as results in [6] suggested this conjecture.

7. CONCLUSIONS

Many parameters play a role in the two methods considered for approximating the blocking probabilities of overflow traffic components. Therefore, the conclusions here cannot be definitive, but are necessarily provisional. Under this restriction the following can be said.

The method of Manfield and Downs is preferable in the case of two traffic streams, neither of which is Poissonian. The method of Akimaru and Takahashi is preferable in the case of two streams one of which is Poissonian, and in the case of more than two streams.

Errors in the order of 5% in the approximations occur.

Two remarks can be made here. First, the extension in the method of Akimaru and Takahashi to the general case of two overflow traffics should perhaps be further studied.

Secondly, an approximation of the equivalent traffic of Wilkinson's ERT has been found which is easier to calculate than the exact value and is more accurate than Rapp's approximation.

8. ACKNOWLEDGEMENT

I am grateful to Ir. B. Sanders of the Dr. Neher Laboratorium of the Dutch PTT who provided me with some important calculation results.

REFERENCES

[1] R.I. Wilkinson — Theories for toll traffic engineering in the USA. Bell System Technical Journal, Vol. 35, No.2, p. 421-514, 1956.

[2] A.A. Fredericks — Congestion in blocking systems - a single approximation technique, Bell System Technical Journal, Vol. 59, No. 6, p. 805-827, 1980.

[3] J. de Boer Limits and asymptotes of overflow curves, Proc. 10th Int. Teletraffic Congress, Montreal, Paper 5.3.4., 1983.

[4] J. Regnier, P. Blondeau, W.H. Cameron Grade of service of a dynamic call-routing system, Proc. 10th Int. Teletraffic Congress, Montreal, Paper 3.2.6, 1983.

[5] E.A. van Doorn On the overflow process from a finite Markovian queue, Performance Evaluation, Vol. 4, No. 4, p. 233-240, 1984.

[6] D.R. Manfield, T. Downs Decomposition of traffic in loss systems with renewal input, IEEE Transactions on Communications, Vol. COM-27, No. 1, p. 44-58, 1979.

[7] H. Akimaru H. Takahashi An approximate formula for individual call losses in overflow systems, IEEE Transactions on Communications, Vol. COM-31, No. 6, p.808-811, 1983.

[8] A. Kuczura The interrupted Poisson process as an overflow process, Bell System Technical Journal, Vol. 52, No. 3, p. 437-448, 1973.

[9] E.W.B. van Marion Influence of holding time distribution on blocking probabilities of a grading, Tele (English edit.), No. 1, p. 17-20, 1968.

APPENDIX

A1. THE PARAMETERS OF AN IPP

An IPP has three parameters denoted in [8] and [7] by λ, ω and γ respectively, where

λ =intensity during on-time of the Poisson process x mean holding time
$1/\gamma$=mean on-time of the Poisson process
$1/\omega$=mean off-time of the Poisson process

In formula (12) of [8] λ, γ and ω are expressed in A and the first three factorial moments $M^{(j)}$ (j=1,2,3) of the number of occupied overflow lines. These factorial moments can be written as

$$M^{(1)} = M$$

$$M^{(2)} = M(M+Z-1)$$

$$M^{(3)} = \frac{2AM(M+Z-1)^2}{A+Z(M+Z-1)}$$

Substituting this in formula (12) of [8], after some algebraic manipulation we get

$$\lambda = \frac{(M+Z-1)\{A(2Z-M-2)+MZ(M+Z-1)\}}{Z(M+Z-1)(M+2Z-2)-AM} \qquad (12)$$

$$\omega = \frac{M}{\lambda}\left(\frac{\lambda-M}{Z-1}-1\right) \qquad (13)$$

$$\gamma = \omega\left(\frac{\lambda}{M}-1\right) \qquad (14)$$

When M,Z and A are known λ can be calculated by (12) and hence also ω and γ.

A2. B_1/B_2 ACCORDING TO MMD

Equations (4.9) and (4.10) of [6] give the mean numbers O_1 and O_2 of calls of stream 1 and 2 respectively, overflowing from the N lines in a time interval wherein all N lines are occupied. Because

$$\frac{B_1M_1}{B_2M_2} = \frac{O_1}{O_2} \qquad (15)$$

it is necessary to evaluate O_1 and O_2. It follows from the above equations that

$$\frac{O_1}{O_2} = \frac{M_1}{M_2} \cdot \frac{\dfrac{\mathcal{L}_1(N)}{1-\mathcal{L}_1(N)} + \dfrac{M_2}{N}}{\dfrac{\mathcal{L}_2(N)}{1-\mathcal{L}_2(N)} + \dfrac{M_1}{N}} \qquad (16)$$

where $\mathcal{L}_i(\cdot)$ is the Laplace transform of the interarrival time distribution of the IPP concerned. Formula (8) of [7] gives

$$\frac{\mathcal{L}_i(N)}{1-\mathcal{L}_i(N)} = \frac{\lambda_i}{N} \cdot \frac{N+\omega_i}{N+\omega_i+\gamma_i} \qquad (i=1,2)$$

Using (13) and (14) we find that the right hand side of this equation is equal to

$$\frac{M_i}{N}\left(1+\frac{(\lambda_i-M_i)(Z_i-1)N}{M_i\{\lambda_i-M_i+(N-1)(Z_i-1)\}}\right) = \frac{M_i}{N}\{1+(Z_i-1)f_i\}$$

Substitution in (16) and use of (15) yields

$$\frac{B_1}{B_2} = \frac{M_1\{1+(Z_1-1)f_1\}+M_2}{M_2\{1+(Z_2-1)f_2\}+M_1}$$

A3. B AND $E_{L+N}(A)/E_L(A)$ FOR N=1,2,3

A and L are the equivalent Poisson traffic and the equivalent number of lines of ERT when the overflow is characterized by M and Z.

a) N=1
The recurrence relation for the Erlang loss formula

$$\frac{1}{E_{L+1}(A)} = 1 + \frac{L+1}{A} \cdot \frac{1}{E_L(A)}$$

leads to

$$\frac{E_L(A)}{E_{L+1}(A)} = \frac{AE_L(A)+L+1}{A} = \frac{M+L+1}{A} \qquad (17)$$

Now

$$Z=1-M + \frac{A}{M+L+1-A}$$

so that

$$\frac{M+L+1}{A} = \frac{M+Z}{M+Z-1}$$

Substitution in (17) gives

$$\frac{E_L(A)}{E_{L+1}(A)} = \frac{M+Z}{M+Z-1}$$

According to (4), and taking the term with j=0 apart, for N=1

$$B^{-1}=1 + \frac{1}{M+Z-1} = \frac{M+Z}{M+Z-1}$$

Hence for N=1

$$B^{-1}=E_{L+1}(A)/E_L(A)$$

b) N=2
Two times application of the recurrence formula gives

$$\frac{E_L(A)}{E_{L+2}(A)} = \left(\frac{M+Z}{M+Z-1}\right)^2 + \frac{Z}{A(M+Z-1)}$$

On the other hand (4) gives

$$B^{-1}=\frac{M+Z-1}{M+Z-1} + \frac{2(Z-1+\lambda-M)}{(M+Z-1)\{M(Z-1+\lambda-M)+2(Z-1)(\lambda-M)\}}$$

and substitution of λ from (12) yields after some algebra

$$B^{-1} = \left(\frac{M+Z}{M+Z-1}\right)^2 + \frac{Z}{A(M+Z-1)}$$

c) N=3
Proceeding in the same way, but with much more algebra, we find

$$\frac{E_L(A)}{E_{L+3}(A)} = \left(\frac{M+Z}{M+Z-1}\right)^3 + \frac{M+3Z(M+Z)}{A(M+Z-1)^2} + \frac{Z(2-M)}{A^2(M+Z-1)}$$

The complicated expression for B^{-1} is not of much use here. Putting it equal to $E_L(A)/E_{L+3}(A)$ we get a cubic equation for A:

$$DA^3+GA^2+HA+P=0 \qquad (18)$$

with

$D=2M$
$G=M(M+Z-1)(4Z-M-6)$
$H=2Z(M+Z-1)^2.\{M^2-M(3Z-2)-6(Z-1)^2\}$
$P=MZ^2(M+Z-1)^3.(2-M)$

Putting

$$A=aZ(M+Z-1)$$

(18) can be simplified to

$$da^3+ga^2+ha+p=0 \qquad (19)$$

with

$d=2MZ$
$g=M(4Z-M-6)$
$h=2\{M^2-M(3Z-2)-6(Z-1)^2\}$
$p=M(2-M)$

For Z=1 the three roots of this equation are $a_1=1$, $a_2=1$, $a_3=(M-2)/2$.
In this case a=1 or A=M is clearly the desired solution, i.e. the smallest root in the interval $[1,\infty)$. It is difficult to see for $Z>1$ whether (19) has one or three real roots. The left hand side is then negative for a=1, namely $-12(Z-1)^2$, and it is $+\infty$ for $a=\infty$. Therefore, we follow the same rule for $Z>1$, i.e. the smallest root in $[1,\infty)$ should be taken.

A4. COMPARISON OF TWO APPROXIMATIONS OF THE EQUIVALENT POISSON TRAFFIC IN ERT

The solution of (18) has been compared with Rapp's approximation

$$A=MZ+3Z(Z-1) \qquad (20)$$

and the exact value of the equivalent Poisson traffic for given M and Z. The comparison has been made for the cases of Table 3 and the results are given in Table 4.
The table shows that (18) gives a uniformly - though sometimes only slightly - better approximation than (20).

M	Z	Equivalent random traffic according to		
		(20)	(18)	Exact
11.4243	1.0796	12.592	12.569	12.569
22.1458	1.1001	24.693	24.676	24.676
21.1551	1.2342	26.977	26.931	26.931
4.1291	2.1006	15.609	14.949	14.877
8.3009	1.8640	20.304	19.861	19.846
6.5201	1.8275	16.384	15.995	15.976
2.0020	2.2032	12.364	11.907	11.694
6.3107	1.9353	17.643	17.110	17.082
13.693	1.6113	25.019	24.801	24.797
4.3745	2.2446	18.200	17.473	17.378
8.6832	2.0293	23.887	23.343	23.320
16.066	1.7000	31.115	30.871	30.868
7.5732	2.0858	22.591	21.985	21.953
11.882	1.9861	29.474	29.019	29.007
19.265	1.7363	37.285	37.056	37.053

Table 4. Comparison of different methods of calculation of the equivalent random traffic

TELETRAFFIC ISSUES in an Advanced Information Society
ITC-11
Minoru Akiyama (Editor)
Elsevier Science Publishers B.V. (North-Holland)

ANALYSIS AND DIMENSIONING OF NON-HIERARCHICAL TELEPHONE NETWORKS

Marc LEBOURGES*, Christopher BECQUE and David SONGHURST

* Centre National d'Etudes des Telecommunications,
Issy-les-Moulineaux, France
British Telecom Research Laboratories, Felixstowe, UK, IP11 8XB

ABSTRACT

British Telecom's strategy for telephone network digitalisation is leading to radical changes in both the structure of the network and the call routing strategy used in it. In particular, mutual overflow routings will be used together with a trunk reservation facility. Analysis methods involving decomposition and moment models are normally used for hierarchical alternative routing networks, but are inadequate for the new strategy. A methodology introduced recently by F Le Gall and J Bernussou [1] is developed and applied to this problem, and found to give satisfactory results for end-to-end blocking.

1 INTRODUCTION

Over the past 3 years British Telecom (BT) has been developing a new strategy for the full digitalisation of the telephone network. This will involve an early move to a fully digital main (trunk) network, but with both analogue and digital local exchanges co-existing for a longer period.

As the strategy has evolved, a range of possible routing schemes have been proposed. The main feature of these schemes is the connection of local exchanges to two 'parent' units, with mutual overflow between these alternative paths. One approach is outlined in Section 2.

The BT Teletraffic Division has been concerned with modelling the performance of proposed network and routing strategies, and with the development of suitable dimensioning algorithms. This work has required research into new analysis techniques. The algorithm presented in Section 5 was developed by Mr M Lebourges, who worked with the Teletraffic Division on a study visit from December 1982 to March 1984, funded by CNET.

2 THE BT NETWORK

2.1 Analogue Network

The existing analogue network is a 4-level hierarchy, in which little use is made of automatic alternative routing (AAR). Most main network (trunk) traffic is carried at the second level usually via either 1 or 2 trunk exchanges. The top 2 levels (the 'Transit Network') were introduced primarily as a means of enabling full subscriber trunk dialling. Below the Transit Network, many trunk exchanges are interconnected by direct routes provided on an economic basis according to the level of traffic.

2.2 Digitalisation Strategy

Many factors have influenced the strategy adopted for digitalisation. The following are of particular relevance:

- Need for rapid implementation
- Interworking with analogue network
- Modularity of digital transmission and switching equipment
- Desire to simplify network planning procedures
- Flexibility to handle an increasingly wide range of services

Underlying all of these is the basic objective of designing a cost-effective and resilient digital network. The main features of the resulting strategy are the concentration of traffic onto fewer and larger routes and trunk switching units, and the widespread use of overflow routing strategies. The definition of these routing strategies is still evolving.

2.3 New Network Structure

The main network will comprise about 50 digital main switching unit (DMSU) sites, fully interconnected by traffic routes. Consideration is being given to the introduction of AAR via a smaller backbone tier in order to improve the efficiency and resilience of this network.

Digital local exchanges (DLE) will generally have two traffic routes: one to the parent DMSU to carry trunk traffic, and one to a digital principal local exchange (DPLE) to carry traffic within the local area. All traffic routes are bothway, and mutual overflow is provided between the two paths for both incoming and outgoing traffic (Figure 1).

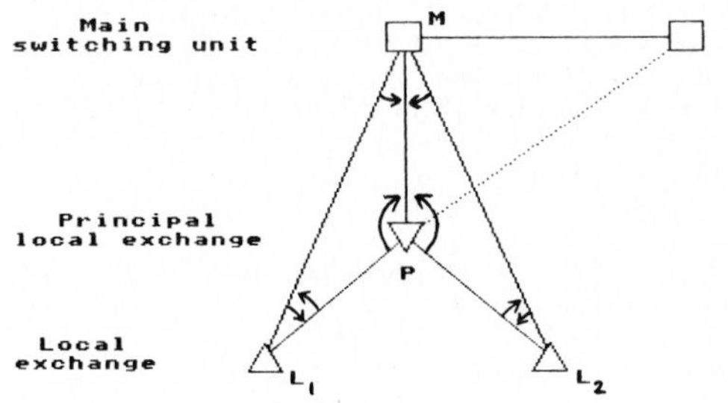

Figure 1 Routing within the DPLE area

For example a call from L_1 to L_2 has the following sequence of alternative paths:

L_1PL_2 L_1PML_2 L_1ML_2 L_1MPL_2

It is likely that a priority reservation system (software-controlled trunk reservation - see [2]) will be used to control overflows within the DPLE area.

The DPLE site will also be connected to a second DMSU, as shown in Figure 1, but the traffic routing strategy to be used for this path has yet to be finalised. Other routes may exist, e.g. direct interconnection of DPLEs, but these will be relatively infrequent. The largest urban areas will use variants of this strategy, but probably retaining the basic features of dual parenting and mutual overflow (see [3] for one example).

Other important features of the new strategy are the use of digital distribution frames at DPLE and DMSU sites to enable rapid re-routing of digital transmission modules, and extensive use of remote concentrator units parented on local exchanges.

3 NETWORK DIMENSIONING

In hierarchical AAR networks the 'cluster engineering' approach to dimensioning is straight-forward. The cluster consists of a final choice circuit group together with all high-usage groups from which it receives overflow traffic. Grade-of-service criteria are applied to each traffic stream offered to the cluster.

With non-hierarchical routing this approach may be difficult or impossible. However in the planned BT network the DPLE-area forms a cluster that can be dimensioned as an entity.

The definition of reference load levels for overload criteria is a significant problem. Analysis of field measurements of traffic flow has suggested the use of 3 load levels with the following form:

1 Normal load
2 Simultaneous overload of all local traffic
3 Simultaneous overload of all trunk traffic

This is more severe than applying overload criteria separately to each circuit group, but less severe than applying a uniform overload of all traffic in the DPLE-area.

Simplified dimensioning algorithms are being developed for interactive operational use. The more sophisticated analysis approach presented in this paper is used to calibrate simpler algorithms, and to analyse the performance of areas of the network under a wide range of traffic conditions. This latter process is being used to provide statistical information on the distribution of end-to-end grade-of-service under realistic ranges of traffic conditions.

4 TELETRAFFIC ASPECTS

4.1 Performance Analysis

The most interesting part of the new strategy from a teletraffic viewpoint is the local network which is represented by the DPLE area shown in Figure 1. Analysis of this structure is not straightforward. Classical decomposition methods work well when the link-by-link analysis can be ordered, but in the case of the DPLE area the mutual overflow arrangement does not allow that ordering. One way of approaching this problem is to carry through the decomposition method and, by assuming the independence of links LM and LP, apply an iteration. A 2-moment algorithm can be envisaged which could be applied to a basic DPLE area. Such an algorithm and its limitations are described below.

4.2 A 2-Moment Method

4.2.1 Analysis of link PM

Evaluation of the PM link requires prior knowledge of the first two moments of all traffic streams offered to it. This includes, for each local exchange, incoming and outgoing trunk traffic streams overflowing from the LM link and incoming local traffic overflowing from LP. Each of these streams is offered to the PM link as part of its alternative path. Given then that these streams have been evaluated a standard 2-moment method (such as Wilkinson's Equivalent Random Traffic (ERT) or an interrupted Poisson process model) can be applied.

4.2.2 Analysis of links LM and LP

Consider an algorithm which applies Wilkinson ERT to the LM and LP links in turn (treating a single local exchange) and so iterates until the mean overflow traffics satisfy a convergence criterion. At each stage of the iteration the algorithm must be able to separate the total non-carried traffic (mean M and variance V) into that traffic which is lost and that which is offered as overflow (m, v) to the other link. Several heuristic formulae have been proposed for splitting means and variances of overflow traffic streams, e.g. by Wallstrom as presented in [4]. Thus the logic of the algorithm would be as follows:

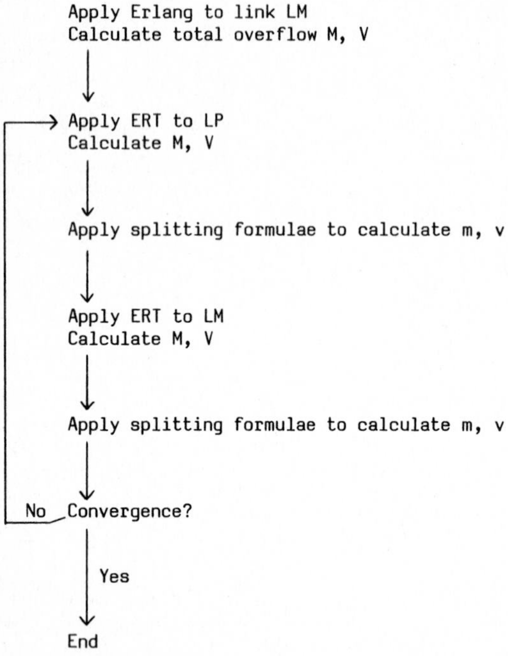
Apply Erlang to link LM
Calculate total overflow M, V
Apply ERT to LP
Calculate M, V
Apply splitting formulae to calculate m, v
Apply ERT to LM
Calculate M, V
Apply splitting formulae to calculate m, v
No
Convergence?
Yes
End

Some initial studies of this type of iteration algorithm suggest that, although computationally it is quite fast (convergence normally within 4 to 6 iterations), it can suffer severe drawbacks.

The predominant assumption that LM and LP are statistically independent leads to the expectation that this method should underestimate stream blockings and that the accuracy would deteriorate at higher loadings. In fact this was found to be the case (by comparison with simulation studies), but with the additional inconsistency that calculated local and trunk traffic stream blocking probabilities often differed despite the fact that they were offered the same circuit availability. The size of these errors was not significantly affected by either a change in the starting point of the iteration or a change in the accuracy of the convergence requirement.

It may be possible to improve this algorithm for the specific routing strategy shown in Figure 1 by taking account of the full availability (to all incoming and outgoing traffic) of all circuits connected to a local exchange. However it is also necessary to model other strategies which do not retain this full availability (e.g. where traffic from P to L does not overflow to PM).

4.3 Further Analysis Problems

The 2-moment method described above considered a very basic DPLE area and has ignored the effects of baulking. In this structure the second-choice path of each traffic stream contains two links in tandem, and local traffic overflows via the high usage link LM. Therefore a practical analysis method should be able to account for baulking effects.

Section 2 mentioned that trunk reservation would be used to control traffic flows. The use of this system to control traffic overflowing to the PM link can be analysed by using an interrupted Poisson process (IPP) model [2]. The use of trunk reservation at the local exchange would complicate the calculation of overflow moments in the iteration, but IPP methods are still feasible [5].

4.4 Application

While moment methods might be applicable as the basis of long-term planning or dimensioning tools, a more consistent and reliable method is required for analytic purposes.

5 THE INVERSE-ERLANG METHOD: CALCULATION OF END-TO-END BLOCKINGS

This method, introduced by Le Gall and Bernussou [1], overcomes most of the teletraffic problems discussed above. First we describe the method, then we adapt it to the DPLE area network and trunk reservation, and finally we give some results.

5.1 Description

The mean carried traffic on a specified trunk group is equal to the summation, over all traffic streams which may use it, of traffic intensity multiplied by some combination of stationary trunk group blocking probabilities. These exact equations can be derived from the transient equations for the mean carried traffic on each trunk group. Approximations are needed to use them: only trunk groups originating from the same node and related by traffic overflow are considered to be dependent. Assuming this, the probabilities described above, and thus the mean carried traffics, are given in terms of time-congestion probabilities of individual and sets of dependent trunk groups. But we need further sets of equations to supplement this set, which is denoted Equations (1).

Two further assumptions are made: to each trunk group (respectively, to each set of dependent trunk groups) corresponds a fictitious Poisson traffic which, when offered to the trunk group (respectively, to a fictitious trunk group whose capacity is the sum of the set), leads to:

- the same time-congestion (respectively joint time-congestion)
- the same mean carried traffic (respectively a mean carried traffic on the fictitious trunk group equal to the sum of the set)

as in the real system.

The first assumption gives Equations (2) and the second Equations (3). Equations (1), (2), (3) make a complete system which can be solved iteratively. These equations will now be developed in full for the DPLE area network.

5.2 Application to the DPLE area with trunk reservation

The DPLE area network has been described in section 2. Let $L_1, \ldots, L_n$ be the local exchanges (generally $n < 15$), P the DPLE, and M the DMSU.

There are bothway Poisson streams between each pair of exchanges, with intensities

AL_{ij} for streams L_i to L_j

ALP_i and APL_i for streams L_i to P and P to L_i

ALM_i and AML_i for streams L_i to M and M to L_i

APM and AMP for streams P to M and M to P

(Note that P is also a local exchange and therefore a source and sink of traffic. Although M is solely a trunk switching unit, it is treated as the source and sink of trunk traffic in and out of the DPLE area.)

The trunk group capacities are NLP_i, NLM_i and NMP

The equations developed below will refer to the routing strategy shown in Figure 1, with the exception that traffic from P to L_i does not overflow onto PM (this overflow was not specified in the strategy when the analysis was undertaken).

In addition, the use of trunk reservation on LP links is modelled. Three classes of traffic are defined, in decreasing order of priority:

1. Traffic offered from P to L_i
2. First-choice traffic offered from L_i to P (i.e outgoing local traffic from L_i)
3. Trunk traffic outgoing from L_i and overflowing from L_iM to L_iP

The Inverse-Erlang method had given accurate end-to-end blocking estimates from the DPLE area without trunk reservation. Thus its general structure was maintained while adapting it to model trunk reservation.

Only the pair of links originating from each local exchange were considered as dependent. In particular the PM link was assumed independent from all others because the number of links overflowing onto it prevented any limited dependence structure being modelled.

Trunk reservation is modelled by dividing the mean carried traffics on appropriate LP and LM links into 3 parts corresponding to priority. Blocking probabilities and fictitious Poisson traffics are divided correspondingly.

The equations can now be presented, with the following additional notation (where the suffix i refers to the local exchange, and k (= 1, 2, 3) the order of priority):

XLP_{ik}, XLM_{ik}, XPM	mean carried traffics
PLP_{ik}, PLM_i, PPM	blocking probabilities
$PLPM_{ik}$	joint blocking probabilities
YLP_{ik}, YLM_i, YPM, $YLPM_{ik}$	fictitious Poisson traffics

Equations (1)

$$\begin{aligned} XLP_{i1} &= APL_i(1-PLP_{i1}) \\ &+ AML_i\,(PLM_i - PLPM_{i1})(1-PPM) \\ &+ \sum_{j\neq i} AL_{ji}\big[(1-PLP_{j2})(1-PLP_{i1}) + \\ &(PLP_{j2}-PLPM_{j2})(PLM_i - PLPM_{i1})(1-PPM)\big] \end{aligned}$$

$$\begin{aligned} XLP_{i2} &= ALP_i\,(1-PLP_{i2}) \\ &+ \sum_{j\neq i} AL_{ij}\,(1-PLP_{i2})(1-PLP_{j1}) \end{aligned}$$

$$XLP_{i3} = ALM_i\,(PLM_i - PLPM_{i3})\,(1-PPM)$$

$$\begin{aligned} XLM_{i1} &= AML_i\,(1-PLM_i) \\ &+ \sum_{j\neq i} AL_{ji}(PLP_{j2} - PLPM_{j2})(1-PLM_i) \end{aligned}$$

$$\begin{aligned} XLM_{i2} &= ALP_i\,(PLP_{i2} - PLPM_{i2})(1-PPM) \\ &+ \sum_{j\neq i} AL_{ij}\big[(PLP_{i2} - PLPM_{i2})(1-PLM_j) + \\ &(PLP_{i2} - PLPM_{i2})(PLM_j - PLPM_{j1})(1-PPM)\big] \end{aligned}$$

$$XLM_{i3} = ALM_i\,(1-PLM_i)$$

$$\begin{aligned} XPM &= (APM + AMP)(1-PPM) + \sum_i \sum_{j\neq i} AL_{ij}\,\cdot \\ &\cdot\big[(PLP_{i2} - PLPM_{i2})(PLM_j - PLPM_{j1})\,(1-PPM)\big] \\ &+ \sum_i \big[ALP_i(PLP_{i2} - PLPM_{i2})(1-PPM) \\ &+ AML_i\,(PLM_i - PLPM_{i1})(1-PPM) \\ &+ ALM_i\,(PLM_i - PLPM_{i3})(1-PPM)\big] \end{aligned}$$

Equations (2)

$$\left.\begin{array}{l} YLP_{i1}, YLP_{i2}, YLP_{i3} \\ NLP_i \text{ and t.r.} \\ \text{parameters} \end{array}\right] \longrightarrow PLP_{i1}, PLP_{i2}, PLP_{i3}$$

$$\left.\begin{array}{l} YLPM_{i1}, YLPM_{i2}, YLPM_{i3} \\ (NLM_i + NLP_i) \text{ and t.r.} \\ \text{parameters} \end{array}\right] \longrightarrow PLPM_{i1}, PLPM_{i2}, PLPM_{i3}$$

$$YLM_i, NLM_i \longrightarrow PLM_i$$

$$YPM, NPM \longrightarrow PPM$$

Note: calculation of blockings on a trunk group using trunk reservation and offered Poisson traffics is a simple adaptation of Erlang's formula (e.g. see [6])

Equations (3)

$$YLP_{ik} = \frac{XLP_{ik}}{1 - PLP_{ik}}$$

$$YLM_i = \frac{\sum_{k=1}^{3} XLM_{ik}}{1 - PLM_i}$$

$$YLPM_{ik} = \frac{XLP_{ik} + XLM_{ik}}{1 - PLPM_{ik}}$$

$$YPM = \frac{XPM}{1 - PPM}$$

Iteration Scheme

The fictitious Poisson traffics are initially set equal to the first-choice traffics offered to the trunk groups. Equations (2) then give blocking probabilities. From there Equations (1) give mean carried traffics, and Equations (3) then give new values of the fictitious Poisson traffics. The iteration can be terminated by a test on the stability of the fictitious Poisson traffics. (In practice the number of iterations has been arbitrarily limited to 10 without damaging the accuracy of the method. Otherwise the analysis of heavily loaded networks can lead to numerous iterations.)

5.3 Accuracy of the Method: Numerical Results

The Inverse-Erlang model has been tested against a network simulation package, developed by the BT Teletraffic Division, on DPLE-area networks with 2, 4, 6 and 9 local exchanges, with and without trunk reservation. Due to the number of streams in these networks, and the wide range of their intensities, simulation studies were expensive in CPU-time (20-30 minutes on an IBM 3081) while the analytical model needed a few seconds only. The accuracy of the model is good, particularly without trunk reservation. The protective effects of trunk reservation tend to be overestimated.

End-to-end loss probabilities of streams using multi-link paths are estimated rather well, but the loss probabilities of P-M and M-P streams are consistently underestimated. Further studies have shown, perhaps surprisingly, that extending the equations in order to remove the assumption of independence of the PM route does not actually improve the results.

The networks studied were generally rather heavily loaded with the worst streams having end-to-end blockings between 6% and 45%. In such cases the method generally gives good results for streams with high blockings, but gives larger relative errors for streams with low blockings.

The following example, a DPLE-area network with 6 local exchanges where trunk reservation is used on 4 of the LP routes, is representative of our results:

i \ j	L_1		L_2		L_3		L_4		L_5		L_6		P		M	
L_1			17.74	3.1 / 4.4	15.12	3.6 / 5.0	10.08	2.3 / 3.3	5.04	3.1 / 4.0	2.02	2.9 / 3.3	50.0	1.2 / 1.7	150.0	3.7 / 2.5
L_2	16.92	3.7 / 3.5			12.69	5.5 / 6.1	8.46	4.3 / 4.3	4.23	5.1 / 5.4	1.69	4.8 / 5.2	44.0	3.2 / 2.9	120.0	6.3 / 4.8
L_3	13.74	2.8 / 2.4	12.09	4.2 / 4.6			6.87	3.4 / 3.5	3.43	4.2 / 4.0	1.4	3.9 / 3.7	37.5	2.3 / 1.6	90.0	2.7 / 2.9
L_4	8.39	4.4 / 3.8	7.38	5.7 / 5.9	6.29	6.1 / 6.2			2.10	5.7 / 5.5	0.84	5.4 / 4.9	25.0	3.8 / 3.0	60.0	4.4 / 4.4
L_5	3.87	2.4 / 2.5	3.40	3.8 / 4.5	2.90	4.2 / 4.8	1.93	2.9 / 3.4			0.39	3.4 / 2.1	12.5	1.8 / 1.5	30.0	5.3 / 5.2
L_6	1.48	2.5 / 2.8	1.30	3.9 / 3.7	1.11	4.3 / 4.3	0.74	3.1 / 3.6	0.37	3.8 / 2.7			5.0	1.9 / 1.4	15.0	2.9 / 3.3
P	55.6	0.6 / 1.0	46.08	2.0 / 3.1	36.88	2.4 / 3.5	21.91	1.2 / 1.9	9.83	1.9 / 2.8	3.69	1.7 / 2.3			250.0	2.7 / 4.7
M	150.0	0.3 / 0.8	120.0	1.3 / 2.2	90.0	2.7 / 2.8	60.0	1.9 / 2.8	30.0	2.7 / 3.6	15.0	2.9 / 3.4	250.0	2.7 / 4.7		

Key

Traffic intensity (Erlangs)	GOS (%) Analytical
	Simulation

The simulation results have confidence intervals generally between 10% and 30% of the estimated mean grades of service, but much larger for very small traffics.

The following parameters define the network:

	L_1	L_2	L_3	L_4	L_5	L_6	P
LM & PM capacity	300	220	150	90	40	20	630
LP capacity	220	200	190	140	80	40	-
t.r parameters (priorities 2, 3)	2, 4	1, 2	0,0	1,1	0,1	0,0	-

Overall, the method is less accurate than the best available algorithms for individual losses on overflow trunk groups in simple hierarchical clusters. However considering the complexity of the problem and the simplicity of this algorithm, the resulting estimates of end-to-end blocking are satisfactorily good.

6 CONCLUSIONS

When alternative routing is organised on a non-hierarchical basis, involving mutual overflow, standard methods of analysis are inadequate since the network cannot be decomposed sequentially. Attempts to adapt decomposition methods by introducing iteration can give poor results owing to the need to assume independence between traffic streams that are actually correlated.

The Inverse-Erlang method described here was found to give satisfactory results, considering that it is able to estimate end-to-end blocking in networks using mutual overflow and trunk reservation.

Although the algorithm is essentially a one-moment model, it is <u>not</u> equivalent to a decomposition model where all traffics are represented by Poisson streams. The use of detailed equations for traffic flow, and the introduction of fictitious Poisson traffics to calculate individual and joint blocking probabilities, ensure that the effects of overflow traffic can be modelled in a reasonable way. This approach is best suited to complex networks where end-to-end traffic streams may be routed over a number of alternative multi-link paths.

Further study of the Inverse-Erlang method is desirable, to investigate its applicability to a wider range of networks, and to gain a better understanding of its accuracy and convergence properties.

ACKNOWLEDGEMENT

Acknowledgement is made to the Director of System Evolution and Standards Department of British Telecom for permission to publish this paper.

REFERENCES

[1] F le Gall, J Bernussou, 'An analytical formulation for grade of service determination in telephone networks' IEEE Trans. Commun., vol COM-31, No 3 pp 420-4, March 1983

[2] D J Songhurst, 'Protection against traffic overload in hierarchical networks employing alternative routing', Proc. Telecommunication Networks Planning Symposium, Paris, pp214-220, 1980

[3] J R Bonser, R H Thompson, 'The analysis of transient performance for nodal failure in an alternative routing network', paper submitted to 11th Int. Teletraffic Congress, Kyoto, 1985

[4] K Lindberger, 'Simple approximations of overflow system quantities for additional demands in the optimisation', Proc 10th Int. Teletraffic Congress, Montreal, paper 5.3.3 1983

[5] J Matsumoto and Y Watanabe, 'Analysis of individual traffic characteristics for queueing systems with multiple Poisson and overflow inputs', Proc. 10th Int Teletraffic Congress, Montreal, paper 5.3.1, 1983.

[6] J M Akinpelu, 'The overload performance of engineered networks with non-hierarchical and hierarchical routing', Proc 10th Int. Teletraffic Congress, Montreal, paper 3.2.4, 1983

TELETRAFFIC ISSUES in an Advanced Information Society
ITC-11
Minoru Akiyama (Editor)
Elsevier Science Publishers B.V. (North-Holland)

DIMENSIONING ALTERNATE ROUTING NETWORKS WITH OVERLOAD PROTECTION

Gyula SALLAI and Zoltan DELY

PKI - Research Institute of the Hungarian PTT
Budapest, Hungary

ABSTRACT

In hierarchical alternate routing networks traffic overload protection methods are to be used for protecting the traffic offered directly to the final routes. The paper presents a concept for direct dimensioning of the optional groups so that the cost and overload performances are in optimum relation. Taking into account the modularity aspects and full-grouping considerations a unified theorem and optimum design criteria are derived, as an extension of the theorem and criteria developed for cost-optimal dimensioning. The presented criteria are available to compile securized THF (tandem, high-usage, full-group) design diagrams to choose the type and size of the optional group concerned under a given nominal load and overload ratio. The cost-optimal and securized solutions and THF diagrams are extensively compared.

INTRODUCTION

In hierarchical alternate routing networks the transversal, optional circuit groups are established generally to provide cost reductions. Normally they are high-usage groups with overflow onto another optional or a final circuit group. The rapidly expanding use of digital switching and transmission facilities necessitates a reconsideration of dimensioning methods taking into account the digital peculiarities and involving the service protection requirements. The impact of the digital peculiarities, as 24 or 30 channel modularity of the circuit groups, preferation of the both-way trunking over one-way trunking, etc. are handled in many papers [1 - 7].

As the minimum permitted size of the digital optional groups is greater than the minimum permitted group sizes in analogue networks, the number of economic optional groups will decrease and the established - final and optional - groups will be larger. This tendency is supported by the change of cost parameters, i.e. by the decrease of the line cost and tandem switch cost due to the digitalization [2, 3, 8]. Thus, in general the economically optimum digital network will have a more vulnerable network structure than the economically optimum analogue network has. Therefore the network dimensioning should involve service protection methods, against traffic overloads and component failures. Regarding the overloads in hierarchical networks, preventive methods are to be used for protecting the traffic offered directly to a final route, as

a) limitation of the marginal load on the final groups [2, 9, 10],

b) splitting of the final groups into an overflow subgroup and a protective subgroup applied to the directly offered traffic portion only [10, 11, 12],

c) forced full-grouping of the optional groups to prevent very peaked traffic from overflowing [2, 4, 6].

Generally, a higher overload protection can be obtained by using high-usage groups of larger size than the size relating to the least-cost solution at the nominal load.

Usually we try to find a least-cost solution to a predetermined combination of the protection methods and use an iterative procedure including sizing and protection analysis phases [9, 13, 14]. In the following we present *an efficiency concept and a procedure for direct sizing of the optional groups where the extra cost due to the scaling up of the optional groups and overload performances at a given overload ratio are in optimum relation.* Taking account of the modularity aspects and full-grouping considerations a unified theorem and optimum criteria are derived. The criteria are available to construct decision diagrams to facilitate the design of optional groups with an efficient overload protection. These "securized" THF diagrams as an efficiently securized version of the cost-optimal THF diagrams provide the classification of the optional group: tandem routing (T), high-usage group with overflow (H) and full-group without overflow (F), and the number of circuit modules optimum to a given nominal offered load, overload ratio and desired traffic performance level calculated from the alternate routing pattern. We point out some features of the securized modular THF diagrams to clarify some alternate routing and overload protection problems.

1. THE PRINCIPLE OF THE SIZING METHOD

Usually, the high-usage groups are dimensioned to least-cost; an optimum number of circuits and overflow probability are given at the nominal load. Overloading such a high-usage group, the overflow probability and the overflow traffic onto the final group are increased. The overload protection from the aspect of the sizing of the optional groups may be defined with the aid of excess overflow traffic, which causes an undesired increase of the grade-of-service over the final group.

Considering a high-usage group with N circuits nominal offered load M and overload ratio $\tau \geq 1$, and denoting the congestion function by $B(N,M)$, the overflow traffic is $U=M.B(N,M)$ under nominal case, $U_\tau=\tau.M.B(N,\tau M)$ under overload condition.

The protection level at overload ratio τ can be defined as

$$S(N) = 1-\frac{U_\tau-U}{\tau M} = \frac{Y_\tau+U}{\tau M} = 1-B(N,\tau M)+\frac{B(N,M)}{\tau}, \quad (1)$$

where $U_\tau-U$ is the excess overflow traffic, $Y_\tau = \tau M-U_\tau$ is the carried traffic of the optional group under overload condition. Obviously $S(N)=1$ if $\tau=1$, for higher τ values $Y_\tau>Y=M-U$, $S(0)=1/\tau$, $S(\infty)=1$ and $\partial S(N)/\partial N=S'>0$. Thus if the size of the concerned optional group is increased, the overload protection is improved, i.e. the excess overflow traffic is decreased (the carried overflow traffic over the high-usage group is increased). This overload protection consideration stimulates to have an optional group with a larger size than that relating to the least-cost solution at nominal load. The overall cost of handling M erlangs by an optional high-usage circuit group of size N and a tandem route of marginal capacity β (*) can be estimated as

$$C(N) = N.c_h + \frac{M.B(N,M)}{\beta}.c_t$$

where c_h is the cost per circuit of the concerned optional high-usage group, c_t is the average cost per circuit in the overflowing network of the high-usage group. The least-cost continuous solution is given at size N_c, where:

$$-M\frac{\partial B(N_c,M)}{\partial N} = c_h\frac{\beta}{c_t}$$

with simple notations:

$$\delta(N_c,M) = L,$$

where δ is called marginal occupancy, $L=\beta c_h/c_t$ is called the desired traffic performance level.

To find the most efficient solution we define a *qualification factor Q*, expressing the ratio of the protection level and the relative cost of handling traffic load M, as:

$$Q(N) = \frac{S(N)}{C(N)/C(0)} = \frac{1-B(N,\tau M)+B(N,M)/\tau}{\frac{N.L}{M}+B(N,M)} \quad (2)$$

Maximizing Q according to N under $\tau>1$, $N_\tau \geq N_c$ is yielded, so that

$$\frac{S'(N)}{S(N)} - \frac{C'(N)}{C(N)} \begin{cases} >0 & N<N_\tau \\ =0 \quad \text{if} & N=N_\tau \\ <0 & N>N_\tau \end{cases}$$

Thus the scaling up of the cost-optimal optional group to N_τ results in an efficient improvement in protection level, the additional circuits provide greater improvement in protection level than the relative increment in the overall cost. If $\tau=1$, then $N_\tau = N_c$ (Fig. 1).

In the followings taking into account the modularity (that the size of the circuit group is to be in multiple of a fixed module size m), the permitted loss probability of the optional groups B_0, as well as the feasible minimum group size N_{min}, we show decision criteria for choosing the type of the concerned optional group (T, H or F) and the optimum number of modules.

* The marginal capacity is the additional traffic that can be offered to the route by adding a circuit under condition of constant blocking. The marginal capacity of a circuit group is a theoretical maximum value, depending on the mean and the peakedness of the offered traffic, which is usually not fully utilized in the course of the dimensioning, but a practical constant value (0.7 ... 0.9) is taken.

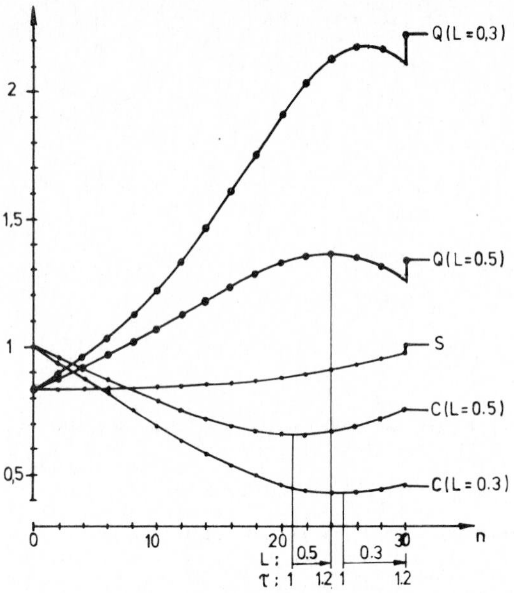

Fig. 1 Protection level S, relative cost C=C(N)/C(0) and qualification factor Q=S/C curves for M=20 erlangs, m=1, τ=1.2 at L=0.5 and 0.3 respectively. In the first case $n\tau$=nc+3=24, high-usage group, in the second case $n\tau$=n0, full group.

2. MODULAR HIGH-USAGE GROUPS

Let m be the module size, n be the number of modules, and from (2)

$$Q_m(n) = Q(n.m) = \frac{M.S_m(n)}{n.m.L+M.B_m(n)}, \quad (3)$$

where $S_m(n)=S(n.m)$, $B_m(n)=B(n.m,M)$. Thus the efficient solution n_τ is given if

$$Q_m(n_\tau-1) < Q_m(n_\tau) \geq Q_m(n_\tau+1). \quad (4)$$

We generalize

$$\delta_m(n,M) = \delta(n.m,M) = -\frac{M}{m}[B_m(n)-B_m(n-1)],$$

the marginal occupancy with module size m [4] in the form of:

$$\delta_m^\tau(n,m)=\delta^\tau(n.m,M)=-\frac{M}{m}\frac{B_m(n)S_m(n-1)-B_m(n-1)S_m(n)}{nS_m(n-1)-(n-1)S_m(n)}$$

involving the improvement of the overload protection.

We obtain from (4), that

$$\delta_m^\tau(n_\tau,M) \geq L > \delta_m^\tau(n_\tau+1,M), \quad (5)$$

i.e. n-th module is to be established, if $\delta_m^\tau(n,M)$, the average occupancy of the circuits in the module corrected by the improvement of the protection level satisfies the desired performance level L. The n is optimum, if the (n+1)th module does not fulfill this condition any more.

Substituting τ=1 the inequality of the least-cost solution with n_c modules is derived as:

$$\delta_m(n_c,M) \geq L > \delta_m(n_c+1,M).$$

Because $\delta_m^\tau(n,M)$ is decreasing versus n, and $\delta_m^{\tau 2}(n,M)>\delta_m^{\tau 1}(n,M)>\delta_m(n,M)$ if $\tau 2>\tau 1>1$, then we have that $n_{\tau 2}\geq n_{\tau 1}\geq n_c$.

3. FULL GROUPING

If the permitted loss probability of an optional group is B_o, then we have an upper limit for the number of circuits and modules, N_o and n_o=-ent($-N_o/m$) respectively. To exploit the permitted increase of the point-to-point congestion we may establish a fully provided group without overflow. These so-called full-groups provide an overload protection for final routes, the peaked traffic being prevented from overflowing, $U=U_\tau=0$. Thus we can define $S(N_o)=S_m(n_o)=1$, and

$$Q_m(n_o) = \frac{M}{n_o m L},$$

(Fig.1). Taking account of the possibility of full-grouping a high-usage group, we obtain the efficient solution, where

$$\max_{n=0,1,..n_o} Q_m(n) = \max\{Q_m(n_\tau),Q_m(n_o)\}.$$

Generalizing the partial average capacity [4]

$$\pi(n.m,n_o.m,M) = \frac{M.B_m(n)}{m.(n_o-n)}$$

in the form of

$$\pi_m^\tau(n,n_o,M) = \frac{MB_m(n)}{m[n_oS_m(n)-n]} = \pi_m[n,n_oS_m(n),M] \quad (6)$$

we obtain that $Q_m(n_o) \geq Q_m(n)$ for all $n < n_o$, i.e. it is efficient to establish full group, if:

$$\pi_m[n,n_oS_m(n),M] \geq L \text{ for all } n<n_o. \quad (7)$$

Introducing

$$\pi_m^\tau{}_{min}(n_o,M) = \min_n \pi_m[n,n_oS_m(n),M] = \pi_m[n_F,n_oS_m(n_F),M]$$

where obviously $n_F<n_oS_m(n_F)<n_o$, we can write (7) as

$$\pi_m^\tau{}_{min}(n_o,M) \geq L. \quad (8)$$

With the parameters n_o and M omitted, if $\tau 2>\tau 1>1$, then $\pi_m^{\tau 2}(n)>\pi_m^{\tau 1}(n)>\pi_m(n)$, thus $\pi_m^{\tau 2}{}_{min}>\pi_m^{\tau 1}{}_{min}>\pi_{m,min}$ and $n_F(\tau 2)<n_F(\tau 1)<n_F(1)$.

We can observe that the function $\pi_m^\tau(n,n_o,M)$ starts from $\alpha_m(n_o,\tau M)$, the average load of the circuits of the full group under overload condition, i.e.

$$\pi_m^\tau(0,n_o,M) = \frac{\tau M}{n_o m} = \alpha_m(n_o,\tau M) = \alpha_m^\tau(n_o,M),$$

and the relationship between $\pi_m^\tau(n)$ and $\delta_m^\tau(n)$ functions, as follows:

$$\pi_m^\tau(n) \gtrless \pi_m^\tau(n-1) \gtrless \delta_m^\tau(n). \quad (9)$$

Consequently if $\pi_m^\tau(0) = \alpha_m^\tau \geq \delta_m^\tau(1)$, then the function $\pi_m(n)$ will monotonously increase and $n_F=0$. In these cases the function $\pi_m^\tau(n)$ dominates over $\delta_m^\tau(n)$ for all n, a full group provides better qualification than a high-usage one. The condition $\alpha_m^\tau \geq \delta_m^\tau(1)$ is obviously found for $n_o=1$ or at extremely large offered load M. For example if m=1 and the traffic load is of random type this is the case when $n_o \leq \tau M+1$.

In other cases, if $\alpha_m^\tau < \delta_m^\tau(1)$, then $\pi_m(n)$ is decreasing to $\pi_m(n_F)$. Here from (9) we find that

$$\delta_m^\tau(n_F) \geq \pi_m^\tau(n_F) > \delta_m^\tau(n_F+1).$$

Since after the intersection $\pi_m^\tau(n)$ is monotonously increasing, dominates over $\delta_m^\tau(n)$, thus it is efficient to scale up a high usage group to satisfy B_o.

In sum, the full grouping involves the high-usage groups of size $n_\tau > n_F(\tau)$, and the ones of size $n_F(\tau)$ for which $\pi_m^\tau(n_F,n_o,M) \geq L$, where $n_F(\tau)$ is decreasing versus τ.

For practical way to full grouping, we express from (7) $\tilde{n}_o$ = ent($\tilde{N}_o/m$), the size of the full group, which would be desirable to the efficient full grouping of a high usage group of size n modules. We obtain:

$$\tilde{N}_o = \frac{1}{S_m(n)}\left(mn+\frac{MB_m(n)}{L}\right)$$

Thus, after determining n_c, the optimum number of high-usage modules, we try to find a number of modules not higher than $\tilde{n}_o$, which satisfies the congestion objective B_o. If the congestion of the circuit group of size $\tilde{n}_o$ modules is greater than B_o, then the efficient full-grouping is not possible.

The cost penalty in case of efficient full-grouping of a cost-effective high-usage group of size n_c modules, i.e. when $Q_m(n_o) \geq Q_m(n_c)$ is given as:

$$\frac{C_m(n_o)}{C_m(n_c)} \leq \frac{1}{S_m(n_c)} \leq \tau .$$

Consequently, choosing the value of τ, the overload ratio we limit the extra cost.

4. TANDEM ROUTING

It is usual to set a lower limit n_{min}, below which an optional group will not be established. This means that we try to find the maximum of $Q_m(n)$ with respect to n=0, n_{min}, $n_{min}+1$, ... n_o and makes necessary the conversion of circuit groups having 1, 2, ... $n_{min}-1$ modules according to (5) and (8) into a high-usage group of n_{min} modules or a full group of $n_o^* = \max(n_o,n_{min})$ modules or the offered traffic will be routed on tandem route (n=0).

To evaluate the tandem routing, the qualifications $Q_m(0) = 1/\tau$, $Q_m(n_{min})$ and $Q_m(n_o^*)$ can be compared.

The tandem route is more efficient than the high-usage group if $Q_m(0) > Q_m(n_{min})$, i.e.

$$L > \frac{\tau M[1-B(n_{min}m,\tau M)]}{n_{min}m} = \eta_m(n_{min},\tau M) = \eta_m^\tau(n_{min},M) \quad (10)$$

where $\eta_m^\tau(n_{min},M)$ means the average occupancy of the circuit group of size $N_{min} = n_{min}.m$ under overload condition, as an extension of $\eta_m(n_{min},M)$, the average occupancy.

When n_{min}=1, then $\eta_m^\tau(1,M) = \delta_m^\tau(1,M)$, otherwise $\eta_m^\tau(n_{min},M) > \delta_m^\tau(n_{min},M)$, i.e. there is a range in L, where the rounding up to n_{min} modules is advantageous. Seeing that $\eta_m^\tau(n_{min},M) > \eta_m(n_{min},M)$, the overload protection stimulates to establish a high-usage group against tandem routing.

Comparing to full group, the tandem routing is more efficient if $Q_m(0) > Q_m(n_o^*)$, i.e.

$$L > \frac{\tau M}{n_o^* m} = \alpha_m(n_o^*,\tau M) = \alpha_m^\tau(n_o^*,M), \quad (11)$$

where $\alpha_m^\tau(n_o^*,M)$ is the average offered traffic under overload condition, as an extension of $\alpha_m(n_o,M)$, the average offered traffic. Because $\alpha_m^\tau(n_o^*,M) = \tau\, \alpha_m(n_o,M)$, the overload protection significantly emphasizes the full-grouping against tandem routing.

The minimal size n_{min} and the permitted loss probability B_o determine a traffic limit M_H so that $B(n_{min}.m, M_H) = B_o$.
So, if $M \leqslant M_H$, then $n_o^* = n_{min}$, otherwise $n_o^* = n_o$.

Summarizing the relations (10) and (11) we can write that the tandem routing is efficient, if

$$L > \max\{\alpha_m(n_o^*, \tau M);\ \eta_m(n_{min}, \tau M)\} \qquad (12)$$

Accordingly, the criterion of tandem routing under enhanced overload protection is equivalent to the criterion calculated at τ times traffic load. The right side of the relation (12) gives a value of $\tau M/N_{min}$ in the domain $M \leqslant M_H$ (otherwise it is less). Hence if $M < N_{min}L/\tau$, then the traffic load M should be tandem routed.

Introducing the limit n_{min}, the criterion to establish full group has been directly effected according to (11). Generally we must find the minimum of $\pi_m^{\tau*}(n) = \pi_m^{\tau}(n, n_o^*, M)$ with respect to $n=0, n_{min}, n_{min}+1, \ldots$. Cancelling the not permitted group sizes instead of (8)

$$\pi_m^{\tau*}{}_{min}(n_o^*, M) \geqslant L$$

is valid, i.e. the criterion to establish full-group against a high-usage group is also influenced. Taking account of the relationship

$$\pi_m^{\tau*}(n_{min}) \gtrless \alpha_m^{\tau}(n_o^*) \gtrless \eta_m^{\tau}(n_{min}) \qquad (13)$$

which can be easily proven, and that $\pi_m^{\tau*}(n, n_o^*, M) = \pi_m^{\tau}(n, n_o, M)$ when $n_o \geqslant n_{min}$ we have, with notation $P=\pi_m^{\tau*}(n_{min})$:

$$\pi_m^{\tau*}{}_{min} = \begin{cases} \alpha_m^{\tau}(n_o^*) & P \geqslant \eta_m^{\tau}(n_{min}) \\ \pi_m^{\tau}(n_{min}) \quad \text{if} & \eta_m^{\tau}(n_{min}) > P > \delta_m^{\tau}(n_{min}) \\ \pi_m^{\tau}{}_{min} & P \leqslant \delta_m^{\tau}(n_{min}) \end{cases} \qquad (14)$$

where the conditions are corresponding to $n_F=0$, $n_F=n_{min}$ and $n_F>n_{min}$, respectively. Additionally, the criterion of tandem routing (12) can be written as

$$L > \begin{cases} \alpha_m(n_o^*, \tau M) & \text{if} \quad n_F=0 \\ \eta_m(n_{min}, \tau M) & \phantom{\text{if}} \quad n_F \geqslant n_{min}. \end{cases}$$

If $n_{min}=n_o^*$, i.e. $M \leqslant M_H$, then we have the case $n_F=0$ and $\pi_m^{\tau*}{}_{min} = \alpha_m^{\tau}(n_{min})$. If $n_{min} < n_o^* = n_o$, i.e. $M > M_H$, then $P=\pi_m^{\tau}(n_{min})$ and $\pi_m^{\tau*}{}_{min} \geqslant \pi_m^{\tau}{}_{min}$.

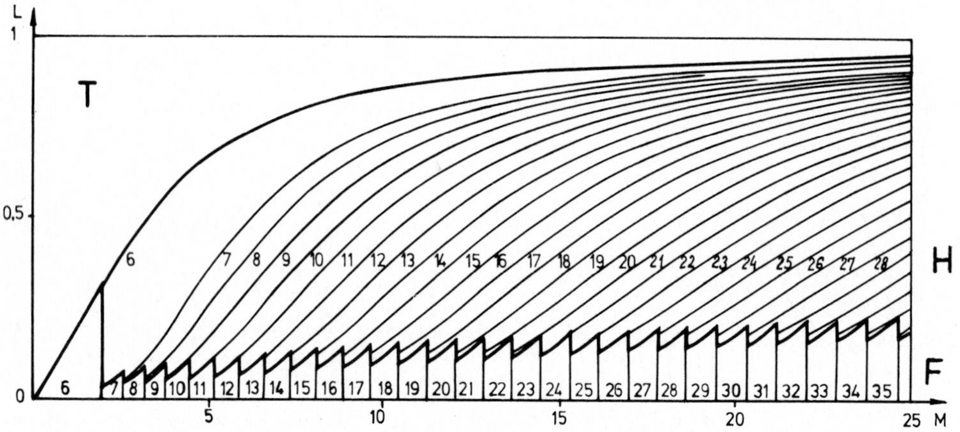

Fig. 2 THF diagram for nominal load (τ=1), with unit size modules (m=1), N_{min}=6, B_o=0.01

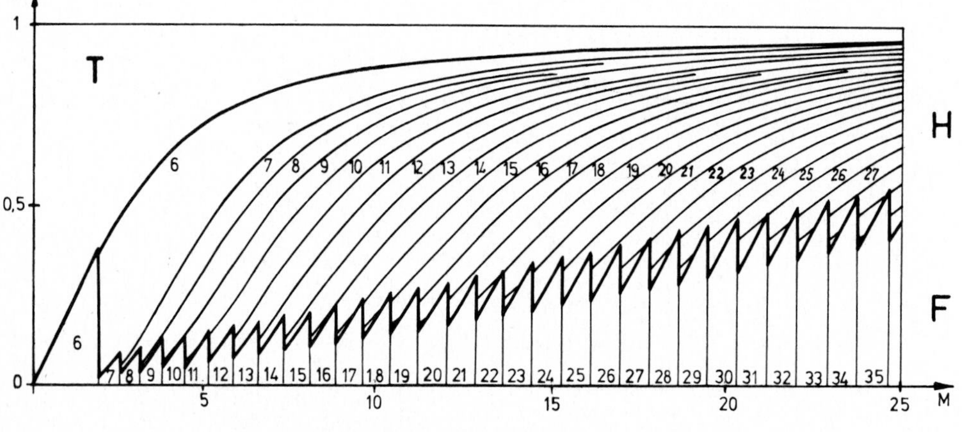

Fig. 3 Securized THF diagram for overload ratio τ=1.2, with unit size modules (m=1), N_{min}=6, B_o=0.01

5. SECURIZED THF DIAGRAMS

As a result of dimensioning taking the protection considerations also into account, the optional groups are not established (tandem routing T), or it is realized as a high-usage group with overflow (H) or as a non-overflowing full-group (F).

Evaluating $\max\{\alpha_m(n_o^*,\tau M);\ \eta_m(n_{min},\tau M)\}$ versus M, a curve $T^\tau(M)$ is derived, which represents maximum overload capability of optional groups with $n \geqslant n_{min}$, taking the value of B_o into account. T(M) is the limit curve of tandem routing. Calculating $\eta_m^{\tau *}{}_{min}(n_o^*,M)$ versus M, a curve $F^\tau(M)$ can be given, which provides capability of the optional groups of size $n \geqslant n_F$ scaled up to full-group of size n_o^*, taking account of the change in overload protection. $F^\tau(M)$ indicates the limit of the full-grouping. Drawing the curves $T^\tau(M)$ and $F^\tau(M)$ on L - M plane, three domains are designated: above $T^\tau(M)$ the domain of T-class, between $T^\tau(M)$ and $F^\tau(M)$ the domain of H-class, below $F^\tau(M)$ the domain of F-class. Also drawing $\delta_m^\tau(n,M)$ for $n = n_{min}+1, \ \dots \ n_F$ in the H-domain a so-called securized THF diagram is compiled. The diagram gives the efficient routing of the offered traffic M, and the number of modules at a given module size and overload ratio τ and at any desired performance level L. Figs. 2-6 present THF diagrams for some practical cases. All diagrams suppose a blocking objective B_o=0.01 and a pure-chance offered traffic M. Figs. 2-3 are related to the unit-size module with N_{min}=6. Figs. 4-6 present diagrams for module size m=30 with n_{min}=1, which are typical to the design of two-way digital optional groups.

For comparison we show diagrams for τ=1 (cost-optimal solution [4]), and τ=1.2, as well as the $F^\tau(M)$ and $T^\tau(M)$ curves for various τ values under m=30 in Fig.6.

We can see that the curves $F^\tau(M)$ have saw-toothed form; a lead is at load M if $N_o = N(M,B_o) = n_o.m$ independently of τ. The lead is increasing with m and τ, and if $\alpha_m^\tau(n_o^*) \geqslant 1$, we have got F-domain at any L in a certain domain of the load M.

The curves $T^\tau(M)$ and $F^\tau(M)$ coincide in certain traffic ranges. From (13) and (14) we obtain that coincidence is at a traffic load M, if and only if

$$\alpha_m^\tau(n_o^*) \geqslant \eta_m^\tau(n_{min}) \tag{15}$$

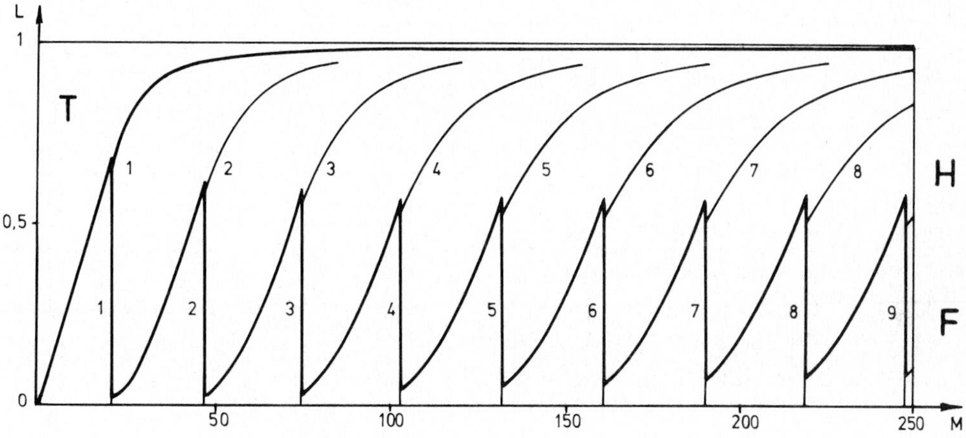

Fig. 4 THF diagram for nominal load (τ=1), module size m=30, n_{min}=1, B_o=0.01

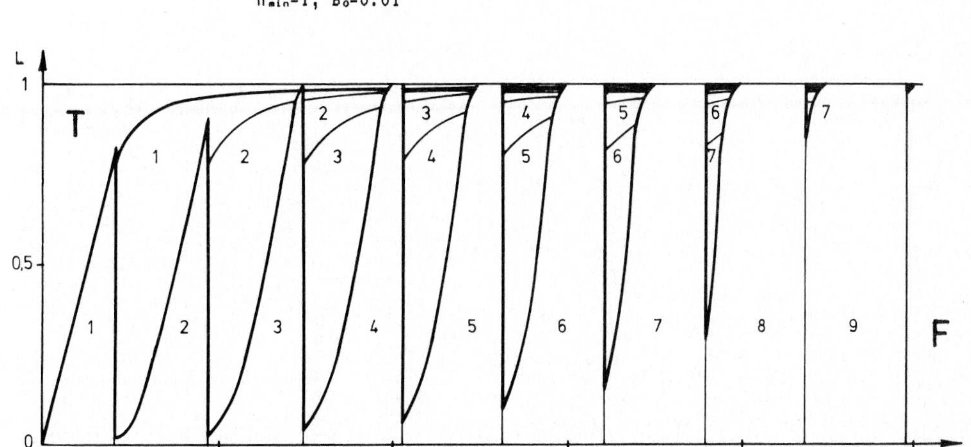

Fig. 5 Securized THF diagram for overload ratio τ=1.2, with module size m=30, n_{min}=1, B_o=0.01

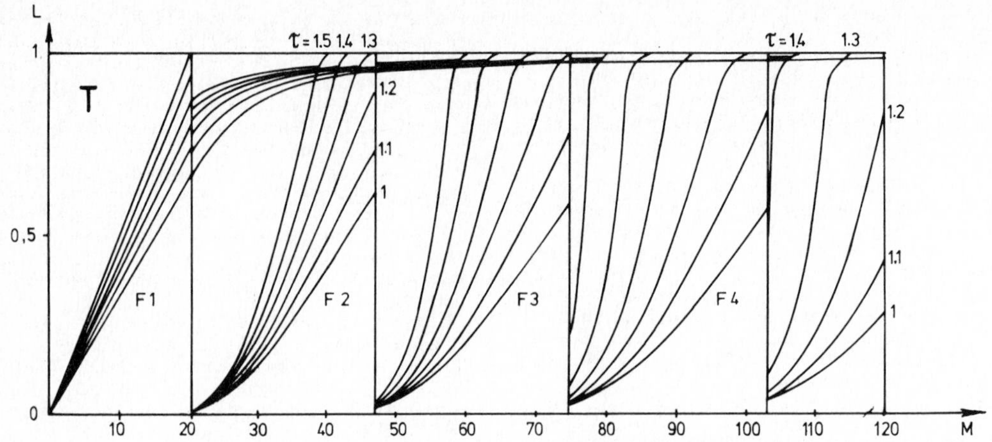

Fig. 6 T^τ and F^τ curves for overload ratios τ=1, 1.1, 1.2, 1.3, 1.4 and 1.5 with module size m=30, n_{min}=1, B_o=0.01

and then $T^\tau(M) = F^\tau(M) = \alpha_m^\tau(n_o^*)$.
(Otherwise $T^\tau(M) = \eta_m^\tau(n_{min}) > \alpha_m^\tau(n_o^*) > F^\tau(M)$.)
The condition (15) can be written as

$$n_{min} \geqslant n_o^*[1-B(n_{min}m,\tau M)]. \qquad (16)$$

Analysing this condition we can define an upper and a lower critical load M_U^τ and M_L^τ resp., so that $T^\tau(M) = F^\tau(M)$ for $M \leqslant M_L^\tau$ and $M \geqslant M_U^\tau$, hence H-domain is found only if $M_L^\tau < M < M_U^\tau$. Condition (16) is obviously satisfied in the range $(0,M_H)$, because $n_o^*=n_{min}$. In general $M_L^\tau \geqslant M_H$, however $M_L^\tau > M_H$ only if $n_{min}+1 \geqslant 1/B(n_{min}\cdot m,\tau M_H)$. If M tends to infinite, $\eta_m^\tau(n)$ tends to 1, $\alpha_m^\tau(n)$ approaches $\tau/(1-B_o)$, consequently M_U^τ exists necessarily. With the aid of the asymptotical features of the congestion function we obtain from (15) approximately $\tau M_U^\tau \cong N(M_U^\tau,B_o)+m$. If m=1 then $M_U^\tau \cong 83$ erlang, if m=30 M_U^τ=285 erlangs are given, under τ=1.2, B_o=0.01. Studying Fig.5 from practical point of view, above 170 erlangs F-domain is found. If τ=1.5 we can consider pure F-domain above 60 erlangs (Fig.6). Fig.6 also shows that the $T^\tau(M)$ curves can be well approximated as:

$$T^\tau(M) \geqslant \min\{\tau M/N_{min};1\} .$$

6. CONCLUSION

The 24 or 30 channel modularity of the circuit groups is an essential feature of digital networks and has significant impact on network planning, mainly due to the security requirements. The paper presented unified criteria for optimum modular engineering an optional circuit group in an alternate routing network, involving overload protection considerations. The developed criteria are the extension of the criteria of the cost-optimal engineering. They are based on an efficiency concept and give an optimum combination of overload protection provided and extra cost required from increasing the size of the optional group concerned. The criteria are valid for random and non-radom traffic load and available to construct securized modular THF design diagrams for calculating the number of circuit modules optimum to a given nominal load and overload ratio under an alternate routing pattern. It is found that main features in the design criteria of cost-optimal dimensioning are held in the design criteria of the generalized, securized dimensioning.

REFERENCES

[1] W.B. Elsner, "Dimensioning trunk groups for digital networks," Proc. 9th ITC, Torremolinos, no. 421, 1979.
[2] R.W. Horn, "End-to-end connection probability," Proc. 9th ITC, Torremolinos, no. 627, 1979.
[3] U. Mazzei, G. Miranda, P. Pallotta, "On a full digital long distance network," Proc. 9th ITC, Torremolinos, no. 426, 1979.
[4] G. Sallai, Z. Dely, "Modular engineering and full grouping theorems for alternate routing networks," Proc. 10th ITC, Montreal, no. 4.3B-2, 1983.
[5] J.P. Farr, "Modular engineering of junction groups in metropolitan networks," Proc. 8th ITC, Melbourne, 1976.
[6] R.N. Rao, "Improved trunk engineering algorithm for high-blocking networks," 9th ITC, Torremolinos, no. 616, 1979.
[7] M. Combes, F. Horn, J. Kerneis, "Simple modular algorithm for dimensioning digital links, Proc. 10th ITC, Montreal, no. 4.3B-4, 1983.
[8] P.A. Caballero, J.M. Silva, "Effects of digitization on the optimum structure of junction network," Proc. 10th ITC, Montreal, no. 4.3A-4, 1983.
[9] J. Berbineau, J.P. Daoudal, "Traffic securization in 'Ile-de-France'," Proc. 1st Int. Netw. Planning Symp. Paris, pp. 173-181, 1980.
[10] G. Sallai, "Efficiency aspects in network securization," Proc. 2nd Int. Netw. Planning Symp. Brighton, pp. 157-162, 1983.
[11] D.J. Songhurst, "Protection against traffic overload in hierarchical networks employing alternate routing," Proc. 1st Int. Netw. Planning Symp. Paris, pp. 214-220, 1980.
[12] A. Jansons, "Optimization of a telecommunication network based on adjustable access possibilities," Proc. 10th ITC, Montreal, no. 4.3B-6, 1983.
[13] P. Lindberg, U. Mocci, A. Tonietti, "A procedure for minimizing the cost of a transmision network under service availability constraints in failure conditions," Proc. 10th ITC, Montreal, no. 2.3-3, 1983.
[14] D.G. Haenschke, D.A. Kettler, E. Oberer, "Network management and congestion in U.S. telecommunication networks," IEEE Trans. on Communication. vol. 29, no. 4, pp. 375-385, 1981.

TELETRAFFIC ISSUES in an Advanced Information Society
ITC-11
Minoru Akiyama (Editor)
Elsevier Science Publishers B.V. (North-Holland)

FUNDAMENTAL TRAFFIC NETWORK PLANNING UNDER UNCERTAINTY

K. Y. CHOW and A. H. KAFKER

AT&T BELL LABORATORIES
HOLMDEL, NEW JERSEY, U.S.A.

Abstract

In this paper, we propose methods that allow network planners to explicitly account for both the stochastic nature of the fundamental traffic network planning problem and their attitude toward the risk of prematurely exhausting a switch. Some innovations to fundamental traffic network planning are introduced: 1) forecast uncertainty is tracked through real data; 2) risk analysis is performed in the rehome selection process; 3) the relief strategy is chosen based on cost, outlays for capital and labor, and risk.

1. Introduction

Fundamental planners must decide when and where to add switches to a telecommunications network. These decisions are typically made three to five years before the switch is installed. Therefore, all plans are based on forecasts and subject to error. In this paper, we propose methods that allow the planners to explicitly account for both the stochastic nature of the problem and their attitude toward the risk of prematurely exhausting a switch.

1.1 Fundamental Planning Problem

The first step in any plan is to identify those switches that require relief and when the additional capacity is required. Figure 1 illustrates the three options available to a planner for each year: 1) continue with the present plans and risk the exhaust, 2) rehome some of the subtending traffic to other switches to defer the exhaust, and 3) build a new switch and rehome some traffic from the exhausting to the new switch. Some combination of these actions will determine the plan.

If the planner decides to rehome some of the subtending offices away from the exhausting switch, then the selection of the offices to be rehomed will determine the probability of attaining the desired relief. This decision depends on the planner's attitude toward the risk of exhausting the switch in each year of the study. The planner will choose the plan that incorporates the best tradeoff between the risk that a switch will exhaust and the cost of implementing the plan.

1.2 Previous Work

Attempts have been made in the past to solve the traffic network planning problem as an optimization problem; however, most of these methods failed to take forecast uncertainty into account. Recently, the problem was formulated as a stochastic dynamic programming problem and it was shown that an inventory-type rehoming policy results in the minimum expected cost[1]. This was the first time that forecast uncertainty was considered in the model. Nevertheless, the conclusion was drawn based on the assumptions that the switch is termination limited and that only one relief switch is involved. One can show that if more relief switches are involved, the inventory-type rehoming policy does not always produce the minimum expected cost. Furthermore, many modern switches are attempt limited. Consequently, the application of this solution procedure is somewhat limited.

1.2.1 New Approach

In this paper, we propose a more general model, which will not have the shortcomings described above, to solve this problem and we use a different criterion for formulating the rehoming strategies. We generate a set of rehoming policies such that the plan cost, the risk of switch exhaust, and the outlays for capital and labor are all acceptable to the network planners. As a result of our method, we bring the following innovations to fundamental traffic network planning:

1) Forecast uncertainty is explicitly tracked through the real data and no assumption on the form of the distribution function is made.

2) For the first time, the network planners are able to plan for switch relief based on their attitudes toward risk.

3) The best relief strategy is chosen based on the plan cost, the outlays for capital and labor, and the risk of a plan rather than only on the expected cost of a plan.

4) There is no need to assume which of the two parameters, terminations and attempts, will determine switch exhaust.

1.3 Outline of Paper

The remainder of the paper is divided into two parts. Section 2 describes the planning procedure that should be followed to take advantage of our methods. This section shows how the planner would use our method to help make the required decisions. Section 3 contains the mathematical basis for our method and the calculations needed to support our method.

2. The Planning Procedure

The planning procedure has three steps: 1) identify switches that will exhaust in each year, 2) specify a course of action, and 3) determine desirable rehomings. Each of these steps will be explained in the following sections.

2.1 Switch Exhaust Forecasting

The first step in solving the traffic network planning problem is to predict a switch exhaust. In the past, either a deterministic forecast (e.g., [3]) or a stochastic forecast with an assumed distribution function (e.g., [2]) switch utilization was used with no effort to track and update the assumed distribution function. In order to more accurately reflect the stochastic nature of the forecasting process, we have devised a method to track and update the distributions associated with switch utilization.

The distributions are constructed based on measurements of the total terminations and the total number of attempts at each switch. These measurements are compared to previous forecasts to derive the error distribution associated with each parameter as a function of time. The actual calculations to perform this construction are provided in Section 3.1.

Using the method described in [3] to forecast switch utilization and then the method described above to track the uncertainty of the forecast, we can construct the probability distribution functions for the total terminations and the total number of attempts. Since a switch is either attempt limited or termination limited, the exhaust probability is equal to the probability that an attempt or termination threshold is exceeded.

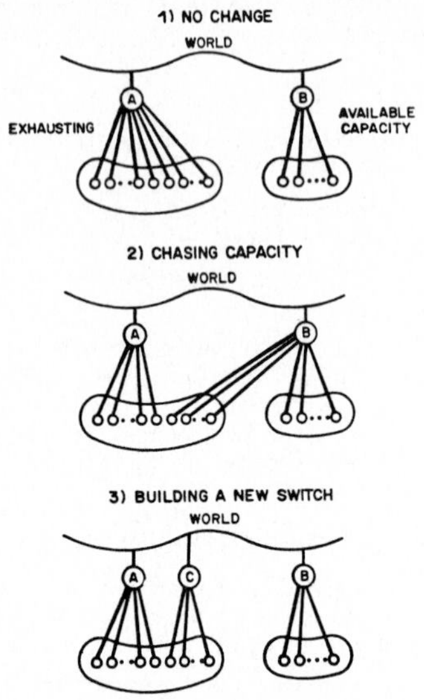

FIGURE 1 THREE PLANNING OPTIONS

2.2 Plan Specification

The second step for solving the switch relief problem is to choose a relief strategy. First, the network planner specifies several alternatives; each consists of the years in which rehomings are desired and if a new switch is to be built, the year in which it will be built. In addition, the planner must specify the candidate relief switches in each year. These relief switches can be used by the planner to provide a desired number of years of deferral for the exhausting switch. The offices to be rehomed in each year are selected by solving an integer linear program which is formulated in Section 3.2, based on the network planner's attitude toward risk, and the constraints on outlays for capital and labor. The cost associated with each alternative is evaluated and the network planner reviews the costs to determine the best relief strategy.

2.3 Rehome Selection Under Uncertainty

An integer linear program similar to the one described in [3] is used to solve the rehome selection problem. The explicit consideration of risk in this process requires several additional inputs by the planner. These inputs reflect the planner's attitude toward risk based on knowledge about the distribution functions of the attempt and termination forecasts. The method for determining these inputs is described in Section 3.3.

One factor that helps determine an acceptable level of risk is the year under consideration. Certainly a planner is more concerned about a switch with exhaust probability 0.7 in the second year than a switch with the same exhaust probability in the tenth year. In the latter case, no immediate action is necessary and the network planner can afford to wait for several years to see whether the switch is actually going to exhaust. Therefore, the acceptable level of risk should be increasing as a function of time.

The network planner is presented, for a year in which rehomings occur, with a list of offices to rehome. The probability of exhaust associated with this set of rehomings is also displayed. Financial constraints may force the planner to reassess the acceptable level of risk. The final choice is made by balancing the constraints against the risk. The planner then proceeds to the next year in which rehomings occur and repeats the process. This procedure enables the planner to formulate a relief strategy that 1) is risk-acceptable, 2) satisfies the financial constraints, and 3) will not exhaust the relief switches.

3. Traffic Network Planning Methods

In this section, we will first describe how forecasts are tracked and how exhaust probabilities are generated from the tracking results. Then, we will explain how to plan for switch relief to account for the information about the probability of exhaust.

3.1 Forecast Tracking and Exhaust Probability Calculation

In the fundamental network planning process, two items are tracked:

a) accuracy of termination forecast, and

b) accuracy of attempt forecast.

Since a switch is either termination limited or attempt limited,[1] the exhaust probability can be calculated based on the forecasts from the exhaust prediction tools in [3] and the information from a) and b). In the following, we will describe the procedure for tracking a forecast and how to calculate the exhaust probability.

3.1.1 Forecast Tracking Method

The objective of the forecast tracking procedure is to determine the error distributions for the items we track. The key in this step is to compare forecasts with measurements. Since forecast errors tend to grow as the length of forecast interval increases, the error distributions are functions of the forecast interval.

We now describe the procedure for calculating error distributions for a forecast. This procedure is used for both the attempt and termination forecasts in the traffic network planning process. Assuming that the tracking procedure is started in Year 0, we first introduce the following notation:

$F_{ij}^k(x), k=1, \cdots, z$: the forecasted value for a particular item x in Year j made in Year i for Switch k. (For example, in Year i, we forecast the number of terminations for z switches in Year j.)

$M_j^k(x), k=1, \cdots, z$: the measurement for a particular item x in Year j for Switch k. (For example, the numbers of terminations for z switches measured in Year j.)

$R_{j\ell}^k(x), k=1, \cdots, z$: the new samples obtained in Year j for the forecast errors with the length of forecast interval equals ℓ.

$E_{j\ell}^k(x), k=1, \cdots, \gamma$: the forecast errors which are used to form the error distribution for a forecast x with interval length ℓ in Year j.

$D_\ell(x,y)$: the error distribution for a forecast x with interval length ℓ and parameter y.

1. In theory, a switch can also be usage limited; however, it rarely happens in practice.

$D_\ell^j(x,y)$: the error distribution for a forecast x with interval length ℓ and parameter y, estimated in Year j.

In Year 0, we start the tracking procedure. In this year, a forecast is made and $\left\{F^k_{oj}(x),\ j=1,\ \cdots\ ,n;k=1,\ \cdots\ ,z\right\}$ are determined. Next year, the measurements are taken and $\left\{M^k_1(x),\ k=1,\ \cdots\ ,z\right\}$ are obtained. Also, a new forecast is made and $\left\{F^k_{1j}(x),\ j=2,\ \cdots\ ,n+1;k=1,\ \cdots\ ,z\right\}$ are determined. Consequently, in this year, the forecast errors, $\left\{R^k_{11}(x),\ k=1,\ \cdots\ ,z\right\}$, can be calculated by comparing the forecasted values, $\left\{F^k_{01}(x),\ k=1,\ \cdots\ ,z\right\}$, with the measurements, $\left\{M^k_1(x),\ k=1,\ \cdots\ ,z\right\}$. Since this is the first time we calculate the error distribution with the length of forecast interval equals 1, and $\left\{R^k_{11}(x)\right\}$ are the only samples we have, $\left\{E^k_{11}(x),\ k=1,\ \cdots\ ,z\right\}$ are the same as $\left\{R^k_{11}(x),\ k=1,\ \cdots\ ,z\right\}$. Then, $D_1^1(x,y)$ is calculated using $\left\{E^k_{11}(x)\right\}$.

In Year 2, both $\left\{F^k_{2j}(x),\ j=3,\ \cdots\ ,n+2;k=1,\ \cdots\ ,z\right\}$ and $\left\{M^k_2(x),\ k=1,\ \cdots\ ,z\right\}$ are determined. Comparing the forecasts made in Year 0 for Year 2, $\left\{F^k_{02}(x),\ k=1,\ \cdots\ ,z\right\}$,[2] and the forecasts made in Year 1 for Year 2, $\left\{F^k_{12}(x),\ k=1,\ \cdots\ ,z\right\}$, with the measurements taken in Year 2, $\left\{M^k_2(x),\ k=1,\ \cdots\ ,z\right\}$, we obtain the new samples for the 2 year forecast errors $\left\{R^k_{22}(x),\ k=1,\ \cdots\ ,z\right\}$, and those for the 1 year forecast errors, $\left\{R^k_{21}(x),\ k=1,\ \cdots\ ,z\right\}$, respectively. Since the error distribution for $\ell = 2$ has not been estimated before, $\left\{E^k_{22}(x)\right\}$ are the same as $\left\{R^k_{22}(x)\right\}$ and $D_2^2(x,y)$ is calculated through $\left\{E^k_{22}(x)\right\}$. To estimate $D_1(x,y)$, we can either use $\left\{R^k_{21}(x),\ k=1,\ \cdots\ ,z\right\}$ directly or update $D_1^1(x,y)$ by incorporating $\left\{R^k_{21}(x)\right\}$ into $\left\{E^k_{11}(x)\right\}$ to form $\left\{E^k_{21}(x),\ k=1,\ \cdots\ ,2z\right\}$. If $D_1(x,y)$ has not changed from Year 1 to Year 2, the latter method will produce a better estimation for $D_1(x,y)$ because we have more samples. Therefore, we have to test whether $D_1(x,y)$ has changed from Year 1 to Year 2 before calculating $D_1^2(x,y)$.

The chi-square test is used to test $D_1(x,y)$. First, we divide $D_1^1(x,y)$ into cells of equal probability and create a histogram. Then, the distribution formed through $\left\{R^k_{21}(x)\right\}$ is tested against $D_1^1(x,y)$. If the test succeeds, $\left\{R^k_{21}(x)\right\}$ are incorporated into $\left\{E^k_{11}(x)\right\}$ to form $\left\{E^k_{21}(x)\right\}$ which are used to construct $D_1^2(x,y)$. Otherwise, the distribution constructed through $\left\{R^k_{21}(x)\right\}$ is used for $D_1^2(x,y)$. By doing this, we can react to the changes in the error distributions and increase the accuracy of error distribution estimation.

From the cases described above, we can therefore summarize the steps to be performed in Year j for estimating $\left\{D_\ell(x,y)\ ,\ \ell=1,\ \cdots\ ,j\right\}$ in the flowchart shown in Figure 2.

1) Take measurements, $\left\{M^k_j(x),\ k=1,\ \cdots\ ,z\right\}$.

2) Calculate $\left\{R^k_{j\ell}(x),\ k=1,\ \cdots\ ,z\right\}$ for $\ell=1,\ \cdots\ ,j$.

$$R^k_{j\ell}(x) = (F^{k\,3}_{j-\ell,j}(x) - M^k_j(x))/F^k_{j-\ell,j}(x)\ .^4$$

3) Calculate $D_j^j(x,y)$.

a) Let $E^k_{jj}(x) = R^k_{jj}(x)$ for k=1, ... ,z.

b) Use $\left\{E^k_{jj}(x),\ k=1,\ \cdots\ ,z\right\}$ to construct $D_j^j(x,y)$.

4) Calculate $D_\ell^j(x,y)$ for $\ell=1,\ \cdots\ ,j-1$.

a) Let X_ℓ be the distribution constructed from $\left\{E^k_{j-1,\ell}(x)\right\}$.

b) Let Y_ℓ be the distribution constructed from $\left\{R^k_{j\ell}(x)\right\}$.

c) Use the chi-square test to test Y_ℓ and X_ℓ.
If "rejected", go to Step d).
Otherwise, go to Step e).

d) Let $\left\{E^k_{j\ell}(x)\right\} = \left\{R^k_{j\ell}(x)\right\}$ and $D_\ell^j(x,y) = Y\ell$. Next ℓ.

e) Let $\left\{E^k_{j\ell}(x)\right\} = \left\{E^k_{j-1,\ell}(x)\right\} \cup \left\{R^k_{j\ell}(x)\right\}$

f) Construct $D_\ell^j(x,y)$ from $\left\{E^k_{j\ell}(x)\right\}$.

3.1.2 Exhaust Probability Calculation

A switch exhausts when the termination or attempt capacity is exceeded. Therefore, the exhaust probability, P(A,j), for a particular switch A in Year j can be expressed as follows:

2. If rehoming is performed in Year 1, $\left\{F^k_{02}(x)\right\}$ should be adjusted to account for the rehoming performed.

3. $F^k_{j-\ell,j}(x)$ should be adjusted to account for the rehomings performed between Year j-ℓ and Year j.

4. Forecast errors are calculated in percentage.

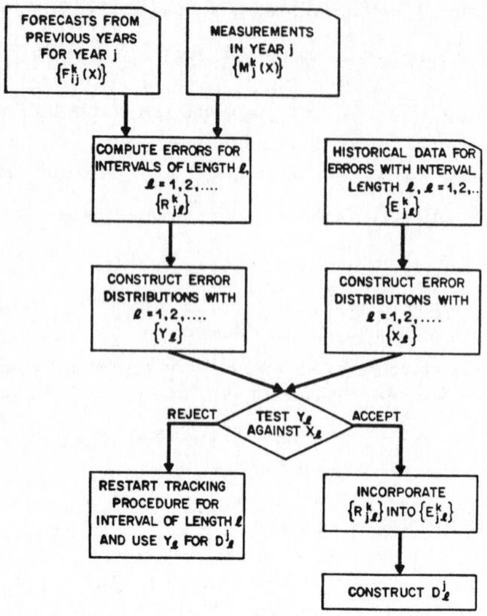

FIGURE 2 TRACKING PROCEDURE IN YEAR j

$$P(A,j) = Prob\left\{U_t(j) > C_t \text{ or } U_a(j) > C_a\right\}$$

$$= Prob\left\{U_t(j) > C_t\right\} + Prob\left\{U_a(j) > C_a\right\}$$

$$- Prob\left\{U_t(j) > C_t \text{ and } U_a(j) > C_a\right\} \qquad (1)$$

where $U_t(j)$ = the termination utilization for Switch A in Year j,

C_t = the termination capacity for Switch A,

$U_a(j)$ = the attempt utilization for Switch A in Year j, and

C_a = the attempt capacity for Switch A.

From Equation (1), it is clear that to calculate P(A,j) we have to know the distribution functions for both $U_t(j)$ and $U_a(j)$. In addition, we also have to know $Prob\{U_t(j) > C_t \text{ and } U_a(j) > C_a\}$. This can be done through numerical integration once we know the distribution functions for both $U_t(j)$ and $U_a(j)$. However, for simplicity, we can skip calculating $Prob\{U_t(j) > C_t \text{ and } U_a(j) > C_a\}$ by using the following approximation:

$$P(A,j) \approx \max\left\{Prob\{U_t(j) > C_t\},\ Prob\{U_a(j) > C_a\}\right\}.$$

In the previous section, we have described how to estimate the error distributions; therefore, the problem here is how to use the error distributions we estimate to construct the distributions for both U_t and U_a. Suppose in Year i, we have estimated the error distributions, $\left\{D^i_\ell(T,y),\ell=1,2,\ldots,n\right\}$ and $\left\{D^i_\ell(A,y),\ell=1,2,\ldots,n\right\}$, for termination and attempt forecasts with forecast interval ranging from 1 to n following the procedure described in Section 3.1.1. Consequently, $D^i_\ell(T,x)$ and $D^i_\ell(A,y)$ represent the probabilities that the ℓ-year termination and attempt forecasts have errors smaller than x and y, respectively. Let $F^i_j(T)$ and $F^i_j(A)$ be the termination and attempt forecasts for Year j forecasted in Year i. Hence, the forecast interval, ℓ, for these two forecasts is equal to (j-i). Since $U_t(j)$ and $U_a(j)$ represent the termination and attempt utilizations in Year j, the ℓ-year termination and attempt forecast errors can be expressed as $\left(F^i_j(T) - U_t(j)\right)/F^i_j(T)$ and $\left(F^i_j(A) - U_a(j)\right)/F^i_j(A)$, respectively. Thus, the distribution functions for $U_t(j)$ and $U_a(j)$ can be derived from the following:

$$Prob\{U_t(j) \le t\} = Prob\left\{F^i_j(T)-U_t(j) \ge F^i_j(T)-t\right\}$$

$$= Prob\left\{\left(F^i_j(T) - U_t(j)\right)/F^i_j(T) \ge \left(F^i_j(T) - t\right)/F^i_j(T)\right\}$$

$$= 1 - D^i_\ell\left[T,(F^i_j(T)-t)/F^i_j(T)\right]$$

$$Prob\{U_a(j) \le a\} = 1 - D^i_\ell\left[A,(F^i_j(A)-a)/F^i_j(A)\right]$$

Therefore,

$$Prob\{U_t(j) > C_t\} = D^i_\ell\left[T,(F^i_j(T)-t)/F^i_j(T)\right]$$

$$Prob\{U_t(j) > C_a\} = D^i_\ell\left[A,(F^i_j(A)-a)/F^i_j(A)\right]$$

The attempt forecast is a multiple-hour forecast. However, in the tracking process, we only track the attempt forecast for the busiest hour and apply the resulting error distributions to all hours in the rehoming selection process. We believe that this will not seriously affect the rehome selection since only the number of attempts in the busiest hour is used to determine whether the attempt capacity has been exceeded.

Also, as described in [3], the attempt forecast is made by forecasting the number of trunks by signaling type terminating on a switch and multiplying these numbers by their associated attempt per trunk ratio. Therefore, in the tracking process, instead of tracking the total number of attempts, we can track the number of attempts by signaling type and convolve the resulting error distributions to form the distribution function for the attempt forecast.

3.2 Rehome Selections and Risk Analysis

For each year in which rehomings are performed, an integer program is solved. The objective of rehome selections is to lower the risk of switch exhaust to the level specified by the planners in the most economical way. We now provide the formulation of the integer program.

Minimize $\sum_{\substack{i\epsilon I \\ j\epsilon J}} C_{ij}X_{ij}$

Subject to

$$\sum_{j\epsilon J} X_{ij} \le 1 \quad \text{for all } i\epsilon I \qquad (1)$$

$$\sum_{\substack{i\epsilon I \\ j\epsilon J}} a_{ik}X_{ij} \ge \bar{a}_k \quad \text{for all } k\epsilon K \qquad (2)$$

$$\sum_{i\epsilon I} b_{ik}X_{ij} \le \bar{b}_k \quad \text{for all } j\epsilon J \text{ and all } k\epsilon K \qquad (3)$$

$$\sum_{\substack{i\epsilon I \\ j\epsilon J}} t_i X_{ij} \geq \bar{t} \tag{4}$$

$$\sum_{i\epsilon I} s_{ij} X_{ij} \leq \bar{s}_j \quad \text{for all } j\epsilon J \tag{5}$$

and,

$$X_{ij} \in (0,1) \tag{6}$$

where

I = the set of offices subtending the exhausting switch,

J = the set of existing relief switches,

K = the set of switch busy hours,

C_{ij} = the cost of rehoming Office i to Switch j, $i\epsilon I$ and $j\epsilon J$,

X_{ij} = the indicator, X_{ij} = (1,0) = (rehome, do not rehome) Office i to Switch j,

a_{ik} = the forecasted number of attempts removed in Hour k from the exhausting switch through rehoming of Office i,

b_{ik} = the forecasted number of attempts to be added to a relief switch in Hour k through rehoming of Office i,

$\bar{a}_k$ = the number of attempts to be removed from the exhausting switch in Hour k such that the resulting exhaust probability is equal to the risk specified by the network planner,

$\bar{b}_{jk}$ = the upper bound on the number of attempts to be added to relief Switch j in Hour k such that the resulting exhaust probability is equal to zero,

t_i = the forecasted number of terminations removed from the exhausting switch through the rehoming of Office i,

s_i = the forecasted number of terminations added to the relief switch through the rehoming of Office i,

$\bar{t}$ = the number of terminations to be removed from the exhausting switch such that the resulting exhaust probability is equal to the risk specified by the network planner,

and

$\bar{s}_j$ = the upper bound on the number of terminations to be added to Switch j such that the resulting exhaust probability is equal to zero.

Equation (1) ensures that each subtending office is rehomed to at most one relief switch. Equations (2) and (4) guarantee that the resulting rehoming policy is acceptable for the network planner. Equations (3) and (5) ensure that the relief switches will not exhaust within the planning horizon.

Before this integer program is solved, the values for $\{\bar{a}_k\}$, $\{\bar{b}_{jk}\}$, $\bar{t}$, and $\{\bar{s}_j\}$ have to be determined. These values depend on the acceptable risk level specified by the planner. Once the acceptable risk level, γ, for a particular year is specified by the planner, $\{\bar{a}_k\}$, $\{\bar{b}_{jk}\}$, $\bar{t}$, and $\{\bar{s}_j\}$ are determined based on the distribution functions for the attempt and termination forecasts.

Since the method for calculating each set of values is the same, we only describe how $\{\bar{a}_k\}$ are calculated. Let F_k be the distribution function for the number of attempts in Hour k and z_k be the smallest number such that $(1-F_k(z_k)) \leq \gamma$. Then, z_k is the attempt capacity needed in Hour k that will insure the risk of switch exhaust to be within the level specified by the planner if no rehoming is performed. Let C_k be the forecasted attempt capacity in Hour k. Consequently, the difference between C_k and z_k is the number of attempts in Hour k that we want to reduce through rehoming. Therefore,

$$a_k = \max\{0, z_k - C_k\} .$$

The integer program can be solved by the method described in [4]. After the integer program is solved, a set of offices to be rehomed away from the exhausting switch to the relief switches or the new switch in a particular year is produced. If the cost and labor associated with rehoming these offices are feasible, the planner can then proceed to the next year in which rehomings will occur. Otherwise, the planner may have to rehome only a subset of the offices.

If the financial constraints force the planner to rehome only a subset of offices, the final choice is made by balancing the constraints against the risk. In this case, the planner has to raise the acceptable risk level and rerun the integer program until the financial constraints are met. However, if the resulting risk of switch exhaust is too high under the existing financial constraints, the planner has a strong case to request more capital.

Based on the discussion above, we can then summarize the steps to be performed in a particular year in which rehomings occur:

1. The planner specifies the acceptable risk level γ.
2. Adjust the attempt and termination forecasts to reflect the rehomings selected in the previous years.
3. Calculate the exhaust probability p.
4. If $p \leq \gamma$, go to next rehoming year. Otherwise, go to next step.
5. Calculate $\{\bar{a}_k\}$, $\{\bar{b}_{jk}\}$, $\bar{t}$, and $\{\bar{s}_j\}$.
6. Solve the integer program and identify the offices to rehome.
7. Calculate the cost and labor associated with rehoming the offices identified in Step 6.
8. If the cost and labor are feasible, go to next rehoming year. Otherwise, go to next step.
9. The planner specifies a higher risk level and repeats steps 5-7 until the cost and labor associated with rehoming are feasible.
10. The planner either adopts the rehoming produced in Step 9 or requests for more capital.

4. Conclusion

Like most of the planning process, forecasting plays an important role in fundamental traffic network planning. Because of the stochastic nature of the forecasting process, one must take forecast uncertainty into account in the planning process. In this paper, we have presented the methods for tracking a forecast and utilizing the tracking results in analyzing the risk associated with a switch relief plan. Although these methods are developed for fundamental traffic network planning, they can be easily modified to apply to many other planning processes.

REFERENCES

[1] Chow, K., David, A., and Ionescu-Graff, A., "Switch Capacity Relief Model, Theoretical Development," 10th International Teletraffic Congress, Montreal, Canada, June 1983.

[2] Coco, R., Farel, R., Potter, R., and Wirth P., "Relationships Between Utilization, Service and Forecast Uncertainty for Central Office Provisioning," 10th International Teletraffic Congress, Montreal, Canada, June 1983.

[3] David, A., and Farber, N., "The Switch Planning System for the Dynamic Nonhierarchical Routing Network," 10th International Teletraffic Congress, Montreal, Canada, June 1983.

[4] Rosenberg, E., "An Integer Programming Approach to Switch Relief in a Telecommunication Network," paper presented at the TIMS/ORSA Conference, Chicago, 1983.

TELETRAFFIC ISSUES in an Advanced Information Society
ITC-11
Minoru Akiyama (Editor)
Elsevier Science Publishers B.V. (North-Holland)

DIGITAL TELEPHONE JUNCTION NETWORK PLANNING SYSTEM CONSIDERING LOGICAL AND PHYSICAL NETWORK ASPECTS

Makiko YOSHIDA and Hiroyuki OKAZAKI

C & C Systems Research Laboratories
NEC Corporation, KAWASAKI, 213 JAPAN

ABSTRACT

This paper describes a junction network optimization technique which takes into account higher order PCM transmission systems. The optimization procedure, composed of logical- and physical-network design parts, is proposed. During the optimization process, data representing both networks are iteratively transferred between these two parts. The proposed method introduces plural hierarchical module sizes, while the conventional modular engineering uses a unique module size. The case study shows that considerable cost saving was provided as compared with the conventional method.

1. INTRODUCTION

The rapid introduction of digital switching and transmission facilities requires a reconsideration of conventional network planning methods. Main problems are as follows.

- Transmission facilities are provided on the basis of a variety of module sizes, e.g., 30, 120, 480 and 1920 channels, because of the PCM hierarchy used in digital networks.
- Junction networks can be viewed from two aspects, i.e. a logical network and a physical network. Diversity in the transmission media, such as cable PCM, fiber optic system, digital microwave system and so on, increases the necessity for a network optimization method which takes into account these two aspects simultaneously.

The logical network is concerned with routing rules and the logical number of circuit groups to be established between exchanges (Fig.1). The physical network corresponds to the physical organization of a network, i.e., the selection of transmission media, their geographical layout and the number of digital carrier systems for PCM hierarchical order (Fig.2).

A logical network dimensioning method, called modular engineering, is known [1]. It takes into account non-linear trunk cost, by constraining the size of circuits to be multiples of a fixed size, and provides network cost saving. However, conventional modular engineering has the following problems.

- Higher PCM hierarchical order cannot be taken into account, because the dimensioning method only uses a unique module size.
- Routing rules between exchange offices remain to be predetermined throughout the dimensioning procedure.
- The method considers the physical network conditions only through the approximate per-trunk cost ratios between alternate and direct routes.

On the other hand, several physical network optimization methods have been reported [2],[3]. However, these papers did not consider logical network optimization importance.

To solve the above mentioned problem, this paper proposes a junction network optimization method which equally considers both logical- and physical-network aspects. The proposed network planning method introduces plural module sizes, i.e., an administrative logical trunk circuit module size and physical module sizes (e.g., 30, 120, 480 and 1920 channels), so that further cost reduction can be obtained by introduction of appropriate PCM hierarchical order systems for each route.

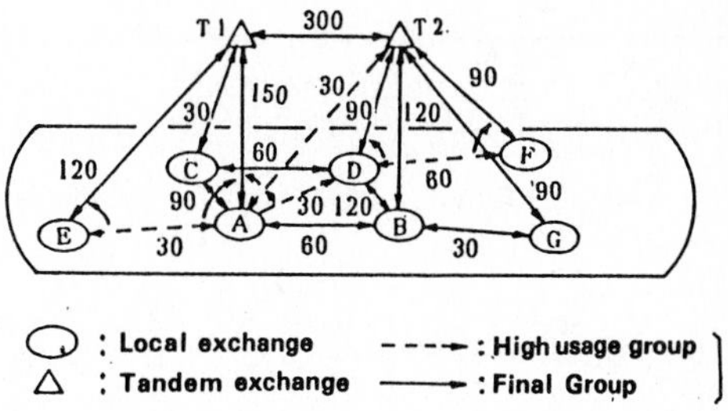

Fig.1 Logical network

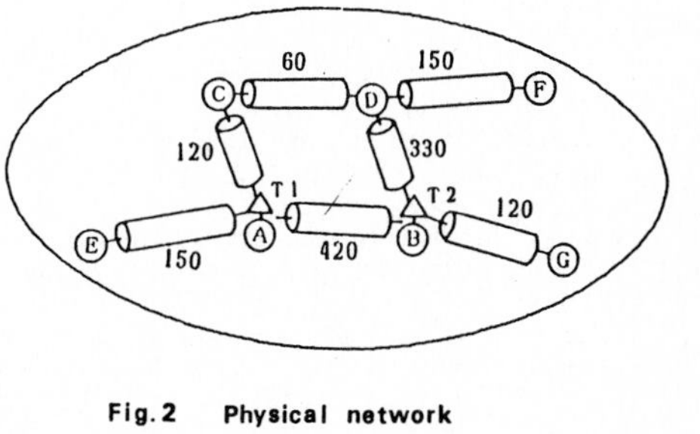

Fig.2 Physical network

2. PLANNING SYSTEM OUTLINE

2.1 Network Configuration

The proposed junction network optimization procedure deals with urban area telephone networks which consist of local offices (LO) and local tandem offices (LMO). As routing methods between local switching offices, the following three routing methods are considered,

(1) Direct routing,
(2) Alternate routing,
(3) Tandem routing.

As shown in Fig.3, The proposed procedure deals with the alternate routing method, in which second choice routes are established between origin offices and destination side tandem exchanges. As a result, there are at most three possible routes in the alternate routing method dealt in this paper, i.e., first- and second-routes having high-usage groups and a third-route as a final route. All possible routing patterns between local switching offices are shown in Fig.4, in which the direct routing is pattern (1), the alternate routing is patterns (2),(3),(4) and (5) ,and tandem routing corresponds to patterns (6) and (7). The circuit group size is constrained to multiples of a fixed module, called a module size in the logical network.

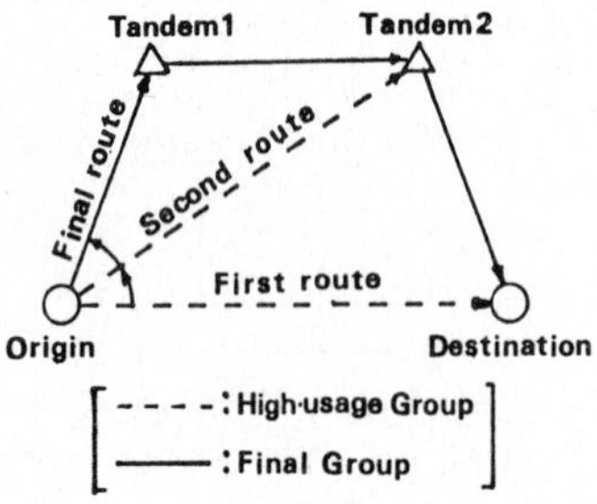

Fig.3 Alternate routing method

As for physical networks, cable PCM and fiber optic systems are candidates for transmission media to use. Transmission system capacities are called module sizes in the physical network.

For simplicity, it is assumed that switching and transmission nodes are located in the same place.

2.2 Input Data

The following input data are necessary for the optimization program,

1. Total number of exchanges included,
2. Individual office ranks (LO or LMO),
3. Individual tandem exchange areas,
4. Number of subscribers in each exchange,
5. Amount of traffic offered between exchanges (Traffic Matrix),
6. Available links to use in physical network and their length,
7. Module sizes in logical- and physical-networks,
8. Grade of service constraint(final group blocking probability),
9. Facility cost data.

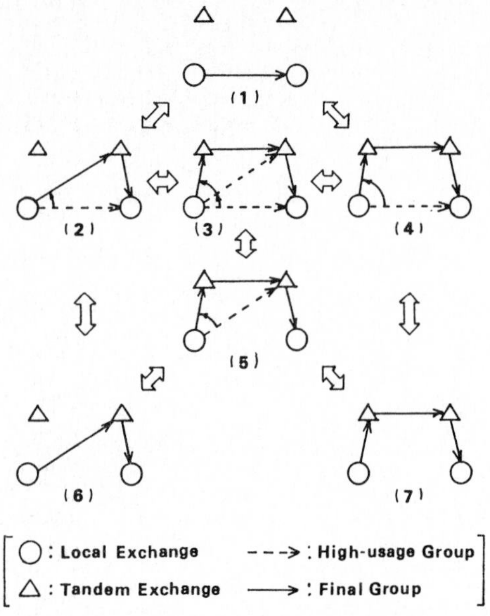

Fig. 4 Possible routing patterns

2.3 Output Data

The optimization program outputs the following items.

1. Routing rules between local offices,
2. Required number of circuit groups between exchanges,
3. Path for individual circuit groups on transmission network,
4. Selection and dimensioning of transmission systems for each route,
5. Required number of multiplexing facilities,
6. Detailed cost tables.

2.4 Network Cost Calculation Method

The cost element for switching, transmission and multiplexing systems used in the optimization program are explained briefly.

2.4.1 Switching Cost : C_E

$$C_E = c_f + c_s \cdot n_s + c_d \cdot n_d \qquad (1)$$

where

n_s : number of subscribers,
n_d : number of digital trunks,
c_f : fixed cost,
c_s : unit cost per subscriber,
c_d : unit cost per digital trunk.

2.4.2 Transmission Cost : C_T

$$C_T = \sum_s [c_t(s) \cdot L + c_r(s) \cdot \lfloor L/d - 1 \rfloor] \cdot n_t(s) \qquad (2)$$

where

L : cable length (km),
d : repeater span (km),
$\lfloor x \rfloor$: minimum integer greater than x,
$n_t(s)$: required number of s th order transmission systems,

$c_t(s)$: unit cost per system per km for s th order transmission line,

$c_r(s)$: unit cost per system for s th order repeater.

2.4.3 Multiplexing Cost : C_M

$$C_M = \sum_s [c_m(s) \cdot n_m(s) + c_d(s) \cdot n_d(s)] \quad (3)$$

where

$n_m(s)$: number of multiplexers which convert from s to (s+1) th order PCM hierarchy,

$n_d(s)$: number of demultiplexers which convert from (s+1) to s th order PCM hierarchy,

$c_m(s)$: unit cost for multiplexer which converts from s to (s+1) th order PCM hierarchy,

$c_d(s)$: unit cost for demultiplexer which converts from (s+1) to s th order PCM hierarchy.

3. OPTIMIZATION METHOD

3.1 Problem Description

In order to make the problem clear, let us generalize a junction network optimization problem. The following parameters are defined:

G(V,U) Network topology, where V is a set of v nodes (switching/transmission nodes) and U is a set of u links (available physical links between nodes).

$a_{i,j}$ Traffic demand originating at node i destined for node j (i,j=1,2,...,v).

$r_{i,j}$ Routing pattern for i,j traffic (Fig.4).

$n_{i,j}$ Number of circuit groups between nodes i,j in the logical network.

B_o Grade of service for the network. This is a bound on final group blocking probability.

$B_{i,j}$ Blocking probability on circuit group i,j.

$p_{i,j}^k$ Circuit assignment rule to the physical network. If a circuit group i,j is routed along a physical link k, $p_{i,j}^k=1$; otherwise, $p_{i,j}^k=0$ (k=1,2,...,u).

$l_{k,s}$ Number of s th order transmission systems on physical link k (s=1,2,3,4).

$m_{i,s}$ Number of s th order multiplexing system in node i (s=1,2,3).

mod_s Module size on the logical network(s=0); or on the physical network (s=1,2,3,4).

$C(n_{i,j}, l_{k,s}, m_{i,s})$ Network cost providing $n_{i,j}$ trunks on switching offices, $l_{k,s}$ transmission systems on physical links and $m_{i,s}$ multiplexing facilities in transmission nodes.

The junction network optimization problem can be written;

Given: G(V,U) and $a_{i,j}$

Minimize: Network cost $C(n_{i,j}, l_{k,s}, m_{i,s})$

With respect to: $r_{i,j}, n_{i,j}, p_{i,j}^k, l_{k,s}$ and $m_{i,s}$

Under constraints:

$$B_{i,j} \leq B_o \quad \text{(for final group)} \quad (C1)$$

$$n_{i,j} = mod_0 \cdot x \ (x\text{:integer}) \quad (C2)$$

$$\sum_s l_{k,s} \cdot mod_s \geq \sum_i \sum_j n_{i,j} \cdot p_{i,j}^k \quad (C3)$$

The implication of the constraints is :

(C1) Final group link blocking probabilities must be no greater than the given grade of service.

(C2) The number of circuit groups in the logical network is constrained to multiples of given module size on the logical network, mod_o.

(C3) Each physical link capacity must be no less than the circuit demand on the link.

Transmission link facility parameter ($l_{k,s}$) and node facility parameter ($m_{i,s}$) are design variables, which have an interrelation with each other. Since it is too complicated to optimize both variables simultaneously, the proposed optimization procedure adopts the following simplified subproblems. First, for the variable $l_{k,s}$, selection of PCM hierarchical order and dimensioning are carried out, on a link by link basis, so that the lowest transmission system cost can be obtained. Next, for variable $m_{k,s}$, a transfer level concept is introduced for simplicity. Transfer level is a certain PCM hierarchical order level, at which through circuit bundles are directly transferred from incoming to outgoing ports in transmission nodes. The level is defined on a circuit-group by circuit-group basis. Then, the number of multiplexing systems is calculated using $l_{k,s}$, $p_{i,j}^k$ and the transfer level. Therefore, networks are optimized with respect to the routing method, circuit demand and circuit assignment to a physical route.

3.2 Optimization Procedure

The proposed design method is composed of logical- and physical-network design parts, as shown in Fig.5. The former part determines traffic routing rules and the number of circuit groups between switching nodes. The latter part optimizes circuit-group assignment and transmission medium selection/dimensioning on the physical network. During the optimization process, the data representing the logical- and physical-network are iteratively transferred between these two parts. The optimization process makes locally optimal networks according to a physical network initializing parameter. The locally optimal solution having the lowest cost is adopted as a final solution.

The junction network optimization proceeds, taking the initiative by the logical network design part. The optimization order is logical network initialization, high-usage group optimization and routing method optimization. In each step, the physical network is modified and optimized by a physical optimization module composed of physical link deletion and route change. Exceptionally, physical network initialization is required after the logical network initialization.

3.2.1 Logical Network Initialization Module

Routing methods between local offices ($r_{i,j}$) are established by judging whether or not high-

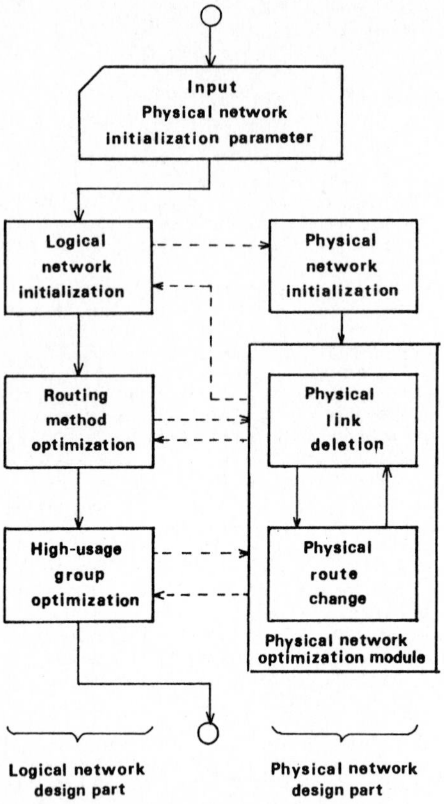

Fig. 5 Optimization procedure

usage/final groups are set up on the first and second routes. For the first route, the decision criterion is shown in Table 1, i.e., offered traffic between exchanges ($a_{i,j}$) is compared with appropriate thresholds [A1,A2]. Next, the number of circuits ($n_{i,j}$)on the first route is determined by a marginal occupancy method [4] in case of high-usage groups using approximate cost ratios; or by Erlang B formula in case of final groups. Then, the traffic offered to the second route is calculated, the link setting for the second route decision is made, and the number of circuits ($n_{i,j}$ for first link of second route) is determined in the same way as for the first route. Finally, the number of final circuit groups ($n_{i,j}$ for third route) is calculated using the Equivalent Random Theory [5].

Table 1 Offered traffic and circuit group on first route

$a_{i,j} < A1$	:	No circuit
$A1 \leq a_{i,j} < A2$	:	High-usage group
$A2 \leq a_{i,j}$	:	Final group

3.2.2 Physical Network Initialization Module

Plural physical networks are set up in order to obtain a global optimum network out of locally optimal solutions. The following shows how to make initial physical networks. A network which consists of all available links given by input data is called a full network. On the other hand, a partial network is defined as a network composed of a subset of available links. First, a partial network is selected according to a physical network initializing parameter. Next, paths for individual circuit groups ($p_{i,j}{}^k$) are set up on the partial network by the shortest-path routing criterion. Then, transmission media selection and dimensioning ($l_{k,s}$) are carried out for individual transmission links. In addition, the number of node facilities ($m_{i,s}$), e.g., multiplexers and demultiplexers, is calculated based on the assumption described in 3.1. The obtained initial physical networks vary in regard to concentrated degree for each link and route length for each traffic flow.

3.2.3 Routing Method Optimization Module

The routing methods determined in the logical network initialization module do not always give the optimum network structure. Therefore, the following routing method perturbation is carried out. The routing method between nodes ($r_{i,j}$) is changed by adding/deleting a first, second or third route. This process is shown in Fig.4. Consider the cases wherein a routing method between a certain pair of local offices is (4) in Fig.4. This routing method can be changed to (1) and (7) by deleting the third and the first route, respectively, or can be changed to (3) by adding the second route. In each of these three networks, (1),(3) and (7), obtained by the perturbation, traffic flow is calculated and circuit dimensioning is performed. Then, the network cost is calculated. The routing method between the node pair which gives the lowest network cost is adopted. This optimization module proceeds repeatedly, as long as the cost is reduced.

3.2.4 High-Usage Group Optimization Module

The number of high-usage circuit groups ($n_{i,j}$) is recalculated, based on actual cost ratios obtained in the physical network design part, instead of approximate cost ratios used in previous modules. The amount of overflow traffic and the required number of final groups are recalculated accordingly. Every time the number of high-usage groups is recalculated, the physical network is modified by the physical route change module. When all high-usage circuits are processed, the network is optimized in physical network optimization module. This procedure is also carried out as long as the network cost decreases.

3.2.5 Physical Link Deletion Module

A link in the physical network is deleted and the circuits traversing the link are set up on a more economical route in this module, as shown in Fig.6. A list of alternate paths for each pair of nodes, when a certain link is deleted, is prepared beforehand. First, a link to delete is selected. Next, delete the link and change the route of circuits traversing the link to several candidate routes. Then, the new network cost is calculated. If the cost is reduced, the route which gives the lowest network cost is selected.

This procedure is executed in expectation of network cost reduction due to transmission system economy of scale, i.e., the transmission cost per channel generally decreases while its capacity increases.

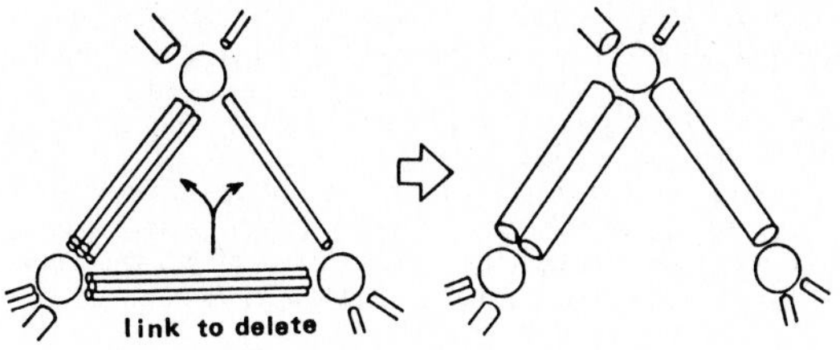

Fig.6 Physical link deletion example

3.2.6 Physical Route Change Module

The physical routes for individual circuit groups are changed to obtain a more efficient network from a view point of the high fill rate at transmission lines, i.e., circuit groups can be packed into physical links effectively. An example of this process is shown in Fig.7. First, a circuit whose physical route is changed is selected. Next, the route is changed onto the candidate routes, which were listed beforehand as described in 3.2.5 and the new network cost is calculated. If the new cost is lower than the old one, the new route is employed.

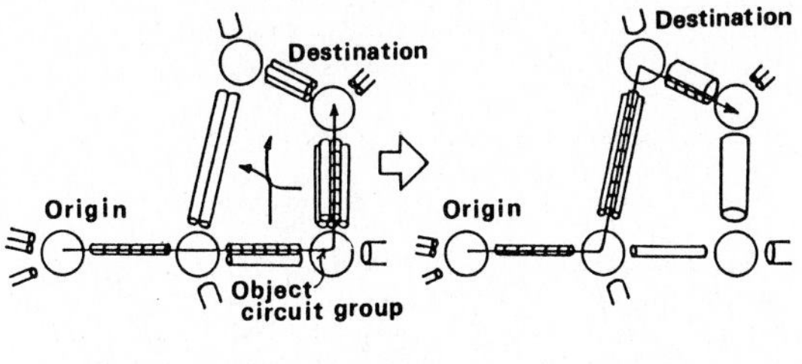

Fig. 7 Physical route change example

4. A CASE STUDY EXAMPLE

The procedure was applied to a hypothetical urban area telephone network with 2 local tandem offices and 8 local offices. Figure 8 shows exchange office layout, distances between exchange offices and the number of subscribers covered by each exchange. Traffic flow matrix statistics are shown in Table 2. Junction network dimensioning and optimization were executed under the conditions shown in Table 3. In the following, the optimization method by modular engineering, with a unique module size (30 ch), is called a conventional method. It is compared with the proposed new method.

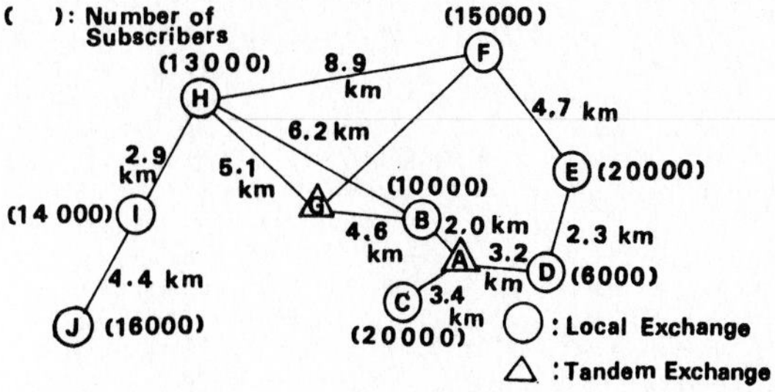

Fig. 8 Studied network configuration

Table 2 Traffic flow matrix statistics

Number of local tandem exchanges :	2
Number of local exchanges :	8
Total traffic flow :	1963.8 erl
Average traffic flow :	35.1 erl
Standard deviation :	26.5 erl
Minimum traffic flow :	5.2 erl
Maximum traffic flow :	115.2 erl

Table 3 Network optimization condition

(1) Grade of service constraints
- Link blocking for first route : 3 %
- Link blocking for second route : 2 %
- Link blocking for third route : 1 %

(2) Circuit establishment thresholds
- [A1,A2] for first route : [5 erl,30 erl]
- [A3,A4] for second route : [5 erl,30 erl]

(3) Module sizes
- Logical network :
  mod_0 = 30
- Physical network :
  mod_1 = 30 (Cable PCM)
  mod_2 = 120 (No use)
  mod_3 = 480 (Fiber optic system)
  mod_4 =1920 (Fiber optic system)

4.1 Network Performance at Each Optimization Step

Figure 9 shows network costs at each optimization step. Network costs optimized by the proposed method are compared with the conventional modular engineering in regard to relative cost. In Fig. 9, network cost, after the logical and physical network initialization (Step 1), obtained by the conventional method is set to be 1.0. Figure 10 shows the ratio for individual routing methods (tandem, alternate or direct routing), and the ratio for number of channels using each PCM hierarchical order at each optimization step.

As shown in Fig. 9, 5.7 % cost reduction, compared with the conventional method with the module size of 30 channels, was obtained by the proposed method. For reference, 12.4 % cost reduction, compared with the conventional analog junction network dimensioning method where no modular engineering is adopted, was also obtained.

Figure 9 shows that the cost reduction effect of the logical network optimization was observed at step 3, i.e., routing method optimization combined with physical network optimization. The ratio for tandem and alternate routing methods increases while that for direct routing method decreases at step 3. As a result, the number of final circuit groups for tandem or alternate routing becomes large, because traffic offered to those groups increases.

On the other hand, the cost reduction effect of the physical network optimization can be directly seen between step 1 and step 2 in Fig. 9. At step 1, it makes little difference in network cost between the two methods whether third order PCM systems are used or not. However, after the physical network optimization, 3.6 % cost reduction in relative cost was achieved. The reason is that the number of physical links decreases and the number of links using third order PCM transmission systems increases due to physical route change and link deletion procedures.

No cost reduction was obtained by the high-usage group optimization. The reason is that the module size in a logical network used in the proposed method equals the lowest physical network module size (30 ch). If module sizes with less than 30 channels were adopted, network cost reduction can be expected by the high-usage group optimization module.

When junction network was optimized only from a logical network aspect using a unique module size, the network extremely tends toward a star like constitution [6]. However, it has a moderate distribution of routing patterns, as shown in Fig.10, because it is optimized from both logical- and physical-network aspects.

As mentioned above, it was clarified that the junction network optimization method, which equally considers both logical- and physical-network aspects, is very important.

4.2 Cost Reduction vs Offered Traffic Increase

Figure 11 shows network cost reduction in

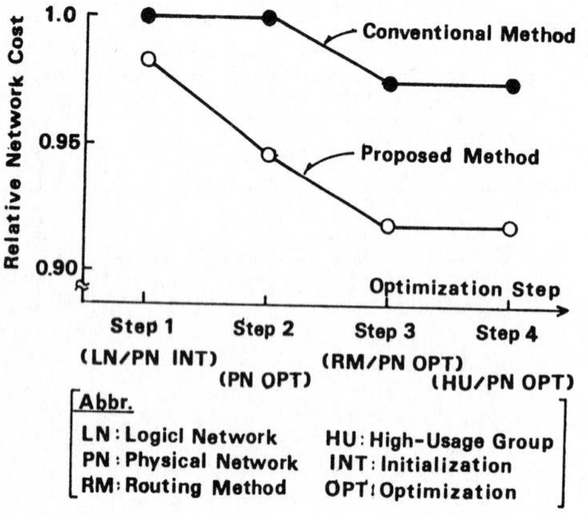

Fig. 9 Network cost at each optimization step

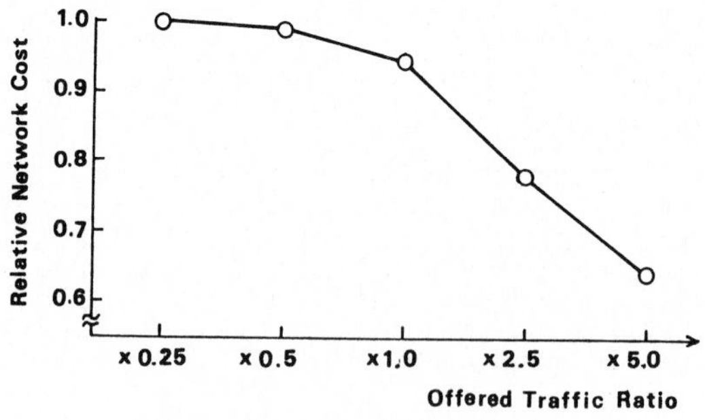

Fig.11 Network cost reduction vs offered traffic increace

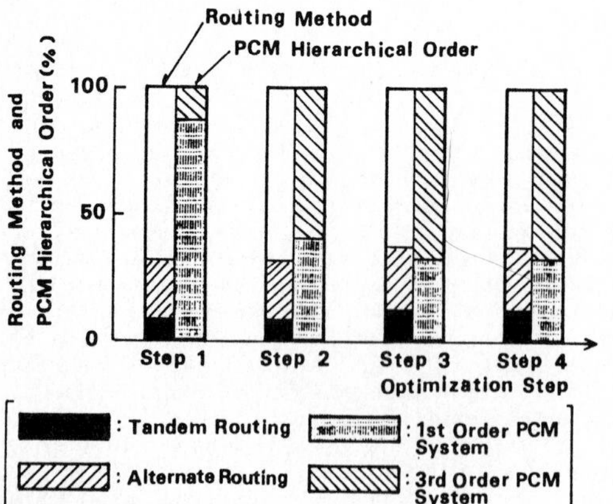

Fig.10 Logical- and physical-network performance at each opimization step

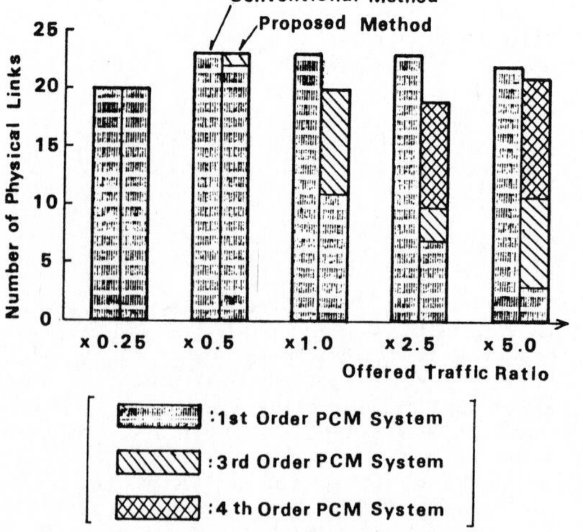

Fig. 12 Selection of transmission systems on physical links

regard to relative cost, when original total offered traffic volume decreased/increased. In this figure, network costs, obtained by conventional modular engineering, are set to be 1.0 at each offered traffic ratio.

Figure 12 shows results of transmission system selection obtained by both conventional and proposed methods. In this case study, a transmission link is considered to be a one-way link.

When offered traffic is light, e.g., the ratios are 0.25 or 0.5, transmission network configuration and obtained network costs were not so different. However, when offered traffic is heavy, e.g., the ratio is greater than 1.0, considerable network cost saving was provided. Under heavy traffic conditions, the number of physical links decreased. This caused the selection of higher order transmission systems on many physical links, as shown in Fig.12. The maximum network cost reduction was 36 % when the ratio is 5.0.

5. CONCLUSION

This paper has shown a new junction network planning system which can take into account large scale transmission facilities in a PCM hierarchy. In order to consider higher order transmission systems, the proposed method introduces plural module sizes, while the conventional modular engineering uses a unique module size.

The proposed method consists of logical- and physical-network design parts. During the optimization process, data representing both networks are iteratively transferred between these two parts. As a result, junction networks can be optimized from both switching- and transmission-network aspects.

Network studies for hypothetical urban telephone network were executed. Case studies show that the proposed method can reduce total network cost form several to more than 30 percent, compared with the conventional method, depending on network scale and/or network traffic load.

ACKNOWLEDGMENT

The authors would like to express their grateful thanks for the valuable discussions with Dr. T. Yamaguchi, Research Manager, and Mr. H. Goto, Manager, Communication Research Laboratory, C&C Systems Research Laboratories.

REFERENCES

[1] W.B.Elsner, "Dimensioning Trunk Groups for Digital Networks," 9th ITC, Torremolinos, 1971.
[2] B.Yaged, "Minimum Cost Routing for Static Network Models," Networks 1, pp.139-172, 1971.
[3] Y.Okano, M.Makino and T.Miki, "Transmission NetworkOptimizationin Junction Networks," Proc. of ICC '84, pp.185-188, Amsterdam, 1984.
[4] C.W.Pratt, "The Concept of Marginal Overflow in Alternate Routing," 5th ITC, New York, pp.51-58, 1967.
[5] R.I.Wilkinson, "Theories for Toll Traffic Engineeringin the U.S.A.," BellSyst. Tech. J., 35, 2, pp.471-514, 1956.
[6] H.Okazaki, "Planning an Urban Area Digital Telephone Junction Network," NEC Res. & Develop., 73, pp 43-52, Apr. 1984.

TELETRAFFIC ISSUES in an Advanced Information Society
ITC-11
Minoru Akiyama (Editor)
Elsevier Science Publishers B.V. (North-Holland)

SIRIUS — JUNCTION NETWORK PLANNING: A DYNAMIC METHODOLOGY

Switched Network/Transmission Network; Long Term/Short Term

VIDAL, D. M. **ABRANTES, L. V.**

T.L.P. — Telefones de Lisboa e Porto
Lisboa, Portugal

ABSTRACT

The development of computer aided junction network planning systems have been directed to the establishement of a fast and accurate forecasting dynamic methodology.

The objective is to define the short, medium and long term networks, which is performed in two complementary phases. Firstly, the long term network and the optimal provisioning periods are established considering the present network as being totally saturated. Secondly, an optimized administration of the existing network (annual) is done in agreement with the first phase, which guarantees an evolution in the horizon network sense.

1. INTRODUCTION

The SIRIUS methodology (figure 1) defines the evolution of the junction network, minimizing the switching costs and the transmission costs during the studied period, in agreement with an optimized administration of the existing network, to satisfy the traffic demand under a prescribed quality of service and the technical plans: routing plan, transmission plan, and numbering plan.

The junction network planning is performed in two complementary phases, with the consequent establishement of the short, medium and long term networks.

The first phase of the SIRIUS methodology, defines the long term network (15 to 20 years) and all the evolution, considering the present network as being totally saturated. Economical optimization criterions are used.

The second phase defines the short term networks (annual) in agreement with an optimized administration of the existing network, in order to delay the switching and transmission equipment acquisition, and an evolution in the horizon network sense.

The global optimization of the junction network is a complex process of difficult treatment. It's possible to divide the study of the junction network into two parts: The functional network involving the logical organization from the view-point of how the traffic is routed; and the physical network that refers to the topological layout. The nexus between functional and physical networks, basically the transmission costs, is not enough to demand the global optimization. So, this problem is solved by treating these two aspects in a separate way, but never loosing the global sense of the process.

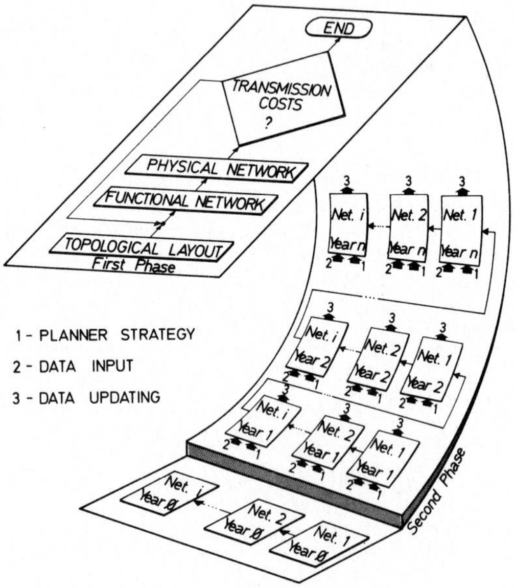

Fig. 1 The SIRIUS methodology

2. THE FIRST PHASE - LONG TERM PLANNING

2.1 Introduction

The long term network and it's evolution are established by optimization of the functional and physical networks separately, taking into account the transmission costs through an iteractive process (figure 2).

The optimization procedure begins with the establishment of the basic topological layout, with a number of routes normally superior to the size of the present network and the characterizing parameters settled within initial bounds.

The next module is the functional network optimization using economical optimization criterions to satisfy the traffic demand under a prescribed quality of service.

In the first iteration of the functional/physical procedure, the shortest distance between each two centers is taken to define any link. The transmission marginal costs are estimated from this distance by an empirical procedure depending on the switching equipment technology. In the next iterations the optimal transmission marginal costs are defined by the transmission module.

The switched marginal costs, the equipment

modularity, the traffic matrix and the imposed restrictions (to the absolete equipment or to force some routing schemes) are also data that supply the functional module. The results are the selection of routing schemes, the number and location of the transit centers and their capacities, the size of the functional routes and the characteristic parameters that define the selected links. This information per link is converted in information per physical route (arc) to supply the physical network optimization.

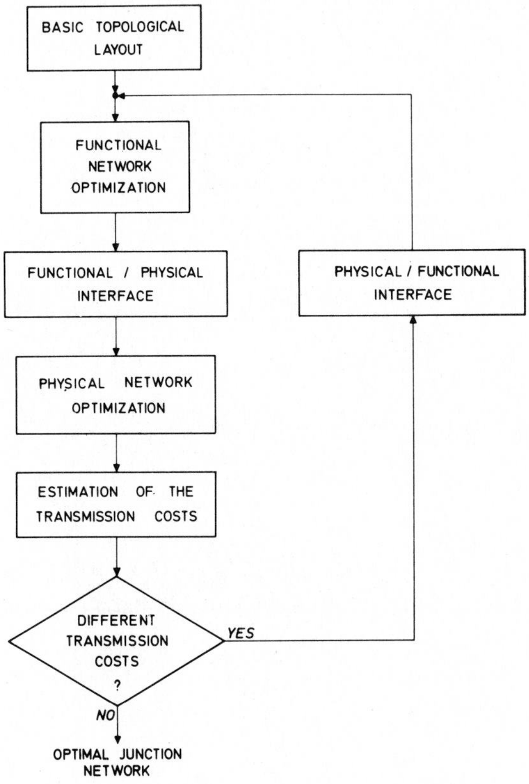

Fig. 2 The first phase: Long term planning.

In the physical module, starting from the basic topological layout, the marginal costs associated with each physical route (arc) and the correspondent transmission equipment are calculated. The planner may impose the transmission equipment or compare some of them in a minimum cost base.

The elimination of each arc is studied considering all the others fixed.

The results are the transmission equipment per arc with the respective year of acquisition, during the studied period, the eliminated arcs. the arcs of integration, the links and the correspondent circuits associated with the final arcs, and the transmission costs of each arc.

Some of these results are integrated in the next module which converts the information per physical route (arc) into information per link, and gives more information about transmission costs to supply the functional network optimization in the next large iteration.

The junction network optimization is concluded when the transmission costs are almost the same in two consecutive large iterations and the optimal network is reached.

2.2 Functional network

2.2.1 Data organization

The information corresponding to input and output data is organized in a logical and systematic sequence. In this way the input data is divided into six groups:

The general data - A general description of the network to be studied is given and the quality service criterions are defined. The running modes are specified, providing the computer program with information about the data flux in the general algorithm. The precision of the optimization algorithms are also specified.

The functional routing data - The information is arranged by functional route with dimension equal to the square of the number of local exchanges. The traffic matrix of the horizon year is estimated by the traffic engineering department and automaticly transfered to this group of information. The planner defines the functional routing schemes, the patterns and the reference of the transit centers allowed or forbidden.

The link data - In this group, every possible link is described and the information is arranged by local exchange to local exchange, local exchange to tandem, tandem to tandem and tandem to local exchange. The minimum and the maximum number of circuits, the initial traffic of the transit links (mean and variance) and availability are specified. The transmission equipment and the marginal transmission costs are estimated by the physical network optimization and automaticly transfered to this group of information. They are not available in the first iteration of the switching/transmission procedure.

The central data - The topological localization, the switching equipment, the minimum and the maximum transit capacity are specified.

The topological layout data - A description of arcs and nodes is given, referring to the distance.

The costs data - The switching costs are given per incoming and outgoing circuit at each local exchange and tandem, in the network. At each tandem the switching costs are also given per erlang of traffic that passes through. The transmission costs are given per circuit and per circuit x km. They are used to initiate the switching/transmission optimization or to make an economic evaluation on new links in other iterations. The equipment modularity is also referred.

In the same way, the output data is divided into three complementary groups:

The functional routing results - The optimized routing plan is obtained from the allowed configurations. The transmission attenuation point to point is also estimated.

The link results - A complete description of the high and low usage links, such as transmission equipment, number of circuits, traffic, congestion, transmission attenuation, costs, etc., is obtained.

The statistical data - Finally, statistics on functional patterns, links, transmission attenuation and costs are presented.

The information about the number of circuits calculated in every link is automaticly transferred to the functional/physical interface.

2.2.2 Algorithm

The general algorithm of the functional network optimization is heuristic, with a modular organization, involving fast and accurate optimization routines (figure 3). A general description of the principal modules is presented:

Data read - This program selects the group of data required by the functional network optimization from the common acess file and makes a complete validity test.

Link marginal costs - The estimation of the link marginal transmission costs is a result of the physical network optimization and is not available in the first iteration of the functional/physical procedure, or when treating new links. The problem of transmission equipment initialization is solved in two elementary operations. Firstly, the shortest way through a minimum number of nodes is calculated to the links. Secondly, with this distance and the switching equipment technology, the optimal transmission equipment is defined by an empirical criterion and the respective marginal transmission costs are estimated.
The marginal switching costs are also estimated.
The present value of annual charges criterion is used to quantitatively measure the economics involved in each alternative, because it's necessary to consider equipment having different characteristics of life, purchasing costs, operation and maintenance expenses.

Initial solution - This module calculates the high or low usage direct routes and the consequent traffic in transit links to initiate the optimization procedure. The Rapp's solution is used for alternate routing. The single routing initialization takes the optimal routing from the allowed configurations, admitting a maximum efficiency in the involved transit links.

Alternative routing optimization - The routing is optimized, taking into account the allowed or defined patterns, by an iterative calculation procedure propose by 'B. Wallström' |6|.

Single routing optimization - A heuristic procedure is used, which calculates the optimal routing in each iteration from the allowed configurations and from the network established in the previous iteration |12|.

The general algorithm has proved that the total cost of the network optimized with mixed routing criterions, practicly doesn't change after one or two iterations.

2.3 Physical network

2.3.1 Data organization

The information is organized in four groups corresponding to different sources:

The existing network data - The present physical structures are described, referring to the distance, the transmission equipment, the corresponding links and the existing number of circuits.

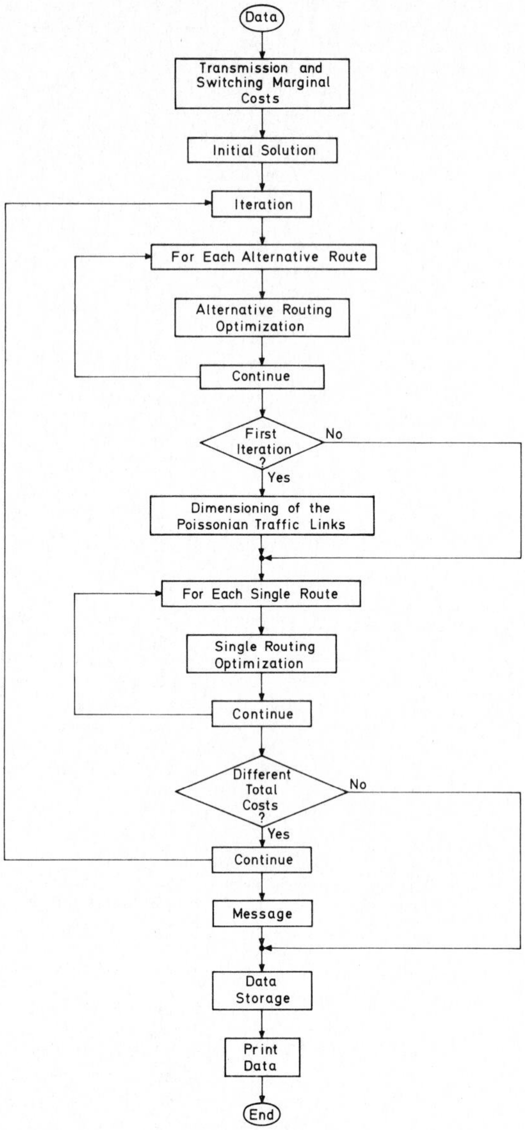

Fig. 3 Functional network optimization

The topological layout data - A description of arcs and nodes is given, referring to the distance and to the number of free ducts in conduits.

The functional network data - Some functional network optimization results are automaticly transferred to this group of information, from the functional/physical interface. They are the physical routes description in the horizon year and the corresponding links referring to the number of circuits.

The physical network data - The planner may

define the security schemes required and restrictions of the physical network optimization, such as to forbid or to fix some arcs and the transmission equipment, or to impose some physical routing schemes.

The forecasting results are given per arc, referring to the transmission equipment, the corresponding investment period, the new links and the number of circuits, the arcs of integration, the increase of circuits, the transmission attenuation and the transmission costs. Finally, statistics on links, arcs, transmission equipment and costs are presented.

The information regarding transmission equipment, new routing schemes for links, and marginal transmission costs are automaticly transferred to the physical/functional interface.

2.3.2 Algorithm

The general procedure for physical network optimization is heuristic with a modular organization and involves fast optimization routines (figure 4). A description of the principal modules is presented:

Optimal transmission equipment and provisioning periods - Economic criterions are used to obtain the optimal transmission equipment, the investment amount, and the optimal provisioning periods from the present till the horizon year satisfying the requested circuits. The demand for circuits in a route is assumed to increase linearly, to determine the economic period of provision, which corresponds to the minimum value of the present worth of all extensions during an unlimited period of time.

Initial solution - An evaluation of the investment costs during the studied periods is done, taking into account the initial topological layout.

Temporary supression - The suppression of one arc involves the change of the corresponding links to other arcs. This integration is done in terms of the minimum cost, which is calculated by the 'Moore - Dijkstra' algorithm |7|.
Starting always from the initial solution, each arc is supressed and the consequent optimal configuration is economically evaluated.

Definitive suppression - After running every arc the new solution is taken, corresponding to the suppressed arc whose configuration produces the minimum investment costs. Just one arc per iteration is eliminated. In this case the network description and the costs are updated.

The procedure stops when the elimination of any arc produces a higher cost configuration.

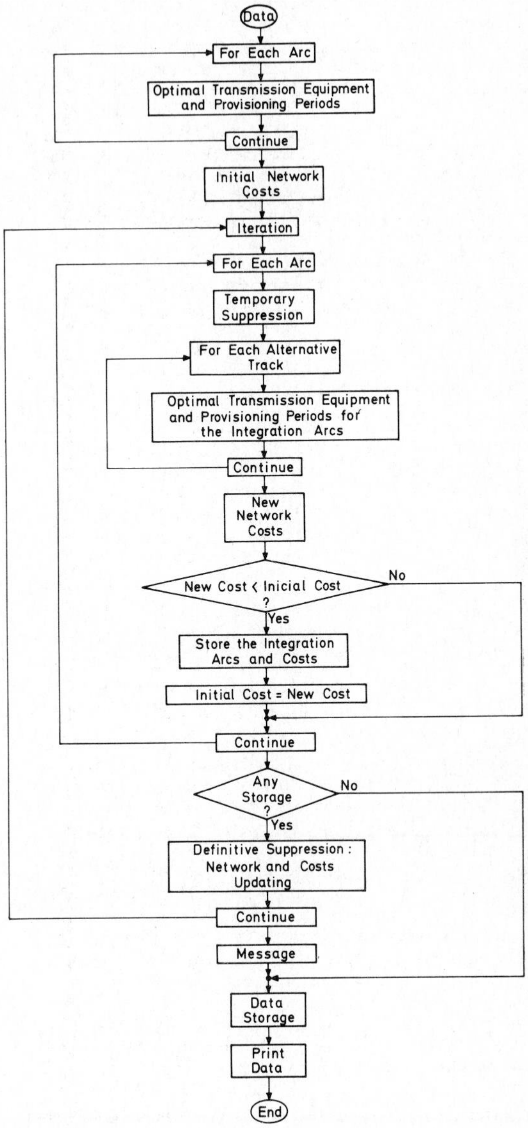

Fig. 4 Physical network optimization

3. THE SECOND PHASE - SHORT TERM PLANNING

3.1 Introduction

In this phase the existing network is taken into account in order to delay the switching and the transmission equipment acquisition, in agreement with an optimized administration of the installed equipment.

When the rerouting problem is not solvable, the equipment acquisition is done in accordance with the first phase although shifted in the time; so, the tendency to the horizon network is obtained.

The developed process for short term planning (annual) permits the simultaneous treatment of several networks with hierarchic dependence for several years. Yearly, from the zero year to the horizon year (figure 1).

Data is organized in six files: the routing plan, the links of 'n - 1' year, the functional network, the physical network and the topological layout.

The forecasting results are divided into eight groups: the initial and final results of functional routing, such as routing schemes and transmission attenuation, the initial and final results of link and sub-link, such as distance, track, traffic, total number of circuits, increase of circuits and category, the results of physical route (arc), such as total and increase number of telephonic and non telephonic circuits, and the results of equipment acquisition.

3.2 Algorithm

The second phase program may run each block in sequence from the 'network 1/year 1' till the 'Last Network/Horizon Year', although the short and medium term blocks are the most important (figure 1).

The running procedure is similar in each block, involving the optimized administration of the existing equipment, the tendency to the horizon network, the planner strategy and the data updating.

A description of the principal modules is presented to the 'Network 1/year 1' block (figure 5):

Routing plan - The hierarquical dependence among sub-networks to be studied is defined, as well the single or alternative routing, the patterns allowed or forbidden and the overflows. The planner may fix some routes but a valid routine confirms if the basic plan is fulfilled.

Initial traffic - It's necessary to take into consideration traffic that overflows from routes in the sub-networks not under study to routes in the current block.

Functional network optimization - The routines are similar to the equivalent program in the first phase.

Physical network optimization - The data in the physical network file is quite similar to the equivalent data described in the first phase, although a reference to the unoccupied number of circuits is done in the present network.

The required circuits in each existing link, coming from the annual demand growth, are integrated in the original routing if the respective capacities are sufficient. On the contrary, it's not possible to make the same routing and so, a rerouting is performed in terms of minimum cost, making use of the unoccupied circuits |8|. The same routing method is used to integrate new links.

If it is possible to find a suitable routing scheme for every circuit comming from the annual demand growth in the current block (sub-network), the problem is solvable and the investment is not necessary.

On the other hand, the problem is not solvable and the program may stop or not, making the investments in accordance to the first phase or to the planner strategy. In this case the planner may force some routing schemes, create or eliminate arcs, force some investments, treat several investment hypotheses and compare some of them in a minimum cost base, or let the program calculate the economical solution.

If an investment proposed by the planner is not necessary, it's automaticly eliminated by the program.

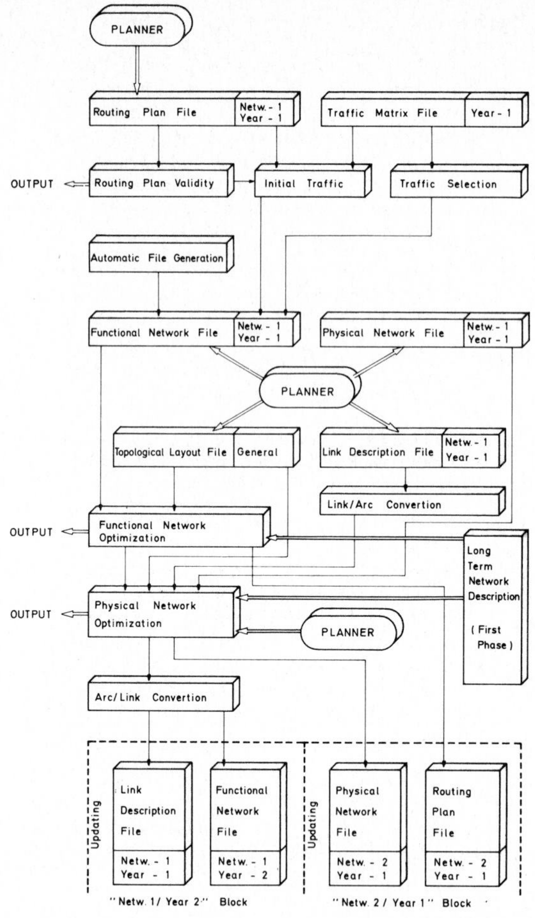

Fig. 5 The second phase: Short term planning (Annual). "Network 1/Year 1" Block.

4. CONCLUSIONS

The SIRIUS methodology has been sucessfully applied to the Lisbon network in the short, medium and long term planning. The complete package of programs, previously described, have about 20000 lines of instructions written in Fortran 77. Input and output data is stored in a work disk, occupying about 100 Mbytes.

The complexity encountered when attempting to carry out long term and short term planning, has been sucessfully reduced by separating the switching and transmission problems, although never loosing the global sense of the process, and by treating several networks with hierarchic dependence for several years (annual).

The SIRIUS methodology has proved to be a fast, manageable and practical method for optimizing junction networks in the long and short term planning.

REFERENCES

|1| STANDARD ELÉCTRICA, S.A., "Telecommunication Planning,"

|2| R. I. Wilkinson, "Theory of toll traffic engineering in the U.S.A.," Bell SysT. Tech. J., vol. 35, nº 3, pp. 421 - 514, 1956.

|3| Y. Rapp, "Planning of Junction Network in a Multi-exchange Area. I. General Principles," Ericsson Tech. 20:1, pp. 77 - 130, 1964.

|4| Y. Rapp, "Planning of Junction Network in a Multi-exchange Area. II. Extensions of the Principles and Applications," Ericsson Tech. 21:2, pp. 187 - 240, 1965.

|5| C. W. Pratt, "The Concept of Marginal Overflow in Alternate Routing," A.T.R. 1 : 1 & 2, pp. 76 - 82, 1967.

|6| B. Wallström, "Methods for Optimizing Alternative Routing Networks," Ericsson Tech. 25:1, pp. 3 - 28, 1969.

|7| M. Gondran, M. Minoux, "Graphes et Algorithmes,"

|8| M. Minoux, "Planification a Court et a Moyen Terme d'un Réseau de Télécommunications," Annales des Télécommunications, tome 29, nºs 11 - 12, pp. 509 - 536, 1974.

|9| M. Minoux, "Multiflots de Coût Minimal avec fonctions de Coût Concaves," Annales des Télécommunications, tome 31, nº 3 - 4, 1976.

|10| Heinrich Kremer, "Ortsnetzplanung," 1963

|11| Ollie Smidt, "Engineering Economics,"

|12| D. Vidal, L. Abrantes, "Planeamento da Rede de Interligação. I - Programas de Longo Prazo," T.L.P., LISBOA, 1983.

TELETRAFFIC ISSUES in an Advanced Information Society
ITC-11
Minoru Akiyama (Editor)
Elsevier Science Publishers B.V. (North-Holland)

STRATEGICAL SYSTEM PLANNING: APPLICATION TO EXCHANGE EVOLUTION

Mario BONATTI - Giorgio GALLASSI and Salvatore CAVALLO

Central Research Laboratories, ITALTEL
Settimo Milanese (Milano), Italy

ABSTRACT

The point of view of a manufacturing company on the planning objectives and criteria for exchange evolution is presented, along with some application examples to system and design choices of UT-LINE electronic digital exchanges family.

A System Challenge Analysis methodology is described (procedures, models and algorithms) and the typologies of the results that can be obtained are shown.

1. INTRODUCTION

1.1. Position

People want new telecommunication services (eg ISDN). Operating companies want new systems in order to offer new services and to adopt new network management technologies (eg DNHR). Competitors offer new systems based on new technologies and implementing new concepts, new capabilities and larger ranges of coherent capacities.

This is the environment in which a manufacturing company takes the very difficult decision of adding a new switching system family for public network (conceived as a coherent set of switching systems covering a broad range of required capacities and capabilities) to his product portfolio (market mix).

The decision to start a development, requiring usually some thousand of man-years, is taken only if market and technology factors exceed a treshold level.

The decision is influenced by all the factors usually considered in medium and long range planning: economical and financial aspects, the company dimension, the R & D resources, new ideas, etc...(Ansoff [1]).

In this context, the strategical plans of the operating company, the philosophies on which they are based and the objectives they pursue are relevant inputs for the planning of a manufacturing company. As the opposite is also true, the resulting plans are temporary equilibria reached in a four part game: subscribers, operating company, the manufacturing company and its competitors in a more or less regulated environment.

The evaluation of the system alternatives conceived as system portfolio alternatives is achieved through System Challenge Analysis techniques, in a System Engineering's working context.

The System Engineering definition of the word "PLANNING" emphasizes: "agreements on the total program of work, consisting of many projects, that the organisation wants to pursue", "extensive background of information, so that an attack of proper breadth and scope may be initiated" (Hall [2]), and selections of the pursued projects according to the forecasted success of the systems they will produce (forecasted system challenge). The challenge of a system, depending on the needs satisfied and on the problems solved is meaured by the contribution that the system will offer to the economical objectives of the company, obtained simulating the fortune of each system or of a system family.

1.2. State of the Art

The evolution of the planning studies is characterized by a transition from static approach methodologies to evolution optimizing methodologies, that define time sequences of optimal choices, as for example, the evolutions of the switching nodes which optimize the transition to Integrated Digital Networks (IDN), or to Dynamical Non Hierarchical Routing techniques (DNHR) or to Integrated Service Digital Networks (ISDN).

As far as the evolution to IDN is concerned, the most studied aspects are the optimal introduction modalities (overlay, replacement) of new systems into the existing network versus the ratio between the costs (provisioning, operating, maintaining) of the new systems and the costs of the existing systems [3] [4]. The choice of the optimal modalities depends on some constraints of a general kind: limitation of the allowable budget for each modality, limitation in the production of the systems (limited systems availability), delayed decisions etc. These constraints modify the optimal solution [5] and oblige the planning partners to define priorities on modernisation actions [6].

A further aspect in this context arises from the possibility of remote switching (optimal assignment of the remote machines to the host machines, coherence between the choices concerning the host machines and the remote machines) [7].

More details on the results of an extensive State of the Art review can be found in [8] along with the consulted bibliography.

1.3. This Paper

In a preceeding paper, presented to the ITC 10 [9], we have illustrated a planning methodology and its applications to the evolution of the district trunking networks.

In this paper we present the different analysis typologies that can be developed during the life cycle of a switching system and some applications to the design choices for the UT LINE (a family of electronic digital exchanges that covers the network needs capabilities from the local to the intertoll exchanges, up to 100K subscribers/60K circuits).

In § 2 we present a set of possible applications In § 3 we explain the procedure, the models and the algorithms supporting the applications. In § 4 some applications referring to UT LINE design are described in detail.

2. SYSTEM CHALLENGE ANALYSIS APPLICATIONS

2.1. The Product Life Cycle

System Challenge Analysis is asked to solve specific problems in different phases of product life cycle.

The typology of the results of System Challenge Analysis depends both on the product life cycle phase involved and on global effort of System Engineering in that phase.

In each phase of the product life cycle, also during the operational life, relevant decisions may be taken among alternatives, that could inpact on system family fortune.

The System Challenge Analysis, therefore, works on different levels of detail and the actually followed process can be rationalized as a top-down process in which the higher level decisions (first phase decisions) are taken simulating the outcomes of the lower level decisions (Hierarchical Planning), as shown in the following table:

PRODUCT LIFE CYCLE	SYSTEM ENGINEERING PROCESS	CHALLENGE ANALYSIS RESULTS
System Definition	Program Planning	Ansoff Response Charts Optimal System Portfolio I Growth Share Matrices I
Exploratory Development	Project Planning I	Evaluation of the Economical Selection Criterion among Architectural Alternatives
System Development	Project Planning II	Evaluation of the Economical Selection Criterion among Subsystem Alternatives
Industrial Engineering Marketing	Production Planning Business Plans	Optimal System Portfolio II Growth Share Matrices II Dynamical Threshold Maps I
Field Tests	System Evolution	
Operational Life Introduction Development Maturity Decline	Current Engineering and Marketing,Operational and Performance Studies	Dynamical Threshold Maps II

2.2. System Definition Phase (Program Planning)

In this phase, Portfolio Analysis Techniques are used to compare alternatives such as existing switching systems improvements, or new systems planning to face transitions such as transition to DNHR, or transition to ISDN, or transition to new technologies.

The Ansoff Response Chart is used to choose the current level of planned results and to show the associate sequence of response costs. In fig. 1a) the trend of the current level of planned results

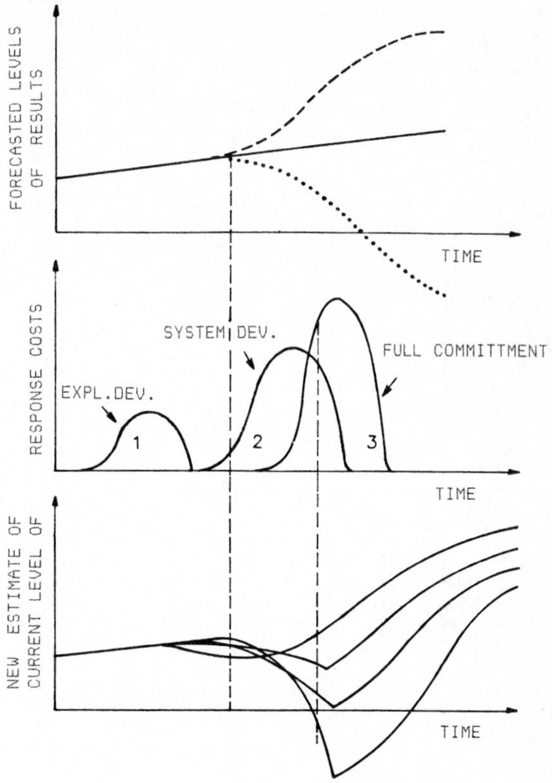

FIG. 1 ANSOFF RESPONSE CHART

(corresponding to the current portfolio) is shown (unbroken line), along with forecasted gains (if a new competitive system family to face new demand were ready at time T1) (broken line) and with forecasted losses (if the opposite is true) (dotted line). In fig. 1b) a possible sequence of response costs is shown (the full commitment costs stand for the cost arising from the total effort of the whole company in supporting the new business). The shapes of the response costs and their time allocations are decisional variables versus which new estimates of current level results can be made (see fig. 1c).

McKinsey and Marakon Matrices are used to show the positions of business based on systems such as large local exchanges with low traffic per source, or medium local exchange with high traffic per source, or medium toll exchanges with medium traffic per circuit, or large toll exchanges with high traffic per circuit.

In fig. 2 (McKinsey matrix) the position of a business is shown versus the industry attractivness (i.e. the relative market share).

In fig.3 (Marakon matrix) the position of a business is shown versus the Business Return on Equity (ROE) and the Growth of Business Assets. A business placed on the diagonal is growing at the same rate of its ROE, and neither generates cash nor requires cash from the company. Business above the diagonal are cash generators, and those below the diagonal are cash consumers.

In these figures circles represent each business, with the area within them proportional to the total sales (or total number of systems installed, or total number of lines installed).

Business as the business S4 in fig. 2 and 3 would be abandoned in a full competitive environment. In a more or less regulated environment of public goods production, the business can be vital for the total performance of the global public utility. It must be sustained.

The McKinsey matrices allow a modeling of the dynamic of a business. In fig. 5d, the current position of the business is represented by a circle. Continuing current strategies would erode the business position as indicated by a square, but committing increased resources should improve the position (star).

BUSINESS STRENGTH	HIGH	MEDIUM	LOW
HIGH	INVESTMENT AND GROWTH S1	SELECTIVE GROWTH	SELECTIVITY
MEDIUM	SELECTIVE GROWTH S3	SELECTIVITY S2	HARVEST/DIVEST S4
LOW	SELECTIVITY	HARVEST/DIVEST	HARVEST/DIVEST

INDUSTRY ATTRACTIVNESS

FIG .2 McKINSEY MATRICE

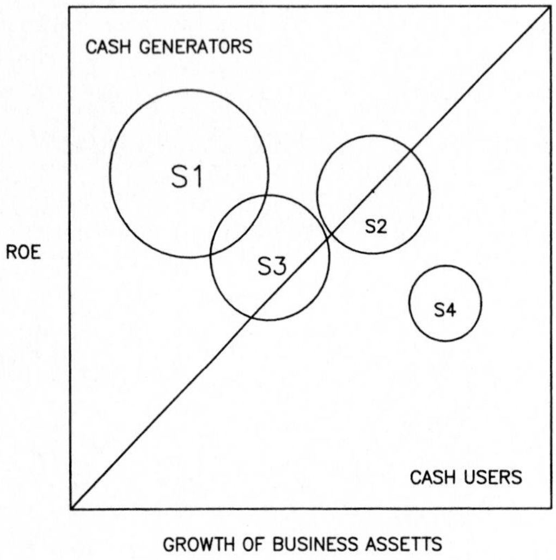

FIG. 3 MARAKON MATRIX

2.3. Exploratory Development (Project Planning I)

Hall recognizes five interrelated functions in this phase: problem definition, selecting objectives, system synthesis (alternatives formulation), system analysis, system choice (selecting the best alternative). The System Challenge Analysis can have a great impact on the system analysis.

As an example, let us choose between the two following system alternatives for a large digital mixed switching exchange

1. alternative - distributed switching network
   - centralized control

2. alternative - distributed switching network
   - distributed control

The 1.st alternative is characterized by the maximum intermediate capacities of the switching network (SI) and of the control (CI), and by the maximum final capacities (SF and CF). The 2.nd alternative is characterized only by SF and CF.

Our interest is firstly put on SF,CF and CI.

Given the costs trends of the switching network and of the two control architectures, and the demand in number of subscribers and in number of circuits, the minimum cost provisioning sequence is calculated for the two alternatives. The two sequences are compared as shown in fig.4a. The fig. 4a can be transformed in the fig. 4b taking into account the variations of operation and maintenance costs, of the costs due to the progress in component technology and in the company production technology, and the demand evolution.

The procedure also gives an answer on the optimal maximum capacities.

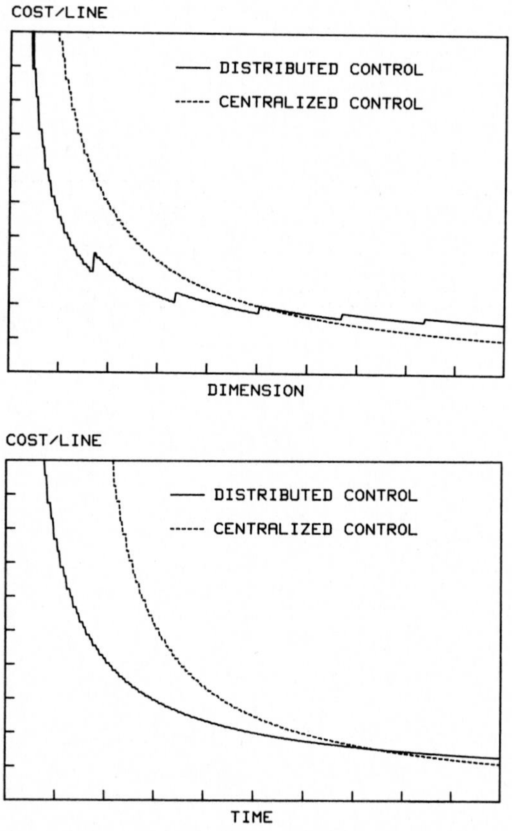

FIG.4 EXPLORATORY DEVELOPMENT AND SYSTEM DEVELOPMENT CHOICES

2.4 System Development (Project Planning II)

"This phase begins only after a decision has been taken that a development project will be undertaken. Operationally this phase is merely recycling of the previous phase, except that all steps are performed in much greater detail" and on lower level alternatives.

Coming back to the example of § 2.3 the attention is now on the maximum intermediate capacities characterizing the distributed switching network.

2.5 Business Plan

In this phase the emphasis is firstly put on the definition of residual alternatives. These alternatives are analysed versus the impact on the optimal company response to the demand using the existing system portfolio: typical results are the Growth Share Matrices shown in fig. 5.

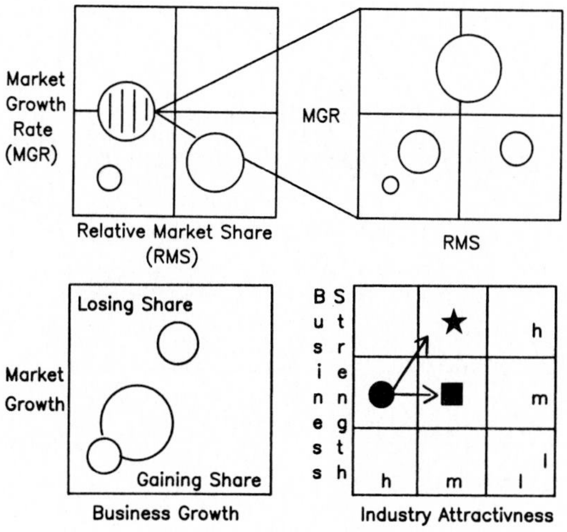

FIG.5 GROWTH SHARE MATRICES

In the fig. 5a the position of a business is shown versus the relative market share enjoyed by the business and the % of growth of the market in the most recent years.

In the fig. 5b the position of a business is analysed in terms of their components.

The fig. 5c shows the position of each business versus the total market growth and the growth rate of the business whith reference to a significant period.

Below the diagonal line are business that have increased sales at a rate higher than their markets, i.e. they have increased their market share. Falling above the diagonal are business that decreased their share of market (Lewis [11]).

In a second step the emphasis is put on the long range requirements of the operating company.

Typical results are the Dynamic Threshold Maps showing (for example) the best sequence of investment typologies (growing/overlay/replacements) which produce an optimal modernization of the network (fig. 6).

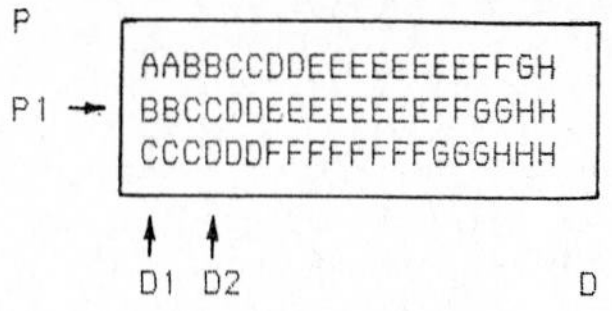

FIG. 6 EVOLUTION STRATEGIES VS. DEMAND DIMENSION D AND THE COST TRADEOFF P

The Dynamic Threshold Maps show that if the trade off between the costs of existing systems and the new system family is P1, the total demand will be satisfied up to the dimension D1 using the strategy B, up to the time dimension D2 using the strategy B etc... An example of strategies could be the distribution of nodes in the following classes of growth: using the exisisting machines, or overlaying the existing machines with the new ones, or replacing overlayed existing machines, or finally immediately replacing existing machines.

3. PROCEDURES, MODELS AND ALGORITHMS

3.1 Procedures

The most important features of the optimisation network model, built up to support the Challenge Analysis studies, are:

- the dynamical integration between the trunking and the switching planning
- the extension to the multifunctional nodes

The models inputs are organized on specialized data bases, one for each data cathegory, i.e.

- the Network Data Base (NDB), containing all the initial configurations of switching nodes and of transmission links
- the Demand Data Base (DDB), containing the trends of the demand matrices, and the modernisation strategies of network that can be chosen by the Operating Company
- the System Scenario Data Base (SDB), containing the technico-economical characteristics of each considered switching and transmission system (capacities, modularities, GOS constraints; cost for each function, maintenance and operational costs, production and operational economical lifes, compatibilities with each other system, etc..)

The Procedures access these Data Bases through a processing procedure (PREP) which controls the coherence between the data and computes all the global variables (inflation effects on cost profiles, GOS constraints, modernisation strategy constraints, etc).

The computational body of the model consists of two procedures

SERENAT which performs the trunking planning

CLEMO which performs the switching planning

The dynamical integration is modeled through a set of rules (IR) concerning reciprocal constraints, reciprocal synergies, priorities, etc.

The Models Outputs are organized on two Data Bases:

- the Exchange Evolution Data Base (EDB) contains the evolution of the switching nodes as the sequence of installation and maintenance actions that satisfies the demand arising from the augmentation of required trunk network. The number and the reusability of the reallocated lines are also recorded for each substituted machine.

- the Link Evolution Data Base (LDB) contains the evolution of each transmission link as the sequence of carriers and of transmission systems to be installed in order to satisfy the trunks demand, arising from the augmentation of required switching capacities.

A Challenge Estimation Procedure (CEP) uses these two data bases

1) to synthetize the data resulting from various simulations, identifying the critical factors in the different scenarios (the multivariate morphological analysis is used).

2) to project the synthetized data from the chosen sample of switching nodes and transmission links to the whole population (the whole network).

A Forecasting Procedure (FORP) to supply the input data bases is under development.

The Procedures Flow is shown in fig.7.

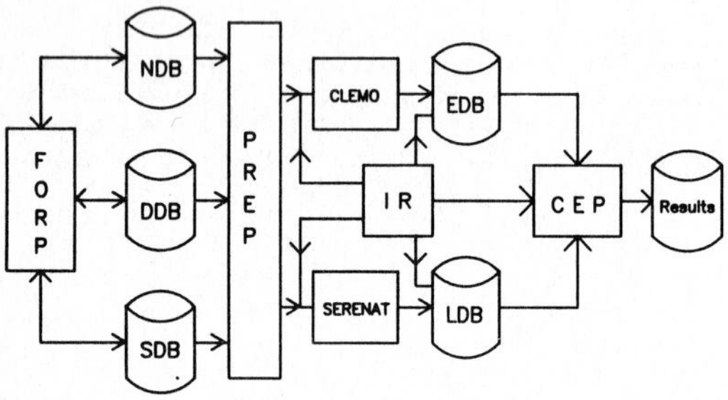

FIG.7 PROCEDURES FLOW

3.2 Models and Algorithms

Given

- the inital state of the switching node
- the demand trend
- the technico-economical system scenario

and the constraints

- maximum budget
- production economical life of each system
- modernisation rules

the CLEMO procedure founds the optimal evolution of each switching node as the minimum cost sequence of installation, maintenance and operation actions needed to satisfy the demand growth during a planning period.

The CLEMO procedure is based on a Branch and Bound Algorithm (CEBB) with the following decision tree:

root	:=	initial state of switching node
leaves	:=	the possible final states of the switching node that will satisfy completely the demand growth
intermediate nodes	:=	all the feasible states in the switching node evolution, defined step-by-step by the alternatives bounded by the preceeding choices.

4. RESULTS TYPOLOGIES

The applications mentioned in § 2 have been and will be achieved by the Central Research Laboratories of ITALTEL to support the life cycle phases of the UT LINE, in particular for the system UT 100, the maximum capacity system of the family with local and transit functions (fig.8,[13]).

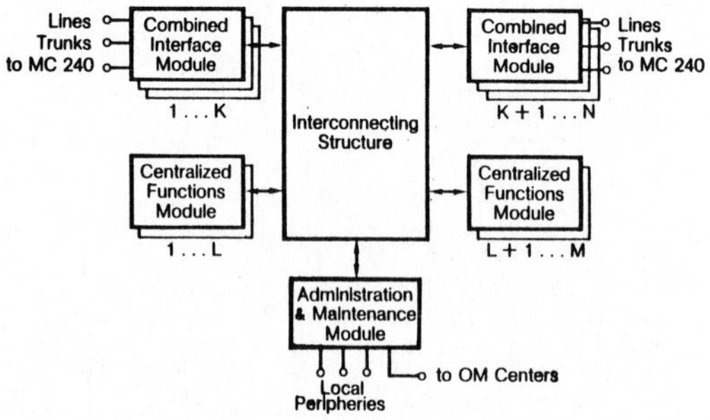

FIG.8 UT SYSTEM ARCHITECTURE

4.1 System Development Phase

In this phase the System Challenge Analysis can offer guidelines regarding

- the impact of system choices on the fortune of the system in the phase of its introduction and of its development (choice significance)
- the discrimination between the alternatives considered.

To assess the choice significance, the effectualness of the optimisation on a component is evaluated taking into account the contribution of this component to the growth of the system (dimensioning rules) and the characteristic of the considered market. The effectualness is measured as the percent decrease of the mean cost per line versus a percent decrease of the component cost.

To compare system alternatives at the subsystem level (see §2.2), each alternative is evaluated versus the trade-off between its cost and its performability (measured by the percentage of the lines which can be installed, satisfying a multipoint service level objective [14]).

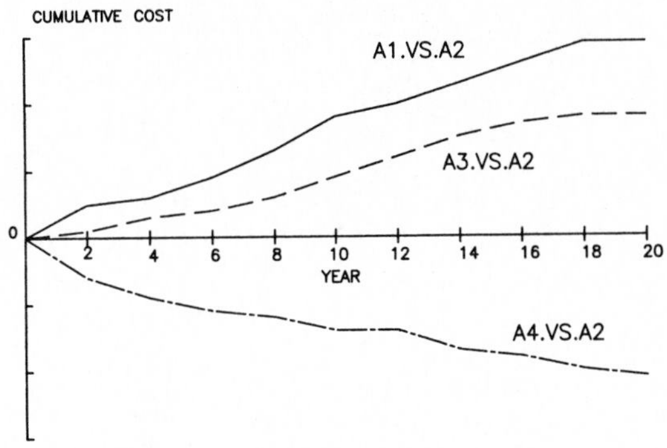

FIG.9 TOTAL INVESTMENT VOLUMES

This trade-off is solved only evaluating the total volume of investments that each alternative needs to satisfy the time dynamics of the considered market.

The fig. 9 shows the compairison among the total volume of needed investments over a 20 years period for the different alternatives regarding the concentration between the incoming lines and the links to the internal switching network (A1 being the alternative with the highest concentration factor and A4 the one with the lowest).

4.2 Operational Life

The most profitable introduction and development modalities of a new digital switching system family into a network depends on a set of factors:

- trade-off between the costs of the new family and the costs of the existing ones (provisioning, operating and maintaining)
- the present configuration of the switching nodes (number, type, capacities of the existing switching machines)
- the demand growth trend
- the planned transmission network modernization (PCM penetration).

The fig.10 shows the Dynamical Threshold Maps for a switching node given its present capacity and an equivalent yearly demand growth rate (in the pair (t1,t2), t1 indicates the introduction year of a new machine and t2 the suppression of the old machine). The effects of the cost trade-off between the new family and the existing electromechanical switching machines are shown (fig. 11a.vs.fig.11c). The transit function is very important for the large capacity switching nodes considered in these applications. In these cases, the number of installed circuits is very high. This characteristic has a great impact on the evolution of the node: the digitalisation process is accelerated (fig.11a.vs.fig.11b).

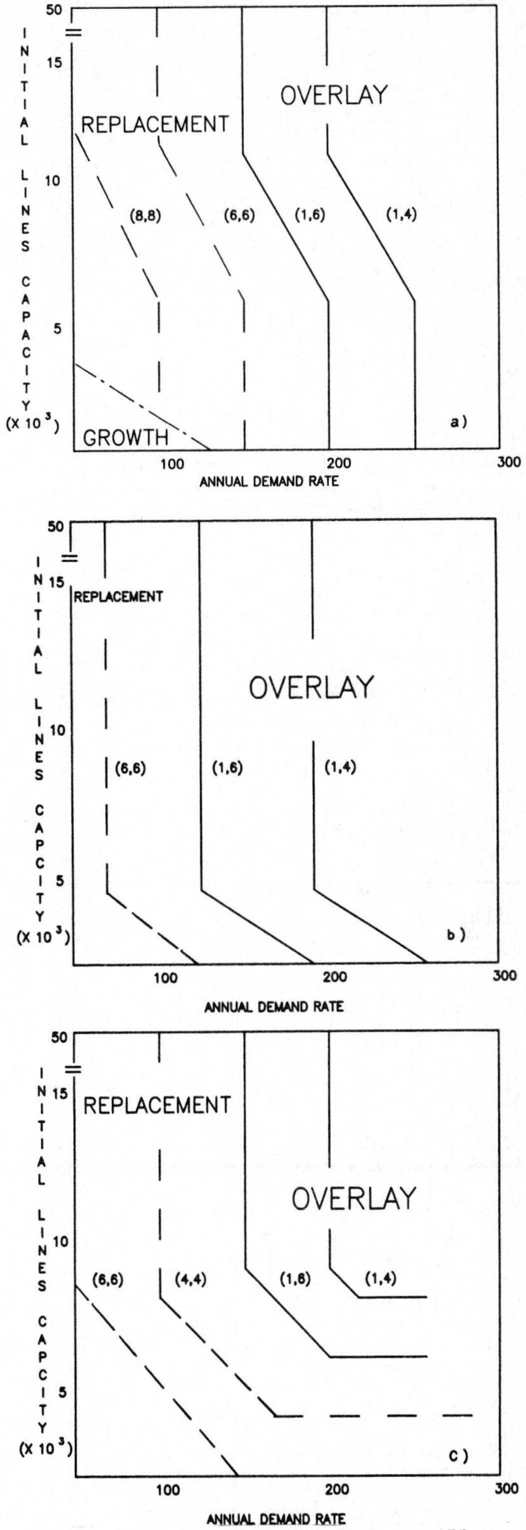

FIG.10 NODE DYNAMICAL THRESHOLD MAPS

4.3 Market Analysis

The time dynamics of the market share of the large capacity mixed exchanges can be characterized by

G : quota of lines growth satisfied by augmenting old machines
O : quota of lines growth satisfied by overlaying new machines to the existing ones
R1: quota of lines growth satisfied by replacing the existing machines with the new ones
R2: quota of lines substituted by new machines.

A quartet of values (G, O, R1, R2) defines a market evolution modalilty during one or more years.

Using the single node Dynamical Threshold Maps and the Demand Growth Scenario, the Market Dynamical Threshold Map of fig. 11 can be defined.

In the single node Maps we never have a coexistence of the four elementary evolution alternatives.

In the Market Maps a coexistence is evident. The difference is due to the presence of high traffic nodes in the Market Demand and of special implementations (containers) in the Existing Network.

A more synthetic and effective presentation is done by the map of fig. 6 where A, B, C, etc stand for specific mixes of evaluation modalities, the rows corresponding to various cost trade-offs and the columns corresponding to the market dimension.

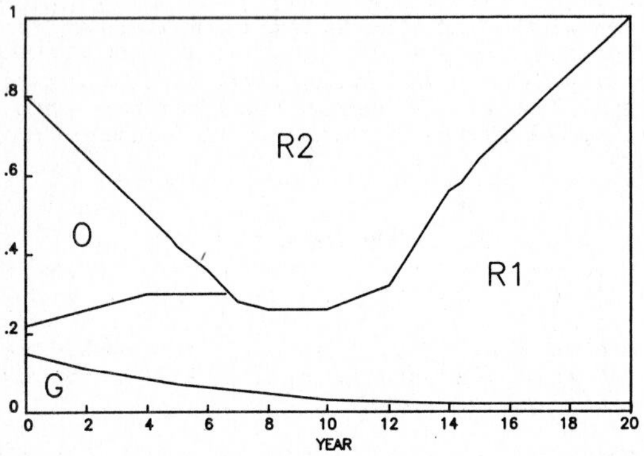

FIG.11 MARKET DYNAMICAL THRESHOLD MAP

REFERENCES

[1] H.I. Ansoff, "Strategic Management", MacMillan Press Ltd, London, 1979

[2] A.D. Hall, "A Methodology For System Engineering", Van Nostrand Co, Princeton, 1962

[3] V. Navarro, P. Santan, C. Tirado, "Economic evaluation of parameters that influence the configuration of mixed analog/digital long distance networks", Proc. Networks'83, Brighton, 1983

[4] H. Okazaki, H. Miyamoto, "A sensitivity study on network planning", Proc. 10th Int. Teletraffic Congress, Montreal, 1983

[5] V.P. Gupta, "Resources constraints modernization planning (RECOMPT)", Proc. Links For The Future, Amsterdam, 1984

[6] K.R. Crooks, "Programming the digital modernization of a national local exchange network", Proc. Networks '83, Brighton, 1983

[7] H.H. Hoang, H.T. Lav, "Otptimizing the evolution of digital switching in a local telephone network", Proc. Links For The Future" Amsterdam, 1984

[8] G. Gallassi, M. Bonatti, "A state of the art review on telecommunications systems strategical planning", Internal Memo ITALTEL/DCRS/PRS/0204/Feb. 85

[9] M. Bonatti, G. Gallassi, P. Camerini, L. Fratta, F. Maffioli, "A dynamic planning method for telecommunication networks and its performance evaluation for district trunk network", Proc. 10th Int. Teletraffic Congress, Montreal, 1983

[10] A.C. Hax, N.S. Majluf, "The use of the industry attractivness-business strength matrix in strategic planning", Interface, Vol. 13, no. 3, pp. 54-71, April 1983

[11] A.C. Hax, N.S. Majluf, "The use of the growth-share matrix in strategic planning", Interface, Vol. 13, no. 1, Feb. 1983

[12] W.W. Lewis, "Planning By Exception", Strategic Planning Associates, Washington, 1977

[13] A. Bovo, A. Bellman, "UT100/60 An electronic digital family of exchanges for large capacity applications", Proc. ISS '84, Florence, May 1984

[14] M. Bonatti, C. Barbuio, G. Cappellini, U. Padulosi, "Models for an effective definition of end-to-end GOS parameters and for their repartition in a IDN network", presented to 11th Int. Teletraffic Congress, Kyoto, 1985

TELETRAFFIC ISSUES in an Advanced Information Society
ITC-11
Minoru Akiyama (Editor)
Elsevier Science Publishers B.V. (North-Holland)

MODULAR DESIGN OF A LARGE METROPOLITAN TELEPHONE NETWORK: A CASE STUDY

Leslie BERRY* and Richard HARRIS**

The University of Adelaide* and Telecom Australia**
Adelaide and Melbourne, Australia

ABSTRACT

We describe the application of an optimisation procedure to a large evolving metropolitan telephone network. This network, proposed for Melbourne in 1990, consists of an analogue component together with a growing digital component superimposed upon it.

The method allows for modular allocation of circuits, nonlinear circuit cost functions and prescribed bounds for origin to destination (OD) grades of service. This work extends the model in [1] and a number of new improvements have been incorporated into the design procedure to assist the network planner.

Results of the application are discussed and are compared with an alternative (more conventional) design approach. This comparison reveals that there are significant cost savings to be made using the new method.

1. INTRODUCTION

At the previous ITC, an interactive computer based system for designing minimum cost circuit switched alternative routing networks was presented [1]. The approach was to solve a nonlinear integer program which gave optimal modular circuit allocations for each link of the network whilst guaranteeing specified upper bounds on the OD grades of service were not violated. Results were given for the application of the system to the metropolitan Adelaide network. This network had 37 nodes, 4 tandem switching centres and 1141 OD pairs. A cost reduction of 35% from a reasonable starting point was achieved.

Whereas the Adelaide network allowed for at most 3 alternative routes, each within 3 fixed routing schemes, the present study considers a very much larger network with up to 15 alternative routes (chains) for each OD pair, and an unrestricted number of routing patterns. For the Melbourne network, considered in this study, the total number of chains increased by a factor of approximately 150.

A number of special features have been included in the extended optimisation system, such as:

(1) A data transformation algorithm which allows the user to input data in a simple format related to a local view of the individual links of the network. This algorithm generates the complete connected graph and permitted chains for an hierarchical network.

(2) A network transformation algorithm which permits complex hierarchical routing structures to be dimensioned by conversion of the network into an equivalent "flow decomposable" form.

(3) An algorithm to determine the optimal cost and mix of transmission media to be used on a link.

(4) An interactive mode which enables planners to make or assess potential changes to circuit allocations on links of the network.

The purpose of the study was to develop a practical dimensioning model for large digital networks incorporating circuit modularities and mixtures of transmission types; and to compare such a model with conventional design systems such as the SWITCHNET system [4] as used in Telecom Australia.

2. NETWORK DESIGN AND OPTIMISATION MODEL

The model used for this study represents an extension to the work presented in [1]. In this section, we shall briefly restate this model, for the sake of completeness, and introduce the new extensions and features.

We consider a full availability network consisting of E exchanges and L directed links connecting these exchanges. All exchanges are divided into nodes which represent the switching functions occurring at that exchange site, and each node performs one of the following three functions: (1)origin for fresh traffic, (2)destination for terminating traffic and, (3)transit (tandem) switching function. From this set of nodes, we determine a set of N origin-destination (OD) pairs and create a traffic demand matrix with elements t^k (k=1,..N) representing the offered traffic for OD pair k; the traffic t^k is assumed to be pure chance. Traffic between each OD pair k may use up to m(k) different routes (chains) and must satisfy specified OD grade of service objectives. The j-th chain for an OD pair k is represented by the symbol R_j^k. An extension to the model of [1] is to allow the model to select the most cost effective transmission media (or a mixture of media types) on a link. Each link i may have up to T(i) different transmission media types available to carry the traffic. The circuits on these links are grouped into their various media types and, corresponding to each type, we specify the permitted circuit module sizes. Details of this new feature and the circuit allocation procedure are described in subsection 5.1 of the paper.

In mathematical terms, the model is a nonlinear integer programming problem and can be formally described as follows:

$$\text{Minimise: } C(\mathbf{n}) = \sum_{i=1}^{L} c_i(n_i)$$

where $C(\mathbf{n})$ = total cost of links in the network

$c_i(n_i)$ = total cost of link i

$$= \sum_{r=1}^{T(i)} c_{ir} n_{ir} \qquad (1)$$

c_{ir} = cost per circuit of link i with transmission media type r;

n_{ir} = number of circuits on link i with transmission media type r;

$n_i = \sum_{r=1}^{T(i)} n_{ir}$ is the total number of circuits on link i.

Subject to:

$$(1-B^k)t^k \leq \sum_{j=1}^{m(k)} h_j^k < t^k \quad \text{for } k=1,\dots,N \qquad (2)$$

$$f_i = \sum_j \sum_k a_{ij}^k h_j^k \quad \text{for } i=1,\dots,L \qquad (3)$$

$$h_j^k \geq 0 \quad \text{for } j=1,\dots,m(k) \text{ and } k=1,\dots,N \qquad (4)$$

$$n_{ir} \geq 0 \quad \text{for } i=1,\dots,L \text{ and } r=1,\dots,T(i) \qquad (5)$$

Integer multiples of 1,15 or 30 according to r.

where B^k = prescribed maximum end to end loss for OD pair k;

h_j^k = traffic carried on chain R_j^k (chain flow represented as $\mathbf{h}$ in vector form);

f_i = total traffic carried on link i;

$$a_{ij}^k = \begin{cases} 1 & \text{if link i is on chain } R_j^k. \\ 0 & \text{otherwise.} \end{cases}$$

In the original continuous model, described in [2], the chain flow vector $\mathbf{h}$ was updated in the optimising process and the number of circuits required on a link was determined from the offered and carried traffic moments. In the present model, the independent variables are the numbers of circuits on "flow assignment" links (c.f. subsection 3.3). Since the network is hierarchical, the process of computing the chain flows can be performed in a single pass through the network, starting with the direct routes and moving through the successive choices until the final choice chains are reached. At this point, circuit requirements are computed in order to satisfy the end to end congestion constraints. At each stage of this dimensioning process, the mean and variance of the offered traffic to link i are computed as:

$$M_i = \sum_j \sum_k a_{ij}^k M_j^k \quad \text{(Mean offered traffic)} \qquad (6)$$

$$V_i = \sum_j \sum_k a_{ij}^k V_j^k \quad \text{(Variance of offered traffic)} \qquad (7)$$

where $M_1^k = t^k$

$$M_{j+1}^k = t^k - \sum_{s=1}^{j} h_s^k \quad \text{for } j=1,\dots,m(k)$$

$$V_{j+1}^k = \frac{M_{j+1}^k}{6}\left\{3-M_{j+1}^k+\sqrt{(3-M_{j+1}^k)^2+12t^k\left\{\frac{M_{j+1}^k}{t^k}\right\}^{.1}}\right\} \qquad (8)$$

The variance of the lost traffic from link i is:

$$v_i = \frac{(M_i-f_i)}{6}\left\{3-(M_i-f_i)+\sqrt{(3-M_i+f_i)^2+12A_i\left\{\frac{1-f_i}{M_i}\right\}^{0.1}}\right\} \qquad (9)$$

where $A_i = V_i + 3\frac{V_i}{M_i}(\frac{V_i}{M_i} - 1)$ is the well-known Rapp formula for the equivalent random traffic. Circuits on the final choice links are computed using Berry's formula:

$$n_i = f_i + A_i\left\{\frac{(M_i - f_i)}{M_i-f_i-1)(M_i-f_i)+v_i} - \frac{M_i}{M_i^2 - M_i + V_i}\right\} \qquad (10)$$

As indicated above, the hierarchical nature of the network enables it to be dimensioned in a single pass. This is achieved by ordering the links in a special sequence, referred to as the "dimensioning sequence", and then, provided the network has been transformed into "flow decomposable" form (c.f. subsection 3.3 for details) the individual chain flows may be determined by applying Wallström's splitting formula to the dimensioning sequence of "flow assignment links" (c.f. subsection 3.3).

3. TRANSFORMATION OF THE NETWORK

3.1 Network description.

The Melbourne network consists of 162 exchanges which can be subdivided into 889 switching nodes. The functional breakup of these nodes is: 280 originating traffic nodes, 326 transit nodes and 283 terminating traffic nodes. This results in 67,517 OD pairs with significant traffic levels between them; of these, approximately 1800 have direct routes linking them; the remaining OD pairs must utilise the overflow routes in order to terminate their traffic demands. There is a large number of routing schemes available, and particular schemes are specified by the input data for the various OD pairs, according to the switching technology available to interconnect them. The maximum number of chains for any OD pair is 15, and since there are approximately

456,000 chains, the average number of choices for an OD pair is 7 chains.

The overflow network consists of 18,073 links and a maximum of two different transmission types are possible for a given link. Voice frequency (VF) analogue circuits may be utilised singly, whilst Pulse Code Modulation (PCM) circuits are typically used in multiples of 15 or 30 circuits. Each link of the network is classified into one of 172 different types for the purposes of determining link costs. Associated with these 172 types, there are 186 different cost curves which specify a media type, termination and repeater equipment, and distance related cost information for the transmission media. The list of media types available for any specific link determines which cost curves will be used in this calculation.

3.2 Conversion to chain format

The size of the Melbourne network in 1990 precluded the possibility of enumerating the 456,000 chains manually! Furthermore, the large number of possible overflow patterns also precluded individually programming these patterns for automatic generation. Thus, a procedure was required to develop an arbitrary network structure from a simplified input specification. The procedure developed was a composite of the process utilised in the SWITCHNET network design system and an algorithm specially written by the authors to produce the required network specification. The data required for the procedure consists of three items for each OD pair and overflow link in the network, viz: (i)the originating or transit node name; (ii)the transit or terminating node name; (iii)the name of the node to receive overflow traffic from this link. (If the link has no overflow, then this name will be left blank.)

This data is clearly much simpler to provide and, in the present instance, it required 85,600 sets of information to be prepared. The first stage in the processing of this data is to number the nodes, commencing with the originating traffic nodes and progressing to the transit nodes and finally numbering the destination nodes. The numbering must be performed in such a way that links of the network **always** join a low numbered node to a higher numbered node, with traffic also flowing in that direction. The purpose of this rule is to prevent loops in the chain generation phase of the process. If the input data cannot satisfy this condition, the input data is rejected and the user must create "dummy nodes" in such a way that this rule can be satisfied. The second stage involves sorting the links into a list for which the origin and transit exchange numbers are sorted into descending order and the destination nodes of the link are sorted with a secondary sort into ascending order. Link numbers are then allocated to all OD pairs starting with 1 and ending with N (even if, physically, the actual direct route does not exist or is not permitted to have circuits). The remaining overflow links are then numbered from N+1 to L, commencing with the lowest origin and ending with the last transit exchange.

The final stage of the algorithm involves joining together of the links to form the legal chains available to each OD pair. This is a complex process since the algorithm produces many invalid chains as well as valid ones. To overcome the validity problem, the overflow (third parameter) is used to determine which chains are correct. The general procedure can be illustrated by the following simple two destination network.

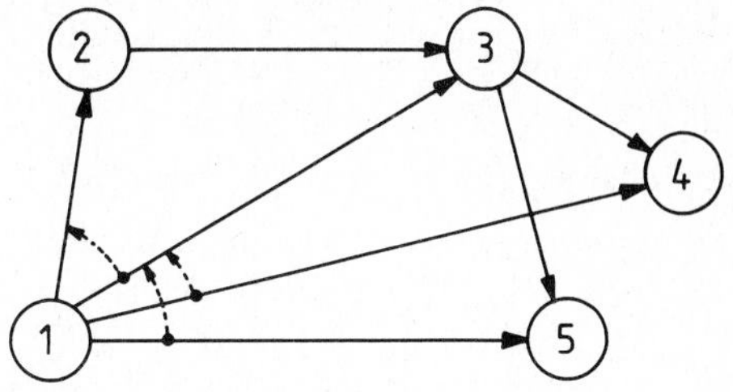

Fig. 1: Simple two destination network.

The procedure outlined above results in the following link list:

Row	Nodes
1	3 4
2	3 5
3	2 3
4	1 2
5	1 3
6	1 4
7	1 5

The list is now expanded by starting at the top and copying each row containing the destination node, from earlier rows which have that destination node as the first node in the row progressively, until the end of the list is reached; e.g. rows 1 and 2 remain unchanged, but the destination of row 3 is node "3" and hence, rows 1 and 2 which have origin node "3" will each be copied to produce 2 new rows (3a and 3b) respectively as follows:

3a	2 3 4
3b	2 3 5

Having completed this expansion we continue to row 4 and repeat the procedure. The final expanded list is:

Row	Nodes
1	3 4
2	3 5
3a	2 3 4
3b	2 3 5
4a	1 2 3 4
4b	1 2 3 5
5a	1 3 4
5b	1 3 5
6	1 4
7	1 5

From this list, the chains for OD pair "1-4" are located in order by searching the list from bottom to top, and matching the origin "1" with the destination "4". The resulting ordered list of chains is given as:

Chain 1: {1, 4}
Chain 2: {1, 3, 4}
Chain 3: {1, 2, 3, 4}

For a network the size of Melbourne, this process takes nearly two hours of CPU time on a Honeywell DPS8 computer and requires a sizeable amount of backing storage. The end result is a list of nodes which specify the individual chains joining all OD pairs of the network, these lists are then converted into link numbers. This completes the conversion of the input data into the network specification required by the model.

3.3 Transformation into "flow decomposable form".

In order to generalise the methods used in [1] so that they may be applied to general hierarchical routing networks, we introduce the concept of a "flow decomposable" network. Given a circuit allocation vector **n**, the vector of chain flows **h** may be estimated from a subset of the circuit allocation vector called the partial circuit vector $\mathbf{n}_p$, provided the network satisfies certain conditions. We call the class of networks that meet these conditions "flow decomposable".

Definition 1: When one or more streams are offered to a single link, the carried traffics on that link are called "single link flows".

In general, the flow on a link forming part of a chain depends on the number of circuits on preceding and succeeding links on that chain. For example, the flows on link IY in Fig.2 depend on the number of circuits on links YJ_1 and YJ_2. However, provided that the final choice routes $IXYJ_1$ and $IXYJ_2$ are dimensioned to give a sufficiently low OD blocking probability, the flows on chains IYJ_1, IYJ_2 can be accurately estimated by determining the single link flows for link IY. With this assumption, from the partial circuit vector (n_1,n_2,n_3) it is easy to compute the chain flows h_1^1, h_1^2, h_2^1, h_2^2 and then the flows on the final routes h_3^1, h_3^2, are determined from OD grade of service values. Then, from the chain flow vector **h**, the chain flow model [2] is used to compute the complete circuit vector **n**.

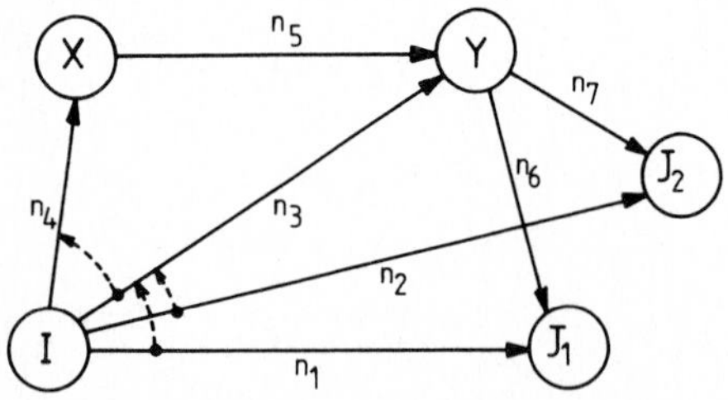

Fig. 2: Single origin - two destination network

It should be noted that the assumption of a single link flow pattern on link IY is restrictive in the sense of limiting the class of feasible chain flows considered during the optimisation.

Definition 2: A link of a network is said to be "chain disjoint" iff for each OD pair using the link, it belongs to only one of the set of permissible chains for that OD pair. That is, for every k and $j_1 \neq j_2$, $i \in R_{j_1}^k \Rightarrow i \notin R_{j_2}^k$.

Definition 3: A link is said to be a "flow assignment link" iff it satisfies the following:
(i) it is chain disjoint;
(ii) the link is carries single link flows.

Definition 4: A network is said to be "flow decomposable" iff for each OD pair, every permissible chain, except the last choice, has exactly ONE flow assignment link.

Fig. 2 is an example of a flow decomposable network, the flow assignment links being IJ_1 and IJ_2 (the direct links) and IY. A further example is given in Fig. 3, the flow assignment links being IJ, IY, XJ, XY, and ZJ.

A simple non flow-decomposable network is illustrated in Fig. 4. For example, there does not exist a chain disjoint link on R_3 as $I\text{-}T_1$ belongs to both R_2 and R_3 and the link $T_1\text{-}J$ belongs to both R_2 and R_4.

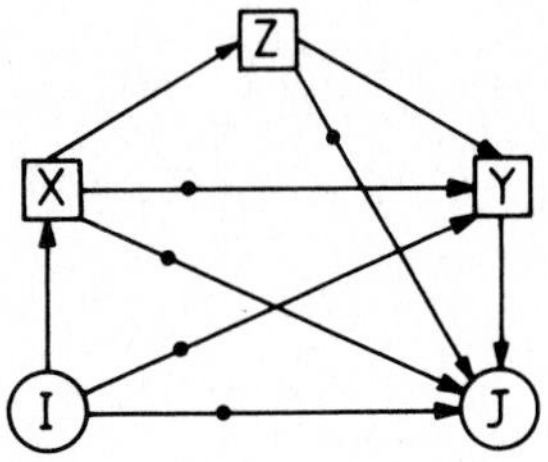

Chains

R_1: I-J
R_2: I-Y-J
R_3: I-X-J
R_4: I-X-Y-J
R_5: I-X-Z-J
R_6: I-X-Z-Y-J

Fig. 3 A flow decomposable network

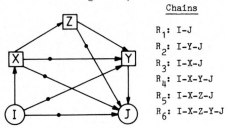

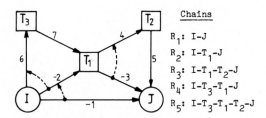

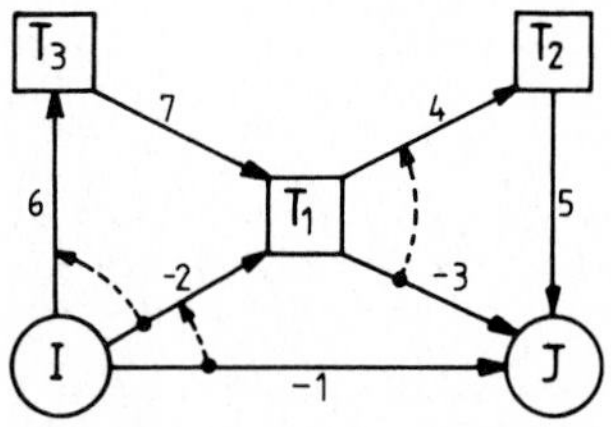

Chains

R_1: I-J
R_2: $I\text{-}T_1\text{-}J$
R_3: $I\text{-}T_1\text{-}T_2\text{-}J$
R_4: $I\text{-}T_3\text{-}T_1\text{-}J$
R_5: $I\text{-}T_3\text{-}T_1\text{-}T_2\text{-}J$

Fig. 4: A non flow-decomposable network

To transform the above network into an equivalent flow decomposable network, we introduce **dummy flow assignment links** labelled "-8, -9 and -10" as shown in Fig. 5. A negative signed link number indicates that the link is not on a final choice chain for any OD pair. The dummy links are introduced to the network at no cost. Direct route -1 and dummy links -8, -9 and -10 control the chain flows on R_1,R_2, R_3 and R_4 respectively.

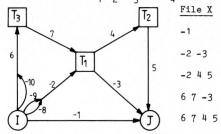

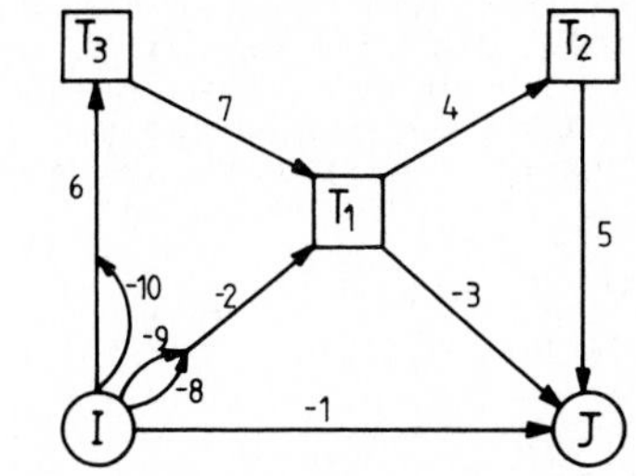

File X

-1

-2 -3

-2 4 5

6 7 -3

6 7 4 5

Fig. 5: An equivalent flow decomposable network

The following algorithm allocates dummy flow assignment links to chains where they are needed, for arbitrary multiple OD networks. The rows of file X contain ordered chain lists for each OD pair (e.g. see Fig. 5) and it is assumed that the links of the network are labelled consecutively from 1 to L (the number of links in the network) together with their appropriate sign. The elements of the file FAL are the flow assignment links for each high usage chain.

Transformation Algorithm

1. Initialise each row of the FAL file to zero.
2. Set NL ← L + 1.
3. Declare each row of the file X unscanned.
4. Is there an unscanned row of file X?
   Yes: Go to step 5.
   No: Change the sign of numbers in FAL and stop.
5. Are all of the elements of the row positive?
   Yes: Declare the row scanned and go to 4.
   No: Go to step 6.
6. Is there a negative number in the selected unscanned row which, for every OD pair, occurs in ONE chain only?
   Yes: Choose the first such negative signed link, declare that link a flow assignment link and write for all rows of file X containing the link, the absolute value of the link number into the corresponding rows of FAL. Declare each row of X containing that flow assignment link scanned. Go to 4.
   No: Is the entry in the corresponding row of FAL positive?
   Yes: Declare the row scanned. Go to 4.
   No: Enter NL in the corresponding row of FAL. Set NEG equal to the first negative number in that row of X and declare the row scanned. For each unscanned row of the same OD pair, whenever both the link number NEG occurs in a row of X, and the corresponding row entry of FAL is zero, enter values in FAL of NL+1,NL+2,... respectively for each such row. For each unscanned row of different OD pairs, whenever both the link number NEG occurs in a row of X, and the corresponding row entry of FAL is zero, enter in FAL the values NL,NL+1,NL+2,... respectively for each such row. (These rows of X contain potential chain disjoint links, and the numbers NL,NL+1,... are dummy flow flow assignment link numbers.)
7. Set NL equal to the next largest unused integer. Go to 4.

The above algorithm is applied to the network once only. The practical implementation was performed efficiently using simple expansion and sorting sequences.

4. DESIGN PROCEDURE OVERVIEW

The network transformation and design procedures have been programmed in FORTRAN for a Honeywell DPS8 computer and currently amount to nearly 8000 lines of documented code.

The system can be thought of in four parts:

(1) Network Transformation.
In this phase, the procedures outlined in subsections 3.2 and 3.3 are applied to the network to develop the complete network specification and the dimensioning sequence for the links.

(2) Initialisation
In this phase, data files are set up which contain information about OD pairs and links in the network. For OD pairs, information is held concerning the offered traffic, current chain flows, overflow chain variances, direct route circuits, cost curve data for the direct routes, and origin and destination node numbers. For links, information is held concerning the mean and variance of the offered traffic and the total carried traffic, cost data and numbers of circuits on the routes, exchange names and overflow node numbers.

(3) Design and Optimisation
This phase may be performed either by automatic optimisation strategies (c.f. Section 5), or by specific changes made through the use of an interactive program which enables a planner to interrogate the network data base and compile a list of alterations based on his knowledge of the network. The lists compiled either by the planner in the interactive phase or by automatic generation, are passed to a main batch processing program which carries out the alterations to the data base. When this has been completed, the program determines the new network cost.

(4) Report
Finally, at the completion of these phases, or a specific optimisation strategy, it is possible to produce full or partial reports for use by technical or non-technical staff for further study or implementation.

5. OPTIMISATION PROCEDURES

5.1 Optimal Transmission Bearer Allocation

Two types of transmission bearers were considered within the network. These were PCM circuits installed in modules of size 1, 15 or 30, and VF circuits installed in modules of unit size. The general forms of the 185 different cost curves are illustrated in Fig. 6 which shows the the cost per circuit as functions of distance between exchanges. The discontinuities in the curves are due to changes in cable diameter and installation of repeater equipment at prescribed distances.

Suppose that a chain flow pattern **h** requires n circuits on a particular link, and that from the cost curves for that link, the cost per circuit is c_1 for a VF bearer and c_2 for a PCM bearer. Let K be the module size for PCM groups. We determine the optimal mix of VF and PCM bearers as follows:

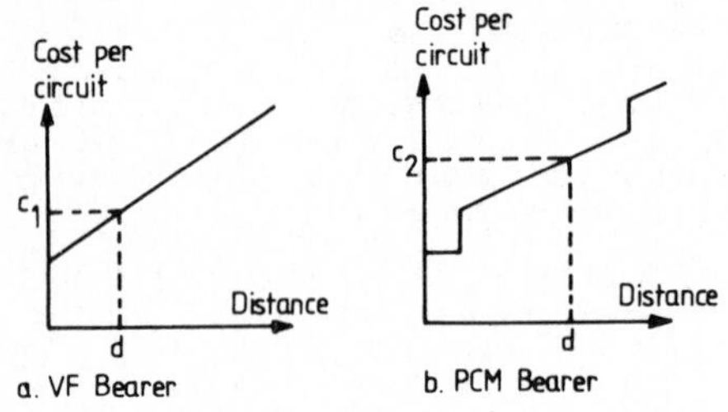

Fig. 6: Typical circuit cost curves

Let x be the number of circuits using VF and therefore, at least n-x circuits must use PCM. The problem may now be formulated:

$$\text{Min. } \{c_1 x + c_2(1+[\tfrac{n-x-1}{K}])K \text{ for } x=0,1,\ldots,n\} \quad (11)$$

where [.] denotes "the greatest integer less than or equal to ."

This optimisation problem can be simplified by defining

$m = (1 + [\frac{n-1}{K}])$, where m is the minimum number of PCM modules to provide n circuits. Then (11) is equivalent to:

$$\text{Min } \{c_2 Km \text{ , } c_2 K(m-j)+(n-(m-j)K)c_1 \text{ for } j=1,2,..m\} \quad (12)$$

A straightforward algebraic argument leads to the following solution:

For the case $c_1 \leq c_2$ allocate n VF circuits.

For the case $c_1 > c_2$, let $r=n-(m-1)K$.

Then, if $c_2 K \leq c_1 r$ then provide m PCM modules of size K and reset n=Km, otherwise provide m-1 modules of size K and r VF circuits. Note that it is necessary to reset n in order to compute correct overflow means and variances.

5.2 Heuristic Strategies

In [1], the concept of **chain critical factors** was introduced.

Definition 5: Chain critical factors, ρ^k_j, are defined by

$$\rho^k_{j+1} = \frac{M^k_{j+2}}{V^k_{j+1} + (M^k_{j+1})^2/V^k_{j+1}} \qquad j=1,..,m(k)-1 \quad (13)$$

From this concept, an heuristic optimisation strategy was developed from observations made on the relative sizes of the chain critical factors in the network. Work in the present study has involved extending this heuristic to more than three chains and to taking incremental chain costs per unit flow into account before applying the strategy.

Definition 6: The incremental chain costs per unit flow ω^k_j are defined by:

$$\omega^k_j = \sum_i a^k_{ij}\, \theta_i \quad (14)$$

where θ_i = incremental cost/flow on link i;

$$= c_i \frac{\Delta n_i}{\Delta f_i} \quad (15)$$

where c_i = cost per circuit on link i for this increment;

Δn_i = permitted change in circuits on i;

Δf_i = change in carried traffic for this change in circuits.

Note that $\theta_i=0$ if the actual number of circuits required is less than the module size which must be allocated because of transmission media constraints, i.e. the addition of a circuit would not affect the actual number of circuits required for the link.

In past studies ([1]), it was shown that the number of circuits required to carry specified minimum mean traffic stream flows was essentially determined by one stream. This stream was designated the **critical stream.** Heuristic methods used in the previous study identified streams which resulted in "wastage" of link resources and marked circuit groups for changes.

The present study network was designed for a 1% end to end congestion probability for all OD pairs. The network was designed firstly using the conventional methods outlined in [4]. The circuit values obtained by this method were then transfered to our model to be used as a starting point for the optimisation procedure. The inital cost of this network was $28,323,000.

The first step in our procedure was to determine the chain flows associated with this network and to assess the end to end grade of service achieved by the conventional model. The network grade of service was found to be better than the nominal 1% value, and hence. an algorithm was constructed to adjust the circuit values on links in order to achieve the nominated grade of service more precisely. This algorithm considered each flow assignment link (in the dimensioning sequence) and determined the maximum number of circuits required on that link for the grade of service standards and the modularity constraints to be satisfied, and reduced the circuit levels to that maximum if they exceeded this limit.

The second procedure used the critical factors to determine circuit reductions on direct routes. If the direct route had four or more circuits and the final choice chain had a high critical factor (>0.2), then, provided the incremental cost per unit flow on the direct route was not the lowest for all chains, a reduction by one circuit was made to that route. This procedure was repeated until no further improvements could be made to the network cost.

The final heuristic method used in the present study examined the chain critical factors and established a "scoring system" which identified circuit groups for changes. The scoring system was applied to flow assignment links and is conducted in the following way:

Scoring System:
Examine the chain critical factors for each OD pair using the flow assignment link and determine an OD score τ^k where:

$$\tau^k = \sum_{j=1}^{m(k)-1} \tau_j^k \qquad (16)$$

$$\tau_j^k = \begin{cases} +1 \text{ if } \rho_j^k \leq 0.1 \\ 0 \text{ if } 0.1 < \rho_j^k < 0.2 \qquad j=2,..m(k)-1 \\ -1 \text{ if } \rho_j^k \geq 0.2 \text{ and } h_{j+1}^k = 0 \end{cases} \qquad (17)$$

Having determined the individual OD pair scores, they are summed to produce a score for the flow assignment link. Generally speaking, a high positive score suggests that circuits should be increased on the flow assignment link, whilst a high negative value suggests a decrease in circuits is warranted. Since the scoring system is only concerned with traffic carrying efficiency and not with cost, it is supplemented with knowledge about the incremental chain costs before alterations are made to the circuit values on affected links.

6. RESULTS

The strategies outlined in the previous section were applied to the Melbourne Network and a 5% reduction to a new total network cost of $26,900,000 has been achieved. This does not represent an optimal solution, as further improvements can still be achieved using the existing strategies. However, note that such reductions are consistent with the improvements shown to be possible from comparisons of the conventional design with continuous forms of the Berry optimising models discussed in earlier papers, (e.g. [3]). It should also be pointed out that improvements of 35% from an initial starting point, reported in [1], have not yet been realised in the present study and this can be attributed to the fact that, in this case, the initial point selected is much closer to an optimal solution. In addition, much of the initial optimisation has concentrated on the elimination of wastage in the network, rather than fine tuning to reach a global optimum. In the continuous model, the optimality conditions are satisfied when the chains with positive flow have equal marginal chain costs per unit flow. For the integer model described in this paper, we have used the incremental chain costs per unit flow as a guide to optimality, since no other optimality conditions have been developed for this type of integer program. Modularity constraints add a further complication to this issue and it is known that such constraints create local minima. Studies are continuing on the question of suitable optimality criteria for the model.

Application of the various optimising strategies described above to the 890 node Melbourne Network generally each required 10-15 minutes of CPU time. The most significant calculation effort is the computation of the chain flow vector from the partial circuit vector $\mathbf{n_p}$. This process takes approximately 1/2 hour CPU time and 2 1/2 hours of IO processing time to complete. Studies are also being carried out to reduce this overall computation time.

7. CONCLUSIONS

In this paper, we have described our ongoing research into improved dimensioning and optimising models based upon the concept of end to end grade of service constraints. We have demonstrated that cost gains of at least 5% are possible over conventionally designed networks (employing marginal occupancy and link by link grades of service constraints). The work also demonstrates that the Berry models can successfully be applied to very large networks (up to 1000 nodes) - despite the large numbers of variables involved.

As a practical tool for our research and for planning engineers , a comprehensive interactive software package has been developed to enable interaction with the design process. Studies of new optimising strategies are planned, and it is hoped that improvements will be found which further reduce the costs of providing large and complex networks of the future.

8. ACKNOWLEDGEMENTS

The permission of the Chief General Manager, Telecom Australia to present this paper is gratefully acknowledged.

9. REFERENCES

[1] L.T.M. Berry, "Optimal Dimensioning of Circuit Switched Digital Networks", Proc. 10th ITC, Montreal, Paper 2.1.6, 1983.

[2] L.T.M. Berry, "A Mathematical Model for Optimising Telephone Networks", Ph.D. Thesis, University of Adelaide, Dec. 1971.

[3] R.J. Harris, "Comparison of Network Dimensioning Models", Proc. 10th ITC, Montreal, Paper 4.3b.3, 1983.

[4] Y.M. Chin, "SWITCHNET - A Generalised Network Dimensioning Aid", Proc. of the Second International Network Planning Symposium, NETWORKS 83, Brighton,U.K.

TELETRAFFIC ISSUES in an Advanced Information Society
ITC-11
Minoru Akiyama (Editor)
Elsevier Science Publishers B.V. (North-Holland)

ADAPTIVE CHANNEL ASSIGNMENTS IN A SMALL CELL MOBILE RADIO COMMUNICATION SYSTEM

Masakazu SENGOKU, Wako KUDO and Takeo ABE

Faculty of Engineering, Niigata University,
Ikarashi, Niigata 950-21, Japan

ABSTRACT

In a small zone system, the algorithm used to make the channel assignment for a call has a great effect on system performance. The efficiency of the channel usage in the dynamic channel assignment method at low blocking probability is superior to a fixed channel assignment method. At high blocking probability, the inequality of the efficiency is reversed. In this paper, the cause of the above results is theoretically studied. And it is shown that an adaptive channel assignment algorithm is necessary to maintain the high efficiency of the channel usage at any blocking probability. And also, it is shown that rearrangement of channels produce a more increased channel occupancy at high blocking probability than a usual dynamic channel assignment algorithm. At last, a simple adaptive channel assignment algorithm is proposed.

1. INTRODUCTION

The frequency bands in mobile radio communication systems are limited. Thus the efficient use of the frequency spectrum is one of the most important problems in the system. In a small zone (cell) system, it is possible to reuse the same channel (frequency or time slots) at the same time in the other zones, and the algorithm used to make the frequency assignment for a call has a great effect on system performance (traffic characteristics). That is, in the system, the telephone traffic-carrying capability varies with the method of channel assignment. The channel assignment methods are roughly classified into two types. One is a fixed channel assignment method and the other is a dynamic channel assignment method. It is known by computer simulations that the efficiency of the channel usage in the dynamic channel assignment method at low blocking probability is superior to a fixed channel assignment method[1]. This means that at high blocking probability, a fixed channel assignment method is superior to a dynamic channel assignment method. We have not seen clearly why the inequality of channel usage of a dynamic channel assignment algorithm and a fixed channel assignment algorithm reverses. In this paper, we study theoretically the cause of the reversion of the inequality. If the blocking probability in the reversion, namely, the reversion point, rises, the efficiency of the channel usage is improved. We show that rearrangement of channels has a great effect on rising the reversion point. And also, we show that an adaptive channel assignment algorithm is necessary to maintain the high efficiency of the channel usage at any blocking probability. Furthermore, a simple adaptive channel assignment algorithm is proposed and a computer simulation results of a mobile radio communication system using this algorithm is presented.

2. THE CHANNEL ASSIGNMENTS

2.1 The Channel Assignment Methods

In a small zone mobile radio communication system, a service area is divided into a number of small zones. The method of the division depends on the property of the service area. If a channel frequency is issued in a zone, it may not be used in some other zones because of interference of the frequency. If the buffer zones of a given zone consist of all zones closer than **n** zones away from it, the buffering system is called the **n**-belt buffering. This means that a channel reuse interval is n+1.

There are some methods of channel assignment in a mobile radio communication system. They are roughly classified two types. One is a fixed channel assignment and the other is a dynamic channel assignment. In a fixed channel assignment method, a subset of channels available to the system is permanently reserved for use within each coverage zone. In the most general form of dynamic channel assignment, any channel can be used in any coverage zone. In the others (this may be contained in the class of the dynamic channel assignment methods), there is a hybrid channel assignment method containing both fixed channels and dynamic channels. And various algorithm of dynamic channel assignments have been proposed.

2.2 The Results of Computer Simulations

We begin with an example of computer simulations. The system adopted here has 61 zones as illustrated in Fig.1. Each zone is a regular hexagon and each edge of the hexagons is 1 km in length. The traffic carried per channel is much concerned in the buffering system. In this computer simulations, 2-belt buffering system is examined. The following assumptions are made.

(1) Telephone traffic distribution is uniform over the service area.

(2) The arrival of initiations (requests for service) forms a Poisson process and the holding time is an exponential random variable with the mean value of 1.5 minutes. And the system is a

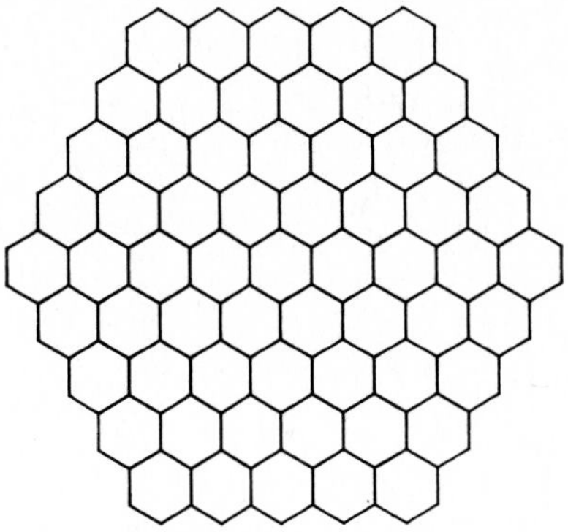

Fig.1 A service area.

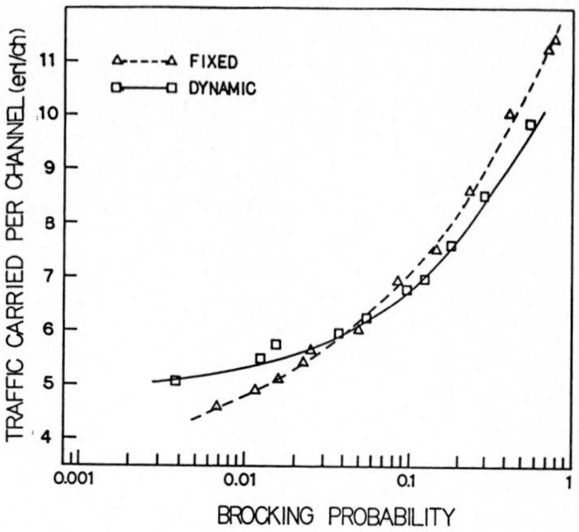

Fig.2 Simulation results.

loss system.

In general, it is not so easy to make a model of the movement of the mobile subscribers in the service area. Let us define the model by the following assumption.

(3) At the time of call attempts, they are moving to uniform direction.

(4) They move at a constant speed in a straight line in a period which has a truncated Gaussian distribution with the mean value of 1 minutes and standard deviation of 1 minute. Then, they change the speed and the direction. The speed distribution of the mobile is as follows. 0km/h is 30%, 15km/h, 30km/h, 45km/h are 20% respectively, and 60km/h is 10%. The direction has the truncated Gaussian distribution with the mean value of zero and standard derivation of 90 degrees.

In this example, the following two channel assignment algorithms are examined.

(a) Fixed channel assignment.

(b) Dynamic channel assignment.

The first available method is used in these algorithms. In the case of fixed channel assignment, the number of channels is 15 per zone and in the case of dynamic channel assignment, the number of channels is 100 in whole system. The results of computer simulations are shown in Fig.2. The axis of ordinate denote the blocking probability and the axis of abscissa denote traffic carried per channel. The traffic carried per channel is calculated by the following equation.

$$a_c = na(1-B)/n_c \qquad (2-1)$$

where a_c : traffic carried,
n : the number of zones,
a : traffic intensity offered in each zone,
B : blocking probability and
n_c : the number of channels in whole system.

From the result of computer simulations in Fig.2, at low blocking probability, the efficiency of channel usage using dynamic channel assignment method is superior to fixed channel assignment method. However, when the blocking probability become high, the traffic-carrying capability using fixed channel assignment method is superior to dynamic channel assignment method. So, two curved lines are intersected in Fig.2. Let us call the point of intersection "reversion point". In Fig.2, the reversion point is at about 4% of blocking probability. In general, a point of the reversion varies with the number of channels and the structure of system. If the number of channels increase, the reversion point shifts to the left (decrease) generally. This reversion of the inequality of channel usage of a dynamic channel assignment and a fixed channel assignment has beenseen in many computer simulations[1][5]. However, we have not seen clearly why the inequality reverses. We will theoretically study the cause at Sec.4.

3. REARRANGEMENTS

If the blocking probability of the reversion of the inequality namely, the reversion point, rises, then the efficiency of the channel usage is improved. In this section, we show that rearrangement of channels has a great effect on rising the reversion point. We begin with introduction of rearrangements of a mobile radio communication system.

3.1 A Formulation of Rearrangement in a Mobile Radio Communication System[5][6]

Rearrangement is a method of breaking through a blocked call by changing routes of calls already in progress[8]. This method is applicable to various systems that adopt a common control switching system. A mobile radio communication system using dynamic channel assignment is a common control system. So, the rearrangement can be applied to the system and by which the performance of the system may be considerably improved.

It is convenient to consider a channel assignment in a small zone mobile radio communication system by a following nonoriented graph G.

$$G = (V, E), \; |V| = n \qquad (3-1)$$

where V and E are the set of vertices and edges respectively.

(i) Calls attempts occur on each vertex.

(ii) n_c channels are assigned one by one to each call on each vertex, where n_c is the number of

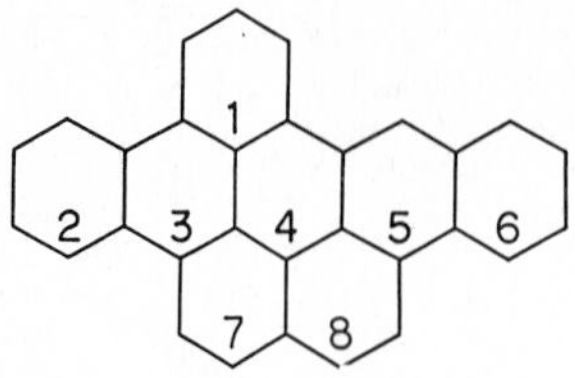

(a) A service area.

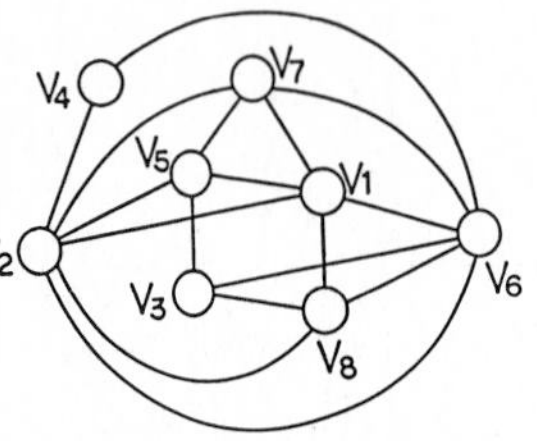

(b) Its graph G.

Fig.3 A service area.

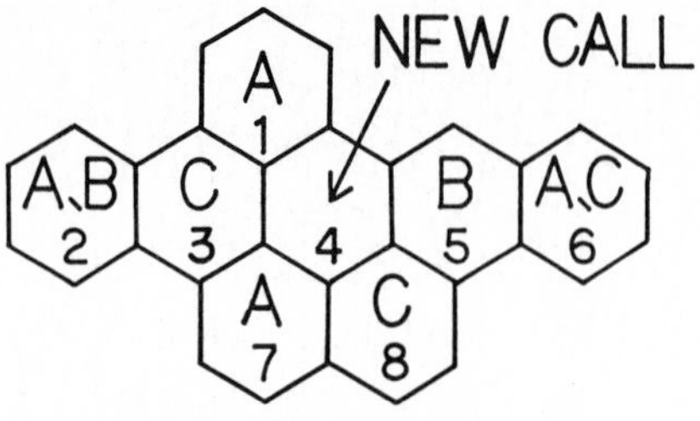

Fig.4 A state of channels.

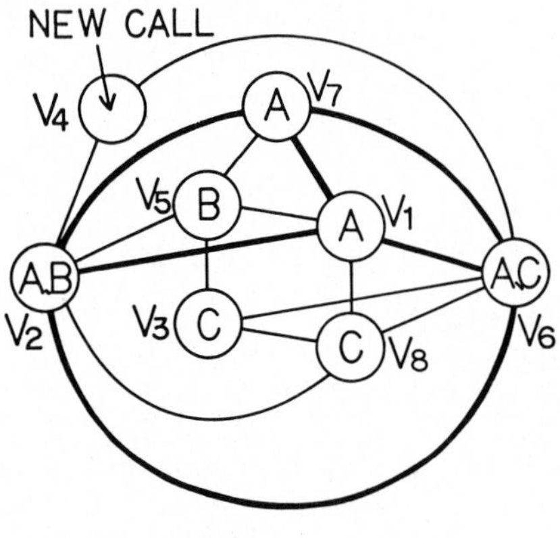

(a) The graph G of Fig.4.

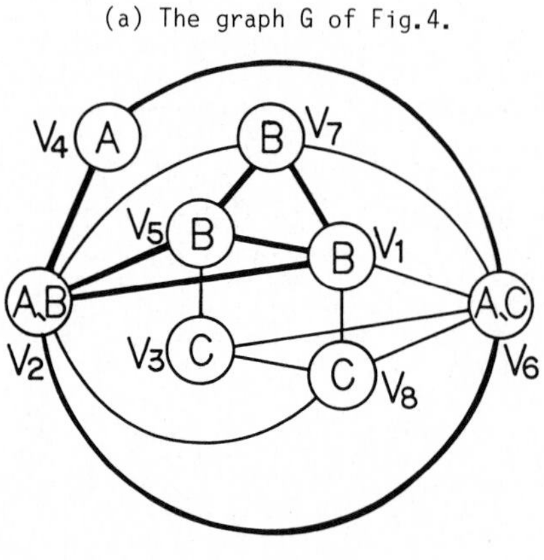

(b) A state of channels after rearrangements.

Fig.5 Achannel assignment with rearrangement.

channel of the system. Only the calls on the vertices which are adjacent can use the same channel. If the n_c+1st channel is required for a new call, the new call is blocked.

In a mobile radio communication system, the vertices $v_1,\ldots,v_n$ of G correspond to the zone $z_1,\ldots,z_n$ of a service area, and an edge $e_k=(v_i,v_j)$ of G represents that a zone z_i does not belong to the buffer zones of z_j. For example, consider a service area of Fig.3(a). If the buffer zones of a zone consist of all zones closer than one zones away from it, the graph G is Fig.3(b).

In this definition of a graph G, G has the following properties.

(a) The vertices which have calls using the same channel form a clique (a complete subgraph).

(b) The calls on vertices in a independent set of vertex can not use the same channel.

From these properties, the rearrangement problem can be considered as the channel assignment problem to cliques in G.

For example, we consider a channel assignment with rearrangement in the case of Fig.4. The system in Fig.4 is 1-belt buffering system, and it has three channels in whole system and let the labels of these channels be A,B and C respectively. The letters in a zone show that a channel of the letter is used in the zone. Suppose that a new call attempt occurs in a zone z_4. The buffer zones of a zone z_4 consist of zones z_1,z_3,z_4,z_5,z_7 and z_8. Since channels A,B and C are used in the buffer zones of z_4, there is no channels for the new call of z_4. However, if some channels are permitted to reassign, the new call may not be blocked. This channel assignment with rearrangement can be shown on graph G. A state before rearrangement is shown in Fig.5(a). In this graph, there are a number of cliques, for example, a channel A is used on a clique $c_A=\{v_1,v_2,v_6,v_7\}$ which is drawn with thick lines. A channel assignment is performed as follows.

(1) A clique c_B that is a set of vertices using the channel B can be enlarged to a set of vertices including v_1. That is to say, the channel B is assigned instead of the channel A since the channel B can be used in v_1. Thus, a clique c_A becomes $\{v_2,v_6,v_7\}$ and a clique c_B becomes $\{v_1,v_2,v_5\}$.

(2) Similarly, the channel B is assigned instead of the channel A to a call on v_7. Therefore, a clique c_A becomes $\{v_2,v_6\}$ and a clique c_B becomes $\{v_1,v_2,v_5,v_7\}$.

(3) After all, a clique c_A can be enlarged to aset of vertices including v_4 and the channel A can be assigned to a new call on v_4.

The thick lines in Fig.5(b) show cliques c_A and c_B.

The rearrangement problem is formulated as follows.

Let a set of maximal cliques in G be Q.

$$Q = \{q_1, q_2, \ldots, q_m\} \quad (3-2)$$

T is a (nxm) matrix as

$$T = \{t_{ij}\}, \; t_{ij} = 1 \text{ for } v_i \in q_j, \; t_{ij} = 0 \text{ for } v_i \notin q_j \quad (3-3)$$

For two vectors

$$X = [x_1, x_2, \ldots, x_m]^T \text{ and } W = [w_1, w_2, \ldots, w_n]^T,$$

the problem becomes :

$$\text{Minimize } Z = \sum_{i=1}^{m} x_i \quad (3-4)$$

subject to a constraint

$$TX \leq W \quad (3-5)$$

$$\sum_{i=1}^{m} x_i \leq n_c \quad (3-6)$$

where $x_i = 0, 1, 2, \ldots,$ and w_i is the number of calls on a vertex v_i.

The solutions X and Z represent the assignment of channels to maximal cliques of G and the minimum number of channels which is necessary for calls of G, respectively. Although this formulation is expressed in a simple form, it is difficult to solve it. Because this problem is equivalent to the well known problem of colorability of a graph. The colorability problem is NP-complete[9]. The NP-complete problem is known as one of the intractable problem. So the rearrangement problem is very difficult to obtain the solution for large system, and we need an approximate algorithm.

3.2 The First Level Rearrangement Algorithm

(A) Generally, it is possible to assume that the origination of calls is at random and the probability of the origination of two or more calls at a time is negligibly smaller than the probability of the origination of no or only one call. Then, the rearrangement problem in a mobile radio communication system is finding an algorithm to assign a channel to one new call under the condition that no call in progress is forced to terminate.

Let C be a subset of cliques in G.

$$C = \{c_1, c_2, \ldots, c_{nc}\} \quad (3-7)$$

where $c_i \in C$ is a clique consisting of vertices which have calls using a channel **i**. When a channel **j** is not used in G, $c_j \in C$ is ϕ . Let a new call attempt occurs in a vertex v_i. If a clique $c_k \in C$ such that $c_k \not\ni v_i$ and

$$\{v_i\} \cup c_k \subseteq q_j, \; ^{\exists} q_j \in Q$$

exists in G, then $c_k \cup \{v_i\}$ is a clique (this operation is called the enlargement of a clique). So, a channel **k** can be assigned to the new call and in this case the rearrangement is unnecessary.

(B) If c_k does not exist in G, the rearrangement is necessary. In this case, if the rearrangement of all calls in progress is permitted, the algorithm of rearrangement is complicated as mentioned above. The algorithm in which candidates of rearrangement are restricted to **r** channels in calls **r**th level algorithm of rearrangement. In this **r**th level algorithm, If r=1,the algorithm is very simple. Let us consider the 1st level algorithm.

Let W be a vector which represents the number of calls (include the new call) on each vertex.

Step 1. Select a channel **k** as a candidate.

Step 2. Select a maximal clique q_j ($q_j \ni v_i$) containing the maximum vertices which use the channel **k** and are adjacent with v_i.

Step 3. Let X_r be a vector in which **j**th component is 1 and the other are zero.

$$X_r = [0, 0, \ldots, 1, 0, \ldots, 0]^T$$

Step 4. Set $W_r = W - TX_r$

Step 5. The components of the W_r are 1 or 0. Reassign the call on vertices of the "1" component of W_r using the (A) technique (the operation of the enlargement of a clique). If all of the above calls can not reassign, return Step 1.

Step 6. Assign the channel **k** to the new call (If all channels are examined, the new call is blocked).

The first level rearrangement does not contain the repetition except the Step 1, and the selection of a candidate channel at Step 1 is not more than n_c times. The worst-case complexity is $O(N \cdot n_c)$, where N is the maximum number of buffer zones of a zone. So, this may be one of the simplest algorithm in this reassignment problem.

An example of the computer simulations is shown in Fig.6. Channel assignment algorithms are fixed channel assignment, dynamic channel assignment with rearrangement and hybrid channel assignment. Assumptions in this simulations are similar to assumptions in Sec.2.2. 15 channels per zone is used in a fixed channel assignment method and 100 channels is used in a dynamic channel assignment method. In the case of hybrid channel assignment, each zone can use 10 fixed channels and 5 dynamic channels, and after the fixed channels can not use, a dynamic channel is assigned to a new call.

From this result, the dynamic channel assignment method with rearrangement is superior to a fixed and a hybrid channel assignment method. And the reversion point shifts to the right, that is, the traffic characteristics using dynamic channel assignment with rearrangement is superior to without rearrangement. Therefore, the traffic characteristics can be considerably improved by rearrangement. However, the channel assignment with rearrangement can't also avoid the reversion. That is, a dynamic channel assignment method with rearrangement is inferior to fixed channel assignment at higher blocking probability. Thus, it is necessary to propose a method not producing the reversion.

4. THE REVERSION OF THE EFFICIENT OF CHANNEL USAGE IN TWO CHANNEL ASSIGNMENT

In this section, we study the reversion of the inequality of channel usage of a dynamic channel assignment and a fixed channel assignment.

The traffic characteristics of the mobile radio communication system using fixed channel assignment method can be calculated by Erlang's B formula with the number of channels and offered traffic intensity. On the other hand, the traffic

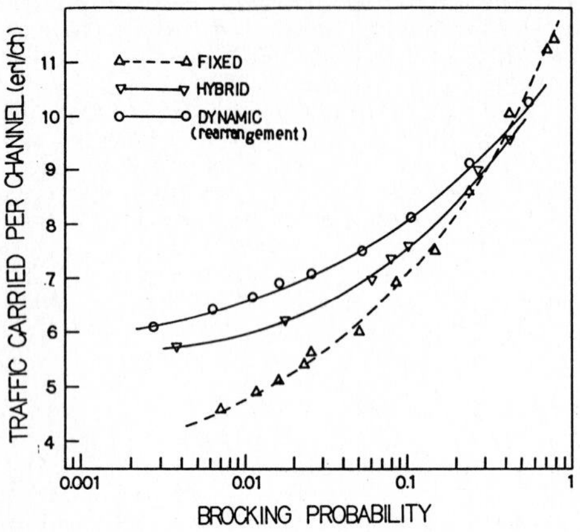

Fig.6 Simulation results.

characteristics of the mobile radio communication system using dynamic channel assignment method have not been obtained analytically. However, the exact solution is known in a system which use only one channel[4]. So, using this solution, it can be known that a method of finding approximately properties of a system with multi-channels[4].

Consider a nonoriented graph G as in Sec.3.1 and define Θ_x and Θ_x^j.

Θ_x : A set of X-cliques of G.

Θ_x^j : A set of X-cliques including v_j of Θ_x in G.

k_m : The number of vertices of maximum cliques of G.

$$\theta_s = \{v_{i1}, v_{i2}, \dots\} \qquad (4-1)$$

Then origination of calls in each zone forms a Poisson process with the coefficient of λ_i and the holding time is an exponential random variable with the mean value of $1/\mu_i$. Therefore, the offered traffic intensity is $a_i=\lambda_i/\mu_i$ (erl), and the traffic carried in zone z_j is

$$a_{c,j} = \frac{\sum_{x=1}^{k_m} \sum_{\theta_s \in \Theta_x^j} a_{i1}\, a_{i2} \cdots a_{ix}}{1 + \sum_{x=1}^{k_m} \sum_{\theta_s \in \Theta_x} a_{i1}\, a_{i2} \cdots a_{ix}} \qquad (4-2)$$

And the blocking probability in the zone z_j is

$$B_j = (a_j - a_{c,j})/a_j \qquad (4-3)$$

The blocking probability in the whole system can be calculate with similar method. But in the case of a system with a large number of channels, the number of available states (that is, cliques) of channel usage becomes very large. So, the equation of states can't be solved, and the general solution can't be found in the system with multi-channels. From this equation of states, we can see the cause of the reversion of the inequality of channel usage of a dynamic channel assignment and a fixed channel assignment. Let us consider by an example. Let us consider graphs in Fig.7. They correspond to the systems of 6 and 7 zones, respectively. The graph G represents the relations between a zone and its buffer zones. In these systems, let the number of channels be two, A and B, and let a dynamic channel assignment method use. For simplicity, assume that the mobile unit does not cross a zone boundary during a communication and a channel can use in any zones. There are many states of channel usage. In these systems, a state is changed when a call arrives or departs.

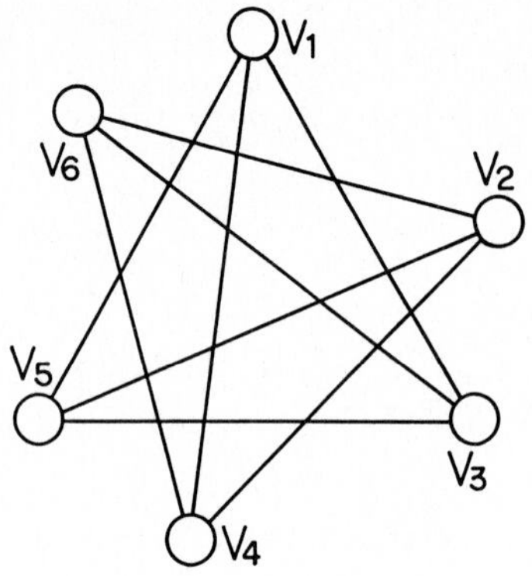

(a) 6 zones system.

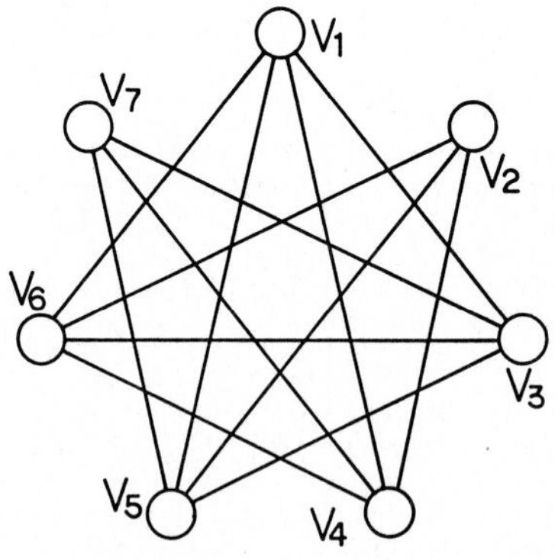

(b) 7 zones system.

Fig.7 Graphs of 6 and 7 zones system.

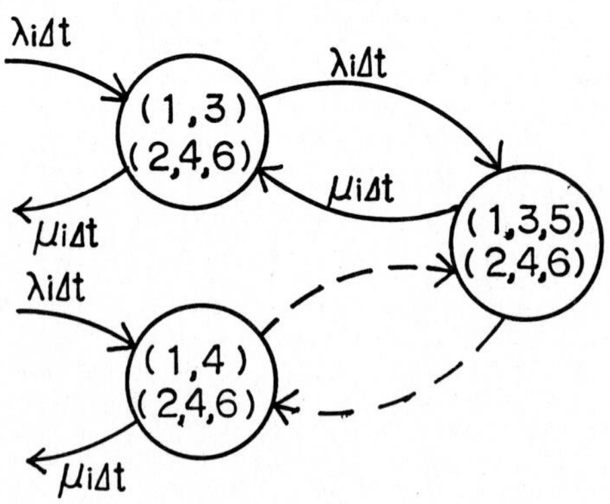

Fig.8 A flow diagram of 6 zones system.

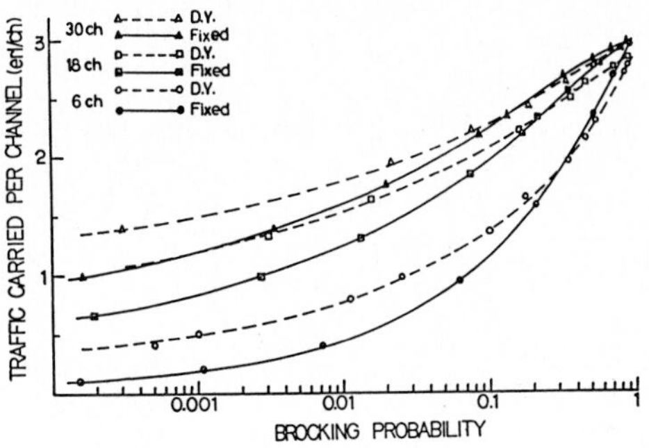

Fig.9 Simulation results (6 zones system).

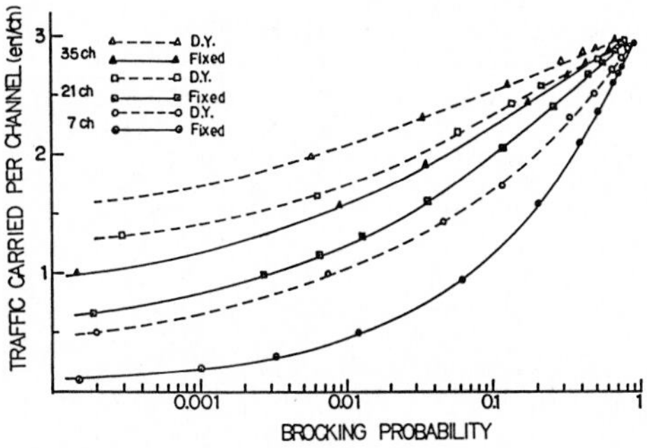

Fig.10 Simulation results (7 zones system).

A part of the state flow diagram of 6 zones system is shown in Fig.8. In 6 zones system and 7 zones system, the maximum number of zones that can use a same channels at the same time is 3. That is, the number of vertices of maximum cliques is 3. The sets of the numbers in the circle in Fig.8 show the numbers of zone using the channel A and channel B. For example, the most right circle shows that the channel A is used at zones z_1, z_3 and z_5, and the channel B is used at zones z_2, z_4 and z_6. An arrow shows next state after Δt, and its subscript shows the flow (transition) probability.

The state of the most right circle shows that of the most efficiency of channel usage. In the fixed channel assignment, the state of the channel assignment is a subset of the state in the most right circle in Fig.8, and it flows to the state of this circle according to the increase of the traffic offered. In the case of dynamic channel assignment, if a state flows smoothly to the state of this most right circle, the traffic characteristics will not be inferior to the traffic characteristics using fixed channel assignment. However, in 6 zones system, the state of (1,4) and (2,4,6) can not flow (shift) to the most right circle directly. If such states do not exist in the system, the system does not produce the reversion point of the inequality of channel usage of a dynamic channel assignment and a fixed channel assignment. The existence of these state is a cause of the reversion. For example, the results of the computer simulations of 6 zones system and 7 zones system is shown in Fig.9 and Fig.10, respectively. The reversion point is produced in 6 zones system and it is not produced in 7 zones system.

Thus, we get the following results:

"The inequality is not reserved in a system with a graph in which a maximal clique is identical with a maximum clique".

In the system of graph (A), the inequality is reserved. However, in the system of graph (B), the inequality is not reserved. And in the system of graph (B), dynamic channel assignment method has higher efficiency of channel usage than a fixed channel assignment method at any blocking probability, since a maximal clique is identical with a maximum clique in the graph of the system.

5. AN ADAPTIVE CHANNEL ASSIGNMENT

In general, it is rare that in the graph of a system a maximal clique is identical with a maximum clique. So, most of the system have a reversion point (blocking probability of the reversion of the inequality of channel usage of a dynamic channel assignment and a fixed channel assignment). Thus, in order to maintain the high efficiency of the channel usage at any blocking probability, we assign a channel for a new call according to traffic offered. That is, an adaptive (to traffic offered) channel assignment algorithm is necessary. The reversion point varies with the zone structure and the number of channels. So, it is difficult to find the reversion point analytically.

In this section, we propose a simple adaptive algorithm. The algorithm consists of two part, that is, a dynamic channel assignment part and a fixed channel assignment part. The choice between two parts depends on traffic offered in the system. The basic idea of the algorithm is using the dynamic channel assignment when offered traffic intensity is low, and when the offered traffic intensity becomes high, the method of the channel assignment change to the method using fixed channel assignment step by step.

For example, let us consider the case increasing traffic offered. In the first, assign the dynamic channel to calls, and at the same time, calculate the blocking probability assuming that the calls are assigned by fixed channel assignment method. And if the blocking probability of a fixed channel assignment is lower than that of a dynamic channel assignment, then a fixed channel assignment is used. In this case, blocking probability is a percentage of the number of blocking calls in the number of arriving calls. So, it is available to compare only the number of blocking calls in spite of blocking probabilities. That is, the channel assignment method is changed to fixed channel assignment step by step when the number of blocking calls using dynamic channel assignment is larger than using fixed channel assignment. If the channel assignment method is changed for all of channels at the same time, the blocking probability is increased rapidly. If a channel is assigned by dynamic channel assignment method, the channel can not be used by fixed channel assignment method. So, it is necessary to wait until the channel is unused.

The results of the computer simulations using the adaptive channel assignment is shown in Fig.11 and Fig.12. The system has 61 zones. It is the same as Fig.1. The system is 1-belt buffering system. The number of channels in Fig.11 and Fig.12 are 18 and 36 in whole system respectively.

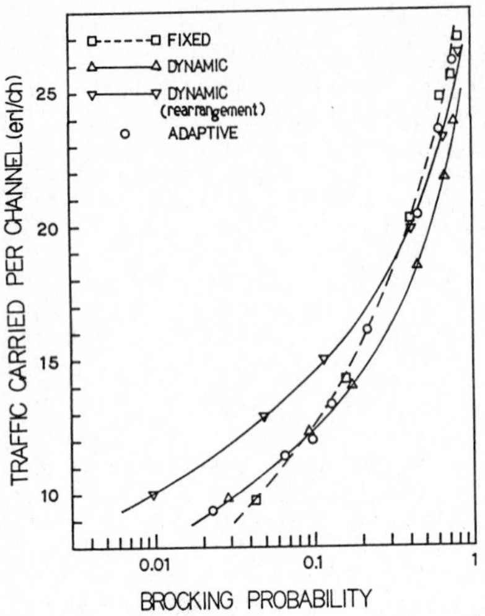

Fig.11 Simulation results (61 zones, 18 channels).

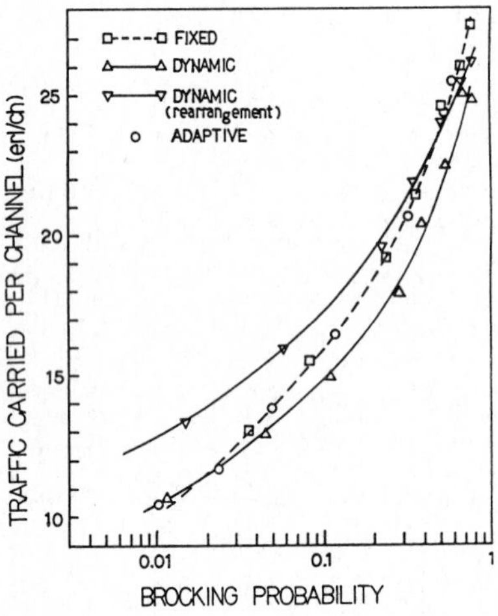

Fig.12 Simulation results (61 zones, 36 channels).

The axis of abscissa denote the blocking probability and the axis of ordinate denote traffic carried per channel. A solid line is the result using dynamic channel assignment, the broken line is the result using fixed channel assignment, and the result using the adaptive channel assignment is shown as "o".

From the results of the computer simulations, the traffic characteristics using an adaptive channel assignment method is similar to that using dynamic channel assignment without rearrangement at low blocking probability, and it is similar to that using fixed channel assignment at high blocking probability. So, the algorithm using adaptive channel assignment is superior to the dynamic channel assignment without rearrangement because of it can prevent the "reverse" at any blocking probability. This adaptive channel assignment is not using rearrangement, but it can be used with rearrangement. The adaptive channel assignment with rearrangement may be more superior to a method without rearrangement and is also able to prevent the reversion. And, the method has the following disadvantage. It must change the channel assignment step by step, so it must spend any time until the channel assignment is completed and it must always calculate the blocking probability using fixed channel assignment method, so this algorithm is more complex than dynamic channel assignment.

6. CONCLUSION

The inequality of the efficiency of the channel usage of a dynamic channel assignment and a fixed channel assignment reverses at some blocking probability in most of mobile radio communication systems. The zone structures in which the reversion does not occur are revealed. And it is shown that rearrangement of channels has a great effect on rising the reversion point. And a simple adaptive channel assignment algorithm in which the reversion does not take place is proposed and the validity of the channel assignment algorithm is confirmed by computer simulations.

7. REFERENCES

[1] Edited by W. C. Jakes, "Microwave Mobile Communications", (Chapter 7, pp.545-), John Wiley & Sons Inc. (1974)
[2] D. C. Cox and D. O. Reudink, "Dynamic Channel Assignment in Two-dimensional Large-scale Mobile Radio Systems", Bell System Technical Journal, 51, 7, pp.1611-1629, (Sep. 1972)
[3] M. Sengoku, K. Ito and T. Matsumoto, "A .po60 Dynamic Frequency Assignment Algorithm in Mobile Radio Communication Systems", Trans. IECE, Japan, Vol. E61, pp.527-533, (July, 1978)
[4] M. Sengoku, "Telephone Traffic in a Mobile Radio Communication System Using Dynamic Frequency Assignments", IEEE, Trans. on Vehicular Technology, VT-29, No.2, pp.270-278, (May, 1980)
[5] M. Sengoku, M. Kurata and Y. Kajitani, "Rearrangements in a Small Cell Mobile Radio System", Proc. IEEE and NTC, International Conference on Communications", Vol. 81CH1648-5, pp.2371-2375, (June, 1981)
[6] M. Sengoku, M. Kurata and Y. Kajitani, "Application of Rearrangement to a Mobile Radio Communication System", Trans. IECE, Vol. J64-B, No.9, pp.978-985, (Japanese)
[7] M. Sengoku, M. Kurata, H. Kariya and T. Abe, "Application of Rearrangement to a Hybrid Channel Assignment Scheme in Mobile Radio Communication Systems", Trans. IECE, Vol. J65-B, No.3, pp.340-341, (1982), (Japanese)
[8] M. Akiyama, "Modern Communication Switching Engineering", Tokyo Denkishoin, (1972), (Japanese)
[9] A. W. Aho, J. E. Hopcroft and J. D. Ullman, "The Design and Analysis of Computer Algorithms", Addison-Wesley, (1975)

TELETRAFFIC ISSUES in an Advanced Information Society
ITC-11
Minoru Akiyama (Editor)
Elsevier Science Publishers B.V. (North-Holland)

TRAFFIC CHARACTERISTICS OF DAMA TASI/DSI SYSTEM

Shohei SATOH, Sen'ichi TANABE and Hiroyuki MOCHIZAI

NEC Corporation
1131-Hinode, Abiko, Chiba, Japan

ABSTRACT

The paper investigates the effect of application of time assignment speech interpolation (TASI) to fully variable demand assignment multiple access (DAMA) satellite communication system. From the traffic engineering point of view, it is regarded as a type of delay-system with grading which has special distributed service discipline. The exact solution is not reached. An approximation method is used to evaluate the DAMA-TASI system. The results show there is no traffic capacity increase of the DAMA-TASI system compared with point-to-multi point variable destination DAMA system.

1. INTRODUCTION

It is widely recognized in normal voice communications that a channel of a telephone circuit carries no talk bursts when another direction channel of the same circuit carries talk bursts. This speech redundancy has been utilized to attain efficient use of sub-marin cables or satellite channels in the well known time-assignment speech interpolation (TASI) system [1] or in digital speech interpolation (DSI) systems [2].

Roughly speaking, the system accommodates ℓ telephone trunks in the capacity of $\ell/2$ telephone channels. The efficient use of channels may cause a service degradation known as freeze-out of speech or speech clipping, because with some probability there is no idle channel in the system when a talk burst arrives. A type of DSI is a speech predictive encoding communication (SPEC) system [3], in which a service degradation is perceived as a degreese of signal-to-noise ratio.

The service degradations have been analyzed for the system which has fixed number of channels. Regarding satellite communication systems, point-to-point pre-assigned systems and point-to-multi-point destination variable systems are application fields of traditional TASI/DSI technique mentioned above.

This paper investigates the possibility of applying TASI/DSI technique to the fully variable demand assignment multiple access (DAMA) satellite communication system. The propagation delay between earth stations requires some contrivances in the system, and raise a new traffic model to be solved.

2. DAMA-TASI SYSTEM

This section presents the proposed demand assignment multiple access system with TASI/DSI (DAMA-TASI in short). Detaile implementations such as signalling sequence signalling traffic are not considered here. Traffic characteristics of speech channel are main concern in this paper.

There are N earth stations and each station has ℓ terrestrial telephone circuits. Of course in the actual system each station does not have the same number of terrestrial circuits, but in this paper we suppose it has, for simplicity. Each station has m-satellite channels which are exclusively reserved to the station, and there are n-common satellite channels which are assigned to stations on demand. The total satellite speech channels are, therefore, $N\cdot m+n$.

When a talk burst arrives on a terrestrial circuit, a satellite channel out of m-reserved channel is assigned to the circuit. The earth station requests one common channel to the control station which manages the assignment of common channels. The control station is assumed for explanation simplicity, though it is possible to conceive the system which has no control station. Time duration between an earth station sends a request signal and the station receives a channel assignment signal is T which includes propagation time of signals and processing time in the control station. The speech clipping will occur, if more than m-1 talk burst arrives in T for this case.

An earth station requests or releases common channels so as to maintain the number of reserved idle channel be

m for $i=0$ to $\ell-m$

$\ell-i$ for $i=\ell-m+1$ to ℓ

where i is the number of terrestrial circuits on which there are talk bursts at the time (See Figure 1).

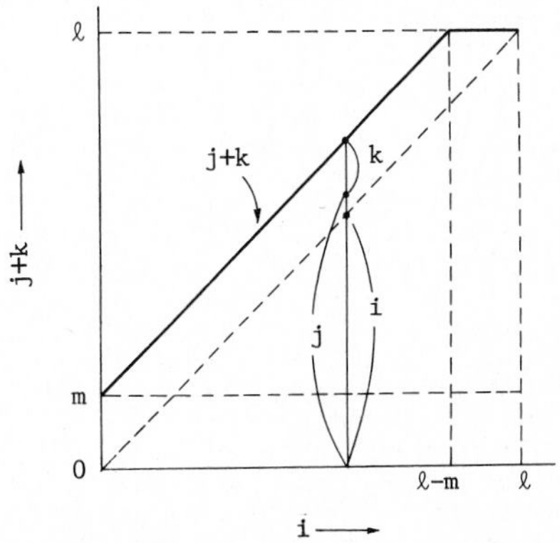

Fig. 1 Explanation of i,j and k

We represent the state of an earth station by three variables i,j and k, explained below.

i ... mentioned above (i=o, ..., ℓ)

j ... the number of satellite channels which are assigned to the station (j=m, ..., ℓ)

k ... the number of satellite channels which are requested by the station but not yet the assignments are received by the station (k=o, ..., ℓ-m)

Figure 2 shows the detail channel request/release operation of an earth station.

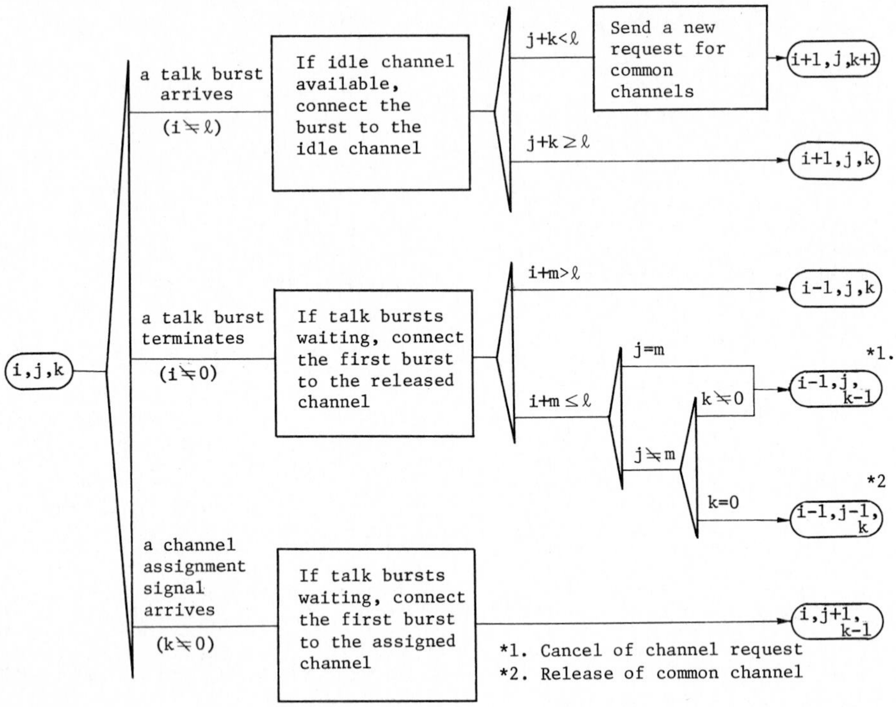

Fig. 2 Channel Request/Release Operation of an Earth Station

3. TRAFFIC MODEL

The DAMA-TASI system is considered as a type of delay-grading system with a special service discipline. Figure 3 shows this grading system. Notations and explanations given in section 2 are also used hereafter. Additional assumptions and remarks are described below.

(1) A time of talk burst follows negative exponental distribution with a mean c. The talk burst vanishes from the system after the time length whether it is served or not.

(2) A length of idle time between successive talk bursts on a terrestrial circuit follows also negative exponential distribution with a mean b=(1-a)/a, where "a" is a occupancy of talk burst on a terrestrial circuit.

(3) In the real situation, all terrestrial circuits of an earth station are not always busy. But here, we assume that they are, because in some occations all busy probability is not so small and noticeable service degradations in established calls are undesirable.

(4) At each earth station, talk bursts are served on first-come-first-served (FCFS) basis. And at the control station, channel assignment signals are processed on FCFS basis, too. Each station has exclusively assigned channels and each station does not release common channels when talk bursts terminate but still there are talk bursts waiting on its other terrestrial circuits (See Figure 2). Because of these, talk bursts at different earth stations are not served on FCFS basis.

The traffic model mentioned above is a complicated one. Exact solution is not given in this paper. We divide the system into two parts. The one is "a earth station" and the other is "a control station or common channels". Though actually these two are not statistically independent, it is assumed they are.

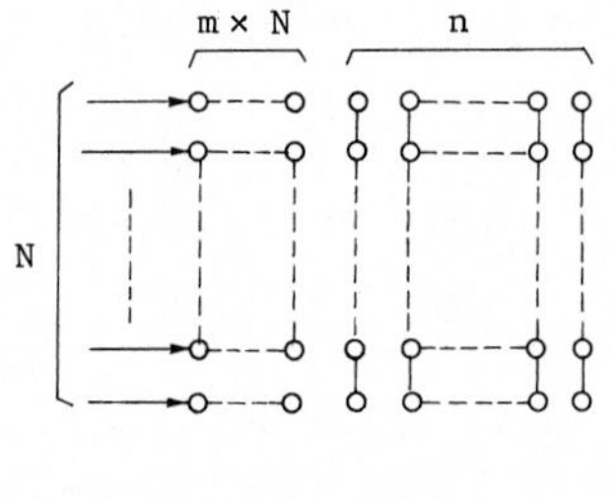

Fig. 3 Delay System with Grading

4. WAITING TIME AT EARTH STATION

An arrived talk burst must wait if there is no idle satellite channel at the station. This case occurs when the exclusively reserved channels are busy and there is no idle common channel or when so many talk bursts arrive in propagation time of channel request or assignment signals.

In this section, a method to derive waiting time distribution at an earth station with constant T is shown, where T is the time duration from a channel request sending to a channel assignment receiving by the earth station. The T includes propagation delays, processing time at control station and waiting time to common channels. The waiting time and the processing time is not constant, but it is assumed that variations of these are so small compared with the propagation time that total time can be regarded constant. The propagation time is about 260 ms if the control station is on a satellite or 520 ms if it is on earth's surface.

Let $P(t;i,j,k)$ be the probability that on earth station is in state (i,j,k) at time t. Figure 2 represents state transition of the earth station. Transition probabilities for talk burst arrival and termination in time dt are

$$P_b = P(t;i,j,k)\cdot(\ell-i)\cdot\frac{1}{b}\cdot dt \tag{1}$$

$$P_c = P(t;i,j,k)\cdot i\cdot\frac{1}{c}\cdot dt. \tag{2}$$

Transition probability Pd for channel assignment signal arrival is the probability that there is a channel request sending at time t-T and the request is not canceled. ($P_d = P_{d1}\cdot P_{d2}$)
The first part of the probability P_d is

$$P_{d1} = \sum_{j+k<\ell} P(t-T;i,j,k)\cdot(\ell-i)\cdot\frac{1}{b}\cdot dt. \tag{3}$$

To obtain the second part of the probability, a first-passage-problem has to be solved, because a channel request is canceled when the number of talk burst termination exceeds the number of talk burst arrival for the first time after the channel request. This problem is simplified as follows.

Consider the time duration T from t-T to T, and the total number of talk burst arrivals and terminations in the duration. The average arrival and termination rate is the same and the sum is

$$a\cdot\ell + a\cdot\ell = 2\cdot a\cdot\ell. \tag{4}$$

The second part of P_d is obtained using the result of first-passage in coin-tossing game [4].

$$P_{d2} = 1 - \sum_{f=0}^{\infty} e^{-2a\ell T}\cdot\frac{(2a\ell T)^{2f+1}}{(2f+1)!}\cdot\sum_{g=0}^{f}\frac{1}{2g+1}\binom{2g+1}{g+1}2^{-(g+1)} \tag{5}$$

Thus, transition probabilities P_b, P_c, P_d are obtained and these are used to get steady state probability

$$P(i,j,k) = P(\infty;i,j,k). \tag{6}$$

The $P(i,j,k)$ is calculated by repetition, i.e., from $P(t;i,j,k)$ to $P(t+dt;i,j,k)$.

Waiting time distribution $W(>t)$ of talk burst is derived from $P(i,j,k)$ considering the waiting more than t occurs in the following case.

(A) $i \geq j$ and

(B) $d \leq i-j$, where d is a sum of h (= number of talk burst termination out of j in time t) and d-h (= a number of channel assignment signals received in time t) and

(C) The length of talk burst which arrives and waits is more than t.

These conclude

$$W(>t) = \left\{\sum_{i=m}^{\ell-1}\sum_{j=m}^{i} P(i,j,k)\left\{\sum_{d=0}^{i-j}\sum_{h=0}^{d}\binom{j}{h}\left(e^{-t/c}\right)^{j-h}\cdot\left(1-e^{-t/c}\right)^{h}\cdot\binom{k}{d-h}\cdot\left(\frac{t}{T}\right)^{d-h}\cdot\left(1-\frac{t}{T}\right)^{k-(d-h)}\right\}\cdot\right\} e^{-t/c}$$

$$\text{where } k = i+m-j \text{ for } i+m<\ell$$
$$= \ell-j \quad \text{for } i+m \geq \ell \tag{7}$$

5. WAITING FOR COMMON CHANNEL

The probability of waiting is taken as a measure to evatuate waiting for common channel and it is aimed to set the probability so small compared with one at mentioned in section 4. Let q_i be the probability that there are i talk burst requests and served by common channels. The q_i is N-convolution of P_i, which is the probability that there are i talk bursts at an earth station requesting and served by common channel. The P_i is derived from binomial distribution as follows.

$$P_i = \begin{cases}\binom{\ell}{i} a^i\cdot(1-a)^{\ell-i} & \text{for } i=0\sim\ell-m-1\\ \sum_{j=\ell-m}^{\ell}\binom{\ell}{j} a^j\cdot(1-a)^{\ell-j} & \text{for } i=\ell-m\end{cases} \tag{8}$$

Probability of waiting for common channel is

$$P = \sum_{i\geq n} q_i \tag{9}$$

6. NUMERICAL EXAMPLE AND EVALUATION

Waiting time distribution for ℓ=10,20,30 and m=2,3,4,5 is calculated according to ε_q. (7), and shown in Figure 4. The mean length c of talk burst is about 1.5 second [1]. Speech clipping more than 50ms is said to be annoying. If to set a GOS that probability of clipping more than 50ms is less than 2 %, this is interpreted as

$$W(>0.05/1.5) \leq 0.02\ . \tag{10}$$

To satisfy this GOS, each station has to have m=4 or 5 exclusively assigned channel.

Probability of waiting P for common channel is shown in Figure 5. Here, we set another GOS of $P \leq 0.5\%$. The total number of satellite channel to satisfy these two GOS's is shown in Table 1 and compared with a point-to-multipoint destination variable system, of which speech clipping probability more than t is given by

$$W(>t)=\sum_{i=m}^{\ell-1} \binom{\ell-1}{i} a^i (1-a)^{\ell-i-1} \cdot e^{-(i+1)}$$

$$\cdot \sum_{j=0}^{i-m} \binom{i}{m} \left(e^{\frac{t}{c}}-1\right)^j \qquad (11)$$

The number of setellite channel per earth station m for the system is also determinated by the same GOS (10) and shown in Table 1.

For all calculated cases, Table 1 shows that point-to-multipoint system gives less number of total channels than DAMA-TASI system.

7. CONCLUSION

The traffic model of DAMA-TASI system is presented as a delay-grading system with distributed service discipline. The waiting time distribution of talk burst (i,e,clipping time distribution) is obtained by an approximation method. Numerical examples show that traffic capacity of the system is no greater than that of point-multipoint variable destination DAMA system.

ACKNOWLEDGEMENT

The authors wish to express their thanks to members of First Switching Division of NEC Co. for discussions and supports.

REFERENCES

[1] Bullington, K and Fraser, J.M.: "Engineering aspect of TASI", Bell Syst. Tech. J., Vol.38, No.2 (March 1959)

[2] Ota, C and Amano K.: "A Digital Speech Interpolation", Trans. of IECE of Japan Vol.56-A, No.8 (August 1973)

[3] Sciulli, J.A and Campanella S.J: "A Speech Predictive Encoding Communication System for Multichannel Telephony", IEEE TRANS., Vol. COM-21, No.7. (July 1973)

[4] Feller, W.: "An Introduction to Probability Theory and Its Applications", Vol.1, John Wiley & Sons (1957)

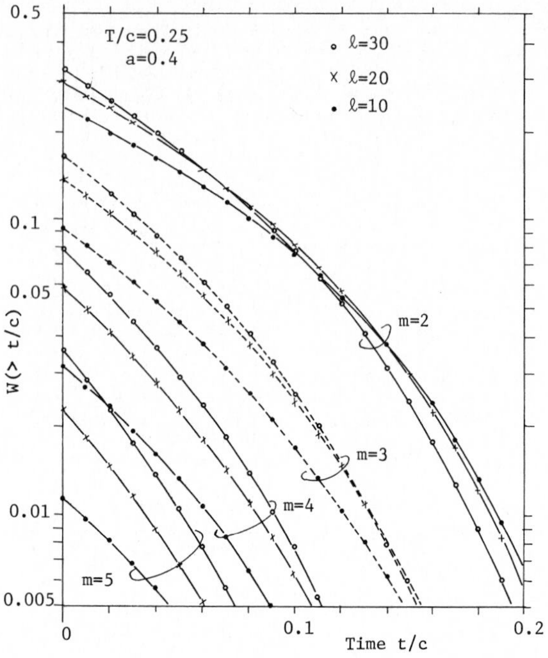

Fig. 4-a Waiting Time Distribution of Talk Burst

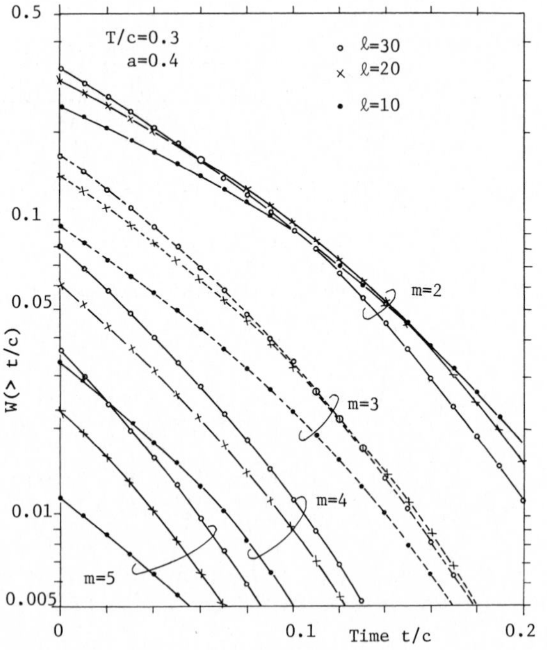

Fig. 4-b Waiting Time Distribution of Talk Burst

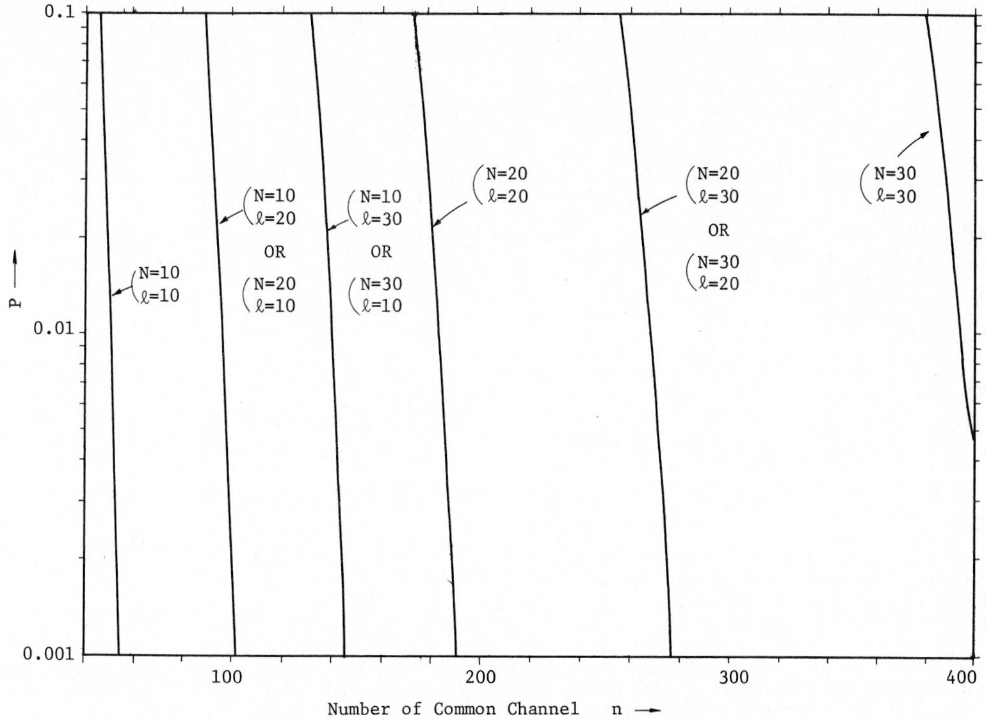

Fig. 5 Probability of Waiting P for Common Channel

Table 1. Require Number of Satellite Channel

	N	10			20			30		
	ℓ	10	20	30	10	20	30	10	20	30
A	m	4	5	5	4	5	5	4	5	5
	n	53	99	143	99	186	272	143	272	400
	N·m+n	93	149	193	179	286	372	263	422	550
B	m	8	14	18	8	14	18	8	14	18
	N·m	80	140	180	160	280	360	240	420	540

A: DAMA-TASI System

B: Point-to-multipoint Destination Variable DAMA System

TELETRAFFIC ISSUES in an Advanced Information Society
ITC-11
Minoru Akiyama (Editor)
Elsevier Science Publishers B.V. (North-Holland)

AN OPTIMAL INSERTION OF A SATELLITE SYSTEM IN A TELEPHONE NETWORK

Milena BUTTO'(*), Giovanni COLOMBO (*), Ivo PILLONI (*), Carlo SCARATI (*)
Lorenzo ZANETTI POLZI (**)

(*)CSELT - Centro Studi e Laboratori Telecomunicazioni S.p.A. -
Via G.Reiss Romoli, 274 - 10148 Torino (ITALY)
(**)SIP/DG - Via Flaminia 169 - ROMA (ITALY)

ABSTRACT

A procedure is reported for the optimal insertion of an SS/TDMA satellite system in a TLC network. After a layered definition of the system functions and cost elements, the satellite network is modelled according to an oriented graph, pointing out the network switching points and sharing levels. According to this model, a procedure is described as a sequence of separated optimization steps linked to each other by the most significant correlation elements. An example is finally presented, allowing a comparison among different system architectures, each of them being defined by the TDMA terminal bit rate, the penetration level of the earth stations, the utilized DA configurations and the eventual presence of switching points in the network.

1. INTRODUCTION

An SS/TDMA system (allowing a rational band use) with a few spot beam antennas is considered.

As far as the access network is concerned, different sharing levels on the trunk groups connecting the telephone exchanges to the related ground station are available. Space resources are managed according to different Demand Assignment (DA) schemes, involving different system utilization levels. Channels efficiency is increased by redundancy reduction equipments (ADPCM, DSI).

Achieving the optimal insertion of the satellite system means defining:

- the traffic load to be shifted on the SS;
- the number of earth stations, the hierarchical level of insertion, their location and the telephone exchanges linked to each of them;
- the TDMA terminal allocation on the frequency band;
- the DA functions and the sharing levels on the access network links.

The optimal satellite dimensioning is obtained on the global satellite system cost which is composed by local access network, earth station and control network costs. The payload cost is not taken into account since it does not substantially affect the system configuration choice and since the satellite capacity is supposed to be fixed. Nevertheless this cost element can be added after the previous optimal insertion has been found, in order to compare different technological solutions for telecommunication networks (satellite, optical fiber, ...).

Only the circuit-switched telephone service is considered. The optimal solution constraint is the quality of service expressed by the end-to-end loss probability.

The satellite network obtained in such a way represents the basic network structure in which the above mentioned satellite facilities can be taken into account.

2. SYSTEM FUNCTIONAL LAYERS AND COST ELEMENTS

Each call engages resources belonging to four different functional layers:

- access network
- earth station
- space frame
- control network.

Each layer, identified with the resources accomplishing the related function, can be modelled by compact blocks, sufficient to point out the most significant cost and operation elements. [1,2,3,4]

Only cost elements allowing a comparative evaluation of the possible system solutions are considered.

Access Network

As far as the full duplex access links are concerned, two sharing levels are foreseen:

- not-shared circuit trunk groups (D-trunks), each one intended for a single exchange-exchange oriented traffic;
- shared circuit trunk groups (S-trunks), each one intended for either exchange-satellite or satellite-exchange traffic.

The first kind of trunk groups involves the same traffic partition on the space channels. On the contrary, the S-trunk coexists with all the DA schemes, but it requires switching functions in the station. The above switching functions can be placed in the exchange where the station is located.

The access network cost is the sum of all the exchange-station link costs C_e.

The sharing level choice leads to one or more trunk groups, composing the exchange-station link. The following linear cost function can be assumed:

$$(1) \qquad C_e = N_e\, c_e + c_{e_o}$$

where N_e is the link capacity, c_e and c_{e_o} the circuit and the fixed costs respectively, depending on the link length.

Earth Station

In principle an earth station can be outlined by four blocks, each one representing a functionally closed equipment set [5].

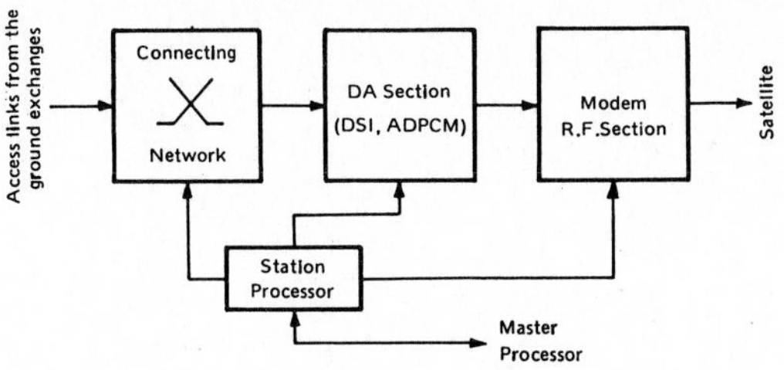

Fig. 1 - Earth station functional blocks

The functions performed by the station processor (SP) depend on the DA schemes assigned to the satellite system. In particular, SP exchanges informations with the Master Processor (MP). The DA functions do not affect the costs of modem and RF sections. In a pure permutative satellite system, SP and Connecting Network can be avoided.

In general, the cost of a single station s can also be expressed in a linear form:

(2) $$C_s = N_s \, c_s + n_{TER} \cdot c_{TER} + c_{so}$$

being N_s the overall access channel number, c_s the channel cost (switching and DA), n_{TER} the number of terminals, c_{TER} the related cost and c_{so} the fixed equipment cost.

According to this model, the DA scheme complexity and the access link sharing level are mainly reflected on the earth station cost.

Space Frame

Every frequency band is univocally associated to a transponder and on it a space frame is defined, filled by all the terminals working on that frequency. [6]

A defined sharing level of the space channels corresponds to every DA scheme. Four DA schemes will be considered:

E-E: end-to-end between exchanges; the satellite accomplishes channel trunk groups, each one intended for a single exchange-exchange oriented traffic (relation);

S-S : end-to-end between stations; the satellite accomplishes channel trunk groups; each one shared by all the relations ending in a couple of terminals;

VD : Variable Destination; the satellite accomplishes channel trunk groups, each one shared by all the relations rising in a terminal and ending in the set of terminals associated to the same transponder;

FV : Variable Origin, Destination and Window (Fully Variable); the satellite accomplishes channel trunk groups each one shared by all terminals of a transponder towards any other system terminal, regardless of the transponder it belongs to.

All these DA schemes can coexist on the satellite. The frame cost elements depend on the DA schemes: they are included in (2) as far as the station functions are concerned, while they contribute to the MP cost for all the centralized functions.

Control Network

A control network is used in order to exchange informations among earth stations, on board processor and a centralized processor unit (MP). All the costs affering the centralized functions are included, for simplicity, in the station cost.

3. SATELLITE NETWORK MODELLING

The satellite network can be modelled according to an oriented graph: the vertices represent the network switching points while the edges stand for the trunks connecting them. More precisely, the directional edges represent monodirectional circuit trunks in the access links and monodirectional channel trunks in the space segment.

If the space full-duplex connection, linking a couple of nodes, is achieved through two monodirectional channel trunks showing a univocal coincidence of the occupancy state, the related graph edge is double oriented (two arrows). This last case holds when the following (necessary) condition is fullfilled: the two interested nodes are homologous (i.e. they represent the same sharing level). In that condition, in fact, the occupancy in the two transmission directions is symmetrical.

In the network, four switching points (graph vertices) are singled out and the following description will be used:

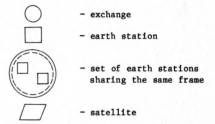

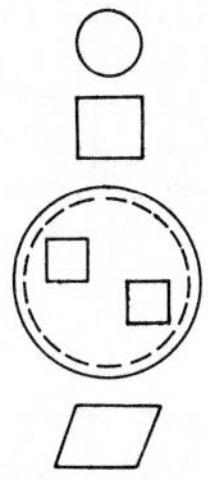

- exchange
- earth station
- set of earth stations sharing the same frame
- satellite

The routing associated to each exchange-exchange traffic flow will be represented by a sequence of numerical labels, each of them being assigned to a graph edge.

With the above description, the network configurations can be modelled according to the four DA schemes.

E-E Configuration

Only the exchanges accessing the satellite system are graph vertices. The stations will not appear as they simply carry out trunk permutations.

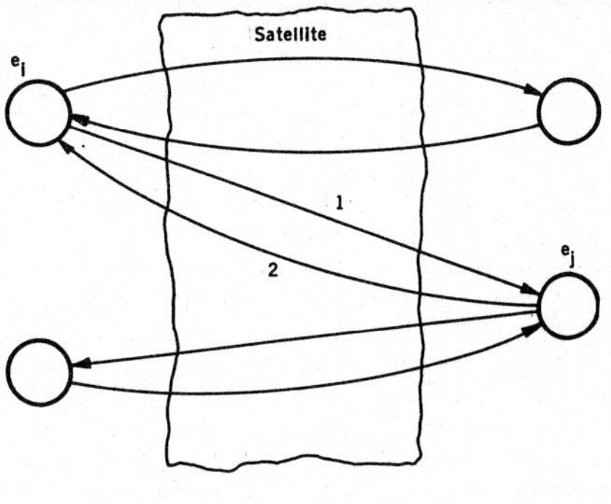

Fig. 2 - Oriented graph for the E-E configuration

Every traffic relation utilizes one only switching segment:

Traffic Relation	Routing
$e_i \rightarrow e_j$	1
$e_j \rightarrow e_i$	2

If N_1 and N_2 are the capacities of the trunks 1 and 2 respectively, each one of the exchanges e_i, e_j will be linked to the related station through two circuit trunk groups of capacity N_1 and N_2, while the two stations are spatially linked through two channel trunk groups each of them of capacity N_1+N_2.

Even if the network nodes are homologous, the related graph edges are not double oriented as they substantially represent the access links. This is why the preceding condition about the bidirectional edges is only necessary.

S-S Configuration

In addition to the exchanges, the earth stations are now included: they perform a switching function.

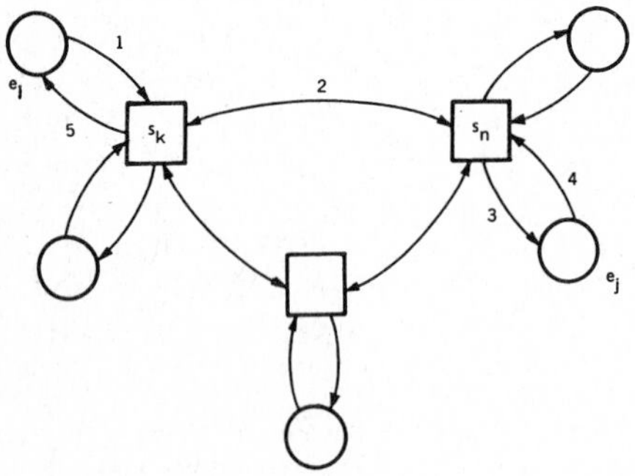

Fig. 3 - Oriented graph for the S-S configuration

A path through the graph is composed by three edges:

Traffic Relations	Routing
$e_i \rightarrow e_j$	1 2 3
$e_j \rightarrow e_i$	4 2 5

The space link between s_k, s_n is composed by two identical channel trunks, both loaded by the traffic flow generated between s_k and s_n in the two switching directions. This means that the station-station links are bidirectional circuit trunk groups, also being the exchange-station links monodirectional trunk groups.

VD Configuration

In this configuration, the frame channel sharing in destination adds another switching point.

A path through the graph is composed by four edges:

Traffic Relations	Routing
$e_i \rightarrow e_j$	1 3 4 5
$e_j \rightarrow e_i$	6 4 3 2

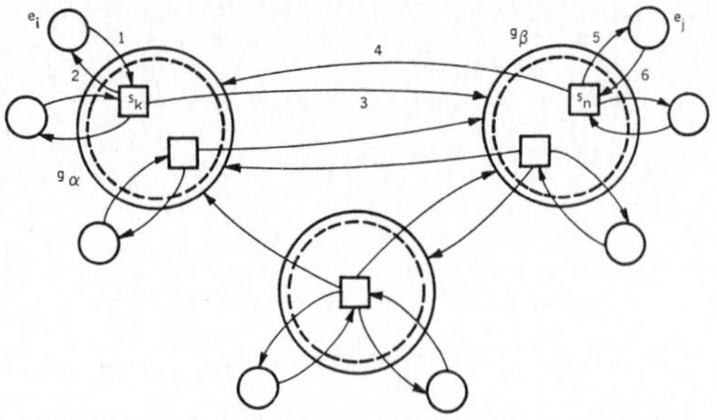

Fig. 4 - Oriented graph for the VD configuration

The channel trunk 3 is loaded by the traffic flow generated between s_k and g_β in the two switching directions; the channel trunk 4 is loaded by the traffic flow generated between s_n and g_α in the two switching directions.

FV Configuration

In this configuration the station-to-station link is guaranteed by a connecting network (a three stage T-S-T arrangement when just a classical S-stage is used on board [7].

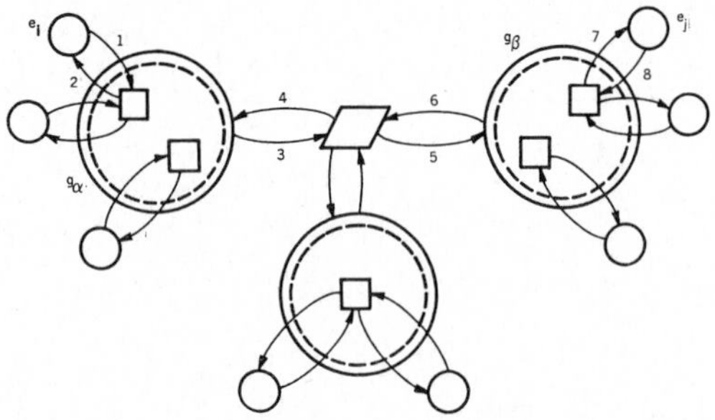

Fig. 5 - Oriented graph for the FV configuration

The path through the graph is composed by six edges but in this case the space channel availability is constrained by an internal blocking due to the particular occupancy states on the up and down links.

Traffic Relations	Routing
$e_i \rightarrow e_j$	1 3 5 6 4 7
$e_j \rightarrow e_i$	8 6 4 3 5 2

Each one of the two trunks (3,4; 5,6) ending in a transponder (g_α; g_β) is loaded with the total traffic rising or expiring in the transponder itself.

The path description pointed out in this paragraph, together with the hypothesis of the occupancy statistical independence among the trunks, allows to calculate the end-to-end blocking probability by the following formula: [8]

$$B(e_i, e_j) = 1 - \prod_{t \in L_{ij}} (1-B_t) \tag{3}$$

where L_{ij} is the set of trunks used by the traffic

relation (e_i, e_j).

The model (3) also holds for the FV Configuration when the internal blocking is considered to be negligible.

4. OPTIMAL SATELLITE INSERTION IN A TLC NETWORK

The optimal satellite network dimensioning involves the solution of a series of correlated problems. Due to the relevance of the overall complexity, the optimum is obtained by sequentially solving separated problems, taking into account every time the most significant correlations.

In the following, the various problems tied to the optimal satellite dimensioning are presented. For each of them the cost function to be minimized and the related constraints are pointed out. No emphasis is put on the algorithms implemented for the solutions: they substantially belong to optimization problem classes, developed in the literature.

4.1 Choice Of The Satellite Traffic Load

The satellite is assumed to be used as a telephone network enlargement to absorb traffic load increases.

We assume that the shift of a traffic amount on the satellite does not produce modifications in the terrestrial network topology and hierarchy.

The choice of the traffic to be shifted on the satellite can be performed according to different criteria. We choose moving on the satellite defined parts of the whole relation load (overflowing by the terrestrial routings).

This choice involves a large access capillarity but does not produce substantial modifications in the terrestrial routing criteria. Besides, DA schemes have to be introduced.

For instance in an assigned load condition, the shifted relations are those suffering the highest end-to-end loss probabilities or those having the highest carried erlang cost.

4.2 Earth Station Location

The access network topology is completely described by both the earth station set I with the related geographical location and the partition on the above stations of the set E of the exchanges accessing the satellite.

The cost elements which contribute to define the earth stations location are the following:

- the station cost C_s, as expressed in (2);
- the access link cost C_e, as expressed in (1).

The optimal station location is determined by the set J E with the related exchange partition such that the following minimum is reached:

$$(4)\quad \min_{(I,\Omega_s)} \sum_{s\in I} \left[C_s + \sum_{e\in\Omega_s} C_e \right] ; \quad \bigcup_{s\in I} \Omega_s \equiv E$$

being Ω_s the set of the exchanges linked to the station s.

The costs appearing in (4) depend on the space segment DA configurations which at this point are not yet known (they will just be determined on the basis of the solution of (5) in section 4.4). The elements creating this dependence are the access trunk capacities and the presence of switching and DA equipments in the station. So an iterative procedure should start; but it can be avoided when the above dependence has not a strong impact on the solutions.

4.3 TDMA Terminal Allocation On The Frequency Band

A multibeam satellite system introduces a number of advantages but, conversely, requires the following problems to be solved:

- how to carry out the connections among the stations through a feasible transponder switching function;
- how to assign the available frequency band to the TDMA terminals (terminal/transponder association).

Solving the first problem means defining, along the frames, the transponder-to-transponder connections which can be settled in a permanent way for Fixed Window DA schemes or re-defined frame by frame (call by call basis) when Variable Window DA schemes are adopted.

For the terminal/transponder association we choose as optimal a criteria giving a uniform channel distribution on the frames by assuming a given mean frame efficiency. Obviously, the number of transponder depends on the TDMA terminal transmission speed.

4.4 Choice Of The Optimal Configuration And Satellite Network Dimensioning

At this point we have already defined the satellite traffic load and the stations location with the related assignment of the available frequency bands.

Now we have to determine the optimal configurations (in the sense of section 3) and, accordingly, the optimal space and access trunk group capacities.

The solution of this problem is carried out by minimizing the global satellite network cost (regardless of the payload cost):

$$(5)\quad C_{NET} = \sum_{s\in I} \left[C_s(\phi) + \sum_{e\in\Omega_s} C_e(\phi) \right]$$

being $C_s(\phi)$ the s-th station cost when the configuration (°) ϕ holds and $C_e(\phi)$ the cost of the access trunks, linking the exchange e with the station s.

The minimum is calculated as a function of both the configuration ϕ and the space-access trunk group capacities, according to the following constraints:

(6) $B(e_i, e_j) < b_{max}$: exchange-to-exchange loss probability not exceeding an assigned value

(7) $\sum_{t\in Hg} M_t(\phi) < T$, $g=1,2,\ldots G$: overall trunk space capacity (inferred by the configuration ϕ on each transponder) not exceeding the frame capacity T. G is the number of transponders, Hg is the set of space trunks belonging to the transponder g.

Note that the objective function (5) is formally identical to the one expressed by (4); nevertheless the two functions are minimized on different sets of independent variables. The optimum

(°) From now on "configuration" stands for "set of the configurations foreseen on the whole network"

of (5) is carried out coherently with the optimal topological choices performed in (4).

The constraint (6), in the hypothesis of statistical independence (°) of the trunk groups forming the path $e_i \to e_j$, can be expressed as:

(8) $B(e_i,e_j) = B_A + B_S(e_i,e_j) \cdot (1-B_A) \leq b_{MAX}$

where B_A and B_s are the blocking probabilities suffered by the (e_i,e_j) traffic relation on the access network and on the space segment respectively. These probabilities can, in turn, be evaluated by the blocking probabilities related to the single trunks, as in (3).

The constraint (8) points out a loss probability partition between terrestrial and space segments. High values of B_A involve low access network costs, but low blocking probabilities on the space path are necessary and consequently, by (7), high sharing levels and station costs are needed.

The optimal partition depends on both the ϕ configurations and the access link costs (both of them are the optimal solutions we are looking for in (5)).

Besides it has to be noted that, due to the choices on the access network sharing levels, these last ones are univocally determined by the DA schemes: this eliminates one problem variable.

Another important consequence of those choices is that the configuration ϕ influences the access trunk capacities only for the number of E-E links it involves on each earth station.

Finally, the constraint which mainly conditions the above optimal loss partition and, consequently, the DA scheme choice, is expressed through (7) by the finite frame capacity.

So we reach the minimum (5) in two steps: in the first one we determine the configurations with the related space trunk capacities, in the second one we determine the optimal dimensioning of the access links.

Configuration choice and related space trunk capacities

The optimal configuration is obtained by fixing for each traffic relation a maximum space segment loss $B_S(e_i,e_j)$ and by satisfying the constraints (7). For optimal configuration we mean the one minimizing the earth station cost (switching network and SP) under the following constraints:

$$(9) \quad \begin{cases} \min\limits_{\phi} \sum\limits_{s \in I} C_s(\phi) \\ \sum\limits_{t \in Hg} M_t(\phi) \leq T \\ B(s_k, s_n) \leq B_S(e_i,e_j);\ s_k,s_n \in I; \\ e_i \in \Omega_{s_k};\ e_j \in \Omega_{s_n} \end{cases}$$

in which the last constraint expresses that the loss of the space path between stations s_k and s_n, must not exceed a maximum value.

The main problem is how to choose the space segment losses $B_S(e_i,e_j)$. At this point, as we do not know anything of the access network, we choose $B_S(e_i,e_j)$ in order to guarantee that an access trunk dimensioning exists such that:

(10) $B(e_i;e_j) = b_{MAX} \qquad e_i,e_j \in E$

Condition (10), of course, does not lead to the minimum of the objective function (5) as the access link costs could impose an unbalancement on the traffic relation losses (always under the constraint (7)): it just allows the space segment choice not to add "a priori" constraints in the optimal network dimensioning. Now, the following statement can be verified:

"A sufficient condition on the value of $B_S(e_i,e_j)$ so that (10) can be fulfilled is:

(11) $B_S(e_i,e_j) = b_s \qquad e_i,\ e_j$

where b_s is chosen in order to verify the constraint (7). Condition (11) becomes even necessary if the network is completely connected (in the sense that every access trunk conveys or receives traffic to or from all the frame segments accessed by its station)"

These results follow by observing that:

- all the traffic relations flowing on the same space path suffer on it the same blocking probability (°);
- every access trunk load is split on or collected from a subset of space paths, the subset depending on both the traffic relations shifted on the satellite and the configuration ϕ;
- every single path occupies four frames due to the bidirectional telephone service: in each of them the constraint (7) must be satisfied.

The above remarks support the criterion according to which the minimum (9) is reached:

a) for a fixed b_s: $0<b_s<b_{MAX}$ the optimal configuration ϕ_o is reached by (9) in which $B_S(e_i,e_j)=b_s \ \forall\ e_i,\ e_j$. Together with ϕ_o the space trunk capacities $M_t(\phi_o)$, among which the frame capacity have been subdivided, are known;
b) as the previous capacities $M_t(\phi_o)$ do not completely fill the frame channels (the higher is the load unbalancement, the lower is the frame filling), they can be re-defined with the criterion of maximizing the frame filling level, keeping fixed the configuration ϕ_o; so some space paths will reduce their loss.

Access link dimensioning

Configuration ϕ_o being fixed, the sharing levels of the access network are univocally assigned. Besides, at the above point b) all the space paths are dimensioned and so their loss $B_S(e_i,\ e_j)$ is known. Now, the only variables to be determined in order to reach the minimum of (5) are the losses of the access links, being known the offered loads to each one of them.

The goal is achieved by solving the following miminum problem:

(°) Acceptable hypothesis for all the sharing levels but the E-E (section 3).

(°) We work in a poissonian environment

$$(12)\quad \begin{cases} \min\limits_{N_e} \sum\limits_{s \in I} \sum\limits_{e \in \Omega_s} C_e(\phi_o) \\ B_A(e_i,e_j) \leq b_{MAX} - B_S(e_i,e_j) \end{cases}$$

As we have already said, the value of b_s imposed in point a) conditions the solutions to (9) and (12). More precisely, a variation in b_s infers opposite cost variations on the satellite system and the access network: the minimum (5) is finally reached for that value of the loss b_s for which the above cost variations compensate with each other.

5. A CASE STUDY

The presented example just shows the procedure functionality: it has not the objective of dealing with real system evaluations. The procedure has been implemented by making use of the non-linear optimization models (branch and bound, conjugate grandient method of Fletcher and Reeves). As far as the end-to-end loss probability evaluation is concerned, the classical analytic models based on the product from (3) have been adopted.

According to the constraints shown in sect. 4.4, the case study refers to a satellite system with a fixed utilizable capacity of 36.000 channels at 64 kbit/s (i.e. 9000 bidirectional telephone circuits). 34 telephone exchanges offer the satellite 4535 erlang split in 618 relations with an average load per relation of 7.34 erlang. The satellite covers the 34 exchanges with 5 spots. The optimal insertion is carried out under the constraint of b_{max}=0.01 exchange-to-exhcange loss probability. No alternative routing is foreseen inside the satellite network.

Starting from the above defined data and constraints, the procedure determines the optimal access network topology, the D.A. configurations and the system dimensioning.

The earth station RF cost is determined by assuming the cost of a kbit/s to be a decreasing function of the TDMA bit rate. Besides the cost of each terminal is proportional to the related RF cost, while the switching function cost (added, when needed in the earth station) is assumed to be a constant.

In Fig. 6 the system cost behaviour is sketched as a function of the TDMA terminal bit rate. More precisely, both the diagram variables are expressed with respect to a reference system configuration.

The overall system cost rate (curve 1) is not too much sensitive versus the TDMA bit rate (at least in the assumed cost hypothesis). The earth station and the access network costs (curves 2 and 3 respectively) are, on the contrary, more sensitive versus the TDMA bit rate: with low speed terminals we have low station costs and so the procedure assigns a great number of earth stations reducing the access network cost; when high speed terminals are adopted, the station cost increases and the procedure assigns a low number of earth stations leading to high access network costs. The earth station cost has a minimum, since at least one station per spot is needed.

In Tab. 1, for each bit rate, the satellite system architecture is summarized: number of stations and satellite transponders in each spot, channel percentage in the E-E and S-S configurations. In the considered transmission speed range (40÷130 Mbit/s) and satellite capacity the procedure succeds in dimensioning the satellite network by making use of just the first two configurations. Fig. 6 and Tab. 1 have been obtained by assigning the space segment a loss probability b_s=0.005 (see section 4.4).

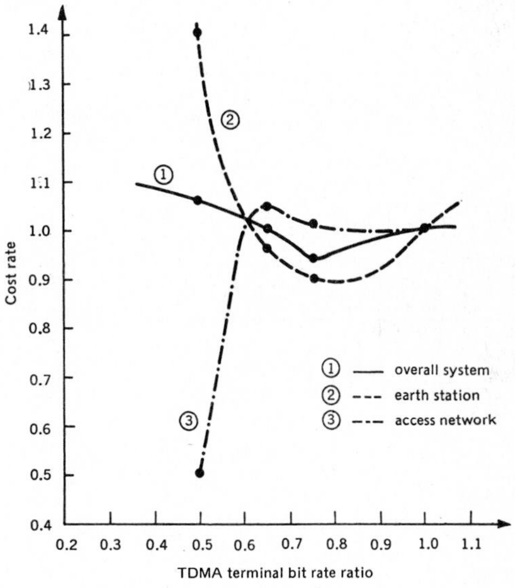

Fig. 6 - Network cost behaviour

TDMA bit rate ratio	SPOT 1		SPOT 2		SPOT 3		SPOT 4		SPOT 5		Channel distribution among system configuration	
	Stat.	Transp.	Stat.	Transp.	Stat.	Transp.	Stat.	Transp.	Stat.	Transp.	E-E %	S-S %
1	2(2)*	1	2 (2)	1	2 (1)	1	1 (1)	1	3 (1)	1	44	56
0.714	2 (0)	1	3 (0)	2	3 (0)	2	3 (0)	1	3 (0)	1	100	0
0.625	2 (2)	1	3 (1)	3	3 (0)	2	3 (0)	1	2 (2)	1	88	12
0.5	4 (2)	2	6 (5)	3	3 (2)	2	2 (1)	1	4 (1)	2	66	34

(*) The number between braeckets indicates the number of stations accomplishing switching functions.

Tab. I - Satellite system architecture.

STATION COST RATE	SPOT 1			SPOT 2			SPOT 3			SPOT 4			SPOT 5			CHANNEL DISTRIBUTION AMONG SYSTEM CONFIGURATION		
	St.	Sw.	SP	St.	Sw.	SP	St.	Sw.	SP	St.	Sw.	SP	St.	Sw.	SP	E-E %	S-S %	VD %
1	2	2	/	2	2	/	2	1	/	1	1	/	3	1	/	44	56	0
0.5	2	2	/	3	3	1	3	3	1	2	1	/	3	1	/	29	59	12
0.25	5	5	3	6	6	3	4	4	2	3	3	1	5	5	/	/	44	56

Tab. II - Influence of the earth station cost on the system architecture

Tab. II shows different system architectures induced by earth station cost variations (due for instance to technological improvements). The reference cost (rate 1) is assigned to the reference TDMA terminal bit rate and this transmission speed is kept constant for the three configurations. The lower the station cost, the greater the number of stations (terminals), all of them sharing only one transponder per spot (being fixed the satellite capacity). The presence of many terminals leads to a high degree of traffic partition, hence higher sharing levels (DA configurations) become necessary.

References

[1] Kota S.L., "Demand Assignment Multiple Access (DAMA) techniques for satellite communications", National Telecommunication Conference, New Orleans (1981)

[2] Preti R., De Padova S. and Puccio A., "Integration of a satellite switched system with the terrestrial network", International Switching Symposium, Montréal (1981)

[3] Perillan L., Rowbotman T.R., "Intelsat VI SS-TDMA system definition and technology development", 5th International Conference on Digital Satellite Communication, Genoa (1981)

[4] Lombard D., Rancy F., "TDMA demand assignment operation in Telecom I business service network", National Telecommunications Conference, New Orleans (1981)

[5] Pontano B.A., Dicks J.L., Colby R.J., Forcina G., Phiel G.F., "The Intelsat TDMA/DSI system", IEEE Journal on selected areas in Comm. Vol. SAC-1 N°1, January 1983

[6] Luvison A., Preti R., Tosalli A., "Analysis of the suitability of Demand Assignment techniques in TDMA Satellite Systems", XXXIII Congress of International Astronautical Federation, Paris-September 1982

[7] Buttò M., Colombo G., Scarati C., "Performance Evaluation of DA/SS-TDMA satellite systems with wide bandwidth traffics", X-th ITC, Montreal, June 1983

[8] Buttò M., Colombo G., Tonietti A., "On point-to-point losses in communication networks", VIII ITC, Melbourne 1976

TELETRAFFIC ISSUES in an Advanced Information Society
ITC-11
Minoru Akiyama (Editor)
Elsevier Science Publishers B.V. (North-Holland)

AN ANALYSIS OF TWO-HOP RANDOM-ACCESS RADIO NETWORKS WITH HIGHER CHANNEL UTILIZATION

Aleksander T. KOZLOWSKI

Institute of Telecommunications
Technical University of Gdańsk
Gdańsk, Poland

ABSTRACT

A two-hop random-access radio network is considered in this paper. It is shown that using error detecting repeaters, the channel utilization increases in comparison with that of a two-hop system using simple repeaters which retransmit packets only. This utilization may be much improved if capture exists in the second-hop channel and different repeaters use packets of different lengths. An algorithm minimizing cost of system realization is obtained and the network structure with minimum cost and maximum channel utilization generated by this algorithm is presented.

1. INTRODUCTION

Random-access radio computer networks like nonslotted or slotted ALOHA have many advantages, see e.g. [1] - [4] . However, they also have a few disadvantages - among the most important are the low channel utilization and relatively high cost of realization, especially in the satellite version. It has been shown, [3] - [4] , that a two-hop random-access radio network partly eliminates these two disadvantages. In this system, users transmit their packets to repeaters over first-hop channels and the repeaters send these packets to the centre via the second-hop channel. This conception is characterised by a smaller cost of system realization and an even higher channel utilization though only if two repeaters are used. The above idea of two-hop systems has been presented under the assumption that simple repeaters are used - repeaters only receive and repeat packets. The situation is quite different if repeaters can detect errors in packets received from users. In this case, packets which are involved in a multiaccess collision in the first-hop channel are not transmitted over the second-hop channel. Thus, the second-hop channel is free from overlapping packets sent by users collaborating with one repeater, and second-hop channel utilization may be improved. This utilization may be greater than ever if the capture effect is assumed in the second-hop channel and different repeaters use packets of different lengths. In addition, it is shown that there are sets of packet lengths and user traffics for which the second-hop channel utilization reaches its maximum value.

The second important problem of such a network is its realization cost. It is obvious that the cost of a satellite channel is much greater than the cost of a terrestrial radio channel with a transmitting range of about 100 miles. Thus, from the cost point of view, the number of repeaters in the system ought to be made as small as possible. On the other hand, the terrestrial radio channel's transmitting range does not allow too large an area to be served. First-hop channel utilization does not allow too many users to be served either. Thus, the number of repeaters is limited by the area covered by the system and by the number of system users located over this area. In this situation, the number of repeaters has to be chosen as a compromise between the system realization cost and the number of system users.

The present paper considers the problems mentioned above. In the first section, a two-hop random-access system with error detecting repeaters is analysed. Next, it is shown that the utilization of a nonslotted random-access channel can be improved if repeaters use packets of different lengths. Finally, an algorithm for two-hop network design is presented. This algorithm minimizes the system realization cost giving a system with minimum cost and maximum utilization.

2. TWO-HOP RANDOM-ACCESS SYSTEM ANALYSIS

Let us consider a two-hop random-access radio system like the two-hop ALOHA network [3] . All system users are divided into K classes marked $\mathcal{N}_1$, $\mathcal{N}_2$, $\mathcal{N}_3$, ... , $\mathcal{N}_K$ in such a way that $\mathcal{N}_1 \cup \mathcal{N}_2 \cup \mathcal{N}_3 \dots \cup \mathcal{N}_K = \mathcal{N}$ and $\mathcal{N}_i \cap \mathcal{N}_j = \{\emptyset\}$ for $i \neq j$, $i,j = 1,2,3, \dots, K$, where $\mathcal{N} = \{1,2,3, \dots, N\}$ and N is the number of users in the system. Users allotted to each class transmit their packets to the centre through their repeater over a first-hop random-access terrestrial radio channel (FHC). The repeater decodes packets received from its users and sends them at random times to the centre on a frequency other than that used at FHC if they were received correctly. Thus, all packets which are involved in a multiaccess collision in FHC will not be transmitted over the second-hop satellite channel (SHC). The powers of the transmitters are chosen in such a way that packets sent by k-th class users cannot disturb packets transmitted from l-th class users during their reception by the l-th repeater. All system users transmit packets of the same length equal to T.

Let A_j denote the event that a j-th system user sends its packet to the first-hop channel in some time interval[x/] and $Pr\{A_j\}$ the probability of this event. It is natural to assume that users transmit their packets independently. Thus $Pr\{A_j \cap A_l\} = Pr\{A_j\} Pr\{A_l\}$ if $j \neq l$. Let us now consider a packet sent by an i-th user allotted to the k-th class to FHC in the interval $\langle t, t+T \rangle$, where t is the beginning moment of the considered packet's transmission. This packet will be transmitted over SHC if this packet is successfully received by the k-th repeater. Transmission over SHC is successful if no other repeater has sent a packet to SHC. This means that zero, two, three, ... users allotted to other classes transmit their packets to their FHC's. Thus, the probability $P_{i,k}$ of successful user-centre transmission of the considered packet is given as follows, [4] :

$$P_{i,k} = Pr\left\{ \bigcap_{\substack{j \in \mathcal{N}_k \\ j \neq i}} \bar{A}_j \right\} \prod_{\substack{l=1 \\ l \neq k}}^{K} \left[1 - \sum_{n \in \mathcal{N}_l} Pr\left\{ \bar{A}_n \right\} Pr\left\{ \bigcap_{\substack{h \in \mathcal{N}_l \\ h \neq n}} A_h \right\} \right] \qquad (1)$$

where $\bar{A}_j$ is the reverse event of the event A_j.

Let us next define the i-th user throughput $S_{i,k} = \lambda_{i,k} T$ as the average number of new packets generated by the i-th user per packet transmission time T, i.e. the input rate normalized with respect to T, where $\lambda_{i,k}$ is the mean i-th user packet rate. Packet retransmissions, which occur in the considered system, increase the mean offered traffic to $G_{i,k}$ packets per packet transmission time T, where $G_{i,k} > S_{i,k}$. Thus, the relationship between the i-th user throughput and traffic is

$$S_{i,k} = G_{i,k} P_{i,k} \qquad (2)$$

Note that successful transmissions over SHC of packets generated by users allotted to different classes are separable events. Thus the k-th class SHC throughput S_k is equal to:

$$S_k = \sum_{i \in \mathcal{N}_k} G_{i,k} P_{i,k} \qquad (3)$$

Similarly, the total system throughput S_T may be given as follows

$$S_T = \sum_{k=1}^{K} S_k = \sum_{k=1}^{K} \sum_{i \in \mathcal{N}_k} G_{i,k} P_{i,k} \qquad (4)$$

We will assume later that:

i/ the moments of each user packet transmission are uniformly distributed over the interval $\langle t-T, t+T \rangle$,

ii/ the number of each class users is large, so each class throughput is assumed to be a Poisson point process,

iii/ each FHC traffic is a Poisson point process.

Using the above assumptions and Eq. (4) we have, [4] :

$$S_T = \sum_{k=1}^{K} G_k e^{-\alpha G_k} \prod_{\substack{i=1 \\ i \neq k}}^{K} \left[1 - \alpha G_i e^{-\alpha G_i} \right] \qquad (5)$$

where

$$\alpha = \begin{cases} 1 & \text{for the slotted protocol} \\ 2 & \text{for the nonslotted protocol} \end{cases}$$

whereas G_k is the k-th class traffic. Let each class of users generates the same traffic $G_k = G/K$, where G is the total system traffic. Thus, the total system throughput is equal to:

$$S_T = G e^{-\frac{\alpha G}{K}} \left[1 - \frac{\alpha G}{K} e^{-\frac{\alpha G}{K}} \right]^{K-1} \qquad (6)$$

Fig. 1 shows the relative changes in the maximum two-hop system throughput S_T in comparison with the maximum star system throughput S_0 versus the number of repeaters K. The maximum two-hop system throughput decreases monotonically with the number of classes, and for more than 30 classes this

x/ this means: i/ in some slot for the slotted multiaccess protocol and ii/ in a time interval $\langle t-T, t+T \rangle$ for the nonslotted multiaccess protocol, where t is the beginning moment of the analysed packet's transmission.

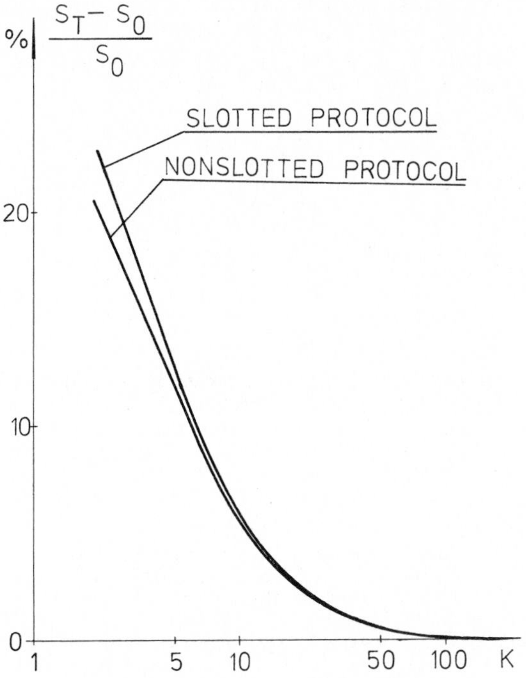

Fig. 1. The relative two-hop system throughput S_T changes versus the number of classes K. S_0 - star system throughput.

throughput is, in practice, equal to the star system throughput. The greatest two-hop system throughput increase takes place for two classes of users. At the same time, the relative increase of this throughput for the slotted protocol is greater than that for the nonslotted protocol, especially when the number of classes is less than 5-6.

Let us now consider the system analysed above assuming capture in the second-hop channel. Suppose that repeater powers are chosen in such a way that the packets transmitted by the k-th repeater always dominate packets sent by an l-th repeater if $k < l$; $k,l = 1, 2, 3, \ldots, K$. Thus, whenever one packet from the k-th repeater interferes with packets from the k+1-th, k+2-th,..., K-th repeaters, we assume that only the first packet may be received correctly at the centre. In other transmission conflict situations this packet is lost. Thus, successful transmission over SHC of a packet sent by k-th class users may be disturbed only by a packet simultaneously transmitted by users allotted to the 1-st, 2-nd, ..., k-1-th class. The probability $P'_{i,k}$ of a successful k-th class packet transmission then is equal to, [4] :

$$P'_{i,k} = Pr\left\{ \bigcap_{\substack{j \in \mathcal{N}_k \\ j \neq i}} \bar{A}_j \right\} \prod_{l=1}^{k-1} \left[1 - \sum_n Pr_l \left\{ A_n \right\} Pr\left\{ \bigcap_{\substack{h \in \mathcal{N}_l \\ h \neq n}} \bar{A}_h \right\} \right] \qquad (7)$$

After an analysis similar to that given in the first part of this section, the total two-hop system throughput with capture S_{TC} may be obtained as follows

$$S_{TC} = \sum_{k=1}^{K} G_k e^{-\alpha G_k} \prod_{i=1}^{k-1} \left[1 - \alpha G_i e^{-\alpha G_i} \right] \qquad (8)$$

It can be shown that the total throughput S_{TC} reaches its maximum value

$$S_{TC_{max}} = \frac{1 - \left(1 - e^{-1}\right)^K}{\alpha} \qquad (9)$$

for $\hat{G}_1 = \hat{G}_2 = \ldots = \hat{G}_K = 1/\alpha$

Fig. 2 shows the maximum two-hop system throughput $S_{CT_{max}}$ versus the number of classes in the system. This throughput

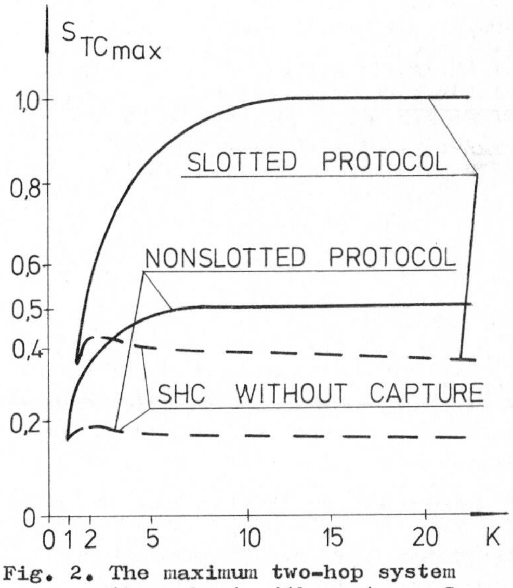

Fig. 2. The maximum two-hop system throughput with capture $S_{CT_{max}}$ versus the number of classes K.

increases monotonically with the number of classes and for 10 classes reaches almost the theoretical limit of channel utilization, equal to $1/\alpha$.

The nonslotted system utilization given above may be much improved if different repeaters use packets of different lengths and if capture in SHC exists. So, let us assume that each packet generated by k-th class users is of a length requiring T_k seconds for its transmission. The packet header is the same length for each packet in the system and requires T_h seconds for its transmission. Thus, $T_k = T_h + T_{d,k}$, where $T_{d,k}$ is the data transmission time for the k-th class packet.

Note, that a k-th class packet will be successfully transmitted via SHC if users allotted to the 1-st, 2-nd,... , k-1-th classes do not transmit exactly one packet during the intervals $\langle t-T_1, t+T_k \rangle$, $\langle t-T_2, t+T_k \rangle$, ... , $\langle t-T_{k-1}, t+T_k \rangle$ respectively, where t is the beginning moment of the considered packet's transmission. Thus, the total data throughput S_{TCD} of a two-hop nonslotted system with capture and different packet lengths, obtained by modifying Eq. (8) , may be given as, [4] :

$$S_{TCD} = \sum_{k=1}^{K} \left(1 - \frac{T_h}{T_k}\right) G_k e^{-2G_k} \prod_{l=1}^{k-1} \left[1 - \left(1 + \frac{T_k}{T_l}\right) S_l e^{-\left(1 + \frac{T_k}{T_l}\right) S_l} \right] \qquad (10)$$

where S_l is the l-th class throughput. It has been shown [4] , that there is a set $\mathcal{T}' = \{T_1', T_2', \dots, T_K'\}$ of packet lengths for which each component in the sum (10) exceeds its equivalent for packets of the same length if the set $\mathcal{G} = \{G_1, G_2, \dots, G_K\}$ is given. In addition, there are values $\hat{T}_1, \hat{T}_2, \dots, \hat{T}_K$ for which these components reach maxima. Using the above packet lengths and traffics giving the maximum total system throughput obtained before a local maximum of total system data throughput for a system with capture and different packet lengths may be obtained.
Let us however find the global maximum data throughput for the system with different packet lengths and capture by using optimization methods. The optimization problem is formulated as follows:

$$S_{TCD}\left\{ \mathcal{T}, \mathcal{G} \right\} \longrightarrow \max \qquad 11a$$

with additional conditions

$$0 \leqslant G_k \leqslant G_{max} \qquad 11b$$

and

$$T_h \leqslant T_k \leqslant T_{max} \qquad 11c$$

for $k = 1,2,3,\dots,K$.

Optimization of the problem formulated above leads to 2K complicated equations, which may be resolved only by using computer methods. A few solutions of the optimization problem are illustrated in Fig. 3.

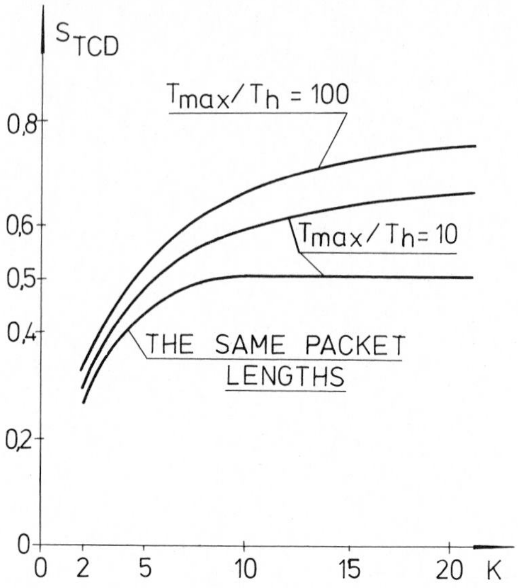

Fig. 3. The total data throughput S_{TCD} for a two-hop system with capture and different packets lengths versus the number of classes K.

Note that the total data throughput given in Fig. 3 increases in comparison with the total data throughput S_{TC} for packets of the same length for all system users. This increase is greater for a high number of classes. However, the throughput S_{TCD} slowly becomes saturated with the number of classes and probably decreases if the number of classes approaches the number of system users. It is worth mentioning that if the number of user classes is more than 10, the throughput S_{TCD} exceeds 0.5, i.e. surpasses the most optimistic results given for the nonslotted random-access system with capture described by Abramson, [2] . The maximum values of the total data throughput S_{TCD} are equal to about 0.65-0.75 and depend on the maximum packet length. In addition, the longer the packet, the greater the improvement in the total data throughput.

3. LOW-COST TWO-HOP SYSTEM DESIGN

The two-hop structure of random-access radio systems allows to be connected
way. Thus, the realization cost of such a system may be different for different system configuration. The problem of system realization cost is especially important when the system serves users

located over a large area. In this case the satellite channel is usually used for connecting users (repeaters) with the centre. Because the cost of a satellite channel is much greater than the cost of a terrestrial radio channel with a range of some tens of miles, as many users as possible ought to be connected with the centre through repeaters. From this conclusion, an algorithm for designing a two-hop system with minimum system realization cost is proposed. This algorithm generates a system structure giving the maximum total data throughput with minimum cost of system realization. A simple example of two-hop system design is now solved using this algorithm. In this example, a system consisting of 2500 users whose location is randomly generated inside a square area is considered. The linear dimensions of this square are 5 times greater than the maximum transmitting range of the terrestrial radio channel. In addition, it is assumed that the cost of a satellite channel exceeds 2000 times the longest terrestrial radio channel cost. The algorithm run is repeated a few times to generate different user locations. The results given by the algorithm are shown in Fig. 4, where the changes in the two-hop system realization cost C in comparison with the cost of a star system C^* are presented versus the number of user classes K. The cost C of two-hop system realization quickly decreases

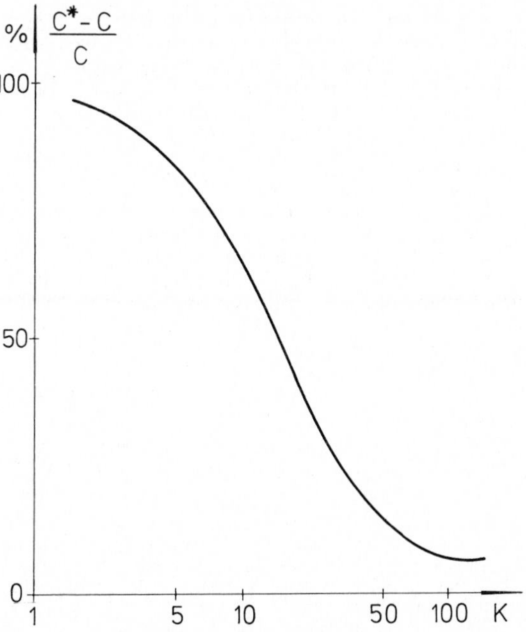

Fig. 4. The relative changes of two-hop system realization cost C with comparison of star system realization cost C^* versus K.

with the number of classes K for K in the range $\langle 2,50 \rangle$. The two-classes system realization cost is close to the cost C^* of a star system. The two-hop system realization cost is lowest for 100-110 classes, and if K increases, this cost slowly increases. It may be predicted that for a number of classes close to the number of system users this cost starts to approach the cost of a star system. It is obvious that the system structures generated by the algorithm are characterized by maximum channel utilization, i.e. the channel utilization changes in the way shown in Fig. 3.

4. CONCLUSIONS

In the present paper the problems of two-hop random-access system utilization and system realization cost are considered. It has been shown that the utilization of a two-hop system may be improved if error detecting repeaters are used, although in this case, the improvement in this utilization is not too great. The utilization of the analysed system increases if capture between different repeaters in the second-hop channel is assumed. If the packet lengths are different for different repeaters, the system utilization of a two-hop nonslotted system is even greater than 0.5. It is worth mentioning that the repeaters considered in our analysis retransmit only user packets after error detection. Thus, each user has to realize the retransmission rules itself - there is end-to-end acknowledgment. In such a case, only packets correctly received by a repeater are sent to the centre via the second-hop channel and the capacity of this channel is not fully utilized. However the utilization of the second-hop channel may be improved if repeaters have packet buffers and realize their own retransmission rules.

In the last part of this paper, an algorithm which minimizes the cost of two-hop system realization is proposed. This algorithm allows systems with a high system utilization and low realization cost to be designed.

REFERENCES

[1] L. Kleinrock, S. Lam, "Packet Switching in a Multiaccess Broadcast Channel: Performance Evaluation", IEEE Trans. on Comm., vol. 23, no. 4 pp. 410-422, 1975.

[2] N. Abramson, "The Throughput of Packet Broadcasting Channels", IEEE Trans. on Comm., vol. 25, no. 1, pp. 117-128, 1977.

[3] I. Gitman, "On Capacity of ALOHA network and some Design Problems", IEEE Trans. on Comm., vol. 23, no. 3, pp. 305-317, 1975.

[4] A. T. Kozlowski, "Channel Utilization of Random-Access Radio Channel with Different Packet Lengths", is proposed to IEEE Trans. on Comm.

TELETRAFFIC ISSUES in an Advanced Information Society
ITC-11
Minoru Akiyama (Editor)
Elsevier Science Publishers B.V. (North-Holland)

CHANNEL UTILIZATION AND BLOCKING PROBABILITY IN A CELLULAR MOBILE TELEPHONE SYSTEM WITH DIRECTED RETRY

Berth EKLUNDH

Ericsson Radio Systems
Lund, Sweden

A B S T R A C T

A Directed Retry facility, which enables subscribers in a Mobile Telephone System to look for free radio channels in more than one cell, is investigated with respect to blocking probability and channel utilization. An iterative procedure is devised, by which the dependencies among cells can be illustrated. This procedure makes use of theories developed for overflow systems in classical telephony, and proves to be very accurate for the situations under study. Analytical results are compared with simulations and good agreement is observed. Results show that a substantial improvement, as compared with systems without a Directed Retry facility, can be achieved as far as carried traffic is concerned. The improvement is accomplished at the expense of those subscribers who cannot make use of the Directed Retry facility due to variations in radio coverage.

I N T R O D U C T I O N

Recent advances in micro electronics have led to proliferation of versatile communication equipment, such as simple video terminals and vehicle mounted or portable telephone sets. Mobile telephones communicate over a radio channel, and the increased number of mobile users has resulted in frequencies available for mobile communication becoming a scarce resource. Recently, a worldwide agreement was made to allocate bandwidth for mobile communications in the 900 MHz band [1]. With currently available technique, this band can accommodate 1000-2000 duplex channels for voice communication.

If no further steps are taken to increase the number of frequencies, only 2000 calls can simultaneously be established in a mobile telephone system which uses these channels for subscriber to subscriber communication. In most systems this figure is too small by far, and a regular pattern has been devised by which the same frequencies can be used at several locations. The pattern has usually a hexagonal structure, and each hexagon is referred to as a cell. A cell is allocated a specific number of frequencies, corresponding to a number of channels, and the same frequencies are repeated in some distant cell which is sufficiently far away for the co-channel interference between the cells to be negligible.

A subscriber, who wants to make a call from his mobile terminal, has to be allocated one of the channels in a cell to be able to complete his call. This is accomplished by a signalling sequence which is sent over a dedicated channel common to all subscribers in several cells. The use of a common signalling channel increases utilization of this channel and leaves other channels free for voice communication. If all voice channels are occupied, the call attempt is blocked. The number of voice channels allocated to each cell is usually rather low, in the order of ten to twenty channels. The channels are futhermore allocated in a fixed manner in most systems, the so called Fixed Channel Assignment [5]. Because of the small number of allocated channels, utilization is quite low for reasonable blocking probabilities.

The low channel utilization is a main concern in cell structured mobile telephone systems, and various suggestions have been made to increase utilization. Some system designs make use of the signalling channels for as large a part of the call as possible, and send digits and similar information over this channel as Ericsson Radio Systems' CMS 8800 and the French-German EC 900 [2,3,4]. Queueing of call requests awaiting free channels is also employed [3]. These techniques increase channel availability by decreasing necessary holding times. Another technique is to increase the number of potential channels that may be used by the subscribers in a cell. This can be accomplished by making larger cells, but such a solution works against the advantages accomplished by structuring the systems into cells, i e to reuse frequencies. A better technique is to create several layers of cell structures with several small cells composing one larger cell and to have common frequencies in the larger cells which can handle overflow from the smaller ones. These systems are referred to as Dynamic Channel Assignment scheme and Hybrid Channel Assignment scheme and have been studied extensively [5]. These schemes can be further improved by moving calls from the larger cell to one of the smaller ones when channels become free, so called channel reassignment. This technique is known from ordinary telephone systems and has proven positive effects [6].

There are some disadvantages with the dynamic channels assignment schemes. Cox [6] has shown that dynamic channel assignment function less satisfactorily than fixed assignment for very high load, and this has been confirmed by Kahwa and Georganas [5] for both dynamic and hybrid assignment schemes. Another disadvantage of dynamic schemes is that the transmitters have to be able to transmit not only the frequencies allocated permanently to that cell, but also any of the frequencies which belong to the pool of dynamic frequencies. Such a facility complicates transmitter engineering and aerial design.

Yet another way to increase channel utilization, which does not suffer from the above-mentioned disadvantages, is to use what is known as Direcetd Retry. The analysis of such a facility is the purpose of this paper.

DIRECTED RETRY

The previously mentioned hexagonal cell structure is an idealization in all practical systems, since radio coverage cannot be made to coincide with the hexagonal form. Several subscribers therefore hear not only the transmitter which covers the cell in which the subscriber is currently located, but also transmitters in some of the neighbouring cells. In a system without Directed Retry, subscribers are only allowed to search for free channels in their "own" cell, i e the cell in which they are situated for the moment. This restriction is relaxed in a system with Directed Retry, and subscribers can look for free channels in one or more of the cells that surround the subscriber's "own" cell. This facility increases the number of potential channels and thereby the channel utilization.

A calling subscriber will make his call attempt in a common channel, shared on a random access basis by all subscribers in a large area, consisting of several cells (the FCC specification [4], for example, specifies a multiple access protocol similar to BTMA (Busy Tone Multiple Access)). If the switching office observes that all voice channels in the subscriber's cell are occupied, it may advise the subscriber (i e the mobile telephone equipment) to check the quality of the channels in neighbouring cells. If the quality is found to be adequate, then the call is directed to that channel (hence the term Directed Retry). If no channel with adequate transmission quality is found, the call attempt is blocked.

Several important issues are of interest in a system with Directed Retry. Some will be addressed in this paper:

- Traffic carried in a system with Directed Retry is greater than in a system without Directed Retry in the case of a given load. How large is the gain, and does it defend the increased complexity that results from the Directed Retry facility?

- Since not all subscribers hear transmitters in more than one cell, some subscribers will experience the system as giving better service, while others will find it to be worse compared with a regular system. How large are these effects, and what measures should be used to describe them?

There are, in principle, two methodologies that can be used to model a cellular system of the above kind in order to find system performance measures: analytical models and simulation. Each of these methods has its merits and drawbacks. These will be discussed in conjunction with the results from the models.

Very little is known about the behaviour of mobile subscribers as far as holding times, mobility and similar measures are concerned. Thus, any elaborate models of these properties cannot be defended but some standard assumptions are adopted in the models described below:

- Calls that originate and terminate in a cell are assumed to form a Poisson process with constant intensity.

- Holding times are assumed to be exponentially distributed.

  These assumptions are commonly accepted in modelling telephone traffic, and there are presently no data available to support any other assumptions.

- A third assumption that has to be made in order to model the Directed Retry function has to do with the likelihood of a subscriber hearing one or more transmitters in cells other than his "own". Two cases can be identified: one case in which Directed Retry can be used for both originating and terminating calls, and one case in which only originating calls can use Directed Retry. It is assumed here that a random subscriber has a constant probability of hearing one extra transmitter. In order to minimize the number of parameters, a probability p_0 is introduced, which in the second case takes care of the probability of hearing one extra transmitter as well as how much of the total traffic in the cell that originates in the cell. This means that if 50% of the traffic in a cell originates in the cell and if 50% of the subscribers can hear one extra transmitter, p_0 should be given the value 0.25. In the first case p_0 illustrates the probability of hearing more than one transmitter only. p_0 could be regarded as an overlap probability, which illustrates how much of the traffic in one cell can influence the surrounding cells. Subscribers are consequently assumed to hear at most one extra transmitter. There is a small probability that a subscriber could hear two or more transmitters, but the situations in which this occurs are rare and are therefore not incorporated in the analysis. Blocking probability for a single cell should be some 5% if the system is properly dimensioned, and overflow to a third cell would in such a case occur with a probability of 0.25% on average. The increase of traffic due to this blocking is considered negligible in the models.

- It is furthermore assumed that calls which originate or terminate in a cell occupy one channel only, i e calling and called subscriber are not located in the same cell. This assumption is of course an approximation, but the effects are not important if the cells of the system are small, in which case only a minor part of the traffic is intra-cell traffic.

ANALYTICAL MODEL

When a subscriber, who wants to make a call, finds all the channels of his "own" cell occupied, he is normally blocked. If there is a Directed Retry function in the system and if the subscriber is able to hear some other transmitter, he is allowed to look for free channels in the cell in which the transmitter he can hear is located. If there are free channels in this cell he may occupy one of these. This will only happen, however, when all channels are busy in his "own" cell and the traffic that flow from one cell to another will have the character of an overflow traffic, which is well known from classical traffic theory [7]. In the same way as one cell may overflow traffic to its neighbouring cells it may receive traffic from these. This matter complicates the analysis.

In the analytical model the channels of a cell are viewed as a grading which accepts traffic directed to and from the cell itself and overflow traffic to and from the surrounding cells. The direct traffic is, as stated above, assumed to be Poisson, but the same assumption cannot be made for the overflow traffic since it has a different character to the Poisson traffic. The overflow traffic is therefore represented by an ERT group with n^* channels offered a traffic of A^* Erlangs. (ERT, Equivalent Random Theory).

Consider a structure of only seven cells organized as in figure 1, and assume the direct traffic of a single cell to be A. Each of the cells at the border is represented by a grading and its surrounding cells by an ERT group. The overflow traffic from the ERT group, and that part of the direct traffic that cannot hear more than one transmitter, i e which cannot make a Directed Retry, is considered as one traffic stream offered to the grading of figure 2 (overflow traffic + $(1-p_o)A$). Another traffic stream consists of the direct traffic that can hear more than one transmitter (p_oA).

It is therefore possible to represent a cell at the border by a grading as in figure 2.

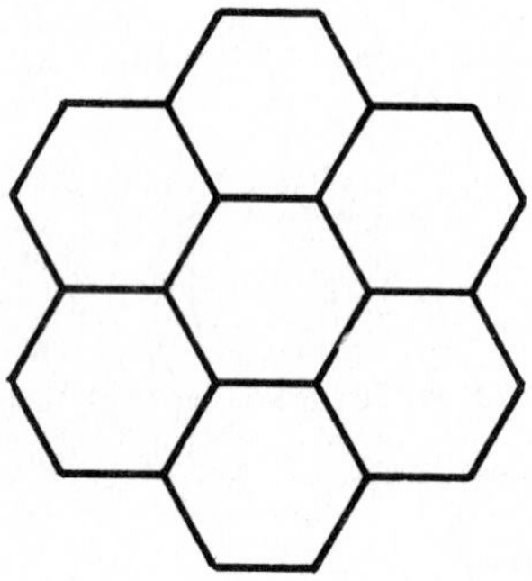

Figure 1.

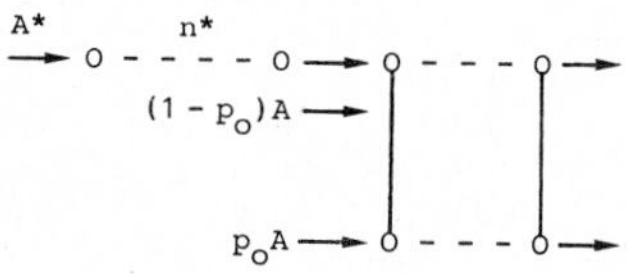

Figure 2.

At this stage A^* and n^* are still unknown, but they will be given values in the iterative scheme that follows below. However, given A^* and n^*, it is possible to calculate with standard methods how much of the traffic p_oA will be blocked from the grading. Let us assume the mean and variance of this traffic to be m and v respectively. If we picture the entire seven cell structure as a system of linked gradings, we arrive at figure 3. Each grading represents a cell which receives two kinds of traffic: one which is allowed to overflow to some other cell and one which is not. Out of the total overflow traffic, 5/6 do not overflow to the centre cell, while 1/6 does. In a symmetric situation the centre cell of figure 1 will in all receive a traffic with mean m and variance v, (6*1/6m, 6*1/6v), since it is surrounded by six cells. The means and the variances of the overflow from the six cells can be added and represented by another ERT group. The centre cell can thus be seen as a structure similar to that of figure 2 and represented in the same way.

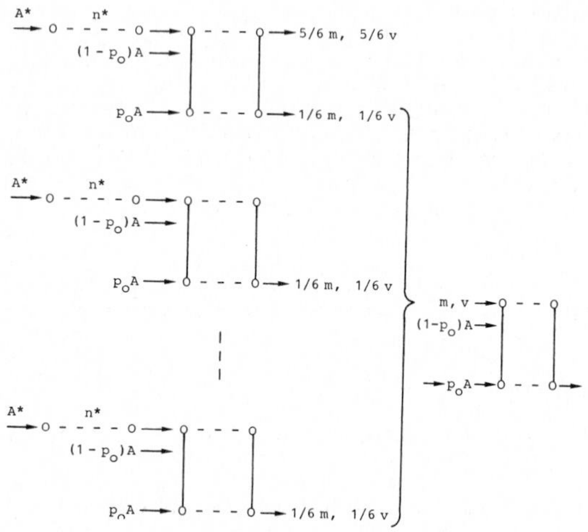

Figure 3.

Generally there are not just seven cells in a structure, but rather an "infinite" number of cells covering an entire country, or several countries. The output from the rightmost grading of figure 3 therefore has the same character as the input to that cell from other cells. An iterative procedure can consequently be applied to figure 3 in which the output of the second grading is fed back to the input of the first and there represented as an ERT group. The iterative procedure is repeated until there is no significant change in overflow between successive iterations. In the models, the mean and the variance of the overflow from the grading is solved "exactly" by numerically solving the system of equations resulting from the state vector representation of figure 2. The "exact" solution technique requires n^* to be an integer, however, and the system is solved for three different integer values in the vicinity of n^* and a Newton interpolation is then employed. Some other method could have been used to represent the overflow traffic [8,9], but ERT seems to give quite accurate results.

From the offered traffic and the blocked traffic it is simple to calculate the blocking probability and the channel utilization.

SIMULATION MODEL

The analytical model described above contains some approximations in addition to the general model assumptions concerning arrival process, holding times and so forth. One approximation is that individual cells are assumed to be independent, while they in reality can be expected to be heavily dependent on each other. A consequence of this is that the true blocking probability for an arriving call which is allowed to make more than one attempt cannot be extracted from the analytical model. An approximation is to let the blocking probability for such a call be the square of the blocking probability for an individual cell.

These considerations, and a general interest in validating the results, suggested the use of a simulation model of the cell structure and the Directed Retry facility. Input data to the simulation model was limited and a Monte Carlo simulation was considered sufficient. The simulation model itself was validated by comparing with known results. As an example, the blocking probability for $p_o = 0$ should be equal to the value obtained from the Erlang B formula. For a more detailed discussion on the use of simulation models see [10] and also [11]

The simulation model differs from the analytical model in a few crucial aspects:

- The cellular pattern covers a limited geographical area, i e it contains a specific number of cells.
- Calls that are blocked in one cell are, if the conditions are fulfilled, directed to one of the neighbouring cells.
- The overflow traffic maintains its true character, and is not, as in the analytical model, described by its first two moments only.

By comparing simulation results and analytical results, some questions can be answered: Does the iterative scheme converge and how well does it describe the interaction among the cells? Can individual cells be considered as independent as far as blocking is concerned? Does the limited number of cells in a "real" situation have any impact on the performance of the system?

RESULTS

Results from the analytical model and the simulation model show that very good agreement can be obtained between the results. Simulation results are extremely costly to produce, however, because of the large number of cells that have to be illustrated in the simulation model. A 20x20 cell pattern has proved to be sufficient to describe an "infinite" cell pattern if the (10,10) cell is used for measurements. But to get statistically significant results for this cell a very large number of calls have to be generated in the entire cell structure. The difficulties in generating statistically significant results are furthermore dependent on the overlap and blocking probabilities since it is indeed the overflow traffic that has to be significant in volume to create noticeable changes in blocking for an individual cell.

The improved service offered to some subscribers is accomplished at the expense of others. It is therefore reasonable to define a new blocking probability which takes into account both the service offered to subscribers who can use the Directed Retry facility and the service offered to those who cannot use it.

Indicate by p_b the total blocking for all subscribers and by p_c the blocking of an individual cell. Define p_b as

$$p_b = (1-p_o)p_c + p_o p_c^2$$

where p_o is the overlap probability as defined earlier.

This definition assumes that the blocking probability for a call in a cell is independent of the arrival process, i e it assumes all attempts are made at a random time. The validity of the assumption that a subscriber at his second attempt will experience a blocking probability equal to p_c will be discussed below.

To increase channel utilization, mobile telephone systems are assumed to work with higher blocking than landbased telephone systems. Figures around 5% have been mentioned. In the case of 10 channels per cell this corresponds to an offered traffic of 6-7 Erlangs, and results are presented for these traffic values. Traffic is assumed to be distributed uniformly among the cells.

Only a minor part of the results presented below have been verified by simulations. This is quite natural, since the aim of the analytical results is to give information that cannot be produced by simulation in a simple manner. The agreement between simulation results and analytical results is so good however, that there is reason to believe that the analytical models give a fair description of the system for more situations than the simulated ones. The simulation results are presented as 5% confidence limits.

Figure 4 shows p_c and p_b as functions of offered traffic for an overlap probability of 0.5, i e $p_o = 0.5$. The figure shows that if p_b is accepted as the blocking probability instead of p_c, traffic could be increased by approximately 20% in the case of a given blocking probability. Subscribers who cannot make more than one attempt because they are unable to hear more than one transmitter will suffer from the increased traffic, and experience a blocking equal to p_c.

In figure 5, p_c is shown as a function of offered traffic for three different overlap probabilities, p_o = 0.1, 0.5 and 0.9. For $p_o = 0.5$, analytical results are compared with simulation results. The agreement is, as may be seen, very good. Besides the effect the overlap probability has on p_c itself, it determines to what extent subscribers are allowed to make more than one attempt, which influences the total blocking p_b.

A similar phenomenon can be observed in figure 6 where p_c is shown as a function of the overlap probability. When overlap is increased from 0 to 1, blocking increases by approximately 50%. Traffic offered is 7.5 Erlang; at other traffic rates, change may be more or less dramatic. p_b is a good estimator of the blocking experienced by a random subscriber provided that p_c describes the probability of being blocked at the second attempt when the first has failed (for which the probability is p_c).

Figure 7 shows calculated and simulated results for the blocking probability with retry as a function of offered traffic. It can be seen that p_c^2 (solid line) describes the probability of blocking with fair accuracy, but also that it is a slight underestimate at high traffic rates. This is a consistent difference, since the probability of being blocked in the vicinity of a cell with all channels occupied ought to be larger than in a random cell. In spite of this, p_b can be assumed to give a good description of the overall blocking probability. For situations with evenly distributed traffic load, Directed retry can indeed increase channel utilization or decrease blocking for a random subscriber. Whether or not the facility should be implemented depends on the effects on unfavoured subscribers and on the cost of increased program complexity in the switching office.

CONCLUSIONS

The analyses of this paper have shown that a substantial gain can be accomplished by the use of Directed Retry. The increased number of channels potentially available to the average subscriber can be used either to decrease blocking or to increase channel utilization. Subscribers that are unable to hear more than one transmitter suffer from the scheme, however, because traffic that flows in from surrounding cells occupies channels in their cell. Blocking in a loss system is relatively insensitive to load increase up to a certain level, at which change is more dramatic. This effect is even more apparent in a system with Directed Retry and dimensioning has to be rather conservative.

It is quite obvious that the overlap probability could be used as a system parameter, by change of transmitter power, but disadvantages, such as increased co-channel interference, may appear.

It has been shown that an iterative analytical model can be used to model the system and that the accuracy of such a model is very good. This is a great advantage since simulation results are very costly to produce due to the large number of cells that have to be illustrated in a simulation model.

REFERENCES

[1] IEEE Transactions on Communications: Joint Special Issue on the 1979 World Administrative Radio Conference (WARC-79), Vol. COM-29, August 1981.

[2] H. Pfannschmidt and G. Rolle, "Architecture and Radio Technology of EC 900 - A New High Capacity Automatic Mobile Telephone System", ISS'84, Florence, Italy, 1984.

[3] H. Auspurg, H. E. Binder, Ph. Dupless and J. Lamy, "Advanced Switching Services in a New Cellular Mobile Communication System (EC 900)", ISS'84, Florence, Italy, 1984.

[4] Federal Communications Commision, OST Bulletin no 53, (EIA CIS-3)

[5] T. J. Kahwa and N. D. Georganas, "A Hybrid Channel Assignment Scheme in Large-Scale, Cellular-Structured Mobile Communication Systems" IEEE Transactions on Communications, COM-26, April 1978.

[6] D. C. Cox and D. O. Reudink, "Increasing Channel Occupancy in Large-Scale Mobile Radio Systems: Dynamic Channel Reassignment", IEEE Trans. on Veh. Tech., VT-22, November 1973.

[7] R. I. Wilkinson, "Theories for Toll Traffic Engineering in the USA", B. S. T. J. 35 (1956):2

[8] A. Kuczura, "The Interrupted Poisson Processes as an Overflow Process", B. S. T. J. 52 (1973):3

[9] B. Wallström and L. Reneby, "On Individual Losses in Overflow Systems", ITC 9, Torremolinos 1979.

[10] B. Eklundh, "Simulation of Queueing Structures - A Feasibility Study", Ph. D. Thesis, Department of Telecommunication Systems, Lund Institute of Technology, Lund, Sweden, 1982.

[11] S. H. Bakry and M. H. Ackroyd, "Teletraffic Analysis for Single-Cell Mobile Radio Telephone Systems", IEEE Transactions on Communications, COM-29, March 1981.

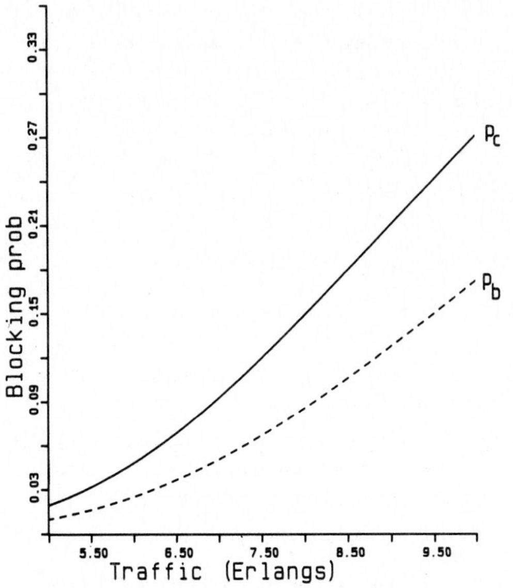

Fig. 4. p_b and p_c versus cell traffic.
p_o = 0.5

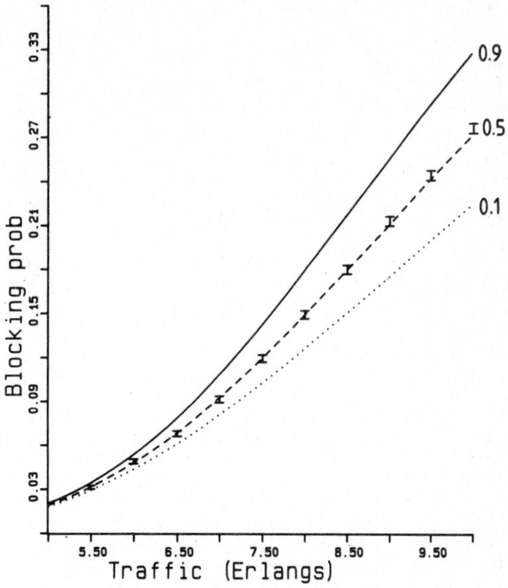

Fig. 5. p_c versus cell traffic.
p_o = 0.1, 0.5 and 0.9

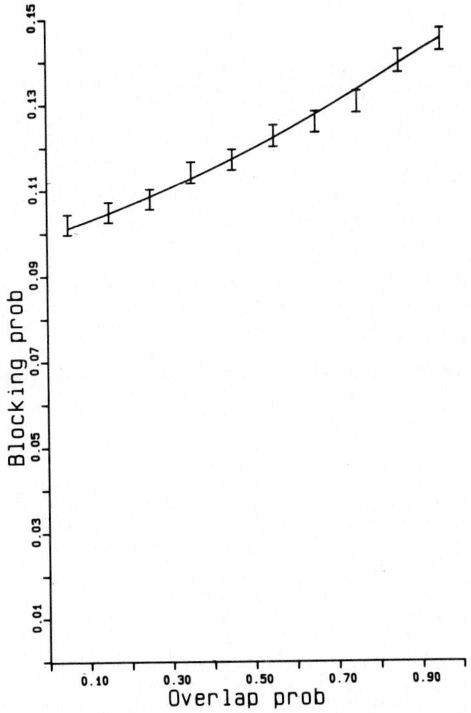

Fig. 6. p_c versus overlap probability.
A = 7.5

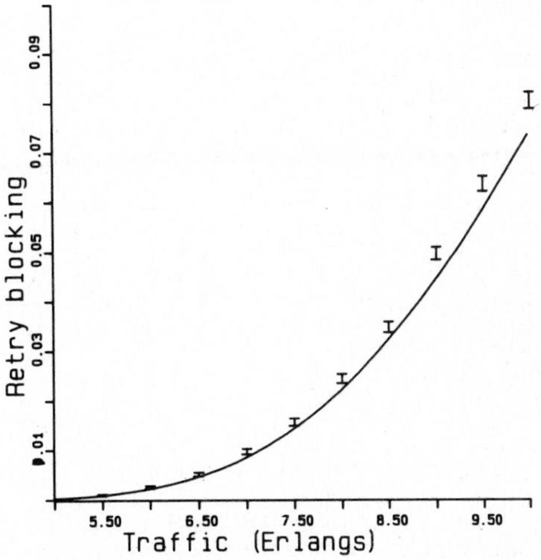

Fig. 7. Blocking probability after retry.
p_o = 0.5

TELETRAFFIC ISSUES in an Advanced Information Society
ITC-11
Minoru Akiyama (Editor)
Elsevier Science Publishers B.V. (North-Holland)

A HEURISTIC PROCEDURE FOR OPTIMIZING A DIGITAL TELEPHONE NETWORK

Ruth KLEINEWILLINGHÖFER-KOPP

Research Institute Of The Deutsche Bundespost
Darmstadt, Germany

ABSTRACT

In this paper a method is given for dimensioning circuit groups in digital networks. The model described includes hierarchical and non-hierarchical networks, bothway operation of the circuit groups, the multiplexing hierarchy of transmission links, and end-to-end blocking constraints. This problem is solved by means of a heuristic procedure which takes into account both all traffic relations and the capacity of the transmission links in each step of the algorithm.

1. INTRODUCTION

An important task in planning the switching network consits in determining the most economical number of circuits between any two switching nodes. Literature gives a variety of dimensioning procedures developed for the application in analog networks. In digital networks, however, there are a number of new technical possibilities which have to be considered in the planning, too. For example, digital networks can assume a hierarchical as well as non-hierarchical structures. Furthermore, bothway operation of the circuit groups is possible and finally the multiplexing hierarchy of the transmission links, which has 30 circuits in the smallest group already, is to be taken into consideration. For the solution of this problem in digital networks a mathematical model is defined and a procedure is given in which, moreover, the grade of service is not defined by values indicating the allowed link blockings on final choice routes, but it is required that the end-to-end blocking value of each traffic relation does not exceed a certain given value. Examples illustrate the method applied and the respective results are discussed.

2. THE MODEL

For the network considered the following is given:

- the set of switching nodes (the set of vertices of the model graph)

$$\mathbb{V} = \{v_i / i = 1,\ldots,N\} \qquad N \in \mathbb{N}$$

- the distances between any to nodes

$$l: \begin{cases} \mathbb{V} \times \mathbb{V} \longrightarrow \mathbb{R}_+ := \{r \in \mathbb{R} / r \geqq 0\} \\ (v_i, v_j) \longrightarrow l(i,j) := l(v_i, v_j) \end{cases}$$

- and the total traffic which is to pass from any switching node to any other switching node

$$AV: \begin{cases} \mathbb{V} \times \mathbb{V} \longrightarrow \mathbb{R}_+ \\ (v_i, v_j) \longrightarrow AV(i,j) := AV(v_i, v_j) \end{cases}$$

As the internal traffic of a switching node is not offered to the network, we set without loss of generality

$$AV(i,i) = 0 \qquad \forall\, v_i \in \mathbb{V}.$$

<u>Definition 1:</u>

A tupel $(v_i, v_j) \in \mathbb{V} \times \mathbb{V}$ is called <u>traffic relation</u> with origin v_i and destination v_j, iff $AV(i,j) > 0$. The set of all traffic relations is given by

$$\mathbb{VB} := \{vb(i,j) := vb(v_i, v_j) / v_i, v_j \in \mathbb{V}, AV(i,j) > 0\}$$

and the quantity by $K := |\mathbb{VB}|$.

The hierarchy of the switching nodes and the method of setting up a call - information assumed to be known - determine the routes (represented as a sequence of nodes) on which the traffic $AV(i,j)$ can pass through the network in order to get from origin v_i to destination v_j.

Knowledge of the traffic routes for all traffic relations implies that also the switching nodes are known which act as transit nodes, and that the set of links, i.e. the edges of the model graph

$$\mathbb{E} := \{e_i / i = 1,\ldots,M\} \qquad M \in \mathbb{N},$$

is given as follows:

> Two nodes v_i and v_j are incident to an edge in $\mathbb{E}$ if and only if there is a traffic route in which v_i is directly followed by v_j or v_j is directly followed by v_i.

For dimensioning a network, the traffic routing strategy applied must be known. However, the model can also be formulated independently of the strategies used since the effect of all strategies is the same in the following respect: Each of them provides a selection of traffic routes and specifies the conditions under which a traffic route is examined as to whether or not it can be occupied. This is referred to as "weight on the traffic routes".

Definition 2:

The set of all possible traffic routes, together with the weight on the routes induced by the routing strategy is called list of traffic routes.

Of course, this list of traffic routes does not include information about the link capacities

$$n_i \in \mathbb{N} \cup \{0\} = \{0,1,2,3,\ldots\}$$

we are looking for. Notice that $n_i = 0$, $i = 1,\ldots,M$, is allowed, i.e. not all edges in $\mathbb{E}$ must necessarily be installed. This is to be decided in consideration of the costs

$$K_i(n_i) \in \mathbb{R}_+ \qquad i = 1,\ldots,M,$$

resulting from the installation of the edge e_i with capacity n_i. In addition, a grade of service is to be provided for which the condition is set up that

> the value of the end-to-end blocking EEB_{ij} of each traffic relation vb(i,j) is not higher than a given parameter $\varepsilon_{ij} \in \mathbb{R}$.

It is well known that EEB_{ij} depends on the link blockings

$$x_i \in [0,1] \subset \mathbb{R} \qquad i = 1,\ldots,M,$$

defining the probability $P(e_i \text{ is busy})$ of all circuits n_i of e_i being busy.

The above conception is now illustrated by the following example.

Example 1:

In a non-hierarchical network with four switching nodes $\mathbb{V} = \{v_1,v_2,v_3,v_4\}$ two traffic relations vb(1,2) and vb(1,3) are given. The list of traffic routes is shown in Table 1.

As can be seen from the table, the routing strategy (alternative with step-by-step routing) defines an order of the traffic routes for each traffic relation induced by the conditions (column 'weight') under which the traffic route is examined with regard to whether or not it can be occupied. The sign (-) indicates that this route is examined first without any conditions. The model graph thus defined is given in Fig. 1.

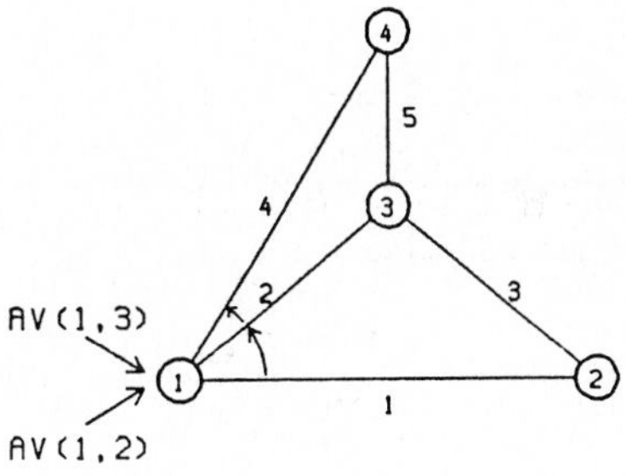

Fig. 1 Example 1

Table 1 List of traffic routes

vb	number	nodes	edges	weight
(1,2)	1	v_1-v_2	e_1	-
	2	v_1-v_3-v_2	e_2-e_3	e_1 is busy
	3	v_1-v_4-v_3-v_2	e_4-e_5-e_3	e_1 and e_2 are busy
(1,3)	1	v_1-v_3	e_2	-
	2	v_1-v_4-v_3	e_4-e_5	e_2 is busy

The list of traffic routes also provides information to determine the EEB_{ij}. For this purpose, we define the event

$$AW_k(i,j) = (\text{traffic route } k \text{ of } vb(i,j) \text{ is offered a traffic}) ,$$

which represents the condition set up in column 'weight'.

Formula for the EEB_{ij}

The end-to-end blocking of a traffic relation vb(i,j) which can make use of L different traffic routes is:

$$EEB_{ij} = 1 - \sum_{k=1}^{L} P\Big((AW_k(i,j)) \text{ and } (\text{traffic route } k \text{ is not busy})\Big).$$

For the calculation of these values we make the usual

Assumption:

$$P((e_i \text{ is busy}) \text{ and } (e_j \text{ is busy})) = P(e_i \text{ is busy}) \cdot P(e_j \text{ is busy}) \quad \forall\, e_i, e_j \in \mathbb{E},\ i \neq j$$

which means that the events are stochastically independent.

Thus, the model is as follows:

Given

- the (undirected) graph $G(\mathbb{V},\mathbb{E})$ with the set of nodes $\mathbb{V}$ and set of edges $\mathbb{E}$
- the traffic matrix $AV = \{AV(i,j)/v_i, v_j \in \mathbb{V}\}$
- the distances $l = \{l(i,j)/v_i, v_j \in \mathbb{V}\}$
- the list of traffic routes
- the grade of service parameters $\varepsilon_{ij} \in [0,1] \subset \mathbb{R} \quad \forall\, vb(i,j) \in \mathbb{VB}$
- the costs for switching and transmission equipment

Required

the number of circuits $n_i \equiv 0 \mod 30$, $\forall\, e_i \in \mathbb{E}$, resulting as the solution of the problem

$$\text{(B)} \qquad \text{Minimize} \quad C = \sum_{i=1}^{M} K_i(n_i) \qquad \text{subject to} \quad EEB_{ij} \leqq \varepsilon_{ij} \quad \forall\, vb(i,j) \in \mathbb{VB}$$

3. THE HEURISTIC PROCEDURE

Since the values $K_i(n_i)$ depend also on the transmission equipment which is partly not known before the transmission network is calculated, the problem (B) can hardly be solved in its given generality. That is why we make, as usual, the following approach

$$K_i(n_i) =: c_i \cdot n_i \qquad i = 1,\dots,M,$$

which, however, requires an adaptation of the factors c_i to the given situation by an iteration between switching and transmission network optimization ([3]).

The functional relations between n_i, x_i and traffic A_i offered to e_i result from the traffic theory applied. If A_i is Poissonian, the relation is given by the Erlang formula

$$x_i = E(n_i, A_i) = \frac{A_i^{n_i}}{n_i!} \Bigg/ \sum_{j=0}^{n_i} \frac{A_i^{j}}{j!} .$$

It is well known that, if the traffic offered has other stochastical attributes (for example overflow traffic), the relations are much more complicated. To dimension a link which is offered non-Poissonian traffic, Wilkinson proposed the "equivalent random method" [5], which is widely used in the procedures for planning a switching network. Applying this method one has to distinguish between links offered Poissonian traffic and those offered overflow traffic. To meet the different dimensioning functions and the end-to-end blocking constraints in the procedure given by Blaauw [2] the whole problem is divided into the dimensioning of the 'high usage links' on the one hand and into the dimensioning of the 'final choice links' on the other. Since the problem dealt with here can hardly be decomposed especially when bothway operation of the links is considered, we suggest now another method:

> Each traffic offered to a link is set to be a Poissonian traffic, with overflow traffic being replaced by a Poissonian traffic with a higher mean.

This is done by applying a function A_E that depends on the traffic offered to the traffic route the overflow traffic comes from, the probability that the traffic overflows, and the blocking of the link to which the overflow traffic is offered.

Although being only a kind of 1-moment-method for dimensioning the links, this approach offers the advantage that no further decomposition of the problem (the network) is required. (When setting up function A_E each overflow traffic was described by an Interrupted Poisson Process [4]. Thus, also the higher moments were taken into account).

Each of the A_i can now be represented as a function of the values given in the traffic matrix and the blocking vector $x = (x_1,...,x_M)$, i.e. $A_i = A_i(AV,x)$. Since the end-to-end blockings are also functions of x, we can use x_i, $i=1,...,M$, as independent variables of our problem (B). The number of circuits is then calculated by applying the Erlang formula for which an 'inverse function' $nn(A_i,x_i)$ of $E(n_i,A_i)$ was determined.

For solving (B), the use of a nonlinear optimization method seems to be obvious, which, however, involves problems arising from the fact that the objective function is 'highly' nonlinear, nonconcave and nonconvex. Moreover, in contrast to the problem solved by Berry [1], we have to choose $n_i \equiv 0 \bmod 30$, which may have the consequence that, when observing this condition, the next point x determined by using the optimization method, leads to higher costs C.

For these reasons, a heuristic procedure has been developed, the main steps of which are indicated in the following:

1. Choose a feasible point $x \in [0,1]^M$, i.e. a point meeting the constraints

   $$EEB_{ij}(x) \leqq \varepsilon_{ij} \qquad \forall\ vb(i,j) \in \mathbb{VB}$$

2. Calculate the 'flow' in the network, i.e. the traffic offered to each link

   $$A_i(AV,x) \qquad \forall\ e_i \in \mathbb{E}$$

3. Calculate $n_i = nn(A_i,x_i) \qquad \forall\ e_i \in \mathbb{E}$

   and $$C = \sum_{i=1}^{M} c_i \cdot n_i$$

   If $n_i \equiv 0 \mod 30 \qquad \forall\ e_i \in \mathbb{E}$

   and x feasible go to step 4.

   If not, choose an appropriate, feasible x and go to step 2.

4. "Heuristic for improvement"

   Examine for each traffic relation vb(i,j) whether a reduction of the costs C can be achieved by another distribution of the traffic AV(i,j) to the different traffic routes.

   If no traffic relation was determined, go to step 5.

   Otherwise choose x as proposed and go to step 2.

5. End with the best solution $x^{(opt)}$, $n^{(opt)}$, $C^{(opt)}$.

A basic problem arising in solving (B) is that the total number of circuits required depends on the flow distribution in the entire network. This is illustrated in the following example.

Example 2:

A traffic relation vb(1,3) with AV(1,3) = 80 Erl. can make use of two routes. The first route uses the edges e_1 and e_2 and the second one the edges e_3 and e_4, the latter being chosen only if edge e_1 is busy. The required grade of service is $\varepsilon_{13} = 0.55$ and the costs per unit are $c_1 = c_2 = 1$ and $c_3 = c_4 = 1.5$.

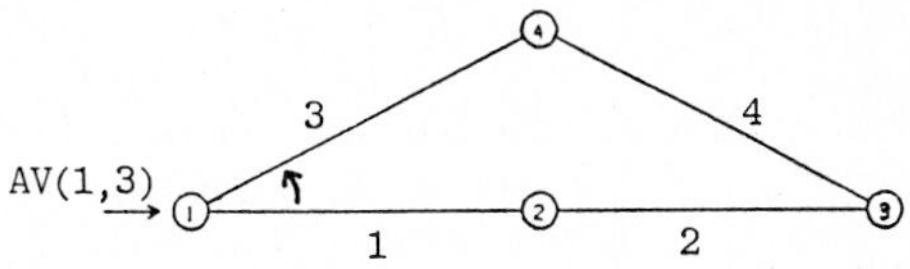

Fig. 2 Example 2

The grade of service is achieved if the traffic is carried only on one route (e.g. the second one, cf. Fig. 3) which has 60 circuits per edge.

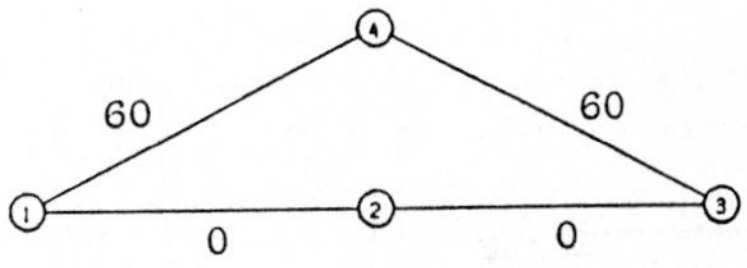

Fig. 3 Feasible solution, C = 180

In contrast, this does not apply to the solution shown in Fig. 4, since $EEB_{13} \simeq 0.66 > \varepsilon_{13}$ although the number of circuits is the same as in Fig. 3.

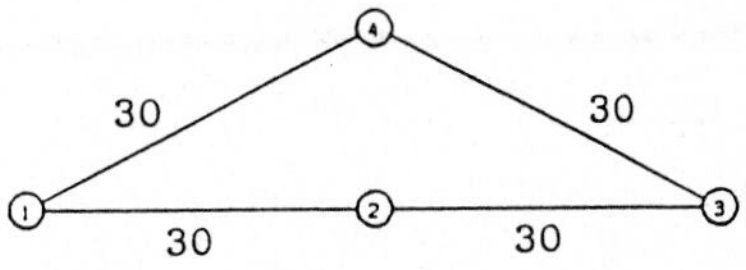

Fig. 4 Unfeasible solution

In a mathematical optimization procedure it is determined by which direction and steplength a given solution is to be modified in order to achieve an improvement of the solution.

Due to the multiplexing hierarchy of the transmission links a "steplength" is given as a number of circuits which in our case is a multiple of 30. Furthermore, the solution can be modified only by adding or removing circuits on at least one edge, i.e. by "rerouting the traffic". Even if the right direction is chosen temporary increases in costs cannot be excluded. Under the condition that exactly 30 circuits per edge are to be installed on the first route, the unfeasible solution in Fig. 4 of example 2 leads to the feasible but more expensive solution in Fig. 5.

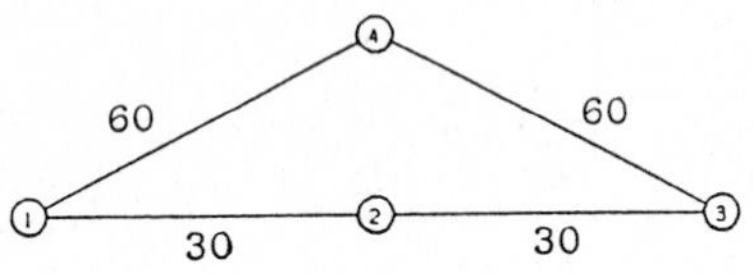

Fig. 5 Feasible solution, C = 240

Another rerouting of 30 circuits from the second to the first route yields the optimum solution.

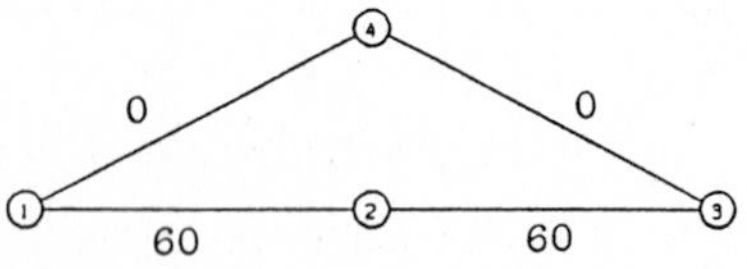

Fig. 6 Optimum solution, C = 120

If the procedure presented here is applied to this example the solution in Fig. 5 is not taken into account. Based on the feasible starting point

$x_1 = x_2 = 1$ and $x_3 = x_4 = 0.2$ (step 1),

the algorithm (step 2 and 3) calculates as the first feasible solution the one shown in Fig. 3. The heuristic for improvement indicates that when using the first route costs can be reduced ("direction"), and determines a total of 60 circuits as one "steplength", which immediately leads to the optimum solution (Fig. 6).

In more extensive examples with several traffic relations, however, more expensive interim solutions (cf. Fig. 5) cannot be avoided due to the mutual interference of the traffic on common edges. That is why the algorithm does not represent a method of descent as far as the costs are concerned, although it allows the effects of the flow distribution on the overall network to be observed quite well.

4. APPLICATIONS

This algorithm was applied to a number of network examples which differed from one another with respect to their structures, the number of nodes (N), edges (M), and traffic relations (K). In these examples the dependence of the solution on the starting point selected was examined. The result was that various solutions were obtained which, however, normally differed only slightly. The time needed for calculating the first feasible solution mainly depends on the number of traffic relations and edges (cf. steps 2 and 3 of the algorithm). In an example with the values K = 1550 and M = 820 not more than 60 secs. were required on an IBM 8031. The calculation time for the remaining iterations is shorter since in these cases only the number of traffic relations whose traffic was rerouted must be taken into account.

5. REFERENCES

[1] L.T.M. Berry, "An Application of Mathematical Programming to Alternate Routing," A.T.R., vol.4, no.2, pp.20-27, 1970.

[2] A.K. Blaauw, "Optimal Design of Hierarchical Networks with Alternate Routing Based on an Overall Grade of Service Criterion," Networks, Paris, pp.116-123, 1980.

[3] K.D. Hackbarth, "A Heuristic Procedure for Calculating Telecommunication Transmission Networks in Consideration of Network Reliability," to appear in Math. Progr. Study, vol.MP 428, 1985.

[4] A. Kuczura, D. Bajaj, "A Method of Moments for the Analysis of a Switched Communication Network's Performance," IEEE Trans. Comm., vol.COM-25, no.2, pp.185-193, 1977.

[5] R.I. Wilkinson, "Theories of Toll Traffic Engineering in the U.S.A.," Bell Syst. Tech. J., vol.35,no.3, pp.421-514, 1956.

TELETRAFFIC ISSUES in an Advanced Information Society
ITC-11
Minoru Akiyama (Editor)
Elsevier Science Publishers B.V. (North-Holland)

END TO END BLOCKING IN TELEPHONE NETWORKS: A NEW ALGORITHM

Jean-Pierre GUERINEAU and Jacques LABETOULLE

Centre National d'Etudes des Télécommunications
CNET/PAA/ATR 38 Rue du Général Leclerc
92131 Issy Les Moulineaux, FRANCE

ABSTRACT

In this paper, we present a new method for calculating point-to-point traffic stream proba bilities within a telephone network. This method takes into account the baulking effects by adding virtual capacities to some of the trunk groups.

The paper firstly reviews other existing methods for calculating point-to-point grades of service and then describes the new algorithm. It concentrates on the differences between the new algorithm and the existing ones and presents a comparison of results.

This new method appears to give good results and is easy to implement. It can be easily introduced in a planning tool for calculating networks in which significant losses are accepted.

1. INTRODUCTION

In the design of telephone networks, planners minimize the cost under constraints of grade of service to obtain alternative routing tables and trunk group sizes. Different parameters have been chosen for use as grade of service criteria such as the loss probability on last choice trunk groups [1], and more recently other methods have been developed taking into account the point-to-point losses [2]-[3].

The method of [1] uses a simple model of the network to compute the loss probability [4] and the algorithm is not time consuming. [4] gives good results when losses are low and this is the case when the planner designs the network under normal conditions. When designing private networks for which high losses have to be accepted, errors become significant and this phenomenon also occurs in the case of link failures and concentration of traffic in one direction.

The method of [2] uses a simplified algorithm to compute the end-to-end loss probability and can be used for large networks. The algorithm of [3] takes into account the blocking probability on a path and not only on a single trunk group, this means that a call blocked on a trunk group of a path has to be removed from the previous trunk groups of its path. This algorithm being based on an iterative method, the computer time is very long which prevents its application to large networks. So several methods have been used to analyse the end-to-end loss probabilities of telephone networks, but their complexity is too high to implement them in a planning tool. We present in this paper, a new algorithm more precise than that described in [2] and which can be easily introduced in a planning tool.

Firstly, we present a brief survey of existing analysis techniques, and then we describe our method. Finally, we give results on several networks and we compare them with exact solutions or simulations and other approximate methods.

2. NETWORK MODEL

In most of the methods, some assumptions have been used for network analysis and a certain number of them are the same in all methods.

- Originating traffics are Poisson.
- Call holding times are independent with exponential distribution function.
- Network is in statistical equilibrium.
- Each exchange has full access to outgoing trunks.
- Call set-up times are negligible.
- Blocked calls are cleared and do not repeat attempt.
- The occupancy distributions of trunk groups are independent.

Some methods take into account the dependence of trunk groups [5] but these methods can be used as references to test validity of approximate methods. They are too complex to deal with large networks and we will not discuss them in this paper.

The main differences between traffic models are the following :

- depending on the method, streams of overflow traffic are described by their moments (one, two or first three moments) and can be modelled as renewal processes [6] or with "equivalent random theory" [7],
- streams of carried traffic are modelled as Poisson processes or as smooth traffics [8],
- some of the methods distinguish the indi vidual streams by apportioning the total overflow.

3. PRESENTATION OF METHODS

3.1. The Simplest Model

Firstly, we present the method of [4] which applies Erlangs formula and the equivalent random theory. It does not take into account the fact that a call which is carried on a given link may be blocked further along its path and so it ignores the effects of baulking (this assumption is valid when the losses are low). The method is well suited to hierarchical networks in which the list of trunk groups can be ordered and the traffic offered on a given link (i,l) can be obtained easily. Then carried and overflow traffic are computed and offered to corresponding trunk groups. So, the main advantage is that the list of trunk groups is explored only once. The point-to-point loss probability is obtained by scanning the list of trunk groups in reverse order. The main assumption made here is that the trunk groups are independent.

It can be seen that it is not necessary to perform any iteration, so this algorithm is fast and can be applied to large networks. In the case of high losses, errors become significant due to the simplifications in the model. A certain number of models have therefore been developed to improve results, by considering losses on a path and they are discussed below.

3.2. Use of a Reduced Offered Traffic

The various models presented in this section take into account the fact that a call going through a trunk group could be blocked further on the path. In general, these methods consider the traffic offered to a trunk group (i,k) to be of the form

$$t = \sum_{j\in\xi} t_{ij} - \tau$$

where ξ is a set of all destination nodes to which calls are sent from i via k and τ is a traffic corresponding to the blocked traffic on the rest of the paths in which (i,k) is included. The methods differ mainly in the way they compute τ and also in the number of moments taken into account.

Katz [9] was the first to introduce this method. To represent routing schemes, a route tree generation (see example in Fig. 1) is build.

In the paper, a load assignement procedure is considered where the mean and variance of offered parcel are adjusted. An adjustment factor is required to account for the increased availability of links seized by calls which fail to reach their destination. The effect of the adjustment is to reduce the offered loads to effective values from which link blocking probabilities may be computed. This adjustment factor is obtained empirically from simulation results on a certain number of networks. Different formulae are employed for the link overflow parameter calculations, depending on the variance to mean ratio of the link offered load. An iteration procedure is performed to obtain point-to-point loss probabilities in which the initial values of these probabilities are set to zero and in each subsequent iteration, calculations are performed to obtain improved estimates of the traffic parameters. These calculations use the network routing plan and the traffic parameters computed in the previous iteration. The main disadvantage of this method is the use of a correction factor which is obtained empirically from simulation results.

The paper [10], using similar criteria, proposes analytic procedure to obtain the adjustment factor. In the paper two philosophies are considered for calculating the traffic offered to a group. In the first one, the traffic offered to a trunk group is the traffic carried by the route plus the traffic lost because of the trunk group congestion, this is the hypothesis of Katz'paper. In the second one, the traffic offered to the trunk group is the traffic carried by the groups of the path except the group under examination. By comparison with simulations, it is shown that the first philosophy generally overestimates losses and the second one underestimates them.

Two papers are presented by Manfield and Downs [11]-[12] in which the main difference lies in the number of moments used, in the first one, one moment is used and in the second three moments are used to describe the overflow traffic, carried traffics are modelled as Poisson processes. A method of [13] is used to represent uniquely the routing scheme for a network with a given route plan and arbitrary form of route control. Imaginary loss nodes are introduced in the route tree in order to distinguish the ways in which calls are lost in the system. Fig.2 shows an example.

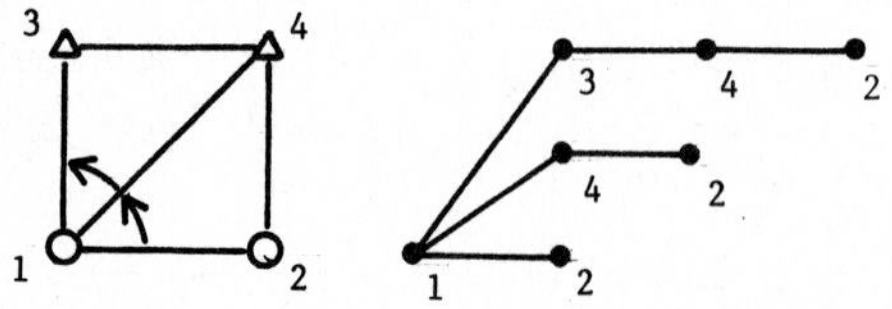

Fig 1. Example of route tree generation

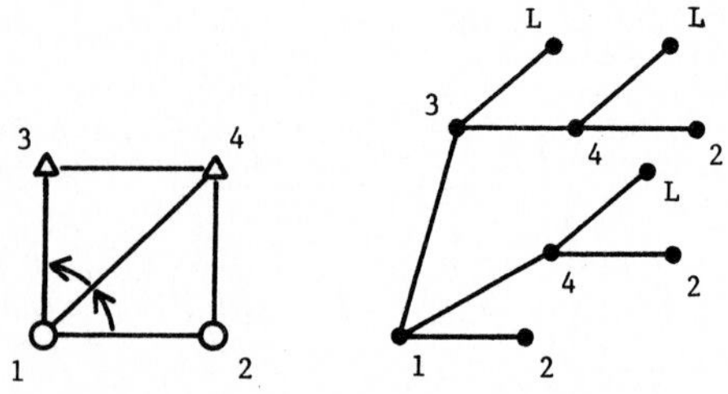

Fig 2. Example of augmented route tree for step-by-step route control.

The paper considers a general link model as a set of limited avalability servers. Considering just the case of a single traffic stream offered to a link, a call which arrives and finds a free server is rejected with a constant baulking probability b according to an independent Bernouilli trial. The offered traffic is decomposed into two processes, one corresponds to a reduced offered traffic a'=a(1-b), the other of intensity "ab" is not accepted on the link. The algorithm is decomposed into two phases. In the first, assuming that the link blocking probabilities and the overflow characteristics of each link are known, the link offered traffics are derived. An important point is that the carried traffic of each link due to one stream of path offered traffic is numerically the same. The second phase consists of link traffic segregation. With the one moment method, results are easy to obtain, but with a three moment method, the carried and overflow traffic belonging to each individual offered stream have to be determined, each stream being modelled as a revewal process by a moment match technique.

3.3. Use of a Fictitious Offered Traffic

In the paper of F. Le Gall and J. Bernussou [14], a one moment model is considered which incorporates the definition of fictitious offered traffic which enables one to take into account the deviation of smooth and peaked traffics from the Poisson. A differential equation can be written for the mean carried traffic in transient state and it can be viewed as a fluid equation. An approximation is made to define a fictitions instantaneous traffic, so that the Erlang formula can be applied even in the transient state. A generalisation is made for networks which require the knowledge of network structure and the knowledge of the routing scheme at each node. A numerical integration of the differential equations until the steady state is reach can be performed. It may be preferable to resolve directly the system for the steady state using relaxation techniques.

4. PRESENTATION OF OUR METHOD

4.1. Aim Objective

When loads are high, the method of [4] is to simple and errors become significant. We can see what happens on a simple example shown in Fig.3.

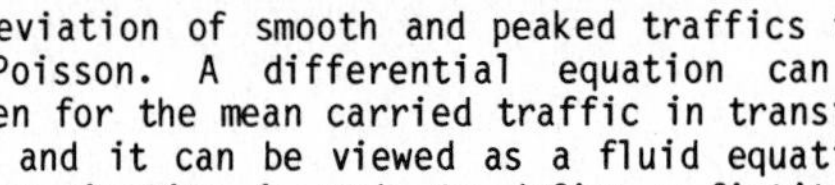

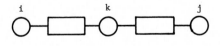

Fig 3. Simple example

We consider a traffic stream which uses links (i,k) and (k,j). In a simple model like one presented in [4], a call carried on link (i,k) and blocked on (k,j) will still be considered as carried by (i,k). So, it can be deduced that $p_2 a_e$ trunks are occupied on (i,k) for calls which were rejected by (k,j), (p_2 being the blocking probability of trunk group (k,j) and a_e the carried traffic on link (i,k) which is offered to (k,j)). So the blocking probability on (i,k) is certainly overestimated by this simple model.

In our method, we propose to take account of calls blocked on a later part of the path. This is done by considering that these calls occupy fictitious trunks on (i,k), so we increase the size of the trunk group by x trunks where $x=p_2 . a_e$. In computing the blocking probabilities, we will have to consider non-integer number of trunks, so adapted formulae should be used.

4.2. Description of the Method

- Model of network.

Consider the assumptions presented in paragraph 2 in which traffic streams are represented by the first two moments. Overflow traffic streams are computed using equivalent random theory and the superposition of carried traffic streams is modelled as a Poisson process. The total offered traffic to a link is the superposition of a Poisson process, and if this link receives overflow traffic, of a renewal process, so using the method of [15], we can decompose the different traffic streams to obtain individual loss probabilities.

- Description of the algorithm.

The algorithm is divided in two phases :

. Knowing the sizes of links, point-to-point losses are computed using the method developed in [4].

. Knowing efficiency rates, additional virtual values are computed.

Let us consider a given trunk group (i,k) noted m : e represents the calculated carried traffic on m. A part of this traffic will be blocked further downstream, and let x_m be this lost traffic. We will increase the size of trunk group m by this same value x_m as in the simple example of Fig 3.

Let ξ be the set of streams offered to link m.

Let $L\tau$ be the set of links occupied by the stream τ, downstream of k.

Let $p_{n,\tau}$ be the loss probability of the stream τ, on link n.

Let $t_{n,\tau}$ be the offered traffic of the stream τ on link n.

We have :

$$x_m = \sum_{\tau\in\xi} \Big(\sum_{L\tau} p_{n,\tau} . t_{n,\tau} \Big) \quad (1)$$

Let $e_{k,\tau}$ be the efficiency rate for each pair of nodes and for each stream.

$$e_{k,\tau} = \frac{\text{carried traffic of stream } \tau \text{ from k to destination of } \tau}{\text{offered traffic of stream } \tau \text{ from k to destination of } \tau}$$

We also have :

$$x_m = \sum_{\tau\in\xi} t_{m,\tau} .(1-p_{m,\tau}).(1-e_{k,\tau}) \qquad (2)$$

It can be seen that when computing losses on a given link, we have to store in memory the offered traffic and the lost traffic corresponding to each stream. In the same way, when computing the point-to-point loss probabilities, we have to store the efficiency rate for each stream.

If we use exactly the same model as in [4] without apportioning the individual traffic streams, the above formulae become simpler : losses and efficiency rates do not depend upon individual streams, so we have :

$$x_m = \sum_{\tau\in\xi} \left(\sum_{n\in L\tau} p_n . t_n \right) \qquad (3)$$

$$x_m = \sum_{\tau\in\xi} t_m .(1-p_m)(1-e_k) \qquad (4)$$

In (4) e_k depends on the origin k and the extremity of the stream, but does not depend on the stream itself.

We have supposed that the network can be ordered. So by scanning the list of links only once, we obtain loss probabilities on each link. In the same way, scanning the list in reverse order we can compute the efficiency rate for each stream and for each node. Then virtual additional sizes are computed for each link, using equation (2). The algorithm then iterates and stops when the differences between loss probabilities of two consecutive iterations are smaller than a given value.

We have checked the convergence of the method for different networks and under several operating conditions : normal load, small overloads, high overloads. Convergence is obtained generally within two or three iterations. To check the validity of the method, we have changed the starting conditions of the iteration by considering some non-zero, initial values of x_i. In every example, the method converges on the same solution, so we can assume that most of networks could be processed with our method.

Finally, the method can be summarized in the flowchart shown in Fig 4.

5. RESULTS AND COMPARISONS

In this paragraph, we present some results obtained on different networks. First, we compare our method with results presented by Manfield and Downs [12] on the same network and we discuss the differences obtained. Then, we give results on a small network with different loads and we compare results with those obtained by a simulation method. Finally, we test the algorithm on a network composed of links in series and we compare with exact results ; the main interest here is to show where the results differ from those of the exact model and to analyse the results to forecast the types of errors due to the method.

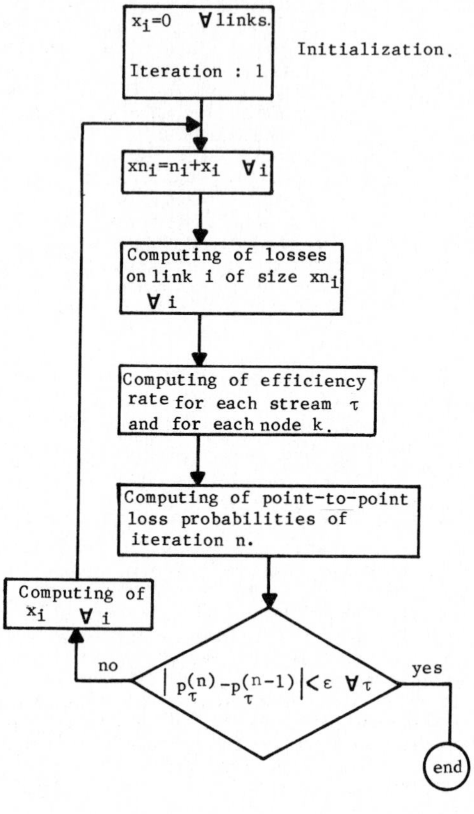

Fig 4. Flowchart of the algorithm

We consider the network of Fig. 5, with step-by-step route control.

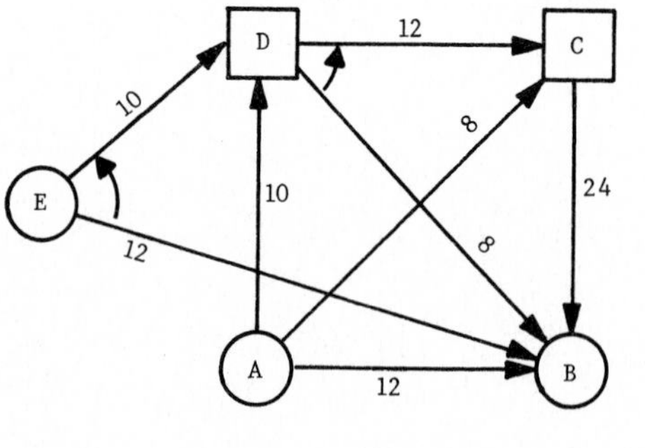

Fig 5. First Network

The number of trunks in each link is indicated adjacent to the link. A and E are origin exchanges, B is the unique destination exchange, C and D are tandem exchanges. It is assumed that there is also external traffic arriving at C and D via B. This traffic represents traffic coming from other parts of the overall network. We assume this traffic is Poisson since it must already have been carried on other links. The alternative routes from the various exchanges are indicated with arrows (an arrow indicates the overflow). From A to B, we have two overflow links : respectively A-C and A-D. From D to B, we have one overflow link : D-C and form E to B, one overflow link E-D.

We compare our method with Manfield and Downs [12], for normal condition in table I, and for overload condition in table II.

stream	offered Traffic	G.O.S (%) Proposed method 1st step	final results	Method of [12]	simulation
A-B	20	3.9	3.7	3.7	4
C-B	8	7.1	7.7	7.7	7.8
D-B	6	6.2	5	5.1	6.7
E-B	16	5.2	4.5	4.5	4.4

Table I - Example 1 - Normal condition

stream	offered traffic	G.O.S Proposed method 1st step	final results	Method of [12]	simulation
A-B	24	13	12.3	11.4	11.7
C-B	8	14.6	22.3	17.6	19.7
D-B	8	23	17.5	24.2	18.9
E-B	20	18.6	14.1	16.6	14.1

Table II - Example 2 - Overload condition

In example 1, we obtain the same results as Manfield and Downs. With our method we obtain final results at second iteration. Compared with simulation results, we have a good precision, except for traffic stream D-B where the relative error on the traffic lost is of 25 %. It can be seen that only one extra iteration improves the results of [4], which correspond to the first iteration. In example 2, it can be seen that errors are of the same order of magnitude as in Manfield and Downs, and compared with simulations results the maximum relative error, obtained on stream C-B, has a value of 14 %. The final result is obtained with 3 iterations and compared to [4], the precision is greatly increased. If we consider the total traffic lost in the network, the grade of service is 14.9 % and the simulation gives 14.5 %. As a result, we can see that errors are of the same order of magnitude as in Manfield and Downs, but the grade of service of the whole network is 14.1 % with their method. So, for this example, our method overestimates the total grade of service, the method of [12] underestimates it. In this example, it was not necessary to split into individual streams because on a given link, we have either a Poisson process or an overflow process but not both.

5.2. Second Network

We will present here results on a small network with a mixing of Poisson and overflow traffic on some links, so it will be possible to check the improvement due to the splitting of streams. The network is represented on Fig 6.

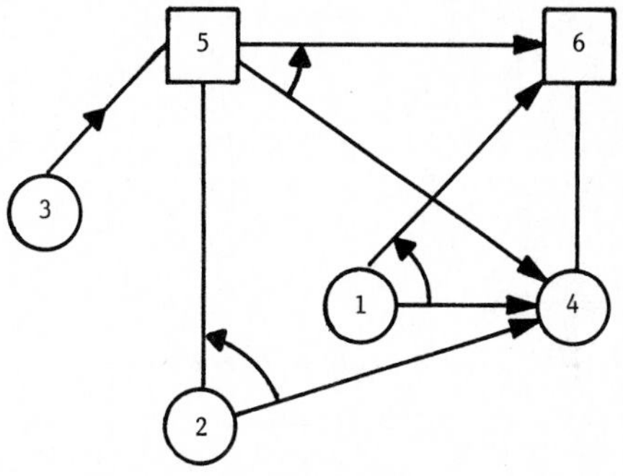

Fig 6. Second Network

Traffics are offered from origin exchanges 1-2-3 to exchange 4. Moreover, two streams are offered from nodes 2 and 3 to the outside of the network, the loss probabilities of these streams on the outside are respectively q_1 and q_2. The alternative routes form the various exchanges are indicated with arrows. We have checked our method with different values of traffic and different sizes of trunk groups. The method has been performed with and without splitting of streams to evaluate the improvement due to the splitting. As reference, a simulation method of QNAP [16] has been used. For example 3 (see table III), we have compared our method with that of [14]. For examples 3, 4, 5, we have q_1=q_2=0.1.

Size of links	Stream	Offered traffic	G.O.S (%) QNAP	Our method without splitting	Our method with splitting	Method of [14]
1-4:4	1-4	5	15.2	16.1	15.9	16.2
1-6:3	2-4	6	15.1	13.2	14.8	19.6
6-4:4	3-4	4	28.5	31.8	31.1	27.1
2-4:5	2-out side	4	29.5	32.9	30.2	25.9
2-5:6						
5-4:4	3-out side	3	27.6	27.7	27.9	28.9
5-6:3						
3-5:8						

Table III - Example 3 - G.O.S. results

Results for example 4 are shown in table IV

Size of links	Stream	Offered Traffic	G.O.S (%)		
			QNAP	Our method without splitting	with splitting
1-4:5	1-4	5	10.6	11	10.9
1-6:3	2-4	4	9.3	7.8	9.1
6-4:3	3-4	4	38	41.1	40.7
2-4:5	2-out	4	24.4	31.6	30.2
2-5:5	side				
5-4:3	3-out	3	29	33.7	33.9
5-6:2	side				
3-5:6					

Table IV - Example 4 - G.O.S results

In example 5, we consider the same network, and we add two streams representing traffics coming of the outside of the network : a stream of 5 to 4 carried on links 5-6 and 6-4 ; a link from 6 to 4 carried on link 6-4.

Size of links	Stream	Offered traffic	G.O.S (%)		
			QNAP	Our method without splitting	with splitting
1-4:4	1-4	5	21.6	21.1	21
1-6:3	2-4	6	16.5	14.9	16.5
6-4:4	3-4	4	33.2	36.2	36.3
2-4:5	5-4	1.2	48	56.7	52.4
2-5:6	6-4	1.2	43.5	39.9	39.5
5-4:4	2-out	4	27.8	32	29.3
5-6:3	side				
3-5:7	3-out	3	25.6	26.2	26.2
	side				

Table V - Example 5 - G.O.S results

The results obtained for tables III, IV, V show a good precision compared to exact results of QNAP. Splitting of streams leads to some improvements of G.O.S, but even without splitting we obtain correct results. The convergence is obtained after 2 or 3 iterations. Also for this example, we decrease error by 50 % from the first iteration to the second or third iteration.

It can be seen that in tables III and IV, the maximum relative error is about 20 % for stream 2 to outside. These examples deal with high overloads and even in these conditions, we obtain correct results for the different streams. In example 5, we have added two small streams to check the robustness of the method when we modify the distribution of streams. For streams 5-4 and 6-4, we have a relative error on the G.O.S of about 10 %.

5.3. Links in series

We consider the system composed of two links in series of Fig. 7. We have three traffic streams :

a stream 1 carried on links N_1 and N_2 of intensity a_{12}
a stream 2 carried on link N_1 of intensity a_1
a stream 3 carried on link N_2 of intensity a_2

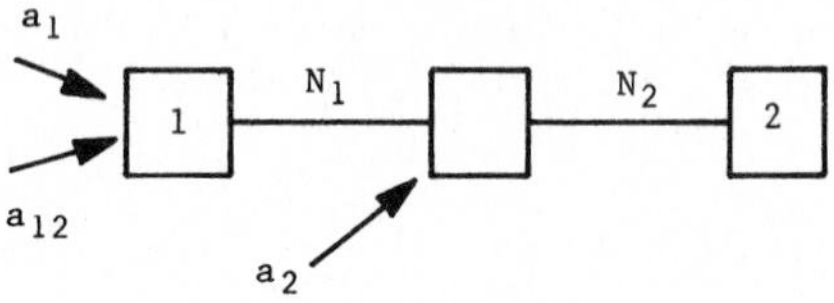

Fig 7 . Links in series

In this example when losses are high, effects due to the dependence of links become significant and we have checked our method to analyse the errors occuring on the G.O.S. Our results have been compared with exact values obtained from QNAP. N_1 and N_2 have been set to small values to compute the Markov system but conclusions can be generalized.

In table VI, we summarize results obtained for example 6, with $N_1=N_2=3$.

a_{12}	a_1	a_2	G.O.S (%)					
			QNAP Flow 1	Flow 2	Flow 3	Proposed method Flow 1	Flow 2	Flow 3
3	0	1	45	-	45	51.9	-	38.6
3	1	0	45	45	-	51	40.3	-
3	0	0	34.6	-	-	44	-	-
3	0.5	0.5	45	33.6	33.6	51.2	31.3	29
1.5	0	0.5	21	-	20.8	27.6	-	18.6
1.5	0.5	0	21	20.8	-	26.7	19.4	-
1.5	0	0	13.4	-	-	21.2	-	-
1.5	0.25	0.25	20.5	17.8	17.8	26.7	14.9	13.8

Table VI - G.O.S - results for links in series

It can be seen that when $a_{12}=3$ and $a_1=a_2=0.5$, the loss probability of stream 1 is overestimated and the losses on streams 2 and 3 are underestimated. In the case of smaller load, we obtain the same kind of results.

The above considerations can enable us to forecast the types of errors in networks. In table V, example 5, we have obtained results which could be forecasted by the above considerations : stream 5-4 is overestimated and stream 6-4 is underestimated, which is consistent with results obtained in table VI.

Using the method of [8] to model smooth traffics, we have applied our method to examples of table VI. We have not found a significant improvement in every case and so, we have decided to keep the original model.

We have also applied our method to examples given in [5] in which to compare the robustness of our method when traffics and number of trunks are modified and we have obtained roughly the same results as Katz [9] and Manfield and Downs [12].

6. CONCLUSION

The method presented in this paper can be applied to hierarchical networks or to networks in which trunk groups can be ordered, with a step-by-step routing control.

Comparisons with other methods show an equivalent precision over a wide range of network types. The use of virtual capacities to take into account the baulking effects is easy to implement and leads to a method which is robust and converges rapidly. An important advantage is the possibility to introduce this algorithm for calculation of point-to-point loss probabilities in a planning tool ; the computing time is doubled for the calculation of the grade of service but the improvement is important and can justify the use of the method.

REFERENCES

[1] C.W. Pratt, "The concept of marginal overflow in alternate routing", Austral. Telecom. Res, vol.1, no.1-2, pp.76-82, 1967.

[2] J. Dressler et al., "COST201 : A European Research project. A Flexible procedure for minimizing the costs of a switched network taking into account mixed technologies and end-to-end blocking constraints", Proc. 10th Int. Teletraffic Congress, Montreal, 1983.

[3] A. Girard and Y. Cote, "Sequential routing for circuit switched networks", I.E.E.E Trans.on.Com., vol.32, no.12, pp.1234-1239, 1984.

[4] B. Camoin and S. Guitonneau, "Network design taking into account breakdowns and traffic overloads", Proc. 9th Int. Teletraffic Congress, Torremolinos, 1979.

[5] D.R. Manfield, "Carried traffic in circuit-switching networks", A.E.U, vol.35, no.9, pp.360-368, 1981.

[6] A. Kuczura and D. Bajaj, "A method of moments for the analysis of a switched communication network's performance", I.E.E.E Trans.on. Com., vol.25, no.2, p.185-193, 1977.

[7] R.I. Wilkinson, "Theories for Toll Traffic Engineering in the U.S.A", B.S.T.J, vol.35, pp.421-514, 1956.

[8] B. Sanders, W.H. Haemers and R. Wilcke, "Simple Approximation techniques for conges tion functions for smooth and peaked traffic", Proc. 10th Int. Teletraffic Congress, Montreal, 1983.

[9] S. Katz, "Statistical Performance analysis of a switched communications network", Proc. 5th Int. Teletraffic Congress, New-York, 1967.

[10] M. Butto, G. Colombo and A. Tonietti, "On Point-to-Point Losses in communications networks", Proc. 8th Int. Teletraffic Congress, Melbourne, 1976.

[11] D.R. Manfield and T. Downs, "On the one-moment analysis of telephone traffic networks", I.E.E.E Trans.on Comm., vol.27, no.8, pp.1169-1174, 1979.

[12] D.R. Manfield and T. Downs, "A moment method for the analysis of telephone traffic networks by decomposition", Proc. 9th Int. Teletraffic Congress, Torremolinos, 1979.

[13] P.M. Lin, B.L. Leon and C.R. Stewart, "Analysis of Circuit Switching Networks Employing Originating Office Control with Spill-Forward", I.E.E.E Trans.on.Comm., vol.26, no.6, pp.754-765, 1978.

[14] F. Le Gall and J. Bernussou, "An Analytical formulation for grade of service determination in telephone Networks", I.E.E.E Trans.on.Comm., vol.31, no.3, pp.420-424, 1983.

[15] H. Hakimaru, H. Takahashi and T. Ikeda, "Optimum design of alternative routing systems with constrained individual call losses", Proc. 10th Int. Teletraffic Congress, Montreal, 1983.

[16] J.J. Guillemaud, D. Chanderisis, M. Veran and P.H. De Rivet, "QNAP" (Queing network Analysis Package) Reference Manual CII Honeywell Bull and INRIA, 1980.

TELETRAFFIC ISSUES in an Advanced Information Society
ITC-11
Minoru Akiyama (Editor)
Elsevier Science Publishers B.V. (North-Holland)

ANALYTICAL MODELS OF A BOC TRUNK PROVISIONING PROCESS

DEBORAH A. ELSINGER and CHARLES D. PACK

BELL COMMUNICATIONS RESEARCH
Red Bank, New Jersey
USA

ABSTRACT

We derive mathematical models for a simplified, generic trunk provisioning process. The models are significantly more general than those of the Trunk Provisioning Operating Characteristic (TPOC) studies, and they build on previous work in traffic models, traffic forecasting, trunk planning, demand servicing, and modular engineering.

The results are closed-form mathematical expressions for the distributions of such quantities as realized trunks, forecasted trunks and blockings, and reserves and shortages of trunks as a function of various provisioning practices and load characteristics. The distributions can be used to describe the tail behavior effects of any combination of provisioning policy and load characteristics. That is, they relate assumptions about forecasting, servicing, and traffic attributes to their effects on fractions of groups in either the reserve or shortage categories. They also can be used to illustrate the limited improvements in these quantities that can be achieved solely by modifying provisioning practices.

The models have been verified via field data.

1. INTRODUCTION

1.1 Simplified, Generic BOC Trunk Provisioning Process

As indicated in [1] and shown in Figure 1, a simplified model for the BOC trunk provisioning process is composed of five generic functions: measuring network traffic (I); projecting (forecasting) traffic (II); engineering (sizing) the network in a static and deterministic fashion (III); expanding (or contracting), in an orderly manner, the network's capacity to account for load dynamics, uncertainty and a variety of costs (IV); and adjusting network capacity on an emergency basis ("demand servicing") (V).

1.2 Previous Work

At previous ITCs and in other publications, we and our former colleagues from AT&T and Bell Laboratories, addressed improvements in these generic planning functions. For example, the work by David and Pack [2] and by Moreland [3] provided the basis for new traffic forecasting algorithms. The efforts of Hill and Neal [4], Neal [5], Elsner [6], and Ash *et al* [7] have significantly improved traffic modeling and engineering procedures, resulting in more accurate trunk estimates and more efficient network utilization. The notion of formalizing and systemizing the planned trunk network capacity function, recognizing load characteristics and costs, was described by Kashper, Pack and Varvaloucas [1] and Kashper and Varvaloucas [8]. Finally, the process of accounting for the statistical variability of traffic measurements in deciding when and how much capacity should be added on an emergency (demand) basis is described in a paper by Szelag [9], using many of the notions in [4,5].

1.3 Some Key Relationships

While this generic process in Figure 1 is almost universally employed, to various degrees of sophistication, the interrelationships among these key functions (I-V) are not well understood. (We will discuss briefly, below, the pioneering work in this area, the "TPOC" models by Franks *et al* [10]). However, for a given *network*, one would expect that such a process would result in the key *average* relationships that are "ideally" depicted in Figure 2. (While the plot is of average busy-season trunks, similar notions would apply for, say, quarterly trunk quantities.) That is, for any base year, there exists an unknown true average busy-season trunk requirement, R_T. We use available busy-season measurements (function I) to estimate R_T by R. Then, using a forecasting process (function II), such as SPA [2,3], traffic is projected. The quantity D is the forecast of R_T made 1-year ago. Thus, ideally, D and R are unbiased estimates of R_T, i.e., $E(D) = E(R) = R_T$, with (typically) $\sigma_D > \sigma_R$. The engineering (III) and capacity expansion (IV) functions are combined, as for example in [1], to produce a biased estimate of trunk requirements. This average bias assures adequate service and a cost effective network evolution, with reasonable network utilization despite forecast uncertainties. This "administered" forecast, made *1 year ago* to correspond to the unbiased quantity D, is labeled A. The number of trunks, I, that are actually in service for a given busy season is clearly a function of R, D, A as well as demand servicing actions (function V) that took effect just prior to the busy season. Therefore, I is often well-approximated by planned action A, but may be modified by demand servicing. Finally, using the common, but often misleading terminology, we have

$$R_T - I = \begin{cases} \text{"reserve capacity" if } R_T < I \\ \text{"shortage" if } R_T > I. \end{cases}$$

Of course, $R_T - I = E(R-I) = E(D-I)$ is really the *mean* quantity that we often try to estimate by $R-I$. The distribution of $R-I$, especially, as it relates to models for the key functions in Figure 1, is the heart of this paper. The quantity $R-I$ is used because, from the models of Neal and others, we can make statistical inferences about $R_T - I$ from the easily-measured $R-I$.

1.4 Objectives and TPOC Comparisons

We provide an integrated, analytical model of the simplified BOC trunk provisioning process depicted in Figure 1. The model clearly illustrates the cause and effect relationships among various load characteristics, provisioning algorithms and practices, and network utilization results. The results can be and have been useful in analyzing existing algorithms or the need for new procedures, evaluating the effectiveness of provisioning policies and/or their implementors, and in assessing the impacts of new techniques on the network.

This work, while motivated by different objectives, is similar in spirit to the pioneering Trunk Provisioning Operating Characteristics (TPOC) models of Franks *et al* [10]. However, our approach is quite different, largely due to a different motivation. TPOC has significant modeling detail for the measurement process (since new measurements and their utilization were key TPOC concerns); we allow much more generality in the provisioning model because our concerns included a more global evaluation of the process. The effect is that, for many statistics of interest, we provide more general analytical results, relative to *published* TPOC material, for evaluating the trunk provisioning process. Some specific comparisons are in Table 1.

Finally, because of the differing motivations, our results will be displayed in a different format, typically trading off fraction of groups with significant reserves versus fraction of groups with significant shortages, whereas TPOC tradeoff curves are usually shown as mean trunk reserves versus fraction of groups with significant shortages.

1.5 Overview of Results

We produce closed-form mathematical expressions for distributions of such quantities as realized trunks and blockings, forecasted loads and trunks, and reserves and shortages of trunks as a function of various provisioning practices and load characteristics. These distributions can be used to describe the tail behavior effects of various assumptions on network service and utilization. In fact, a major application is the quantification of the limited benefits of simplistic modifications in provisioning practices (augmentation/disconnect policies) without significant new technologies in forecasting, and in network flexibility, robustness, and control.

Several examples are presented. In one case we show that for an *average* network reserve capacity of 10% (i.e., $E(I-R) \approx .1E(R)$), the fraction of trunk groups with reserve in excess of $3\sigma_{R-R_T}$ may exceed 40 to 50%. The models have been verified via field data.

II. MODELS OF THE PROVISIONING PROCESS

Traditionally, the statistics $R-I$ (*estimated* required trunks minus in-service trunks for a busy season) and $.01-\bar{B}$ (*objective* average blocking - *estimated* average blocking for the busy season) have been used as measures of *network* utilization and service, respectively. However, as we pointed out in Section 1.3, these statistics, by themselves, do not account for the dynamic nature of the provisioning process nor the relationships among the key ingredients. That is, they measure the *effects* of various complex events and practices, but do not hint at or allow analysis of the cause and effect relationships. Thus, they provide no clues as to how to make things better or, even, what is realistically achievable.

2.1 Trunking Models and Distributions

The model for the trunk difference, $R-I$, builds on the relationships depicted (idealistically) in Figure 2. While the figure suggests that $E(D) = E(R) = R_T$, it is not a necessity for our model; however, it is true that $E(R) \approx R_T$ (or can be made so), and SPA [2,3] produces essentially unbiased demand forecasts ($E(D) \approx R_T$). Except for demand servicing activity, which should be less frequent based on [8] and [9], it is usually true that $E(A) \approx I$. In fact, as we will show, if average reserve ($E(I-R)$) is sufficiently large, then the demand servicing component is negligible.

2.1.1 Model

Algebraic decomposition results in the following identity:

$$R-I = (R-R_T) + (R_T-D) + (D-A) + (A-I), \quad (2.1)$$

where $(R-R_T)$ is the trunk estimation error [4,5], (R_T-D) is the demand forecast error [2,3], and $(D-A)$ is essentially a trunk bias term (*given* D) resulting from the usual augmentation and disconnect practices. The TPOC analyses described a family of simple policies for producing A from D [10], whereas the TIP algorithms systematized and customized practices based on forecast and load characteristics, group type, and a variety of costs [1]. The quantity $(A-I)$ is a measure of trunks added in demand servicing (in excess of A) prior to the imminent busy season. It is assumed here that A is usually implemented well in advance (1-2 months) of the busy season. It must be emphasized that the four terms in (2.1) are not independent. The model (2.1) is valid on a single trunk group level. However, for convenience and statistical reliability, we will later assume that we are modeling an arbitrary (random) trunk group in some large population, e.g., a trunk group category such as primary high usage groups or network finals.

2.1.2 Distributions

Given (2.1), one could take several approaches in deriving a distribution for $R-I$. We will illustrate two.

Normal Approximation: We note that if $E(R) \approx E(D) \approx R_T$ and we make the *restrictive* assumption that $(D-A)$ and $(A-I)$ are deterministic (this can be thought of as having totally *mechanized*, nonrandom conversions of D to A and A to I), then

$$N_{R-I} = \frac{R-I-[(D-A)+(A-I)]}{\sqrt{\sigma^2_{R-R_T}+\sigma^2_{R_T-D}}} \quad (2.2)$$

is approximately normally distributed with mean 0 and variance 1. Of course, this requires the reasonable assertions that *previous* forecast error (R_T-D) and *current* measurement error $(R-R_T)$ are independent and approximately normally distributed.

More General Approximation:

We assume that A and I are related by

$$I = max(A, r) \quad (2.2)$$

where r is the peak prebusy-season trunk *estimate*. (Recall R is a busy-season estimate. We ignore the rare event when I<A.) By definition $r \le R$; we further assume that r is uniformly distributed on [0,R]. This latter assumption is important only if $A < R$ and is quite reasonable if $A >> 0$. Clearly, (2.2) captures the essential relationship between planned and demand servicing and is similar to the model in [1].

Next, for this analysis we choose a relationship between A and D that captures the essential effects of most provisioning practices, allows us to obtain some closed form expressions, and which we have shown, in independent studies, is a fairly robust one as far as obtaining good quantitative results. We assume that $Z = (A-D)/D$ is distributed as

$$P[Z \le z] = P[A \le D(z+1)] = 1-e^{-\lambda z} \quad (2.3)$$

(of course, it is here that we are assuming that (2.1) models an arbitrary group in some population for which (2.3) is, on the whole, valid). The distribution is appealing for two reasons: (i) as is typical, most groups have $A \gtrsim D$, while a few have $A >> D$, and (ii) $1/\lambda$ is a convenient (approximate) parameterization of mean percent reserve capacity because, as can be shown, $E(I-R)\ /\ E(R) \approx E((A-D)/D)$. With these assumptions, we can rewrite (2.1) as

$$R-I = \delta + \epsilon + D - \max(D(Z+1), r),$$

$$= \min(R_T + \epsilon - (Z+1)(R_T - \delta), R_T + \epsilon - r), \quad (2.4)$$

where $\delta = R_T - D$ and $\epsilon = R - R_T$. Thus, the normalized, complementary distribution of $R-I$ is

$$\bar{F}_{R-I}(x\sigma_{R-R_T}) = P\left[R-I > x\sigma_{R-R_T}\right]$$

$$= P\left[\epsilon+\delta(Z+1) - ZR_T > x\sigma_{R-R_T},\ R_T+\epsilon-r > x\sigma_{R-R_T}\right].$$

Now, conditioning on ϵ and using the fact that r is uniform on $[0, \epsilon + R_T]$, we have

$$\bar{F}_{R-I}\ (x\sigma_{R-R_T}) = \int_{\max(x\sigma_{R-R_T} - R_T,\ -R_T)}^{\infty} \bar{F}_\delta\ N_1\, dF_\epsilon(s).$$

where

$$N_1 = \left[(x\sigma_{R-R_T} - s + ZR_T)/(Z+1)\right] \frac{\min(R_T + s - x\sigma_{R-R_T},\ s + R_T)}{R_T + s}$$

Finally, from (2.3) we get

$$\bar{F}_{R-I}\ (x\sigma_{R-R_T}) = \int_{\max(x\sigma_{R-R_T} - R_T,\ -R_T)}^{\infty} \int_0^{\infty} \bar{F}_\delta N_2\ \lambda e^{-\lambda z}\ dz dF_\epsilon(s)$$

where

$$N_2 = \left[(x\sigma_{R-R_T} - s + zR_T)/(z+1)\right] \min\left(1 - \frac{x\sigma_{R-R_T}}{R_T + s},\ 1\right) \quad (2.5)$$

F_δ is a quantification of forecast error relating to a particular forecast process, such as discussed in [1, 2, 3, 8, 10]. For our discussion and examples later we will assume F_δ *is* $N(0, \sigma^2_{R_T - D})$. F_ϵ is derived in [4,5]. Thus, given λ and the true requirement R_T, the characterization is complete. From [2.5], we derive that

$$\frac{E\ (I - R)}{R_T} = \left[\int_0^{\infty} \left(\bar{F}_{R-I}(x) - F_{R-I}(-x)\right) dx\right] / R_T. \quad (2.6)$$

Numerical studies of (2.6) allowed us to quantify the tradeoff between planned and demand servicing stated in (2.2) and to clarify the difference between $1/\lambda$ and average reserve. For example, as $1/\lambda$ increases, the role of demand servicing approaches 0. However, it can be shown that, even under reasonable assumptions, the contribution of demand servicing activity, prior to the busy season, to reserve capacity is nearly always minimal. That is, even for $1/\lambda = 0$, with typical assumptions about $\sigma^2_{R-R_T}$ and $\sigma^2_{D-R_T}$, demand servicing contributes less than 1% to the expected reserve. (Note that unanticipated demand occurring *during* the busy season may result in augmentation *after* the busy season and hence is not reflected in busy-season I). It can be shown that, for $1/\lambda = 0$,

$$\rho = E\left(\frac{I - R}{R_T}\right) \quad (2.7)$$

$$\leq \tfrac{1}{2}\left[\frac{\sigma^2_{R_T - D}}{R_T^2} + \frac{\sigma^2_{R - R_T}}{R_T^2} + \frac{\sigma^2_{R - R_T}\,\sigma^2_{R_T - D}}{R_T^4}\right].$$

Some examples are given in Table 2.

Thus, for most cases of interest, the demand servicing contribution is less than 1% and $E\ (R - I)/R_T\ \approx\ 1/\lambda$.

2.2 Blocking Distributions

In this section, we develop a model for observed blocking. The supplementary notations required are:

- $\hat{B}$ - the busy-season average observed blocking estimated from traffic measurements
- B_T - the true busy-season average blocking
- $\bar{a}$ - the mean offered load
- z - the peakedness of the offered load
- ϕ - the day-to-day variation exponent, which indicates the level of day-to-day variation (low, medium, or high)

Using (2.2), (2.3) and letting $\delta = R_T - D$ and $\epsilon = R - R_T$, we can write

$$I = \max\left(D(Z + 1), r\right)$$
$$= \max\left((R_T - \delta)\ (Z + 1), r\right).$$

Then, the distribution of the number of trunks in service is

$$P\left[I \leq x\right] = F_I(x) = P\left[\ max(R_T - \delta)\ (Z + 1), r) \leq x\right]$$
$$= P\left[\ (R_T - \delta)\ (Z + 1) \leq x, r \leq x\right]$$

Since we assume r, for a given R (or ϵ), to be uniformly distributed over the range $(0, R_T + \epsilon)$, and ϵ and δ to be independent, we can condition on ϵ to get, for $x \geq 0$,

$$F_I(x) = \int_{s = -R_T}^{\infty} P\left[\delta \geq R_T - \frac{x}{Z + 1}\right] \frac{\min(x, R_T + s)}{R_T + s}\ dF_\epsilon(s)$$

Finally, if we let $\bar{F}_\delta(x)$ denote the complement of $F_\delta(x)$, it follows that

$$F_I(x) = \int_{s = -R_T}^{\infty} \int_{z = 0}^{\infty} \bar{F}_\delta\left(R_T - \frac{x}{z + 1}\right) \lambda e^{-\lambda z}\ dz \quad (2.8)$$

where

$$M = \min\left(\frac{\mu_s R_T + \lambda}{R_T + \lambda}\right) dF_\epsilon\ (\lambda).$$

To obtain the corresponding distribution of observed blocking, $F_{\hat{B}}(x)$, we first condition on I and use the two-parameter beta distribution, $F_{\hat{B}}(x \mid I)$, derived in [5]. Then,

$$F_{\hat{B}}(x) = \int_0^{\infty} F_{\hat{B}}(x \mid I)\ dF_I. \quad (2.9)$$

In effect, this is the same as offering traffic, parameterized by $\bar{a}$, z, and ϕ, to a random I, characterized by $F_I(x)$, instead of to R_T as in [5]. The distribution of δ and ϵ are the same as assumed in Section 2.1, above.

III. EXAMPLES

3.1 Standardization and Banding

3.1.1 Standardization

For ease of graphical illustration, we will plot each trunk group distribution in units of σ_{R-R_T}, as shown in (2.5). In this way, the basic graph will be the same for all examples, independent of σ_{R-R_T}, group size (R_T), day-to-day variation, peakedness, and module size. The drawback to this standardization is that the normalizing factor σ_{R-R_T} is largest for inefficient trunk groups (small R_T, large z, large ϕ, etc.), *seeming* to cause some counterintuitive results. For example, for a fixed *average* reserve capacity, large groups seem to have more mass in the "reserve" tails than small groups. In fact, it is really the abscissa units that are different.

3.1.2 Banding

To assist in quantifying the fraction of trunks or groups in the tails of a distribution, we use a "banding" concept, as in [5]. That is, we plot on each curve the 1st, 10th, 90th, and 99th quantile points of the measurement error distribution $f_{R-R_T}(x)$ or $f_B(x \mid R_T)$, as appropriate. As in [5], the values to the left of x_{01} are assumed to correspond to reserves (Band 4) and to the right of x_{99} are shortages (Band 5); $x_{01} < x < x_{10}$ is Band 2, $x_{10} < x < x_{90}$ is Band 1, and $x_{90} < x < x_{99}$ is Band 3. We use slightly different quantiles for the blocking distributions. Also, in order to better interpret the results, we assume that the curves represent a population, e.g., category, of similar trunk groups rather than a single trunk group. Hence, the ordinate is in units of "fraction of trunk groups", while the abscissa is, of course, in units of σ_{R-R_T}.

3.2 Examples

We now illustrate via examples the results of the models for both trunks required and blocking. We will vary the key parameters in ways that might, for example, correspond to the various provisioning policies, types of trunk groups, levels in hierarchy, and load estimates. For example, it is hardest to estimate and forecast load for small, tandem grade of service (GOS) groups receiving overflow. These groups also have the highest peakedness and day-to-day variation.

3.2.1 Trunk Group Examples

The legend to the plots in Figures 3 - 8 is as follows. The subtitle indicates the values of peakedness (z), day-to-day variation ($\phi = L$ *or* H *for Low or High*), the relationship between σ_m ($=\sigma_{R-R_T}$) *and* σ_f ($=\sigma_{D-R_T}$), and the modular sizing rule assumed (1, 12, or 24). The solid curve is always $f_{R-R_T}(x)$; its associated Bands (4, 2, 1, 3, 5) are delimited by solid vertical lines. The other plots are the new distribution (2.5) for different valves of R_T. The examples show that, for a reserve capacity of about 10 percent, Band 4 and Band 5 mass range from about 30-70 percent and 3-10 percent, respectively. The explanation is simple. Large positive (negative) forecast errors combined with large positive (negative) measurement errors make it likely that Band 4 (Band 5) events occur. The reserve capacity cannot reduce the uncertainty (spread) in the measurements and forecasts; rather it provides (indicates preferences) for the reserve over the shortage case.

Example 1: Typical Case:

Figure 3 illustrates a typical case in which Band 4 percentages range from 33 to 64, while Band 5 percentages range from 7 to 3 for an average reserve capacity of 10%

Example 2: Effects of Day-to-Day Variation: $\phi = H$

We now use the same parameters as in the first example, but increase the level of day-to-day variation to "high." As shown in Figure 4, the increase in ϕ tends to reduce the mass in Band 4, while increasing the mass in Band 5. The interpretation is that, since σ_m is greater in this case, the tails of the original distribution are greater - providing more allowance in the bands associated with the solid curve.

Example 3: Effects of Reserve Capacity:

Figure 5 corresponds to the first example, except the $1/\lambda$ takes on value .05, illustrating the significant impact of reserve capacity requirements on band allowances, especially Band 4 and Band 5.

Example 4: Effects of Forecast Variation:

Examination of Figures 6 and 7 seems to indicate that assumptions concerning forecast error have more impact on Band 5 mass than on Band 4, especially for $\lambda = 10$. That is, for typical levels of reserve capacity (10-15 percent), the Band 4 mass is more sensitive to λ than to σ_f, while the Band 5 mass is strongly dependent on both parameters.

Example 5: Effects of Modular Sizing:

Figure 8 suggests that banding allowances are not strongly dependent on the module size. That is, σ_m seems to capture the essence of the effects of modular sizing.

3.2.2 Blocking Example

It has been shown that the distribution of measured blocking (which results from having a finite number of hours in a finite study period and from properties of the blocking estimator) has two components; one is discrete, the probability of no blocking, and the other is continuous. The continuous component can be approximated by a two-parameter beta distribution. This blocking distribution for R_T is shown by the solid curve in Figure 9. Figure 9 compares the R_T distribution to the new in-service distribution (dotted curve) for a particular combination of R_T, ϕ, z, λ, and forecast and measurement errors.

For this example, $R_T = 100$ trunks, $z = 1.0$, $\phi = 1.5$ (L), $\lambda = 10$, $\sigma_m / R_T = .025$, and $\sigma_f / R_T = .05$. For internal company reasons in this example, we used (approximate) quantiles 0, 1, 97, 2, and 0 as the reference points for the solid curves (rather than those for the trunking distribution). Using the new model, the new masses are 30-17-49-3-1; the large shift to the left is mainly due to the 10 percent reserve capacity.

IV. SUMMARY

We use a simplified model of a trunk provisioning process to derive analytical expressions for various quantities of interest. These statistics can be used to analyze planning algorithms, provisioning practices, engineering effectiveness, network utilization and service, and certain new technologies. We illustrate the essential tradeoffs between measures of reserve capacity and shortages as a function of various traffic characteristics, forecasting algorithms, and provisioning practices. The examples given show, for reasonable levels of *average* reserve capability (10-15%) and other typical parameters, that the fraction of groups (or trunks within a category of groups) seem to exhibit extraordinary quantities of "reserve" capacity (40-50%) at the same time as there are some groups (3-5%) with significant trunk shortages. Attempts to decrease the reserves "tail" by simple provisioning practices quickly result in lengthening the shortage tail. Thus, something more fundamental such as forecast quality or more efficient network utilization (e.g., [7]) is required to change the basic tradeoffs. While these basic tradeoffs have been observed in practice and the models have been proven accurate in many test cases, the examples are not intended to reflect current BOC conditions or provisioning practices.

Our models improve upon the TPOC results in many respects (See Table 1). However, there are some qualifications to the results derived herein:

1. The simplifying assumptions (2.2) and (2.3) should be evaluated for relevance based on local conditions and practices. For example, while the notion is not recommended (8), one could implement the *average* reserve capacity by trying to inflate each group forecast by the same percentage, i.e., A = k D. However, the analysis of Section II could still be easily carried out in that case.
2. We have not tested the models in networks that have topologies that are constantly changing, either in a planned or unplanned fashion. The problem has less to do with the model than with the fact that trunk groups are constantly coming and going, making the specification of key network parameters, such as in (2.5), very difficult to define and interpret. Thus, for example, our models might be generally applicable to the dynamic, nonhierarchical networks in [7]; however, (2.2) may not directly apply and other quantities may need to be redefined (perhaps with difficulty).

V. ACKNOWLEDGEMENTS

We would like to thank G. C. Varvaloucas and C. R. Szelag for their valuable contributions to this work.

REFERENCES

[1] A. Kashper, C. D. Pack, G. C. Varvaloucas, "Minimum Cost Multiyear Trunk Provisioning," ITC10, Montreal, 1983.

[2] A. J. David and C. D. Pack, "The Sequential Projection Algorithm: A New and Improved Traffic Forecasting Procedure," ITC9, Torremolinos, 1979.

[3] J. P. Moreland, "A Robust Sequential Projection Algorithm for Traffic Load Forecasting," *BSTJ*, 61, January 1982, pp. 15-38.

[4] D. W. Hill and S. R. Neal, "Traffic Capacity of a Probability Engineered Group," *BSTJ*, 55, September 1976, pp. 831-842.

[5] S. R. Neal, "Effective Network Administration in the Presence of Measurement Uncertainty," ITC9, Torremolinos, 1979.

[6] W. B. Elsner, "Dimensioning Trunk Groups for Digital Networks," ITC9, Torremolinos, 1979.

[7] G. R. Ash *et al*, "Intercity Dynamic Routing: Architecture and Feasibility," ITC10, Montreal, 1983.

[8] A. Kashper and G. C. Varvaloucas, "Trunk Implementation Plan for Hierarchical Networks," *BSTJ*, 63, January 1984, pp. 57-88.

[9] C. R. Szelag, "Trunk Demand Servicing in the Presence of Measurement Uncertainty," *BSTJ*, 59, July 1980, pp. 845-860.

[10] R. L. Franks, *et al*, "A Model Relating Measurement and Forecast Errors to Provisioning of Direct Final Trunk Groups," ITC8, Melbourne, 1976. See, also, *BSTJ*, February 1979.

Table 1: *TPOC Comparison*

CP/DE	All Analytic	All Groups	High level Measurement Error Model (function I, Fig. 1)	General Forecast (II)	General Engineering Parameters (III)	General Provisioning Practices (IV)	Demand Servicing Model(V)
TPOC	Analysis and Simulation	Direct Finals	Detailed Measurement Model	Equilibrium Growth or Simulation	Limited Parameters or Simulation	Special Equilibrium Add/Disconnect or Simulation	Not Included

R_T	D/D Var	σ_{R-R_T} / R_T	$\sigma_{R_T-D} / \sigma_{R-R_T}$	bound on ρ
10	Low	.08	2	.95%
100	Low	.025	2	.094%
10	High	.084	2	1.1%
10	High	.084	4	1.8%
10	High	.084	10	3.9%

Table 2: Examples of Bound on Contribution of Demand Servicing to Average Reserve Capacity.

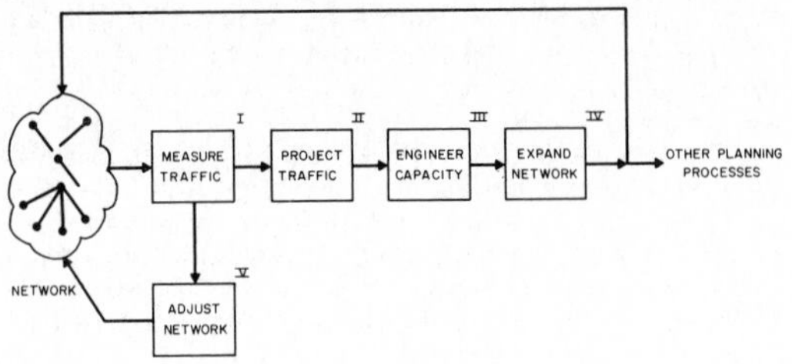

FIGURE 1 SIMPLIFIED, GENERIC TRUNK PROVISIONING PROCESS

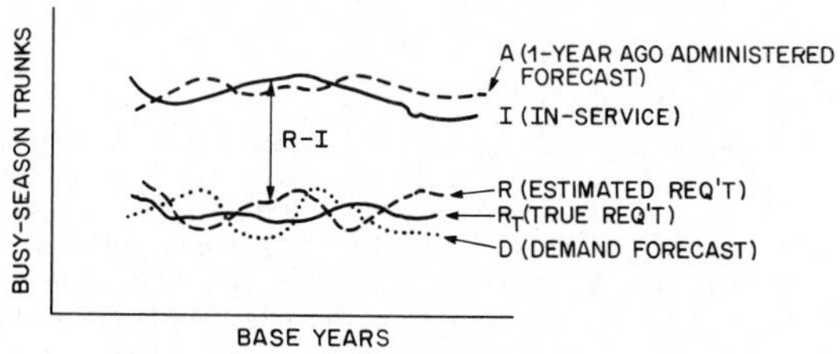

FIGURE 2 RELATIONSHIPS AMONG PROVISIONING STATISTICS

FIGURE 3 TRUNK DEVIATION (R-I) DISTRIBUTIONS

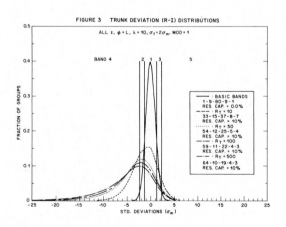

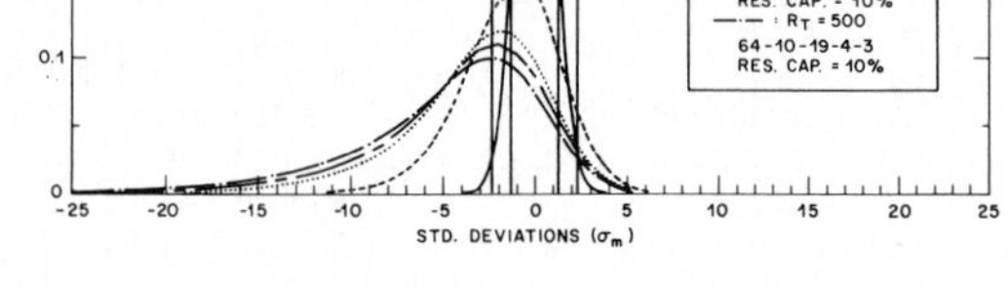

FIGURE 4 TRUNK DEVIATION (R-I) DISTRIBUTIONS

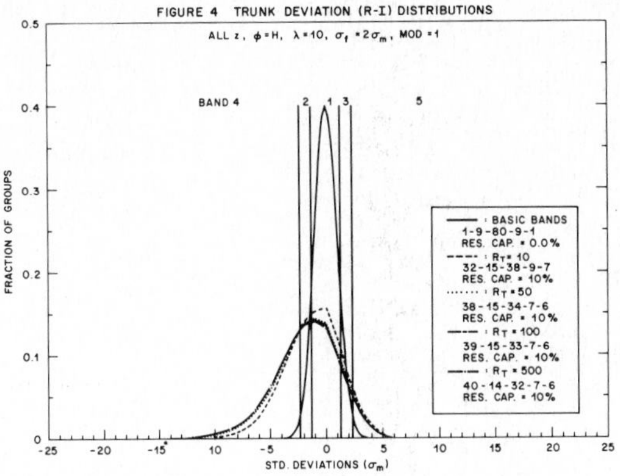

FIGURE 7 TRUNK DEVIATION (R-I) DISTRIBUTIONS

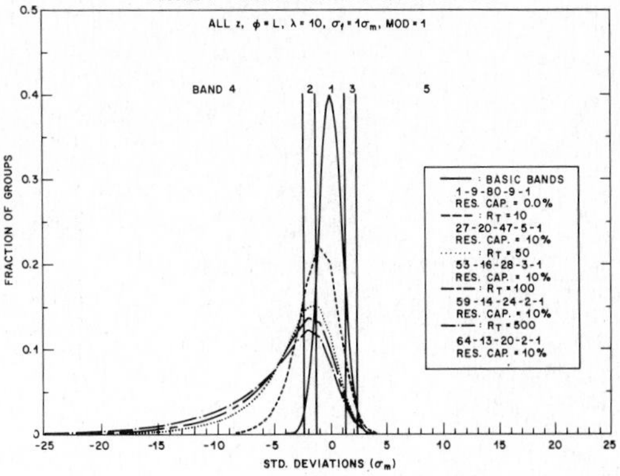

FIGURE 5 TRUNK DEVIATION (R-I) DISTRIBUTIONS

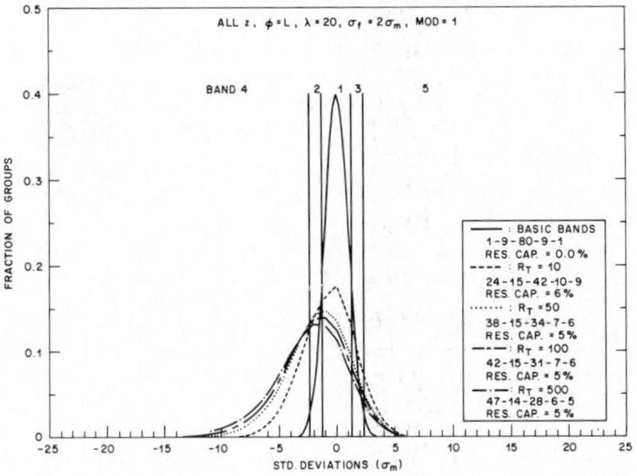

FIGURE 8 TRUNK DEVIATION (R-I) DISTRIBUTIONS

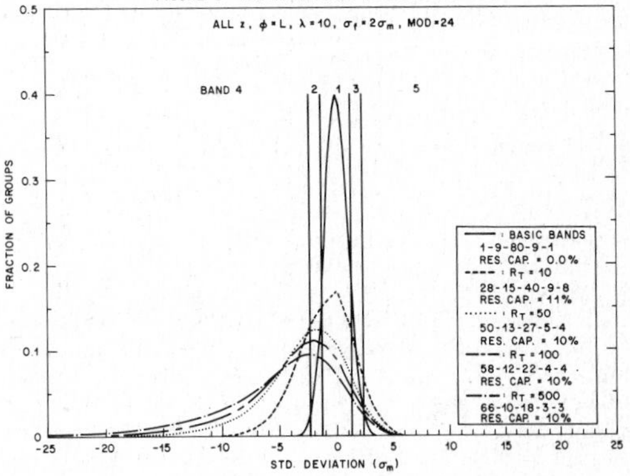

FIGURE 6 TRUNK DEVIATION (R-I) DISTRIBUTIONS AND BANDS

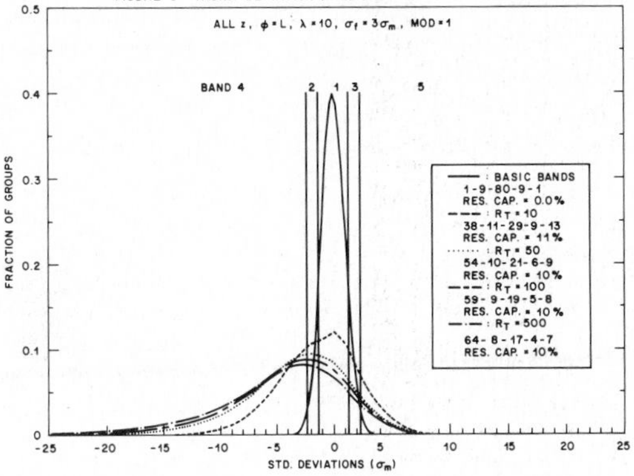

FIGURE 9 BLOCKING DISTRIBUTIONS

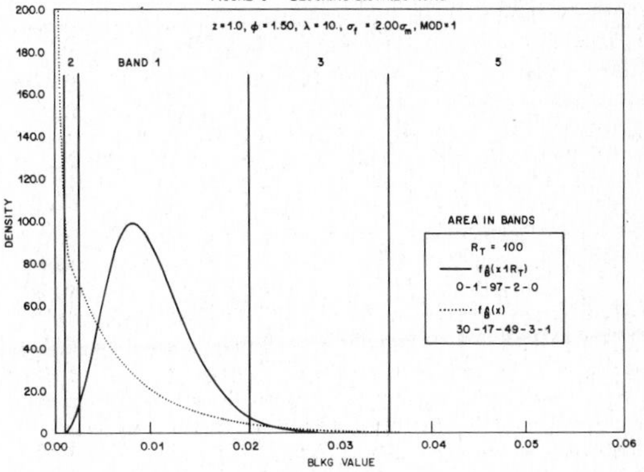

TELETRAFFIC ISSUES in an Advanced Information Society
ITC-11
Minoru Akiyama (Editor)
Elsevier Science Publishers B.V. (North-Holland)

OPTIMIZATION OF DIGITAL NETWORK STRUCTURES

André H. ROOSMA

Dr Neher Laboratories, Netherlands PTT
Leidschendam, Netherlands

ABSTRACT

With the introduction of new technologies in telecommunication networks it becomes important to determine the optimal structure of these networks in terms of the origin-destination routings and the presence of trunk groups. Algorithms given here stem from either of two approaches. One is based on traffic-carrying-efficiency formulas, as used in network sizing. The other focusses more on the combinatorics of modular engineering and reformulates the problem as a mixed integer linear program.
Numerical results affirm our idea that the second approach performs best for large module-sizes. Under reasonable assumptions concerning costs etcetera, networks tend to be fully interconnected and overflow decentralized if the module-size is small, less connected and more backbone-oriented if the module-size is large. Optimal hierarchy of nodes often appears to be non-transitive, as opposed to common practice in networks.

1. INTRODUCTION

In teletraffic engineering much attention has been paid to telecommunication-network performance evaluation, network sizing and various types of routing strategies (e.g. [1],[2],[3]). In all these studies, the network structure, in terms of the presence of trunk groups between pairs of nodes and the possible routes between two nodes, is usually assumed to be fixed. Since alteration of the network structure could imply reallocation of installed switching and transmission equipment this assumption is generally justified.
With the introduction of new technologies, such as digital switching and optical transmission, the allocation problem is there and the way is open to a more radical change in network structures.
In many studies questions have been raised about the optimality of the number of hierarchical levels in the network (e.g.[5]), the presence of direct high-usage links, and so on. All these questions could be dealt with if there would be a clear view on optimal structures for telecommunication networks. This paper aims at being a first step towards the acquisition of such a view.

1.1. Preliminaries, assumptions and scope

In the sequel the words 'origin', 'destination', 'tandem' will be used to denote the respective nodes involved in (handling) certain node-to-node traffic.

Tandem nodes are supposed to be ordinary network nodes as well, with originating and terminating traffics etcetera.

The usual assumptions will be made, like:
- offered traffic is Poisson;
- blocked calls are cleared;
- independence of trunk group occupations (and - blockings);
- nodes are non-blocking.

The major part of the paper is devoted to algorithms and principles for the optimization of the structure of networks. Though the algorithms deal primarily with circuit- switched telecommunication (telephone or integra- ted services) networks, many other networks can be dealt with in a similar way. The approach will be applicable to both unidirectional (one-way trunking) networks and bidirectional (two-way) networks.

2. PROBLEM FORMULATION

The problem of the optimal structure can roughly be formulated as follows:
Given a set of nodes $\{1,2,...,n\}$, matrices of node-to-node offered traffic A_{ij}, costs per trunk c_{ij} and initial costs of a trunk group between two nodes C_{ij}, and maximal blocking values for the node-to-node traffic Bmax.
The problem is to find the matrix of trunk group sizes N (with elements N_{ij}), and the set of routing paths for each traffic-relation (including overflow strategies). These should be optimal in terms of the costs of the network.

It is also possible to separate the network structure problem and the network sizing problem, replacing the matrix of trunk group sizes by an interconnection-matrix Y (with elements 1/0 - trunk group present/ not present). The optimal trunk group sizes and associated network costs could be calculated by a network sizing algorithm that allows for a general routing strategy (as in [2],[3]).

In practice there will be some more constraints on the solution set. There can be constraints on the reliability of origin-destination connections (which may be improved by multi-path routing). The set of switching centres that can be given a function as tandem switches can be constrained to be a subset of the set of nodes. Furthermore, in

a digital switching and transmission environment, the trunk group sizes will necessarily be a multiple of a certain module-size m.
The network improvement problem can be described as above, with some constraints on the presence or size of trunk groups.

The difficulty of the problem is grounded in the combination of stochastic offered traffic, combinatorial allocation aspects and the large scale of most networks. Probably this is the reason why the problem has not been treated so far in this general form. Some aspects have been discussed in [4] and [5]. However, these studies assume hierarchical network structures, an assumption we will not make. Valuable work has been done in [2], though the network structure problem is not solved integrally there.

3. ALGORITHMS

To solve the network-structure problem described, two approaches have been investigated. One focusses on the network efficiency in dealing with the stochastic traffic, the other is more devoted to the combinatorial aspect of the problem. As a consequence, it is to be expected that the first approach gives very satisfactory results when the module-size is small as compared to the node-to-node traffic intensities. The second approach will perform relatively better if the module-size is large. In both approaches a rough network sizing method, based on a one-moment analysis, is incorporated.

3.1. Traffic-Carrying-Efficiency Formulas.

In the first approach, to each origin-destination pair is assigned another node, serving as a tandem-switch. The tandem-switch is selected on the basis of network costs: various prospective nodes are tested and the one minimizing the total network cost is selected. This involves only the evaluation of 'triangles', consisting of origin-, destination- and tandem-node.
Though relatively simple, this approach appears to be quite powerful. In the algorithm, one only has to evaluate 'triangles', but the resulting routing paths may consist of any number of links. Network hierarchy (or non-hierarchy) is established in a natural way and the trunk groups are dimensioned optimally.

The outline of the algorithm is as follows:
First, the network is dimensioned with all traffic routed along the most direct routes (if possible: a direct link from origin to destination). This will yield a certain (probably high) total network cost.
Iteratively, an origin-destination relation is selected and for various prospective tandem nodes the reduction in total network costs, involved in using that node as a tandem for the relation, is computed. These computations are based on the efficiency-formulas of Pratt (see [1]), stating for a simple triangle network:

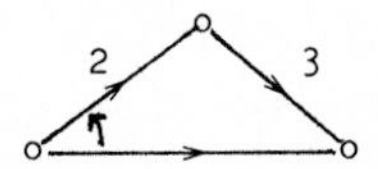

$$c1/H1 \;=\; c2/\beta 2 + c3/\beta 3$$

where H = marginal occupancy = $(\partial Y/\partial N)_A$

β = marginal capacity = $(\partial A/\partial N)_B$

Given the β's of the links from origin to tandem and from tandem to destination (the prospective overflow links), the resulting H of the direct link is used to compute the number of trunks N of the direct link. The overflowing traffic is computed and the overflow links dimensioned on a final trunk group grade of service criterion. In this way each prospective tandem involves a prospective reduction in total network costs.
The one yielding highest reduction in total network costs is selected and the network is updated accordingly (including the addition of the overflowing traffic to the direct traffic on the overflow links).
Once an origin-destination pair is treated and a tandem switch has been determined, no overflow on the direct high-usage trunk group between this node-pair is allowed anymore. Only if it is optimal not to let the traffic overflow, the direct trunkgroup is still 'final' and can be used in an overflow route for other traffic streams. In this way convergence to a feasible solution is guaranteed.

In this approach the combinatorial problem is not integrally solved but approximatingly treated by selecting the origin-destination pairs in a certain order. Different ways of determining this order have been investigated. Numerical experiments on networks of 4, 12 and 19 nodes showed that it is optimal to select the pairs in the order of increasing traffic, starting with the (smallest) traffic streams to or from the smallest switching centre (in terms of total incoming and outgoing traffic). A motivation for this is that, generally, large switching centres will be more likely to be selected as tandem switches. So, first eliminating small switching centres as possible tandem switches will hardly obstruct optimality.

3.2. A Linear Programming Model

A second approach which has been explored is to model the problem in such a way that it can be written as a mixed-integer linear programming (MILP) model (a kind of multicommodity flow / allocation model). Here, the difficulty is in the stochastic character of the traffic.
MILP-models known have no capability of dealing with stochastic variables, like the traffic in this case. However, the following appeared to be a rather good approach for eliminating this difficulty, specially for relatively large module-sizes.
One can derive the extra number of Erlangs of traffic that can be offered at a certain blocking to a trunkgroup when an extra ℓ-th module is installed. This will be called the marginal capacity of that module. For a module-size of 30 and a blocking of 0.01, the marginal capacities of a first until tenth module are given in table 1 as an example.
It will be clear that the sizing is based on a maximal trunk group blocking criterion. The trunk group blockings can be chosen in such a way that the end-to-end blocking constraint will be satisfied.
The approach implies that for traffic with other stochastic characteristics the method needs only marginal modification.

TABLE 1. Marginal capacity/ module (module-size = 30, blocking = 0.01).

module	capac.	module	capac.
1	20.3	6	28.8
2	26.6	7	29.0
3	27.8	8	29.2
4	28.3	9	29.2
5	28.6	10	29.3

In this approach we have chosen for the concept of traffic sharing, rather than overflow. Certain portions of a node-to-node offered trafficstream are routed along different paths. For more details we refer to the literature on this subject (e.g. [2]).
If only routes of at most two links are considered, let D^{ℓ}_{ij} be 0/1 variables denoting the presence of the ℓ-th module on link ij , a^{ℓ} the capacity of an ℓ-th module on a link, and M_{ikj} the part of the traffic from i to j that is routed via k.
Now, the problem can be formulated as:

$$\text{minimize: } \Sigma_{ij\ell}\, c_{ij}\, D^{\ell}_{ij} + \Sigma_{ij}\, C_{ij}\, D^{1}_{ij} \qquad (a)$$

subject to:

$$\Sigma_k M_{ikj} \geqslant A_{ij} \qquad (b)$$

$$M_{ikj} \geqslant 0 \qquad (c)$$

$$\Sigma_k M_{kij} + \Sigma_k M_{ijk} \leqslant \Sigma_{\ell}\, a^{\ell}\, D^{\ell}_{ij} \qquad (1)\ (d)$$

$$D^{\ell}_{ij} = 0/1 \qquad (e)$$

$$D^{\ell}_{ij} \geqslant D^{\ell+1}_{ij} \qquad (f)$$

This problem is closely related to the well-studied class of multicommodity flow /allocation problems. Efficient methods to solve this problem, making use of its specific structure, are under investigation. However, the problem can already be treated using known techniques (cf. [6], [7], [8] and [9] where similar problems are solved, using (generalized) Benders' decomposition). If the network is large, column-generators and the like, can be used to save computer-memory. Feasibility of solutions is always guaranteed in this approach.
The network improvement problem can be dealt with by adding only a set of restrictions on the trunk group sizes. This implies no essential modification of the algorithm, which is very useful with regard to the evolution of a present network to the optimal one in future.

To allow for routes of more than two links, the M_{ikj}'s should be interpreted less strictly, relaxing the condition: $M_{ikj} \geqslant 0$ (1c)
to: $\Sigma_k M_{kij} + \Sigma_k M_{ijk} \geqslant 0$ (1c').
This is best illustrated by the following example (see illustration below).
Routing the traffic from 1 to 2 via 3 and 4 can be established by routing the traffic from 1 to 2 via 3 and routing all traffic on 1-3 via 4 (so, all D_{12}'s = 0 , as well as D_{13}'s = 0).

Then, $M_{133} + M_{132} \leqslant \Sigma_{\ell} D^{\ell}_{13} = 0$ (1d),
implies $M_{133} \leqslant -M_{132} = -A_{12} < 0$
This approach guarantees that no 3-links route is token, if also some 2-links route has free space (shortest path property).

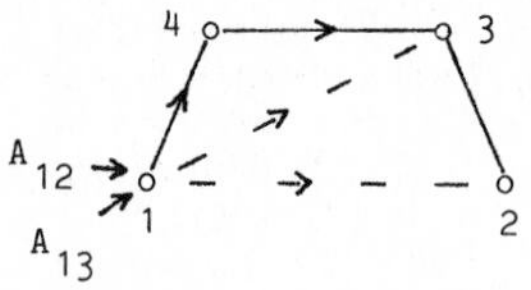

$M_{133} = -A_{12}$
$M_{143} = A_{12}+A_{13}$
$M_{132} = A_{12}$

3.3. Heuristics Based On The MILP-Formulation

For medium-size networks (ca.20 nodes) it is possible to solve the MILP problem (1), utilizing the specific structure of the restrictions. For very large networks however, the branching proces involved in the search for the integer D's may grow very time-consuming. In this case, the MILP problem (1) is approximated by a series of LP problems. For each LP problem there is a rather simple solution that is generally very close to optimal.
Therefore, introduce a slack-variable Y_{ij}, the not yet offered traffic, in the lefthandside of (1b), thereby relaxing this restriction:

$$\Sigma_k M_{ikj} + Y_{ij} \geqslant A_{ij} \qquad (2a)$$

$$Y_{ij} \geqslant 0 \qquad (2b)$$

Define:

a_{ij}= marg.capac. of next module on i-j,
$= a^{\ell}$, with ℓ such that $D^{\ell}_{ij} = 0,\ D^{k}_{ij} = 1,\ k<\ell$;

$$c'_{ij} = \begin{cases} c_{ij} + C_{ij} & \text{if } D^{1}_{ij} = 0; \\ c_{ij} & \text{otherwise.} \end{cases}$$

One step of the heuristic iterative algorithm is as follows:

Let a set of M's, Y's and D's be given, fulfilling (2a,b, 1c'-f) (initially: M=0, D=0, Y=A).
Add one module (the 'next' D_{ij}:=1) to the relation i-j that maximizes:

$$a_{ij}\, Y_{ij} \,/\, c'_{ij}$$

Regard the D's as fixed and solve the LP problem (3) given by (2a,b, 1c',d) and the minimizing criterion:

$$\min\ \Sigma_{ij}\, c'_{ij}\, Y_{ij}$$

In this way modules are added iteratively, on the place where they (at that stage in the algorithm) are most useful, until all traffic can be offered (Y's = 0).
The process can be accelerated by adding more than one module at a time, at such places that the addition of one will not affect the 'usefulness' of the others.
This process may yield a solution which is sub--optimal, but numerical results are encouraging.

A heuristic 'solution' to problem (3) can be found from the previous one as follows:
On adding a module to the relation i-j, the M_{ijj}

can be increased by $\min(Y_{ij}, a_{ij})$ and Y_{ij} decreased by this amount. If a_{ij} was greater than Y_{ij}, some room is left over on link i-j. This can be used by Y_{ik}'s or Y_{kj}'s for which there is also spare capacity on k-j or i-k, respectively.
In using this spare capacity for each of these Y's there will be involved a reduction of potential costs:

$$c_{ik}' \; Y_{ik} \; / \; a_{ik} \quad \text{(or a similar expression for kj)}$$

Now, fill the spare capacity with those Y's that give the largest reduction in potential costs (reducing the particular Y's and increasing the M_{ijk} or M_{kij} involved).

To fasten the process, an initialization procedure can be used. It is easy to see that direct modules (i.e. on the direct link) that are completely filled, will seldomly be non-optimal. Even the strategy which installs modules for half the module-size of traffic on direct links has proven to perform well, at uniform costs per trunk and small initial costs per trunk group. Such a strategy could be justified by the fact that even for an efficiency of almost 100 % at the tandem links, the costs per Erlang traffic on tandem and direct route are about the same.

3.4. Other Heuristics

Some other approaches towards heuristic algorithms have been investigated. The algorithm based on traffic efficiency formulas has been modified to deal with traffic sharing in stead of overflow. To this end, the efficiency of a direct module and modules along an alternate route is measured using a cost/benefit analysis, the benefit defined as the sum of the traffic that can be removed from the set of not-yet-offered--traffic and part of the empty capacity that is introduced (which might be used by other tandem traffic). After initialization (see previous paragraph), for each origin-destination relation the optimal route is selected, and the traffic placed on this route. The sequence in which the relations are treated is in principle the same as in the algorithm, based on the traffic carrying efficiency formulas.

4. RESULTS

The algorithms derived have been applied to various networks. The first being a 4-node network on which a lot of other research had been done, and for which there were good indications as to what the optimal structure and trunk group sizes would be. The aggregated input-data for this network are given in table 2.
The application of the algorithms to networks from the operational field gives a good idea of the performance of the algorithms on larger networks. The aggregated input-data of real 12-node and 19-node networks are given in tables 3 and 4 respectively.
The results are given in tables 5 and onward.

In the tables the following notations are used:

TCE = Traffic Carrying Efficiency algorithm;
MILP = Mixed Integer Linear Program (see text, only allowing routes of 2 links);
H1# = heuristic, based on MILP-formulation, as described in par. 3.3, version 1.#;
H2# = heuristic, based on combination of ideas from the TCE approach and traffic sharing as described in par. 3.4, version 2.#;
DO = routing traffic Direct Only; no overflow, no sharing;

(The heuristics allow routes of any length, the various versions differ in initialization and some other minor aspects.)
(Heuristic vers.1.2 is a bidirectional version);

I : total number of modules in the network (equivalent to the total costs);
II : maximum number of modules at one node (origin. + termin.);
III : connectivity (relative number of relations with direct trunk group);
IV : number of relations with more than one routes;
V : number of relations using a tandem node;
VI : maximum usage of any one tandem node (in number of relations);

The measures II, III, V, VI are related to the hierarchy of the networks created. If II is relatively large, and/or VI is large (specially as compared to V) the routing is more centralized, backbone-oriented.

TABLE 2. 4-NODE NETWORK - AGGREGATED INPUT-DATA

number of nodes:	4
number of traffic-relations:	12
total traffic:	355 Erl.
average relation:	ca. 30 Erl.
largest node (origin.+termin.traffic):	225 Erl.
smallest node (origin.+termin.traffic):	120 Erl.
initial costs per trunk group:	0
costs per trunk (circuit; channel):	1
blocking criterion:	1 %

TABLE 3. 12-NODE NETWORK - AGGREGATED INPUT-DATA

number of nodes:	12
number of traffic-relations:	132
total traffic:	3487 Erl.
average relation:	ca. 26 Erl.
largest node (origin.+termin.traffic):	1274 Erl.
smallest node (origin.+termin.traffic):	242 Erl.
initial costs per trunk group:	0
costs per trunk (circuit; channel):	1
blocking criterion:	1 %

TABLE 4. 19-NODE NETWORK - AGGREGATED INPUT-DATA

number of nodes:	19
number of traffic-relations #0:	336
total traffic:	5722 Erl.
average relation:	ca. 17 Erl.
largest node (origin.+termin.traffic):	1638 Erl.
smallest node (origin.+termin.traffic):	165 Erl.
initial costs per trunk group:	0
costs per trunk (circuit; channel):	1
blocking criterion:	1 %

In tables 5, 6 and 7 the performance of the algorithms is compared for the 4-, 12- and 19-node networks respectively. A problem here is that one would wish to derive absolute lower bounds to the costs. As stated before, research is continuing on efficient MILP-algorithms. The one used in the results presented here is a modified general-use algorithm. It optimizes continuously on the basis of the full set of columns of the problem, which appeared to be rather time-consuming for these problems. It was manually stopped after a certain computer time (minutes for the 12-node network and more than an hour for the 19-node network), so there is no absolute guarantee of optimality as yet.

TABLE 5. COMPARISON OF METHODS - AGGREGATED RESULTS 4-NODE NETWORK

module-size: 1						
	I	II	III	IV	V	VI
TCE	470	329	12/12	6	6	6
MILP	488	298	12/12	3	3	2 *)
H10	491	302	12/12	3	3	3
H11	488	298	12/12	3	3	2 *)
H13	489	299	12/12	2	2	1
H21	489	299	12/12	3	3	2
D O	491	300	12/12	0	0	0
*) can be considered the true optimum for traffic sharing						
module-size: 15						
	I	II	III	IV	V	VI
TCE	35	23	11/12	6	7	4
MILP	35	22	11/12	4	5	2
MILP	35	22	12/12	5	5	3
H10	37	24	12/12	5	5	3
H11	36	22	12/12	2	2	2
H21	36	22	12/12	4	4	2
(the MILP-algorithm gives more total-cost--identical solutions here)						
module-size: 30						
	I	II	III	IV	V	VI
TCE	19	12	11/12	5	6	3
MILP	18	11	10/12	3	5	3
H10	19	14	10/12	4	4	4
H11	20	12	12/12	1	1	1
H13	20	12	12/12	1	1	1
H21	20	13	11/12	5	6	3

From the results, it appears that some heuristics (like version 1.1) tend to underestimate the combinatorial aspects, since they do not combine as much tandem traffic on a few nodes as the MILP does. Version 1.0 seems to centralise too much resulting in networks with higher costs than, e.g., version 1.1. The optima seem to be rather flat. This may induce other factors, like maintainability, reliability and ease of planning to become more important. Then, a solution given by, e.g., version 1.1 may become more desirable than an 'exact optimum', as deliverd by a sophisticated MILP-algorithm.

TABLE 6. COMPARISON OF METHODS - AGGREGATED RESULTS 12-NODE NETWORK

module-size: 30						
	I	II	III	IV	V	VI
TCE	176	66	79/132=.60	20	73	21
MILP	174	62	85/132=.64	33	63	36
H10	181	68	78/132=.59	62	92	42
H11	178	56	105/132=.80	32	49	13
H13	174	54	104/132=.80	32	53	15
H14	181	66	85/132=.64	64	80	43
H15	180	56	111/132=.84	40	48	12
H21	177	64	79/132=.60	59	83	38

TABLE 7. COMPARISON OF METHODS - AGGREGATED RESULTS 19-NODE NETWORK

module-size: 30						
	I	II	III	IV	V	VI
TCE	306	92	159/342=.46	37	220	75
MILP	308	108	133/342=.39	41	220	127
H10	327	111	143/342=.42	168	261	125
H11	326	78	221/342=.65	82	147	39
H14	322	101	140/342=.41	169	263	100
H21	316	118	143/342=.42	144	254	125
(the MILP and H14 use node 16 most as tandem, all others use node 3 most.)						

To demonstrate the applications for the algorithms, the effect of centralization on the costs of the networks was studied. For this purpose, the set of potential tandem switches in the 19-node network was restricted to predetermined subsets of 4 and 2 nodes. Results are given in table 8.
The algorithms have also been used to study the effect of larger or smaller module-size and the effect of bidirectional (two-way) trunk groups. Some results of the former study are presented in table 5.

TABLE 8. INFLUENCE OF THE NUMBER OF POTENTIAL SWITCHING CENTRES - AGGREGATED RESULTS 19-NODE NETWORK

module-size: 30						
	I	II	III	IV	V	VI
TCE(19)	306	92	159/342=.46	37	220	75
TCE(4)	314	125	143/342=.42	43	242	119
TCE(2)	315	157	113/342=.33	37	266	148
H10(19)	327	111	143/342=.42	168	261	125
H10(4)	317	131	146/342=.43	108	248	185
H10(2)	324	132	151/342=.44	73	230	155
H14(19)	322	101	140/342=.41	169	263	100
H14(4)	318	128	143/342=.42	87	244	153
H14(2)	318	144	143/342=.42	66	238	192
H21(19)	316	118	143/342=.42	144	254	125
H21(4)	314	153	128/342=.37	95	264	196
H21(2)	317	171	122/342=.35	73	267	214

5. CONCLUSIONS

From the results some general conclusions can be derived on the optimal structures of networks. For module-sizes that are small as compared to the offered traffic intensities and rather uniform costs per trunk and low initial costs per trunk group, the optimal structure of a network will be almost fully interconnected. In this case a decentralised overflow-strategy will be slightly better than a centralised one (typically 2% lower in cost).
For large module-sizes, small switching centres will be connected only to one or a few large switching centres nearby (high traffic intensities, low costs).
In both cases, the optimal number of potential tandem-switches will depend largely on the structure of the costs and traffic intensities. However, the optima seem to be rather flat with respect to the centralization of overflow. This suggests that taking reliability constraints into account will not result in highly increased costs of the networks.
Network hierarchy appears to be a fuzzy concept, since the hierarchy of nodes often appeared to be non-transitive (node 1 higher than node 2, concerning most of its outgoing links, 2 higher than 3, and 3 higher than 1).
By its character, the first algorithm (the TCE algorithm, as described in par. 3.1.) will always give a link-hierarchical solution. This is a quite attractive side-effect regarding eventual planning and evaluation processes. The algorithms based on traffic sharing are interesting because of the fact that there is no culmination of peakedness of traffic in the network.

6. SUGGESTIONS FOR FURTHER INVESTIGATIONS

The results give enough reason for further development of the algorithms presented. Combination of the various approaches should lead to better heuristics.
A version of the first algorithm that will allow for mutual overflow is under investigation. Link hierarchy will no longer prevail in solutions created by such an algorithm.
The problem could also be treated as a multi-stage dynamic programming problem. A first stage could be to establish the hierarchy in the routing, with some prospective tandem-switches per relation. Perhaps this could be done by means of gravitation-methods. The second stage, then being the allocation of trunk groups and tandem-switches, could be treated by a simplified version of one of the algorithms presented. The more precise network sizing could be viewed upon as a third stage. The first stage would reduce the problem thus significantly (specially for large module-sizes as compared to traffic intensities), that an efficient MILP-algorithm would be able to treat even large networks in an acceptable time in the second stage.

If the reliability of connections has to be taken into account, the second approach described is very worthwile. Only one set of inequalities is added to the MILP-formulation of the problem. In the first approach more than one tandem switch could be selected and the overflowing traffic shared over the two alternate routes.
The way in which this influences the optimal network structure should be investigated.

ACKNOWLEDGEMENT

Helpful discussions with L.Rob van Vliet are gratefully acknowledged. Rob van Vliet and Anton J. Goos (both with DNL) are also thanked for modification of a MILP computer package for implementation of the present problem.
Eric A.van Doorn (Twente Univ.of Techn.) is acknowledged for his helpful comments on a first version of the manuscript.

REFERENCES

[1] C.W. Pratt, "The concept of marginal overflow in alternate routing", Proc. 5th Int. Teletraffic Congress, New York, 1967, pp.51-58.
[2] A. Girard, Y. Cote, Y. Ouimet, "A comparative study of non-hierarchical alternate routing", Telecommun.Networks Planning Conf., Brighton, March 1983, pp.70-74.
[3] F. Le Gall, "One moment model for telephone traffic", Appl.Math.Modelling, 1982, Vol.6, pp.415-423.
[4] P.A. Caballero, J.M. Silva, "Effects of digitization on the optimum structure of junction network", Proc. 10th Int.Teletraffic Congress, Montreal, 1983, sess.4.3A, paper 4.
[5] G. Sallai, Z. Papp, "A statistical method for optimizing hierarchical network structure", Telecommun.Networks Planning Conf., Paris, Sept./Oct.1980, pp.131-137.
[6] J.A. Ferland, A. Girard, L. Lafond, "Multicommodity flow problem with variable arcs capacities", J.Opl.Res.Soc. Vol.29, 5, 1978, pp.459-467.
[7] H.H. Hoang, "Topological optimization of networks: A nonlinear mixed integer model employing generalized Benders' decomposition" IEEE Trans.Autom.Control, Vol. AC-27, No.1, Febr.1982.
[8] M. Minoux, J.Y. Serreault, "Subgradient optimization and large scale programming: An application to optimum multicommodity network synthesis with security constraints", RAIRO Operations Research, Vol.15, No.2, May 1981, pp.185-203.
[9] G. Cote, M.A. Laughton, "Large-scale mixed integer programming: Benders-type heuristics", Eur.Jl.Opl.Res., Vol.16, June 1984, pp.327-333.

Postscript.

At the last moment before release of this paper, a collegue of the author, Mr. P.A. Bruijs, achieved preliminary results by means of a more sophisticated MILP-algorithm of his. Allowing for routes of any length, it found a solution with total number of modules 170 for the 12-node network, and within 15 minutes computer-time a solution with total number of modules 299 for the 19-node network (both for module-size = 30 and no restrictions on the tandem switching centres). These results are extremely interesting since they can compete with the solutions obtained by the TCE-algorithm, based on (optimal) overflow. Research will continue in this field, taking into account also the transmission aspects and reliability constraints so as to integrate the transmission and switching optimization processes (which in common practice have been separated up to now).

TELETRAFFIC ISSUES in an Advanced Information Society
ITC-11
Minoru Akiyama (Editor)
Elsevier Science Publishers B.V. (North-Holland)

ENGINEERING TRAVERSE TRUNKS IN A TRANSIT NETWORK

Makoto YOSHIDA and Noriyuki IKEUCHI

Musashino Electrical Communication Laboratory, NTT
Tokyo, Japan

ABSTRACT

A new network design method called the "Skewer Connection Method" is proposed for constructing future networks which must meet the rapidly growing demand for new service traffic. In this method, traffic from one center to another is carried by a tandem route switched through a third same-level center like a skewer. The method makes use of the traffic bundling effect which enables the number of direct trunks to be reduced. The economic advantage, i.e., cost saving, of this method is analyzed by derived formulas and numerical examples. It is shown that the skewer connection method is advantageous when: i) the node-to-node traffic is small, ii) the nodes are located far apart, and iii) skewer tandem node cost is small. This method improves existing hierarchical networks and enhances the possibilities of constructing even higher-quality networks in the future.

1. INTRODUCTION

Most existing public telephone networks have hierarchical structures with an alternate routing scheme, in which overflow traffic from a direct traverse trunk is routed to a second choice path, if such exists. This alternate routing scheme results in economical network engineering and operation. However, predicted cost reductions in network components through rapidly changing technology coupled with high growth in traffic are now provoking studies on new network structures and trunk engineering methods.

The above trends suggest that future network nodes will be densely connected to each other with direct trunks; i.e., quasi-complete graph networks ([1] - [3]). However, applying this type of structure to the ever-growing large-scale transit networks will require highly sophisticated network design and administration methods. To make both network facilities and administration as economical as possible, simpler regularized topologies for practical networks are desirable. As one such topology, a hierarchical mesh structure constructed according to the following rules is presented:

i) Transit centers closely related to each other in terms of both traffic flow and geographical location are connected in a mesh, and
ii) Traffic to/from other areas flows via a specified transit center.

This paper proposes the "skewer connection method", a new engineering method for using direct traverse trunks suitable to the above type of hierarchical mesh structure. In this method, small traffic loads which cannot afford direct traverse trunks to their destined center are carried by a tandem route switched through a third center on the same level like a skewer. The method is compared with a conventional alternate routing method mainly from an economical viewpoint to clarify its area of application. For this purpose, we have formulated and closely examined the advantages of the skewer connection and conventional alternate routing methods over the method in which only direct final trunks are utilized.

2. DESIGN METHODS AND NETWORK MODEL

2.1 Network Design Methods

The following four basic network design methods using direct trunks in different ways are considered:

Method A: Traffic to/from other areas is carried only by basic trunks via higher level transit centers (no direct trunks).

Method B: Traffic to/from other areas is carried only by direct trunks used as final routes.

Method C: Traffic to/from other areas is first offered to direct trunks used as high-usage routes, then overflows to basic trunks if necessary.

Method D: Traffic to/from other areas is carried by direct trunks used as final routes with skewer connections.

In terms of network costs, Method C is generally advantageous over Methods A and B because of the increased trunk efficiency provided by alternate routing. Method D is more economical than Method B because of the increased efficiency obtained from the traffic bundling by the skewer connection. The merits and demerits of Methods C and D cannot be discussed by simple qualitative observations, however; it is necessary to compare them quantitatively. In Section 3, the cost advantages of the alternate routing and skewer connection methods over Method B are examined. A hybrid method which uses both alternate routing and skewer connection is not discussed in this paper.

2.2 Network Model

An essential model of a hierarchical mesh structure which can be analyzed easily with no loss in generality is shown in Fig.1. This structure is a part of a large transit network consisting of higher-level RC's (Regional transit Centers) and lower-level TC's (Transit Centers). The TC's in an RC area are fully interconnected by direct final trunks because these nodes are located in short distance and have large traffic demands between one another. It is assumed that both RC-RC and RC-TC basic trunks provide high trunk efficiency. The above three methods B, C and D differ only in how the T_1-T_2 or T_1-T_3 traffic in Fig.1 is carried. Therefore, in Section 3, the design methods are compared in terms of the relative network (switching and transmission) cost required for carrying these two traffic. The network cost realized by each design method was formulated in Table 1 using the following parameters:

$a_1(a_2)$: node-to-node traffic for T_1-T_2 (T_1-T_3),

a_0 : background traffic for T_2-T_3,

C_s : switching cost per erlang*

C, c_i : transmission cost of one trunk,

α : ATC (additional trunk capacity) of basic trunks,

$n_1(n_2)$: the numbers of T_1-T_2 (T_1-T_3) direct traverse trunks,

$a_1'(a_2')$: overflow traffic from T_1-T_2 (T_1-T_3) direct traverse trunks,

$N(a)$: the numbers of trunks required for offered load "a" under the constraints of link blocking probability b_0.

Trunk dimensioning methods employed are:

1) The number of high usage (direct traverse) trunks n_1 and n_2 in the alternate routing method are engineered with the so-called ECCS method**.
2) The number of final direct traverse trunks is calculated under the constraints of the link blocking b_0 (set 0.01 in the numerical examples in this paper).
3) The grade of service for calculating the number of basic trunks accepting overflow traffic from high usage routes is the same as the above 2). This number is usually calculated by using the mean and the variance

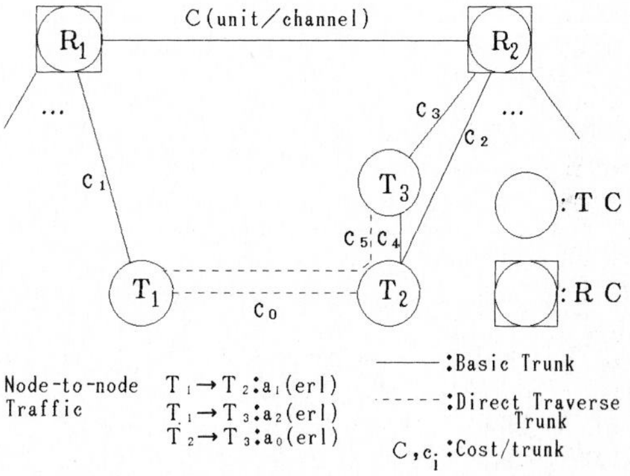

Fig.1 Network model

---

* Switching cost is assumed to be proportional to switched traffic. For example, this cost function is suitable for building block structure switching systems.

** ECCS method: economical hundred call seconds method. This is called the LTC (last trunk capacity) method in Japan.

Table 1 Comparison of network resources (except common resources)

Resources \ Method			(a) Basic trunk only	(b) Direct final trunk only	(c) Alternate routing	(d) Skewer Connection
Switching cost		T_2	0	0	0	$a_2 \times C_s$
		R_1	$(a_1+a_2)\times C_s$	0	$(a_1'+a_2')\times C_s$	0
		R_2	$(a_1+a_2)\times C_s$	0	$(a_1'+a_2')\times C_s$	0
Transmission Cost	Basic trunk	R_1-R_2	$\frac{(a_1+a_2)}{\alpha}\times C$	0	$\frac{(a_1'+a_2')}{\alpha}\times C$	0
		T_1-R_1	$\frac{(a_1+a_2)}{\alpha}\times c_1$	0	$\frac{(a_1'+a_2')}{\alpha}\times c_1$	0
		R_2-T_2	$\frac{a_1}{\alpha}\times c_2$	0	$\frac{a_1'}{\alpha}\times c_2$	0
		R_2-T_3	$\frac{a_2}{\alpha}\times c_3$	0	$\frac{a_2'}{\alpha}\times c_3$	0
		T_2-T_3	$N(a_0)\times c_4$	$N(a_0)\times c_4$	$N(a_0)\times c_4$	$N(a_0+a_2)\times c_4$
	Direct trunk	T_1-T_2	0	$N(a_1)\times c_0$	$n_1\times c_0$	$N(a_1+a_2)\times c_0$
		T_1-T_3	0	$N(a_2)\times c_5$	$n_2\times c_5$	0

of traffic. However, for simplicity, only the mean value is used in the numerical examples in Section 3.

3. COMPARISON OF DESIGN METHODS

Let us examine the economic advantages (cost savings) of alternate routing (Method C) and skewer connection (Method D) over the method in which only direct final trunks are used (Method B). Since the difference among Methods B, C and D results from the different methods of trunk operation used (i.e., direct final, high usage and skewer connection), the advantage of Method C defined in this paper can be called cost saving by high usage trunking, and the advantage of Method D can be called cost saving by skewer connection. These cost savings are expressed as follows:

Cost saving by high usage trunking $H=X_b-X_c$, (1)

and

Cost saving by skewer connection $S=X_b-X_d$, (2)

where X_i = network cost designed by method i.

3.1 Economic Advantages of High Usage Trunking (H)

Many papers have dealt with alternate routing (e.g., Refs [4] - [6]). This section focuses on the cost saving H achieved through using direct traverse trunks as high usage trunks, taking a cost ratio (defined here as direct route cost / alternate route cost) and traffic offered as parameters. The relation H>0 is easily deduced, considering that high usage trunks are optimally engineered by the ECCS method.

Using terms in Table 1, cost saving H is calculated as:

$$H=\sum_{i=1}^{2} H_i \tag{3}$$

$$H_i=\{N(a_i)-n_i\}C(i)-\frac{a_i'}{\alpha}\{C+c_1+c_{i+1}+2\alpha C_s\}$$

$$=W_i[K_i\{N(a_i)-n_i\}-\frac{a_i'}{\alpha}] \tag{4}$$

where

$$K_i=C(i)/W_i, \tag{5}$$

$$W_i=C+c_1+c_{i+1}+2\alpha C_s, \tag{6}$$

$C(1)=c_0$ and $C(2)=c_5$.

The above value K_i (i=1,2) is the cost ratio of direct route / alternate route for traffic a_i, and W_i corresponds to the alternate route cost. The term H_i is the cost saving for traffic a_i by high usage trunking. Since H_1 and H_2 have the same form, the characteristics of H can be clarified by examining either H_1 or H_2. For this purpose, the general expression H_i (Eq.(4)) is used.

Cost saving H_i is proportional to the cost of the alternate route W_i. The term H_i depends on traffic a_i and cost ratio K_i when W_i is fixed. The characteristics of H_i vs. a_i are shown in Fig.2. The oscillation of cost curves is due to the discreteness of the numbers of trunks. From observing Fig.2, it can be said that:

a) When traffic is fixed, the larger the cost ratio is, the larger the cost saving is. In other words, high usage trunking is effective in cases where the direct route cost is not much cheaper than the alternate route cost (see Appendix 1).

b) For a cost ratio in excess of roughly 1/2, the larger the offered traffic is, the more the cost saving is. In contrast, for a cost ratio of less than roughly 1/2, the cost saving disappears as the offered traffic increases (see Appendix 2).

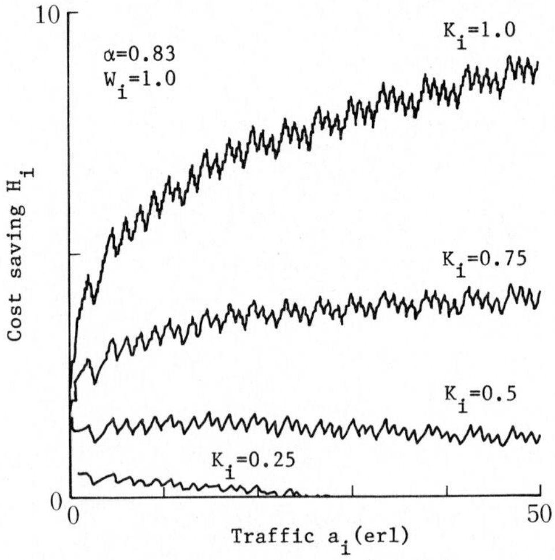

Fig.2 Cost saving by high usage trunking

3.2 Economic Advantages of Skewer Connection (S)

Using the factors in Table 1, cost saving by skewer connection S (Eq.(2)) is expressed as follows:

$$S=N(a_2)c_5+\{N(a_1)-N(a_1+a_2)\}c_0$$

$$+\{N(a_0)-N(a_0+a_2)\}c_4-a_2C_s$$

$$=Pc_0+Qc_4-a_2C_s-N(a_2)\Delta c, \tag{7}$$

where

$$c_5=c_0+c_4-\Delta c \quad (\Delta c\geq 0),$$

$$P=N(a_1)+N(a_2)-N(a_1+a_2), \tag{8}$$

and $Q=N(a_0)+N(a_2)-N(a_0+a_2)$. (9)

From the above equations, cost saving S is characterized as the trade-off between the positive terms (Pc_0 and Qc_4) and the negative terms (a_2C_s and $N(a_2)\Delta c$).

Both of the two positive factors (Pc_0 and Qc_4) signify a reduction in the number of direct final trunks. This reduction is due to the increased trunk efficiency obtained from combining traffic parcels which are too small to justify sending them by direct routes into a larger traffic parcel which may be sent by a direct route economically.

In order to analyze the above cost saving factors in detail, we first discuss P and Q. The term P is a function of traffic a_1 and a_2. Figure 3 shows a relationship between a_2 and P when a_1 is fixed. The oscillation of the curves is due to the discreteness of the numbers of trunks, as mentioned in Sect.3.1. From this figure, it is understood that the value of P becomes larger as a_2 increases, but levels off when a_2 is large.

The former characteristic is analytically proven by:

$$\frac{\partial P}{\partial a_2} = \frac{\partial N(a_2)}{\partial a_2} - \frac{\partial N(a_1+a_2)}{\partial (a_1+a_2)} = \frac{1}{f(a_2)} - \frac{1}{f(a_1+a_2)} > 0 \qquad (10)$$

where f(a) means the ATC for traffic a.

The latter characteristic is confirmed by:

$$\lim_{a_2\to\infty} P = N(a_1) - \lim_{a_2\to\infty}\{N(a_1+a_2)-N(a_2)\} = N(a_1) - \frac{a_1}{\lim_{a_2\to\infty} f(a_2)} = N(a_1) - a_1 \qquad (11)$$

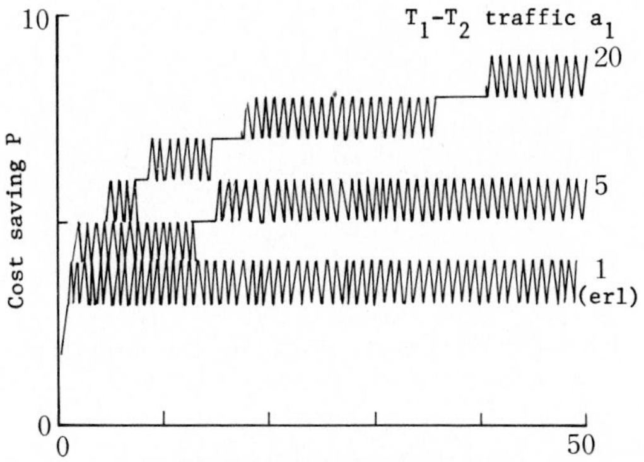

Fig.3 Cost saving P vs. traffic a_2

It is easily understood that a similar relationship holds between a_1 and P when a_2 is fixed. Furthermore, the relationship between Q and a_0 or a_2 is similar to that between P and a_1 or a_2. Therefore, the total value of the positive factors (Pc_0+Qc_4) also increases as a_2 increases, but levels off in a very large a_2 region.

The negative factors have the following characteristics. One of the negative factors, a_2C_s, is the skewer connection switching cost. This is incurred by traffic a_2 switched at node T_2. This factor increases linearly as a_2 increases. The other negative factor, $N(a_2)\Delta c$, is the cost saving in Method B (positive factor for Method B), obtained from the reduction in T_1-T_2 route length or transmission facility cost at T_2 node.

Considering all the above positive and negative factors, cost saving by skewer connection is characterized as a convex function of traffic a_2 and disappears in a large a_2 region. These characteristics are illustrated in Fig.4.

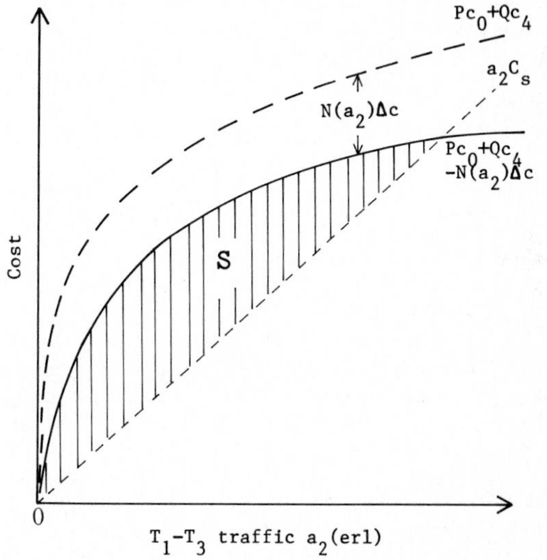

Fig.4 Components of cost saving S

3.3 Comparison between Methods C and D

Based on the discussions in Sections 3.1 and 3.2, in which cost parameters are fixed, this section compares the two cost savings H and S, varying the values of cost parameters.

Let us suppose that $c_0=C$, $c_5=c_0+c_4$ and $c_1=c_2=c_3=c_4=c_f$ (set to a fixed value), then

$$A = W\sum_{i=1}^{2}[K_i\{N(a_i)-n_i\}-a_i'/\alpha], \qquad (12)$$

$$S=PC+Qc_f-a_2C_s, \qquad (13)$$

where $W=C+2c_f+2\alpha C_s$, $K_1=C/W$ and $K_2=(C+c_f)/W$.

When the transmission cost C is large, cost saving H proportional to W becomes large. Furthermore, H does not level off no matter how large the traffic a_2 is, because large C yields large cost ratios K_1 and K_2. On the other hand, when C is small, cost saving H becomes small and levels off in a large a_2 region. In the same manner, cost saving S by skewer connection increases as C increases, but it always disappears in a large a_2 region regardless of the value of cost parameter C. These relationships are illustrated in Fig.5.

From the above discussion, it can be said that the skewer connection method is more favorable than alternate routing when:

a) The transmission cost of R_1-R_2 (or T_1-T_2) link (which is proportional to the node-to-node distance) is large, and the traffic from T_1 to T_3 is small, and
b) Cost/erl of a skewer tandem switching system is relatively small, and transmission multiplexing/demultiplexing cost located at a skewer tandem center is small.

4. CONCLUSION

A hierarchical mesh structure with a "skewer connection" design method was proposed as a favorable transit network structure for the future. The skewer connection method was found to be advantageous when: i) the node-to-node traffic is small, ii) the nodes are located far apart, and iii) skewer tandem node cost is small.

The trunk engineering process of the skewer connection method is easier than that of conventional alternate routing. However, if this method is applied throughout a large-scale network, the final result will be a kind of non-hierarchical network having a "chaotic" quasi-complete graph, whose design and operation would be quite complicated. Therefore, the skewer connection should be utilized in an economically feasible portion of a network, instead of being spread over an entire network, in conjunction with the alternate routing which has a wide application area.

REFERENCES

[1] M.Imase, K.Okada and H.Ichikawa, "Graph theoretical approach to a highly reliable network," Trans. IECE Japan (Section B), vol. J66-B, no. 3, pp. 337-344, 1983.
[2] Y.Tanaka and M.Akiyama, "Constitution of non-hierarchical multi-link communication network," Trans. IECE Japan (Section B), vol. J66-B, no. 12, pp. 1494-1501, 1983.
[3] A.J.Fischer, D.A.Garbin, T.C.Harris and J.E.Knepley, "Large scale communication networks - design and analysis," Omega, Int. J. of Mgmt Sci., vol. 6, no. 4, pp. 331-340, 1978.
[4] Y.Rapp, "Planning of junction networks in a multi-exchange area I. General principles," Ericsson Technics, vol. 1, pp. 77-130, 1964.
[5] C.J.Truitt, "Traffic engineering techniques for determining trunk requirements in alternate routing trunk networks," Bell Syst. Tech. J., vol. 33, no. 2, pp. 277-302, 1954.
[6] S.Abe and H.Akimaru, "An optimum design of multi-stage alternative routing network system," Trans. IECE Japan (Section E), vol. J66-E, no. 7, pp. 435-441, 1983.

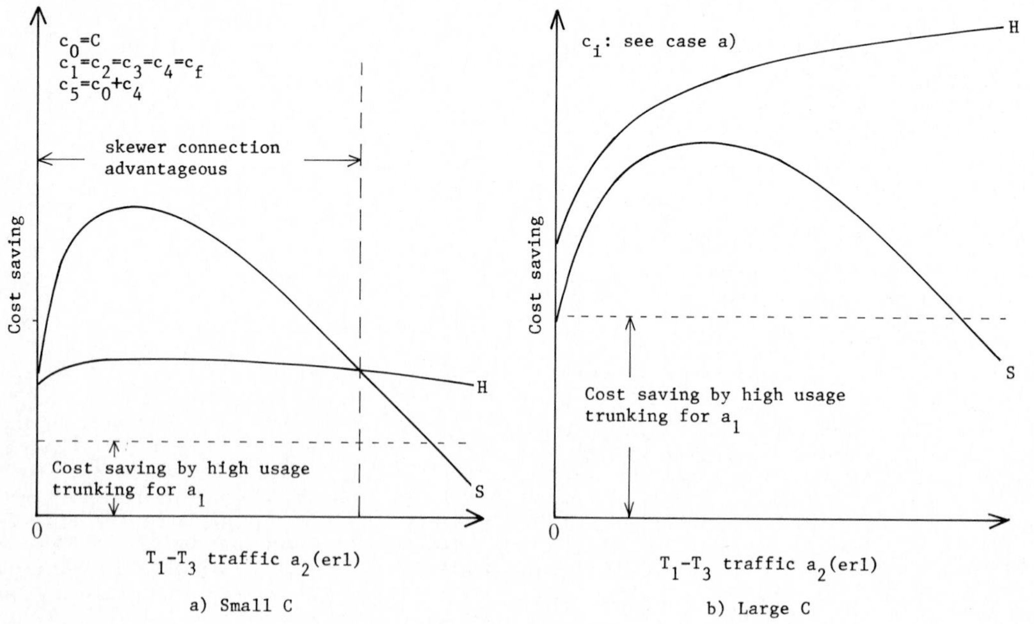

Fig.5 Comparison of cost savings H and S

APPENDIX 1

Let $H(K)$ be the cost saving by high usage trunking when the cost ratio is K. Assuming $K_1<K_2$,

$$H(K_2)-H(K_1)=[K_2N-\{aE(n_2,a)/\alpha+K_2n_2\}] -[K_1N-\{aE(n_1,a)/\alpha+K_1n_1\}] =(K_2-K_1)N-a\{E(n_2,a)-E(n_1,a)\}/\alpha -K_2n_2+K_1n_1. \quad \text{(A-1)}$$

Since $\alpha K=-\dfrac{\partial\{aE(n,a)\}}{\partial n}$,

$$\alpha K_1(n_1-n_2)<aE(n_2,a)-aE(n_1,a)<\alpha K_2(n_1-n_2).$$

Thus,

$$H(K_2)-H(K_1)>(K_2-K_1)N-K_2(n_1-n_2)-K_2n_2+K_1n_1 =(K_2-K_1)(N-n_1)>0, \quad \text{(A-2)}$$

where N : the number of direct trunks required to carry traffic "a" within a given link blocking probability objective,

n_i : the number of high usage trunks required when cost ratio is K_i,

$E(n,a)$: link blocking probability when traffic "a" is offered to n trunks.

APPENDIX 2

Cost saving by high usage trunking (H) is equal to zero iff no traffic is desired to be offered to basic trunks; i.e., the number of high usage trunks n is equal to the value of N in Appendix 1. This condition is:

$$\frac{aE(N-1,a)-aE(N,a)}{\alpha}>K \quad \text{(A-3)}$$

and $E(N, a)=b_0$. (A-4)

From the above relationships,

$$K<\frac{b_0}{\alpha}\left(\frac{N}{1-b_0}-a\right). \quad \text{(A-5)}$$

By using the following approximate formula of Wilkinson for $b_0=0.01$,

$$N=0.970a+1.752\sqrt{a}+2.277 \quad (a>6.2), \quad \text{(A-6)}$$

Ineq.(A-5) is expressed as follows:

$$0.0202a-1.770\sqrt{a}+(100\alpha K-2.3)<0. \quad \text{(A-7)}$$

There exists "a" which satisfies Ineq.(A-7) iff

$$\alpha K<0.41. \quad \text{(A-8)}$$

Thus, when $\alpha=0.83$,

$$K<0.5. \quad \text{(A-9)}$$

TELETRAFFIC ISSUES in an Advanced Information Society
ITC-11
Minoru Akiyama (Editor)
Elsevier Science Publishers B.V. (North-Holland)

TELETRAFFIC PROBLEMS IN CELLULAR MOBILE RADIO SYSTEMS

Neil MACFADYEN and David EVERITT

British Telecom Research Laboratories
Felixstowe, UK
IP11 8XB

ABSTRACT

The TACS cellular mobile radio system has many areas of teletraffic interest. This paper considers 3 of these: the random-access signalling; the benefits available from off-air call setup; and the effects of handover between cells.

1 INTRODUCTION

The UK has chosen TACS as its new national cellular mobile radiotelephone system. Because more than one company will be operating this in public service, the Teletraffic Division of British Telecom has undertaken a detailed study of the characteristics of the air interface and the performance of different system facilities, in order to enable informed agreement upon detailed standards. This paper presents some of the more interesting results of the study.

The first area considered is the random-access signalling by the mobiles. This has interesting overload behaviour, displaying a severe hysteresis loop, and requires extensive control measures to stabilise the system against a drift into unacceptable congestion.

The second area we consider here is that of speech channel capacity. In particular, we look at the effects of handover, and the possible benefits due to off-air call-setup. In either case the effects require more than mere analysis for a decision as to their importance, and raise important questions as to service policy.

2 ANALYSIS OF SIGNALLING CAPACITY

In the TACS system, signalling between the base and the mobile stations is effected through an overlay structure on the voice channels when these are allocated. On the other hand, replies to paging messages or initial attempts at system access are made on a dedicated signalling channel known as the Reverse Control Channel (RCC) and answered on the corresponding system controlled Forward Control Channel (FCC), and this gives rise to interesting behaviour under high traffic.

The random-access signalling is essentially a non-persistent proxy CSMA-CD with limited perseverance. The FCC is a continuous 8 kbit/s data stream, every 11th bit of which (the Busy/Idle bit) is set if and only if there is an effective transmission on the RCC then in progress; by listening to which a mobile station can determine the busy/idle status of the RCC and hence regulate its own transmissions. This control can however break down during times of high traffic, for the B/I bit remains unset whenever 2 or more access attempts clash, and so opens wider the window for further clashes.

The access procedure then follows the standard CSMA-CD rules, with the obvious modifications. The interesting features arise because

a) There is an unguarded period of nominally 56 bits before the B/I bit is set for a successful access, and messages involved in a clash persist for some 115 bits before abandonment;

b) The repeat attempts are limited, to N times of finding the channel busy, and/or M times of involvement in call-clash;

c) There is an overall timeout on the access procedure.

The behaviour of the land station is complementary to that of the mobile, and is of no present interest. Full details of the interface specification can be found in [1].

We consider now the ideal behaviour of the RCC under the assumptions that all messages are of a fixed duration H, and that the entire stream of access attempts is effectively Poisson with rate λ. The unguard duration will be denoted B; the call-clash time of 115 bits will be the time unit; and we shall ignore the granularity of the channel and, in the first instance, the timeouts. Set

$\Pr\{\text{an attempt sees busy on arrival}\} = p(\lambda)$
$\Pr\{\text{an attempt collides with another}\} = q(\lambda)$
$\Pr\{\text{an attempt arrives to a free RCC}\} = \pi$

We assume that q is independent of the order of the attempt. Then the total attempt rate λ is related to the fresh intent rate Λ by

$$\lambda = \Lambda \sum_{i=0}^{M} \sum_{j=0}^{N} \binom{i+j}{i} p^j q^i$$

and the stream of losses - i.e the signalling grade of service - satisfies

$$\beta = p^{N+1} \sum_{i=0}^{M} \binom{N+i}{i} q^i + q^{M+1} \sum_{j=0}^{N} \binom{M+j}{j} p^j$$

But $p(\lambda)$ is just the guarded throughput: hence

$$\pi \lambda (H - B) \ell^{-\lambda B} = \Lambda(1 - \beta)(H - B)$$

Furthermore

$$\pi = 1 - \pi \lambda [R(\lambda) + H\ell^{-\lambda B}]$$

where $R(\lambda)$ is the mean time spent in call-clash by an attempt which arrives to a genuinely free RCC. It is not difficult to show, by considering the duration of a clash period, that this is simply

$$R(\lambda) = (1 - \ell^{-\lambda B}) \ell^{\lambda}/\lambda - B\ell^{-\lambda B}$$

Finally, the rate at which clash periods start is clearly $\lambda\pi (1 - e^{-\lambda B})$; so that since the mean number of accesses contributing to a clash period can be shown to be $(1 + e^{\lambda})$, we have

$$q = \pi (1 - \ell^{-\lambda B})(1 + \ell^{\lambda})$$

Starting with the overall attempt-rate λ, it is now possible to use the equations above to solve successively for π, q, p, β and Λ.

The form of the curves of β against Λ depends critically on the values of M and N. When M is 15 or lower, these are of typical Erlang shape; but as M is raised further the curve bends back upon itself, thus forming the physically unrealisable portion familiar to random-access theory. In this system, however, the limitation on the number of repeat attempts by any one source makes the curve turn over yet again, forming an S-shape (see Figure 1) with a stable upper branch. (The dashed curve in this Figure is the contribution of call-clash to the GOS; the solid loop is that of the remainder, the channel-busy condition.)

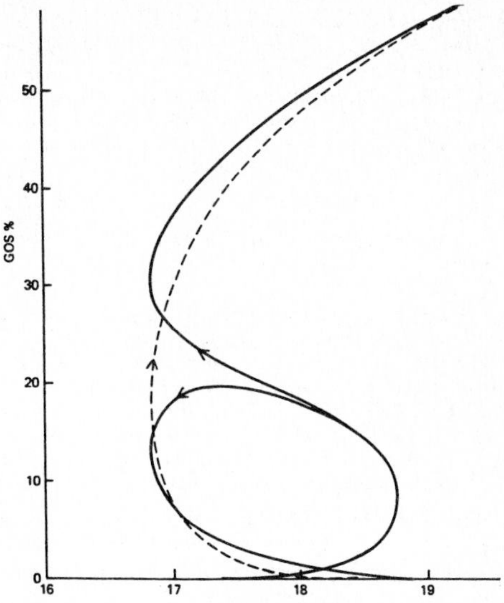

Figure 1

TACS random-access signalling GOS as a function of offered load in KBH access-intents (KBHAI): analytic results
For call-clash maximum M = 30
Channel-busy maximum N = 30

This corresponds to the existence of a hysteresis loop, so that over a certain range of attempt-rate there are 2 possible solutions, representing low and high congestion operating regimes respectively. Outside this range, only a single solution exists; and we observe that there is no possibility of total system lockup as in systems with unbounded repeats. Figure 2 illustrates this for N = 10 and differing M.

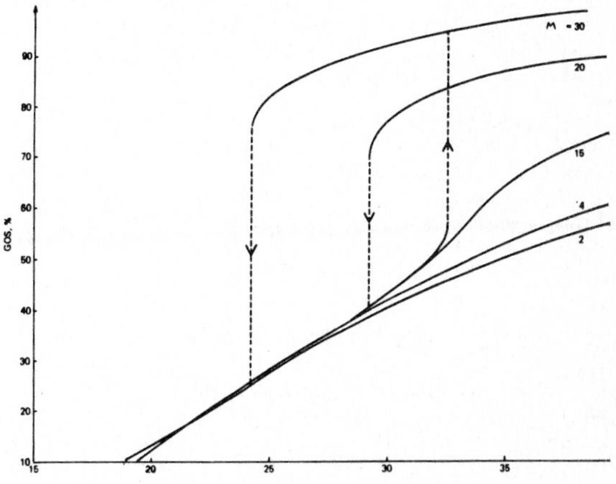

Figure 2

TACS random-access signalling GOS: implications of analysis
For N = 10 and M varied

The entry-threshold of this loop is insensitive to the value of M, but increasing that parameter increases the size of the loop and lowers the exit-point. Raising the value of N on the other hand makes the loops narrower but located at lower values of GOS or of intent-rate. In addition, there is a tendency for the loop to appear at lower values of M. For N=M=30, near the lower exit point of the loop it is possible to operate at a GOS value as low as 1% or as high as 35%.

In any practical system we would expect to operate well below these limits. If the signalling GOS is set at not more than 1%, the existence of the hysteresis loop leads to well-defined recommendations for the values of N and M; conversely, for any fixed values of these which allow hysteresis there is a maximum access-intent rate (or, correspondingly, GOS) at which it is safe to operate.

2.1 Effect of Timeouts

Since the operating curves of the RCC have infinite or negative slope close to the regions of interest, it does not at once follow that a conclusion that relatively long timeouts have negligible effect is in fact justified, and so a quantitative study is necessary.

We assume that the individual repeats in a single call string all see the system in equilibrium, so that they are effectively independent. An analysis of the validity of this leads to the gratifying conclusion that this assumption is indeed reasonable where it matters. Let P_{ij} be the probability that the access timer is still within time after a total of i collisions and j busies; then since the most recent event was a clash with probability i/(i+j), the entire previous analysis remains valid except that the loss probability β must now be replaced by

$$\beta = \sum_{j=0}^{N} \sum_{i=0}^{M} \binom{i+j}{j} q^i p^j \times \left\{ \left(iP_{i-1,j} + jP_{i,j-1} \right) / (i+j) - P_{i,j} \right\}$$

$$+ p^{N+1} \sum_{i=0}^{M} \binom{N+i}{i} q^i P_{i,N}$$

$$+ q^{M+1} \sum_{j=0}^{N} \binom{M+j}{j} p^j P_{M,j}$$

It remains to determine P_{ij}. We do this by approximating the sum of the (uniform) randomisation intervals involved by a Normal random variate; once again, this is not good for small values of (i+j), but in that region the remaining factors in the expression above make the fact irrelevant. By using the usual Hastings approximation for the Gaussian the problem is then easily solved numerically.

The results of this are interesting. Imposing a timeout has a beneficial effect on the grade of service: it tends to suppress the hysteresis loop, and fewer intents are lost. The effects are however very small at realistic levels of N or M; only when the effective timeout becomes as low as 2 seconds or so do they make any difference to the access rates which see a GOS of 1%. Figure 3 illustrates this.

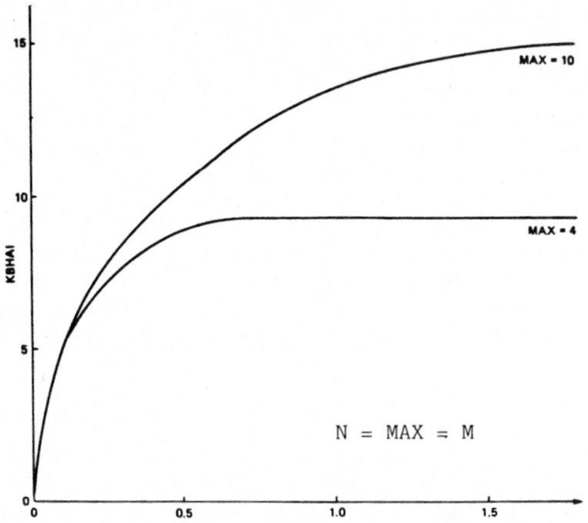

Figure 3

Capacity of RCC for 1% GOS in presence of timeout (seconds)

2.2 Drift into Congestion

If the presence of an effectively infinite number of mobiles is assumed, then for infinite M and N the RCC will always eventually drift into congestion. Since manually-reinitiated sequences of access attempts, which serve to multiply these control parameters, cannot be prevented, it is of interest to study the expected stable life of the system before this occurs.

So label the system state by the number of intents still in the backlog, and consider the set of instants at which the RCC genuinely becomes free: then with the standing assumptions on Poissonality of arrivals, these points determine an embedded Markov chain. So too does the subset of instants at which a successful attempt ends. Decompose the transition probability matrix between these former points into the disjoint probabilities S of success and F of failure.

In order to examine these probabilities, we need to determine the arrival-rate of access-attempts conditional upon a clash period being in progress with k attempts in the retry phase. Let the mean retry rate of a single backlogged attempt be μ, and set $\Lambda_k = \Lambda + k$ with $\Lambda_{-2} = \Lambda_{-1} = \Lambda_0 = \Lambda$. We then approximate the system by assuming that if the system state is k, and a clash period is in progress (i.e. after the first interrupt), the total arrival-rate is Λ_{k-2}. This corresponds to (a) ignoring finite-source effects, and (b) assuming that on average precisely 2 attempts overlap simultaneously. It is not easy to decide which way this approximation will be in error. Using the easily-proved result that the probability that a clash period ends without any further interrupt by a fresh call attempt is

$$g(\lambda,f) = \{\ell^{\lambda} - (\ell^{\lambda} - 1)(1 - f)\}^{-1}$$

where f is the probability that any given call attempt is a fresh access, not a retry, we can then write down an expression for $F_{i\ i+n}$ by decomposing it into a sum of 4 terms according to whether the initial and first interrupting calls are fresh or repeats. This then gives rise to the convenient recursive expressions

$$F_{i,i+n} = g_{i+n} G_{i,n}$$

with

$$G_{i,0} = \frac{i\mu}{\Lambda_i}\left(1 - \ell^{-B\Lambda_{i-1}}\right)\frac{(i-1)\mu}{\Lambda_{i-1}}$$

$$G_{i,1} = \left(1 - g_i\right)G_{i,0}$$

$$+ \frac{i\Lambda\mu}{\Lambda_i}\left\{\left(1 - \ell^{-B\Lambda_i}\right)/\Lambda_i + \left(1 - \ell^{-B\Lambda_{i-1}}\right)/\Lambda_{i-1}\right\}$$

$$G_{i,2} = \left(1 - g_{i+1}\right)G_{i,1} + \left(\Lambda/\Lambda_i\right)^2\left(1 - \ell^{-B\Lambda_i}\right)$$

$$G_{i\ n+1} = \left(1 - g_{i+n}\right)G_{i,n} \qquad n \geqslant 2$$

Here g_k is just $g(\Lambda_{k-2}, \Lambda/\Lambda_{k-2})$.

The term S_{ij} which represents successful accesses can be written down immediately:

$$S_{ij} = \frac{\Lambda}{\Lambda_i}\ell^{-\Lambda_i B}\frac{[\Lambda(H - B)]^n}{n!}\ell^{-\Lambda(H-B)}$$

$$+ \frac{i\mu}{\Lambda_i}\ell^{-(\Lambda_i-\mu)B}\frac{[\Lambda(H - B)]^n}{n!}\ell^{-\Lambda(H-B)}$$

Here we have set j = i + n, and the second term appears only for n > 0.

The transition matrix between instants at which the RCC clears after a successful access is then clearly just $P = (1-F)^{-1}S$, which solves the problem of finding the transition probabilities between points of the embedded Markov chains.

To solve the actual time-dependent drift problem, we observe that when the system is not in congestion, the mean time between successful accesses must perforce be the mean time $1/\Lambda$ between fresh arrival attempts. The transition probabilities in time $t = n/\Lambda$ are therefore well approximated (for large n) by the elements of P^n. The numerology of all this is straightforward. The method adopted was to truncate the system state space at the value of 50 and make this state absorbing, which is quite adequate when the retry failure rate μ is reasonably fast.

Figure 4 shows the results of this. The descending curve is a plot of the median time-to-congestion, in hours; the ascending one, the probability that the system will become locked up within one hour. No attempt should be made to extrapolate to shorter times, since the approximations are then invalid. We observe that, with no controls upon congestion, the drift is a most important factor: if a probability of lockup within the hour as high as 1% is acceptable, then the RCC capacity is only some 8,500 busy-hour access intents, which represents a de-rating of its nominal maximum possible rate (based on the usual operating-characteristic arguments) by a factor of more than 2. A requirement for a MTBF of 1000 hours, or say 4 years at 1 busy-hour per day, would limit the capacity to only 6,500 BHAI.

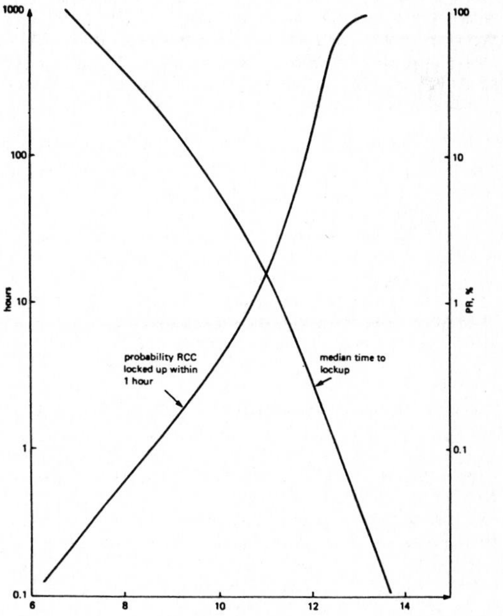

Figure 4

RCC drift into congestion for unbounded perseverance

An analysis for the case of limitations on the number of access attempts in a sequence is considerably more difficult. In practice, TACS has sufficent controls available to it of various kinds that the drift can be detected and reversed at an early stage; the results above then indicate that if the full capacity of the RCC is to be realised we can expect these to be used rather frequently.

3 SPEECH CHANNEL CAPACITY

3.1 Effects of Off-Air Call Setup

A significant gain in capacity may be possible if the speech channels do not have to be allocated before the called subscriber answers a mobile originated call - an arrangement known as Off-Air Call Setup, or OACSU. TACS supports this protocol, which involves the queueing of calls awaiting answer, and therefore raises new questions of system dimensioning and performance.

Such a system can be modelled very simply by an M/M/N/S queue, where N denotes the number of speech channels available in any cell, and S the number of queue places available for calls still awaiting either answer or a free speech channel. An investigation of the adequacy of this, bearing in mind the expected absence of very short calls under OACSU, has indicated that it is surprisingly robust. One possible pair of dimensioning criteria is then to provide that number of queueing places such that the probability of a new call finding all these full and all speech channels busy (the loss GOS) is limited, and also that the 90%-point of its overall delay distribution is lower than (say) 1/10 of its mean conversational holding time, or some 10 seconds. We do not consider this totally satisfactory, however, since it can lead in small channel-groups to very long delays for the calls which do have to wait.

We therefore consider the effects of the additional requirement, that the conditional mean delay too shall be limited to this value. At this point however another consideration enters: in an OACSU system, the delay and loss are experienced not only by the calling but also the called subscriber, and it is not clear that conventional requirements are adequate. This may be especially true if the called number is a PBX. This suggests that, while a 10% GOS without OACSU may be reasonable, if OACSU is implemented a better GOS should be required. If we choose therefore to compare OACSU systems at 10% GOS with non-OACSU at 2%, it is then by no means always true that OACSU leads to increased capacity, as the Table below shows. This give the total (speech + setup) traffic capacity of 4 system designs: pure loss or queueing + loss, in each case with and without OACSU, and uses the tripartite dimensioning criterion above. To aid comparisons, the capacities given are in terms of Erlangs of carried traffics.

Pure Loss			Loss and Delay		
N	On-Air	OACSU	S	On-air	OACSU
6	3.4	2.7	0	3.4	2.7
12	8.6	7.6	1	9.3	8.4
20	15.8	15.2	3	17.2	17.5
30	25.3	25.3	6	27.3	29.6

The only free parameter here is N: the queue limit S is determined by the dimensioning algorithm.

It is clear from these figures that OACSU leads to a genuine increase in traffic capacity only with channel groups larger than a certain GOS dependent minimum, unless it is considered acceptable to throw appreciable loss or delay onto the called subscriber as well as the calling. Since the inconvenience of a failed call is, with OACSU, seen by both parties, it would however logically seem necessary to count it twice instead of once only. It is indeed not clear that even so we are not seriously under-representing the total annoyance caused.

3.2 Analysis of Handover Behaviour

Handover is the dynamic assignment of a call in progress to a new speech channel (usually in a different cell) when the mobile station has moved and reception is no longer adequate. If we consider the balanced situation where the mean hand-in and hand-out rates coincide, it is not immediately clear in which direction the effects will lie: since the handover both decreases the effective mean holding time in a cell while at the same time keeping the traffic constant, hence reducing queueing delays; and also degrades the service seen by fresh calls by assuming priority over them in the assignment of channels. It is therefore necessary to examine this in detail.

We model the system as a multiserver 2-priority queue where the high priority calls represent the hand-in traffic in a cell. Both streams are assumed to be Poissonian, and the effects of the hand-outs are accounted for by reducing the overall mean holding-time. This of course corresponds to the number of handovers which a call experiences being geometrically distributed. The fresh traffic is assumed to see a finite maximum queue length, whereas the hand-ins see one which is effectively infinite.

The model can be analysed by a method due to Basharin [2], which we can only outline here. Let p_{ij} be the probability that there are i high and j low priority customers in the queue simultaneously, and all servers busy; and let q_k be the probability that there are k calls in service and none in the queue. Let the high (low) priority calling-rates be λ_1 (λ_2), and the mean holding-time $1/\mu$. Then we can write down immediately

$$q_k = q_0 \, (\Lambda/\mu)^k / k!$$

$$p_{0,0} = q_n$$

where Λ is the overall arrival-rate $(\lambda_1 + \lambda_2)$. If we now introduce the vectors

$$\underset{\sim}{p}_i = (p_{i0}, p_{i1}, \ldots p_{iS})$$

the equilibrium equations for the p_{ij} become

$$(\underset{\sim}{p}_0, \underset{\sim}{p}_1, \ldots) \, Q = 0$$

where the transition matrix Q is infinite-dimensional and block-partitioned, with each block a (m + 1) x (m + 1) square matrix. In this form, Q is actually tridiagonal, with

$$\begin{aligned} Q_{ii} &= B - (\lambda_1 + n\mu)I \qquad (i > 1) \\ Q_{i,\, i+1} &= \lambda_1 I \\ Q_{i,\, i-1} &= n\mu I \end{aligned}$$

where I is the unit matrix. The matrix B satisfies

$$B_{i,\, i+1} = \lambda_2 = -B_{ii}$$

with

$$B_{m+1,\, m+1} = 0$$

and all other entries vanishing.

To solve these equations we assume a relation of the form

$$p_i = p_0 \, S^i;$$

which results in a matrix quadratic equation for the unknown S

$$\lambda_1 I + S[B - (\lambda_1 + n\mu)I] + S^2 \mu = 0$$

In the case in hand, S is upper triangular with $S_{ij} = s_{j-i}$ for j<m+1, and this equation is not difficult to solve. We are then in possession of all the information needed to analyse the system of interest.

The Table below shows some of the results from this model, for a loss GOS of 10% and the delay criteria discussed earlier, for various proportions of handover traffic (which has been assumed to be balanced). All Figures are in erlangs of total (ie., speech + setup) offered traffic, and assume 15% of ineffective traffic.

N	S	p(d)	f: 0	f: 0.3	f: 0.5
6	0	-	4.5	4.1	0.5
12	1	.12	12.1	11.9	1.7
20	2	.21	22.5	22.1	21.9
30	3	.36	48.9	48.2	47.7
40	4	.36	48.9	48.2	47.7
48	5	.43	59.5	58.9	58.3

The column labelled p(d) gives the probability that a fresh call has to queue at all, when there is no handover (f=0).

These figures show that balanced handover has little effect upon the system capacity. Overall, it slightly reduces the waiting time (to a p(d) value of 0.33 for instance when N = 48 and f = 0.5), but at the same time makes the loss GOS rather worse - and it is that effect which dominates. The net result is slightly to lower the capacity. The case of unbalanced handover is rather different: an excess of hand-ins causes the performance to deteriorate rather quickly. The effect is less pronounced with a delay than a loss system, which is an argument in favour of the former architecture.

4 CONCLUSIONS

The signalling structure of TACS is extremely rich and the presence of a hysteresis loop with a flip into congestion must be taken into account when determining the operational capacity. In practice however these effects occur at rather high loadings, and there are sufficient effective controls available to make the signalling channel virtually transparent under normal operating conditions.

The use of Off-Air Call Setup and queueing facilities can raise the nominal capacity of a system considerably. They should not however be used without serious consideration of their effects on the called party, and may not be as advantageous as expected.

The effects of handover on speech channel capacity in a properly dimensioned system are quite negligible, provided that there is a reasonable balance between hand-ins and hand-outs. When that is not the case, it can lead to significant performance degradation.

ACKNOWLEDGEMENT

Acknowledgement is made to the Director of System Evolution and Standards Department of British Telecom for permission to publish this paper.

REFERENCES

1 United Kingdom Total Access Communications System: Mobile Station - Land Station Compatibility Specification (1983)

2 G P Basharin, 'Poisson Service Systems with Priorities and Limited Waiting Capacity', Proc 5th Int. Teletraffic Congress, New York (1967)

TELETRAFFIC ISSUES in an Advanced Information Society
ITC-11
Minoru Akiyama (Editor)
Elsevier Science Publishers B.V. (North-Holland)

TRAFFIC POLICIES IN CELLULAR RADIO THAT MINIMIZE BLOCKING OF HANDOFF CALLS

Edward C. Posner and Roch Guérin

Department of Electrical Engineering, California Institute of Technology
Pasadena, CA 91125 USA

ABSTRACT

In cellular radio, mobile subscribers are successively served by transmitters associated with the different cells through which they pass. Since a cellular system is only allocated a fixed number of channels within a given cell, it is possible for a subscriber to enter a cell where all channels are already busy. This would result in a disconnection in the middle of a call. Several schemes are investigated to greatly reduce the frequency of such occurrences. All of the schemes are based on allowing such "handoff calls" to be given priority, to queue, or a combination of both. Literal and numerical expressions of blocking probabilities and waiting times are obtained, and the influence of each method on other types of calls is estimated. It is found that significant decrease in blocking of handoff calls can be obtained, paid for only by a small penalty for other calls. The method, therefore, seems to be an easy and efficient way of improving the perceived service quality of cellular radio.

1. INTRODUCTION

We consider a cellular radio system with n channels in a given cell. For this cell, the following assumptions are made:

- Two types of traffic are handled by the cell. The *handoff calls* enter the cell while already being served by another site. They are handed off to the site of the cell they enter. The *originating calls* correspond to calls initiated or initially received by customers located in the cell. Once a call is in progress, its original type (handoff or originating) cannot be determined.
- Memoryless service time, on the average equal to $1/\mu$ sec. The service time refers to the entire call duration, assumed independent of the number of handoffs to other cells as the mobile drives throughout the service area.
- Memoryless handoff rate on the average equal to h per second per call. The call duration in a given cell will, therefore, be taken memoryless. As vehicle speed and direction as well as the distance from the cell boundaries usually change during a call, this approximation is not too inaccurate and might even be very close to reality on the average for hand-held units.
- Memoryless arrival rate for originating calls, on the average equal to λ per second.
- Memoryless arrival rate for handoff calls, on the average equal to γ per second.

• Our goal is to derive methods of decreasing the blocking probability of handoff calls without increasing the blocking probability of originating calls too noticeably. The main reason for this goal is that handoff calls represent calls already in progress, and the blocking of one of them implies an interruption in the middle of a conversation. Therefore, in order to preserve a certain perceived service quality, this type of occurrence should be made as rare as possible. Several possibilities arise.

• The first investigated is the blocking of newly-originating calls as soon as only r channels among the initial n are still free. This allows us to build a "guard band" before handoff calls will be blocked, which will happen only when all n channels are occupied. Clearly, the influence on the blocking probability of originating calls will mainly depend on the value of r. We would like to keep it relatively low while obtaining a substantial decrease in the blocking probability of handoff calls.

• The second possibility is to allow handoff calls to be queued, hoping for a short average delay which would not make the interruption too perceptible.

• The third possibility is to combine the two above solutions, keeping a certain number of guard channels for the handoff calls, while also allowing them to be queued if no channel is available. The allowed queue size can either be taken infinite or finite and equal to some number L.

• We will briefly describe the first possibility and then directly study the third one, since the second corresponds to the special case $r = 0$. (The first possibility is also a special case of the third one, namely $L = 0$.) What is meant by *description* is the state probabilities of the cell, a state being the number of customers in service or waiting for service in the cell. We will then derive the expression of meaningful parameters such as blocking probabilities, probability of delay, average delay, etc.

2. GUARD CHANNELS

• First define some useful notation:

$$\eta = \mu + h \qquad a = \frac{\alpha}{\eta}$$

$$\alpha = \lambda + \gamma \qquad b = \frac{\gamma}{\eta}$$

• If we decide to keep r guard channels to protect handoff calls from high blocking, basic queueing theory can be shown to give:

State probabilities:

$$0 \le i \le n - r \qquad P(i) = \frac{a^i}{i!} P(0)$$

$$n - r \le i \le n \qquad P(i) = \frac{a^{n-r} b^{i-(n-r)}}{i!} P(0)$$

where

$$P(0) = 1 \Big/ \left[\sum_{i=0}^{n-r-1} \frac{a^i}{i!} + a^{n-r} \sum_{i=n-r}^{n} \frac{b^{i-(n-r)}}{i!} \right]$$

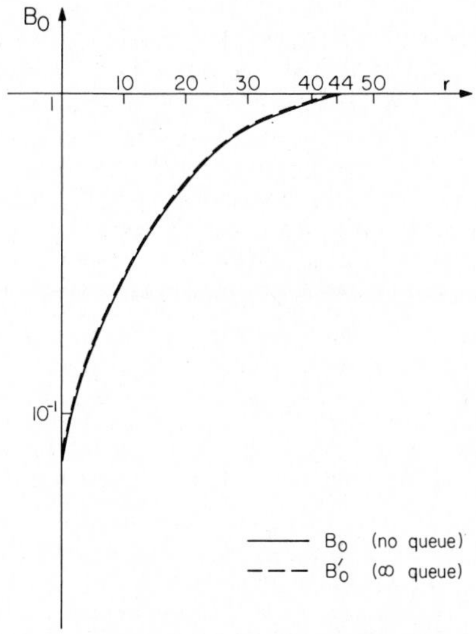

Fig. 1 - Blocking probabilities for originating calls in cases of no queues or infinite queues.

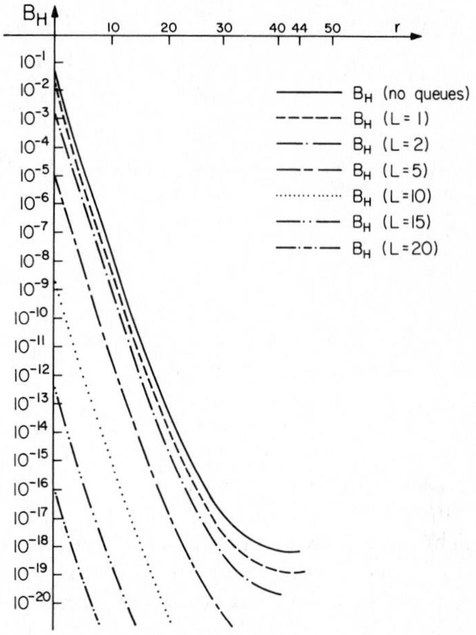

Fig 2.- Blocking probabilities for handoff calls in cases of no queues and finite length queues

The blocking probability of originating calls is then given by:

$$B_0 = P(i \ge n - r)$$

$$\Rightarrow B_0 = \left[a^{n-r} \sum_{i=n-r}^{n} \frac{b^{i-(n-r)}}{i!} \right] \times P(0) \; ; \qquad (1)$$

Similarly for handoff calls we have:

$$B_H = P(i \ge n) \Rightarrow B_H = \frac{a^{n-r} \times b^r}{n!} \times P(0) \; ; \qquad (2)$$

Plots of these two probabilities as functions of r can be found in Figs. 1 and 2, and will be discussed later.

3. QUEUEING OF HANDOFF CALLS AND GUARD CHANNELS

• We first assume that infinite queues are allowed for handoff calls. The state balance equations can then easily be derived using the following state diagrams:

– $0 \le i \le n-r-1$ $\qquad P(i+1) = \frac{a}{i+1}\, P(i)$

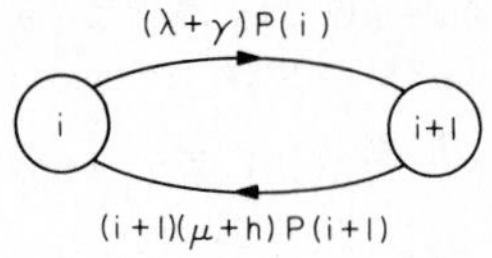

– $n-r \le i \le n-1$ $\qquad P(i+1) = \frac{b}{i+1}\, P(i)$

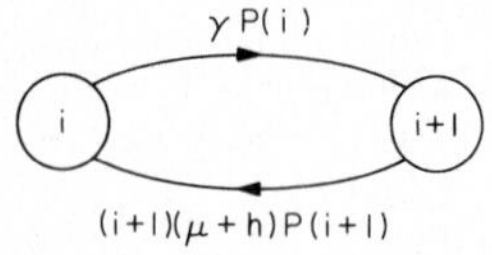

– $n \le i$ $\qquad P(i+1) = \frac{b}{n}\, P(i)$

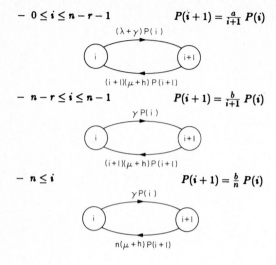

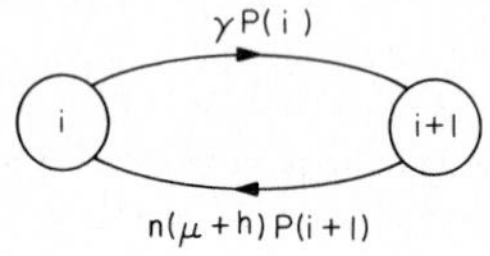

$$\Rightarrow \begin{cases} 0 \le i \le n-r, & P(i) = \frac{a^i}{i!} \times P(0) \\ n-r \le i \le n, & P(i) = \frac{a^{n-r} b^{i-(n-r)}}{i!} \times P(0) \\ n \le i, & P(i) = \frac{a^{n-r} b^{i-(n-r)}}{n! n^{i-n}} \times P(0) \end{cases}$$

where $P(0)$ is given by:

$$P(0) = 1 \Bigg/ \left[\sum_{i=0}^{n-r-1} \frac{a^i}{i!} + a^{n-r} \sum_{i=n-r}^{n-1} \frac{b^{i-(n-r)}}{i!} + \frac{a^{n-r} b^r}{(n-1)!(n-b)} \right]$$

• We can again compute the blocking probability for originating calls:

$$B_0' = P\{i \ge n-r\}$$

$$\Rightarrow B_0' = \left[\left(\frac{a}{b}\right)^{n-r} \sum_{i=n-r}^{n-1} \frac{b^i}{i!} + \frac{a^{n-r} b^r}{(n-1)!(n-b)} \right] \times P(0) \qquad (3)$$

We can remark that due to the fact that handoff calls are allowed to queue, B_0' is slightly greater than the previous B_0, even in the case $r = 0$ (handoff calls have priority). E.g.,

$$B_0 \simeq 6.46 \times 10^{-2} \text{ versus}$$

$$B_0' \simeq 7.79 \times 10^{-2} \text{ for } r = 0\,.$$

$$(a = 40\,,\ b = 8\,,\ n = 44)$$

B_0' is always larger than B_0 (except that for $r = n$ they are both equal to 1). Since handoff calls are queued, there is no blocking probability, but we can compute an equivalent expression which would here be the probability of being delayed:

$$P\,\{\text{handoff call is delayed}\} = P\{i \ge n\} = P_H\{>0\}$$

$$\Rightarrow P_H\{>0\} = \frac{a^{n-r} b^r}{(n-1)!(n-b)} \times P(0)\,; \qquad (4)$$

We can also obtain the probability of being delayed more than a certain time t:

$$\begin{aligned} P_H(>t) &= P_H(>0) \times P_H(\text{delay } > t \,/\, \text{delayed}) \\ &= P_H(>0) \times e^{-t[n(\mu+h)-\gamma]} \\ &= P_H(>0) \times e^{-t\eta(n-b)} \end{aligned}$$

$$\Rightarrow P_H(>t) = \frac{a^{n-r} b^r}{(n-1)!(n-b)} \times P(0) \times e^{-t\eta(n-b)}\,; \qquad (5)$$

Eqn. (5) gives the expression of the average delay for handoff calls:

$$W_H = \frac{a^{n-r} b^r}{(n-1)!(n-b)^2\,\eta} \times P(0) \qquad (6)$$

• However, we should remark that the average delay given delayed, that is, the average delay encountered by customers that will have to queue, is independent of r and equal to:

$$D_H = \frac{1}{\eta} \times \frac{1}{n-b} \qquad (7)$$

This tells us that increasing the number of guard channel will only decrease the number of handoff calls that have to be queued, without reducing the average delay for those actually queueing. A heuristic explanation for this can be obtained by remembering that once a call is in process, it is impossible to determine its previous type (handoff or originating). Therefore, once we are in the situation where all servers (channels) are busy, we have no information on the possible value of r, and our waiting time will only depend on the residual service time of the customers in service. These are clearly independent of r because of the memorylessness.

If we take again: $a = 40$, $b = 8$, $n = 44$ and choose $\eta \simeq 0.012$ ($\mu \simeq .5/min$, $h \simeq .2/min$), we get:

$$D_H \simeq 2.31 \text{ sec}$$

On the average the delay introduced in the handoff process by queueing, using the above parameters, is greater than 2 seconds and therefore perceptible when experienced.

• We will now investigate the case with finite queue size equal to some number L. The results will only be the truncation of the previous expressions and a blocking probability for handoff calls will again appear. The only change in the state probabilities is that they are equal to zero for i strictly bigger than $n + L$. This will introduce a modification in the expression of $P(0)$, which can be obtained without recalculation (truncation).

$$P_L(0) = 1 \Bigg/ \left[\sum_{i=0}^{n-r-1} \frac{a^i}{i!} + \left(\frac{a}{b}\right)^{n-r} \sum_{i=n-r}^{n-1} \frac{b^i}{i!} + \frac{a^{n-r}b^r}{(n-1)!} \times \frac{1-(\frac{b}{n})^{L+1}}{n-b} \right] ;$$

The other state probabilities are obtained simply by replacing $P(0)$ by $P_L(0)$. This gives a blocking probability for originating calls equal to:

$$B'_{L0} = \left[(\frac{a}{b})^{n-r} \sum_{i=n-r}^{n-1} \frac{b^i}{i!} + \frac{a^{n-r}b^r}{(n-1)!} \times \frac{1-(\frac{b}{n})^{L+1}}{n-b} \right] \times P(0) ; \quad (8)$$

As a check, Eqn. (8) reduces to Eqn. (3) if we let $L \to \infty$ and to Eqn. (1) if $L = 0$. Similarly, for handoff calls we have:

$$B'_{LH} = P\{i = n + L\}$$

$$\Rightarrow B'_{LH} = \frac{a^{n-r}b^{L+r}}{n!\, n^L} \times P(0) ; \quad (9)$$

Again we can check that Eqn. (9) goes to 0 if $L \to \infty$ and reduces to Eqn. (2) if $L = 0$. We can also derive the probability for a handoff call to be delayed:

$$P\{\text{handoff call is delayed}\} = P\{n \le i < n + L\}$$

$$= P_{LH}(> 0)$$

$$\Rightarrow P_{LH}(> 0) = \frac{a^{n-r}b^r}{(n-1)!} \times \frac{1-(\frac{b}{n})^L}{(n-b)} \times P(0) ; \quad (10)$$

Eqn. (10) gives 0 for $L = 0$ and reduces to Eqn. (4) if we let $L \to \infty$, as must be.

Finally, we have the probability that a handoff call is delayed more than t:

$$P_{LH}(> t) = P_{LH}(> 0) \times e^{-t\eta(n-b)}$$

$$\Rightarrow P_{LH}(> t) = \frac{a^{n-r}b^r}{(n-1)!} \times \frac{1-(\frac{b}{n})^L}{n-b} \times P(0) \times e^{-t\eta(n-b)} ; \quad (11)$$

Eqn. (11) gives 0 for $L = 0$ and reduces to Eqn. (5) as $L \to \infty$, as again must be. From Eqn. (11) we can obtain the average delay for handoff calls:

$$W_{LH} = \frac{a^{n-r}b^r}{(n-1)!(n-b)^2\eta} \times \left(1 - (\frac{b}{n})^L\right) \times P(0) ; \quad (12)$$

Again Eqn. (12) equals 0 if $L = 0$ and reduces to Eqn. (6) if $L \to \infty$. Due to the truncation property, the average delay given delayed is again independent of r and still equal to:

$$D_{LH} = D_H = \frac{1}{\eta(n-b)}$$

• To illustrate all the above formulas, we will plot them as functions of r in Figures 2, 3, and 4.

• The first encouraging conclusion we can draw from the observation of the curves is that allowing the handoff calls to queue has only a small (deleterious) influence on the blocking probability of originating calls. There is therefore no heavy penalty for the system in having this feature present. Practically, we can see that if we allow a queue of length $L = 2$ and take a guard band of $r = 2$ channels, we have the following results (assuming the previously used parameters):

$$B_{2H} \simeq 9.65 \times 10^{-5}$$

while we had before:

$$B_H \simeq 6.46 \times 10^{-2} .$$

We have a clearly tolerable average waiting time $W_{2H} \simeq 8 \times 10^{-3}$ sec but we shall remember that we still have: $D_{2H} \simeq 2.3$ sec, but with a probability of being delayed equal to $P_{2H}(> 0) \simeq 3.46 \times 10^{-3}$, which is small.

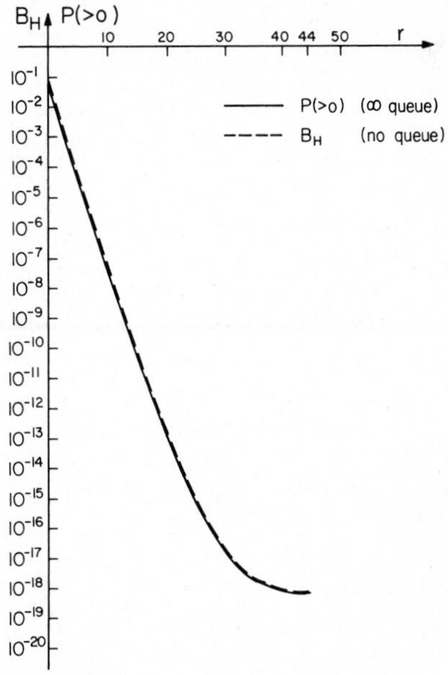

Fig. 3 - Probability of delay (infinite queue) and blocking probability (no queue) for handoff calls

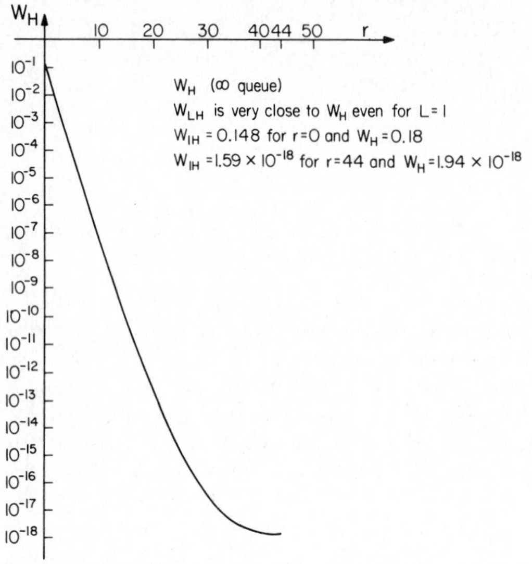

Fig. 4 – Average waiting time for handoff calls in case of infinite queue

On the other hand, we had $B_0 \simeq 6.46 \times 10^{-2}$ with no queue allowed and now we have $B'_{20} \simeq 10^{-1}$ which is a rather small increase. Therefore, this seems a good trade to take since we have a rather good improvement for handoff calls, which is only paid by a small degradation of the service quality for the originating calls.

4. POSSIBLE GENERALIZATION

• A natural follow-on to the above method is to allow both types of calls to be queued while keeping a certain type of priority for the handoff calls, either through guard channels or through a classical preassigned priority classification. This design concept is however more complicated than the previous one, and also doesn't seem to provide as many advantages. Note that the penalty inflicated to the originating calls when using the method described in Section 3 is not too high.

• In the case of finite queue lengths, we have derived recursive formulas for the state probabilities, making the computation of the different system parameters (blocking probabilities, average waiting time, etc.) possible. We shall not report on this here.

5. CONCLUSION

• The method described in Section 3 is a relatively simple and useful way to improve the service quality of a cellular system by protecting calls from being disconnected in the middle of a conversation. This goal is achieved without too severe a penalty on grade service for newly orginating calls. Moreover, the method used is based on techniques familiar to classic telephony. It is therefore a possible solution at a rather low investment cost to the problem of improving the perceived quality of cellular service. This type of improvement might be rather critical in a start-up market, where customers don't necessarily have any particular level of confidence in the cellular concept.

TELETRAFFIC ISSUES in an Advanced Information Society
ITC-11
Minoru Akiyama (Editor)
Elsevier Science Publishers B.V. (North-Holland)

CAPACITY OF A PSTN-MOBILE TELEPHONE NETWORK IN VIEW OF THE SPECIAL REQUIREMENTS FROM A MOBILE SUBSCRIBER

Mats Düring

Telefonaktiebolaget LM Ericsson
Stockholm, Sweden

ABSTRACT

In this report a model is derived for the voice channel allocation in a mobile network using small cell technique with overlaid cells, rearranging and two type of priorities. Both analytical formulas and simulation results are obtained.

1. INTRODUCTION

A project has been started at ERICSSONs telecommunication department with the intention to get a total overview of the capacity of a mobile network including the special requirements from a mobile subscriber (such as mobility).

This project will try to create a model of the PSTN (Public Switched Telephone Network) - MOBILE network and using that to get knowledge about the function of the system i.e. what is the behaviour of the mobile subscriber and his influence on the system, what is the capacity of the system from a traffic point of view, how to dimension etc. The model will of course also make use of traffic measurement results to make it more consistent with reality. The project is going to be divided into two main parts, one concerning the processor capacity and one concerning the switching capacity (including link capacity) observing that they of course influence each other.

This paper will concentrate on the part of the network called the MSC area and specially on the voice channel capacity for a specific way of distributing voice channels among certain areas called cells.

When a mobile network is going to be dimensioned from the radio-channels point of view the major problem is, if the network is of such a size that it could not be expected to be served by the radio-channels made available by the administration, how to increase the total number of available channels. A number of techniques have been proposed and implemented. The simplest and probably the first suggestion was to arrange the total number of frequencies into a number of groups. These groups are then allocated to well defined areas called cells in a way that it is possible to reuse these groups a certain distance away ruled by the reuse formula.

The reuse formula tell us how to minimize the co-channel interference. This technique is called fixed channel assignment. Other techniques use the dynamic or hybrid channel assignment schemes where the last one is a mixture of fixed and dynamic. In this scheme a number of fixed channels and channels common to a number of cells are allocated, see reference (4).

The scheme used here is a type of hybrid assignment with rearrangement, see reference (2) and (3). This technique is here called overlaid cells with rearranging. The model is also extended with two kind of priorities.

Part A of the analysis gives a short description of the mobile network restricting it to the part of the network of interest to us in this paper. In part B the model is derived using this interpretation of the system.

The derivation of a model for our system could be done using either analytical tools, simulation or a combination. This paper will use the last. In a forthcoming paper these results will also be compared with traffic measurements from actual working systems.

2. GENERATION OF THE MODEL

Today, relatively little is known about the behaviour of mobile networks. The intention with the project is as mentioned above to derive knowledge of a mobile network including mobile subscriber behaviour such as mobility, holdingtimes, calling patterns and so on. Because of subscriber mobility in the network there are other special requirements e.g. roaming, handoff and other signalling functions (using e.g. CCITT No.7), see figure 1 and 2.

2.1 Modelling

The modelling is done stepwise. First formulas are derived without these special requirements such as mobility (see definitions below). The next step is to expand the model with the mobility requirement. After this it is possible to see what effect the mobility has on the capacity.

This paper concentrates on the first step and the special area of the network called the MSC area.

2.1.1 The intepretation

The approach to the problem was in this case first of all to create a descriptive model of the network. This is mainly done by first reading the functional descriptions to see how the system actually works.

This basic description is in the project as well as in this paper based on AXE 10. The next thing is to translate to traffic terms, and delete things not of importance or negligible.

2.1.2 The creation

Now this interpretation was used to generate the analytical formulas and the simulation program. The results from the simulation program were then checked against the analytical formulas. Also traffic measurements from actual working systems are investigated.

The results from the traffic measurements as well as the analytical formulas and simulations with mobility parameters will be presented in a forthcoming paper.

2.2 Part A, the network model

A short description of a PSTN-MOBILE network is as follows.

a) The Mobile Services Switching Centre (MSC) area;
b) The Public Land Mobile Network (PLMN);
c) The interconnection of the MSC area, PLMN and PSTN networks.

One reason for dividing the model into these three parts is the possibility to see the effect that mobile telephony has on ordinary telephony and vice versa.

2.2.1 The total network

Figure 1 and 2 indicates a network including the MSC, PLMN and PSTN parts.

The three basic actions that generate load on such a system is:

a) Ordinary calls;
b) Customers moving around;
c) Calls with movement.

In other words this is the main difference between ordinary and mobile telephony.

2.2.2 The interconnection of the MSC area, PLMN and PSTN

The main purpose with this section is to find out how the traffic is flowing through the exchange based on the number of traffic types possible to have in the system.

2.2.3 The PLMN network

This section involves the specific mobile network parameters such as roaming, interrogation, call to visiting subscribers and the signalling influence on the link load.

2.2.4 The MSC area

As mentioned above this report will deal with the MSC area part and specially the voice channel capacity.

An MSC area is defined as the part of the PLMN network covered by an MSC. An MSC area may comprise several traffic cells. A traffic cell area is the part of the MSC area covered by a base station. Every mobile in a traffic cell area can be reached by the radio equipment of the base station.

The communication channels are the link between the mobiles and the rest of the network. The channels could be of two types, namely radio-channels and link-channels.

This model is divided into four main parts namely, the traffic cell, the communication channels, the customers and the traffic flow of customers.

<u>Customers</u>

A customer is recognized by its class which among other things describes the priority of this customer (two priorities here).

<u>Traffic flow of customers</u>

The traffic flow or the traffic types is TCELLoutgoing, TCELLincoming, TCELLintra or TCELLinter. Note that TCELLinter traffic need allocation of 2 channels in a TCELL.

The radio-channels

The radio-channels are the connection between the mobiles and the exchange and they are of two basic types. On one hand the control channel(s) (TCC) which are allocated to a traffic cell and serves as a common resource for the different signalling needs between the mobile and the base station. On the other hand the voice channels (TVC) which are the actual communication resource between the mobile subscribers and the rest of the network. If all voice channels are busy when a call arrives the call is lost.

The traffic cells

Because of the reason mentioned in the introduction it is essential to increase the number of channels made available from the administration. In this paper we will define the traffic cells (usually a hexagonal structure) in the following manner:

The MSC area is divided into a number of traffic cells (TCELLs). The MSC area has also a number of radio-channels allocated to it. These radio-channels is divided into two set of channels. The first set for the primary cells and the second set for the secondary cells, see below and figure 2.

Each set of radio-channels is divided into a number of groups (e.g. 7 or 21). The result is that each group has a unique set of radio-channels.

Now, each TCELL, primary or secondary, is given one of the groups, from the primary set if it is a primary cell, from the secondary if it is a secondary cell. The criteria when distributing these groups among the TCELLs is ruled by the reuse formula. The reuse formula tell us how to minimize the co-channel interference.

2.2.4.1 The traffic model for the MSC area

Let us now describe our model of the MSC area in a number of steps as follows.

STEP1

First of all let us assume that we have distributed the groups in the first set of radio-channels among the primary TCELLs as mentioned above. This kind of distribution is called FIXED channel assignment and means that this group of channels in the TCELL is exclusively used inside this TCELL and could NOT be used by another TCELL within the reuse distance.

STEP2

The groups in the second set of radio-channels is distributed among the secondary TCELLs i.e. are overlaid the primary TCELLs. In other words inside a secondary TCELL there are a number of primary TCELLs, see figure 2. The consecuence is that the channels in a secondary TCELL is a common pool of channels for a number of primary TCELLs.

STEP3

We now combine this system with rearrangement, which means that when a call in a certain primary TCELL using a fixed channel terminates another call in the same primary TCELL using a common channel is moved to the fixed channel.

STEP4

The next step involves priorities for different kind of users. The model will be defined for two kind of users with two kind of priorities which means that from the common pool a number of channels will be assigned exclusively for the customers with highest priority.

2.3 Part B, formulas

2.3.1 Analytical formulas

2.3.1.1 Presumptions

We assume that all the calls in our model is generated according to a Poission process. We also assume that the holding times are exponentially distributed and that mobility (handoff and roaming) is not included.

Our desire is to derive Grade of Service (GOS) and state probability formulas.

This is the basis from which we then are able to derive the offered and carried traffic corresponding to a given set of GOS values. The set of GOS values will be explained in detail later but now we will define it as a set of values which correspond to a PARTICULAR installation of our system.

Since the system is working as a loss system the GOS is equal to probability of congestion. The TCELL voice channel system in the MSC area is as a consequence of the definitions above described as follows.

```
          n(1)          m          l
A(1)
----->    0>...0>  0>...0>  I-0 ...I-0
B(1)      I    I   I    I   I      I
----->    0>...0>  0    0   I      I
                   I    I   I      I
             .     I    I   I      I        1
                   I    I   I      I
          n(r)     I    I   I      I
A(r)               I    I   I      I
----->    0>...0>  0>   0>  I-0 ...I-0
B(r)      I    I   I    I
----->    0>...0>  0... 0
          ___  ____  ____  ___  ____

          n(1)
A(1)
----->    0>...0>  0>...0>  I-0 ...I-0
B(1)      I    I   I    I   I      I
----->    0>...0>  0    0   I      I
                   I    I   I      I
             .     I    I   I      I        k
          n(r)     I    I   I      I
A(r)               I    I   I      I
----->    0>...0>  0>   0>  I-0 ...I-0
B(r)      I    I   I    I
----->    0>...0>  0... 0
```

NOTE that r is the number of primary TCELLS within the reuse distance, n is equal to the number of channels in each primary TCELL, m plus l is equal to the number of channels in the secondary TCELL, and k is the number of secondary TCELLs or reuse groups.

The channels in the third group can only be selected by the A traffic i.e. the traffic with the highest priority. Rearrangement implies that when there are calls in the second group corresponding to a particular first group this first group is always full. And further on if there is a call in the third group corresponding to a particular first group this first group AND the second group is full. I.e. when a call terminates in either a first OR second group a call, if any, is moved from the highest possible group.

The formulas for the state probabilities and GOS will be derived in a number of steps as follows.

STEP 1

Let's start with ONE type of traffic offered to ONE group inside ONE cell.

```
                 n           m           l
             A

1     ------>  0>... 0>   0>... 0>   0....0
```

The state probability formula is of course

$$Pk = \frac{\frac{A^k}{k!}}{\sum_{k=0}^{n+m+l} \frac{A^k}{k!}}$$

and the formula for the congestion

$$B1(1,n+m+l) = \frac{\frac{A^{(n+m+l)}}{(n+m+l)!}}{\sum_{k=0}^{n+m+l} \frac{A^k}{k!}}$$

The carried traffic in the first group is

$$A(n) = n - \frac{(n+1, m+l) \times E(n+m+l, A)}{A^{(m+l)} \times E(n, A)} \times (n - A(1 - E(n, A)))$$

$$\textit{where } (n+1, n+l) = (n+1) \times (n+2) \ldots (n+1+m+l-1)$$

STEP 2

Let us now add another type of traffic B which can only select channels in the primary and secondary group.

```
                   n           m           l
            A1

1     ------>   0>... 0>   0>... 0>   0....0
                I     I    I     I
            B1  I     I    I     I
      ------>   0>... 0>   0 ... 0
```

The equation for the state probabilities for this system is:

For $0 \le k \le n$

$$Pk = \frac{(A1+B1)^k}{k!} \times P_0$$

and $n < k \le n+m$

$$Pk = \frac{(A1+B1)^n \times (A1+B1)^{(k-n)}}{k!} \times P_0$$

and finally $n+m < k \le n+m+l$

$$Pk = \frac{(A1+B1)^n \times (A1+B1)^m \times A1^{(k-n-m)}}{k!} \times P_0$$

The normalization condition is of course

$$1 = \sum_{k=0}^{n+m+l} Pk$$

and we reach the formula for the call congestion that the A and B traffic experiences

$$BA1(1,n+m+l)=\frac{\dfrac{(A+B)^{(n+m)}\times A^{l}}{(n+M+l)!}}{\sum_{k=0}^{n+m}\dfrac{(A+B)^{k}}{k!}+\sum_{k=n+m+1}^{n+m+l}\dfrac{(A+B)^{(n+m)}\times A^{(k-n-m)}}{k!}} \qquad (1)$$

$$BB1(1,n+m+l)=\frac{\sum_{k=n+m}^{n+m+l}\dfrac{(A+B)^{(n+m)}\times A^{(k-n-m)}}{k!}}{\sum_{k=0}^{n+m}\dfrac{(A+B)^{k}}{k!}+\sum_{k=n+m+1}^{n+m+l}\dfrac{(A+B)^{(n+m)}\times A^{(k-n-m)}}{k!}}$$

STEP 3

The next step in the process is to add one group (TCELL) to the previous one.

In the following we will calculate the call congestion, separately for the A and B traffic, in the FIRST group of the grading.

```
                  n1          m          l
         A1

1     ------>   0>... 0>   0>... 0>   I-0...I-0
                I     I    I     I    I     I
         B1     I     I    I     I    I     I
      ------>   0>... 0>   0     0    I     I
                           I     I    I     I
                           I     I    I     I
         A2       n2       I     I    I     I
1     ------>   0>... 0>   0>... 0>   I-0...I-0
                I     I    I     I
         B2     I     I    I     I
      ------>   0>... 0>   0 ... 0
```

As mentioned in reference (2) this formula can be derived in terms of the nominator TA(r, n+m+l) and the denominator N(r,n+m+l) where r is the number of groups in the grading and n, m, l are the number of channels in the primary, secondary and thirdly group.

The derivation is done in the following way:

When we add one group T(2,n+m+l) will contain thoose states which correspond to congestion for the A1 and B1 traffic due to that a number of the m and l channels is occupied by the A2 and B2 traffic.

It is now possible to derive the T(2,n+m+l) formula for the A and B traffic in terms of T(1,n+m+l).

First the A traffic

TA(2,n+m+l) contains three terms:

One that expresses the states of the second group when it does not effect the first group.

$$TA'(2,n1+m+l)=\sum_{k=0}^{n2}\frac{(A2+B2)^{k}}{k!}\times TA(1,n1+m+l)$$

One that expresses the states of the second group when it does effect the first group but only through the secondary group.

$$TA''(2,n1+m+l)=\sum_{k=1}^{m}\frac{(A2+B2)^{(n2+k)}}{(n2+k)!}\times TA(1,n1+m+l-k)$$

One that expresses the states of the second group when it does effect the first group both through the secondary and thirdly group.

$$TA'''(2,n1+m+l)=\sum_{k=1}^{l}\frac{(A2+B2)^{(n2+m)}A2^{k}}{(n2+m+k)!}\times TA(1,n1+l-k)$$

The B traffic

TB(2,n1+m) will also contain three terms.

$$TB'(2,n1+m)=\sum_{k=0}^{n2}\frac{(A2+B2)^{k}}{k!}\times TB(1,n1+m)$$

$$TB''(2,n1+m)=\sum_{k=1}^{m}\frac{(A2+B2)^{(n2+k)}}{(n2+k)!}\times TB(1,n1+m-k)$$

$$TB'''(2,n1+m)=\sum_{k=1}^{l}\frac{(A2+B2)^{(n2+m)}A2^{k}}{(n2+m+k)!}\times TB(1,n1)$$

STEP 4

We now continue to add groups to our grading and end up with the following

```
               n(1)          m         l
        A(1)
       ----->  0>...0>  0>...0>  I-0 ...I-0
1       B(1)   I     I  I     I  I       I
       ----->  0>...0>  0     0  I       I
                     .  I     I  I       I
                     .  I     I  I       I
                     .  I     I  I       I
               n(r)     I     I  I       I
        A(r)            I     I  I       I
       ----->  0>...0>  0     0  I-0 ...I-0
r       B(r)   I     I  I     I
       ----->  0>...0>  0...  0
```

The recursion formula for the nominator now read as follows

$$TA\,(r,n(r-1)+m+l) = TA'\,(r,n(r-1)+m+l) + TA''\,(r,n(r-1)+m+l) + TA'''(r,n(r-1)+m+l)$$

and

$$TB\,(r,n(r-1)+m) = TB'\,(r,n(r-1)+m) + TB''\,(r,n(r-1)+m) + TB'''(r,n(r-1)+m)$$

The same expression is derived for the denominator and we are now in the position to calculate BA(r) and BB(r) as

$$BA\,(r,n(r-1)+m+l) = \frac{TA\,(r,n(r-1)+m+l)}{NA\,(r,n(r-1)+m+l)}$$

$$BB\,(r,n(r-1)+m) = \frac{TB\,(r,n(r-1)+m)}{NB\,(r,n(r-1)+m)}$$

As mentioned in description of the flow of customers it is possible to have a 2 channel call, because the call can both originate and terminate in a single primary TCELL. Instead of using formula(1) as the basis of our calculation we could use the modified Erlang formula developed in reference(1) which takes into account 2 channel calls as well.

2.3.2 Simulation

A simulation program has been developed and used for the above system. The type of simulation used was process simulation. The results from the simulation was checked against the theoretical results and the correspondence was very good.

The next step will be, as said above, to build in other things like mobility. The results here however are without any extras such as mobility.

3 CAPACITY

The main purpose with this paper is to from the total number of channels given by the administration to the MSC aera, which are

$$NTOT = NPRIM + NSEC = \sum_{i=1}^{r} n(i) + NSEC$$

and with reuse of the reuse groups

$$NTOTR = \sum_{j=1}^{k}\sum_{i=1}^{r} n(i) + \sum_{j=1}^{k} (m(j)+l(j))$$

and with rearrangement obtain the optimum disposition of channels among the three type of groups. I.e. to maximize the carried traffic. This optimum disposition is given for each set of Grade of Service values (congestion) that the administration has specified. The set consists of GOS values for the A and B traffic per group (TCELL). The results presented here, figure 3 and 4, is calculated in terms of fixed channel assignment i.e. only primary cells.

4 CONCLUSION

The results from the analytical formulas and simulations for a system with overlaid cells and rearrangement show that this system has a capacity that exceeds the capacity of fixed channels assignment systems. The increase of capacity depends on the specific implementation of the system. But so far investigated an increase of 35% offered traffic for the specific implementation in figure 3 has been calculated. Also a substantial gain of increased number of subscribers is accomplished for the system in figure 4.

When priority traffic is used, if this amount of traffic is small compared to the ordinary traffic, a relatively large decrease in GOS is achieved without affecting the ordinary traffic to a large extent. If the parameters used in reference(2) is applied to the formula derived here the results are the same.

REFERENCES

(1) M.H. ACKROYD and S.H. BAKRY, Teletraffic Analysis for Single Cell Mobile Radio Telephone Systems. IEEE Transactions on Communications Vol COM-29 No. 3 March 1981

(2) L. RENEBY, Rearranging of Calls in Telecommunication Systems. Nordic Teletraffic Seminar (NTS5) Trondheim 1984

(3) D.C. COX and D.O. REUDINK, Increasing Channel Occupancy in Large Scale Mobile Radio Systems; Dynamic Channel Reassignment. IEEE Trans. on Veh. Tech, VT-22, November 1973

(4) T.J. KAHWA and N.D. GEORGANAS, A Hybrid Channel Assignment Scheme in Large-Scale, Cellular-Structured Mobile Communication Systems. IEEE Trans. on Comm. COM-26, April 1978

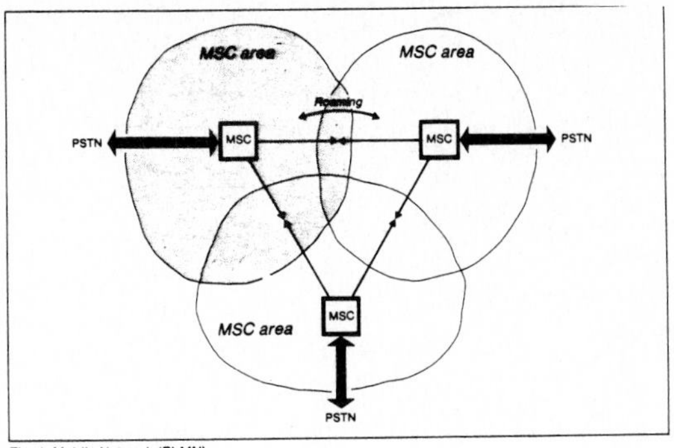
MSC area
MSC area
Roaming
PSTN
MSC
MSC
PSTN
MSC area
MSC
PSTN

Fig. 1 Mobile Network (PLMN)

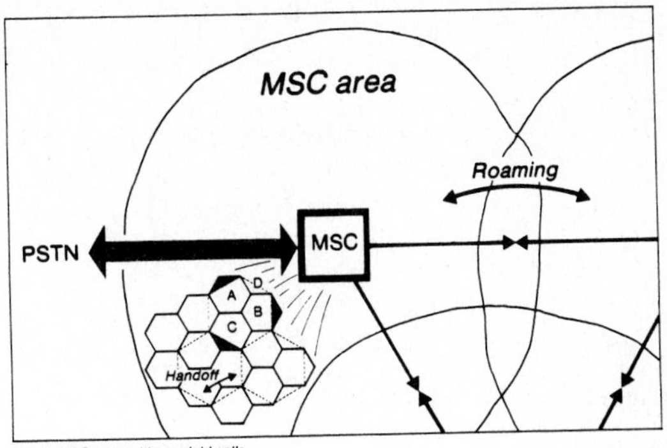
MSC area
Roaming
PSTN
MSC
A
B
C
D
Handoff

Fig. 2 MSCarea with overlaid cells

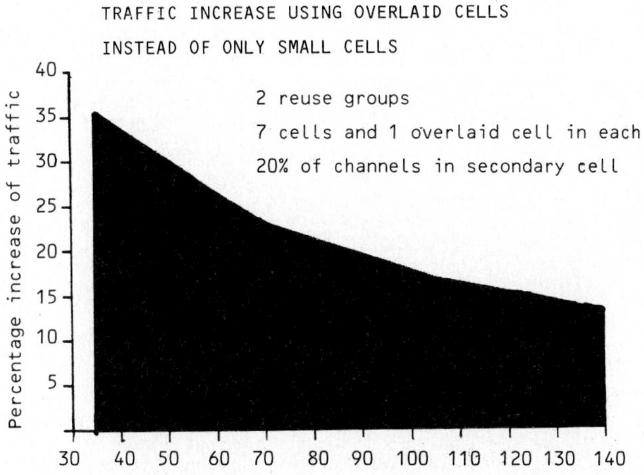
TRAFFIC INCREASE USING OVERLAID CELLS
INSTEAD OF ONLY SMALL CELLS
2 reuse groups
7 cells and 1 overlaid cell in each
20% of channels in secondary cell
Percentage increase of traffic
40
35
30
25
20
15
10
5
30
40
50
60
70
80
90
100
110
120
130
140
Number of available channels before reuse

Fig.3

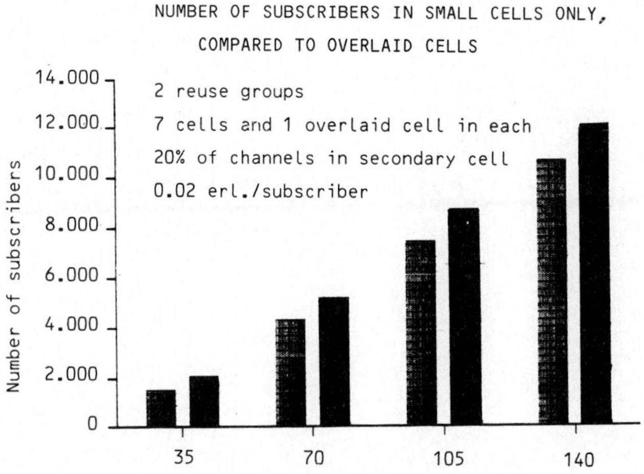
NUMBER OF SUBSCRIBERS IN SMALL CELLS ONLY,
COMPARED TO OVERLAID CELLS
2 reuse groups
7 cells and 1 overlaid cell in each
20% of channels in secondary cell
0.02 erl./subscriber
Number of subscribers
14.000
12.000
10.000
8.000
6.000
4.000
2.000
0
35
70
105
140
Number ot available channels before reuse

Fig.4

TELETRAFFIC ISSUES in an Advanced Information Society
ITC-11
Minoru Akiyama (Editor)
Elsevier Science Publishers B.V. (North-Holland)

TRAFFIC MODELING OF A CELLULAR MOBILE RADIO SYSTEM

Oscar AVELLANEDA, Raj PANDYA* and George BRODY***

BNR
**Ottawa, Canada*
***Richardson, Texas*

ABSTRACT

Cellular mobile radio systems being developed and implemented in North America, Europe and Japan use the frequency reuse concept whereby the same set of frequencies can be used in noncontiguous cells. However, calls in progress may have to be handed off from one RF channel to another as the mobile unit travels across cell boundaries during a call in progress. Further, the operations for setting up, monitoring, coordinating and controlling calls in a cellular mobile radio are radically different from those in a land-based system.

This paper presents the results of a simulation analysis to estimate the performance of the dynamic load-sharing algorithm which is required to ensure negligible probability of call cutoff due to unsuccessful handoffs. It also describes a traffic model to assess the performance of the data link that carries the control messages between the MTX and the cell site controllers.

1. INTRODUCTION

Cellular mobile systems represent the latest technology for providing reliable public mobile telephone service to a large number of customers while providing service quality and features comparable to those of the public switched telephone network (PSTN).

The principle of cellular systems is to divide a large service area into cells with diameters from 2 to 20 kilometers, each of which has a number of radio frequency (RF) channels. Transmitters in adjacent cells operate on different frequencies to avoid interference. However, since the transmitter power in adjacent cells is kept relatively low, cells that are sufficiently far apart can reuse the same set of frequencies.

The coverage range and capacity of the cellular system is potentially unlimited. As the market grows, additional cells can be added and as traffic demand increases in a given area, cells can be split or sectored to accommodate the additional traffic.

The Mobile Telephone Exchange (MTX) in a cellular system allows calls in progress to continue uninterrupted while the mobile moves from one cell to another. The MTX automatically transfers or hands off the call to a different free channel in the adjacent receiving cell.

The operations for setting up, monitoring, coordinating and controlling calls in a cellular mobile systems are radically different from those in a land-based system. In addition, innovative algorithms for dynamic load-sharing between adjacent cells are needed to ensure negligible probability of call cutoff due to unsuccessful handoff.

To meet all these requirements, the MTX must have a high degree of intelligence, design sophistication, and reliability. DMS-MTX*, a cellular mobile system with these characteristics is described in [1] and [2], and its basic architecture is illustrated in Fig. 1.

The traffic parameters that have bearing on the performance and design of a cellular mobile system include:

- blocking on the RF channels,
- blocking on the cellular-to-PSTN circuits,
- probability of a call encountering unsuccessful handoff,
- throughput and delay associated with the data link, and
- call setup delays between the MTX and mobile units.

The blocking performance of the RF channels and the PSTN-connecting trunks can be modelled using classical traffic models [3]. The performance with respect to the remaining three performance parameters for the DMS-MTX system was investigated in detail using computer simulation models and/or analytical models.

In this paper, we present the results on the handoff performance in the presence of the unique dynamic load-sharing strategy implemented in the DMS-MTX system. We also describe a traffic model and provide results on the delay and throughput performance of the high-level data link control HDLC data link that carries the control messages between the MTX and the ce' site controllers.

The next section provides a brief description of the dynamic load-sharing algorithm and the cellular system traffic simulator used for assessing the handoff performance. Results on

*DMS-MTX is a Trade Mark of Northern Telecom Ltd.

the sensitivity of handoff performance to traffic and system parameters are also presented.

Section 3 describes the traffic models used for the HDLC data link and provides results on the throughput and transfer times.

Conclusions of the analysis are summarized in Section 4.

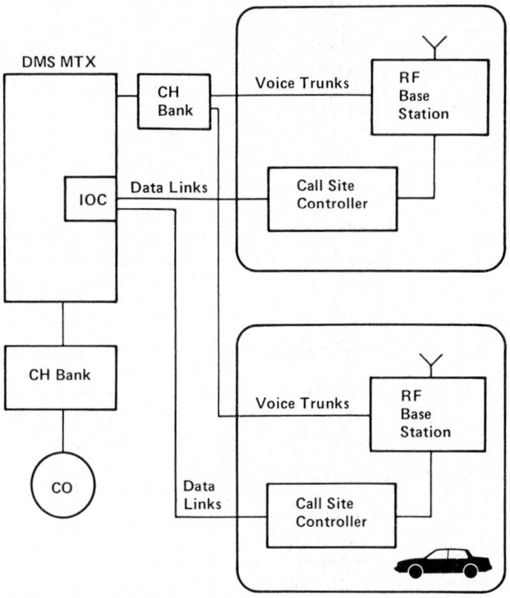

CH Bank: Channel Bank
IOC : Input—Output Controller
CO : Central Office/Local Exchange
RF : Radio Frequency
MTX : Mobile Telephone Exchange

Fig. 1 DMS—MTX cellular mobile radio system

2. DYNAMIC LOAD-SHARING ALGORITHM AND HANDOFF PERFORMANCE

2.1 Introduction

One of the unique characteristics of cellular mobile radio systems is their ability to hand off calls in progress between cells. In the DMS-MTX system, when a mobile user moves out from a cell site, he will reach a signal level which has been selected as the beginning of the handoff area. This level is called the RSSI (Received Signal Strength Indicator) and it is used to indicate when a mobile user is a potential candidate for handoff. When conditions are suitable, the call will be transferred from one cell to the next. Call handoff will be more frequent in mature systems where cell sizes may be quite small.

To avoid blocking or loss of calls during handoff, DMS-MTX gives priority to calls requiring handoff. This is achieved by using the dynamic load-sharing algorithms (directed retry and directed handoff) which reserve a given number of voice channels for calls being handed off into the cell.

2.2 The Handoff Algorithm

As we mentioned before, a key feature of cellular systems is the ability to handoff a mobile unit from one cell to another during a call in progress. Basically there are two reasons why a call is handed off: The RSSI from the mobile unit is becoming too weak to maintain satisfactory service; or the traffic load must be balanced among the adjacent cells. The handoff algorithm consists of two parts:

- computing which cells are to receive the request for RSSI message, i.e., which cells are potential candidates for receiving the call being handed off, and
- ranking the responses of these cells from best to worst.

In order to perform these tasks, the handoff process determines in which of the following four states the cell is:

- Normal: originations, terminations and handoff are possible
- Directed Retry: only handoffs into or out of the cell are allowed
- Directed Handoff: only handoff requests due to weak RSSI are accepted. Cell also attempts to handoff calls in progress to the adjacent cells even though they may not be potential candidates for handoff in the 'Normal' state; or
- Unavailable: Cell is either down or has all its channels busy.

To compute which cells are to receive the request for RSSI message, the handoff process obtains the current state of all the adjacent cells. A request for RSSI is sent to cells which are in normal or directed retry state and to directed handoff state only if the reason for the handoff is weak RSSI. The ordering (from best to worst) of the responses (only cells with better RSSI will respond) is done as follows:

- First, normal cells according to RSSI
- Next, directed retry or directed handoff according to channel availability.

Based on the above information, the MTX will choose a suitable cell for handing off the call.

2.3 Dynamic Load-Sharing Algorithm

Dynamic load-sharing algorithms are the means to avoid the blockage of calls or the loss of calls during handoff, and they take advantage of the overlapping cell coverage areas generally designed into a multicell system. This load-sharing is performed on a cell-by-cell basis to help maintain some open voice channels for calls requiring handoff due to a weak signal strength. There are two stages in the algorithm for load-sharing:

- Directed retry -

  routing of calls attempting originations or page responses to a cell site OTHER than the one normally chosen.

- Directed handoff -

  transfer of some calls in progress to an adjacent cell even though they would not have been handed off under normal state.

If a given cell is in directed retry state and all the adjacent cells are also in directed retry or directed handoff state, the system will respond with a reorder for originations and a release for a page response.

2.4 Cellular System Traffic Simulator

In order to assess the handoff and dynamic load-sharing algorithms, different systems were modelled and studied with the help of the Cellular System Traffic Simulator (CSTS). This computer-aided simulation tool is a comprehensive call-by-call simulator for mobile systems. The simulation's key parameters are specified by the user through a set of input files. Some of these parameters are: the length of the simulation run, the number of cells and their coverage areas, the average holding time, and the traffic distribution. The cellular system's geographic service area is overlaid with a rectangular grid used in locating the positions of cell site antennas and mobile units with calls currently in progress. The typical statistics collected are: breakdown of call setups according to call type and call disposition; breakdown of calls according to whether they were completed normally, blocked on handoffs or have left the service area; blocking on the RF channels; and the land line trunks, etc. This software tool has been used extensively for analyzing the performance of actual systems during their planning stages.

2.5 The Test System

The results presented in this paper are based on a reference 13-cell mature system. This system is depicted in Fig. 2. It may be considered as an isolated piece of a larger system which has been fragmented for study purposes. The parameters considered in the specification of this system are summarized below:

- Holding times

  Cellular mobile radio systems will have relatively short holding times, ranging from 70 s to 140 s. The mean call holding time was set at 120 secs.

- Traffic mix

  Normally the Mobile to Land (M-T-L) traffic will be greater than the Land to Mobile (L-T-M) traffic and the Mobile to Mobile (M-T-M) traffic will be relatively low. A distribution of 65% M-T-L, 30% L-T-M and 5% M-T-M traffic was assumed.

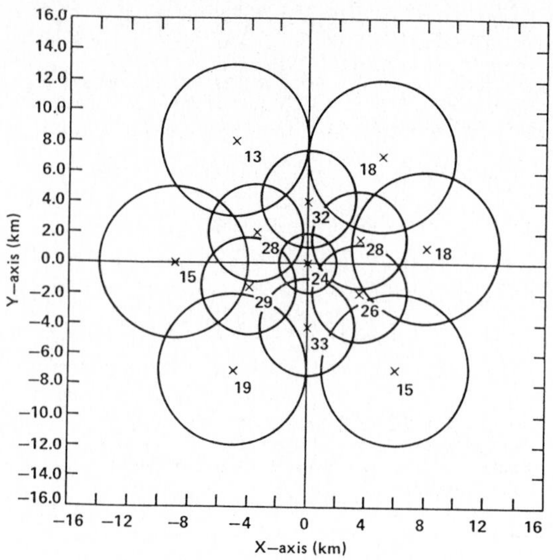

Fig. 2 Cell layout (mature system)

- Mobile traffic distribution

  This is an important factor for the design of cellular systems. Mobile traffic is usually highly concentrated within certain sectors of a metropolitan area. As an illustration, Fig. 3 is a three-dimensional depiction of mobile traffic density for the reference system studied.

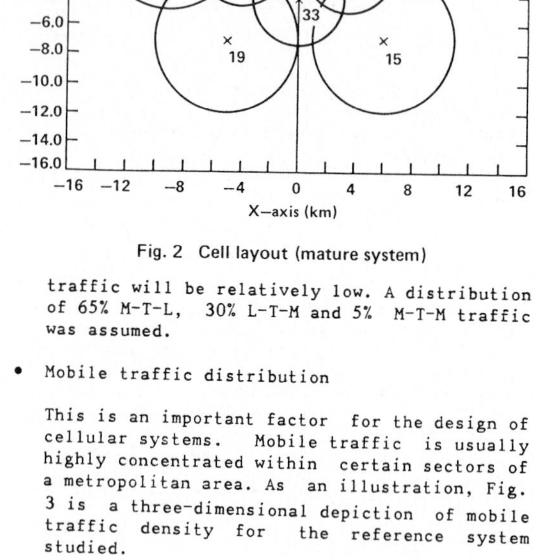

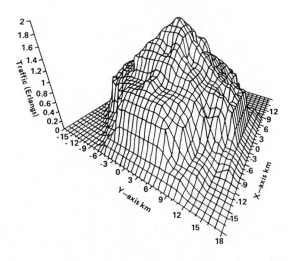

Fig. 3 Mobile traffic distribution (Mature system)

- Contours

  They are defined as circles along which the received radio signal strengths are equal to given signal thresholds. The handoff contour at -85 dBm identifies when a handoff attempt should be initiated. The bad service contour characterizes the minimum RSSI required to maintain adequate service. Mobile units will hit this contour only if they cannot be handed off to another cell in spite of continued attempts by the system to do so.

- Voice channel grade of service

  The RF channels were dimensioned for 2% blocking under normal load and no dynamic load-sharing. The number of RF channels in each cell is presented in Fig. 2.

- Cell coverage

  For cellular applications, the coverage is defined as the location of the handoff contour. It is assumed that the small cells have handoff radii equal to 2 km, the medium ones have handoff radii equal to 3.2 km and the large cells have handoff radii equal to 5 km.

- Cell overlapping

  This is a very important factor for handoff reliability because a layout with a small amount of overlap has less flexibility in handling shifts in mobile user density. The classic hexagonal minimum coverage layout results in only 5.7% overlap between any two cells if the hexagons are replaced by circles. This amount of overlap is not enough for reliable handoffs. On the other hand, excessive overlap may have negative cochannel interference effects, not to mention the extra cost of either increased transmitter power or antenna height.

- Mobile user speed

  Mobile units are assumed to have Gaussian distributed speeds with means that vary according to specific sectors of the metropolitan area. Mobile units are assumed to travel in the small cell (cell number 1) with an average speed of 30 km/hr with a standard deviation of 20 km/hr. In the system periphery, mobile units have an average speed of 90 km/hr with standard deviation of 30 km/hr.

2.6 Results

The improvement in handoff performance provided by the dynamic load-sharing algorithm is illustrated by Fig. 4 where the probability of a subscriber receiving unsatisfactory handoff service (RSSI below -97 dBm) is shown against % overload in traffic. Similarly, Fig. 5 represents the overall blocking perceived by the subscribers as a function of percent overload. Again one can observe that the dynamic load-sharing algorithm provides lower blocking and results in a 10 - 20% increase in traffic handling capacity.

The effect of reserving RF channels is illustrated in Fig. 6 where the following parameters are plotted against the number of RF channels reserved for handoffs:

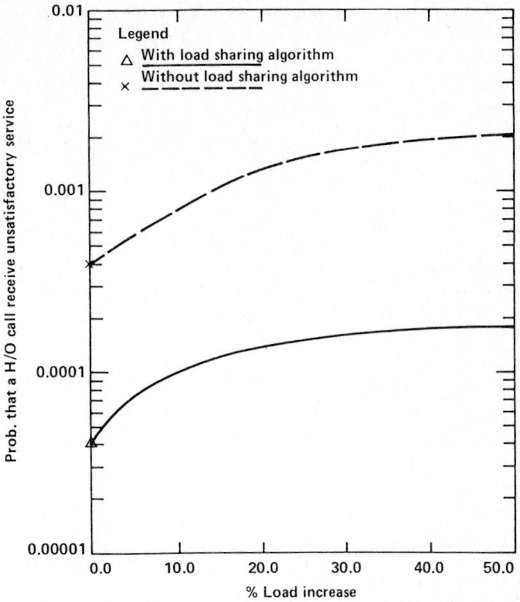

Fig. 4 Effect of load sharing on handoff performance

- probability that a handoff attempt has to wait for a free RF channel,

- overall system blocking perceived by subscribers, and

- average blocking for calls in a cell.

This scheme works in a manner similar to alternate routing where, as you decrease the

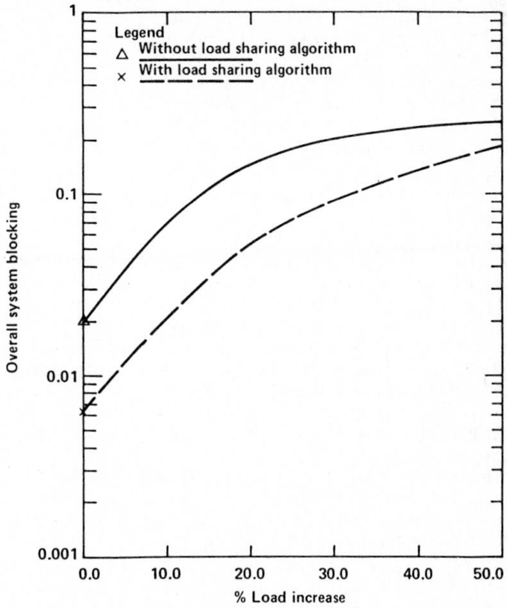

Fig. 5 Effect of load sharing algorithm on overall system blocking

number of first choice trunks (increase in reserved RF channels), the blocking on the first route increases but the overall blocking is much lower. As the number of reserved channels increases, the blocking of calls within a cell increases. However, the overall system blocking curve exhibits a minimum so that the number of reserved channels can be chosen to minimize the overall system blocking. The probability that a handoff attempt coming into a cell will have to wait for a free channel decreases with the increase in number of reserved channels.

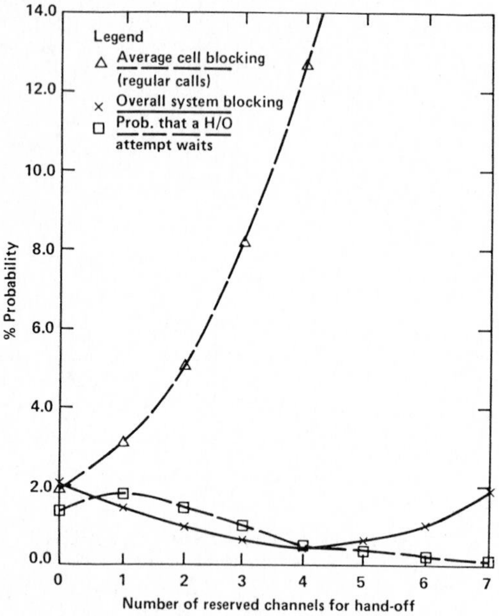

Fig. 6 Effect of reserved channels on system performance

3. HDLC DATA LINK PERFORMANCE

3.1 Introduction

A high-capacity cellular mobile system, such as DMS-MTX, requires a sophisticated centralized control to coordinate the actions of the switching network, the cell site controllers (CSC) and, the mobile units. This coordination is accomplished via a data transmission path, which is established between the MTX and the mobile unit. This path consists of dedicated high-level data link control (HDLC) links between the MTX and the CSCs and the radio data channel from the CSCs to the mobile units. Characteristics of the HDLC procedure may be found in [4].

In this section we describe the traffic model to assess the performance of the HDLC protocol under asynchronous balanced response mode (ABM). Two effects may impact performance: sequence-number starvation and error-recovery procedures. Transfer time of messages and throughput are the two primary performance criteria of the link. We follow the analysis presented in [5] to obtain estimates of the link throughput and the transfer time of messages. These two parameters are considered in the specification of the data link so that a uniform and satisfactory overall performance is attained.

3.2 HDLC Traffic Model

A schematic representation of the model underlying the performance of the data link is shown in Fig. 7. Messages to be transmitted from HDLC CSC to HDLC IOC (input/output controller) or vice versa, are stored in the send buffer of the sending station, where they have to wait for transmission. The following assumptions are considered in the model:

- full-duplex point-to-point link connecting the MTX and each CSC
- channel produces statistically independent bit errors with probability BER;
- Poisson arrival process of rates λ_1(CSC) and λ_2 (MTX);
- fixed-length information frames (I-frames) and
- no processor/buffer limitations.

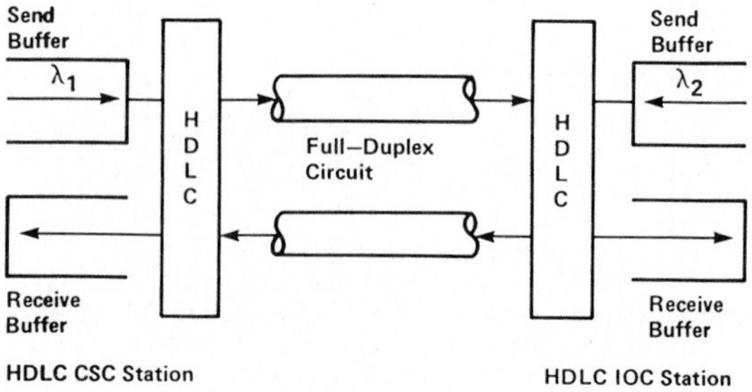

Fig. 7 Structure of the data link model

The results presented in [5] are based on the concept of virtual transmission time which can be regarded as the equivalent of service time. Since the protocol ensures transmission of the frames in order of arrival, the service time is either the single information frame transmission time or the recovery plus transmission time (in the case of errors). Hence the expected transfer time E[T] is determined by using the well-known results from M/G/1 queuing systems.

$$E[T] = \frac{\lambda E[T_0^2]}{2(1 - \lambda E[T_0])} + E[T_0] + t_1 \qquad (1)$$

where $E[T_0]$ and $E[T_0^2]$ denote the first two moments of the virtual transmission time and t accounts for the processing as well as the propagation times. Since $\rho = \lambda T_0$, while the bit error probability is equal to zero, the expression (1) becomes the M/D/1 formula

$$E[T] = \frac{\rho E[T_0]}{2(1-\rho)} + E[T_0] + t_1 \qquad (2)$$

The maximum information throughput (bits/unit of time) of the data link is given by

$$H = \frac{L}{T_0}$$

where L is the average length of a message.

In our specific case, the modulus is equal to 8, the bit error probability is 10^{-6}, the processing time is 50 ms and it is assumed that the message length is 200 bits.

3.3 *Messages Carried By The HDLC Data Link*

The following types of messages travel between the MTX and the CSCs:

- Load Messages

  They are exchanged for downloading programs and data from the MTX to the CSC over the data link.

- System Maintenance Messages

  These messages are exchanged between the MTX and the CSC for conducting diagnostics, maintenance functions, and for reporting or querying of alarm conditions at the CSC.

- Man/Machine Messages

  These messages are exchanged for testing and querying the status of the HDLC controllers, etc. .

- Operational Measurements Messages

  Each Operational Measurements record contains a given number of messages. They are normally collected for statistical purposes.

- Call Processing Messages

  These messages serve as instructions to the CSC mobile for originations, terminations, or handoffs.

All these messages have different lengths and also different generation rates. For instance, the loading messages do not normally contribute to the loading of the link since they are exchanged during low traffic hours. The call processing message traffic is a function of the number of cells in the system, as well as the number of adjacent cells for handoff purposes. The first one is due to the fact that for land-to-mobile calls, the MTX does not know the location of the mobiles and it pages all the cells.

3.4 *Results*

Fig. 8 provides the relationship between the useful channel load and the transfer time. The useful channel load represents the channel utilization for successful transmission.

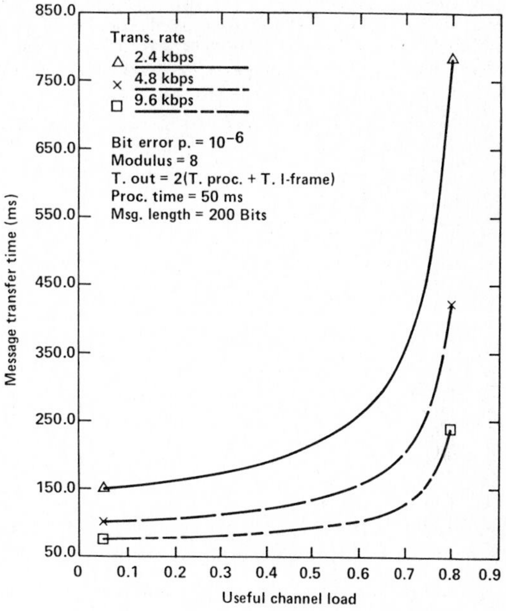

Fig. 8 Message transfer time vs useful channel load

The effect of system size and channel occupancy on the maximum allowable loading on the channel is illustrated in Fig. 9. The drop in allowable loading with increase in the number of

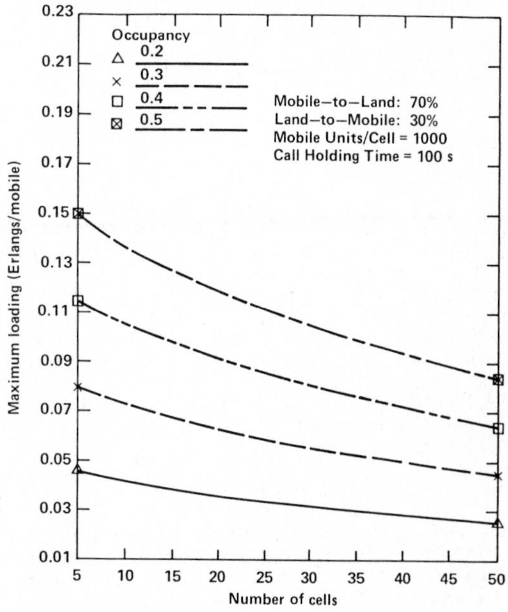

Fig. 9 Maximum loading vs number of cells

cells is caused by the fact that for a land-to-mobile call it is assumed that all the cells in the system are paged to locate the mobile. If the paging is done on zone basis, this effect will be eliminated. If the percentage of land-to-mobile calls increases, this effect will become more noticeable.

Fig. 10 provides results on the effect of system size and traffic loading on message transfer times.

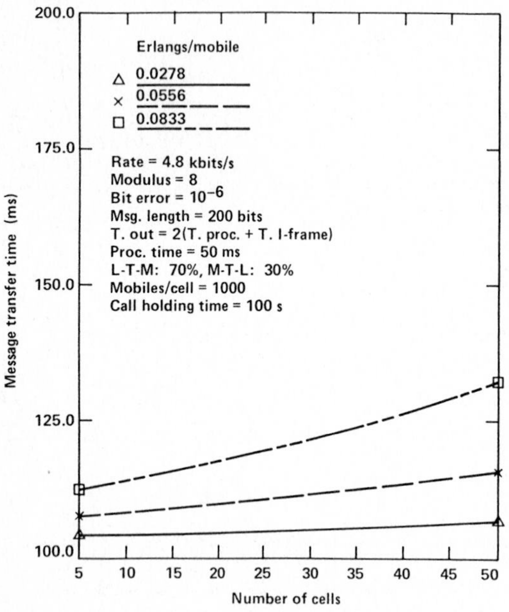

Fig. 10 Mean transfer time vs number of cells

4. SUMMARY AND CONCLUSION

In this paper we initially introduced the performance analysis of the dynamic load-sharing algorithm for DMS-MTX. This was investigated using a comprehensive call-by-call simulation tool for cellular mobile systems. Some representative results have been presented for a theoretical, mature system. These results indicate that the load-sharing algorithm makes the system behave as an alternative routing or progressive grading system. We found that the load-sharing algorithm increases the effective capacity of the system by 10% to 20% .

Finally we described the performance of the HDLC data link which carries the control information between the switch and the cell site peripherals. Representative results of the message transfer delay as well as occupancies have been presented.

5. REFERENCES

[1] Brody, G. Patel, R., and Rowland, R., "Application of Digital Switching in a Cellular Mobile Radio System", International Switching Symposium, May 7-11, 1984, Florence, Italy.

[2] Ma, F.K. and Chau, S. "DMS-MTX Turnkey System for Cellular Mobile Radio Application", IEEE 1984 Vehicular Technology Conference, May 21-23, Pittsburgh, Pennsylvania, U.S.A.

[3] Pandya, R.N. and Brown, D.M., "Performance Modelling for an Automated Public Mobile Telephone System", International Communications Conference, June 13-17, 1982, Philadelphia, U.S.A.

[4] Data Communications - High Level Data Link Control Proc. - Elements of Procedure (Independent Numbering). International Standard ISO 43 35, 1976.

[5] Bux, W., Truong, H.L., High Level Data Link Control Traffic Considerations", Proc. of the 9th Internat. Teletraffic Congress, Torremolinos, 1979.

TELETRAFFIC ISSUES in an Advanced Information Society
ITC-11
Minoru Akiyama (Editor)
Elsevier Science Publishers B.V. (North-Holland)

APPROXIMATIONS FOR A SINGLE SERVER QUEUE WITH A BRANCHING POISSON ARRIVAL PROCESS

Masakiyo MIYAZAWA

Dept. of Information Sciences, Science University of Tokyo,
Noda City, Chiba 278, Japan

ABSTRACT

Approximations are discussed for a single server queue with a Branching (Clustered) Poisson arrival process. In this queue, customers arrive in the system in clusters which form a Poisson process. For simplicity, we assume that the size of the cluster is deterministic and the arrival interval of customers in the same cluster is a constant length. We also assume that the service time of customers are i.i.d.. Interesting feature of this model is the dependency in the arrival process, which causes the difficulty in analysis. In this paper, a new type of approximation formulas for the mean number of customers in the system are obtained by using Benés' formula for a queue with a general input. It is shown that those approximations have good asymptotic properties. Also, their numerical values are compared with simulations and other approximations.

1. INTRODUCTION

This paper considers approximation formulas for a single server queue with a branching Poisson arrival process and the i.i.d. (independently identically distributed) service times. We denote this queue by BP/GI/1. We are only concerned with the mean number of customers in the system in the steady state.

In the present paper, a branching Poisson process means a special case of a Poisson cluster process such that each cluster is composed of finite number of points. We call the first arriving customer in each cluster a parent and call other customers in the same cluster children. We assume that parents form a stationary Poisson process with a parameter λ and the interarrival times of customers in the same cluster is a constant length a. For simplicity, we assume further that the number of children in the same cluster is a constant number k. Thus, the arrival rate of our input process is $\lambda(k+1)$.

A BP/GI/1 queue appeared in teletraffic problems (cf. [7]). But, it is also interesting as a queue with a non renewal input. Its arrival process is very different from a GI-type input such as in a GI/GI/1 queue. In general, the analysis of this type of a queue is very difficult since we need very complex state space to apply Markov process to it. So approximations for BP/GI/1 are needed. For it, Murao [7] tried to get an approximation by replacing the input process with a GI-type input by matching the distribution of the interarrival time. This is an only result as far as the author knows. In this paper, we propose two types of approximations. The first one is a simple approximation by using a batch arrival queue, but it is only efficient for small a. The second one is an approximation by using Benés' formula for a queue with general input (cf. [1]).

This paper is composed of five sections. The basic properties of BP/GI/1 is considered in Section 2. Approximation formulas are obtained in Section 3. Numerical values of various approximations and simulations are compared in Section 4. And, in the final section, some concluding remarks are given.

2. BP/GI/1 QUEUE

Firstly we mention the notation and basic assumptions for our BP/GI/1 queue. The service discipline of the queue is allowed to be arbitrary if it satisfies that the server is busy when customers are in the system. It is convenient to use a point process to describe the arrival process of customers (cf. [4] and [5]). Let $\{N(.)\}$ be a stationary point process. Then, N(A) denotes the number of points in the set A for any Borel set A in the real line, and the joint distribution of $N(A_i+t)$ for i=1,...,n dose not depend on t, where $A+t = \{x;x=a+t \text{ for any } a\in A\}$. We denote the arrival process of customers of BP/GI/1 by $\{N(.)\}$. As introduced in Section 1, we assumed that $\{N(.)\}$ is a stationary Poisson cluster process, where each cluster contain k+1 points (one parent and k children) and the intervals of points in the same cluster is a. We denote the arrival rate of parents by λ, and so $EN((0,1])=\lambda(k+1)$, where E denotes an expectation. All informations on a point process is given by its gfl. (generating functional). For $\{N(.)\}$, we can easily get its gfl.. But we can also obtain our interested characteristics by an elementary calculation, and so we do not use gfl.. Let S_n be the service time of the n-th customers, where the first customer after time 0 is numbered 1. We assume that S_n is i.i.d. and independent of the arrival process. The distribution of S is denoted by G, where G is assumed to have the first two moments, and the suffix of S is abbreviated. The abbreviation of suffixes is always done in the later if there is no confusion. The mean and variance G are denoted by m and σ^2, respectively. Related to N(.) and S_n, we introduce the count N(t) and the total work load K(t) for any t>0 by

$$N(t) = N((0,t]) \quad (1)$$

$$K(t) = S_1+\dots+S_{N(t)} \quad (2)$$

Those two quantities can be easily handled in the theory of point processes. On the other hand, we can also consider the arrival process by the sequence of interarrival times $\{T_n\}$, where T_n denotes the time between the arrival epochs of the n-th and (n+1)-th customers. $\{T_n\}$ is stationary under the condition that the 1-st customer arrives at time 0. In general, this sequence is correlated and its joint distributions are very complex. But, its one dimentinal distribution F is very simple and given as follows.

$$F(t) = \begin{cases} 1 - e^{-\lambda(k+1)t} & (t \leq a) \\ 1 - \frac{1}{k+1} e^{-\lambda(ka+t)} & (t > a) \end{cases} \qquad (3)$$

In this paper, we are only concerned with the queueing process in the steady state. So we assume that $\rho = \lambda(k+1)m < 1$. It is well known that there exists stationary queueing processes such a queue length process under this assumption. We denote them by $\{L(t)\}$, $\{V(t)\}$, $\{W_n\}$ for the number of customers in the system, the virtual waiting time, and the actual waiting time, respectively. Note that W_n is stationary as a sequence. See [6] for the details of them.

The purpose of this paper is to give approximations for EL in BP/GI/1, where the variable of L is omitted since $\{L(t)\}$ is stationary. As easily seen, the arrival process of BP/GI/1 equals to the compound Poisson process with a constant size k+1 when a=0, and it equals to the Poisson process with a parameter $\lambda(k+1)$ when $a=+\infty$. Hence, in those two extrimal cases, EL is well known and given as follows.

Lemma 2.1 In BP/GI/1, we have, for a=0,

$$EL = \frac{\rho}{2(1-\rho)}(k+2+\rho(\delta^2-1)) \qquad (4)$$

and, for $a=+\infty$,

$$EL = \frac{\rho}{2(1-\rho)}(1+\rho(\delta^2-1)) \qquad (5),$$

where $\delta = \sigma/m$.

By the continuity of queues (cf. [2]), EL converges to those two cases as a tends to 0 or to infinity, respectively, since the arrival process converges to the compound Poisson or Poisson process. This asymptotic property is prefer to hold for approximations of EL.

3. APPROXIMATIONS

In the following subsections, we consider various approximations for EL of BP/GI/1.

3.1 GI-type approximation

Murao [7] proposed the approximation by GI/GI/1 queue, where the distribution of T is assumed to be F. Since EL of GI/GI/1 is not known, she considered the case that S is exponentially distributed, i.e., BP/M/1. Denote the Laplace transform of F by $\psi(s)$ and the root in the unit circle of the equation

$$z = \psi(\mu(1-z))$$

by α. Then, Murao's approximation of EL for BP/M/1 is given by

$$EL(MUR) = \frac{\rho}{1-\alpha} \qquad (6).$$

EL(MUR) converges to the exact value as a tends to infinity but converges to 2(k+1)/(k+2) times of the exact value as a tends to 0. Thus, this approximation is not good for small a. Our numerical test shows that it is not so good even for large a. Murao pointed out that this approximation ignores the correlation of $\{T_n\}$ and so it is only applicable when the correlation is small.

3.2 Batch-type approximation

When a is sufficiently small against m, we can expect the service of customers is done like as in a batch arribal M/G/1 queue with high probability. So we consider modify EL of the batch arrival queue. Note that the difference of total waiting time of customers in the same cluster (or batch) between the batch arrival queue and BP/GI/1 is ak(k+1)/2 if there is no idle time during their service. By Lemma 2.1, we give the approximation by the batch arrival queue by

$$EL(BAT) = \frac{\rho}{2(1-\rho)}(k+2+\rho(\delta^2-1)) - \lambda ak(k+1)/2 \qquad (7)$$

Our numerical results show that this approximation is very good for a less than m/2.

3.3 New-type approximation

The above two approximations is simple but very restrictive in applications. Here we consider an approximation app icable for all a. First, one may think about the heavy traffic or diffusion approximation (cf. [2]). In those approximations, N(t) or L(t) is approximated by a diffusion process. There are two not good points of those approximations. One is that it is good only for a high traffic case, and the other is that Var(N(t)) or Var(K(t)) is approximated by a linear function of t even though it is not linear. This non-linearity is essential in BP/GI/1 since Var(N(t))/t equals to λ as t tends to 0 and to $\lambda(k+1)$ as t tends to infinity. Hence, we need to consider an approximation in which this non-linearity is effective. For this purpose, we use the interesting result obtained by Benés [1]. His result is given for the virtual waiting time process in G/G/1, where G/G/1 denote a single server queue with a stationary input. He was only concerned with the process starting time 0 under the condition that the system is empty at time 0, which is not stationary. It is easy to extend his result to the stationary virtual waiting time process $\{V(t)\}$ since we know the existence of it. Thus, the following theorem is slightly different from his result itself (cf. also p.29 of [3]).

Theorem 3.1 (Benés [1]) In G/G/1, if the traffic intensity $\rho < 1$ then we have, for any $w > 0$,

$$\int_0^w P(V(0)>u)du = (1-\rho)\int_0^{+\infty} P(0<K(u)-u<w \mid V(0)=0)du \qquad (8).$$

In particular,

$$EV(0) = (1-\rho)\int_0^{+\infty} P(0<K(u)-u \mid V(0)=0)du \qquad (9).$$

From the well known invariance relations (cf. [6]),

$$EL = \lambda_a(EW+m) \qquad (10)$$

$$EV = \lambda_a(mEW + E(S^2)/2) \qquad (11),$$

we have

$$EL = (EV - \lambda_a E(S^2)/2)/m + \rho \qquad (12),$$

where $\lambda_a = \lambda(k+1)$. Hence, from (9) and (12), we can get EL if the integration in (9) is obtained. In general, this integration is very difficult since the term conditioned by V(0)=0 is contained.

In BP/GI/1, we approximate

$P(0<K(u)-u|V(0)=0)$ by $P(0<K(u)-u|N((-b,0])=0)$, where b is a nonnegative constant, which is determined later. Then we can calculate this approximate value in principle since parent customers in BP/GI/1 form a stationary Poisson process. However, its expression is very complicated and its integration is difficult even numerically. So we give a further approximation for it by assuming the normal distribution with the same mean and variance. This two steps of approximations are essential here. Thus we need to evaluate the mean and variance of $K(u)-u$ conditioned by $N((-b,0])=0$. Let

$$N_b(u) = N(u)-M(b,u)$$
$$K_b(u) = S_1+\dots+S_{N_b(u)},$$

where $M(b,u)$ is the number of customers arrived in the interval $(0,u]$ whose parent or some of other children arrived in the interval $(-b,0]$. The following lemma is easily obtained from Poisson properties of the parent customer process.

Lemma 3.1 In BP/GI/1, we have

$$E(K(u)|N((-b,0])=0) = mEN_b(u) \quad (13)$$
$$E((K(u)-mEN_b(u))^2|N((-b,0]=0) = \sigma^2 EN_b(u)+m^2 Var(N_b(u)) \quad (14).$$

The calculation of $E(N_b(u))$ and $Var(N_b(u))$ are not difficult if we note that only cutomers arrived in the interval $(-ka,0]$ affect the arrival process after time 0. The results are given in the next lemma.

Lemma 3.2 In BP/GI/1, we have, for $j=1,2,\dots$, if $0<u-ja\leq \max(0,a-b)$, then

$$EN_b(u) = \lambda(u(k+1)-bj_k(2k-j_k+1)/2) \quad (15)$$
$$Var(N_b(u)) = \lambda\{((2j_k+1)k-j_k^2+j_k+1)u -aj_k(j_k+1)(3k-2(j_k-1))/3 -bj_k((j_k+1)(2j_k+1)+6j_k(k-j_k))\} \quad (16),$$

and, if $\max(0,a-b)<u-ja\leq a$, then

$$EN_b(u) = \lambda(j_k+1)(u+a(k-j_k)-b(2k-j_k)/2) \quad (17)$$
$$Var(N_b(u)) = \lambda\{(j_k+1)(6(j_k+1)u +a(6k(j_k+1)-2j_k(4j_k+5)) -b(6k(j_k+1)-j_k(4j_k+5))/6\} \quad (18),$$

where $j_k = \min(j,k)$.

The expression of (15)~(18) may seem to be complex, but they are linear function of u for each interval. Thus, the right hand sides of (13) and (14) are also linear in each sub-intervals. We denote them $c(t)$ and $d(t)$, respectively. Then, the approximation for EV is given by

$$EV(app) = (1-\rho)\int_0^{+\infty}\int_{p(u)}^{+\infty}\frac{1}{\sqrt{2\pi}}e^{-y^2/2}dydu = \frac{1-\rho}{\sqrt{2\pi}}\int_0^{+\infty}up'(u)e^{-p^2(u)/2}du \quad (19),$$

where

$$p(u) = \frac{u-c(u)}{d(u)} \quad (20).$$

Hence, our approximation for EL is given by,

$$EL(app) = \frac{EV(app)}{m} + \rho(1-\delta^2) \quad (21).$$

We remark that EL(app) is exact for a=0 and $a=+\infty$, i.e., its values equal to those of M/GI/1 and the batch arrival M/GI/1, respectively. This fact easily proved if we note that $p(u) = const\times\sqrt{u}$. Further, EL(app) depends continuously on a and therefore EL(app) satisfies the asymptotic property discussed in the end of Section 2. Now we are assuming that b is constant with respect to a but arbitrary. We need to determine a suitable b, which is an approximated value for the minimum time s such that $N((-s,0])=0$ implies $V(0)=0$. By considering this property of b and numerical tests, we propose the following value for it.

$$b = \rho^2 m/(1-\rho) \quad (22)$$

In the next section, we consider this case and the case of b=0. Except a very light traffic case, b of (22) gives fairly good approximations for all a.

4. NUMERICAL CONSIDERATION

We compare numerical values of approximations discussed in the previous section with simulation results. We have performed simulations for various distributions of the service time. But we consider only BP/M/1 with 3 children here since other cases are not so much different from this case. We consider the cases that m=0.1, a=0.001, 0.01, 0.05, 0.1, 0.5, 1, 2 and 10. Results are given in Tables 1. In BP/GI/1, there is no exact result except the extrimal cases, whose values are given in Table 2. Therefore, reliability of simulation results is important. Our simulation has been done in personal computers (PC-9801F of NEC) and it is programed by language C (Lattice C). We used two series of random numbers, one for the arrival process and the other for the service time, and those random numbers, which are generated by multiplicable congruential method, are chosen so to satisfy several statistical tests for about 150000 runs. We determine the length of simulation runs by counting the number of busy cycles until the rate of the length of 95% reliable interval to the mean busy cycle is less than 1 plus 2.5%.

In Table 1, (sim), (appl), (app2), (MUR) and (BAT) denote the simulation result of EL, EL(app) of b given by (22), EL(app) of b=0, EL(MUR) and EL(BAT), respectively. To calculate the values of EL(appl) and EL(app2), we use numerical integrations by simple rectangle approximation. The reason why we give (app2) is that this case disregards the condition V(0)=0 and it is one of the worst selections of b. Comparing Table 1 with Table 2, we see that our simulations gives very good values in the extrimal cases. From Table 1, we can see the following points.

(i) (BAT) gives very good values for a less than m/2 and not so bad even a less than m.

(ii) (appl) is fairly good for all a except ρ = 0.1 and 0.9, whose cases are not so bad.

(iii) (app2) behaves like (appl) for ρ less than 0.5 and so it is also fairly good. But, for ρ greater than 0.5, (app2) is not good.

(iv) (MUR) is not good except the case that a is about 0.1.

The reason why (appl) and (app2) is less good for ρ =0.1 seems to be due to the assumption of normal distribution. And its improvement seems to be difficult unless we remove the assumption. On the other hand, for ρ =0.9, we may improve (appl) by chosing more suitable b.

5. CONCLUDING REMARKS

In this paper, we proposed two approximations EL(BAT) and EL(app) for BP/GI/1. EL(BAT) is very simple but appricable only for small a. On the other hand, EL(app) is complicated but applicable for all a. Actually, it takes longer time (more than 5 hours in the worst case) to compute EL(appl), where we use FORTRAN on a personal computer. Nevertheless, the author hope this type of approximation will be used furthermore since it has a possibility appricable to other queueing models. For example, we can apply this method to GI/GI/1 queue. In this case, we put b=0, then we have

$$EN_0(u) = \lambda u \tag{23}$$

$$Var(N_0(u)) = \lambda^3 \beta^2 u \tag{24},$$

where $1/\lambda$ and β^2 are the mean and variance of the interarrival time T. Hence, we obtain

$$EL(app) = \frac{\rho(\rho E(S^2)/m^2 + (\lambda\beta)^2 - 1)}{2(1-\rho)} + \rho \tag{25},$$

where $\rho = \lambda m$. This EL(app) is known as one of good approximations for EL (cf. [6]). We are only concerned with the mean of L, but it is noted that our approximation is possible also for higher moments of L.

Finally, the author thanks Mr. H. Miyazawa, M. Sasaki and T. Sumino of my laboratory for performing simulations and some of calculations.

Table 1 Numerical results of BP/M/1 (k=3, ES=0.1)

ρ/a	0.001	0.01	0.05	0.1	0.5	
0.1	0.2776	0.2645	0.2148	0.1741	0.1125	(sim)
	0.2661	0.2286	0.1530	0.1234	0.1111	(appl)
	0.2733	0.2381	0.1556	0.1241	0.1113	(app2)
	0.4411	0.4117	0.2948	0.1952	0.1117	(MUR)
	0.2763	0.2627	0.2028	0.1277		(BAT)
a =	1	2	10	(MRE)		
	0.1113	0.1115	0.1111	-		(sim)
	0.1111	0.1111	0.1111	(29.2%)		(appl)
	0.1113	0.1111	0.1111	(28.7%)		(app2)
	0.1111	0.1111	0.1111			(MUR)
ρ/a	0.001	0.01	0.05	0.1	0.5	
0.3	1.0680	1.0289	0.8814	0.7461	0.4824	(sim)
	1.0593	0.9403	0.8297	0.7365	0.4626	(appl)
	1.0670	1.0371	0.9173	0.8015	0.4779	(app2)
	1.7015	1.5888	1.1536	0.7828	0.4314	(MUR)
	1.0669	1.0264	0.8464	0.6214		(BAT)
a =	1	2	10	(MRE)		
	0.4391	0.4255	0.4289	-		(sim)
	0.4312	0.4286	0.4286	(8.6%)		(appl)
	0.4335	0.4286	0.4286	(7.4%)		(app2)
	0.4286	0.4286	0.4286			(MUR)
ρ/a	0.001	0.01	0.05	0.1	0.5	
0.5	2.4814	2.4124	2.1870	1.9462	1.3516	(sim)
	2.4873	2.3633	1.8317	1.8173	1.3636	(appl)
	2.4974	2.4751	2.3791	2.2680	1.6564	(app2)
	3.9701	3.7093	2.2794	1.9014	1.0083	(MUR)
	2.4925	2.425	2.125	1.75		(BAT)
a =	1	2	10	(MRE)		
	1.1514	1.0481	1.0022	-		(sim)
	1.1459	1.0813	1.0000	(16.9%)		(appl)
	1.2983	1.2983	1.0677	(22.6)		(app2)
	1.0001	1.0000	1.0000			(MUR)
ρ/a	0.001	0.01	0.05	0.1	0.5	
0.7	5.8231	5.7751	5.3075	5.0763	3.9745	(sim)
	5.8188	5.6904	5.1203	4.4592	3.5604	(appl)
	5.8319	5.8183	5.7586	5.6853	5.1531	(app2)
	9.2636	8.6600	6.4543	4.6209	2.3587	(MUR)
	5.8228	5.7283	5.3083	4.7833		(BAT)
a =	1	2	10	(MRE)		
	3.4559	2.8578	2.3236	-		(sim)
	3.2652	2.8011	2.3219	(12.2%)		(appl)
	4.6102	3.8299	2.4042	(34.0%)		(app2)
	2.3335	2.3333	2.3333			(MUR)
ρ/a	0.001	0.01	0.05	0.1	0.5	
0.9	22.483	22.363	21.906	21.443	19.175	(sim)
	22.479	22.348	21.755	21.027	15.974	(appl)
	22.493	22.489	22.469	22.444	22.251	(app2)
	35.731	33.421	25.218	18.559	9.1417	(MUR)
	22.486	22.365	21.825	21.15		(BAT)
a =	1	2	10	(MRE)		
	18.402	16.402	12.924	-		(sim)
	12.731	12.365	10.608	(30.8%)		(appl)
	22.006	21.529	18.308	(41.7%)		(app2)
	9.0009	9.0000	9.0000			(MUR)

MRE = Maximum relative error

Table 2 Two extrimal cases of BP/M/1 (k=3, ES=0.1)

ρ/a	0 (Batch arrival)	$+\infty$ (Poisson)
0.1	0.2777	0.1111
0.3	1.0714	0.4286
0.5	2.5	1.0
0.7	5.8333	2.3333
0.9	22.5	9.0

REFERENCES

[1] V.E. Benés, "General stochastic processes in the theory of queues," Addison-Wesley, New York, 1963.

[2] P. Billingsley, "Convergence of Probability Measures," John Wiley & Sons Inc., New York, 1968.

[3] A.A. Borovkov, "Stochastic Processes in Queueing theory," Nauka, Moscow (1972), English translation, Springer-Verlag, Berlin, 1976.

[4] D.J. Daley and D. Vere-Johnes, "A summary of point processes, " Stochastic Point Processes ed. by P.W.A. Lewis, John Wiley & Sons Inc., New York, 1972.

[5] M. Miyazawa, "A formal approach to queueing peocesses in the steady state and their applications," J. Appl. Prob. 16, 332-346, 1977.

[6] M. Miyazawa, "Approximations of the steady state distributions in queues," Proc. of the international semiar "Fundamental of teletraffic theory", Moscow, 1984.

[7] H. Murao, "On approximations for the queue with branching Poisson input," (in Japanese) Spring meeting of Oper. Res. Soc. of Japan, 1975.

TELETRAFFIC ISSUES in an Advanced Information Society
ITC-11
Minoru Akiyama (Editor)
Elsevier Science Publishers B.V. (North-Holland)

REFINING DIFFUSION APPROXIMATIONS FOR GI/G/1 QUEUES: A TIGHT DISCRETIZATION METHOD

Toshikazu KIMURA

Tokyo Institute of Technology, Tokyo, Japan

ABSTRACT:

This paper provides an effective method of refining diffusion approximations for the queue length distribution of a GI/G/1 queue with finite or infinite waiting space: A tight discretization method is developed for a class of GI/G/1 queues to obtain a discrete queue length distribution at arbitrary moments from a continuous probability distribution of the approximating diffusion process. By tight we mean that there exists a queueing system yielding the discretized queue length distribution as its exact solution. To see the quality of diffusion approximations with tight discretization, they are numerically compared with previously established diffusion approximations, various heuristic approximations and the exact results for some particular cases.

1. INTRODUCTION AND SUMMARY

Diffusion approximations for queues have been now recognized as basic approximation models in teletraffic theory as well as queueing theory. There are a number of interesting applications of diffusion approximations to general teletraffic models such as single server queues (11), many server queues (10,13) and networks of queues (20,27). Diffusion approximations also can be applied to the analysis and synthesis of systems with more complicated service mechanism, e.g., priority queues (1) and queues with removable server (12,18).

In heavy traffic situations, diffusion approximations can be often identified and justified by heavy traffic limit theorems: For a queueing characteristic process, e.g., the queue length process, heavy traffic limit theorems assure that the process in an unstable queue, when appropriately scaled and translated, converges weakly to a Brownian motion process; see (29). This implies that the heavy traffic limit theorems provide not only useful descriptions of unstable queues but also useful approximations of stable queues, especially, in heavy traffic. However, in moderate traffic, the limiting diffusion process is sometimes inappropriate as an approximation of the underlying process. In general, diffusion approximations become inaccurate as the traffic becomes light.

Another serious drawback of diffusion approximations is that they are not consistent with available exact results for particular cases. For example, consider a GI/G/1 queue. For the queue length process in the GI/G/1 queue, several different diffusion approximations have been proposed by many researchers; see (28,31). However, none of the approximations for the queue length distribution are consistent with any exact results, even with the M/M/1 queue. For the mean queue length, some of the diffusion approximations fail the consistency check with the M/G/1 queue, i.e., they do not coincide with the Pollaczek-Khintchine formula; see (31). The lack of consistency makes diffusion approximations unreliable with respect to the accuracy. Whitt (31) illustrated the need to refine diffusion approximations by consistency checks with bounds.

The purpose of this paper is to provide an effective method of refining diffusion approximations for the queue length distribution of a GI/G/1 queue with finite or infinite waiting space in such a way that the refined approximations are consistent with exact results for certain basic queueing systems. For ease of exposition, we consider the standard GI/G/1/N model with a single server, $N - 1$ waiting places $(1 \leq N \leq \infty)$, the FCFS (first-come first-served) discipline and i.i.d. (independent and identically distributed) service times (with a general distribution) that are independent of a renewal arrival process. It should be, however, noted that our method in this paper applies to any other queueing systems whenever their arrival and service processes do not depend on the queue length process; cf. (14). One may readily apply this method to bulk queues (2), queues with random service interruptions (3), etc.

Although a few refining methods of the diffusion approximations have been proposed so far, they only dealt with either modifications of the boundary conditions of the limiting diffusion process (8) or modifications of only the mean queue length (24); see also (9, Chapter 4). In this paper, we propose a <u>tight discretization method</u> for obtaining a "discrete" queue length distribution at arbitrary moments from the "continuous" probability distribution of the diffusion process. By <u>tight</u> we mean that there exists a queueing system yielding the discretized queue length distribution as its exact solution. We call this system a <u>base</u> of discretization.

The rest of this paper is organized as follows: In Section 2, we first focus on the GI/G/1/N queue $(N < \infty)$. From a partial differential equation with certain boundary conditions, we derive a distribution function of a diffusion process approximating the queue length process in the GI/G/1/N queue, assuming that the system is in equilibrium state. Using a tight discretization method, we obtain a diffusion approximation for the queue length distribution of the system. In Section 3, we apply the discretization method to the GI/G/1 queue with unlimited waiting room, obtaining a simpler diffusion approximation for the queue length distribution. A class of PH/M/1 queues of which interarrival-time distribution can be characterized by its first two moments is used as the base of discretization. Diffusion approximations for various congestion measures of the GI/G/1 queue, e.g., the mean waiting time and the coefficient of variation of interdeparture times, can be explicitly derived from the approximate queue length distribution. To improve the accuracy of these diffusion approximations, we develop a heuristic modification of the diffusion parameters. In Section 4, to see the quality of the diffusion approximations with tight discretization, they are numerically compared with previously derived diffusion approximations, various heuristic approximations such as the Kraemer and Langenbach-Belz (21) approximation and the exact results for some particular cases.

2. GI/G/1 QUEUE WITH LIMITED WAITING ROOM

2.1. System Descriptions

In this section we focus on a GI/G/1 queueing system with limited waiting room. The system can be specified by the following assumptions: Customers individually arrive at the system according to a recurrent process. Let u_n $(n = 1,2,\dots)$ be the interarrival time between the $(n-1)^{st}$ and n^{th} arriving customers; $\{u_n; n \geqq 1\}$ is a sequence of nonnegative i.i.d. random variables with a cdf (cumulative distribution function) $A(t)$ $(t \geqq 0)$. The maximum number of customers allowed in the system is N $(\geqq 1)$ including a customer in service. If a customer arrives when the waiting room is fully occupied, he cannot enter the system and so that he is overflowed. Assume that overflowed customers are cleared, namely, they leave the system and have no effect upon it. Customers allowed to enter the system are served in order of arrivals. Let v_n $(n = 1,2,\dots)$ be the service time of the nth entering customer; $\{v_n; n \geqq 1\}$ is a sequence of nonnegative i.i.d. random variables with a cdf $B(t)$ $(t \geqq 0)$ and is independent of $\{u_n\}$. For convenience, let u and v denote generic interarrival and service times, respectively.

2.2. Diffusion Model

Let $Q(t)$ be the number of customers either waiting or being served at time t $(\geqq 0)$ in the system. We approximate the discrete-valued process $\{Q(t); t \geqq 0\}$ by a time-homogeneous diffusion process $\{X(t); t \geqq 0\}$ with state space $I = [0, N)$, using the fact that the process $Q(t)$ apart from boundary behavior follows the same probability law just as with the case $N = \infty$: Let $dX(h)$ be the increment in the diffusion process $X(t)$ accrued over a time interval of length h (> 0), i.e., $dX(h) = X(h) - X(0)$. Then, the diffusion process $X(t)$ in the interval $I^o = (0, N)$ can be characterized by the diffusion parameters

$$b \equiv \lim_{h \downarrow 0} \frac{1}{h} E[dX(h) \mid X(0)] = \lambda - \mu, \tag{1}$$

and

$$a \equiv \lim_{h \downarrow 0} \frac{1}{h} E[\{dX(h)\}^2 \mid X(0)] = \lambda c_a^2 + \mu c_s^2, \tag{2}$$

where $\lambda = 1/E(u)$ and $\mu = 1/E(v)$ and c_a (c_s) is the coefficient of variation (standard deviation devided by mean) of u (v); see [11].

To obtain the approximating diffusion process for the process $Q(t)$ in the GI/G/1/N queue, we need appropriate boundary conditions at $X = 0$ and N. We simply add a reflecting barrier at $X = N$, while we use a certain sticky barrier at the origin, taking into account the probability mass at the boundary. This mass corresponds to the probability that the system is empty. Note that these two boundaries act as impenetrable barriers and so that the process $X(t)$ is conservative, i.e., $P\{X(t) \in I \mid X(0) = x_0\} = 1$ for all $t \geqq 0$ and $x_0 \in I$.

Let $P(x,t \mid x_0)$ $(x, x_0 \in I)$ be the conditional cdf of $X(t)$ initially starting from $X(0) = x_0$, i.e., $P(x,t \mid x_0) = P\{X(t) \leqq x \mid X(0) = x_0\}$. Then, it is known that the cdf $P(x,t \mid x_0)$ satisfies the partial differential equation

$$\frac{\partial P}{\partial t} = \frac{1}{2} a \frac{\partial^2 P}{\partial x^2} - b \frac{\partial P}{\partial x}, \qquad x \in I^o \tag{3}$$

with the initial condition

$$P(x,0 \mid x_0) = U(x - x_0), \tag{4}$$

where $U(x)$ is the unit step function degenerated at $x = 0$; see [11]. In what follows, we will restrict our attention to the equilibrium cdf $P(x) = \lim_{t \uparrow \infty} P(x,t \mid x_0)$ whose existence is assured because the process $X(t)$ is conservative. From (3), we obtain the ordinary differential equation

$$\frac{1}{2} a \frac{d^2P}{dx^2} - b \frac{dP}{dx} = 0, \qquad x \in I^o. \tag{5}$$

The associated boundary conditions are

$$P(0) = p_0, \tag{6a}$$

and

$$P(N) = 1, \tag{6b}$$

where p_0 is the equilibrium probability that the system is empty. Solving (5) together with (6), we have

$$P(x) = \begin{cases} p_0 + (1 - p_0) \dfrac{1 - e^{\gamma x}}{1 - e^{\gamma N}}, & \rho \neq 1 \\ p_0 + (1 - p_0) \dfrac{x}{N}, & \rho = 1, \end{cases} \tag{7}$$

for $x \in I$, where $\rho = \lambda/\mu$ is the traffic intensity and

$$\gamma = \frac{2b}{a} = \frac{2(\rho - 1)}{\rho c_a^2 + c_s^2}. \tag{8}$$

To identify the cdf $P(x)$ completely, we need to specify the empty probability p_0. It is, however, difficult to obtain an exact expression of p_0 for a general GI/G/1/N queue. Hence, we will use an appropriate approximation for p_0. To obtain simple approximations for p_0, it is convenient to use an intimate relation between diffusion models for a two-stage cyclic queue and the GI/G/1/N queue: Consider a cyclic queueing system which consists of two sequential stages; see **Figure 1**. This system has been analyzed by several researchers (e.g., (5,23)) as a model of multiprogrammed computer systems. We suppose that there are N ($\geq$ 1) programs in the central processing unit (CPU) - data transfer unit (DTU) cycle. Each program goes through both stages in sequence and then returns to the first stage. This process continues indefinitely. Programs in the CPU and DTU queues are served according to the FCFS discipline.

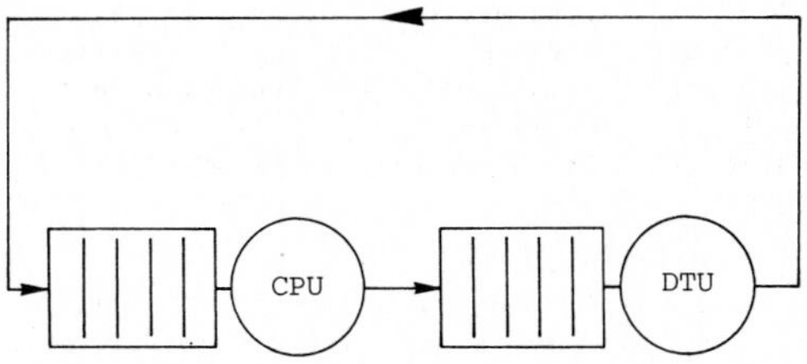

Figure 1. Cyclic queueing system.

If we assume here that (i) the sequence of DTU service or auxiliary memory access and transfer times is i.i.d. as with $\{u_n\}$; (ii) the sequence of CPU service or processing times is also i.i.d. as with $\{v_n\}$; and (iii) CPU and DTU processing times are mutually independent, then the queue length process at the CPU stage can be approximated by a diffusion process with the same diffusion parameters as with X(t); see (6,7,8): Gaver and Shedler (6,7) used the same boundary conditions as (6), while Gelenbe (8) adopted a different kind of boundaries with jumps, both obtaining approximations for the CPU utilization $1 - p_0$; see also (16,17). Let π_0 be an approximation for p_0. Then, their results are rewritten as:

(a) <u>Gaver and Shedler (6) approximation</u>:

$$\pi_0 = \begin{cases} \dfrac{1-\rho}{1-\rho\tilde{\rho}^N}, & \rho \neq 1 \\ \dfrac{1}{1+\dfrac{2N}{c_a^2+c_s^2}}, & \rho = 1; \end{cases} \tag{9}$$

(b) <u>Gaver and Shedler (7) approximation</u>:

$$\pi_0 = \begin{cases} \dfrac{1-\tilde{\rho}^N-\rho(1-\tilde{\rho}^{N-1})}{(1+\rho)(1-\tilde{\rho}^N)-\rho(1-\tilde{\rho}^{N-1})}, & \rho \neq 1 \\ \dfrac{1}{N+1}, & \rho = 1; \end{cases} \tag{10}$$

(c) <u>Gelenbe's (8) approximation</u>:

$$\pi_0 = \begin{cases} \dfrac{1-\rho}{1-\rho^2\tilde{\rho}^{N-1}}, & \rho \neq 1 \\ \dfrac{1}{2}\,\dfrac{1}{1+\dfrac{N-1}{c_a^2+c_s^2}}, & \rho = 1, \end{cases} \tag{11}$$

where

$$\tilde{\rho} = e^{\gamma} = \exp\{\frac{2(\rho-1)}{\rho c_a^2 + c_s^2}\}. \tag{12}$$

From some numerical comparisons for particular cases in (6,7,8), we see that all of the approximations perform well, providing accuracy sufficient for most engineering applications. In particular, the approximation (c) is better than (a) when N is small, and the approximation (b) is more accurate than (a) and (c) for the case $c_a > 1$. It is worth noting that all of the approximations give the exact result for $N = \infty$, i.e., $\lim_{N\uparrow\infty} \pi_0 = 1 - \rho$ if $\rho < 1$. However, there is a serious deficiency in these approximations: The approximations with $\tilde{\rho}$ in (12) do not coincide with any exact results, even with the M/M/1/N queue. To deal with this deficiency, Gaver and Shedler (7) developed a distribution-dependent modification of $\tilde{\rho}$, based on Wald's identity. Let $\alpha(s) = E(e^{-su})$ and $\beta(s) = E(e^{-sv})$ for Re $s \geq 0$. Then, under a weak assumption on u, they proposed

$$\tilde{\rho} \simeq \beta(\bar{s}), \tag{13}$$

where $\bar{s}$ is the unique positive solution of the equation

$$\alpha(-s)\beta(s) = 1. \tag{14}$$

When u and v are exponentially distributed, we have $\tilde{\rho} = \rho$, so that all of the approximations for the M/M/1/N queue yield the exact result

$$\pi_0 = p_0 = \begin{cases} \dfrac{1-\rho}{1-\rho^{N+1}}, & \rho \neq 1 \\ \dfrac{1}{N+1}, & \rho = 1. \end{cases} \tag{15}$$

It should be, however, noted that such consistency cannot be observed for other systems, even by the use of the approximation (13).

2.3. A Tight Discretization Method

From the <u>continuous</u> cdf P(x) of the diffusion process X(t), we will derive an approximate <u>discrete</u> distribution of the process Q(t) in the GI/G/1/N queue. Let $p_k = \lim_{t\uparrow\infty} P\{Q(t) = k \mid Q(0)\}$ and $\tilde{p}_k$ be its diffusion approximation (k = 0,...,N). We say that an approximate queue length distribution $\{\tilde{p}_k\}$ is <u>tight</u> if there exists a queueing system for which $\tilde{p}_k = p_k$ for all $k \geq 0$. We call this system a <u>base</u> of discretization.

Before introducing our discretization method for P(x), we show a typical discretization method adopted by many researchers so far:

$$\begin{aligned} \tilde{p}_0 &= P(0) \\ \tilde{p}_k &= P(k) - P(k-1), \quad k = 1,\dots,N, \end{aligned} \tag{16}$$

and hence

$$\tilde{p}_k = \frac{(1-p_0)(1-\tilde{\rho})\tilde{\rho}^{k-1}}{1-\tilde{\rho}^N}, \quad k = 1,\dots,N \tag{17}$$

with the approximation $p_0 \simeq \pi_0$. As shown in Section 2.2, the approximate distribution (17) is a tight one whose base is the M/M/1/N queue if we approximate $\tilde{\rho}$ by (13). However, if we use the usual definition (12) for $\tilde{\rho}$, (17) is no longer a tight approximation. In what follows, we assume that $\tilde{\rho}$ is defined by (12) as usual.

Now suppose that we have an exact queue length distribution for a particular queueing system. Using this distribution, we develop a discretization method yielding a tight approximation whose base coincides with this system. Let $\{p_k^*\}$ and $P^*(x)$ be the exact queue length distribution at arbitrary moments and the cdf via diffusion approximation for the base system, respectively. Similarly, we attach the asterisk * to any parameters of the base system. An essential idea in our discretization method is to shift base points of the discretization from the integer points in I^o, depending upon the distribution $\{p_k^*\}$: Define the following sequence $\{x_k;\ k = 0,\ldots,N\}$ as base points of the discretization; see **Figure 2**:

$$x_0 = 0, \qquad x_N = N,$$

and

$$P^*(x_k) = \textstyle\sum_{i=0}^{k} p_i^*, \qquad k = 1,\ldots,N-1, \tag{18}$$

from which

$$x_k = \begin{cases} \dfrac{1}{\gamma^*}\ln \Theta_k, & \rho \neq 1 \\[2ex] \dfrac{N}{1-p_0^*}\textstyle\sum_{i=1}^{k} p_i^*, & \rho = 1, \end{cases} \tag{19}$$

for $k = 0,\ldots,N$, where

$$\Theta_k = 1 - \frac{1-\exp(\gamma^* N)}{1-p_0^*}\textstyle\sum_{i=1}^{k} p_i^*, \tag{20}$$

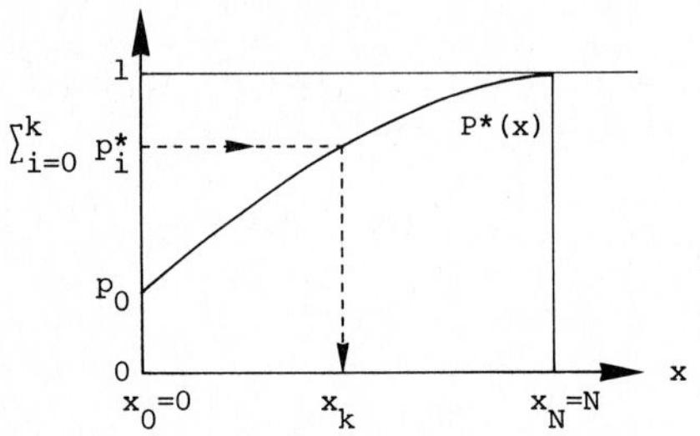

Figure 2. Base points of discretization.

Using these base points, we discretize the cdf $P(x)$ as

$$\begin{aligned} \tilde{p}_0 &= P(0) \\ \tilde{p}_k &= P(x_k) - P(x_{k-1}), \qquad k = 1,\ldots,N. \end{aligned} \tag{21}$$

From (7), (19) and (21), we obtain

$$\tilde{p}_k = \begin{cases} \dfrac{1-p_0}{1-\exp(\gamma N)}(\Theta_{k-1}^d - \Theta_k^d), & \rho \neq 1 \\[2ex] \dfrac{1-p_0}{1-p_0^*}p_k^*, & \rho = 1, \end{cases} \tag{22}$$

for $k = 1,\ldots,N$, where

$$d = a^*/a. \tag{23}$$

Here we need to approximate the empty probability p_0 for the system to be approximated. The easiest way of such an approximation is to set $p_0 \simeq p_0^*$, from which it can be proved that (22) is a tight approximation, i.e., $\tilde{p}_k^* \simeq p_k^*$ for all k. Although the approximation $p_0 \simeq \pi_0$ does not provide a tight approximation, it may be still accurate enough for practical applications.

3. GI/G/1 QUEUE WITH UNLIMITED WAITING ROOM

In this section we consider the case $N = \infty$, i.e., the standard GI/G/1 queue with unlimited waiting room. Applying the tight discretization method to the GI/G/1 queue, we give a simple diffusion approximation for the queue length distribution. In particular, we use the GI/M/1 queue as a base of discretization, obtaining explicit approximations for several queueing characteristics and their asymptotic properties in heavy traffic.

3.1. Diffusion Model

The diffusion process $X(t)$ approximating $Q(t)$ in the GI/G/1 queue has the same diffusion parameters as with the case $N < \infty$ and the state space $I = [0, \infty)$. The equilibrium cdf of $X(t)$, $P(x)$, also satisfies the ordinary differential equation (5) with the boundary conditions

$$P(0) = 1 - \rho, \tag{24a}$$

and

$$P(\infty) = 1; \tag{24b}$$

see (11). It should be noted here that the boundary condition at the origin does not contain any undetermined parameters; cf. (6a). The condition (24a) reflects the exact empty probability of the GI/G/1 queue, which is valid even for stationary non-renewal arrival processes; see [4, (4.2.3)].

To assure the existence of the cdf $P(x)$, we assume $\rho < 1$, or equivalently, $b < 0$ in (5). Solving (5) together with (24) yields

$$P(x) = 1 - \rho\exp(\gamma x), \qquad x \geq 0. \tag{25}$$

If we apply the tight discretization method (21) with (19) to (25), then we obtain the base points of discretization as

$$x_k = \frac{1}{\gamma^*}\ln\Big(1 - \frac{1}{\rho}\textstyle\sum_{i=1}^{k} p_i^*\Big), \qquad k \geq 0, \tag{26}$$

and hence the approximate queue length distribution is

$$\begin{aligned} \tilde{p}_0 &= 1 - \rho \\ \tilde{p}_k &= \rho^{1-d}\Big\{\Big(1-\sum_{i=0}^{k-1} p_i^*\Big)^d - \Big(1-\sum_{i=0}^{k} p_i^*\Big)^d\Big\}, \quad k \geq 1. \end{aligned} \tag{27}$$

Remark 1. If we apply the usual discretization method (16) to (25), then we have

$$\begin{aligned} \tilde{p}_0 &= 1 - \rho \\ \tilde{p}_k &= \rho(1-\tilde{\rho})\tilde{\rho}^{k-1}, \qquad k \geq 1, \end{aligned} \tag{28}$$

which coincides with Kobayashi's [20] diffusion approximation.

In what follows we will investigate detailed properties of the approximation (27) in which the GI/M/1 queue is assumed as the base of discretization. It is well known that the GI/M/1 queue length distribution at arbitrary moments is

$$p_0 = 1 - \rho$$
$$p_k = \rho(1-\omega)\omega^{k-1}, \qquad k \geq 1, \tag{29}$$

where $\omega \in (0, 1)$ is the solution of the equation for z:

$$z = \alpha(\mu(1-z)). \tag{30}$$

To compute the root ω, it is convenient to use the recursive formula

$$z_0 = \tilde{\rho}^*$$
$$z_{n+1} = \alpha(\mu(1-z_n)), \qquad n \geq 0, \tag{31}$$

which yields the unique fixed point $z_\infty = \omega$ if $\rho < 1$; cf. (28) for the initial point of the recursion. Substituting (29) for $\{p_k^*\}$ in (26), we obtain

$$x_k = -\frac{(1+\rho c_a^2)\ln\omega}{2(1-\rho)}k, \qquad k \geq 0, \tag{32}$$

from which we see that the value of the kth base point x_k is proportional to k, and so that the length of the interval (x_{k-1}, x_k) is independent of k. From (27) and (29), we obtain

$$\tilde{p}_0 = 1 - \rho$$
$$\tilde{p}_k = \rho(1-\omega^d)\omega^{(k-1)d}, \qquad k \geq 1, \tag{33}$$

for which

$$d = \frac{\rho c_a^2 + 1}{\rho c_a^2 + c_s^2}. \tag{34}$$

Remark 2. Since diffusion approximations are two-moment approximations, we need either to replace or to approximate the cdf A of the GI/M/1 queue by an appropriate distribution which can be completely characterized by its first two moments. Of course, when $c_a = 0$, we use the degenerate distribution, i.e., $A(t) = U(t - 1/\lambda)$ and hence $\alpha(s) = \exp(-s/\lambda)$. When $c_a > 0$, it is often convenient to use distributions with exponential building blocks; see Whitt (30). There are two cases, depending on whether c_a ia less than one or greater than one. Here we use a gamma distribution (Γ) when $0 < c_a \leq 1$ and a mixture of two exponentials with balanced means (H_2^b) when $c_a > 1$. The Laplace-Stieltjes transforms of A are, when $0 < c_a \leq 1$,

$$\alpha(s) = \left(\frac{\kappa\lambda}{s+\kappa\lambda}\right)^\kappa \tag{35}$$

with

$$\kappa = 1/c_a^2, \tag{36}$$

and when $c_a > 1$,

$$\alpha(s) = \sum_{i=1}^{2}\frac{q_i\nu_i}{s+\nu_i} \tag{37}$$

with

$$q_i = \frac{1}{2}\left\{1 \pm \sqrt{\frac{c_a^2-1}{c_a^2+1}}\right\}, \quad \nu_i = 2\lambda q_i, \quad i = 1,2. \tag{38}$$

These distributions are also useful for the distribution-dependent modification of $\tilde{\rho}$ in (13).

From the approximate queue length distribution (33), we can derive various queueing characteristics for the GI/G/1 queue: Let EW be the mean waiting time (until beginning service) in the GI/G/1 queue. Then, with the aid of Little's formula, we have

$$EW = \frac{1}{\lambda}\sum_{k=1}^{\infty}(k-1)\tilde{p}_k = \frac{\omega^d}{\mu(1-\omega^d)}. \tag{39}$$

Let c_d be the coefficient of variation of interdeparture times. Marshall (25) showed the following relation between EW and c_d:

$$c_d^2 = c_a^2 + 2\rho^2 c_s^2 - 2\lambda(1-\rho)EW. \tag{40}$$

The relation (40) was first used together with an approximation for EW to approximate departure process in networks of queues by Kuehn (22). Following Kuehn (22), we combine (39) and (40) to obtain

$$c_d^2 = c_a^2 + 2\rho^2 c_s^2 - \frac{2\rho(1-\rho)\omega^d}{1-\omega^d}. \tag{41}$$

These results can be utilized in decomposition-approximations for networks of queues. In particular, a software package called QNA (Queueing Network Analyzer) has been developed to calculate approximate congestion measures for open non-Markovian networks of queues; see (32). The approximation (41) combined with the previous results for superposition process provide a new approximation in QNA-like softwares.

We now investigate asymptotic properties of the GI/M/1-base approximation in heavy traffic. First, we characterize heavy traffic limits of ω and its derivative:

Lemma 1. $$\lim_{\rho\uparrow 1}\omega = 1. \tag{42}$$

Lemma 2. $$\lim_{\rho\uparrow 1}\frac{d\omega}{d\rho} = \frac{2}{1+c_a^2}. \tag{43}$$

(Since proofs of these lemmas are straightforward, we omit the proofs.) From Lemmas 1 and 2, we immediately have

Theorem. For the GI/M/1-base diffusion approximation,

(a) $$\lim_{\rho\uparrow 1} x_k = k, \tag{44}$$

(b) $$\lim_{\rho\uparrow 1} 2\mu(1-\rho)EW = c_a^2 + c_s^2. \tag{45}$$

From (44) we see that the base points in our tight discretization method asymptotically coincide with the usual (i.e., integer-valued) base points as the traffic becomes heavy. Also, we see from (45) that the heavy traffic limit of EW is consistent with Kingman's (19) result.

3.2. Heuristic Modification

The approximate distribution (33), of course, gives the exact solution for the GI/M/1 queue. However, for the M/G/1 queue, it does not give the exact solution even for the mean waiting time. In this subsection, we will modify (33) both to be consistent with the Pollaczek-Khintchine formula and to be still tight for the GI/M/1 queue.

For the above purpose, we modify the

diffusion parameters through d: Let $\tilde{d} \equiv \tilde{d}(c_a^2, c_s^2)$ denote a modification of d. Obviously, we must have

$$\tilde{d}(c_a^2, 1) = 1. \tag{46}$$

To let (39) be exact for the M/G/1 queue, we need

$$\omega^{\tilde{d}(1,\, c_s^2)} = \frac{\rho(1 + c_s^2)}{2 - \rho(1 - c_s^2)}. \tag{47}$$

Taking into account $\omega = \rho$ for the M/M/1 queue, we propose the following modification $\tilde{d}$ satisfying (46) and (47) simultaneously:

$$\omega^{\tilde{d}} = \frac{\omega(c_a^2 + c_s^2)}{c_a^2 + 1 - \omega(1 - c_s^2)}, \tag{48}$$

i.e.,

$$\tilde{d}(c_a^2, c_s^2) = \log_\omega\{\frac{\omega(c_a^2 + c_s^2)}{c_a^2 + 1 - \omega(1 - c_s^2)}\}. \tag{49}$$

Substituting (48) into (39) yields

$$EW = \frac{\omega(c_a^2 + c_s^2)}{\mu(1 + c_a^2)(1 - \omega)}, \tag{50}$$

which is exact both for the GI/M/1 and M/G/1 queues.

4. NUMERICAL COMPARISONS

To see the quality of our diffusion approximations with tight discretization, we compare them with previously established diffusion approximations, heuristic approximations and the exact results for some particular cases. Since it is difficult to obtain exact results for queues with finite waiting space, we restrict our comparisons to GI/G/1 queues with unlimited waiting room.

Table 1 compares diffusion approximations for the queue length distribution $\{p_k\}$ of the $M/E_2/1$ queue (ρ = 0.6, 0.9). Four different diffusion approximations are considered, namely, the approximation (33), its modification (33) with (48), Gelenbe's [9, (4.26a)] approximation and Kobayashi's [20, (2.9)] approximation (= (28)). In our two approximations we use the M/M/1 queue as the base of discretization, i.e., $\omega = \rho$ in (33). Since all of these approximations give the exact result for the empty probability, $\tilde{p}_0 = p_0 = 1 - \rho$ is excluded from the comparisons.

Table 1 shows that the modified approximation is much better than the others in moderate traffic. In heavy traffic, our and Kobayashi's approximations are quite accurate. In particular, it is interesting that the M/M/1-base approximation (33) performs about the same as Kobayashi's approximation. Gelenbe's approximation performs very poorly, especially, for the probability p_1. This seems to be due to the boundary condition of the approximating

Table 1. A comparison of diffusion approximations for the queue length distribution of the $M/E_2/1$ queue (ρ = 0.6, 0.9).

Approximations	p_1	p_2	p_3 (× 10)	p_4 (× 10)	p_5 (× 10)	p_{10} (× 10)
ρ = 0.6						
Gelenbe [9]	.173	.220	1.065	.514	.249	.00655
Kobayashi [20]	.310	.150	.724	.350	.169	.00445
M/M/1-base	.315	.150	.712	.339	.161	.00392
Mod. M/M/1-base	.282	.149	.791	.419	.222	.00922
Exact	.276	.154	.817	.425	.220	.00800
ρ = 0.9						
Gelenbe [9]	.061	.112	.968	.839	.727	.356
Kobayashi [20]	.120	.104	.900	.780	.677	.331
M/M/1-base	.120	.104	.901	.781	.677	.331
Mod. M/M/1-base	.116	.101	.881	.767	.668	.335
Exact	.110	.101	.893	.780	.679	.338

Table 2. A comparison of approximations for the mean queue length of $E_m/E_2/1$ queues (m = 2,4,9 and ρ = 0.6, 0.9).

Approximations	$E_2/E_2/1$		$E_4/E_2/1$		$E_9/E_2/1$	
	ρ=0.6	ρ=0.9	ρ=0.6	ρ=0.9	ρ=0.6	ρ=0.9
Gelenbe [9]	.600	4.275	.488	3.263	.425	2.700
Kobayashi [20]	.349	3.841	.248	2.833	.193	2.275
M/M/1-base	.337	3.837	.238	2.830	.186	2.272
GI/M/1-base	.282	3.739	.166	2.698	.110	2.128
Mod. GI/M/1-base	.393	3.953	.262	2.907	.195	2.333
Kimura [15]	.351	3.900	.237	2.872	.183	2.316
KL [21]	.403	3.976	.242	2.873	.155	2.249
Page [26]	.409	3.978	.277	2.929	.203	2.346
Exact	.378	3.923	.237	2.853	.162	2.261

diffusion process; see (16) for a similar situation and its modification.

Table 2 compares the diffusion approximations with several heuristic approximations for the mean queue length of $E_m/E_2/1$ queues (m = 2,4,9 and ρ = 0.6,0.9). Three heuristic two-moment approximations are added here for comparisons, namely, the approximations of Kimura (15), Kraemer and Langenbach-Belz (21) (abbreviated by KL) and Page (26). The modified M/M/1-base approximation is excluded from the comparisons because it is less accurate than the approximation without modification.

From Table 2 we see with a little surprise that the M/M/1-base approximation is better than the GI/M/1-base approximation. The modified GI/M/1-base approximation is better than that without modification, but it is sometimes less accurate than the M/M/1-base approximation. A major reason for these results may be that the GI/M/1-base approximations tend to be too sensitive to the arrival distribution form. However, for systems with high-variable interarrival-time distributions, the GI/M/1-base approximations might be more accurate than the M/M/1-base approximation. As shown in Table 1, Kobayashi's approximation performs about the same as the M/M/1-base approximation, both providing accuracy comparable to the heuristic approximations. Gelenbe's approximation is less accurate than the others for all cases in the table.

REFERENCES

(1) Abdou, E., "Diffusion Approximation for Preemptive-Resume Queueing System," Proc. of the 10th International Teletraffic Congress, Montreal, 1983.

(2) Chiamsiri, A.B. and M.S. Leonard, "A Diffusion Approximation for Bulk Queues," Mgmt. Sci. 27, 1188-1199 (1981).

(3) Fischer, M.J., "An Approximation to Queueing Systems with Interruptions," Mgmt. Sci. 24, 338-344 (1977).

(4) Franken, P., Konig, D., Arndt, U. and Schmidt, V., Queues and Point Processes, Akademie-Verlag, Berlin, 1981.

(5) Gaver, D.P., "Probability Models for Multiprogramming Computer Systems," J. ACM 14, 423-438 (1967).

(6) Gaver, D.P. and G.S. Shedler, "Processor Utilization in Multiprogramming Systems via Diffusion Approximations," Opns. Res. 21, 569-576 (1973).

(7) Gaver, D.P. and G.S. Shedler, "Approximate Models for Processor Utilization in Multiprogrammed Computer Systems," SIAM J. Comput. 2, 183-192 (1973).

(8) Gelenbe, E., "On Approximate Computer System Models," J. ACM. 22, 261-269 (1975).

(9) Gelenbe, E. and I. Mitrani, "Analysis and Synthesis of Computer Systems," Academic Press, New York, 1980.

(10) Halachmi, B. and W.R. Franta, "A Diffusion Approximation to the Multi-Server Queue," Mgmt. Sci. 24, 522-529 (1978).

(11) Heyman, D.P., "A Diffusion Model Approximation for the GI/G/1 Queue in Heavy Traffic," Bell System Tech. J. 54, 1637-1646 (1975).

(12) Kimura, T., "Optimal Control of an M/G/1 Queueing System with Removable Server via Diffusion Approximation," European J. Opnl. Res. 8, 390-398 (1981).

(13) Kimura, T., "Diffusion Approximation for an M/G/m Queue," Opns. Res. 31, 304-321 (1983).

(14) Kimura, T., "A Unifying Diffusion Model for State-Dependent Queues," Research Report on Information Sciences, No. B-132, Tokyo Institute of Technology, 1983.

(15) Kimura, T., "A Two-Moment Approximation for the Mean Waiting Time in the GI/G/s Queue," Research Report on Information Sciences, No. B-154, Tokyo Institute of Technology, 1984.

(16) Kimura, T., "Diffusion Approximation for a Tandem Queue with Blocking," Math. Oper. u. Stat., Ser. Optimization 16, 000-000 (1985).

(17) Kimura, T., K. Ohno and H. Mine, "Diffusion Approximation for GI/G/1 Queueing Systems with Finite Capacity: II - The Stationary Behaviour," J. Opns. Res. Soc. Japan. 22, 301-320 (1979).

(18) Kimura, T., K. Ohno and H. Mine, "Approximate Analysis of Optimal Operating Policies for a GI/G/1 Queueing System," Memoirs of the Faculty of Engineering, Kyoto University. 42, 377-390 (1980).

(19) Kingman, J.F.C., "The Heavy Traffic Approximation in the Theory of Queues," Proc. Symposium on Congestion Theory, pp. 137-159, University of North Carolina Press, Chapel Hill, 1965.

(20) Kobayashi, H., "Application of the Diffusion Approximation to Queueing Networks I: Equilibrium Queue Distributions," J. ACM 21, 316-328 (1974).

(21) Kraemer, W. and M. Langenbach-Belz, "Approximate Formulae for the Delay in the Queueing System GI/G/1," Proc. of the 8th International Teletraffic Congress, pp. 235-1/8, Melbourne, 1976.

(22) Kuehn, P.J., "Approximate Analysis of General Queueing Networks by Decomposition," IEEE Trans. Commun. COM-27, 113-126 (1979).

(23) Lewis, P.A.W. and G.S. Shedler, "A Cyclic-Queue Model of System Overhead in Multiprogrammed Computer Systems," J. ACM 18, 199-220 (1971).

(24) Marchal, W.G., "An Approximate Formula for Waiting Times in Single Server Queues," AIIE Trans. 8, 473-474 (1976).

(25) Marshall, K.T., "Some Inequalities in Queueing," Opns. Res. 16, 651-665 (1968).

(26) Page, E., Queueing Theory in OR, Butterworth, London, 1972.

(27) Reiman, M.I., "Open Queueing Networks in Heavy Traffic," Math. Opns. Res. 9, 441-458 (1984).

(28) Shanthikumar, J.G. and J.A. Buzacott, "On the Approximations to the Single Server Queue," Int. J. Prod. Res. 18, 761-773 (1980).

(29) Whitt, W., "Heavy Traffic Limit Theorems for Queues: A Survey," Mathematical Methods in Queueing Theory, pp. 307-350, Lecture Notes in Economics and Mathematical Systems, No. 98, Springer-Verlag, New York, 1974.

(30) Whitt, W., "Approximating a Point Process by a Renewal Process, I: Two Basic Methods," Opns. Res. 30, 125-147 (1982).

(31) Whitt, W., "Refining Diffusion Approximations for Queues," Opns. Res. Letters 1, 165-169 (1982).

(32) Whitt, W., "The Queueing Network Analyzer," Bell System Tech. J. 62, 2779-2815 (1983).

TELETRAFFIC ISSUES in an Advanced Information Society
ITC-11
Minoru Akiyama (Editor)
Elsevier Science Publishers B.V. (North-Holland)

THE RELAXATION TIME OF SINGLE SERVER QUEUEING SYSTEMS WITH POISSON ARRIVALS AND HYPEREXPONENTIAL/ ERLANG SERVICE TIMES

Julian Keilson †, Fumiaki Machihara ‡ and Ushio Sumita †

† Graduate School of Management
University of Rochester
Rochester, New York 14627
U.S.A.

‡ Musashino Electrical Communication
Laboratory
Nippon Telegraph and Telephone
Public Corporation, Musashino-shi, Tokyo, Japan

ABSTRACT

The relaxation time required to justify use of steady state information for M/G/1 systems is described. The presentation, based on a more extensive theoretical study [7], is oriented to the needs of telecommunication practice. Hyperexponential and Erlang service time distributions are discussed in detail with theoretical and numerical results provided. The understanding and quantification of the system relaxation time requires a knowledge of the structure of the busy period transform in the complex plane.

0. INTRODUCTION

The relaxation time of a system is a measure of the time required for the system to reach ergodicity from an arbitrary initial state. For many of the stochastic models employed in telephony, only the ergodic distribution is available. Use of ergodic information, however, can only be justified when underlying parameters such as traffic intensity remain constant over a period long enough for the ergodic distribution to set in. Such parameters are seldom constant in time and a knowledge of the system relaxation time is essential before ergodic information can be employed for planning purposes. The importance of the relaxation time has often been ignored in design practice.

For a system modeled by a finite Markov chain which is reversible in time, (Keilson [3], Kelly [9]), e.g. a birth-death process, the notion of relaxation time is well understood. (See, for example Keilson [3].) For such a chain $J(t)$ with transition probability matrix $\underline{\underline{P}}(t)$ of order $N+1$ and ergodic probability vector $\underline{e}^T$, one has

$$(0.1) \qquad \underline{\underline{P}}(t) - \underline{1}\underline{e}^T = \sum_{i=1}^{N} \exp(-\gamma_i t)\underline{\underline{J}}_i$$

where $\underline{1}$ is a column vector having all elements equal to one and $0 < \gamma_1 < \gamma_2 < \ldots < \gamma_N$. Furthermore the matrices $\underline{\underline{J}}_i$ are of rank one and satisfy $\underline{\underline{J}}_i\,\underline{\underline{J}}_j = \delta_{ij}\,\underline{\underline{J}}_i$ where $\delta_{ij} = 1$ if $i = j$ and $\delta_{ij} = 0$ if $i \neq j$. Clearly the exponential decay rate γ_1 dominates the speed of the convergence of $\underline{\underline{P}}(t)$ to its ergodic matrix $\underline{1}\underline{e}^T$. The relaxation time of $J(t)$ may then be defined to be $|\gamma_1|^{-1}$ (see [3], §8.10). Let $P_0(t) = P[J(t) = 0 \mid J(0) = 0]$ and define $\pi_0(s) = \mathcal{L}\{P_0(t)\} = \int_0^\infty e^{-st}P_0(t)dt$. One then easily sees from (0.1) that

$$(0.2) \qquad \mathcal{L}\{\frac{d}{dt}P_0(t)\} = s\pi_0(s) - 1 = -\sum_{i=1}^{N} \frac{\gamma_i}{s+\gamma_i}\,\underline{1}_0^T \underline{\underline{J}}_i\,\underline{1}_0$$

where $\underline{1}_0^T = (1,0,\ldots,0)$. Hence the relaxation time of $J(t)$ is the reciprocal of the magnitude of the first singularity of $s\pi_0(s)$ on the negative real axis.

In a recent paper by Keilson, Machihara and Sumita [7], this relaxation time has been extended to ergodic M/G/1 systems. Let $N(t)$ be the number of customers in system at time t and define $E(t) = P[N(t) = 0 \mid N(0) = 0]$ with the Laplace transform $\varepsilon(s) = \mathcal{L}\{E(t)\}$. Since the system is assumed to be ergodic, one has $\lim_{t\to\infty} E(t) = \lim_{s\to 0+} s\varepsilon(s) = e_0 > 0$. If the system is exponentially ergodic [4, 5] so that $e^{\theta t}(E(t) - e_0) < M$ for some $\theta > 0$ and for sufficiently large t, then there exists $\xi^* < 0$ such that

$$(0.3) \qquad \xi^* = \inf\{\xi : s\varepsilon(s) \text{ is regular in } D(\xi)\}$$

where $D(\xi) = \{s \mid Re(s) > \xi\}$. The natural candidate for the relaxation time is then

$$(0.4) \qquad T_{\text{rel}} = |\xi^*|^{-1}.$$

The relaxation time T_{rel} of (0.4) is directly related to the spectral structure of $\sigma_{BP}(s) = E[e^{-sT_{BP}}]$ where T_{BP} denotes the server busy period. From the Takács functional equation, one has

$$(0.5) \qquad \sigma_{BP}(s) = \alpha(s + \lambda - \lambda\,\sigma_{BP}(s))$$

where λ is the Poisson arrival intensity, $\alpha(s) = E[e^{-sT}]$, and T is the random service time. On the other hand, it can be readily seen that $\frac{d}{dt}E(t) = -\lambda E(t) + \lambda \int_0^t E(t-x)s_{BP}(x)dx$, and hence $\varepsilon(s) = (s + \lambda - \lambda\,\sigma_{BP}(s))^{-1}$. Combining this with (0.5), one obtains

$$(0.6) \qquad 1 - s\varepsilon(s) = \frac{\rho\sigma_{BP}^*(s)}{(1-\rho) + \rho\sigma_{BP}^*(s)}.$$

Here $\sigma_{BP}^*(s) = (1 - \sigma_{BP}(s))/sE[T_{BP}]$. It has been shown in [7] that under certain conditions all finite singular points of $\sigma_{BP}(s)$, and hence those of $\sigma_{BP}^*(s)$ are branch points. Then the denominator of (0.6) has no zeros. Consequently ξ^* of (0.3) also satisfies

$$(0.7) \qquad \xi^* = \inf\{\xi : \sigma_{BP}(s) \text{ is regular in } D(\xi)\},$$

i.e., ξ^* is also the first singularity of $\sigma_{BP}(s)$ on the negative real axis. In [7], the regularity structure of $\sigma_{BP}(s)$ has been studied at length. The results have then been applied to analyze the relaxation time of M/G/1 systems.

It should be emphasized that for an M/G/1 system the passage time to the idle state from an initial state can be arbitrarily long, in contrast to the situation for a finite Markov chain. The relaxation described in this paper is that for a system starting in the idle state.

The goal of this paper is to present the relaxation time results for single server queueing systems in a concrete way oriented to the needs of telecommunication practice. To do so, we will describe the relaxation time of the following queueing systems:

(i) M/M/1 (K);
(ii) M/G/1 with hyperexponential service times;
(iii) M/G/1 with Erlang service times.

In Section 1, the main results of Keilson, Machihara and Sumita [7] are summarized for the reader's convenience. The relaxation time of M/M/1 (K) systems is then analyzed in Section 2. An M/M/1 (K) system is an ergodic time reversible finite Markov chain and its spectral structure has been found by Ledermann and Reuter [10]. Here we provide an independent derivation based on the monotone matrix approach of Keilson and Kester [6]. In Sections 3 and 4, the relaxation time of M/G/1 systems with hyperexponential and Erlang service times is discussed based on the results of [7]. Section 5 is devoted to numerical examples. The behavior of the relaxation time for the queueing systems above is explored numerically by varying underlying system parameters.

1. Spectral Structure of M/G/1 Systems

The key results of Keilson, Machihara and Sumita [7], which play an important role in subsequent sections of this paper will now be summarized. Consider an M/G/1 queueing system governed by Poisson arrival intensity λ and service time c.d.f. $A(\tau) = P[T_s \le \tau]$ having Laplace - Stieltjes transform $\alpha(s) = E[e^{-sT_s}] = \int_0^\infty e^{-s\tau} dA(\tau)$. The busy period and its transform are denoted by T_{BP} and $\sigma_{BP}(s) = E[e^{-sT_{BP}}]$ respectively. It is assumed that:

(AS1) The service time c.d.f. $A(\tau)$ is absolutely continuous with p.d.f. $a(\tau)$.

(AS2) The system is ergodic, i.e. $0 < \rho < 1$ where $\rho = \lambda E[T_s]$.

(AS3) There exists s^* such that $-\infty < s^* < 0$ and $\alpha(s)$ is regular in the right half plane $D(s^*)$ bounded by s^*.

(AS4) $\lim_{s \to s^*+} (1 + \lambda \frac{d}{ds} \alpha(s)) < 0$.

(AS5) $\alpha(s)$ is regular at $s = \infty$ and vanishes there.

When T has bounded support, $\alpha(s)$ is entire and $s^* = -\infty$. This must be treated as a limiting case.

The two functions

$$\varsigma(s) = s + \lambda - \lambda\, \sigma_{BP}(s) \tag{1.1}$$

and

$$\Psi(s) = s + \lambda\, \alpha(s) - \lambda \tag{1.2}$$

play an important role. In particular we note from (0.5) that the Takács equation takes the form

$$s = \Psi(\varsigma(s)). \tag{1.3}$$

The regularity structure of $\sigma_{BP}(s)$, established in [7] is described next.

Theorem 1.1

(a) As a function of real u, $\Psi(u)$ is strictly convex in the interval (s^*, ∞), and has a unique minimum point ς_0 with $s^* < \varsigma_0 < 0$.

(b) Let $s_0 = \Psi(\varsigma_0)$. Then $s_0 < 0$ and $s_0 = \inf\{\xi : \sigma_{BP}(s)$ is regular in $D(\xi)\}$.

(c) s_0 is a branch point of $\sigma_{BP}(s)$ with multiplicity two.

(d) Let P be the set of finite singular points of $\sigma_{BP}(s)$. Then every $s \epsilon P$ is a branch point of $\sigma_{BP}(s)$ and P is bounded.

The relaxation time of such M/G/1 systems in the sense of Theorem 1.1 is then defined as

$$T_{\text{rel}} = |s_0|^{-1}. \tag{1.4}$$

When the service time p.d.f. $a(\tau)$ is fixed, the relaxation time becomes a function of λ. To emphasize the dependence on λ, we write $T_{\text{rel}}(\lambda), \varsigma_0(\lambda), s_0(\lambda)$ and $\rho(\lambda)$. In [7], it has been shown that:

Theorem 1.2

Let $\mu_k = E[T_S^k], k = 1, 2, \ldots,$ and $\sigma^2 = \text{Var}[T_S]$.

(a) $T_{\text{rel}}(\lambda) = [\int_{\varsigma_0(\lambda)}^\infty h(\lambda, y) dy]^{-1}$ where $h(\lambda, y) = 1 - \lambda \int_0^\infty e^{-y\tau} \tau a(\tau) d\tau$.

(b) $T_{\text{rel}}(\lambda)$ is monotonically increasing in $\lambda, 0 < \lambda < \mu_1^{-1}$ and $T_{\text{rel}}(\lambda) \longrightarrow +\infty$ as $\lambda \longrightarrow \mu_1^{-1}$.

(c) $\frac{T_{\text{rel}}(\lambda)}{\mu_1} \sim \frac{2\rho(\lambda)}{(1-\rho(\lambda))^2} \left[1 + \frac{\sigma^2}{\mu_1^2} + \frac{1-\rho(\lambda)}{\rho(\lambda)^2} \cdot \frac{\mu_3}{\mu_1 \mu_2}\right]$ as $(1 - \rho(\lambda)) \to 0+$ for fixed service time distribution.

Let X_1 and X_2 be positive random variables with p.d.f.'s $a_1(\tau)$ and $a_2(\tau)$ respectively. Local ordering (a refinement of stochastic ordering) of X_1 and X_2 is then defined by (see e.g. Keilson and Sumita [8])

$$X_1 \prec_\ell X_2 \overset{\text{def}}{\Leftrightarrow} \frac{a_1(\tau)}{a_2(\tau)} \downarrow \text{ in } \tau \text{ where defined.} \tag{1.5}$$

The following theorem has been also given in [7].

Theorem 1.3

Let T_{rel_1} and T_{rel_2} be the relaxation times of two M/G/1 systems with the service time T_{S_1} and T_{S_2} respectively having the same arrival intensity λ. If $T_{S_1} \prec_\ell T_{S_2}$, then $T_{rel_1} \leq T_{rel_2}$.

2. The Relaxation Time of M/M/1 (K) Systems

Consider an M/M/1 (K) system with Poisson arrivals of rate λ, service rate μ, and system capacity K. Let $N(t)$ be the number of customers in system at time t and let $p_{mn}(t) = P[N(t) = n \mid N(0) = m]$. Clearly $N(t)$ is an ergodic time reversible Markov chain on $\mathcal{N} = \{0, 1, 2, \ldots, K\}$ and the transition probability matrix $\underline{\underline{P}}(t) = [p_{mn}(t)]$ has a spectral representation as in (0.1). For M/M/1 (K) systems, the decay rates γ_i $i = 1, 2, \ldots, K$ can be explicitly obtained. One has

$$\gamma_i = (\sqrt{\lambda} - \sqrt{\mu})^2 + 2\sqrt{\lambda\mu}\left\{1 - \cos\left(\frac{\pi i}{K+1}\right)\right\}, 1 \leq i \leq K. \tag{2.1}$$

This result was first found by Ledermann and Reuter [10]. A simple independent derivation is provided here.

Let $\nu = \lambda + \mu$. Following the uniformization procedure (see, e.g. Keilson [3]), one sees that

$$\underline{\underline{P}}(t) = e^{-\nu t(\underline{\underline{I}}_K - \underline{\underline{a}}_K)} = \sum_{k=0}^{\infty} e^{-\nu t}\frac{(\nu t)^k}{k!}\underline{\underline{a}}_K^k \tag{2.2}$$

where $\underline{\underline{I}}_K$ is the identity matrix of order $(K+1)$ and $\underline{\underline{a}}_K$ is a stochastic matrix of order $(K+1)$ given by

$$\underline{\underline{a}}_K = \begin{pmatrix} \mu/\nu & \lambda/\nu & & & \underline{0} \\ \mu/\nu & 0 & \lambda/\nu & & \\ & \ddots & \ddots & \ddots & \\ & & \mu/\nu & 0 & \lambda/\nu \\ \underline{0} & & & \mu/\nu & \lambda/\nu \end{pmatrix}. \tag{2.3}$$

Let $\lambda_i(\underline{\underline{a}}_K)$ be the eigenvalues of $\underline{\underline{a}}_K$, $i = 1, \ldots, K$, excluding the trivial eigenvalue 1. One then sees easily from (0.1) and (2.2) that

$$\gamma_i = \nu(1 - \lambda_i(\underline{\underline{a}}_K)), \quad i = 1, 2, \ldots K. \tag{2.4}$$

To find $\lambda_i(\underline{\underline{a}}_K)$, we let $\underline{\underline{t}}_K$, as in Keilson and Kester [6], be the square lower triangular matrix of order $(K+1)$ with 1's on and below the main diagonal and 0's elsewhere. Then $\underline{\underline{t}}_K$ is nonsingular and its inverse is $\underline{\underline{t}}_K^{-1}$ with 1's on the main diagonal, elements -1 on the adjacent diagonal below the main diagonal, and 0's elsewhere. One then finds that $\underline{\underline{t}}_K^{-1}\,\underline{\underline{a}}_K\,\underline{\underline{t}}_K$ has the block form

$$\underline{\underline{t}}_K^{-1}\,\underline{\underline{a}}_K\,\underline{\underline{t}}_K = \begin{pmatrix} 1 & \vdots & \underline{c}^T \\ \cdots & \vdots & \cdots\ \cdots\ \cdots \\ \underline{0} & \vdots & \underline{\underline{\delta}}_K \end{pmatrix} \tag{2.5}$$

where $\underline{\underline{\delta}}_K$ is tridiagonal with

$$\underline{\underline{\delta}}_K = \begin{pmatrix} 0 & \lambda/\nu & & & \underline{0} \\ \mu/\nu & 0 & \lambda/\nu & & \\ & \ddots & \ddots & \ddots & \\ & & \mu/\nu & 0 & \lambda/\nu \\ \underline{0} & & & \mu/\nu & 0 \end{pmatrix}. \tag{2.6}$$

Let $\underline{\underline{D}} = \text{diag}\{\theta, \theta^2, \ldots, \theta^K\}$ be the diagonal matrix of order K whose j-th diagonal element is θ^j where $\theta = \sqrt{\mu/\lambda}$. The matrix $\underline{\underline{\delta}}_K$ can then be converted to a symmetric matrix via the similarity transform

$$\underline{\underline{\delta}}_K^* = \underline{\underline{D}}^{-1}\,\underline{\underline{\delta}}_K\,\underline{\underline{D}} = \frac{\sqrt{\lambda\mu}}{\nu}\underline{\underline{B}}_K \tag{2.7}$$

where

$$\underline{\underline{B}}_K = \begin{pmatrix} 0 & 1 & & & \underline{0} \\ 1 & 0 & 1 & & \\ & \ddots & \ddots & \ddots & \\ & & 1 & 0 & 1 \\ \underline{0} & & & 1 & 0 \end{pmatrix}. \tag{2.8}$$

Since eigenvalues are invariant under a similarity transform, the eigenvalues of $\underline{\underline{a}}_K$ are those of $\underline{\underline{B}}_K$ multiplied by the constant $\sqrt{\lambda\mu}/\nu$. Let $g_{K+1}(x) = \det(\underline{\underline{B}}_K - x\underline{\underline{I}}_K)$ be the characteristic polynomial of $\underline{\underline{B}}_K$, with $g_0(x) \stackrel{\text{def}}{=} 0$ and $g_1(x) \stackrel{\text{def}}{=} 1$. One then sees that

$$g_{K+2}(x) = -x\, g_{K+1}(x) - g_K(x)\ , \ K \geq 0. \tag{2.9}$$

Hence for the generating function $\gamma(u,x) = \sum_{K=0}^{\infty} g_K(x)u^K$, one obtains

$$\gamma(u,x) = u \cdot (u^2 + xu + 1)^{-1}. \tag{2.10}$$

This, in turn, implies that

$$g_{K+1}(x) = S_K(-x)\ , \ K \geq 0\ , \tag{2.11}$$

where $S_K(x)$ is the Chebyshev polynomial of the second kind, see e.g. Abramowitz and Stegun [1]. If one denotes the j-th zero of $g_{K+1}(x)$ by $x_j^{(K)}$, $1 \leq j \leq K$, one has

$$x_j^{(K)} = -2\cos\left(\frac{\pi j}{K+1}\right)\ , \ 1 \leq j \leq K. \tag{2.12}$$

It should be noted that $x_{K+1-j}^{(K)} = -x_j^{(K)}$. By letting $i = K+1-j$, one finds from (2.4) that

$$\gamma_i = \nu\left\{1 - \frac{\sqrt{\lambda\mu}}{\nu} \cdot 2\cos\left(\frac{\pi i}{K+1}\right)\right\} \tag{2.13}$$

which coincides with (2.1).

We note that, when $K = 1$, $\gamma_1 = \nu$. As $K \to \infty$, the spectrum becomes more dense. At the limit, the spectrum is continuous, extending over the interval $[-(\sqrt{\lambda} + \sqrt{\mu})^2, -(\sqrt{\lambda} - \sqrt{\mu})^2]$, see [10]. The next theorem is immediate from (2.1).

Theorem 2.1

Let $T_{rel}(\lambda,\ \mu,\ K)$ be the relaxation time of an M/M/1 (K) system with Poisson arrival rate λ, service rate μ and system capacity K. Then

$$T_{rel}(\lambda,\ \mu,\ K) \tag{2.14}$$
$$= \frac{1}{(\sqrt{\lambda}-\sqrt{\mu})^2 + 2\sqrt{\lambda\mu}\left\{1-\cos\left(\frac{\pi}{K+1}\right)\right\}}.$$

It should be observed that $T_{rel}(\lambda,\ \mu,\ K)$ is monotonically increasing in K, and is symmetric in λ and μ. In particular, $T_{rel}(\lambda,\ \mu,\ 1) = 1/\nu$ and, if $\rho = \lambda/\mu < 1$, $T_{rel}(\lambda,\ \mu,\ \infty) = 1/(\sqrt{\lambda}-\sqrt{\mu})^2$. The system capacity K has its greatest influence on the relaxation time when λ is close to μ. For $\lambda = \mu = \eta$, one has

$$T_{rel}(\eta,\eta,K) = \left[2\eta\left\{1-\cos\left(\frac{\pi}{K+1}\right)\right\}\right]^{-1} \tag{2.15}$$
$$\sim \frac{(K+1)^2}{\eta\pi^2} \text{ as } K \to \infty.$$

3. The Relaxation Time of M/G/1 Systems with Hyperexponential Service Times

The results described in Section 1 can be developed more explicitly when service time distributions belong to certain classes. In this section M/G/1 queueing systems with hyperexponential service time distributions are discussed. Service time distributions of Erlang type will be treated in Section 4.

The hyperexponential case is of special importance since the confluence of N different M/M streams with parameter sets (λ_i, μ_i) is equivalent to an M/H stream with $\lambda = \sum_{i=1}^{N} \lambda_i$, $a_T(\tau) = \sum_{i=1}^{N} \frac{\lambda_i}{\lambda}\mu_i e^{-\mu_i\tau}$.

Let the service time p.d.f. $a(\tau)$ be of the form

$$a(\tau) = \sum_{i=1}^{N} p_i\ \theta_i\ e^{-\theta_i\tau} \tag{3.1}$$

where $p_i \geq 0, \sum_{i=1}^{N} p_i = 1$ and $0 < \theta_1 < \cdots < \theta_N < \infty$. The Laplace transform $\alpha(s)$ is then given by

$$\alpha(s) = \sum_{i=1}^{N} p_i \frac{\theta_i}{s+\theta_i} \tag{3.2}$$

so that $s^* = -\theta_1$. The function $\Psi(s)$ of (1.2) takes the form

$$\Psi(s) = s + \lambda \sum_{i=1}^{N} p_i \cdot \frac{\theta_i}{s+\theta_i} - \lambda. \tag{3.3}$$

To find the relaxation time, one has to solve the equation $\frac{d}{ds}\Psi(s) = 0$ in the interval $(s^*,0)$, i.e.,

$$\frac{d}{ds}\Psi(s) = 1 - \lambda\sum_{i=1}^{N} p_i \frac{\theta_i}{(s+\theta_i)^2} = 0. \tag{3.4}$$

The solution ς_0 in $(s^*,0)$ may be found via the bisection method. The relaxation time T_{rel} is then given by

$$T_{rel} = |s_0|^{-1}\ ;\ s_0 = \Psi(\varsigma_0). \tag{3.5}$$

We have seen in Theorem 1.2(b) that the relaxation time of an M/G/1 system increases as λ increases when the service time distribution is fixed. We have also seen in Theorem 1.3 that for fixed λ the relaxation time increases as the service time distribution increases in local ordering. In what follows, we elaborate this result further in the context of hyperexponential service times.

Theorem 3.1

Consider two M/G/1 systems having the same arrival intensity λ and the service time H_1 and H_2 with p.d.f. $h_1(\tau) = \sum_{i=1}^{N} p_i\ \theta_i\ e^{-\theta_i\tau}$ and $h_2(\tau) = \sum_{i=1}^{N} q_i\ \theta_i\ e^{-\theta_i\tau}$ respectively. The corresponding relaxation times are denoted by T_{rel_1} and T_{rel_2}. If $p_i/q_i \leq p_{i+1}/q_{i+1}$, $i = 1, 2, \cdots, N-1$, then $T_{rel_1} \leq T_{rel_2}$.

Proof

Let E_θ be an exponential random variable with p.d.f. $a_\theta(\tau) = \theta e^{-\theta\tau}$. It can be readily seen that E_θ decreases locally as θ increases, i.e., $\theta_1 < \theta_2 \Rightarrow E_{\theta_1} \succ_\ell E_{\theta_2}$. The condition of the theorem implies that $p \succ_\ell q$. Hence from Theorem 4.5 of Keilson and Sumita [8] we see that $H_1 \prec_\ell H_2$. The theorem now follows from Theorem 1.3. ■

Theorem 3.1 states that if the mixing probabilities are locally ordered, so are the random service times. The relaxation time is then increased accordingly. We next show that the relaxation time is decreased if the spectral interval of the service time is shifted toward the right enough to leave no overlap.

Theorem 3.2

Consider the two M/G/1 systems in Theorem 3.1 where $h_1(\tau) = \sum_{i=1}^{M} p_i\ \alpha_i\ e^{-\alpha_i\tau}$ and $h_2(\tau) = \sum_{j=1}^{N} q_j\ \beta_j\ e^{-\beta_j\tau}$. If $0 < \alpha_i < \beta_j$ for $1 \leq i \leq M$ and $1 \leq j \leq N$, then $T_{rel_1} \geq T_{rel_2}$.

Proof

Let $H_1 = W_1\ E_1$ and $H_2 = W_2\ E_2$, where W_1 and W_2 are mixing variates independent of the exponential random variables E_1 and E_2 of mean one. Then $W_1 \succ_\ell W_2$ so that $H_1 \succ_\ell H_2$ [8]. ■

When the service time density is completely monotone e.g. is in the subclass of hyperexponential densities, the first singularity of $\sigma_{BP}(s)$ on the negative real axis and indeed the entire Riemann structure of $\sigma_{BP}(s)$ can be found. It has been shown in Keilson [2] that for completely monotone service time densities the busy period density $s_{BP}(\tau)$ is also completely monotone with a representation $s_{BP}(\tau) = \int_A^B \theta\, e^{-\theta\tau} w(\theta) d\theta$. For our hyperexponential service time density, one has $0 < A$ and $B < \infty$. The functions $\varepsilon(s)$ and $\sigma_{BP}(s)$ are regular at all s away from a branch cut running along the negative s- axis from $-B$ to $-A$. The relaxation time is $T_{\mathrm{rel}} = A^{-1}$. The reader is referred to [7] for more detailed discussions.

4. The Relaxation Time of M/G/1 Systems with Erlang Service Times

Consider next M/G/1 queueing systems with Erlang service time distribution, conventionally denoted by $M/E_k/1$. Here E_k is the sum of k independent and identical exponential random variables each with parameter $k\theta$ so that $E[E_k] = 1/\theta$. The distribution function $A_k(\tau)$ and the Laplace transform $\alpha_k(s)$ are given by

$$A_k(\tau) = 1 - \sum_{r=0}^{k-1} e^{-k\theta\tau}\frac{(k\theta\tau)^r}{r!} \tag{4.1}$$

and

$$\alpha_k(s) = \left(\frac{k\theta}{s+k\theta}\right)^k. \tag{4.2}$$

Since $\Psi(s) = s+\lambda\left(\frac{k\theta}{s+k\theta}\right)^k - \lambda$, the equation $\frac{d}{ds}\Psi(s) = 0$ can be explicitly solved. It is found that the solutions of this equation are a set of $(k+1)$ points uniformly distributed over a circle in the complex plane. These map into $(k+1)$ other points also distributed about a circle which are branch points for $\sigma_{BP}(s)$. One then has

$$T_{\mathrm{rel}}(k) = \left[k\theta + \lambda - (k+1)(\lambda\mu^k)^{\frac{1}{k+1}}\right]^{-1}. \tag{4.3}$$

It has been shown in [7] that $T_{\mathrm{rel}}(k)$ is strictly decreasing in $k, k = 1,2,\cdots$. Hence among $M/E_k/1$ queueing systems with fixed mean service time $1/\theta$, M/M/1 has the largest relaxation time and M/D/1 the smallest. Note that M/D/1 lies in the closure of our class since $\lim_{k\to\infty}\alpha_k(s) = e^{-s/\theta}$.

5. Numerical Exploration of Relaxation Times

In this section, the relaxation time of the single server queueing systems discussed above is explored numerically. For convenience, a hyperexponential random variable with p.d.f. $a(\tau)$ given in (3.1) is denoted by $H(\underline{p}, \underline{\theta})$ where $\underline{p} = (p_1, \cdots, p_N)$ and $\underline{\theta} = (\theta_1, \cdots, \theta_N)$. Similarly an Erlang random variable having the Laplace transform $\alpha_k(s)$ of (4.2) is deonted by $E_k(\theta)$. All figures are given at the end of this section.

In Figures 5.1 (a) and (b), the relaxation times of M/M/1 (K) systems are plotted as a function of K, when $\mu = 1$ is fixed and $\lambda = 0.1$, 0.25, 0.5, 0.75 and 0.9. It can be seen that the relaxation time increases as K or λ increases when the other two parameters are fixed. For $\lambda = 0.1$ or 0.25, the relaxation time converges to its limiting value quickly after an intial period of rise, say $1 \le K \le 15$. The effect of K on the relaxation time becomes more significant as λ increases. When $\lambda = 0.9$, the relaxation time is still growing even at $K = 250$.

In Figure 5.2, the relaxation times of M/G/1 systems with service times $H_i(\underline{p}_i, \underline{\theta}), 1 \le i \le 3$, are depicted for $0.1 \le \lambda \le 1.2$, where $\underline{p}_1 = (0.1, 0.2, 0.3, 0.4)$, $\underline{p}_2 = (0.25, 0.25, 0.25, 0.25)$, $\underline{p}_3 = (0.4, 0.3, 0.2, 0.1)$ and $\underline{\theta} = (1, 2, 3, 4)$. We note that $\underline{p}_1 \succ_\ell \underline{p}_2 \succ_\ell \underline{p}_3$ and the relaxation times are ordered as shown in Theorem 3.1. Figure 5.3 illustrates Theorem 3.2 numerically. Here three service times are chosen to be $H_i(\underline{p}, \underline{\theta}_i)$, $4 \le i \le 6$, where $\underline{p} = (0.5, 0.5)$, $\underline{\theta}_4 = (2, 2.1)$, $\underline{\theta}_5 = (2.2, 2.3)$ and $\underline{\theta}_6 = (2.4, 2.5)$. Figure 5.4 shows an example where the relaxation times of two M/G/1 systems with hyperexponential times can cross over each other as λ changes. Here the two service times $H_i(\underline{q}, \underline{\theta}_i)$, $i = 7,8$, are chosen with $\underline{q} = (\frac{1}{3}, \frac{1}{3}, \frac{1}{3})$, $\underline{\theta}_7 = (1, 3, 5)$ and $\underline{\theta}_8 = (1.5, 1.6, 1.7)$. One finds that :

X	$E[x]$	$\mathrm{Var}[x]$	$E[x]\left(1+\frac{\mathrm{Var}[x]}{E^2[x]}\right)$
H_7	0.511	0.506	1.501
H_8	0.627	0.395	1.257

Since $E[H_7] < E[H_8]$, the relaxation time for H_8 dominates that for H_7 when λ is small. As the systems approach saturation, however, this inequality is reversed. This is so because the term for the first order asymptotic expansion of the relaxation time associated with H_7 is larger than that associated with H_8. (See Theorem 1.2 (c).) It is again observed in Figures 5.2 through 5.4 that the relaxation time increases as λ increases.

In Figures 5.5 (a) and (b), the relaxation times of M/G/1 systems with Erlang service times $E_\theta(k)$ are plotted as a function of k when $\theta = 1$. The five curves correspond to $\lambda = 0.1$, 0.25, 0.5, 0.75 and 0.9. In all cases, the relaxation time is monotonically decreasing in k and converges to the relaxation time of the corresponding M/D/1 system rather quickly.

Finally Figures 5.6 and 5.7 demonstrate the asymptotic approximation of the relaxation time for two M/G/1 systems. In Figure 5.6, the service time is $H(\underline{p}, \underline{\theta})$ where $\underline{p} = (0.25, 0.25, 0.5)$ and $\underline{\theta} = (0.5, 1, 2)$. Only the first order approximation of Theorem 1.2 (c) is employed here. In Figure 5.7, the service time is $E_\theta(k)$ with $\theta = 1$ and $k = 5$. Both the first order and the second order approximations are tested. Note that the mean service time is one for the two M/G/1 systems. In each case the first order approximation provides a relative error of around 10% at $\lambda = 0.8$. In Figure 5.7, the second order approximation becomes better than the first order approximation for $\lambda \ge 0.6$, and provides a relative error of 3% at $\lambda = 0.8$.

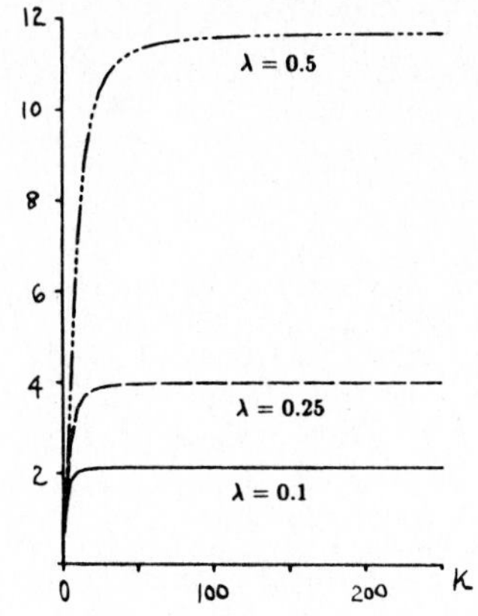

Figure 5.1(a). T_{rel} of M/M/1(K).

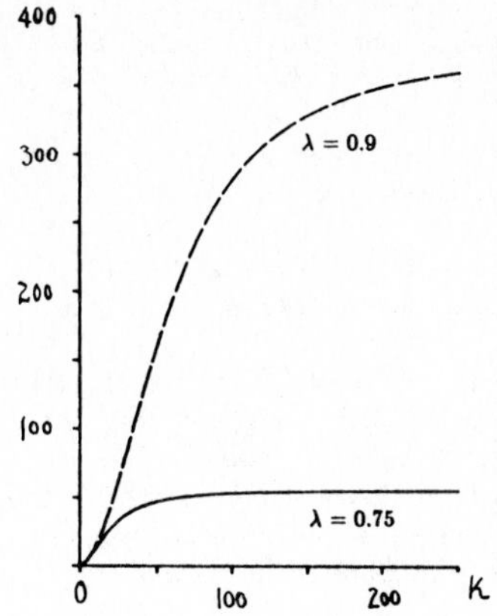

Figure 5.1(b). T_{rel} of M/M/1(K).

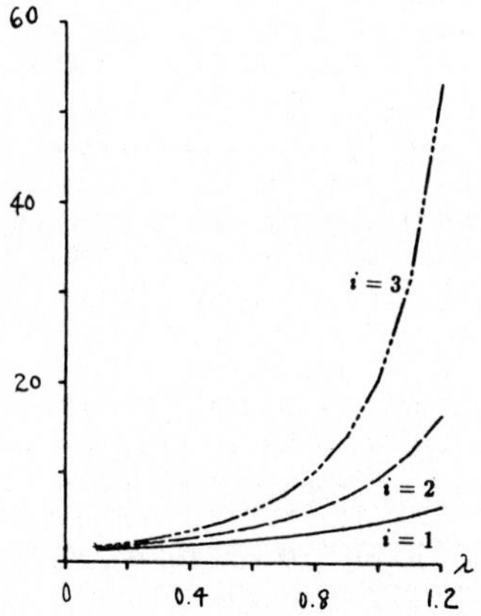

Figure 5.2. T_{rel} **of M/G/1 with Hyperexponential Service Times** $H_i(\underline{p}_i, \underline{\theta})$, $i = 1,2,3$.

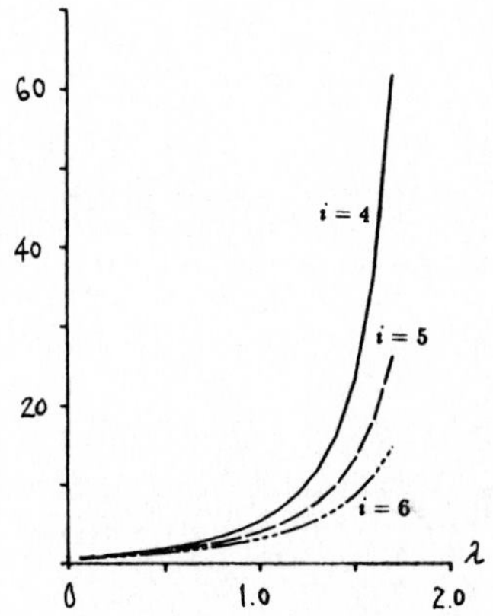

Figure 5.3. T_{rel} **of M/G/1 with Hyperexponential Service Times** $H_i(\underline{p}, \underline{\theta}_i)$, $i = 4,5,6$.

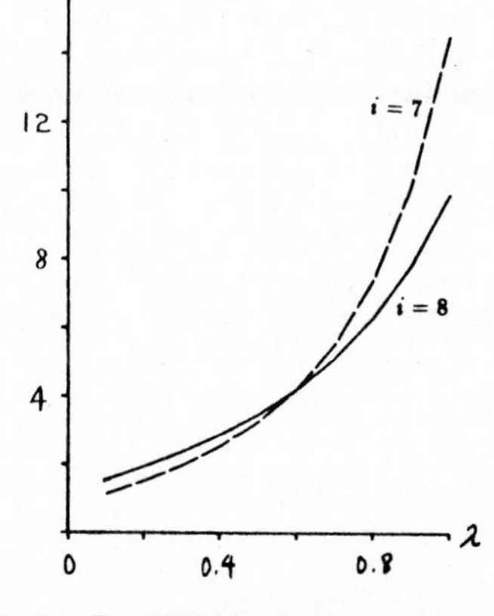

Figure 5.4. T_{rel} **of M/G/1 with Hyperexponential Service Times** $H_i(\underline{q}, \underline{\theta}_i)$, $i = 7,8$.

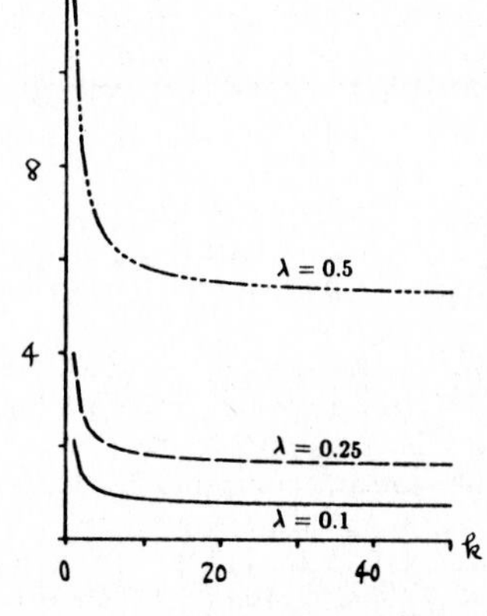

Figure 5.5(a). T_{rel} **of M/G/1 with Erlang Service Times** $E_\theta(k)$.

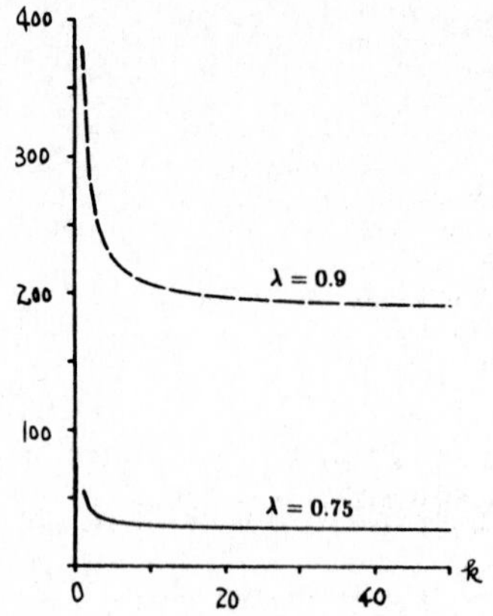

Figure 5.5(b). T_{rel} of M/G/1 with Erlang Service Times $E_\ell(k)$.

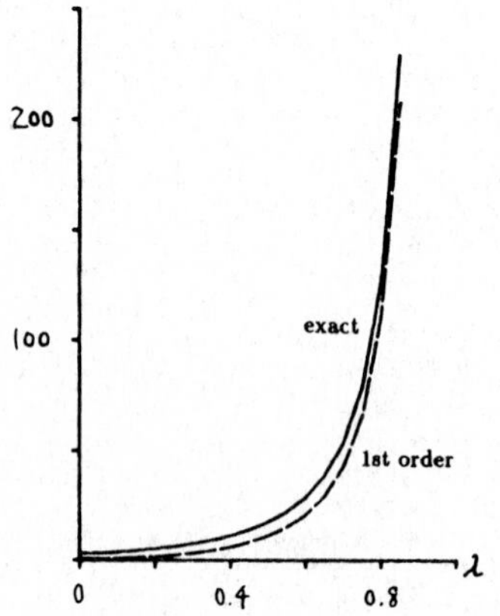

Figure 5.6. Asymptotic Approximation of T_{rel} with $H(p, \ell)$.

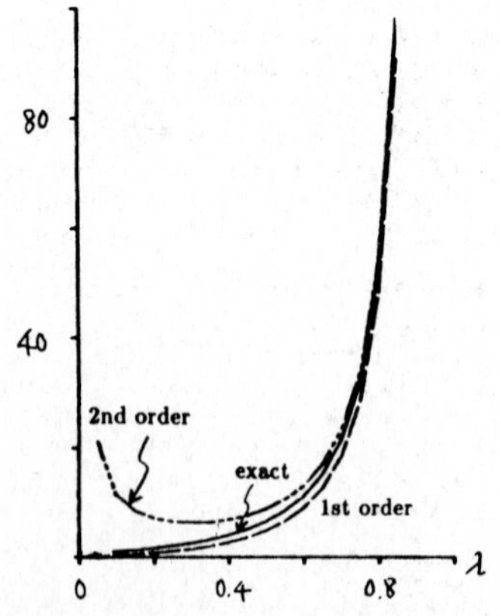

Figure 5.7. Asymptotic Approximation of T_{rel} with $E_\ell(k)$.

Acknowledgement

The authors wish to thank Caron Clair for her hard work and editorial contributions. Some technical support was provided by Masaaki Kijima and Yasushi Masuda.

References

[1] M. Abramowitz and I.A. Stegun (1965), *Handbook of Mathematical Functions*, Dover, New York.

[2] J. Keilson (1978), "Exponential Spectra as a Tool for the Study of Server Systems with Several Classes of Customers," Journal of Applied Probability, 15, pp. 162-170.

[3] J. Keilson (1979), *Markov Chain Models - Rarity and Exponentiality*, Springer-Verlag, New York.

[4] J. Keilson and H. Callaert (1978), "On Exponential Ergodicity and Spectral Structure for Birth-Death Processes I," *Stochastic Processes and Their Applications*, Vol. 1, No. 2, pp. 187-216.

[5] J. Keilson and H. Callaert (1979), " on Exponential Ergodicity and Spectral Structure for Birthe-Death Processes II," *Stochastic Processes and Their Applications*, Vol. 1, No. 3, pp. 217-235.

[6] J. Keilson and A. Kester (1977), " Monotone Matrices and Monotone Markov Processes, " *Stochastic Processes and Their Applications*, Vol. 5, No. 3, pp. 231-241.

[7] J. Keilson, F. Machihara and U. Sumita (1984), "Spectral Structure of M/G/1 Systems - Asymptotic Behavior and Relaxation Time - ," Working Paper Series QM8414, Graduate School of Management, University of Rochester.

[8] J. Keilson and U. Sumita (1982), "Uniform Stochastic Ordering and Related Inequalities," *The Canadian Journal of Statistics*, Vol. 10, No. 3, pp. 181-198.

[9] F.P. Kelly (1979), *Reversibility and Stochastic Networks*, John Wiley & Sons, New York.

[10] W. Ledermann and G.E.H. Reuter (1954), " Spectral Theory for the Differential Equations of Simple Birth-and-Death Processes," *Philos. Trans. Royal Soc. A. 246*, pp. 321-369.

TELETRAFFIC ISSUES in an Advanced Information Society
ITC-11
Minoru Akiyama (Editor)
Elsevier Science Publishers B.V. (North-Holland)

APPROXIMATIONS FOR BURSTY (AND SMOOTHED) ARRIVAL QUEUEING DELAYS BASED ON GENERALIZED PEAKEDNESS

A. E. ECKBERG, Jr.

AT&T - Bell Laboratories
Holmdel, New Jersey, USA 07733

ABSTRACT

The *peakedness* characterization of a teletraffic stream has long been recognized as an extremely useful one for the analysis of loss systems. In this paper, we present a preliminary investigation into the utility of this characterization for the analysis of delay systems as well. Restricting attention, for simplicity, to single server queueing systems G/G/1 with FCFS service, we derive an explicit approximation to the waiting time distribution, parameterized by quantities which can be easily computed from the *peakedness functional* of the arrival stream and the first two moments of the service time. The peakedness functional characterization allows arbitrary stationary arrival streams to be treated, and avoids focusing attention on aspects of arrival streams which in many circumstances may be misleading, such as interarrival time statistics. The accuracy of this approximation is assessed by several examples, and rules of thumb are given predicting its degree of accuracy, or lack thereof, in potential applications.

1. INTRODUCTION AND SUMMARY

1.1 Introduction

The performance analysis of modern telecommunication systems invariably involves predicting queueing delays at various points in the system in terms of either measured or assumed traffic and system parameters. Examples are packet switching systems for voice and/or data where transport and switching delays can significantly affect the viability of service. More often than not, system complexity causes the arrival streams at queues to have complex structures which can be expected to have non-negligible effects on the magnitudes of queueing delays. For example, both data and packetized voice traffic sources often are "bursty," and this burstiness tends to increase delays; on the other hand, data links often "smooth" a packet process, and this smoothing tends to decrease delays at subsequent switches.

While recent advances in the formulation and analytic solution, as well as effective numerical solutions, of quite elaborate queueing models allow many complex systems to be almost exactly modeled and analyzed, the application of such models is often inappropriate due to the time and effort that is required. Moreover, the dimension of the parameter space associated with such a model is often so large that it is difficult, if not impossible, to estimate appropriate values of model parameters from traffic measurements.

Thus, there is the need for simple modeling and approximation approaches that capture the most important effects of traffic characteristics on queueing delays. Such simple methods would prove extremely useful in deriving "quick and dirty" approximations for the early performance assessment of alternative system designs, and in understanding performance sensitivities to various traffic characteristics so that appropriate traffic measurements can be planned. Another important benefit derived from simple modeling and approximation methods is new insight into decomposition approaches for the analysis of large systems.

In this paper we consider only cases where service requirements of individual arrivals at a service system are mutually independent, and also independent of the arrival stream itself. Thus it is meaningful to focus attention on characterizing the arrival stream, which can be mathematically modeled as a point process. Two basic approaches can be taken to describe such a process. In one approach, attention is focused on the sequence of interarrival times, $\{T_i, i \geqslant 1\}$, separating successive arrivals to the system (if arrivals can occur in batches, then some of the T_i's may equal zero). In the second approach, attention is focused on the counting process, $\{N_t, t \geqslant 0\}$, associated with the point process; N_t is a piece-wise constant random process which increases at arrival epochs by integer quantities of magnitude equal to the number of arrivals at those epochs.

These are clearly two equivalent descriptions of an arrival stream when a complete characterization is available. However, the differing focuses of these two descriptions give rise to different approximation philosophies when a complete characterization of an arrival stream is not available, e.g., when the nature of the stream can only be approximately described in terms of measured traffic quantities. For example, when attention is focused on interarrival times, it is inviting to assume that successive interarrival times are independent, i.e., that the stream is a renewal process, since this assumption tremendously reduces the complexity of the arrival stream characterization and the analysis of its interactions with service systems. However, in many cases this independence of interarrival times is not present, and to assume its presence will result in significant characterization inaccuracies, regardless of how accurately the interarrival time distribution may have been determined. The difference between these two approaches has been pointed out in [1], where partial characterization methods based on $\{T_i\}$ are called stationary-interval methods, while those based on $\{N_t\}$ are called asymptotic methods.

1.2 Peakedness as an arrival stream characterization

Peakedness is a traffic characteristic which was originally introduced to aid in the approximate analysis of complex loss systems. More recently, it has been shown in [2] that a generalized concept of peakedness is equivalent to a complete second order point process characterization of an arrival stream. No specific assumptions need be made about the structure of the stream (e.g., Poisson, renewal, orderly, etc.) except that it is stationary. The arrival stream is characterized in terms of a *peakedness functional*, $Z[\cdot]$, which takes *complementary holding time distributions* as arguments, and maps them into *peakedness values*. For a given complementary holding time distribution G, the peakedness value $Z[G]$ is defined as $\lim_{t \to \infty} Var[K_t]/E[K_t]$, where K_t is the number of busy servers in a fictitious infinite server group, with holding time distribution $1-G$, to which the arrival stream is hypothetically offered. The intuitive concept is that, if a given complementary holding time distribution characterizes the "reaction time" of the arrival stream with a system, then the resulting peakedness value is a potentially useful measure of stream variability with respect to that system. It is often convenient to restrict the domain of $Z[\cdot]$ to exponential holding time distributions, and this defines an *exponential peakedness function*, $z_{\exp}(\beta)$, mapping the holding time rate, β, i.e., the reciprocal of the mean holding time, to a peakedness value.

The reader is referred to [2] for a summary of the origins of the peakedness concept and for details on representations and properties of the peakedness functional used in this paper. Particularly noteworthy is the fact that the pair $(\lambda, Z[\cdot])$, where $\lambda = \lim_{t\to\infty} E[N_t]/t$ is the intensity of the arrival stream, provides a complete second order characterization of the counting process N_t. Equivalent second order characterizations are the pairs $(\lambda, V(\cdot))$, $(\lambda, k(\cdot))$, and $(\lambda, U(\cdot))$, where $V(\cdot)$ is the *variance-time curve*, $k(\cdot)$ is the *covariance density*, and $U(\cdot)$ is the *expectation function* of the counting process.

The characterization in terms of the expectation function is particularly relevant in many situations; this function is the generalization of the renewal function associated with a renewal process, and is defined as:

$$U(x) = \text{the expected number of arrivals following, and no later than a time } x \text{ from, an arbitrarily chosen arrival, for } x \geq 0$$
$$= 0, \text{ for } x < 0 \qquad (1.1)$$

(At first glance it may appear that $U(\cdot)$ provides only a *first order* characterization of the stream; however, it is due to the fact that time in (1.1) is being measured from an arbitrarily chosen arrival that $U(\cdot)$ provides second order information.) Defining $\hat{U}(\cdot)$ to be the Laplace-Stieltjes transform of $U(\cdot)$:

$$\hat{U}(s) = \int_{0-}^{\infty} e^{-sx} dU(x) \qquad (1.2)$$

it can be shown that the value of the exponential peakedness function at the holding time rate β is given by

$$z_{\exp}(\beta) = 1 + \hat{U}(\beta) - \lambda/\beta \qquad (1.3)$$

Finally, a parameter that is often relevant in describing an arrival stream is the quantity V/M, defined as the variance-to-mean ratio for the number of arrivals occurring in a very long interval of time, i.e., $V/M = \lim_{t\to\infty} Var[N_t]/E[N_t]$. It can be shown that

$$V/M = 2 \lim_{\beta\to 0} z_{\exp}(\beta) - 1 \qquad (1.4)$$

$$= 1 + 2 \lim_{\beta\to 0} [\hat{U}(\beta) - \lambda/\beta] \qquad (1.5)$$

(Equation (1.5) is an illustration of second order information contained in the expectation function.)

In many cases, it is of interest to estimate the value of $z_{\exp}(\beta)$, for an actual arrival stream, for a very small value of β; for such cases, (1.4) provides a convenient method. For larger values of β, (1.3) can be used, based on the following procedure for estimating $\hat{U}(\beta)$:

i. Set $i = 1$.

ii. Generate a pseudo-random integer-valued variable K_i with a "suitably large" mean value.

iii. Generate an exponentially distributed pseudo-random variable E_i with mean $1/\beta$.

iv. Allow K_i arrivals to occur, and select the next arrival. Set a timer to the value E_i, and record the number of subsequent arrivals that occur before the timer expires; denote this as $N(E_i)$.

v. Increment i, and loop back to step ii.

It can be easily seen that the sample mean of the quantities $N(E_i)$ generated via the above procedure will be an unbiased estimate of $\hat{U}(\beta)$, as long as the mean of the quantities K_i is sufficiently large (the K_i's simply provide a convenient method for "selecting an arbitrary arrival.").

1.3 The issue of appropriate holding time distributions for peakedness determination

We have stated that $(\lambda, Z[\cdot])$ provides a complete second order characterization of the arrival stream. However, except for those cases where the structure of the arrival stream is extremely well understood, it may be impossible to exactly determine the peakedness functional, i.e., the *mapping* $Z: G \to$ *peakedness value* $= Z[G]$; one must focus instead on determining one or more peakedness values directly. This is especially the case when measurements must be taken on an actual arrival stream. Moreover, if we are interested in using peakedness to approximately characterize the interaction of the arrival stream with a given service system, then it is clear that for many G's the resulting $Z[G]$ will bear little relation to the problem at hand. It is in part the determination of an appropriate G that makes the use of the peakedness concept somewhat of an art.

For loss systems there is an obvious G that makes intuitive sense; simply using the complementary distribution of the servers' holding times results in a peakedness value that may be used in such computational procedures as the equivalent random method [3] and Hayward's approximation [4]. However, for delay systems, the appropriate choice of G is not so obvious, due principally to the fact that G should approximately characterize the "reaction time" of the arrival stream with the delay system, and it is essentially this reaction time that we are attempting to determine by analyzing the delay system. Specific questions that need answers are:

- When there is access to only a single peakedness value, at what G should $Z[\cdot]$ be evaluated to provide maximum information relevant to quantifying delays?
- How can the entire peakedness functional $Z[\cdot]$, or the values $Z[G_i]$ at several G_i's, best be utilized in quantifying delays?
- How much information relevant to quantifying delays is contained in the peakedness functional, and how well can delays be approximated in terms of a single $Z[G]$ as compared with other simple delay approximations?

Addressing these questions for general delay systems, with arbitrary numbers of servers and arbitrary service disciplines, is well beyond the scope of this study. The objective of this paper is to address these questions for the simplest class of queues with general stationary input: G/G/1 queues with FCFS service discipline (we do *not* restrict attention to queues with renewal input, i.e., GI/G/1). It is hoped that insights gained from this simpler study will eventually impact on the issues of more general delay systems.

The essence of the peakedness-based delay approximation that will be derived below is that in many cases an appropriate choice for G is

$$G(x) = P[W > x | W > 0] \qquad (1.6)$$

where W is the waiting time in the queue. Thus, answers to the first and second questions above are that if only a single peakedness value is available, it should be with respect to the best *a priori* estimate of the conditional delay distribution (e.g., a *design objective* delay distribution may suffice), while if several peakedness values are available, they should be with respect to a set of complementary distributions that span likely delay distributions. The answer to the third question will be provided by some of the examples later in this paper.

1.4 Outline of the remainder of the paper

The remainder of this paper is organized as follows. In Section 2 we briefly consider the GI/M/1 queue, and show that: i) $z_{\exp}(\cdot)$ uniquely determines the waiting time distribution; and ii) knowledge of $z_{\exp}(\beta)$ provides a better approximation to the waiting time distribution than does knowledge of the interarrival time variance, whenever $\beta^{-1} \geq E[W|W>0]$. In Section 3 we derive an approximation to the waiting time distribution in terms of the function $z_{\exp}(\cdot)$ and the first two moments of service time. This approximation is then tested on several examples in Section 4, where it is shown that in many cases the approximation yields better results than would be obtained from complete knowledge of the interarrival time distribution and an assumption of renewal process arrivals. Finally, Section 5 contains some concluding remarks.

2. OBSERVATIONS ON THE GI/M/1 WAITING TIME

A convenient starting point in the investigation of the effects of peakedness on delays is the solution to the GI/M/1 queue (renewal input, exponential service time distribution), where the existence of an exponential delay distribution is well known:

$$P[W \leqslant x] = 1 - \omega\, e^{-\mu(1-\omega)x},\ x \geqslant 0 \tag{2.1}$$

where μ is the service rate and, with $\phi(\cdot)$ denoting the Laplace-Stieltjes transform of the interarrival time distribution, ω is the unique real number in the interval (0,1) which satisfies

$$\omega = \phi(\mu(1-\omega)) \tag{2.2}$$

Since for a renewal process the expectation function is just the renewal function, for which the Laplace-Stieltjes transform is known to be $\phi(\cdot)/[1-\phi(\cdot)]$, it is seen from (1.3) that

$$\phi(s) = \frac{\lambda + s[z_{\exp}(s) - 1]}{\lambda + s z_{\exp}(s)} \tag{2.3}$$

It follows that the waiting time distribution is just

$$P[W \leqslant x] = 1 - (1-\beta/\mu)\, e^{-\beta x},\ x \geqslant 0 \tag{2.4}$$

where β satisfies

$$\beta = \frac{\mu - \lambda}{z_{\exp}(\beta)} \tag{2.5}$$

Thus, knowledge of $z_{\exp}(\cdot)$ allows exact determination of the waiting time distribution in GI/M/1!

Now, if only a single peakedness *value*, $z_0 = z_{\exp}(\beta_0)$, were available, how well could the delay distribution be determined? It can be shown, e.g., using explicit results in [5], that the function $z_{\exp}(\cdot)$ can be bounded sharply (i.e., the bounds can be attained) in terms of λ, β_0, and z_0 as follows:

$$z_1(\beta;\lambda,\beta_0,z_0) \leqslant z_{\exp}(\beta) \leqslant z_2(\beta;\lambda,\beta_0,z_0),\ \text{for } 0 \leqslant \beta \leqslant \beta_0 \tag{2.6a}$$

$$z_2(\beta;\lambda,\beta_0,z_0) \leqslant z_{\exp}(\beta) \leqslant z_1(\beta;\lambda,\beta_0,z_0),\ \text{for } \beta \geqslant \beta_0 \tag{2.6b}$$

where, with x_0 being the unique root in $(0,\infty)$ to

$$x_0 = \frac{\lambda+\beta_0 z_0}{\lambda\beta_0}\left[1 - e^{-\beta_0 x_0}\right] \tag{2.7}$$

the functions $z_1(\cdot)$ and $z_2(\cdot)$ are

$$z_1(\beta;\lambda,\beta_0,z_0) = \frac{\lambda x_0}{1-e^{-\beta x_0}} - \lambda/\beta \tag{2.8}$$

$$z_2(\beta;\lambda,\beta_0,z_0) = \left\{1 - \left[\frac{\lambda+\beta_0(z_0-1)}{\lambda+\beta_0 z_0}\right]^{\beta/\beta_0}\right\}^{-1} - \lambda/\beta \tag{2.9}$$

These sharp bounds on $z_{\exp}(\cdot)$ will now yield sharp bounds on β, via (2.5), in terms of λ, β_0, and z_0. Clearly, a better approximation will result from knowledge of $z_{\exp}(\beta_0)$ the closer β_0 is to the quantity $\mu(1-\omega)$, the reciprocal of the conditional mean $E[W|W>0]$.

A set of sharp bounds analogous to (2.6) can be derived for $z_{\exp}(\cdot)$ in terms of λ and σ^2, the variance of the interarrival times. It is shown in [5] that these bounds are not as tight as those in (2.6), in particular for the range $\beta \geqslant \beta_0$. We thus conclude that in the GI/M/1 queue, a single peakedness value provides more information relevant to quantifying delays than does the variance of interarrival times, as long as the peakedness value results from a β_0 no larger than $1/E[W|W>0]$.

3. *A PEAKEDNESS-BASED DELAY APPROXIMATION FOR G/G/1 QUEUES WITH FCFS SERVICE*

In this section we derive an approximation for the delay distribution for G/G/1 queues with FCFS service in terms of the peakedness functional of the arrival stream. This approximation results from a series of simplifying assumptions, the first of which is that the complementary delay distribution is approximately exponential, i.e.

$$P[W>x] = \alpha\, e^{-\beta x},\ \text{for } x \geqslant 0 \tag{3.1}$$

This approximation is at least partly justified by the facts that i) with few exceptions (3.1) will be asymptotically true as $x \to \infty$, and ii) the range of x-values over which (3.1) is nearly correct increases as the server utilization, ρ, increases. We next introduce probabilistic arguments and further simplifying approximations to determine the parameters α and β.

First note that from (3.1)

$$\alpha/\beta = E[W] \tag{3.2a}$$

$$= \tau_r\, E[S_a] + \tau_1\, E[Q_a] \tag{3.2b}$$

$$= \tau_r\, P[W>0] + \tau_1\, E[Q_d] \tag{3.2c}$$

where S_a and Q_a denote the number of jobs in service (0 or 1) and in queue, respectively, just prior to an arbitrary arrival, Q_d denotes the number of jobs left in queue by an arbitrary job departing the queue (including those that do not wait in the queue), τ_1 is the mean service time of an arbitrary job, and τ_r is the mean remaining service time for the job in service at an arbitrary arrival if $S_a>0$. Equation (3.2c) follows from (3.2b) by the equivalence of the queue distributions at arrivals to, and departures from, the queue; see [3].

We now note that Q_d equals the number of arrivals to the queue during the waiting time of an arbitrary arrival, and introduce the approximation that

$$E[Q_d|W] = U(W) \tag{3.3}$$

where $U(\cdot)$ is the expectation function of the arrival process. For renewal arrivals this approximation is exact, and for many arrival processes and queueing systems this approximation will be quite good; however, there are situations (one will be illustrated in an example) where this estimate of $E[Q_d|W]$ is systematically biased, and thus will yield a poor approximation. (For example, for streams characterized by a slowly varying instantaneous arrival rate, some values of which may produce server utilizations close to or greater than one, for large values of W it will typically be that $E[Q_d|W] \gg U(W)$; also, for certain arrival streams with finite source characteristics, it may be that $E[Q_d|W] \ll U(W)$ for large values of W.) Accepting (3.3) as a reasonable approximation for most cases, we now have

$$E[Q_d] = E[U(W)] = \alpha\, \hat{U}(\beta) \tag{3.4}$$

from (3.1) and an integration by parts.

Combining (3.2c) and (3.4), and noting that $P[W>0]=\alpha$, we find that β can be determined via

$$\beta = \frac{1}{\tau_r + \tau_1 \hat{U}(\beta)} \tag{3.5}$$

or, by making use of (1.3),

$$\beta = \frac{\mu - \lambda}{\tau_r/\tau_1 + z_{\exp}(\beta) - 1} \tag{3.6}$$

where $\mu = 1/\tau_1$ is the service rate. Finally, we need an approximate expression for τ_r, and a simple approximation for which the accuracy tends to increase with increasing server utilization is just

$$\tau_r = \frac{\tau_2}{2\,\tau_1} \tag{3.7}$$

where τ_2 is the second moment of the service times; (3.7) is exact when the arrival stream is Poisson.

To determine α, we use the fact that

$$P[W>0] = 1 - \frac{1}{E[\#\ \textit{jobs served in a busy period}]} \tag{3.8}$$

Similarly to (3.4), we approximate

$$E[\#\ \textit{jobs served in a busy period}|B] = 1 + U(B) \tag{3.9}$$

where B denotes the duration of a busy period. If we were to assume that the busy periods are approximately exponentially distributed, with parameter γ, then

$$E[\#\ \textit{jobs served in a busy period}] = 1 + \hat{U}(\gamma) \tag{3.10}$$

Observing that for both the GI/M/1 and the M/G/1 queues the ratio of the mean duration of a busy period to the mean conditional waiting time is just τ_1/τ_r, where τ_r is given in (3.7), we are now motivated to use $\beta\tau_r/\tau_1$ for γ in (3.10). This leads to the following approximation for α:

$$\alpha = \frac{\hat{U}(\beta\tau_r/\tau_1)}{1+\hat{U}(\beta\tau_r/\tau_1)} \tag{3.11}$$

or

$$\alpha = 1 - \frac{\beta\tau_r/\tau_1}{\beta\tau_r z_{\exp}(\beta\tau_r/\tau_1)/\tau_1 + \lambda} \tag{3.12}$$

Thus, the final delay approximation is (3.1) with α and β computed via (3.6), (3.7), and (3.12). This approximation is exact for all GI/M/1 queues, since all of the approximating assumptions above are exact in these cases. Moreover, it can be seen that, while (3.1) is not exact for M/G/1 queues in general, it does result in exact values for $P[W>0]$ and $E[W]$ (recall that $z_{\exp}(\cdot)=1$ for Poisson arrivals).

4. *EXAMPLES OF THE PEAKEDNESS-BASED DELAY APPROXIMATION*

In this section we summarize conclusions drawn from several examples that have been used to test the accuracy of the delay approximation derived in Section 3. These examples by no means give a complete evaluation of the accuracy of the approximation, but do serve to illustrate cases where it can be expected to be good, and other cases where it cannot. The examples also serve as an illustration of how easily the exponential peakedness function, $z_{\exp}(\cdot)$, can be determined. To conserve space, numerical results are not given here; rather we summarize the conclusions we have drawn from examination of many numerical examples.

4.1 The MR(2)/M/1 queue

A fairly general class of stationary arrival processes is the class of Markov-renewal processes governed by the evolution of an n-state Markov chain, which we denote as the class of MR(n) processes. A typical such process is completely described by an $n\times n$ matrix of (usually defective) distribution functions:

$$A(x) = [A_{j,k}(x)] \tag{4.1}$$

where, if t_i denotes the ith arrival epoch, X_t denotes the Markov chain state at t, and $T_i=t_i-t_{i-1}$ denotes the ith interarrival time, then

$$A_{j,k}(x) = P[X_{t_{i+1}}=k,\ T_{i+1}\leqslant x \mid X_{t_i}=j;\ X_{t_m},\ m\leqslant i-1;\ T_m,\ m\leqslant i] \tag{4.2}$$

The ordinary renewal process is a special case where $n=1$; and by using an $n>1$ and appropriate $A_{j,k}$'s, various types of dependences between interarrival times may be modeled.

A quantity of basic interest is the matrix of Markov chain transition probabilities at arrival epochs:

$$P = [P_{j,k}],\quad P_{j,k} = \lim_{x\to\infty} A_{j,k}(x) \tag{4.3}$$

The invariant vector, π, of P, defined as

$$\pi = \pi P,\quad \sum_i \pi_i = 1 \tag{4.4}$$

is the vector of Markov chain state probabilities at an arbitrary arrival. It is convenient to define the matrix of Laplace-Stieltjes transforms of the components of A:

$$\Phi(s) = [\phi_{j,k}(s)],\quad \phi_{j,k}(s) = \int_{0-}^{\infty} e^{-sx}dA_{j,k}(x) \tag{4.5}$$

and the Laplace-Stieltjes transform of the *Markov-renewal kernel*

$$H(s) = [H_{j,k}(s)] = \Phi(s)[I-\Phi(s)]^{-1} \tag{4.6}$$

Although an MR(n) is in general not a renewal process, one can partially characterize it by its interarrival time distribution, which is easily seen to have the Laplace-Stieltjes transform

$$\phi(s) = \sum_{j,k} \pi_j P_{j,k}\, \phi_{j,k}(s) \tag{4.7}$$

Also, the expectation function of the process can be shown to have the Laplace-Stieltjes transform

$$\hat{U}(s) = \sum_{j,k} \pi_j H_{j,k}(s) \tag{4.8}$$

and the exponential peakedness function can be obtained via (1.3).

It is of interest to compare the quality of the peakedness-based delay approximation with an approximation that uses $\phi(s)$ in a GI/M/1 assumption, i.e., via (2.1) and (2.2) (the latter would be a typical approximation approach if one were concentrating on interarrival time statistics).

The solution of the MR(n)/M/1 queue has been known for some time, (see, e.g., [6]), and in these test examples we have focused on the special case where $n=2$, where it can be shown that the complementary delay distribution is of the form

$$P[W>x] = \alpha_0 e^{-\beta_0 x} + \alpha_1 e^{-\beta_1 x},\ x\geqslant 0 \tag{4.9}$$

Moreover, there is a well defined procedure for obtaining the quantities α_i and β_i.

A considerable number of MR(2)/M/1 examples have been considered numerically, and the above mentioned two approximations have been compared with the exact solutions, with the following conclusions derived:

- Uniformly over all examples considered, the peakedness-based approximation yields more accurate estimates of the mean waiting time and the asymptotic decay rate of the delay distribution that does the renewal-process-based approximation.
- For most examples considered, the peakedness-based approximation also yields a more accurate estimate of $P[W>0]$.
- The accuracy of the peakedness-based approximation for mean delay increases with the server utilization, and is typically within 10% of the true mean for server utilizations $\rho \geqslant .50$, and much better than this for $\rho \geqslant .75$.
- On the other hand, the amount of error in the renewal-process-based approximation may be arbitrarily large, depending on the characteristics of the MR(2)/M/1 queue being approximated.

4.2 The PCP/G(B)/1 queue

To investigate the quality of the approximation when the service times are not exponentially distributed, a PCP/G(B)/1 queue was considered. In this system, the arrivals constitute a Poisson Cluster Process (PCP) with clusters of arrivals being spawned by an underlying Poisson process, and each cluster consisting of a random number of equally spaced arrivals. Individual clusters will typically overlap. The PCP is characterized by: λ', the rate at which *clusters* originate; δ, the (deterministic) spacing between arrivals in a cluster; $g(\cdot)$, the probability generating function of the number of arrivals in a cluster; and n_1 and n_2, the first and second moments of the number of arrivals in a cluster. The service times in the queueing system are characterized by the first two moments τ_1 and τ_2, and in addition it is assumed that the service times are bounded from below by the spacing of arrivals in a single cluster (thus, the "B" in the notation G(B) denotes "bounded service"):

$$P[\textit{service time} \geqslant \delta] = 1 \tag{4.10}$$

From a generalization of an observation made by Farel ([7]), it can be shown that the mean waiting time in this model is given by

$$E[W] = \frac{\rho\tau_2/2\tau_1 + [\tau_1+(\rho-1)\delta]\dfrac{n_2-n_1}{2n_1}}{1-\rho} \tag{4.11}$$

where $\rho = \lambda' n_1\tau_1$ is the server utilization.

For the PCP it can be shown that the expectation function is just

$$U(x) = \lambda' n_1 x + \sum_{i=1}^{\infty} u(x-i\delta)\, P[K\geqslant i] \tag{4.12}$$

where the random variable K is defined as the number of subsequent arrivals in the same cluster as an arbitrarily chosen arrival, and $u(\cdot)$ is the unit step function:

$$u(x) = 0,\ x<0 \qquad = 1,\ x\geqslant 0 \tag{4.13}$$

It then follows from a length-biasing argument that

$$\hat{U}(s) = \frac{\lambda' n_1}{s} + e^{-s\delta}\left[\frac{1-e^{-s\delta}+[g(e^{-s\delta})-1]/n_1}{(1-e^{-s\delta})^2}\right] \quad (4.14)$$

and $z_{\exp}(\cdot)$ can be obtained via (1.3).

The peakedness-based approximation was tested on this example for a wide range of model parameter values, with the same basic conclusions as were derived in Section 4.1, i.e., that the approximation is quite good, especially as the server utilization increases.

4.3 A packetized voice example

The final example that we discuss is the application of the peakedness-based delay approximation to the prediction of delays that would be experienced at a single trunk in a packet switching network over which the packets associated with many packetized voice virtual circuits are being transmitted. This situation may be modeled as a PCP/D/1 queueing system, but the picture is somewhat different from that in Section 4.2 in that now the (deterministic) service times are very much less (typically .02 as large) than the inter-packet spacing within a packet cluster.

The peakedness-based delay approximation was calculated and compared with results obtained via a simulation. For server utilization up to $\rho = .75$ the agreement was very good; however, for higher utilizations, the approximation rapidly lost accuracy.

There is an explanation for this behavior of the approximation: at high utilizations the validity of (3.3) in the approximation derivation becomes more questionable. In this queueing system, large delays typically occur when a large number of clusters happen to be overlapping; thus, if an arrival experiences a large delay, the mean number of arrivals during its waiting time is larger than $U(W)$. That is, there is positive correlation between an arrival's waiting time and the number of subsequent arrivals during its waiting time which is not captured entirely by the functional relationship $U(W)$.

5. CONCLUSIONS

In this paper we have demonstrated that a simple delay approximation based on generalized peakedness provides an effective way to approximately predict delays. However, the problem to which the approximation is being applied should be scrutinized for phenomena such as was illustrated in Section 4.3, which can reduce the accuracy of the approximation. The accuracy of the delay approximation seems to depend critically on the validity of the approximation (3.3); if this is approximately true, then the delay approximation can be expected to yield accurate results, especially for relatively large server utilizations. Equation (3.3) is a poor approximation when the arrival stream infrequently enters periods of sustained instantaneous arrival rates which push the server utilization close to or greater than one.

Finally, it has also been shown that the value of $z_{\exp}(1/E[W|W>0])$ is the most relevant peakedness information, from a single peakedness value, for predicting delays.

6. REFERENCES

[1] W. Whitt, "Approximating a Point Process by a Renewal Process, I: Two Basic Methods," *Operations Research,30* (1982), pp. 125-147.

[2] A. E. Eckberg, "Generalized Peakedness of Teletraffic Processes," *Proceedings, 10th International Teletraffic Congress*, Montreal, 1983.

[3] R. B. Cooper, *Introduction to Queueing Theory*, Macmillan, 1972.

[4] A. A. Fredericks, "Congestion in Blocking Systems - A Simple Approximation Technique," *Bell System Technical Journal, 59* (1980), pp. 805-827.

[5] A. E. Eckberg, "Sharp Bounds on Laplace-Stieltjes Transforms, with Applications to Various Queueing Problems," *Mathematics of Operations Research, 2* (1977), pp. 135-142.

[6] E. Cinlar, "Queues with Semi-Markovian Arrivals," *Journal of Applied Probability,4* (1967), pp. 365-379.

[7] R. A. Farel, unpublished work.

TELETRAFFIC ISSUES in an Advanced Information Society
ITC-11
Minoru Akiyama (Editor)
Elsevier Science Publishers B.V. (North-Holland)

QUEUES WITH SEMI-MARKOV INPUT AND SERVICE TIMES

R. G. ADDIE

Telecom Australia Research Laboratories
Melbourne, Australia

ABSTRACT

The concept of a semi-Markov process is introduced and proposed as a method of modelling a queueing arrival process with non-independent inter-arrival times. A formula for the stationary distribution of delay experienced by signals whose arrival times form such a process, when passing through a first-in-first-out queue, is given in terms of a certain Matrix Wiener-Hopf factorization. This formula applies to a large class of semi-Markov models for the arrival process. The service times of signals passing through the queue may be taken to be independently and identically distributed, with a distribution which has a rational Fourier transform, or to form a semi-Markov sequence in a manner analogous to the sequence of inter-arrival times. An algorithm for Wiener-Hopf factorization is applied to solve a specific example problem using the given formula.

1. INTRODUCTION

The concept of a semi-Markov process generalizes that of a renewal process. It allows one to model arrival processes subject to clustering of arrivals and other phenomena resulting from non-independence of inter-arrival times. This type of traffic can be expected in a packet switched data network, for example, because the data sources have this characteristic and because the processing of data traffic by a node in a data network introduces some degree of correlation into the data stream.

In section 2, the concept of a semi-Markov process is introduced, including an example. In section 3, we give a general formula for the distribution of delays experienced by signals forming such a process when they pass through a queueing system. This formula is expressed in terms of a matrix Wiener-Hopf factorization of a certain matrix function which summarizes, in a certain way, all the details concerning the input process and the distribution of processing times of signals.

An algorithm from [1] is used in section 4 to compute the delay distribution corresponding to the example given earlier. Concluding remarks are presented in section 5.

Although queues with semi-Markov inputs and service times have been studied before, [2] - [3], and matrix Wiener-Hopf factorization has been proposed as a method of solution, [4] - [6], the results of the present paper are new in one important respect. The factorization concept of [4] - [6] does not lend itself to computation, whereas the factorization presented here is readily computed in a large class of cases. The method of expressing the solution in terms of a Wiener-Hopf factorization leads to a more general result than those given in [2] - [3]; in [2], the service times must form an independently, identically distributed sequence of negative-exponentially distributed random variables, which are also independent from the arrival process; in [3] the inter-arrival time and service time processes must also be independent from each other and, in addition, the arrivals must form a Poisson process. Despite the greater generality of the present method, because an algorithm for Wiener-Hopf factorization exists, computation of the solution remains quite convenient and efficient.

2. SEMI-MARKOV PROCESSES

The concept of a semi-Markov process was introduced in [7] and [8]. A semi-Markov process is a special type of "marked point process" on the real line; by a marked point process we mean a point process in which with each point is associated a "mark", or "state", taking values in a certain state space. In general this state space may be infinite, but in this paper we shall always assume that it forms a finite set, referred to henceforth as the state space of the process.

In order for a marked point process to form a semi-Markov process we require that the bivariate process of marks and inter-arrival times, associating each inter-arrival time with the mark of the point at the end of the inter-arrival interval, forms a Markov process with stationary transition probabilities. Furthermore the conditional probability of any future event, given the most recent mark and inter-arrival time, is required to depend on the value of the most recent mark only.

We shall usually discuss a semi-Markov point process in terms of the corresponding inter-arrival time process, $(X_n,Y_n)_{n\geq 0}$ say, where Y_n

denotes the time between the $n-1^{st}$ and the nth points and X_n denotes the mark at the nth point.

Example

We may loosely describe the example we wish to investigate as follows. Signals arrive at a single-server queue according to a Poisson process with intensity α_i while the "state" of the arrival process is i. Furthermore, the state of the arrival process changes, from i to j say, with probability p_{ij}, after each signal has been generated. We wish to determine the stationary distribution of delay experienced by signals passing through this system.

To be a little more specific, we suppose that the arrival times of signals form a semi-Markov process, as described above, for which

$$P\{X_2=j,Y_2\leq x|X_1=i\} = p_{ij}(1-e^{-\alpha_j x}),$$

$1 \leq i,\ j \leq n,\ x \in \mathbb{R}$. We assume, in addition, that signals have random service times which are independent of each other and of the arrival process, and have an exponential distribution with mean $1/\mu$. We will usually assmume that changes of state of the X process take place much less often than signal arrivals. This justifies the loose description of the process given initially.

We shall see later how the stationary distribution of waiting times of signals passing through this system may be calculated.

The process $(X_n)_{n\geq 0}$, of a semi-Markov process $(X_n,Y_n)_{n\geq 0}$, forms a Markov process, with stationary transition probabilities, in its own right. Let us suppose, to be specific, that the state space for this process is $\{1,...,k\}$ and its transition matrix is $P = (p_{ij})$. From the theory of stationary Markov chains, [9], we know that there exists a stationary probability distribution, $\pi = (\pi_1,...,\pi_k)$, say, for this process. We assume, in addition, that this is the only stationary probability distribution for the process $(X_n)_{n\geq 0}$, and that $\pi_i \neq 0$, $i = 1,...,k$. We denote by $E_\pi\{Z\}$ the expected value of any random variable Z defined in terms of $X_1,Y_1,X_2,...$, when X_1 has the distribution π.

The main purpose of this paper is to show how we can express and calculate the joint stationary distribution of delays and signal state for signals passing through a queue, such as in the above example. We shall denote this vector function by $\lambda(x) = (\lambda_1(x),...,\lambda_k(x))$ with the understanding that $\lambda_i(x)$ is the stationary probability that a signal arrives when $X_n = i$ and is delayed for less than or equal to time x before service begins, assuming that a unique joint stationary probability distribution for these quantities exists.

Any vector of right continuous, non-decreasing functions, $f(x) = (f_1(x),...,f_k(x))$, such that

$$\lim_{x\to-\infty} f_i(x) = 0, \quad i = 1,...,k,$$

$$\lim_{x\to\infty} f_i(x) = p_i, \text{ such that } \sum_{i=1}^{k} p_i = 1,$$

will be referred to as a vector distribution function. Note that in this paper we use row vectors exclusively.

3. FORMULA FOR THE STATIONARY DELAY DISTRIBUTION

The formula for λ which we derive in this section is stated in Theorem 3. It is given in terms of the Fourier-Stieltjes transform of λ; the Fourier-Stieltjes transform of such a function is defined in §3.1. In the next two sections we outline the derivation of this Theorem. For the reader mainly interested in the formula, these sections may be skimmed.

3.1 An Algebra and a Vector Space of Transition Functions

In order to be able to express the formula for λ we need to define a type of transition function which incorporates information about the service time distribution of signals as well as the arrival process. The following function turns out to be the appropriate one:

$$Q(x) = (q_{ij}(x)) \tag{1a}$$

where

$$q_{ij}(x) = P\{S_{n-1}-Y_n \leq x,\ X_n = j \mid X_{n-1} = i\}, \tag{1b}$$

$x \in \mathbb{R}$, in which S_n denotes the service time of the nth signal, and $(X_n,Y_n)_{n\geq 0}$ is the inter-arrival process. The transition probability matrix for $(X_n)_{n\geq 0}$ is then given by $p_{ij} = \lim_{x\to\infty} q_{ij}(x) - \lim_{x\to-\infty} q_{ij}(x)$, $0 \leq i,\ j \leq k$.

For the results of this section to hold, all that it is necessary to assume of the process $(S_n)_{n\geq 0}$ is that the process $(X_n,S_{n-1}-Y_n)_{n\geq 0}$ forms a semi-Markov sequence, as defined in [3]. That is, $(X_n,S_{n-1}-Y_n)_{n\geq 0}$ must form a Markov process with stationary transition probabilities in which the conditional probability of any event defined in terms of future values of $X_n,S_{n-1}-Y_n$, given the past of the process, depends on the current value of X_n only. This is true, for example, if the process $(S_n)_{n\geq 0}$ forms a sequence of independent, identically distributed random variables, each of which is independent of the process $(X_n, Y_n)_{n\geq 0}$.

The general name for a transition function such as Q is a semi-Markov transition function. A product of semi-Markov transition functions, generalizing convolution of distribution functions, is defined by $Q^{(2)} = Q^{(1)}Q^{(0)}$ with $Q^{(2)} = (q_{ij}^{(2)})$, where

$$q_{ij}^{(2)} = \sum_{\ell=1}^{k} q_{i\ell}^{(1)} * q_{\ell j}^{(0)}, \qquad (2a)$$

in which "*" denotes convolution, as defined for distribution functions, i.e.,

$$q_{i\ell}^{(1)} * q_{\ell j}^{(0)}(y) = \int_{-\infty}^{\infty} q_{i\ell}^{(1)}(y-x)\,dq_{\ell j}^{(0)}(x) . \qquad (2b)$$

By including matrix functions which have function entries which are right continuous but not necessarily increasing, the set of all semi-Markov transition functions of dimension k, say, may be embedded in a linear algebra (also termed a ring, see [10], p152, for a definition), which we denote by $\mathbf{B}(k)$. The identity element of this algebra, denoted by E, is the matrix whose diagonal entries are unit step functions, and whose off-diagonal entries are zero functions (the identity matrix is denoted by I). In addition to being a matrix of right continuous functions, we require of any element, e.g. $Q = (q_{ij})$, of $\mathbf{B}(k)$, that

$$|Q|_{\mathbf{B}(k)} = \sup_{i=1}^{k} \sum_{j=1}^{k} \int_{-\infty}^{\infty} |dq_{ij}(x)| < \infty ; \qquad (3)$$

this ensures that the product of any two elements of $\mathbf{B}(k)$, as at (2a-b), is well-defined. Furthermore, two transition functions, Q and R say, which differ by a constant matrix, in which case all of the Stieltjes integrals at (3) take the value zero, so $|Q-R|_{\mathbf{B}(k)}=0$, will be identified, as elements of $\mathbf{B}(k)$.

The reader familiar with the theory of signed measures will recognise that $\mathbf{B}(k)$ is isomorphic to the algebra of $k\times k$ matrices of signed measures. A proof that $\mathbf{B}(k)$ forms an algebra, in fact a complete normed algebra, is given in [12].

The set of transition functions satisfying the weaker condition

$$\sup_{x\in\mathbf{R}} \int_{a+x}^{b+x} |dq_{ij}| < \infty , \; i,j = 1,\dots,k, \; a,b \in \mathbf{R},$$

forms a vector space containing $\mathbf{B}(k)$, which we denote by $\mathbf{V}(k)$. Again, any two functions which differ by a constant matrix are regarded as representing the same element of $\mathbf{V}(k)$. The product of an element $Q = (q_{ij})$, of $\mathbf{B}(k)$, and an element $R = (r_{ij})$, of $\mathbf{V}(k)$, is the element $S = (s_{ij})$ of $\mathbf{V}(k)$ defined by

$$s_{ij}(x) = \sum_{\ell=1}^{k} \int_{-\infty}^{\infty} dq_{i\ell}(y)\big(r_{\ell k}(x-y)-r_{\ell k}(-y)\big).$$

A similar definition applies to the product RQ, with $R \in \mathbf{V}(k)$, $Q \in \mathbf{B}(k)$, and to the product fQ, when f is a vector function and $Q \in \mathbf{B}(k)$. In the latter case the result is a vector function.

We say a certain transition function Q(x) is <u>concentrated on the interval A</u> in $\mathbf{R}$ if Q(x) is constant on each interval in $\mathbf{R}\backslash A$ and $\lim_{x\to y-} Q(x) = Q(y)$ for any $y \notin A$. (The latter condition ensures that if A does not include its left endpoint, e.g. $A = (y,z)$, then Q(x) does not have a jump there.) This same terminology will be used for vector functions, with exactly the same definition. A transition function (vector function) is said to be <u>non-decreasing</u> if each element of the defining matrix (vector) is a non-decreasing function.

The <u>Fourier-Stieltjes transform</u> of an element of $\mathbf{B}(k)$ is defined to be $\hat{Q}(\theta) = (\hat{q}_{ij}(\theta))$, where

$$\hat{q}_{ij}(\theta) = \int_{-\infty}^{\infty} e^{i\theta y}\,dq_{ij}(y) , \quad \theta \in \mathbf{R}.$$

The Fourier-Stieltjes transform of a vector function is defined analogously.

The Fourier-Stieltjes transform of the product of two elements, $Q = (q_{ij})$, $R = (r_{ij})$ say, of $\mathbf{B}(k)$ satisfies the equation

$$(QR)\hat{}\,(\theta) = \hat{Q}(\theta)\hat{R}(\theta), \quad \theta \in \mathbf{R},$$

where the product on the right hand side is defined to equal the matrix $S(\theta) = (s_{ij}(\theta))$, where

$$\hat{s}_{ij}(\theta) = \sum_{\ell=1}^{k} \hat{q}_{i\ell}(\theta)\hat{r}_{\ell j}(\theta) , \quad \theta \in \mathbf{R}.$$

An analogous equation holds when Q is replaced by a vector function.

3.2 Canonical Factorization

A factorization

$$E-Q = (E-Q_+)(E-Q_-) ,$$

of transition functions in $\mathbf{B}(k)$ is said to be <u>left-canonical</u> (c.f. [11]), if:

(i) Q_+ is concentrated on $(0,\infty)$ and there exists a transition function $\tilde{Q}_+ \in \mathbf{V}(k)$ concentrated on $(0,\infty)$ such that $(E+\tilde{Q}_+)(E-Q_+) = E$,

(ii) Q_- is concentrated on $(-\infty,0]$ and there exists a transition function $\tilde{Q}_- \in \mathbf{V}(k)$ concentrated on $(-\infty,0]$ such that $(E-Q_-)(E+\tilde{Q}_-) = E$.

Intuitively, a left-canonical factorization is one in which the left factor is concentrated on $[0,\infty)$, together with its inverse, and the right factor is concentrated on $(-\infty,0]$, together with its inverse. Also, as a normalization condition, it is required that $\lim_{x\to 0-} Q_+(x) = Q_+(0)$.

If the definition of canonical factorization given in [11], for example, were applied in the present context, the terms $\tilde{Q}_+$ and $\tilde{Q}_-$ would also be required to belong to $\mathbf{B}(k)$. For this reason, the type of canonical factorization defined here should be termed improper. It is the use of such an "improper" type of canonical factorization which allows us to solve the problem at hand in an effective way.

The type of factorization defined in [4] and [6] allows a solution for the problem at hand which is identical in form to that given below. It is probable, in fact, that the factorizations are identical in most cases. Given that this is the case, the contribution of the following theorem is an alternative means of characterising the factorization in question which has the advantage that a ready means of computing the factors is available for a large class of cases.

The rather technical condition on $P\{Y_2>0|X_1=i,X_2=j\}$ may be unnecessary, but the currently existing proof of the existence of a canonical factorization requires this to prove that $\tilde{Q}_+ \in \mathbf{B}(k)$.

Theorem 1

If $E_\pi\{S_1-Y_2\} < 0$, and $P\{Y_2>0|X_1=i,X_2=j\} > 0$, $0 \le i, j \le k$, then there exists a unique left-canonical factorization

$$E - Q = (E - Q_+)(E - Q_-),$$

where Q_+, $\tilde{Q}_+$, Q_- and $\tilde{Q}_-$ are all non-decreasing, and $\tilde{Q}_+ \in \mathbf{B}(k)$. Furthermore the unique stationary joint distribution of queueing delay and signal state is given by

$$\lambda(x) = \pi(I - \hat{Q}_+(0))(E + \tilde{Q}_+(x)),$$

with the understanding that $\tilde{Q}_+$ is represented by a transition function $\tilde{Q}_+(x)$ such that $\tilde{Q}_+(-0) = -E$.

Remark

The condition that $\tilde{Q}_+(-0) = -E$ ensures that $\lambda(x) = 0$, $x < 0$; if any other representation of $\tilde{Q}_+$ were used, $\lambda(x)$ as defined here would not satisfy the condition $\lim_{x\to-\infty}\lambda(x) = 0$, and hence could not be a vector distribution function.

Proof

It follows readily from the results in [12], that in the present case a canonical factorization with the required properties, and a stationary distribution $\lambda(x)$, exist, and both are unique. The formula for $\lambda(x)$ may be proved as follows.

Let D_n denote the delay experienced by the nth signal. Then

$$D_{n+1} = \max\{0, D_n+S_n-Y_{n+1}\}, \quad n > 0, \qquad (4)$$

(c.f. [13], eq. (2)).

Define the operator Π on vector functions of the form $f(x) = (f_1(x),\ldots,f_k(x))$ by $\Pi f = g = (g_1,\ldots g_k)$, where

$$g_i(x) = \begin{cases} \lim_{y\to-\infty} f_i(y), & x < 0, \\ f_i(x) - f_i(0), & x \ge 0, \end{cases}$$

whenever the limit exists. This operator is the natural generalization in this context of the operator denoted in the same way in [13].

In order to solve for the stationary joint distribution of X_n and D_n we now transform (4) into an equation in terms of vector distributions. Let $\lambda^{(n)}$ denote the vector distribution for (X_n, D_n), i.e. $\lambda^{(n)}(x) = (\lambda_1^{(n)}(x),\ldots,\lambda_k^{(n)}(x))$, where

$$\lambda_i^{(n)}(x) = P\{X_n=i, D_n \le x\}.$$

Then, since X_n, D_n are defined in terms of the past of the semi-Markov sequence $(X_n, S_{n-1}-Y_n)$, the vector distribution, $\eta^{(n+1)}$, of $(X_{n+1}, D_n+S_n-Y_{n+1})$, is given by $\lambda^{(n)}Q$, i.e. $\eta^{(n+1)}(x) = (\eta_1^{(n+1)}(x),\ldots,\eta_k^{(n+1)}(x))$, where

$$\eta_i^{(n+1)}(x) = \sum_{j=1}^{k} \int_0^\infty \lambda_j^{(n)}(dy) q_{ji}(x-y).$$

Also, if (X,V) is any pair of random variables with X in $\{1,\ldots,k\}$ and V in $\mathbf{R}$ and with joint distribution described by the vector distribution function $\eta(x) = (\eta_1(x),\ldots,\eta_k(x))$, then $(X_n, \max\{0,V_n\})$ has the vector distribution function $\Pi\eta$.

Using both of these facts we see that (4) leads to the following equation in vector distributions :

$$\lambda^{(n+1)} = \Pi(\lambda^{(n)}Q) ,$$

and hence, the stationary vector distribution, λ, for (X_n, D_n) must satisfy the equation

$$\lambda = \Pi(\lambda Q) . \tag{5}$$

Now (5) is equivalent to

$$\Pi(\lambda(E-Q)) = 0 \tag{6}$$

together with the requirement that the vector distribution function $\lambda(x)$ is concentrated on $[0,\infty)$. The choice

$$\lambda(x) = \pi(I-\hat{Q}_+(0))(E+\tilde{Q}_+(x))$$

is concentrated on $[0,\infty)$. To see that (6) holds, observe that

$$\lambda(E-Q) = \pi(I-\hat{Q}_+(0))(E-Q_-) \tag{7}$$

is concentrated on $(-\infty,0]$, so $\Pi(\lambda(E-Q))$ is concentrated on $\{0\}$. Furthermore, using the properties of canonical factorization, and the fact that

$$\hat{f}(0) = \lim_{x\to\infty} f(x) - \lim_{x\to-\infty} f(x) ,$$

$$\hat{R}(0) = \lim_{x\to\infty} R(x) - \lim_{x\to-\infty} R(x) ,$$

for any vector function, f, and transition function, R, we see that

$$\lim_{x\to\infty} \Pi(\lambda(E-Q))(x) - \lim_{x\to-\infty} \Pi(\lambda(E-Q))(x)$$

$$= \lim_{x\to\infty} \lambda(E-Q)(x) - \lim_{x\to-\infty} \lambda(E-Q)(x)$$

$$= \pi(I-\hat{Q}_+(0))(I+\hat{Q}_-(0)), \text{ by (7)},$$

$$= \pi(I-\hat{Q}(0))$$

$$= \pi(I-P)$$

$$= 0 ,$$

since π is a stationary distribution for the process $(X_n)_{n\geq 0}$. It follows that (5) holds also.

Notice, now, that the factor $(I-\hat{Q}_+(0))$ equals

$$I - \lim_{x\to\infty} Q_+(x) + \lim_{x\to-\infty} Q_+(x)$$

$$= \left[I + \lim_{x\to\infty} \tilde{Q}_+(x) - \lim_{x\to-\infty} \tilde{Q}_+(x) \right]^{-1} ,$$

and hence, recalling that $\lim_{x\to-\infty} \tilde{Q}_+(x) = 0$,

$$(I - \hat{Q}_+(0))(I + \tilde{Q}_+(x))$$

tends to the limit I as $x \to \infty$. This ensures that $\lim_{x\to\infty}\lambda(x) = \pi$. It follows from (5) (since each side of the equation must have the same total variation) that each component of $\lambda(x)$ is monotonic, and hence non-decreasing, and by the choice of representation of $\tilde{Q}_+$, $\lim_{x\to-\infty}\lambda_i(x) = \lambda_i(0-) = 0$, $i = 1,\dots,k$. Thus $\lambda(x)$ is a vector distribution. This concludes the proof.

3.3 Wiener-Hopf Factorization of Matrix Functions

Suppose $\Phi(\theta)$ is a $k\times k$ matrix of functions,

$$\Phi(\theta) = (\phi_{ij}(\theta)),$$

defined on the real line. Typically, $\Phi(\theta)$ will be the Fourier transform of a function in **B**(k). A <u>(left) Wiener-Hopf factorization</u> of $\Phi(\theta)$ is a factorization

$$\Phi(\theta) = \Phi_+(\theta)\,\Phi_-(\theta), \quad \theta \in \mathbf{R},$$

in which

(a) $\Phi_+(\theta)$ is analytic and bounded in $\mathrm{Im}(\theta) > 0$ and continuous in $\mathrm{Im}(\theta) \geq 0$, and $\Phi_-(\theta)$ is analytic and bounded in $\mathrm{Im}(\theta) < 0$ and continuous in $\mathrm{Im}(\theta) \leq 0$, and

(b) $\Phi_+^{-1}(\theta)$ is analytic and bounded in $\mathrm{Im}(\theta) > \varepsilon$ for any $\varepsilon > 0$ and $\Phi_-^{-1}(\theta)$ is analytic and bounded in $\mathrm{Im}(\theta) < -\varepsilon$ for any $\varepsilon > 0$.

Note: condition (b) is somewhat weaker than the usual requirement (as in, e.g., [14]) that condition (a) apply to the inverses of the factors as well as to the factors themselves. For this reason, the type of Wiener-Hopf factorization defined here should be termed "improper".

<u>Theorem 2</u>

Taking Fourier Transforms of a canonical factorization of transition functions gives rise to a Wiener-Hopf factorization, in which $\lim_{x\to+\infty}\Phi_+(ix) = I$.

<u>Proof</u>

The proof may be based on the proof of Theorem 29.2 of [15] .

This result provides us with a practical means of obtaining a canonical factorization. It would be nice to have a general result which ensures that, in the context of Theorem 1, every Wiener-Hopf factorization arises as the Fourier Transform of a canonical factorization, or by modifying such a factorization by multiplying the factors by constant matrices. Such a result is not true, in general. It is possible, however, to give simple conditions, satisfied in all reasonable cases, under which a Wiener-Hopf factorization of a rational matrix function does take this form; for example, if each of the functions $\det(\Phi_+(\theta))$ and $\det(\Phi_-(\theta))$ have at most one zero on the real line, and when this zero occurs it has multiplicity one, then the Wiener-Hopf factorization $\Phi(\theta) = \Phi_+(\theta)\Phi_-(\theta)$ arises from a canonical factorization in this way.

The following result follows from Theorem 1, by taking Fourier transforms.

Theorem 3

If Q is as defined at (1a-b), and

$$I - \hat{Q}(\theta) = \Phi_+(\theta)\Phi_-(\theta), \quad \theta \in \mathbf{R},$$

is a Wiener-Hopf factorization which corresponds to a canonical factorization, the Fourier transform of λ may be expressed as

$$\hat{\lambda}(\theta) = \pi\Phi_+(0)\Phi_+^{-1}(\theta) .$$

4. EXAMPLE CALCULATION

Suppose, making the example introduced in section 2 more specific, that $m = 3$, $\alpha_1 = .5$, $\alpha_2 = .75$, $\alpha_3 = 1.0$, $\mu = 1.0$, and

$$(p_{ij}) = \begin{pmatrix} .95 & .05 & 0.0 \\ .05 & .9 & .05 \\ 0.0 & .05 & .95 \end{pmatrix}$$

Then $\pi = (1/3, 1/3, 1/3)$ and

$$\hat{\Phi}(\theta) \overset{\Delta}{=} I - \hat{Q}(\theta) =$$

$$\begin{pmatrix} 1 + \frac{.95\mu\alpha_1}{(\omega-\mu)(\omega+\alpha_1)} & \frac{.05\mu\alpha_2}{(\omega-\mu)(\omega+\alpha_1)} & 0 \\ \frac{.05\mu\alpha_1}{(\omega-\mu)(\omega+\alpha_2)} & 1 + \frac{.9\mu\alpha_2}{(\omega-\mu)(\omega+\alpha_2)} & \frac{.05\mu\alpha_3}{(\omega-\mu)(\omega+\alpha_2)} \\ 0 & \frac{.05\mu\alpha_2}{(\omega-\mu)(\omega+\alpha_3)} & 1 + \frac{.95\mu\alpha_3}{(\omega-\mu)(\omega+\alpha_3)} \end{pmatrix},$$

where $\omega = i\theta$. Using the algorithm of [1] we find

$$\Phi(0)\times\Phi_+^{-1}(\theta) = \frac{(\mu-\omega)/\mu}{(1-10.365\omega+29.977\omega^2-26.156\omega^3)} \times$$

$$\begin{pmatrix} 1-8.467\omega+14.117\omega^2 & .308\omega-1.438\omega^2 & .185\omega-.161\omega^2 \\ .385\omega-1.608\omega^2 & 1-7.520\omega+10.581\omega^2 & .798\omega-1.276\omega^2 \\ .471\omega-.729\omega^2 & 1.586\omega-2.865\omega^2 & 1-4.742\omega+5.279\omega^2 \end{pmatrix}$$

whence

$$\hat{\lambda}(\theta) = \frac{(\mu-\omega)/\mu}{(1 - 10.365\omega + 29.977\omega^2 - 26.156\omega^3)} \times$$

$$(1/3-2.54\omega+3.93\omega^2,\ 1/3-1.87\omega+2.09\omega^2,\ 1/3-1.12\omega+.981\omega^2) .$$

By calculating a partial fraction representation of this vector function, and using well known Laplace transform inversion formulae, we find that the stationary marginal distribution of delay is

$$P\{\text{delay} \le x\} = 1-.551e^{-.1607x}-.128e^{-.4231x}-.053e^{-.5623x}.$$

5. CONCLUDING REMARKS

We have presented a fairly general method for determining stationary distributions of delay in queueing systems in which the stream of inputs forms a semi-Markov process and the service times of signals are independently and identically distributed, and are independent from the inter-arrival process. Without significant changes, the same method is applicable to systems in which, in addition, the successive service times of inputs to the queue form a "semi-Markov sequence" (as in [3]).

What makes the described method an effective tool for solving such problems is the availability of an algorithm for Wiener-Hopf factorization of a large class of rational matrix functions [1]. A FORTRAN subroutine has been developed for computing the joint stationary distribution of state and delay for a queue in which both the service times and the interarrival times depend on the underlying state. This subroutine is able to solve problems in which the number of states is less than or equal to 10. Further work is required to determine numerical properties and the limits of applicability of such algorithms.

ACKNOWLEDGEMENT

The permission of the Chief General Manager, Telecom Australia, to present this paper, is gratefully acknowledged.

REFERENCES

[1] R. Addie, "Factorization of Rational Matrix Functions", Telecom Research Laboratories Report, in preparation.

[2] E. Cinlar, "Queues with semi-Markovian arrivals", J. Appl. Prob. Vol. 4, pp 365-379, 1967.

[3] M. R. Neuts, "The single server queue with Poisson input and semi-Markov service times", J. Appl. Prob. Vol. 3, pp 202-230, 1966.

[4] E. Arjas, "On the use of a fundamental identity in the theory of semi-Markov queues", Adv. Appl. Prob. Vol. 4, pp 271-284, 1972.

[5] L. Takacs, "A storage process with semi-Markov input", Adv. Appl. Prob. Vol. 7, pp830-844, 1975.

[6] E. Arjas and T. P. Speed, "Topics in Markov additive processes", Math. Scand. Vol 33, pp 171-192, 1973.

[7] P. Levy, "Processus semi-Markovien", Proc. Int. Congress Math. 3 (Amsterdam), pp416-426, 1954.

[8] W. L. Smith, "Regenerative stochastic processes", Proc. R. Soc. (London) Vol. A232, pp 6-31, 1955.

[9] K. L. Chung, "Markov Chains with stationary transition probabilities", 2nd ed., Springer, Berlin, 1967.

[10] M. A. Naimark, "Normed algebras", 3rd American edition, Wolters-Noordhoff Publishing, Groningen, The Netherlands, 1972.

[11] E. L. Presman, "Factorization methods and boundary problems for sums of random variables given on Markov chains", Izv. Akad. Nauk. SSR, Ser. Mat., Tom 33, No. 4, 1969; English translation, Math. USSR Izvestiya, Vol. 3, No. 4, pp 815-852, 1969.

[12] R. Addie, "Discrete time Markov-additive processes", PhD thesis, Monash University, 1984.

[13] J. F. C. Kingman, "On the algebra of queues", J. Appl. Prob. Vol. 3, pp 285-326, 1966.

[14] I. C. Gohberg and M. G. Krein, "Systems of integral equations on a half line with kernels depending on the difference of the arguments", Uspehi Mat. Nauk Vol. 13, No. 2 (80) pp 3-72, 1958; English translation, Am. Math. Soc. Transl. (Series 2) Vol. 14, pp 217-287, 1960.

[15] G. Doetsch, "Introduction to the theory and application of the Laplace transform", Springer-Verlag, Berlin, New York, 1974.

TELETRAFFIC ISSUES in an Advanced Information Society
ITC-11
Minoru Akiyama (Editor)
Elsevier Science Publishers B.V. (North-Holland)

SELECTED TRAFFIC PROBLEMS ASSOCIATED WITH THE RESTRUCTURING OF THE TELEPHONE INDUSTRY IN THE UNITED STATES

R.R. MINA

FEDERAL COMMUNICATIONS COMMISSION, USA. *

ABSTRACT

This paper deals with selected traffic engineering problems associated with the restructuring of the telephone industry in the United States resulting from the settlement of the U.S. Department of Justice antitrust case against AT&T. It focuses on the requirement that the local operating companies provide access to multi-interexchange carriers that is "equal in type, quality and price" to that provided to AT&T. It provides an optimum cost equal access trunk provisioning plan based on high-usage alternate route techniques.

BACKGROUND AND INTRODUCTION

The settlement of the AT&T antitrust case marks a new era in the U.S. telecommunications industry. Briefly, the settlement agreement divests AT&T of any ownership interests in the companies that provide local exchange service. The agreement breaks up the former, vertically -integrated Bell System that was largely owned by AT&T,

into one company providing long distance service (AT&T) and twenty-two local exchange companies (LECs) providing local service (known as the Bell Operating Companies, or BOCs). The BOCs each serve a separate region of the country, whereas AT&T competes with several companies to provide long distance service to customers throughout the country. Because of the importance of the divestiture as an historic event in global telecommunications, I thought it would be appropriate to record in the pages of the proceedings of the Eleventh ITC, as an addendum to my paper, an overview of that agreement. It is known as the Modified Final Judgment (MFJ).

Local Access and Transport Area (LATA)

In addition to divesting the LECs from AT&T, the MFJ called for the separation of "exchange" and "interexchange" telecommunications functions. This partitioning was accomplished by dividing the country into about 200 exchange areas, called LATAs (Local Access and Transport Areas). The LECs are responsible for the transport of traffic between customers within the LATA (intra-LATA), while the Interexchange Carriers (ICs) are responsible for the transport of traffic between LATAs (inter-LATA). The LECs are also responsible for the transport of inter-LATA traffic from the customers to the ICs at an IC's point-of-presence (PoP) within the LATA.

LATAs were designed according to guidelines contained in the MFJ. They were formulated on the concept of demographic areas which have common social, economic and other interests. Most states contain several LATAs but sparsely populated states consist of a single LATA. One of the largest LATAs, for example, is the New York metropolitan LATA. It includes over 300 end offices serving over six million subscribers' lines and is served by four National Plan Area (NPA) codes.

LATA networks are generally 2-level hierarchical networks which are independent sub-sections of the 5-level North American Network. The end offices are the lower level and the access tandem(s) are the higher level. The access tandem (AT) concentrates the originating traffic and distributes the terminating traffic for a number of end offices in an area. In large LATAs where more than one AT is required, the LATA is divided into sectors and each sector is served by a sector tandem. Figure 1 illustrates the trunking arrangement in a LATA with double tandems. The backbone final route chain (shown by double lines)

---

* The views expressed in this paper are those of the author and do not necesarily reflect the views of the Federal Communications Commission.

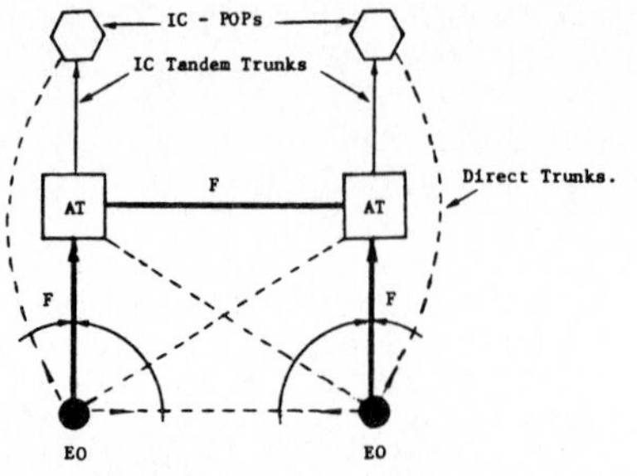

Fig. 1 Illustrating 2 - Level Hierarchical Routing in a LATA with two Access Tandems.

consists of three final trunk groups; two interconnecting the end offices with their home access tandem(s) and one group interconnecting the two tandems. High-usage (HU) trunk groups with alternate routing to the finals interconnect the end offices with each other and with the tandem other than their home tandem. In the restructured networks, the ICs connecting trunks may consist of direct (high-usage) trunks (shown by dotted lines) terminating at the end offices and/or tandem trunks (shown by solid lines) terminating at the tandems.

The MFJ called also for providing equal dialing parity to all interexchange carriers. This has been accomplished by updating the generic program in large end offices of the electronic type to route toll calls dialed with the prefix "1" or "0" to a particular interexchange carrier based on the customer's presubscription to that carrier. Equal dialing parity was therefore provided without the need to revise the National Plan Area (NPA) code. In addition to presubscribing, a customer can route his call to a particular carrier by dialing a 5-digit (10XXX) carrier code. The generic also provides other features such as the separation of intra- and inter-LATA traffic and routing of traffic to its destination at its origin.

FORMULATION OF THE PROBLEM

There are many problems which have become apparent since the divestiture. The most obvious concerns of the policy makers in the new environment have related to promoting competition and economic efficiency. These concerns, however, must be balanced against the technical realities of the network. At this stage, the efficiencies emanating from technical solutions involving the structure and functions of the network have received little attention. This paper, therefore, will focus on such technical solutions in a network which is undergoing a transition from a monolithic structure with a single service provider to one geared towards several service providers. It is specifically directed toward the problem of access, that is the means by which the facilities of competing interexchange carriers may be efficiently interconnected with the networks of the LECs.

In a competitive marketplace, it is expected that the demand handled by the new competing carriers will gradually increase, and that handled by the former service provider will gradually decrease at the same rate. The demand is dynamically controlled by the customer's assignment of the carrier either by prior presubscription or by dialing the carrier's access code.

Distribution of the demand among the carriers is therefore unpredictable and the forecast of the access trunk requirements is consequently uncertain. This can lead to heavy unnecessary expenditures by both the LECs and the ICs unless a plan is provided to reduce the cost penalty resulting from shifting the demand from one carrier to another. In the meantime, the former dominant carrier will be left with access trunk capacity in excess of its requirement at no extra cost to it because the access tariff is based on the "minutes of use".

It is the specific intent of this paper to:

(1) Develop an optimum trunking plan to provide to multi-interexchange carriers access that is "equal in type, quality and price" to that provided to AT&T without augmenting the quantities of toll connecting trunks required in the former Bell network.

(2) Provide the basis for an algorithm for dimensioning the carrier's direct and tandem access trunks based on equal cost per unit of traffic and equal grade-of service.

The solution suggested by this paper is to segregate a common pool of access trunk capacity to be used by all carriers according to their share of the demand. This can be accomplished by converting the toll connecting trunk groups to alternate routing. The former toll connecting trunk groups provided access for intercity calls. They were engineered as full (non-alternate route) trunk groups.

In this paper I am not claiming a novel method or theory for solving this problem; rather, I am adapting the alternate route techniques to multiple-carrier operation. Those techniques have been promoted in this Congress and are widely used for the optimization of communication networks. Because the interest in this paper may extend beyond this Congress to those who are not familiar, or do not have access to those techniques, I will review in more detail the relevant formulas for solving the problem at hand.

In Section 1, the theories and formulas for sizing the trunk groups using alternate route trunking are reviewed. Section 2 provides an algorithm for the implementation of an equal access plan that fully complies with the MFJ.

1. THEORY OF ALTERNATE ROUTING

The principle of alternate routing is applied in communication networks by providing a first choice high-usage route and a second-choice alternate route when the call fails to find an idle trunk in the first route. The network optimization process involves an evaluation of the trade-offs between switching and transmission costs. Finding the proper balance will lead to the least-cost network.

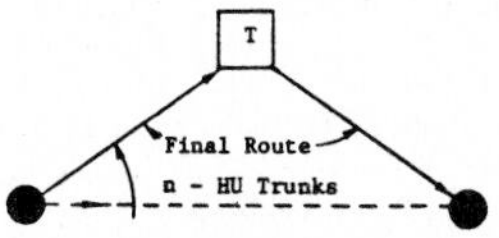

Fig. 2 - Basic Alternate Route Triangle.

Alternate route engineering determines the optimal quantity of trunks in the HU group that results in the least-cost network. The solution to this problem was first published by Truitt [1] and was further extended by many other authors [2 to 7]. Traditionally, trunk engineering of alternate route networks was based on that model, known as the "cost ratio" method.

Figure 2 shows the basic alternate route triangle.

It can be shown that the load carried by the last incremental trunk in a truck group consisting of n trunks, to which a Poisson load with mean $\underline{a}$ is offered is given by:

$$\alpha(n) = aB(n-1, a) - aB(n,a) \quad \ldots \quad (1)$$

where $B(n,a) = \frac{a^n}{n!} \Big/ \sum_0^n \frac{a^i}{i!}$

is the Erlang -B function,

The optimum n of the number of HU trunks denoted by n* is that for which (n*) is close to the economic load, known as the "economic CCS" (ECCS). The ECCS is given by the relation:

$$ECCS = K/R \quad \ldots \quad (2)$$

where K is the "marginal capacity" of the alternate route to traffic overflowing from the HU group and R is the "cost ratio" of a final trunk to a HU trunk.

In terms of the ECCS, $\alpha(n^*)$ is given by:

$$\alpha(n^*) \geq ECCS > \alpha(n^* + 1) \quad \ldots \quad (3)$$

Determination of the ECCS depends on the marginal capacity K and the cost ratio R. Crucial inaccuracies, however, have been found in practice in arriving at their values. The marginal capacity K was assumed to be a constant. Arbitrary values around 28 CCS were used. It has been shown lately [8] that K is a specific derivative that is a function of the blocking, the "background" traffic and its peakedness and the overflow traffic from the first route and its peakedness. Formulas for computing the true value of the marginal capacity K were provided [8]. For low blocking, Dayem (Appendix A of [8]) provided the following formula which is valid for most cases of interest:

$$K = (1-B)^{-1} \quad \ldots \quad (4)$$

This shows that the marginal capacity K approaches one (1) erlang (36 CCS) at B.01. The accuracy of the process depends to a large extent on the accuracy of the marginal capacity K. In the restructured network, large capacity, highly efficient final groups will be required at each end office.

2. THE EQUAL ACCESS PLAN.

This section describes the equal access plan that meets the MFJ requirements to provide to all carriers access that is "equal in type, quality and price" to that of AT&T. In general, the plan consists of providing at each end office a common pool of traffic capacity to carry, initially, the full load of the small-volume carriers and the overflow load of high-volume carriers. This will be accomplished by converting the toll connecting trunk groups from full (non-alternate route) groups to alternate routing. The common pool capacity will be provided in the final route which also

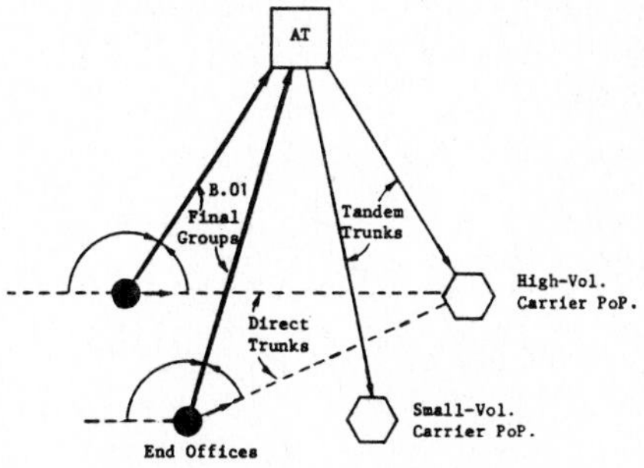

Fig. 3 illustrating Alternate Route Model for Interexchange Carriers Connecting Trunks.

carries the office "background" traffic. The direct and tandem trunks will be optimized for least-cost network based on the marginal capacity and the cost ratio using modular trunk engineering. It has been shown [9, 10] that there are cost savings in using trunk module sizes of 12 and 24 for one-way and two-way trunks, respectively.

Figure 3 illustrates the alternate routing model in an area with several interexchange carriers. The PoPs of the dominant carrier are connected to the end offices by direct (high-usage) trunks and to the tandem by tandem trunks. The PoPs of small-volume carriers are connected only to the tandem. Each end office is connected to its home tandem by a "final" trunk group. It is required to develop an algorithm for sizing the trunks in this model.

Sizing of the Direct Trunks.

With reference to the alternate route triangle in figure 2, it can be shown that the traffic carried by the last incremental module, in a trunk group consisting of r modules of size m is given by:

$$\alpha(rm) = aB(rm-m, a) - aB(rm, a) \quad \ldots \quad (5)$$

The optimum number of modules r, denoted by r* that determines the optimum size of the direct trunks is that for which $\alpha(r^*m)$ is close to the economic load. The economic load, EL, in this case is given by:

$$EL = m\,K/R \quad \ldots \quad \ldots \quad (6)$$

where K and R have the same meaning as before.

In terms of the economic load, the optimum number of modules r* is given by the relation:

$$\alpha(r^*m) \geq mK/R > \alpha(r^*m + m) \quad \ldots \quad (7)$$

The mean of the traffic offered to the alternate route, i.e. the additional traffic capacity required to be provided in the final group is given by:

$$y(r^*m, a) = a\,B\,(r^*m, a) \quad \ldots \quad (8)$$

In order to provide capacity in the common pool in addition to that required for optimum overflow load, it will be necessary to increase the overflow load by reducing the optimum direct modules by one or two modules.

This process will also be used, in a similar way, to establish new direct groups for the new carriers. In order to provide additional capacity in the common pool, then the direct group will have one module less than r.

The process for sizing the final group and tandem trunks are well known [11].

CONCLUSION

The process presented here provides a method for optimizing the direct and tandem trunks for all carriers that results in the least-cost network without augmenting the quantities of access trunks used by a single-service provider. It is based on the use of large size final routes whose marginal capacity approaches one (1) erlang at B.01 and constant cost ratio R. It requires only the forecast loads used in the former network to determine the dimensions of the carriers access trunks.

This process satisfies the MFJ's three-phase equal access stipulations:

1. Equal type - is ensured by providing access to all carriers based on the direct-alternate route model. All carriers contribute to and share a common pool of access traffic capacity. Access of high-volume carriers consists of tandem and direct trunks, if economically justified. Access of small-volume carriers consists, initially, of tandem trunks only.

2. Equal quality - is ensured by equal grade-of-service provided by the common final route. The quality of transmission is controlled by the tariffed type of interconnection and the use of digital facilities and digital tandem switching.

3. Equal Price - is controlled by using final routes with high marginal capacity and a uniform cost ratio R for all carriers. This will result in equal cost per minute-of-use.

4. Last, but not least, the carrier access charge tariff can be based on the cost of the direct-alternate route model.

Acknowledgment.

I wish to thank P.L. Wynns for encouraging me to prepare this paper, and to thank J.M. Kraushaar and J. Windhausen for their contributions.

REFERENCES

1. Truitt, C.J. - "Traffic Engineering Techniques for Determining Trunk Requirements in Alternate Routing Networks", B.S.T.J., Vol. 33, pp. 277-302, March 1954.

2. Wilkinson, R.I. - "Theories for Toll Traffic Engineering in the USA", B.S.T.J., Vol. 35, pp. 44-514, 1956.

3. Pratt, C.W. - "The Concept of Marginal Overflow in Alternate Routing", Fifth ITC, New York, 1967.

4. Rapp, Y. - "Planning of Junction Network in a Multiexchange Area", Ericsson Tech., Vol. 20, No. 1, 1964.

5. Asgersen, C. - "A New Danish Traffic Routing Plan with Single Stage Alternate Routing and Consistent Use of Service Protection Finals for First Routed Traffic to Tandem Offices", Fifth ITC, New York, 1967.

6. Valenzuela, J.G. - "Program for Optimizing Networks with Alternate Routing", Sixth ITC, Munich, 1970.

7. Harrington, J.S. - "Equivalent High Usage Circuits and their Application in the Analysis and Design of Local Networks Employing Alternate Routing", Eighth ITC, Melbourne, 1973.

8. Rao, R.N. - "Improved Trunk Engineering Algorithm for High Blocking Hierarchical Networks", Ninth ITC, Torremolinos, Spain, 1979.

9. Levine, S.W. and Wernander, H.A. - "Modular Engineering of Trunk Groups for Traffic Requirements", Fifth ITC, New York, 1967.

TELETRAFFIC ISSUES in an Advanced Information Society
ITC-11
Minoru Akiyama (Editor)
Elsevier Science Publishers B.V. (North-Holland)

ADDENDUM *

OVERVIEW OF THE MODIFIED FINAL JUDGMENT

The Modified Final Judgment, approved by Judge Greene on August 24, 1982, settled the antitrust case filed by the U.S. Department of Justice against AT&T almost eight years earlier. Besides terminating the antitrust action, the agreement provided for several fundamental changes in AT&T's corporate structure. Before turning to summarize the terms of that agreement, however, it will be helpful to trace briefly the pre-divestiture environment.

For most of the last fifty years, AT&T has dominated the U.S. telecommunications market as a vertically integrated, privately-owned monopoly. AT&T owned 100% of the Western Electric Company, which manufactured and sold telephone products and equipment to its subsidiaries. AT&T and Western Electric together owned 100% of Bell Laboratories. AT&T Long Lines provided nearly 100% of the U.S. and international long distance service. Finally, AT&T owned a controlling, and often a complete, interest in 20 of the 22 Bell Operating Companies (BOCs) that provide local telephone service within their exchange areas.

Because of AT&T's monopoly control, AT&T has been regulated by both federal and state governments. In 1934, Congress passed the Communications Act, thereby creating the Federal Communications Commission (FCC). Congress gave to the FCC the authority to regulate all "interstate" communications services for the purpose of making available, "so far as possible, to all the people of the United States a rapid, efficient, Nation-wide, and world-wide wire and radio communication service with adequate facilities at reasonable charges..." (Section 1). State governments retain the authority to regulate all intra-state communications, and they do so in a variety of ways depending on local circumstances. The FCC, however, may preempt state regulation of an intra-state activity if the activity is connected to an interstate activity and the state's regulation conflicts with federal law or policy.

Over the last thirty years, the FCC and the U.S. courts have opened the gates to competition to AT&T in both the telephone equipment market and the long distance market. As far back as 1949, the U.S. Department of Justice maintained that AT&T's monopoly over the provision of customer premises equipment (CPE) was unjustified. In that year the Justice Department filed an antitrust suit against AT&T, seeking to force AT&T to relinquish its Western Electric stock, and seeking generally to separate telephone manufacturing from the provision of telephone service. The antitrust case resulted in a settlement agreement between AT&T and Justice, known as the 1956 Consent Decree. The Decree allowed AT&T to keep Western Electric, but prohibitted Western Electric from manufacturing any equipment other than that used by the Bell System, and restricted AT&T to providing common carrier communications services.

Despite the Justice Department's failure to separate telephone manufacturing from telephone service, the FCC and the courts have gradually permitted competition in the telephone equipment market. In the "Hush-a-phone" decision in 1956 and the Carterphone decision in 1968, customers were allowed to attach non-AT&T telephones to the network. These decisions paved the way for non-AT&T companies to begin manufacturing and marketing telephones in competition with Western Electric. The rapid growth of alternative sources of telephones and other customer premises equipment (CPE) allowed the Commission to deregulate CPE in the 1979 Second Computer Inquiry Decision. AT&T may now market CPE without filing tariffs as long as it does so under a separate subsidiary. The Computer II decision forced AT&T to split into two companies: AT&T-Information Systems (ATTIS), which markets telephone equipment and so-called "enhanced services", and AT&T-Communications (ATTCOM), which supplies basic transmission services.

The growth of competition in the long distance market has proceeded more slowly than competition in the CPE market. Beginning with the Commission's approval of an MCI application to provide private line service in competition with AT&T in 1969, the Commission and the courts have gradually expanded the opportunities for competition into the entire long distance market. Competition was hindered, however, by AT&T's ownership of the local exchange companies, with which the competing carriers had to interconnect in order to allow customers to obtain access to their services. Two dangers to competition were identified: first, it was feared that AT&T would subsidize its competitive long distance services with profits from its local service, over which

---

* Compiled by John Windhausen

AT&T retained a virtual monopoly; second, it was feared that the local exchange companies would discriminate in favor of AT&T Long Lines, giving AT&T higher quality connections to their local switches than they would give to the competitors. Because of AT&T's potential ability to use its ownership of the BOCs to engage in these forms of anti-competitive conduct, the Justice Department once again filed an anti-trust action against AT&T, seeking this time to separate the BOCs from AT&T, and seeking generally to separate local service from long-distance service.

The settlement agreement (MFJ) that resulted from this anti-trust action modifies the 1956 Consent Decree in several ways. AT&T retains its ownership of Western Electric and Bell Laboratories but AT&T is no longer restricted by the terms of the 1956 Consent Decree to providing common carrier communications services. The MFJ frees AT&T to enter the data communications market, the computer market, and all forms of information services except electronic publishing, which the judge found to be an infant industry that would be too easily dominanted by AT&T. Western Electric, Bell Labs, AT&T Information Systems, and AT&T International together are now known as AT&T Technologies.

The most dramatic provisions of the MFJ call for the complete divestiture of AT&T's ownership interests in the Bell Operating Companies. The divestiture, which went into effect on January 1, 1984, forced AT&T to relinquish $100 billion in fixed plant, or about 2/3 of its total fixed assets. Judge Greene has also ruled that AT&T may no longer use the "Bell" name and logo. The twenty-two Bell Operating Companies have been combined into seven independent regional holding companies (know as regional Bell Operating Companies, or RBOCs). Each BOC is limited to providing local exchange service within its area.

The divestiture alone does nothing to prevent the BOCs from discriminating against other long distance carriers, however. The MFJ solved this problem by requiring the BOCs to provide equal access to all long distance carriers upon bona fide request. Specifically, the agreement calls for the BOCs to provide access that is "equal in type, quality, and price" to that provided to AT&T. The MFJ requires the BOCs to offer equal access from end offices serving at least one-third of its access lines by September 1, 1985, and requires the BOCs to offer equal access to all of its access lines by September 1, 1986.

Prior to the MFJ, competing interexchange carriers obtained a variety of different types of access connections that were inferior to that of AT&T. As equal access is put into effect the competing carriers will obtain the same high quality connections as AT&T. The MFJ also provides that the BOCs will collect access charges from the interexchange carriers at rates set by tariffs filed with the FCC. As the competing carriers obtain the same high quality connetions as AT&T, they will also pay the same high access charges.

The MFJ is a complex and detailed document, of which only the main points have been presented here. The growth of competition makes telecommunications a dynamic industry, and Judge Greene and the FCC will continue to monitor the terms of the agreement to determine how well it withstands the tests of an innovative marketplace. Whether competing carriers will ever succeed in posing a significant challenge to AT&T's continued dominance is yet to be seen, but the MFJ at least establishes the basic groundwork for a fully competitive environment.

TELETRAFFIC ISSUES in an Advanced Information Society
ITC-11
Minoru Akiyama (Editor)
Elsevier Science Publishers B.V. (North-Holland)

Class of Service Analysis of Traffic Variations at Telephone Exchanges

Vladimir A. Bolotin

Bell Communications Research
Red Bank, New Jersey, U.S.A.

Abstract

Class of service analysis of subscriber line traffic identifies linear functional relations between a peak traffic value and the average busy season (ABS) value. For the highest day (HD),

$$HD\ value = ABS\ value + (ABS-to-HD\ Increment),$$

where, in a given area at a given period of day, ABS-to-HD increment is a linear function of ABS traffic per subscriber line. This is apparently true of any other peak statistic of interest, such as THD (ten highest day average), 4HD (fourth highest day), OAM (once-a-month), etc.

The introduction of the ABS-to-PEAK increment instead of the PEAK/ABS ratio simplifies and improves peak period estimation procedures for traffic engineering. In particular, the increment method has made it possible to estimate HD usage and calling rate for all Bell Operating Companies.

1. Overview

This traffic variability analysis continues a series of subscriber traffic studies described earlier at ITC9 [1] and ITC10 [2]

An extreme traffic value, e.g. the highest day (HD) CCS/Line or HD calling rate, is commonly estimated as

$$HD\ value = (ABS\ value) \times (HD/ABS\ ratio). \quad (5)$$

As shown in this paper, the HD/ABS ratio can be adequately analyzed only in rare cases of homogeneous subscriber traffic (e.g. purely residential, purely individual business, etc.). In general, this ratio is an empirically derived characteristic of day-to-day traffic variability, which is difficult to describe analytically.

To develop an analytical model of variability, we examine the major factor behind the differences in day-to-day traffic variability among telephone exchanges. This factor is subscriber class of service mix. In a given area at a given period of day, class of service mix is the most significant factor that defines how large is the deviation of extreme traffic values from the average busy season value.

We introduce (see sections 2 and 3) a **new day-to-day variability characteristic of the exchange traffic - "ABS to HD Increment,"** so that

$$HD\ value = (ABS\ value) + (ABS-HD\ Increment). \quad (8)$$

Analysis of the ABS to HD increment is based on linear functional relation (described in section 2) between ABS traffic and subscriber class of service mix. The same linear relation was found to exist between HD traffic and class of service mix.

In section 4, we further show that the HD traffic is a linear function of the ABS traffic and the ABS-HD increment is also a linear function of the ABS traffic:

$$ABS-HD\ Increment = (Z-1) \cdot a + S\ . \quad (13)$$

Equation (13) suggests how **typical variability characteristics of the total telephone area can be found and used for quantification of day-to-day traffic variability at each single exchange.** First, the coefficients Z and S may be found by simple linear regression with large amount of input data collected at all exchanges in the given area. Then, ABS-HD increment value for a particular exchange is found by formula (13), in which a is the ABS value (CCS/Line, Calls/Line) at this exchange. The HD value is obtained by adding the ABS-HD increment to the ABS value - formula (8). This method is explained in detail in section 5.

The ABS-HD increment analysis reveals how the exchange-to-exchange variability splits in two parts (see section 5). The larger part, described by equation (13), is due to the class of service variability among exchanges. This part is easily defined and can be used for prediction of extreme traffic for any exchange with given class of service mix or given ABS traffic per line. The uncertainty of this prediction is to be attributed to the second part, which is mostly due to year-to-year variability of the extreme traffic at the exchange. In this sense, the highest day traffic would be predicted with almost all possible accuracy.

Estimates of the coefficients Z and S in function (13) were obtained by linear regression for all Bell Operating Companies (BOCs) in the U.S.A. The estimates are based on observations at more than 800 exchanges during 5-6 years (section 6).

Section 7 presents a more detailed analysis of relations between the ABS-HD increment and ABS usage or calling rate depending on class of service mix and on time of day.

The accuracy improvement achieved by the new method is illustrated in section 8.

Time consistent busy hour values are assumed throughout this paper. It can be shown that the principles apply equally well to bouncing busy hour statistics used in extreme value engineering [3], [4].

The analysis described in this paper is applicable not only to the highest day traffic but to any other extreme value traffic statistic of interest, such as THD (ten highest day average), 4HD (fourth highest day), OAM (once-a-month)[4] etc.

2. *Class of Service Decomposition of Telephone Exchange Load*

Consider a telephone area and an exchange within it. Let there be r classes of service in the area with the traffic intensity per line a_i in the $i-th$ class. Vector $\mathbf{a}=(a_1,a_2,\ldots,a_r)$ **is the area class of service traffic vector.** Let n_i be the number of lines in the $i-th$ class and n be the total number of lines at the exchange. Denote $\mathbf{n}=(n_1,n_2,\ldots,n_r)$. In the previous studies[2], we characterized the (total) exchange traffic intensity A by its mean value

$$E(A) = \mathbf{n}\cdot\mathbf{a} = \sum_{i=1}^{r} n_i a_i \,, \qquad (1)$$

and by its variance $V(A)$, the latter being a measure of the residual variation after the class of service variation among exchanges is removed. Correspondingly, **the exchange traffic intensity per line** a has the mean value

$$E(a) = E\left(\frac{A}{n}\right) = \frac{\mathbf{n}}{n}\cdot\mathbf{a} = \mathbf{x}\cdot\mathbf{a} \,, \qquad (2)$$

where $\mathbf{x}=(x_1,x_2,\ldots,x_r)$ **is the exchange class of service mix vector:** $x_i=n_i/n$.

We have shown[2] that, with the time of day and the area given, most of the exchange-to-exchange traffic variation is attributed to class of service mix differences among exchanges. Moreover, the residual variation, i.e. the variance $V(a)$, is small enough to allow reasonably accurate estimation of ABS exchange load on the basis of the exchange class of service mix $\mathbf{x}$ and the area class of service loads $\mathbf{a}$. The residual variability is mostly year-to-year variability of the ABS load.

The class of service traffic vector in formulae (1) and (2) may refer to the average busy season load, to the highest day (HD) load, to the ten highest day average load etc. (The only difference is that the year-to-year variability of the HD load is somewhat larger than that of the ABS load.) To be specific, we shall reserve notation a for the ABS traffic, and use h for the HD traffic:

$$E(H) = \mathbf{n}\cdot\mathbf{h} = \sum_{i=1}^{r} n_i h_i \,, \quad E(h) = \mathbf{x}\cdot\mathbf{h} \,, \qquad (3)$$

where h is the highest day exchange traffic intensity per line.

The class of service decomposition (1)-(3) is equally applicable to subscriber call volume (calling rate). In this case, vectors $\mathbf{a}$ and $\mathbf{h}$ would describe ABS and HD call volume instead of ABS and HD traffic intensity.

For class of service analysis of day-to-day traffic variation we shall express the r.v. a and h as

$$a = \mathbf{x} \cdot \mathbf{a} + \epsilon_a , \quad h = \mathbf{x} \cdot \mathbf{h} + \epsilon_h , \qquad (4)$$

where ϵ_a, ϵ_h are r.v. with zero mean values and the variances $V(a)$ and $V(h)$.

3. *ABS-HD Increment vs. HD/ABS Ratio*

A peak traffic value, e.g. the high day (HD) CCS/Line or HD calling rate, is commonly estimated as

$$HD\ value = (ABS\ value) \times (HD/ABS\ ratio) \ . \qquad (5)$$

To analyze the HD/ABS ratio, we assume that the area and time of day are given, so that the area class of service load vectors **a** and **h** are constants. Using (4) we express the HD/ABS ratio as

$$HD/ABS\ Ratio = \frac{\mathbf{x} \cdot \mathbf{h} + \epsilon_h}{\mathbf{x} \cdot \mathbf{a} + \epsilon_a} \ . \qquad (6)$$

For one exchange, the class of service mix vector **x** is given, and the variability of ratio (6) is the residual variability (not related to class of service) expressed by the random terms ϵ. For the total population of exchanges, the class of service component **x** is a random variable, too, and the fraction (6) has much larger total variability. A typical example of this phenomenon is shown in Fig. 1. For the total population, the full range of the measured HD/ABS ratio values in Fig. 1 is from 1.05 (for purely business exchanges) to 1.50 (for purely residential exchanges). For a given exchange, i.e. for each fixed class of service mix (fixed percentage of residential lines), the ratio has a much narrower range; variation within this range is attributed mostly to year-to-year differences as explained in section 5.

For traffic engineering applications, it may be feasible to quantify the residential and business HD/ABS ratio for each telephone area at each of the three busy hour time periods - morning, afternoon and evening. But it is much more difficult to quantify the HD/ABS ratio as a function of class of service mix - the relation suggested by the tentatively drawn curve in Fig. 1, apparently a non-linear function. A major difficulty lies in the fact that, according to formula (6), the HD/ABS ratio is a ratio of random variables.

Consequently, day-to-day traffic variability may be adequately described by the HD/ABS ratio, but only in rare cases of homogeneous subscriber traffic (e.g. purely residential, purely individual business, etc.). In general, this ratio is a characteristic of day-to-day traffic variability that is difficult to describe analytically.

As we show in section 4, functional relation between HD traffic and ABS traffic is essentially linear. Therefore, instead of the HD/ABS ratio, we introduce a new variability characteristic, which is also linearly related to class of service mix. We define the new variability characteristic as the difference between the HD and the ABS values:

$$ABS-HD\ Increment \equiv h-a = \mathbf{x} \cdot (\mathbf{h}-\mathbf{a}) + \epsilon_h - \epsilon_a . \qquad (7)$$

Then, instead of (5) we shall use

$$HD\ value = ABS\ value + ABS-HD\ Increment \ . \qquad (8)$$

Linear relations of this type result in sums and mixtures of random variables. These are easier to analyze

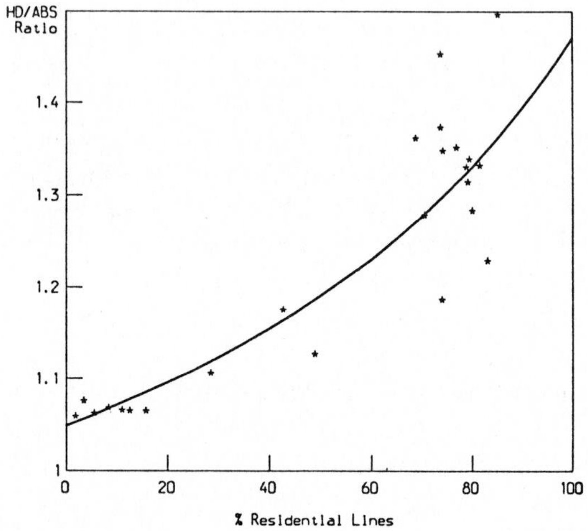

Fig. 1. Example of HD/ABS Ratio Variability

than relations in the form of ratios (6). In particular, the non-linear function for the HD/ABS ratio will be reduced to a linear function for the ABS-HD increment, as explained in detail in section 5.

4. *ABS-HD Increment as a Linear Function of ABS Traffic*

In many practical applications it is enough to consider only two classes of service, residential and business (the latter includes all non-residential services). Consider a_r, a_b, h_r, h_b, where a is ABS traffic value, e.g. CCS/Line, h is the high day CCS/Line, r stands for "residential", b stands for "business". The following analysis shows that the HD traffic and the ABS-HD increment are linear functions of the ABS traffic.

With the proportion of residential lines equal to x_r, formulae (4) for a and h, will be written:

$$a = a_r x_r + a_b(1-x_r) + \epsilon_a , \qquad (9a)$$

$$h = h_r x_r + h_b(1-x_r) + \epsilon_h . \qquad (9h)$$

Elimination of x_r shows that the HD and ABS traffic values are linearly related:

$$h = Z \cdot a + S - Z \cdot \epsilon_a + \epsilon_h = Z \cdot a + S + \epsilon , \qquad (10)$$

where

$$Z = \frac{h_b - h_r}{a_b - a_r} , \quad S = \frac{h_r a_b - h_b a_r}{a_b - a_r} . \qquad (11)$$

Formula (7) is now

$$ABS-HD\ Increment = (Z-1) \cdot a + S + \epsilon . \qquad (12)$$

The residual ϵ describes year-to-year variability of the increment about its mean value (shown in Fig. 2, section 5). The mean value is a good estimate of the ABS-HD increment engineering value:

$$ABS-HD\ Increment = (Z-1) \cdot a + S . \qquad (13)$$

Special cases of (13) are discussed in section 7.

Equation (13) quantifies day-to-day traffic variability at any exchange. The coefficients Z and S may be found by a simple linear regression with large amount of data from all exchanges in the given area. An example of the new method is described in the next section.

5. *ABS-HD Increment Method: An Example*

Fig. 2 shows the relation between HD and ABS usage as a function of the class of service mix for two classes of service, residential and non-residential (business). These data describe morning traffic in a telephone area.

The mean, median and other percentiles were obtained from the raw data which consisted of 220 points observed in 1978-82 at 59 exchanges in one Bell Operating Company. The x-axis variable is expressed both in ABS usage and in proportion of residential lines. These are linearly related - see section 4, formula (9a).

Single observation points shown in Fig. 2 represent typical year-to-year variation exemplified by 17 exchanges. A vertical line with 4 or 5 points on it represents the observed range of year-to-year HD load variability at one exchange. For most of the 17 exchanges, the range of the year-to-year variation in any given exchange is on the order of the variation range shown by the percentiles for the entire 59 exchange sample. (Only one exchange out of 17 has a much narrower range, with the increment value under 0.3 CCS/Line.)

We conclude that the high day usage at an exchange can be predicted on the basis of the high day usage and class of service mix data for other exchanges in the telephone area. The inaccuracy (uncertainty) of this prediction should be attributed mostly to the variation of the highest day load from year to year. In this sense, the highest day traffic would be predicted with as much accuracy as feasible. Because of this fact, the class-of-service data can be used not only for default engineering but as an important component in forecasting, along with historical data available for the given exchange.

Fig. 3 shows the median and the quantiles of the HD/ABS ratio calculated from Figure 2 data as

$$HD/ABS\ Ratio = 1 + \frac{ABS-HD\ Increment}{ABS\ Value} . \qquad (14)$$

Since the ABS CCS/Line always depends on class of service mix[2], the HD/ABS ratio also does. It is clear that the increment method automatically accounts for this factor.

Suppose that we tried to do this analysis in terms of the HD/ABS ratio instead of ABS-HD increment. The linear-fractional fitting curve (6) should be considered (instead of the straight line) - a very significant complication in comparison with the linear regression. The comparative analysis of the 17 exchanges would not show that their year-to-year variation is essentially the

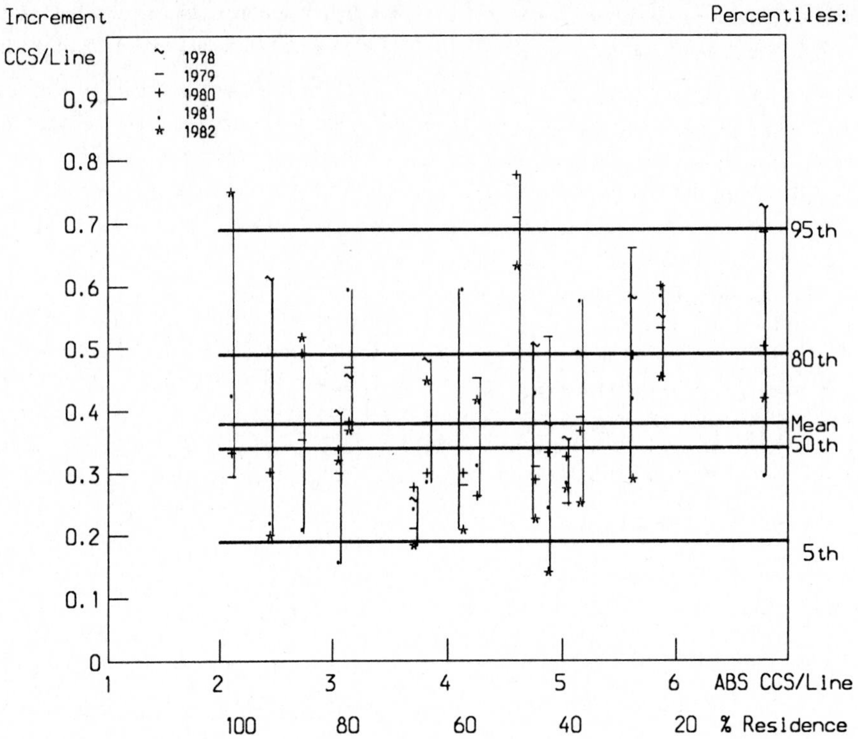

Fig. 2. ABS-HD Increment vs. Class of Service Mix

same. This fact would be masked by different HD/ABS ratio ranges at different class of service mixes as shown in Fig. 3.

Fig. 4 describes the subscriber calling rate increment for the same population of exchanges.

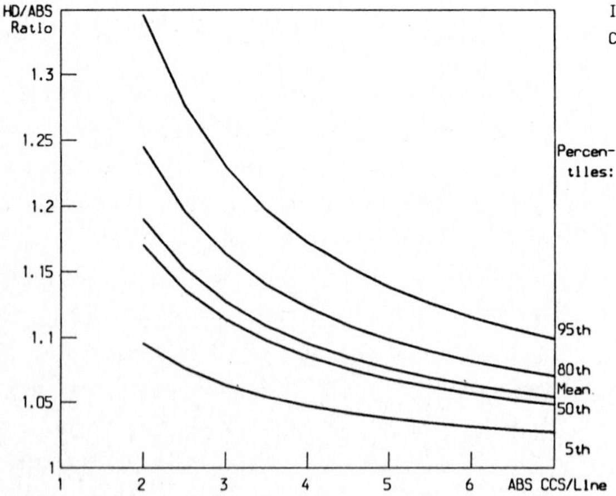

Fig. 3. HD/ABS Ratio for ABS-HD Increment in Fig. 2

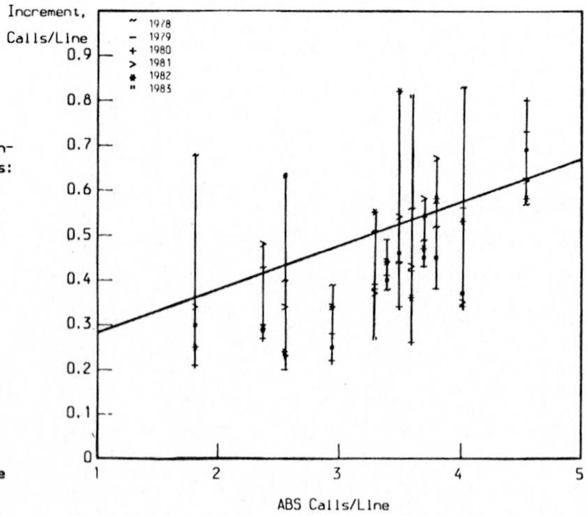

Fig. 4. Calling Rate ABS-HD Increment

6. *HD Traffic Estimation by the Increment Method - All Bell Operating Companies*

Estimates of the coefficients $(Z-1)$ and S in function (13) were obtained by linear regression for all Bell Operating Companies in the U.S.A. Sample estimates are given in Tables I and II (only 3 out of the 17 companies are shown in each table).

Table I. Sample ABS-HD CCS/Line Increment

Bell Companies	Morning			Afternoon			Evening		
	S	Z-1	n	S	Z-1	n	S	Z-1	n
1	.37	.00	210	.45	.00	139	.02	.16	57
2	.72	-.04	200	.55	-.01	322	.18	.09	72
3	.28	.04	278	.21	.07	82	.06	.12	68

n - sample size (number of exchange-years); S in CCS/Line

Table II. Sample ABS-HD Calling Rate Increment

Bell Companies	Morning			Afternoon			Evening		
	S	Z-1	n	S	Z-1	n	S	Z-1	n
1	.18	.10	210	-.01	.20	159	.01	.24	33
2	.31	.11	194	.31	.10	321	.05	.25	66
3	.20	.17	297	-.05	.26	85	-.13	.34	63

n - sample size (number of exchange-years); S in Calls/Line

Among the 17 companies, the sample size for each estimation (a fixed period of day for a company) varies between 15 and 278 ("exchange-years" points). The total volume of data used for each complete table is over 4000 points observed at more than 800 exchanges.

With the values of Z and S based on the data from all exchanges in the area, the ABS-HD increment and the HD value for a particular exchange is found by formulae (13) and (8), in which a is the ABS value (CCS/Line, Calls/Line) at this exchange.

7. *Types of Relation between ABS-HD Increment and ABS Traffic*

This section provides additional analysis of the increment behaviour depending on class of service settings and on time of day.

Generally, in the daytime (morning and afternoon)

$$\frac{h_r}{a_r} > \frac{h_b}{a_b} \text{ and } a_b - a_r > 0\,. \tag{15}$$

Therefore $S>0$. (See formulae (11) and (13)). These relations can be applied to the evening traffic, too, if the residential subclass with custom calling services, in particular call-waiting service, substitutes for the business class. (Residential subclasses have been discussed in[2].)

An extreme type of the increment behaviour is observed when $h_r{:}a_r = h_b{:}a_b$, so that $S=0$, and

$$HD/ABS\ Ratio = Z = const. \tag{16}$$

In particular, for the evening time the business class should be replaced by the custom calling service residential class, which has almost the same HD/ABS ratio as the regular residential class. According to our comprehensive results on traffic in all Bell Operating Companies (see section 6), this case is almost never applicable to usage (CCS/Line), but is applicable to some of the afternoon and evening calling rates.

Another, more typical, extreme in the increment behaviour is represented by the morning usage (CCS/Line) example described in section 5 (Fig. 2). In this case, $Z-1 = 0$, and by (11) and (13)

$$ABS-HD\ Increment = h_r - a_r = h_b - a_b = const. \tag{17}$$

Here, the ABS-HD increment (not the HD/ABS ratio) is the same both for the residential and business class. Consequently the value of the usage increment is independent of the ABS CCS/Line value. This is the case in most of the morning and about half of the afternoon usage data, which show small values of $(Z-1)$. In these cases, formula (17) is clearly applicable.

Intermediate cases, with non-zero values of both $Z-1$ and S, are observed in a smaller part of the afternoon usage cases, in almost all evening usage cases and in all morning calling rate cases. Formula (13) is applicable in these cases.

8. *Accuracy of HD Load and Calling Rate Estimation by the Increment Method*

The morning usage increment for Company 2 (see Table I) is $0.72-0.04\cdot(CCS/Line)$. By formula (14), $HD/ABS\ Ratio = 0.96 + \dfrac{0.72}{ABS\ CCS/Line}$. This relation is drawn in Fig. 5.

The histogram in Fig. 5 shows the HD/ABS ratio for all 200 sample points (200 exchange-years). Each of 10 histogram bars represents about 10% of the sample points. The horizontal line designates the average HD/ABS ratio, computed for all exchanges used in the study (with the morning, afternoon and evening busy hour).

The average company HD/ABS ratio (1.17) reflects the actual situation very poorly. At least for the 4 right histogram bars, i.e. for 40% of exchanges, the difference between the company average and exchange actual HD/ABS ratio is about 8.5%. That would mean an overestimation of HD usage by 8.5 % if the company average were used. For another 20% of exchanges there will be overestimation, too, by 3-4%. On the other hand, for predominantly residential exchanges, with ABS usage under 3 CCS/Line, the HD usage would be underestimated by 8% or more. This could cause serious service problems.

The increment based curve by formula (14) follows the histogram closely: the HD/ABS ratio based on the increment estimates the actual ratio correctly.

A similar analysis shows that for calling rate the overestimation and underestimation errors are about the same as for usage.

As seen in Fig. 5, when the observed value of ABS usage increases from predominantly residential (2-3 CCS/Line) to predominantly business (6-8 CCS/Line), the HD/ABS ratio changes from 1.32 to 1.05. For calling rate the HD/ABS estimate may be as large as 1.50 and as small as 1.16.

In the afternoon traffic, this class of service phenomenon is usually less pronounced than in the morning traffic because the actual range of the ABS values is narrower in the afternoon. For the same Company 2, the observed afternoon values of ABS calling rate are mostly in the range from 1.5 to 2.5 with the corresponding HD/ABS ratio in the range from 1.31 to 1.23. The total range, including a small number of unusual exchanges, is from 1.36 to 1.18. For afternoon usage, the basic range of the HD/ABS ratio is narrower, only 1.21 to 1.15, the total range being between 1.26 to 1.10.

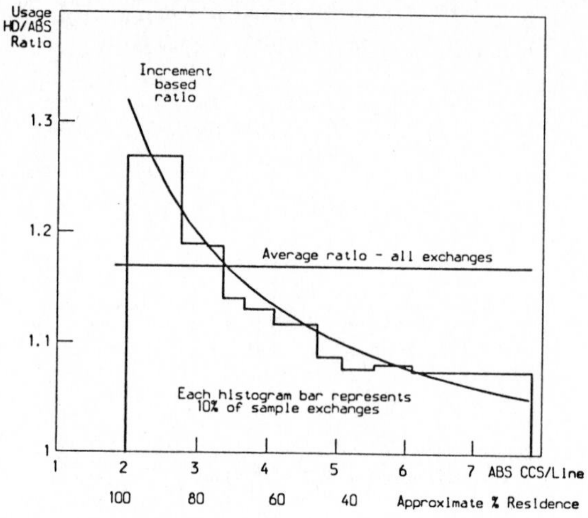

Fig. 5. HD/ABS Ratio Estimation by Increment

The evening traffic is predominantly residential. Therefore the HD/ABS ratio does not change noticeably within the range of the observed evening ABS usage and calling rate. The ABS usage range itself can be larger depending on the penetration of special features, as explained in our ITC10 paper[2].

9. *Acknowledgment*

I am grateful to Mr. J.G. Kappel who contributed a great deal to this study providing valuable advice and criticism.

REFERENCES

1. Bolotin, V.A., Prokopovich, V.S. *Characterizing Traffic Variation.* Proc. 9th International Teletraffic Congress, Torremolinos, pp. 131.1-131.5, 1979.

2. Bolotin, V.A., Kappel, J.G. *Bell System Traffic Usage by Class of Service.* Proc. 10th International Teletraffic Congress, Montreal, 2.4-1, 1983.

3. Barnes, D.H. *Observations of Extreme Value Statistics in Small Switching Offices.* Proc. 9th International Teletraffic Congress, Torremolinos, pp. 311.1-311.5, 1979.

4. Friedman, K.A. *Precutover Extreme Value Engineering of a Local Digital Switch.* Proc. 10th International Teletraffic Congress, Montreal, 1.4-1, 1983.

TELETRAFFIC ISSUES in an Advanced Information Society
ITC-11
Minoru Akiyama (Editor)
Elsevier Science Publishers B.V. (North-Holland)

TRAFFIC ADMINISTRATION IN THE BRAZILIAN NETWORK

Carlos Alberto NUNES, Ricardo CONDE* and Paulo Cesar CARVALHO*

Telecomunicações Brasileiras S.A. - TELEBRÁS
Telecomunicações de Minas Gerais S.A. - TELEMIG*
BRAZIL

ABSTRACT

The Telecommunication Company's final objective is to render good service to its users, in an economic manner, that will lead to reasonable rates and an adequate return on investments made.

In order to attain this objective, plant equipment must be as suitable as possible to the volume of traffic. Thus, it is essential that, within the Company, there be a system of traffic administration, so that, by means of monitoring the central switching offices, it may be possible to detect distortions from the optimal conditions and allow the same to be corrected.

The implementation of a systematic traffic administration policy in Brazil, went into effect from 1979 on, coordinated by the telecommunications holding company, TELEBRÁS. The results already obtained on a national scale, and by TELEMIG in particular, has made a marked contribution to improving the quality of the service.

1. INTRODUCTION

The telecommunications organization in Brazil is composed of 28 state companies, responsible for local and intrastate services, and another company, EMBRATEL in charge of the interstate services.

These companies constitute a system led by Telecomunicações Brasileiras S.A. - TELEBRÁS, a holding company, directly connected with the Ministry of Communications.

Figure 1 shows this system within the framework of the Brazilian telecommunications system.

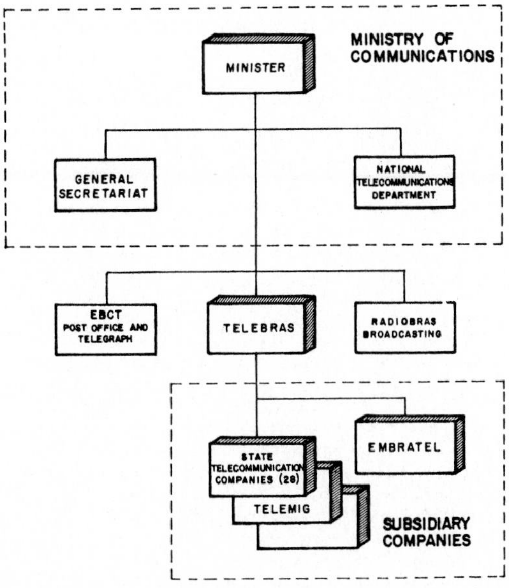

Figure 1 - The Brazilian Communications System

Upon setting up TELEBRÁS in 1972, Brazil had only 2 380 000 telephones (2.4 telephones per hundred inhabitants). At this time special stress was placed upon an increase in the number of main stations. Figure 2 shows the growth in telephone density in the Brazilian network during the period of 1972 to 1984.

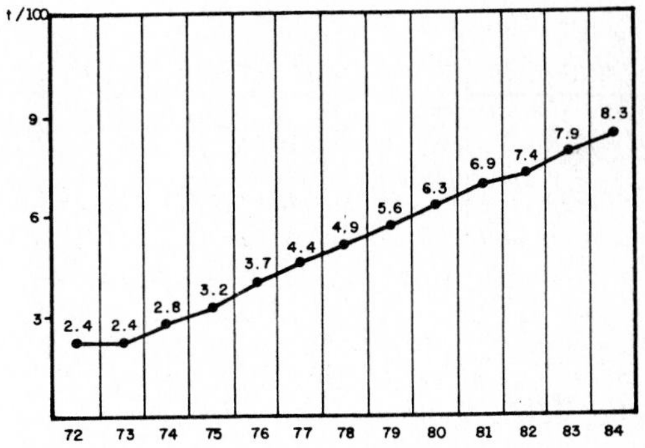

Figure 2 - Telephones per 100 Inhabitants

In 1978, when a total of 5 552 000 telephones was reached (4.9 telephones per hundred inhabitants) the quality of the service rendered became a serious problem. The rate of dial tone delay longer than 3 seconds was 95%, the completion rate was 36.3%, and the rate of equipment blockage and failure was 23.6%.

On the basis of these observations, it was determined that there was a need for a systematic effort aimed at improving the quality of the service, as well as to utilize the plant equipment more efficiently, through the introduction of new operational methods and procedures.

One of the responses to this need was development and implementation of the Traffic Administration System (TAS) in the operating companies.

Traffic Administration and the Program for Reduction of Loss due to the Called Subscriber (LCS), in addition to other programs, were responsible for improving the quality of the Brazilian network service, as can be seen by the changes in some of the indicators shown on the graphs in figures 3 to 8.

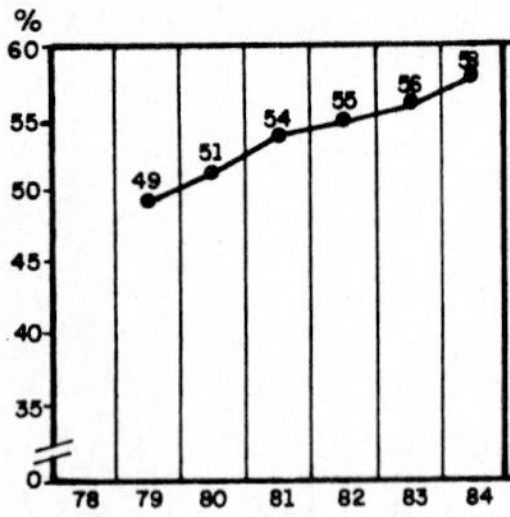

FIG. 3 - LOCAL COMPLETION RATE

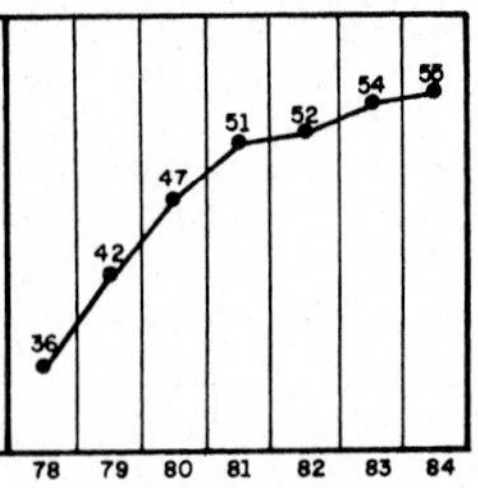

FIG. 4 - DDD COMPLETION RATE

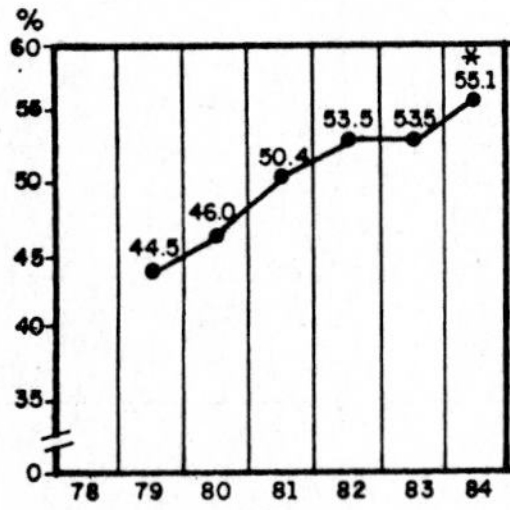

* JAN - SEP 1984

FIG. 5 - IDD COMPLETION RATE ORIGINATED IN USA

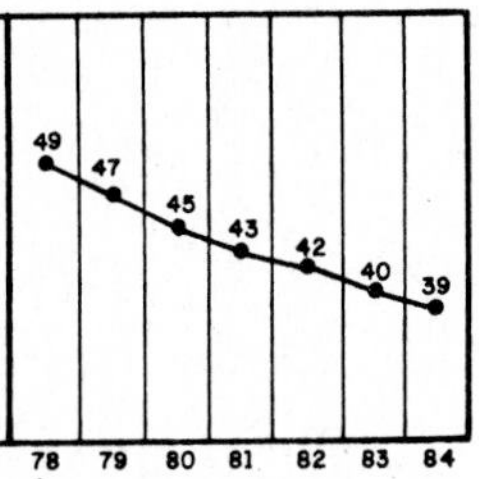

FIG. 6 - LOSS DUE TO THE CALLED SUBSCRIBER (L C S)

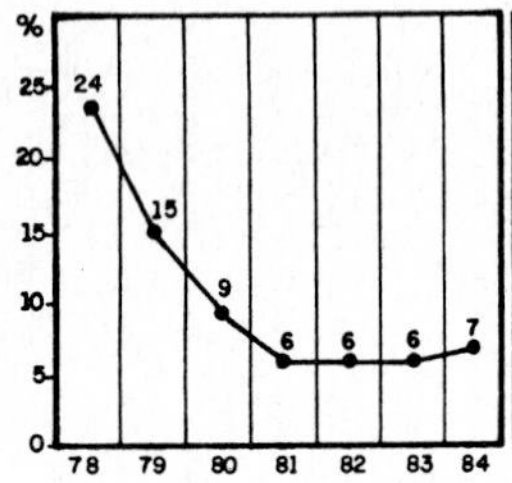

FIG. 7 - BLOCKAGE AND EQUIPMENT FAILURE RATE

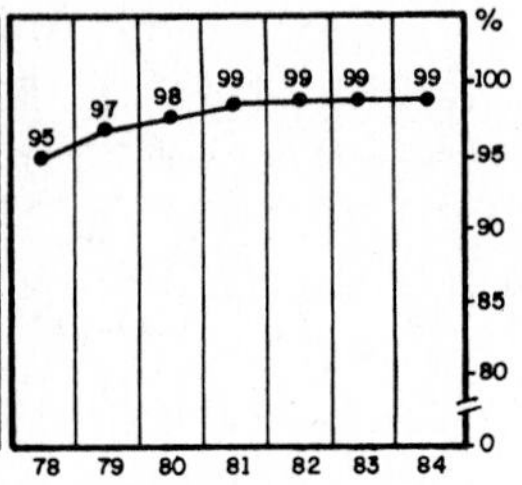

FIG. 8 - DIAL TONE DELAY

2. THE TRAFFIC ADMINISTRATION SYSTEM IN BRAZIL

2.1 Concept

TAS's objective is to monitor and control telephone system traffic flow in order to assure:
- good traffic flow, in relation to the number of facilities;
- efficient utilization of existing traffic facilities;
- maximization of traffic flow capacity;
- quality as uniform as possible in the service rendered to the various users, from the point of view of traffic flow;

As to philosophy, TAS can be divided into two phases:

A preventive phase carried out on the basis of short term (1 to 2 year) traffic forecasts, with the objective of foreseeing congestion and under-utilization of equipment, far enough in advance to program preventive action in order to avoid these problems materializing.

A corrective phase, that fundamentally encompasses action required to eliminate the actual congestion verified.

2.2 TAS organization

TAS, according to the characteristics of its activities, is organized in two operational groups:

First Group - Encompasses the job of continually monitoring the system, and actions of a preventive and corrective nature.

This group includes the following activities:
- acquiring, checking and summarizing traffic data and quality of service indicators;
- forecasting short term (1 to 2 years) traffic data;
- analysing measured traffic data and forecasts;
- assignment of main stations and subscriber's telephone numbers;
- evaluation of the degree of traffic balance between switching offices, and its effects;
- definition and implementation of corrective actions to solve problems that either actually exist or are foreseen;
- determination of switching office traffic load capacities.

TAS, in order to make its implementation process easier, was divided into the following sets of functions:
- main station administration;
- traffic administration at local central offices;
- traffic administration at toll center offices;
- traffic administration at toll switchboards;
- traffic administration in trunk groups.

Second Group - This group includes activities that constitute network management. Network management is the group of activities carried out in real time, with the objective of assuring efficient utilization of existing facilities, when a traffic overload or equipment failure takes place, and maximizing traffic flow capacity in the telephone network.

Networks with alternate routes, and common control equipment, although very efficient when operating under design conditions, suffer rapid deterioration when submitted to traffic overloads. This makes an efficient system of monitoring and control convenient, not to say necessary, so that it may be possible to minimize undesirable effects of most serious traffic overloads and equipment failures. The vital role that the telecommunications network plays, in a country during emergencies, justifies the investments necessary to implement a Network Management System.

At present, only EMBRATEL operates a Network Management System in Brazil, called SSCTC, illustrated in Figure 9. SSCTC is basically a telesupervision and telecontrol system.

The telesupervision function is performed by analysing the conditions of various supervision points in the toll center offices, and by centralized data processing, in the control center. In this manner, it is possible to gain a rapid cognizance of the performance of equipment at these toll center offices and the conditions of traffic flowing through them.

Through the telecontrol function, protective and expansive actions are carried out, such as rerouting, blocking, and announcement machine operation.

At present the SSCTC supervises around 250 000 points at the 33 EMBRATEL toll center offices.

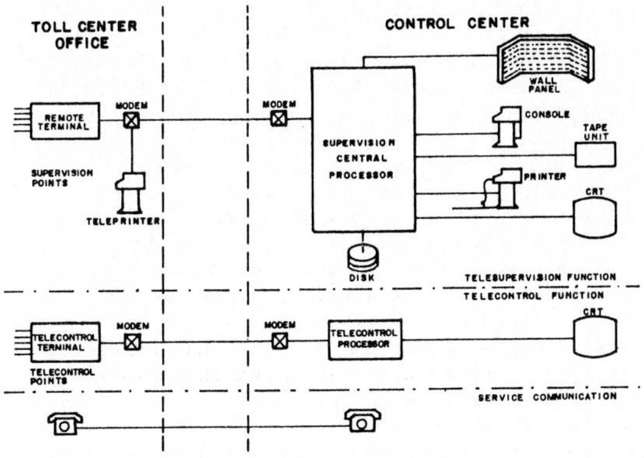

Figure 9 - The EMBRATEL SSCTC System

2.3 Supervision Resources

TAS was developed in such a way, that all functions, except network management, could be implemented by utilizing the conventional supervision resources at the mechanical-electrical central offices, that constitute almost the entire Brazilian network.

Nevertheless, in order to keep abreast of developments, both in technology and operational procedures, stress has been placed on the development and production of modern traffic supervision systems, by domestic manufacturers.

The main one of these systems, SITASU, was developed by a state operator company, Telecomunicações do Rio de Janeiro S.A. - TELERJ.

SITASU, as it was conceived utilizes ICUP technology, and in addition to this, incorporates the most recent developments in the theory of repeated calls, in the theory of renovation, and in applied statistics. SITASU, in the functions of traffic supervision and switching, can be organized on up to 3 levels: remote unit, regional center, and system center. The remote unit level can cover up to 16 000 points of data acquisition and 4 000 points for blocking (Figure 10). The Regional Center can take care of up to 16 remote units, and the System Center can take care of up to 16 Regional Centers.

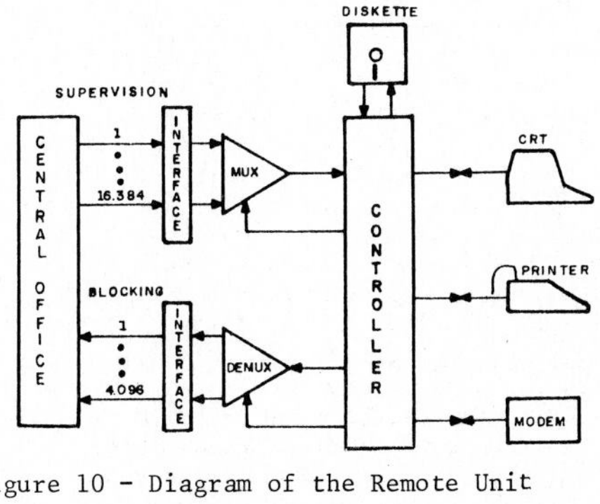

Figure 10 - Diagram of the Remote Unit

A summary of SITASU's main new characteristics is as follows:
- intelligent remote unit, capable of complete local data processing, including finding the busy hour;
- measuring of efficient traffic in trunks and common control equipment, besides the traditional measurement of occupancy traffic;
- detection and blocking of killer switches, through algorithms based on the renovation theory. In table 1, some typical time lags between failures and automatic blocks are shown, for switches whose efficiency is nil, belonging to groups with normal traffic;

SWITCH	TIME
Marker	30 seconds
Register	15 minutes
Junctor	15 minutes

Table 1 - Time Lag for Automatic Block

- precise probability model, for the random variables involved, in order to determine exceptions per facility group. As a result, besides confirming the killer effect, it is possible to identify the "killer efficient", a junctor that systematically disconnects the call during the conversation;
- non-parametric probability model, to determine the time basis for measuring traffic, in order to assure a previously specified precision. The SITASU time basis for data acquisition is programmable, as shown below:

  Events: 25 ms, 250 ms, 1 s, and 5 s;
  Times: 1 s, 5 s, 20 s, and 50 s;

  For equipment with a very rapid switching time, it is possible to employ a processor to supervise switches of up to a level of 4 ms occupation time and 4 ms idle time.

In addition to SITASU, the following supervision equipment can be mentioned:
- Portable Traffic Recorder: equipment used to measure and process traffic data of up to 500 points individually, utilizing the ICUP technique;
- Automatic Answer Time Recorder: equipment used to measure the answer time in manual services. It collects and processes data in up to 256 trunks. Basically, it gives the rate at which calls are answered, in a shorter time than a programmable set reference time.

The automatic toll ticketing system is also utilized as a supervision resource for DDD calls. A program is put into effect every month, that allows all calls to be recorded, whether completed or not. The uncompleted calls are detailed, in regard to the following reasons for not being completed: line busy, no answer, equipment blockage, and failure and others. The data obtained from this program is also utilized for study of subscriber behavior.

2.4 Training Resources

To back up implementation of TAS, in the operating companies, a training program was developed based on the floowing resources:
- Traffic Theory Course: a course developed for personalized instruction, involving telephone

traffic theory and its application. This course is aimed at university level personnel, with training in the field of mathematics;
- Introduction to Traffic Theory: similar to the previous course, that takes up telephone traffic theory only superficially, with greater emphasis on practical applications;
- Traffic Seminar: the Traffic Seminar is a resource that is being utilized for publicizing, on a national level, studies, experiences and developments in connection with the traffic sphere, as well as the results obtained from implementing TAS. Three seminars have been held since 1978, with a fourth scheduled for 1986.

3. THE TELEMIG TRAFFIC ADMINISTRATION SYSTEM

3.1 Concept

Telecomunicações de Minas Gerais S.A. - TELEMIG, one of the 28 state companies, operates in the State of Minas Gerais, located in the Southeastern Region of the Country, which also includes the states of São Paulo and Rio de Janeiro, among others. These three states are the economically most important in the Country. The TELEMIG telecommunications system has about one million telephones (7.30 telephones per hundered inhabitants), and serves 635 municipalities.

From the point of view of marketing, for TELEMIG, TAS represents the management of its main product — Teletraffic. Implementation of TAS, in a systematic manner, took place from 1980 on, in relation to the various sets of functions already described, except in regard to network management, which the Company plans to implement in the near future, utilizing the SITASU system. Prior to this, although traffic administration was carried out, it was in a rather inefficient manner. TAS was implemented in accordance with the TELEBRÁS holding company, taking TELEMIG's peculiar characteristics into consideration.

3.2 Organization

Geographically, TELEMIG is divided into seven operational areas, called Operational Regions. Each Operational Region is subdivided into districts, making a total of 20, and these in turn are divided into subdistricts. The seven Operational Regions, together with the staff, composed of the Marketing, Data Communications, and Operational Planning and Engineering Departments, report to the Operations Director's Office.

TAS activities are decentralized at the regional, district and operational unit levels. An operational unit can be understood as the portion of the system, such as toll center offices, switchboard centers, operations centers, etc., where each set of traffic administration functions is carried out. Traffic Managers are assigned to each one of these levels, who are organized on the basis of rank, to carry out their duties (Figure 11), under the direction of the Traffic Division, that reports to the Operational Planning and Engineering Department. This Division is also in charge of the physical control of trunk facilities.

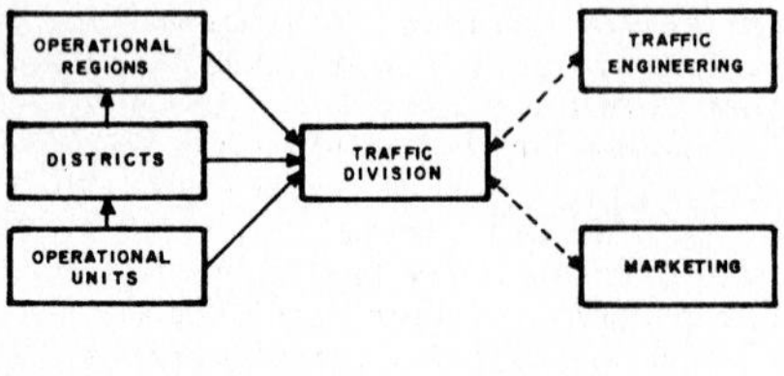

Figure 11 - Decentralized Traffic Administration Organization

The gratifying results obtained from this policy of TAS decentralization at TELEMIG, is evidenced by the fact that a traffic doctrine has spread throughout the whole operational sphere, with the equipment maintenance, commercial and marketing sectors engaged in carrying out these activities. Furthermore, this process made it feasible to obtain the human resources needed, mostly personnel who can give only part of their time to these duties, and whose help would be difficult to obtain for a centralized staff.

In Figure 11 we also stress the two main TAS interfaces: the fields of Traffic Engineering and Marketing. At TELEMIG, Traffic Engineering reports to the Technical Planning Department, and Technical Planning in turn reports to the Technical Director's Office. Its principal interface with Traffic Administration is in respect to network expansion, and the solution of problems of traffic congestion. On the other hand, the Marketing Department interfaces with Traffic Administration, in the solution of network underutilization problems, increasing utilization of the services offered and proposing new ones.

3.3 The traffic administration process at TELEMIG

Figure 12 illustrates the traffic administration process employed by TELEMIG, as well as the interfaces.

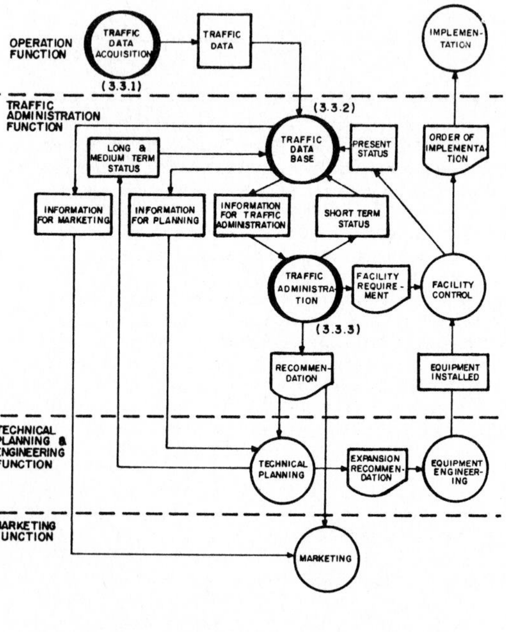

Figure 12 - Traffic Administration at TELEMIG

3.3.1 Data Acquisition

The first step in the process consists in acquiring traffic data through supervision equipment installed at local central offices, toll center offices, tandems, PBX, and toll switchboards. These measurements are carried out every month, during 5 working days at the busy hour, except at local central offices with less than a thousand main stations, at which measurements are taken quarterly. Based on this data, the representative traffic figures are found for traffic during the month (VRM), the year (VRA), yearly movable traffic (VRAM), and the yearly forecast (VRAP), that will serve as a basis for Traffic Administration to decide on actions to be put into effect. These concepts are defined below:

VRM - second highest figure during the 5 measuring days;

VRA - second highest VRM for the traffic year, that covers a period of 12 consecutive months, in regard to which the traffic data is consolidated;

VRAM - second highest VRM for the last 12 months;

VRAP - VRA forecasts for the next traffic year.

The busy hour, during which this information is obtained, is found every year.

Table 2 shows the distribution of the several types of supervision equipment used by TELEMIG. It can be seen that the great majority of the supervision equipment is not automatic, and that it monitors central offices with cross-bar or cross-point technology. Installation of the first SPC central offices at TELEMIG is scheduled for 1986.

CENTER OFFICE \ MONITORING EQUIPMENT	MECHANICAL-ELECTRICAL COUNTER	PAPER TAPE	AUTOMATIC	TOTAL
LOCAL	192	23	7	222
TOLL	12	11	2	25
TANDEM	2	-	2	4
PBX	-	-	(*)	(*)
SWITCHBOARD	-	-	9	9

(*) The PBX do not possess permanent equipment and are monitored by portable equipment on a rotation basis.

Table 2 - Supervised Central Offices, According to Type of Monitoring Equipment

Data obtained by nonautomatic equipment is processed by computerized systems, generating reports that, when added to those supplied by automatic equipment, are fed to a Traffic Data Base. The automatic monitoring equipment is SITASU type or similar. TELEMIG is carrying out a plan to install this new equipment and substitute conventional equipment.

3.3.2 Traffic Data Base

The Traffic Data Base constitutes the second step in the process. This is a base implanted in the Company's main computer, utilizing Data Bank software, that allows on-line access by the various sectors. As shown in figure 12, this Base is fed by traffic data and information in regard to the network's physical configuration, both now and in the future, on a short, medium, and long term basis. Furthermore, by means of monthly updating, it supplies historical data to carry out traffic forecasts.

Fundamentally therefore, the data fed to the Base are the representative figures for average traffic; the figures found for congestion, the number of facilities installed and in service, their traffic capacities, and the estimated traffic figures, and the respective number of facilities during the next 7 years, as well as 3 five year periods, besides the monitored limits of traffic.

The main benefits observed as a result of Base utilization are:

a) uniformity in the data available at the several Company sectors;
b) management reports, containing information on the state of the network, are issued;
c) it supplies information, on which Administration and Traffic Engineering can act;
d) reliable traffic data is quickly available;
e) it provides an overall view of the network.

As an example of Base potential and flexibility, examples of some of the answers it can supply are given below:

a) traffic intensity in trunks;
b) total number of facilities programmed for a given central office;
c) over engineered routes;
d) under engineered routes;
e) alterations programmed in the network for a given date;
f) number of outgoing routes from a given central office.

3.3.3 Traffic Administration

The third step is traffic administration itself, that consists of the decision-making process in regard to traffic flow control, based on the continuous monitoring of the system in operation, as described in items 3.3.1 and 3.3.2. This process aims at achieving efficient utilization of the system's traffic facilities. This objective can only be reached by means of a conscious effort by all the technical and operational sectors involved.

The decision-making process at TAS has given top priority to preventive, rather than corrective administration, seeking to act before overloads occur in the system. For the purpose of controlling traffic flow, in the various sets of functions, TAS utilizes parameters set according to the type of facility, described as follows:

a) Trunk Facilities - upper (UL) and lower (LL) limits of control are utilized for each route, for which the traffic figures indicate normal conditions. Otherwise, measures are taken to increase trunk group capacity, or to provide available facilities for future rearrangements. These limits are set according to the route grade of service. At TELEMIG, the following equation is used:
UL = 3 GOS and LL = 0.1 GOS.
b) Local Central Offices Facilities and Main Stations - the parameter utilized for Traffic Administration in local central offices and Main Station Administration is load capacity, that represent the largest possible number of main stations that can be connected by the end of the next traffic year. Load capacity is figured according to the amount of traffic forecast, and the grade of service rendered by the various facilities at the central office. Should this load capacity be less than the physical capacity, it would hinder the assignment of all of the central office main stations, until the recommended measures are

put into effect (for instance: rearrangements, extensions). In addition to this, a balance of traffic and main stations, among basic groups of subscribers, is aimed at, through assignment of telephone numbers.

c) Toll Center Office Facilities - Traffic Administration at toll center offices employs load capacity as a parameter, which is the same as the one utilized for traffic administration at local central offices. In the case of toll centers, load capacity is the maximum incoming traffic that can be handled by the center office without congestion in its common control equipment.

d) Toll Switchboard Facilities - the toll switchboard trunk facilities are evaluated by means of the same UL and LL parameters as those described in "a". Furthermore, the manual service answer time is also evaluated, so as to fill the proper number of positions.

Figures 13 and 14 show the preventive and corrective flows employed by Traffic Administration at TELEMIG. As is shown in Figure 13, preventive administration is based on the VRA forecasts for 1 and 2 year terms. These forecasts are analysed, which makes it possible to foresee future congestion and slack in the facilities. Facilities in these conditions are considered exceptions. From this phase on, the process is centralized, and fundamentally consists of consolidating, on a company-wide basis, all the regional results and analyses, the Series of Preventive Administration Meetings, and the solutions decided upon. These solutions, that consist of the basic proposals made by Traffic Administration, along with the Traffic Engineering programs, constitute the Traffic Expansion Plan. Execution of this Plan is subject to the limitations imposed by the Company's investment budget.

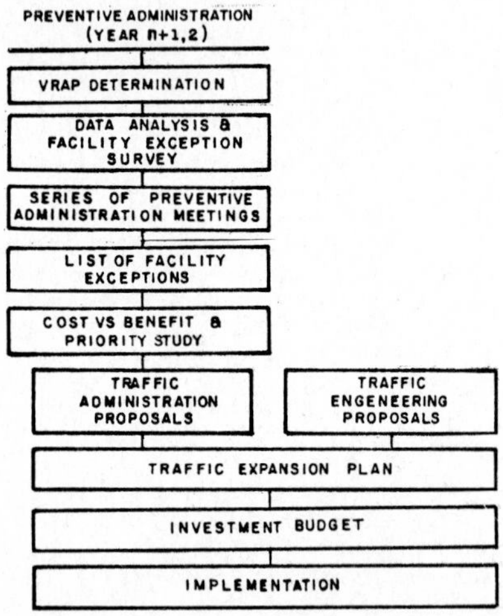

Figure 13 - Preventive Traffic Administration

The figure for Corrective Administration shows that the monthly analysis, carried out by the several decentralized levels of Traffic Administration, generates a list of facilities that are exceptions. This list, for management purposes, comprises a consolidated Exception Report by the staff, and which, besides exceptions, gives the figures for performance of the various TAS sets of functions, and the overall figures for completed calls in the TELEMIG network. Measures to be taken by Corrective Administration, when they require financial resources, are fitted within the limits for year n.

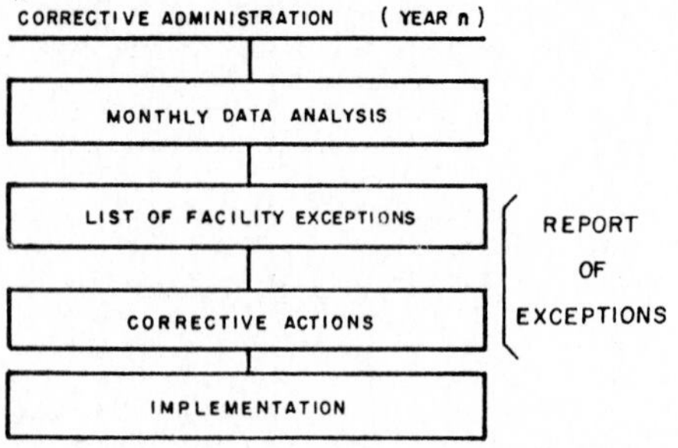

Figure 14 - Corrective Traffic Administration

3.4 Human resources involved

Decentralization of activities was the main device found to engage the largest number of personnel in the Traffic Administration process. Consequently, at present 1.1% of the total TELEMIG work force of 7 697 employees is involved in these activities. Nevertheless, only 20% of this personnel work for Traffic Administration full time.

An important factor in this process is training. Training has been provided by means of courses on Telephone Traffic Theory, and Introduction to Telephone Traffic Theory, described in item 2.4. Besides these, TELEMIG used its own resources to develop courses such as the Telephone Traffic Administration Course, and the training module designed for Company managers who work in this field.

A group that includes all the Traffic Managers within the decentralized organization, called the Traffic Administration Group, engages in exchanging accounts of experiences among the managers, and evaluation of the overall results obtained by Traffic Administration at TELEMIG. This group meets twice a year.

Another event which has contributed to training and developing personnel is the TELEBRÁS System Traffic Seminar.

3.5 Operational results

The results seen in the present stage of Traffic Administration have indicated that the policy has been a success. The measures implemented by this system have contributed to improving the quality of the service rendered by the TELEMIG network, as shown in figure 15.

It shows that the completion rate has grown progressively. The present completion rate, around 57%, represents a very gratifying result, when compared to international standards.

In regard specifically to TAS sets, of functions, there are figures that allow its performance to be evaluated. Among others, we stress the rate of congested DDD routes.

Figure 16 shows the grouth of this figure, and figure 17 shows the relative distribution of congested DDD routes per traffic flow. It shows that out of a total of 98 congested routes, in December 1984, 85% have a traffic flow of 5 erlangs. These low capacity routes serve small places (50 to 300 main stations) with a high unsatisfied demand. The solution of these cases normally would require substitution of the transmission system, involving significant investments with low returns. Therefore, the problem of these cases are being solved progressively. This is not the case for medium and high capacity routes, for which the equipment is modular and the short term returns on the investment are higher.

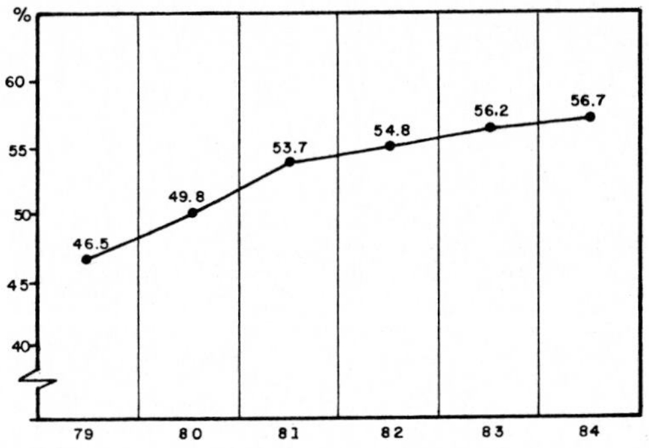

Figure 15 - Completion Rate in the TELEMIG Network

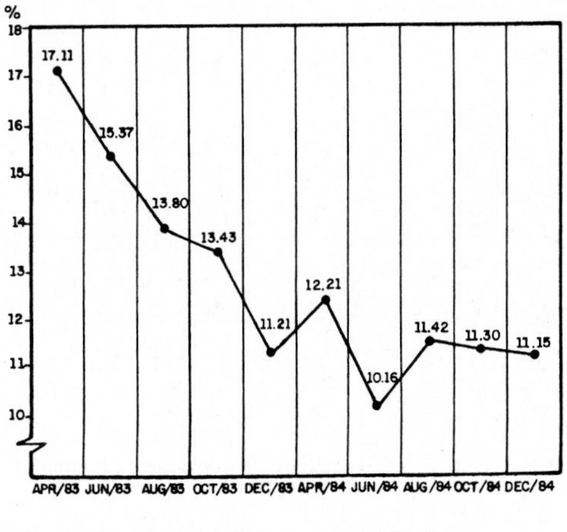

Figure 16 - Rate of Congested DDD Routes

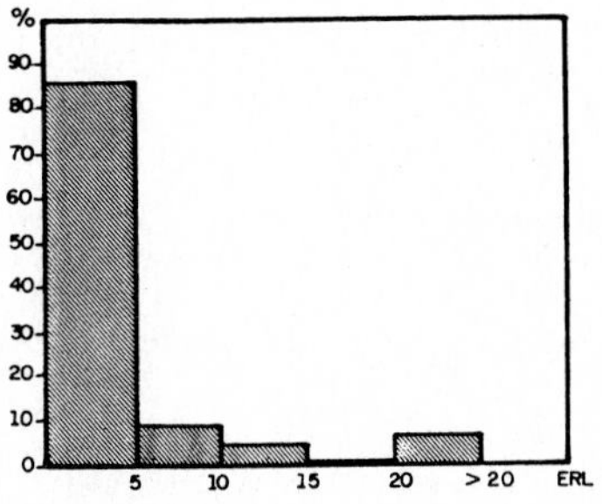

Figure 17 - Relative Distribution of Congested DDD Routes per Traffic Flow - Dec. 1984

The proportion of TELEMIG participation in traffic expansion, within the total amount invested, is growing every year, showing the effort the Company is making to solve these cases, and its recognition of what it means in relation to its operational income. In 1985, for instance, TELEMIG will invest approximately 5% of its total investments in traffic expansion, without taking into consideration the expansion of its facilities, together with the projects for expansion of the number of main stations.

On the other hand, Traffic Administration measures are not limited to network expansion alone. Slack facilities are frequently utilized for solving cases of congestion.

4. CONCLUSION

This paper gives priority to stressing the process of development and implementation of Traffic Administration in the Brazilian network, and at TELEMIG in particular.

The operational results that can be reaped from a traffic administration system, fully justify its introduction.

In Brazil, in addition to the highly satisfactory results reported in this paper, we have the fact that a new teletraffic approach has been spread throughout the whole corporation.

REFERENCES

(1) "Política de Administração de Tráfego", Telebrás Practices System, 1981
(2) E.A.O. Ramalho, T.C. Lustosa, "Gerência de Rede na Embratel", Proc. 3rd Telebrás System Teletraffic Seminar, Brasília, pp. 167-180, 1983
(3) Traffic Administration, Proc. 1st, 2nd, 3rd Telebrás System Teletraffic Seminar, 1978, 1980, 1983
(4) G. Nunes, "Bases de Tempo para Medição de Tráfego", Telebrás Review, pp. 20-26, March 1984
(5) S.W. Lewis, "ICUP Tracks Killer Trunks, Nips Traffics in the Bud" Telephone Engineer Management, USA, 1975.

TELETRAFFIC ISSUES in an Advanced Information Society
ITC-11
Minoru Akiyama (Editor)
Elsevier Science Publishers B.V. (North-Holland)

BENEFITS OF EXTREME VALUE ENGINEERING FOR A LOCAL DIGITAL SWITCH

Gillian WOODRUFF, D.Murray BROWN**, Kalyan BASU*** and William ROBINSON**

* BNR, Ottawa, Canada
** Bell Canada, Toronto, Canada
*** BNR, Atlanta, U.S.A.

ABSTRACT

The development and application of extreme value engineering (EVE) procedures for specific switching systems have been introduced in several previous ITC papers [1][2][3][4]. EVE is a method of estimating traffic loads for the purpose of forecasting traffic-sensitive equipment requirements and bases load estimates on the distribution of daily peak hour loads.

This paper reviews general advantages of EVE and expected EVE benefits for a generic local digital switch. Although there are several inherent advantages with EVE, the benefits are shown to be very dependent on the system architecture, traffic environment, and current administration system. For several reasons, new systems may be best engineered according to EVE.

1. *INTRODUCTION*

Traffic load estimates used for switching equipment provisioning play a crucial role in determining future service provided to subscribers. As inputs to a forecasting procedure, these estimates must be both accurate and stable to minimize the extent of over- or under-engineering in the switching office.

Traditionally, Bell Canada has used time-consistent engineering (TCE) procedures because of the simplicity of TCE traffic data collection requirements. TCE procedures were developed at a time when data had to be collected and analyzed manually. The TCE method requires traffic values from the same hour each day, called the time-consistent busy hour (TCBH), for a three month duration (not necessarily consecutive), called the busy season. The average busy season busy hour (ABSBH) load was suitable for traffic engineering purposes, due to both the stability of the ABSBH load for forecasting, and the graceful service degradation displayed by electromechanical switching components.

Present and future switching technologies have components with steep load/service relationships, and are therefore sensitive to infrequent peak loads. These components must be provisioned according to service objectives at the forecasted peak load to ensure that the blocking or delay remains within reasonable limits. For this purpose the TCE method requires a stable estimate of the high day busy hour (HDBH) load. However, the existing HDBH load estimation procedure is not widely or consistently used and is often replaced by more subjective methods of questionable accuracy. Also, the concept of a true busy hour is not justified in many cases at present, and traffic is expected to become less predictable in the future.

Extreme value engineering (EVE) is a method which can alleviate several of the inherent disadvantages of TCE in the new traffic environment. The EVE method can be applied to daily peak hour load data. Furthermore, EVE lends itself to a computerized administration system. Several ITC papers [1][2][3][4] have discussed the development and application of EVE procedures for specific systems.

The general question of when it is beneficial to adopt EVE has not yet been addressed in the literature. This important information is the starting point for those who may be contemplating a changeover to EVE, but are unsure of the concept and its potential benefits or drawbacks. The major goals of this paper are to summarize

- generally known facts about EVE,
- specific cost and service benefits expected with EVE in the Bell Canada local digital switch (LDS) environment, and
- system and environmental characteristics which would lead to a decision whether to adopt EVE.

In order to assess specific LDS cost and service benefits expected with EVE it was necessary to compare the implications of provisioning according to both TCE and EVE. This comparison was carried out by

- developing a database to provide information on expected traffic characteristics,
- developing EVE-based procedures appropriate for the comparison,
- evaluating the overall capital cost and peak service expected from application of TCE and EVE, and
- identifying the impact expected with EVE on both implementation effort and on-going administration.

This paper concentrates on the details of these four steps, which illustrate many of the considerations which should be made when identifying EVE benefits. Section 2 gives an overview of the LDS architecture assumed, and the type and nature of traffic data collected. Section 3 presents the TCE and EVE procedures for the LDS, and a general description of EVE. Section 4 summarizes the benefits identified with EVE for the LDS. Aspects of EVE implementation are discussed in Section 5, and the dependence of EVE benefits on specific system characteristics is outlined in Section 6. Section 7 concludes the paper.

2. DESCRIPTION OF THE LOCAL DIGITAL SWITCH

2.1 Architecture

The architecture assumed in the analysis is shown in Figure 1. Lines terminate on local line concentrators (LLCs) at the host switch, or on remote line concentrators (RLCs). These line concentrators (LCs) are connected to the network time-switches of the host via 2, 3 or 4 digital links of 30 voice channels each. The LCs have both a physical line termination capacity, and a traffic (usage) capacity dependent on the number of links to the host and the LC blocking requirements.

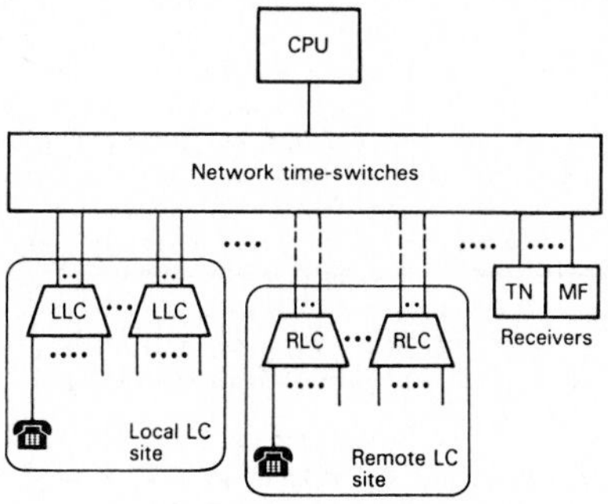

Figure 1: LDS Architecture

The primary difference between the LLC and the RLC is cost. The links from RLC-to-host are longer, and cost an order of magnitude more than the links from LLC-to-host. The basic LLC and RLC costs are approximately equal. The LCs are provisioned at each site, with the same number of links for administration reasons, according to the least expensive configuration which satisfies both physical and traffic requirements.

Tone (TN) receivers for lines and multifrequency (MF) receivers for trunks reside on peripheral modules with direct access to the network. All receivers of one type are accessed as one group and are provisioned according to their allocated delay requirements. The central time-switched network is assumed to contribute negligible blocking so that it is always provisioned according to physical termination requirements. The host CPU is attempt-sensitive and an accurate forecast of its exhaust date is required. However, this analysis assumes an environment where CPU capacity is not reached.

In summary, the components of the LDS affected on a year-to-year basis by the traffic engineering procedure are assumed to be

- local line concentrators (LLCs),
- remote line concentrators (RLCs),
- tone (TN) receivers, and
- multifrequency (MF) receivers.

These components are considered in terms of capital cost and service for the comparison of TCE and EVE.

2.2 Sample Traffic Characteristics

In order to examine the types of traffic behavior expected with these components, hourly traffic data was collected during the busy seasons of two sample LDS offices, one primarily business and one residential. These offices consisted of 5 LC sites totalling 56 LCs, and provided a sufficient database for the analysis.

Figures 2 and 3 and Table 1 summarize characteristics observed from the data. Figure 2 is a plot of the busy season coefficient of variation (CV), defined as the ratio of the standard deviation to the mean, of the TCBH loads vs. the CV of the daily peak hour loads. The graph shows that the TCBH loads were more variable than the peak hour loads, agreeing with similar observations from other sources [3] [5], and suggesting that estimates derived from peak hour loads should be more stable than those from TCBH loads.

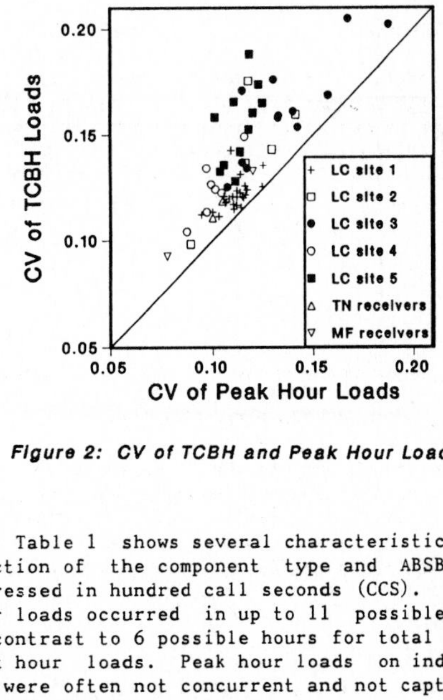

Figure 2: CV of TCBH and Peak Hour Loads

Table 1 shows several characteristics as a function of the component type and ABSBH load, expressed in hundred call seconds (CCS). LC peak hour loads occurred in up to 11 possible hours, in contrast to 6 possible hours for total LC site peak hour loads. Peak hour loads on individual LCs were often not concurrent and not captured by a total site load measurement.

The peak hour loads which did not occur in the TCBH are referred to as side hour peak loads (SHPLs), and were observed during 35%-85% of the busy season days. There was a wide variation in the amount by which SHPLs exceeded TCBH loads of the same day. All receiver and total LC site peak hour loads stayed fairly close to TCBH

	ABSBH LOAD (CCS)			
	0-300	300-1500	1500-2500	2500-60000
COMPONENTS	RECEIVERS	RESIDENTIAL LCs	BUSINESS LCs	LC SITES
# HOURS OF PEAK LOAD	5-7	5-11	4-7	4-6
% DAYS WITH SHPLs	40-50	40-85	40-65	35-60
% SHPLs EXCEEDING 130% of TCBH LOAD	0-5	5-40	0-7	0-7
MAXIMUM % DIFFERENCE OF SHPL AND TCBH LOAD	15-80	35-120	15-60	15-65

Table 1: Sample Data Characteristics

loads, but individual LC peak hour loads could deviate widely from the TCBH loads, especially with residential subscribers - maximum deviations sometimes exceeded 100%. These observations imply that the use of TCBH loads for provisioning components, such as LCs, serving a small subscriber group, is not enough to guarantee that service at the peak load will remain within acceptable bounds.

Figure 3 shows the CV of the peak hour loads vs. the ABSBH load for individual LCs and the total LC site loads. Similar results can be obtained for TCBH loads, illustrating that the CV of total LC site loads is not equal to that of individual LCs.

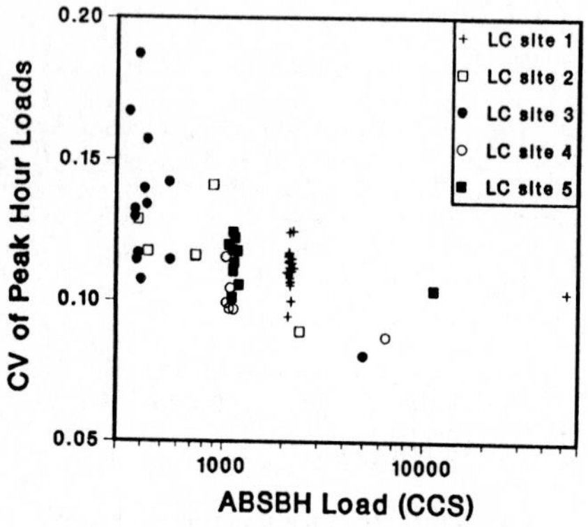

Figure 3: Dependence of LC Peak Hour Load Variation on Load Size

3. *DESCRIPTION OF TCE AND EVE FOR THE LDS*

3.1 Purpose

The LDS traffic behavior observed in Section 2.2 suggests that potential problems with TCE may be greatest for LC provisioning. Capital cost and peak service with both TCE and EVE in the Bell Canada LDS environment were estimated in order to assess the magnitude of potential TCE problems, and to determine whether EVE may provide some improvement. Section 3.2 describes the TCE methods assumed in the analysis. Section 3.3 introduces the EVE concept and the procedures developed for this comparison.

3.2 TCE for the LDS

3.2.1 TCE Reference Loads

The TCE method for LC and receiver provisioning requires estimates of two reference loads:

ABSBH - the average TCBH traffic over the busy season, and
HDBH - the highest TCBH traffic of the year.

The TCBH is determined retroactively as that hour which yields the highest average load over the busy season. The HDBH estimate is not the actual observed HDBH value, a very unstable quantity, but is based on the only appropriate statistical method used in Bell Canada. This estimate is given by

$$\text{HDBH} = \text{ABSBH} \times (1.68R - .68) \quad \text{for } R < 1.3,$$

where R is an estimate of the ratio of the average of the 10 highest busy hour loads to the ABSBH load.

3.2.2 LC Provisioning with TCE

The LCs are provisioned on the basis of the reference load estimates calculated for the <u>total LC site</u>, due to the prohibitive administration effort which would be involved in tracking the TCBH of individual LCs. The sum of the individual LC ABSBH or HDBH loads is assumed to equal the total site ABSBH or HDBH load, respectively.

The LC capacity is dependent on the blocking requirements and the LC load/service relationship, the latter being dependent on the number of links to the host. The requirements are

- 1.9% blocking at ABSBH load, and
- 9% blocking at HDBH load.

The least expensive link configuration is calculated by first determining, for each configuration, the maximum number of lines which the LC can support to satisfy the above service requirements and not exceed the physical limitation. Each configuration is either average traffic-, peak traffic-, or physically-limited. The configuration corresponding to the least expensive cost per line is then chosen, with the restriction that the same number of links per LC must be provisioned at each site.

Figure 4 shows the ranges of HDBH CCS/line values which would produce peak traffic- and physically-limited LLC and RLC configurations. The CCS/line bands producing traffic-limited configurations are narrower for the LLC than the RLC, due to the lower LLC link cost.

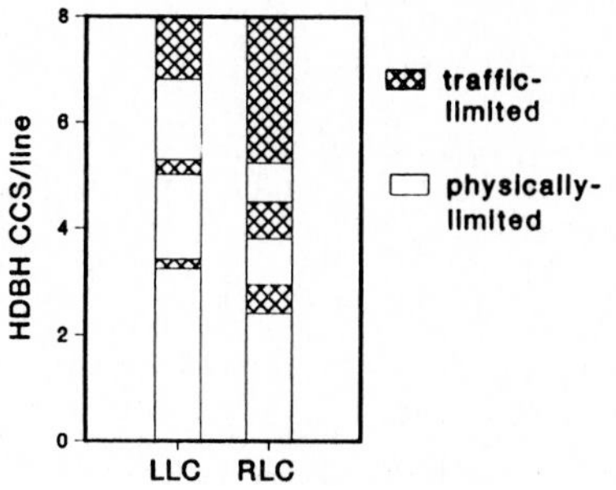

Figure 4: Dependence of LC Capacity Limitation on HDBH Load per Line

3.2.3 Receiver Provisioning with TCE

The TN and MF receivers are provisioned according to delay requirements at both ABSBH and HDBH loads, with none of the complications of LC provisioning.

3.3 EVE

3.3.1 General Description of EVE

EVE is a method for estimating "extreme" loads, such as those achieved or exceeded once per day, week, month, or year. This estimation is accomplished by making use of

- a set of several peak loads, each reported over a certain time interval (e.g., day or week), to estimate their
  - average (APK), and
  - standard deviation (SD), and
- the underlying distribution of the peak loads, described by APK and SD. Appropriate distributions are the Gumbel distribution [1], and the normal distribution raised to an appropriate power [2][3]. Distributions requiring higher order moments are not attractive since a primary objective is to minimize the complexity of the procedure adopted.

Any EVE load estimate can be expressed as

APK + c x SD,

where c is a constant depending on the extremity of the estimated load (e.g., weekly, monthly, yearly) and the theoretical distribution [2]. Key advantages of EVE are:

- Stability of the extreme load estimates for forecasting. All estimates are based on APK and SD, derived from the entire data set.
- Accuracy of the extreme load estimates, dependent on the closeness of the assumed and actual peak load distribution.
- Flexibility for adaptation to specific applications. EVE enables the use of any
  - extreme load estimate,
  - peak load duration (e.g., 1/2 hour, 1 hour),
  - peak load reporting interval (e.g., 1 day, 1 week), and
  - data set size (e.g., 1 month, 3 months).
- Suitability for a computerized provisioning tool [5]. EVE requires a minimal database when storing the values APK and SD as moving averages, and lends itself to statistical techniques for automatically rejecting outlying data [4]. Widespread automation of these procedures ensures a consistent application of the EVE concept.
- All-hours monitoring without all hours of traffic data. Potential EVE service objectives, such as "blocking should not exceed 5% more than once per month on average" are easily compared to monitored service.
- Ability to track traffic variability. EVE provides better service protection for subscribers experiencing highly variable traffic loads. Load variability estimates could be extrapolated for different traffic sizes and mixes.

3.3.2 EVE for the LDS

3.3.2.1 EVE Reference Loads

In order to compare the effectiveness of TCE and EVE for provisioning peak traffic-limited components of the LDS, suitable EVE procedures were developed for this analysis. The peak EVE reference load chosen for direct comparison with TCE was the highest peak hour load of the year (YRPK).

The YRPK load was assumed to be estimated at the end of each month using the equation

YRPK = APK + 3.27 x SD,

and averaged over the three highest months. This relationship is based on

- a least-squares fit to the data, and
- hypothesis tests to determine a good peak hour load distribution,

and can be interpreted as the maximum peak hour load expected in 66 days (approximately a busy season) assuming a Gumbel distribution [1].

3.3.2.2. EVE Service Objectives

For comparison with TCE, meaningful EVE service objectives were defined. Three different definitions were examined:

- EVE1: "Equal numbers" EVE

  The same numerical service objectives were used for the similar TCE and EVE loads. For example, the TCE 9% blocking objective at the HDBH load would translate into an EVE 9% blocking objective at the YRPK load. This addresses the question of whether TCE provides its objective peak service, and whether EVE could come closer to the objective.

- EVE2: "Equal capital cost" EVE

  The EVE service objectives were determined such that the same capital cost requirements as with TCE would result. This addresses the question of whether equipment redistribution with EVE would result in an overall improvement in peak service.

- EVE3: "Equal peak service" EVE

  The EVE service objectives were chosen such that the same overall peak service as with TCE would result. In this context, peak service was defined as the service (blocking or delay) level exceeded once, on average, during the year. This addresses the question of whether any capital savings could be realized by a more equitable allocation of equipment.

3.3.2.3 LC Provisioning with EVE

The sum of the YRPK estimates of each LC is generally greater than the YRPK estimate of the total LC site. The reasons for this are both the non-coincident hours in which the LC loads peak, and the higher variation of smaller LC loads as exhibited in Figure 3. The optimum LC provisioning method would be to track the traffic of each LC separately using EVE, as outlined in [4].

However, for this analysis, it was of interest to compare the performance of TCE and EVE on an equal basis, with LCs provisioned according to total site load measurements. Accordingly, suitable empirical equations, similar to those in [4], were developed to infer individual LC YRPK loads from the total site YRPK load estimate, and to estimate the LC line capacity. The least expensive LC link configuation was determined in the same manner as with TCE.

3.3.2.4 Receiver Provisioning with EVE

The peak traffic-limited TN and MF receivers were assumed to be provisioned in a manner analogous to TCE, using determined EVE YRPK delay requirements and the estimated YRPK load.

4. BENEFITS OF EVE FOR THE LDS

4.1 Offices Affected by EVE

The current global LDS environment considered for the overall capital cost and peak service assessment consisted of 33 offices, covering a total of 33 LLC sites, 20 RLC sites, and 326000 lines. General TCE data available for these offices indicated, by using the TCE methods outlined in Section 3.2, the following:

- Almost all LC sites were limited by their HDBH, rather than ABSBH, service requirements.
- Only 50% and 20% of the offices were limited by the HDBH service requirements for MF and TN receivers, respectively. This was due to the predominantly small office sizes in the environment considered.

These observations suggest that significant benefits with EVE will probably lie with LC, rather than receiver, provisioning.

4.2 Benefits of EVE for LCs

Based on Figure 4, 5% of LLC lines and 35% of RLC lines are currently traffic-limited (rather than physically-limited) in Bell Canada. However, only 10% of the total LDS lines are served by RLCs, bringing the total number of lines served by traffic-limited LCs to 8%. This implies that any capital cost advantages with EVE would be insignificant relative to total capital investment. Trends towards a higher incidence of RLCs may increase the importance of EVE. However, decreasing transmission costs or higher link capacities would cause a trend towards physically-limited peripherals which do not require traffic engineering.

The database described in Section 2.2 consisted of LC traffic values from almost the same number of lines as those served by traffic-limited LCs. A one-to-one correspondence (with appropriate CCS/line scaling) was assumed between recorded LC traffic and traffic from a peak traffic- limited LC site with similar characteristics. EVE and TCE procedures applied to the data determined LC line capacities and expected traffic-sensitive costs. Peak service with EVE and TCE was evaluated from the traffic data with adjustments to balance and place traffic at the required LC line capacity.

Unlike EVE, TCE consistently underestimated both the observed YRPK and HDBH loads of LCs, due to the incorrect assumption that HDBH loads of each LC sum to the total LC site HDBH load. This problem has the potential to lead to subjective over-provisioning. The service deterioration expected with TCE is indicated in both Table 2, which shows the peak blocking objectives determined for EVE2 and EVE3 (as defined in Section 3.3.2.2), and Figure 5, which shows the average number of days per busy season on which the traffic-limited LC peak blocking was higher than 9%.

METHOD	PEAK BLOCKING OBJECTIVE
TCE	9 %
EVE1	9 %
EVE2	16.5%
EVE3	18.7%

Table 2: EVE and TCE Peak Blocking Objectives

Figure 5 indicates that, unlike TCE, EVE met its peak service objective. The peak service benefits with EVE are high for traffic-limited LCs due to the following factors:

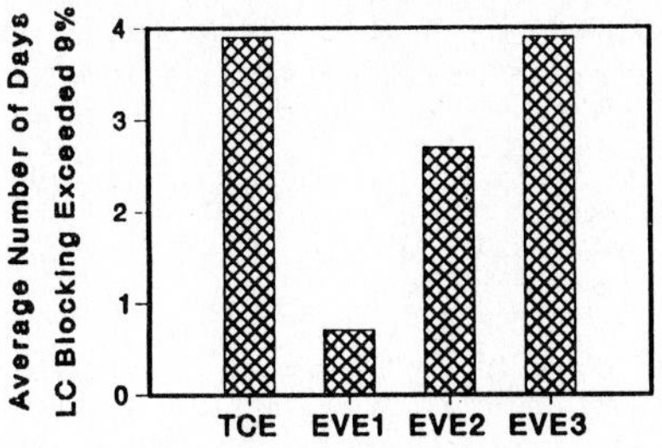

Figure 5: Average Number of Days Traffic-Limited LC Blocking Exceeded 9%

- The LCs have steep load/service relationships.
- The LC traffic can be highly variable throughout the day and from day-to-day.
- With EVE, individual LC variability can be measured or approximated.

The estimated LC cost differences between EVE and TCE are shown in Table 3. Total cost includes fixed line and office costs for all LDS offices. Tables 2 and 3 and Figure 5 indicate that a slight capital cost saving with EVE may be possible with an equipment reallocation (or equivalently, some improvement in peak service may be possible for the same cost). However, these results may be misleading because

- additional peak load estimation error due to year-to-year variation and forecasting may eliminate EVE2 and EVE3 differences,
- LC equipment must be provisioned modularly so that, in practice, extra capital investment in spare capacity is necessary, and
- in practice, TCE is not applied strictly as assumed in this analysis. High peak load underestimation penalties cause a tendency to over-provision.

4.3 Benefits of EVE for Receivers

No particular capital cost or peak service benefits with EVE for receivers could be identified for the following reasons:

- Both TCE and EVE appeared to meet peak service objectives.
- Relatively small office sizes yield shallow load/service relationships for receivers, making average traffic-limitations common, and imposing low penalties on peak load underestimation.
- Receiver traffic was not as variable either throughout the day or from day-to-day, and is not split into higher variability groups.
- Receiver cost is small compared to LC cost, making the penalties with over-provisioning small.

COST	DEVIATION FROM TCE		
	EVE1	EVE2	EVE3
TRAFFIC-SENSITIVE COST, TRAFFIC LIMITED LC SITES	4.6%	0.0%	-1.7%
TOTAL COST, ALL SITES	0.5%	0.0%	-0.2%

Table 3: Expected Capital Cost Differences Between EVE and TCE

5. IMPLEMENTATION OF EVE

The previous sections have discussed potential benefits with EVE for the LDS in terms of peak service and capital cost. However, EVE introduction would have a major effect on the day-to-day administration of traffic data. This section summarizes

- the areas of on-going administration which would benefit from EVE introduction, and
- the implementation effort required to introduce a new procedure such as EVE.

The extent to which on-going operations may benefit from EVE depends largely on the existing administration and whether a new concept in administration is desired. The flexibility of EVE application and the suitability of EVE for a computerized system open doors to many data analysis techniques which would previously have been impossible to apply on a large scale. The largest benefit would probably be apparent in switching from a manual to computerized system, as described in [5], resulting in less manual supervision and reporting, and a sharing of resources. However, existing large-scale computerized systems incorporating TCE could also benefit from the introduction of EVE procedures by

- eliminating the need to manually review and track potential busy hours and busy season months,
- streamlining operations by providing a consistent data analysis tool with the ability to indicate traffic level, variability and forecasts in an appropriate report format,
- providing statistical identification, reporting, and rejection of faulty data,
- gearing service reports more closely to service objectives,
- providing load balancing procedures according to EVE reference loads, rather than the ABSBH load, to equalize the peak service experienced by individual groups, and
- receiving fewer complaints from subscribers who suffer service penalties due to the TCE method.

Because EVE represents a new direction in traffic data analysis, there is a great deal of effort required for

- understanding the EVE concept and its potential applications,

- developing EVE procedures suitable for the specific system, requiring
  - a traffic database,
  - TCE-EVE load conversion,
  - peak hour load distribution characterization,
  - suitable EVE reference loads,
  - EVE service requirements,
  - tests to detect invalid data,
  - capacity tables,
  - load balancing procedures,
  - service measurement indices, and
  - suitable forecasting methods,
- specifying the content and format of EVE reports,
- educating provisioners and administrators on the use of EVE concepts and reports,
- rewriting and distributing provisioning documentation, and
- developing the software support system.

The above requirements represent a high implementation cost, much of which is necessary for the introduction of any new system, independent of EVE. However, the high changeover costs for an existing large-scale computerized system using TCE are likely to outweigh the potential benefits of EVE.

6. *SUMMARY OF EVE BENEFITS*

The advantages of EVE over TCE can be divided into those which are system-independent or dependent.

The intrinsic system-independent advantages include

- more accurate and stable load estimates,
- all-hours monitoring without the collection of all hours of data,
- suitability for a computerized system providing a consistent provisioning tool, and
- flexibility in application.

Whether any large service or cost benefits exist with EVE is very system- and environment-dependent:

- Components characterized by steep load/service relationships are usually limited by their peak service requirement and require an accurate peak load estimate. There is a tendency to overprovision these components, which should be reduced to some extent using EVE.
- EVE is ideal for components subjected to highly uncertain or variable traffic. With EVE, peak loads are captured regardless of when they occur and their variability is measured.
- Attempt-sensitive components with a short holding time, such as processors, often require a shorter measurement interval than one hour since they are sensitive to peaks, and are therefore good candidates for EVE.
- Only components which have the potential to be traffic-limited could gain from application of EVE. As demonstrated in this paper, factors influencing the type of limitation include
  - architecture (distribution and modularity),
  - cost distribution, and
  - traffic levels.

A changeover to EVE on an existing large computerized traffic administration system would not likely be justified by service or capital cost benefits alone, and large savings in administration cost would not be expected. However, new systems may be best engineered according to EVE, due to

- the introduction of new services, expected to cause an unpredictable and highly variable traffic environment,
- an increased incidence of remote concentrators, which have a reasonable likelihood of being traffic-limited,
- a trend towards peak-sensitive components, including processors, and
- the processing power now available for traffic data analysis.

7. *CONCLUSIONS*

This paper has illustrated the major factors which should be considered if an adoption of EVE for switch component provisioning is contemplated. The assessment of cost and service benefits with EVE based on sample data is very difficult, and overall benefits can only be accurately quantified after EVE is introduced in the field. Introduction of EVE on an existing large-scale system would probably not be justified due to the high cost of changeover. However, the benefits with EVE identified in this paper warrant serious consideration of adopting EVE for new systems. The extent of benefits reaped is largely dependent on characteristics of both the system and traffic environment.

REFERENCES

[1] D.H. Barnes, "Extreme Value Engineering of Small Switching Offices," Proc. 8th Int. Teletraffic Congress, Melbourne, 1976.

[2] K.A. Friedman, "Extreme Value Analysis Techniques," Proc. 9th Int. Teletraffic Congress, Torremolinos, 1979.

[3] D.H. Barnes, "Observations of Extreme Value Statistics in Small Switching Offices," Proc. 9th Int. Teletraffic Congress, Torremolinos, 1979.

[4] K.A. Friedman, "Precutover Extreme Value Engineering of a Local Digital Switch," Proc. 10th Int. Teletraffic Congress, Montreal, 1983.

[5] D.H. Barnes, J.J. O'Connor, "Small Office Network Data System", Bell Syst. Tech. J., vol. 62, no. 7, pp. 2397-2431, September 1983.

TELETRAFFIC ISSUES in an Advanced Information Society
ITC-11
Minoru Akiyama (Editor)
Elsevier Science Publishers B.V. (North-Holland)

DEVELOPMENT OF TRAFFIC DATA COLLECTION AND ANALYSIS SYSTEM FOR VARIOUS ESS TYPES IN KOREA

Keesoo HONG and Jinho KIM

Research Center, KTA
Seoul, Korea

ABSTRACT

The traffic data in telecommunications network operation can be used for the following purposes :

- efficient use of existing facilities
- keeping up the grade of service level for subscribers
- planning and engineering of telecommunications network

Not only for the purposes described above, but for the automatic network management, the Centralized Traffic Management System(CTMS) has been developed.

In this article, we will introduce the CTMS architecture including the traffic data interface units of which primary functions are as follows ;

- temporarily storing the traffic data from the ESS
- communication with the host computer to send the traffic data or receive messages.

1. INTRODUCTION

In general, the method of traffic measurement and data collection for the electromechanical switching systems is independent of the exchange type.

On the other hand, the electronic switching system(ESS) measures the traffic using the internal software in its own way, so an appropriate system for traffic data collection from the various types of ESS should be developed.

The centralized Traffic Management System (CTMS) was developed to meet the needs described above in Korean situation, which could collect traffic data from the ESS and process it to control and manage the telephone network.

The CTMS performs the functions as follows ;

- traffic data collection from the different types of ESS in Korea through the data line
- maintaining the central office equipment data
- storing the traffic data for further applications.

The CTMS consists of two basic parts ; host computer system and traffic data interface units for each types of ESS. The data transfer interface processor(DTIP) was developed for the ESS which records the traffic measurement results on the magnetic tape.

On the other hand, the pollable data unit (PDU) was also developed for the ESS which produces the traffic measurement results on the local printer.

Both interface units, which were added to the relevent ESS, are under the control of the commands from the host computer system. Figure 1 illustrates the overall CTMS configuration.

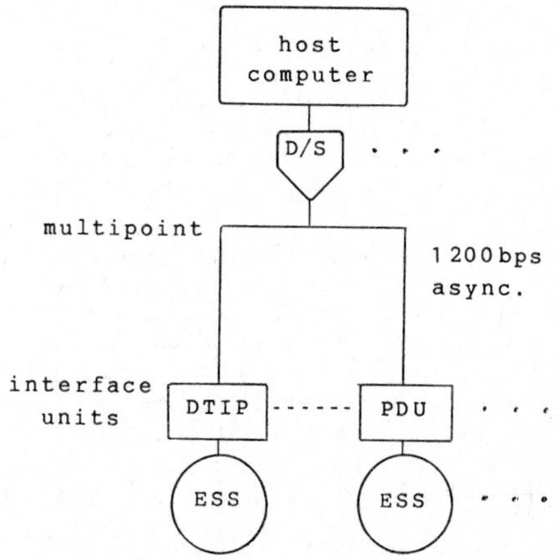

Fig. 1 The CTMS Configuration

2. INTERFACE UNITS

The interface units temporarily store the traffic data from the ESS and send it to the host computer after receiving the appropriate commands. As described above, two kinds of interface units (DTIP and PDU) were developed according to the traffic data collection method from the ESS.

2.1 Data Transfer Interface Processor(DTIP)

As shown in Figure 2, the DTIP catches the traffic data when it flows from the magenetic tape controller to magnetic tape unit in the ESS. The DTIP stores the traffic data until it receives the commands, "send", from the host computer.

The ESS(M10CN) has two CPUs(A and B) to operate in load sharing mode, and both CPUs measure the traffic and transmit it to MTU simultaneously. Accordingly, the DTIP was designed to use two processors to operate independently.

The DTIP hardware was configured using the Z80A microprocessor and its family.

The DTIP investigates the line status between MTC and MTU to catch the traffic data. Communications with the host computer is accomplished through the data set on the idle condition of the line after checking out the status between two processors in DTIP. As the interface units are connected to the host computer in multipoint mode,

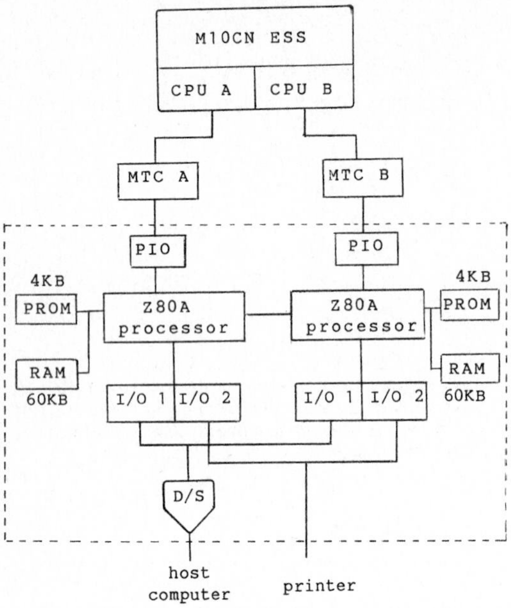

Fig. 2 DTIP Configuration

each unit has its own site code(8 bits) to distinguish the messages from the host computer.

The software of DTIP was designed to operate in interrupt mode with its four main routines as follows;

- data collection routine ; catches the traffic data between the MTC and MTU, and stores it in RAM
- communication routine ; receives the commands from the host computer and performs its function after checking out the site code predefined
- system initialization routine ; keeps up the system parameters and controls the printer for the special messages from DTIP or host computer
- self-diagnostic routine ; investigates the PROM status.

Figure 3 shows the DTIP software structure.

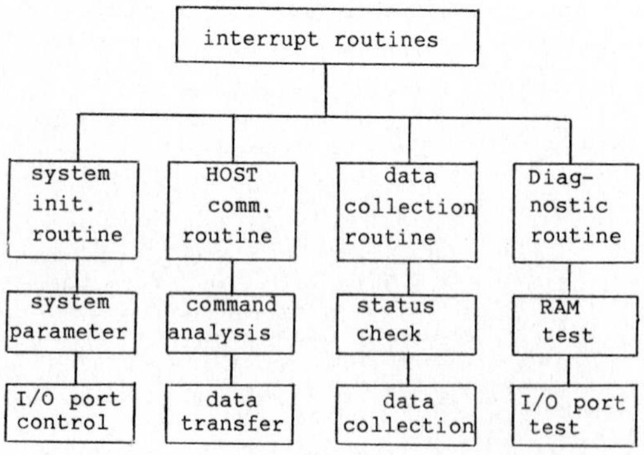

Fig. 3 DTIP S/W Structure

2.2 Pollable Data Unit(PDU)

For the ESS(NO.1A) which has its own printer channel for the traffic data, the PDU collects the traffic data and transmits it to host computer. The PDU hardware also used the Z80A microprocessor and its family as illustrated in Figure 4.

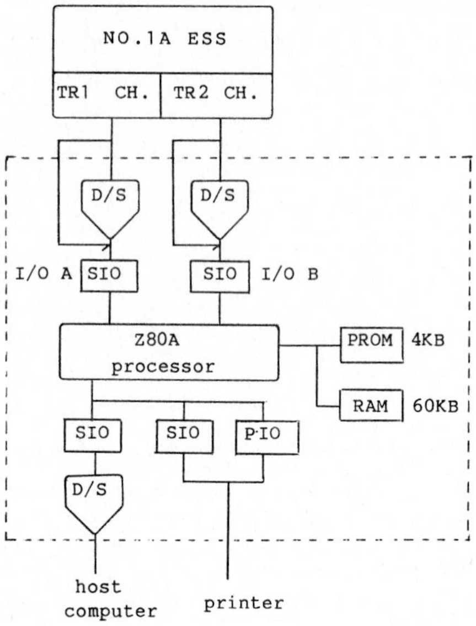

Fig. 4 PDU Configuration

From the I/O channel A and B, the PDU receives the traffic data or other messages from the ESS. These two channels can be connected to the PDU through the data set or directly using EIA 25 pin connector. The PDU has two switches(8 bits), one for its site code and the other for its transmission speed control to the host computer.

The software of PDU was designed to operate in interrupt mode with its four main routines as follows ;

- I/O control routine ; receives the traffic data from the channel of the ESS and stores it in the buffer. When the messages are not for traffic data, this routine sends them to the printer-buffer for output on local printer.
- communication control routine ; receives commands from the host computer and performs its function after checking out the site code. If the messages are for local printer, this routine just forwards them to the printer-buffer.
- printer control routine ; investigates the three independent buffer areas(I/O A, I/O B and printer buffer) and prints out the contents of the buffer on local printer not-overlappingly.
- self-diagnostic routine ; investigates the PROM every 25 minutes. If the error is detected after comparing the checksum of PROM, it generates a sound for warning.

Figure 5 illustrates the PDU software structure.

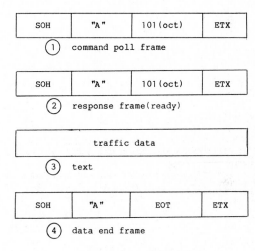

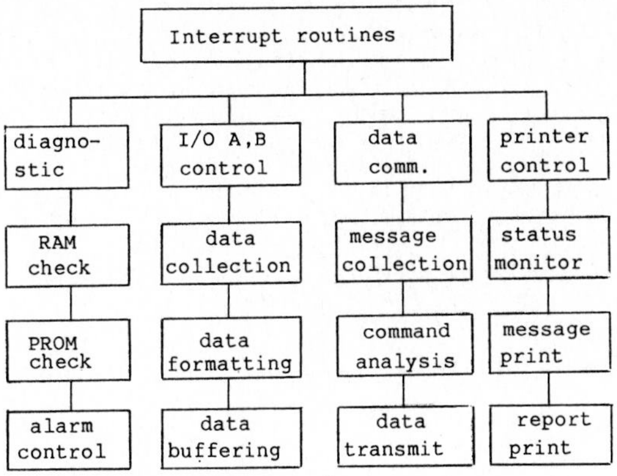

Fig. 5 block diagram of PDU software

3. CTMS ARCHITECTURE

The computer system controls the interface units which were connected to the ESS through data lines in multipoint mode. VAX11/780 system has been used as the CPU with 6MB main memory, 900MB disk storage, 3 M.T units, 8 multiplexors and the other peripheral devices.

3.1 Interface Unit Control Technique

One I/O channel in CTMS computer has multi-connection to several sets of interface units which have their own inherent site codes.

When the command poll has been sent through the channel in CTMS computer, the interface unit detects it and compares the site code fields. If the site code in the command poll is consistent with that of the interface unit, it activates the request-to-send(RTS) signal "ON"to send response message.

SOH	site code	command/ response	ETX

message format

One of the ASCII character A to Z can be used as a site code of interface units, and 0 to 9 for the command and response code.

The implemented command class is as follows;
- transmission request or cancellation of the traffic data stored in the interface units
- initialization of interface units
- timeout control of data line
- start or end of transmission
- status request for interface units, etc.

Data communications between the computer system and the interface units is achieved in handshaking mode. A command of CTMS computer is usually followed by a response from the interface units.

If the CTMS computer could not receive the response from the interface units, it retransmits the command every 1 second.

The command is aborted after 3 times of retransmission.

3.2 Traffic Data Polling

On-line data polling is accomplished by the scheduling table in the CTMS computer. The invoked scheduling program analyzes the scheduling table every quarter hour and the format is as follow ;

device type	up/ down	command-1	command-2

secheduling table format

The number of tables which the computer system retains, is the number of sites multiplied by the number of quarter hours.

Device type field of the format displays classification of interface units, and up/down field indicates the status of the units. Command -1 and command-2 fields represent the contents of commands. Among these, the data polling commands are expressed in 100(oct) plus the number of traffic data classifications predefined.

On receiving the data polling command from the host computer, the interface unit transmits the traffic data which has been stored in its buffer area. If there is no traffic data stored, it transmits not-ready response frame to the host computer. Followings are the message frame formats in transmission sequence.

SOH	"A"	101(oct)	ETX

(1) command poll frame

SOH	"A"	101(oct)	ETX

(2) response frame(ready)

traffic data

(3) text

SOH	"A"	EOT	ETX

(4) data end frame

message frame formats

4. DATA ANALYSIS SOFTWARES

This part describes the software systems, which uses the informations in the common data base to analyze the traffic data for the operations and maintenance of the exchanges.

The period of data analysis can be determined according to its operational purpose.
- 30sec. or 5min. ; network control
- 15min. or hours ; on-line reporting
- weeks or months ; network administration
- weeks or months or years ; network engineering/planning

In this paragraph, the following software systems will be described.
- on-line report generation and distributor
- detailed reporting system
- planning and engineering reporting system
- load balance analysis system

4.1 On-line Report Generation and Distributor

The traffic data collected by the data polling software is stored on addressed memory and then forwarded to mass storage(magnetic disk). The reporting software analyzes the traffic data using the exchange equipment informations in common data base(CDB), and generates the exception reports to be printed out on the interface unit's printer.

The items analyzed by this software are as follows ;

- abnormal holding time(suspect usage)
- overflows on trunk group, junctor, signaling devices beyond the threshold.
- other informations needed to maintain the exchange facilities.

4.2 Detailed Reporting System

This software analyzes the traffic data collected weekly or monthly and the following reports can be produced.

- busy hour traffic calculation
- traffic flow diagram
- day to day traffic variation in a week
- number of required trunks
- traffic per subscriber in busy hour
- local and toll traffic
- traffic on common devices
- number of non-completed calls
- other informations

4.3 Planning and Engineering Reporting System

Using the annually accumulated data processed by the detailed reporting system, the traffic variation during a year can be produced to determine the busy season traffic. This system makes it possible for the engineer to calculate the seasonal or safety factors, so as to adjust the efficient number of the network operation facilities.

4.4 Load Balance Analysis System

The traffic congestion on the subscriber concentrator occurs occasionally upon the subscriber's behavior. This system analyzes the subscriber's line concentration circuit traffic of the specified period, and makes it possible to keep the traffic balance among the subscribers module Figure 6 shows the overall software structure of CTMS.

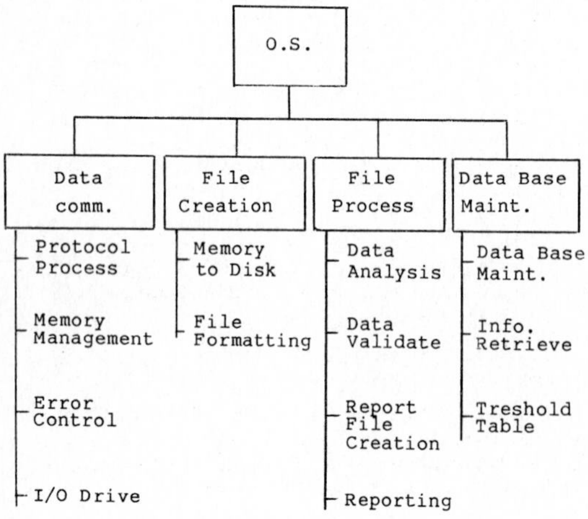

Fig. 6 CTMS Software Structure

5. CONCLUSION

The efficient administration and management of telephone network could be accomplished since the CTMS provides the sufficient informations for the short and long term study of traffic patterns in Korea.

For the automatic control of the toll telephone network, a specialized system with the graphical display capabilities of the network status, should be developed to collect and process the traffic data on shorter interval (at least less than 5 minutes).

REFERENCES

(1) "Microprocessor Applications Reference Book, Vol.1".
Zilog, July, 1981. Zilog, Inc.
(2) "Total Network Data System", The BSTJ. Sep. 1983, Vol.62, No.7, Part 3
(3) "Bell System Practices", AT&TCo standard, ADDENDUM 231-070-515, Issue 1, Dec, 1980
(4) "Model 9700 Digital Tape Tranport manual" - Kennedy co.
(5) 160 ITT 52013 xxxx-DEBE, BTM
160 ITT 52122 xxxx-EDBE, BTM
160 ITT 52122 xxxx-ECBE, BTM
160 ITT 52222 xxxx-ECBX, BTM

TELETRAFFIC ISSUES in an Advanced Information Society
ITC-11
Minoru Akiyama (Editor)
Elsevier Science Publishers B.V. (North-Holland)

RELATIONS AMONG THE MOMENTS OF QUEUE LENGTH DISTRIBUTION IN THE SYSTEM M/G/1 WITH UNIT-SERVICE AND BULK-SERVICE

Yo MURAO and Gisaku NAKAMURA

Shibaura Institute of Technology, Ohmiya, Japan
Shinshu University, Nagano, Japan

ABSTRACT

The queueing system M/G/1 with unit-service and bulk-service is analyzed from a new point of view. Some relations among the moments of queue length distribution are first found, and the n-th order moment, $n \geq S$, is derived by using the k-th order moments, $k < n$, where S is the maximum batch size of bulk-service queue. It is noted that the above derivation is made without the zeros of the denominator of probability generating function for queue size.

1. INTRODUCTION

Queueing systems are analyzed for the various congestion problems occured in the telecommunication system and the computer system. The present paper investigates the congestion characteristics of bulk-service queueing system M/G/1(bulk), which includes the ordinary unit-service system M/G/1 as a special case, from a new point of view. The relations are found among the moments of queue length distribution, and the n-th order moment, $n \geq S$, is expressed by the k-th moments, $k < n$, where S is the maximum batch size.

It is noted that the formulas are given by the same expressions for both the bulk-service system and the unit-service system.

The moments of queue length distribution, such as the mean and the variance, are usually obtained by differentiating the probability generating function of queue length, after the derivation of the probability generating function with the aid of zeros of its denominator. The present investigation is motivated to find a new approach in which no zeros are required.

2. DESCRIPTION OF QUEUEING SYSTEM

In the system M/G/1(bulk), calls (or jobs) randomly arrive at the system with rate λ. The service times are identically distributed, mutually independent random variable with distribution B(t). It is assumed that

$$\int_0^\infty t^n \, dB(t) = b_n < \infty, \quad n \geq 1. \tag{1}$$

The calls are served by a single server in batches, according to the following discipline. Let s and S be non-negative integers satisfying $0 \leq s \leq S < \infty$. The server commences his services only if there are at least s calls in the queue. If the number of calls in the queue is from s to S (s and S are included), the entire calls are served in a batch, whereas, if it is greater than S, only S calls are taken into service. This discipline is called a general bulk-service rule [1], and the system includes many typical bulk-service queueing systems as special cases.

If $s = 0$ and $S > 0$, the system is the classical bulk-service system discussed by Bailey [2], if $s = 1$ and $s \leq S$, the system discussed by Le'Gall [3], and if $0 < s = S$, the system by Fabens [4]. If $s = S = 1$, the system is reduced to the ordinal (unit-service) system M/G/1, and if $s = 0$ and $S = 1$, the system M/G/1 with enforced idle time [5]. Since the paper investigates the general bulk-service system, the ordinal system M/G/1 is naturally included.

3. ANALYSIS

Let t_k and t'_k be the epochs at which the k-th service commences and finishes, respectively. Put q_k be the queue length at t_k (q_k does not include the calls that are about to be served), and r be the number of arrivals during the k-th service. Consider

$$q_k + r,$$

then there are three cases:

(i) $q_k + r \geq S$

(ii) $S > q_k + r \geq s$

(iii) $s > q_k + r \geq 0$

In the first case, S calls are served in the $(k+1)$-st service, which commences at $t'_k = t_{k+1}$. The queue length is expressed as

$$q_{k+1} = q_k + r - S \tag{2}$$

In the second case, the entire calls are served in the $(k+1)$-st service, which commences at $t'_k = t_{k+1}$. The queue length is 0, that is

$$q_{k+1} = 0. \tag{3}$$

In the third case, the $(k+1)$-st service is postponed until the queue length attained to s. Thus, in this case, $t_{k+1} > t_k$, and the entire calls are served,

$$q_{k+1} = 0. \tag{4}$$

Define $U(q_k, r)$ the number of calls having the following two properties: (1) the calls are served in the $(k+1)$-st service, (2) the calls arrive at the system until the epoch t'_k. Then,

$$U(q_k, r) = \begin{cases} S, & \text{for (i)}, \\ q_k + r, & \text{for (ii)}, \\ q_k + r, & \text{for (iii)}. \end{cases} \quad (5)$$

Thus, defining

$$d(q_k, r) = S - U(q_k, r), \quad (6)$$

Eqs.(2), (3), and (4) are unified in a single form:

$$q_{k+1} = q_k + r - S + d(q_k, r). \quad (7)$$

From Eqs.(5) and (6), the relation

$$\{q_k + r - S\}^i \{d(q_k, r)\}^j = (-1)^i \{d(q_k, r)\}^{i+j}, \; j > 0 \quad (8)$$

is obtained and, by taking n-th power of both hand sides of Eq.(7),

$$\{q_{k+1}\}^n = \{q_k + r - S\}^n + \sum_{j=1}^{n} \binom{n}{j} \{q_k + r - S\}^{n-j} \{d(q_k, r)\}^j = \{q_k + r - S\}^n - \{-d(q_k, r)\}^n, \quad n > 0 \quad (9)$$

holds.

Now, consider the Taylor's expansion of $f(x + \eta)$, then, with the aid of Eq.(9),

$$f(x + q_{k+1}\zeta) = f(x + [q_k + r - S]\zeta) - f(x - d(q_k, r)\zeta) + f(x) \quad (10)$$

is derived. Put

$$f(x) = e^{-x},$$

then Eq.(10) leads to

$$e^{-q_{k+1}\zeta} = e^{-(q_k + r - S)\zeta} - e^{d(q_k, r)\zeta} + 1. \quad (11)$$

Put $e^{\zeta} = 1 + \theta$, and denote

$$(x)_n = x(x-1)(x-2)\cdots\cdots(x-n+1),$$

then the left hand side of Eq.(11) is written as

$$(1+\theta)^{-q_{k+1}} = 1 + \sum_{n=0}^{\infty} \frac{(-q_{k+1})_{n+1}}{(n+1)!} \theta^{n+1} \quad (12)$$

and the first and second terms of the right hand side are written as

$$(1+\theta)^{-(q_k + r - S)} = 1 + \sum_{n=0}^{\infty} \frac{(-\{q_k + r - S\})_{n+1}}{(n+1)!} \theta^{n+1} \quad (13)$$

$$(1+\theta)^{d(q_k, r)} = 1 + \sum_{n=0}^{\infty} \frac{(d(q_k, r))_{n+1}}{(n+1)!} \theta^{n+1}. \quad (14)$$

Since the coefficients of θ^{n+1} ($n \geq 0$) in Eq.(12) must coincide with the sum of the coefficients of θ^{n+1} in Eqs.(13) and (14), it is given that

$$(-1)^{n+1} q_{k+1}(q_{k+1}+1)(q_{k+1}+2)\cdots(q_{k+1}+n) = (-1)^{n+1}(q_k+r-S)(q_k+r-S+1)(q_k+r-S+2)\cdots\cdot(q_k+r-S+n) - \{d(q_k,r)\}\{d(q_k,r)-1\}\{d(q_k,r)-2\}\cdots\cdot\{d(q_k,r)-n\},$$
$$\text{for } n \geq 0. \quad (15)$$

Since

$$d(q_k, r) = 0 \text{ or } 1 \text{ or } 2 \text{ or } \cdots\cdot \text{or } S,$$

the relation

$$d(q_k, r)\{d(q_k, r) - 1\}\cdots\cdots\{d(q_k, r) - n\} = 0$$

must hold for $n \geq S$. Thus

$$q_{k+1}(q_{k+1}+1)(q_{k+1}+2)\cdots\cdot(q_{k+1}+n) = (q_k + r - S)(q_k + r - S + 1)(q_k + r - S + 2)\cdots\cdot(q_k + r - S + n), \quad n \geq S \geq 1 \quad (16)$$

is derived. Using the well-known formula

$$t(t+1)(t+2)\cdots\cdot(t+n-1) = \sum_{m=1}^{n} (-1)^{m+n} s(n, m) t^m, \quad \text{for } n \geq 1,$$

where $s(n,m)$ is the first kind of Stirling's number, Eq.(16) becomes

$$\{q_{k+1}\}^{n+1} + \sum_{m=1}^{n} (-1)^{m+n+1} s(n+1, m)\{q_{k+1}\}^m = \{q_k\}^{n+1} + \sum_{j=0}^{n} \binom{n+1}{j} (r - S)^{n+1-j} \{q_k\}^j + \sum_{m=1}^{n} (-1)^{m+n+1} s(n+1,m)\{q_k\}^m + \sum_{m=1}^{n} (-1)^{m+n+1} s(n+1,m) \times \left[\sum_{j=0}^{m-1} \binom{m}{j} (r - S)^{m-j} \{q_k\}^j\right],$$
$$\text{for } n \geq S \geq 1. \quad (17)$$

Now, assume that the queue is in statistical equilibrium, then

$$E[\{q_{k+1}\}^n] = E[\{q_k\}^n] \equiv E[q^n], \quad \text{for } n \geq 1$$

is satisfied. By taking the expectation for both sides of Eq.(17), it is obtained that

$$\sum_{j=0}^{n} \binom{n+1}{j} E[(r-S)^{n+1-j}]\, E[q^j] + \sum_{m=1}^{n} (-1)^{m+n+1} s(n+1,m) \times \sum_{j=0}^{m-1} \binom{m}{j} E[(r-S)^{m-j}] E[q^j] = 0,$$

$$\text{for } n \geq S. \qquad (18)$$

Eq.(18) is expressed in a slightly modified form:

$$\binom{n+1}{n} E[r-S]E[q^n] + \sum_{j=0}^{n-1} \binom{n+1}{j} E[(r-S)^{n+1-j}] \times E[q^j] + \sum_{m=1}^{n} (-1)^{m+n+1} s(n+1,m) \sum_{j=0}^{m-1} \binom{m}{j} \times E[(r-S)^{m-j}]E[q^j] = 0, \quad n \geq S. \qquad (19)$$

By exchanging the order of summations on the third term of the left hand side in Eq.(19), it is obtained that

$$\binom{n+1}{n} E[r-S]E[q^n] + \sum_{j=0}^{n-1} \binom{n+1}{j} E[(r-S)^{n+1-j}] \times E[q^j] + \sum_{j=0}^{n-1} E[q^j] \sum_{m=j+1}^{n} (-1)^{m+n+1} \times s(n+1,m) \binom{m}{j} E[(r-S)^{m-j}]$$

$$= \binom{n+1}{n} E[r-S]E[q^n] + \sum_{j=0}^{n-1} E[q^j] \times \{ \binom{n+1}{j} \times E[(r-S)^{n+1-j}] + \sum_{m=j+1}^{n} (-1)^{m+n+1} \times s(n+1,m) \binom{m}{j} E[(r-S)^{m-j}] \}$$

$$= \binom{n+1}{n} E[r-S]E[q^n] + \sum_{j=0}^{n-1} E[q^j] \times \sum_{m=j+1}^{n+1} (-1)^{m+n+1} s(n+1,m) \binom{m}{j} E[(r-S)^{m-j}]$$

$$= 0, \quad \text{for } n \geq S. \qquad (20)$$

Thus, the n-th order moment of queue length distribution $E[q^n]$ is derived as

$$E[q^n] = \frac{\sum_{j=0}^{n-1} E[q^j] \sum_{m=j+1}^{n+1} (-1)^{m+n+1} s(n+1,m) \binom{m}{j} E[(r-S)^{m-j}]}{(n+1)\{E[S]-E[r]\}}$$

$$\text{for } n \geq S \geq 1. \qquad (21)$$

Eq.(21) shows that the n-th order moment, $(n \geq S)$, of queue length distribution is expressed by the k-th order moments, $(n > k \geq 1)$. Then, by substituting the i-th moments, $(1 \leq i \leq S-1)$, into the j-th moments, $(S \leq j \leq n-1)$, recurrently, any moment the order of which is equal to or greater than S is principally expressed only the first $S-1$ moments.

4. EXAMPLES

In this section, special cases when $S=1$ and $S=2$ are considered. Applying the formula derived in the previous section, the moments of queue length distribution are explicitly calculated. It is shown that the moments obtained by the formula coincide with ones by differentiating the generating functions in the ordinary method.

4.1 M/G/1 system

Putting $S=1$ in Eq.(21), the n-th order moment $(n \geq 1)$ of queue length distribution in the system M/G/1 with unit-service is obtained.

For the sake of notational abbreviation, let

$$\begin{cases} E[q^n] = Q_n, \\ E[r^n] = R_n, \\ E[q^o] = Q_o = 1. \end{cases} \qquad (22)$$

Then, for $n=1$ and $n=2$,

$$Q_1 = \frac{\sum_{m=1}^{2} (-1)^{m+2} s(2,m) \binom{m}{0} E[(r-1)^m]}{2(1-R_1)} = \frac{(R_1-1)+(R_2-2R_1+1)}{2(1-R_1)} = \frac{R_2-R_1}{2(1-R_1)} \qquad (23)$$

$$Q_2 = \frac{\sum_{j=0}^{1} Q^j \sum_{m=j+1}^{2} (-1)^{m+1} s(3,m) \binom{m}{j} E[(r-1)^{m-j}]}{3(1-R_1)} = \frac{3Q_1(R_2-1)+R_3-R_1}{3(1-R_1)} \qquad (24)$$

are obtained after a little calculations. It is easily verified that Q_1 and Q_2 coincide with the first and second order moments of queue length distribution derived by the ordinary method.

Let P_r, $(r=0,1,2,\cdots\cdots)$, be the probability that r calls arrive at the system during a service interval (t_k, t_k'), $(k=0,1,2,\cdots\cdots)$, $K(z)$ be the probability generating function of $\{P_r\}$, and $\beta(s)$ be the Laplace-Stieltjes transform of service time, i.e.,

$$P_r = \int_0^{\infty} \frac{e^{-\lambda t}(\lambda t)^r}{r!} dB(t) \qquad (25)$$

$$K(z) = \sum_{r=0}^{\infty} P_r z^r = \int_0^{\infty} e^{-\lambda t(1-z)} dB(t) = \beta[\lambda(1-z)]. \qquad (26)$$

Thus, using Eq.(1), the moments of the number of arrivals are given by

$$R_1 = E[r] = \sum_{r=0}^{\infty} r\, P_r = K'(z)\big|_{z=1} = \lambda b_1 \quad (27)$$

$$R_2 = E[r^2] = \sum_{r=0}^{\infty} r^2 P_r$$

$$= K''(z)\big|_{z=1} + K'(z)\big|_{z=1}$$

$$= \lambda^2 b_2 + \lambda b_1 \quad (28)$$

$$R_3 = E[r^3] = \sum_{r=0}^{\infty} r^3 P_r$$

$$= K'''(z)\big|_{z=1} + 3K''(z)\big|_{z=1} + K'(z)\big|_{z=1}$$

$$= \lambda^3 b_3 + 3\lambda^2 b_2 + \lambda b_1. \quad (29)$$

Since the probability generating function of queue length distribution at the epochs just after the services commence is given by

$$G(z) = \frac{(1-\lambda)(1-z)}{K(z)-z},$$

the first and second moments are calculated as

$$Q_1 = \frac{\lambda^2 b_2}{2(1-\lambda b_1)} \quad (30)$$

$$Q_2 = Q_1 \frac{\lambda^2 b_2 + \lambda b_1 - 1}{1-\lambda b_1} + \frac{\lambda^3 b_3 + 3\lambda^2 b_2}{3(1-\lambda b_1)}. \quad (31)$$

The well known Pollaczek-Khintchine formula gives the probability generating function of the number of calls in the system at the epochs just after the services commence (which is also the number of calls in the queue at the epochs just before the services commence), and also gives the mean number of calls. Denote the mean number of calls in the system just after the epochs by m_1. Then, by Eqs.(27) and (30), it is obtained that

$$m_1 = Q_1 + R_1 = \frac{\lambda^2 b_1}{2(1-\lambda b_1)} + \lambda b_1. \quad (32)$$

It is easily seen that the first moment Q_1 given by Eq.(23) coinsides with one derived from the Pollaczek-Khintchine formula. Next, denote the variance of calls at the same epochs by σ^2. Then, by the definition,

$$\sigma^2 = Q_2 - Q_1^2 + (R_2 - R_1^2). \quad (33)$$

Substituting Eqs.(27), (28), (30), and (31) into the right hand side of Eq.(33), it is shown that the second moment given by Eq.(24) coinsides with one given by the Pollaczek-Khintchine formula.

4.2 M/D/1 system with enforced idle time

Consider the queueing system M/D/1 with enforced idle time, which is a special case of the system M/G/1 with the same property investigated by Powell et al.[5]. Let w_n be the number of services during waiting time. Thus, a call is served at the $(n+1)$-st service after its arrival. In this section, the stochastic property for w_n is studied instead of one for waiting times. It is only based on the theoretical simplicity.

Let g_n be the probability with which the number of calls is in the queue at the epochs just after the services commence, and $G(z)$ the generating function of $\{g_n\}$. Let $W(z)$, $G(z)$, and $K(z)$ be the probability generating functions for $\{w_n\}$, $\{g_n\}$, and $\{P_r\}$, respectively, then

$$W(z) = \sum_{n=0}^{\infty} w_n z^n = [G(z) - (1-\lambda)]/\lambda \quad (34)$$

$$G(z) = \sum_{n=0}^{\infty} g_n z^n = \frac{(1-\lambda)(1-z)}{K(z)-z} \quad (35)$$

$$K(z) = \sum_{r=0}^{\infty} P_r z^r = e^{-\lambda(1-z)} \quad (36)$$

are derived. By the ordinary calculations, the mean and the variance for w_n are

$$\bar{w} = \lim_{z\to 1} \frac{dW(z)}{dz} = \frac{\lambda}{2(1-\lambda)} \quad (37)$$

$$\sigma_w^2 = \lim_{z\to 1} \frac{d^2W(z)}{dz^2} + \bar{w} - (\bar{w})^2$$

$$= \frac{\lambda^2(2+\lambda)}{6(1-\lambda)^2} + \frac{\lambda}{2(1-\lambda)} - \frac{\lambda^2}{4(1-\lambda)^2} \quad (38)$$

By putting $b_n = 1$, $(n = 1,2,\cdots\cdots)$, in Eqs.(30) and (31), the first and the second moments of queue length distribution are given

$$Q_1 = \frac{R_2 - R_1}{2(1-R_1)} = \frac{\lambda^2}{2(1-\lambda)} \quad (39)$$

$$Q_2 = \frac{3Q_1(R_2-1) + R_3 - R_1}{3(1-R_1)}$$

$$= \frac{\lambda^3(2+\lambda) + 3\lambda^2(1-\lambda)}{6(1-\lambda)^2} \quad (40)$$

for the epochs just after the services commence in the system M/D/1 with enforced idle time. Using the relation between $W(z)$ and $G(z)$, which is given by Eq.(34), and Q_1 and Q_2 obtained by Eqs.(39) and (40), respectively, it is derived that

$$\bar{w} = \frac{Q_1}{\lambda} = \frac{\lambda}{2(1-\lambda)} \quad (41)$$

$$\sigma_w^2 = \frac{Q_2}{\lambda} - Q_1^2 = \frac{\lambda^2(2+\lambda)}{6(1-\lambda)^2} + \frac{\lambda}{2(1-\lambda)} - \left\{\frac{\lambda}{2(1-\lambda)}\right\}^2 \quad (42)$$

These results coincide with the $\bar{w}$ and σ_w^2 given by Eqs.(37) and (38). The higher order moments of w_n are usually derived by differentiating W(z) for many times (the number of times depends on their orders).

It is noted that the ordinal calculation of the k-th order moment ($k \geq 3$), which is made by differentiating W(z), is quite troublesome. The present method is tractable to obtain Q_n, ($n \geq 3$), in the explicit form, because the recurrent computation is efficiently made.

4.3 M/G/1 system with bulk-service

Consider the queueing system M/G/1 with bulk service, the batch size of which is 2. First, the second order moment is derived. Putting S = 2 and n = 2 in Eq.(21),

$$Q_2 = \frac{R_3 - 3R_2 + 2R_1 + 3Q_1(R_2 - 2R_1)}{3(2-R_1)} = \frac{R_2 - 2R_1}{2 - R_1} Q_1 + \frac{R_3 - 3R_2 + 2R_1}{3(2-R_1)} \quad (43)$$

is obtained with some calculation, Next, if $\Pi(z)$ and $\Phi(z)$ are the probability generating functions of queue length at the epochs just before the services complete, and just after the services commence, respectively, they are given by

$$\Pi(z) = \sum_{n=0}^{\infty} \pi_n z^n = \frac{K(z)\sum_{n=0}^{1}(z^2 - z^n)\pi_n}{z^2 - K(z)} = \frac{K(z)(2-\lambda)(z-1)}{z^2 - K(z)} \cdot \frac{z - z_1}{1 - z_1} \quad (44)$$

$$\Phi(z) = \sum_{n=0}^{\infty} \phi_n z^n = \Pi(z)/K(z) \quad (45)$$

where K(z) is the probability generating function of $\{P_r\}$ given by Eq.(26), and z_1 in Eq.(44) is a zeros of the denominator of $\Pi(z)$ within the unit circle.

By substituting $\Pi(z)$ in the right hand side of Eq.(45),

$$\Phi(z) = \frac{2 - R_1}{1 - z_1} \cdot \frac{(z-1)(z-z_1)}{z^2 - K(z)} = k \cdot \frac{(z-1)(z-z_1)}{A(z)} \quad (46)$$

where

$$k = \frac{2 - R_1}{1 - z_1},$$

$$A(z) = z^2 - K(z). \quad (47)$$

The derivatives of A(z) are written as

$$\begin{cases} A'(z) = 2z - K'(z), \\ A''(z) = 2 - K''(z), \\ A'''(z) = -K'''(z), \end{cases}$$

and, for z = 1,

$$\begin{cases} A'(1) = 2 - R_1, \\ A''(1) = 2 - R_2 + R_1, \\ A'''(1) = -(R_3 - 3R_2 + 2R_1). \end{cases} \quad (48)$$

From Eq.(46), the first derivative of $\Phi(z)$ is

$$\Phi'(z) = k \cdot \frac{\{2z - 1 - z_1\}A(z) - (z-1)(z-z_1)A'(z)}{\{A(z)\}^2}$$

and

$$\Phi'(1) = k \cdot \frac{2A(z) - (z-1)(z-z_1)A''(z)}{2A(z)A'(z)}\bigg|_{z=1} = k \cdot \frac{2A'(1) - (1-z_1)A''(1)}{2\{A'(1)\}^2}$$

Since $Q_1 = \Phi'(1)$, Q_1 is derived as

$$Q_1 = \frac{1}{1 - z_1} - \frac{2 - R_2 + R_1}{2(2 - R_1)}. \quad (49)$$

Next, the second derivative is also calculated, but the precise form is to complicate to write down. So, the main results are only shown.

$$\Phi''(1) = \frac{2 - R_1}{1 - z_1} \cdot \frac{1}{6(2-R_1)^3} \cdot B'''(z)\big|_{z=1} \quad (50)$$

where

$$B'''(1) = -6(2-R_1)(2-R_2+R_1) + 2(1-z_1)(2-R_1)(R_3 - 3R_2 + 2R_1) + 3(1-z_1)(2-R_2+R_1)^2, \quad (51)$$

$$\Phi''(1) = \frac{1}{1-z_1} \cdot \frac{2 - R_2 + R_1}{2 - R_1} + \frac{R_3 - 3R_2 + 2R_1}{3(2-R_1)} + \frac{(2-R_2+R_1)^2}{2(2-R_1)^2}. \quad (52)$$

$$Q_2 = \Phi''(1) + Q_1$$

$$= \frac{R_2 - 2R_1}{2 - R_1} Q_1 + \frac{R_3 - 3R_2 + R_1}{3(2 - R_1)} \qquad (53)$$

From Eqs.(43) and (53), it is shown that the second order moment derived by Eq.(21) coincides with one obtained in a usual way. The derivation in this subsection is seem to be easier than the ordinal method.

5. CONCLUSION

The simple relation among the moments of queue length distribution in the M/G/1 with unit-service and bulk-service are derived without use of zeros of the denominator of the generating function. By the relation the k-th order moment is expressed by j-th order moments ,($j = 1, 2, \cdots, k-1$), recurrently. It is necessary, however, to obtain the first $S-1$ moments in the other methods.

Some applications are shown for the ordinal M/G/1 system, the M/D/1 system with enforced idle time, and the M/G/1 system with bulk-service.

REFERENCES

[1] M.F. Neuts,"A General Class of Bulk Queues with Poisson Input," *Ann. Math. Statist.*, Vol. 38, pp.759-770, 1967.

[2] N.T.J. Bailey,"On Queueing Processes with Bulk Service," *J. Roy. Statist. Soc.*, Ser.B, Vol. 16, pp.80-87, 1954.

[3] Le'Gall,"Les systems avec ou sans attente et les processus stochastiques," Dunod, Paris, 1962.

[4] A.J. Fabens,"The Solution of Queueing and Inventory Models by Semi-Markov Processes," *J. Roy. Statist. Soc.*, Ser.B, Vol. 23, pp.113-117, 1961.

[5] B.A. Powell and B. Avi-Itzhak,"Queueing Systems with Enforced Idle Time," *J. Opns. Res. Soc. of Amer.*, Vol. 15, pp.1145-1156, 1967.

TELETRAFFIC ISSUES in an Advanced Information Society
ITC-11
Minoru Akiyama (Editor)
Elsevier Science Publishers B.V. (North-Holland)

APPROXIMATE ANALYSIS FOR BULK QUEUEING SYSTEM WITH COMPOSITE SERVICE DISCIPLINE

Kazuhiro Ohtsuki*, Toshio Shimizu** and Toshiharu Hasegawa*

*: Dept. of Appl. Math. & Physics, Kyoto University, Japan

**: Application System Development Section
Matsushita Electric Industrial Co., LTD., Japan

ABSTRACT

In this paper, we consider a transportation type bulk arrival bulk service queueing system with composite bulk service discipline as a fundamental research on future communication networks. In this system, customers arrive in group and customers in a group have an identical attribute. From the head of the queue, customers with the same attribute are served within a finite bulk size at the same time.

For such bulk service queueing system, the average queue length of customers in the system is approximately analyzed. In our approach, each class of customers with the same attribute is assumed to form their own queue. An approximation method for obtaining queue length of each queue just before the service occasion epoch, where probability generating function approach and embedded Markov chain method are utilized. Numerical examples show our approximation results are well verified by the simulation ones.

1. INTRODUCTION

This paper considers a queueing system where customers arrive in group according to compound Poisson process and form a single queue. Customers in a group have an identical attribute. Waiting customers are served by a single server in a bulk service fashion according to the predetermined discipline. From the head of the queue, customers with the same attribute are served within a finite bulk size at the same time. Customers in a group may not be served simultaneously, then a group is divided into two or more subgroups. Customers having arrived in different groups can be served in the same bulk if they have an identical attribute. A service is carried out during a time interval between service occasion epochs. The time intervals between service occasion epochs are independently and identically distributed with an arbitrary distribution.

This kind of queueing system can be applied to analyse some of stochastic behaviors in communication system as a mathematical model. Examples are following. In a packet switching network, each switching node assembles several packets having the same attribute as a common destination into a frame with fixed length. And gate nodes perform almost the same role to connect multiple local area networks through terrestrial global network or satellite communication network, and in the case of ISDN, all external interface will play the same role. Even in satellite communication system terrestrial stations can be considered to have a similar discipline to increase the utilization factor in the future. In those cases, a customer, an attribute, a server and a service time are regarded as a packet, a destination or a kind of packet, a switching equipment for framing and a time interval of sending a frame, respectively. And a service occasion epoch is regarded as a time epoch that a gate opens, or as a time boundary point of a slot in a synchronized system (in this case, the time intervals of service occasion epochs are assumed to obey a constant distribution).

The already published works for bulk service queueing systems are almost concerned with the systems where all customers have an identical attribute, called HBQS(homogeneous bulk queueing system). Bailey [1] derived the equilibrium distribution of queue length in a HBQS, where customers arrive individually, by the embedded Markov chain method. Jaiswal [4] solved the same problem as the one of Bailey except that the maximum number of customers to be served at the same time is not constant and obtained time-dependent solution by phase method [5] Miller [6] studied group arrival and group service, namely $M^x/G^y/1$ queue. Many researchers (Bhat [2] Cohen [3] Neuts [7] et al.) considered various versions of HBQS. Watanabe et al. [8] proposed Exclusive Group Service (EGS) discipline in $M^x/G^y/1$ type bulk queueing system, where arriving groups have different attributes though customers in group have an identical attribute. Our model proposed here has a composite bulk service discipline which allows different groups to have the same attribute.

For such bulk service queueing system, the average queue length of customers in the system is approximately analyzed. In our approach, each class of customers with the same attribute is assumed to form their own queue. There exist as many queues as the number of different attributes. An approximation method for obtaining queue length of each queue just before the service occasion epoch, where probability generating function (p.g.f.) approach and embedded Markov chain method are utilized. Our approximation results are well verified by the simulation results but have tendency to underestimate the average number of customers.

2. MODEL

Let us consider a transportation type bulk arrival bulk service queueing system model under composite bulk service discipline.

Customers arrive in group according to compound Poisson process, that is, the groups of customers arrive according to Poisson process and the sizes of groups, the number of customers in a group, are independent and identically distributed with an arbitrary distribution. Furthermore all customers of a group have the same attribute. The service, which is in batch of fixed capacity, begins just after a service occasion epoch and the time intervals of service occasion epochs are independently and identically distributed with an arbitrary distribution. The server is able to accommodate as many customers as possible within the capacity if at the back there are the groups of customers whose attributes are the same as the first one in the queue. If there are no customers in the queue at the time of completing the service, next service does not start as soon as the customers arrive at the system but starts after a next service occasion epoch.

In Fig.1, a symbol stands for a customer and the same symbol means an identical attribute. The number figured in a symbol is the group number arriving the node. Fig.1-(a) shows a situation just before service starting. The queue consists of four groups and the number of customers of the head group is 2. Customers of the head group in the queue have a claim to be served at first within the bulk size which is 4 in this figure. If the server can serve more customers yet, he accommodates the subsequent customers who have the same attribute as the head group. Then, the system state changes as Fig.1-(b). In our approach, each class of customers with the same attribute is assumed to form their own queue (see Fig.2). There exist as many queues as the number of different attributes. The service discipline is put another way as follows. Customers in the queue in which the earliest arrival group have joined

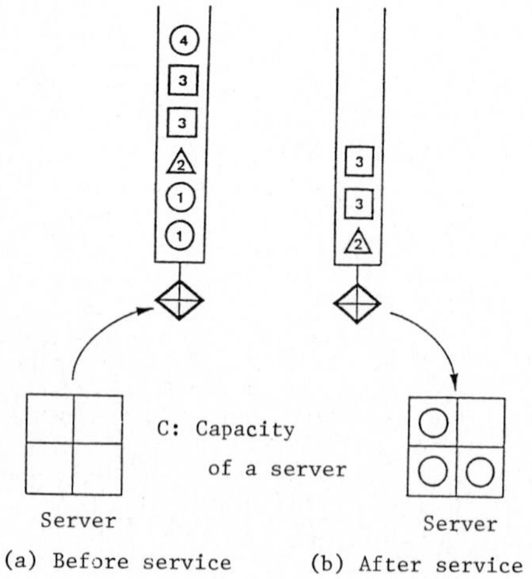

Fig. 1 Bulk queueing system model with composite service discipline.

are served within the capacity according to FCFS discipline. Note that FCFS discipline only applies to the customers (or groups) who have the same attribute. For example, in the case of Fig.2, group 4 is served earlier than group 2 and 3, although he arrived later.

Following notations are employed for the parameters of this queueing model.

n: The number of attributes.

λ_i: The arrival rate (Poisson rate) of groups of customers whose attribute is i in queue i. (i=1,2,...,n)

λ: The total arrival rate (Poisson rate) of groups of customers. $\lambda = \Sigma \lambda_i$.

b_k^i: The probability that a group of customers whose attribute is i consists of k customers at its arrival, where $k \geq 1$.

c: The capacity of a server.

V(t): The probability distribution function of the time intervals of service occasion epochs.

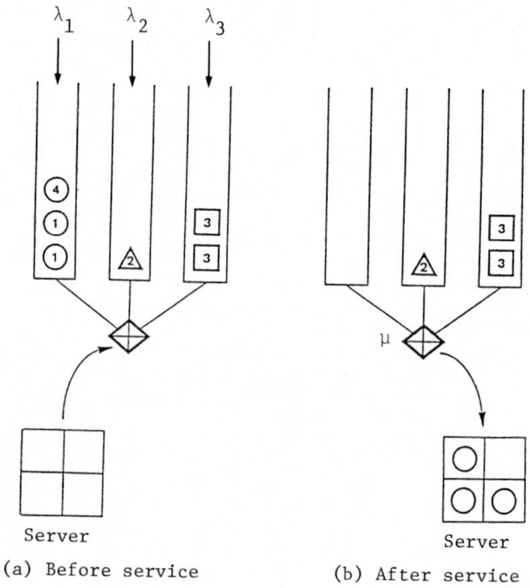

Fig. 2 Equivalent model for the bulk service queueing system model described in Fig.1.

3. ANALYSIS

3.1. Approximate Analysis for Mean Queue Length

We propose approximation method of obtaining a probability generating function for queue length of customers immediately before the beginning of service under a composite bulk service discipline. The outline of our method is as follows. We pay attention to the behavior of customers who have the same attribute, that is, we analyse the average length of tagged queue under the assumption that q, the probability that customers in the tagged queue, if any, are served is obtained. Let us introduce the following notations (random variables) to formulate our queueing model. We assume that queue i is formed by only customers whose attribute is i and has an infinite buffer.

V_j : The time interval of the j-th service occasion epoch and j+1-st one. we call V_j as j-th time interval.

X_j^i: The number of arriving customers at queue i during the time interval V_j.

W_j^i: The number of waiting customers in queue i immediately before the j-th service occasion epoch.

Y_j^i: The number of customers in queue i who are accommodated during the j-th time interval.

B_k^i: The size of k-th arriving customer-group in queue i. Let us call B_k^i as the customer-group size in queue i.

In what follows, we pay attention to queue 1 (i.e. customers of attribute 1) and analyse the queue length of queue 1. To make simple, the above random variables are rewritten as follows.

$$X_j = X_j^1,\ W_j = W_j^1,\ Y_j = Y_j^1,\ B_k = B_k^1.$$

Accordingly, the following recurrence relations can be provided for the queue length W_j of customers and the number of served customers Y_j, as shown in Fig. 3.

$$W_{j+1} = W_j - Y_j + X_j\ , \tag{1}$$

$$Y_j = \begin{cases} \min[W_j, c] & \text{(the case where customers can be served)} \\ 0 & \text{(otherwise).} \end{cases} \tag{2}$$

Eq.(2) shows that the number of served customers, Y_j, depends on whether customers in queue 1 can be served or not rather than the queue length W_j. Therefore, it is necessary to grasp not only the situation of queue 1 but also that of other queue. We must know which head customers of each queue arrive earliest of

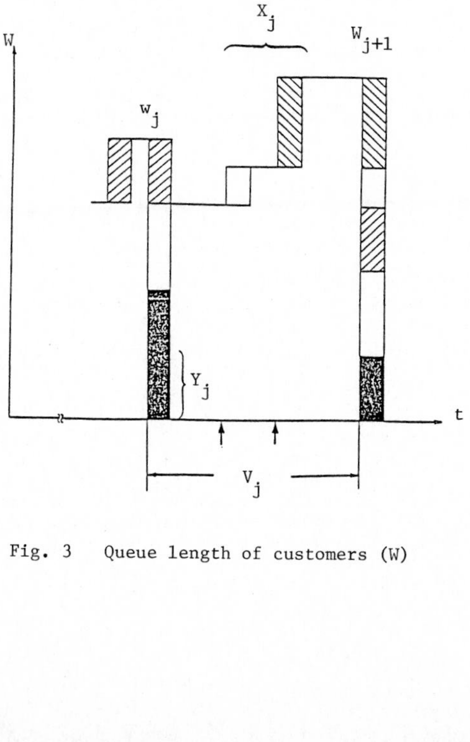

Fig. 3 Queue length of customers (W)

all in order to analyze this system exactly. Thus, an exact analysis is extremely difficult for the queueing model like this. Then approximation method for queue length W_j are proposed by using the probability that customers in queue 1 are able to be collected.

If subscript j of the above random variables is omitted, it means that j need not be considered.

First, the probability p_k that k customers arrive at queue 1 during a current service is,

$$p_k = \Pr[X_j=k] = \Pr[X=k]$$

$$= \sum_{k=0}^{\infty} \Pr[X=k|V=t]dV(t)$$

$$= \int_0^{\infty} \sum_{n=0}^{\infty} \exp(-\lambda_1 t)\frac{(\lambda_1 t)^n}{n!}\, b_k^{n*} dV(t) \tag{3}$$

where $b_k^{n*} = \Pr[B_1+B_2+...+B_n=k]$.

The probability generating function, p.g.f., P(z) of X is derived from both the p.g.f. B(z) of B, i.e., the size of a customer group at its arrival time, and V(t).

$$P(z) = \sum_{k=0}^{\infty} p_k z^k$$

$$= \int_0^{\infty} \exp(-\lambda_1 t[1-B(z)])dV(t). \tag{4}$$

$$B(z) = \sum_{k=0}^{\infty} b_k z^k\ . \qquad (|z| \leq 1) \tag{5}$$

According to Eqs.(1) and (3), the following equation is held.

$$\Pr[W_{j+1}=k \mid W_j=n,\ Y_j=m] = \Pr[x_j=k+m-n] = p_{k+m-n} \quad (m \leq n). \tag{6}$$

Multiplying by z^k on both sides of the above equation and adding over k, we have,

$$\sum_{k=0}^{\infty} \Pr[W_{j+1}=k \mid W_j=n,\ Y_j=m]z^k = P(z)z^{n-m}. \tag{7}$$

Furthermore, the following equation is obtained, if W_j and Y_j are removed.

$$\sum_{k=0}^{\infty} \Pr[W_{j+1}=k]z^k$$

$$= P(z) \sum_{n=0}^{\infty} \{\Pr[W_j=n,\ Y_j=\min(n,c)]z^{n-\min(n,c)} + \Pr[W_j=n,\ Y_j=0]z^n\}$$

$$= P(z) \sum_{n=0}^{\infty} \{\Pr[W_j=n,\ SV(j)]z^{n-\min(n,c)} + \Pr[W_j=n,\ NSV(j)]z^n\}\ . \tag{8}$$

Where, SV(j) and NSV(j) denote the event that the customers in queue 1 can be and can not be served during the j-th time interval, respectively. We introduce the following assumptions in analyzing the model approximately. The probability that the service is available at the j-th service occasion epoch, q, is independent of the queue length W_j, and identical. The approximation method for q is explained in Section 3.2.

$$\Pr[W_j=n,\ SV(j)] = q \cdot \Pr[W_j=n]\ . \tag{9}$$

$$\Pr[W_j=n,\ NSV(j)] = (1-q)\Pr[W_j=n]\ . \tag{10}$$

Introducing Eqs.(9) and (10), Eq.(8) is rewritten as follows.

$$\sum_{k=0}^{\infty} Pr[W_{j+1}=k]z^k$$

$$= P(z) \sum_{n=0}^{\infty} \{q \cdot Pr[W_j=n]z^{n-\min(n,c)} + (1-q)Pr[W_j=n]z^n\}$$

$$= P(z)\{q \sum_{n=0}^{c-1} Pr[W_j=n] + q \sum_{n=c}^{\infty} Pr[W_j=n]z^{n-c} + (1-q) \sum_{n=0}^{\infty} Pr[W_j=n]z^n\} . \quad (11)$$

Defining $w_k=\lim(j\to\infty)Pr[W_j=k]$ from the steady state assumption, the p.g.f. R(z) is given as follows.

$$R(z) = \sum_{n=0}^{\infty} w_n z^n = A(z)/D(z) , \quad (12)$$

where,

$$A(z) = q \sum_{n=0}^{c-1} w_n(z^c-z^n) , \quad (13)$$

and

$$D(z) = z^c\{P^{-1}(z)-(1-q)\}-q . \quad (14)$$

The c unknown variables w_n (n=0,1,...,c-1) can be determined from the facts that R(z) is analytical in $|z|\le 1$ and that D(z)=0 has the c zeroes in $|z|\le 1$ by Rouche's theorem. If we assume that D(z)=0 has 1 and ξ_i (i=2,3,...,c) as its simple zeroes, w_n (n=0,1,...,c-1) are obtained as the solution of the following linear equations.

$$\dot{A}(1) = \dot{D}(1) . \quad (15a)$$

$$\sum_{n=0}^{c-1} w_n(\xi_i{}^c-\xi_i{}^n) = 0 \quad (i=2,3,..,c) . \quad (15b)$$

Eq. (15a) is derived by the equation lim (z →1)R(z) = 1.

Then, the average queue length of customers in queue 1, E[L1] is obtained as follows.

$$E[L1] = \dot{R}(1) = \frac{\ddot{A}(1)\dot{D}(1) - \dot{A}(1)\ddot{D}(1)}{2\{\dot{D}(1)\}^2} , \quad (16)$$

where,

$$\dot{A}(1) = q \sum_{n=0}^{c-1} w_n(c-n) , \quad (17a)$$

$$\ddot{A}(1) = q \sum_{n=0}^{c-1} w_n\{c(c-1)-n(n-1)\} , \quad (17b)$$

$$\dot{D}(1) = c\cdot q-\dot{P}(1) , \quad (17c)$$

and

$$\ddot{D}(1) = c(c-1)q - 2c\dot{P}(1) + 2\{\dot{P}(1)\}^2 - \ddot{P}(1) . \quad (17d)$$

3.2. Approximation method for q

Introducing the probability, q that customers in queue 1 can be served, the approximate probability generating function R(z) for the queue length of customers in queue 1 can be obtained in Section 3.1. Whether customers in queue 1 can be served depends on not only the state of queue 1 but also the states of other queues. So it is very difficult to calculate q exactly. Then, we propose the following approximation equation for q using c, b_i(the mean group size of customers in queue i), λ_i and μ (the mean service rate).

$$q = (\mu - \sum_{i=2}^{n} \frac{\lambda_i b_i}{c})/\mu = 1 - \sum_{i=2}^{n} \frac{\lambda_i b_i}{c\mu} . \quad (18)$$

This equation is derived under the assumption that a server drops in queue 1,

$$(\mu - \sum_{i=2}^{n} \frac{\lambda_i b_i}{c})$$

times during μ times service occasions and can collect customers in queue 1.

4. NUMERICAL EXAMPLES AND EVALUATIONS

We now evaluate the approximation method by comparing them with simulations.

Table 1 shows the validity of our simulation experiment, where simulation results are compared with exact theoretical values the case that the number of attributes is equal to one (i.e., HBQS case), and the time intervals of service occasion epochs are constantly distributed.

The following numerical results are derived in the case that the sizes, B^i of customer groups at their arrival time are according to a geometric distribution with the mean equal to b_i (=1/p), i.e.,

$$b_k{}^i = Pr[B^i=k] = p(1-p)^{(k-1)} \quad (k=1,2,...).$$

At first, we show the numerical results in the case where the time intervals of service occasion epochs are according to an exponential distribution with parameter and the number of attributes is equal to two.

Fig. 4 shows the relationship between the average queue length of customers in queue i, E[Li] (i=1,2) and the Poisson arrival rate of groups of customers whose attribute is 1, λ_1 in the case where λ_2=0.1, b_1=b_2=3.0, μ = 1.0 and c = 5. Lines and symbols present analytical results and simulation results respectively. From this figure, we find that the theoretical results are well verified by the simulation results. Further it is found that E[L1] is affected by the change of λ_1 but E[L2] is not so much. Namely, in such a case, the mean queue length of one queue is not much

E[L] \ λ	0.1	0.2	0.3	0.4	0.5	0.6	0.7	0.8	0.9
Simulation	0.45	0.98	1.61	2.40	3.33	4.78	7.03	11.4	24.3
Exact	0.45	0.98	1.60	2.37	3.38	4.80	7.06	11.4	24.1

Table 1 Comparison between simulation and exact.

μ = 1.0, b = 3.0, c = 3 .

influenced by the arrival rate of group-customers in other queue.

Fig. 5 shows the relation between E[L1] and c in the case where μ = 1.0, λ_1= λ_2= 0.1 and b_1= b_2 = 5.0. As illustrated in this figure, it is recognized that if c is sufficiently larger than b_1 (=b_2), E[L1] changes little. That is because all waiting customers in the same queue are considered to be served at once.

Next, we consider the case where the service occasion epochs have constant time intervals and the number of attributes is three.

In Fig. 6 the relationship between E[L1] and λ_1 are shown with λ_1=λ_2=λ_3, c=5 and μ=1.0 under various value of b_1 (=b_2=b_3). It is confirmed that slight increase of the bulk size of arriving customers causes the large queue length in the case of heavy traffic. Furthermore, even if the mean arrival number per time (i.e. $\lambda_i \cdot b_i$) is the same, the larger the bulk size becomes, the more the queue length increases.

Fig. 7 shows the mean queue length of each queue as a factor of b_1. Similarly to Fig. 4, we can say that the mean queue length of one queue is not so much influenced by the mean customer-group size in other queue.

Fig. 8 shows the mean queue length for various λ_1 in the case where $\lambda_1 b_1$ is fixed to 0.4, b_2=b_3= 4.0, c=3, μ=1.0. The fitness of our appoximation is not good for the case when the average bulk size of another attribute is too large.

In general, our approximation results are well verified by the simulation results shown in these figures but has tendency to underestimate the simulation ones, which is caused by the approximation of the probability, q_i, that customers in queue i, if any, can be served next derived. In eq. (18), we assumed $\lambda_i b_i/c$ as the mean frequency of service per time for queue i, but this value is smaller than exact one because a server does not always serve c customers. Therefore, the value, q of our method is greater than the real one, then mean queue length is underestimated.

Many examples show our approach provides approximation results with rather good accuracy, and, in general, has the similar tendency with these examples.

5. CONCLUSION

In this paper, we consider a transportation type bulk service queueing system with the composite service discipline as a fundamental research on future communication networks. In our model, waiting customers which have the same attribute can be served at the same time. Then, the number of the waiting customers with the same attribute is the key factor for performance of the systems. In our approach, each class of customers with the same attribute is assumed to form their own queue. An approximation method for obtaining queue length of each queue just before the service occasion epoch, where probability generating function approach and embedded Markov chain method are utilized. Many examples show that our proposed method provides high approximation accuracy evaluating by the comparison with the results of simulations. With the proposed method, it has become possible to evaluate the characteristic quantities of the system considered.

Our proposed method can be applied to analyse some of stochastic behaviors in communication system.

References

[1] N.T.J. Bailey, "On queueing process with bulk service," J. of Royal Stat. Soc., Ser. B 16, 1954, pp.80-87.

[2] U.N. Bhat, "On single-server bulk-queueing process with binomial input," Oper. Res., 12, 1964, pp.527-533.

[3] J.W. Cohen, "The Single Server Queue," North-Holland, Amsterdam, 1969.

[4] N.K. Jaiswal, "Bulk-service queueing problem," Oper. Res., 8, 1960, pp.139-143.

[5] N.K. Jaiswal, "Time-dependent solution of the bulk-service queueing problem," Oper. Res., 8, 1960, pp.773-781.

[6] R.G. Miller, "A contribution to the theory of bulk queues," J. of Royal Stat. Soc., Ser. B 21, 1959, pp.320-337.

[7] M.F. Neuts, "Queues solvable without Rouche's theorem," Oper. Res., 27, 1979, pp.767-781.

[8] M. Watanabe, T. Shimizu, Y. Takahashi and T. Hasegawa, "Approximate analysis for exclusive bulk service queues (in Japanese)", Kokyuroku, 490, Research Inst. Math. Science, Kyoto University, 1983, pp.227-247.

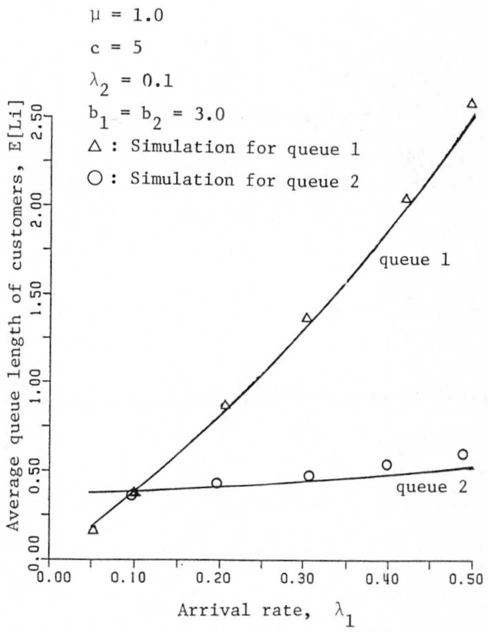

Fig. 4 Average queue length characteristics for various λ_1.

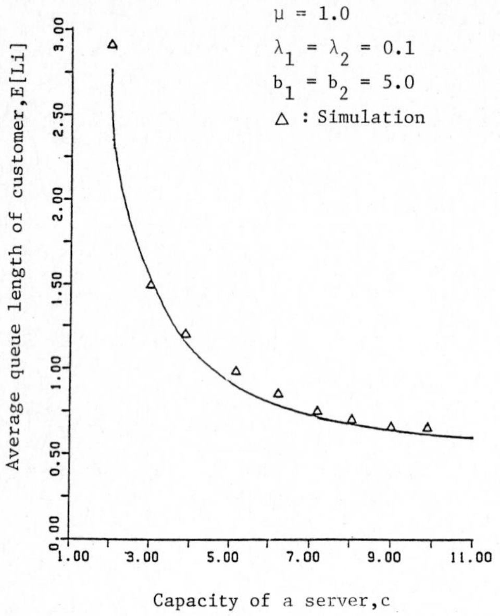

Fig. 5 Average queue length characteristics for various capacity of a server.

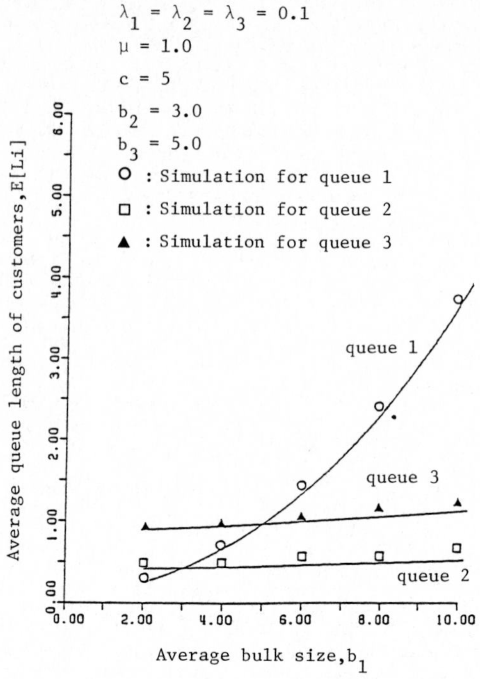

Fig. 7 Average queue length characteristics for various b_1.

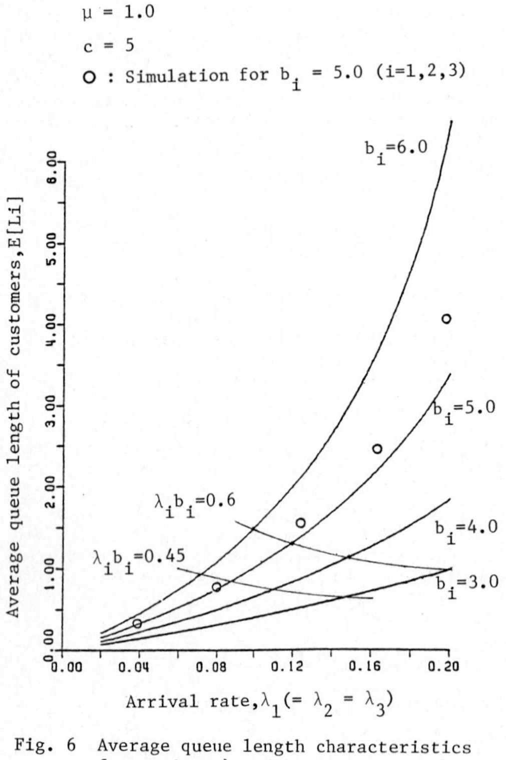

Fig. 6 Average queue length characteristics for various λ and b, where $\lambda_1=\lambda_2=\lambda_3$, and $b_1=b_2=b_3$.

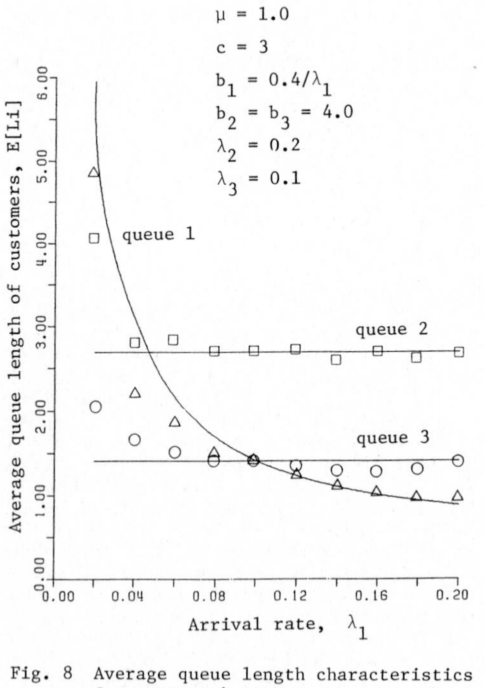

Fig. 8 Average queue length characteristics for various λ_1, where b_1 is inversely proportional to λ_1.

TELETRAFFIC ISSUES in an Advanced Information Society
ITC-11
Minoru Akiyama (Editor)
Elsevier Science Publishers B.V. (North-Holland)

MODELING OF COMPOUND TRAFFIC STREAMS IN COMPUTER COMMUNICATION NETWORKS

H.-Dieter SUEDHOFEN   Peter F. PAWLITA

Tech. Univ. Aachen   Siemens AG
Aachen, FRG   Munich, FRG

ABSTRACT

Computer communication network models usually include a submodel representing several simultaneous user traffic streams. This paper analyzes improvements of accuracy by realistic models of user traffic streams. Based on measured user interactions, influences of empirical and usual theoretical input traffic models on system performance measures are compared, including models of compound traffic streams and small user populations. The analysis comprises

- deviations of queueing measures from usual "ideal" models,
- impact of modeling depth in different applications.

The factors on which the deviations depend are identified, as well as certain areas where simple models cannot be accepted.

1 INTRODUCTION

In computer communication networks a certain number of users concurrently demand transmission, processing, and storage facilities. Performance evaluation of computer communication networks is generally based on modeling of user traffic, the communication network itself, and the end systems (computer and terminal facilities), in terms of queueing theory. By "user traffic" we mean any stochastic process of user requests (and responses) at a shared resource.

Realistic modeling of user traffic is a key factor for successful performance evaluation of communication networks or network components. The interactive user is an important but uncertain factor in a network. Unfortunately, modeling of user behavior seems to be less developed than modeling the operation of transmission or processing facilities. This is due mainly to the lack of measurements. Uncertainty about realistic modeling leads to generous, frequently unreasonable safety margins ("over-dimensioning") at best; but misestimation of user traffic can also lead to network project failures at worst because queueing systems usually are very sensitive to type and parameter variations of the queueing process.

The modeling depth depends on the modeling objective: the usual practice in design and development of real networks or network components is to introduce several modeling simplifications in order to reduce the complexity of mathematical analysis or simulation. The errors caused by such modeling simplifications are only partly known. Thus, there is a strong demand for empirically based and practically applicable user models [15]. Network developers and designers need practical decision aids to choose models and modeling depth for a given problem, if given error limits are not to be exceeded.

In this paper we investigate the influence of simplifications of input traffic on estimated system behavior. It will be shown how performance predictions depend on the modeling depth of input traffic of a - preferably small - group of customers. In chapter 2 we introduce a model for single user and user population input traffic. The interarrival times of both input processes will be modeled by hyperexponential distributions, whose parameters are fitted to the given first three ordinary moments. The statistical data for the empirical models originate from a measurement project of which selected results were presented in [14],[16],[18]. The project comprised measurements in the following typical TP systems:

System measured	Abbreviaton
• Time-sharing system, predominantly with technical/scientific tasks	TST
• Time-sharing system, commercial	TSC
• Banking transaction system	BNK
• On-line data collection system	DCL
• Inquiry-response data base system, 2 different user populations	INQ
• Tourist reservation system	TOU

In chapter 3 the impact of simplifications of some input processes will be derived directly from queueing theory. In chapter 4 we will analyze the influence on models based on measurements. For this we will introduce models of measured user interactions as appropriate traffic sources in computer communication networks. In chapter 5 applications of the investigation results will be discussed.

2 MODELING OF INPUT TRAFFIC

The common assumption regarding the aggregate arrivals of user demands to a communication system is that they form a Poisson process. This is allowed only for a great number of subscribers like in telephone systems. But very frequently the population size associated to a node in a computer communication network is comparatively small. Albin [1] showed that for the superposition of up to 1024 i.i.d. renewal processes as input to an exponential server the expected number in the system can be significantly greater than for Poisson input of the same rate. Beyond this Bux [3] showed that for a reliable performance prediction of open single server queueing systems one must consider moments and the form of the distribution function for service and interarrival times as well.

In this paper we investigate input processes of interactive users to computer communication systems. Thus an appropriate description of the interarrival time distribution is important for exact user traffic modeling. Any distribution function with rational Laplace transform can be modeled arbitrarily close by a phase type distribution with equal mean introduced by Bux and Herzog [2]. This distribution is a simplification of the general phase type distribution presented by Cox [8] and a special form of a general mixture of Erlang distributions (cf.fig 1). It has less parameters and therefore the complexity of the state space is reduced in analytical calculations. On the other hand with a mixture of Erlang distributions one can reach a better mapping of real system variables onto model parameters. It is therefore easy to model the system behavior over a broad range by adapting the appropriate model parameter while for adjusting the phase type distribution of Bux and Herzog one has to establish a completely new solution. A second major reason in favor of the Erlang mixture is that it includes the important hyperexponential distributions.

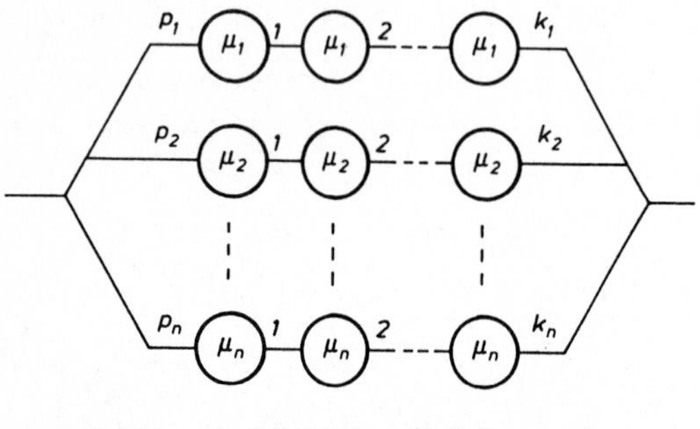

p_i weighting probabilities ($i=1,2,\dots,n$)
k_i number of exponential phases in branch i
μ_i transition rate in single phase in branch i

Fig. 1: Mixture of Erlang distributions

Measurements in several computer communication systems (see above) have shown that the distribution of interarrival times between successive demands generated by a single user have a squared coefficient of variation c_A^2 significantly greater than 1 (cf. tab. 1). The simplest distribution function to model $c_A^2 > 1$ is the hyperexponential distribution with parameters p, λ_1, λ_2. Often the parameters are fitted to mean, variance and the so-called "balanced means" condition, i.e. $p/\lambda_1=(1-p)/\lambda_2$ [1],[12]. This fixes the normalized third moment to $m_3=E(t_A^3)/6E(t_A)^3=c_A^2(c_A^2+1)/2$.

system	$\bar{a}$	c_A^2	m_3	p	λ_1	λ_2
DCL	36.0	1.66	2.86	0.958	0.0316	0.00740
INQ	28.6	5.11	28.50	0.982	0.0434	0.00302
user 1	47.6	2.82	10.30	0.983	0.0240	0.00252
user 2	25.0	5.86	39.50	0.986	0.0492	0.00285
TOU	67.9	5.52	15.90	0.883	0.0324	0.00287

Tab. 1: Selected measurement results and parameters of the hyperexponential model

As we will see in chapter 3, the influence of the third moment on system performance is not negligible. Moreover the measurement results indicate that the contributions of the branches are rather "unbalanced". We therefore determine the parameters of the interarrival pdf $a_1(t)$ of a single user by its mean $\bar{a}$, its squared coefficient of variation c_A^2, and its normalized third moment m_3 (cf. tab.1).

$$\lambda_1 = 2 / [x - \sqrt{x^2-4y}] \tag{1}$$

$$\lambda_2 = 2 / [x + \sqrt{x^2-4y}] \tag{2}$$

$$p = (\lambda_2/\lambda - 1)/(\lambda_2/\lambda_1 - 1) \tag{3}$$

with

$$\lambda = 1 / \bar{a} \tag{4}$$

$$x = \frac{2\bar{a}}{c_A^2-1}\left[m_3 - \frac{c_A^2+1}{2}\right] \tag{5}$$

$$y = \frac{2\bar{a}^2}{c_A^2-1}\left[m_3 - \left(\frac{c_A^2+1}{2}\right)^2\right]. \tag{6}$$

Given the complementary interarrival time distribution $A_1(t)$ of a single user, the pdf $a_N(t)$ of the aggregate input process of N users is [9]

$$a_N(t) = -\bar{a}_N \frac{d}{dt}\left(A_1(t)\left\{\frac{1}{\bar{a}_1}\int_t^\infty A_1(u)du\right\}^{N-1}\right) \tag{7}$$

with mean $\bar{a}_N = \bar{a}_1/N$. This yields for hyperexponentially distributed interarrival times

$$a_N(t) = \frac{\lambda^{N-1}}{N}\sum_{i=1}^{N} f(i)(g(i))^2 e^{-g(i)t} \tag{8}$$

with a j-th ordinary moment of

$$\overline{a_N^j} = \frac{j!\lambda^{N-1}}{N}\sum_{i=1}^{N} f(i)/g(i)^{j-1} \tag{9}$$

$$f(i) = \binom{N}{i}[p/\lambda_1]^{N-i}[(1-p)/\lambda_2]^i \tag{10}$$

$$g(i) = (N-i)\lambda_1 + i\lambda_2 \, . \tag{11}$$

In our following investigations we will model this aggregate point process again with a hyperexponential distribution whose parameters fit the first three ordinary moments of the interarrival times. This is in better agreement with the measurement results [18] than the approximation formula for the squared coefficient of variation given by Sevcik [17]

$$c_N^2 = 1 + \sum_{i=1}^{N} (\lambda_i/\lambda)^2 (c_i^2 - 1) = 1 + \frac{c_1^2 - 1}{N} \quad (12)$$

which in addition neglects the impact of the third moment.

3 INFLUENCE OF INPUT TRAFFIC

3.1 Basic considerations

In computer communication networks the behavior of interactive users has considerable impact especially on systems near to the user interface, like communication processors. To model the input traffic of such a single server system one has two basic choices

a) a closed queueing system with finite population,
b) an open queueing system with "infinite" population.

In closed queueing systems a user cannot generate a new demand, as long as his last one is not yet finished by the server. According to this feedback the system is always stable and the influence of the input process on system performance is only small and mainly based on the total number of users and their individual, average "think" time, termed "user time" in the following. Analytical analysis of closed queuing systems compared to open ones, is rather complicated, especially for the FCFS scheduling discipline (First Come First Served) with arbitrary service time distributions.

The popularity of open queueing systems is based on a wide-spread availability of models. They are useful as known submodels in series or networks of queues, and for studies of maximum throughput. Their performance is very sensitive to changes of

a) model parameters (e.g. mean values of service or interarrival time)
b) model processes (e.g. distribution types), and
c) combinations of both.

This implies two consequences. First, inaccurate estimation of process type or parameters can cause considerable misestimation of performance measures. This is especially true for the frequent (!) case of queueing processes with distributions of interarrival or service times with a squared coefficient of variation $c^2>1$. An example of sensitivity was given in [16]. Secondly, simplifications of models lead to unintentional changes of process type and result in a different system behavior. It is this second point on which we focus. Due to mathematical tractability any queueing analyst will use simple Markov type models if at all possible. In this chapter we directly derive some influences of simplification from queueing theory. To quantify the deviations, we define a sensitivity factor

$$S_X(A,B) = (X_B - X_A) \,/\, X_A \quad (13)$$

with X : quantity under consideration (e.g mean waiting time)
and A,B: types of models (e.g. A=M/M/1,B=G/M/1).

This sensitivity factor, here refering to input process variation, depends on several other quantities, e.g. the utilization ρ. It expresses the deviation of a quantity X, e.g. mean waiting time, in a model of type B compared to a reference model A. We mainly investigate changes of the input process in its type or its modeling depth. Of course, similar considerations apply to queueing sensitivity with respect to changes of service time distributions or to variations of both processes.

3.2 Open queueing systems

We choose the classical M/M/1 queueing system for our reference model because it implies the strongest simplifications in assuming exponential distributions for both interarrival and service times. The impact of the service time distribution has been studied successfully with the M/G/1 queueing system. Compared to M/M/1 the mean waiting time is increased by a factor of $(c_S^2+1)/2$, dependent on the squared coefficient of variation of the service time. To study the influence of the input process the dual system G/M/1 seems appropriate. The formulas for the mean values of variables describing the system performance, e.g. waiting time, delay, number in system, are similar to M/M/1, but with a quantity σ in "most" places of the utilization ρ. The mean waiting time e.g. is $W = \sigma/(1-\sigma)/\mu$. Given the Laplace transform $A^*(s)$ of the interarrival time distribution, σ is determined by:

$$\sigma = A^*(\mu - \mu\sigma). \quad (14)$$

For the $H_2/M/1$ queueing system with parameters ρ, λ_1, λ_2 of the input process and parameter μ of the output process, the solution of equ.(14) for σ is

$$\sigma = [\mu+\lambda_1+\lambda_2 - \sqrt{(\mu+\lambda_1-\lambda_2)^2 - 4\mu\rho(\lambda_1-\lambda_2)}]/2\mu. \quad (15)$$

In the following we will discuss the deviation of mean waiting time of a $H_2/M/1$ compared to a M/M/1 queueing system with the same arrival rate $\lambda=1/\bar{a}$, and an identical exponential service process (cf.fig 2). We derive this deviation in terms of the mean interarrival time $\bar{a}$, the squared coefficient of variation c_A^2, the normalized third moment m_3, and the utilization ρ. First of all this deviation depends on the absolute value of c_A^2, and in the limit for $\rho \to 1$ it is fixed to $(c_A^2-1)/2$. But its course over ρ strongly varies with the value of the normalized third moment m_3 (cf. fig.2). The range of values for m_3 is bounded below by m_{3min} and separated into two regions by its value for the "balanced means" condition m_{3b}:

$$m_{3min} = \left(\frac{c_A^2+1}{2}\right)^2 \; ; \quad m_{3b} = c_A^2 \, \frac{c_A^2+1}{2}. \quad (16)$$

Smaller values of m_3 lead to a distinct maximum of deviation

$$S_{\overline{W}max}\ (\ M/M/1\ ,\ H_2/M/1\ ) = \frac{r[c_A^2,m_3]}{s[c_A^2,m_3]} \quad (17)$$

at

$$\rho_{max} = \frac{s[c_A^2,m_3]}{r[c_A^2,m_3]} \cdot \frac{2m_3-3c_A^2+1}{2c_A^2-1} \quad (18)$$

with

$$r[c_A^2,m_3] = 4m_3^2 - 4m_3(3c_A^2-1) + (c_A^2+1)^2(2c_A^2-1) \quad (19)$$

and

$$s[c_A^2,m_3] = 4m_3^2 - (c_A^2 + 1)^2. \quad (20)$$

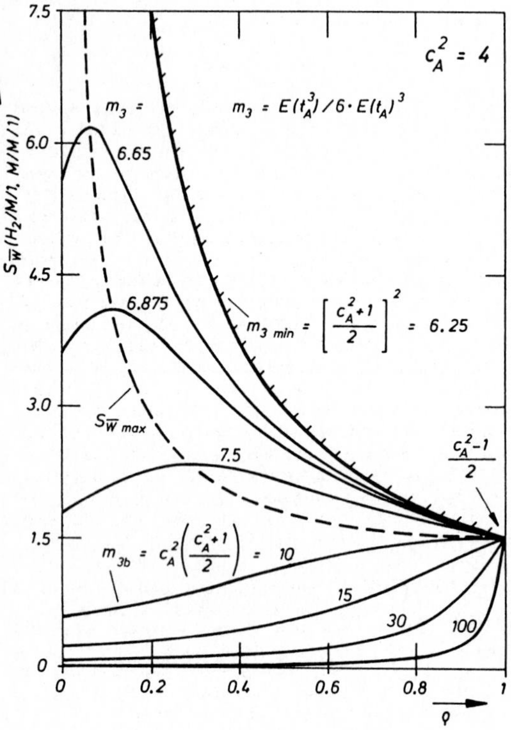

Fig. 2: Deviation of mean waiting time $\overline{W}$ of $H_2/M/1$ vs. $M/M/1$ for constant mean $\overline{a}=1$, constant squared coefficient of variation $c_A^2=4$, and with the normalized third moment m_3 as parameter

With greater values of m_3 the deviation between $H_2/M/1$ and $M/M/1$ decreases. This is due to the fact that for increasing values of m_3 the branch probabilities tend to 1 and 0 resp., so that there is one dominating exponential branch.

The set of curves is qualitatively equal for all values of c_A^2. But the interval (0,1) of values ρ covers different sections of their course. The upper value $\rho=1$ is fixed and the lower value $\rho=0$ varies with c_A^2. This results for $c_A^2<3$ into a third region, where the maximum deviation occurs for $\rho=0$, while the absolute maximum occurs for a theoretical $\rho<0$. This is true for all curves with $m_3>(3c_A^2-1)/2$.

To give a numerical example of the described coherence we consider two different values of m_3, keeping the mean $\overline{a}=1$ and the squared coefficient of variation $c_A^2=4$ constant. For $m_3=15$ the parameters of the H_2-input process are $p=0.9663$, $\lambda_1=1.2963$, and $\lambda_2=0.1322$; for $m_3=6.875$ we get $p=0.6752$, $\lambda_1=6.638$, and $\lambda_2=0.3615$. For $\rho=0.4$ the deviation in the mean waiting time of $H_2/M/1$ and $M/M/1$ is only 12% for $m_3=15$, but 345% for $m_3=6.875$. This shows the strong impact of the third moment of H_2-input processes on system performance.

3.3 Compound traffic streams

The input traffic generated by a single user is rather bursty. Therefore a Poisson process very often is not an accurate model of the aggregate arrival stream of a homogeneous population with a small number of users with identical behavior. This especially applies to heterogeneous populations of users with different interarrival and service time distributions. In chapter 4 we will investigate a model with a compound input traffic stream generated by a heterogeneous user population.

3.4 Main influence factors

The sensitivity of system performance regarding the input process depends on the following important factors:

- mean interarrival time,
- higher (2nd, 3rd) ordinary moments,
- composition and size of population.

Because the service time distribution has a main influence on system performance itself, we compare in the following only models with equal service parameters.

4 INFLUENCE ON MODELS BASED ON MEASUREMENTS

We now turn to concrete influences of input traffics by applying results of the measurements cited in the introduction. First we investigate a model for a multiple step transaction system. Then we analyse a compound arrival stream in a system with a heterogeneous user population. At last we consider a model for a buffer with limited capacity and a non-Poisson input process.

4.1 Model of a multiple step transaction system

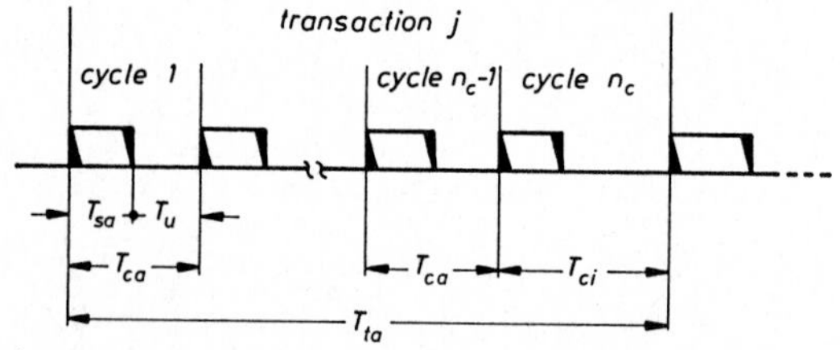

start of input- / end of output-transmission

T_{ta}: *transaction interarrival time*

T_{sa}: *system answer time*

T_u: *user time*

T_{ca}: *interarrival time of cycles during the active period*

T_{ci}: *interarrival time between last cycle of the current transaction and first cycle of the following transaction*

Fig. 3: Definition of multiple step transactions

The dialog of an interactive user with a computer system is cyclical. A multiple step transaction consists of several dialog cycles, which logically belong to the same task (cf.fig.3). The interarrival time T_{ta} between succeeding transactions of the same user consists of two sections. An "active" period of comparatively small cycle interarrival times T_{ca} while the user continuously interacts with the system, and one longer dialog cycle T_{ci} at the end of a transaction, when the user changes his task. For a small group of N users in a multiple step transaction system we introduce the following model (cf.fig.4).

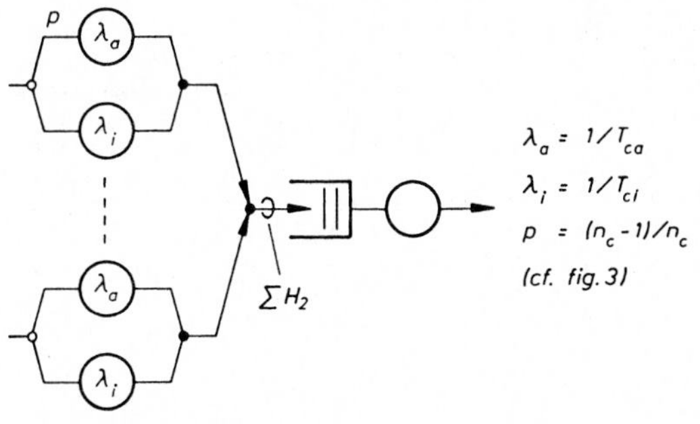

Fig. 4: Model of input traffic for a multiple step transaction system

We validate the model parameters with typical measurement results of two different systems (cf.tab.2).

INQ	p	λ_a	λ_i	TOU p	λ_a	λ_i
	0.747	0.063	0.021	0.645	0.043	0.006
N	$\bar{a}$	c_A^2	m_3	$\bar{a}$	c_A^2	m_3
1	23.78	1.645	2.175	73.18	2.690	4.008
3	7.93	1.203	1.358	24.39	1.921	2.753
5	4.76	1.113	1.189	14.64	1.586	2.115
7	3.40	1.077	1.125	10.45	1.414	1.772
10	2.38	1.053	1.083	7.32	1.281	1.503
15	1.59	1.034	1.053	4.88	1.179	1.305
20	1.19	1.025	1.039	3.66	1.130	1.216

Tab. 2: Model parameters based on measurements

Tab.2 further shows the development of the characteristics of the aggregate point process as the number of users increases. Especially for the TOU-system we observe that even for a number of users greater than 20 the squared coefficient of variation differs significantly from 1.

To investigate the sensitivity of system performance due to changes in the arrival process, we compare simulation results for the mean waiting time $\overline{W}$ of a $\sum H_2/H_2/1$ (cf.fig.4) to a $M/H_2/1$ system and a $\sum H_2/M/1$ to a M/M/1 system (cf.tab.3). We consider both cases because the squared coeficient of variation c_S^2 of the service time has in most cases a stronger influence on system behavior than any parameters of the input process except the mean arrival rate. Both service processes have equal mean service rate $\mu=N\lambda/\rho$ depending on the number of users and the utilization under consideration. For the hyperexponential service time distribution we assume a squared coefficient of variation $c_S^2=2$.

INQ		$\sum H_2/H_2/1$ – $M/H_2/1$	$\sum H_2/M/1$ – $M/M/1$	TOU	$\sum H_2/H_2/1$ – $M/H_2/1$	$\sum H_2/M/1$ – $M/M/1$
ρ	N			N		
0.5	3	0.027	0.079	3	-	-
	5	0.005	0.048	5	0.125	0.200
	7	-0.004	0.034	7	0.081	0.147
	10	-	-	10	0.043	0.106
0.7	3	0.051	0.084	3	-	-
	5	0.019	0.050	5	0.154	0.213
	7	0.007	0.036	7	0.114	0.158
	10	0.002	0.025	10	0.075	0.113
	15	-	-	15	0.040	0.077
	20	-	-	20	0.025	0.058

Tab. 3: Sensitivity factor $S_{\overline{W}}(\sum H_2/\cdot/1, M/\cdot/1)$ in a multiple step transaction system

As expected the difference in system performance for the original input stream and Poisson arrivals decreases with an increasing number of users an rises with greater values of the utilization ρ. The rate of convergence however is very different in both systems. While in the INQ-system the deviation of the mean waiting time is already less than 1% for small user numbers, it remains significantly greater in the TOU-system even for user numbers greater than 20. This is in part certainly due to the influence of the higher order moments. On the one hand the squared coefficient of variation c_A^2 is essentially greater in the TOU-system. On the other hand the value of the normalized third moment m_3 lies in the unfavourable region of great deviations to M/M/1, while in the INQ-system $m_3 \approx m_{3b}$ as for the "balanced means" case (cf.fig.2).

4.2 Compound traffic model

We consider a computer system with two user groups, distinct in size and service demands (cf.tab.4). One group of interactive users with a fixed size of 16 has only small service demands and an "empirical" behavior based on measurement results. The N_2 users of the second group rarely request long intervals of service, like in file transfer or cad/cam applications. The interarrival times of demands of a single user will be modeled by H_2-distributions for each group. Regarding the service time we have three models:

- H_2-distribution different for each group ($2H_2/2H_2/1$),
- exponential distribution different for each group ($2H_2/2M/1$),
- H_2-distribution equal for each group ($1H_2/1H_2/1$).

Arrival	p	λ_1	λ_2		
Group 1	0.9697	0.0444	0.0037		
Group 2	0.8700	0.0185	0.0005		
Service (2H2)	p	μ_1	μ_2	(2M)	μ
Group 1	0.1127	3.381	26.6190		15.0
Group 2	0.9976	0.1101	0.0013		0.1

Tab. 4: Parameters of the compound traffic model

In the third case the service time distribution is the statistical mixture of the service time distributions of the first case, weighted with their frequency of occurence. The arriving user demands will not be distinguished regarding their service, so that we can neglect this distinction for the arrivals as well and model only the superposition of both input streams.

ρ	N_2		$2H_2/2H_2/1$ $\overline{T}$	$2H_2/2M/1$ $\overline{T}$	$2H_2/H_2/1$ $\overline{T}$
0.14	3	T	6.913	1.733	1.438
		T1	6.648	1.512	1.436
		T2	17.140	13.330	1.577
0.17	4	T	9.177	2.272	1.964
		T1	8.809	1.980	1.957
		T2	23.420	13.560	2.217
0.21	5	T	10.820	3.092	2.490
		T1	10.400	2.718	2.485
		T2	23.840	14.780	2.643
0.24	6	T	13.580	3.791	3.051
		T1	13.070	3.344	3.048
		T2	27.020	15.510	3.133

Tab. 5: Mean delay $\overline{T}$ in compound traffic models

In tab. 5 the mean delay for a single user demand is entered for the different service models. The results evidently indicate that only a clear distinction of the users in service yields correct estimation of system behavior for each group. This implies the necessity of separate modeling of each arrival stream too. The results for the exponential service model differ only quantitatively and reflect at least qualitatively the correct system behavior for the single user groups.

Furthermore we see that only with an appropriate modeling depth we perceive the drastical deterioration of system performance for the interactive users due to the increasing size of the second user group.

4.3 Buffer Model

Next, we analyze the following buffer model important in data communication (cf. fig. 5).

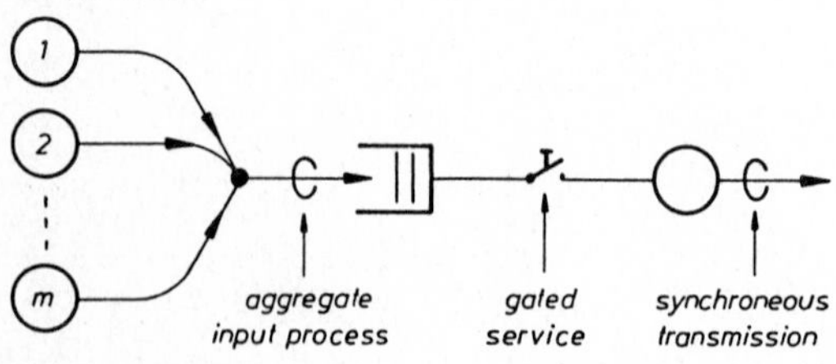

Fig. 5: Buffer model

Given a buffer of limited length L, a character arrival stream and synchronous output transmission, which means that characters can only be removed from the buffer and be transmitted at equal periods T_s. Such a system is frequently illustrated as a model with a "gate" between queue and server. For Poisson input overflow probabilities and queueing delay were studied in [10] and [6]. Under heavy load conditions combined with large buffer size the system behaves similar to a M/D/1/L system.

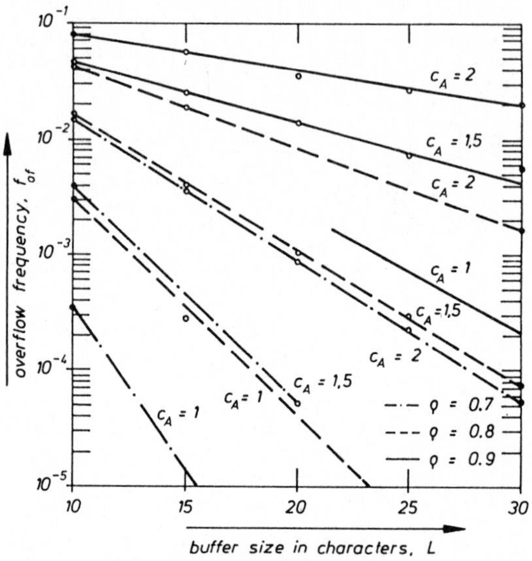

Fig. 6: Simulation results of overflow frequency

Clearly, in case of a small group of terminals as traffic input sources, the Poissonian supposition is too optimistic. Here we refer to the traffic measurement results of aggregate teletypewriter traffic given in [14]. The coefficient of variation c_A of intercharacter times considering 9 resp. 4 terminals amounts to about 1.5 resp. 2.0. By approximating the measured interarrival times with a H_2-distribution the simulation results of overflow frequencies shown in fig. 6 were achieved. As a basis of comparison the values for Poisson input calculated by Chu [6] are also entered.

Increasing c_A from 1.0 to 1.5 and 2.0, the curves in fig.6 show a tremendous increase of overflow frequency f_{of}. Usually, network designers would be interested in systems with overflow probabilities in the range of 10^{-4} and lower. As an example, for a given buffer size L = 20 and ρ= 0.7, the overflow frequency rises by about 3 orders of magnitudes, if c_A increases from 1.0 to 2.0 . In this case the sensitivity factor S_{of} would be greater than 10^3 according to equ. (13) with Poisson input in the reference system A. Varying c_A from 1.0 to 2.0 the sensitivity factors of the mean waiting time for large buffer sizes lie between about 2 (ρ=0.6) and 3 (ρ=0.8) (cf. fig. 7). In addition, we studied the buffer model given in [7] with two heterogeneous user traffics, one being Poisson and one being compound Poisson. Preliminary results of this investigation again show the following:

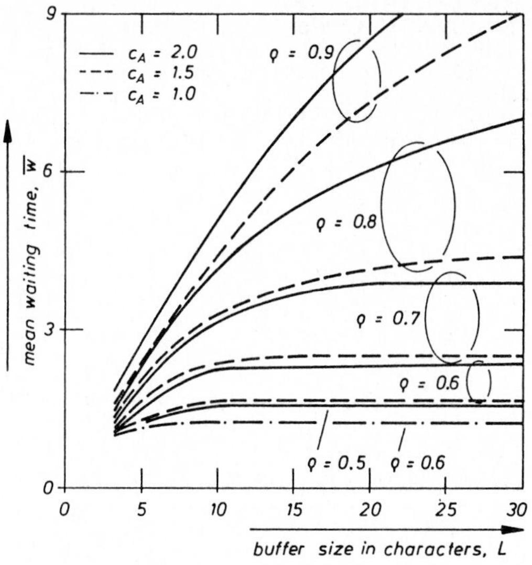

Fig. 7: Mean waiting time of the buffer model

by changing the dominating compound Poisson process to a process with H_2-distribution with a coefficient of variation c_A=1.5 the overflow frequency is increased by 2 orders of magnitude. Thus, we can draw the following conclusion: queueing systems with limited waiting capacity are extremely sensitive to input process type variations, especially with respect to overflow probabilities. Moreover, in case of a small user population and thus possibly a high coefficient of variation the usual Poisson-oriented models should not be used!

5 CONCLUSIONS

Empirically-based information on the sensitivity of queueing network models regarding the input traffic process is important to improve the quality of performance evaluation. The achieved results are valuable for a better modeling of:

- small user populations, e.g. clusters, exogenous traffic in a backbone network, LAN access,
- interactive users with a bursty arrival pattern as in transaction or data base systems,
- shared line configurations like buffers, multiplexors, or concentrators
- heterogeneous populations, e.g. for the user profile at the periphral interface of ISDN.

As it is shown in [11] modeling of the ISDN requires the consideration of various distinct traffic types. Models with mixed traffic on the ISDN basic access lines are a typical case of small and heterogeneous population, because only one multifunctional terminal or a small number of terminals can be connected.

Knowing the sensitivity of models under consideration allows a specific selection and provides a better understanding of the obtained results.

Acknowledgement

The measurement project was supported by the Deutsche Forschungsgemeinschaft, Bonn, the Bundesministerium für Forschung und Technologie, Bonn, and the Siemens AG, Munich.

REFERENCES

[1] Albin, S.L., On approximations for superposition arrival processes in queues, Management Sci. 28(1982), 126-137

[2] Bux, W., Herzog, U., The phase concept: approximation of measured data and performance analysis, in: [5],23-38

[3] Bux, W., Single server queues with general interarrival and phase-type service-time distributions - computational algorithms, Proc. 9.ITC, Torremolinos, 1979

[4] Buzen, J.P., Goldberg, P.S., Guidelines for the use of infinite source queueing models in the analysis of computer system performance, Proc. AFIPS Conf. 43 (1974 NCC), 371-374

[5] Chandy, K.M., Reiser, M., Proc. Int. Symp. on Computer Performance, Measurements and Evaluation, Amsterdam 1977

[6] Chu, W.W., Buffer behavior for Poisson arrivals and multiple synchronous constant outputs, IEEE Trans. Computers, C-19(1970), 530- 534

[7] Chu, W.W., Buffer behavior for mixed input traffic and single constant output rate, IEEE Trans. Comm., COM-19(1972), 230- 235

[8] Cox, D.R., A use of complex probabilities in the theory of stochastic processes, Proc. Camb. Phil. Soc., 51(1957), 313-19

[9] Cox, D.R., Renewal Theory, Methuen & Co. Ltd London 1967

[10] Dor, N.M., Guide to the length of buffer storage required for random (Poisson) input and constant output rates, IEEE Trans. Electronic Computers, 16(1967),683-684

[11] Hofstetter, H., Weber, D., Traffic models for large ISDN-PABX's, Proc. 11.ITC, Kyoto 1985

[12] Kuehn, P., Tables on Delay Systems, University of Stuttgart, 1976

[13] Kuehn, P., Approximate analysis of general queueing networks by decomposistion, IEEE Trans.Comm.27(1979),113-126

[14] Pawlita, P.F., Suedhofen, H.-D., User behavior in teleprocessing networks: analytical models based on empirical data, Proc. 10.ITC, Montreal 1983, Session 3.3

[15] Pawlita, P.F., Suedhofen, H.-D., Role of traffic measurements in information networks, Computer Performance 5(1984),135-143

[16] Pawlita, P.F., Interactive users in network performance modeling, Proc. ECOMA-12, Munich 1984, 56-67

[17] Sevcik, K.C., et al., Improving approximations of aggregate queueing network subsystems, in:[5],1-22

[18] Suedhofen,H.-D., Benutzerverhalten in Fernverarbeitungssystemen : Meßergebnisse zur Tageszeitabhängigkeit charakteristischer Grössen von Einzelbenutzern und Benutzerbündeln, Interner Bericht, RWTH Aachen, Dezember 1984 (in German)

TELETRAFFIC ISSUES in an Advanced Information Society
ITC-11
Minoru Akiyama (Editor)
Elsevier Science Publishers B.V. (North-Holland)

MULTI-SERVER SYSTEM WITH BATCH ARRIVALS OF QUEUEING AND NON-QUEUEING CALLS

Yoshitaka TAKAHASHI and Tsuyoshi KATAYAMA

Musashino Electrical Communication Laboratory, NTT
Tokyo, Japan

ABSTRACT

A multi-server system with batch arrivals of queueing and non-queueing types of calls is presented, and is analyzed under Markovian assumptions. A recursive scheme for the steady-state probabilities is obtained by using the generating function. Traffic measures, e.g., loss probability, mean waiting time, and the probability that the number of calls in the system is zero, are represented by only a finite set of the steady-state probabilities in spite of the infinite state space. Some numerical examples are shown. Traffic characteristics are also discussed.

1. INTRODUCTION

Multi-server systems with single and/or batch arrivals of queueing and non-queueing calls are common in practice. A prime example is the system in which voice and data traffic share the same group of communication channels [1]. Another example is the storage equipment in the Facsimile Intelligent Communication System (FICS) [2].

Assuming Poisson arrivals and exponential service times for both types of calls, the system (hereafter denoted by M_1,M_2 /M/S/S,∞) has been analyzed independently by Cohen [3] and Helly [4]. Pratt [5] analyzed a more general system which allows different service rates ($M_1,M_2/M_1,M_2$/S/S,∞). For a large number of servers (S), however, exact analysis is not practical, although Bhat and Fischer [1] provided an approximation by using the results of M_1,M_2/M/S/S,∞ [3,4].

Making the assumption of Poisson arrivals in the case of traffic input in bursts incurs an underestimation for the traffic measures, e.g., the loss probability or the probability of delay. In an analysis of message packetization, batch Poisson arrivals rather than Poisson arrivals were adopted [6]. Similarly, multi-address calls in the FICS require multi-units of resources, so batch Poisson arrivals are assumed when designing the FICS [2,7].

In this paper, a multi-server system with batch Poisson arrivals of two types of calls ($M_1^{[X]},M_2^{[Y]}$/M/S/S,∞) is studied. Type (1) calls arriving to find that they cannot find idle servers must abandon the system. Type (2) calls arriving to find that they cannot find idle servers enter an infinite waiting room.
The system considered is an extension of the delay system with batch arrivals, $M^{[X]}$/M/S [8,9,10,11], and the loss system with batch arrivals, $M^{[X]}$/M/S/S [12,13], as well as the loss and delay combined system with Poisson arrivals, M_1,M_2/M/S/S,∞ [3,4].

2. QUEUEING MODEL ($M_1^{[X]},M_2^{[Y]}$/M/S/S,∞)

The system under consideration is shown in Fig.1 and is characterized by the following assumptions:

i) There are two types of calls(1),(2). Calls belonging to each type are offered in batches to a service system having identical S servers in parallel.

ii) Interarrival times of batches of type (i) are independently and exponentially distributed with a mean of $1/\lambda_i$ (i=1,2). Arrival processes of type (1) and type (2) are independent of each other.

iii) The numbers of calls included in an arriving batch (called "batch sizes") of type (1) and (2) are i.i.d. random variables with arbitrary distributions $\{g_j\}$, $\{h_j\}$, respectively.

iv) Service times for individual calls belonging to both types are exponentially distributed with a mean of $1/\mu$.

v) When an arriving batch of type (1) is larger than the number of idle servers, calls in the batch are rejected under the following strategies:
Whole Batch Acceptance Strategy (WBAS) --- calls in the batch are totally rejected;

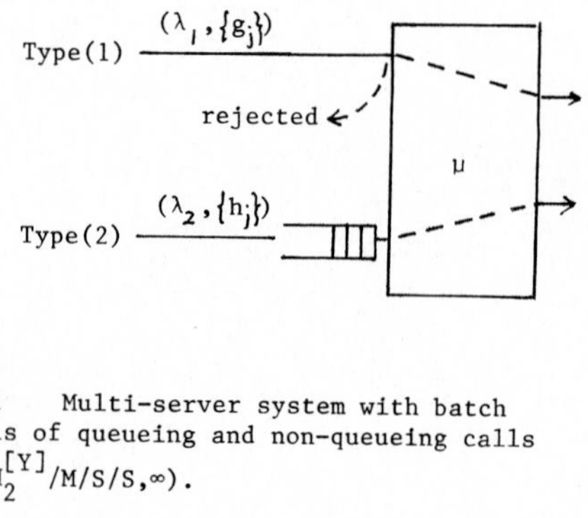

Fig.1 Multi-server system with batch arrivals of queueing and non-queueing calls ($M_1^{[X]},M_2^{[Y]}$/M/S/S,∞).

Partial Batch Acceptance Strategy (PBAS) --- the batch fills idle servers and the remaining calls in the batch are rejected. Whichever strategy is assumed, rejected calls depart immediately and no longer affect the system.

vi) When an arriving batch of type (2) is larger than the number of idle servers, the batch fills the idle servers and the remaining calls of the batch enter a waiting room, whose capacity is infinite. These calls are served individually, based on the FIFO discipline between arrival batches and on the SIRO discipline in the same batch.

Remark: Batch arrival systems had been treated only under PBAS, until Manfield et al. [6] distinguished between WBAS and PBAS.

3. ALGORITHMS FOR THE STEADY-STATE PROBABILITIES

For simplification, the following notations are introduced:

$$a_i \triangleq \lambda_i/\mu, \quad \bar{g}_i \triangleq \sum_{k=i}^{\infty} g_k, \quad \bar{h}_i \triangleq \sum_{k=i}^{\infty} h_k,$$

$$g \triangleq \sum_{k=1}^{\infty} kg_k, \quad h \triangleq \sum_{k=1}^{\infty} kh_k,$$

$$\alpha(i,j) \triangleq \sum_{k=i}^{j} g_k, \quad \rho_i \triangleq a_i h/S.$$

The system state is represented by the number of calls belonging to both types in the system. The steady state exists if and only if $\rho_2 < 1$.

This paper deals only with the steady state. Let $\{p_j\}$ denote the steady-state probabilities. The following steady-state equations can be written in the usual manner:

Under PBAS,

for $0 \leq n < S$,

$$(\lambda_1 + \lambda_2 + n\mu)p_n = \sum_{i=0}^{n-1} \lambda_1 g_{n-i} p_i + \sum_{i=0}^{n-1} \lambda_2 h_{n-i} p_i + (n+1)\mu p_{n+1}, \qquad (1a)$$

for $n=S$,

$$(\lambda_2 + S\mu)p_S = \sum_{i=0}^{S-1} \lambda_1 \bar{g}_{S-i} p_i + \sum_{i=0}^{S-1} \lambda_2 h_{S-i} p_i + S\mu p_{S+1}, \qquad (2a)$$

for $n>S$,

$$(\lambda_2 + S\mu)p_n = \sum_{i=0}^{n-1} \lambda_2 h_{n-i} p_i + S\mu p_{n+1}. \qquad (3)$$

The empty sum ($\sum_{i=0}^{-1}$, etc.) is zero.

Under WBAS, the steady-state equations are the same for $n>S$, but

for $0 \leq n < S$,

$$\left(\sum_{i=1}^{S-n} \lambda_1 g_i + \lambda_2 + n\mu\right)p_n = \sum_{i=0}^{n-1} \lambda_1 g_{n-i} p_i + \sum_{i=0}^{n-1} \lambda_2 h_{n-i} p_i + (n+1)\mu p_{n+1}, \qquad (1b)$$

for $n=S$,

$$(\lambda_2 + S\mu)p_S = \sum_{i=0}^{S-1} \lambda_1 g_{S-i} p_i + \sum_{i=0}^{S-1} \lambda_2 h_{S-i} p_i + S\mu p_{S+1}. \qquad (2b)$$

The normalizing condition is given by

$$\sum_{j=0}^{\infty} p_j = 1. \qquad (4)$$

In the same manner as Abol'nikov [10] and Cromie et al.[11], summing Eq.(1a) or Eq.(1b) for n=0 to n=j, and summing Eq.(3) for n=j to n=∞, the following equations are obtained:

For $0 \leq j \leq S$,

under PBAS,

$$jp_j = a_1 \sum_{i=0}^{j-1} \bar{g}_{j-i} p_i + a_2 \sum_{i=0}^{j-1} \bar{h}_{j-i} p_i; \qquad (5a)$$

under WBAS,

$$jp_j = a_1 \sum_{i=0}^{j-1} \alpha(j-i, S-i) p_i + a_2 \sum_{i=0}^{j-1} \bar{h}_{j-i} p_i. \qquad (5b)$$

For $j>S$,

$$Sp_j = a_2 \sum_{i=0}^{j-1} \bar{h}_{j-i} p_i. \qquad (6)$$

Let us define C_j by

$$C_j \triangleq p_j/p_0. \qquad (7)$$

From Eqs (5)-(6), the following recursive scheme for $\{C_j\}$ is obtained:

$$C_0 = 1. \qquad (8)$$

For $1 \leq j \leq S$,

under PBAS,

$$C_j = \left(\sum_{i=0}^{j-1} [a_1 \bar{g}_{j-i} + a_2 \bar{h}_{j-i}] C_i\right)/j; \qquad (9a)$$

under WBAS,

$$C_j = \left(\sum_{i=0}^{j-1} [a_1 \alpha(j-i, S-i) + a_2 \bar{h}_{j-i}] C_i\right)/j. \qquad (9b)$$

For $j>S$,

$$C_j = a_2 \sum_{i=0}^{j-1} \bar{h}_{j-i} C_i / S. \qquad (10)$$

Equations (9)-(10) represent the flow balance between the states $\{0,1,2,\ldots,j-1\}$ and $\{j,j+1,\ldots\}$.

Manfield et al.[6] and Kabak [8,9] provided the recursive scheme for the finite queueing system ($M^{[X]}/M/S/K$). Applying their approach to the system under consideration, a truncated error is involved, because the system ($M_1^{[X]}, M_2^{[Y]}/M/S/S,\infty$) has an infinite waiting room. On considering the z-transform of $\{C_j\}$, however, algorithms for the steady-state probabilities can be obtained without any truncated error. Let C(z) be the z-transform of $\{C_j\}$, i.e.,

$$C(z) \triangleq \sum_{j=0}^{\infty} C_j z^j.$$

Using Eqs (8)-(10), denoting by H(z) the generating function for the batch size distribution of type (2) $\{h_j\}$, it follows after some calculations:

Under PBAS,

$$C(z) = \frac{\sum_{j=0}^{S-1}(S-j)C_j + a_1 \sum_{j=0}^{S-1}\left(\sum_{i=1}^{S-j} \bar{g}_i z^i\right) C_j z^j}{S - a_2 z(1-H(z))/(1-z)}; \qquad (11a)$$

Under WBAS,

$$C(z) = \frac{\sum_{j=0}^{S-1}(S-j)C_j + a_1\sum_{j=0}^{S-1}(\sum_{i=j+1}^{S}\alpha(i-j,S-j)z^i)C_j z^j}{S - a_2 z(1-H(z))/(1-z).} \quad (11b)$$

Taking into account the normalizing condition (4), and the definitions of $\{C_j\}$,

$$C(1) = 1/p_0 .$$

From the above equation and Eq.(11), using L'Hospital's rule, the representations of p_0 are obtained by a finite subset of $\{C_j\}$:
Under PBAS,

$$p_0 = S(1-\rho_2)/\sum_{j=0}^{S-1} [S-j+a_1\sum_{i=1}^{S-j} \bar{g}_i]C_j ; \quad (12a)$$

Under WBAS,

$$p_0 = S(1-\rho_2)/\sum_{j=0}^{S-1}[S-j+a_1\sum_{i=j+1}^{S}\alpha(i-j,S-j)]C_j . \quad (12b)$$

Therefore, the recursive scheme (8)-(10) for $\{C_j\}$, definitions of $\{C_j\}$ (Eq.(7)), and Eq.(12) give the steady-state probabilities $\{p_j\}$. Moreover, the generating function for the steady-state probabilities, P(z), can be represented by a finite number of the probabilities $\{p_0,p_1,\ldots,p_S\}$. In fact, multiplying both sides of Eq.(11) by p_0, it follows that:
Under PBAS,

$$P(z) = \frac{\sum_{j=0}^{S-1} [(S-j) + a_1(\sum_{i=1}^{S-j} \bar{g}_i z^{i+j})]p_j}{S - a_2 z(1-H(z))/(1-z)} ; \quad (13a)$$

Under WBAS,

$$P(z) = \frac{\sum_{j=0}^{S-1}[(S-j)+a_1(\sum_{i=j+1}^{S}\alpha(i-j,S-j)z^{i+j})]p_j}{S - a_2 z(1-H(z))/(1-z)} . \quad (13b)$$

4. TRAFFIC MEASURES

4.1 Loss Probabilities

Since loss probability for an arbitrary batch is generally different from that for an arbitrary call, we must distinguish between these concepts. Let us call the former "batch loss probability" and the latter "call loss probability". The batch loss probability is defined as the probability that an arbitrary batch will be either partly or wholly rejected. The call loss probability is defined as the probability that an arbitrary individual call will be rejected.

By the batch Poisson assumption, it is known from [15] that:

$$\Pr\left\{\begin{array}{l}\text{the number of calls in the system}\\ \text{immediately before the arrival of}\\ \text{an arbitrary batch}\end{array}\right\} = p_k .$$

The batch loss probability under the condition that arriving batch size is i , $\hat{B}_i$, is given by

$$\hat{B}_i = \sum_{k=(S-i+1)^+}^{\infty} p_k , \quad (14)$$

where $(a)^+ \triangleq \max(a,0)$.

Equation (14) is deduced from the fact that, given that the arriving batch of size i finds k calls in the system, the batch will be rejected under either PBAS or WBAS, if $k > (S-i+1)^+$. By using the total probability rule, the batch loss probability $\hat{B}$ is given as follows:

$$\hat{B} = \sum_{i=1}^{\infty} g_i\hat{B}_i . \quad (15)$$

Substituting Eq.(14) into Eq.(15), it follows after elementary calculations:

$$\hat{B} = 1 - \sum_{k=0}^{S-1}\sum_{i=1}^{S-k} g_i p_k , \quad (16)$$

or

$$\hat{B} = \bar{p}_S + \sum_{k=0}^{\infty} \bar{g}_{S-k+1}\, p_k , \quad (17)$$

with $\bar{p}_S \triangleq \sum_{k=S}^{\infty} p_k = 1 - \sum_{k=0}^{S-1} p_k$.

As Manfield et al.[6] pointed out, the call loss probability is not quite so easily obtained. Considering an arbitrary call, it is known from [14] that

$$\Pr\{\text{arbitrary call arrives in batch of size } i\} = ig_i/g . \quad (18)$$

Denoting by B_i the call loss probability under the condition that the arbitrary call arrives in a batch of size i, and by B the call loss probability, from Eq.(18), we have

$$B = \sum_{i=1}^{\infty} ig_iB_i/g . \quad (19)$$

Noting that the service order in the same batch is random, it follows that:
Under PBAS,

$$B_i = \bar{p}_S + \sum_{k=(S-i+1)^+}^{S-1} (i+k-S)p_k/i ; \quad (20a)$$

Under WBAS,

$$B_i = \sum_{k=(S-i+1)^+}^{\infty} p_k . \quad (20b)$$

Equation (20a) is obtained from the following facts: When the number of calls in the system immediately before its arrival (denoted by k in this paragraph) is more than S-1, the arbitrary call belonging to the batch of size i is always rejected; When k < S, it is rejected with probability (i+k-S)/i under PBAS. Also, equation (20b) is derived from the fact that the arbitrary call is always rejected when $k > (S-i+1)^+$ under WBAS.

Substituting Eq.(20) into Eq.(19), and changing the summation order, the resultant expression for the call loss probability is given as follows:
Under PBAS,

$$B = \bar{p}_S + \sum_{k=0}^{S-1}\sum_{i=S-k+1}^{\infty} (i+k-S)g_ip_k/g ; \quad (21a)$$

Under WBAS,

$$B = \bar{p}_S + \sum_{k=0}^{S-1}\sum_{i=S-k+1}^{\infty} ig_ip_k/g . \quad (21b)$$

It should be noted that all the loss probabilities obtained here are represented by a finite number of the steady-state probabilities $\{p_0,\ldots,p_S\}$. The batch/call loss probabilities under the condition that arriving batch size is i, $\hat{B}_i/B_i$ (Eq.(14)/Eq.(20)) appear to be new measures. These probabilities will be used for designing the FICS [7].

4.2 The Probability of Delay

Let W_1 and W be the time until the first call in the batch or an arbitrary call in the batch enters the service, respectively. Define $W_1^c(0)$, $W^c(0)$ by

$$W_1^c(0) \triangleq \Pr(W_1 > 0) \,, \tag{22}$$

$$W^c(0) \triangleq \Pr(W > 0) \,. \tag{23}$$

Burke [14] shows, using a result from renewal theory, that the probability b_n of an arbitrary call being in the n-th position of its batch is given by

$$\begin{aligned} b_n &= \textstyle\sum_{j=n}^{\infty} h_j/h \\ &= \bar{h}_n/h \,. \end{aligned} \tag{24}$$

The arbitrary call is held up when the number of calls in the system immediately before its arrival is more than S-1, or when it finds k calls in the system immediately before its arrival and the position of its call in the batch is larger than S-k. Using the $\{b_n\}$, it follows that:

$$W^c(0) = \textstyle\sum_{k=S}^{\infty} p_k + \sum_{k=0}^{S-1} \sum_{i=S-k+1}^{\infty} b_i p_k \,. \tag{25}$$

Substituting Eq.(24) into Eq.(25) , changing the summation order, and using Eq.(6) , it follows that

$$W^c(0) = (1-\textstyle\sum_{i=0}^{S} p_i)/\rho_2 \,. \tag{26}$$

For $W_1^c(0)$, it is obvious that

$$W_1^c(0) = \textstyle\sum_{i=S}^{\infty} p_i = \bar{p}_S = 1-\sum_{i=0}^{S-1} p_i \,. \tag{27}$$

4.3 Mean Queue Length

Denoting by $E(L_q)$ the mean queue length, it follows that:

$$\begin{aligned} E(L_q) &= \textstyle\sum_{j=S}^{\infty} (j-S)p_j \\ &= P'(1) + \textstyle\sum_{j=0}^{S-1} (S-j)p_j - S \,. \end{aligned} \tag{28}$$

From Eq.(13), we have:
Under PBAS,

$$\begin{aligned} E(L_q) &= (S(1-\rho_2))^{-1} \\ &\cdot(\textstyle\sum_{j=0}^{S-1}[j(S-j)+a_1\sum_{i=1}^{S-j}(i+j)\bar{g}_i]p_j+a_2[h+H''(1)/2]) \\ &+ \textstyle\sum_{j=0}^{S-1} (S-j)p_j - S \,; \end{aligned} \tag{29a}$$

Under WBAS,

$$\begin{aligned} E(L_q) &= (S(1-\rho_2))^{-1} \\ &\cdot(\textstyle\sum_{j=0}^{S-1} [j(S-j) + a_1\sum_{i=1}^{S-j}(i+j)\alpha(i,S-j)]\, p_j \\ &+ a_2[h+H''(1)/2]) + \textstyle\sum_{j=0}^{S-1} (S-j)p_j - S \,. \end{aligned} \tag{29b}$$

The mean waiting time for an arbitrary call, E(W) is followed by Little's formula:

$$E(W) = E(L_q)/(\lambda_2 h) \,. \tag{30}$$

4.4 The LST of the Waiting Time Distribution

To find the LST of the waiting time distribution $W^*(s)$, the following random variable N is introduced. Let N be the random variable that is in position in the queue (including those being served) immediately after an arrival. By Eqs (24) and (6), it follows that for k>S,

$$\begin{aligned} \Pr(N=k) &= \textstyle\sum_{i=0}^{k-1} b_{k-i}p_i \\ &= \textstyle\sum_{i=0}^{k-1} \bar{h}_{k-i}p_i/h \\ &= p_k/\rho_2 \,. \end{aligned} \tag{31}$$

Conditioning on N, and using the total probability formula,

$$\begin{aligned} W^*(s) &= \textstyle\sum_{k=S+1}^{\infty} \Pr(N=k)(S\mu/(s+S\mu))^{k-S} \\ &+ \textstyle\sum_{k=0}^{S} \Pr(N=k) \,. \end{aligned} \tag{32}$$

From Eq.(31), equation (32) is given as follows:

$$\begin{aligned} W^*(s) &= \rho_2^{-1}\cdot(S\mu/(s+S\mu))^{-S} \\ &\cdot[P(S\mu/(s+S\mu))-\textstyle\sum_{i=0}^{S}p_i(S\mu/(s+S\mu))^{i}] \\ &+ 1 - \rho_2^{-1}(1-\textstyle\sum_{i=0}^{S} p_i). \end{aligned} \tag{33}$$

Let us recall the representation of P(z), i.e., Eq.(13). Any n-th moment of the waiting time can be expressed by a finite number of the steady-state probabilities $\{p_0,\dots,p_S\}$ as well as the probability of delay.

5. SPECIAL MODELS

5.1 The Case of $\lambda_2 = 0$

When λ_2 is equal to 0, the model considered in this paper is reduced to the loss system with batch arrivals ($M^{[X]}/M/S/S$). No explicit solution under WBAS has been found. Mejzler [12] gave an explicit solution for the number of calls in the system under PBAS. However, its numerical calculation is difficult because of the inclusion of the derivatives. Kabak [8,9] and Fujiki et al.[13] have provided the numerical solution under PBAS, and Manfield et al.[6] has provided under WBAS. Under both strategies, the steady-state equations (1), (2), and (3) are consistent with those obtained by [6,12,13]. Noting $\bar{p}_S = p_S$ in this case, it is verified that Eq.(17) and Eq.(21) are consistent with the batch and call loss probabilities obtained by Manfield et al.[6], and Fujiki et al.[13].

5.2 The Case of $\lambda_1 = 0$

The considered model here is reduced to the delay system with batch arrivals ($M^{[X]}/M/S/\infty$), if $\lambda_1 = 0$. No explicit solution for this delay

system has been found yet. However, Abol'nikov [10], and Cromie et al.[11] provided the numerical solution. It can be seen that Eqs (26) and (27) are consistent with the "probability of delay" formulas obtained by Cromie et al.[11]. The mean queue length $E(L_q)$ in Eq.(29) is reduced to the following:
Under both strategies,

$$E(L_q) = \frac{\sum_{j=0}^{S-1} j(S-j)p_j + a_2(h+H''(1)/2)}{S(1-\rho_2)} + \sum_{j=0}^{S-1} (S-j)p_j - S \,. \qquad (34)$$

On the other hand, substituting $a_1 = 0$ into Eq.(13), and taking the limit as $z \to 1$, we have

$$1 = P(1) = \sum_{j=0}^{S-1} (S-j)p_j/(S(1-\rho_2)) \,. \qquad (35)$$

Substituting Eq.(35) into Eq.(34), it follows that

$$E(L_q) = (S(1-\rho_2))^{-1} \cdot \left(\sum_{j=0}^{S-1} j(S-j)p_j + (S\rho_2)^2 + a_2H''(1)/2 - S\rho_2(S-1)\right). \qquad (36)$$

Equation (36) is consistent with the result by Abol'nikov [10].

5.3 The Case of Single Arrivals

The model here is reduced to the loss and delay combined system with single arrivals $(M_1, M_2/M/S/S,\infty)$, if $g_1 = h_1 = 1$. It can be easily verified that the steady-state equations (1), (2), and (3) are consistent with those by Cohen [3], and Helly [4].

6. NUMERICAL EXAMPLES

Numerical results are provided in some special cases. Some traffic characteristics are also discussed. Let ρ be the total offered traffic intensity, i.e.,

$\rho \triangleq (\lambda_1 g + \lambda_2 h)/(S\mu)$.

Denote by Z the random variable with the distribution $\{z_i\}$, and denote by E(Z) and Var(Z) the mean, or the variance of the random variable Z, respectively. To see the influence of the batch size distribution on the traffic measures, consider the following three discrete distributions $\{z_i\}$.

I) Unit distribution:

$$z_i \triangleq \begin{cases} 1, & i=10, \\ 0, & i \neq 10, \end{cases}$$

$E(Z) = 10,\ Var(Z) = 0$.

II) Binomial distribution:

$$z_i \triangleq {}_{20}C_i (1/2)^i (1/2)^{20-i} \,, \quad 0 < i \leq 20,$$

$E(Z) = 10,\ Var(Z) = 5$.

III) Bi-polar distribution:

$$z_i \triangleq \begin{cases} 9/14, & i = 1, \\ 5/14, & i = 15, \end{cases}$$

$E(Z) = 10,\ Var(Z) = 45$.

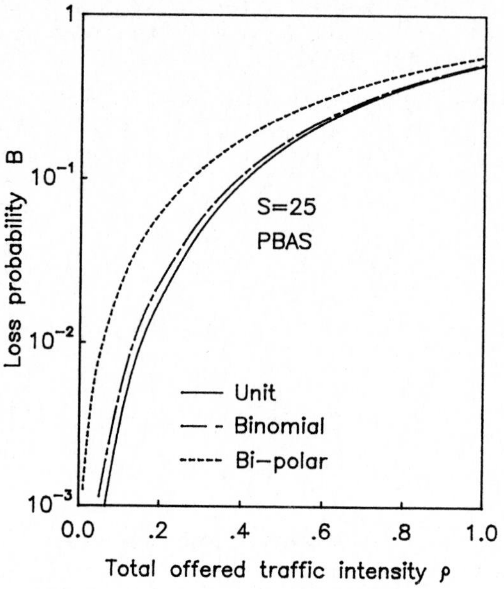

Fig.2 Relation of loss probability with total offered traffic under PBAS.

Throughout the following special cases, it is assumed that the batch sizes of type (1) and type (2) have the same distribution ($g_i = h_i$, $i \geq 1$) and $\lambda_1 = \lambda_2$. This situation can be seen in the FICS ; see [2,7]. For simplicity, let $\mu = 1$.

Figure 2 gives the graph of the call loss probability B vs. ρ for the system with S = 25,

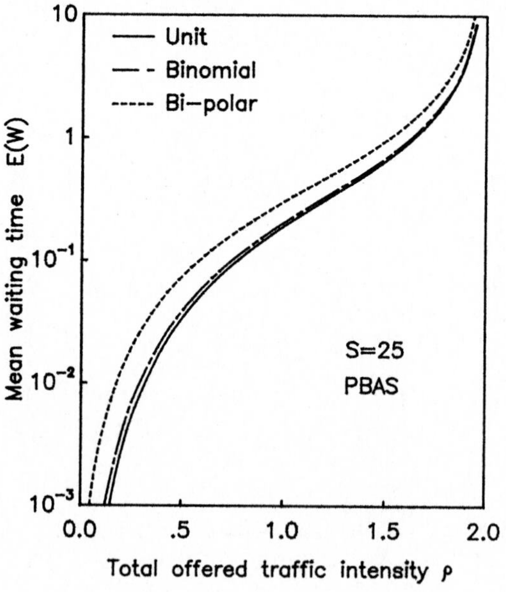

Fig.3 Relation of mean waiting time with total offered traffic under PBAS.

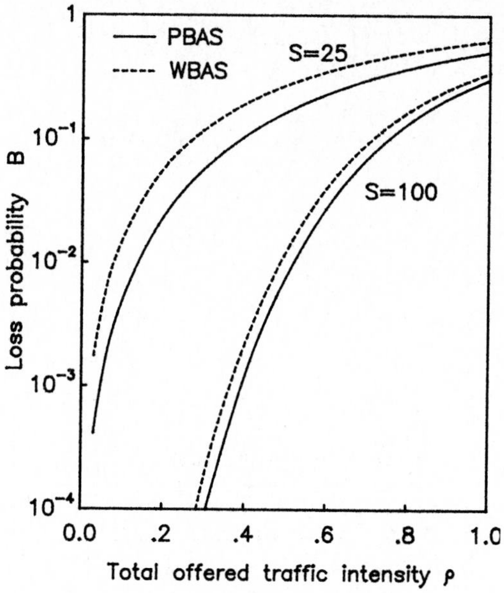

Fig.4 Relation of loss probability with total offered traffic under PBAS/WBAS.

and PBAS. Figure 3 gives the mean waiting time E(W) as in Fig.2. From these figures, it is observed that the loss probability or the mean waiting time increases as the variance of the batch size is larger, regardless of the maximum batch size. Recall that the maximum sizes of the unit, the binomial, and the bi-polar batch size distribution are given by 10, 20, and 15, respectively. It should be noted that the

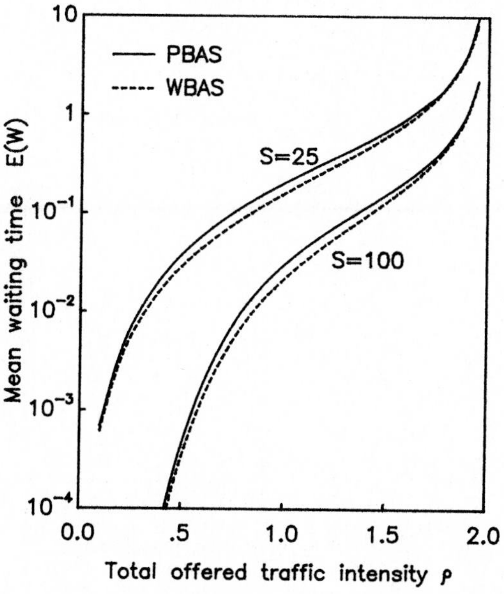

Fig.5 Relation of mean waiting time with total offered traffic under PBAS/WBAS.

influence of the arrival batch size variance on the traffic measures cannot be negligent.

For dimensioning the telecommunication resources (trunk-circuit, memory, etc.) with batch arrivals of calls, it is important to measure (at least) the variance of batch size as well as the batch arrival rate.

The call loss probability B and the mean waiting time E(W) under PBAS and WBAS are shown for the offered traffic intensity in Fig.4 and Fig.5, respectively. In these figures, binomial batch size distribution is assumed and number of servers (S) is either 25 or 100. The loss probability under WBAS is higher than that under PBAS, and the mean waiting time under WBAS is smaller than that under PBAS. This is verified intuitively from the definition of the strategies. For practical design, the choice between PBAS and WBAS depends on the service quality to be satisfied.

Figure 6 shows the loss probability as a function of the number of servers (S) under PBAS and WBAS. The offered traffic intensity is 1.9, and unit batch size distribution is assumed. When S is less than the mean batch size (i.e., $S < 10$), the loss probability is always 1 under WBAS, and less than 1 under PBAS. This is because of the definition of strategies and constant batch size. In the case of $S > 10$, however, the difference between PBAS and WBAS

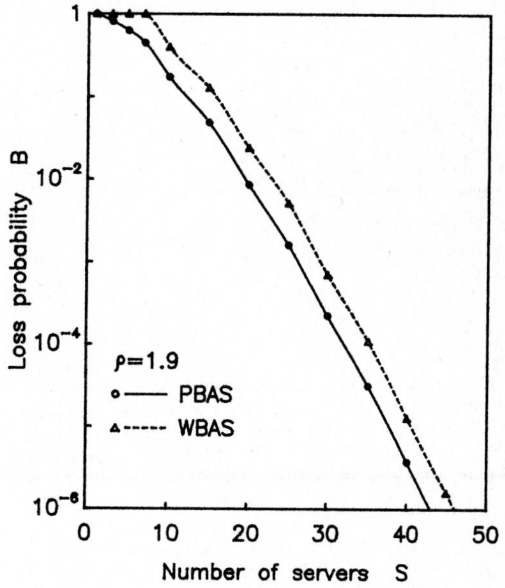

Fig.6 Relation between loss probability and number of servers under PBAS/WBAS.

does not depend on the number of servers, and it seems to remain constant.

Through these examples, it is numerically seen that the calculation speed is about the same as with Erlang B formula.

7. CONCLUSION

A multi-server system with batch arrivals of queueing and non-queueing calls is presented and analyzed under Markovian assumptions ($M_1^{[X]}$, $M_2^{[Y]}/M/S/S,\infty$). A general numerical approach is provided, i.e., the steady-state probabilities p_j are obtained without any truncated error, combining the recursive scheme with the generating function. New traffic measures for the non-queueing calls, i.e., the batch/call loss probabilities given the arriving batch size, are introduced under both PBAS and WBAS. These quantities will be useful for practical traffic design in the FICS. Since main traffic measures can be represented by a finite set of the steady-state probabilities $\{p_0, p_1, \ldots, p_S\}$, programming for the traffic measures is very simple and the calculation speed is about the same as with Erlang B formula. From the numerical examples, it is seen that:

i) The influence of the arrival batch size variance on the traffic measures is worse than that of the maximum batch size. The loss probability and the mean waiting time increase as the variance is larger.

ii) The difference between PBAS and WBAS is not a function of the number of servers (S), if S is larger than the mean batch size, with the total offered traffic intensity being constant.

For further study, the following topics remain:

i) A generalization to the system which allows different service rates ($M_1^{[X]}$, $M_2^{[Y]}/M_1, M_2/S/S,\infty$);

ii) An expansion to the non-Markovian system (e.g., with general service time distributions, general interarrival time distributions).

ACKNOWLEDGEMENT

The authors would like to express their thanks to Prof. Masaya Fujiki of Nihon University for his helpful advice and comments. They also wish to thank Kunio Kodaira, Chief of Teletraffic Section in Musashino Electrical Communication Laboratory, NTT, and Konosuke Kawashima, staff engineer of the same section, for their guidance and encouragement.

REFERENCES

[1] U. N. Bhat and M. J. Fischer, "Multichannel queueing systems with heterogeneous classes of arrivals," Naval Res. Logist. Quart., vol. 23, no. 2, pp. 271-283, 1976.

[2] Y. Takahashi and T. Yokoi, "A traffic analysis for storage capacity with multi-address calls in the FICS," Nat. Conv. Rec. IECE Japan, no. 1587, 1982. (In Japanese)

[3] J. W. Cohen, "Certain delay problems for a full availability trunk group loaded by two traffic sources," Commun. News, vol. 16, no. 3, pp. 105-113, 1956.

[4] W. Helly, "Two doctrines for the handling of two-priority traffic by a group of N servers," Oper. Res., vol. 10, no. 2, pp. 268-269, 1962.

[5] C. W. Pratt, "A group of servers dealing with queueing and non-queueing customers," Aust. Telecommun. Res., vol. 5, no. 2, pp. 34-41, 1971.

[6] D. R. Manfield and P. Tran-Gia, "Analysis of a finite storage system with batch input arising out of message packetization," IEEE Trans. Commun., vol. 30, no. 3, pp. 456-463, 1982.

[7] T. Endo, M. Sawai, and F. Adachi, "Traffic design and service quality in FICS-2," Review of the ECL, NTT, Japan., vol. 31, no. 4, pp. 13-20, 1985.

[8] I. W. Kabak, "Blocking and delays in $M^{(n)}/M/c$ bulk queueing systems," Oper. Res., vol. 16, no.4, pp. 830-840, 1968.

[9] I. W. Kabak, "Blocking and delays in $M^{(x)}/M/c$ bulk arriving queueing systems," Management Sci., vol. 17, no. 1, pp. 112-115, 1970.

[10] L. M. Abol'nikov, "A multichannel queueing system with group arrival of demands," Eng. Cybern., vol. 4, pp. 39-48, 1967.

[11] M. V. Cromie, M. L. Chaudhry, and W. K. Grassmann, "Further results for the queueing system $M^X/M/c$," J. Opl Res. Soc., vol. 30, no. 8, pp. 755-763, 1979.

[12] D. Mejzler, "A generalization of Erlang's loss formulas in queueing," J. Appl. Prob., vol. 5, no.1, pp. 143-157, 1968.

[13] M. Fujiki and A. Toda, "An analysis of a loss system with batch arrivals," Nat. Conv. Rec. IECE Japan, no. 1606, 1979. (In Japanese)

[14] P. J. Burke, "Delays in single-server queues with batch input," Oper. Res., vol. 23, no.4, pp. 830-833, 1975.

[15] R. W. Wolff, "Poisson arrivals see time averages," Oper. Res., vol. 30, no. 2, pp. 223-231, 1982.

TELETRAFFIC ISSUES in an Advanced Information Society
ITC-11
Minoru Akiyama (Editor)
Elsevier Science Publishers B.V. (North-Holland)

THE STABILITY OF TELEPHONE TRAFFIC INTENSITY PROFILES AND ITS INFLUENCE ON MEASUREMENT SCHEDULES AND DIMENSIONING

Asko PARVIALA

Helsinki Telephone Company
Helsinki, Finland

ABSTRACT

The non-continuous traffic measurements base on the assumption that the busy time of the year is stable and can be defined in advance. To test the assumption, a large analysis was made in Helsinki. Comparisons of the yearly intensity profiles of 2,728 circuit groups show that a non-continuous two-week measurement yields a traffic value in average at least 7.6 % lower than the real top two-week value. The dispersion around the average is high. A 90 % reliability requirement thus brings an intensity underestimate of at least 20 % in average. - If the two-week intensity values are based on the mean of highest daily full-hour values, instead of a mean day based on quarter-hours, the traffic values are identical, but the data handling is minimized. Thus the exchange processor will be able to collect the values continuously, to give reports immediately and to store and to handle profile values of up to two years.

1. INTRODUCTION

When scheduling traffic measurements there are three main questions to be answered:

- at what time of the year and
- in which hours of day should the measurements be carried out,
- how will the intensity values be processed.

If the variations of traffic are regular, it will be sufficient to measure the most highly loaded time intervals within a year and a day. The use of such non-continuous measurements, scheduled in advance, will save a lot of measurements and data processing work compared to continuous measurement. The same equipment can be used for several circuit groups by transferring it from place to place, according to the traffic load on the circuit groups. This method has been motivated in conventional exchanges where the traffic measurements have traditionally been made using transportable equipment and the data have mainly been analyzed manually.

If the profile of the traffic intensity varies from year to year and the daily values have no fixed high-load times, the measurements should be spread over the entire time period during which top values can occur.

This paper presents the results of extensive measurements on a large number of traffic routes. Some conclusions on appropriate measurement arrangements are made, too.

2. TRAFFIC ROUTES AND MEASUREMENTS IN HELSINKI

The network of the Helsinki Telephone Company (HTC) covers of the capital of Finland and the surroundings with a radius of about 50 km. The population is 950,000. Altogether 510,000 main telephone lines (January 1984) are connected to 150 local exchanges. These are connected to each other and to five higher level and 20 lower level transit exchanges via 1,080 trunk circuit groups. There are still 2,272 internal circuits groups to be measured.

The measured circuit groups have in average 62 circuits. The circuit groups are loaded up to 72 % of their total traffic capacity, calculated according to the nominal congestion. The average of the highest two-week values is 29.1 Erl per circuit group. A negligible part of the circuit groups is severly overloaded, caused mainly by unexpected growth of the traffic and revealed immediately by the effective measurement method.

During the observation period 1 October 1982 - 1 October 1984, two-week values for 2,742 circuit groups were measured using the AUTRAX system (used in Helsinki since autumn 1977 /Parviala 80/) or other similar methods and 151 routes by other methods. The routing was changed during the observation period on 14 circuit groups. Part of the remaining 2,728 routes were extended during the observation period without changing their task.

The data are in our AUTRAX system collected every full hour circuit per circuit into the central processing unit and aggregated there to produce values for each circuit group. In addition to routine traffic measurements, a statistical analysis has been carried out /Lehtinen 79/.

The circuit groups were divided into separate categories depending on the measured traffic intensity or growth characteristics. The number of circuits was not used for categorizing, because some of the circuit groups were over-dimensioned or extended during the two-year period and the routes generally have sufficient capacity.

The nominal congestion B - calculated by means of the Erlang B-formula - is given in the Finnish Telecommunications Standards /PR 79/ as follows:

- B = 1 % for trunk circuit groups between exchanges (but B = 2 % for groups with less than 10 circuits)
- B = 0,1 - 0,5 % for internal circuit groups of exchanges, with a total max. 1.5 %.

3. THE ANNUAL PROFILES

3.1. The print-out of the annual intensity profile

For each of the 2,728 circuit groups in this study, the annual profile has been plotted out with a two-week resolution over a period of two years. These data are available in 52 memory locations per circuit group stored in the AUTRAX system. A two-week intensity value is obtained as a mean of the busy hour values registered on a full-hour basis on each working day, Fig. 1.

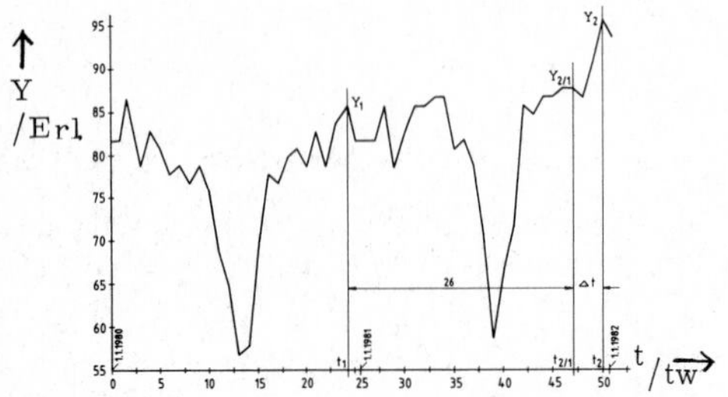

Fig. 1. Two-year profile (1 Jan 1980 - 31 Dec 1981) of one circuit group with reference points. Time is measured in units of two-weeks (tw). The summer holidays are causing characteristic gaps.

In order to determine the measurement weeks for non-continuous traffic measurement in the second year, the time t_1 for the highest intensity Y_1 in the first year is defined. The measurement "window" $t_{2/1}$ is then 26 weeks later with a traffic value $Y_{2/1}$. But the second year has its real highest intensity Y_2 at another time t_2. The scheduling error is given in two-week units:

$$\Delta t = t_2 - t_{2/1} = t_2 - t_1 + 26 \qquad (1)$$

and the percentage error in traffic value due to the timing error is

$$y = \frac{Y_2 - Y_{2/1}}{Y_2} \qquad (2)$$

The circuit groups with $Y_1 < Y_2$ are called here the growing category, groups with $Y_1 > Y_2$ the diminishing category, and groups with $Y_1 = Y_2$ the staying category. (Profiles in "growth categories" above do not absolutely have corresponding growing/diminishing time series trends.)

The timing errors and the intensity error for the observed circuit groups are given in Fig. 2.

Some concentrations towards short timing errors and low intensity errors can be seen. However, for a remarkable number of circuit groups the variability is large which is inconvenient from the view point of non-continuous measurements.

3.2. The timing

Some figures can be extracted from the whole material:

- the high-load times of the second year follow those of the first year in 11.1 % of the circuit groups
- the timing error is less than or equal to 2 two-weeks on either direction on 30.8 % of the circuit groups
- the measurement must last for 6 two-week periods, both before and after to secure the registration of high-load time in 50 % of the circuit groups.

When the material is categorized according to the intensity level of the circuit groups (Table I), it can be seen that

- for large circuit groups the timing errors are slightly smaller than in small or medium sized groups
- the increase or decrease does not influence the timing error very much.

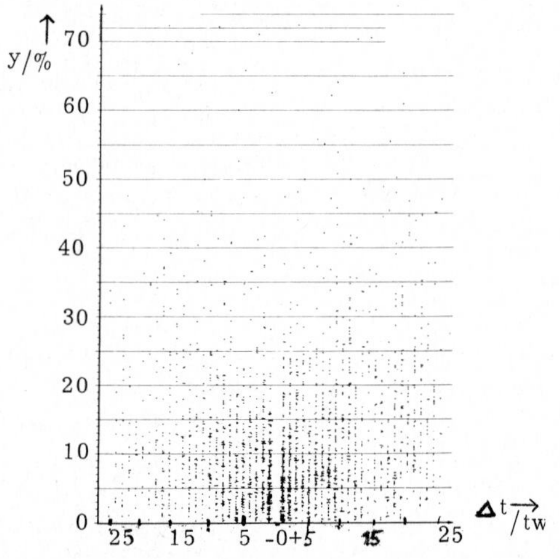

Fig. 2. Timing errors Δt and intensity errors y % for a sample of 2,728 circuit groups when the time of the measurement in the second year is determined on the basis of the first year high load.

Table I: Proportion of circuit groups on which the high-load time occurs exactly one year later or with a deviation of 4 or 12 weeks, categorized according to size and growth.

	Total	Small 10 Erl	Medium 10-100Erl	Big 100 Erl	Increasing	Decreasing	Staying
Circuit groups	2728	1054	1564	110	1445	1156	127
Two-weeks within a difference 0	11,2 %	10,0 %	11,8 %	14,5 %	9,6 %	9,0 %	48,8 %
-2...2	30,8 %	29,6 %	30,9 %	40,9 %	30,2 %	28,2 %	62,2 %
-6...6	50,7 %	50,7 %	50,5 %	54,5 %	46,9 %	53,2 %	72,4 %

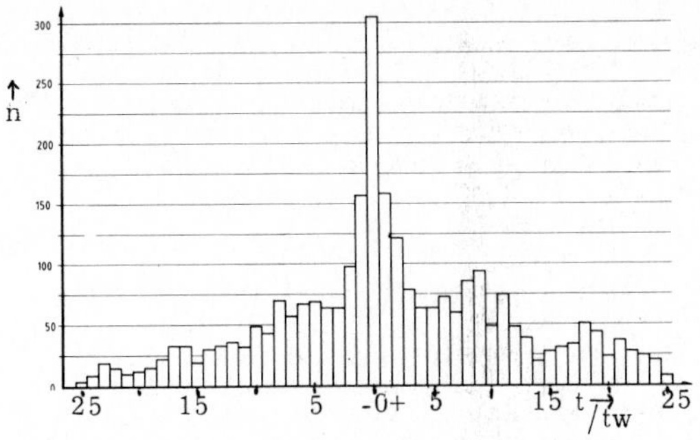

Fig. 3. Distribution of the timing error on 2,728 circuit groups.

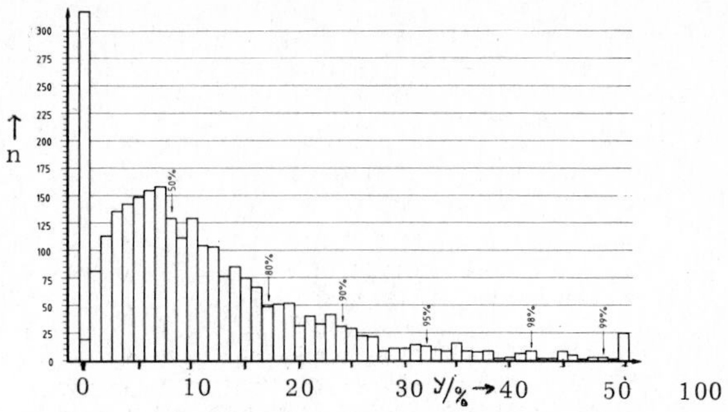

Fig. 4. 2,728 circuit groups classified according to the percentage intensity error caused by timing errors.

3.3. The error of traffic value due to timing error

Quite a large number (27 %) of the circuit groups are in the intensity error classes 2 to 6.9 %, Fig. 4, but much higher percentage errors occur. Some observations about Table II can be made:

- in addition to those 305 circuit groups which have no timing error there are also 13 circuit groups with zero intensity error (but with a timing error);
- the intensity error 7.6 % is weighted by traffic values. The intensity errors concentrate on small circuit groups
- a demand for a realiability of 95 % on all kinds of circuit groups leads to intensity errors of high percentages
- on large circuit groups the intensity error is below 10 % if a reliability of 85 % is enough

Fig. 5 shows that the intensity error grows quite fast near the mean level (7.6 %), when the timing deviates from zero. (The irregularities are great, too.) This is not very significant, if the timing is more or less faulty.

3.4. Errors based on a common profile

The examinations described above were made on a per circuit group basis, i.e. comparisons were made between the first and second year profiles of each circuit group. The search for the high-load time of the first year for each individual circuit group requires too much work in daily practice.

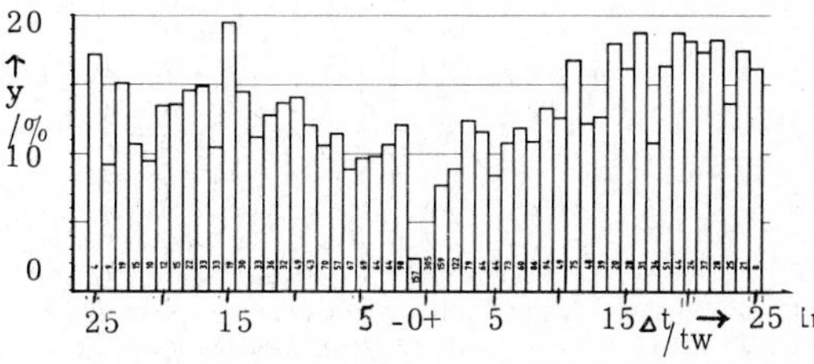

Fig. 5. Intensity error y % compared with timing error Δt.

The high-load time of the first year is in many ways easier to define for the sum traffic of all circuit groups together, using two-week identical summing of profiles, Fig. 6. (Such search for the high-load time could be in the conventional exchanges made by registering the power consumption.)

A comparison between the errors based on the common profile and the corresponding values in the Tables I, II indicates the following about the sum profile vs. individual profiles:

- the timing is less fitting (7.6 % vs. 11.2 %);
- the intensity error increases slightly, when high reliability (e.g. 95 %) is called for (38.9 % vs. 31.9 %).

Table II: Upper limit of the intensity error class for a cumulative proportion of circuit groups, categorized according to size.

	Total	Small 10 Erl	Medium 10-100Erl	Big 100 Erl	Increasing	Decreasing	Staying
Circuit groups	2728	1056	1564	110	1445	1156	127
Weighted mean error in the intensity value	7,6 %	13,7 %	7,8 %	5,2 %	8,1 %	7,4 %	3,3 %
Cumulative proportion of circuit groups							
50 %	7,9 %	12,9 %	6,9 %	3,9 %	8,9 %	7,9 %	0,0 %
80 %	16,9 %	22,9 %	17,9 %	7,9 %	18,9 %	15,9 %	8,9 %
90 %	23,9 %	30,9 %	23,9 %	13,9 %	25,9 %	21,9 %	12,9 %
95 %	31,9 %	37,9 %	34,9 %	17,9 %	33,9 %	27,9 %	16,9 %
98 %	41,9 %	47,9 %	40,9 %	26,9 %	44,9 %	40,9 %	18,9 %

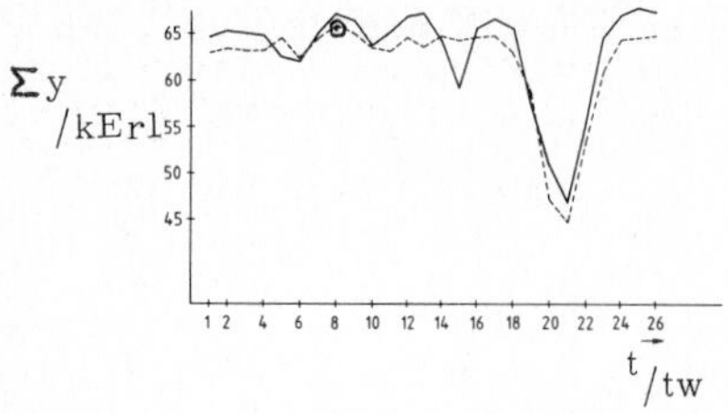

Fig. 6. Two-week identical sum profiles of all 2,728 circuit groups over two years:
------ first year 1 Oct 1982-1 Oct 1983
——— second year 1 Oct 1983-1 Oct 1984
⊙ top value of the first year, end of January

3.5. Earlier studies on yearly profiles

At the 1st ITC, reference was made to "the rather inconsiderable ... month-to-month variations" from September to June in Copenhagen /Brockmeyer 55/. A later thorough Swedish study in two long-distance circuit groups revealed the uncertainty in discovering the high intensity time of the year /Elldin 60/.

At the 4th ITC the CCITT studies were referred as follows: "The busy season was liable to vary from year to year." "... the wrong choice of the busy season might result in a serious underestimate (exceeding 33 %), and the only certain solution to this difficulty was to measure for a much longer period, such as 13 weeks, and to extract from these results the maximum 10-day sample." /Wright 64/

In those days measurements needed separate equipment and also a considerable amount of manual work. Thus only a few circuit groups were studied completely. The measurement methods since that time have been developed and changed completely.

3.6. Conclusions concerning yearly profiles

The measurements in the 2,728 circuit groups in the network in Helsinki show that
- no common high-load time of the year can be shown, but a time with low traffic lasting several weeks is quite clearly seen;
- the times with high load taken from the sum profiles or from circuit-group-identical profiles of earlier years are not of much help in finding the real high-load time, even for longer measurement periods;
- in individual circuit groups the most loaded two-week periods appear at arbitrary times;
- inaccurate timing of the non-continuous measurements causes an underestimate mean of the real intensity: 7.6 % for circuit-group-identical profiles and 8.0 % for yearly sum profiles;
- since there is a considerable deviation from that mean, a reliability of e.g. 90 % necessitates a corrective increase in the intensity values, obtained from non-continuous measurements, of 24 % for circuit-group-identical profiles and of 27 % for yearly sum profiles, leading to an overdimensioning by the same percentages in average.

4. THE TWO-WEEK INTENSITY VALUES

In order to describe the yearly intensity profiles, two-week values (= 10 consecutive working days) obtainable in several ways have been used.

4.1. Some definitions of the busy hours

The CCITT recommendation E 500 /CCITT 84/ defines the "time-consistent busy hour" (TCBH), which can be found by adding the carried traffic values observed during the same quarter-hour on each of the (10) consecutive days. The TCBH is then the four consecutive quarter-hours (of that added-up day) which together yield the largest sum of observed intensity values.

In Helsinki, a good amount of circuit groups was studied in 1983 by making a search for the TCBH timing. Since the observed 783 circuit groups carry traffic from both business and private users, the busy hours occur in the morning (9.30 to 11.15), in the afternoon (14.15 to 15.45) and in the evening (18.00 to 19.45), Fig. 7. (The telephone call charge is the same day and night, having no effect in this respect.) During nearly every quarter-hour of a day there are circuit groups with busy hour just then.

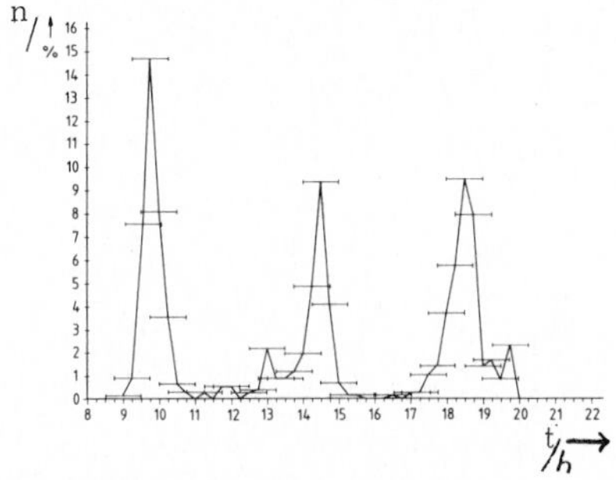

Fig. 7. Timing of busy hours on quarter-hour basis for 783 circuit groups 12-23 Jan 1983.

If the measurements are concentrated on an "in advance chosen measurement hour" (ACMH), e.g.9.45 to 10.45 in the morning, the real busy hour is found only in one of seven circuit groups in area networks like the one in Helsinki. The majority of the circuit groups are measured outside their busy hour, resulting in gross underestimates. The ACMH presents a method by which the traffic values could be obtained with minimal work: one is content with one traffic value daily, measured during a fixed hour. A picture of the traffic situation in alternative routes during that hour is obtained, but it has little to do with optimal dimensioning in networks with non-coincident busy hours /Parviala 83/.

Some telecommunications administrations have enhanced the significance of the ACMH values by prolonging the measurement time from one hour to 2-3 hours in the morning and maybe in the evening, too. The busy hour is calculated from this material later. The result is more representative if such a "post-selected busy hour" (PSBH) is processed from material of the whole day and evening, maybe also the night, like the CCITT recommends by its TCBH. An abundant amount of data is then collected for processing afterwards, mostly as many as 96 traffic values a day.

From the viewpoint of measurements and processing of the material, another method has been developed in Helsinki. This easier method will be described in the following.

4.2. Average daily peak full hour

If the busy hours are chosen from profiles of consecutive days on quarter-hour basis, their average traffic value is higher than the TCBH. But if the daily busy hour is chosen on a full-hour basis (e.g. 10.00-11.00), the average is very close to the TCBH.

This average daily peak full-hour method (ADPH) makes it easier to process the data:

- for every day only one traffic value is chosen on the basis of logical reasoning: the full-hour traffic values of a day are compared with each other and the highest one is stored as representing the daily busy-hour value
- a moving average for the peak full hours of 10 consecutive working days (two weeks) is calculated for use as statistical material or printed out for maintenance purposes.

These logical reasonings are easy enough for an exchange control processor to make beside of other operations and collecting of traffic values. The traffic value print-out is available immediately either daily, in two-week periods or once a year according to need and without separate processing. An alarm of any unusual change in the traffic can be included in these functions /Lehtinen 79/.

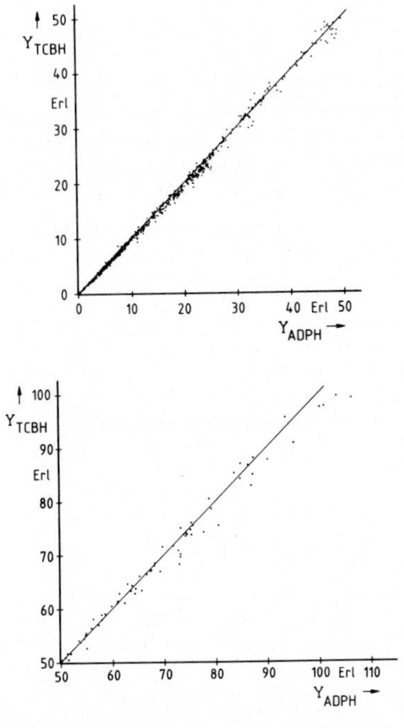

Fig. 8. Simultaneous intensity measurements with TCBH and ADPH methods in 679 circuit groups from 29 Sept to 5 Oct 1979.

The advantages of the ADPH method are obvious, if the traffic values are similar to those obtained with the TCBH method. In order to verify the similarity, the AUTRAX processor was programmed to print out parallel traffic intensity values Y in erlangs using both the TCBH and the ADPH method simultaneously in 679 circuit groups, Fig. 8. A linear regression with a high correlation yields the formula

$$Y_{TCBH} = Y_{ADPH} - 0.8 \text{ Erl}$$

with a calculated dispersion of 1.4 Erl. At low traffic values - below 15 Erl - there is no systematic difference between the results given by the two measurement methods, and it is thus more accurate to say that

$$Y_{TCBH} = Y_{ADPH}$$

When the TCBH method requires that the data are collected every quarter-hour, the data transmission in polling systems like AUTRAX loads the central processor to the extreme and the processor will have no time to perform the other important supervision tasks like surveillance of the individual circuits and reporting. In such systems continuous measurements are not possible beside other tasks, but the TCBH measurement must be planned for a fixed period of time (two weeks) and other supervision activities must be interrupted for that time. These restrictions do not apply to the collecting of data with the ADPH method once an hour only, thus the continuous measurement does not disturb the functions for supervision or printing.

4.3. Earlier studies on busy hours

The non-coincidence, i.e. the daily difference in the occurrence of the busy hours in different circuit groups, was discovered quite early and analyzed in the 1930's /Langer 39/. But other experts had differing opinions: "Experience shows that the busy hour occurs between 9.00 and 11.00 am approximately and that the traffic carried during this period is fairly constant ..." "The hour 9.30-10.30 am has accordingly been chosen as standard busy hour ... (for) traffic measurements ..." /Brockmeyer 55/.

In Helsinki, the studies were carried out by B. Ahlstedt in 1962-63. He measured traffic values during 90 weeks in two circuit groups on a quarter-hour basis /Ahlstedt 64/. The measurements showed that in the circuit group with business traffic, 63 % of the busy hours were between 9.30-11.15 am, and in the circuit group with domicile traffic, 82 % of the busy hours were between 5.45-7.15 pm. Another of Ahlstedt's studies dealt with daily profiles /Ahlstedt 67/, indicating that the daily measurement period must have a length of several hours.

The time-consistent busy hours tend to give 5-10 % lower traffic estimates than the post-selected ones, as was stated in the 1970's /Bear 76/ with reference to earlier studies /Povey 67/.

5. THE MEASUREMENT METHOD AND THE DIMENSIONING

The actual CCITT recommendation E 500 /CCITT 84/ is mainly based on the average of busy hour intensities from 30, 10 and 5 most loaded days of the year. The measurements should be continuous both daily and yearly. - To find out these days, the measurement data will always have to be processed afterwards.

Non-continuous measurements have been accepted as an alternative method in spite of its smaller supervision costs but higher network costs. As the non-continuous measurements may miss the high-load times of the year, the measured intensity values should be increased in accordance with a given correction method described in E 500.

The CCITT recommends the use of the methods in international routes, but for national circuit groups, too, when they carry international traffic.

The use of a daily fixed measurement hour (ACMH), instead of searching for the typical busy hour of the circuit groups with the TCBH or ADPH method, results in too low an intensity value in most circuit groups.

In practice, some telecommunications administrations may use non-continuous measurements without correcting the results. This is perhaps not too hazardous if the nominal congestion is low enough and the forecasting methods are good: the dimensioning may result in a sufficient number of circuits /Parviala 76/. But a more exact measurement method yielding higher intensity values allows higher nominal congestions in order to keep the number of circuits unchanged. The better allocation of augmentation investments leads to lower total network costs.

The collecting of data with the modern processor-controlled measurement methods is mostly not as difficult as the processing, reporting and interpretation of the data, for the latter functions need both processor and human capacity /Palanivelu 84/. Therefore a measurement method which at the origin of traffic data makes analyses and decides which data are worth storing for processing and which can be neglected daily, is superior in practice.

6. CONCLUSIONS

The study of 2,728 national circuit groups shows that the time of high load varies even on the same circuit group from year to year. The yearly non-continuous measurements endanger the finding of that time. No general high-load times were found, but the whole year should be measured. The non-continuous measurements give traffic intensity values which are 7-20 % too low in average, but the underestimate still grows at higher reliability demands.

The described average daily peak full-hour method (ADPH) combines the advantages of the other traffic measurement methods, bud does not have their disadvantages:
- the daily busy hours are revealed on full-hour basis and can be printed out daily;
- the ADPH two-week intensity value can be printed out as a moving average or as a value for a fixed two-week period;
- long time series, up to two years, can be stored in the processor, while the pruning is effective: out of 24 intensity values, the one to be stored can be chosen daily;
- the ADPH value corresponds to the TCBH value recommended by the CCITT; thus the same dimensioning rules can be applied;
- the ADPH intensity values can be used for indicating unusual traffic situations and for giving an early warning of a coming traffic growth.

The non-continuous yearly measurements were motivated in the time of the conventional exchanges when one tried to make effective use of the few expensive separate measurement devices. Daily non-continuous measurements were motivated by the fact that the treating of the traffic data required a vast amount of human work.

Today's processor-controlled collecting of traffic data with the ADPH method is capable of performing continuous measurements for days and years, of eliminating useless data and of printing out refined results. These can hardly be surpassed by the non-continuous measurements in economy or accuracy.

ACKNOWLEDGEMENTS

The statistical material presented here has been extracted with the AUTRAX system by P. Lehtinen, Dr. Techn., and refined by Y. Viinikka, M.Sc. and K. Laaksonen, B.Sc.

REFERENCES

/Langer 39/ Langer M: Die Schwankungen des Fernsprechverkehrs und der Leistung der Betriebsmittel in den Wählerämtern. II: Verkehrsteilung und Verkehszusammenfluss. Siemens Technische Mitteilungen Band Fg 3 (1939) Sept Heft 1. pp. 7-17

/Brockmeyer 55/ Brockmeyer E: A survey of traffic-measuring methods in the Copenhagen telephone exchanges. 1st ITC Copenhagen (1955) and Teleteknik Engl. ed. (1957) 1 pp. 92-105

/Elldin 60/ Elldin A, Tånge I: Långtidsmätning av telefontrafik. Tele (Stockholm) (1960) 1 pp. 197-215

/Ahlstedt 64/ Ahlstedt B V M: Duration of the Congestion State in Local Telephone Traffic. 4th ITC, London (1964)

/Wright 64/ Wright EPG: Conclusions and Recommendations of Working Party 5 of CCITT Study Group XIII. 4th ITC, London (1964)

/Ahlstedt 67/ Ahlstedt B: Load Variance and Blocking Duration in Automatic Telephone Traffic; Load duration. 5th ITC New York (1967) and Sähkö-Electricity in Finland (1967) 10

/Povey 67/ Povey J A: A study of traffic variations and a comparison of post-selected and time-consistent measurements of traffic 5th ITC, New York (1967)

/Bear 76/ Bear D: Principles of telecommunication traffic engineering. Peregrinus/IEE (1976)

/Parviala 76/ Parviala A: Accuracy requirements concerning routine traffic measurements with regard to service level objectives in telephone networks and to certain error and cost factors. ITC8, Melbourne (1976) and Sähkö-Electricity in Finland 50 (1977) 5-6 pp. 187-194

/Lehtinen 79/ Lehtinen P. I: Extreme Value Control and Additive Seasonal Moving Average Model for the Evaluation of Daily Peak Hour Traffic Data. 9th ITC, Torremolinos (1979)

/PR 79/ Puhelinverkkojen rakennemääräykset (Finnish National Telephone Standards) in Finnish and in Swedish only, PTT Helsinki (1979)

/Parviala 80/ Parviala A: The Autrax System as part of the HTC's traffic control routines. Sähkö-Electricity in Finland 53 (1980) 12 pp. 375-379

/Lehtinen 82/ Lehtinen P I: Measuring schedules. Paper read at the Finnish Engineers Schooling Centre (INSKO) in a course of teletraffic measurements (in Finnish only) course no. 62, Helsinki (1982)

/Lehtinen 83/ Parviala A, Lehtinen P: Abundant traffic data leads to better definition of busy hour. Telephony June 27 (1983) pp. 54-62

/Parviala 83/ Parviala A: The suboptimal routing practice with non-coincident busy hours. 10th ITC, Montreal (1983)

/CCITT 84/ Recommendation E 500: Measurement and recording of traffic. Red Book, Vol II. 3 "Network management and traffic engineering." ITU (1985)

/Palanivelu 84/ Palanivelu V: Teletraffic measurement. Telecommunication J. 51 (1984) 8 pp. 453-460

Appendix

The situation of the busy hour is shown in Fig. 7, but it does not describe the busy hour stability. The important question of the stability of busy hours was recently investigated in Helsinki in a separate study.

The traffic in quarter-hours in a local transit exchange with 10,300 circuits divided into 115 circuit groups was observed during four periods (August and November 1984, February and May 1985) each period five days long. Some temporary results are given below.

The busy hour varied so that it remained unchanged on quarter-hour basis during consecutive periods (three months) in 11...18 circuit groups (10...14 %). In no case did the busy hour remain unchanged during all four periods.

The busy hour kept its time of the day (morning, afternoon or evening) in 76 circuit groups (66 %), but leapt to other times of the day (from morning to afternoon or to evening, or vice versa) in 39 circuit groups (34 %).

On the bases of the different traffic profiles, the traffic intensity values depend on the measuring method. The following definitions of methods were used to find the busy hour intensity out of (10) consecutive days measured traffic values.

1. The continuous measurements (with post-selected busy hour):

<u>TCBH, the Time Consistent Busy Hour:</u> The four consecutive highest quarter-hour traffic values taken from a mean day consisting of the traffic values observed during the same quarter-hour on all days.

<u>ADPFH, Average of Daily Peak Full-Hours:</u> the average of the highest traffic values of full hours (ex. 10.00-11.00) of all days.

<u>ADPQH, Average of Daily Peak Quarterly defined Hours:</u> the average of the highest traffic values of four consecutive quarter-hours of all days.

2. The non-continuous measurements

<u>FDMP, Fixed Daily Measurement Period:</u> the daily measurements cover a few (here 2) hours from which the busy hour is post-selected similarly as in the TCBH.

<u>FDMH, Fixed Daily Measurement Hour:</u> the daily measurements are restricted to one hour only.

The measurement time in the FDMP and FDMH methods is defined during earlier continuous measurement per circuit group <u>individually</u> or for the sum traffic of the whole network in question <u>commonly.</u>

The results for 115 circuit groups are given below in Fig. 9.

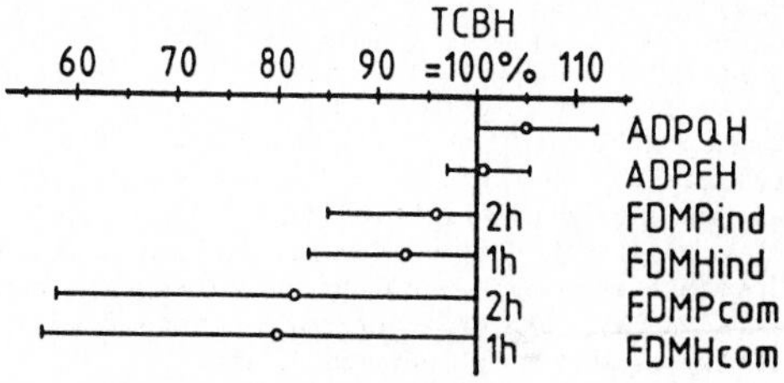

Fig. 9. Traffic values measured on 115 circuit groups simultaneously with different methods (see text), percentages compared to those given by the TCBH method. The means and the variations (up to about 90 % of cases) are given as an average of three measurement periods compared to one period (August 1984).

As a result of the study of 115 circuit groups in Helsinki it can be stated that:

- a 100 % intensity value, equal to TCBH, can be reached with every method, depending on the traffic profile;

- the ADPQH gives intensities over 100 %, here around 105 %;

- the ADPFH corresponds to the TCBH with a moderate variation, as was stated in Fig. 8, too;

- the individually defined non-continuous measurement time gives intensities around 95 %; prolongment of the measurement time from 1 to 2 hours gives a slightly better result;

- the commonly defined non-continuous measurement timing yields quite low traffic values (around 80 %) with large variations. Prolongment of the measurement time does not help notably.

TELETRAFFIC ISSUES in an Advanced Information Society
ITC-11
Minoru Akiyama (Editor)
Elsevier Science Publishers B.V. (North-Holland)

RELATIONSHIP OF SOME TRAFFIC PARAMETERS BASED ON AGT MEASURED DATA

Lansun LEE

Alberta Government Telephones (AGT)
Edmonton, Alberta, Canada

ABSTRACT

This paper describes the development of the relationship of some traffic parameters. The development was the result of the need:
(a) to estimate the mobile busy hour traffic loads with only the average billed minutes per mobile station per month known, and
(b) to equate the grade of service based on Extreme Value Engineering (EVE) so that based on Time Consistent Busy Hour (TCBH) concept. The relationship developed is a set of simple ratios. It has been shown to fulfill the above need, and has been applied to the Aurora-400 automated mobile telephone system operating in Alberta, Canada.

1. INTRODUCTION

In Alberta, Canada, the province-wide automated mobile telephone system is the Aurora-400 system which was designed, developed, and provided by NovAtel Communications Limited. It is operated by the AGT Mobile Communications Division [1] . To engineer the system for operation, the busy hour traffic load is required. Initially, the only available historical traffic parameter is the average billed minutes per mobile station per month. Therefore, it is necessary to develop a relationship for translating the average billed minutes per mobile station per month into the busy hour load.

The most common busy hour loads used in AGT at this time are Average Busy Season (ABS), 10-High Day (10HD), and High Day (HD). All are based on Time Consistent Busy Hour (TCBH) traffic data, which are measured at the same hour of every business day excluding holidays and abnormal days. ABS is the average hourly load of the 3 highest months in a TCBH data year which has 12 consecutive months not necessarily coincident with the calendar year; 10HD is the average hourly load of the 10 highest days in the TCBH data year; and HD is the highest hourly load in the TCBH data year. It is AGT's objective to provide the automated mobile telephone system with equivalent land-line grade of service as perceived by the subscribers. The grade of service may be expressed as the traffic congestion probability for a given busy hour. As the grade of service based on TCBH is generally not easily understood by the subscribers, grade of service based on Extreme Value Engineering (EVE) is therefore considered [2,3,4,5] . However, in AGT, satisfactory grade of service has been well established for TCBH, but has not been defined for EVE. Thus, it is desirable that the automated mobile telephone system grade of service based on EVE can be equated to the grade of service based on TCBH. Hence, in developing the relationship to translate the average billed minutes per mobile stations per month into busy hour load, the busy hour should include not only EVE, but TCBH also.

This paper describes the development of this relationship.

2. THE TRAFFIC PARAMETERS

To engineer the automated mobile telephone system for operation on TCBH basis, the required busy hour load traffic parameters are ABS, 10HD and HD. The latter two are used for engineering the system components which are sensitive to traffic peaks.

To engineer the system on EVE basis, only the peak busy hour loads are considered. The return period must still be selected [2] . A return period is a time interval during which a probability p of the observed peak busy hour loads will exceed a specified value, called the extreme value of the observed peak busy hour loads (EV(0,p)). For a given congestion probability, a system component is dimensioned in accordance with its EV(0,p). To estimate its EV(0,p), the extreme value distribution function $F_X(x)$ for describing the observed peak busy hour loads X may be assumed to be the normal distribution raised to the 6th power [4]. That is,

$$F_X(x) = \left[\frac{1}{\sqrt{2\pi}} \int_{-\infty}^{x} \exp(-y^2/2)dy \right]^6 \tag{1}$$

where

$$y = \frac{x - \mu}{\sigma} \tag{2}$$

is the reduced variate with μ and σ relating to the observed peak busy hour loads average A(0) and standard deviation S(0) as follows:

$$A(0) = \mu + 1.267\sigma \tag{3}$$

and

$$S(0) = 0.642\sigma \tag{4}$$

For

$$x = EV(0,p) \tag{5}$$

then

$$p = 1 - F_X(x) \tag{6}$$

With (2),(3),(4) and (5),

$$EV(0,p) = A(0) + L \cdot S(0) \tag{7}$$

where

$$L = \frac{y - 1.267}{0.642} \tag{8}$$

Numerical values of L for a practical range of p are given in Table 1. If the system component is split into n groups with $EV(0,p)_i$ as the extreme value of the observed peak busy hour loads for group i, it is well known that

$$\sum_{i=1}^{n} EV(0,p)_i = f \cdot EV(0,p) \qquad (9)$$

where f is the splitting factor having a value of unity or larger due to potential lack of time coincidence of peak busy hour loads of the groups [5] . Since f = 1.0 if EV(0,p) approaches ABS or ∞, it is proposed that

$$f = \left(\frac{EV(0,p)}{ABS}\right)^{k\left(\frac{n-1}{EV(0,p)}\right)} \qquad (10)$$

where k is a constant to be determined from the observed data. In AGT, two return periods have been selected: a year, and the high month of a year. The high month is defined as the month of the year which has the highest average daily peak busy hour load per mobile station. For the return period of a year, the observed peak busy hour loads are the monthly peak busy hour loads. Therefore, to use (7), A(0), S(0) and EV(0,p) are respectively the average A(M), standard deviation S(M) and extreme value EV(M,p) of the monthly peak busy hour loads. For the return period of the high month of a year, the observed peak busy hour loads are the daily peak busy hour loads. Therefore, to use (7), A(0), S(0) and EV(0,p) are respectively the average A(D), standard deviation S(D) and extreme value EV(D,p) of the daily peak busy hour loads of the high month. For system components sensitive to traffic peaks, p is chosen to be 1%; otherwise, p is chosen to be about one divided by the number of observed peak busy hour loads in the return period.

Thus, in summary, to engineer the automated mobile telephone system for operation on EVE basis, it can be seen from (7) and (10) that the required busy hour load traffic parameters are ABS, A(M), S(M), A(D) and S(D).

3. THE RELATIONSHIP

It should be noted that, among the required busy hour load traffic parameters, the standard deviations S(M) and S(D) are respectively related to the averages A(M) and A(D) by their coefficients of variation. That is, if CV(M) is the coefficient of variation of the monthly peak busy hour loads,

$$CV(M) = \frac{S(M)}{A(M)} \qquad (11)$$

and if CV(D) is the coefficient of variation of the daily peak busy hour loads of the high month,

$$CV(D) = \frac{S(D)}{A(D)} \qquad (12)$$

Therefore, to translate the average billed minutes per mobile station per month into busy hour load, it is not necessary to have a direct relationship for the standard deviations if the coefficients of variation are known.

Thus, there remains to be established a direct relationship for translating the average billed minutes per mobile station per month into ABS, into 10HD, into HD, into A(M), and into A(D). The simplest relationship may be a ratio. For convenience, the denominator of the ratio is chosen to be the average daily total traffic load for the high month (ADT(HM)). Thus, the required ratios are:

$$R_{ABS} = \frac{ABS}{ADT(HM)} \qquad (13)$$

$$R_{10HD} = \frac{10HD}{ADT(HM)} \qquad (14)$$

$$R_{HD} = \frac{HD}{ADT(HM)} \qquad (15)$$

$$R_{A(M)} = \frac{A(M)}{ADT(HM)} \qquad (16)$$

and

$$R_{A(D)} = \frac{A(D)}{ADT(HM)} \qquad (17)$$

For the ith month, let N(i) be the number of observed days, W(i) be the average billed minutes per mobile station per month, and ADT(i) be the month average daily total traffic load in erlangs per mobile station. Then

$$ADT(i) = \frac{W(i)}{60\ N(i)} \qquad (18)$$

Thus, ADT(HM) on a per mobile station basis is ADT(i) for the high month.

4. THE MEASURED DATA

To estimate the coefficients of variation in (11) and (12) and the ratios of (13) to (17), measured data are needed. Initially, such measured data are not available from the automated mobile telephone system. However, as the system is part of the AGT network, the coefficients of variation and the ratios should be similar to those in the rest of the network. Therefore, the 1982-1983 historical measured data from 9 of the Class 5 Central Offices (CO's) located in Calgary, Alberta, Canada were used. Each of the measured data records contains one month's traffic information on one type of equipment in a CO. The information includes the appropriate TCBH loads, hourly load for 24 hours for each business day, daily total traffic load for each business day, and the number of main stations (MS) for the month. With holidays and abnormal days excluded, peak busy hour load for each day and that for the month can be identified. Thus, for each type of equipment, the following data are available on a per MS basis:

- ABS, 10HD, and/or HD
- A(D) and S(D)
- ADT(HM)
- A(M) and S(M)

These can be used to calculate CV(M), CV(D), R_{ABS}, R_{10HD}, R_{HD}, $R_{A(M)}$ and $R_{A(D)}$. Hence, with various types of equipment in the 9 CO's, there are 2 sets of coefficients of variation and 5 sets of ratios.

For a coefficient of variation, it is possible to have a value from 0 to a very large number. Therefore, to describe a set of coefficients of variation, gamma distribution is assumed,

$$F_G(g) = \frac{v^{u+1}}{\Gamma(u+1)} \int_0^g z^u e^{vz} dz \quad (19)$$

where the parameters u and v relate to the average $E\{G\}$ and 2nd moment $E\{G^2\}$ as follows [6] :

$$u = \frac{2(E\{G\})^2 - E\{G^2\}}{E\{G^2\} - (E\{G\})^2} \quad (20)$$

and

$$v = \frac{E\{G\}}{E\{G^2\} - (E\{G\})^2} \quad (21)$$

It is not difficult to compute the average, the 2nd moment, and the standard deviation for each of the 2 sets of coefficients of variation. The computed averages and standard deviations are listed in Table II. Thus, using (19) to (21), gama distribution can be calculated for each set. Fig. 1 compares the calculated gamma distribution with the distribution of the measured data for CV(M), and shows good agreement between the two. With the average and standard deviation, normal distribution can also be calculated for each set of coefficients of variation. In Fig. 1, the normal distribution for CV(M) is plotted as a dotted curve. It is seen that the normal distribution may serve as an approximation. Similar results have been obtained for CV(D). For the ratios in (13) to (17), they may have a value of up to 1. Thus, a set of ratios may be assumed to be distributed in accordance with the beta distribution,

$$F_B(b) = \frac{\Gamma(d+c+2)}{\Gamma(d+1)\,\Gamma(c+1)} \int_0^b z^d (1-z)^c dz \quad (22)$$

where the parameters c and d relate to the average $E\{B\}$ and 2nd moment $E\{B^2\}$ as follows [6] :

$$c = \frac{E\{B\} + E\{B\} \cdot E\{B^2\} - 2 \cdot E\{B^2\}}{E\{B^2\} - (E\{B\})^2} \quad (23)$$

and

$$d = \frac{2(E\{B\})^2 - E\{B^2\} - E\{B\} \cdot E\{B^2\}}{E\{B^2\} - (E\{B\})^2} \quad (24)$$

Again, it is not difficult to compute the average, the 2nd moment, and the standard deviation for each of the 5 sets of ratios. Table II lists the computed averages and standard deviations. Thus, using (22) to (24), beta distribution can be calculated for each set. Fig. 2 compares the calculated beta distribution with the distribution of the measured data for R_{ABS}, and shows good agreement between the two. Also, normal distribution has been calculated for each set using the corresponding computed average and standard deviation. In Fig. 2 the normal distribution for R_{ABS} is plotted as a dotted curve, and is seen to be a good approximation. For the other 4 sets of ratios, similar results have been obtained.

To determine the constant k in (10), measured data from the Line Link Frames of #5 XB in one of the 9 CO's called Huntington Hills has been used. It has been found that k approximately equals 2.3.

5. APPLICATIONS

To translate the average billed minutes per mobile station per month into various busy hour loads, consider Table III which lists W(i) for May 1983 to April 1984 in Alberta, Canada. According to (18), each W(i) can be converted to ADT(i) as shown in Table III. From Table III, it is seen that the expected high month is January 1984 with

$$ADT(HM) = 0.1894 \text{ erlangs} \quad (25)$$

since, according to (17), it gives the highest A(D) with constant $R_{A(D)}$.

Thus, based on the average ratios and coefficients of variation listed in Table II, and using (7), (11) to (17) and (25), the traffic per mobile station can be calculated as follows:

$$ABS = 0.0176 \text{ erlangs} \quad (26)$$

$$10HD = 0.0189 \text{ erlangs} \quad (27)$$

$$HD = 0.0203 \text{ erlangs} \quad (28)$$

$$EV(M,p) = 0.0210 + 0.0010L \text{ erlangs} \quad (29)$$

$$\text{and } EV(D,p) = 0.0192 + 0.0013L \text{ erlangs} \quad (30)$$

where L depends on p as shown in Table I.

To equate the grade of service based on EVE to that based on TCBH, consider the group of voice channels in a Local Mobile Centre of the automated mobile telephone system [7] . The group of voice channels may be regarded as a system component not too sensitive to traffic peaks. Therefore, in AGT, p is chosen to be 0.08 for the return period of a year and 0.05 for the return period of the high month of a year. Thus, substituting L with the corresponding value from Table 1, (29) and (30) become respectively,

$$EV(M,0.08) = 0.0225 \text{ erlangs} \quad (31)$$

and

$$EV(D,0.05) = 0.0215 \text{ erlangs} \quad (32)$$

Dividing these by (26) results in,

$$\frac{EV(M,0.08)}{ABS} = 1.278 \quad (33)$$

and

$$\frac{EV(D,0.05)}{ABS} = 1.222 \quad (34)$$

According to the previous paper [7] , terminating call blocking probability for total traffic offered to the group of voice channels can be calculated. Let

P_{ABS} = the terminating call blocking probability for ABS offered to the group,

P_M = the terminating call blocking probability for EV(M,0.08) offered to the group, and

P_D = the terminating call blocking probability for EV(D,0.05) offered to the group.

Then, with (33) and (34), relationship among P_{ABS}, P_M, and P_D can be established. Fig. 3 illustrates such a relationship for the mixed traffic as shown. From Fig. 3, it is seen that, for example, $P_{ABS} = 0.020$ implies $P_D = 0.037$ for the 4 voice channels group and 0.053 for the 9 voice channels group. This means that peak busy hour traffic congestion, as usually perceived by the subscribers, varies with the group size of the voice channels if the grade of service is based on TCBH. Such variation can be eliminated if traffic congension probability based on EVE is a constant. As satisfactory grade of service for EVE has not yet been defined in AGT, the relationship such as shown in Fig. 3 helps to equate the grade of service based on EVE to that based on TCBH.

6. CONCLUSION

In conclusion, based on the AGT measured data, relationship among various busy hour load traffic parameters has been developed. The relationship has been shown useful for translating average billed minutes per mobile station per month into various busy hour loads. It has also been shown useful for equating the grade of service based on EVE to that based on TCBH. Indeed, the developed relationship has been applied to the Aurora-400 automated mobile telephone system operating in Alberta, Canada.

ACKNOWLEDGEMENT

The author would like to thank Mr. F. Wissinger for his effort in analyzing the measured data, the AGT Priority Council for the interest on this subject, Mr. L.W. Mills and Mr. D.F. Baillie for their encouragement in having this paper presented to ITC-11, and Mrs. J.L. Young for typing this paper.

REFERENCES

[1] J. Pulford, "Aurora System is built to grow", Telephone Engineer & Management, 1 August, 1984.

[2] D.H. Barnes, "Extreme Value Engineering of Small Switching Offices", ITC-8, Melbourne, Australia, November, 1976.

[3] D.H. Barnes, "Observations of Extreme Value Statistics in Small Switching Offices", ITC-9, Torremolinos, Spain, October, 1979.

[4] K.A. Friedman, "Extreme Value Analysis Techniques", ITC-9, Torremolinos, Spain, October, 1979.

[5] K.A. Friedman, "Precutover Extreme Value Engineering of a Local Digital Switch", ITC-10, Montreal, Quebec, June, 1983.

[6] A. Papoulis, "Probability, Random Variables, and Stochastic Process", McGraw-Hill Book Co., New York, 1965

[7] L. Lee, "A Solution to a Queueing Problem in the Aurora-400 System", Proceedings of the Third International Seminar on Teletraffic Theory, Moscow, USSR, June,1984.

Table I Value of L for a given p in equation (7)

p	L
0.001	3.612
0.002	3.327
0.003	3.151
0.004	3.021
0.005	2.923
0.006	2.838
0.007	2.766
0.008	2.704
0.009	2.647
0.01	2.596
0.02	2.248
0.03	2.032
0.04	1.871
0.05	1.743
0.06	1.635
0.07	1.541
0.08	1.458
0.09	1.382
0.10	1.314

Table II Averages and standard deviations of coefficients of variations and of ratios

	Average	Standard Deviation
CV(M)	0.0489	0.0274
CV(D)	0.0660	0.0267
R_{ABS}	0.0927	0.0125
R_{10HD}	0.1000	0.0177
R_{HD}	0.1071	0.0187
$R_{A(M)}$	0.1108	0.0169
$R_{A(D)}$	0.1015	0.0116

Table III Measured valued of W(i) and the calculated values of ADT(i)

i	W(i)	ADT *
May 1983	116	0.0879
June 1983	133	0.1008
July 1983	113	0.0856
Aug. 1983	151	0.1143
Sept.1983	171	0.1295
Oct. 1983	154	0.1167
Nov. 1983	127	0.0962
Dec. 1983	152	0.1152
Jan. 1984	250	0.1894
Feb. 1984	213	0.1614
Mar. 1984	145	0.1098
Apr. 1984	126	0.0955

* Although the number of days varies from month to month, it is assumed that every month has 22 days.

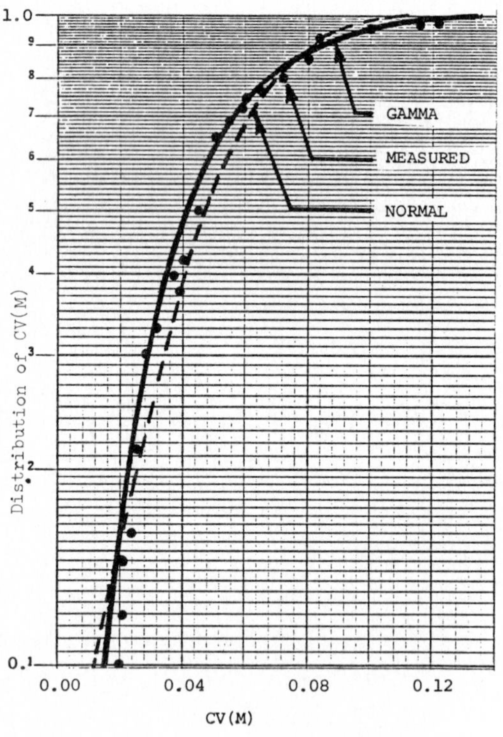

Fig. 1. Distribution of CV(M) with average 0.0489 and standard deviation 0.0274.

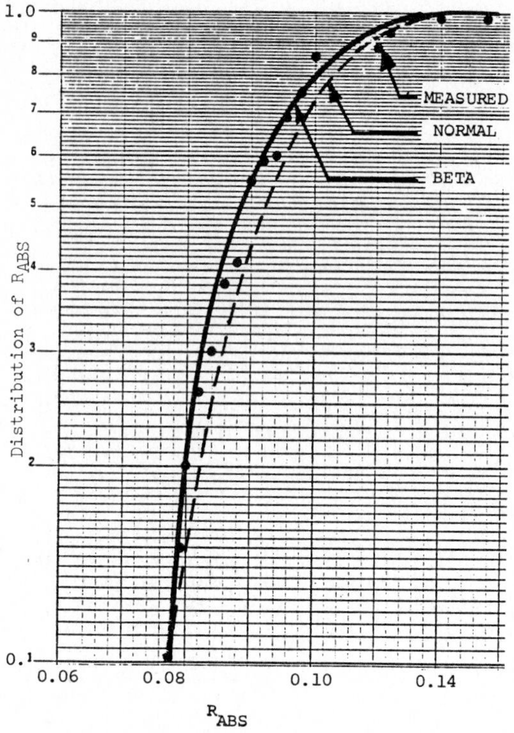

Fig. 2. Distribution of R_{ABS} with average 0.0927 and standard deviation 0.0125.

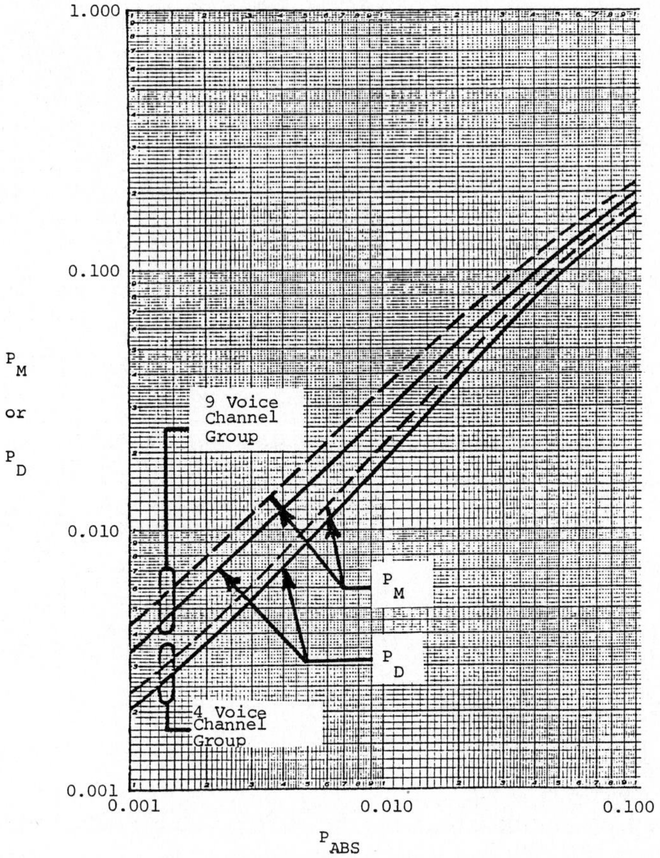

Fig. 3. Relationship of P_{ABS}, P_M and P_D for the case where

Number of queue positions = 2,
Queue time out = 120 sec,
Average service time = 180 sec/call,
0+ calls = 10% of total traffic,
1+ calls = 60% of total traffic, and
Terminating Calls = 30% of total traffic

TELETRAFFIC ISSUES in an Advanced Information Society
ITC-11
Minoru Akiyama (Editor)
Elsevier Science Publishers B.V. (North-Holland)

AN ANALYSIS OF TRAFFIC VARIATIONS IN THE FRENCH TELEPHONE NETWORK

Annie PASSERON and Simone RIVAT

CENTRE NATIONAL D'ETUDES DES TELECOMMUNICATIONS, PAA/ATR
Issy Les Moulineaux, France.

ABSTRACT

This paper deals with a statistical study on traffic measurement data. The study is especially concerned with traffic variation analysis with the goal of comparing different possible ways of choosing representative values for use in final group engineering.

Analysed data concern outgoing traffic in transit centres and were recorded by the French traffic supervisor in 1982 and 1984. They mainly take the form of moving daily peak hour trunk group offered loads.

Statistical tests are first used to check normal and gamma distribution hypotheses for traffic load variations. We then compare the relative merits of different possible methods for trunk group sizing.

1. INTRODUCTION

The general goal of this study is to compare different possible ways of choosing representative values from traffic measurements, for both telephone traffic administration and final group sizing.

More precisely, the study consists in the statistical analysis of a set of traffic data which is large enough to be representative of a national long distance telephone network. This analysis is carried out with two aims. The first is to judge the existing French procedure for choosing representative values in relation to sizing criteria. The second is to simulate other possible methods in order to be able to propose a different procedure which would lead both to improved daily grades of service and decreased network costs.

The existing process for trunk provisioning was defined several years ago when very few traffic measurements could be made and when customers had rather homogeneous needs. Traffic characteristics were modelled simply, resulting in the following system :
each month at the time consistent busy hour [1], five hourly load measurements are made on all trunk groups ; the second highest value is kept as the monthly representative value ; the yearly representative value is the second highest monthly value in the year. The latter is thus a sort of busy season representative value. In the sizing process, the Erlang formula and Wilkinson's ERT are applied to the yearly representative values of offered traffic for a given blocking probability on final groups (currently 1%).

In fact, trunk dimensioning is planned once a year using forecasts of yearly representative values on groups at the end of the study period and is monitored each month using the latest traffic measurements in conjunction with the above sizing models. The performance of the existing network dimensioning method may be directly monitored from measurements on final groups which constitute last choice routes for calls trying to reach their destination. These final groups therefore constitute an important class of trunk groups.

This process may be considered inadequate in the present network for essentially two main reasons.

Firstly, the magnitude of season to season, day to day or hour to hour variations seems to be increasing while the position of the post selected busy hour varies within a wide interval. For instance, for a set of 88 trunk groups, Table I gives the width of the period of the day containing the post selected busy hour during each of 11 working days in May 1984. We exclude busy hours in off-peak call charging periods. The busy hour appears to fluctuate widely. This effect is accentuated for small trunk groups or trunk groups going to transit centres.

Table I : Width of the busy period.

Number of hours	<2	2 to 3	3 to 6
Number of groups (%)	15	18	67

Secondly, the use of electronic exchanges should permit more sophisticated traffic measurements. A statistical study on detailed traffic data has therefore become necessary to improve network administration and trunk provisioning procedures.

This paper deals only with hourly traffic loads and is not concerned with traffic variations inside the hour.

Analysed data are more detailed than those used in the current trunk provisioning method as described above. They are extracted from traffic records made by the new French traffic supervisor which is now being brought into service. These records only concern trunk groups outgoing from transit centres towards either subscriber centres or other transit centres. In the French network organization, the former are in fact final choice groups while some of the latter may overflow on to a special transit centre but only for security reasons. In the planning process, they are all sized for a grade of service of 1 % blocking probality.

Before comparing sizing methods on these data, a preliminary statistical analysis of traffic load variations is made leading, in particular, to theoretical distribution functions for traffic offered to trunk groups.

2. AVAILABLE DATA

2.1. The French Long Distance Network Supervision Centre

After some preliminary trials [8], since 1982, there exists in the French telephone network a supervision centre, whose purpose is to monitor the grade of service of traffic outgoing from regional transit centres.

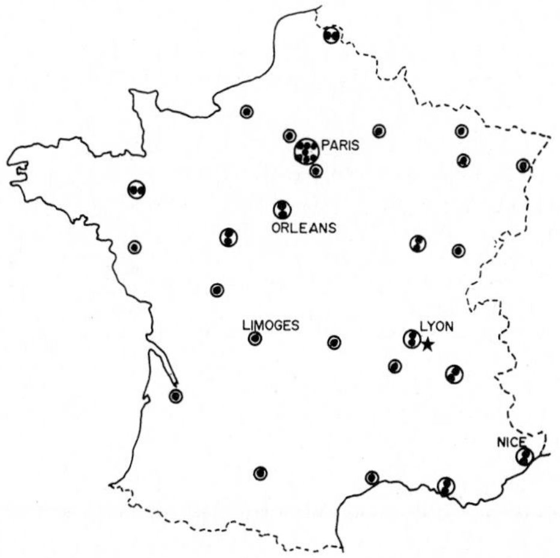

⊛ Number of supervised transit centres in a town.
★ Supervision centre

Fig.1 : Centres connected to the supervision centre in mid 1984.

At mid 1984, about 40 transit centres (see figure 1) were connected to the supervision centre by dedicated data links accounting for almost all space division regional transit centres and including one time division transit centre. Some 33 % of trunks of all incoming and outgoing transit trunk groups of the whole national network were observed. Ultimately, all regional transit centres will be connected, i.e. about 60 centres, and the supervisor will be used for long distance network management and for supplying data for national traffic data bases.

In each supervised centre a terminal station collects traffic measurements describing the state of outgoing trunk groups and some register groups, and sends these data to the supervision centre every fifteen seconds. Measurements are supposed to be made every working day, all day long from about 9 a.m. to about 11 p.m..

For each trunk group, raw data include :

- carried load, number of seizures, number of successful seizures in every four minute period,
- test of all trunks busy state every fifteen seconds,
- number of trunks in service every eight minutes.

As data are recorded on a magnetic tape (1 tape per day), off-line data processing may be performed. Each day, for the normal charging period of the day and for the remainder of the day, the traffic load Moving Daily Peak Hour (MDPH) is calculated, to within an accuracy of 4 minutes, for each individual trunk group and for each centre. Load curves for different trunk groups or centres or groups of them may also be established.

2.2 Data Volume

Data were collected during the whole of 1982 and a large part of 1984. Data for 1983 were very incomplete. In fact, the system is still being brought into service and is not yet completely operational. Due to data transmission errors and equipement failures, measurements are available for less than 17 working days each month.

During 1982 an 1984, some centres were newly connected while others were closed down (reflecting the fast modernization and digitalization of the French network). Overall, we dispose of observations for 58 % of working days in 1982 and 49 % in 1984, but not for all centres.

After data validation (for zero data, clock coherence and observations breaks), 57 % of possible daily measurements from all centres could be kept in 1982 (53 % in 1984), among which 83 % (76 % in 1984) were complete, i.e. all day long, for all trunk groups.

For this study we selected a set of consistent data large enough to be representative of the French network.

2.3. Data Used In The Study

We analysed only hourly loads on final groups outgoing from three regional transit centres, located in Limoges, Nice and Orléans in 1982, and from one, located in Limoges in 1984.

We mainly took into account MDPH loads which are defined during the normal charging period of the day for all avaiblable daily measurements.

Table II shows how much data we obtained throughout 1982 for the above three centres. Trunk groups which were not present during the whole year do not appear.

Data are quite evenly distributed from January to November (5 to 14 days per month) and among different days of the working week.

Table II : Studied data for 1982.

Centre	Number of groups		Maximum number of days
	total	to transit centres	
Limoges	93	28	111
Orléans	79	32	99
Nice	76	31	97

During 1984 Orléans and Nice transit centres were replaced by time division systems and data from Limoges were kept for only 62 days.

Raw data from May 1984 have also been analysed with the aim of making some rough comparisons between MDPH loads and hour by hour daily peak hour (HDPH) loads (i.e. loads measured between 8 am and 9 am, 9 am and 10 am, etc.).

By using an hourly carried load, Te, and the corresponding number of all trunks busy states, NOT, observed in 240 fifteen second cycles it is easy to calculate the offered load to be studied

$$T_0 = T_e/\left(1-NOT/240\right)$$

NOT was never very high and was often zero, trunk groups tending to be overdimensionned. For some groups (60 % of studied trunk groups) it was not possible to get this information and we made the approximation :

$$T_0 = T_e$$

3. TRAFFIC LOAD DISTRIBUTION

It is well known that measured busy hour loads vary from day to day and studies aiming to analyse and model this variability are far from being a recent phenomenon. In [10], for instance, day to day variations were studied during a busy season of 6 weeks on a set of data on Time Consistent Busy Hour (TCBH) loads concerning some 20 trunk groups. A gamma distribution was fitted to the traffic load data and tested graphically. In [6], combined normal distributions were tested graphically for a set of measurements made during two consecutive years and concerning hour by hour loads.

In this section we try to update these results using recent data. The analysis is made for each available trunk group and concerns MDPH loads. In order to throw light on the best choice of yearly representative values, variations are studied inside the whole year and inside a pratical busy season.

3.1. Histograms

Figures 2 and 3 show two typical shapes of the histograms we plotted for MDPH offered load for each of the above trunk groups.

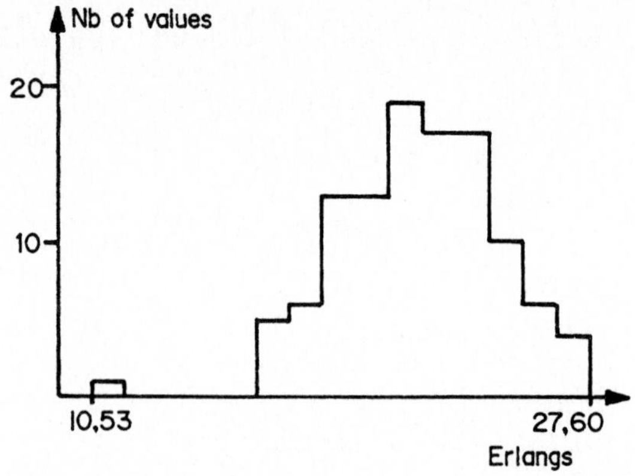

Fig.2 : Histogram of MDPH loads in 1982 for a trunk group from Limoges to another transit centre.

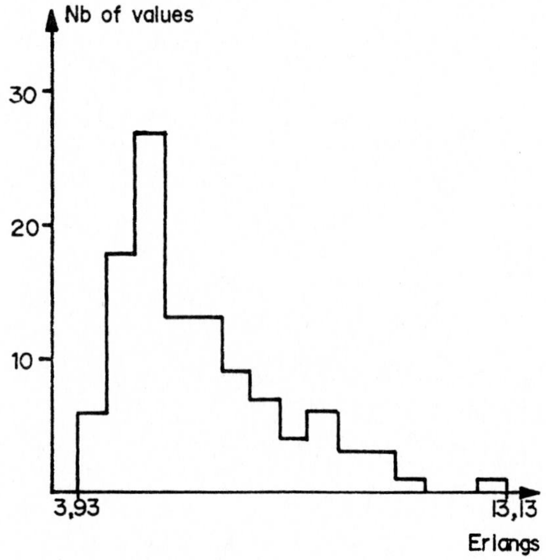

Fig.3 : Histogram of MDPH loads in 1982 for a trunk group from Limoges to a subscriber centre.

The histogram of Figure 2 is rather symmetrical and unimodal ; it looks like a normal distribution. The histogram of Figure 3 is not really symmetrical. It is unimodal and could be approximated by a gamma distribution.

In the case of Limoges in 1982, for example, over all groups, the mean load varies from 1.16E to 119.46E and the coefficient of variation lies between 0.10 and 0.32. All data are, of course, positive.

3.2. Choice Of Distribution Functions

3.2.1. Normal Distribution [4]

This distribution is symmetrical and unimodal. The probability density function is :

$$f(x) = \frac{1}{\sqrt{2\pi}\sigma} \exp\left[-\frac{1}{2}\left(\frac{x-m}{\sigma}\right)^2\right]$$

It is defined for real values of x. Given a sample of n values Xi,

$$\overline{X} = \sum_{i=1}^{n} X_i \text{ is a maximum likelihood unbiased}$$

estimator of m and

$$s^2 = \frac{1}{n-1} \sum_{i=1}^{n} (X_i-\overline{X})^2 \text{ is an unbiased}$$

estimator of σ^2.

The normal distribution can be used as an approximation for other distributions. A known distribution may be replaced by a normal distribution with the same expected value and standard deviation (method of moments). This distribution is easy to use in calculations.

3.2.2. Gamma Distribution Or Type III Of Pearson's System [4]

The probability density function is defined for positive x and (if only two parameters, α and β, are used) is of the form :

$$g(x) = \frac{x^{\alpha-1}\exp(-x/\beta)}{\beta^{\alpha}\cdot\Gamma(\alpha)} \qquad (\alpha > 0 \; ; \; \beta > 0)$$

where $\Gamma(\alpha)$ is the gamma function.

To estimate α and β, the method of moments may be used leading to the following simple formulae.

$$\begin{cases} \tilde{\alpha}.\tilde{\beta} = \overline{X} \\ \tilde{\alpha}.\tilde{\beta}^2 = m_2 \end{cases}$$

where $\overline{X}$ and m_2 are the arithmetic mean and second central moment of a sample of n values X_i.

However, these estimators may be less accurate than maximum likelihood estimators, $\hat{\alpha}$ and $\hat{\beta}$, for which the following approximation may be used if $\hat{\alpha}$ is large enough :

$$\begin{cases} \hat{\alpha}.\hat{\beta} = \overline{X} \\ n^{-1} \sum_{i=1}^{n} LnX_i - Ln\overline{X} = Ln\,(\hat{\alpha} - 0.5) - Ln\,\hat{\alpha} \end{cases}$$

For comparison, we used both estimation methods in this study.

The gamma distribution has a single mode if $\alpha > 1$, and, as α increases, it tends to the normal distribution.

3.2.3. Kolmogorov-Smirnov Distribution Test

This test is used to test gamma and normal distribution hypotheses for traffic loads. Given a sample of n ($n > 35$) independent observations the Kolmogorov-Smirnov test tests whether or not the data may be considered as a random sample of observations from a specified distribution.

For the test, the largest absolute deviation between the sample cumulative distribution and the theoretical distribution is computed. This deviation is compared to a value which is exceeded with a probalitity of 0.10, under the specified hypothesis. This value is computed using the Kolmogorov-Smirnov limiting distribution.

3.3. Test Results

Table III shows results for data on trunk groups observed in the whole year 1982.

Table III : Acceptance (% of trunk groups) for 1982 data.

origin centre	normal	gamma
Limoges	71	75
Orléans	67	67
Nice	86	86

In fitting the gamma distribution, in particular for trunk groups going to transit centres, we sometimes find very high values of α. In this case we use the normal approximation. In all cases α is never less than 10. Both estimation methods give almost identical parameters (α,β) on our specific data (except for a few small trunk groups), which leads us to select the simple method of moments.

For some trunk groups, only one of the two tested distributions is accepted. For instance, the gamma distribution may be accepted for very small trunk groups when the normal distribution hypothesis is rejected. The reason may be that data are near zero but do not include negative values for which the theoretical normal distribution is defined.

Results are not significantly different if we distinguish trunk groups going to transit centres from those going to subscriber centres. In general, if both normal and gamma distributions are not accepted, the coefficient of variation is quite high. The rejection decision does not appear to depend on the level of the traffic load.

For Limoges in 1982 and in 1984, we have tried to define a practical busy season by looking for a period of four consecutive months for which the mean load is the highest in the year for most trunk groups. In fact, such a busy season is only obvious for trunk goups going to subscriber centres. Table IV shows that statistical tests give better results with data taken only in the busy season than with data taken in the whole year. The above comments on gamma fitting are still valid for these data.

Table IV : Acceptance (% of trunk groups) for Limoges in the busy season.

year	normal	gamma
1982	89	88
1984	93	94

In conclusion, the normal approximation for busy season data appears to be acceptable for most trunk groups. Because of its remarkable properties this distribution is used in what follows.

4. FINAL GROUP SIZING

4.1. Different Possible Methods

The objective is to compare different methods for final group sizing. All these methods need traffic data which should be chosen so as to represent the highest loads to be carried during the reference period (one year, for instance). After sizing, a final group is considered to offer a good service if it always meets a given grade of service, estimated, for example, by the proportion of blocked calls, day after day, with the possible exception of a few busy hours. For each sizing method, the above set of traffic data allows us to simulate, for a large number of trunk groups the number of days or hours in a whole reference period for which different blocking levels are exceeded. We suppose that the blocking function which gives the blocking probability on a group from its number of trunks and its offered load is the Erlang function.

Following [2] and [9], we distinguish three approaches as follows.
a) A sufficiently high reference load is chosen and the group is sized to a given blocking probability (using the blocking function).
b) The group is sized for an average blocking objective in a busy season. The distribution function of the load and the blocking function are used.
c) The hourly load which will not be exceeded more than once in a given period is estimated by using an extreme value distribution. The group is sized for this reference load.

The first approach includes both the existing French process, as detailed in the introduction, and the CCITT recommendations [1], i.e. the reference load is the mean of the 30 highest TCBH loads in the year. The second is used in North America [5]. The third belongs to Extreme Value Engineering (EVE).

Taking into account results of the preceding section, we analyse data only on trunk groups for which the normal distribution was accepted.

4.2. Comparison Between CCITT And Existing French Procedures

The comparison is made using MDPH data instead of TCBH data. From MDPH data we define an equivalent French Monthly Representative Value (MRV) based on order statistics. For instance, if 5 MDPH values are available in a month, we keep the second highest, but if 11 MDPH values are available, we keep the fourth highest. A preliminary comparison was made between these supervisor MRVs and the current TCBH MRVs on Limoges data during the whole of 1982.

Depending on the size of each trunk group, the supervisor MRVs are between 4 % and 25 % higher than those of the TCBH. As the CCITT definition is also applied to MDPH loads by taking the mean of the 30. [n/250] highest of n available MDPH data (n = 111 for Limoges, cf Table II), the comparison is thought to be meaningful.

Table V gives global results of the comparison for data in 1982. With the CCITT method the Yearly Representative Value (YRV) is exceeded, on average, on 11 days per year, i.e. 4 % of working days. With the French method the YRV is exceeded, on average, on 37 days per year, i.e. 15 % of working days.

Table V : Undervalue of French YRV (in %) compared to CCITT YRV.

Limoges	Orléans	Nice	Total
8.7	9.8	9.6	9.3

The average traffic load undervalue of 9.3 % leads to an average undervalue in the number of circuits of 8.5 % for a blocking probability equal to 1 %.

The mean of the 3 highest data in the year for each trunk group is, on average, 23 % higher than the French YRV.

The normal distribution modelling allows us to approximate each of the above YRVs with a function the expected value, m, and the standard deviation σ:

$$YRV = m + K.\sigma,$$

of which an unbiased estimator is

$$\bar{X} + K.s \;(\bar{X} \text{ and } s \text{ are defined above}).$$

K is computed from the normal distribution function applied to our data.

On average, we find :

K = 1.73 for the CCITT YRV and
K = 1.04 for the French YRV.

Such a model could be used to compare both methods on a new small sample of data or to predict the performance of these methods.

4.3. Comparison Between CCITT Recommendations And Average Blocking Method

Only the Limges data from 1982 are used.

Average blocking, $\bar{B}$, on a trunk group of N circuits is calculated as follows [5] :

$$\bar{B} = \int_0^{\infty} E[a,N]f(a)da,$$

where f(a) is the normal probability density function of the distribution of loads offered to the trunk group and E[a,N] is the Erlang blocking function.

For each trunk group, f(a) is estimated for the month where the average load is highest. Sizing is then made successively with the CCITT YRV and for an average blocking level of 1 %. Roughly speaking, if the coefficient of variation is lower than 0.15, the average blocking criterion leads to fewer trunks than the CCITT criterion ; the opposite is true if this coefficient is greater than 0.15.

The average blocking method seems better suited to traffic variability and, when used over the highest month of a year, sizing is, on average, only 4.2 % higher than that of CCITT recommendations.

4.4. First Comparisons With EVE

Limoges data from May 1984 are used for this comparison.

In telecommunications two extreme value distributions have already been proposed. In [7], the classical extreme value distribution function (also called the Gumbel distribution) is used :

$$\Pr[X \leq x] = \exp[-e^{-\beta(x-\alpha)}]$$

This is a limiting distribution of the greatest value in a random sample of infinite size. In [3], the normal distribution power 6 is proposed. In order to test these distributions we first calculate the HDPH of 7 typical trunk groups using the traffic profiles of several working days. Figure 4 shows, for instance, the traffic profile observed on May 10, 1984 on a medium size trunk group.

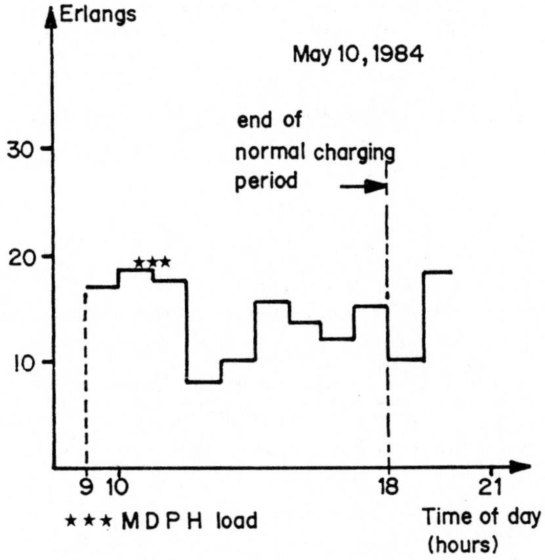

Fig.4 : Hourly loads on a trunk group From Limoges to another transit centre

These data are used to estimate the parameters of the above distributions. In EVE, the load used for sizing is that which will be exceeded once on average during a return period (usually 20 working days), as calculated from an extreme value distribution. After estimation of parameters by the method of moments, the first distribution leads to the return period load :

$$r_{20} = \bar{X} + 1.866\ s$$

and the second gives :

$$r'_{20} = \bar{X} + 1.744\ s.$$

By taking a return period equal to the number of available data we could check the number of times the return period load is exceeded for each of the two distributions. However, due to the small values of the coefficient of variation it is not possible to draw any positive conclusions. Further studies are necessary but as, on average for the 7 trunk groups, the two definitions of r_{20} differ by only 1.4 %, we make the arbitrary choice of the normal 6 modelling. We come to the same conclusions if we calculate a return period load using MDPH data for May 1984.

In comparing the return period load, for hour by hour data (a), the return period load for MDPH data (b) and the MDPH French MRV (c) for May 1984 we find that (c) is always the lowest, that (a) is, on average, 10.9 % higher than (c) and that (b) is on average, 17.8 % higher than (c).

Lastly, we compare the return period load r_{20} calculated from MDPH data over 4 months to the equivalent CCITT load calculated from the same data. We find practically the same loads, which is not suprising since the CCITT load was modelled by :

$\bar{X} + 1.73\ s$ and r_{20} is calculated with
$\bar{X} + 1.744\ s$

Another reason is that both loads have similar definitions on MDPH data : In 1982 on our data the CCITT load was exceeded on 4 % of working days on average and r_{20} is theoretically the load which will be exceeded, on average, once in 20 working days.

5. CONCLUSIONS

Data recorded by the French long distance supervision centre have supplied rich information on traffic load variations. Daily peak hour data may be defined on trunk groups for each working day with a precision of 4 minutes. When, ultimately all regional transit centres are connected, the supervisor could be used to provide data for trunk provisioning and for validating traffic variation models in addition to its primary function of traffic supervision.

From the present study it appears that the normal approximation for traffic variations within the year is acceptable and is as good as an approximation by the gamma distribution. Moreover, the normal approximation is even better for data in a practical busy season of 4 consecutive months.

Concerning YRVs used for trunk group sizing, we first showed that, with our data, MDPH loads are between 4 % and 25 % higher than TCBH loads. If considering only MDPH data, the French method leads to a YRV which is exceeded, on average, on 15 % of working days while the CCITT YRV would be exceeded on just 4 % of working days. An average blocking method applied to a busy season (in the study we took the highest month in a year) leads to nearly the same sizing as the CCITT method while closely fitting traffic variability and being more easy to apply. Only busy season measurements are needed and sizing may be realized by use of the normal distribution. EVE applied to a period of 4 months gives similar results but in this field further studies are necessary.

Concerning traffic monitoring and monthly representative values we have found tentative relations between engineering with hour by hour data, MDPH data and French MRV data. This analysis requires further study using additional data.

REFERENCES

[1] CCITT Yellow Book, Vol II, Fascicule II.3, Recommendation E 500, 1980.

[2] A.C. COLE, " A study of traffic flow in relation to CCITT recommendations for the measurement of telephone traffic flow, call attempts and seizures", 10th Int. Teletraffic Congress, Montréal, 1983.

[3] K.A. FRIEDMAN, "Extreme value analysis techniques", 9th Int. Teletraffic Congress, Torremolinos 1979.

[4] N. JOHNSON and S. KOTZ," Distributions in statistics - continuous univariate distributions - 1", Wiley Series in Probability and Mathematical statistics, New-York, 1970.

[5] A. KASHPER, S.M. ROCKLIN and C.R. SZELAG, "Effects of day to day load variation on trunk group blocking". Bell Syst. Tech. J., vol 61, n°2, pp 123-135, 1982.

[6] P. LE GALL, "Les variations du trafic et la qualité de service", Commutation et Electronique, n°7, pp 44-62, 1964.

[7] P. LEHTINEN, "Extreme value control and an additive seasonal moving average model for the evaluation of daily peak hour traffic data", 9th Int. Teletraffic Congress, Torremolinos, 1979.

[8] M. PEYRADE, A. SCHLATTER, A. SPIZZICHINO, "Etude du trafic téléphonique à l'aide du superviseur du réseau interurbain", "La mesure dans les télécommunications" Congress, Lannion, 1977.

[9] J. RUBAS, "Estimation of reference load from daily traffic distributions", 10th Int. Teletraffic Congress, Montréal, 1983.

[10] R.I. WILKINSON, "A study of load and service variations in toll alternate route systems", 2nd Int. Teletraffic Congress, The Hague, 1958.

TELETRAFFIC ISSUES in an Advanced Information Society
ITC-11
Minoru Akiyama (Editor)
Elsevier Science Publishers B.V. (North-Holland)

CHARACTERISTICS OF NON-VOICE APPLICATIONS OVER THE INTERNATIONAL PUBLIC SWITCHED TELEPHONE NETWORK

Keizo OHNO, Tatsu HIROSE and Masao KOJIMA

Kokusai Denshin Denwa Co., Ltd. (KDD)
Tokyo, Japan

ABSTRACT

This paper describes the investigation results of non-voice applications over the international public switched telephone network. The results have indicated that the current international public switched telephone network is widely used for carrying the non-voice traffic. The ratio of non-voice traffic to the total international subscriber dialling (ISD) traffic to certain countries exceeds 80% in the number of calls. In addition non-voice traffic has a significant difference in its characteristics, such as 24-hour traffic profile and duration of call-holding time, compared with the ordinary voice traffic.

1. INTRODUCTION

The present telephone network is capable of providing a bearer service for non-voice applications including data and facsimile. The occupancy rate of non-voice application traffic in the total traffic over the telephone network has been increasing recently and further increase is anticipated.

The non-voice call is generally considered to be placed between terminals on non-attendant mode of operation at least at the called end, and characteristics of the non-voice call differ from those of voice call which is usually placed considering convenience of a called party. Increase of the non-voice calls may therefore entail special considerations for telephone network designing practice. KDD has been checking the characteristics of non-voice traffic .

2. MEASUREMENT PROGRAM

The investigation was carried out on outgoing telephone traffic from Japan at the KDD's international gateway located in Shinjuku, Tokyo. A total of 14 countries in various time zones were selected for measurement considering the possible influence of time difference to the characteristics(1),(2). About 15 percent of the total calls destined for the countries were picked up on each circuit group observed. The samples were collected by automatic observation equipment developed by KDD. The equipment is capable of distinguishing the non-voice calls, such as data and facsimile from ordinary voice calls with the accuracy of more than 93.9 percent in average. Observation is carried out fully automatically and observation result is stored into the magnetic tape.

Since the equipment does not monitor a call in full duration, the detailed record of calls completed on the same day was utilized to obtain statistics on call duration. The samples which did not have the corresponding completed calls in the record were deleted from the data for analysis.

Traffic data on outgoing international telex calls to the same countries were also utilized to confirm the similarity between non-voice telephone traffic and telex traffic in 24-hour profile.

3. MEASUREMENT RESULTS AND ANALYSIS

3.1 Mean call duration

Table 1 shows mean call duration of non-voice and voice calls, which is sorted by area zone. There is a significant difference in call durations between non-voice and voice traffic, while relatively small deviation by destination is recognized in each traffic. The mean call duration of non-voice traffic is within three minutes in most cases, which is about one half or even one third of the voice traffic. Difference in basic unit for charging may have caused the difference in call duration between ISD and operator assisted calls both in non-voice and voice traffic.

Note: Basic charging unit is 6 seconds for ISD, while 1 minute for operator assited call exceeding 3 minutes.

Table 1 Average Call Duration

Country	Average Call Duration (sec)			
	ISD		Operator assisted	
	non-voice	voice	non-voice	voice
A (in Asia)	134	386	158	478
B (in Europe)	148	456	229	592
C (in N America)	172	555	286	732
D (in Oceania)	156	440	160	629
E (in Asia)	134	297	223	453
F (ditto)	147	296	255	409
G (ditto)	112	266	205	429
H (in M East)	171	670	120	752
I (in Europe)	165	467	196	507
J (ditto)	181	483	176	515
K (ditto)	150	412	243	561
L (ditto)	152	615	246	557
N (in S.America)	199	648	189	567
M (in N America)	156	545	306	678
Average of total monitored calls	156	403	237	540
Deviation by country	22	130	52	107

3.2 Ratio of non-voice traffic

Tables 2 and 3 respectively show the ratio of non-voice traffic to the total traffic in the number of calls completed and in the volume of traffic carried, each classified by destination and category of calls (ISD, operator assisted), which may be characterized by the following:

(1) The current international public switched telephone network is greatly utilized for carrying the non-voice traffic, while the ratio of non-voice traffic to the total traffic varies widely according to the countries.

(2) Non-voice traffic is carried mainly by ISD calls, and operator assisted calls are rarely utilized for non-voice traffic.

(3) The ratio of non-voice traffic destined for the countries, where data and/or facsimile terminals seem to be widely used, is relatively high. More than 80 percent of ISD calls destined for some European countries are of non-voice traffic, which is equal to 50 to 60 percent in terms of traffic volume mainly due to the difference in call duration between non-voice and voice traffic and in the proportion of ISD calls to operator assisted calls (about 70:30 in Japan).

Note: Most of the non-voice calls are facsimile.

Table 2 Ratio of Non-Voice Call

Country	Ratio of Non-Voice Call (%)		
	ISD	Operator assisted	Total
I (in Europe)	83.3	17.6	73.0
L (ditto)	83.1	25.3	70.6
K (ditto)	73.7	18.5	64.1
H (in M East)	76.4	9.4	62.6
C (in N.America)	75.3	14.5	57.1
M (ditto)	75.0	6.8	57.1
B (in Europe)	65.1	8.5	56.0
A (in Asea)	59.1	13.9	49.8
J (in Europe)	62.5	10.0	47.7
D (in Oceania)	56.5	7.4	47.1
N (in S.America)	72.7	4.7	45.3
E (in Asea)	43.3	8.3	36.1
G (ditto)	37.1	4.6	23.8
F (ditto)	17.9	1.4	10.5
Average of total monitored calls	61.6	8.9	45.9
Deviation by country	18.2	6.3	18.1

Table 3 Ratio of Non-Voice Traffic Volume

Country	Ratio of Non-Voice TrafficVolume (%)		
	ISD	Operator assisted	Total
I (in Europe)	63.7	7.6	47.4
L (ditto)	55.0	13.0	39.4
K (ditto)	50.0	8.9	37.0
H (in M East)	45.2	1.6	30.8
B (in Europe)	37.7	3.5	27.9
C (in N.America)	48.6	6.2	26.6
M (ditto)	46.2	3.2	26.4
J (in Europe)	38.4	3.7	24.7
A (in Asea)	33.4	4.5	23.2
D (in Oceania)	31.5	2.0	22.0
N (in S.America)	44.9	1.6	21.2
E (in Asea)	25.6	4.3	18.3
G (ditto)	19.9	2.3	9.7
F (ditto)	9.7	0.9	4.8
Average of total monitored calls	38.3	4.1	22.3
Deviation by country	13.9	3.3	10.7

3.3 24-hour traffic profile

1) Three countries in different time zones are picked up as examples to show 24-hour traffic profile:

Country A is an Asian country and the time difference is -1 hour (one hour behind Japan standard time)

Country B is an European country with the time difference of - 8 hours (summer time).

Country C is located in North America and has a wide range of time difference within the country : -13 hours on the east coast and -16 hours on the west coast (summer time).

2) 24-hour traffic profiles on the above countries are shown in Figures 1 - 4.

(1) Figure 1 indicates the relations between the number of non-voice calls and that of total calls. Solid line shows the number of total calls, which is expressed by hour-by-hour concentration ratio, and dotted line shows the proportion of non-voice calls within the total calls.

(2) Figure 2 indicates the relations between the volume of non-voice traffic and that of total traffic. Solid line shows the volume of non-voice traffic, which is expressed by hour-by-hour concentration ratio, and dotted line shows the proportion of non-voice traffic within the total traffic.

(3) Figure 3 indicates the relations between the number of non-voice calls and that of voice calls. Solid line shows the number of non-voice calls and dotted line shows the number of voice calls, each of which is expressed by hour-by-hour concentration ratio.

(4) Figure 4 indicates the relations between the number of non-voice ISD calls and that of telex calls. Solid line shows the number of non-voice ISD calls and dotted line shows the number of telex traffic, each of which is expressed by hour-by-hour concentration ratio.

3) Following characteristics are pointed out from the figures:

(1) The difference in profile of non-voice traffic and voice traffic become apparent when there is much time difference between originating and terminating countries. In case of countries A and B, peak hours of both traffic are almost the same, but in case of country C, there appears two peaks, one (earlier one) for voice and the other (latter one) for non-voice.

(2) There is a slight difference in profile of the number of calls and that of the volume of traffic. This will be mainly because of the difference in call duration between non-voice traffic and voice traffic.

(3) Non-voice traffic tends to cover non-busy hours for voice traffic to countries with much time difference where business hours are not overlapped. This contributes to flattening the traffic profile and facilitates to increase the efficiency of circuit group during the non-busy hours for such countries. Country C gives a good example for this.

(4) It should also be noted that non-voice traffic may sharpen the peak of the profile in case of short overlapping of business hours existing between two countries. This may affect the dimensioning of network and require additional circuits only to cover quite a short period of time, without effective alternative routing arrangements. Country B is a good example for this. The concentration ratio exceeds 20% in traffic volume as well as in the number of calls.

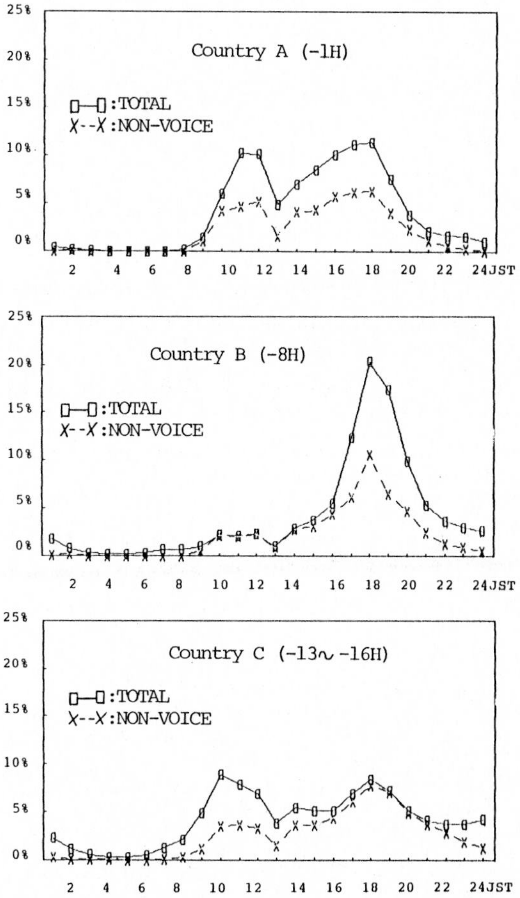

Figure 1 24-hour profile of the number of non-voice calls and total calls (expressed by concentration ratio)

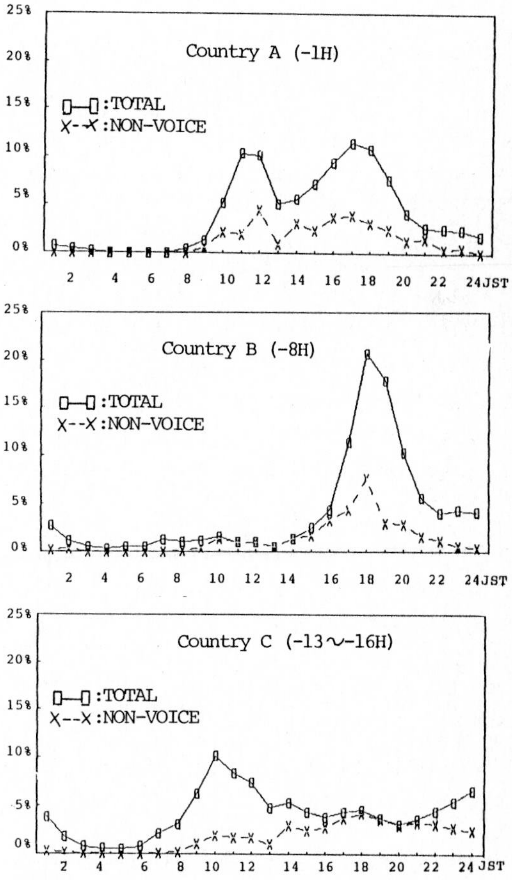

Figure 2 24-hour profile of the volume of non-voice traffic and total traffic (expressed by concentration ratio)

(5) It is interesting to note that non-voice traffic has a similar profile to that of telex traffic [3]. The peak hours of non-voice traffic come between 17:00 and 18:00 JST not only to country B but also country C, no matter what time it is in the terminating country. This period is the time to close offices in Japan, but it is in the middle of business hour in country B and late in the evening or at midnight in country C. This fact also suggests that non-attendant mode of operation is highly utilized for non-voice traffic at least at the receiving end.

4. CONCLUSION

The investigation results have indicated that non-voice traffic has a significant difference in its characteristics compared with the ordinary voice traffic, and 24 hour profile of total telephone traffic is considerably influenced by the volume of non-voice application traffic. Therefore, much more attention should be paid to the non-voice applications and it would be desirable to study further on its influence upon dimensioning of network and method for forecasting two types of traffic in the international public switched telephone network.

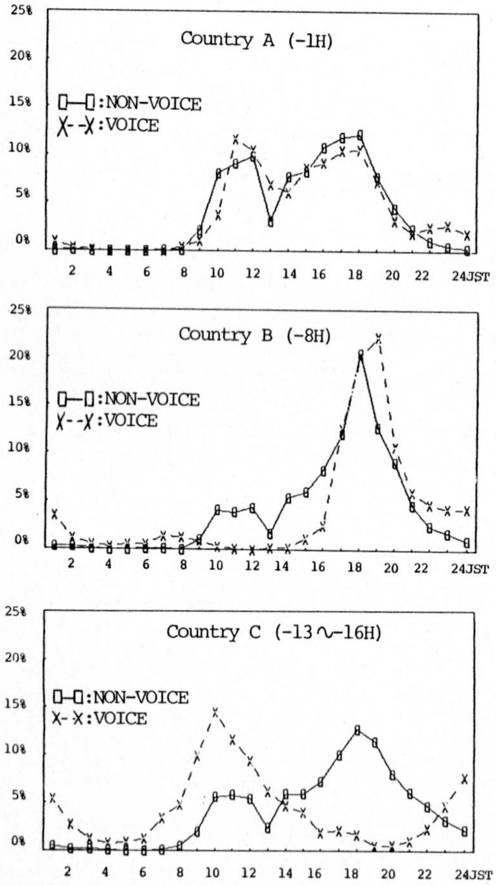

Figure 3 24-hour profile of the number of non-voice calls and voice calls (expressed by concentration ratio)

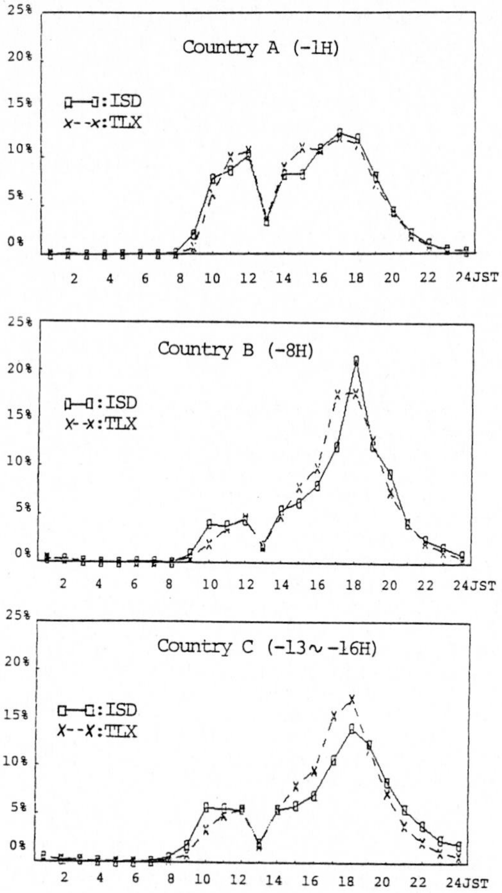

Figure 4 24-hour profile of the number of non-voice ISD calls and telex calls (expressed by concentration ratio)

5. ACKNOWLEDGMENT

The authors would like to express their gratitude to Dr. Sei-ichi Inoue, Director Operations and Network Administration Department, for supporting this work.

REFERENCE

(1) CCITT E series of Recommendation, Supplement No.8, Red Book " Non-voice traffic on the telephone network"

(2) T. Ohta: " Network efficiency and network planning considering telecommunication traffic influenced by the time difference ", 7th ITC, Stockholm, No.425 (1973.6)

(3) N. Hattori and K. Yamada: " Traffic characteristics of the international telex calls ", Proc, 7th ITC, Stockholm, No.443 (1973.6)

TELETRAFFIC ISSUES in an Advanced Information Society
ITC-11
Minoru Akiyama (Editor)
Elsevier Science Publishers B.V. (North-Holland)

FIELD DATA ANALYSIS FOR TRAFFIC ENGINEERING

Eiji TAKEMORI, Yukihiro USUI and Jun MATSUDA

Musashino Electrical Communication Laboratory, NTT
Tokyo, Japan

ABSTRACT

Traffic models, traffic load estimation methods and subscriber behavior are investigated based on field traffic data collected by detailed traffic measuring equipment developed at NTT.

First, inter-arrival time and holding time distributions are identified as exponential and hyper-exponential distributions, respectively. Next, the estimation accuracy of the traffic load for each destination is analyzed through investigation of day-to-day fluctuation characteristics of the ratio and the average holding time of calls for each destination. Finally, subscriber behavior is also analyzed through investigation of the ringing tone trunk and the busy tone trunk holding time distributions as well as the characteristics of repeated calls.

The analysis results in this paper are valuable fundamental materials for construction of communication networks.

1. INTRODUCTION

Recently, the traffic structures of telephone networks have been changing because of the various types of calls. Furthermore, NTT has experimentally offered various new services of INS (Information Network System) such as digital communications and broad-band communications since September of 1984, and this will undoubtedly accelerate the change in traffic structures. Under these conditions, detailed traffic measurements and analyses are indispensable for planning, designing and management of networks. At NTT, D-type Traffic Measuring Equipment (DTME; Appendix 1) has been developed in order to collect detailed traffic data from an electronic exchange.

In this paper, we investigate traffic models, traffic load estimation methods and subscriber behavior based on field traffic data collected by DTME. With regard to traffic models, V. B. Iversen analyzed inter-arrival time and holding time distributions [1], [2]. However, he did not describe a method of identifying distributions. We identify the distributions by means of minimum AIC (An Information Criterion) method [3], and refer to their effects on network dimensioning. For traffic load estimation, the fluctuation characteristics of the ratio and the average holding time of calls for each destination are investigated. With regard to subscriber behavior, the holding times of the ringing tone trunk and the busy tone trunk are analyzed as well as the characteristics of repeated calls.

2. VERIFICATION OF TRAFFIC MODELS

Network dimensioning is carried out by using queueing models with the assumed inter-arrival time and holding time distributions of calls. The more closely the assumed distributions represent the actual traffic characteristics of networks, the more properly the network dimensioning can be carried out. Analyzing field traffic data by means of minimum AIC methods (Appendix 2), we identified inter-arrival time and holding time distributions of arriving calls to a trunk group, and verified usually assumed distributions. From the viewpoint of ease in analysis, we considered exponential, gamma and hyper-exponential distributions as candidate distributions.

Furthermore, we analyzed the short time fluctuation characteristics of calls arriving at an exchange. Short time fluctuation has an important effect on the design of common equipment in an exchange.

2.1 Inter-arrival Time Distribution

We analyzed inter-arrival time distributions of calls for each destination for 24 cases. Part of the results are shown in Table 2.1. Fourteen cases were identified as an exponential distribution, 2 cases as a gamma distribution and 2 cases as a hyper-exponential distribution. The remaining 6 cases were not identified as any particular distribution, i.e., it was equally possible for them to be exponential or hyper-exponential distribution. In all 24 cases, the coefficient of variation ranges from 0.89 to 1.18. This means that the inter-arrival time distributions are close to an exponential distribution. Further, we tested the Poisson characteristics for the number of arriving calls for the cases identified as an exponential distribution (Appendix 3), and the hypothesis that it is a Poisson distribution was not rejected in most cases. Figure 2.1 shows a typical example of inter-arrival time distribution.

2.2 Holding Time Distribution

We analyzed holding time distributions in the busiest hour for several trunk groups. The result was that all cases were identified as a hyper-exponential distribution (Table 2.2). Figure 2.2 shows a typical example of holding time distribution.

From the above results, it has turned out that the actual traffic characteristics are closer to an M/H_2 model than the M/M model which is usually used. Thus, in order to dimension the alternating routes exactly, a theory on overflow from $M/H_2/S/S$ such as the well-known theory on overflow from M/M/S/S is needed.

2.3 Short Time Fluctuation of the Number of Arriving Calls

It is necessary for the design of common equipment in an electronic exchange to know the short time fluctuation of the number of arriving calls. Though Poisson arrival is usually assumed for the call arrival process, it is possible that the actual fluctuation may be different. We measured the number of calls arriving at an electronic exchange in the busiest hour every 20 seconds, and evaluated the fluctuation.

First, we compared the coefficient of variation for the field data with that of Poisson distribution. The coefficient of variation in each measurement is shown in Fig. 2.3. From the figure, we found that it was about 10-20 percent larger than that of Poisson fluctuation at office A, and rather smaller at office B.

Next, we considered an event which occurs with a probability of 1/180 (i.e., one sample per hour) in order to check the spread of the distribution. Assuming Poisson fluctuation, the

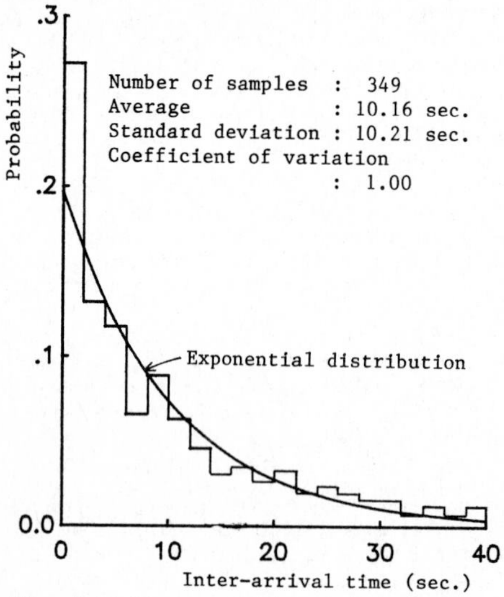

Fig. 2.1 Inter-arrival time distribution.

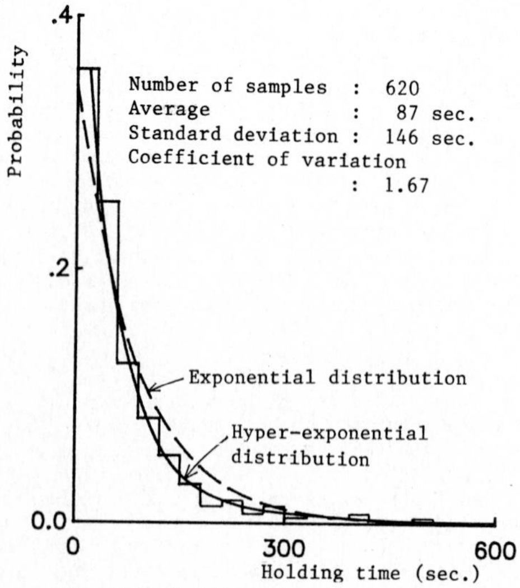

Fig. 2.2 Holding time distribution.

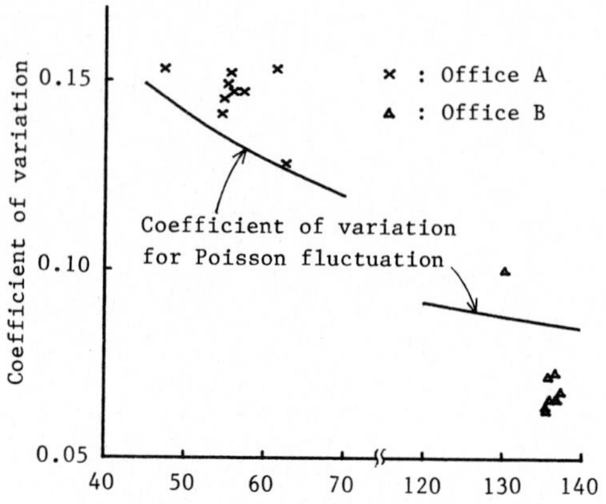

Fig. 2.3 Short time fluctuation of the number of arriving calls.

Table 2.1 Inter-arrival time distribution.

	No. of samples	Mean (sec.)	CV	A I C		
				Exp.	Gamma	Hyper-exp.
1	191	18.63	0.89	* 0	3.73	-
2	158	20.05	1.15	0.91	4.27	0
3	242	14.70	1.04	* 0	14.27	1.61
4	391	9.16	0.91	2.04	* 0	-
5	376	8.70	0.99	* 0	18.82	-
6	464	7.74	1.08	0.46	30.77	0
7	349	10.16	1.00	* 0	1.90	1.99
8	420	8.52	1.02	* 0	21.54	1.79
9	876	4.10	0.98	* 0	57.55	-
10	873	4.10	1.04	0	71.42	0.91
11	1654	2.17	1.07	5.32	295.23	* 0
12	1900	1.89	1.02	0	271.38	0.87

CV = Coefficient of variation
Exp. = Exponential
* = Identified distribution

Table 2.2 Holding time distribution.

	No. of samples	Mean (sec.)	CV	A I C		
				Exp.	Gamma	Hyper-exp.
1	957	167	1.98	412.32	247.15	* 0
2	104	154	1.29	2.32	3.21	* 0
3	178	148	1.70	31.51	21.79	* 0
4	482	158	2.02	235.36	146.03	* 0
5	1165	135	1.64	224.31	164.80	* 0
6	435	116	1.34	36.91	17.54	* 0
7	883	114	1.65	157.66	103.12	* 0
8	620	87	1.67	107.92	93.55	* 0

CV = Coefficient of variation
Exp. = Exponential
* = Identified distribution

probability that a value greater than $u=m+2.54\sqrt{m}$ (m : mean) will occur must be 1/180. We counted the number of samples with values greater than u in an hour, and found that, on the average, the number was about two at office A and one at office B. Thus, fluctuations larger than that of Poisson arrival should be taken into consideration at office A.

Moreover, we investigated the maximum value U of samples in an hour. We indicate its deviation by k defined as follows:

$$k = (U - m)/\sqrt{m}. \quad (2.1)$$

Figure 2.4 shows k obtained from the actual data as well as the expectation of k when assuming Poisson fluctuation. For the actual data, k was about 3.1 at office A and 2.6 at office B, on the average, while, for Poisson fluctuation, k is about 2.6. Thus, fluctuation about 1.2 times as large as Poisson fluctuation should be taken into consideration at office A while Poisson fluctuation will do at office B even in considering the maximum value.

From the above results, it is clear that short time fluctuations in the busiest hour must be investigated carefully at each office.

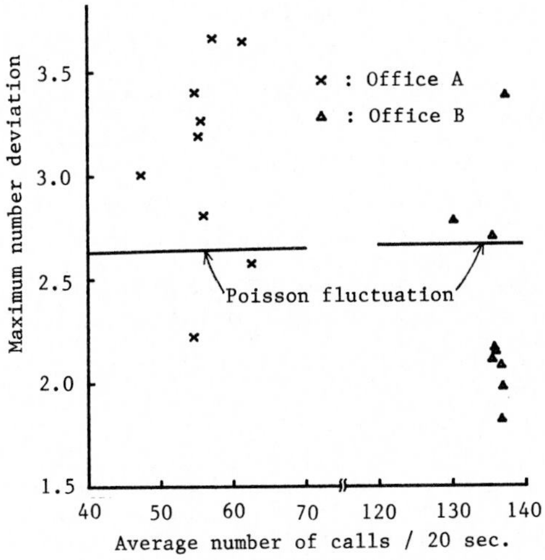

Fig. 2.4 Deviation of the maximum number of arriving calls.

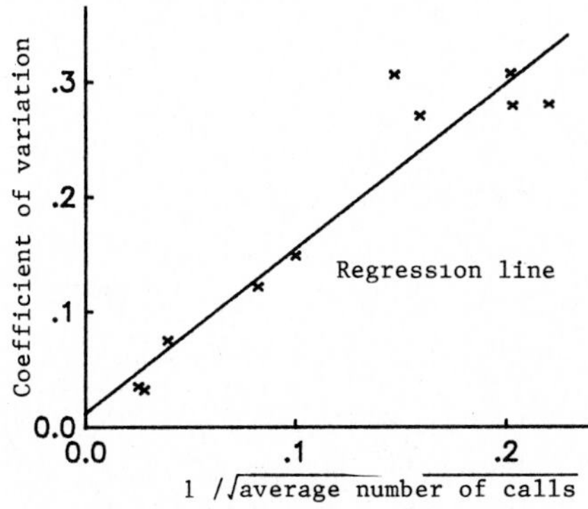

Fig. 3.1 Ratio fluctuation of calls for each destination.

3. ESTIMATION METHOD FOR TRAFFIC LOAD TO EACH DESTINATION

At NTT, network dimensioning is carried out based on the traffic load for each destination. This traffic load is estimated from the total exchange traffic load, the average holding time and the number of calls for each destination. That is, Y_i, the traffic load for destination i is estimated as follows:

$$Y_i = T\, r_i\, h_i / \textstyle\sum_j r_j\, h_j, \quad (3.1)$$

where T : the total traffic load,
r_i : the ratio of the number of calls for destination i to the total number of calls, and
h_i : the average holding time of calls for destination i.

However, it is not practical to measure the average holding time and the number of calls for each destination every day, when considering the load on the CC (Central Control equipment) of an exchange and the traffic data processing center. From this point of view, sampling measurements are adopted practically. Thus the day-to-day fluctuation influences the accuracy of estimation by Eq. (3.1). Fluctuation characteristics and their effects are investigated below.

3.1 Fluctuation Characteristics of the Ratio of Calls to Each Destination

We analyzed day-to-day fluctuation characteristics of the ratio of calls for each destination (the number of calls for a certain destination / the total number of calls). It is thought that the more calls a destination has, the smaller the day-to-day fluctuation will be. We investigated the relationship between the average number of calls and the coefficient of variation in the ratio of calls. The result is shown in Fig. 3.1. This figure shows that CV_i (the coefficient of variation in the ratio of calls) and n_i (the average number of calls for destination i) have the following relationship:

$$CV_i = k/\sqrt{n_i} + \ell \quad (k, \ell : \text{const.}). \quad (3.2)$$

Therefore, letting h_i be the average holding time for destination i, the error of traffic

Table 3.1 Number of circuits.

Correct No. of circuits	Offered traffic load	Est. of CV	95 % interval (Up. / Low.)	No. of circuits
12	5.87	0.13	7.37	15
			4.37	10
30	20.3	0.08	23.5	34
			17.2	27
48	36.1	0.06	40.3	53
			31.9	44
60	46.9	0.06	52.4	66
			41.4	54
90	74.7	0.05	82.0	98
			67.4	83
120	103	0.04	111	129
			94.9	112

Est. of CV = Estimation of coefficient of variation
Up. = Upper bound
Low. = Lower bound
Loss probability : 0.01

load for destination i is evaluated by

$$E_i = (k/\sqrt{n_i}+\ell)n_i h_i = (k\sqrt{n_i}+\ell\, n_i)h_i . \quad (3.3)$$

In the actual data, ℓ is about 1/100 to 1/30 of k. Therefore, it turns out that E_i is nearly proportional to $1/\sqrt{n_i}$.

Table 3.1 shows the number of circuits necessary to satisfy a loss probability of 0.01 taking this fluctuation in the ratio of calls into account. Assuming an average holding time of 150 seconds, we derived the average number of calls, and estimated the coefficient of variation by using Fig. 3.1. Next, assuming that the ratio of calls for a destination obeys a normal distribution, we derived the upper and lower bounds of a 95 percent confidence interval for the traffic load estimated by using the ratio of calls on one day of a year. For these values of traffic, we dimensioned the number of circuits necessary to satisfy a loss probability of 0.01. From Table 3.1, it turns out that the accuracy of the ratio has a large effect on the number of circuits in routes.

3.2 Fluctuation Characteristics of the Average Holding Time

The average holding time of sample j in a day i, x_{ij}, is written as follows:

$$x_{ij} = X + d_i + s_{ij} = x_i + s_{ij}, \quad (3.4)$$

where X : the average of each day's average holding time,

d_i : the day-to-day fluctuation of each day's average holding time, $\sum_i d_i = 0$, and

s_{ij} : the fluctuation in each day's average holding time due to sampling, $\sum_j s_{ij} = 0$.

The coefficient of variation of x_{ij} then becomes

$$D = \sqrt{V[x_{ij}]}/E[x_{ij}] = \sqrt{E[d_i^2]+E[s_{ij}^2]}/X. \quad (3.5)$$

From field data, D^2 can be regressed as follows (Fig. 3.2):

$$D^2 = a / n + b, \quad (3.6)$$

where n : the number of calls in the period of measurement, and

a, b : coefficients of linear regression.

$$\sqrt{E_j[s_{ij}^2]} / x_i = CV_i/\sqrt{n}, \quad (3.7)$$

where CV_i : the coefficient of variation of holding time in a day i.

Therefore, from Eqs (3.4)-(3.7),

$$D_2^2 = E[d_i^2]/X^2 = a/n+b-(CV)^2/n, \quad (3.8)$$

where we approximated that x_i=X and CV_i=CV for all i. Then, D_2, the coefficient of variation by d_i, is obtained as follows:

$$D_2 = \sqrt{V[x_i]} / E[x_i] = \sqrt{E[d_i^2]} / X$$
$$= \sqrt{a / n + b - (CV)^2 / n} . \quad (3.9)$$

Figure 3.3 shows D^2 and D_2^2 versus n. As can be seen from the figure, for small n, the fluctuation due to sampling, $D^2 - D_2^2$, is large compared with the day-to-day fluctuation D_2^2, and the total fluctuation becomes large, while the total fluctuation is small for large n. Therefore, we must sample many calls in order to correctly ascertain the average holding time.

4. ANALYSIS OF SUBSCRIBER BEHAVIOR

A call becomes incomplete for various reasons. It may meet congestion in the network or in the called subscriber line. The called party may be absent. In these cases, how does the calling subscriber behave? How long does he listen to the busy tone or the ringing tone? How long does he wait before repeating the call? How many times does he repeat the call? Such types of subscriber behavior cause ineffective processing of an exchange or ineffective holding of circuits. We investigated these kinds of behavior by using field traffic data.

4.1 Ringing Tone Trunk Holding Time

Ringing tone trunk holding time distribution of ineffective calls is shown in Fig. 4.1. The average holding time is 24.0 seconds, with a maximum holding time of about 2 minutes. The distribution is nearly uniform; about 25 percent of the subscribers hold the trunk longer than 8 seconds.

4.2 Busy Tone Trunk Holding Time

Busy tone trunk holding time distribution is shown in Fig. 4.2. About 80 percent of the subscribers hold the trunk less than 8 seconds. The average holding time is 17.2 seconds, and the maximum holding time is about 5 minutes.

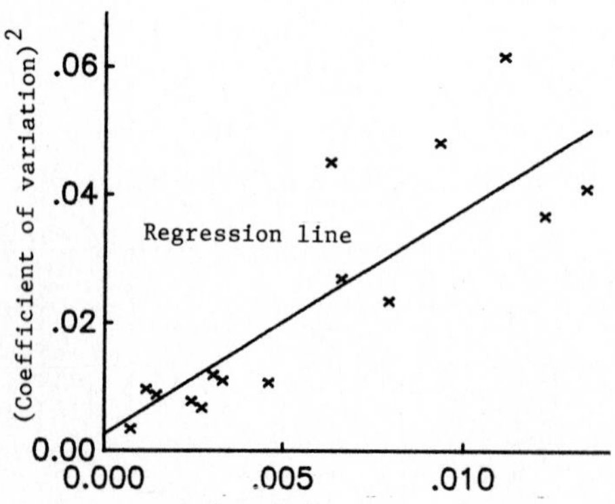

Fig. 3.2 Fluctuation of average holding time (1).

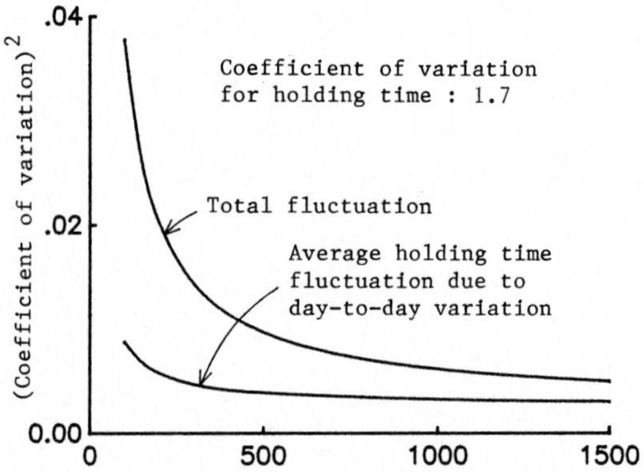

Fig. 3.3 Fluctuation of average holding time (2).

These results were obtained by analyzing incoming calls. About 25 percent of incoming calls become incomplete. About 80 percent of the incomplete calls are a result of the called subscriber failing to answer and about 20 percent of them are a result of the called subscriber line being busy.

4.3 Subscriber Retrial Behavior

When an initial call attempt is not completed, a subscriber may either abandon or repeat his call. This section describes retrial probabilities, retrial time interval distributions, and completion probabilities.

The overall completion probability, the completion probability of the first attempts, the average retrial probability and the completion probability of reattempts are shown in Table 4.1.

The completion probability of the initial attempt is 76.0 percent, a figure higher than the overall completion probability. Of the subscribers whose call attempts are not completed, 52.2 percent attempt retrials, but only 27.1 percent of reattempts are completed. The reattempt completion probability of 27.1 percent is remarkably low compared with the completion probability for the initial attempt. The completion probability of the initial attempt and the overall completion probability are almost equal to Liu's results, but the average retrial probability in NTT is smaller than his result [4].

Whether reattempts are completed or not depends on the retrial time interval. In Fig. 4.3, the retrial time interval distribution is

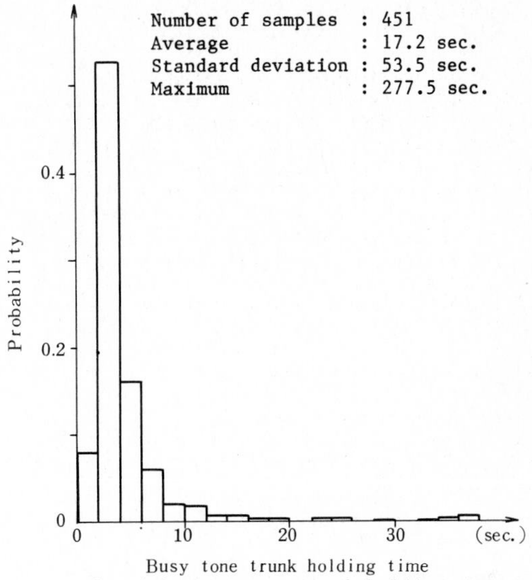

Fig. 4.2 Busy tone trunk holding time distribution.

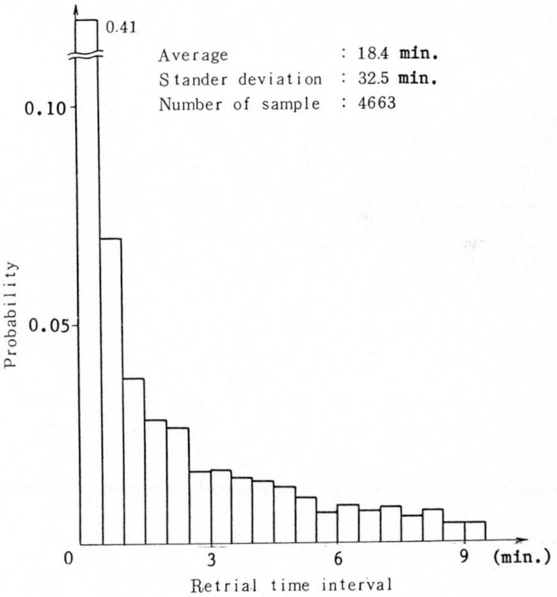

Fig. 4.3 Retrial interval distribution.

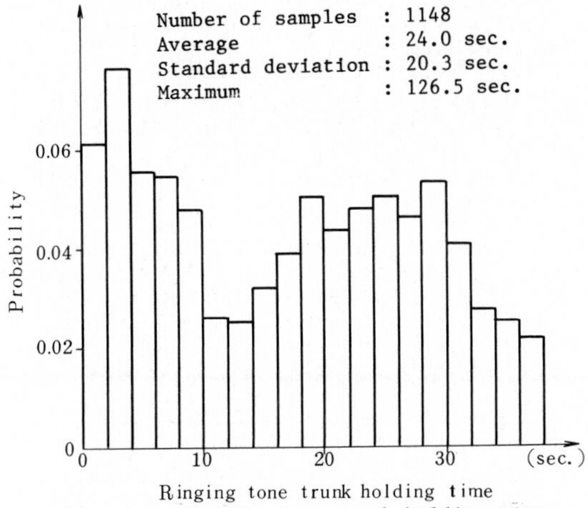

Fig. 4.1 Ringing tone trunk holding time distribution.

Table 4.1 Competion and retrial probabilies.

(a) Overall completion probability	(b) Initial completion probability	(c) Average retrial probability	(d) Completion probability of reattempts	(e) Ineffective call ratio	(f) Ineffective load ratio
67.4%	76.0%	52.2%	27.1%	12.4% *	4.7%

* (e) = (1 - (a)) x (c) x (1 - (d))

shown. It is found that about 40 percent of the reattempts are made within 30 seconds, and it can be thought that many of them inevitably fail again.

The reattempt completion probability depends on the reasons the call was not completed and on the the length of the retrial time interval. The relationships are shown in Fig. 4.4. Reattempts following SB (Subscriber line busy) have a higher completion probability than those following SNA (Subscriber no answer). The reattempt completion probability following SNA is very low regardless the length of the retrial interval.

The completion probability and the retrial probability at each attempt level is shown in Fig. 4.5. It can be seen that the completion probability decreases as the number of retrials increases.

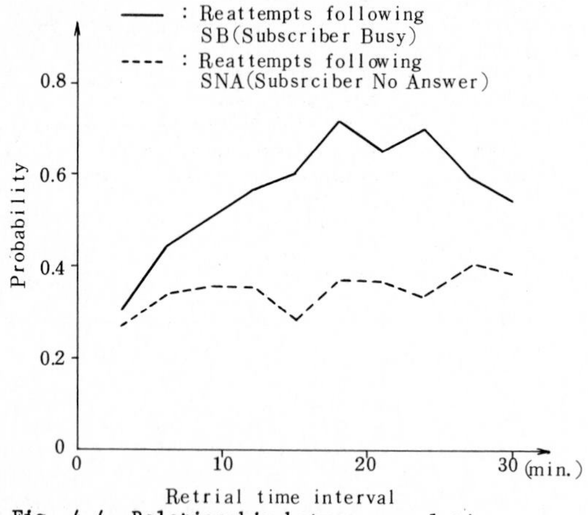

Fig. 4.4 Relationship between completion probability and retrial time interval.

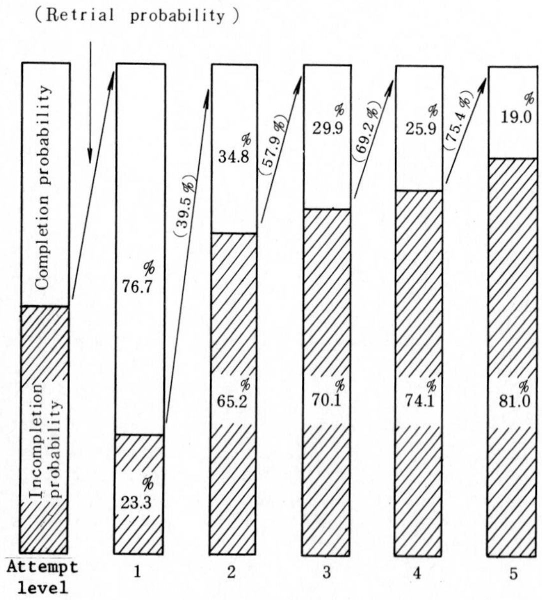

Fig. 4.5 Completion and retrial probability at each attempt level.

Unsuccessful attempts and holding time due to reattempts amount to 12 percent and 4.7 percent, respectively (Table 4.1). This is a significant load on the network resources. To minimize this load, several services (e.g., transfer service) have been introduced.

5. CONCLUSIONS

We have shown the traffic characteristics obtained by analyzing field traffic data. The traffic model was found to be closer to M/H_2 than to M/M. For traffic load estimation, fluctuations were evaluated and their effects on the estimation accuracy were discussed. Subscriber behavior related to ineffective use of network resources was also investigated.

NTT has experimentally offered INS services since September of 1984. Through the use of DTME, INS traffic data is now being collected and analyzed. We plan to continue investigating the traffic characteristics of INS in detail, as these are still relatively unknown.

6. ACKNOWLEDGEMENTS

The authors would like to thank Yohnosuke Harada, engineer of the Switching Systems Section, Musashino ECL, NTT, for his great help in collecting field traffic data. The authors also express their gratitude to Kunio Kodaira, Chief of the Teletraffic Section, and all the members of the same Section for their thoughtful discussions.

7. REFERENCES

[1] V. B. Iversen, "Analysis of Real Teletraffic Processes Based on Computerized Measurements," Ericsson Tech., vol. 29, no. 1, pp.3-64, 1973.

[2] V. B. Iversen, "Analysis of Traffic Processes Based on Data Obtained by the Scanning Method," Teleteknik, vol. 17, no. 2, pp.44-57, 1973.

[3] H. Akaike, "Canonical Correlation Analysis of Time Series and the Use of An Information Criterion," in: R. K. Mehra and D. G. Lainiotis, eds., System Identification, pp.27-96, Academic Press, New York, 1976.

[4] K. S. Liu, "Direct Distance Dialing: Call Completion and Customer Retrial Behavior," Bell System Technical Journal, vol. 59, no. 3, pp.295-311, 1980.

APPENDIX 1. DTME (D-type Traffic Measuring Equipment)

DTME is connected to the memory bus of an electronic exchange (Fig. A1.1). It monitors the execution of the call processing program by watching the address bus, and collects data being written in the specific main memory area of an exchange. Thus, traffic data can be collected continuously even during busy periods.

Traffic data consists of detailed data (Fig. A1.2) and macro data such as the traffic load and the number of calls. DTME sends traffic data to the data processing center after editing the collected data for every call. This reduces the amount of data transferred to the data processing center.

APPENDIX 2. Minimum AIC method

For the obtained data, the parameters of a model can be determined by the maximum likelihood method. The minimum AIC (An Information Criterion) method is used for selecting the best model among the models whose parameters have been determined in this way.

We first define the following notations:

θ : parameter of a model,
x : obtained data,
$L(\theta|x)$: likelihood function,
P : dimension of θ,
$\hat{\theta}$: maximum likelihood estimator of θ.

Then AIC is defined as the following:

$$AIC = -2 \log L(\hat{\theta}|x) + 2P, \qquad (A2.1)$$

where $L(\hat{\theta}|x)$ is called the maximum likelihood function, which represents the suitability of a model. The larger $L(\hat{\theta}|x)$ is, the higher the suitability of a model is. The maximum likelihood function tends to have a larger value when the number of parameters increases. Thus, a good model is obtained when the number of parameters is increased.

On the other hand, Akaike [3] has advocated that a good model is the one in which the suitability is high and the number of parameters is small. AIC is a measure based on this idea.

The first term of Eq.(A2.1) is the measure of suitability of the model and the second term is the penalty for the increase in parameters. Thus, a model having small AIC is the one in which the suitability is good and the number of parameters is small. The minimum AIC method is the method which selects the model having a minimum AIC.

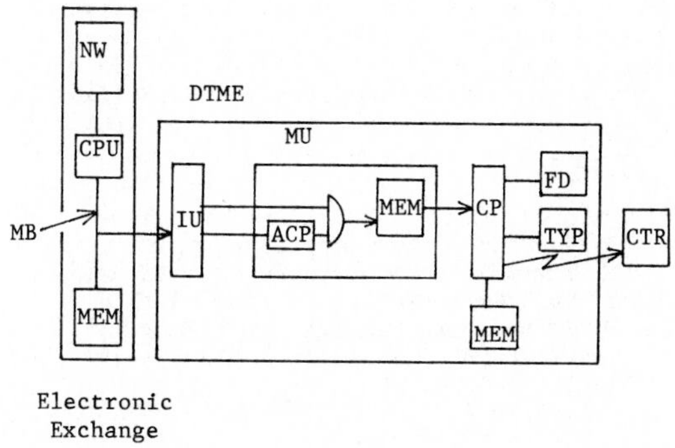

NW : Network	CPU: Central Processing Unit
MB : Memory Bus	MEM: Memory
IU : Interface Unit	ACP: Address Comparator
CP : Control Processor	FD : Floppy Disk
TYP: Typewriter	CTR: Center
MU : Measurement Unit	

Fig. A1.1 D-type traffic measurement equipment.

Originating Subscriber ID
Present State Number
Next State Number
Time
Trunk Number
Subscriber Class

Fig. A1.2 Detailed traffic data.

APPENDIX 3. Chi Square Test for a Poisson Distribution

A chi square test is used to check whether the obtained samples contradict a hypothesis. If $X_1, X_2, \ldots, X_n$, are sampled from a normal distribution $N(m,s^2)$, C^2 defined by (A3.1),

$$C^2 = \sum_{i=1}^{n} (X_i - \bar{X}) / s^2 , \qquad (A3.1)$$

where $\bar{X}$ is a sample mean,

obeys a chi square distribution with a degree of freedom n-1.

On the other hand, a Poisson distribution has the following characteristics:

i) A Poisson distribution can be approximated by a normal distribution when the mean value is greater than 4.

ii) Poisson distribution has the same variance as the mean value.

Consequently, the value,

$$C^2 = \sum_{i=1}^{n} (X_i - \bar{X}) / \bar{X} , \qquad (A3.2)$$

is used to judge whether the hypothesis that mean = variance is rejected at a significance level.

TELETRAFFIC ISSUES in an Advanced Information Society
ITC-11
Minoru Akiyama (Editor)
Elsevier Science Publishers B.V. (North-Holland)

ANALYSIS OF QUEUEING SYSTEMS VIA MULTI-DIMENSIONAL ELEMENTARY RETURN PROCESS

Haruhisa TAKAHASHI AND Haruo AKIMARU

Toyohashi University of Technology, Toyohashi 440, Japan

ABSTRACT

This paper proposes a refinement method for the multi-dimensional diffusion approximation with instantaneous reflection on the boundaries. To this end we define and discuss a multi-dimensional elementary return (diffusion) process which is a natural extension of the one-dimensional elementary return process. The forward equation for the steady-state density functions is derived. Applying this process, an approach for modeling some large queueing systems is shown. Then generating functions for two-station queueing networks and preemptive-resume priority queues are derived by solving a functional equation in two variables. Numerical examples are shown for the preemptive-resume queueing systems. It is also shown that an empty queue probability for these models can be exactly given as well as by one-dimensional case.

1. INTRODUCTION

We consider an n-dimensional diffusion process Y_t that arises in conjunction with large queueing systems such as networks of queues, preemptive-resume queueing systems, etc.. Its state space consists of the interior and boundaries of the parallelotope of R^n (i.e., $0 \le x_i \le M_i$, $i=1,2,\ldots n$). On the interior of this state space, Y_t behaves like an ordinary n-dimensional diffusion process (Brownian motion with drift). Whenever Y_t reaches one of the (n-1)-dimensional hyperplanes (for example, $x_i=0$) it remains there and behaves as an ordinary (n-1)-dimensional diffusion process until an exponentially distributed finite time lapses or until one of the (n-2)-dimensional hyperplanes is reached. If the exponentially distributed finite sojourn time lapses before the process hitting the (n-2)-dimensional hyperplane, Y_t jumps instantaneously to a point on the interior of the state space with $x_i=1$. Otherwise an (n-2)-dimensional hyperplane is reached and the process stays there for an exponentially distributed finite time to act as (n-2)-dimensional Brownian motion. In both cases the process then starts from scratch. The behavior of Y_t on the lower dimensional hyperplane is defined in the same way. The process thus defined is a natural extension of one-dimensional elementary return process [4].

Several works have been devoted to the diffusion approximation for large queueing systems (mainly, for open queueing networks). Kobayashi [10] wrote out the forward equation for open queueing networks with the rough boundary condition that the approximating process must be restricted to the nonnegative orthant. Harrison and Reiman [8] showed the precise boundary conditions for a tandem queue which the reflected Brownian motion should obey. It was also shown that their reflected Brownian motion satisfys the heavy traffic limit theorem [1].

Although the heavy traffic limit theorem was proved for the reflected Brownian motion, it has the disadvantage of failing to approximate the behavior of the light traffic queues. In fact it is known that a serious deviation from exact values arises in a middle or a light traffic condition[10]. With this motivation refinement techniques of the diffusion approximation of queues have been proposed by some researchers[6], [10]. Gelenbe [6] proposed a refinement technique by properly modeling the effect of empty queue probability applying one-dimensional elementary return process defined by Feller. In his approach, the behavior of the light traffic queues can be approximated in a smaller error than by using the reflected diffusion approximation. Our primary goal here is to propose a refinement technique of the multi-dimensional diffusion approximation by extending his work to a multi-dimensional case and to derive the stationary forward equations of the multi-dimensional elementary return process as a model of more general large queueing systems than queueing networks.

In Section 2 and 3 we define the multi-dimensional elementary return process and derive the corresponding stationary forward equations. Using the result derived in Section 3, we then derive in Section 4 the partial differential equations for queueing network with two stations. It is also shown that the empty queue probability for this model can be exactly modeled as well as one-dimensional case of Gelenbe [6]. In Section 5, we discuss a generating function of the two-station queueing networks by analyzing the Laplace transformation of the partial differential equations derived in Section 4. In Section 6 the modeling technique is applied to the preemptive-resume priority queues with feedback. In Section 7 calculation methods for mean queue size of preemptive-resume queueing systems are given and some numerical examples are shown to compare with the exact values. A general notation that the infinitesimal volume elements of R^n is denoted $dx = dx_1 dx_2 \ldots dx_n$ shall be used through the paper.

2. DEFINITION OF THE PROCESS

We shall define a multi-dimensional elementary return process Y_t or $Y(t)$, $t \ge 0$. General case shall be discussed where the state space S^n of the process is the parallelotope of R^n

$$0 \le x_i \le M_i, \quad L_i > 0 \ (i=1,2,\ldots n).$$

Note that M_i may be infinite. S^n consists of its interior and the lower dimensional cells (hyperplanes). The number of 0-cells (vertexes) of S^n is 2^n. Generally the number of (n-k)-cells of S^n is $2^k\binom{n}{k}$, k=1,2,...n and sum to 3^n. We number these cells through so that the (n-k)-cells are denoted by C_i, ($\Sigma_{j=1}^{k-1} 2^j\binom{n}{j} +1 \le i \le \Sigma_{j=1}^{k} 2^j\binom{n}{j}$) for k>0 and by C_0 for k=0. Note that C_0 represents the interior of S^n. Suppose C_j is an m-cell, (0<m≤n), then we write C_j >($\not>$) C_i if C_i is (not) an (m-1)-cell and (or not) contained in the boundary of C_j. Let π^i be the natural projection map from R^n to R^{n-k} which is the sub-space of R^n parallel to the (n-k)-cell C_i. The index of the coordinate axis on $\pi^j(R^n)$ which is orthogonal to $\pi^i(R^n)$ is denoted i*j if C_j>C_i.

On the interior of S^n, Y_t behaves as an n-dimensional Brownian motion which has the (constant coefficient) forward operator

$$L_0^* = \sum_{\phi\phi'=1}^{n} \frac{1}{2} v^0_{\phi\phi'} \frac{\partial^2}{\partial x_\phi \partial x_{\phi'}} - \sum_{\phi=1}^{n} m^0_\phi \frac{\partial}{\partial x_\phi}$$

where the covariance matrix ($v^0_{\phi\phi'}$) is assumed to be non-negative definite. Whenever Y_t reaches an (n-1)-cell C_i, (1≤ i ≤ 2 $\binom{n}{1}$) at time t_1, it stays there for a finite sojourn time T_i and behaves as an (n-1)-dimensional Brownian motion with the (constant coefficient) forward operator

$$L_i^* = \sum_{\phi\phi'\neq i*0} \frac{1}{2} v^i_{\phi\phi'} \frac{\partial^2}{\partial x_\phi \partial x_{\phi'}} - \sum_{\phi\neq i*0} m^i_\phi \frac{\partial}{\partial x_\phi}$$

where the covariance matrix ($v^i_{\phi\phi'}$) is assumed to be non-negative definite. If Y_t reaches on an (n-2)-cell C_j < C_i at t_1+T_i, it stays there for an finite sojourn time T_j and behaves as an (n-2)-dimensional Brownian motion which has the (constant coefficient) forward operator

$$L_j^* = \sum_{\phi\phi'\neq i*0,\ j*i} \frac{1}{2} v^j_{\phi\phi'} \frac{\partial^2}{\partial x_\phi \partial x_{\phi'}} - \sum_{\phi\neq i*0,\ j*i} m^j_\phi \frac{\partial}{\partial x_\phi}$$

with the non-negative definite covariance matrix ($v^j_{\phi\phi'}$). The behavior of Y_t and the forward operator on the lower dimensional cells are defined in the same way except for the 0-cells (vertexes). If one of the 0-cells is reached then Y_t stays there for an exponentially distributed finite sojourn time. The finite sojourn time T_k on C_k (0 ≤ k ≤ 3^n-1) is exponentially distributed if any boundary cell C_i of C_k (i.e., C_i < C_k) is not reached. As soon as the exponential sojourn time lapses before the process hitting the boundaries, a jump in the direction of x_{k*m}-axis occurs to the interior of C_m from C_k, (C_m > C_k) according to some probability density function over 0 < x_{k*m} < M_{k*m}. Let λ_{km} be the positive constant which represents the rate at which jumps occur to a point on the interior of C_m from C_k, (C_m > C_k). Then

$$\text{Prob}\{T_k > t \mid \text{any } C_i < C_k \text{ is not reached}\} = \exp(-t\, \Sigma_m \lambda_{km})$$

where the summation in the exponential function is taken over all m such that C_m > C_k. For our purpose, let the density function for the jump be the Dirac delta function $\delta[x_{k*m}-M(k,m)]$ in which M(k,m) is defined as follows: let

$$\text{sgn}(k,m) = \begin{cases} +1 & \text{if } x_{k*m} = 0 \text{ for } x\in C_k \\ -1 & \text{if } x_{k*m} = M_{k*m} \text{ for } x\in C_k \end{cases} \tag{1}$$

then

$$M(k,m) = \tfrac{1}{2}[1+\text{sgn}(k,m)] + \tfrac{1}{2}[1-\text{sgn}(k,m)](M_{k*m}-1).$$

After the jump, the process starts from scratch. The process Y_t thus defined is the Markov process because there are no point with memory. The notation (1) shall be used hereafter.

3. STATIONARY FORWARD EQUATION

Let the alternative representation of the process be

$$Y(t) = (k, X^k)_t$$

where k represents the index of the cell on which the process sojourns and $X^k = \pi^k(Y)$. We assume the process has the transition probability density defined by

$$p_t^{ij}(x^i,y^j)\,\Delta y^j$$
$$= \text{Prob}\{(k,X^k)_t=(j,y^j),\ y^j\in\Delta y^j \mid (k,X^k)_0=(i,x^i)\}$$

in which Δy^j represents the small volume element on C_j. We decide $p_t^{ij}(x^i,y^j)$ represents the transition probability if C_j is a 0-cell.

Let x^k denote $\pi^k x$. Let ψ_k, k=0,....,3^n-1 be the set of continuous functions f_k: $\pi^k(C_k)$-R^1 that are twice continuously differentiable in x^k except at z^k whose component x_i of R^n ,if contained ,is 1 or M_i-1 for some i. Let $f_k(x^k) \in \psi_k$ be the density function on C_k that satisfys

$$\sum_{k=0}^{3^n-1} \int_{\pi^k C_k} f_k(x^k)\, dx^k = 1 .$$

For the density function f_k, we define the transition operator by

$$(T_t^* f)_j(z^j) = \sum_k \int_{\pi^k C_k} f_k(x^k)\, p_t^{kj}(x^k,z^j)\, dx^k$$

where $f = (f_0,\ldots,f_{3^n-1})$ and $z^j \in \pi^j C_j$. Then the forward operator of the process is defined by [9]

$$A^* = \lim_{t\to 0^+} \frac{T_t^* - I}{t} \tag{2}$$

where I represents the unit operator on $\psi_0 \times \psi_1 \times\cdots\times \psi_{3^n-1}$. The stationary forward equation of the process can be represented in the form

$$(A^* f)_i = 0 , \quad i=0,1,\ldots,3^n-1 . \tag{3}$$

It will be convenient to define the differential operators

$$J^i_\phi[f_i(\cdot)] = -\frac{1}{2}\sum_{\phi'} v^i_{\phi\phi'} \frac{\partial f_i(x^i)}{\partial x_{\phi'}} + m^i_\phi f_i(x^i)$$

and

$$H^{ik}[f_i(\cdot)] = \lim_{x_{k*i}\to N(k,i)} J^i_{k*i}[f_i(\cdot)]$$

where N(k,i)= (1/2)[1 - sgn(k,i)] M_{k*i}, i=1,...,3^n-1, and ϕ , ϕ' represent indexes of the components of x^i and the summation is taken over all such indexes. The following theorem gives the concrete form of the equation (3). We shall maintain all of the notation established earlier.

Theorem. Assume that Y_t has a stationary density function $f_k(\pi^k x)$, k=0,1,...,3^n-1. Then it is the solution of the equations

$$-L_0^* f_0(x) = \sum_{i=1}^{2n} \lambda_{i0} f_i(x^i)\, \delta(x_{i*0}-M(i,0))$$

....

$$-L_k^* f_k(x^k) = \sum_{m \text{ s.t. } C_m > C_k} \{ -\text{sgn}(k,m) H^{mk}[f_m(x^m)] - \lambda_{km} f_k(x^k)\} + \sum_{j \text{ s.t. } C_k > C_j} \lambda_{jk} f_j(x^j)\, \delta(x_{j*k}-M(j,k))$$
$$(1 \le k \le 3^n-2^n-1)$$

....

$$0 = \sum_{m \text{ s.t. } C_m > C_k} -\text{sgn}(k,m) H^{mk}[f_m(x^m)] - \lambda_{km} f_k$$
$$(3^n-2^n \le k \le 3^n-1)$$

with boundary conditions

$$\lim_{x_\phi \to 0} f_k(x^k) = \lim_{x_\phi \to L} f_k(x^k) = 0$$

where x_ϕ is the component of x^k, $0 < k \leq 3^n-1$.

Remark. If all L_i are finite and there are no absorbing states, Y_t is positive recurrent and has a stationary density.

Proof. See [11].

4. TWO-STATION QUEUEING NETWORKS

4.1. Stationary forward equation

In this section we present the forward equation of the elementary return process which approximates two-station queueing networks. Let the mean rate and the variance of exogenous interarrivals of jobs to the queue Q_i be denoted by λ_i and a_i, i=1,2, respectively. Service times at each station are i.i.d. with mean $1/\mu_i$ and variance S_i. Having completed services at each station, jobs proceed to the other station for services with probability γ_i, with probability β_i jobs rejoin the same station, and with probability $1-\gamma_i-\beta_i$ jobs leave the system.

The process behaves as an ordinary two-dimensional diffusion process approximating the busy period queue sizes on the interior of R_+^2. Whenever it reaches x_i-axis, i=1,2, it remains there and behaves as an ordinary one-dimensional diffusion process until an exponentially distributed finite time with mean $1/\lambda_{3-i}+\gamma_i\mu_i$ lapses or until the origin is reached. The process is there approximating the queue size behavior of Q_i conditional on the other queue being empty. If the exponentially distributed finite time lapses before the process hitting the origin, it jumps instantaneously to a point with the same x_i-coordinate and with $x_{3-i}=1$, i=1,2. Otherwise the origin would be reached and the process stays there for an exponentially distributed finite time, after which jumps occur to a point (1,0) or (0,1) with the rate λ_1 or λ_2, respectively. In both cases the process then starts from scratch. The exponential sojourn times of the process on the x_i-axis, i=1,2, or at the origin approximates the periods of one empty queue or of both queues empty by supposing the arrival processes to be the Poissonian with the rate λ_1, λ_2 when either queue is empty.

The stationary forward equations of the two-dimensional elementary return process are given in section 2. These are the natural extension of one-dimensional elementary (or instantaneous) return process which is proposed in [4] and applied to approximate computer system models in [6]. The partial differential equations for the queue size equilibrium density read as follows:

$$\frac{1}{2}A_1\frac{\partial^2 f_0}{\partial x_1^2} + B\frac{\partial^2 f_0}{\partial x_1 \partial x_2} + \frac{1}{2}A_2\frac{\partial^2 f_0}{\partial x_2^2} - C_1\frac{\partial f_0}{\partial x_1} - C_2\frac{\partial f_0}{\partial x_2}$$

$$= -d_2 f_1(x_1)\delta(x_2-1) - d_1 f_2(x_2)\delta(x_1-1) \qquad (4)$$

$$-\frac{1}{2}D_i\frac{\partial^2 f_i}{\partial x_i^2} + E_i\frac{\partial f_i}{\partial x_i}$$

$$= -H_{3-i}(x_i) - d_{3-i}f_i(x_i) + \lambda_i f_3\delta(x_i-1), \quad (i=1,2) \qquad (5)$$

$$K_1 + K_2 + (\lambda_1+\lambda_2)f_3 = 0 \qquad (6)$$

$$f_0(x_1,0) = f_0(0,x_2) = 0$$

$$f_1(0) = f_2(0) = 0$$

where

f_0 : density function on the interior of R_+^2,

f_i, i=1,2: density function on the interior of R^+,

f_3 : probability mass at the origin,

$$H_{3-i}(x_i) = \lim_{x_{3-i}\to 0} -\frac{1}{2}A_{3-i}\frac{\partial f_0}{\partial x_{3-i}} - \frac{1}{2}B\frac{\partial f_0}{\partial x_i} + Cf_0$$

$$K_i = \lim_{x_i-0} -\frac{1}{2}D_i\frac{\partial f_i}{\partial x_i} + E_i f_i, \qquad (i=1,2)$$

$$A_i = \lambda_i^3 a_i + \mu_i^3 S_i(1-2\beta_i) + \mu_{3-i}\gamma_{3-i}(1-\gamma_{3-i} + \gamma_{3-i}\mu_{3-i}^2 S_{3-i}) + \mu_i\beta_i(1-\beta_i+\beta_i\mu_i^2 S_i)$$

$$B = -\Sigma_{i=1}^2 [\mu_i^3 S_i\gamma_i + \mu_i\beta_i\gamma_i(1-\mu_i^2 S_i)]$$

$$C_i = \lambda_i - \mu_i + \beta_i\mu_i + \gamma_{3-i}\mu_{3-i}$$

$$D_i = \lambda_i^3 a_i + \mu_i^3 S_i(1-2\beta_i) + \mu_i\beta_i(1-\beta_i+\beta_i\mu_i^2 S_i)$$

$$E_i = \lambda_i - \mu_i + \beta_i\mu_i$$

$$d_i = \lambda_i + \gamma_{3-i}\mu_{3-i}, \qquad (i=1,2).$$

Notice that f_0 is a two-dimensional probability density function on the interior of R_+^2 and f_i, i=1,2, is a one-dimensional probability density function on the interior of R^+.

The parameters of the forward equations are interpreted as the mean and (co-)variance of the infinitesimal increment of the process, and called drift and diffusion parameters, respectively. Let Q_i, i=1,2, be the number of customers in the station 1 at time t. Then the drift and diffusion parameters on the interior of R_+^2 are given by

$$\lim_{t\to 0^+}\frac{1}{\Delta t}E[Q_i(t+\Delta t)-Q_i(t) \mid Q_1>0, Q_2>0] = C_i$$

$$\lim_{t\to 0^+}\frac{1}{\Delta t}\mathrm{var}[Q_i(t+\Delta t)-Q_i(t) \mid Q_1>0, Q_2>0] = A_i, \qquad i=1,2,$$

and

$$\lim_{t\to 0^+}\frac{1}{\Delta t}\mathrm{cov}[Q_1(t+\Delta t)-Q_1(t), Q_2(t+\Delta t)-Q_2(t) \mid Q_1>0, Q_2>0] = B.$$

The parameters C_i, A_i, i=1,2, and B can be determined directly through the central-limit-theorem type argument[8]. Similarly we have

$$\lim_{t\to 0^+}\frac{1}{\Delta t}E[Q_i(t+\Delta t)-Q_i(t) \mid Q_j=0, i\neq j, Q_i\neq 0] = E_i$$

$$\lim_{t\to 0^+}\frac{1}{\Delta t}\mathrm{var}[Q_i(t+\Delta t)-Q_i(t) \mid Q_j=0, i\neq j, Q_i\neq 0] = D_i, \qquad i=1,2$$

The parameters E_i, D_i, i=1,2, can be determined applying the central-limit-theorem type argument under the condition that $Q_j=0$, $i\neq j$, i,j=1,2,. We assume hereafter that the infinitesimal covariance matrix defined by

$$\begin{pmatrix} A_1 & B \\ B & A_2 \end{pmatrix}$$

is positive definite.

4.2. Functional equation

In this section we show the stationary functional equation for the generating function, and discuss the traffic intensity (or the utilization factor). Let define the Laplace transformations

$$L_0(s_1,s_2) = \int_0^\infty\int_0^\infty f_0(x_1,x_2)\exp(-s_1x_1 - s_2x_2)\,dx_1dx_2$$

$$L_i(s_i) = \int_0^\infty f_i(x_i)\exp(-s_ix_i)\,dx_i\,, \quad (i=1,2)$$

and

$$F_i(s_{3-i}) = \int_0^\infty H_i(x_{3-i})\exp(-s_{3-i}x_{3-i})\,dx_{3-i}\,, \quad (i=1,2)$$

Taking the two-dimensional Laplace transformation of (4), we have

$$[\tfrac{1}{2}A_1s_1^2 + Bs_1s_2 + \tfrac{1}{2}A_2s_2^2 - C_1s_1 - C_2s_2]\,L_0(s_1,s_2) + F_1(s_2) + F_2(s_1)$$
$$= -d_1L_2(s_2)\exp(-s_1) - d_2\,L_1(s_1)\exp(-s_2). \quad (7)$$

Furthermore Laplace transformation of (5) and use of (6) yield

$$F_{3-i}(s_i) = [\tfrac{1}{2}D_is_i^2 - E_is_i - d_{3-i}]\,L_i(s_i) - \lambda_i f_3(1-\exp(-s_i)), \ (i=1,2) \quad (8)$$

As will be shown in section 5, L_0, L_1, L_2 are analytic at the origin. Thus we obtain the power series expansion for L_0, L_i, i=1,2

$L_0(s_1,s_2) = a_0 + a_1s_1 + a_2s_2 + a_3s_1s_2 + \ldots..$

$L_1(s) = b_0 + b_1s + \ldots..$,

and

$L_2(s) = h_0 + h_1s + \ldots..$.

Substitute (8) into (7) and substitute the power series expansion for L_0, L_1, L_2 into the equation thus derived to get the coefficients of the s_1^2, s_2^2, s_1s_2, s_1 and s_2. Then we obtain

$$\tfrac{1}{2}A_1a_0 - C_1a_1 + \tfrac{1}{2}D_1b_0 - E_1b_1 + \lambda_1f_3/2 = -\tfrac{1}{2}d_1h_0$$
$$\tfrac{1}{2}A_2a_0 - C_2a_2 + \tfrac{1}{2}D_2h_0 - E_2h_1 + \lambda_2f_3/2 = -\tfrac{1}{2}d_2b_0 \quad (9)$$
$$Ba_0 - C_1a_2 - C_2a_1 = d_2b_1 + d_1h_1$$

and

$$-C_1a_0 - \lambda_1f_3 = E_1b_0 + d_1h_0 \quad (10)$$
$$-C_2a_0 - \lambda_2f_3 = E_2h_0 + d_2b_0.$$

Notice that

$$a_0 = L_0(0,0) = \int_0^\infty\int_0^\infty f_0(x_1,x_2)\,dx_1\,dx_2$$

$$b_0 = (\text{or } h_0 =)\ L_i(0) = \int_0^\infty f_i(x_i)\,dx_i\,,\ i=1 \text{ (or 2)},$$

and thus

$$a_0 + b_0 + h_0 + f_3 = 1\,. \quad (11)$$

Notice that the mean $E[x_1]$ is given by

$$E[x_1] = \int_0^\infty x_1 \int_0^\infty f_0(x_1,x_2)\,dx_2 + f_1(x_1)\ dx_1$$
$$= -\frac{\partial L_0}{\partial s_1}(0,0) - \frac{\partial L_1}{\partial s_1}(0) = -a_1 - b_1\,, \quad (12)$$

and similarly $E[x_2]$ is given by

$$E[x_2] = -a_2 - h_1. \quad (13)$$

Let the stationary probability that the process sojourns on $\{(x_1, x_2) \mid x_i>0\}$, i=1,2, be

$$\rho_1 = 1 - h_0 - f_3\,,\quad i=1$$
$$\rho_2 = 1 - b_0 - f_3\,,\quad i=2\,.$$

Then from (10) and (11)

$$\rho_i = \frac{\lambda_i(1-\beta_{3-i}) + \lambda_{3-i}\gamma_{3-i}}{\{(1-\beta_1)(1-\beta_2) - \gamma_1\gamma_2\}\,\mu_i}\,,\quad i=1,2.$$

This coincides with the traffic intensity at each station derived from the use of traffic equation [8]. Thus we have obtained the empty queue probability exactly. Notice that Gelenbe [6] introduced the elementary returning boundary to model the empty queue probability (or the utilization factor) in one-dimensional case. Thus we can see that our process is a natural extension of one-dimensional case of Gelenbe in this respect too.

5. GENERATING FUNCTION

5.1. Solution for the special case of B=0.

In this section, we seek the generating function of the diffusion model of two station queueing networks. The main mathematical technique used in this section is so called Riemann-Hirbert boundary value problems. The technique developed can be applied to another diffusion model of queueing systems directly as will be done in the next section.

Rewriting (7) and (8), we have

$$-R(s_1,s_2)L_0(s_1,s_2)$$
$$= G_1(s_1,s_2)L_1(s_1) + G_2(s_1,s_2)L_2(s_2) - g(s_1,s_2)f_3 \quad (14)$$

where

$$R(s_1,s_2) = \tfrac{1}{2}A_1s_1^2 + Bs_1s_2 + \tfrac{1}{2}A_2s_2^2 - C_1s_1 - C_2s_2,$$

$$G_1(s_1,s_2) = d_2\exp(-s_2) + \tfrac{1}{2}D_1s_1^2 - E_1s_1 - d_2\,,$$

$$G_2(s_1,s_2) = d_1\exp(-s_1) + \tfrac{1}{2}D_2s_2^2 - E_2s_2 - d_1$$

$$g(s_1,s_2) = \lambda_1[1-\exp(-s_1)] + \lambda_2[1-\exp(-s_2)]\,.$$

Notice that from (14) to find $L_0(s_1,s_2)$, it is enough to find L_i, i=1,2. We assume through this subsection that B=0.

The following observation will enable us to determine the generating functions L_i, i=1,2. The right-hand side of (14) vanishes whenever $R(x,y) = 0$, provided that $L_0(x,y)$ is finite. Thus for a complex variable w define

$$R_1(x,w) = \tfrac{1}{2}A_1x^2 - C_1x + w = 0, \qquad \text{Re } x \geq 0,$$

$$R_2(y,w) = \tfrac{1}{2}A_2y^2 - C_2y - w = 0, \qquad \text{Re } y \geq 0,$$

so that

$$R(x,y) = R_1(x,w) + R_2(y,w) = 0\,. \quad (15)$$

In what follows, , let $\sqrt{z}$ represent the principal value of the complex function $z^{0.5}$ (for a complex number z s.t. $-\pi < \arg z \leq \pi$), It follows that:

i) $R_1(x,w)$ has exactly one zero

$$x = K_1(w) = [\,C_1 + \sqrt{C_1^2 - 2A_1w}\,]/A_1$$

with multiplicity one for Re w ≤ 0, in Re x $\geq$ 0,

ii) $R_2(y,w)$ has exactly one zero

$$y = K_2(w) = [C_2 + \sqrt{C_2^2 + 2A_2w}\,]/A_2,$$

with multiplicity one for Re w ≥ 0 in Re y ≥ 0. The functions $K_1(w)$ and $K_2(w)$ clearly satisfy (15). It can be seen that $K_1(w)$ is regular for Re w <0 and continuous for Re w $\leq$ 0, and that $K_2(w)$ is regular for Re w >0 and continuous for Re w $\geq$ 0. For Re w = 0 , $(x,y) = (K_1(w),K_2(w))$ is a zero of the kernel R(x,y) and hence it follows from (14):

$$G_1(K_1(w),K_2(w))L_1(K_1(w)) + G_2(K_1(w),K_2(w))\cdot$$
$$L_2(K_1(w),K_2(w)) = g(K_1(w),K_2(w))f_3,\ \text{Re } w = 0. \quad (16)$$

Our problem has almost been reduced to what is known as Riemann boundary value problem [1]: let Φ be a smooth contour, and let Φ^+ (Φ^-) be the left (right) domain of Φ w.r.t. the positive (or counter-clockwise) direction on Φ . Determine a function f(.) satisfying boundary conditions defined on the contour Φ of the form

$$f^+(t) = G(t)f^-(t) + e(t)\,,\quad G(t)\neq 0, (t\ \ \Phi)$$

with

$$f^+(t) = \lim_{\substack{z \to t \\ z \in \Phi^+}} f(z),$$

$$f^-(t) = \lim_{\substack{z \to t \\ z \in \Phi^-}} f(z),$$

where G(.) and e(.) are functions defined on Φ, satisfy the Hölder condition [1] on Φ, and do not vanish everywhere on Φ. Furthermore the following conditions are incurred;

i) f(z) is regular for $z \in \Phi^+$ (Φ^-) and is continuous for $z \in \Phi \cup \Phi^+$ ($\Phi \cup \Phi^-$),

ii) $f(z) \to c$ as $|z| \to +\infty$, where c is a constant.

Notice that in our problem c=0.

The form of solution of the Riemann boundary value problem depends on what is known as "the index" of the problem. This is denoted by and is defined by

$$\chi = \frac{1}{2\pi i}\int_{\Phi} d \ \mathrm{Log}\, G(t)$$

= { increment of the argument of G(t) when t traverses in the positive direction } /2π where $i = \sqrt{-1}$.

To reduce our problem to the Riemann boundary value problem, rewrite (16) as

$$L_1(K_1(w)) = G(w)L_2(K_2(w)) + e(w)f_3,$$

where

$$G(w) = -G_2(K_1(w),K_2(w))/G_1(K_1(w),K_2(w))$$

and

$$e(w) = g(K_1(w),K_2(w))/G_1(K_1(w),K_2(w)).$$

Furthermore set $\Phi = \{ w;\ \mathrm{Re}\ w = 0 \}$, and set $\Phi^+ = \{ w;\ \mathrm{Re}\ w < 0\}$, $\Phi^- = \{ w;\ \mathrm{Re}\ w > 0 \}$. It is easily seen that G(.) and g(.) have finite derivatives everywhere on Φ. Since

$$\lim_{w \to \infty, w \in \Phi} G(w) = (D_2/D_1)(A_1/A_2)$$

and

$$\lim_{w \to \infty, w \in \Phi} e(w) = 0,$$

G(w) and e(w) satisfy the Hölder condition on Φ and there exists a solution for the boundary value problem. For the analysis of the problem we have to investigate the index. From the definition the index of G(w) is given by the increment of the argument of $G_2(K_1(w),K_2(w))$ subtracted by the increment of the argument of $G_1(K_1(w),K_2(w))$ and devided by 2π when w traverses Φ from $-i\infty$ to $+i\infty$. Since the first term of G_1, i.e., the exponential term, vanishes when $w \to \pm i\infty$ and since the square root represents the principal value, we have

$$\arg[G_1] = \lim_{w \to +i\infty} \arg G_1(K_1(w),K_2(w)) - \lim_{w \to -i\infty} \arg G_1(K_1(w),K_2(w))$$

$$= \lim_{w \to +i\infty} \arg \tfrac{1}{2}D_1K_1^2(w) - \lim_{w \to -i\infty} \arg \tfrac{1}{2}D_1K_1^2(w)$$

$$= -\pi .$$

Similarly for $-G_2(K_1(w),K_2(w))$, the increment of the argument on Φ is $-\pi$. Thus $\chi = 0$.

If $\chi = 0$, then there exists the unique solution of the Riemann problem [1]. From [1] the solution of the problem is given by:

$$f(w) = f_3\ \Psi(w) \exp \Gamma(w)$$

$$= \begin{cases} L_1(K_1(w)) , & \mathrm{Re}\ w < 0 \\ L_2(K_2(w)) , & \mathrm{Re}\ w > 0 \end{cases} \tag{17}$$

where

$$\Psi(w) = \frac{1}{2\pi}\int_{-i\infty}^{+i\infty} e(it) \exp[-\Gamma^+(it)] \frac{dt}{it-w},$$

$$\Gamma(w) = \frac{1}{2\pi}\int_{-i\infty}^{+i\infty} \frac{\mathrm{Log}\, G(is)}{is - w} ds,$$

and where

$$\Gamma^+(it) = \tfrac{1}{2}\mathrm{Log}\{ G(it)\} + \Gamma(it).$$

Notice that f_3 can be determined from (10) and (11). Thus the generating function has been found;

$$\begin{aligned} L_1(s_1) &= f(-\tfrac{1}{2}A_1s_1^2 + C_1s_1) , \quad \mathrm{Re}\ s_1 > 0, \\ L_2(s_2) &= f(\tfrac{1}{2}A_2s_2^2 - C_2s_2) , \quad \mathrm{Re}\ s_2 > 0, \end{aligned} \tag{18}$$

with f(.) given in (17). All the equilibrium moments of (x_1,x_2) can now be computed through numerical computation.

5.2 Approximate solution for general case.

To seek the solution for general case, i.e. for $B \neq 0$, we have to determine the functions $K_1(.)$ and $K_2(.)$ on complex w-plane with the following property:

(i) $R(K_1(w),K_2(w)) = 0$, for $w \in \{ w;\ \mathrm{Re}\ w=0\}$

(ii) $K_1(w)$ is analytic for Re w < 0 and continuous for Re w ≤ 0, and $K_2(w)$ is analytic for Re w > 0 and continuous for Re w ≥ 0,

(iii) $K_1(w)$ is multiplicity one for Re w ≤ 0 in Re $K_1(w) \geq 0$, $K_2(w)$ is multiplicity one for Re w ≥ 0 in Re $K_2(w) \geq 0$, and

$$\lim_{w \to \pm i\infty} \mathrm{Re}\ K_i(w) = +\infty ,\ i=1,2.$$

First consider the asymptotic solution for $w \to \pm i\infty$. Let

$$\begin{aligned} s_1 &\sim a_1(-w)^{\gamma} \\ s_2 &\sim b_1 w^{\gamma} , \quad (w \to \pm i\infty), \end{aligned} \tag{19}$$

where $a_1 > 0$, $b_1 > 0$ and $0 < \gamma \leq 1/2$ are real constants. To determine these constants, substitute (19) into

$$R(x,y) \sim \tfrac{1}{2}A_1x^2 + Bxy + \tfrac{1}{2}A_2y^2 = 0.$$

Then we obtain for Re $w \to \pm i\infty$

$$\begin{aligned} &\tfrac{1}{2}A_1a_1^2 \cos(\mp\pi\gamma) + Ba_1b_1 + \tfrac{1}{2}A_2b_1^2 \cos(\pm\pi\gamma) = 0 \\ &\tfrac{1}{2}A_1a_1^2 \sin(\mp\pi\gamma) + \tfrac{1}{2}b_1^2 \sin(\pm\pi\gamma) = 0 \end{aligned} \tag{20}$$

Since the infinitesimal covariance matrix is positive definite, we obtain from (20)

$$a_1 = \sqrt{(A_2/A_1)}\ b_1$$

$$0 < \gamma = (1/\pi) \cos^{-1}[\sqrt{B^2/(A_1A_2)}\,] \leq 1/2 .$$

Now let determine the solution for (i), (ii), (iii) of the form

$$\begin{aligned} K_1(w) &= a_1(z_0 - w)^{\gamma} + \sum_{i=-\infty}^{0} a_i(z_0 - w)^i \\ K_2(w) &= b_1(z_1 + w)^{\gamma} + \sum_{i=-\infty}^{0} b_i(z_1 + w) , \end{aligned} \tag{21}$$

assuming that $K_1(0)=K_2(0)=0$. To show the existence of the solution (21), we employ the well known " theorem for the coincidence of limiting function" in analytic function theory: suppose a sequence $f_n(z)$, n=0,1,... of analytic function on a simply connected domain D be uniformly bounded on any bounded region in D. If $f_n(z)$ converges to f(z) at z_i of D with z_i converging to z of D as $i \to \infty$, then $f_n(z)$ weakly uniformly converges to f(z) analytic in D.

Now consider a sequence w_i, (i=1,2) which converges to zero with Re w_i =0. Define function sequences

$$K_1^n(w) = a_1(z_0-w)^{\gamma} + \Sigma_{i=-n}^{0} a_i(z_0-w)^i ,$$
$$K_2^n(w) = b_1(z_1+w)^{\gamma} + \Sigma_{i=-n}^{0} b_i(z_1+w)^i , \tag{22}$$

where z_0, z_1, a_1 are arbitrary positive real numbers (and thus b_1 is a positive real number), and where the complex numbers a_i, b_i, $-n \le i \le 0$ are determined by solving a system of simultaneous quadratic equations:

$$R(K_1^n(w_j), K_2^n(w_j)) = 0, \quad j=1,2,\ldots,2n,$$
$$K_1^n(0) = K_2^n(0) = 0 \quad . \tag{23}$$

Then $K_1^n(w)$ is analytic for Re $w \le 0$ and $K_2^n(w)$ is analytic for Re $w \ge 0$ for arbitrary n. Furthermore $K_i^n(w)$ for any n is bounded on any bounded region in Re $w \le 0$ and Re $w \ge 0$ for i=1,2, respectively. Since $K_1(0)=0$, and $K_2(0)=0$, the infinite series in (21) converges in Re $w \le 0$ and in Re $w \ge 0$ for i=1,2, respectively. Thus $K_i^n(w)$, i=1,2 is uniformly bounded on any bounded region in Re $w \le 0$ for i=1 and on that region in Re $w \ge 0$ for i=2. From the theorem for coincidence of limitting function, we can conclude that $K_i^n(w)$ converges to an analytic function $K_i(w)$ uniformly (in weak sense) in Re $w \le 0$ for i=1 and in Re $w \ge 0$ for i=2. On the other hand, since $R(K_1(w),K_2(w))$ is analytic on a strip includes the imaginary axis and vanishes at infinitely many point w_j, j=1,2,... on the imaginary axis, it is identically zero on the strip. Thus we have obtained a solution for (i), (ii), (iii).

For the sake of numerical evaluation, we may make an approximation of (21) by using the truncated series (22), e.g., for n=10. The equation (23) for the coefficients is rewritten in the form

$$Ax = g(x) \tag{24}$$

where the nonlinear term g(.) is a quadratic function and $x=(a_0,\ldots,a_n,b_0,\ldots,b_n)^t$. A solution for the nonlinear function (24) can be numerically computed by so-called Picard iteration: i.e.,

$$x^{k+1} = A^{-1}g(x^k), \quad k=0,1,2,\ldots, \tag{25}$$

assuming A is nonsingular. It is convenient to set $a_1=(2A_1)^{\gamma}/A_1$, $z_0= C_1^2/(2A_1)$, $z_1= C_2^2/(2A_2)$, and $x^0 = (a_0,0,\ldots,b_0,0\ldots,0)^t$ with $a_0 = C_1/A_1$ and with $b_0 = C_2/A_2$.

Repeating the same discussion as subsection 5.1, we obtain

$$L_1(s_1) = f(K_1^{-1}(s_1))$$
$$L_2(s_2) = f(K_2^{-1}(s_2)) \tag{26}$$

where f(.) is given by (17) and K_1, K_2 is replaced by (21).

6. APPROXIMATION OF GI_1 GI_2/G_1 $G_2/1$ PRIORITY QUEUES

Consider preemptive-resume queueing systems (which may have inter-queue feedback structure) with two types of customers (see Figure 1). Type 1 customers are given preemptive-resume priority over type 2 customers and both type of customers are served by a single server. External customers arrive at each queue according to a renewal process whose interarrival times have mean $1/\lambda_i$ and variance a_i, i=1 for high priority and i=2 for low priority. The service times for the class i are i.i.d. with mean $1/\mu_i$ and variance S_i. The feedback probability from priority i to priority j is given by P_{ij}.

Repeating the same discussion as in section 4, we obtain diffusion equations (4), (5), (6) or (14) as an approximate model where the coefficients are given as follows:

$$\begin{aligned}
A_1 &= \lambda_1^3 a_1 + \mu_1^3 S_1(1-2P_{11}) \\
&\quad + \mu_1 P_{11}(1-P_{11}+P_{11}\mu_1^2 S_1) \\
A_2 &= \lambda_2^3 a_2 + \mu_1 P_{12}(1-P_{12}+P_{12}\mu_1^2 S_1) \\
B &= -[\mu_1^3 S_1 P_{12} + \mu_1 P_{11} P_{12}(1-\mu_1^2 S_1)] \\
C_1 &= \lambda_1 - \mu_1 + P_{11}\mu_1 \\
C_2 &= \lambda_2 + P_{12}\mu_1 \\
D_1 &= \lambda_1^3 a_1 + \mu_1^3 S_1(1-2P_{11}) \\
&\quad + \mu_1 P_{11}(1-P_{11}+P_{11}\mu_1^2 S_1) \\
D_2 &= \lambda_2^3 a_2 + \mu_2^3 S_2(1-2P_{22}) \\
&\quad + \mu_2 P_{22}(1-P_{22}+P_{22}\mu_2^2 S_2) \\
E_1 &= \lambda_1 - \mu_1 + P_{11}\mu_1 \\
E_2 &= \lambda_2 - \mu_2 + P_{22}\mu_2 \\
d_1 &= \lambda_1 + P_{21}\mu_2 \\
d_2 &= \lambda_2 + P_{12}\mu_1 .
\end{aligned} \tag{27}$$

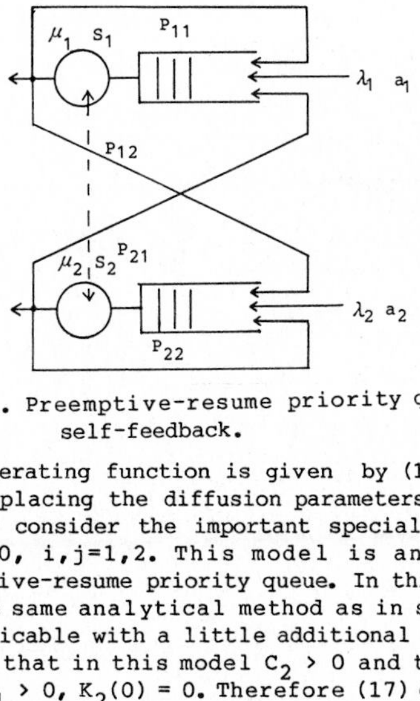

Figure1. Preemptive-resume priority queues with self-feedback.

The generating function is given by (17) and (21) with replacing the diffusion parameters by (27).

Now consider the important special case with $P_{ij} = 0$, i,j=1,2. This model is an ordinary preemptive-resume priority queue. In this case B=0 and the same analytical method as in section 5.1 is applicable with a little additional discussion. Notice that in this model $C_2 > 0$ and thus $K_1(0) = 2C_2/A_1 > 0$, $K_2(0) = 0$. Therefore (17) detremines $L_2(s_2)$ for Re $s_2 \ge 0$ uniquely but $L_1(s_1)$ only for Re $s_1 \ge K_1(0)$. To determine $L_1(s_1)$ for $0 \le$ Re $s_1 \le K_1(0)$, we need the other branch of $K_1(w)$, i.e.

$$T_1(w) = [\, C_1 - \sqrt{C_1^2 - 2A_1 w}\,]/A_1 \, .$$

Since $(T_1(w),K_2(w))$ and $(K_1(w),K_2(w))$ for $w \in [0, C_1^2/(2A_1)]$ are both zeros of the kernel $R(s_1,s_2)$, Re $s_1 \ge 0$, Re $s_2 \ge 0$, we have two expressions: for $w \in [\,0, C_1^2/(2A_1)\,]$

$$L_1(K_1(w)) = G(w)L_2(K_2(w)) + e(w)f_3 \tag{28}$$

and

$$L_1(T_1(w)) = G'(w)L_2(K_2(w)) + e'(w)f_3 \qquad (29)$$

where G(.) and e(w) are defined in section 5,

$$G'(w) = -G_2(T_1(w),K_2(w))/G_1(T_1(w),K_2(w))$$

and

$$e'(w) = g(T_1(w),K_2(w))/G_1(T_1(w),K_2(w)) .$$

From (17) for Re $w \geq 0$, the righthand sides of (28), (29) are known, therefore the lefthand sides are given. Hence for $s_1 \in [0, C_1]$, $L_1(s_1)$ is determined from (29) and for $s_1 \in [C_1, 2C_1]$ it is determined from (28).

From the above discussion, we conclude that the generating function is analytic at the origin. Hence we have again the relation (9),(10),(11) between the coefficients of Taylor series with B=0. Since the simultaneous equation (10) and (11) for a_0, b_0, h_0 with f_3 asssumed to be known are linearly independent we obtain

$$f_3 = 1 - \lambda_1/\mu_1 - \lambda_2/\mu_2 , \qquad (30)$$

under the assumption that stationary distribution exists. The empty queue probability is exactly modeled again.

7. NUMERICAL CALCULATION

In this section we show the method of numerical calculation and numerical examples for the mean queue length of the preemptive-resume priority queue as an example. To evaluate the mean queue length, it is enough to evaluate b_0 (or h_0) and b_1 (or h_1). Then from (10),(30),(9),(12),(13) the mean queue length of both high and low priority queues can be obtained. From (17),(18)

$$b_0 = f(0)$$

$$b_1 = f_3[\Psi'(0) \exp \Gamma(0) + \Psi(0) \Gamma'(0) \exp \Gamma(0)] \qquad (31)$$

where ' represents the derivative w.r.t. w.

The method of singular integration [2],[7] is applied to evaluate the integrals appearing in (31). The 16-point Gaussian numerical integration method is applied dividing the integration interval [-iM, iM] with M being sufficiently large into nine sub-intervals. Since the function Log G(w) in these integrals has to be evaluated for many points and used over and over, tables of the function can be constructed. The numerical solution was rapid in computer time. The solution took over about 6 seconds of a middle class computer (MELCOM 800 III, 1.7 MIPS).

In order to evaluate the accuracy of our approximation method we have compared it numerically in Table 1 with the exact results of M/G/1 priority queues. It is remarkable that our approximation method gives exact value in the Markovian case.

Table 1. Mean queue length of M/G/1 preemptive-resume priority queues.

$S_1=S_2=0.5$

ρ_2	$\rho_1=0.1$ Exact	Diff	$=0.3$ Exact	Diff	$=0.5$ Exact	Diff
0.1	0.123	0.095	0.166	0.156	0.256	0.206
0.45	0.883	0.758	1.569	1.408	7.650	7.425
0.65	2.152	1.972	10.40	10.17		
0.85	13.13	12.90				

$S_1=S_2=1.0$

ρ_2	$\rho_1=0.1$ Exact	Diff	$=0.3$ Exact	Diff	$=0.5$ Exact	Diff
0.1	0.126	0.126	0.174	0.174	0.275	0.275
0.45	1.011	1.011	1.877	1.877	9.900	9.900
0.65	2.629	2.629	13.56	13.56		
0.85	17.19	17.19				

$S_1=S_2=1.5$

ρ_2	$\rho_1=0.1$ Exact	Diff	$=0.3$ Exact	Diff	$=0.5$ Exact	Diff
0.1	0.130	0.158	0.182	0.217	0.294	0.344
0.45	1.139	1.264	2.186	2.346	12.15	12.38
0.65	3.106	3.286	16.71	16.94		
0.85	21.25	21.49				

8. CONCLUSIONS

For queueing networks, it is known that the diffusion process with instantaneous reflection on the boundaries is a heavy traffic model [1]. We have shown that the multi-dimensional elementary return diffusion process models empty queue probability of queueing networks and so those in light traffic conditions. We have also shown that the process can be applied to models not only of queueing networks but also of more general queueing systems such as priority queues. In fact it is directly applicable as a model of the coupled processor[2] and the solution of the preemptive-resume priority queues is available.

REFERENCES

[1] Coffman E.G. and Reiman M.I. Diffusion Approximations for Computer/Communication Systems. International Workship on Appl. Math. and Performance Reliability Models of Computer Communication Systems, Univ. of Pisa. 1983, 169-188.

[2] Cohen, J.W. and Boxma, O.J. Boundary Value Problems in Queueing System Analysis. North-Holland, Amsterdam.New York.Oxford , 1983.

[3] Cox, D.R. and Miller, H.D. The Theory of Stochastic Processes. Chapman and Hall Ltd.,New York, 1965.

[4] Feller, W. Diffusion processes in one dimension. Trans. Amer. Math. Soc. 77, 1954, 1-31.

[5] Foschini, G. J. Equiliblia for Diffusion Models of Pairs of Communicating Computers-Symmetric Case. IEEE Trans. Infor. Theory, IT-28, 1981, 273-284.

[6] Gelenbe, E. On Approximate Computer System Models. J. Assoc. Comput. Mach., Vol22, 1975, pp. 261-269.[7] Harrison, J.M., (1973) The Heavy Traffic approximation for Single Server Queues in Series. J. Appl. Prob. Vol.10, pp.613-629.

[7] Gakhov, F.D., Boundary Value Problems. Pergammon Press, oxford, 1966.

[8] Harrison, J.M. and Reiman, M.I. On the distribution of multi-dimensional reflected Brownian mption. SIAM J. Appl. Math. Vol. 41, No.2, 1981, pp345-361.

[9] Ito, K. Probability theory, Iwanami, Tokyo. 1952.

[10] Kobayashi, H. Application of the Diffusion Approximation to Queueing Networks I: Equilibrium Queue Distributions. J. Assoc. Compute. Mach., Vol.21, 1974, 316-328.

[11] Takahashi, H. and Akimaru, H., Approximate Queueing System Models via Multi-Dimensional Elementary Return Process. On queueing theory and its application, Research Institute for Mathematical Sciences, Kyoto Univ. Vol.519, 1984.

ACKNOELEDGEMENT

The authors are grateful to J.W. Cohen for helpful discussion. This work has been performed under the support of the Scientific Research Funds of the Ministry of Education of Japan.

TELETRAFFIC ISSUES in an Advanced Information Society
ITC-11
Minoru Akiyama (Editor)
Elsevier Science Publishers B.V. (North-Holland)

APL SOFTWARE DEVELOPMENT FOR CENTRAL CONTROL TELEPHONE SWITCHING SYSTEMS VIA THE ROW-CONTINUOUS MARKOV CHAIN PROCEDURE †

Ushio Sumita and J. George Shanthikumar

Graduate School of Management
University of Rochester
Rochester, New York 14627, U.S.A.

School of Business Administration
University of California at Berkeley
CA 94720, U.S.A.

ABSTRACT

We consider a central control telephone switching system having S processors each of which is connected to a line. A stream of calls are generated by a Poisson process with parameter λ. A call finding all S lines busy upon its arrival enters a pool of unsatisfied calls. These milling calls generate a stream of retrial requests from the first-attempt request stream. This milling situation is important in telephony where customers receiving a busy signal will be likely to repeat their requests until they are successful. In this paper, a bivariate Markov chain is developed, describing such milling phenomena. Based on the row-continuous Markov chain procedure of Keilson, Sumita and Zachmann (1981), numerical procedures are then developed for calculating various performance measures of interest. An APL computer software is also given with full details. A numerical example illustrates the use of the software.

0. Introduction

In the ordinary queueing system, customers that arrive while the service system is entirely occupied may be held in a queue. A waiting customer can be accepted as soon as the service system becomes available to him. In communication systems, however, one often encounters another type of queueing situation where the customer can only call in and test the state of the system. The customer enters the service if possible. When the service system is occupied, the customer cannot wait and enters a pool of milling unsatisfied customers generating a stream of repeat requests for service.

This milling situation is important in telephony where customers receiving a busy signal will likely repeat their requests until they are successful. Such queueing systems with retrials have been studied extensively, see e.g., Cohen [1], Keilson, Cozzolino, and Young [3], Kosten [5], Riordan [6], Syski [7], Wilkinson [8], and others.

In this paper, we consider many-server systems with no waiting, Poisson arrivals, exponential service, and different exponential interarrival distributions for first attempts and repeat attempts. The possibility that a milling customer may give up and leave the system is also considered. As we will see, the model can be formulated as a bivariate Markov chain in continuous time with desired skip-free property. The row-continuous Markov chain procedure developed by Keilson, Sumita, and Zachmann [4] is therefore applicable. The main purpose of the paper is to develop an efficient computer software for evaluating various probabilistic entities of interest using the row-continuous Markov chain procedure.

In Section 1, the underlying model is described. Although a central telephone control switching system is discussed for concreteness, the software is applicable to other similar models. In Section 2, relevant results on the row-continuous Markov chain procedure are summarized for the reader's convenience from Keilson, Sumita, and Zachmann [4]. A computer software is developed in Section 3 using APL as a programming language. A numerical example is also given.

† This work has been partially supported by GTE Laboratories, Waltham, Massachusetts.

1. Model Description

We consider a central control telephone switching system having S processors each of which is connected to a line. A stream of calls are generated by a Poisson process with parameter λ. Service time of each call is exponentially distributed with parameter μ. A call finding all S lines busy upon its arrival enters a pool of unsatisfied calls. These milling calls generate a stream of retrial requests from the first-attempt request stream. The milling process is described by two parameters (θ, γ). A milling call tries again with probability θ, and gets discouraged leaving system with probability $1-\theta$. Cycle time between attempts is exponentially distributed with parameter γ. All random variables are indepedent of each other. Let

(1.1) $\quad M(t)$ = number of busy lines at time t,

and

(1.2) $\quad R(t)$ = number of milling calls at time t.

A call finding $M(t) = S$ and $R(t) = R$ upon its arrival is lost. Then the bivariate process $(M(t), R(t))$ becomes a finite Markov chain on $\mathbf{M} \times \mathbf{R} = \{(m,r) : 0 \le m \le S, 0 \le r \le R\}$ governed by the transition rates given in Figure 1.

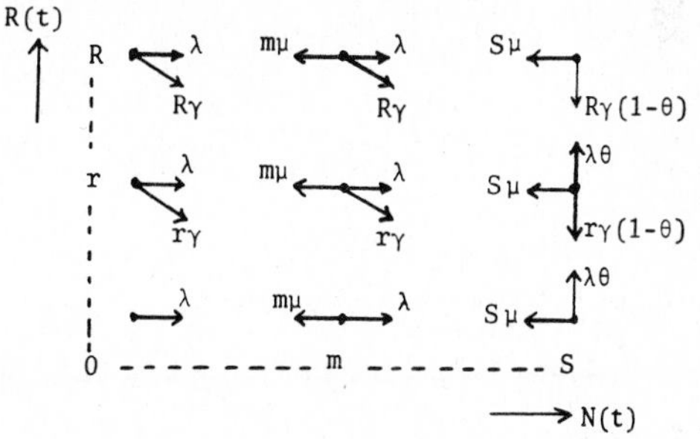

Figure 1: transition rates of $(M(t), R(t))$

This model can be analyzed via the row-continuous Markov chain procedure developed by Keilson, Sumita, and Zachmann [4]. Various probabilistic entities of interest can be numerically evaluated via the procedure as we will see.

2. Row Continuous Markov Chains

In this section, we describe some results from Keilson, Sumita, and Zachmann [4]. The discussion is restricted to the results which are directly relevant to the model given in Section 1.

Consider a bivariate Markov process $\underline{N}(t) = (M(t), R(t))$ on $\mathbf{N}$=$\mathbf{M} \times \mathbf{R}$ where $M(t)$ on $\mathbf{M} = \{m : 0 \le m \le S\}$ and $R(t)$ on $\mathbf{R}$ = $\{r : 0 \le r \le R\}$. Suppose that $\underline{N}(t)$ is governed by the transition rate matrix $\underline{\underline{\nu}} = (\nu_{(m,r)(m',r')})$ which is an $N \times N$ matrix where $N = (S+1) \times (R+1)$. This is equivalent to saying that the dwell time of $\underline{N}(t)$ at (m,r) is exponentially distributed with parameter $\nu_{(m,r)} = \sum_{(m',r')} \nu_{(m,r)(m',r')}$. This process $\underline{N}(t)$ then visits the state (m', r') next with probability $\nu_{(m,r)(m',r')}/\nu_{(m,r)}$.

We assume that $R(t)$ is skip free in the row directions, i.e.,

(2.1) $\nu_{(m,r)(m',r')} = 0$ if $|r' - r| > 1$.

2.1 Structure of $N(t) = (M(t), R(t))$

Because of the skip-free property in the row directions, the structure of $\underline{N}(t)$ is of a generalized birth-death process type in the following sense. Let

(2.2) $\underline{\underline{\nu}}_r^+ = (\nu_{r:mn}^+);\ \nu_{r:mn}^+ = \nu_{(m,r)(n,r+1)},$

(2.3) $\underline{\underline{\nu}}_r^0 = (\nu_{r:mn}^0);\ \nu_{r:mn}^0 = \nu_{(m,r)(n,r)},$

(2.4) $\underline{\underline{\nu}}_r^- = (\nu_{r:mn}^-);\ \nu_{r:mn}^- = \nu_{(m,r)(n,r-1)}.$

We note that $\underline{\underline{\nu}}_r^+$, which is an $(S+1)\times(S+1)$ matrix, represents the transitions from the $r-th$ row of the state space $\mathbf{N}=\mathbf{M}\times\mathbf{R}$ to the $(r+1)-st$ row. Similarly $\underline{\underline{\nu}}_r^0$ represents the transitions within the $r-th$ row and $\underline{\underline{\nu}}_r^-$ from the $r-th$ row to the $(r-1)-st$ row. The bivariate process $\underline{N}(t)$ is then governed by

$$(2.5)\qquad \underline{\underline{\nu}} = \begin{pmatrix} \underline{\underline{\nu}}_0^0 & \underline{\underline{\nu}}_0^+ & & & \underline{\underline{0}} \\ \underline{\underline{\nu}}_1^- & \underline{\underline{\nu}}_1^0 & \underline{\underline{\nu}}_1^+ & & \\ & & \ddots & & \\ & & \underline{\underline{\nu}}_{R-1}^- & \underline{\underline{\nu}}_{R-1}^0 & \underline{\underline{\nu}}_{R-1}^+ \\ \underline{\underline{0}} & & & \underline{\underline{\nu}}_R^- & \underline{\underline{\nu}}_R^0 \end{pmatrix}$$

$$(S+1)(R+1)\times(S+1)(R+1).$$

One easily sees that $\underline{\underline{\nu}}$ has the structure of the matrix version of the birth-death process. Let the diagonal matrix $\underline{\underline{\nu}}_{Dr}$ be defined by

$$(2.6)\qquad \underline{\underline{\nu}}_{Dr} = \mathrm{diag}\{\sum_n (\nu_{r:mn}^+ + \nu_{r:mn}^0 + \nu_{r:mn}^-)\},$$

where $\nu_{0:mn}^- = 0$ and $\nu_{R:mn}^+ = 0$. We further define

$$(2.7)\qquad \underline{\underline{\nu}}_D = \begin{pmatrix} \underline{\underline{\nu}}_{D0} & & & \underline{\underline{0}} \\ & \underline{\underline{\nu}}_{D1} & & \\ & & \ddots & \\ \underline{\underline{0}} & & & \underline{\underline{\nu}}_{DR} \end{pmatrix}.$$

The infinitesimal generator $\underline{\underline{Q}}$ of $\underline{N}(t)$ is then given by

(2.8) $\underline{\underline{Q}} = -\underline{\underline{\nu}}_D + \underline{\underline{\nu}}.$

Let $\underline{\underline{P}}(t) = (P_{(m,r)(m',r')}(t))$ be the transition probability matrix of $\underline{N}(t)$, i.e., $P_{(m,r)(m',r')}(t) = P(\underline{N}(t) = (m',r')|\underline{N}(0) = (m,r))$. One then has

(2.9) $\underline{\underline{P}}(t) = e^{\underline{\underline{Q}}t}$ and $\frac{d}{dt}\underline{\underline{P}}(t) = \underline{\underline{P}}(t)\underline{\underline{Q}}.$

By treating each row as a probabilistic entity, we decompose the state probability vector $\underline{p}^T(t)$, i.e.,

$$(2.10)\qquad \underline{p}^T(t) = (\underline{p}_0^T(t), \underline{p}_1^T(t), \dots \underline{p}_R^T(t))$$

where $\underline{p}_r^T(t)$ is an $(S+1)$ vector whose $m-th$ component is $P(\underline{N}(t) = (m,r))$. It can be readily seen that $\underline{p}^T(t) = \underline{p}^T(0)\underline{\underline{P}}(t)$ and the forward equations are obtained from (2.9) as $\frac{d}{dt}\underline{p}^T(t) = \underline{p}^T(t)\underline{\underline{Q}}$. From (2.5), (2.7), and (2.8), one then sees that

$$(2.11)\qquad \frac{d}{dt}\underline{p}_0^T(t) = -\underline{p}_0^T(t)\underline{\underline{\nu}}_{D0} + \underline{p}_0^T(t)\underline{\underline{\nu}}_0^0 + \underline{p}_1^T(t)\underline{\underline{\nu}}_1^-,$$

$$(2.12)\qquad \begin{aligned} \frac{d}{dt}\underline{p}_r^T(t) &= -\underline{p}_r^T(t)\underline{\underline{\nu}}_{Dr} + \underline{p}_r^T(t)\underline{\underline{\nu}}_r^0 \\ &\quad + \underline{p}_{r-1}^T(t)\underline{\underline{\nu}}_{r-1}^+ + \underline{p}_{r+1}^T(t)\underline{\underline{\nu}}_{r+1}^-, \\ &\quad 1 \le r \le R-1. \end{aligned}$$

$$(2.13)\qquad \frac{d}{dt}\underline{p}_R^T(t) = -\underline{p}_R^T(t)\underline{\underline{\nu}}_{DR} + \underline{p}_R^T(t)\underline{\underline{\nu}}_R^0 + \underline{p}_{R-1}^T(t)\underline{\underline{\nu}}_{R-1}^+.$$

The matrix version of the birth-death process type structure can be seen explicitly in (2.11) through (2.13).

2.2 Row-wise First Passage Time Distributions

Let $s_{r:mn}^+(t)$ be the $n-th$ component of the vector density for the first passage time from (m,r) to $\{(n,r+1) : n \in \mathbf{M}\}$. The downward first passage time density $s_{r:mn}^-(t)$ is defined similarly. Let

(2.14) $\underline{\underline{s}}_r^+(t) = (s_{r:mn}^+(t))$

(2.15) $\underline{\underline{s}}_r^-(t) = (s_{r:mn}^-(t))$

which are matrix $p.d.f.'s$. Then

(2.16) $\underline{\underline{s}}_r^+ = \int_0^\infty \underline{\underline{s}}_r^+(t)dt$ and $\underline{\underline{s}}_r^- = \int_0^\infty \underline{\underline{s}}_r^-(t)dt$

are stochastic matrices. The corresponding matrix Laplace transforms are defined by

$$(2.17)\qquad \begin{aligned} \underline{\underline{\sigma}}_r^+(s) &= \int_0^\infty e^{-st}\underline{\underline{s}}_r^+(t)dt \\ \underline{\underline{\sigma}}_r^-(s) &= \int_0^\infty e^{-st}\underline{\underline{s}}_r^-(t)dt. \end{aligned}$$

The associated moment matrices are also defined by

$$(2.18)\qquad \underline{\underline{\mu}}_r^+ = \int_0^\infty t\underline{\underline{s}}_r^+(t)dt \text{ and } \underline{\underline{\mu}}_r^- = \int_0^\infty t\underline{\underline{s}}_r^-(t)dt.$$

We note that

$$(2.19)\qquad \underline{\underline{s}}_r^+ = \underline{\underline{\sigma}}_r^+(0);\ \underline{\underline{\mu}}_r^+ = -\frac{d}{ds}\underline{\underline{\sigma}}_r^+(s)|_{s=0}$$

and

$$(2.20)\qquad \underline{\underline{s}}_r^- = \underline{\underline{\sigma}}_r^-(0);\ \underline{\underline{\mu}}_r^- = -\frac{d}{ds}\underline{\underline{\sigma}}_r^-(s)|_{s=0}.$$

$(\underline{\underline{s}}_r^+)_{mn}$ is the probability that the process reaches the $(r+1)st$ row for the first time at $(n, r+1)$ given that the process starts at (m,r). $\underline{\underline{\mu}}_r^+$ is the corresponding first moments and hence

$$\sum_{n=0}^{S+1} (\underline{\underline{\mu}}_r^+)_{mn}$$

gives the mean first passage time of $\underline{N}(t)$ from (m,r) to any state in the $(r+1)-st$ row.

To evaluate $\underline{\underline{s}}_r^+, \underline{\underline{s}}_r^-, \underline{\underline{\mu}}_r^+$, and $\underline{\underline{\mu}}_r^-$ efficiently, we introduce a lossy process $\underline{N}_r^*(t)$ on $\{(m,r) : m \in \mathbf{M}\}$ obtained by censoring transitions going out of the $r-th$ row. The infinitesimal generator $\underline{\underline{Q}}_r^*$ of $\underline{N}_r^*(t)$ is given by

$$(2.21)\qquad \underline{\underline{Q}}_r^* = -\underline{\underline{\nu}}_{Dr} + \underline{\underline{\nu}}_r^0.$$

Hence, $\underline{N}_r^*(t)$ has the transition probability matrix

$$(2.22)\qquad \underline{\underline{P}}_r^*(t) = [P_{(m,r)(n,r)}^*(t)] = e^{\underline{\underline{Q}}_r^* t}.$$

We note that $\underline{\underline{P}}_r^*(t)$ is strictly substochastic and $\underline{\underline{P}}_r^*(t) \to \underline{\underline{0}}$ as $t \to \infty$. From the forward equations $\frac{d}{dt}\underline{\underline{P}}_r^*(t) = \underline{\underline{P}}_r^*(t)\underline{\underline{Q}}_r^*$ and the initial condition $\underline{\underline{P}}_r^*(0) = \underline{\underline{I}}$, one finds that

(2.23) $\underline{\underline{\pi}}_r^*(s) = \int_0^\infty e^{-st}\underline{\underline{P}}_r^*(t)dt = (s\underline{\underline{I}} - \underline{\underline{Q}}_r^*)^{-1}.$

To go up from the $r-th$ row to the $(r+1)-st$ row, the process $\underline{N}(t)$ may go up directly, or go down to $(r-1)-st$ row first, and then climbs up twice to the $(r+1)-st$ row through the $r-th$ row. One sees that

$$(2.24)\qquad \begin{aligned} \underline{\underline{s}}_r^+(t) &= \underline{\underline{P}}_r^*(t)\underline{\underline{\nu}}_r^+ \\ &\quad + \underline{\underline{P}}_r^*(t)\underline{\underline{\nu}}_r^- * \underline{\underline{s}}_{r-1}^+(t) * \underline{\underline{s}}_r^+(t), \\ &\quad 1 \le r \le R-1, \end{aligned}$$

where $a(t) * b(t)$ denotes the matrix convolution, i.e., $(\underline{\underline{a}}(t) * \underline{\underline{b}}(t))_{mn} = \sum_k \int_0^t a_{mk}(t-x)\, b_{kn}(x)dx$. By taking the Laplace transform on both sides of (2.24), we obtain

$$\underline{\underline{\sigma}}_r^+(s) = \underline{\underline{\pi}}_r^*(s)\underline{\underline{\nu}}_r^+ + \underline{\underline{\pi}}_r^*(s)\underline{\underline{\nu}}_r^-\underline{\underline{\sigma}}_{r-1}^+(s)\underline{\underline{\sigma}}_r^+(s), \quad 1 \leq r \leq R-1. \tag{2.25}$$

From (2.21) and (2.23) and by setting $s = 0$, Equation (2.25) leads to

(2.26) $(\underline{\underline{\nu}}_{Dr} - \underline{\underline{\nu}}_r^0 - \underline{\underline{\nu}}_r^-\underline{\underline{s}}_{r-1}^+)\underline{\underline{s}}_r^+ = \underline{\underline{\nu}}_r^+.$

Let

$$\begin{aligned} \underline{\underline{\theta}}_r^+ &= \underline{\underline{\nu}}_{Dr}^{-1}\underline{\underline{\nu}}_r^+; \\ \underline{\underline{\theta}}_r^0 &= \underline{\underline{\nu}}_{Dr}^{-1}\underline{\underline{\nu}}_r^0; \\ \underline{\underline{\theta}}_r^- &= \underline{\underline{\nu}}_{Dr}^{-1}\underline{\underline{\nu}}_r^-. \end{aligned} \tag{2.27}$$

We note that $\underline{\underline{\theta}}_r = \underline{\underline{\theta}}_r^+ + \underline{\underline{\theta}}_r^0 + \underline{\underline{\theta}}_r^-$ is a stochastic matrix. The first factor in (2.26) can be rewritten as $\underline{\underline{\nu}}_{Dr}(\underline{\underline{I}} - \underline{\underline{\theta}}_r^0 - \underline{\underline{\theta}}_r^-\underline{\underline{s}}_{r-1}^+)$. Since $\underline{\underline{\theta}}_r$ and $\underline{\underline{s}}_{r-1}^+$ are stochastic, the inverse of the above matrix exists and one has

(2.28) $\underline{\underline{s}}_r^+ = (\underline{\underline{I}} - \underline{\underline{\theta}}_r^0 - \underline{\underline{\theta}}_r^-\underline{\underline{s}}_{r-1}^+)^{-1}\underline{\underline{\theta}}_r^+, 1 \leq r \leq R-1.$

Since

$$\begin{aligned} \underline{\underline{\sigma}}_0^+(s) &= \underline{\underline{\pi}}_0^*(s)\underline{\underline{\nu}}_0^+ \\ &= (s\underline{\underline{I}} + \underline{\underline{\nu}}_{D0} - \underline{\underline{\nu}}_D^0)^{-1}\underline{\underline{\nu}}_0^+ \\ &= (s\underline{\underline{\nu}}_{D0}^{-1} + \underline{\underline{I}} - \underline{\underline{\theta}}_0^0)^{-1}\underline{\underline{\theta}}_0^+, \end{aligned}$$

it can be readily seen that

(2.29) $\underline{\underline{s}}_0^+ = (\underline{\underline{I}} - \underline{\underline{\theta}}_0^0)^{-1}\underline{\underline{\theta}}_0^+.$

Hence $\underline{\underline{s}}_r^+$ can be computed recursively by (2.28) starting with (2.29). Similarly for the downward first passage times one has

$$\underline{\underline{\sigma}}_r^-(s) = \underline{\underline{\pi}}_r^*(s)\underline{\underline{\nu}}_r^- + \underline{\underline{\pi}}_r^*(s)\underline{\underline{\nu}}_r^+\underline{\underline{\sigma}}_{r+1}^-(s)\underline{\underline{\sigma}}_r^-(s), \quad 1 \leq r \leq R-1, \tag{2.30}$$

and $\underline{\underline{\sigma}}_R^-(s) = \underline{\underline{\pi}}_R^*(s)\underline{\underline{\nu}}_R^- = (s\underline{\underline{I}} + \underline{\underline{\nu}}_{DR} - \underline{\underline{\nu}}_R^0)^{-1}\underline{\underline{\nu}}_R^-$.
This then leads to the recursion formulae

$$\underline{\underline{s}}_r^- = (\underline{\underline{I}} - \underline{\underline{\theta}}_r^0 - \underline{\underline{\theta}}_r^+\underline{\underline{s}}_{r+1}^-)^{-1}\underline{\underline{\theta}}_r^-, \quad 1 \leq r \leq R-1, \tag{2.31}$$

starting with

$$\underline{\underline{s}}_R^- = (\underline{\underline{I}} - \underline{\underline{\theta}}_R^0)^{-1}\underline{\underline{\theta}}_R^-. \tag{2.32}$$

By differentiating (2.25) and (2.30) with respect to s and then setting $s = 0$, the moment recursion formulae can be found:

$$\begin{aligned} \underline{\underline{\mu}}_r^+ &= (\underline{\underline{I}} - \underline{\underline{\theta}}_r^0 - \underline{\underline{\theta}}_r^-\underline{\underline{s}}_{r-1}^+)^{-1} \\ &\quad \times (\underline{\underline{\nu}}_{Dr}^{-1} + \underline{\underline{\theta}}_r^-\underline{\underline{\mu}}_{r-1}^+)\underline{\underline{s}}_r^+, \\ &\quad 1 \leq r \leq R-1. \end{aligned} \tag{2.33}$$

$$\underline{\underline{\mu}}_0^+ = (\underline{\underline{I}} - \underline{\underline{\theta}}_0^0)^{-1}\underline{\underline{\nu}}_{D0}^{-1}\underline{\underline{s}}_0^+. \tag{2.34}$$

$$\begin{aligned} \underline{\underline{\mu}}_r^- &= (\underline{\underline{I}} - \underline{\underline{\theta}}_r^0 - \underline{\underline{\theta}}_r^+\underline{\underline{s}}_{r+1}^-)^{-1} \\ &\quad \times (\underline{\underline{\nu}}_{Dr}^{-1} + \underline{\underline{\theta}}_r^+\underline{\underline{\mu}}_{r+1}^-)\underline{\underline{s}}_r^-, \\ &\quad 1 \leq r \leq R-1. \end{aligned} \tag{2.35}$$

$$\underline{\underline{\mu}}_R^- = (\underline{\underline{I}} - \underline{\underline{\theta}}_R^0)^{-1}\underline{\underline{\nu}}_{DR}^{-1}\underline{\underline{s}}_R^-. \tag{2.36}$$

Upward and downward first passage times over a number of rows can be evaluated by differentiating

$$\begin{aligned} \underline{\underline{\sigma}}_{rr'}(s) &= \underline{\underline{\sigma}}_r^+(s)\underline{\underline{\sigma}}_{r+1}^+(s)\cdots\underline{\underline{\sigma}}_{r'-1}^+(s) \\ &= \underline{\underline{\sigma}}_{r,r'-1}(s)\underline{\underline{\sigma}}_{r'-1}^+(s) \end{aligned} \tag{2.37}$$

and

$$\begin{aligned} \underline{\underline{\sigma}}_{r'r}(s) &= \underline{\underline{\sigma}}_{r'}^-(s)\underline{\underline{\sigma}}_{r'-1}^-(s)\cdots\underline{\underline{\sigma}}_{r+1}^-(s) \\ &= \underline{\underline{\sigma}}_{r',r+1}(s)\underline{\underline{\sigma}}_{r+1}^-(s) \end{aligned} \tag{2.38}$$

where $r < r'$. One finds that

(2.39) $\underline{\underline{\mu}}_{rr'} = \underline{\underline{\mu}}_{r,r'-1}\underline{\underline{s}}_{r'-1}^+ + (\prod_{j=m}^{r'-2}\underline{\underline{s}}_j^+)\cdot\underline{\underline{\mu}}_{r'-1}^+$

and

(2.40) $\underline{\underline{\mu}}_{r'r} = \underline{\underline{\mu}}_{r',r+1}\underline{\underline{s}}_{r+1}^- + (\prod_{j=r+2}^{r'}\underline{\underline{s}}_j^-)\cdot\underline{\underline{\mu}}_{r+1}^-.$

2.3 Erogodic Probabilities

When $\underline{\underline{\nu}}_r^-$ are invertible, the ergodic probabilities $\lim_{t\to\infty}\underline{p}_r^T(t) = \underline{e}_r^T$ can be efficiently evaluated. By letting $t \to \infty$ in (2.12), one finds that

$$0 = -\underline{e}_r^T(\underline{\underline{\nu}}_{Dr} - \underline{\underline{\nu}}_r^0) + \underline{e}_{r-1}^T\underline{\underline{\nu}}_{r-1}^+ + \underline{e}_{r+1}^T\underline{\underline{\nu}}_{r+1}^-.$$

Hence when $\underline{\underline{\nu}}_r^-$ are invertible for $r = 1, 2, \ldots, R$, one has

$$\underline{e}_{r+1}^T = \{\underline{e}_r^T(\underline{\underline{\nu}}_{Dr} - \underline{\underline{\nu}}_r^0) - \underline{e}_{r-1}^T\underline{\underline{\nu}}_{r-1}^+\}(\underline{\underline{\nu}}_{r+1}^-)^{-1}, \quad 1 \leq r \leq R-1. \tag{2.41}$$

Similarly from (2.11), one has

(2.42) $\underline{e}_1^T = \underline{e}_0^T(\underline{\underline{\nu}}_{D0} - \underline{\underline{\nu}}_0^0)(\underline{\underline{\nu}}_1^-)^{-1}.$

If $\underline{e}_0^T$ is found, one can calculate $\underline{e}_r^T$ by (2.41) and (2.42) recursively. It has been shown (Keilson [2]) that $\underline{e}_0^T$ is a left eigen vector, apart from a multiplicative factor, of the matrix $\underline{\underline{\beta}}_0$ with eigenvalue one, where

(2.43) $\underline{\underline{\beta}}_0 = \underline{\underline{\nu}}_{D0}^{-1}(\underline{\underline{\nu}}_0^0 + \underline{\underline{\nu}}_0^+\underline{\underline{s}}_1^-)$,

and $\underline{a}_0^T = \underline{a}_0^T\underline{\underline{\beta}}_0 \Rightarrow \underline{e}_0^T = K\underline{a}_0^T$. The constant K can be found by normalization, i.e., calculate $\underline{a}_r^T$ starting with $\underline{a}_0^T$ by (2.41) and (2.42). Then $K = (\sum_{r=0}^R \underline{a}_r^T\underline{1})^{-1}$ and $\underline{e}_r^T = K\underline{a}_r^T$.

2.4 Summary of the procedure

We now summarize the procedure compactly.

(S1) Find $\underline{\underline{\nu}}_r^+, \underline{\underline{\nu}}_r^0, \underline{\underline{\nu}}_r^-$, and $\underline{\underline{\nu}}_{Dr}$.

(S2) $\underline{\underline{\theta}}_r^+ = \underline{\underline{\nu}}_{Dr}^{-1}\underline{\underline{\nu}}_r^+; \underline{\underline{\theta}}_r^0 = \underline{\underline{\nu}}_{Dr}^{-1}\underline{\underline{\nu}}_r^0; \underline{\underline{\theta}}_r^- = \underline{\underline{\nu}}_{Dr}^{-1}\underline{\underline{\nu}}_r^-.$

(S3) $\underline{\underline{s}}_r^+ = (\underline{\underline{I}} - \underline{\underline{\theta}}_r^0 - \underline{\underline{\theta}}_r^-\underline{\underline{s}}_{r-1}^+)^{-1}\underline{\underline{\theta}}_r^+, 1 \leq r \leq R-1,$

starting with

$$s_0^+ = (\underline{\underline{I}} - \underline{\underline{\theta}}_0^0)^{-1}\underline{\underline{\theta}}_0^+.$$

(S4) $\underline{\underline{s}}_r^- = (\underline{\underline{I}} - \underline{\underline{\theta}}_r^0 - \underline{\underline{\theta}}_r^+\underline{\underline{s}}_{r+1}^-)^{-1}\underline{\underline{\theta}}_r^-, 1 \leq r \leq R-1$

starting with

$$\underline{\underline{s}}_R^- = (\underline{\underline{I}} - \underline{\underline{\theta}}_R^0)^{-1}\underline{\underline{\theta}}_R^-.$$

$$\begin{aligned} \underline{\underline{\mu}}_r^+ &= (\underline{\underline{I}} - \underline{\underline{\theta}}_r^0 - \underline{\underline{\theta}}_r^-\underline{\underline{s}}_{r-1}^+)^{-1} \\ &\quad \times (\underline{\underline{\nu}}_{Dr}^{-1} + \underline{\underline{\theta}}_r^-\underline{\underline{\mu}}_{r-1}^+)\underline{\underline{s}}_r^+, \\ &\quad 1 \leq r \leq R-1, \end{aligned} \tag{S5}$$

starting with

$$\underline{\underline{\mu}}_0^+ = (\underline{\underline{I}} - \underline{\underline{\theta}}_0^0)^{-1}\underline{\underline{\nu}}_{D0}^{-1}\underline{\underline{s}}_0^+.$$

The $m-th$ component of $\underline{\underline{\mu}}_r^+\underline{1}$ is the mean first passage time of the process $\underline{N}(t)$ from (m, r) to any state in the $(r+1) - st$ row, where $\underline{1}$ is a vector having all components equal to one.

(S6) $\underline{\underline{\mu}}_{0r} = \underline{\underline{\mu}}_{0,r-1}\underline{\underline{s}}_{r-1}^+ + (\prod_{j=0}^{r-2}\underline{\underline{s}}_j^+)\cdot\underline{\underline{\mu}}_{r-1}^+$

starting with

$$\underline{\underline{\mu}}_{01} = \underline{\underline{\mu}}_0^+.$$

The $m-th$ component of $\underline{\mu}_{0r}\underline{1}$ is the mean first passage time of the process $\underline{N}(t)$ from $(m,0)$ to any state in the $r-th$ row.

$$(S7)\qquad \begin{aligned}\underline{\mu}_r^- &= (\underline{I} - \underline{\theta}_r^0 - \underline{\theta}_r^+ \underline{s}_{r+1}^-)^{-1}\\ &\times (\underline{\nu}_{Dr}^{-1} + \underline{\theta}_r^+ \underline{\mu}_{r+1})\underline{s}_r^-,\\ &1 \le r \le R-1,\end{aligned}$$

starting with

$$\underline{\mu}_R^- = (\underline{I} - \underline{\theta}_R^0)^{-1} \underline{\nu}_{DR}^{-1} \underline{s}_R^-.$$

$$(S8)\quad \underline{\mu}_{Rr} = \underline{\mu}_{R,r+1} \cdot \underline{s}_{r+1}^- + (\prod_{j=r+2}^{R} \underline{s}_j^-)\underline{\mu}_{r+1}^-.$$

$$(S9)\quad \underline{\beta}_0 = \underline{\nu}_{D0}^{-1}(\underline{\nu}_0^0 + \underline{\nu}_0^+ \underline{s}_1^-)$$

$$(S10)\quad \underline{a}_0^T = \underline{a}_0^T \underline{\beta}_0$$

$$(S11)\qquad \begin{aligned}\underline{a}_{r+1}^T &= \{\underline{a}_r^T(\underline{\nu}_{Dr} - \underline{\nu}_r^0)\\ &- \underline{a}_{r-1}^T \underline{\nu}_{r-1}^+\}(\underline{\nu}_{r+1}^-)^{-1},\\ &1 \le r \le R-1,\end{aligned}$$

starting with $\underline{a}_0^T$ and

$$\underline{a}_1^T = \underline{a}_0^T(\underline{\nu}_{D0} - \underline{\nu}_0^0)(\underline{\nu}_1^-)^{-1}.$$

$$(S12)\quad \underline{e}_r^T = K\underline{a}_r^T \text{ where } K = (\sum_{r=0}^{R} \underline{a}_r^T \underline{1})^{-1}.$$

3. Software Development

We now develop a computer software for evaluating the ergodic probabilities and the mean first passage times for the model described in Section 1. APL is used as a programming language. It should be noted from Figure 1 that the bivariate process is skip free in both directions. The row structure can be characterized by the following transition rate matrices.

$$(3.1)\qquad \underline{\nu}_r^+ = \begin{pmatrix} & & \underline{0} & & & \\ 0 & - & - & - & - & 0 & \lambda\theta \end{pmatrix} \qquad 0 < r < R-1$$

$$(3.2)\qquad \underline{\nu}_r^0 = \begin{pmatrix} 0 & \lambda & & & & \underline{0}\\ \mu & 0 & \lambda & & & \\ & 2\mu & 0 & \lambda & & \\ & & \ddots & \ddots & \ddots & \\ & & & \ddots & \ddots & \lambda\\ \underline{0} & & & & S\mu & 0 \end{pmatrix} \qquad 1 \le r \le R-1$$

$$(3.3)\qquad \nu_r^- = \begin{pmatrix} 0 & r\gamma & & & \underline{0}\\ & \ddots & \ddots & & \\ & & \ddots & \ddots & \\ & \underline{0} & & 0 & r\gamma\\ & & & & r\gamma(1-\theta) \end{pmatrix} \qquad 1 < r < R$$

(3.4) $\underline{\nu}_{Dr} = \text{diag}\{\lambda + r\gamma + j\mu, (0 \le j \le S-1), S\mu + r\gamma(1-\theta) + \lambda\theta(1-\delta_{rR})\}$,

where $\delta_{rR} = 1$ if $r = R$ and $\delta_{rR} = 0$ otherwise. We note that $\underline{\nu}_r^+$ and $\underline{\nu}_r^0$ are independent of r and $\underline{\nu}_r^- = r \times \underline{v}$ where $\underline{v}$ is also independent of r. The transition rate matices characterizing the column structure are given by:

$$(3.1)'\qquad {}^c\underline{\nu}_m^+ = \begin{pmatrix} \lambda & & & & \\ \gamma & \lambda & & & \underline{0}\\ & 2\gamma & \lambda & & \\ \underline{0} & & \ddots & \ddots & \\ & & & R\gamma & \lambda \end{pmatrix} \qquad 0 \le m \le S-1$$

(3.2)′

$${}^c\underline{\nu}_m^0 = \delta_{Sm} \times \begin{pmatrix} 0 & \lambda\theta & & & \underline{0}\\ \gamma(1-\theta) & 0 & \lambda\theta & & \\ & \ddots & \ddots & \ddots & \\ & & \ddots & \ddots & \lambda\theta\\ \underline{0} & & & R\gamma(1-\theta) & 0 \end{pmatrix} \qquad 0 \le m \le S,$$

(3.3)′ ${}^c\underline{\nu}_m^- = m\mu \times \underline{I}, 1 \le m \le S$

(3.4)′ ${}^c\underline{\nu}_{Dm} = \text{diag}\{\lambda + j\gamma + \delta_{Sm} \cdot (j\gamma(1-\theta) - \lambda\theta) + m\mu,$
$(0 \le j \le R-1), \lambda + R\gamma + \delta_{Sm} \cdot R\gamma(1-\theta) + m\mu\},$
$0 \le m \le S.$

The column structure is more attractive than the row structure for evaluating the ergodic probabilities since ${}^c\underline{\nu}_m^-$ is invertible while $\underline{\nu}_r^-$ is not. Hence the procedure described in 2.3 is applicable only to the column structure.

The APL programs for calculating the ergodic probabilities and the row-wise mean first passage times are given next with detailed explanations. Relevant minor subroutines are described in 3.1. The main program is given in 3.2, and a numerical example in 3.3.

3.1 Minor subroutines

```
      ∇NIP[⎕]∇
    ∇ R←NIP X
[1]   X
[2]   R←⎕
    ∇
```

This program first displays the question stored in the argument X in alphanumeric mode and then accepts the input from the user.

```
      ∇DIAG[⎕]∇
    ∇ Z←MT DIAG A
[1]   Z←(⍴MT)⍴(,1=MT)\A
    ∇
```

The first argument MT must be a matrix having elements either 0 or 1. The second argument A is a numeric vector where the length of A must be equal to the number of elements of MT having the value 1. The program returns the matrix where those 1-elements of MT are replaced by the values of A in the order from left to right and then up to down. This program is convenient to set up transition rate matrices.

```
      ∇LEFT[⎕]∇
    ∇ Z←LEFT A;CT;W
[1]   CT←10*-8;W←(1↑⍴A)⍴1
[2]  LP:Z←Z÷+/Z←W+.×A
[3]   →((W∧.=Z)∨CT>+/|Z-W)⍴0
[4]   W←Z
[5]   →LP
    ∇
```

This program calculates the left eigenvector of the matrix A associated with eigenvalue one, where A should be entered as the argument. The power method is employed and the program terminates when the difference of the two consecutive vectors falls below 10^{-8} in terms of the absolute norm.

```
      ∇SMS[⎕]∇
    ∇ Z←SMS;DD;K
[1]   Z←(1+⍳K)∘.=⍳K←R+1
[2]   Z←(Z×LM×TH)+VP←GM×(1-TH)×((⍳K)∘
     .=1+⍳K)DIAG⍳K-1
[3]   VP←VP+(IDT←(⍳K)∘.=⍳K)×LM
[4]   DD←IDT DIAG 1÷(S×MU)++/Z
[5]   Z←(█IDT-DD+.×Z)+.×DD×S×MU
    ∇
```

The matrix ${}^c\underline{s}_S^-$ is calculated by this program. Through the first two lines the matrix ${}^c\underline{\nu}_S^0$ of (3.2) is stored in Z. The lines [3] and [4] set up the matrices ${}^c\underline{\nu}_m^+$ of (3.1) and $({}^c\underline{\nu}_{DR})^{-1}$ of (3.4) and store them in VP and DD respectively. The matrix ${}^c\underline{s}_S^-$ is obtained in the line [5], reading (S4) of 2.4 in terms of the column structure. The matrix VP is not used in this program but in $\nabla SM1$ which refers to ∇SMS as a subroutine. It is more convenient to set up VP here than in $\nabla SM1$. The identity matrix of the size $(R+1) \times (R+1)$ is also stored in IDT for later use.

```
      ∇SM1[⎕]∇
    ∇  Z←SM1;I;L
[1]     Z←SMS;I←S-1
[2]    LOOP:L←IDT DIAG 1÷(I×MU)++/VP
[3]     Z←(⌹IDT-L+.×VP+.×Z)+.×L×I×MU
[4]     →(1≤I←I-1)/LOOP
    ∇
```

This program returns the matrix ${}^c\underline{\underline{s}}_1^-$ which will be used subsequently in the program ∇ERG for evaluating the ergodic probabilities. The initial matrix ${}^c\underline{\underline{s}}_S^-$ is first evaluated by the program ∇SMS and stored in Z. Then ${}^c\underline{\underline{s}}_m^-$ is calculated recursively using (S4) of (2.4) in the column structure. In the line [2], $({}^c\underline{\underline{\nu}}_{Dm})^{-1}$ is stored in L where the matrix ${}^c\underline{\underline{\nu}}_m^+$ of (3.1), which is independent of m, has been stored in VP in the subroutine ∇SMS. The line [3] is the recursion formula (S4). We note that ${}^c\underline{\underline{\theta}}_m^0 = 0$ for $0 \leq m \leq S-1$.

```
      ∇MATTH[⎕]∇
    ∇  MATTH M;VM1
[1]     →(M=R)/L1
[2]     DA←ID DIAG 1÷+/VZ+VP+VM1←VM×M
[3]     TP←DA+.×VP;TZ←DA+.×VZ;TM←DA+.×
      VM1
[4]     →0
[5]    L1:DA←ID DIAG 1÷+/VZ+VM1←VM×M
[6]     TZ←DA+.×VZ;TM←DA+.×VM1
    ∇
```

The matrices $\underline{\underline{\theta}}_r^+$, $\underline{\underline{\theta}}_r^-$, and $\underline{\underline{\theta}}_r^0$ of (S2) for the row structure are calculated in this program. The argument M is the row index. The matrix $\underline{\underline{\nu}}_{Dr}^{-1}$ is first stored in DA, and then $\underline{\underline{\theta}}_r^-$ in TM, $\underline{\underline{\theta}}_r^0$ in TZ, and $\underline{\underline{\theta}}_r^+$ in TP. The matrices $\underline{\underline{\nu}}_r^+$, $\underline{\underline{\nu}}_r^0$, and $\frac{1}{r}\,\underline{\underline{\nu}}_r^-$ of (3.1) through (3.3) must be generated outside this program.

```
      ∇UPSPE[⎕]∇
    ∇  UPSPE;J;TZ;TP;TM;MX
[1]     MUP←(R,⍴SUP←((S+1),S+1)⍴(S⍴0),1
      )⍴0
[2]     MATTH 2-J←2
[3]     MUP[1;;]←(⌹ID-TZ)+.×DA+.×SUP
[4]    LOOP:MATTH J-1
[5]     MX←⌈/+/MUP[J;;]←(⌹ID-TZ+TM+.×
      SUP)+.×(DA+TM+.×MUP[J-1;;])+.×
      SUP
[6]     J←J+1
[7]     →(MX≥10*6)/EXIT
[8]     →(R≥J)/LOOP
[9]    EXIT:STOP←J-1
    ∇
```

This program calculates $\underline{\underline{\mu}}_r^+$ of (S5) for the row structure. We note from Figure 1 (or (S3)) that $\underline{\underline{s}}_r^+$ has all elements equal to zero except the $(S+1,\ S+1)$ element having the value 1. This matrix is stored in SUP. Using the program $\nabla MATTH$, the matrices $\underline{\underline{\theta}}_r^+$,$\underline{\underline{\theta}}_r^0$, and $\underline{\underline{\theta}}_r^-$ are first calculated. Then in the line [4], $\underline{\underline{\mu}}_r^+$ is evaluated and stored in the three dimensional array MUP. The program terminates if the maximum mean first passage time exceeds 10^6. If this occurs at the $r^* - th$ row, $\underline{\underline{\mu}}_r^+ = \underline{\underline{0}}$ for $r > r^*$.

```
      ∇DNWARD[⎕]∇
    ∇  DNWARD;INV;J;TZ;TP;TM;MX
[1]     SDN←MDN←(R,(S+1),(S+1))⍴0
[2]     MATTH 1+J←R-1
[3]     SDN[R;;]←(INV←⌹ID-TZ)+.×TM
[4]     MDN[R;;]←INV+.×DA+.×SDN[R;;]
[5]    LOOP:MATTH J
[6]     SDN[J;;]←(INV←⌹ID-TZ+TP+.×SDN[J
      +1;;])+.×TM
[7]     MX←⌈/+/MDN[J;;]←INV+.×(DA+TP+.×
      MDN[J+1;;])+.×SDN[J;;]
[8]     J←J-1
[9]     →(MX≥10*6)/EXIT
[10]    →(1≤J)/LOOP
[11]   EXIT:STOP←J+1
    ∇
```

The matrices $\underline{\underline{s}}_r^-$ and $\underline{\underline{\mu}}_r^-$ are calculated and stored in SDN and MDN respectively in this program. Since no special structure such as $\underline{\underline{s}}_r^+$ above is present, the recursion formulae (S4) and (S7) must be employed. The algorithm terminates when the maximum mean first passage time exceeds 10^6, leaving $\underline{\underline{s}}_r^-$ and $\underline{\underline{\mu}}_r^-$ as zero matrices beyond that point.

```
      ∇PARAM[⎕]∇
    ∇  PARAM
[1]     SPACE,'NUMBER OF SERVERS
                  :  ',,'I5'⍕S
[2]     SPACE,'NUMBER OF MAXIMUN MILLIN
      G CUSTOMERS  :  ',,'I5'⍕R
[3]     ' '
[4]     SPACE,'ARRIVAL RATE :  ',(,'F5.
      2'⍕LM),'    SERVICE RATE
      :  ',,'F5.2'⍕MU
[5]     SPACE,'RETRIAL RATE :  ',(,'F5.
      2'⍕GM),'    RETRIAL PROBABILITY
      :  ',,'F5.2'⍕TH
[6]     ' '
    ∇
```

```
      ∇PTHEAD[⎕]∇
    ∇  PTHEAD;H1;H2;L1;L2
[1]     H1←' R \ S            '
[2]     L1←'--------          '
[3]     H2←,(((S+1),4)⍴' '),('I3'⍕(0,⍳S
      )),((S+1),3)⍴' '
[4]     L2←,((S+1),10)⍴' --------- '
[5]     ' '
[6]     SPACE,H1,H2
[7]     SPACE,L1,L2
    ∇
```

```
      ∇FMT[⎕]∇
    ∇  X FMT Y;ZZ;I;F
[1]     ZZ←(I←1)↑⍴Y
[2]    LP:→(∧/0=⍴F←(WD,DC)⍕Y[I;])/SKIP
[3]     SPACE,((IWD,0)⍕X[I]),'    |   ',
      F
[4]    SKIP:→(ZZ≥I←I+1)/LP
    ∇
```

The program $\nabla PARAM$ displays the parameters and $\nabla PTHEAD$ provides the headings. The program ∇FMT is the format function.

3.2 The main program

```
      ∇MAIN[⎕]∇
    ∇  MAIN;⎕IO;DNF;DNON;UPF;UPON;ID;
      DA;EG;GM;LM;MU;R;S;TH;VM;VP;VZ;
      DC;SPACE;WD;IWD
[1]     ⎕IO←1;SPACE←3⍴' ';WD←10;IWD←5
[2]     IWD←5;WD←10
[3]     INP
[4]     PREP S+1
[5]     ERG
[6]     PTERG
[7]     UPRES
[8]     PTUP
[9]     DNRES
[10]    PTDN
    ∇
```

The main program consists of the eight major subprograms. ∇INP accepts relevant parameter values from the user. Based on those values, the transition rate matrices are prepared in $\nabla PREP$. ∇ERG computes the ergodic probabilities and $\nabla PTERG$ displays the results at the terminal. $\nabla UPRES$ and $\nabla DNRES$ calculate the mean upward and downward first passage times respectively. $\nabla PTUP$ and $\nabla PTDN$ are the corresponding printing functions. Further explanations of the subroutines now follow.

```
      ∇INP[⎕]∇
    ∇  INP
[1]     S←NIP 'ENTER THE NUMBER OF SERV
      ERS.'
[2]     R←NIP 'ERTER THE MAXIMUM NUMBER
       OF MILLING CUSTOMERS.'
[3]     LM←NIP 'ENTER THE ARRIVAL RATE
      LAMBDA.'
[4]     MU←NIP 'ENTER THE SERVICE RATE
      MU.'
[5]     GM←NIP 'ENTER THE RETURN RATE G
      AMMA.'
[6]     TH←NIP 'ENTER THE RETRIAL PROBA
      BILITY THETA.'
    ∇
```

This program accepts relevant parameter values from the user.

```
      ∇PREP[⎕]∇
    ∇ PREP K;V
[1]   VM←GM×V←(1+⍳K)∘.=⍳K
[2]   VM[K;K]←GM×1−TH
[3]   VP←(K,K)⍴((¯1+K*2)⍴0),LM×TH
[4]   VZ←(V×LM)+MU×((⍳K)∘.=1+⍳K)DIAG⍳
      K−1
[5]   ID←(⍳K)∘.=⍳K
    ∇
```

The transition rate matrices $\underline{\underline{\nu}}_r^+$, $\underline{\underline{\nu}}_r^0$, and $\frac{1}{r}\underline{\underline{\nu}}_r^-$ in (3.1) through (3.3) are prepared and stored in ∇P, ∇Z, and VM respectively in this program. We note that both $\underline{\underline{\nu}}_r^+$ and $\underline{\underline{\nu}}_r^0$ are independent of r. Furthermore $\frac{1}{r}\underline{\underline{\nu}}_r^-$ is also independent of r.

```
      ∇ERG[⎕]∇
    ∇ ERG;IDT;VV;VP;J
[1]   EG←((S+1),(R+1))⍴0
[2]   EG[1;]←LEFT(IDT DIAG 1÷+/VP)+.×
      VP+.×SM1
[3]   EG[¯1+J←3;]←(EG[1;]+.×VV←IDT
      DIAG+/VP)÷MU
[4]  LOOP:EG[J;]←((EG[J−1;]+.×VV+IDT×
      (J−2)×MU)−EG[J−2;]+.×VP)÷MU×J−1
[5]   →((S+1)≥J←J+1)/LOOP
[6]   EG←⍉⊖EG÷+/+/EG
    ∇
```

The ergodic probabilities are calculated in this program using Equations (S9) through (S12) in terms of the column structure. The subroutine $\nabla SM1$ returns the intial matrix ${}^c\underline{\underline{s}}_1^-$. Then the lefteigen vector $\underline{a}_0^T$ of $\underline{\underline{\beta}}_0$ in (S9) associated with eigen value one is stored in the first row of the $(S+1)\times(R+1)$ matrix EG. We note that the ergodic probabilities are calculated using the column structure. Therefor the transition rate matrices ${}^c\underline{\underline{\nu}}_m^+$, ${}^c\underline{\underline{\nu}}_m^0$, and ${}^c\underline{\underline{\nu}}_m^-$ in (3.1) through (3.3) and are different from those established in $\nabla PREP$ where the row structure is taken. Hence the variables VV and VP here are localized with respect to ∇ERG. The line [4] is the recursion formula and the normalization of (S12) is done in the line [6].

```
      ∇PTERG[⎕]∇
    ∇ PTERG
[1]   ' '
[2]   'ALIGN PAPER AND HIT RETURN'
[3]   ⍞
[4]   2 1 ⍴' '
[5]   PARAM
[6]   SPACE,' ERGODIC PROBABILITY AT
     ( R , S )'
[7]   SPACE,34⍴'−'
[8]   PTHEAD
[9]   DC←5
[10]  (⌽0,⍳R)FMT EG
    ∇
```

This program prints out the ergodic probabilities.

```
      ∇UPRES[⎕]∇
    ∇ UPRES;MUP;STOP;I;J;SUP;MON
[1]   UPSPE
[2]   UPF←UPON←(R,S+1)⍴R−I←R+2−J←2
[3]   UPON[R;]←UPF[R;]←+/MON←MUP[1;;]
[4]  LP:UPF[I←I−1;]←+/MUP[J;;]
[5]   UPON[I;]←+/MON←(MON+.×SUP)+SUP+
     .×MUP[J;;]
[6]   →(STOP≥J←J+1)/LP
    ∇
```

The mean upward first passage times from (m,r) to any state in the $(r+1)-st$ row as well as those from $(m,0)$ to any state in the r-th row are calculated in this program and stored in UPF and $UPON$ respectively. The first row of the matrices corresponds to the highest row R of the model, and the last row to the lowest row 0. The matrices $\underline{\underline{\mu}}_r^+$ are calculated and stored in MUP in the program $\nabla UPSPE$. The matrix $\underline{\underline{s}}_r^+$ which is independent of r is also stored in SUP in the subroutine. Equation (S6) is used for evaluating $\underline{\underline{\mu}}_{0r}$. The row sums of $\underline{\underline{\mu}}_r^+$ and $\underline{\underline{\mu}}_{0r}$ are the desired results.

```
      ∇PTUP[⎕]∇
    ∇ PTUP
[1]   3 1 ⍴' '
[2]   'ALIGN PAPER AND HIT RETURN.'
[3]   ⍞
[4]   SPACE,' MEAN FIRST PASSAGE TIME
      ( ONE STEP UPWARDS )'
[5]   SPACE,46⍴'−'
[6]   PTHEAD
[7]   DC←1
[8]   (⌽0,⍳R−1)FMT UPF
[9]   3 1 ⍴' '
[10]  'ALIGN PAPER AND HIT RETURN'
[11]  ⍞
[12]  SPACE,' MEAN FIRST PASSAGE TIME
      ( FROM ZERO TO R MILLING CUSTOM
     ERS )'
[13]  SPACE,61⍴'−'
[14]  PTHEAD
[15]  (⌽⍳R)FMT UPON
    ∇
```

This program prints out the results from $\nabla UPRES$. The following two functions are the counterparts of $\nabla URES$ and $\nabla PTUP$ for the downward mean first passage times.

```
      ∇DNRES[⎕]∇
    ∇ DNRES;SDN;MDN;STOP;I;MRN;SRN;J
[1]   DNWARD
[2]   DNF←DNON←(R,S+1)⍴R−1+I←R−2−J←1
[3]   DNF[1;]←DNON[1;]←+/MRN←MDN[R;;]
[4]   SRN←ID
[5]  LOOP:DNF[J←J+1;]←+/MDN[I;;]
[6]   DNON[J;]←+/MRN←(MRN+.×SDN[I;;])
     +(SRN←SRN+.×SDN[I+1;;])+.×MDN[I;
     ;]
[7]   →(STOP≤I←I−1)/LOOP
    ∇

      ∇PTDN[⎕]∇
    ∇ PTDN
[1]   3 1 ⍴' '
[2]   'ALIGN PAPER AND HIT RETURN.'
[3]   ⍞
[4]   SPACE,' MEAN FIRST PASSAGE TIME
      ( ONE STEP DOWNWARDS )'
[5]   SPACE,48⍴'−'
[6]   PTHEAD
[7]   DC←3
[8]   (⌽⍳R)FMT DNF
[9]   3 1 ⍴' '
[10]  'ALIGN PAPER AND HIT RETURN.'
[11]  ⍞
[12]  SPACE,' MEAN FIRST PASSAGE TIME
      (FROM MAX TO R MILLING CUSTOMER
     S )'
[13]  SPACE,60⍴'−'
[14]  PTHEAD
[15]  (⌽0,⍳R−1)FMT DNON
    ∇
```

3.3 An example

To illustrate how to use this software, a simple numerical example is given.

```
      MAIN
ENTER THE NUMBER OF SERVERS.
⎕:
      5
ENTER THE MAXIMUM NUMBER OF MILLING CUSTOMERS.
⎕:
      20
ENTER THE ARRIVAL RATE LAMBDA.
⎕:
      2.0
ENTER THE SERVICE RATE MU.
⎕:
      1.0
ENTER THE RETURN RATE GAMMA.
⎕:
      3.0
ENTER THE RETRIAL PROBABILITY THETA.
⎕:
      0.7

ALIGN PAPER AND HIT RETURN
```

NUMBER OF SERVERS : 5
NUMBER OF MAXIMUM WILLING CUSTOMERS : 20

ARRIVAL RATE : 2.00 SERVICE RATE : 1.00
RETRIAL RATE : 3.00 RETRIAL PROBABILITY : 0.70

ERGODIC PROBABILITY AT (R , S)

R \ S	0	1	2	3	4	5
6	0.00000	0.00000	0.00000	0.00000	0.00000	0.00001
5	0.00000	0.00000	0.00000	0.00001	0.00002	0.00005
4	0.00000	0.00001	0.00004	0.00008	0.00015	0.00024
3	0.00002	0.00010	0.00025	0.00049	0.00078	0.00111
2	0.00022	0.00082	0.00173	0.00270	0.00356	0.00421
1	0.00259	0.00752	0.01188	0.01390	0.01378	0.01249
0	0.12967	0.25934	0.25817	0.16908	0.07959	0.02537

MEAN FIRST PASSAGE TIME (ONE STEP UPWARDS)

R \ S	0	1	2	3	4	5
7	4334030.0	4333107.0	4329341.0	4312360.6	4228374.4	3775704.7
6	558604.9	558413.1	557717.0	554894.8	542228.7	479783.7
5	78888.4	78842.4	78696.1	78169.5	76048.0	66568.7
4	12338.7	12325.7	12290.1	12178.1	11778.3	10177.4
3	2167.7	2163.3	2153.1	2125.4	2039.2	1734.3
2	436.0	434.3	430.7	422.6	400.9	334.2
1	103.1	102.2	100.7	97.8	91.3	74.3
0	29.6	29.1	28.3	27.1	24.7	19.5

MEAN FIRST PASSAGE TIME (FROM ZERO TO R WILLING CUSTOMERS)

R \ S	0	1	2	3	4	5
8	4334406.9	4334406.4	4334405.6	4334404.4	4334402.0	4334396.7
7	558702.1	558701.6	558700.9	558699.6	558697.3	558692.0
6	78918.5	78918.0	78917.2	78916.0	78913.6	78908.3
5	12349.8	12349.3	12348.5	12347.3	12344.9	12339.7
4	2172.3	2171.8	2171.1	2169.8	2167.5	2162.2
3	438.1	437.6	436.8	435.6	433.2	428.0
2	103.8	103.3	102.6	101.3	99.0	93.7
1	29.6	29.1	28.3	27.1	24.7	19.5

MEAN FIRST PASSAGE TIME (ONE STEP DOWNWARDS)

R \ S	0	1	2	3	4	5
20	0.017	0.017	0.017	0.017	0.018	0.047
19	0.018	0.018	0.018	0.018	0.019	0.052
18	0.019	0.019	0.019	0.019	0.020	0.055
17	0.020	0.020	0.020	0.020	0.021	0.058
16	0.021	0.021	0.021	0.021	0.022	0.061
15	0.022	0.022	0.022	0.022	0.024	0.065
14	0.024	0.024	0.024	0.024	0.026	0.069
13	0.026	0.026	0.026	0.026	0.028	0.074
12	0.028	0.028	0.028	0.028	0.030	0.079
11	0.030	0.030	0.030	0.030	0.033	0.086
10	0.033	0.033	0.033	0.034	0.037	0.093
9	0.037	0.037	0.037	0.037	0.041	0.102
8	0.042	0.042	0.042	0.042	0.046	0.112
7	0.048	0.048	0.048	0.048	0.053	0.126
6	0.056	0.056	0.056	0.056	0.063	0.143
5	0.067	0.067	0.067	0.068	0.076	0.165
4	0.083	0.083	0.084	0.085	0.096	0.198
3	0.111	0.111	0.112	0.114	0.130	0.247
2	0.167	0.167	0.168	0.173	0.197	0.338
1	0.334	0.335	0.339	0.351	0.394	0.572

MEAN FIRST PASSAGE TIME (FROM MAX TO R WILLING CUSTOMERS)

R \ S	0	1	2	3	4	5
19	0.017	0.017	0.017	0.017	0.018	0.047
18	0.034	0.034	0.034	0.036	0.068	0.099
17	0.053	0.053	0.056	0.088	0.123	0.154
16	0.073	0.077	0.109	0.145	0.180	0.212
15	0.099	0.133	0.170	0.206	0.241	0.272
14	0.158	0.196	0.234	0.270	0.305	0.337
13	0.226	0.264	0.302	0.338	0.373	0.405
12	0.299	0.337	0.375	0.411	0.446	0.478
11	0.376	0.415	0.453	0.489	0.524	0.555
10	0.460	0.499	0.536	0.573	0.608	0.639
9	0.551	0.590	0.627	0.663	0.698	0.730
8	0.650	0.689	0.726	0.762	0.797	0.829
7	0.759	0.797	0.835	0.871	0.906	0.938
6	0.880	0.919	0.956	0.992	1.027	1.059
5	1.016	1.055	1.093	1.129	1.164	1.195
4	1.174	1.212	1.250	1.286	1.321	1.353
3	1.359	1.398	1.435	1.472	1.507	1.538
2	1.588	1.627	1.664	1.701	1.736	1.767
1	1.896	1.935	1.972	2.008	2.043	2.075
0	2.411	2.450	2.488	2.524	2.559	2.590

ACKNOWLEDGMENT

The work has been motivated by GTE Laboratories, Waltham, Massachusetts. The model formulation has been done through the discussions with A. Hakib and A. Kooharian of GTE Laboratories and J. Keilson of the University of Rochester. The editorial contribution of Caron Clair is gratefully acknowledged. The authors wish to thank GTE Laboratories for its generous support.

REFERENCES

[1] J. W. Cohen (1957), "On the Fundamental Problem of Telephone Traffic Theory and the Influence of Repeated Calls," *Phillips Telecomm. Rev.*, *18*, 49-100.

[2] J. Keilson (1969), "On the Matrix Renewal Function for Markov Renewal Processes," *The Annals of Math. Statist.*, *40*, 1901-1907.

[3] J. Keilson, J. Cozzolino, and H. Young (1968), "A Service System with Unfilled Requests Repeated," *Operations Research*, *16*, 6, 1126-1137.

[4] J. Keilson, U. Sumita, and M. Zachmann (1981), "Row Continuous Finite Markov Chains - Structure and Algorithms," LIDS-P-1078, Laboratory for Information and Decision Systems, Massachusetts Institute of Technology.

[5] L. Kosten (1947), "On the Influence of Repeated Calls in the Theory of Probabilities of Blocking," *DeIngenieur* , *59* (in Dutch).

[6] J. Riordan (1962), *Stochastic Service Systems*, Wiley, New York.

[7] R. Syski (1960), *Introduction to Congestion Theory in Telephone Systems*, Oliver and Boyd, Edinburgh.

[8] R. I. Wilkinson (1956), "Theories for Toll Traffic Engineering in the U.S.A.," *Bell Syst. Tech. J.*, *35*, 421-514.

TELETRAFFIC ISSUES in an Advanced Information Society
ITC-11
Minoru Akiyama (Editor)
Elsevier Science Publishers B.V. (North-Holland)

WAITING TIMES IN A M/M/N/S_1/S_2/HOL-NONPRE QUEUEING SYSTEM

Ralf LEHNERT

TE KA DE Fernmeldeanlagen, Division of Philips Kommunikations Industrie
Nürnberg, F.R.G.

ABSTRACT

In this paper a solution is given for the waiting times in a Markovian queueing system with two priority levels and N servers. The server hunting strategy for high-priority calls is nonpreemptive, that is high-priority calls have to wait until a server becomes idle. The queueing disciplines for both priority levels are FCFS. We assume that calls of both types are associated to separate queues which are bounded to lengths S_1 and S_2, respectively. Furthermore we assume a common exponential service rate for both types of calls. The analysis is done in the steady state of the system.
In the paper a solution is given for the expectation and the complementary distribution function of the waiting time TW for both high- and low-priority calls.
The accuracy of the method is shown by comparison with simulation results.

1. INTRODUCTION

In the past, many single-server priority queueing systems have been analysed, reflecting the strong interest in getting information on the performance of multiprogramming and time-sharing computer systems, e.g. [COHE82], [CONW67],[KLEI76]. For multi-server systems very few solutions exist, e.g. [GAMB83], [KAWA83],[TAYL80]. To our knowledge there is no solution for the waiting time distribution in the case with bounded queues.
An important case is a queueing system with two priority classes, (approximately) Markovian arrival processes and negative exponentially distributed service times. There are numerous applications for queueing systems with two priority classes, e.g. a trunking system where mobile-to-fixed-station calls are prioritized over fixed-to-mobile-station calls. In real life systems queue lengths are usually bounded, especially when delay time bounds are given.

In the paper we derive an analytical solution for the complementary distribution function (CDF) of the waiting time TW for both high- and low-priority calls.

The outline of the paper is as follows: After the definition of the queueing system (sec. 2) we calculate the state probabilities of the system (sec. 3). In sec. 4 the CDF of the waiting time of high-priority calls is calculated. In sec. 5, the central part of the paper, we present an analytical solution for the CDF of the waiting time of low-priority calls. Sec. 6 concludes the paper with some numerical results and a comparison with simulated values.

2. THE QUEUEING SYSTEM

The queueing model is shown in fig. 1. There are two independent traffic sources for high-priority calls (HPCs) and low-priority calls (LPCs). The interarrival times of both traffic streams are assumed to be negative exponentially distributed with arrival rates λ_1 and λ_2, respectively.
The traffic streams are fed into separate queues, which are bounded to S_1 and S_2. Within each queue the queueing discipline is FCFS.
There are N servers. When a server becomes idle, at first the HPC-queue is inspected; if it is empty, calls from the LPC-queue are selected. This is the well-known head-of-the-line strategy [KLEI76].
LPCs in service are not preempted by HPCs.
The service times are assumed to be negative exponentially distributed and common for both HPCs and LPCs. The server hunting strategy is organized such that an idle server is always selected whenever calls are queued.

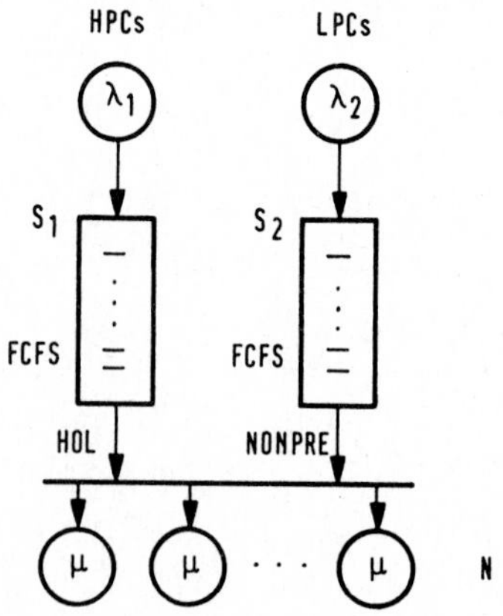

Fig. 1 The queueing model

3. STATE PROBABILITIES

Because of the memoryless processes in the queueing system the state space is rectangular with a dimension (S_1+1, S_2+1).

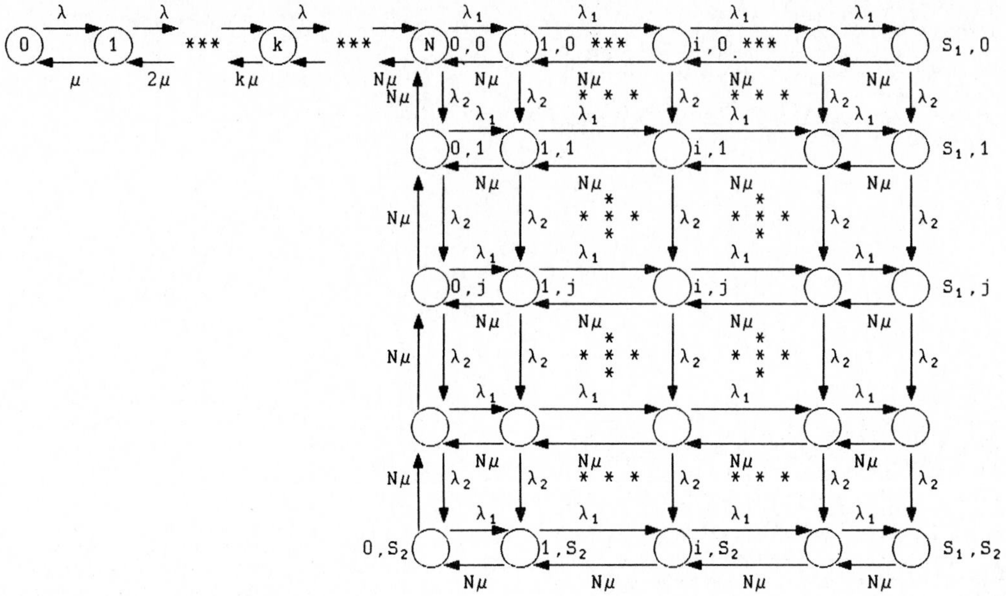

Parameters

$\lambda=\lambda_1+\lambda_2$	total arrival rate		
μ	service rate of one server		
N	no. of servers		
S_1	maximal length of high-priority queue		
S_2	maximal length of low-priority queue		

State probabilities

P_k	$k = 0\ldots N-1$	probability of k servers busy
P_{ij}	$i = 0\ldots S_1$, $j = 0\ldots S_2$	probability that i calls are waiting in the high-priority queue and j calls are waiting in the low-priority queue and all servers are busy

Fig. 2 State space of the queueing system $M/M/N/S_1/S_2$/HOL-NONPRE

At the state (0,0) the vector of the state probabilities (SP) P_k k=0, ... N-1 is appended, which represents the probabilities of k busy servers when both queues are empty. The state space is shown in fig. 2. Because of the steady state Markovian processes all transition probabilities are known.

From the state space we can formulate the equations of statistical balance:

Let

$$\rho = \lambda / \mu \quad , \quad \rho' = \rho / N\mu \quad ,$$
$$\rho_1 = \lambda_1 / N\mu \quad , \quad \rho_2 = \lambda_2 / N\mu \quad ,$$

then

$$(\rho+k)P_k - \rho P_{k-1} - (k+1)P_{k+1} = 0 \qquad k = 1\ldots N-1 \tag{1a}$$
$$(\rho'+1)P_{0,0} - P_{1,0} - P_{0,1} - \rho' P_{N-1} = 0 \tag{1b}$$
$$(\rho'+1)P_{i,0} - \rho_1 P_{i-1,0} - P_{i+1,0} = 0 \qquad i = 1\ldots S_1-1 \tag{1c}$$
$$(\rho_2+1)P_{S1,0} - \rho_1 P_{S1-1,0} = 0 \tag{1d}$$
$$(\rho'+1)P_{0,j} - \rho_2 P_{0,j-1} - P_{0,j+1} - P_{1j} = 0 \qquad j = 1\ldots S_2-1 \tag{1e}$$
$$(\rho_1+1)P_{0,S2} - \rho_2 P_{0,S2-1} - P_{1,S2} = 0 \tag{1f}$$
$$(\rho'+1)P_{i,j} - \rho_1 P_{i-1,j} - P_{i+1,j} - \rho_2 P_{i,j-1} = 0 \qquad i = 1\ldots S_1-1 \; ; \; j = 1\ldots S_2-1 \tag{1g}$$
$$(\rho_1+1)P_{i,S2} - \rho_1 P_{i-1,S2} - P_{i+1,S2} - \rho_2 P_{i,S2-1} = 0 \qquad i = 1\ldots S_1-1 \tag{1h}$$
$$(\rho_2+1)P_{S1,j} - \rho_2 P_{S1,j-1} - \rho_1 P_{S1-1,j} = 0 \qquad j = 1\ldots S_2-1 \tag{1i}$$
$$P_{S1,S2} - \rho_1 P_{S1-1,S2} - \rho_2 P_{S1,S2-1} = 0 \tag{1k}$$

The system of equations can be solved by the well-known Gauss-Seidel elimination method. In systems with long queues S_1 or S_2 we have large matrices which may become computationally complex. We therefore investigated different sparse matrix methods, and solution programs, such as [ZLAT81]. Finally we implemented an algorithm, which makes efficient use of the band-structure of the matrix. With this algorithm the CPU-time needed to compute all state probabilities of a queueing system with e.g. 389 equations is less than 0.5 sec on a VAX-11/780 computer.

By summing up the state probabilities at the vertical or horizontal end of the state space, we get the blocking probability B_1 at the high-priority and B_2 at the low-priority queue, respectively.

$$B_1 = \sum_{j=0}^{S_2} P_{S1,j} \tag{2a}$$

$$B_2 = \sum_{i=0}^{S_1} P_{i,S2} \tag{2b}$$

4. WAITING TIME OF HIGH-PRIORITY CALLS

HPCs waiting in the high-priority queue are not influenced by arriving or waiting LPCs. This is because of the overall Markovian behaviour of the system and the nonpreemptive queueing discipline.

Therefore we can calculate the waiting time TW_1

of a HPC arriving at the state (i,x) by a i - times convolution of the total service time TH_t. TH_t is given by the output process of all N servers being active; it has a mean value of $E\{TH_t\} = 1/N\mu$.

So we get the waiting time of all accepted HPCs with mean

$$E\{TW_1\} = \sum_{i=0}^{S_1-1} \sum_{j=0}^{S_2} P_{ij} \cdot \frac{i+1}{N\mu(1-B_1)} \qquad (3)$$

and CDF

$$F^c_{TW1}(t) = \frac{e^{-tN\mu}}{1-B_1} \cdot \sum_{i=0}^{S_1-1} \sum_{j=0}^{S_2} P_{ij} \cdot \sum_{m=0}^{i} \frac{(tN\mu)^m}{m!} \qquad (4)$$

The results are similar to a M/M/N/S-system without priority but with the difference that all state probabilities also depend on the arrival rate λ_2.

5. WAITING TIME OF LOW-PRIORITY CALLS

LPCs proceed through the queueing system in the following way:
A LPC arrives with probability $P(i_s,j_s)$ at the system's state (i_s,j_s). This results in a transition to the new state (i_s,j_s+1). Now the waiting time TW_2 begins; it ends when this LPC arrives at the state (0,0) which means that it enters service.
At first, the LPC has to wait until the high-priority queue becomes empty. Then a server must become idle before the LPC can proceed from state (0,j) to state (0,j-1) (LPC-down-step).
In the meantime HPCs may arrive and start a new random walk in the high-priority queue.
In this case the total time for a LPC down-step is lengthened by the random walk cycle. The procedure of random walk cycles and LPC down-steps repeats, until this LPC has reached state (0,0).
Because of the FCFS queueing discipline we have to take into account only the LPCs which are in front of the test-LPC; other LPCs which arrive later do not influence the behaviour of the test-LPC.

From the system's behaviour the computation procedure can be derived directly:

At first we have to compute the probability $P_2(b,i)$ that starting from state (i,j) there are b-1 HPC down-steps and 1 LPC down-step to proceed to state (0,j-1). During this random walk we must take into account that paths beyond the maximal queue length S_1 are not allowed. The probability for these fictive paths is included in the blocking probability B_1.

A LPC arriving at state (i_s,j_s) has to go through j_s random walks; the first one starting in column i_s of row j_s+1; the other ones starting in column 0 of the rows $1...j_s-1$.
We call $P_3(b,i_s,j_s)$ the probability that b down-steps occur on that path.

Weighting $P_3(b)$ with the state probabilities that are allowed at the arrival of a LPC, we get a probability $P_4(b)$ that b down-steps occurs between an arrival of a LPC and the beginning of its service.

As all the down-steps are exponential phases with equal service rate $N\mu$, we are now able to calculate the CDF of TW_2 via weighted Erlang-distributions where b is the number of phases.

5.1 Step Probabilities of the Random Walk

A random walk of a LPC occurs in one row of the two-dimensional state space; it includes the transition to the next (upper) row at column i=0. There is a transition in upward direction, when an input of a HPC occurs before an output of a HPC or a LPC from the server pool and a transition in downward direction, when the server pool is faster.
Because of the overall Markovian system both the input and the output process are independent negative exponentially distributed with constant intensity λ_1 and $N\mu$. So the step-up probability p_u is

$$p_u = \int_{\tau=0}^{\infty} f_i(\tau) \cdot F^c_o(\tau)\, d\tau = \lambda_1/(\lambda_1+N\mu) \qquad (5a)$$

and the step-down probability p_d

$$p_d = \int_{\tau=0}^{\infty} f_o(\tau) \cdot F^c_i(\tau)\, d\tau = N\mu/(\lambda_1+N\mu). \qquad (5b)$$

There is one exception: at state (S_1,x) $x=0...S_2$, there is $p_u = 0$ and $p_d = 1$.

5.2 Number of Paths in the Random Walk

Fig. 3 shows a possible path of a random walk with a starting-point i=1, a HPC queue length S_1=5, k=2 touches of the bound S_1, b=10 down-steps and a total of n=17 steps. The ending-point at "altitude" -1 denotes that the transition of a LPC into service is included in the random walk.

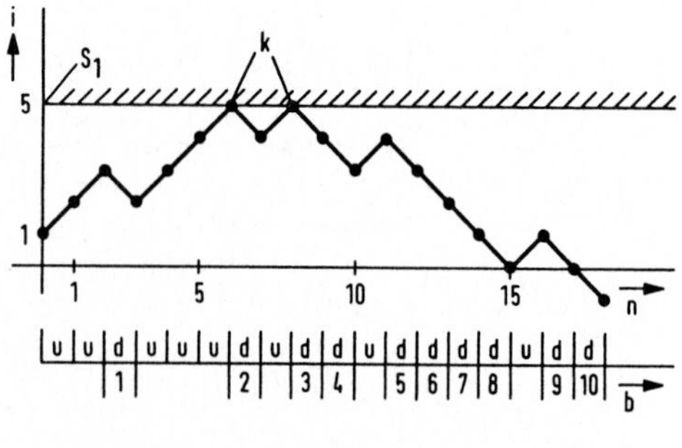

Fig. 3 Typical path of a random walk

There are $M(i,S_1,k,n)$ possible paths of length n to go from starting-point $i=0...S_1$ to the ending-point -1 under the condition that the upper bound S_1 is touched k times.
M can be computed very effectively by methods of coding theory.

5.3 Random Walk

The probability P_1 for a LPC to select a definite path of the random walk depends on the number of up-steps N_u, down-steps N_d and k.

$$P_1 = p_u^{Nu} \cdot p_d^{Nd-k} \cdot 1^k \qquad (6)$$

Rewriting eq. 6 in n and i we have

$$P_1(n,i,S_1,k) = p_u^{(n-i)/2} \cdot p_d^{(n+i)/2-k} \, . \qquad (7)$$

Now we can calculate the probability that any path of length n starting at i occurs, regardless of k :

$$P_2(n,i,S_1) = \sum_{k=0}^{k_{max}} P_1(n,i,S_1,k) \cdot M(i,S_1,k,n) \qquad (8)$$

$$k_{max} = n-(2S_1-i)-1$$

It can be shown easily that P_2 is normalized to one.

Because only the down-steps contribute to the waiting time TW_2, we rewrite P_2 in $P_2(b,i,S_1)$. Obviously, the number b of down-steps is b = (n+i+1)/2.

Now we can interprete P_2 as the number of down-steps which occur when there is a transition from state (i,j) to state (0,j-1), i=0...S_1, j=1...S_2.

5.4 Chaining of Random Walks

A LPC which arrives at state (i_s,j_s) initiates a transition of the system's state to (i_s,j_s+1). From (i_s,j_s+1) there is a first random walk with starting-point i = i_s ending at $(0,j_s)$. Then the next random walk starts with i=0 and ends at state $(0,j_s-1)$. This procedure repeats j_s-1 - times until state (0,0) has been reached. Fig. 4 shows an example of this chaining of random walks for $i_s=2$, $j_s=2$ in the state space of a $S_1=5$, $S_2=4$, N=4 system.

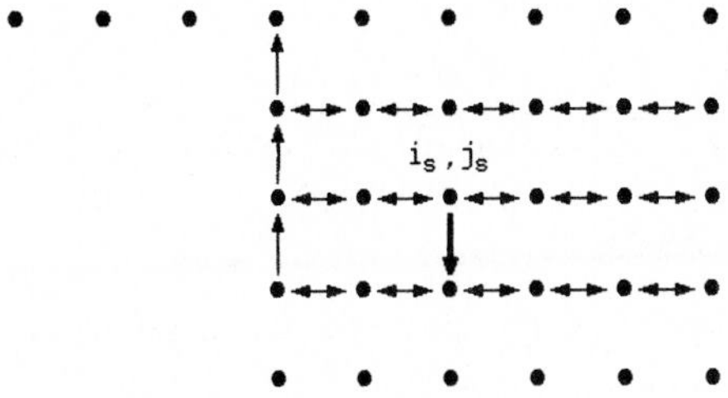

Fig. 4 Chain of random walks in a state space $S_1=5$, $S_2=5$, N=4. Starting point (2,2)

Because of the Markov property of the queueing system all random walks are independent; so the number of down-steps on a chain of random walks is the sum of the number of steps of the involved random walks. This is the convolution of $P_2(i_s)$ with the j_s-times convolution of $P_2(0)$.

We calculate the probability function (PF) $P_3(b,i_s,j_s,S_1)$ that b down-steps occur:

$$P_3(b,i_s,j_s,S_1) = \qquad (9)$$

$$\mathrm{conv}\{P_2(b,i_s,S_1),\mathrm{conv}^{j_s-1}\{P_2(b,0,S_1),P_2(b,0,S_1)\}\}$$

P_3 is normalized to one.

5.5 Probability Function for the Number of Down-Steps

By weighting P_3 with the system state probabilities we get a total PF $P_4(b)$

$$P_4(b) = \sum_{i_s=0}^{S_1} \sum_{j_s=0}^{S_2-1} P_3(b,i_s,j_s,S_1) \cdot P_{is,js} \qquad b=1\ldots\infty \qquad (10)$$

$$= \sum_{k=0}^{N-1} P_k \qquad b=0$$

which gives the probability that b down-steps occur for any LPC that is accepted by the system. The row $j_s=S_2$ is excluded, because LPCs which arrive at these states are blocked.

5.6 Complementary Distribution Function

All down-steps occur when all servers are busy. So, during the chain of random walks, a LPC "sees" a constant service rate $N\mu$. Therefore we finally calculate the CDF as the b-times convolution of exponential service phases weighted with $P_4(b)$.

$$F^c_{TW2}(t) = \frac{e^{-N\mu t}}{1-B_2} \cdot \sum_{b=0}^{\infty} P_4(b) \cdot \sum_{r=0}^{b-1} \frac{(N\mu t)^r}{r!} \qquad (11)$$

The normalization condition is introduced by the denominator $1-B_2$, which regards the omission of row S_2 in PF $P_4(b)$.

As we can see, this stepwise analysis does not give a compact formula for the CDF of TW_2. But the solution procedure is straightforward and computational tractable. For instance, the CDFs of a system with $S_1=S_2=10$, N=5 and $A_1=A_2=2$ can be computed in less than one minute.

We further see that the traffic intensity of LPCs does not explicitely appear in eq. 11, its influence is hidden in the state probabilities P_{ij}. The traffic intensity of HPCs is introduced by the step-probabilities of eq. 5.

6. NUMERICAL EXAMPLES

We will now give some numerical examples and compare the results with results from [TAYL80] and simulation runs.

Fig. 5 shows the PF $P_4(b)$ in a queueing system with symmetric offered traffic $A_1=A_2=2$, queue lengths $S_1=5$, $S_2=15$ and N=5 servers.

We see an excellent agreement between our calculation and the simulation results.

Fig. 6 shows the CDFs of the waiting times of this system.

We have chosen the parameters so that the blocking probabilities are below 1 percent in order to be able to compare the results with the formulas of [TAYL80], which are valid for a

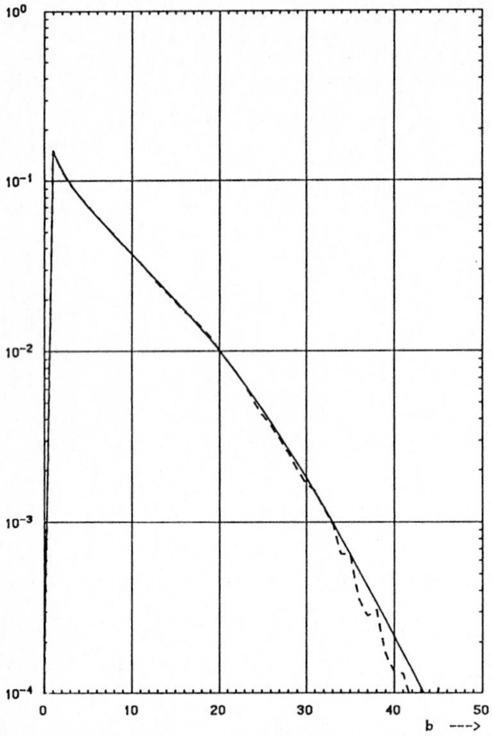

Fig. 5 Queueing system M/M/5/5/15/HOL-NONPRE
Probability function $P_4(b)$ (eq. 10)
$A_1=2$ $A_2=2$
Solid line : calculation
Dashed line: simulation

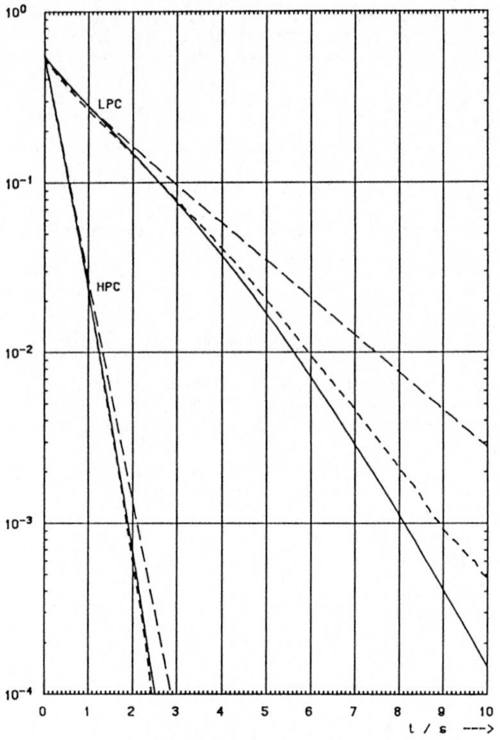

Fig. 6 Queueing system M/M/5/5/15/HOL-NONPRE
CDF of waiting Times TW_1 and TW_2
$A_1=2$ $A_2=2$
Solid lines : calculation
Short dashed lines: simulation
Long dashed lines : calculation from [TAYL80] (infinite queues)

system with infinite queues.

We see an excellent agreement of the TW_1 curves. Because of the small blocking probability B_1 this is true also for the [TAYL80] curve.
The TW_2 curves show that in the infinite queue system there are bigger probabilities for long random walks and therefore there is a bigger coefficient of variation. We think that the deviation of the simulated from the calculated TW_2 is induced by inaccuracies (correlations) in the random number generators of the simulation; the means $E\{TW_2\}$ of calculation and simulation, however, agree very well.

7. CONCLUSION

We presented a method to calculate the complementary distribution function of the waiting time of low-priority calls in a $M/M/N/S_1/S_2$/HOL-NONPRE queueing system.
The method is straighforward and can easily be implemented. It is very accurate with respect to the mean value and also the probability function of the number of down-steps occuring during a random walk induced by arriving high-priority calls.
We think this method can be extended to the cases

- more priority classes
- different service rates per priority class.

ACKNOWLEDGEMENT

The author thanks Dipl.-Ing. H. Jüchter for many helpful discussions and his computing efforts.

REFERENCES

COHE82 Cohen, J.W.: The single server queue. North Holland Publishing Company, Amsterdam, 1982

CONW67 Conway, R.W., et al: Theory of scheduling. Addison Wesley, Reading, Massachusetts, 1967

DAVI66 Davies, R.H.: Waiting-time distribution of a multi-server, priority queueing system. Op. Res. 14, 1968, pp. 133-136

GAMB83 Gambe, E.: Overview in the general area of basic traffic theory and queueing. Proc. 10th Int. Teletraffic Congress, Montreal, 1983, session 4.1, paper no.1

KAWA83 Kawashima, K.: Efficient numerical solutions for a unified trunk reservation system with two classes. Rev. of the El. Comm. Lab. (Japan) 31, 1983, pp. 419-429

KLEI76 Kleinrock, L.: Queueing systems vol. II, Computer applications. John Wiley, New York, 1976

TAYL80 Taylor, I.D.S., Templeton, J.G.C.: Waiting time in a multi-server cutoff-priority queue, and its application to an urban ambulance service. Op. Res. 28, 1980, pp. 1168-1188

ZLAT81 Zlatev, et al: Y12M, Solution of large and sparse systems of linear algebraic equations. Lecture notes in computer science 121, Springer, Berlin, 1981

TELETRAFFIC ISSUES in an Advanced Information Society
ITC-11
Minoru Akiyama (Editor)
Elsevier Science Publishers B.V. (North-Holland)

M/G/1 QUEUE WITH N-PRIORITIES AND FEEDBACK: JOINT QUEUE-LENGTH DISTRIBUTIONS AND RESPONSE TIME DISTRIBUTION FOR ANY PARTICULAR SEQUENCE

B. FONTANA & C. DIAZ BERZOSA

STANDARD ELECTRICA, S.A. RESEARCH CENTER
MADRID, SPAIN

ABSTRACT

This paper considers a queueing system consisting of an M/G/1 queue with N non preemptive priorities where, after a service completion, several new tasks can be queued again with the same or different priority (multiple feedback).

Some performance measures of this model are studied. The results include formulae for the response time distribution of any particular sequence of tasks and for some stationary queue length distributions. These distributions turn out to have a product form. These results enable us to estimate the behavior of the actual processor systems.

1. INTRODUCTION

Actual distributed systems (telecommunication systems, computer centers and networks, local area networks, etc) can generally be characterized as consisting of a set of both hardware resources and processors and a set of tasks, jobs, or messages competing for and accessing those resources. For many types of systems the key resources are the processors; within a processor, task requests appear in sequences. Each sequence is defined by the number and arrangement of tasks, and the priority and processing time of each task. Because there are multiple tasks competing for the processor. it is therefore natural to represent the processor by a queue model. The purpose of the model is to predict the performance by estimating characteristics of the processor occupancy, the queue length and the queueing delays. Analytic performance models are queue models for which these characteristics may be found analytically.

Regarding the structure and behavior of these processors, we consider a single server queue with N non preemptive priorities, FCFS discipline and multiple feedback (after a service completion, several tasks can be queued again with the same or different priority). For each priority, the external arrival process obeys a Poisson law but the service times are governed by a general distribution. We think the above queue model provides insight into, the key parameters affecting the system performance, with sufficient realism.

On the other hand, from a customer's point of view possibly the most important performance measure is the response time, defined as the time spent by a particular sequence of tasks from its arrival to its service completion. Clearly, response time expectations are important in projecting distributed systems, and a typical request of Administrations (as it is reflected in the CCITT recommendations) is that at least 95% of all executed jobs do not exceed a given value of time in travelling through the system; this is a question about the distribution of response time.

During the last years, analytic performance modeling has been an extremely useful tool and an important effort has been made in looking for analytical solutions for the distribution of response time in several queueing models (see e.g. /1/ to /4/), but the results obtained are quite limited in that they determine the response time distribution of a particular sequence in complex actual processors.

In this paper we obtain an exact solution for the Laplace-Stieltjes transform (LST) of the response time distribution for a particular sequence, with any number of tasks processed in any priority in which the processing times are random variables distributed according to different general laws in the described queueing model. Furthermore, this exact solution for LST of the response time distribution has a simple product-form expression which can be easily inverted by numerical methods.

In Section 2 we give a more detailed description of the model, and we study the stationary conditions and carried traffics in Section 3. In Section 4 we define the "engaged time", an important stochastic process in queueing systems; this process allows us to derive, in Section 5, the generating functions (GF) of the queue length distributions at departure, output and random instants and, in Section 7, the LST of response time distribution for any particular sequence. In Section 6, we deal with an extension of the "Arrival Theorem". Finally. in Section 8, the numerical results for response time distribution are compared with those obtained from an actual processor simulation.

2. MODEL DESCRIPTION

Consider a single server queue with N non preemptive priorities and multiple interqueue feedback. Tasks arrive to each priority class according to a Poisson stream with mean rate $\eta_i (i=1,.,N)$. The service times of the i-th class are identically and independently distributed random variables with a distribution function $S_i(t)$ whose average value is h_i.

The assumptions of the priority discipline are: a) a task of the i-th class, if present, is always taken for service prior to a task of the j-th class, if i<j and b) the service of any task is continued to completion and it cannot be preempted by the arrival of a task with lower priority index.

Within each priority class, the service discipline is the well-known first come-first served (FCFS) discipline.

The feedback assumptions are: after being served an i-th class task, immediately $\mathbf{m}=(m_1,.,m_N)$ tasks are generated and queued with probability $f^{i)}(\mathbf{m})$ at the first,.,N-th priority, respectively. The probability that an i-th class task leaves the system is $f^{i)}(\mathbf{o})$. These events are independent of any other event involved.

3. CARRIED TRAFFIC AND STATIONARY CONDITIONS

Before presenting the main results of the model, we study the carried traffic for each priority, the total carried traffic and the system stationary conditions.

We introduce the vector and matrix notations: $\boldsymbol{\eta}=(\eta_1,.,\eta_N)$, $\boldsymbol{\eta}^*=(\eta^*_1,.,\eta^*_N)$, $\mathbf{n}=(n_{ij})$ for $(i,j=1,.,N)$, where n_{ij} is the mean number of immediately generated tasks to the j-th queue by a task from the i-th queue, and η^*_i can be interpreted as the relative arrival rate to the i-th priority class (including feedbacks). With these considerations, the following equation can be written:

$$\boldsymbol{\eta}^*=\boldsymbol{\eta}+\boldsymbol{\eta}^*\mathbf{n} \tag{1}$$

If $\mathbf{I}$ is the NxN unit matrix:

$$\boldsymbol{\eta}^*=\boldsymbol{\eta}(\mathbf{I}-\mathbf{n})^{-1} \tag{2}$$

The carried traffic by the i-th priority class, ρ^*_i, is given by:

$$\rho^*_i=\eta^*_i h_i \tag{3}$$

and therefore the total carried traffic, ρ^*, is the sum of the above traffics.

Let a_{ij} be the element of i-th row and j-th column of the matrix $(\mathbf{I}-\mathbf{n})^{-1}$, which can be interpreted as the average number of tasks generated (after all possible feedbacks) from a fresh task in i-th priority to the j-th priority.

The stationary conditions are:

$$\left.\begin{array}{lll} 1)\ a_{ii}\geq 1 & i=1,.,N & \\ 2)\ a_{ij}\geq 0 & i,j=1,.,N; & i\neq j \\ 3)\ \rho^*<1 & & \end{array}\right\} \tag{4}$$

Evidently, the values n_{ij} can be obtained from the GF $F^{i)}(\boldsymbol{\beta})$ $\boldsymbol{\beta}=(\beta_1,.,\beta_N)$ of $f^{i)}(\mathbf{m})$, in the usual way.

4. ENGAGED-TIME PROCESS

In this section, the "engaged-time" process will be studied. This important stochastic process in queueing system was introduced in /3/, and it will be used to study the stationary queue length probabilities and response time distribution.

We define:

The "engaged time in j-th priority due to an i-th priority task" is the period of time which begins when an i-th priority task starts its service and ends when the server is free, or it initiates service of a k-th priority task (k>j).

Let ${}_{i,j}B(\mathbf{n}^{j+1};t)$ be the density function for the engaged time in j-th priority due to an i-th priority task during which $\mathbf{n}^{j+1}=(n_{j+1},.,n_N)$ new tasks (either due to the Poissonian or the feedback processes) have been generated at the (j+1)-th,.,N-th priority.

Then, following the same reasoning as in /3/, it is obtained

$${}_{i,j}B(\mathbf{n}^{j+1};t)=\sum_{\forall\mathbf{s}(o)}\int_o^t a^{i)}(\mathbf{s}(o);x_o)dx_o$$

$$\prod_{k=1}^{j}\Big(\sum_{\forall\mathbf{s}(k)}\int_o^t {}_{k,j}B^{*m_{o,k})}(\mathbf{s}(k);x_k)dx_k\Big)$$

$$\text{for }\sum_k x_k=t\ ;\quad o\leq k\leq j \tag{5}$$

where ${}_{k,j}B^{*m_{o,k})}(.;.)$ denotes the $m_{o,k}$-th iterated convolution of ${}_{k,j}B(.;.)$ with itself, $\mathbf{s}(o)$ is a N-tuple whose n-th component is $m_{o,n}$ and $\mathbf{s}(v)$, $1\leq v\leq j$ is an (N-j)-tuple whose n-th component is $m_{v,j+n}$. All elements $m_{v,u}$ are integer values ($o\leq m_{v,u}$) which verify the condition:

$$\sum_{v'} m_{v',u}=n_u \text{ for } j<u\leq N,\ o\leq v'\leq j.$$

$a^{i)}(\mathbf{m};x)$ is the density function for the duration of a service of i-th priority in which $\mathbf{m}=(m_1,.,m_N)$ tasks are offered to first,.,N-th priority either due to the Poissonian process or the feedback generated after completion of that service.

The GF and LST $\overline{A}^{i)}(\boldsymbol{\beta};s)$, where $\boldsymbol{\beta}=(\beta_1,..\beta_N)$, can be deduced given that $a^{i)}(\mathbf{m};x)$

is the convolution of two independent processes: the feedback process at the end of the service of i-th priority and the Poisson process during the duration of that service. If $\bar{S}_i(s)$ is the LST of $S_i(t)$ with $Re(s)\geq o$, then $\bar{A}^{i)}(\mathbf{ß};s)$ is given by:

$$\bar{A}^{i)}(\mathbf{ß};s)=F^{i)}(\mathbf{ß})\bar{S}_i(s^*) \quad ; \ |ß_j|\leq 1 \qquad (6)$$

where $s^*=s+(\mathbf{e}-\mathbf{ß})\mathbf{\eta}^T$, and $\mathbf{e}$ is the N row vector in which all components are 1.

The GF and LST ${}_{i,j}\bar{B}(\mathbf{ß}^{j+1};s)$, with $|ß_v|\leq 1$, $Re(s)\geq o$ and $\mathbf{ß}^{j+1}=(ß_{j+1},.,ß_N)$, is:

$${}_{i,j}\bar{B}(\mathbf{ß}^{j+1};s)=\bar{A}^{i)}({}_{1,j}\bar{B}(\mathbf{ß}^{j+1};s),\ldots,$$
$${}_{j,j}\bar{B}(\mathbf{ß}^{j+1};s),\mathbf{ß}^{j+1};s) \qquad (7)$$

these engaged times are obtained as:

* for $i\leq j$; the j roots with the smallest absolute value of the above system of equations, and
* for $j<i\leq N$; they are calculated substituting the above roots in (7).

5. STATIONARY QUEUE LENGTH PROBABILITIES

The key is the stationary queue length probabilities at departure points (points in which the service of a task and the feedback mechanism are just finished). These probabilities can be obtained because a discrete time imbedded Markov chain is generated if the queue process is observed only at those points.

Once these probabilities have been evaluated, it is possible to obtain the stationary joint queue length probabilities at a random and output instants.

To characterize these stationary probabilities, we follow the same technique given in /5/.

5.1 AT DEPARTURE POINTS

Consider the instants $t=t_o,t_1,.$ as the departure points defined above. Let $(\mathbf{Q})_n=(Q_1,.,Q_N)_n$ (n=o,1,.) denote the number of tasks in the first,.,N-th priority at instant t.

Let $q(\mathbf{m})$ be the stationary probability associated with the queue length at t_o ,t_1,. instants, being $\mathbf{m}=(m_1,.,m_N)$; $m_1,.,m_N\geq o$.

$$q(\mathbf{m})=\lim_{n\to\infty} P[(\mathbf{Q})_n=\mathbf{m}] \qquad (8)$$

With the present assumptions, the process $[(\mathbf{Q})_n]_{n\geq o}$ is a Markov chain. If $q(\mathbf{m})$ exists, the following equations must be verified:

$$q(\mathbf{m})=q(\mathbf{o})\sum_{1\leq j\leq N} a^{j)}(\mathbf{m};o)\eta_j/\eta+$$
$$+\sum_{1\leq i\leq N}\ \sum_{\forall \mathbf{s}_i} q(\mathbf{s}_i)a^{i)}(\mathbf{r}_i;o)$$
$$\sum_{\forall\mathbf{m}} q(\mathbf{m})=1 \qquad (9)$$

where $\mathbf{s}_i=(o,.,k_i,.,k_N)$ and $\mathbf{r}_i=(m_1,.,m_{i-1},k'_i,.,k'_N)$, and $k_i+k'_i=m_i+1$ with $1\leq k_i\leq m_i+1$ and $k_v+k'_v=m_v$ with $o\leq k_v\leq m_v$ $(i<v\leq N)$ (10) and:

$$\eta = \mathbf{e\eta}^T \qquad (11)$$

Let $Q(\mathbf{ß})$ be the GF of $q(\mathbf{m})$, then from (9), we deduce:

$$\mathbf{q}^-(\mathbf{ß})[\mathbf{a}(\mathbf{ß})]^T=Q(\mathbf{o})\mathbf{\eta}[\mathbf{b}(\mathbf{ß})]^T/\eta \qquad (12)$$

$\mathbf{q}^-(\mathbf{ß})$ is a N row vector whose i-th element is $Q(\mathbf{ß}_i)-Q(\mathbf{ß}_{i+1})$, where $(\mathbf{ß}_j)=(o,.,ß_j,.,ß_N)$ with $(\mathbf{ß}_1)\equiv(\mathbf{ß})$ and $(\mathbf{ß}_{N+1})\equiv(\mathbf{o})$; $[\mathbf{a}(\mathbf{ß})]^T$ is a column vector whose i-th component is $1-\bar{A}^{i)}(\mathbf{ß};o)/ß_i$ and $[\mathbf{b}(\mathbf{ß})]^T$ is a column vector whose i-th component is $\bar{A}^{i)}(\mathbf{ß},o)-1$.

Substituting $ß_i$ by ${}_{i,j}\bar{B}(\mathbf{ß}^{j+1};o)$ (i=1,.,j) in (12) for j=1,.,N-1, a system of N-1 equations is obtained. From this system and the formula (12) the following matricial expression is derived:

$$[\mathbf{q}^-(\mathbf{ß})]^T=Q(\mathbf{o})[\mathbf{B_T}(\mathbf{ß})]^{-1}\mathbf{B}(\mathbf{ß})\mathbf{\eta}^T/\eta \qquad (13)$$

where $\mathbf{B_T}(\mathbf{ß})$ is a NxN triangular matrix defined by:

$$(B_T(\mathbf{ß}))_{i,j}=o \quad \text{for } i>j$$
$$(B_T(\mathbf{ß}))_{i,j}=1-{}_{j,i-1}\bar{B}(\mathbf{ß}^i;o)/ß_j \quad \text{for } 1<i\leq j$$
$$(B_T(\mathbf{ß}))_{1,j}=1-{}_{j,o}\bar{B}(\mathbf{ß};o)/ß_j\equiv$$
$$\equiv 1-\bar{A}^{j)}(\mathbf{ß};o)/ß_j \qquad (14)$$

and $\mathbf{B}(\mathbf{ß})$ is a NxN matrix defined by:

$$(B(\mathbf{ß}))_{i,j}={}_{j,i-1}\bar{B}(\mathbf{ß}^i;o)-1 \quad \text{for } i>1$$
$$(B(\mathbf{ß}))_{1,j}={}_{j,o}\bar{B}(\mathbf{ß};o)-1\equiv\bar{A}^{j)}(\mathbf{ß};o)-1 \qquad (15)$$

From (12), we finally obtain the GF of the queue lenght probabilities, $Q(\mathbf{ß})$, at departure points.

$$Q(\mathbf{ß})=Q(\mathbf{o})\mathbf{e}[I+[\mathbf{B_T}(\mathbf{ß})]^{-1}\mathbf{B}(\mathbf{ß})]\mathbf{\eta}^T/\eta \qquad (16)$$

$Q(\mathbf{o})$ can be obtained taking into account that $Q(\mathbf{e})=1$, then:

$$Q(\mathbf{o})=\eta/\lim_{(\mathbf{ß})\to(\mathbf{e})} \mathbf{e}[I+[\mathbf{B_T}(\mathbf{ß})]^{-1}\mathbf{B}(\mathbf{ß})]\mathbf{\eta}^T \qquad (17)$$

As can be seen, equation (16) is a matrix product formula.

5.2 AT RANDOM INSTANTS

A fundamental function to evaluate the response time distribution of any particular linear sequence is the joint mixed distribution function $R^{i)}(\mathbf{n};x)$ that at a random instant, an i-th priority task is in service, there are $\mathbf{n}=(n_1,..,n_N)$ task in first,..,N-th priority (not including the task in service) respectively, and the time needed to finish the current service is less than or equal to x.

To obtain this distribution, we first evaluate the joint mixed distribution that at instant t between two departure points, the server is busy with a task of i-th priority, the length of current service time is less than or equal to t', the time needed to complete this service is less than or equal to y, and the number of tasks present in first,..,N-th priority (not including the task in service), is $\mathbf{m}=(m_1,..,m_N)$.

Let $P^{i)}(\mathbf{m};t',y;t)$ be this distribution function and let $p^{i)}(\mathbf{m};t',y;t)$ denote a function such as:

$$P^{i)}(\mathbf{m};t',y:t)=\int_o^{t'}dx\int_o^{y}p^{i)}(\mathbf{m};x,z;t)dz \qquad (18)$$

Following the same methodology as in /5/ and using the queue length probabilities $q(\mathbf{m})$ at departure points, we obtain:

$$p^{i)}(\mathbf{m};t',y;t)=Q(\mathbf{o})\eta_i e^{-\eta(t-t'+y)}\,\Omega(\mathbf{m};t'-y)\frac{dS_i(t')}{dt'}$$

for $o\le y\le t'<t+y;\ t\ge o$ (19)

where $\Omega(\mathbf{m};x)$ is the probability of $\mathbf{m}=(m_1,..,m_N)$ arrivals to first,..,N-th priority during the time interval x given by the Poisson law.

$$p^{i)}(\mathbf{m};t',y;t)=\sum_{\forall \mathbf{s}_i} q(\mathbf{s}_i)\Omega(\mathbf{r}_i;t)\frac{dS_i(t')}{dt'}$$

for $t'=t+y;\ y\ge o;\ t\ge o$ (20)
with conditions defined in (10).

Otherwise:

$$p^{i)}(\mathbf{m};t',y;t)\equiv o \qquad (21)$$

The probability that a time t the system is empty $P_o(t)$, is given by:

$$P_o(t)=q(\mathbf{o})e^{-\eta t} \qquad (22)$$

Let $\bar{P}^{i)}(\boldsymbol{\beta};w_1,s;w_2)$ be the GF in $\mathbf{m}$ and LST in t', y and t; with $\boldsymbol{\beta}=(\beta_1..,\beta_N)$, and $|\beta_i|\le 1;\ Re(w_1),Re(s),Re(w_2)\ge o$.

It follows from (19), (20), and (21):

$$\bar{P}^{i)}(\boldsymbol{\beta};w_1,s;w_2)=[\bar{S}_i(w_1+w_2+(\mathbf{e}-\boldsymbol{\beta})\boldsymbol{\eta}^T)-\bar{S}_i(s+w_1)][\eta_i Q(\mathbf{o})/(\eta+w_2)+(Q(\boldsymbol{\beta}_i)-Q(\boldsymbol{\beta}_{i+1}))/\beta_i]/(s-w_2-(\mathbf{e}-\boldsymbol{\beta})\boldsymbol{\eta}^T) \qquad (23)$$

The N above equations can be written as the following matrix expression:

$$[\mathbf{p}(\boldsymbol{\beta};w_1,s;w_2)]^T=Q(\mathbf{o})\mathbf{D}(w_1,w_2;\boldsymbol{\beta};s)[\mathbf{I}/(\eta+w_2)+[\mathbf{C}(\boldsymbol{\beta})]^{-1}\mathbf{B}(\boldsymbol{\beta})/\eta]\boldsymbol{\eta}^T/(s-w_2-(\mathbf{e}-\boldsymbol{\beta})\boldsymbol{\eta}^T) \qquad (24)$$

where $\mathbf{D}(w_1,w_2;\boldsymbol{\beta};s)$ is a NxN diagonal matrix whose i-th diagonal element is $\bar{S}_i(w_1+w_2+(\mathbf{e}-\boldsymbol{\beta})\boldsymbol{\eta}^T)-\bar{S}_i(s+w_1)$ and $\mathbf{C}(\boldsymbol{\beta})$ is a NxN triangular matrix defined by $(\mathbf{C}(\boldsymbol{\beta}))_{i,j}=o$ for i>j and $(\mathbf{C}(\boldsymbol{\beta}))_{i,j}=\beta_j-{}_{j,i-1}\bar{B}(\boldsymbol{\beta}^i;o)$ for $i\le j$.

Now, we are interested in the stationary joint probability $R^{i)}(\mathbf{n};x)$ previously defined. Applying the key renewal theorem /6/, this probability is given by:

$$R^{i)}(\mathbf{n};x)=1/L\int_o^x dy\int_o^{\infty}dt'\int_o^{\infty}p^{i)}(\mathbf{n};t',y;t)dt \qquad (25)$$

where L is the average time between two successive departures (mean recurrence time) of the Markov renewal process in stationary state.

Thus, the GF and LST of $R^{i)}(\mathbf{n};x)$ is:

$$\bar{R}^{i)}(\boldsymbol{\beta};s)=\bar{P}^{i)}(\boldsymbol{\beta};o,s;o)/L \qquad (26)$$

Taking into account the value $\bar{P}^{i)}(\boldsymbol{\beta};w_1,s;w_2)$ the following matrix expression is obtained:

$$[\mathbf{r}(\boldsymbol{\beta};s)]^T=P_o\mathbf{D}'(\boldsymbol{\beta};s)[\mathbf{C}(\boldsymbol{\beta})]^{-1}\mathbf{E}(\boldsymbol{\beta})\boldsymbol{\eta}^T/(s-(\mathbf{e}-\boldsymbol{\beta})\boldsymbol{\eta}^T) \qquad (27)$$

where $[\mathbf{r}(\boldsymbol{\beta};s)]^T$ is a column vector whose i-th component is $\bar{R}^{i)}(\boldsymbol{\beta};s)$, $\mathbf{D}'(\boldsymbol{\beta};s)=\mathbf{D}(o,o;\boldsymbol{\beta};s)$ and $\mathbf{E}(\boldsymbol{\beta})$ is a NxN matrix defined by ${}_{j,i-1}\bar{B}(\boldsymbol{\beta}^i;0)-1$ for i>j and β_j-1 for $i\le j$.

This GF plays an important role in the evaluation of response time or delays because it represents the queue state from the viewpoints of the system and the external Poissonian arriving processes.

P_o is obtained applying the key renewal theorem to (22):

$$P_o=Q(\mathbf{o})/(L\eta) \qquad (28)$$

otherwise, P_o can also be given by:

$$P_o=1-\rho^* \qquad (29)$$

The GF $P(\boldsymbol{\beta})$ for the stationary queue length distribution for the number of tasks of each priority (either waiting or being served) in the system at a random instant, $P(\mathbf{m})$, can be obtained as:

$$P(\boldsymbol{\beta})=P_o+\boldsymbol{\beta}[\mathbf{r}(\boldsymbol{\beta};o)]^T \qquad (30)$$

5.3 AT OUTPUT INSTANTS

The "output instant" is defined as the instant in which the service of a task has just finished and the feedback has not been initiated.

Let $r(\mathbf{m})$ be the stationary joint probability of having $\mathbf{m}=(m_1,..m_N)$ tasks of the first,.,N-th priority respectively, at output instants. Then:

$$r(\mathbf{m})=Q(\mathbf{o}) \sum_{1\leq j\leq N} \int_o^{\infty} \Omega(\mathbf{m};x)dS_j(x)\eta_j/\eta+$$

$$+\sum_{1\leq i\leq N} \sum_{\forall \mathbf{s}_i} q(\mathbf{s}_i) \int_o^{\infty} \Omega(\mathbf{r}_i;x)dS_i(x) \quad (31)$$

with conditions defined in (10).

The GF for the stationary joint queue length, $R(\mathbf{ß})$, at these points with $|ß_j|\leq 1$ is :

$$R(\mathbf{ß})=Q(\mathbf{o})\mathbf{s}(\mathbf{ß})[C(\mathbf{ß})]^{-1}\mathbf{E}(\mathbf{ß})\eta^{\mathbf{T}}/\eta \quad (32)$$

where $\mathbf{s}(\mathbf{ß})$ is a N row vector whose i-th element is $\bar{S}_i((\mathbf{e}-\mathbf{ß})\eta^{\mathbf{T}})$.

As can be seen, $R(\mathbf{ß})$ is also a matrix product formula.

6. QUEUE LENGTH PROPERTY

It is known that in the M/G/1 the stationary queue length probabilities at random, departure and output instants are the same (see. e.g. /7/).

In /5/ we extend this property for the probabilities of the total number of tasks in the system at random and output instants in the following cases:

a) M/G/1 queue with Bernoulli feedback.
b) M/G/1 queue with two non preemptive priorities in which no more than a task is queued again after a service completion and the probability that a task leaves the system is the same independently of the priority in which was processed.

After some lengthly mathematically treatment which we omit for reasons of conciseness, we have obtained the following result:

"In an M/G/1 queue with N non preemptive priorities and single feedback (no more than one task is queued again after a service completion) and $f^{i)}(\mathbf{o})=\eta L$ (i=1,.,N), independently of service priority, the stationary queue length distribution for the total number of tasks in the system at random and output instants are equal".

This is a natural extension of result obtained by us in /5/.

7. RESPONSE TIME DISTRIBUTION FOR ANY PARTICULAR SEQUENCE

Assume that a particular sequence arrives to the system at a random instant x and requires k services. The length of those services are independent random variables with distributions functions $H_1(t),.,H_k(t)$ and will be processed in priorities $i_1,.,i_k$ respectively. Denote by Z the time needed by this sequence to leave the system. Let $\mathbf{M}=(M_1,.,M_N)$ be the number of customer in the first,.,N-th priority respectively, immediately before the time Z+x (not including the particular requisition). Let us define the stationary distribution:

$$U(\mathbf{i_k};\mathbf{n};\mathbf{H_k};t)=\text{Prob}[Z\leq t,\mathbf{M}=\mathbf{n}] \quad (33)$$

where $\mathbf{i_k}=(i_1,.,i_k)$; $\mathbf{H_k}=(H_1,.,H_k)$ and $\mathbf{n}=(n_1,.,n_N)$. Then:

$$U(\mathbf{i_k};\mathbf{n};\mathbf{H_k};t)=\int_o^t dx \sum_{\forall \mathbf{s}(o)}$$

$$\int_o^x dU(\mathbf{i_{k-1}};\mathbf{s}(o);\mathbf{H_{k-1}};x_o) \prod_{j=1}^{i_k} (\sum_{\forall \mathbf{s}(j)}$$

$$\int_o^x {}_{j,i_k-1}B^{*m_{o,j})}(\mathbf{s}(j);x_j)dx_j) \sum_{\forall \mathbf{s}(i_k+1)}$$

$$\int_o^x \Omega(\mathbf{s}(i_k+1);x_{i_k+1})dH_k(x_{i_k+1})$$

$$\text{for } \sum_u x_u=x \text{ ; } o\leq u\leq i_k+1 \quad (34)$$

where $\mathbf{s}(v)$, $v=o,i_k+1$, is an N-tuple whose j-th component is $m_{v,j}$ and $\mathbf{s}(v)$, $1\leq v\leq i_k$, is an $(N-i_k+1)$-tuple whose j-th component is $m_{v,j+i_k-1}$. All elements $m_{v,u}$ are integer values $(o\leq m_{v,u})$ which verify the following conditions:

$\sum_v m_{v,i_k}=n_{i_k}$ $(1\leq v\leq i_k+1)$; $\sum_v m_{v,u}=n_u$ for $i_k<u\leq N$, $o\leq v\leq i_k+1$ and $m_{i_k+1,j}=n_j$, $1\leq j<i_k$.

Let $\bar{U}(\mathbf{i_k};\mathbf{ß};\mathbf{H_k};s)$ be the LST in t and GF in $\mathbf{n}$ of above stationary probability, with $\mathbf{ß}=(ß_1 ..,ß_N)$; $|ß_1|..,|ß_N|\leq 1$; $Re(s)\geq 0$ and $\bar{H}_k(s)$ the LST of the particular service requisition distributed according with $H_k(t)$.

It can be easily seen from (34) that:

$$\bar{U}(\mathbf{i_k};\mathbf{ß};\mathbf{H_k};s)=\bar{H}_k(s^*)\bar{U}(\mathbf{i_{k-1}};\mathbf{b_k};\mathbf{H_{k-1}};s) \quad (35)$$

where $\mathbf{b_k}$ is a N-tuple whose j-h component is ${}_{j,i_k-1}\bar{B}(\mathbf{ß}^{\mathbf{i_k}};s)$ for $1\leq j\leq i_k$ and $ß_j$ for $i_k<j\leq N$.

Formula (35) is a recurrence relation which is characterized when $\bar{U}(\mathbf{i_1};\mathbf{ß};\mathbf{H_1};s)$ is known.

$$U(\mathbf{i_1};\mathbf{n};\mathbf{H_1};t)=\int_o^t [P_o dH_1(x)\Omega(\mathbf{n};x)+\sum_{p=1}^{N}$$

$$\sum_{\forall \mathbf{s}(-1)} \int_o^x dR^{p)}(\mathbf{s}(-1);x_o) \sum_{\forall \mathbf{s}(o)} D^{p)}(\mathbf{s}(o);x_o)$$

$$\prod_{j=1}^{i_1} \Big(\sum_{\forall \mathbf{s}(j)} \int_o^x {}_{j,i_1-1}B^{*r_j)}(\mathbf{s}(j);x_j)dx_j \Big)$$

$$\sum_{\forall \mathbf{s}(i_1+1)} \int_o^x \Omega(\mathbf{s}(i_1+1);x_{i_1+1})dH_1(x_{i_1+1})]$$

for $\sum_u x_u = x;\quad o \le u \le i_1+1$ (36)

where $\mathbf{s}(v)$, $v=-1,o,i_1+1$ is an N-tuple whose j-th component is $m_{v,j}$ and $\mathbf{s}(v)$, $1\le v\le i_1$ is an $(N-i_1+1)$-tuple whose j-th component is $m_{v,j+i_1-1}$. All elements $m_{v,u}$ are integer values $(o \le m_{v,u})$ which verify the following conditions:

$m_{i_1+1,j}=n_j$ for $1\le j<i_1$ and $\sum_v m_{v,i_1}=n_{i_1}$ $(o\le v\le i_1+1)$; $\sum_v m_{v,u}=n_u$ $(-1\le v\le i_1+1; i_1<u\le N)$.

$D^{p)}(\mathbf{s}(o);x_o)$ is the probability that $m_{o,j}$ $(j=1,..,N)$ tasks appear at j-th priority during the time interval x_o, even due to the fresh arrival process or the feedback produced by an p-th priority tasks. Its expression is:

$$D^{p)}(\mathbf{s}(o);x_o)=(\Omega(.;x_o)*f^{p)})(\mathbf{s}(o)) \quad (37)$$

where * means convolution of the dotted variables of $\Omega(.;x_o)$ with $f^{p)}(.)$.

We define r_j as $m_{-1,j}+m_{o,j}$ for $1\le j<i_1$ and $r_{i_1}=m_{-1,i_1}$.

Taking LST and GF in (36), it is obtained

$$\bar{U}(\mathbf{i_1};\mathbf{ß};\mathbf{H_1};s)=\bar{H}_1(s^*)[P_o+\mathbf{f}(\mathbf{b'}_{\mathbf{i_1}})$$

$$[\mathbf{r}(\mathbf{b}_{\mathbf{i_1}};s^{**})]^{\mathbf{T}}] \quad (38)$$

where $\mathbf{b'}_{\mathbf{i_1}}$ is an N row vector whose j-th component is ${}_{j\ i_1-1}\bar{B}(\mathbf{ß}^{\mathbf{i_1}}\ s)$ for $1\le j<i_1$ and $ß_j$ for $i_1\le j\le N$; $\mathbf{b}_{\mathbf{i_1}}$ is a vector with the same components as $\mathbf{b'}_{\mathbf{i_1}}$ except the i_1-th component whose is ${}_{i_1\cdot i_1-1}\bar{B}(\mathbf{ß}^{\mathbf{i_1}};s)$; $s^{**}=s+(e-\mathbf{b'}_{\mathbf{i_1}})\mathbf{\eta}^{\mathbf{T}}$, and $\mathbf{f}(\mathbf{b'}_{\mathbf{i_1}})$ is a row vector whose j-th component is $F^{j)}(\mathbf{b'}_{\mathbf{i_1}})$.

With these results, if $V(\mathbf{i};t)$, $\mathbf{i}=(i_1,..,i_k)$ is the distribution function of the response time for the particular sequence. Its LST is:

$$V(\mathbf{i};s)=\prod_{n=o}^{k-1} \bar{H}_{k-n}(s^*_n)[P_o+\mathbf{f}(\mathbf{ß^{k)}}_{\mathbf{i_1-1}}(s))$$

$$[\mathbf{r}(\mathbf{ß^{k)}}_{\mathbf{i_1}}(s);s^{**}_k)]^{\mathbf{T}}] \quad (39)$$

where: $s^*_n=s+(e-\mathbf{ß^{n)}}(s))\mathbf{\eta}^{\mathbf{T}}$; being $\mathbf{ß^{n)}}(s)=(ß^{n)}_1(s),..,ß^{n)}_N(s))$, with $ß^{n)}_i(s)$ defined through the following recurrence expression:

$$ß^{n)}_j(s)={}_{j,i_{k-n+1}-1}\bar{B}(ß^{n-1)}{}_{i_{k-n+1}}(s),..,ß^{n-1)}_N(s);s) \quad \text{if } j\le i_{k-n+1}$$

$$ß^{n)}_j(s)=ß^{n-1}{}_j(s) \quad \text{if } j>i_{k-n+1}$$

for $1\le n\le k$, where: $ß^{o)}_j(s)\equiv 1; j=1,..,N$ (40)

and: $\mathbf{ß^{k)}}_{\mathbf{i_1-1}}(s)$ is a N row vector whose j-th component is $ß^{k)}_j(s)$ if $1\le j<i_1$ and $ß^{k-1)}_j(s)$ if $i_1\le j\le N$ and $\mathbf{ß^{k)}}_{\mathbf{i_1}}$ has the same components as $\mathbf{ß^{k)}}_{\mathbf{i_1-1}}$ except the i_1-th component whose is $ß^{k)}_{i_1}(s)$, and:

$$s^{**}_k=s+(e-\mathbf{ß^{k)}}_{\mathbf{i_1-1}})\mathbf{\eta}^{\mathbf{T}} \quad (41)$$

As can be seen, the LST for the response time distribution of any type of linear sequence with k tasks in an M/G/1 queue, with any number of non-preemptive priorities and multiple feedback laws, is a product of k+1 LST, k of them defined by the distributions of the services to be performed by each particular task of the sequence.

This result can be considered as an extension of two previous results:

a) LST for the response time distribution for a sequence with k different service requisitions in an M/G/1 queue with Bernoulli feedback.
b) LST for the response time distribution for a sequence with k different service requisitions in an M/G/1 queue with two non-preemptive priorities and multiple feedback (see /4/).

Both results have also a product formula.

8. NUMERICAL EXAMPLE

In order to give an analytical tool for system engineering groups which are interested in finding delay values in actual processor systems, we first study an example.

Let us consider a processor with two non preemptive priorities, to which there arrive ten types of sequences with different arrival rate. Each sequence is defined by the number or tasks (from 1 to 18) and the number of branches (from 1 to 4). The processing time of each task is constant and different from any other task. This processor was modeled for calculating the queue parameters defined in Section 2 with the following criteria:

- the carried traffic for each priority, the carried traffic for i-th to j-th priority, the total carried traffic;
- the first and second moments of the service time in each priority; and
- the maximum number of tasks due to feedback in each priority

are equal in the model and the processor.

The sequence studied is constituted by three tasks processed in 1, 2, and 2 priorities with processing times: 0.225 msec., 3.34 msec. and 3.868 msec., respectively.

To obtain numerical results for the response time distribution in the theoretical model and in the actual system, two computer programs (numerical and simulation) were written.

Due to the fact that the theoretical results are expressed in terms of LST,and these are given in function of the GF and LST of the "engaged time" which are the roots of the systems of equation given in (7), two main numerical problems arise: inversion of LST and calculation of zeros of complex system of equations. For the first problem, we use the Gaussian quadrature formula developed by Piessens for the Bromwich integral /8/ and the zeros were computed with the desired precision using Newton's method. The matrix inversion of (27) was made by the diagonalization method. The total accuracy obtained was at least of four significant digits.

The simulation program was written using the regenerative technique /9/ at a 90% confidence interval.

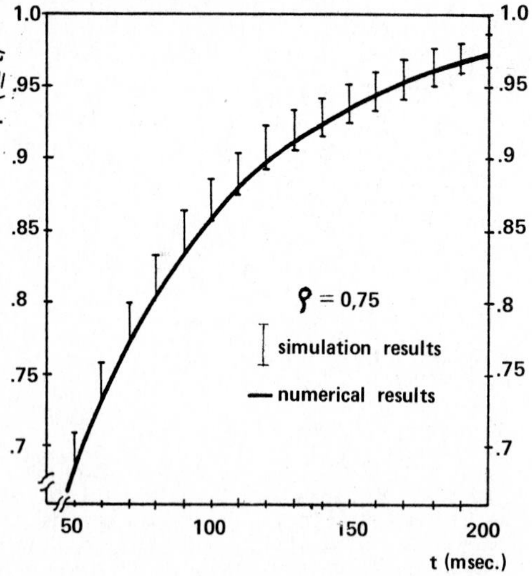

Fig. 1 Waiting Time Distribution

Figure 1 shows a comparison between simulation and numerical results for the waiting time distribution (without the service time) of the particular sequence described above, when the total carried traffic by the system is 0.75 erlangs. The agreement is in general good: there are a few discrepancies at lower probabilities but in specification range the numerical values obtained from the model are within the confidence intervals.

9. CONCLUSIONS

This paper deals with the response time distribution in an M/G/1 queue with N non-preemptive priorities and multiple feedback. The main results can be summarized as follows

- We have obtained the generating functions for the stationary joint queue length probabilities at departure, output and random instants.
- The Laplace-Stieltjes transform for the distribution of the time needed to perform a sequence with any particular services has been found.

It is important to notice the above results have matrix and product form, in spite of the dependence that exists in the model studied in the paper.

A computer program has been written to compute the formulae obtained and the results were compared with those from a simulation. Good agreement was obtained and therefore it can be deduced that the model studied in this paper represents in a satisfactory way the characteristics (non Poissonian input of tasks, non-preemptive priorities, different service time, etc) of an actual processor system.

REFERENCES

/1/ N.K. Jaiswal, "Performance Evaluation Studies for Time Sharing Computer Sistems", Performance Evaluation, Vol.2, No.4, Dec. 1982.

/2/ S.F. Yashkov, "Some Remarks on Processor-Sharing Sistems", Proc. 3th Int. Seminar on Fundamentals of Teletraffic Theory, Moscow, June 1984.

/3/ B. Fontana, "Queue with two Priorities and Feedback: Joint Queue Length Distribution and Response Time Distributions for Specific Sequences", 10th ITC, Montreal, 1983.

/4/ B. Fontana, "M/G/1 Queue with two Non-Preemptive Priorities and Feedback: Response Time Distribution for any Particular Sequence", 3th Seminar on Fundamentals of Teletraffic Theory, Moscow, June 1984.

/5/ B. Fontana & C. Diaz Berzosa, "Stationary Queue Length Distribution in an M/G/1 Queue with two Non-Preemptive Priorities and General Feedback", Performance of Computer-Communication Systems. Proc. IFIP WG 7.3 TC6 2nd Int. Symp., Zurich, March 1984 (North-Holland).

/6/ E. Cinlar, "Introduction to Stochastic Processes", Prentice Hall, Englewood Cliff, 1975.

/7/ E. Gelenbe & I. Mitrani, "Analisis and Synthesis of Computer Systems", Academic Press Inc., London, 1980.

/8/ R. Piessens, "Some Aspects of Gaussian Quadrature Formulae for the Numerical Inversion of the Laplace Transform", Computer Journal, Vol.14, No.4, November 1971.

/9/ D.L. Iglehart & G.S. Shedler, "Regenerative Simulations of Response Times in Networks of Queues", J. ACM, Vol.25, No.3, July 1978.

TELETRAFFIC ISSUES in an Advanced Information Society
ITC-11
Minoru Akiyama (Editor)
Elsevier Science Publishers B.V. (North-Holland)

TRAFFIC FORECASTING WITH MINIMUM DATA

Flavian AUBIN and Beatriz CRAIGNOU

Direction Générale des Télécommunications
Service du Trafic, Equipement et Planification
Paris, France

Centre National d'Etudes des Télécommunications
Division Architecture et Trafic dans les Réseaux
Issy-Les-Moulineaux, France

ABSTRACT

This paper presents a method for estimating volumes and direction of different types of traffic for short or medium term forecasting at the various aggregation levels.

The model may be applied with either rough or detailed data. It is particularly well adapted to networks undergoing a rapid development since it takes into account modifications to the number of lines by differentiating subscriber categories whose traffic evolves in a quite different manner, in volume as well as in direction.

Time series, while not necessary to apply the model, are used to determine values of its different parameters.

1. INTRODUCTION

Traffic forecasting constitutes an essential stage in the management and planning of telecommunications networks.

Good forecasting is always desired by the planner but he does not always have sufficient knowledge of past and present network conditions to accurately estimate future traffic volumes at a given horizon year. This is the case in networks undergoing rapid growth or offering new services whose characteristics are imperflectly known.

As it is expensive to observe all the lines of a subscriber network and to mesure all traffic flows, forecasting must be based on mathematical models using statistical values.

Forecasts are realized here by a model which calculates traffic per line category in conjunction with a traffic growth model.

Traffic forecasting processes have to be consistent at every hierarchical and geographical level of the network.

In France we distinguish the main exchange, the local area containing one or several main exchanges and secondary and primary transit zones. Compatible forecasts must be determined for the whole network.

The model has been tested for total and long distance traffic forecasts (cf. 5.2.).

2. TRAFFIC REPRESENTATION AND MINIMUM DATA

In this paper T denotes a total traffic.

Traffic measurements are made on subscriber groups or on trunk groups. Integer $j \in J1$ will denote a type of traffic evaluated on subscriber groups, $j \in J2$ a type of traffic evaluated on trunk groups. So, the same type of traffic will be denoted by two different integers according to the place of measurement.

Relationships between types of traffic may be represented by a directed tree. Let us consider traffic on subscriber groups. We associate j=0 to the total incoming and outgoing traffic, a non-zero even integer to outgoing traffics, an odd integer to incoming traffics. Figure 1 shows the corresponding directed traffic tree.

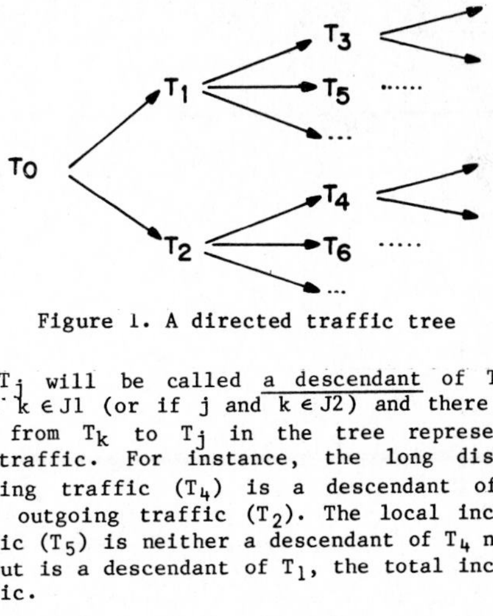

Figure 1. A directed traffic tree

T_j will be called <u>a descendant</u> of T_k if j and $k \in J1$ (or if j and $k \in J2$) and there is a path from T_k to T_j in the tree representing the traffic. For instance, the long distance outgoing traffic (T_4) is a descendant of the total outgoing traffic (T_2). The local incoming traffic (T_5) is neither a descendant of T_4 nor of T_6, but is a descendant of T_1, the total incoming traffic.

This representation is useful when T_0 is not known. In particular, for long distance traffic forecasts it is not necessary to know the volume of other types of traffic.

NC categories are distinguished. Here, "category" means a category of subscribers or a category of lines. Categories are denoted by the subscript i.

To each category i and type of traffic j is associated an analytical component of T, represented here by the mean type j traffic per subscriber (or per line) of category i, t_{ij}.

There is also an associated set of intervals, each one defined by a lower bound (LB) and an upper bound (UB) on t_{ij}. If t_{ij} is given in erlangs,

$$0 \leq LB_j(i) \leq t_{ij} \leq UB_j(i) \leq 1$$

These intervals may, or may not, overlap.

Categories may be defined, for instance, such that for T_0, the intervals associated with $t_{i,0}$ constitute a partition of $[0,1]$. Another definition will allow lines dedicated exclusively to incoming traffic to constitute a category, the same applying to lines dedicated exclusively to outgoing traffic, and the intervals of these two categories may coincide.

The minimum input data required concerns :

- the number of categories to be considered and, for each one, the set of bounds on at least one type of traffic j,

- present and forecast number of lines per exchange and average percentage of lines for each category at studied area level,

- for every exchange, the estimated traffic volume T_j per studied type j. This estimation may be deduced from the number of subscribers using values obtained for other reference exchanges.

Any other information only contributes to refine forecasts (for instance, traffic distribution per type, percentage of subscribers in each categorie at exchange level, ...).

3. NUMBER OF LINES VS MEAN TRAFFIC

Experience shows that mean telephone traffic per line diminishes when the number of lines increases. Figures 2 and 3 compare number of lines and mean traffic per line in France in the last decade.

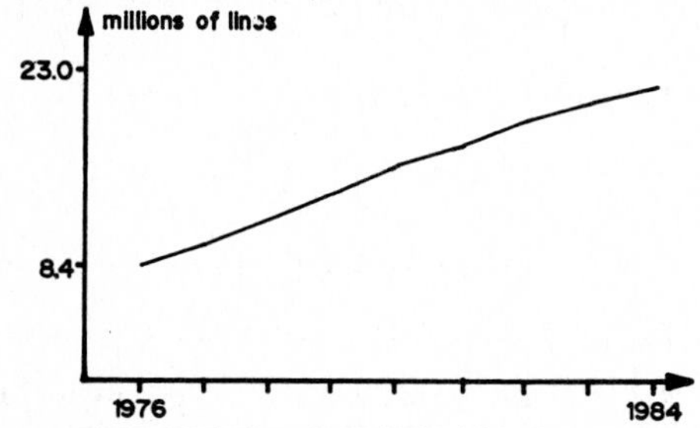

Figure 2. Number of lines in France

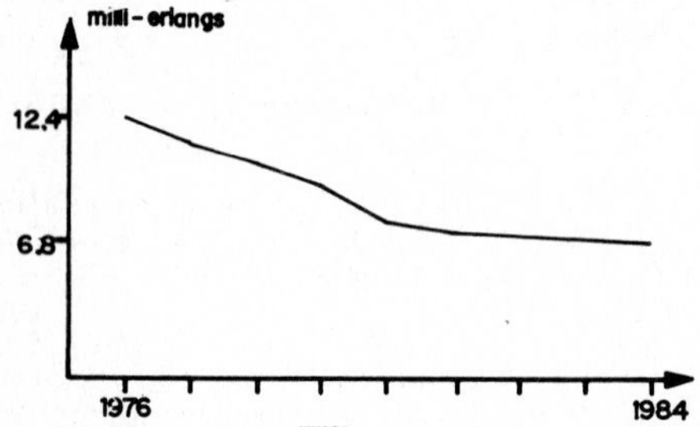

Figure 3. Long distance mean outgoing traffic per line

At present the number of lines is increasing slowly and mean telephone traffic per line is expected to level out in a few years time at a value perhaps a little higher than todays value.

The same phenomenon applies to traffic generated by new services. As far as the recently offered videotex service (teletel and electronic directory) is concerned, the number of terminals increased in 1984 by 300 %, but the mean traffic per terminal decreased by 23 %.

Now, if we consider two categories, residential and business subscribers, mean traffic per subscriber may be low for both categories in rural areas, while in medium size towns and in the outskirts of big cities, it may be much higher for the second category. In big cities, the difference between mean traffics per subscriber may become smaller for both categories, as shown in figure 4.

Finer categories may be considered, according to socio-economic criteria for instance.

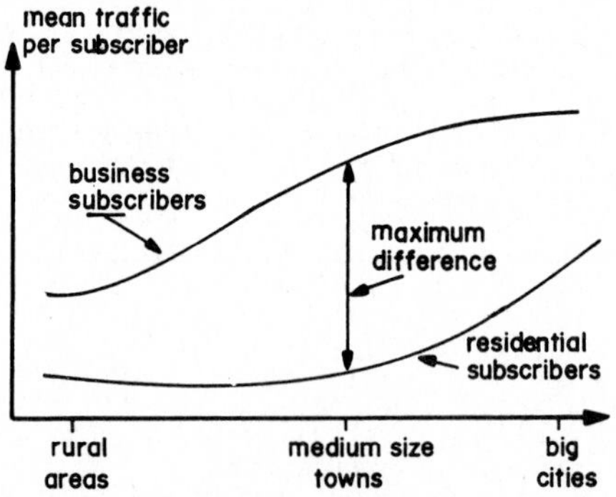

Figure 4. Variations of mean traffic per subscriber

4. TRAFFIC DECOMPOSITION

4.1. General Principle

Let T_j be the total j traffic.

For every exchange and each category i, traffic T_j will be split into a set of analytical components t_{ij}. These components will be determined by resolution of the following system of equations S, where N represents the number of subscribers of the urban area where the exchange is located. Constraints on t_{ij} according to the definition of categories (cf. 2.) are given by eq. (3).

$$S \begin{cases} T_j = \sum_{i=1}^{NC} n_i \, t_{ij} & (1) \\ t_{ij} / t_{i-1,j} = F_{ij}(N), \; i>1 & (2) \\ LB_j(i) \leq t_{ij} \leq UB_j(i) & (3) \end{cases}$$

Ratios $F_{ij}(N)$ are previously determined as it is explained in the following paragraph.

4.2. Estimation of Traffic Ratios between Categories

The values of F_{ij}, eq.(2), are estimated first at national level, then at studied area level and exchange level.

At national level, the values of F_{ij} may be deduced from traffic values defining categories or estimated from data corresponding to different zones. In all cases, national values of F_{ij} become implicit values for other levels.

At studied area level, implicit values are corrected if enough values of traffic per category are provided, while at exchange level these values are ratified or modified to satisfy constraints introduced by the definition of categories.

In the absence of traffic measurements and with no precise ideas about the ratios, we take F_{ij} equal to a constant value, a priori

$$F_{ij}(N) = \frac{LB_j(i)+UB_j(i)}{LB_j(i-1)+UB_j(i-1)}$$

With more precise ideas or sufficient measurements, F_{ij} may be given by a monotonic growing function like

(A) $F_{ij}(N) = a_{ij}\ N^{b_{ij}}$

or a function having a maximum, like

(B) $F_{ij}(N) = a_{ij} \exp \dfrac{-(\log N - \log M)^2}{b_{ij}} + E_{ij}$

where M represents the number of subscribers of a medium town of the country or the considered zone, and E_{ij} the lower bound of values $t_{i,j}/t_{i-1,j}$ corresponding to at least 3 urban areas of large, medium and small size, respectively.

$F_{ij}(N)$ varies according to socio-economic population structures, the choice of categories, call charging structures and so on.

The final form is determined from observation of traffic per category in different zones. The same applies for parameters a_{ij}, b_{ij} and E_{ij}. At regional level they are calculated by the least squares method and the form giving the best correlation coefficient is chosen.

4.3. Resolution of System S

a) Generalities

The system S is solved for the first j=k such that T_k is known (cf. section b) ; values t_{ik} and F_{ik} are determined for every categorie i considered.

The analytical components t_{ij}, the ratios F_{ij} and the set of bounds $LB_j(i)$ and $UB_j(i)$ corresponding to each descendant T_j of T_k are then evaluated. This evaluation needs the proportion P_{ijk} of trafic of type k which is also of type j, per category i.

If T_j is a descendant of T_k, $P_{ijk}=t_{ij}/t_{ik}$. Otherwise, $P_{ijk}=0$ (4)

P_{ijk} may be obtained from intermediate relations ; we have, for instance, $P_{i,4,0} = P_{i,4,2} \times P_{i,2,0}$.

The a priori value of P_{ijk} is obtained from measurements and knowledge of traffic volumes. As an example, let the total outgoing and incoming traffic, T_0, be given. Then,

- $P_{i,2,0}=0$ and $P_{i,1,0}=1$ for incoming dedicated lines ; $P_{i,2,0}=1$ and $P_{i,1,0}=0$ for outgoing dedicated lines ; and with a lack of measurements, take $P_{i,2,0}=P_{i,1,0}=0.5$ for lines carrying incoming and outgoing traffic.

- Or, if i=1 corresponds to residential subscribers and i=2 correspond to business subscribers, $P_{1,2,0} > 0.5$, $P_{2,2,0} < 0.5$, and so on.

- With sufficient measurements, $P_{i,2,0}$ is the corresponding average value.

P_{ijk} may also be calculated in a more sophisticated manner, taking into account the size of the local area expressed by its number of subscribers, NA, and the number of subscribers in the country, NN. For instance, if T_6 represents the local outgoing traffic for the considered area and T_2 the first traffic known ; we have

$$P_{i,6,2} = \left(\frac{NA}{NN}\right)^{g_i},\ g_i > 0, \text{ for all } i.$$

Parameter g_i is estimated by the least squares method from provided values corresponding at least to two local areas of different size.

If T_j is a descendant of T_k, F_{ij} may be obtained from the previously calculated values of F_{ik} and P_{ijk} :

$$F_{ij} = F_{ik} \times P_{ijk}/P_{i-1,j,k} \qquad (5)$$

The set of traffic bounds per category are such that, if T_j is a descendant of T_k

$$\left.\begin{aligned} LB_j(i) &= LB_k(i) \times P_{ijk} \\ UB_j(i) &= UB_k(i) \times P_{ijk} \end{aligned}\right\} \qquad (6)$$

b) Calculations at Exchange Level

Let T_k be the first known traffic.

At exchange level, the analytical components t_{ik} of T_k are determined by the iterative resolution of system S, according to the following algorithm.

Calculations are performed for the initial and preceeding years.

The superscript m on a variable indicates its value at iteration m. Iterations are limited to a previously given number, maxm.

Step 1. m := 0. For i = 2 to NC, let R^m_{ik} be equal to $F_{ik}(N_Y)$, the implicit value of F_{ij}, where N_Y is the number of lines of the area where the exchange is located, at year Y.

Step 2. m := m+1. For i=1 to NC calculate mean traffic values t_{ik} and ratios R_{ik} as follows :

$$R^{m}_{1,k} = 1, \quad G^{m}_{1,k} = 1,$$

for i≠1, $G^{m}_{i,k} = \prod_{c=2}^{i} R^{m-1}_{ck}$

$$E^{m} = T_k \Big/ \sum_{i=1}^{NC} n_i \, G^{m}_{ik}$$

if m < maxm,

$$t^{m}_{ik} = \min\left(UB_k(i), \max(LB_k(i), E^{m} x G^{m}_{ik})\right)$$

if m = maxm, $t^{m}_{ik} = E^{m} \times G^{m}_{ik}$

for i≠1, $R^{m}_{ik} = t^{m}_{ik} / t^{m}_{i-1,k}$

Step 3. If t^{m}_{ik} equals one of its bounds, for at least one category i, and if m < maxm go to Step 2. Should it be otherwise, set ml=m and if, and only if, ml=maxm redefine bounds of categories by

$$LB^{ml}_k(i) = \min\left(LB_k(i), t^{ml}_{ik}\right)$$

$$UB^{ml}_k(i) = \max\left(UB_k(i), t^{ml}_{ik}\right)$$

Step 4. Determine definitive values of mean traffics, ratios and bounds, for i=1 to NC, by

$$t_{ik} = t^{ml}_{ik}, \quad F_{ik} = R^{ml}_{ik},$$

$$LB_k(i) = LB^{ml}_k(i), \quad UB_k(i) = UB^{ml}_k(i)$$

Values for t_{ij}, F_{ij} and a possible modification of $LB_j(i)$ and $UB_j(i)$ may be obtained from the corresponding values for T_k, if T_j is a descendant of T_k, using equations (4), (5) and (6).

5. TRAFFIC GROWTH

5.1. A Model of Mean Traffic Growth per Category

Elementary traffic values t_{ij} are extrapolated by a growth model using historical values and the increase of subscribers within the studied area. For new exchanges, values from old ones given as references are taken into account.

A growth factor G_{ijp} of mean traffic per subscriber (or per line) t_{ij}, during period p going from initial year Y_0 to future year Y_f, is given by

$$G_{ijp} = E_{ijp} \, K^{p}_{ij} \left(1 + \Theta_p\right)^{\alpha_{ij}}$$

where E_{ijp} is a coefficient taking into account call charge modifications, publicity campaigns, the introduction of new services, and so on, occuring during the whole period p ; K_{ij} denotes annual trends of traffic growth ; Θ_p is the growth rate of the number of subscribers in the studied area and period p ; α_{ij} is such that $0 < \alpha_{ij} < 0{,}5$; p is expressed in years or fraction of year.

Growth factors G_{ijp} are calculated at national level from measurements of traffic on a representative sample. If measurements, or reliable samples, are not available, the growth rate of "financial" traffic (number of basic charge units) per category - if it is known - may be used as the general traffic growth. If sufficient data are avaible for the preceeding years G_{ijp} is calculated by exponential smoothing. K_{ij} and α_{ij} may be deduced from G_{ijp}.

At regional and exchange levels, G_{ijp} may be fitted to historical traffic growth if the traffic segmentations are identical for past and present years. In particular, trends K_{ij} deduced from traffic observation at exchange level have priority.

The confrontation of parameters at different levels is quite useful. Differences must be explained and implicit values may be improved.

5.2. Mean Traffic Extrapolation

Let t^{0}_{ijs} denote the mean traffic of type j per subscriber of category i calculated for exchange s at the initial year Y_0 and G_{ijp} its associated growth factor for the period p going from the initial year to the target year Y_f.

The corresponding mean traffic per subscriber at year Y_f is then

$$t^{f}_{ijs} = t^{0}_{ijs} \times G_{ijp} \qquad (7)$$

Forecast mean traffics per subsc iber are calculated for every exchange in service at year Y_0, even if it will be replaced or its service area modified at year Y_f.

5.3. Experimental Results

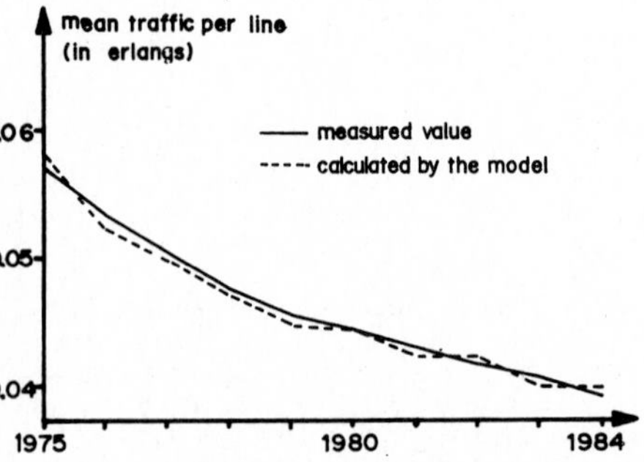

Figure 5. Evolution of mean traffic per line in the last decade in France (not including the region of Paris)

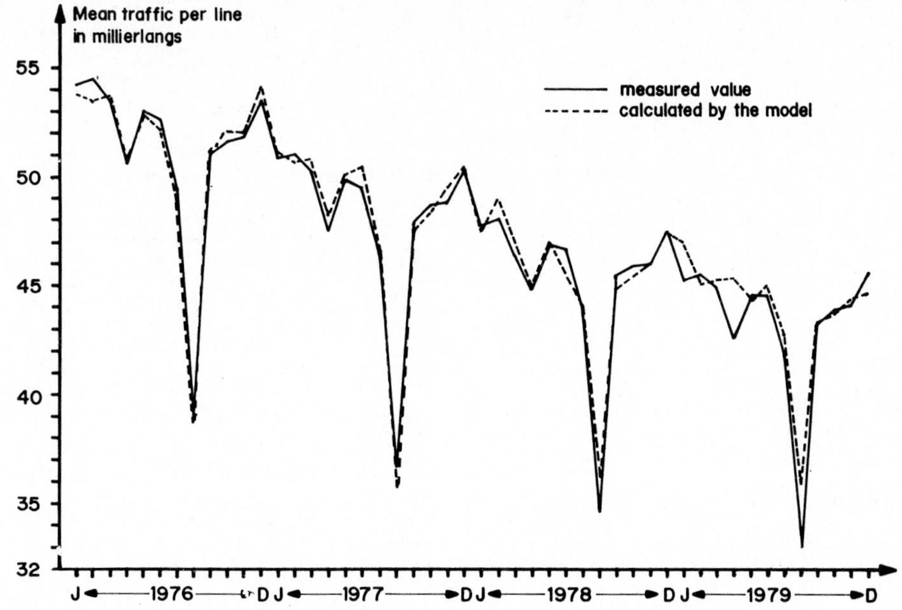

Figure 6. Monthly evolution of mean traffic per line from January 1976 to December 1979 in France (not including the region of Paris)

This model has been tested in France to forecast total and long distance traffic. Calculations have been made for the last ten years, for every month and every year, taking into account seasonal effects. Figures 5 and 6 compare observed values and values obtained by the model.

6. TRAFFIC VOLUME FORECASTING

6.1. At Exchange Level

Let n_{is} denote the expected number of subscribers of category i, for exchange s, at the future year Y_f, supposing its service area to be non modified during the whole period p.

Forecast volume of traffic of type j for exchange s is then

$$T^f_{js} = \sum_{i=1}^{NC} \left[t^f_{ijs} \times n_{is} + \sum_{b \varepsilon B} \left(t^f_{ijb} \, \Delta n_{ib} \right) \right] \qquad (8)$$

where B is the set of exchanges b whose service area is modified by the transfer of Δn_{ib} of its subscribers to exchange s.

Forecast traffic for new exchanges n, in service at year Y_f is calculated by eq. (8) with $t^f_{ijn}=0$ and using t^f_{ijb}, the mean forecast traffic per subscriber for exchange b, whose subscribers behaviour serves as a reference for the subscribers to be connected to exchange n, represented here by Δn_{ib}.

If Δn_{ib} is not known, it will be assumed that $\Delta n_{ib} = \Delta n_b \times n_{ib}/n_b$, where n_b represents the total number of subscribers of exchange b, Δn_b the total number of transferred subscribers.

Forecast traffic for exchange b, from which a part, Δn_{ib}, or all of its forecast n_{ib} subscribers will be transferred to another exchange is then

$$T^f_{jb} = \sum_{i=1}^{NC} t^f_{ijb} \left(n_{ib} - \Delta n_{ib} \right)$$

6.2. At Higher Levels

Forecast type j traffic volume at area level is the sum of forecast total traffic volume of all exchanges s located within the area :

$$T_j = \sum_s T_{js}$$

At transit zones level, forecast volume of traffic may be calculated in a similar manner by addition of forecast volumes at the immediate lower level. However, these values have to be consistent with forecast traffic volume obtained by traffic apportionment at higher levels.

6.3. On Trunk Groups

On trunk groups, growth of some types of traffic may be slightly less than growth of the same type of traffic at the exchange level.

Let $j \in J1$ and $j' \in J2$ correspond to the same type of traffic measured on subscriber groups and on trunk groups respectively.

Then, $T_{j'} = T_j^{\alpha}$

The value of depends on the representative traffic value on trunk groups calculation method.

The representative value which is generally an extreme value, tends towards the mean value as traffic growths.

7. TRAFFIC BY DIRECTION FORECASTING

7.1. Matrices Construction

The key to traffic distribution is given by an initial matrix, drawn up from measurements or using gravity or community of interests models. In the latter cases the matrix may be improved by taking into account economic or administrative relations.

The gravity model determines total traffic between an origin σ and a destination d by

$$T(\sigma,d)= C\ T(\sigma)^{\alpha}\ T(d)^{1-\alpha}/D(\sigma,d)^{\beta},\ 0\leq k,\alpha,\beta\leq 1$$

$T(\sigma,d)$ is a function of the total traffic of ends $T(\sigma)$ and $T(d)$ and the distance between them, $D(\sigma,d)$. Parameter C describes economic or administrative relations between the cities where exchanges are located. C, α and β are estimated by statistical regression.

When enough measurements are provided a method based on factorial analysis [1], allows us to complete missing data by an iterative process described by the following algorithm (where the number of iterations is limited to Z).

Let $(t(a,b)_z)$ be the matrix of traffic flows generated for all exchanges a,b of the studied area, at iteration z.

Step 1. z := 0. Initialize missing traffics at the values given by the gravity model, for instance.

Step 2. z := z+1. Perform a factorial analysis of matrices $(t(a,b)_{z-1})$ and determine the minimum number of x(z) factorial axes required to reconstitute the matrix structure.

Replace missing traffics by the projections of their previous values on the hyperplane determined by the x(z) axes.

Step 3. If the new values of missing traffics differ from their previous values by more than a given ε and if $z<Z$, go to Step 2.

Otherwise, these last values of missing traffics will not be further modified.

The matrix so completed, as matrices obtained in another way, should be modified using Kruithof's method [2] to fit previously determined global outgoing or incoming traffic values.

As for traffic by volume, traffic by direction forecasts are obtained at the various aggregation levels. Thus, elementary traffic flows matrices (between main exchanges), regional traffic matrices (between local areas or transit zones) and national traffic matrices (between primary transit zones) have to be drawn up.

Traffic by direction is, in general, well known at exchanges level within local areas, which make easier matrices construction.

Within a transit zone, traffic flows between neighboring local areas may be determined from measurements, but for distant local areas measurements are not always made. Traffic on transit trunk groups is known in volume but is only given by direction for some particular flows (on direct transversal trunk groups). Un example of this situation is given in figure 7.

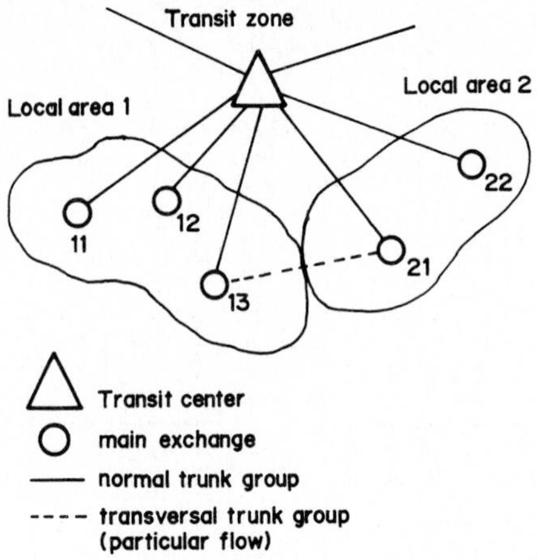

Figure 7. Traffic flows between local areas

So, at regional levels, traffic by direction is not always known. Then, a model (gravity or another) have to be used to build or complete the matrix from any kind of available data.

7.2. Matrices Forecasting

To ensure consistency between matrices drawn up and projected at every level, two methods are proposed.

A. Ascending method :

- From outgoing and incoming traffics at every exchange derive initial matrices between main exchanges, between local areas, between secondary and primary transit zones.
- Determine traffic evolution at every level.
- Project initial matrices to obtain forecast matrices satisfying traffic evolution at every level.

B. Descending method :

- Forecast total traffic (long distance, international, etc) at national level.
- Split total traffic at every level in such a way that, if $T_{j,0}$ denotes total type j traffic at a given level, for the country C, primary transit zones (or main regions) R, secondary transit zones (or sub-regions) D, local areas A and exchanges s,

$$T_{j,C} = \sum_{R} T_{j,R},\quad T_{j,R} = \sum_{D} T_{j,D}$$

$$T_{j,D} = \sum_{A} T_{j,A},\quad T_{j,A} = \sum_{s} T_{j,s}$$

- Determine traffic flows and matrices at every level.

To be consistent, traffic matrices are thus first projected to the horizon period and then modified to match forecast traffic volumes obtained at a higher level. In fact, each traffic volume becomes a constraint for the matrix at a lower level to be added to the constraints that might be already imposed to apply the projection method.

Introduction of these constraints avoids deviations caused by a blind use of any projection such as kruithof's method.

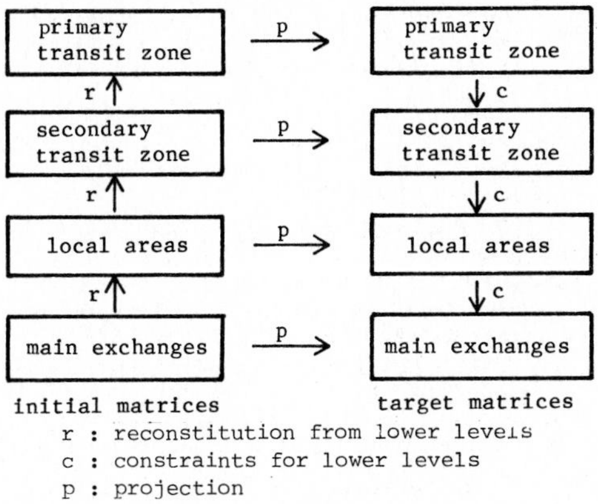

Figure 8. Ascending and descending traffic matrices forecasting method

8. CONCLUSION

Traffic forecasts are realized by the model in a quite satisfactory manner when good measurements are available and subscriber lines are finely segmented. Nevertheless, and this constitutes its interest, it is easily adaptable and allows us to obtain good results when measurements have not been made or are not reliable at studied area level.

Consideration of types of traffic per category allows a better understanding of the network evolution and better forecasts may be expected.

The proposed traffic matrix projection method based on statistical values provides a solution when measurements are missing.

The main characteristic of the method is its modularity, so that different traffic forecasting methods or matrix projection methods may be integrated according to the quantity and quality of available data.

REFERENCES

[1] Diday E. et Collaborateurs. Optimisation en classification automatique. Tomes 1 et 2. INRIA. F-78150 Rocquencourt, 1979.

[2] Kruithof J. Telefoonverkeersrekening - De Ingenieur, 52, 8, E15-E25, 1937.

TELETRAFFIC ISSUES in an Advanced Information Society
ITC-11
Minoru Akiyama (Editor)
Elsevier Science Publishers B.V. (North-Holland)

A METHOD FOR QUANTIFYING ERRORS IN FORECASTS

Robert WARFIELD and Mark ROSSITER

Telecom Australia Research Laboratories
Melbourne, Australia

ABSTRACT

The error in a forecast of a traffic vector is defined as the difference between the forecast vector and the true vector. Considering that the true value of the traffic vector is not available in practice, it is not obvious how to quantify the error vector. Given a forecast, which was made in the past, of a vector that has since been measured, the available data consists of: a forecast vector, a measured vector, and the parameters of the measurement. A method for estimating the size of the error vector from the available data is presented in this paper.

1. INTRODUCTION

The overall objective of this study is to develop a method for placing confidence limits on forecasts of traffic vectors. For the study reported here, it is assumed that there is a routine procedure in use for forecasting traffic vectors. The basic assumption is made that the errors made in past forecasts are typical of the errors that are still being made. Therefore, if the error distribution of old forecasts can be quantified by comparing them with present traffic vectors, this information can be used to calculate confidence limits for new forecasts.

A forecast traffic vector will, of course, differ from the true vector. The true traffic vector will never be available, though if one waits until the year to which the forecast applied, a reasonable approximation to the true vector can be obtained by measurement.

The problem is to compare the measurement, which involves some error, with the forecast, and hence quantify the error in the forecast.

The problem is cast as one of Statistical Inference. The error vector is modelled as a random variable with a multivariate distribution of known form. Inferences are made about the unknown parameters of the distribution.

In this paper, forecast errors are modelled as having a distribution of the same form as the asymptotic distribution of measurement errors. That is, forecast errors are assumed to be distributed Normally, with variance proportional to the true value of traffic intensity. Thus the error vector distribution is characterized by a single parameter, which is termed the "equivalent measurement duration". This terminology was introduced in [1], and similar work was reported in [2]. More details are reported in [3].

The likelihood function of the unknown parameters of the error distribution is found, and from this the Maximum Likelihood estimators are derived. It is found that the Maximum Likelihood Estimator of the equivalent measurement duration of the forecast error is biased. A correction is introduced to reduce the bias. The variance of the estimator is also investigated. In order to verify the predicted properties of the estimators, the algorithms are applied to practical data.

1.1 Notation

In studying forecast errors, the following notation is used:

- μ is an n-dimensional vector of teletraffic intensities. The particular example considered is the vector of offered traffic intensities from a given origin to a number of destinations.
- X is a forecast of the vector μ.
- Y is a measurement of the vector μ. It is subject to measurement error, arising from the finite observation period.

To denote the year to which the vector refers, a value is shown in parentheses. For example,

- X(k) is the forecast of the traffic vector at year k.

To denote individual elements of a vector, the index is shown as a subscript. For example,

- μ_i is the i'th element of the traffic vector μ, for some i in the range $1 \leq i \leq n$. This would represent, in the case of a traffic dispersion vector, the traffic offered to the i'th destination.

To describe the distribution of a random variable, the notation

$$\sim N\text{ (mean, variance)}$$

is used to indicate "has a normal distribution with the specified mean and variance".

2. BACKGROUND

The basis for the work reported in this paper is given in [1] and [3]. For convenience, a brief summary of those results is presented here. Note that in [1], [2], and [3], a Bayesian approach was used. However, in estimating equivalent measurement duration later in this paper a more conventional approach using methods of Classical Statistical Inference is used.

In [1], [2], and [3], a Kalman Filter approach to forecasting and measurement is used. At each stage, information from all past measurements is combined with the current measurement and then projected forward to form a forecast.

The procedure uses well known results for the asymptotic distribution of a measurement of offered traffic with continuous recording (see [4]), namely

$$Y_i \sim N(\mu_i, \frac{2\mu_i}{T_m}) \qquad (1)$$

where

T_m is the equivalent measurement duration of the distribution of Y. It is is simply the measurement duration, measured in units of mean holding times.

Forecast errors are modelled using the concept of equivalent measurement duration. For present purposes, this leads to the model that the forecast has the distribution

$$X_i \sim N(\mu_i, \frac{2\mu_i}{T_f}) \qquad (2)$$

where

T_f is the equivalent measurement duration of the forecast.

At each stage the past forecast and the present measurement are combined to estimate the traffic.In [1], the estimator used is:

$$\mu' = \frac{T_f}{T_f + T_m} X + \frac{T_m}{T_f + T_m} Y \qquad (3)$$

The one-year-ahead forecast is then found by multiplying the present estimator by a growth matrix, which is assumed to be known. The equivalent measurement duration of this new forecast is then given, approximately, by:

$$\frac{1}{T_f(k+1)} \doteq \frac{g}{T_f(k) + T_m(k)} + \frac{1}{f} \qquad (4)$$

where

g is a parameter derived from the growth matrix. For practical purposes it can be taken to be the growth factor of the total traffic, which is assumed to be constant from year to year. The assumption of constant g is for simplicity only - if growth factors for each year are available they can be used, with a corresponding increase in the complexity of the equations.

f is a parameter which accounts for the introduction of errors through the forecasting process. It is interpreted as the equivalent measurement duration of a forecast one year ahead based on perfect knowledge of the traffic at the base year.

In the following, the approximate nature of the equation (4) for the evolution of equivalent measurement duration is ignored - the equation is treated as exact. This allows answers to be obtained, although their accuracy is questionable. The final test of the answers obtained is not in the rigour of their derivation, but in their applicability to data which is generated by simulation or experiment.

If the value of f is known, then it is straightforward to predict the equivalent measurement duration of any forecast by using the equations above. Knowing the equivalent measurement duration of a forecast allows a confidence interval to be placed on each element of the forecast traffic vector.

However, f cannot be estimated directly, since it is never possible to base a forecast on perfect knowledge of the base year traffic. Rather, the accuracy of an actual forecast, which is based on imperfect knowledge of the base year traffic, is estimated, and then the corresponding value of f is determined by the procedure explained below.

If there is a measurement at year 0, and no further measurement till year K, then the equivalent measurement duration of the forecast for year 1 is given by

$$\frac{1}{T_f(1)} = \frac{g}{T_f(0) + T_m(0)} + \frac{1}{f} \qquad (5)$$

and, by induction, the equivalent measurement duration of the forecast for year K is

$$\frac{1}{T_f(K)} = \frac{g^{K-1}}{T_f(1)} + (1+g+\ldots+g^{K-2}) \frac{1}{f} \qquad (6)$$

Substituting for $T_f(1)$, this becomes

$$\frac{1}{T_f(K)} = \frac{g^K}{T_f(0) + T_m(0)} + (1+g+\ldots+g^{K-1}) \frac{1}{f} \qquad (7)$$

This equation relates $T_f(K)$ to g, f, $T_f(0)$, and $T_m(0)$. If $T_f(K)$ is estimated and it is desired to calculate the corresponding value of f, then a value of $(T_f(0) + T_m(0))$ is needed. The value of $T_m(0)$ is assumed known, however $T_f(0)$ is unknown. Normally, $T_m(0)$ will be greater than $T_f(0)$ by a considerable factor (a measurement is much more accurate than a forecast). For this purpose it is assumed that the equivalent measurement duration of the old forecast is equal to the equivalent measurement duration of the new forecast, that is

$$T_f(0) \quad = T_f(K) \tag{8}$$

Equality between the equivalent measurement durations of the two forecasts could arise, for example, if they were based on measurement data of the same duration and age. In any case, the true value of $T_f(0)$ is not critical in the calculations. If greater accuracy were needed, $T_f(0)$ could be estimated directly by comparing the old forecast X(0) with the old measurement Y(0).

$$\frac{1}{T_f(K)} = \frac{g^K}{T_f(K) + T_m(0)} + \frac{g^K - 1}{g - 1}\,\frac{1}{f} \tag{9}$$

The application of the equation above is to find a value of f that corresponds to an estimated value of $T_f(K)$. That is, if $T_f(K)$ is estimated, the corresponding value of f can be computed by solving the above equation. Then the corresponding values of $T_f(k)$ for k greater than zero can be calculated recursively using (4).

3. ESTIMATION OF EQUIVALENT MEASUREMENT DURATION

The data available is taken to be a forecast and a measurement of the same traffic vector, $\mu(K)$. All quantities below relate to the year K. For brevity this is not shown explicitly.

It is assumed that the forecast is unbiased and has errors in each component that are independent and distributed normally with equivalent measurement duration equal to T_f, which is to be estimated. That is,

$$X_i \sim N(\mu_i, \frac{2\,\mu_i}{T_f}) \tag{10}$$

The measurement is taken to be unbiased, with individual components independent, and having an equivalent measurement duration of T_m, which is known, that is

$$Y_i \sim N(\mu_i, \frac{2\,\mu_i}{T_m}) \tag{11}$$

The forecast and the measurement errors are assumed to be independent.

For very small mean values, measurement and forecasting errors cannot be assumed to have a Normal distribution of the form given above. A Normal distribution of errors implies that there is a positive probability of a negative traffic. This probability is usually negligible, but it becomes unacceptably large for very small mean values.

From the assumptions above, the joint probability density function of X and Y can be written as a function of μ, T_m, and T_f , as follows:

$$f(X,Y;\mu,T_m,T_f) = \{ \frac{1}{2\,\pi} \}^n$$

$$\prod_{i=1}^{n} \{ \sqrt{\frac{T_m}{2\mu_i}}\ \exp(\frac{-1}{2}\ \frac{T_m\,(Y_i - \mu_i)^2}{2\,\mu_i})$$

$$\sqrt{\frac{T_f}{2\mu_i}}\ \exp(\frac{-1}{2}\ \frac{T_f\,(X_i - \mu_i)^2}{2\,\mu_i}) \ \} \tag{12}$$

From the probability density function, the log-likelihood function of the unknown parameters, μ and T_f , can be seen to be

$$\ln L(T_f, \mu) = \frac{1}{2} \sum_{i=1}^{n} [-2 \ln (4\,\pi\,\mu_i) + \ln (T_f) - \frac{T_f\,(X_i - \mu_i)^2}{2\,\mu_i} + \ln (T_m) - \frac{T_m\,(Y_i - \mu_i)^2}{2\,\mu_i} \ ] \tag{13}$$

As an initial attempt to estimate the unknown parameters, Maximum Likelihood Estimators can be found by taking partial derivatives of the log-likelihood with respect to the unknowns, and equating to zero (see [5]). This gives the following necessary conditions for the Maximum Likelihood Estimators:

$$\frac{1}{\hat{T}_f} = \frac{1}{n} \sum_{i=1}^{n} \frac{(X_i - \hat{\mu}_i)^2}{2\,\hat{\mu}_i} \tag{14}$$

and

$$(\hat{T}_f + T_m)\,\hat{\mu}_i^{\,2} + 4\,\hat{\mu}_i - \hat{T}_f\,X_i^2 - T_m\,Y_i^2 = 0 \tag{15}$$

This quadratic in $\hat{\mu}_i$ can be solved to give

$$\hat{\mu}_i = \frac{-2 + \sqrt{4 + (T_m + \hat{T}_f)(T_m Y_i^{\,2} + \hat{T}_f X_i^{\,2})}}{T_m + \hat{T}_f} \tag{16}$$

The positive root is chosen because the negative root would always give a non-positive value for the estimator of μ_i.

The system of equations relates $\hat{T}_f$ and $\hat{\mu}_i$, and can be solved iteratively for them.

An alternative estimator for μ may be obtained by combining the forecast and measurement in inverse proportion of their variance. This leads to an estimator of the same form as that given in the previous Section, with T_f replaced by its estimate - that is

$$\mu_i' = \frac{X_i \hat{T}_f + Y_i T_m}{\hat{T}_f + T_m} \quad (17)$$

The two estimates μ of are approximately equal for large traffics. For example, if

$X_i = 10,\quad Y_i = 9,$

$\hat{T}_f = 20,\quad T_m = 100,$

then

$\hat{\mu}_i = 9.16$, and $\mu_i' = 9.17$.

However, there is a significant difference for small traffics. If $\hat{T}_f$ and T_m are as before and

$X_i = 0.01,\quad Y_i = 0.02,$

then

$\hat{\mu}_i = 0.008$, and $\mu_i' = 0.018$.

Note that the maximum likelihood estimate does not lie between the measurement and forecast. This is because, with the model used, decreasing the mean decreases the variance. The more sharply peaked distribution for $\mu = 0.008$ has a higher probability density function for the given measurement and forecast. Thus it appears that the Maximum Likelihood Estimator is biased for small values and that it would be better to use the previously described estimator, μ'. Therefore the alternative estimator is used henceforth.

3.1 True Traffic Vector Known

To investigate the distribution of the estimator of T_f, we start by assuming μ is known. For this case, the quantity

$$T_f \sum_{i=1}^{n} \frac{(X_i - \mu_i)^2}{2\,\mu_i}$$

is Chi-squared with n degrees of freedom. Hence,

$$E\left\{ \frac{1}{n} \sum_{i=1}^{n} \frac{(X_i - \mu_i)^2}{2\,\mu_i} \right\} = \frac{1}{T_f} \quad (18)$$

and

$$\mathrm{Var}\left\{ \frac{1}{n} \sum_{i=1}^{n} \frac{(X_i - \mu_i)^2}{2\,\mu_i} \right\} = \frac{2}{n\,T_f^2} \quad (19)$$

Now let S be defined by

$$S \triangleq \frac{1}{T_f} \quad (20)$$

and $\hat{S}$ by

$$\hat{S} \triangleq \frac{1}{n} \sum_{i=1}^{n} \frac{(X_i - \mu_i)^2}{2\,\mu_i} \quad (21)$$

From the above, $\hat{S}$ is an unbiased estimator of S. It is well known (see, for example, [5]) that, for an unbiased estimator,

$$E\{1/\hat{S}\} \doteq \frac{1 + \mathrm{Var}(\hat{S}) / S^2}{S} \quad (22)$$

$$= \frac{1 + 2/n}{S} \quad (23)$$

Hence,

$$E\left\{ \frac{n}{n+2} \frac{1}{\hat{S}} \right\} \doteq \frac{1}{S} \quad (24)$$

$$= T_f \quad (25)$$

Therefore, for the case of known μ, the quantity

$$\frac{n}{n+2} \left\{ \frac{1}{n} \sum_{i=1}^{n} \frac{(X_i - \mu_i)^2}{2\,\mu_i} \right\}^{-1}$$

is approximately unbiased as an estimator of T_f.

3.2 Unknown Traffic Vector

It is useful to determine the expected value of the quantity

$$\sum_{i=1}^{n} (X_i - \mu_i')^2$$

as follows. Substituting for μ'_i,

$$E\left\{ \sum_{i=1}^{n} (X_i - \mu_i')^2 \right\} =$$

$$E\left\{ \sum_{i=1}^{n} \left[\frac{T_m}{T_m + T_f} (X_i - Y_i) \right]^2 \right\} \quad (26)$$

$$= \left[\frac{T_m}{T_m + T_f} \right]^2 \sum_{i=1}^{n} \mathrm{Var}(X_i) + \mathrm{Var}(Y_i) \quad (27)$$

since X and Y have the property that

$$E\{X_i\} = E\{Y_i\} = \mu_i \quad (28)$$

and they are independent. Therefore,

$$E\left\{ \sum_{i=1}^{n} (X_i - \mu_i')^2 \right\} =$$

$$= \frac{T_m}{T_f} \frac{1}{T_m + T_f} \sum_{i=1}^{n} 2\,\mu_i \quad (29)$$

Using this result, it follows that

$$E\{ \sum_{i=1}^{n} \frac{(X_i - \mu_i')^2}{2\,\mu_i} \} = \frac{n}{T_f}\,\frac{T_m}{T_m + T_f} \quad (30)$$

This equation can be rearranged to give

$$\frac{(T_m + T_f)\,\frac{1}{n}\,E\{ \sum_{i=1}^{n} \frac{(X_i - \mu_i')^2}{2\,\mu_i} \} - 1}{T_m - T_f} = \frac{1}{T_f} \quad (31)$$

This result suggests that if the estimator of T_f were iteratively adjusted until

$$\frac{1}{\hat{T}_f} = \frac{\frac{T_m + \hat{T}_f}{n} \sum_{i=1}^{n} \frac{(X_i - \mu_i')^2}{2\,\mu_i} - 1}{T_m - \hat{T}_f} \quad (32)$$

and an estimator of S were taken as

$$\hat{S} = 1/\hat{T}_f \quad (33)$$

then the resulting value of $\hat{S}$ would be a good estimator for S. This general strategy will be developed later in this paper, however an obstacle must first be overcome: although the numerator in the summation term above involves the estimator of μ, the denominator involves the true value of μ, which is not available. An approximately unbiased estimator of the quantity

$$\frac{1}{\mu_i}$$

can be obtained, as follows:

$$E\{1/\mu_i'\} \doteq \frac{1 + \mathrm{Var}(\mu_i') / \mu_i^2}{\mu_i} \quad (34)$$

and

$$\mathrm{Var}(\mu_i') = \frac{2\,\mu_i}{T_f + T_m} \quad (35)$$

(see [1] or [3]). Hence,

$$E\{ \frac{1}{\mu_i'} \} \doteq \frac{1}{\mu_i} (1 + \frac{2}{\mu_i (T_f + T_m)}) \quad (36)$$

For values of μ_i greater than 1, and for realistic values of T_m and T_f, the factor on the right of the equation above is very close to unity. For example, with

$T_m = 100$, $T_f = 20$

and

$\mu_i = 10$,

$$E\{ \frac{1}{\mu_i'} \} \doteq \frac{1}{10} (1 + \frac{1}{600})$$

$$= 0.10017$$

However, for

$\mu_i = 0.1$,

$$E\{ \frac{1}{\mu_i'} \} \doteq \frac{1}{.1} (1 + \frac{1}{6})$$

$$= 11.6667$$

As a simple numerical illustration of the problem, take the data of the last example and consider,

$$\mathrm{Var}(\mu_i') = \frac{2\,\mu_i}{T_f + T_m}$$

$$= .001666$$

Three typical values of the estimator are chosen as: the mean, mean plus one standard deviation, and mean minus one standard deviation. These three values, and their inverses are:

$\mu_+' = .141,\quad 1/\mu_+' = 7.101$

$\mu_0' = .100,\quad 1/\mu_0' = 10.000$

$\mu_-' = .059,\quad 1/\mu_-' = 16.899$

and the average of these three inverses is equal to 11.333, rather than 10.000.

An approximate correction can be effected by solving the equation

$$\frac{1}{\mu_i'} = \frac{1}{\mu_i} (1 + \frac{2}{\mu_i (T_f + T_m)}) \quad (37)$$

for μ_i in terms of μ_i'. This equation comes from the expression for the expected value of $1/\mu'_i$, replacing the expectation with the actual value. Solving the equation in this way leads to

$$\mu_i^* = \frac{1}{2} (\mu_i' + \sqrt{\mu_i'^2 + \frac{8\,\mu_i'}{T_m + T_f}}) \quad (38)$$

It has been found that using this expression to estimate $1/\mu_i$ gives a lower bias than that of the uncorrected estimator. To illustrate this, the previous numerical example is taken up again:

$\mu_+' = .141,\quad \mu_+^* = .156,\quad 1/\mu_+^* = 6.415$

$\mu_0' = .100,\quad \mu_0^* = .115,\quad 1/\mu_0^* = 8.730$

$\mu_-' = .059,\quad \mu_-^* = .073,\quad 1/\mu_-^* = 13.749$

and the average of the inverses of the μ^* values is equal to 9.631, which is an improvement over the uncorrected figure (in fact, it is a slight over-correction).

4. COMPUTATION OF ESTIMATES

Using the results above, a computer program was developed to estimate μ and T_f from a forecast, X, a measurement, Y, and the equivalent measurement duration T_m . The algorithm used is iterative, and begins with a guess of the value of T_f. After the initial guess is made, the steps described below are followed until the resulting estimates become stable from one iteration to the next.

Step 1 :

The traffic vector is estimated from

$$\mu' = \frac{T_f}{T_f + T_m} X + \frac{T_m}{T_f + T_m} Y \qquad (39)$$

Step 2:

The corrected estimator of $1/\mu_i$ is found from

$$\mu_i^* = \frac{1}{2} (\mu_i' + \sqrt{\mu_i'^2 + \frac{8 \mu_i'}{T_m + T_f}}) \qquad (40)$$

Step 3:

A new estimate of S is calculated using

$$\hat{S} = \frac{(T_m + \hat{T}_f) \frac{1}{n} E\{ \sum_{i=1}^{n} \frac{(X_i - \mu_i')^2}{2 \mu_i^*} \} - 1}{T_m - \hat{T}_f} \qquad (41)$$

and the previous estimate of T_f is replaced by

$$\hat{T}_f = \frac{1}{\hat{S}} \qquad (42)$$

Once the process of repeating the 3 steps above has converged, we have an estimator of S that has been "corrected" for bias, even though some approximations have been used. To obtain an estimator of T_f, it would be desirable to apply a correction as dealt with in Section 3.1 in the case of a known traffic vector. However, it would be difficult to obtain an expression for the variance of the estimator arrived at by the iterative process described above. Instead, for simplicity, the same correction factor is used as the one derived in Section 3.1, namely

$$T_f^* = \frac{n}{n+2} \hat{T}_f \qquad (43)$$

Simulation studies have indicated that this correction factor, though not rigorously justified, does improve the bias of the estimator of T_f.

To give some ability to predict the variability of the estimators, the results of Section 3.1 on the variance of the estimators were pressed into service. The applicability of these results was investigated using simulation studies.

This procedure was tested using data generated by simulation. Values of X and Y were constructed from a random number generator and using specified values of μ, T_m, and T_f.

As an example of the simulation results, the following case is presented.

$T_m = 100$, $T_f = 20$, $n = 100$

Values were chosen for the "true" traffic vector; to give a spread of values the formula used was

$\mu_i = i$ for $i = 1,\ldots,100$

Twenty-five pairs of X and Y vectors were generated using the distributions assumed in the model. For each pair T_f was estimated, with the following results:

The cell frequencies for the batch of 25 simulated experiments are given in Table 1.

Table 1. Histogram of Estimated T_f Values

Cell	Number of Occurrences
< 16	1
16-18	7
18-20	7
20-22	4
22-24	4
>24	2

The statistics for the same batch of 25 experiments are given in Table 2.

Table 2. Statistics from a Sample Simulation Run

	$\hat{T}_f$	$\hat{S}$
Sample Mean	19.88	0.0505
True Value	20.00	0.0500
Sample Std. Dev.	3.3	0.008
Predicted Std. Dev.	2.8	0.007

5. CONFIDENCE INTERVALS

In this Section we describe the calculation of confidence intervals on forecasts. Two cases need to be considered: a priori confidence intervals, and a posteriori intervals calculated conditional on an observation. Firstly, suppose we have some forecast vector X for future traffic, and the equivalent measurement duration T_f

of X is known (having been calculated from the evolution of equivalent measurement duration). Then a 100(1-α)% confidence interval for μ_i is

$$X_i \pm \Phi^{-1}(1-\alpha/2)\sqrt{\frac{2X_i}{T_f}} \qquad (44)$$

where Φ is the standard normal distribution function. Strictly speaking, this interval is approximate since the standard deviation of X_i is unknown.

In the second instance, we may wish to find a confidence interval for μ_i, given a forecast X_i, with equivalent measurement duration T_f, and an independent measurement Y_i, with equivalent measurement duration T_m. Now, omitting the subscript i, we note that the quantity

$$R_{Y/X} \triangleq \frac{\sqrt{T_m}\,(Y-\mu)}{\sqrt{T_f}\,(X-\mu)}$$

has a Cauchy distribution, being the ratio of two independent standard normal random variables (see [5]). Furthermore, by solving a quadratic in μ it can be shown that the inequality

$$|R_{Y/X}| < a \qquad (45)$$

holds if and only if μ lies in the interval with endpoints given by

$$\frac{(T_mY - a^2T_fX) \pm a\sqrt{T_fT_m}\,|X-Y|}{T_m - a^2T_f}$$

provided the constant a satisfies the inequality

$$0 < a < \sqrt{\frac{T_m}{T_f}} \qquad (46)$$

Hence, a 100(1-α)% confidence interval for μ can be constructed by taking

$$a = \tan[(1-\alpha)\pi/2]$$

provided the confidence level 1-α satisfies

$$1-\alpha < (2/\pi)\ \mathrm{artan}\,[\sqrt{(T_m/T_f)}]$$

Of course, one sided confidence intervals can also be calculated using this Cauchy property.

For confidence intervals centered on

$$\mu_i' = \frac{T_fX_i + T_mY_i}{T_f + T_m}$$

we can use the fact, which is easy to show, that μ_i' is independent of $X_i - Y_i$, hence

$$\frac{(T_f+T_m)\ (\mu_i'-\mu_i)}{\sqrt{T_fT_m}\ (X_i-Y_i)}$$

has a Cauchy distribution. This quantity may be used to construct confidence intervals with length proportional to the difference between the forecast and the observation. From sample calculations, intervals constructed in this way appear to have more satisfactory properties than those constructed using (45). Further studies are needed to determine which is more acceptable in practice.

6. CONCLUSIONS

A method for quantifying the error in a forecast traffic vector has been proposed. Tests using sample data indicate that the proposed estimator of equivalent measurement duration is aceptable in terms of bias and variance. A method for finding confidence limits for a forecast traffic vector has also been proposed. Further empirical studies are needed to evaluate the practical applicability of the proposals.

ACKNOWLEDGEMENT

The permission of the Chief General Manager, Telecom Australia, to present this paper is gratefully acknowledged. The authors also acknowledge the contribution of M.R. Dix, particularly the development and running of the computer programs used.

REFERENCES

[1] Warfield R.E. "Optimal Planning and Processing of Teletraffic Measurements", Ninth International Teletraffic Congress, Torremolinos, Spain, 1979

[2] Mariñ J.I. "A Mathematical Model for Traffic Matrices Calculation" Ninth International Teletraffic Congress, Torremolinos, Spain, 1979)

[3] Warfield R.E. "A Study of Traffic Measurement and other Estimation Problems" Ph. D. Thesis, 1980, University of New South Wales, Sydney, Australia.

[4] Riordan J. "Telephone Traffic Time Averages", Bell System Technical Journal, Vol. 30, 1951, pp 1129-1144.

[5] Mood A.M., Graybill F.A., and Boes D.C., "Introduction to the Theory of Statistics", McGraw-Hill, 1974.

TELETRAFFIC ISSUES in an Advanced Information Society
ITC-11
Minoru Akiyama (Editor)
Elsevier Science Publishers B.V. (North-Holland)

INTERNATIONAL TRAFFIC VOLUME FORECASTING

A. LEWIS AND J.M. CLAUDIUS

TELEGLOBE CANADA

ABSTRACT

A new model for forecasting telephone traffic volumes between two countries has been developed with the form.

Volume = a.(#of telephone sets in less developed country) +b.(phone density in less developed country) +c.(#of telephone sets in more developed country) +d.

Comparisons with time series have been performed; the results showed that, in general, the above formula gives more accurate forecasts. This paper also provides methods for estimating statistical data when missing and computing forecasts when data is unavailable (using grouping).

1. NEED

Well proven methods exist for preparing forecasts of international telephone traffic between countries with established relations and a moderate level of telecommunications development.

These methods have been observed to not perform satisfactorily when one of the countries involved has a relatively low level of telephone development. Thus, it becomes necessary to find a simple and accurate method of estimating traffic volumes for this case. Here is a simple and exact consideration: to have a telephone conversation, telephone sets are required. This fact gives the hypothesis that the traffic volumes are closely related to the number of telephones. This holds particularly true in those countries where the telephone equipment is a limiting factor.

In the following pages, this consideration will be used to quantify the traffic flow in order to forecast a future demand.

2. CHOICE OF PARAMETERS

The availability of statistical data is a very important consideration, in fact even the International Telecommunication Union (ITU) yearbook of statistics [1] is far from being complete. However, there are a few items well described for most countries including many of those with low levels of telephone development. These are:

- the number of telephone stations (sets),
- the number of mainlines,
- the number of inhabitants,
- telephone mainlines per 100 inhabitants,
- telephone stations (sets) per 100 inhabitants.

In a study called "Some Aspects of Trunk Traffic Forecasting" by K. Mirski [2], the effect of telephone density on the number of calls generated by one subscriber has been demonstrated.

The model should thus make use of both the number of telephones and the density; available statistical data includes telephone lines and telephone sets as well as densities for both lines and sets. Since a priori there was no basis for deciding which to use, both were statistically tested. A linear regressionwas performed for 44 countries usingeach time the following 8 available independent variables:

- number of telephones in country x,
- number of lines in country x,
- number of telephones per 100 residents in country x,
- number of lines per 100 residents in country x,
- number of telephones in Canada,
- number of lines in Canada,
- number of telephones per 100 residents in Canada,
- number of lines per 100 residents in Canada.

Since the objective was a model for forecasting traffic between Canada and other countries, naturally Canadian statistics are involved.

In all 44 cases, the dependent variable was paid minutes of telephone traffic between country x and Canada.

Stepwise, multiple regression analysis was used involving re-examination at every stage of the variables incorporated nto the model. This process is continued until no more variables will be admitted to the equation and none are rejected. Since not all of the 8 variables are present at the final stage, it was possible to detect the variables most frequently used in all 44 countries. From 8 variables, the number boiled down to four, i.e.:

- the number of telephone sets in country x: N_x
- the number of telephone sets in Canada: N_c
- the phone density in country x: D_x
- the phone density in Canada: D_c

The next step was to test the strength of the relation between the independent variable paid minutes and each of the four variables listed above. The results showed the probability that these variables taken individually have no effect upon the paid minutes:

- for N_x the probability is: .35

  for N_c : .33

  for D_x : .34

  for D_c : .4

From the fact that these are averages (for 44 countries) one must understand that small changes indicate an important difference, and therefore we excluded the phone density in Canada from the model. The three others have been kept.

This decision appears intuitively correct, since Canada has a very high level of telephone development (about 2 telephones per 3 inhabitants) and thus the density has probably little limiting effect on calling patterns.

The results now indicate a model of the form:

- paid minutes = f (N_x, N_c, D_x);

we found that the best combination of these variables is the linear combination:

- paid minutes= $aN_x + bN_c + cD_x + d$.

To confirm this finding many non-linear equations were statistically tested; some were more precise inside the interval of known data (i.e., a higher R^2) but none gave better forecasts, i.e., predictions beyond the known data.

3. COMPARISON WITH TIME SERIES:

In order to prove the importance of such a result, a regression analysis was derived using time as the independent variable as is normal practice, and a comparison was done using the correlation coefficient and the forecasts as criteria.

First, recall that adding independent variables to a model can only increase R^2 (correlation coefficient) and never reduce it. Since R^2 often be made large by including a large number of variables, there exists a modified measure that recognizes the number of independent variables in the model. The "adjusted coefficient of multiple determination", denoted R^2_a is defined:

$$R^2_a = 1-\frac{(n-1)}{(n-p)}\frac{SSE}{SSTO}$$

SSE = Sum of Squares Error
SSTO= Sum of Squares Total
where: n = number of observations
p-1 = number of ind. variables.

R^2_a may become smaller when another independent variable is introduced into the model. The average R^2 and R^2_a for the 44 countries are listed below:

	R^2	R^2_a
quadratic with time	.899	.824
new model	.899	.775

From R^2 one can see that the regressions equivalently fit the given points. It does not provide infomration about how good would be an eventual forecast. R^2_a confirms this equivalency by a being being lower in the new mode.

To test if the variables are part of a better model, the forecasts generated by the new formula are compared with the ones coming from the time model in the following way.

Values for the number of phones in country x (N_x) for the number of phones in Canada (N_c) and for the phone density in country x (D_x), were derived for 1982 using the values of 1975 to 1981; each variable had its own quadratic regression with time. These values (for 1982) were plugged in the equation:

$$PD\ MIN = a\ N_x + bN_c + cD_x + d$$

where a, b, c & d were given by the regression analysis using known data of 1975 to 1981. Then a forecast was found for 1982 using the quadratic with time based on 1975 to 1981. Each forecast was compared with the known value of 1982.

The average relative errors for 44 countries were found to be:

quadratic with time: 16.8%
best forecast for 15 countries

new model : 15.1%
best forecast for 27 countries

(for more details, see Table I).

Since the quadratic with time performed poorly compared to the new model, a linear fit with time was tested but the results were worse than with the quadratic.

4. GROUPING

For several developing countries, statistical figures are either inexistant or found to be inhomogenous, thus making a statistical analysis meaningless. In other cases, volumedata is unavailable due to the introduction of a new route. Sometimes also a whole new telephone system is being established and needed past statistical data is obviously inexistant.

In these cases the forecast can be computed using parameters derived for a group of countries having a similar telephone development.

In this study we used both density and number of telephone sets as parameters for grouping. The reasons for using both factors are as follows: on one hand the number of telephones gives a quantitive use of these telephones. On the other hand the phone density is a qualitative variable which provides an indication about the phoning habits of a subscriber but is unable to solely quantify the demand. It follows that since these factors are complementary, they will together describe the traffic accurately.

Many relations between these two variables were tested for developing countries and eventually a useful relation was found:

$$\sqrt{(D_x + 1) \cdot N_x} \qquad [1]$$

where D_x = # of ph./100 res. in country x

N_x = # of telephones in country x.

This equation represents the geometric mean of $(D_x + 1)$ and N_x.

Forty-four countries were first ranked using equation [1], and then ranked following the paid minutes to Canada recorded in 1982. R^2 (correlation coefficient) between the two ranks was found to be .85.

Then the 44 countries were clustered into 4 groups and constants found for the average equation for each group. These were then used to produce forecasts for each country for comparison with actual values and forecasts developed individually for each country.

The results are given in Table II. Naturally, individual forecasts perform better then forecasts using group constants but these done with group constants are sufficiently good to be used when individual data is lacking.

5. ESTIMATING THE INDEPENDENT VARIABLES

The ease of finding the data needed is the first constraint to this estimate, precision being the second. Two methods are proposed:

1. Obtain from the local telecommunication carrier of the country data for the numbers of phones and population. Then by dividing the number of phones by the number of 100 inhabitants, one gets the phone density. Note that one should look for consistancy between obtained data and past statistics if available.

2. CCITT GAS 5 has published a study [3] which shows a correlation between the GNP (gross national product) and the telephone density. The logarithmic regression was:

$$\log(\text{Tp. dens.}) = 1.303 \log(\text{GNP}) - 3.353$$

Multiplying the phone density by the number of hundred inhabitants gives the number of telephones. To find the GNP, one can consult World Bank publications. The number of inhabitants is also provided in these publications.

6. CONCLUSION

The scarceness of statistical data for countries having a developing telecommunication system, calls for a simple, naturalistic forecasting method. Statistical and comparative analysis showed that, for these countries, the new model is a slight improvement over time series when statistical data exists.

A grouping method has been used to provide forecasts where past statistical data is not available. Although of lower accuracy, this method will give a general indication of future demand.

Another possible application of this model is in long term forecasting but only preliminary work has been done on this.

REFERENCES

[1] "Yearbook of Common Carrier Telecommunications Statistics" - ITU, Geneva - published annually.

[2] K. Mirski, "Some Aspects of Trunk Traffic Forecasting" - Proc. 10th Int. Teletraffic Congress, Montreal.

[3] CCITT GAS 5, "Methods Used in Long Term Forecasting of Telecommunications Demand and Required Resources" - ITU, Geneva, 1983.

TABLE I

FORECASTING FOR 1982

COUNTRY	ACTUAL VALUE	NEW MODEL FORECAST	RELATIVE ERROR	TIME SERIES FORECAST	RELATIVE ERROR
Angola	1 930	1 847	4%	2 390	24%
Argentina	650 398	646 945	0.5%	728 196	12%
Botswana	6 900	5 605	18%	8 427	22%
Brunei	34 024	38 292	12%	39 382	13%
Colombia	536 969	600 374	12%	753 555	40%
Congo	1 461	904	38%	995	32%
Costa Rica	256 069	230 601	10%	312 132	22%
Cuba	95 035	89 725	5%	86 161	9%
Djibouti	1 488	408	72%	229	85%
Ecuador	226 439	257 391	13%	257 209	13%
El Salvador	154 178	224 767	46%	199 869	30%
Fiji	359 312	429 517	19%	448 672	25%
Ghana	12 799	9 811	23%	9 670	24%
Hungary	365 322	445 800	22%	455 536	25%
India	810 866	774 480	4%	772 234	5%
Indonesia	205 080	194 112	5%	197 825	4%
Kenya	244 989	251 158	2%	254 081	4%
Kuwait	149 904	148 199	1%	150 532	0.4%
Madagascar	2 303	2 846	23%	2 313	22%
Malawi	12 043	13 068	8%	13 083	9%
Malaysia	720 820	911 053	26%	926 723	29%
Mauritius	33 146	35 596	7%	35 852	8%
Morocco	168 535	169 236	0.3%	174 199	3%
Neth. Ant.	48 126	40 756	51%	40 101	52%
Niger Rep.	4 729	3 413	28%	4 862	3%
Pakistan	337 077	459 924	36%	422 754	25%
Panama	111 107	118 344	6%	125 095	13%
Papua N.G.	20 464	20 266	1%	24 632	20%
Paraguay	21 901	22 283	2%	22 558	3%
Poland	345 943	320 750	7%	327 409	5%
Rhodesia	43 570	40 552	7%	43 898	0.7%
Rwanda	1 910	895	47%	1 296	32%
Seychelles	3 257	3 090	5%	3 238	0.6%
South Africa	826 203	931 722	13%	945 002	14%
Sri Lanka	77 845	77 033	1%	89 108	14%
Swaziland	10 367	10 575	2%	10 383	0.1%
Syrian Afr.	92 948	89 733	3%	88 098	5%
Tanzania	87 126	88 954	2%	105 077	21%
Thailand	106 454	92 729	13%	97 664	8%
Togo	6 696	4 930	26%	4 745	29%
Tonga	4 472	2 877	35%	4 117	8%
Tunisia	83 497	87 875	2%	85 155	2%
Uganda	7 829	8 006	2%	7 722	1%
Zambia	23 246	24 537	5%	28 038	21%
AVERAGE			15.1%		16.8%

TABLE II

GROUPING AND FORECASTS FOR 1982

COUNTRY	GROUP	ACTUAL VALUE	FORECAST	RELATIVE ERROR	ERROR OF INDIVIDUAL FORECAST
Angola	1	1 930	5 005	159%	4%
Argentina	4	650 398	428 804	34%	0.5%
Botswana	1	6 900	3 427	50%	18%
Brunei	2	34 024	58 289	71%	12%
Colombia	4	536 969	307 009	43%	12%
Congo	1	1 461	3 528	141%	38%
Costa Rica	3	256 069	250 329	2%	10%
Cuba	3	95 035	201 331	111%	5%
Djibouti	1	1 488	2 259	51%	72%
Ecuador	3	226 439	150 380	33%	13%
El Salvador	3	154 178	74 278	52%	46%
Fiji	3	359 312	866 571	75%	19%
Ghana	1	12 799	5 789	54%	23%
Hungary	4	365 322	342 256	6%	22%
India	4	810 866	177 422	78%	4%
Indonesia	3	205 080	117 098	43%	5%
Kenya	3	244 989	126 663	48%	2%
Kuwait	3	149 904	275 348	83%	1%
Madagascar	1	2 303	4 049	75%	23%
Malawi	1	12 043	4 797	60%	8%
Malaysia	4	720 820	190 326	73%	26%
Mauritus	2	33 146	50 497	52%	7%
Morocco	3	168 635	95 531	43%	0.3%
Neth. Ant.	3	84 126	159 371	89%	51%
Niger Rep.	1	4 729	1 877	60%	28%
Pakistan	3	337 077	94 867	71%	36%
Panama	3	111 107	188 040	69%	6%
Papua N.G.	2	20 464	38 563	88%	1%
Paraguay	2	21 901	46 059	110%	2%
Poland	4	345 943	496 153	43%	7%
Rhodesia	3	43 570	124 838	186%	7%
Rwanda	1	1 910	1 251	34%	47%
Seychelles	1	3 257	5 752	76%	5%
South Afr.	4	826 203	521 802	37%	13%
Sri Lanka	2	77 845	38 662	50%	1%
Swaziland	1	10 367	3 892	62%	2%
Syrian Ara.	3	92 948	236 225	154%	3%
Tanzania	2	87 126	38 761	55%	2%
Thailand	3	106 454	146 861	38%	13%
Togo	1	6 696	1 923	71%	26%
Tonga	1	4 472	2 068	53%	35%
Tunisia	3	83 497	125 881	51%	2%
Uganda	1	7 829	4 176	46%	2%
Zambia	2	23 246	35 210	51%	5%

TELETRAFFIC ISSUES in an Advanced Information Society
ITC-11
Minoru Akiyama (Editor)
Elsevier Science Publishers B.V. (North-Holland)

TRAFFIC INTEREST MATRICES- A FORECASTING SCHEME FOR DEVELOPING NETWORKS

Herbert Leijon

Telefonaktiebolaget LM Ericsson
Stockholm, Sweden

ABSTRACT

The planning of a telecommunications network should be based upon a sound traffic forecast. A reliable traffic interest matrix is then needed but is however difficult to arrive at since recorded traffic data may be incomplete, of varying quality and perhaps not relevant for the future situation. The methodology presented here concentrates on the construction of the present traffic interest matrix, and it is hypothetical insofar as it builds up the matrix from assumed traffic characteristics but at the same time it utilizes available recorded traffic data as far as possible. It works stepwise, with correction of assumed model parameter values between the steps, and it takes conceivable future changes of traffic characteristics into consideration. The scheme has a modular structure, i.e. the models are replaceable.

1. INTRODUCTION

When planning a telecommunications network for any future point of time T, the forecasted traffic interest matrix is needed. An element of the matrix $A_{kl}(T)$ should preferably denote the individual traffic interest from any traffic area k to any traffic area l. A commonly used forecasting scheme is based upon the assumed knowledge of the present traffic interests $A_{kl}(0)$, the present subscriber distribution $n_k(0)$, and a reliable forecast of the future subscriber distribution $n_k(T)$. Furthermore, such a forecast should be made for each class of subscribers separately, the total forecast then being the aggregate of the separate ones.

Much work has been spent on the study of traffic growth models, less on the study of the present traffic interests $A_{kl}(0)$. The preparation of such a matrix offers, however, in practise great difficulties. The existing network contains usually a mixture of different types of analogue equipment, in many cases both crossbar and step-by-step systems. The network losses are often quite high, indicating also high rates of repeated call attempts. Especially in step-by-step networks, such repeated call attempts cause abnormal holding times and considerable additional inefficient traffic load on the interconnecting routes. There are no or very limited possibilities for traffic or call dispersion measurements; neither may the recorded route traffics be used for the calculation of traffic interests, since they carry not only inefficient traffic, but also an anonymous mixture of calls of different origins and destinations.

Even if we after all had a method for derivation of the present traffic dispersion from the traffic records, such a matrix would still not be a relevant basis for a sound forecast, since the future network is supposed to offer an improved service and less inefficient traffic compared to the present one, to show a changed traffic profile, and maybe be subject to changed subscriber behaviour due to changed tariff policy, etc.

Summarizing these obstacles, we find that $A_{kl}(0)$ is,

i) generally impossible or difficult to obtain from traffic records,

ii) of varying quality: some values will be most uncertain, others will be missing,

iii) not relevant for the future situation.

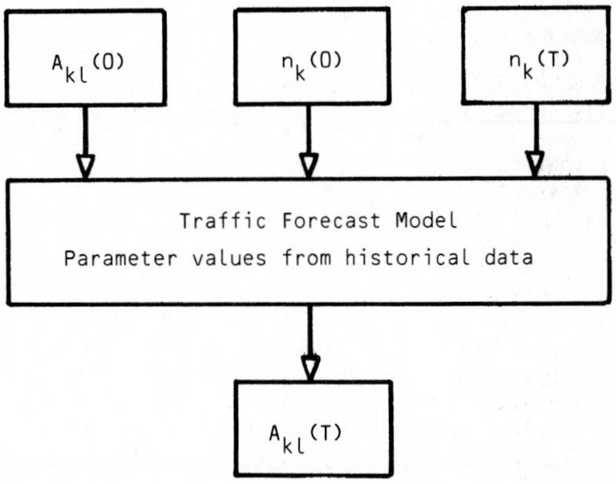

Fig. 1 Traditional forecasting scheme.

What we really need is a method that utilizes available recorded data as far as reasonably possible but is not being absolutely dependent on a complete supply of such data. This implies a considerable amount of individual judgement and decision making, i.e. the model must be mixed.

The main idea of the scheme presented here is then to define traffic parameter values that can be checked against traffic records in order to ensure, as far as possible, that they do not disagree with the present traffic situation. The parameters must be suitable as a basis for the future traffic interest forecast, which means that the values must be possible to update according to expected changes of subscriber behaviour and network quality.

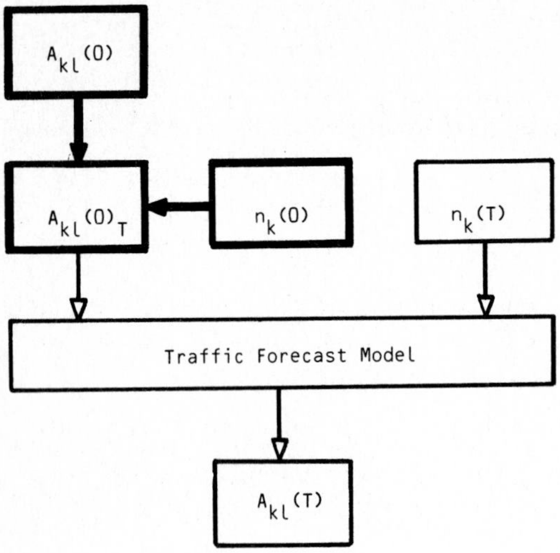

Fig. 2 Mixed model.

2. BASIC PARAMETERS

2.1 Definitions

A Total traffic
a Traffic per subscriber line
y Call intensity
h Holding time
D Dialling time per digit
B Congestion level
R Routing vector
d Dispersion factor
W Traffic interest weight
n No. of subscriber lines

Subscripts:
b,c Subscriber class no.
k,l Traffic area no.
u,v Exchange area no.
r Route no.
o Originating
t Terminating
· Total amount
0 Present time
T Future time
* Recorded quantities

2.2 Subscriber Classes

A number of subscriber classes should be defined. A subscriber class should be reasonably homogenous as concerns traffic level and subscriber behaviour. It must of course also be possible to estimate the present and future distribution of the number of subscribers per class. Examples of subscriber classes are:

a) Residentials, high and middle class
b) Residentials, lower class
c) Single business lines of various kinds
d) Lines to small PBXes
e) Lines to larger PBXes
f) Coin boxes
g) Data users, switched lines
h) Data users, leased lines

2.3 Traffic Areas and Exchange Areas

An area where a telecommunications network exists is divided into a number of exchange areas. Traffic records are related to these exchange areas. In favourable cases, we may know some present traffic interests between exchange areas $A_{uv}(0)$, and also the number of subscribers per class b in each area, $n_{bu}(0)$. For planning purposes, however, we need to forecast the future traffic interests between traffic areas $A_{kl}(T)$ rather than $A_{uv}(T)$. Furthermore, we want to make separate forecasts for different subscriber classes and then aggregate those into a total forecast.

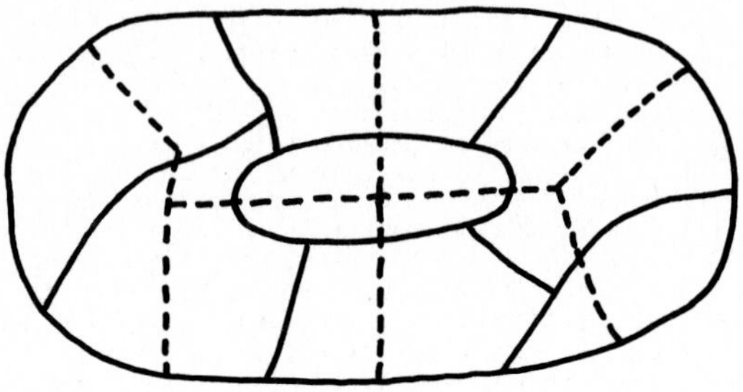

Fig. 3 Traffic areas and Exchange areas

This means that we should divide the entire area into traffic areas. Since we have a need to translate forth and back between exchange areas and traffic areas during the forecasting process in regard both to the number of subscribers per class and to the traffic interests, each traffic area should be relatively homogenous from subscriber class point of view.

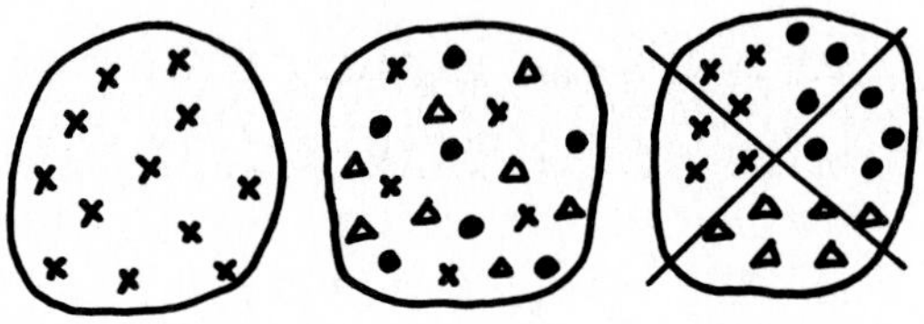

(a) (b) (c)

Fig. 4 Traffic areas.
(a) One subscriber class: suitable traffic area.
(b) Several classes, but well mixed: suitable traffic area.
(c) Non-suitable traffic area.

If this is fulfilled, we can always quite simply calculate

$$n_{bu} = \sum_k n_{ku} \cdot n_{bk}/n_k$$

and

$$n_{bk} = \sum_u n_{ku} \cdot n_{bu}/n_u$$

since

$$n_{bku} = n_{ku} \cdot n_{bk}/n_k$$

and

$$n_{bu} = \sum_k n_{bku}$$

where

n_{bu} = no. of subscribers of class b in exchange area u

n_{bk} = no. of subscribers of class b in traffic area k

etc.

2.4 Traffic Records

At least parts of the following traffic records are usually available:

For exchanges,

a) Total originating and terminating traffics $A_o^*(0)$ resp. $A_t^*(0)$

b) Total no. of carried originating and terminating calls $y_o^*(0)$ resp. $y_t^*(0)$

For traffic routes,

c) Total carried traffics $A_r^*(0)$

d) Total no. of carried calls $y_r^*(0)$

e) Congestion level $B_r^*(0)$

Our matrix for present traffic interests between exchange areas contains for the moment then only the total originating and terminating traffics, except for traffic cases when register control and end-to-end signalling is employed, where we may have records or estimates of the corresponding traffics between exchange areas $A_{uv}^*(0)$.

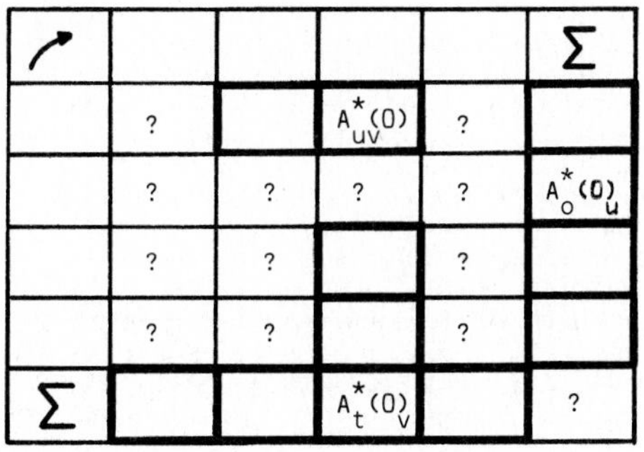

Fig. 5 Traffic records related to the traffic matrix. ▭ = Known values.

Furthermore, we will see that $\sum A_o^*(0)$ usually is much greater than $\sum A_t^*(0)$. The difference is mainly due to dialling traffic for calls that fail before reaching the terminating exchange, and it is the step-by-step calls that by far play the dominating role for the occurrence of this ineffective traffic. Irrespective of the network losses and of the rate of re-attempts, we do not expect that dialling traffic is going to load the future interconnecting network. This traffic should therefore be removed from the observed originating traffic values.

What we can do so far is the following:

i) Define $A_{u.}^*(0)' = A_o^*(0)$ and $A_{.v}^*(0)' = A_t^*(0)$ as start values in the matrix, and subtract the known $A_{uv}^*(0)$-values from this matrix, thus obtaining the new totals

$$A_{u.}^*(0)'' = A_{u.}^*(0)' - \sum_v A_{uv}^*(0)$$

$$A_{.v}^*(0)'' = A_{.v}^*(0)' - \sum_u A_{uv}^*(0)$$

ii) Adjust $A_{u.}^*(0)''$ into $A_{u.}^*(0)'''$ so that $\sum A_{u.}^*(0)''' = \sum A_{.v}^*(0)''$. That may be done in a simple way, by calculating each originating traffic as

$$A_{u.}^*(0)''' = A_{u.}^*(0)'' \cdot [\sum A_{.v}^*(0)'' / \sum A_{u.}^*(0)'']$$

or on a somewhat more individualistic basis, e.g. by first calculating the overall quantity of inefficient traffic per originating call

$$h_i = [\sum A_o^*(0) - \sum A_t^*(0)] / \sum y_o^*(0)$$

or preferably, if $y_{uv}^*(0)$ corresponding to the known $A_{uv}^*(0)$-values also are known,

$$h_i = [\sum A_o^*(0) - \sum A_t^*(0)] / [\sum y_o^*(0) - \sum y_{uv}^*(0)]$$

and then adjusting each originating traffic as

$$A_{u.}^*(0)''' = A_{u.}^*(0)'' - y_o^*(0)_u \cdot h_i$$

resp.

$$A_{u.}^*(0)''' = A_{u.}^*(0)'' - [y_o^*(0)_u - \sum_v y_{uv}^*(0)] \cdot h_i$$

iii) Now we add back the $A_{uv}^*(0)$-values to the matrix again and accept the new originating traffics as totals:

$$A_{u.}^*(0) = A_{u.}^*(0)''' + \sum_v A_{uv}^*(0)$$

$$A_{.v}^*(0) = A_{.v}^*(0)'' + \sum_u A_{uv}^*(0)$$

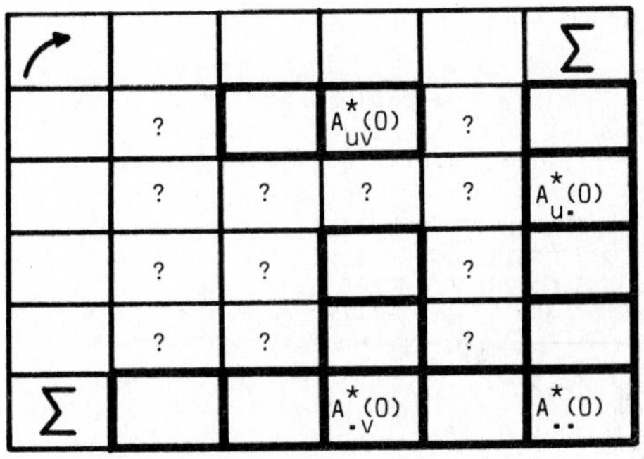

Fig. 6 Adjusted and restored matrix of inter-exchange traffics. ▭ = Known values.

2.5 Forecast Parameters

Our aim is to forecast the future traffic interests between traffic areas, $A_{kl}(T)$. It is of course valuable from planning point of view to have the possibility of separate forecasts for different kinds of traffic, e.g. data traffic on leased lines, business-to-business traffic, etc. But besides that, the final forecast is much more reliable if it is the aggregate of separate ones. Another point is, that a forecast of total originating and terminating traffics, $A_{k.}(T)$ resp. $A_{.l}(T)$, generally is more accurate than the point-to-point forecast, $A_{kl}(T)$. The ideal forecast is then the following:

i) Originating and terminating traffics per subscriber class and traffic area are forecasted, $A_{bk.}(T)$ resp. $A_{b.l}(T)$.

ii) These are aggregated, giving total originating and terminating traffics per traffic area, $A_{k.}(T)$ resp. $A_{.l}(T)$.

iii) Independently of the total traffic forecasts, the point-to-point traffics between subscriber classes are forecasted, $A_{bkcl}(T)$.

iv) These are aggregated, giving the point-to-point traffics for all subscribers, $A_{kl}(T)$.

v) The originating and terminating traffic forecasts $A_{k.}(T)$ resp. $A_{.l}(T)$ are trusted and thus distributed over the matrix, using the separate point-to-point forecast values $A_{kl}(T)$ as distribution factors.

We need consequently such traffic forecast parameters as can be checked against available traffic records, be adapted to the future conditions, and in combination with the subscriber distribution data can be used for calculation of the desired traffic quantities. Three such forecast parameters are central for the proposed scheme:

i) $a_{b.}$ = total originating traffic per subscriber line in subscriber class b. The property of this parameter is that it is relatively universal, i.e. it varies not too much between different places of similar character and stage of development, and it is also rather stable over time.

ii) d_{bc} =traffic dispersion factor, shows how the originating traffic per subscriber of class b is spread over all classes. $\sum_c d_{bc} = 1$. The property of the parameter is a little less universality than that of the first one, i.e. it is more locally influenced and it's values change also more with the development of the area.

iii) W_{bkcl} =traffic interest weight. The parameter corresponds to the tendency that a subscriber of class b and belonginging to traffic area k has to call a subscriber of class c due to the fact that the latter belongs to area l. For example, a high class residential subscriber might have a clear tendency to call small shops, provided that these shops are situated in the same area or in the city center, but a diminutive tendency to do the same if they are far away or are situated in a lower class residential district.

This parameter is of course of completely local character, and it's values may also change considerably with the development of the area. Fortunately, the individual weights can be taken as very round figures without causing serious errors in the aggregated traffic quantities.

3. FORECASTING PROCEDURE

3.1 Calculations for the Present Point of Time

The goal is to find realistic present values of the forecast parameters $a_{b.}(0)$, $d_{bc}(0)$ and $W_{bkcl}(0)$ for well-defined traffic areas. The following procedure could be applied:

a) We collect the parts of the following data that are available:

$A^*_r(0)$ = Route traffics

$y^*_r(0)$ = Carried call intensities on the routes

$B^*_r(0)$ = Congestion level on the routes

$A^*_o(0)$ = Originating exchange traffics

$A^*_t(0)$ = Terminating exchange traffics

$y^*_o(0)$ = Originating carried call intensities

$y^*_t(0)$ = Terminating carried call intensities

$A^*_{uv}(0)$= Exchange-to-exchange traffics

$y^*_{uv}(0)$= Exchange-to-exchange call intensities

$R_{uv}(0)$= Routing vector for step-by-step traffic

b) We define subscriber classes and traffic areas, which implies that the following relation matrices should be prepared:

$n_{bk}(0)$ = No. of subscribers of class b in area k

$n_{ku}(0)$ = No. of subscribers in traffic area k that are connected to exchange area u

Because of the homogenity principle applied to the choice of traffic areas, $n_{bu}(0)$ can be derived from these relation matrices.

c) In section 2.4 was shown how the recorded data could be used for a partial preparation of the traffic matrix $A^*_{uv}(0)$ after the removal of estimated inefficient traffic. Since the point-to-point traffics in the matrix will be used as check values during the calculation of forecast parameter values, some kind of confidence intervals should be attached to them. The size of a confidence interval depends of course on how the particular exchange-to-exchange traffic value was derived. This is best exemplified by some examples:

i) Say that the traffic from one exchange to another is carried on a direct low-loss route where it is properly recorded. The meter shows, say 100 erl. If we consider the possible deviation from the true mean value being at most 5%, then $A^*_{uv}(0)_{min}$=95 erl., $A^*_{uv}(0)_{max}$=105 erl.

ii) Take now the case when alternative routing is employed. Say that we have recorded the carried traffic on the direct high-usage route = 80 erl. and the congestion level on the same route = 20%. If we suppose that we have estimated the point-to-point congestion to about 5% e.g. by using a traffic route tester, then we may calculate the total traffic arriving to the terminating exchange as $80\cdot[1-0.05]/[1-0.20]$ =95 erl., i.e. 80 erl. goes via the high-usage route and 15 erl. via the tandem network. But the 15 erl.-figure is highly uncertain. Say that there is a possible deviation of 60% or 9 erl. Therefore, we may put $A^*_{uv}(0)_{min}$ =86 erl., and $A^*_{uv}(0)_{max}$ =104 erl.

iii) Cases where a great part or the whole traffic is routed via the tandem network may give rise to such uncertainty of estimated point-to-point traffics that the value of such estimates is doubtful.

d) Now we determine the originating traffic per subscriber line in each subscriber class, $a_{b.}(0)$, in the following way:

Solve the equation system

$$\sum_b n_{bu}(0) \cdot a_{b.}(0) = A^*_{u.}(0)$$

$$u = 1,2,\ldots U$$

where U = no. of exchange areas.

If there are in all S subscriber classes, we will get $\binom{U}{S}$ sets of solutions.

Since the assumption that the originating traffic per subscriber belonging to a particular class is constant irrespective of the exchange area of course can not be absolutely true, and the "known" data $n_{bu}(0)$ and $A^*_{u.}(0)$ furthermore are more or less uncertain, some of the sets will look a bit strange, as they will comprise also extreme values, e.g. negative values and very high values as well. Fortunately, extremely low and extremely high values generally belong to the same sets. What we do is to remove those sets from the lot. From the remaining acceptable sets we calculate the most likely values of $a_{b.}(0)$. There are several possibilities to do that. The simplest way is to consider each class b separately and estimate $a_{b.}(0)$ as the median of all accepted values. Another way is to apply the method of least squares to each class individually, or to consider all classes simultaneously. The flexibility of the so composed set may be increased by determining also a confidence interval for each $a_{b.}(0)$-value. Again, there are several possibilities. A statistically calculated 95% confidence interval could be used, but also a fixed percentage around the chosen value, or maybe the whole range of values from the different sets.

e) We will need the terminating traffic per subscriber line in each class $a_{.c}(0)$ as a check value when we determine the traffic dispersion factors, so we repeat the procedure as per d) above, but now solving the equation system

$$\sum_{c} n_{vc}(0) \cdot a_{.c}(0) = A^{*}_{.v}(0)$$

$$v = 1,2,\ldots U$$

Again, sets containig extreme values are rejected, and representative $a_{.c}(0)$-values and confidence intervals are calculated from the remaining ones.

f) Now we come to the delicate problem of determining the traffic dispersion factors $d_{bc}(0)$. The definition of d_{bc} is: The proportion of the originating traffic per subscriber line of class b that terminates among class c subscribers. Consequently, $\sum_c d_{bc} = 1$, and in the $d_{bc}(0)$ -matrix, we set the values row by row. We will understand the idea by imagining that Fig. 7 is a picture shown on a visual data screen.

To our guidance, our earlier determined $a_{b.}(0)$ -values are shown to the left. The matrix on the top right should be filled up by us, from experience and through local knowledge, and by reasoning, row by row. At the extreme bottom, the earlier determined $a_{.c}(0)$ -values with their confidence intervals are displayed.

When we have set all $d_{bc}(0)$ -values, our computer calculates the resulting values

$$a'_{.c}(0) = [\sum_{b} n_{b}(0) \cdot a_{b.}(0) \cdot d_{bc}(0)] \;/\; n_{c}(0),$$

appearing immediately below the $d_{bc}(0)$ -matrix.

Next step is to compare these resulting $a'_{.c}(0)$ -values with the check values $a_{.c}(0)$ displayed further below, and to decide whether the observed differencies can be accepted or not. If not, the $d_{bc}(0)$ -matrix is revised, which is quite simple because e.g. high $a'_{.c}(0)$ -values relate to high $d_{bc}(0)$ -factors, etc.

g) The traffic distribution weight W_{bkcl} is defined as a measure of the tendency that a subscriber of class b and belonging to traffic area k has to call a subscriber of class c, due to the fact that that subscriber belongs to traffic area l. Therefore, each pair of b,c -values can be treated separately in the process of setting the $W_{bkcl}(0)$ -values. Furthermore, a very limited set of round values can be used, e.g. three values 1,2or3. In that case, 1 ="Low", 2 ="Normal" and 3 ="High". There may of course be reason to use a finer scale, e.g. five values 1,2,3,4 or 5. In that case 1 ="Very low", 2 ="Low", 3 ="Normal", 4 ="High" and 5 ="Very high".

Again, let us imagine that we are looking at the data display. If we set a pair of b,c -values, a matrix filled with 3:s appears. The 3:s are default values, which will be used if we do not set other values.

h) All basic traffic parameters now having been determined, we can calculate

$$A_{bkcl}(0) = [a_{b.}(0) \cdot d_{bc}(0) \cdot n_{bk}(0)] \cdot$$
$$\cdot [n_{cl}(0) \cdot W_{bkcl}(0)]/[\sum_{l} n_{cl}(0) \cdot W_{bkcl}(0)]$$

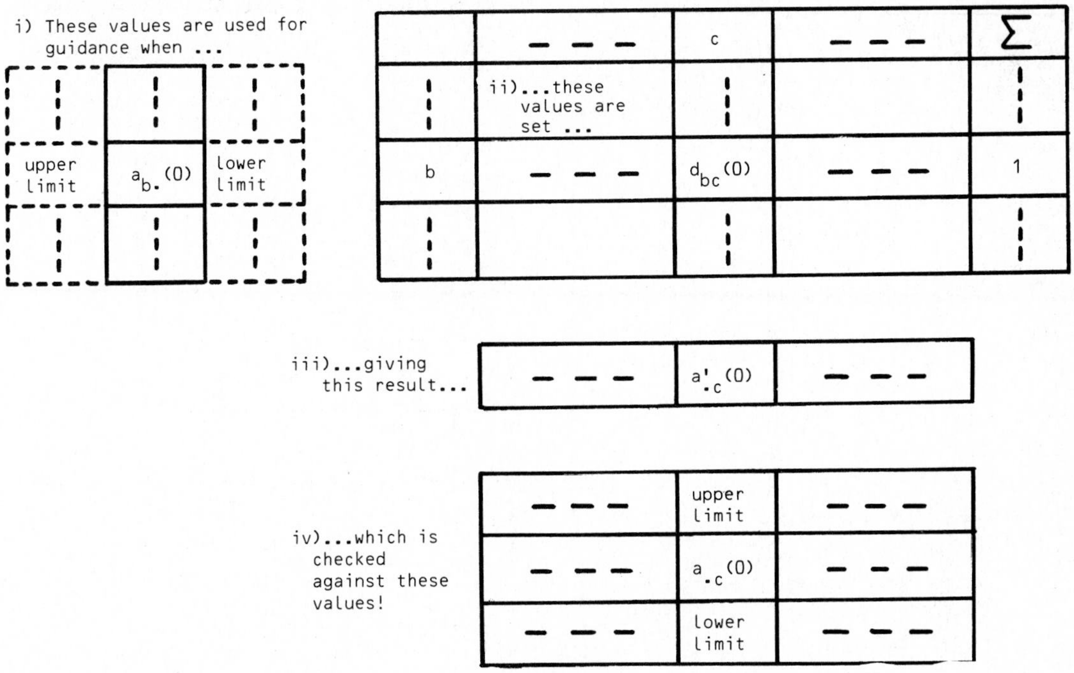

Fig. 7 Setting and checking of traffic distribution factors.

$$A_{kl}(0) = \sum_b \sum_c A_{bkcl}(0)$$

and

$$A_{uv}(0) = \sum_k \sum_l A_{kl}(0)\cdot[n_{ku}(0)\cdot n_{lv}(0)]/[n_k(0)\cdot n_l(0)]$$

The values $A_{uv}(0)$ can be checked against the known $A^{*}_{uv}(0)$ -values (if any). But the recorded route traffics $A^{*}_{r}(0)$ should also be utilized for checking!

b =..., c =...

k \ l	1	2	3	4
1	3	3	3	3
2	3	3	3	3
3	3	3	3	3
4	3	3	3	3

k \ l	1	2	3	4
1	5	5	3	1
2	4	3	3	3
3	3	3	3	3
4	3	3	3	3

We change some of the values....

Fig. 8 Setting of $W_{bkcl}(0)$ -values.

i) Calculation of hypothetical route traffic values: Suppose that totally y call attempts are made in exchange u in order that some traffic shall reach exchange v. Suppose also that the calls are set up in a step-by-step part of the network, via routes nos. 1, 2, 3, 4 and 5 in that order, where route no. 1 is the outgoing route from exchange u, and route no. 4 is the incoming route to exchange v. The congestion levels are B_1, B_2, B_3 and B_4, resp. It can then be shown, that about $y\cdot[1-B_1]\cdot B_2$ calls are carried by route 1 but rejected by route 2. What happened on route 1 was that next digit was dialled, and after that the call was lost. To go one step further: $y\cdot[1-B_1]\cdot[1-B_2]\cdot B_3$ calls are carried by route 1, and accepted by route 2, but rejected from route 3. For these calls, 2 digits were dialled after the acceptance on route 1.

By proceeding with the same kind of reasoning and neglecting some disturbing factors like strange subscriber behaviour etc., but starting from the terminating exchange instead, we may calculate, approximately, the contribution from traffic case uv to the carried traffic on route r as

$$\begin{aligned} A_{ruv}(0) = y_{uv}(0)\cdot[&D\cdot B_2/[1-B_2]\cdot[1-B_3]\cdot[1-B_4] \\ &+ 2\cdot D\cdot B_3/[1-B_3]\cdot[1-B_4] \\ &+ 3\cdot D\cdot B_4/[1-B_4] \\ &+ 3\cdot D + h_t(0)] \end{aligned}$$

where

D = Dialling time per digit
$h_t(0)$ = Remaining holding time for calls that reach the terminating exchanges
$y_{uv}(0)$ = No. of calls from exchange u that reach exchange v

The general calculation procedure may be as follows:

i) Estimate $h_t(0) = [\Sigma A^{*}_t(0)]/[\Sigma y^{*}_t(0)]$

ii) Estimate $y_{uv}(0) = A_{uv}(0)/h_t(0)$

iii) Estimate the contribution of traffic on route r from traffic case uv as

$A_{ruv}(0) = f[A_{uv}(0), R_{uv}(0), B_{uv}(0)]$

where

$A_{uv}(0)$ is calculated earlier
$R_{uv}(0)$ is the routing vector, telling us which routes are used, and in what order
$B_{uv}(0)$ are the congestion levels for the routes defined in the routing vector. Should the individiual congestion values not be known, one way is to use an estimated common average value instead.

iv) The $A_{ruv}(0)$ -values can now be aggregated into total route traffics:

$$A_r(0) = \sum_u \sum_v A_{ruv}(0)$$

j) It is now time to check our hypothetical values $A_{uv}(0)$ and $A_r(0)$ against the traffic records $A^{*}_{uv}(0)$ resp. $A^{*}_r(0)$. Too large differences indicate that, primarily, our $W_{bkcl}(0)$ -values should be revised. However, smaller deviations can be neglected, since the whole business up to this point has been quite a tricky one.

We conveniantly let our computer prepare two lists to be shown on the data display, one list displays bad route traffic cases, and the other one shows the bad exchange-to-exchange traffic cases. The computer should pick the worst case first, then the next to worst one, etc.

Now, large absolute deviations are more serious than small ones, but on the other hand, large relative deviations are also more serious than small ones. Therefore, we must make a compromise between these two principles. Furthermore, we should also make an allowance for certain reasonable variations around the recorded value, say p % for route traffics. As an example, the expression

$$[\,|A^{*}_r(0) - A_r(0)| - A^{*}_r(0)\cdot p/100]^2 / A^{*}_r(0)$$

could be used to find the worst case for the route traffic list. Fig 9 shows how this list could be designed. The exchange-to-exchange traffic list is prepared analogously. One way of calculating the confidence intervals for exchange-to-exchange traffics was described in section b) above.

u=.., v=.., $A^{*}_r(0)$=.., $A_r(0)$=...

b	c	k	l	W_{bkcl}	Contribution to $A_r(0)$

Fig. 9 List for route traffic check.

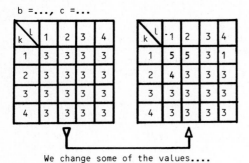

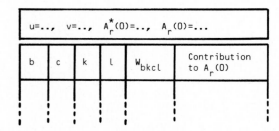

3.2 Calculations for the Future Point of Time

The main task of this paper is to illustrate a way of finding the present values of some traffic parameters that are important for the future traffic forecast. Therefore, only some brief comments will be made as to how these parameters should be updated in order to be relevant for that future situation.

The traffic profile for the hours of the day is often quite deformed in older networks. For example, the present busy hour to all hours traffic ratio may be, say 1/12, while if the network should work at a good quality of service level, the value of the same ratio would be, say 1/8. This indicates that the traffic parameters must be revised. Dependent on how much we know or can reasonably believe about the present conditions, such a revision could be done in several ways. Two examples follow.

First example:

i) Adjust the $A_{uv}(0)$ -values individually, for expected changes of the traffic profiles.

ii) Calculate new $A_{u.}(0)$ -values.

iii) Calculate new sets of $a_{b.}(0)$, corresponding to the sets that were accepted before.

iv) Calculate the new $a_{b.}(0)$ -values, but do not change $d_{bc}(0)$ or W_{bkcl}!

Second example:

i) Revise the $a_{b.}(0)$ -values directly.

After that, $a_{b.}(0)$, $d_{bc}(0)$ and $W_{bkcl}(0)$ should be updated to the expected future conditions of development and subscriber characteristics, $a_{b.}(T)$, $d_{bc}(T)$ resp. $W_{bkcl}(T)$. Combining them with the subscriber forecast $n_{bk}(T)$, the future traffic interest matrices are calculated.

4. CONCLUSIONS

The forecasting scheme presented here has the following properties:

i) Recorded data are used as far as reasonably possible for the calculation of forecast parameter values.

ii) The forecaster's experience and local knowledge is used for setting hypothetical values of the remaining forecast parameters.

iii) The hypothetical values are utilized to calculate quantities that can be checked against recorded data.

iv) Where serious deviations are obtained, the forecast parameter values are revised through decisions made by the forecaster.

v) The calculations are based upon simple and replaceable algorithms, suitable for computer applications. Personal computers may very well be used.

vi) The scheme works stepwise. Judgement and decision-making are essential elements of each step. The scheme is thus best suited for interactive use.

vii) The sensitivity of the forecast due to variations of the basic parameter values is therefore easily investigated.

ix) The forecast is likewise easily updated when more traffic data are being collected.

TELETRAFFIC ISSUES in an Advanced Information Society
ITC-11
Minoru Akiyama (Editor)
Elsevier Science Publishers B.V. (North-Holland)

THE MANAGEMENT OF THE UNCERTAINTIES IN SHORT TERM NETWORK PLANNING

Maurizio FILIPPINI and Umberto TRIMARCO

SIP Società Italiana per l'Esercizio delle Telecomunicazioni
Rome, Italy

ABSTRACT

When dealing with uncertainties in traffic forecasting or facilities failures, the solutions consisting in provisioning stand-by elements generally do not allow to determine fixed correlations between costs and expected benefits.

The aim of the paper is to give elements for the decision, putting the grade of service achievable in several conditions in relation to the probability the events will occur and to the costs of the needed interventions.

With special attention to the long distance network, the proposed methodology allows an easy extension of the optimization concepts to the short term planning, by evaluating the costs we have to face when sizing the network for marginal traffic increases.

1. INTRODUCTION

The target of the researches in the field of network optimization was, in the past years, mainly the most economic way of routing traffic from one node to another.

The results were then:
- optimizing algorithms, able to match minimum cost and fixed grade of service constraints concerning final trunk-groups;
- new concepts (1) and computer tools (2) on grade of service goals, suitable to face the problem involved by the growth and the digitalization of the network.

A few years experience showed that the mere economic optimization did not give the solution to the problem: too many uncertainties are present in a system as complex as long distance network is (trunks failures, forecast errors, facilities supplying or installation delays, traffic overloads).

Further studies have been carried out aiming at redudancy or stand-by trunks optimization: in any case the expected grade of service should fall into the interval bounded by the upper planned level and the lower degraded one (3).

Yet the way of growth of the long distance network looks like the worst enemy of a correct traffic routing: the coming into operation of every new trunk is the result of a complex action. Several components of different kinds are involved, belonging to different regional operation centres, supplied by different factories, and even, in some contries, operated by different companies.

During the time needed to bring the network up to the planned size, final trunk groups in hierarchical networks are loaded by more traffic than we might have forecast.

The common opinion, shared by a large number of planners, is that only the dynamic management of a non hierarchical network would enable the operating companies to route the traffic according to the best trade-off between costs and quality of service.

Yet in many countries, whose size in longitude does not involve more than one hour, hierarchical routing patterns are going to remain competitive for about the next ten years, waiting for:
- full switching modernization
- new district areas structure
- traffic growth.

The above statement, beyond the trivial consideration that SPC switching machines are needed, is based on first evaluations applied to the evolution of the Italian network structure: non hierarchical routing proved to be more economical (about 8% less expensive) if no traffic link between any two districts is less than 10 Erl large. That will bring other elements to change the structure by merging more present district areas (5); the planned merging will be insufficient nevertheless, if traffic growth does not help.

Some further steps in the optimization of a hierarchical network are then an attractive target in order to improve both the grade of service and the control of traffic evolution (6).

In our paper the correlation between improved grade of service goals and cost increases is studied in order to give elements for the decision to the company managers.

The approach consists in characterizing traffic values with their probability distribution: the composition of several matrices with equiprobable values allows us to build up a series of interventions on the network. The correlation is then found between the costs of the interventions and the probability that traffic demand will occur.

This simple and global approach needs a suitable tool to optimize the marginal increases, taking into account existing network or planned increases for more probable values.

In the following sections the tool is described and the global correlation for various levels of uncertainty is shown.

2. THE OPTIMIZATION TOOL

Generally speaking, an optimization algorithm is designed for the most economic sizing of a network in a given time and referred to the grade of service goals.

When we have to plan a new network the goal is reached. But when we deal with increases of an existing network, characterized by its peculiar redudancies or typical design criteria, the mere optimization algorithm fails.

In fact, when one considers the basic equation in Pratt's algorithm (7):

$$C_1/H = C_2/\beta_2 + C_3/\beta_3$$

where

H is the marginal occupancy and

β the marginal capacity (fig. 1)

it is easy to see that low values in the final trunks of the former network, force traffic routing via direct trunks. On the contrary, low marginal capacities mean unloaded trunks thus calling for more traffic routed via final trunk groups.

Thus in the short term planning the result will be a pendulum effect in sizing the final trunk groups.

In other words the equilibrium condition reached in the optimization is unstable: in order to force the stability of the solution an eurhistic methodology is here proposed (fig. 2).

The iterative process is initialized by fixing an arbitrarily wide variation field for every marginal capacity β_i of the final trunk groups.

If the resulting size R of the i-th final trunk group is less than the existing one E, we have to force marginal capacity β_i to the centre point of the above interval; the interval itself has to be furtherly narrowed by raising the lower value β_m up to the marginal capacity resulting when loading the existing trunk group with new traffic.

On the contrary if R > E, we have to:

- keep the former variation field if $\beta > \beta_M$
- narrow the interval by putting $\beta_M = \beta$ (resulting from E trunks loaded with new traffic).

Before proceeding to the following iterations the actual β_i is computed for R trunks and real traffic, and the new variation field is determined.

By this way a fast convergence can be reached, taking into account the existing trunks.

In order to exploit existing capacities, another concept has been introduced into the short term planning procedure: in the first sizing the cost of the marginal trunk in a high usage trunk group is kept equal to that of the existing ones. The resulting oversize N'_o of the group will be reduced by the following iterations, as the optimum size is found by minimizing the total cost function (fig. 3):

$$C_t(A,i) = C_e N_e + C_n i + C_2\,(A - Y(N_e+i)) + C_3 Y\,(N_e+i)$$

where

$C_t(A,i)$ is the total cost

C_e is the cost per existing trunk $(0<n\leq N_e)$

C_n is the cost per new trunk $(i>0)$

C_2 and C_3 are the costs per erlang met with in the branches 2 and 3

A is the offered traffic from I to J

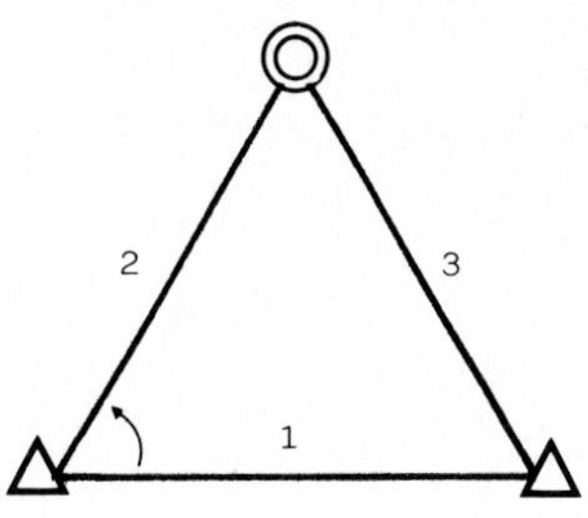

Fig. 1 Basic alternate routing

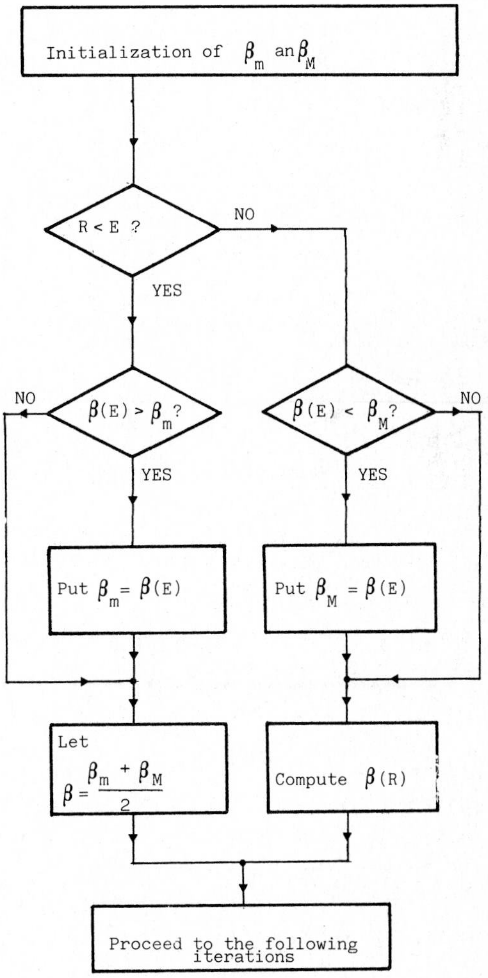

Fig; 2 Eurhistic method for marginal capacity

Y (N) is the traffic routed by a group of N trunks

Decreasing i from $(N'_o - N_e)$ to 1 the condition $(\partial C_t / \partial i)_A = 0$ gives us the value of i that, minimizing C_t^A, will exploit existing capacities at the best (fig. 4).

3) THE MANAGEMENT OF FORECASTING UNCERTAINTIES

The problem of uncertainty management may be faced in two steps:

- as no management can be conceived on the unknown events, firstly we have to examine the nature and the variation field of uncertain elements;
- the second step consists in the development of a flexible structure, able to deal with the largest amount of probable values in the above elements.

When we focus our attention on the planning of a large and interconnected network, as the long distance one is, the uncertain elements mainly are:

- traffic forecasts referring to links deeply influenced by seasonal peak effects or by hardly predictable growth-rate;
- facility provisioning time: important facilities, requiring external works, such as new radio or cable links and buildings for transmission or switching plants, may occur later than in the scheduled year.

The final effect of the resulting delays in trunks availability, or of the mismatching between digital and analogue facilities, is the overload of the network actually operating.

In order to give useful elements for the decision on the entity of the interventions, the uncertainty range of the traffic forecast must be bounded within determined edges.

In the proposed aproach, the traffic matrix should not be compiled in a deterministic way, but, on the contrary, each value should be accompanied by its probability distribution. This approach, of course, has a real meaning and offers a useful aid only when the forecast is the result of a large spread of values.

According to the above method, a series of matrices, each characterized by the cumulative probability that traffic demand will not be larger, may be available.

By means of the eurhistic method shown above we can plan, beyond the network related to the maximum probability matrix, the increases suitable to satisfy the marginal demand. Beside that the grade of service achievable when higher traffic demand occurs can be evaluated.

The investments needed in the focused cases, each related to an assigned probability of the demand and to a degraded, but known and determined, grade of service, are the elements for the final decision on the composition of the program.

A similar approach can be used when dealing with the degradation of the network availability due to the probable lack of some important facility: the optimization may be forced to route the traffic via the existing facilities beyond the economic limit. We can thus grant, even in degraded conditions, the achievement of an assigned grade of service.

The improvements introduced in the short term planning procedure give us the opportunity for prolonging the utilization of the hierarchical

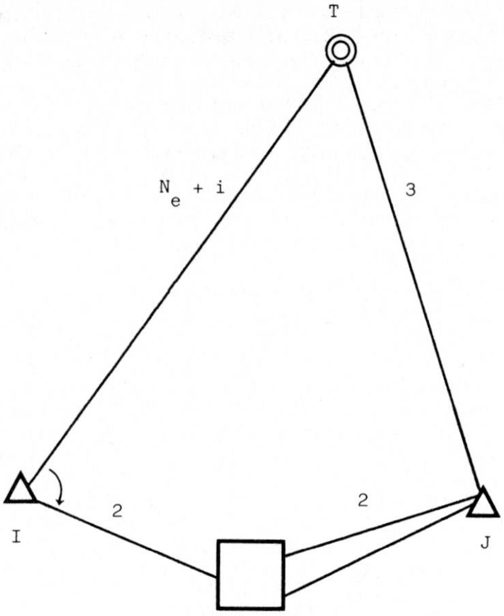

Fig. 3 Determining the optimum size of the higt usage trunk group IT

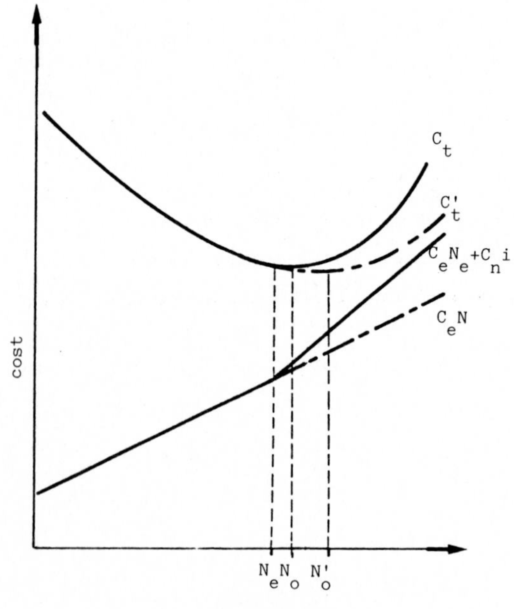

Fig. 4 The influence of the new capacity in the optimization

network. To this purpose the shown arrangement on costs per existing trunks is very useful.
The constrained optimization will determine an extra cost related to the improvement in grade of service we want have, even in case of lacking facilities.

In both cases a global correlation (fig.5) is found between grade of service, probability of the demand, technical resources needed and investments. According to the desired (or minimum allowed) goal is then easy to determine the financial resources involved by the alternative decisions.

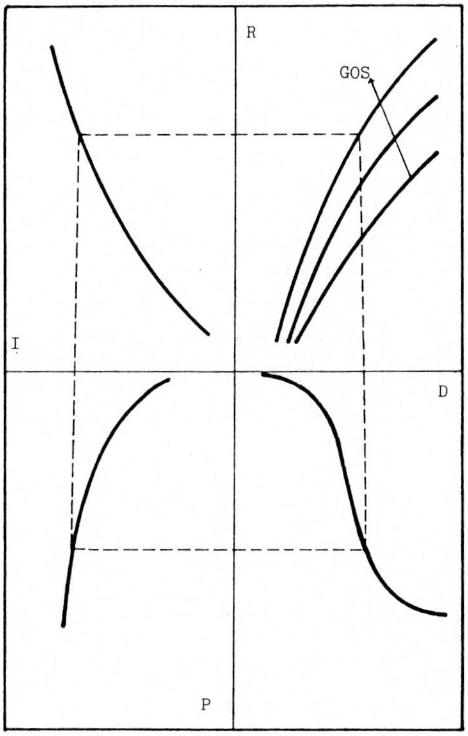

Fig. 5 Correlation between Probability (P) of the Demand (D), technical Resources (R) Grade of Service (GOS) and Investments (I)

References

(1) R.W. HORN - End to end connection probability - The next major engineering issue - 9th ITC Torremolinos, Oct. 1979.
(2) M. GAVASSUTI, G. GIACOBBO SCAVO, U. TRIMARCO - A computer procedure for optimal dimensioning of evolving toll networks under constraints - 10th ITC Montreal. Jun. 1983.
(3) P. LINDBERG, U.MOCCI, A. TONIETTI - Cost 201: A european research project - A procedure for minimizing the cost of trasmission network under service availability constraints in failure conditions - 10th ITC Montreal. Jun.1983.
(4) G.R. ASH, A.H. KAFKER, K.R. KRISHNAN - Intercity dynamic routing architecture and feasibility - 10th ITC, Montreal. Jun. 1983.
(5) U. MAZZEI, G. MIRANDA, P. PALLOTTA. On a full digital long-distance network - 9th ITC, Torremolinos, Oct. 1979.
(6) A.N. KASHPER, G.C. VARVALOUCAS - Trunk implementation plan for hierarchical networks - AT&T Bell Laboratories Technical Journal Jan. 1984.
(7) C.W.PRATT - The concept of marginal overflow in alternate routing - 5th ITC New York, Jun. 1967

TELETRAFFIC ISSUES in an Advanced Information Society
ITC-11
Minoru Akiyama (Editor)
Elsevier Science Publishers B.V. (North-Holland)

A NEW TYPE AGGREGATION METHOD FOR LARGE MARKOV CHAINS AND ITS APPLICATION TO QUEUEING NETWORKS

Yukio TAKAHASHI

TOHOKU UNIVERSITY
Sendai, Japan

ABSTRACT

In some queueing networks we encounter in traffic models, each node is connected with only a few adjacent nodes. Tandem queueing networks are typical examples of them. In such a network, stochastic behavior of a node might be expected to be approximately independent of that of nonadjacent nodes, though it may depend highly on the behavior of adjacent nodes.

Here, a new aggregation method is proposed to approximately calculate the stationary probabilities of large Markov chains derived from such queueing networks. It exploits the nearly independent feature between nonadjacent nodes, and makes it possible to approximately calculate the stationary marginal probabilities with relatively small computational burden.

1. INTRODUCTION

If a queueing network does not have a product form solution, it is usually very difficult to calculate its stationary probabilities and its performance measures exactly. A usual approach is to form a Markov chain which describes the stochastic behavior of the network and to numerically solve the set of linear equations of the stationary probabilities of the chain. If the network has r nodes and if J_k states are required to represent the state of Node k, then we need

$$J = \prod_{k=1}^{r} J_k$$

states to represent the state of the whole network. J grows very rapidly as r increases. So the naive Markov chain approach fails even for networks with more than a few nodes.

In order to overcome this difficulty, a number of approximation methods have been proposed. For example, [9] and [1] proposed relatively simple approximation methods for open networks with blocking and [10] did so for open networks without blocking.

In most of the approximation methods, the idea of node-by-node decomposition is used. Each node of the network is approximated by a suitable standard queueing model and the performance measures are calculated as if these approximate queues are mutually independent.

Such decomposition methods enable us to compute performance measures with relatively small computational burden. However, error assessment is not easy for the results. In some cases they may give a fairly accurate solution but in some cases they may not. We have no means to estimate the accuracy of the results in specific cases. This is a major drawback of the decomposition methods.

In this paper, another approximation method is proposed for networks with a Markov chain representation. This method is based on the aggregation technique for large scale Markov chains. First we choose a set of aggregated states of the chain and form a set of equations for the corresponding stationary probabilities. This is done by adopting the usual aggregation technique. Generally the set of equations so formed contains variables other than the ones first chosen. To make the set of equations solvable, we have to introduce some assumptions on the excess variables. Usually this is done by assuming some statistical independence among nodes. Then we can reform the set of equations so that it contains only the variables first chosen. The resultant equations may not be linear. But a suitable iterative method seems to work well for solving the set of equations.

The merit of this method is in the flexibility in the choice of the aggregated states and in the independence assumption imposed. If we want a more accurate solution, we may choose more variables first and impose a weaker assumption. If we want a rough result with less computational burden, then we may choose less variables and assume more strict independence on the node. The accuracy of the solution, of course, greatly depends on the closeness of the independence assumption to the real situation. However, if we adopt multiple levels of assumptions and if the discrepancy among the results is small, then we might as well conclude that the results are sufficiently accurate.

In the next section, we investigate the method using an example of the application to a tandem network, and in Section 3, we present some numerical results and give some comments on the applicability of the method.

2. A NEW AGGREGATION METHOD

In order to make the investigation clearer, we shall use an example of a tandem network with 5 nodes.

2.1 A tandem network example

Let us consider the tandem network with 5 nodes designated in Fig.1. Customers arrive at the network through a phase type renewal process. Node k (k=1,2,...,5) has s_k servers for service and m_k buffers for customers waiting for their service. We consider the following simple blocking rule. All servers at Node k are

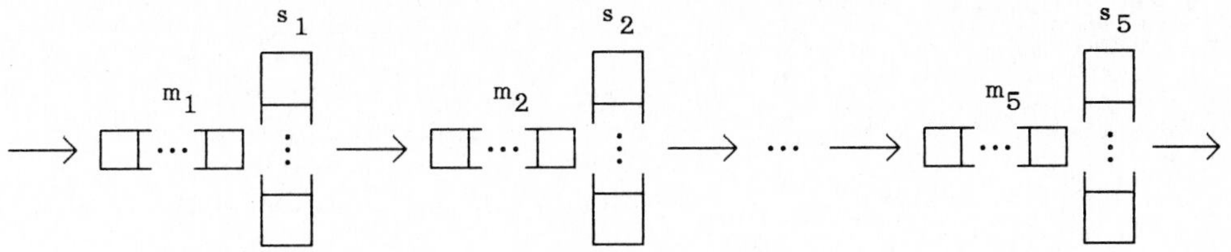

Fig.1. A tandem network model

blocked and stop their service when Node k+1 becomes full. If the service for a customer at Node k+1 is completed and a vacancy appears at the buffer, then the servers at Node k are immediately unblocked. Arriving customers are lost if Node 1 is full.

If the service time distributions are of phase types, then the state of the network can be represented by a quintuple

$$(i_1, i_2, i_3, i_4, i_5),$$

where i_k represents the state of Node k. If we number the states of the network in lexicographic order, then the infinitesimal generator of the Markov chain which represents the stochastic behavior of the network takes the form

$$\begin{aligned} Q = {} & Q_1 \otimes I_2 \otimes I_3 \otimes I_4 \otimes I_5 \\ & + Q_{12} \otimes I_3 \otimes I_4 \otimes I_5 \\ & + I_1 \otimes Q_{23} \otimes I_4 \otimes I_5 \\ & + I_1 \otimes I_2 \otimes Q_{34} \otimes I_5 \\ & + I_1 \otimes I_2 \otimes I_3 \otimes Q_{45} \\ & + I_1 \otimes I_2 \otimes I_3 \otimes I_4 \otimes Q_5, \end{aligned} \tag{1}$$

where Q_{kj} is a $(J_kJ_j)\times(J_kJ_j)$ matrix governing the state transitions caused by a service completion at Node k, Q_1 is a $J_1\times J_1$ matrix governing the state transitions caused by an arrival of a customer, and Q_5 is a $J_5\times J_5$ matrix governing the state transitions caused by a service completion at Node 5. I_k is the identity matrix of order J_k, and $\otimes$ represents the Kronecker product operation of matrices.

Note that

$$\begin{aligned} & Q_1e_1 = 0, \quad Q_5e_5 = 0 \\ & Q_{k,k+1}(e_k \otimes e_{k+1}) = 0, \ k=1,\dots,5, \end{aligned} \tag{2}$$

where e_j is the column vector of order J_j with all entries equal to 1.

The stationary probability vector x satisfies the equations

$$\begin{aligned} xQ &= 0 \\ xe &= 1 \end{aligned} \tag{3}$$

where e is a column vector with all entries equal to 1.

2.2 Some notations

We denote the entry corresponding to the state (i_1,i_2,i_3,i_4,i_5) of the stationary probability vector x by $x(i_1,i_2,i_3,i_4,i_5)$ and their marginals by

$$\begin{aligned} x^k(i_k) &= \sum_{\substack{i_\ell \\ \ell\neq k}} x(i_1,i_2,i_3,i_4,i_5) \\ x^{kj}(i_k,i_j) &= \sum_{\substack{i_\ell \\ \ell\neq k,j}} x(i_1,i_2,i_3,i_4,i_5) \\ x^{kjm}(i_k,i_j,i_m) &= \sum_{\substack{i_\ell \\ \ell\neq k,j,m}} x(i_1,i_2,i_3,i_4,i_5). \end{aligned} \tag{4}$$

Row vectors having these marginal probabilities as entries are denoted by x^k, x^{kj}, and x^{kjm} respectively. If we put

$$\begin{aligned} z_k &= e_1\otimes\dots\otimes e_{k-1}\otimes I_k\otimes e_{k+1}\otimes\dots\otimes e_5 \\ z_{kj} &= e_1\otimes\dots\otimes e_{k-1}\otimes I_k\otimes e_{k+1}\otimes \\ &\quad \dots\otimes e_{j-1}\otimes I_j\otimes e_{j+1}\otimes\dots\otimes e_5 \\ z_{kjm} &= e_1\otimes\dots\otimes e_{k-1}\otimes I_k\otimes e_{k+1}\otimes \\ &\quad \dots\otimes e_{j-1}\otimes I_j\otimes e_{j+1}\otimes \\ &\quad \dots\otimes e_{m-1}\otimes I_m\otimes e_{m+1}\otimes\dots\otimes e_5, \end{aligned} \tag{5}$$

then x^k, x^{kj}, and x^{kjm} are represented as follows:

$$\begin{aligned} x^k &= xz_k \\ x^{kj} &= xz_{kj} \\ x^{kjm} &= xz_{kjm}. \end{aligned} \tag{6}$$

We also denote conditional probabilities as follows.

$$\begin{aligned} x^{j|k}(i_j|i_k) &= \frac{x^{kj}(i_k,i_j)}{x^k(i_k)} \\ x^{m|jk}(i_m|i_k,i_j) &= \frac{x^{kjm}(i_k,i_j,i_m)}{x^{kj}(i_k,i_j)} \,. \end{aligned} \tag{7}$$

2.3 The first stage approximation

For the first stage approximation, we will take $\{i_k\}$, $i_k=1,\dots,J_k$, $k=1,\dots,5$, as aggregated states. Namely, we take marginal probabilities $x^k(i_k)$ as variables to be determined.

In order to derive equations for the variables $x^k(i_k)$, we will post multiply z_k to the first equation of (3). Then from (1) and (6), we have

$$
\begin{aligned}
0 &= xQz_1 = x^1Q_1 + x^{12}Q_{12}(I_1 \otimes e_2)\\
0 &= xQz_2 = x^{12}Q_{12}(e_1 \otimes I_2) + x^{23}Q_{23}(I_2 \otimes e_3)\\
0 &= xQz_3 = x^{23}Q_{23}(e_2 \otimes I_3) + x^{34}Q_{34}(I_3 \otimes e_4)\\
0 &= xQz_4 = x^{34}Q_{34}(e_3 \otimes I_4) + x^{45}Q_{45}(I_4 \otimes e_5)\\
0 &= xQz_5 = x^{45}Q_{45}(e_4 \otimes I_5) + x^5Q_5.
\end{aligned} \tag{8}
$$

These equations contains vectors $x^{k,k+1}$'s other than x^k's. So we have to make some assumption which relates $x^{k,k+1}$'s to x^k's. The most natural one is the following.

Assumption A1

We assume that for any possible choice of k, i_k, i_{k+1}

(9a) $$x^{k|k-1}(i_k|i_{k-1}) = x^k(i_k)$$

or equivalently,

(9b) $$x^{k,k+1} = x^k \otimes x^{k+1}.$$

This assumption asserts that the stochastic behavior of successive nodes is mutually independent. However it does not mean that

(10) $$x(i_1,i_2,i_3,i_4,i_5) = \prod_k x^k(i_k).$$

Hence, the independence assumption (9) is not so strict one.

Inserting (9) to (8), we have the desired set of equations for x^k's.

(11a) $$0 = x^1Q_1 + (x^1 \otimes x^2)Q_{12}(I_1 \otimes e_2)$$

(11b) $$0 = (x^1 \otimes x^2)Q_{12}(e_1 \otimes I_2) + (x^2 \otimes x^3)Q_{23}(I_2 \otimes e_3)$$

(11c) $$0 = (x^2 \otimes x^3)Q_{23}(e_2 \otimes I_3) + (x^3 \otimes x^4)Q_{34}(I_3 \otimes e_4)$$

(11d) $$0 = (x^3 \otimes x^4)Q_{34}(e_3 \otimes I_4) + (x^4 \otimes x^5)Q_{45}(I_4 \otimes e_5)$$

(11e) $$0 = (x^4 \otimes x^5)Q_{45}(e_4 \otimes I_5) + x^5Q_5.$$

Together with the normalization constraints

(11f) $$x^k e_k = 1, \quad \text{for } k=1,\ldots,5,$$

these equations uniquely determine the marginal probabilities $x^k(i_k)$.

Most of the equations in (11) are nonlinear equations. However, if we regard, for example, x^1 and x^3 in (11b) being known, then (11b) is a set of linear equations for variables $x^2(i)$'s. So, the set of equations (11) can be solved numerically using a suitable iterative method as usually done in the aggregation/disaggregation method.

2.4 The second stage approximation

For the second stage approximation, we will take pairs (i_k, i_{k+1}) of states of successive nodes. So, we will take $x^1(i_1)$'s and $x^{k|k-1}(i_k|i_{k-1})$'s as variables to be determined. Here we don't take $x^{k-1,k}(i_{k-1},i_k)$'s as variables because some of them are related as

(12) $$x^{k-1,k}(e_{k-1} \times I_k) = x^k = x^{k,k+1}(I_k \times e_{k+1}).$$

As we have done in the first stage approximation, we can derive a set of equations for these variables by post multiplying $z_{k-1,k}$ to the first equation of (3) as follows.

$$
\begin{aligned}
0 &= xQz_1 = x^1Q_1 + x^{12}Q_{12}(I_1 \otimes e_2)\\
0 &= xQz_{12} = x^{12}(Q_1 \otimes I_2 + Q_{12})\\
&\quad + x^{123}(I_1 \otimes Q_{23})(I_1 \otimes I_2 \otimes e_3)\\
0 &= xQz_{23} = x^{123}(Q_{12} \otimes I_3)(e_1 \otimes I_2 \otimes I_3)\\
&\quad + x^{23}Q_{23}\\
&\quad + x^{234}(I_2 \otimes Q_{34})(I_2 \otimes I_3 \otimes e_4)\\
0 &= xQz_{34} = x^{234}(Q_{23} \otimes I_4)(e_2 \otimes I_3 \otimes I_4)\\
&\quad + x^{34}Q_{34}\\
&\quad + x^{345}(I_3 \otimes Q_{45})(I_3 \otimes I_4 \otimes e_5)\\
0 &= xQz_{45} = x^{345}(Q_{34} \otimes I_{45})(e_3 \otimes I_4 \otimes I_5)\\
&\quad + x^{45}(Q_{45} + I_4 \otimes Q_5).
\end{aligned} \tag{13}
$$

In this case, the most natural assumption will be the one given below.

Assumption A2

We assume that for any possible choice of k, i_{k-1}, i_k, i_{k+1}

(14) $$x^{k+1|k-1,k}(i_{k+1}|i_{k-1},i_k) = x^{k+1|k}(i_{k+1}|i_k).$$

This assumption A2 asserts that the stochastic behavior of Node k-1 and Node k+1 is independent under the condition of the state of Node k being i_k. Strictly speaking, the assumptions A1 (which was imposed in the first stage approximation) and A2 here cannot be compared directly. A1 does not automatically imply A2. However, intuitively, we may consider that A2 is weaker than A1 and that the resultant solution is closer to the real situation.

Using the relation (14), we can reform the set (13) of equations to the one for variables $x^1(i_1)$'s and $x^{k|k-1}(i_k|i_{k-1})$'s. The resultant set of equations is slightly more complicated than (11) to write down in a vector and matrix form, but is not difficult to write a computer program for implementation of an iterative calculation.

Table 1. Numerical results of the application to a tandem network example

		loss prob.	L_1	L_2	L_3	L_4	L_5
Case 1 $\lambda = 1$	1st stage approximation	.2425	2.309	1.184	2.283	1.120	2.132
$\mu_1 = \mu_3 = \mu_5 = .5$	2nd stage approximation	.2640	2.404	1.479	2.353	1.298	2.093
$\mu_2 = \mu_4 = 1$	exact	.2661	2.407	1.465	2.347	1.280	2.078
Case 2 $\lambda = 1$	1st stage approximation	.3589	2.739	2.426	2.209	1.990	1.669
$\mu_1 = \mu_3 = \mu_5 = .5$	2nd stage approximation	.3612	2.753	2.434	2.204	1.973	1.651
$\mu_2 = \mu_4 = .5$	exact	.3580	2.746	2.429	2.204	1.980	1.659
Case 3 $\lambda = 1$	1st stage approximation	.4248	2.944	2.777	2.071	2.239	1.438
$\mu_1 = \mu_3 = \mu_5 = .5$	2nd stage approximation	.4263	2.952	2.782	2.098	2.194	1.417
$\mu_2 = \mu_4 = .4$	exact	.4273	2.950	2.784	2.098	2.205	1.425

Model: A tandem network with 5 nodes. Every node has two servers and a buffer of size 2. Customers arrive through a Poisson process with rate λ, and are served at Node k subjecting to the exponential distribution with rate μ_k. If Node k+1 is full, then the servers at Node k are blocked and stop their service. Arriving customers are lost if Node 1 is full. L_k is the mean number of customers at Node k in the steady state.

3. NUMERICAL EXAMPLES AND SOME COMMENTS

3.1 Numerical examples

This approximation method was tested with the tandem network used in the preceding section to investigate the method. The model tested is as follows. Each node of the network has two servers and a buffer of size 2. Namely, $s_k = m_k = 2$ for all k from 1 through 5. Customers arrive at the network through a Poisson process with rate λ, and are served at Node k subjecting to the exponential distribution with rate μ_k.

Table 1 summarizes the results of three cases with different parameter sets. For each case, the values of the loss probability and of the mean number L_k of customers at Node k, k= 1,...,5, are calculated from the results of the first and the second stage approximations. They are close with each other. This seems to indicate that those values are fairly accurate. In fact, they are very close to the exact values calculated by the aggregation/disaggregation method. Especially, the values by the 2nd stage approximation are very accurate. The maximum relative error is less than 1.5%.

An intuitive reasoning of the accuracy of the results by the 2nd stage approximation is that the stochastic behavior of nonadjacent nodes is nearly independent though that of adjacent nodes is highly dependent. This must be a great hint for applications of this method to the analyses of queueing networks.

The number of equations to be solved in the 1st stage approximation is 25, and that in the 2nd stage approximation is 125, while a system of 3125 equations have to be solved in the exact calculation by the aggregation/disaggregation method.

3.2 Some comments on the applicability

The aggregation/disaggregation method makes it possible for us to calculate exact solutions of fairly large Markov chains (see, e.g., [6], [8], [4] and [5]). However, it still has a dimensionality problem that the computation is at least linear order of the number of the states. So, even using the aggregation/disaggregation method, practically we cannot treat Markov chains having more than 100000 states.

Another type of aggregation method was developed by Courtois in [2]. If it is applied successfully, it saves a lot of computing time as shown in [3]. However, for the approximate solution obtained being accurate, the Markov chain has to have a special structure. This requirement restricts the applicability of the method strictly.

The approximation method proposed here has much more flexibility. The condition requested is far weaker than that of Courtois's method. The level of approximation can be chosen arbitrarily. Furthermore, by comparing the results for two different levels of approximations, we can roughly estimate the accuracy of the solutions. These seems to show a wide applicability of this method.

REFERENCES

[1] O. J. Boxma and A. G. Konheim, "Approximate analysis of exponential queueing systems with blocking," Acta Inform. vol.15, pp.19-66, 1981.

[2] P. J. Courtois, "Decomposability -- Queueing and Computer System Approximations," Academic Press, New York, 1977.

[3] H. Morimura, "Some applications of the recent techniques in the theory of queues," Research Reports on Information Sciences B-107, Tokyo Institute of Technology, Tokyo, 1982.

[4] P. J. Schweitzer, "Aggregation methods for large Markov chains," Mathematical Computer Performance and Reliability (eds., G. Iazeolla, P. J. Courtois and A. Hordijk), North-Holland, Amsterdam, pp.275-286, 1984.

[5] L. P. Seelen, "An algorithm for Ph/Ph/c queues," Research Report 131, Dept. of Actuarial Sciences and Econometrics, Vrije Univ., Amsterdam, 1984.

[6] Y. Takahashi, "A lumping method for numerical calculations of stationary distributions of Markov chains," Research Reports on Information Sciences B-18, Tokyo Institute Technology, Tokyo, 1974.

[7] Y. Takahashi, "Weak D-Markov chain and its application to a queueing network," Mathematical Computer Performance and Reliability (eds., G. Iazeolla, P. J. Courtois and A. Hordijk), North-Holland, Amsterdam, pp.153-165, 1984.

[8] Y. Takahashi and Y. Takami, "A numerical method for the steady-state probabilities of a GI/G/c queueing system in a general class," J. Operations Research Society of Japan, vol.19, pp.147-157, 1976.

[9] Y. Takahashi, H. Miyahara and T. Hasegawa, "An approximation method for open restricted queueing networks," Operations Research, vol.28, no.3, Part I, pp.594-602, 1980.

[10] W. Whitt, "The queueing network analyzer," Bell System Tech. J., vol.62, pp.2779-2815, 1983.

TELETRAFFIC ISSUES in an Advanced Information Society
ITC-11
Minoru Akiyama (Editor)
Elsevier Science Publishers B.V. (North-Holland)

Approximate Analysis of Queueing Networks with Nonpreemptive Priority Scheduling

Satoru IKEHARA and Masahiro MIYAZAKI

Yokosuka Electrical Communication Laboratory, NTT,
Yokosuka-shi, 238 Japan

Abstract

An asymptotic approximation analysis method is proposed here for evaluating closed type queueing network models with nonpreemptive priority scheduling. By limiting the number of classes to two, service station with nonpreemptive priority scheduling are separated into real servers, which process high priority service requests, and shadow servers, which process low priority service requests. This is done to reduce the system models to BCMP queueing network models. Service rates for real servers and shadow servers are asymptotically determined using virtual holding and super request concepts.

Comparison among exact solutions (numerical analysis), simulation results and the results of our analysis, shows that our method produces better approximations than those using other scheduling disciplines.

This method can also be applied to models with an plural number of preemptive and nonpreemptive priority scheduling service stations.

1. Introduction

Multiclass models [1∿3], such as those that execute conversational and batch jobs concurrently have been analyzed by queueing networks [4∿6] under the same priority assumption for every kind of job. But recently developed theory dose not allow solution of these models when priority service disciplines are present. Many approximation techniques [7∿16] have been proposed for central server models with priority. Most of these dealt with preemptive priority. Only a few studies [7, 15] analyzed nonpreemptive priority by flowequivalent methods which use aggregation of portions of the model. Though aggregation is computationally advantageous, it does not capture the interaction between individual portions.

Priority scheduling has been studied using techniques other than approximation. These can be seen in the studies of combined interactive and batch jobs, where jobs are scheduled by dynamic priority discipline [17], in an analysis of central server models, where optimal scheduling was determined [18], and in the analysis using the idea of the first passage [19]. Unfortunately, these analytical techniques have a disadvantage: the Markovian balance enquation must be solved directly. Therefore, if the scale of the model is increased even slightly, calculation becomes almost impossible.

This paper proposed an approximate analytical method for queueing network models with nonpreemptive priority scheduled service stations. For the preemptive method, Reiser proposed approximation analysis to evaluate central server models [8]. In his model, contentions between different types of requests are considered only at one service station (CPU); other stations (Terminals) are free from such a contention. Sevcik extended Reiser's approximation technique for terminal stations which show such a contention [9].

This paper extends Servcik's real and shadow server method further, to evaluate nonpreemptive priority models using a Virtual holding concept. High and low priority requests affect each other at a nonpreemptive priority scheduled station. Therefore, mutual influences must be considered. The new method separates the nonpreemptive priority scheduled stations into real servers and shadow servers for high priority and low priority requests, respectively. The original server is replaced with these two servers to reduce models to BCMP queueing networks. Service rates for real servers and shadow servers are determined using virtual holding and super request concepts. These concepts help derive the relation between server utilization factors for the original server and replaced servers. Service rates for real and shadow servers are calculated using this relation and the BCMP theorem.

Comparisons with exact solutions for a simple model and simulation results for an actual system model showed that this method is sufficiently useful for evaluating the influences of a scheduling method on system performance to replace simulations.

2. Analytical Model

Consider a BCMP queueing network model restricted in the following manner:

① Two kinds of requests exist; high and low priority requests.

② Routing is a decision sequence made upon job arrival and after each service completion. Such a transition conforms to a closed chain. The population of high priority requests in the first chain, and that of low priority requests in the second chain are represented by N_1 and N_2, respectively.

③ Requests are scheduled by a nonpreemptive priority discipline at a service station, i. High priority calls and low priority calls are respectively scheduled by FCFS (First Come First Served) discipline.

④ The number of servers for nonpreemptive priority scheduled station i is assumed to be 1. Server holding time for both high priority requests and low priority requests is assumed to be distributed exponentially. If station i is shorted, this model reduces to a BCMP queueing model and can directly be analyzed by an ordinary program package [6].

3. Analysis

3.1 Virtual Holding and Server Separation

The service rates at the nonpreemptive priority scheduled station i are represented by μ_1 and μ_2 for high and low priority requests. Hereafter μ_{i1}, μ_{i2}, ρ_{i1}, ρ_{i2}, ... are simply denoted as μ_1, μ_2, ρ_1, ρ_2, ... abbreviating the suffix i for station i. Similarly, server utilzation factors are represented by ρ_1 and ρ_2. Throughputs can then be expressed by $\lambda_1 = \mu_1 \rho_1$ and $\lambda_2 = \mu_2 \rho_2$.

At this station, high priority and low priority requests affect each other and cause time delays to arise. These delays can be represented by virtual server holding time Δh_1 and Δh_2. These service time inflations are added to the original server holding time h_1 ($= 1/\mu_1$) and h_2 ($= 1/\mu_2$). That is, the average server holding times are considered to have increased to h_1' ($= h_1 + \Delta h_1$) and h_2' ($= h_2 + \Delta h_2$). consequently, the original service station, i, can be replaced by stations i_1 and i_2, as shown in Fig.1. Stations i_1 and i_2 execute high and low priority calls, respectively.

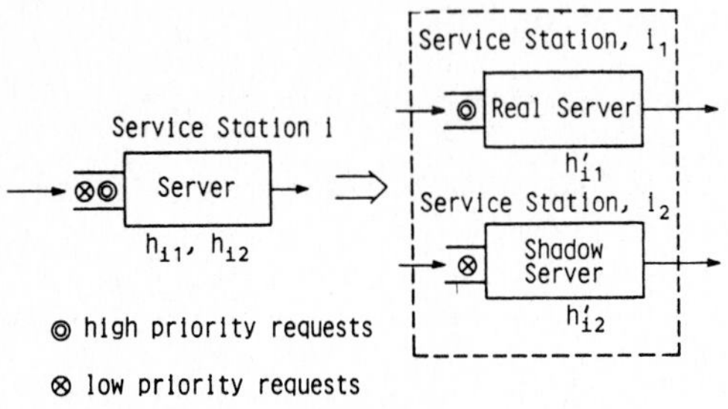

Fig.1 Service Station i Replacement by Real Server and Shadow Server

Server holding times for the new stations i_1 and i_2 are not exponentially distributed. If one assumes that they are exponentially distributed, and that virtual holding time is not much than the original service time, the new model reduced to a BCMP queueing model and can be analyzed by an ordinary program package, such as QNET 4 [5] and QSEC [6]. This approximation yields better results for smaller values of Δh_1 and Δh_2.

3.2 Average Service Time for Real Server and Shadow Server

Service time inflation for a high priority request waiting for the completion of a low priority request execution at original station i is represented by the virtual holding time of a high priority request at replaced station i_1. Therefore, the server utilization factor, ρ_1', for station i_1 will be larger than ρ, for station i by as much as the virtual holding time by a low priority request. Letting this defference be ρ_x:

$$\rho_1' = \rho_1 + \rho_x \ . \tag{1}$$

The average service time for station i_1 is expressed by h_1'. Since the throughput for the real server coincides with that of high priority requests for the original station,

$$\rho_1'/h_1' = \rho_1/h_1 \ . \tag{2}$$

Substituting Eq.(1) into Eq.(2),

$$h_1' = h_1(1 + \rho_x/\rho_1) \ . \tag{3}$$

Similarly, let the virtual holding time for the shadow server be ρ_y. The utilization factor, ρ_2', and the average service time. h_2', for the station, i_2, can be represented by

$$\rho_2' = \rho_2 + \rho_y \quad \text{and} \tag{4}$$

$$h_2' = h_2(1 + \rho_y/\rho_2) \ . \tag{5}$$

Thus, both h_1' and h_2' are expressed using ρ_x and ρ_y.

3.3 Probability of Virtual Holding

Here, balance equations for the original station, i, will be solved to yield ρ_x and ρ_y.

(1) Super Request Approximation for High Priority Requests

In considering the high priority requests that continuously hold the station, i, we note that they can be gathered to from a single super request. The queueing system for high priority requests can then be reduced to a finite (one) population model where state transitions between idle states and busy periods take place iteratively.
In the analysis for preemptive priority scheduling [8, 9], there is the assumption* that the population for high priority requests (N_1) is sufficiently large in contrast to 1. The same assumption holds here.

The average idle time for a super request can be represented by $1/\lambda_1$. Where, λ_1 represents an arrival rate for an original high priority call.

Average service time (busy period) for a super request can be approximately derived as follows. The probability that a high priority request arrives at station i when there are no other high priority requests already at station i is $1 - \rho_1 - \rho_x$. That is, under the assumption $N_1 >> 1$, the probability can be approximated by the probability that no high priority request exists at station i in equilibrium. This means that a super request (busy period) is an aggregate of the amount of $1/(1 - \rho_1 - \rho_x)$ high priority requests. Thus, the average service time, W_i, and service rate, μ_1^*, for a super request can be represented by

---

* Reiser and Sevcik represented the service rate for the shadow server by $\mu_2' = \mu_2(1 - \rho_1)$ for preemptive cases without assumption. This relation, however, can be derived from the assumption that $N_1 >> 1$.

$$\left.\begin{array}{l} W = h_1/(1 - \rho_1 - \rho_x) \quad \text{and} \\ \mu_1^* = 1/W \;. \end{array}\right\} \qquad (6)$$

(2) Balance Equations and Solutions

When solving balance equations for station i alone, the population of low priority requests, N_2, is assumed to be much larger than one. High priority requests are dealt with as super requests, as was just defined.

State variables, k and n, which represent the state of the station, i, can be defined

$$k = \begin{cases} 0\text{: There is no high priority request at the station,} \\ 1\text{: High priority requests are waiting (Virtual holding state),} \\ 2\text{: High priority request is being served.} \end{cases}$$

n: Number of low priority requests at the station.

If $k = 0$ or 2, then $0 \le n \le N_2$, and if $k = 1$ then $1 \le n \le N_2$. The probability that station i will take the state (k, n) in equilibrium, is defined as P(k, n). The normalization condition is represented by

$$P(0,0) + P(2,0) + \sum_{k=0}^{2} \sum_{n=1}^{N_2} P(k,n) = 1 \qquad (7)$$

The state transition diagram for station i is shown in Fig.2. From the definitions, the server utilization factors, ρ_1 and ρ_2, and ρ_x and ρ_y, can be written as follows;

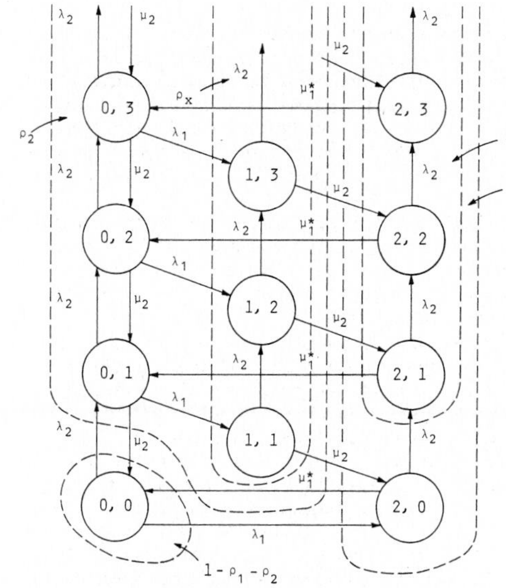

Fig.2 State Transition Diagram for Service Station i (Approximated bt $N_2 >> 1$)

$$\rho_1 = \sum_{n=0}^{N_2} P(2,n) \;, \qquad (8)$$

$$\rho_2 = \sum_{k=0}^{1} \sum_{n=1}^{N_2} P(k,n) \;, \qquad (9)$$

$$\rho_x = \rho_2 - \sum_{n=1}^{N_2} P(0,n) = \sum_{n=1}^{N_2} P(1,n) \quad \text{and} \qquad (10)$$

$$\rho_y = \rho_1 - P(2,0) = \sum_{n=1}^{N_2} P(2,n) \;. \qquad (11)$$

There are $3N_2 + 2$ states in Fig.2. For each of these, the following balance equation can be derived under the super request assumption

$$(\lambda_1+\lambda_2)\, P(0,0) = \mu_2\, P(0,1)+\mu_1^*\, P(2,0) \;, \qquad (12)$$

$$\left.\begin{array}{l} (\lambda_2+\mu_2)\, P(1,1) = \lambda_1\, P(0,1) \;, \\ (\lambda_2+\mu_2)\, P(1,2) = \lambda_1\, P(0,2)+\lambda_2\, P(1,1) \;, \\ \quad \vdots \\ \mu_2\, P_i(i,N_2) = \lambda_1\, P(0,N_2)+\lambda_2\, P(1,N_2-1) \;, \end{array}\right\} \qquad (13)$$

$$(\mu_1^*+\lambda_2)\, P(2,0) = \lambda_1\, P(0,0)+\mu_2\, P(1,1) \;. \qquad (14)$$

By summing up both sides of Eq.(13) and substituting into Eq.(10), ρ_x can be derived as

$$\rho_x = \frac{\lambda_1 h_2}{1+\lambda_1 h_2} \rho_2 \;. \qquad (15)$$

From Eq.(12), the first equation of (13), and Eq.(14), P(2,0) can be derived as

$$P(2,0) = \frac{\lambda_1(\lambda_1+2\lambda_2+\mu_2)(1-\rho_1-\rho_2)}{\lambda_1\, \mu_1^*+(\lambda_2+\mu_1^*)(\lambda_2+\mu_2)} \;. \qquad (16)$$

Substituting Eq. (16) in Eq.(11), ρ_y can be represented as

$$\rho_y = \lambda_1 \left\{\frac{1}{\mu_1} - \frac{(\lambda_1+2\lambda_2+\mu_2)(1-\rho_1-\rho_2)}{\lambda_1\, \mu_1^* +(\lambda_2+\mu_1^*)(\lambda_2+\mu_2)}\right\} . \qquad (17)$$

3.4 The Relation Between Average Service Time and Utilization Factor

By substituting Eq.(15) in Eq.(3), the average service time for the real server, h_1', is derived as

$$h_1' = h_1\left(1 + \frac{h_2\, \rho_2}{h_1 + h_2\, \rho_1}\right) . \qquad (18)$$

Similarly, substituting Eq.(17) in Eq.(5), the average service time for the shadow server can be represented as

$$h_2' = h_2\left[1 + \frac{\mu_1\, \rho_1}{\rho_2} \left\{ \frac{1}{\mu_1} - \frac{(\mu_1\, \rho_1+2\mu_2\, \rho_2+\mu_2)(1-\rho_1-\rho_2)}{\mu_1\, \mu_1^*\, \rho_1+\mu_2(\mu_2\, \rho_2+\mu_1^*)(1+\rho_2)} \right\}\right] . \qquad (19)$$

Here, μ_1^* (= 1/W) can be expressed from Eqs.(6) and (15) as

$$\mu_1^* = \mu_1\left(1 - \rho_1 - \frac{\rho_1\, \rho_2\, h_2}{h_1+\rho_1\, h_2}\right) . \qquad (20)$$

Since the values for h_1 (= $1/\mu_1$) and h_2 (= $1/\mu_2$) were given as model description parameters, when ρ_1 and ρ_2 are determined the average service time, h_1' and h_2', for the real and shadow server can be calculated from Eqs.(18) ~ (20).

3.5 Asymptotic Approximation Method

An asymptotic approximation analysis can be established to yield the values for ρ_1 and ρ_2. The initial values for $\rho_1^{(0)}$ and $\rho_2^{(0)}$ can be chosen arbitrarily so that the relation $0 \leq \rho_1(0) + \rho_2(0) \leq 1$ holds. However, for quick convergence, it is better to use network analysis results, assuming PS (Processor Sharing) scheduling as the intial values for $\rho_1^{(0)}$ and $\rho_2^{(0)}$.

The analytical procedure can be described as follows

Step 1: The iteration counter, k, is initialized to zero. Solving the model (where station i is assumed to be scheduled by PS) using BCMP theorem yields values for $\rho_1^{(k)}$ and $\rho_2^{(k)}$, where $k = 0$.

Step 2: After substituting $\rho_1^{(k)}$ and $\rho_2^{(k)}$ in Eqs.(18) and (19), h_1' and h_2' are then calculated.

Step 3: Through the BCMP theorem, a solution is possible for the new model, where station i is replaced by the real server, the service rate of which is h_1', and the shadow server, the service rate of which is h_2'. This yields the utilization factors, $\rho_1'^{(k+1)}$ and $\rho_2'^{(k+1)}$, for the real server and the shadow server, respectively.

Step 4: From Eqs.(1), (4), (15) and (17), $\rho_1^{(k+1)}$ and $\rho_2^{(k+1)}$ can be calculated* using $\rho_1'^{(k+1)}$ and $\rho_2'^{(k+1)}$ given by Step 3.

Step 5: If $\rho_1^{(k+1)}$ and $\rho_2^{(k+1)}$ coincide to a sufficient degree of accuracy with $\rho_1^{(k)}$ and $\rho_2^{(k)}$, then the analysis is complete. Otherwise, increase k to k + 1, and return to Step 2.

4. Case Studies and Considerations

4.1 Comparison with Exact Solution (Two Service Stations)

4.1.1 Model Descriptions

In the two station model shown in Fig.3, there is no waiting time at station NO.2 since the number of servers is infinite. Thus, the

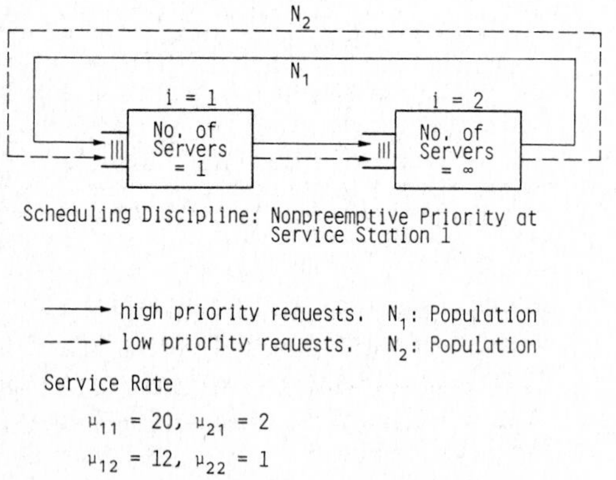

Fig.3 Example with 2 Service Stations

* Numerical calculation, such as a two-dimensional binary search, will be necessary for the solution.

model coincides with a finite source model with N_1 input lines for high priority requests and N_2 input lines for low priority requests.

The balance equation, the number of which is $2N_1 N_2 + N_2 + 1$, can be solved numerically; approximation results can be compared with exact solutions.

4.1.2. Comparison of Results

The responce time characteristics are compared in Figs.4 and 5. Since the responce time for Station 2 is constant, only the responce time for Station 1 is shown in Figs.4 and 5.

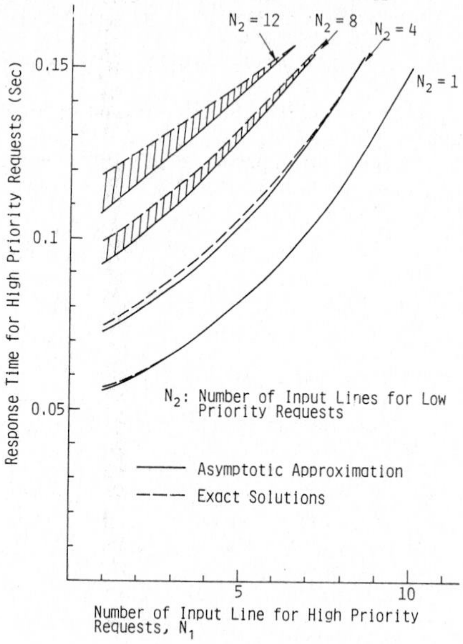

Fig.4 Response Time for High Priority Requests

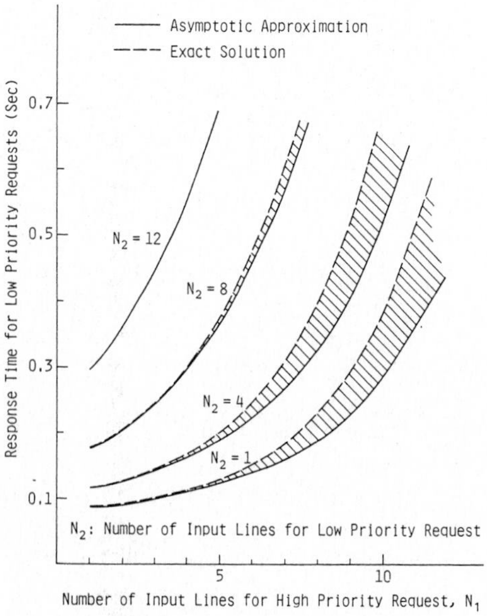

Fig.5 Response Time for Low Priority Requests

With respect to these figures, the following considerations are possible.

i) Analytical errors of the response time for high priority requests, T_1, increase with N_2. Similarly, analytical errors concerning the response time for low priority requests, T_2, increase with N_2.

ii) Analytical errors concerning response time for high priority calls, T_1, decrease as the population N_1 increases. The same applies for analytical errors concerning the response time, T_2, for low priority requests.

That is, virtual holding for low priority requests increases with increase in high priority request population. Virtual holding for high priority requests also increases, with increase in low priority request population. These increase the coefficients of variation for the real and the shadow servers. However, service time was approximated by exponential distribution. This is the reason why phenomena similar to i) are caused and why the results turn out very favorable

Trend ii) indicates that the trends in i) will be weakened with an increase of N_1 or N_2. That is, if N_2 is fixed and N_1 is increased, the portion of high priority execution increases, and virtual holding for every high priority request decreases. Thus, service time for the real server shows a close exponential distribution. If N_1 is fixed and N_2 is increased, the throughput for low priority requests increases, virtual holding for every low priority request decreases, and service time for the shadow server approaches a exponential distribution.

In addition to what was just mentioned, Fig. 3 shows that the analytical errors increase with a decrease in the service rate, μ_{i1} and μ_{i2}, for Station 1.

4.2 Comparison with Simulation Results (5 station model)

4.2.1 Model Description

A queueing network model with five service stations is shown in Fig.6. Parameter values for this model are listed in Table 1. This is the system model for a scientific computation system with two kinds of processors, namely a CPU and CCP (Communication Control Processor).

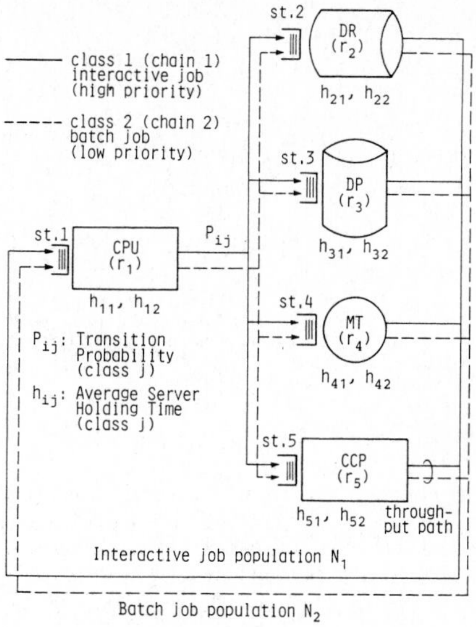

Fig.6 System Model with Interactive and Batch Jobs

The results of the approximation analysis are compared with the simulation results by GPSS (General Purpose Simulation System). These results were also compared with the results for other three scheduling methods. The following schedulling methods were selected for the two processors (CPU, CCP).

① FCFS (First Come First Served): by simulation.

② PS (Processor Sharing): by BCMP queueing theory.

③ Preemptive Priority Scheduling: by Sevcik's approximation and simulation.

④ Nonpreemptive Priority Scheduling: by the authors' approximation and simulation.

Other stations (NOs.2, 3 and 4) are scheduled by FCFS.

4.2.2 Analytical Results and Considerations

Results are shown in Figs.7, 8. Observations and considerations are listed here.

(1) Comparison of Approximation and Simulation Results

These results showed the present analysis to be very useful as an evaluation method to take the place of simulation. From Fig.7, the

Table 1. Parameter Value Table

	Device Name	No. of Unit (r_i)		Average Server Holding Time h_{ij}	Average No. of visits to Station i/Job	Transition Probability P_{ij}
Service Station 1	CPU	1	class 1	13.9	23.3	
			class 2	21.2	809	
Service Station 2	Drum	5	class 1	17.2	11.5	0.494
			class 2	17.2	251	0.310
Service Station 3	Disk Pack	14	class 1	61.7	8.4	0.361
			class 2	61.7	410	0.506
Service Station 4	Magnetic Tape	9	class 1	55.9	2.4	0.102
			class 2	55.9	147	0.182
Service Station 5	Communication Control Processor	1	class 1	92.8	1.0	0.043
			class 2	2280	1.0	0.0012

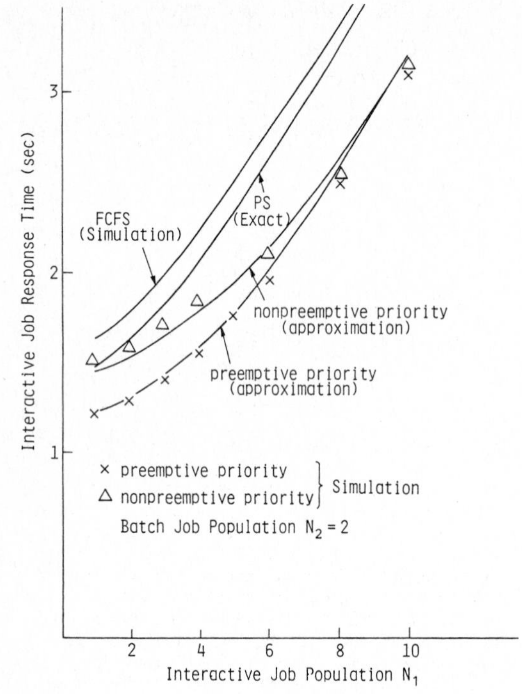

Fig.7 Comparison of Response Time with Simulation

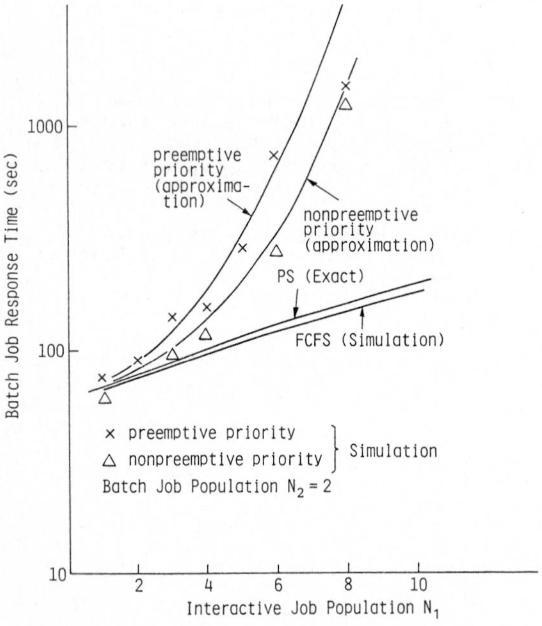

Fig.8 Comparison of Response Time with Simulation

same trends for high priority requests as in the previous example can be observed.

In addition, the present analysis required 5∿7 iterations for the solutions to converge to a three figure order. Two seconds were used for every iteration by computer (3 MIPS). However, simulation required 3600 seconds per data point.

(2) Comparison among Several Scheduling Methods

It can be seen in Figs.7, 8 that the present analytical results with nonpreemptive priority scheduling are closer to those for FCFS scheduling in a small N_1 region. However, in a large N_1 region, the results approach those of preemptive priority scheduling.

The reason for this is that if population N_1 is small, virtual holding for low priority requests is also small, and it is easy to provide the service with low priority requests. However, if population N_1 is large, high priority requests consecutively hold the server in many cases. The chance that a low priority request can be provide for the server then decreases rapidly.

4.3 Approximation Errors in this Analysis

The analysis described above employed the following approximations to solve the model defined in Section 2.

① The service times for the real server and the shadow server were approximated by exponential distributions. (§3.1)

② Assuming that the population for high priority requests, N_1, is sufficiently larger than one, high priority requests were approximately represented by super requests. (§3.3 (1))

③ Assuming that the population for low priority requests, N_2, is also sufficiently larger than one, the arrival rate of low priority requests at station i was considered to be independent of the state of the total system. (§3.3(2))

In an initial considering of the influence from ①, the service time for the real server is the sum of the original service time for a high priority request at station i and the waiting time for the completion of low priority request execution, namely a virtual holding. The same goes for the service time for the shadow server. Subsequently, the coefficient of variation will be larger than that of an exponential distribution. Nevertheless, the service time is approximated as in ①. Therefore, response time will be underestimated by approximation analysis. This underestimation tends to grow according to an increase in the virtual holding. The phenomenon shows that the greater the service time, h_{i1}, or the population, N_1, for high priority calls, the larger the aproximation error for the response time, T_2, for a low priority call will be. On the contrary, the more service time, h_{i2}, or the population, N_2, for a low priority request grow, the greater the error for high priority calls will be.

The greater the populations, N_1 and N_2, for high and low priority requests, the better approximations ② and ③ hold. Subsequently, the approximation errors caused from ② and ③ can be expected to decrease, according to the increases in N_1 and N_2.

In summary, the following considerations can be made.

ⓐ As population N_1 increases, the approximation error for the response time, T_1, decreases. And as population N_2 increases, the approximation error for the response time, T_2, decreases.

ⓑ As service time h_{i1} and h_{i2} decrease, response time T_1 and T_2, for low priority and high priority requests, decrease.

These considerations can be made certain from the case studies.

5. Conclusion

An asymptotic approximation analysis method has been proposed here for evaluating queueing models with nonpreemptive priority scheduled stations. This method dealt closed type queueing network models which exexute two kinds of calls, where the workloads are different from each other.

The important aspects of this analysis are as follows:

① Nonpreemptive priority scheduled stations were separated into real servers, only for high priority requests, and shadow servers, only for low priority requests. The original stations were replaced by these two kinds of stations.

② The relations between average service time for the replaced stations, and server utilization factors for high and low priority requests at the original station, were derived using virtual holding and super requests handling concepts.

③ Combining reduced model ① and relation ②, the average service time for both the real and the shadow server were asymptotically determined.

When the population for each chain is small, a direct numerical analysis for balance equations is possible. For large populations, however, the numerical analysis becomes too expensive. Thus, in this analysis, the populations for both high and low priority requests were assumed to be much larger than 1.

However, case studies showed that, even for small populations, if the workload per request was small enough, the approximation holds well. Consequently, it became clear that this analysis was useful as an evaluation technique for comparing the performance of nonpreemptive priority scheduling with that for FCFS, PS and other scheduling methods.

This analysis could easily be extended to queueing network models which include both preemptive priority scheduled stations and nonpreemptive priority scheduled stations.

References

[1] J.E.Neilson, "An analytic performance model of a multiprogrammed batch-time shared computer," Proc. of the ISCPME (1976), 59-70.

[2] C.A.Rose, "A calibration-prediction technique for estimating computer performance," Proc. AFIPS, NCC (1977), 813-818.

[3] M.A.Diethelm, "An empirical evaluation of analytical models for computer system performance prediction," Proc. of the ISCPME (1977), 1 -160.

[4] F.Baskett, K.M.Chandy, R.R.Muntz and F.G. Palacios, "Open closed and mixed networks of queues with different classes of customers," J. of ACM, Vol.22, No.2, (1975), 248-260.

[5] M.Reiser and H.Kobayashi, "Queueing networks with multiple closed chains; theory and computational algorithms," IBM J. of Res. and Dev., Vol.19, No.3 (1975), 283-294.

[6] S.Ikehara and S.Yamada, "BCMP queueing network analysis program QSEC," ECL Tech. J., Vol.29, No.5 (1980), 1051-1078, in Japanese.

[7] C.H.Sauer and K.M.Chandy, "Approximate analysis of central server models," IBM J. Res. Develop., 19, 3, pp.301-313, 1975.

[8] M.Reiser, "Interactive modeling of computer system," IBM SYST. J., Vol.15, No.4, 1976, 309-329.

[9] K.C.Sevcik, "Priority scheduling disciplines in queueing network models of computer systems," Information Processing 77, IFIP, North-Holland, Amsterdam, 1977.

[10] B.Meister, "A tandem queueing system with priorities," Computing, 19, pp.203-208, 1978.

[11] M.Reiser, "A queueing network analysis of computer communication networks with window flow control," IEEE Trans. Commun., COM-27, 8, pp.1199-1209, 1979.

[12] R.J.Morris, "Priority queueing network," Bell Syst. Tech. J., 60, 8, pp.1745-1769, 1980.

[13] J.S.Kaufman, "Approximate analysis of priority scheduling disciplines in queueing network models of computer systems," Proc. ICCC '82, pp.955-961, 1982.

[14] W.Schmitt, "Approximate analysis of Markovian queueing networks with priorities," Proc. ITC10, Session 1.3, 3, 1983.

[15] W.M.Chow and P.S.Yu, "An approximation technique for central server queueing models with a priority dispatching rule," Performance Evaluation, 3, 1, pp.55-62, 1983.

[16] A.Kumar, "Equivalent queueing networks and their use in approximate equilibrium analysis," Bell Syst. Tech. J., 62, 10, pp.2893-2910, 1983.

[17] C.E.Landwehr, "An endogeneous priority model for local control in combined batch-interactive computer system," Proc. of the ISCPME, 282-295, 1976.

[18] J.R.Spirn, "Multi-queue scheduling of two tasks," Proc. of the ISCPME, 102-108, 1976.

[19] Y.T.Wang, "Analysis of some overload control strategies for queueing systems with delayed feedback," Performance Evaluation, 1, 4, pp.305-319, 1981.

TELETRAFFIC ISSUES in an Advanced Information Society
ITC-11
Minoru Akiyama (Editor)
Elsevier Science Publishers B.V. (North-Holland)

TRANSIT DELAY DISTRIBUTIONS IN PRIORITY QUEUING NETWORKS

Paul J. KUEHN and Wolfgang SCHMITT

Institute of Communications Switching and Data Technics
University of Stuttgart, Fed.Rep.of Germany

Department of Communications
University of Siegen, Fed.Rep.of Germany

ABSTRACT

The paper presents an analysis procedure for transit delay distributions of closed queuing networks with preemptive or nonpreemptive priorities. The procedure is developed in two steps: at first, the decomposition of the priority queuing network with P classes of priorities into P queuing networks, one for each priority class, and, secondly, the analysis of the transit delay distribution function for a single-priority class network. The first step is based on a decomposition technique for priority stations into load- and queue length-equivalent single-class stations with state-dependent service rates. In the second step, such networks are analyzed by a first passage time method exactly as well as approximately by network transformations. The procedure involves several approximative assumptions which are validated by computer simulations.

1. INTRODUCTION

1.1 Problem

Transit delays are of primary interest for applications of interactive computer systems (terminal-I/O), paged computer systems (disk-I/O), computer communications networks (packet delays), or signalling networks based on the common channel interoffice signalling system (call setup delays). Modelling of such cases leads to closed or open queuing networks.

Queuing networks have received, therefore, much attention in research during recent years. Most research has been concentrated on average delays of product-form queuing networks [1,2], decomposition and aggregation techniques [3,4,5] in case of networks with a complex structure or general service centers. More recently, first approaches have been made towards multi-class queuing networks with priorities [6,7,8,9,10].

Transit delay distributions have been analyzed only recently in case of closed Markovian queuing networks [11,12,13,14,15]. The distribution of transit delays allows a much deeper insight into the network's behaviour as, e.g., for the analysis of the percentiles in case of terminal response times, post-dialling call delays, dialtone delays, packet-acknowledgement delays, etc. This analysis, however, is more difficult and requires time-dependent processes even in the stationary case.

Models of the real world of computer systems and communications switching control lead often to priority queuing networks which have to be analyzed with respect to the average delays and with respect to the distribution of transit delays such as response times, cycle times, or flow times.

An example of the class of models treated in this paper is shown in Fig. 1.

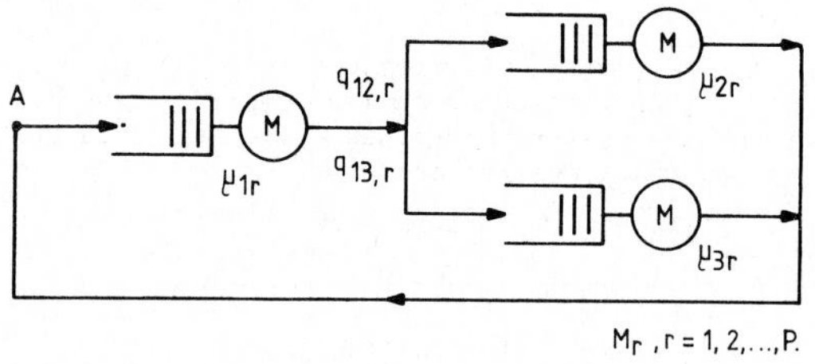

Fig. 1. Multi-Class Priority Queuing Network

The central server type model of Fig. 1 consists of 3 Markovian service stations. There are P classes of customers with populations M_r for class r, r = 1,2,..., P. The service rates μ_{ir} depend on the station number i and on the priority class r. Each class of customers is routed independently according to a class-individual probabilistic routing matrice $Q_r = (q_{ij,r})$, where $q_{ij,r}$ denotes the probability of a class-r customer being routed to station j after leaving station i. Within each station or, at least within one of the stations, customers are scheduled for service according to a preemptive or nonpreemptive priority discipline. The queue discipline within each station is FIFO for each class; other disciplines could be principally included, as well.

We are interested in the transit delay distribution of a priority r-test customer, i.e. the distribution of the time elapsing betweeen two successive passages of that r-test customer through the control point A, also called the cycle time of a class r-customer. Similar problems arise in the analysis of the transit delay for the passages of an r-test customer between two arbitrary points within the network.

1.2 Outline of the Analysis

The analysis is based on some recent results for both, the transit delay analysis for single-class networks and decomposition techniques for multi-class priority networks. The two solution steps (1) and (2) are outlined in Chapter 2 and 3:

11 th International Teletraffic Congress (ITC), Kyoto/Japan, September 4 - 9, 1985.

(1) Decomposition of the P-class network into P single-class networks for each of the priority classes. This decomposition provides equivalent service centers for each considered priority class. The equivalent service centers own a state-dependent service rate which reflects the influence of all other priority class customers. This approximate decomposition method has been validated for average queuing delays.

(2) Transit delay analysis of the equivalent class r-network. The transit delay distribution analysis rests on the method of first passage times. The fate of a class r-customer is explicitly considered and described by backward-type differential equations. This method allows also for state-dependent service rates.

Note that the method of step 2 is principally also applicable to solve the given problem of transfer delays within multipriority networks. The state description technique, however, requires multiple variables per station so that the resulting systems of differential equations are too complex to solve.

In case of more complex network structures, both steps (1) and (2) can be combined with aggregation methods to replace a considered station's complete environment by one composite server.

In the paper we develop both steps (1) and (2) separately. Approximate methods will be validated by simulation results.

The combination of both steps to the transit delay analysis of priority class queuing networks will be exemplified by some network examples in Chapter 4. Since the procedure involves several approximative assumptions, the analytical results are validated by computer simulations.

2. DECOMPOSITION OF PRIORITY QUEUING NETWORKS

2.1 Decomposition Principle

The method we will apply to decompose the priority stations of a queuing network with two priority classes is described comprehensively in [8] and [10]. Hence, concerning this method, we confine ourselves to a summary of the essentials.

The decomposition of a two-class priority queuing model into two single-class queuing models without a priority discipline is best explained by reference to Fig. 2, where:

λ_r	arrival rate
μ_r	service rate
x_r	number of class-r customers in the system
r = 1,2	class index; r = 1: high priority class r = 2: low priority class
$\mathbb{N}$	set of positive integers
:=	equals by definition

We call these single-class queuing models "Virtual Server" models. This is due to the fact that each class is assumed to receive its service from a dedicated Virtual Server, whose service time distribution function is chosen such, that the priority mechanism is appropriately taken into account. Replacing the original server by these

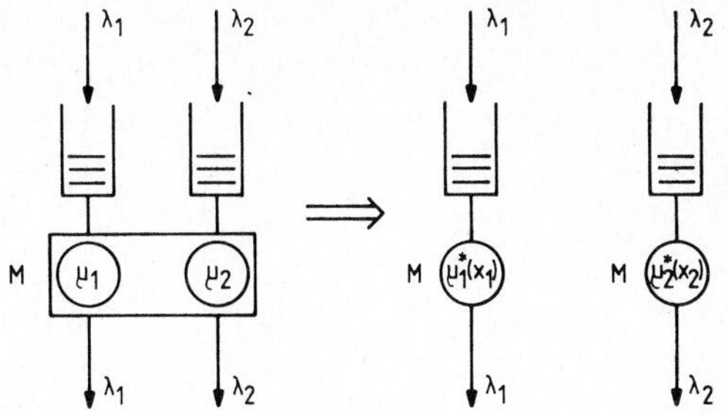

Fig. 2. Decomposition of a priority queue into Virtual Server models with state-dependent reduced service rates

Virtual Servers, a two-class priority queuing network is transformed into a queuing network without a priority discipline.

An important feature of the Virtual Server models presented in Fig. 2 are the state-dependent reduced service rates $\mu_r^*(x_r)$ (r=1,2) defined by Eqs. (2.1a-b):

$$\mu_r^*(x_r) := c_r(x_r)\cdot\mu_r, \qquad (2.1a)$$

$$c_r(x_r) := \text{prob(class r-customer in service} \mid x_r \text{ class r-customers in system)}$$

$$x_r \in \mathbb{N};\ r=1,2\ . \qquad (2.1b)$$

It has been proved in [9] that this choice of the service rates provides an exact description of the marginal distributions $p_r(x_r)$ (r=1,2) of system states in the priority queuing models M/M/1/PRE and M/M/1/NONPRE. Concerning the global behaviour of traffic flow, our approach is still an approximation. Particularly, the output streams are in general not Poisson processes.

We call our decomposition method the state-dependent reduced occupancy approximation and refer to it as to the s d r o a .

2.2 Test-Bed Examples for Decomposition

In this section we illustrate the application of our decomposition method by means of two test-bed queuing networks.

First, we consider the cyclic queuing model •/M/1/PRE → •/M/1/PRE (see Fig. 3a). Decomposition of the priority stations leads to two single-class queuing models depicted in Fig. 3b. In accordance with Eqs. (2.1a-b), the service rates $\mu_{ir}^*(x_{ir})$ (i=1,2; r=1,2) are determined by the following equations:

$$\mu_{i1}^* = \mu_{i1} \quad , \ i=1,2 \qquad (2.2a)$$

$$\mu_{i2}^*(x_{i2}) = c_{i2}(x_{i2})\cdot\mu_{i2} \quad , \ i=1,2 \qquad (2.2b)$$

where

$$c_{i2}(x_{i2}) = \text{prob}(X_{i1}=0 \mid X_{i2}=x_{i2}),\ x_{i2}=0,1,\ldots,M_2;\quad i=1,2. \qquad (2.2c)$$

The random variable X_{ir} denotes the number of class-r customers at station i and M_r denotes the network population of class r-customers.

In the present case the calculation of the unknown reduction coefficients $c_{i2}(x_{i2})$ can be carried out exactly by setting up the global balance equations and solving them (see [7]). For instance let

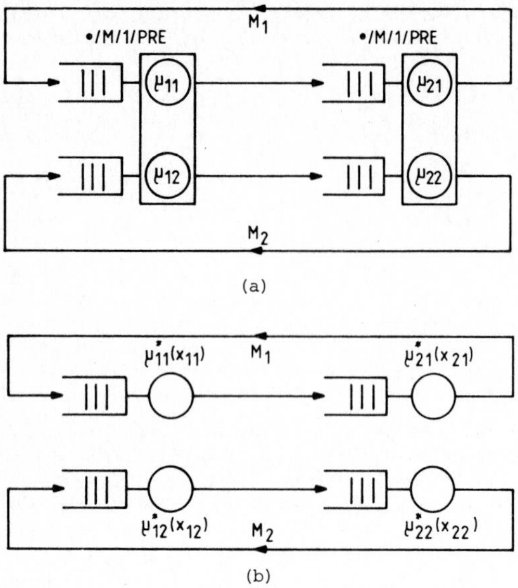

Fig. 3. (a) Cyclic queuing model •/M/1/PRE → •/M/1/PRE

(b) Decomposition into two single-priority networks

$M_1 = M_2 = 2$

$\mu_{11} = 0.2, \quad \mu_{12} = 1.0,$

$\mu_{21} = 0.1, \quad \mu_{22} = 0.1.$

Then we get

$c_{12}(1) = 0.11475, \quad c_{12}(2) = 0.07018,$

$c_{22}(1) = 0.50820, \quad c_{22}(2) = 0.09091.$

It must be mentioned here that the exact values for the coefficients $c_{i2}(x_{i2})$ lead to an exact description of the marginal distribution $p_{i2}(x_{i2})$ of the cyclic queuing model •/M/1/PRE → •/M/1/PRE. For a proof see [10].

Unfortunately, an exact calculation of the reduction coefficients is only possible for priority networks of little complexity. If this requirement will not be fulfilled, we rely on appropriate approximation techniques (see [10]). In order to distinguish this approach from the s d r o a we call it the modified s d r o a and refer to it as to the m s d r o a .

We have applied such an approximation technique to decompose our second test-bed queuing network for preemptive priorities with a structure as depicted in Fig. 1. The procedure to calculate the reduction coefficients for this network type is described in section 3.2 of [10]. Here, we give only some numerical values for the reduction coefficients $c_{i2}(x_{i2})$.

Let $M_1 = M_2 = 2$, $\mu_{ij} = 1.0$ ($i=1...3$; $r=1,2$),

$q_{12,r} = q_{13,r} = 0.5$ ($r=1,2$).

Then, $c_{12}(1) = 0.25609 \qquad c_{12}(2) = 0.18085$

$c_{22}(1) = 0.51291 \qquad c_{22}(2) = 0.42324$

$c_{32}(1) = 0.51291 \qquad c_{32}(2) = 0.42324$

Similar results have been obtained for nonpreemptive priorities, see [9]. The decomposition principle has been validated by computer simulations for a wide parameter range which has shown an acceptable accuracy.

3. TRANSIT DELAY ANALYSIS IN SINGLE-CLASS MARKOVIAN NETWORKS

3.1 Transit Delay as First Passage Time Problem

We consider a class of Markovian queuing models which can be discribed by a Markov chain with an enumerable set of states and a continuous time parameter. The behavior of the Markov chain can be described by the well-known Chapman-Kolmogorov relation from which two sets of differential equations for the transition probabilities can be derived, the Kolmogorov forward equations and backward equations [16].

The waiting time or the response time of a customer can be considered as "life times" T of one or several tagged (test) customers within a properly defined set S of states. The life time terminates when the test customer leaves S for the first time entering a "taboo" set $H = \overline{S}$; his life time is equal to the "first passage time" to H. The life time process can be considered as a special process with "absorbing" states in H. This modified process can be constructed from the system state transition probabilities under the condition that states in H are excluded. The state of the modified process must be specified such that all effects which may influence the life time T of the test customer directly or indirectly are reflected properly.

The general procedure of this life time process analysis has been outlined in [15]. Here, we refer only to the main results.

Let $w(t|i)$ denote the conditional complementary life time distribution function (df) where the considered life time of a test customer has started at initial state i. Then, the life time process is described by the set of Kolmogorov backward-type equations

$$\frac{d}{dt} w(t|i) = -q_i w(t|i) + \sum_{j \neq i} q_{ij} w(t|j) \qquad (3.1a)$$

where

$$q_i = \sum_{j \neq i} q_{ij} + \varepsilon_i \quad , \; i,j \in S. \qquad (3.1b)$$

Within Eqs. (3.1a,b) q_{ij} denotes the instantaneous rate of transitions from state i to state j in S, q_i the rate for transitions from i to any other state in S or H, and ε_i the life time terminating rate from state i into H. These rates are found by considering the underlying life time process.

From Eqs. (3.1a), a set of linear equations can be derived for the ordinary k-th moments of the conditional life time $m_i^{(k)}$:

$$q_i m_i^{(k)} - \sum_{j \neq i} q_{ij} m_j^{(k)} = k m_i^{(k-1)} , \qquad (3.2)$$

where $m_i^{(0)} = 1, \quad i,j \in S, \; k \in \mathbb{N}.$

The total complementary life time df is composed from the initial state distribution $\Pi(i)$ and the conditional life time df's $w(t|i)$ according to

$$W(t) = \sum_{i \in S} \Pi(i) \cdot w(t|i) \qquad (3.3)$$

3.2 Exact Analysis

The exact analysis of the life time process will be demonstrated for the example of a cycle time analysis of a closed queuing network with two stations, as shown in Fig. 3b, where the service rate $\mu_i(x_i)$ of station i depends on the number of customers x_i in that station, i = 1,2. The queue discipline in each station is FIFO.

The exact result of the cycle time df for this model is known only in the simpler case of constant service rates, see [11,13,15].

Now we consider the case where the instantaneous service rate of a server may depend on the actual number of customers in that station. Combinatorial methods for the cycle time analysis are not adequate since in this case succeeding customers in a queue behind the test customer may influence his cycle time even if overtaking is not possible. The state description has now to be augmented.

Let $\zeta(t) = (i,j,k)$ define the state of cycle time process, where

k indicates the station where the test customer is currently located in

i = number of customers in station 1 consisting of the test customer and his predecessors in line (k = 1)
or
total number of customers in station 1 (k = 2)

j = total number of customers in station 2 (k = 1)
or
number of customers in station 2 consisting of the test customer and his predecessors in line (k = 2).

The state transition diagram is shown in Fig. 4 for the special case of M = 3 (extension to general M is straightforward).

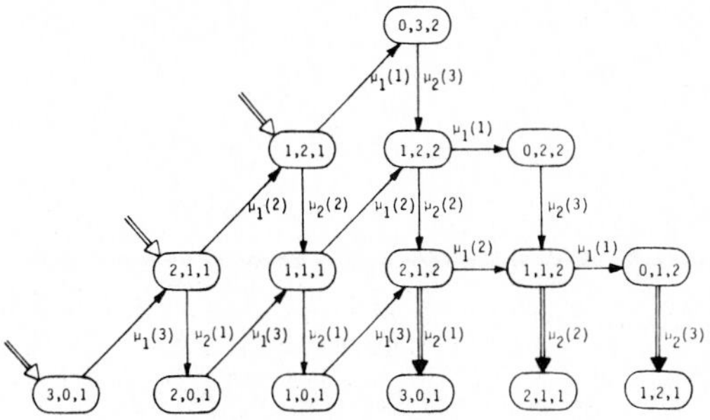

Fig. 4. State transition diagram for the cycle time for a cyclic queuing system with two stations and state-dependent service rates.

The arrows at the left hand side indicate the initial states a test customer meets on the arrival at station 1. The bold arrows at the lower right indicate the termination of the cycle time, i.e. the transitions into the absorbing space H.

The exact analysis of the cycle time df proceeds as follows:

- Definition of the set of $M\cdot(M+1)$ differential equations, one for each state of S according to Eq. (3.1).
- Calculation of the arrival state distribution for the arrival of a test customer at station 1 according to [17,18].
- Numerical solution of the system of differential equations either in the time domain (e.g. by a Runge-Kutta method) or in the Laplace-domain (i.e., solving for the corresponding eigenvalues).
- Calculation of the complementary df of the cycle time according to Eq. (3.3).

For larger populations M, larger queuing network structures, or both, the exact solution of the df requires an enormous amount of computing. Instead of the full df, we may be interested in the moments of lower order especially the second moment m_2 or the coefficient of variation c (the first moment is already known from the theory of product-form queuing networks [1,2]).

The solution for the ordinary moments requires only the solution of a set of linear equations according to Eq. (3.2), one set for each order k, k = 1,2,.... In the particular case of Fig. 4, each set can be solved recursively: starting with the right hand side column, all quantities are obtained by proceeding bottom-up, column-by-column from right to left.

From the lower conditional moments (or unconditional total moments), say k = 1,2 or k = 1,2, and 3, we can approximately construct the df. In Section 2 we will mainly concentrate on the coefficient of variation of cycle times only.

3.3 Approximate Analysis

To further reduce the computational amount, several approximate methods can be applied. Two different methods have been developed so far:

(a) Network transformations

(b) Independent flow time approximations

a) Network Transformations

Closed product-form queuing networks can be transformed into a cyclic queuing network of two stations (as shown in Fig. 3b) consisting of a considered station and a "composite station" by use of the so-called Norton Theorem for product-form queuing networks [3,5]. This transformation however, is only exact with respect to the throughput rate and state probabilities and, therefore, for the average cycle time. The higher moments and the df of the cycle time are only approximations.

This method has proved to be quite accurate for the following cases, see [15].

- Aggregation of several tandem stations into one composite station
- Aggregation of several parallel stations with equal flow times into one composite station.

By this method, certain types of networks can be structurally reduced step by step to facilitate the numerical computations. However, subnets with cyclic paths should not be aggregated into a composite station; although the average time is exact, the second moment can be quite underestimated.

b) Independent Flow Time Approximation (IFTA)

This method is applied to the original network structure. The flow times a test customer may observe in each of the passed stations are analyzed under an independently calculated arrival state distribution. The total flow time can be found by a summation of all component flow times, e.g., by convolutions of their respective df's.

First results of this method are quite encouraging [19]. This method will be reported by a forthcoming paper.

4. RESULTS AND VALIDATION

4.1 Test-Bed Networks

In the following a number of examples will be reported to show how priorities influence the cycle time in queuing networks and how accurate the analysis method works. The examples refer to three test-bed networks:

Test-Bed Network 1: Cyclic queuing network with two stations and two classes of nonpreemptive priorities (structure as in Fig. 3a).

Test-Bed Network 2: as 1, but with preemptive priorities.

Test-Bed Network 3: Central server queuing network model (CSM) with three stations and two classes of preemptive priorities (see Fig. 1).

4.2 Results

a) Results for Test-Bed Network 1

This model has been analyzed for a fixed population $M_1 = M_2 = 2$. The service rates are:

$\mu_{11} = 0.2$ $\quad\quad$ μ_{12} variable

$\mu_{21} = 0.1$ $\quad\quad$ $\mu_{22} = 0.1$

The results are given for the average cycle times t_{C1} and t_{C2} and for the cycle time coefficients of variation c_1 and c_2 for high priority and low priority customers, respectively (The average cycle times are exact values). All results are compared to simulations (given in brackets).

μ_{12}	t_{C1}	t_{C2}
1.0	27.35 (27.41 ± 0.15)	81.16 (81.71 ± 0.88)
0.1	30.08 (30.08 ± 0.11)	75.96 (76.27 ± 0.63)
0.01	109.4 (109.4 ± 1.13)	223.0 (223.4 ± 1.96)
0.001	903.5 (902.9 ± 8.10)	2023. (2024. ± 18.5)
0.0001	8806. (8802. ± 90.0)	20020 (20039 ± 212.)

μ_{12}	c_1	c_2
1.0	.6275 (.5846 ± .0028)	.7378 (.6434 ± .0044)
0.1	.6072 (.5763 ± .0023)	.6628 (.6300 ± .0048)
0.01	.8906 (.9111 ± .0074)	.6673 (.6266 ± .0076)
0.001	1.100 (1.097 ± .0060)	.7033 (.6957 ± .0057)
0.0001	1.129 (1.127 ± .0059)	.7067 (.7038 ± .0046)

Table 1. Results for test-bed network 1

b) Results for Test-Bed Network 2

In case of preemptive priorities, class 1-customers are not affected by class 2-customers. Therefore, for class 1 the results of the single-class network applies which are known explicitly [11,13, 15].

Subsequently, we give the results for class 2-customers for various populations. The service rate parameters are as in a).

b1) $M_1 = 2, \quad M_2 = 2$

μ_{12}	t_{C2}	c_2
1.0	173.3 (173.4 ± 2.27)	.8468 (.7721 ± .0149)
0.1	190.9 (188.8 ± 1.84)	.7585 (.7059 ± .0176)
0.01	435.8 (435.4 ± 11.0)	.6196 (.6113 ± .0170)
0.001	3516. (3599. ± 239.)	.7076 (.7368 ± .0607)

b2) $M_1 = 2, \quad M_2 = 5$

μ_{12}	t_{C2}	c_2
1.0	355.9 (357.0 ± 4.13)	.6756 (.6005 ± .0118)
0.1	367.0 (367.1 ± 5.98)	.6212 (.5652 ± .0066)
0.01	899.9 (902.7 ± 17.3)	.4532 (.4452 ± .0107)
0.001	8750. (8423. ± 459.)	.4504 (.4314 ± .0243)

b3) $M_1 = 5, \quad M_2 = 2$

μ_{12}	t_{C2}	c_2
1.0	1665. (1729. ± 113.)	.9584 (.8998 ± .0463)
0.1	1684. (1710. ± 105.)	.9436 (.8696 ± .0355)
0.01	1896. (1928. ± 83.5)	.8170 (.7885 ± .0404)
0.001	4743. (4869. ± 304.)	.6151 (.5871 ± .0677)

b4) $M_1 = 5, \quad M_2 = 5$

μ_{12}	t_{C2}	c_2
1.0	3245. (3407. ± 294.)	.8079 (.6667 ± .0295)
0.1	3257. (3345. ± 219.)	.7991 (.6665 ± .0351)
0.01	3436. (3408. ± 171.)	.6995 (.6470 ± .0371)
0.001	10020 (9716. ± 480.)	.4428 (.4596 ± .0532)

Table 2. Results for test-bed network 2

c) Results for Test-Bed Network 3

As in case b), the class 1-customers are not affected by class 2-customers. The results for the average cycle time are exactly known [1,2]; results for the cycle time coefficient of variation have been validated in [15], so that we can concentrate on the low priority class only.

We note, that the cycle time analysis has been based on the additional approximation of network transformation: both I/O-stations of the CSM for class 2-customers have been aggregated to one composite server. We consider only those cycles starting at station 1.

The parameters of the CSM are as follows:

$M_1 = 2 \quad , \quad M_2 = 2$ and 5

$\mu_{11} = \mu_{21} = \mu_{31} = \mu_{22} = \mu_{32} = 1,\ \mu_{12}$ variable

$q_{12,1} = q_{12,2} = 0.5$

c1) $M_1 = 2$, $M_2 = 2$

μ_{12}	t_{C2}		c_2	
10.	4.549	(5.757 ± 0.032)	.669	(.7218 ± .0078)
1.0	9.480	(10.76 ± 0.132)	.717	(.7277 ± .0099)
0.1	74.35	(73.07 ± 2.025)	.722	(.7254 ± .0265)
0.01	737.0	(730.4 ± 61.67)	.707	(.7353 ± .1026)

c2) $M_1 = 2$, $M_2 = 5$

μ_{12}	t_{C2}		c_2	
10.	6.680	(4.873 ± 0.090)	.502	(.8804 ± .0159)
1.0	18.70	(19.43 ± 0.153)	.569	(.5969 ± .0096)
0.1	185.1	(185.4 ± 6.620)	.460	(.4748 ± .0164)
0.01	1842.	(1805. ± 175.0)	.447	(.4433 ± .0406)

Table 3. Results for test-bed network 3

4.3 Validation

The presented analysis method for transit delays in priority queuing networks rests on various approximation hypothesises:

(1) decomposition of the priority network into networks without priorities

(2) aggregation of subnets into composite stations.

Approximation (1) has been carried out with respect to state distributions; the application of that principle to transit delay df's is one source of error.

Approximation (2) has been applied to reduce the complexity in the analysis. This method does work under certain conditions as mentioned in Section 3.3. A further reduction of errors will be obtained through the advanced analysis procedure IFTA.

For all these sources of errors, most analysis results are within a 10 % range of the simulation results which is good enough in most cases of applications. Variations in the population size, service rates, priority schedule, or network structure indicate that the analysis method applies to a wider range of applications.

CONCLUSION

This paper presents - for the first time - an analysis method for transit delay distributions of Markovian priority queuing networks. This problem could be solved exactly in principle using the same method as for single-class networks, see [15]. To reduce the computational complexity, the problem has been simplified by decomposition of the multi-class priority network into equivalent single-class networks and aggregation of subnetworks into composite server stations. The results indicate that the method applies to a wider range of parameters as population size, service rates, priority schedule ,or network structure. The method can be applied to quite realistic cases where priorities are used to meet specific real-time percentile criterions as in case of dial-tone or acknowledgement delays in telecommunication systems, or in case of turn-around delays in interactive computer systems.

ACKNOWLEDGEMENT

The programming and computing efforts of Mr. Stefan Hohl are greatly appreciated.

REFERENCES

[1] Baskett, F., Chandy, K.M., Muntz, R.R., Palacios, F.G., Open,Closed and Mixed Networks of Queues with Different Classes of Customers. J.ACM 22 (1975), 248-260.

[2] Reiser, M., Lavenberg, S.S., Mean Value Analysis of Closed Multichain Queuing Networks. J.ACM 27 (1980), 313-322.

[3] Chandy, K.M., Herzog, U., Woo, L., Parametric Analysis of Queuing Networks. IBM J. Res. and Develop. 19 (1975), 36-42.

[4] Kuehn, P.J., Approximate Analysis of General Queuing Networks by Decomposition. IEEE Trans. on Commun. COM-27 (1975), 113-126.

[5] Kritzinger, P.S., van Wyk, S. Krzesinski, A.E., A Generalization of Norton's Theorem for Multiclass Queuing Networks. Int. Journ. on Performance Evaluation 2 (1982), 99-107.

[6] Kaufman, J.S., Approximate Analysis of Priority Scheduling Disciplines in Queuing Network Models of Computer Systems. Proc. 6th Int. Conf. on Computer Commun., London (1982), 955-961.

[7] Morris, R.J.T., Priority Queuing Networks. Bell System Technical Journal 60 (1981), 1745-1763.

[8] Schmitt, W., Approximate Analysis of Markovian Queueing Networks with Priorities. 10th International Teletraffic Congress, Montreal (1983), Congressbook, paper 1.3-3.

[9] Schmitt, W., Traffic Analysis of Priority Queueing Networks. 39th Report on Studies in Congestion Theory, Institute of Switching and Data Technics, University of Stuttgart (1984).

[10] Schmitt, W., On Decompositions of Markovian Priority Queues and Their Application to the Analysis of Closed Priority Queuing Networks. Proc. Performance '84, Elsevier Science Publishers B.V., North Holland (1984), 393-407.

[11] Chow, We-Min, The Cycle Time Distribution of Exponential Cyclic Queues. J.ACM 27 (1980), 281-286.

[12] Daduna, H., Passage Times for Overtake-Free Paths in Gordon-Newell Networks. Adv. Appl. Prob. 14 (1982), 672-686.

[13] Schassberger, R., Daduna, H., The Time for a Round-Trip in a Cycle of Exponential Queues. J.ACM 30 (1983), 146-150.

[14] Boxma, O.J., The Cyclic Queue with one General and one Exponential Server, Adv. Appl. Prob. 15 (1983).

[15] Kuehn, P.J., Analysis of Busy Periods and Response Times in Queuing Networks by the Method of First Passage Times. Proc. Performance '83, North Holland Publ. Comp. (1983), 437-455.

[16] Syski, R., Markovian Queues. Symp. on Congestion Theory. The Univ. of North Carolina Press, Chapel Hill (1964), 170-227.

[17] Lavenberg, S.S., Reiser, M., Stationary State Probabilities at Arrival Instants for Closed Queuing Networks with Multiple Types of Customers. J. Appl. Prob. 17 (1980), 1048-1061.

[18] Sevcik, K.C., Mitrani, I., The Distribution of Queuing Network States at Input and Output Instants, J.ACM 28 (1981), 358-371.

[19] Hohl, S., Kuehn, P.J., Approximate Analysis of Flow Time Processes in Queueing Networks by the Method of Independent Flow Time Approximation. Monograph Univ. of Stuttgart (1985).

TELETRAFFIC ISSUES in an Advanced Information Society
ITC-11
Minoru Akiyama (Editor)
Elsevier Science Publishers B.V. (North-Holland)

ANALYSIS OF A QUEUEING NETWORK WITH POPULATION SIZE CONSTRAINTS

Olav ØSTERBØ

Norwegian Telecommunication Administration,
Oslo , Norway

ABSTRACT

A queueing model for a virtual circuit in a packet network with end-to-end window flow control presented. First we examine the steady state balance equations, and the solutions of the equations for some special cases are given. These solutions suggest product form approximation in the general case, and approximative formulas for end-to-end performance are derived. Some numerical examples are also presented to test the approximation.

1. INTRODUCTION

The purpose of this paper is to study the performance of a virtual circuit in a packet network with end-to-end window flow control. A virtual circuit with end-to-end window flow control operates by limiting the numbers of packets (from the virtual circuit) in transit within the network. The maximum number of packets within a virtual circuit is called the window size.

In this paper we examine a simplified protocol described by:

A. A packet arriving at a VC when the window is full, (ie when the number of packets in the VC have reached the window size) is placed in an external queue (window queue). The first time the number of packets get less then the window size, a packet is removed from the window queue and transmitted over the circuit.

B. A packet arriving at a VC when the number of packets are less then the window size is transmitted immediately.

Shortly we can say that an end-to-end window mecanism removes possible bottlenecks from the transport network to the outstanding window queue. Therefore an important portion of the total delay for a packet is the waiting times in the wndow queue. Also the maximum input rate of packets to a VC is reached when the mean waiting time in the window queue approaches infinity.

In the litterature packet networks usually are modelled as

(i) open queueing networks
(ii) closed queueing networks [1] [2]

For networks without flow control the open queueing model is applied. However, for networks with end-to-end window flow control the closed queueing model can only give the limit performance of the VCs. (To use a closed model one has to assume that each VC is full, ie the number of packets in transit have reached the window size). Recently Dao [3] has given some approximative formulas for the mean delay in a packet network with end-to-end flow control. His analysis is based on the idea that a VC may be viewed as a multiserver queue where the window size is the number of servers. However, the exactness of the formulas derived is not properly discussed.

The paper is organized as follows: In section 2 we analyse a single VC modelled as a series of queues with a window queue in front to handle packets arriving at the circuit when the window is full. Finally we examine some numerical examples in section 3.

2. ANALYSIS OF A QUEUEING NETWORK FOR A SINGLE VIRTUAL CIRCUIT

2.1 Description of the queueing model

In this section we analyse a single VC modelled as a series of queues $Q_1, \ldots, Q_N$ with a window queue Q_B in front handle packets arriving the circuit when the window is full. (See fig 1.)

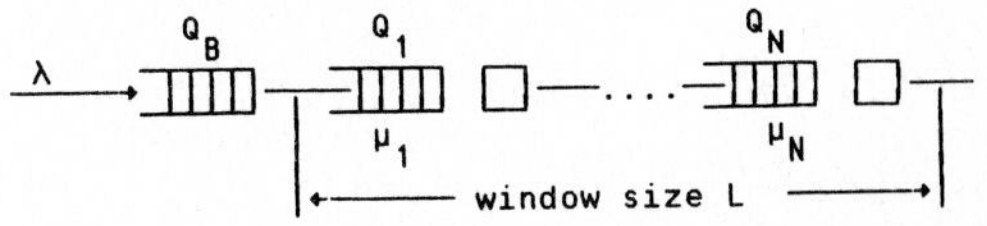

Fig 1.1 The queueing model of a single VC.

Let us now state precisely the assumptions:
Consider a queueing system consisting of N single server queues $Q_1, \ldots, Q_N$. The dynamic in the system is described by:

- An arriving packet (customer) joins Q_1 immedeately if the number of packets (customers) present in the system are less than the <u>window size</u> L.

- An arriving packet (customer) joins the window queue Q_B if the number of packets (customers) present have reached the window size L.

- When a packet (customer) leaves the system (ie Q_N) a packet (customer) in the window queue Q_B (if there are any) immediatly joins queue Q_1

- Further we assume packets (customers) to arrive at the system according to a Poisson process with rate λ, and the service times in $Q_1,\dots,Q_N$ are independent, neg.exp. distributed with parameter $\mu_r, r=1,\dots,N$.

2.2 Analysis of the steady state balance equations.

For the queueing system above we define the stochastic variable:

M_B - the number of packets (customers) in the window queue Q_B

M_r - the number of packets (customers) in queue Q_r, $r=1,\dots,N$

$(M_B, M_1,\dots,M_N)$ will be a multi dimentional Markov process and we define the steady state probabilities

$$p(i,j) = P(M_B=i, M=j) \quad \text{where}$$

$$M = (M_1,\dots,M_N) \text{ and } j = (j_1,\dots,j_N)$$

Let us also define the sets:

$$A(k)= \{j;\ j_1+\dots+j_N=k\}$$

and

$$B(k)= \{j;\ j_1+\dots+j_N\leq k\}$$

(Obvious $B(k)=A(0)\cup\dots\cup A(k)$.)

To get the balance equations on compact form we also define the operators $T_{lk}, T_{.k}, T_{l.}$ acting on the state vector $j = (j_1,\dots,j_N)$:

$$T_{lk}j = \begin{cases}(j_1,\dots,j_l-1,\dots,j_k+1,\dots,j_N) & \text{if } l<k\\ (j_1,\dots,j_k+1,\dots,j_l-1,\dots,j_N) & \text{if } l>k\end{cases}$$

$$T_{l.}j = (j_1,\dots,j_l-1,\dots,j_N)$$

$$T_{.k}j = (j_1,\dots,j_k+1,\dots,j_N)$$

Hence the steady state balance equations reads:

$$\{1+\sum_{r=1}^{N}\delta_{j_r}\varrho_r^{-1}\}p(0,j)=\delta_{j_1}p(0,T_{1.}j) + \qquad j\varepsilon B(L-1)$$
$$\sum_{r=1}^{N-1}\delta_{j_{r+1}}\varrho_r^{-1}p(0,T_{r+1r}j)+\varrho_N^{-1}p(0,T_{.N}j) \qquad \text{(i}$$

$$\{1+\sum_{r=1}^{N}\delta_{j_r}\varrho_r^{-1}\}p(0,j)= \delta_{j_1}p(0,T_{1.}j) + \qquad j\varepsilon A(L)$$
$$\sum_{r=1}^{N-1}\delta_{j_{r+1}}\varrho_r^{-1}p(0,T_{r+1r}j)+\delta_{j_1}\varrho_N^{-1}p(1,T_{1N}j) \qquad \text{(ii)}$$

(2.1)

$$\{1+\sum_{r=1}^{N}\delta_{j_r}\varrho_r^{-1}\}p(i,j)= p(i-1,j) + \qquad i>0,\ j\varepsilon A(L)$$
$$\sum_{r=1}^{N-1}\delta_{j_{r+1}}\varrho_r^{-1}p(i,T_{r+1r}j)+\delta_{j_1}\varrho_N^{-1}p(i+1,T_{1N}j) \qquad \text{(iii)}$$

where $\varrho_r = \lambda/\mu_r \quad ,r=1,\dots,N$

and $\delta_{j_r} = \begin{cases}0 & j_r=0\\ 1 & j_r>0\end{cases}$

The solution of (2.1) cannot be found by using the well known technique of partial balance equations [4], and therefor the solution cannot be of product form. The equations (2.1)(ii) and (iii) can be combined by introducing the generating function

$$P(x,j)=\sum_{i=0}^{\infty}x^i p(i,j) \qquad ,j\varepsilon A(L)$$

giving for $j\varepsilon A(L)$

$$\{1-x+\sum_{r=1}^{N}\delta_{j_r}\varrho_r^{-1}\}P(x,j)-\sum_{r=1}^{N-1}\delta_{j_{r+1}}\varrho_r^{-1}P(x,T_{r+1r}j)-$$
$$\delta_{j_1}\varrho_N^{-1}x^{-1}P(x,T_{1N}j)=\delta_{j_1}\{p(0,T_{1.}j)-\varrho_N^{-1}x^{-1}p(0,T_{1N}j)\} \qquad (2.2)$$

(2.2) is a linear system of equations and determine $P(x,j)$ uniquely in term of the right hand side. Together with (2,1)(i) and the usual demand that $P(x,j)$ is analytic inside $|x|<1$, determine $P(x,j)$, $j\varepsilon A(L)$ and $p(0,j)$, $j\varepsilon B(L-1)$ up to a constant, and the constant is of course found by the fact that the probabilities sum unity ie.

$$\sum_{j\varepsilon A(L)}P(1,j) + \sum_{j\varepsilon B(L-1)}p(0,j) = 1$$

In the general case it seems difficult to solve (2.2) explicitly. However since we primarily are interested in the moments of M_B, we therefore suggest a Taylor serie of $P(x,j)$ about $x=1$.

$$P(x,j)=\sum_{k=0}^{\infty}P^k(j)\ (x-1)^k \qquad j\varepsilon A(L)$$

Inserting in (2.2) and collecting equal potens of $(x-1)$ give the following equations for $P^k(j)$, $k=0,1,\dots$:

$$D\{P^0(j)\}=\delta_{j_1}\{p(0,T_{1.}j)-\varrho_N^{-1}p(0,T_{1N}j)\} \qquad \text{(i)}$$

(2.3)

$$D\{P^k(j)\}=P^{k-1}(j)+(-1)^k\delta_{j_1}\varrho_N^{-1}\{\sum_{n=0}^{k-1}(-1)^n P^n(T_{1N}j)-p(0,T_{1N}j)\} \qquad \text{(ii)}$$

Where D{ } is the "closed chain" operator given by

$$D\{P(j)\}=\{\sum_{r=1}^{N}\delta_{j_r}\varrho_r^{-1}\}P(j)-\sum_{r=1}^{N-1}\delta_{j_{r+1}}\varrho_r^{-1}P(T_{r+1r}j)-P(T_{1N}j)$$

(2.4)

Let us examine the equations (2.3) somewhat closer. If $P_H^k(j)$ satisfy (2.3), then we can add a closed chain solution $c_k\varrho_1^{j_1}\ldots\varrho_N^{j_N}$ where c_k is arbitrary. Hence the general solution of (2.3) must be of the form:

$$P^k(j) = P_H^k(j) + c_k\varrho_1^{j_1}\ldots\varrho_N^{j_N} \quad (2.5)$$

where c_k is an arbitrary constant. However the sum of the closed chain operator is zero for arbitrary P(j) ie. $\sum_{j\varepsilon\bar{A}(L)}D\{P(j)\}=0$. So summing the equation (2.3)(ii) gives an extra equation to determine the c_k:

$$\sum_{j\varepsilon\bar{A}(L)}\{P^0(j)-\delta_{j_1}\varrho_N^{-1}P^0(T_{1N}j)\}=-\delta_{j_1}\varrho_N^{-1}\sum_{j\varepsilon\bar{A}(L)}p(0,T_{1N}j) \quad \text{(i)}$$

and for k=1,2...

$$\sum_{j\varepsilon\bar{A}(L)}\{P^k(j)-\delta_{j_1}\varrho_N^{-1}P^k(T_{1N}j)\}= \quad (2.6)$$

$$(-1)^k\delta_{j_1}\varrho_N^{-1}\sum_{j\varepsilon\bar{A}(L)}\{\sum_{n=0}^{k-1}(-1)^nP^n(T_{1N}j)-p(0,T_{1N}j)\} \quad \text{(ii)}$$

From (2.3) and (2.6) we can determine $P^k(j)$ successively starting with k=0 ((2.3)(i) and (2.6)(i)). In the general case it seems difficult to get explicit expressions for $P^k(k)$. In the next section we apply the methods described above on the special case L=1 and L=2, N=2 , but first we examine the consequences of summing the equations (2.3)(i) for $P^0(j)$. The "extra" equation for p(0,j) is

$$\Delta(L-1) = \sum_{j\varepsilon A(L-1)}\{p(0,j)-\varrho_N^{-1}\,p(0,T_{.N}j)\} = 0 \quad (2.7)$$

However if we sum (2.1)(i) over jεA(k), k=1,2,...,L-1 we get

$$\Delta(k) = \Delta(k-1)\ ,\quad k=1,2,\ldots,L-1$$

and since $p(0,0,\ldots,0)=\varrho_N^{-1}p(0,0,\ldots,1)$ we have $\Delta(0)=0$ and therefore $\Delta(k)=0$, $k=0,1,\ldots,L-1$.

Therefore will a solution of (2.1)(i) also satisfy (2.7), so the "extra" equation is dependant of the equations (2.1)(i) and gives no "extra" information about p(0,j), jεA(L-1).

2.3 The case L=1

In this special case the equations (2.2) is easy to solve explicitly, and the result is:

$$P(x,0,\ldots,j_r=1,\ldots,0)= \quad (2.8)$$

$$p(0,0,\ldots,0)(1-\frac{1}{x})\frac{\varrho_1^{-1}\ldots\varrho_{r-1}^{-1}(1-x+\varrho_{r+1}^{-1})\ldots(1-x+\varrho_N^{-1})}{(1-x+\varrho_1^{-1})\ldots(1-x+\varrho_N^{-1})-\frac{\varrho_1^{-1}\ldots\varrho_N^{-1}}{x}}$$

Taking the limit x→ 1 gives

$$P(1,0,\ldots,j_r=1,\ldots,0)=\frac{\varrho_r}{1-(\varrho_1+\ldots+\varrho_N)}p(0,0,\ldots,0)$$

and normalization gives $p(0,0,\ldots,0)=1-(\varrho_1+\ldots+\varrho_N)$. Summing over the states jεA(1) gives the z-transform for the numbers in the window queue:

$$P(x)=\sum_{j\varepsilon\bar{A}(1)}P(x,j) \quad (2.9)$$

$$=\{1-(\varrho_1+..+\varrho_N)\}(1-\frac{1}{x})\frac{(1-x+\varrho_1^{-1})\ldots(1-x+\varrho_N^{-1})}{(1-x+\varrho_1^{-1})\ldots(1-x+\varrho_N^{-1})-\frac{\varrho_1^{-1}\ldots\varrho_N^{-1}}{x}}$$

which is Pollaczek-Khinchine formula for the number in a M/G/1-queue

2.4 The case N=2, L=2

The case N=2, L=2 represents the simplest "non trivial" case. By this example we demonstrate the last method described in section 2.3, and also we might hope that this example may suggest approximative solutions for the general L and N

Let us denote p_0 the probability that the system is emty (ie. p_0= p(0,0,0)). From (2.1)(i) and (ii) we solve for $p(0,j_1,j_2)$:

$$p(0,0,1)=p_0\varrho_2$$

$$p(0,1,0)=p_0(\varrho_1+\frac{(\varrho_1\varrho_2)^2}{\varrho_1+\varrho_2+2\varrho_1\varrho_2})$$

$$p(0,0,2)=p_0(\varrho_2^2+\frac{\varrho_1\varrho_2^3}{\varrho_1+\varrho_2+2\varrho_1\varrho_2}) \quad (2.10)$$

$$p(0,1,1)=p_0(\varrho_1\varrho_2+\frac{(1+\varrho_1)\varrho_1\varrho_2^3}{\varrho_1+\varrho_2+2\varrho_1\varrho_2})$$

Hence the solution for p(0,j) for jεB(L-1) can not be of product form for general L and N. A particular solution of (2.3)(i) is:

$$P_H^0(2,0) = 0$$

$$P_H^0(1,1) = \frac{\varrho_1\varrho_2^3}{\varrho_1+\varrho_2+2\varrho_1\varrho_2} \quad (2.11)$$

$$P_H^0(0,2) = \frac{\varrho_2^4}{\varrho_1+\varrho_2+2\varrho_1\varrho_2}$$

The general solution is therefore

$$P^0(j_1,j_2)=P_H^0(j_1,j_2) + c_0\varrho_1^{j_1}\varrho_2^{j_2}\ ,\ j_1+j_2=2$$

The equation (2.6)(i) determines c_0 to be

$$c_0 = \frac{s_1}{s_1 - s_2} - \frac{\varrho_2^2}{\varrho_1 + \varrho_2 + 2\varrho_1\varrho_2} \quad \text{where}$$

$$s_0 = 1 \;,\; s_1 = \varrho_1 + \varrho_2 \;,\; s_2 = \varrho_1^2 + \varrho_1\varrho_2 + \varrho_2^2$$

The final result for $P^0(j_1,j_2)$ reads:

$$P^0(2,0) = p_0\left(\frac{s_1}{s_1 - s_2}\varrho_1^2 - \frac{(\varrho_1\varrho_2)^2}{\varrho_1 + \varrho_2 + 2\varrho_1\varrho_2}\right)$$

$$P^0(1,1) = p_0\frac{s_1}{s_1 - s_2}\varrho_1\varrho_2 \tag{2.12}$$

$$P^0(0,2) = p_0\frac{s_1}{s_1 - s_2}\varrho_2^2$$

p_0 is found so that

$$\sum_{j_1+j_2=2} P^0(j_1,j_2) + \sum_{j_1+j_2\leq 1} p(0,j_1,j_2) = 1 \quad \text{giving}$$

$$p_0 = \frac{\frac{1}{s_2} - \frac{1}{s_1}}{\frac{S_2}{s_2} - \frac{S_1}{s_1}} \tag{2.13}$$

where $S_0 = s_0$, $S_1 = s_0 + s_1$, $S_2 = s_0 + s_1 + s_2$.

In a quite similar way we determine $P^1(j_1,j_2)$ from (2.3)(ii) and (2.6)(ii), and for the mean number in the window queue Q_B we get:

$$E[M_B] = \sum_{j_1+j_2=2} P^1(j_1,j_2) = \tag{i}$$

$$\frac{p_0}{(s_1 - s_2)^2}\{s_2^2 s_1 + 2s_1 s_2\varrho_1\varrho_2 - s_1^2 2(\varrho_1\varrho_2^2 + \varrho_1^2\varrho_2)\} + \varepsilon$$

For the mean number in Q_1 and Q_2 we also get

$$E[M_1] = p_0\{\varrho_1 + \frac{s_1}{s_1 - s_2}(2\varrho_1^2 + \varrho_1\varrho_2)\} - \varepsilon \tag{ii}$$

(2.14)

$$E[M_2] = p_0\{\varrho_2 + \frac{s_1}{s_1 - s_2}(2\varrho_2^2 + \varrho_1\varrho_2)\} \tag{iii}$$

and the probability that the window is full p_L:

$$p_L = \sum_{j_1+j_2=2} P^0(j_1,j_2) = \frac{1}{\frac{S_2}{s_2} - \frac{S_1}{s_1}} - \varepsilon \tag{iv}$$

and ε is defined to be

$$\varepsilon = p_0 \frac{(\varrho_1\varrho_2)^2}{\varrho_1 + \varrho_2 + 2\varrho_1\varrho_2} \tag{v}$$

The maximum throughput for the system is found when $E[M_B] \to \infty$ or $p_L \to 1$ ie. when

$$s_1 = s_2$$

solving for λ_{max} we get

$$\lambda_{max} = \frac{\mu_1^{-1} + \mu_2^{-1}}{(\mu_1^{-1})^2 + (\mu_1^{-1})(\mu_2^{-1}) + (\mu_1^{-1})^2} \tag{2.16}$$

Hence we get the closed chain throughput when $\lambda \to \lambda_{max}$, as we expected.

We close this section by examine the magnitude of ε defined in (2.14)(iv). First we observe that $\varepsilon \to 0$ when $(\varrho_1,\varrho_2) \to (0,0)$ and $(\varrho_1,\varrho_2) \to (\varrho_{1max},\varrho_{2max})$

(where $\varrho_{imax} = \frac{\lambda_{max}}{\mu_i}$, i=1,2) and the maximum value ε_{max} is

$$\varepsilon_{max} = 0.0086 \quad \text{for } \varrho_1 = \varrho_2 = 0.477$$

On the other hand if we approximate the probabilities $p(0,j_1,j_2)$ by a product solution ie.

$$p(0,j_1,j_2) = p_0\,\varrho_1^{j_1}\varrho_2^{j_2} \quad , \; j_1 + j_2 \leq 2 \tag{2.17}$$

we get (2.14) with $\varepsilon = 0$, so the magnitude of ε describes the distinction between the exact solution and a approximative one by assuming product form solution for $p(0,j)$, $j\varepsilon B(L-1)$. Since $0 \leq \varepsilon \leq 0.0086$ the two solutions are almost indistinguishable from a practial point of view. Encouraged by this observation we apply the product approximation for general L and N.

2.5 Product approximation for general L and N

We assume that $p(0,j)$ can be approximated by:

$$p(0,j) = p_0\,\varrho_1^{j_1}\cdots\varrho_N^{j_N} \quad \text{for } j\varepsilon B(L) \tag{2.18}$$

So (2.3)(i) becomes $D\{P^0(j)\} = 0$, giving

$$P^0(j) = c_0\,\varrho_1^{j_1}\cdots\varrho_N^{j_N} \quad \text{for } j\varepsilon A(L) \tag{2.19}$$

Let us also define the sums:

$$s_L = s_L(\varrho_1,\ldots,\varrho_N) = \sum_{j\varepsilon A(L)} \varrho_1^{j_1}\cdots\varrho_N^{j_N} \tag{2.20}$$

Then (2.6)(i) gives

$$c_0 = p_0\frac{s_{L-1}}{s_{L-1} - s_L}$$

and p_0 is determined so that

$$\sum_{j\varepsilon A(L)} P^0(j) + \sum_{j\varepsilon B(L-1)} p(0,j) = 1 \quad \text{giving}$$

$$p_0 = \frac{\frac{1}{s_L} - \frac{1}{s_{L-1}}}{\frac{S_L}{s_L} - \frac{S_{L-1}}{s_{L-1}}} \qquad S_L = \sum_{k=0}^{L} s_k \tag{2.21}$$

From (2.19) we get the probability that the window is full, p_L:

$$p_L = \frac{1}{\frac{S_L}{s_L} - \frac{S_{L-1}}{s_{L-1}}} \tag{2.22}$$

The mean number in the queues Q_r, r=1,...,N is found by:

$$E[M_r] = \sum_{j\in A(L)} j_r\, P^0(j) + \sum_{j\in B(L-1)} j_r\, p(0,j) =$$

$$\frac{\varrho_r}{1-\varrho_r}\left\{ 1 - p_L\left(\frac{d_L^{(r)}}{s_L} - \frac{d_{L-1}^{(r)}}{s_{L-1}}\right)\right\} \quad , r=1,\dots,N \qquad (2.23)$$

where $d_L^{(r)}$, r=1,...,N , is the sum

$$d_L^{(r)} = d_L^{(r)}(\varrho_1,\dots,\varrho_N) = \sum_{j\in A(L)} j_r\, \varrho_1^{j_1}\cdots\varrho_N^{j_N} \qquad (2.24)$$

To get the mean number in the window queue Q_B we must solve for $P^1(j)$. For k=1 and using the product approximation (2.18), the equations for $P^1(j)$, (2.3)(ii) and (2.6)(ii) are:

$$D\{P^1(j)\} = \qquad (i)$$

$$\frac{p_0}{s_{L-1} - s_L}\{s_{L-1}\varrho_1^{j_1}\cdots\varrho_N^{j_N} - \delta_{j_1} s_L\, \varrho_1^{j_1-1}\varrho_2^{j_2}\cdots\varrho_N^{j_N}\}$$

$$(2.25)$$

$$\sum_{j\in A(L)}\{ P^1(j) - \delta_{j_1}\varrho_N^{-1}P^1(T_{1N}j)\} = -p_0\frac{s_L s_{L-1}}{s_L - s_{L-1}} \qquad (ii)$$

For N=2 a particular solution of (2.25)(i) is

$$P_H^1(j_1,j_2) = \frac{p_0}{s_L - s_{L-1}}\{\delta_{j_1} s_L \sum_{i=0}^{j_1-1} i\varrho_1^i\varrho_2^{L-i} - s_{L-1}\sum_{i=0}^{j_1} i\varrho_1^i\varrho_2^{L+1-i}\} \qquad (2.26)$$

So the general solution of (2.15)(i) is

$$P^1(j_1,j_2) = c_1\varrho_1^{j_1}\varrho_2^{j_2} + P_H^1(j_1,j_2)$$

From (2.25)(ii) we determine c_1:

$$c_1 = \frac{p_0}{(s_L - s_{L-1})^2}\{s_{L-1}s_L + s_L(b_L - b_{L-1}) - s_{L-1}(b_{L+1} - b_L)\} \qquad (2.27)$$

where $b_L = b_L(\varrho_1,\varrho_2) = \sum_{j_1+j_2=L} j_1\, j_2\, \varrho_1^{j_1}\varrho_2^{j_2}$

Hence we get:

$$E[M_B] = \sum_{j_1+j_2=L} P^1(j_1,j_2) =$$

$$\frac{p_L}{s_L - s_{L-1}}\left\{ s_L + 2b_L - \frac{s_{L-1}b_{L+1}}{s_L} - \frac{s_L b_{L-1}}{s_{L-1}}\right\} \qquad (2.28)$$

The sum b_L can also be written

$$b_L = \frac{1}{2}\left(L^2 s_L - dd_L^{(1)} - dd_L^{(2)}\right)$$

where $dd_L^{(r)} = \sum_{j_1+j_2=L} j_r^2\, \varrho_1^{j_1}\varrho_2^{j_2}$, r=1,2

For general L we let

$$b_L = \frac{1}{2}\left(L^2 s_L - \sum_{r=1}^{N} dd_L^{(r)}\right) \qquad (2.29)$$

and

$$dd_L^{(r)} = dd_L^{(r)}(\varrho_1,\dots,\varrho_N) = \sum_{j\in A(L)} j_r^2\, \varrho_1^{j_1}\cdots\varrho_N^{j_N} \qquad (2.30)$$

We must emphasize that (2.28) only is derived for N=2, and therefore we regard the generalisation (2.29) and (2.30) as an approximation. The maximum throughput for the system is reached when $p_L \to 1$, ie. if λ_{max} is the solution of the equation

$$s_{L-1}(\varrho_1,\dots,\varrho_N) = s_L(\varrho_1,\dots,\varrho_N)$$

giving

$$\lambda_{max} = \frac{s_{L-1}(\mu_1^{-1},\dots,\mu_N^{-1})}{s_L(\mu_1^{-1},\dots,\mu_N^{-1})} \qquad (2.31)$$

Which is the throughput in a N-node cyclic network. When $\lambda = \lambda_{max}$ (2.23) gives:

$$E[M_r] = \frac{d_L^{(r)}}{s_L} \quad \text{and} \quad \sum_{r=1}^{N} E[M_r] = L \qquad (2.32)$$

We therefore conclude that the product approximation gives the correct closed chain solution when $\lambda \to \lambda_{max}$.

We close this section by listing the formulas for the mean values derived. (The mean queueing times follows directly from Littles formula.)

$$W_r(L,\lambda,\varrho_1,\dots,\varrho_N) = \frac{E[M_r]}{\lambda}$$

$$E[M_r] = \frac{\varrho_r}{1-\varrho_r}\left\{ 1 - p_L\left(\frac{d_L^{(r)}}{s_L} - \frac{d_{L-1}^{(r)}}{s_{L-1}}\right)\right\} \qquad r=1,\dots,N$$

$$W_B(L,\lambda,\varrho_1,\dots,\varrho_N) = \frac{E[M_B]}{\lambda} \qquad (2.33)$$

$$E[M_B] = \frac{p_L}{s_L - s_{L-1}}\left\{ s_L + 2b_L - \frac{s_{L-1}b_{L+1}}{s_L} - \frac{s_L b_{L-1}}{s_{L-1}}\right\}$$

where

$$p_L = \frac{1}{\dfrac{S_L}{s_L} - \dfrac{S_{L-1}}{s_{L-1}}} \qquad (2.34)$$

and

$$b_L = \frac{1}{2}\left(L^2 s_L - \sum_{r=1}^{N} dd_L^{(r)}\right)$$

and the sums s_L, S_L, $d_L^{(r)}$ and $dd_L^{(r)}$

$$s_L = \sum_{j\in A(L)} \varrho_1^{j_1}\cdots\varrho_N^{j_N} \quad , \quad S_L = \sum_{k=1}^{L} s_k$$

$$d_L^{(r)} = \sum_{j\in A(L)} j_r\, \varrho_1^{j_1}\cdots\varrho_N^{j_N} \qquad (2.35)$$

$$dd_L^{(r)} = \sum_{j\in A(L)} j_r^2\, \varrho_1^{j_1}\cdots\varrho_N^{j_N} \qquad r=1,\dots,N$$

In Appendix A we derive some recursion formulas for the sums in (2.34), and from these formulas s_L, S_L, $d_L^{(r)}$ and $dd_L^{(r)}$ are easily computed. The time required is the same as for a closed chain with the corresponding N and L.

2.6 The accuracy of the approximations for N=2

In section 2.4 we examined the difference between the exact and approximative formulas ε and we found ε_{max} =0.0086 for L=N=2. For small L and N the balance equations (2.1) are easy to solve numerically. We have solved (2.1) using Gauss-Seidels method [5] for N=2 and L=3,4,5 and computed the differance between the exact and the approximative formulas in section 2.5 . Let

$E[M]=E[M_1]+E[M_2]$, and denote $\varepsilon_B=E[M_B]^{ex}-E[M_B]^{ap}$ and $\varepsilon=E[M]^{ex}-E[M]^{ap}$ the difference between the exact and approximative mean values. In Table 2.1 the result is given. It seems that approximatly

$$\frac{\max(\varepsilon_B)}{L} \approx \frac{\max(\varepsilon)}{L} \approx 0.005 \qquad (2.35)$$

We therefore conclude that the accuracy of the approximative formulas seems to be very close to the exact values, and from a practical point of view they are indistinguishable.

Table 2.1 The maximum difference between the exact and the approximative mean values for N=2 and L=2,3,4,5.

L	$\max(\varepsilon_B)$	$\max(\varepsilon)$
2	0.0086	0.0086
3	0.014	0.015
4	0.018	0.020
5	0.022	0.025

3 SOME NUMERICAL RESULTS

Below two test networks A (N=2), B (N=5) are given. As for the case N=2 ,L=2 the maximum difference between exact and approximative solutions occurs when all the service times have equal means.

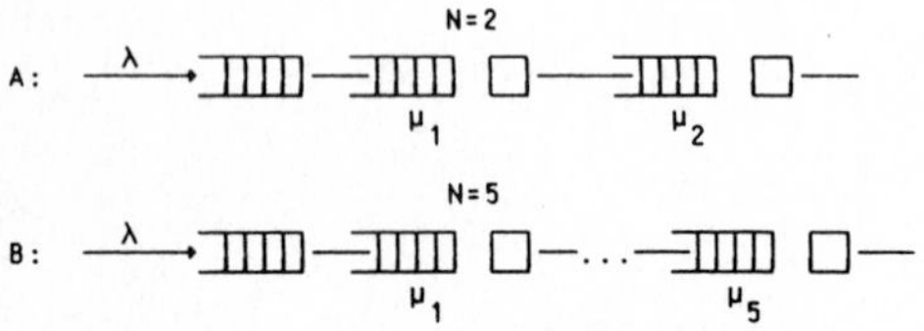

Fig 3.1 The test networks studied. The arrival streams are assumed to be Poissoian, and the service times are independent identically neg.exp. distributed so that the delay for zero load is unity.

For case A we present both exact and approximative calculations (Fig 3.2). For window size L=2 the exact results is given in section 2.4 and for L=3,4,6 the exact curves are obtained by solving the balance equations (2.1) iterative. As pointed out in section 2.6 the error introduced by the product approximation is neglectable for N=2. The product approximation gives a little to high values for W_1 (Fig 3.2 (i)), for W_2 the results are almost identical (Fig 3.2 (ii)), and for W_B the values are little below the exact ones (Fig 3.2 (iii)). For the total delay W^T the errors in W_1 and W_B cancel (Fig 3.2 (iv)).

In Fig 3.3 the total delay W^T for case B (N=5) is given by using the formulas (2.33)-(2.35). Because of the increase in the number of equations (to solve numerically), we have tested the approximations by simulations, and runs are carried out for the loads equal 0.1, 0.5, 0.75 and 0.9 of the maximal throughput.

The agreement between the simulated values and the calculated curves is quite well. However the difference is not neglectable as for case A (N=2). Nevertheless we conclude that the approximative formulas (2.33)-(2.35) describe end-to-end performance of a VC with sufficient accuracy.

Fig 3.2 (v) and Fig 3.3 illustrate that the end-to-end performance of a single VC strongly depends on:

(i) - the window size L

(ii)- the numbers of nodes a VC passes through N

We see that the window size will limit the throughput strongly (Fig 3.2 (iv) and Fig 3.3), and the dependence of the (of window size) will increase with the numbers of nodes N.

4 REMARK

The result for a single VC may be extended to cover networks with many active VCs. The main assumtion (approximation) is:

- When considering a fixed VC we assume the other VCs as open (ie. without window flow control limitations).

By redefinting the service times one can use tne formulas for a single VC.

REFERENCES

[1]' M. Reiser, "A Queueing Network Analysis of Computer Communication Network with Window Flow Control",IEEE Trans. on com., vol. com-27, no.8, pp.1199-1209, 1979.

[2] S.S. Lam, J.W. Wong, "Queueing Network Models of Packet Switching Networks", Performance Evaluation, vol.2, no.3, pp.161-180, 1982.

[3] M. Dao, "Modelisation d'un Reseau a Commutation de Paquets et de son Controle de Flux de Bout en Bout", Proc. Intern. Seminar on Mod. and Perf. Eval. Methodology vol.II, pp.183-201, Paris 1983.

[4] F.P. Kelly, "Reversibility and Stochastic Networks", Wiley, New York, 1979.

[5] G. Dahlquist, Å. Bjørck, "Numerical Methods" Prentice-Hall, Inc., Englewood Cliffs, New Jersey, 1974.

APPENDIX A

For the sums s_L, S_L, $d_L^{(r)}$, $dd_L^{(r)}$ in (2.35) we get the following recursion formulas:

$$d_L^{(r)} = \varrho_r(d_{L-1}^{(r)} + s_{L-1}) \quad \text{(A1)}$$

$$dd_L^{(r)} = \varrho_r(dd_{L-1}^{(r)} + 2d_{L-1}^{(r)} + s_{L-1}) \quad \text{(A2)}$$

$$\sum_{r=1}^{N} d_L^{(r)} = L s_L \quad \text{(A3)}$$

and

$$S_L = S_{L-1} + s_L \quad \text{(A4)}$$

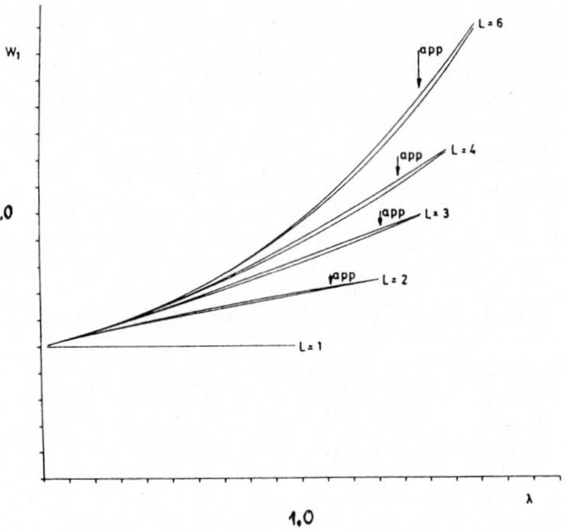

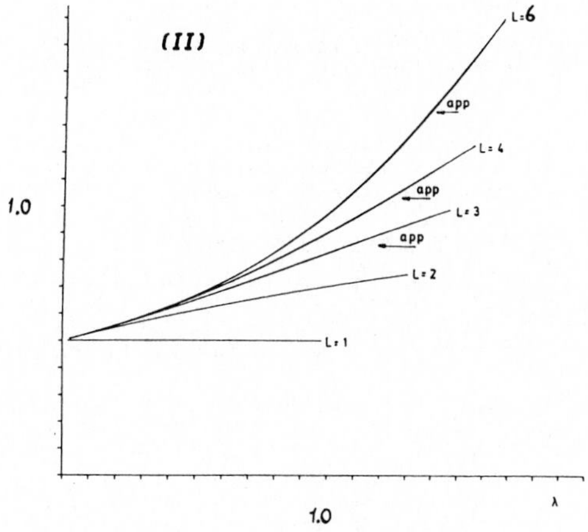

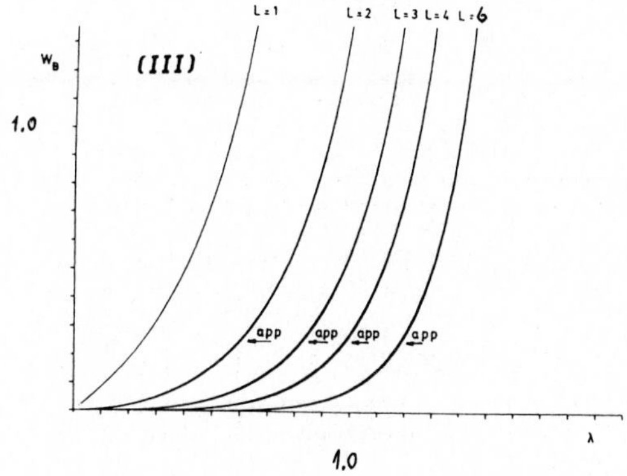

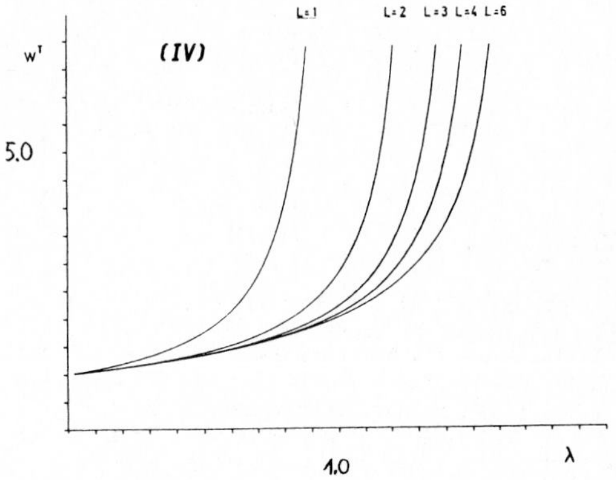

Fig 3.2
The mean queueing time in queue Q_1, W_1 -(i)
The mean queueing time in queue Q_2, W_2 -(ii)
The mean waiting time in the window queue Q_B, W_B-(iii)
The total mean delay W^T-(iv)
for network A with window size L=1,2,3,4,6.
The curves marked "app" is the approximative ones based on the product approximation in section 2.5.

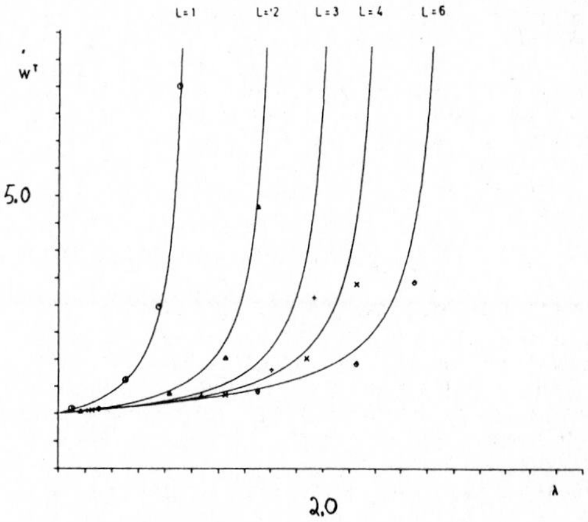

Fig 3.3 The mean total delay W^T for the network B with window size L=1,2,3,4,6. The simulated values are marked o - L=1, Δ - L=2, + - L=3, x - L=4, and ◇ - L=6.

TELETRAFFIC ISSUES in an Advanced Information Society
ITC-11
Minoru Akiyama (Editor)
Elsevier Science Publishers B.V. (North-Holland)

AN ADAPTIVE SHORT-TERM TRAFFIC FORECASTING PROCEDURE USING KALMAN FILTERING

Prosper CHEMOUIL, Bruno GARNIER

Centre National d'Etudes des Télécommunications
Issy-les Moulineaux, France

ABSTRACT

This paper deals with short-term traffic forecasting methods in telephone networks. Analysis in traffic observations shows that there exist substantial variations within a season. Then traffic routing patterns could be updated during the year according to seasonal load variations in order to take advantage of the non-coincidence of the traffic flows. A preliminary step to implement multi season traffic routing is however to determine reliable and precise forecasting models. In this paper attention is focused on this last item and an adaptive forecasting method is proposed using Kalman filtering.

1. INTRODUCTION

This paper addresses the problem of short-term traffic forecasting, which is of major importance in improving planned and trunk demand servicing for telephone networks.

In the french long distance telephone network, dimensioning and traffic routing patterns are determined once a year from a Yearly Representative Value (YRV) of the load offered to each trunk group. This YRV is derived from measurements of carried traffic on trunks. Each month, carried traffic is measured on trunk groups during five days at the mean busy hour; the second highest value of the five ones is chosen as the Monthly Representative Value (MRV). Finally the YRV is defined as the second highest value of the twelve MRVs.

At present the deployment of SPC networks is making possible the traffic routing to be more dynamic. This allows to take advantage of the non-coincidence of traffic variations that exist in the french telephone network. But the determination of multi-season traffic routing patterns first requires the development of accurate short-term forecasting procedures (one month to one year).

Several methods have been proposed to short-term traffic forecasting :

• The Box-Jenkins approach [1], [2], is based on the autoregressive integrated moving average (ARIMA) approach. It consists in modelling the traffic load behaviour and in identifying the model parameters. This approach was found to give accurate results insofar as the time-series remain stable. However it has three major drawbacks :

* first a lot of data is needed to fit the model (around 60 data corresponding to a five-year period of monthly observations).

* second though the modelling is of general form, identification may lead to different models according to the time-series.

* finally, as the model parameters are tuned only for the fitting period, the filter cannot adapt to sharp variations which may affect the traffic load (for example when a new trunk group is created). Only fitting the model again is appropriate but it would require another five-year period of observation to get a new good model.

• The Holt-Winters method [2], [3], is based on exponential smoothing. In this approach a 14th order model is derived by considering variables as the mean deseasonalized level, the trend factor and 12 seasonal factors (in case of monthly observations). These 14 state variables are updated at each observation using three smoothing constant parameters. They are determined in order to minimize the mean square error of the one-step ahead prediction. One main interest of the Holt-Winters approach lies in the uniqueness of the model. This procedure gives good results and can be implemented in a simple automatic way. But it suffers the following drawbacks :

* a great number of observations (about 24 to 30 data) is needed to tune the three smoothing parameters.

* The smoothing parameters would not be reliable anymore in case of break in the time-series. Forecasts could be therefore inaccurate.

• The constant Kalman filter which has been recently proposed [4], [5] is based upon exponential smoothing as the Holt-Winters method. Due to the markovian properties of the associated stochastic model, Kalman filtering provides a simple and recursive solution that determines the optimal linear estimation of the traffic load. This method does not require a lot of initial data to fit the smoothing constants, but it needs a good knowledge of the statistical parameters of the model and observation noises (mean and variance). These parameters can be tuned in two ways : First, in a deterministic view, they are fixed in order to get a good response-time of the filter. Second initial data are used to identify means and variances of the process noises. The two methods become nevertheless inaccurate as soon as there is an important change in the time-series. This is the main drawback of the constant Kalman filter.

All the above methods are unable to improve the forecasts in case of break in the time-series which can occur very often on trunk groups when traffic is removed. The main reason is that constant filters are set up according to a long history and cannot be fitted again with few data.

It is then necessary to design self-tuning filters which are fitted along with time. As Kalman filtering exhibits very nice properties, it is proposed in this paper to make adaptive the Kalman filter.
The principle of the method is to identify the means and the variances of the noises which are the only parameters of the model that are undefined. The parameter estimation is then converted to an optimization problem which is solved using the maximum a posteriori approach. Consequently, the smoothing parameters of the Kalman filter are tuned at each observation and changes in the statistics are therefore taken into account. If the time-series is stable the smoothing parameters tend to their optimal constant value.

The first step in developing forecasting methods is the description of traffic behaviour. Three classes of traffic behaviour are found :

- a seasonal behaviour where substantial variations exist according to the months. See Figs. 1 and 2.

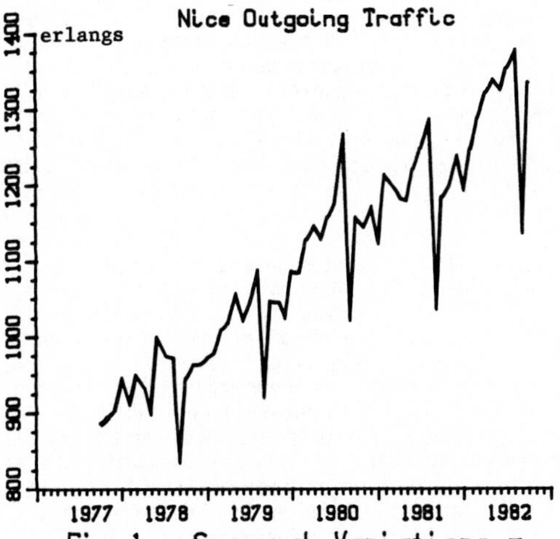

Fig.1 - Seasonal Variations -

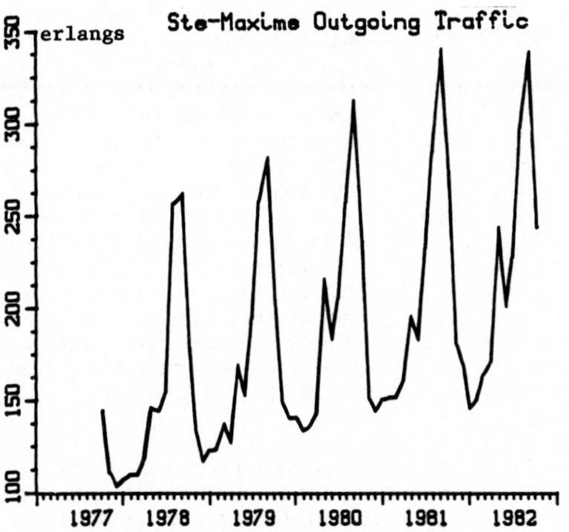

Fig.2 - Seasonal Variations -

- a linear trend behaviour in which no particular seasonal variation is detected. Variations around the linear trend are small (see Fig. 3).

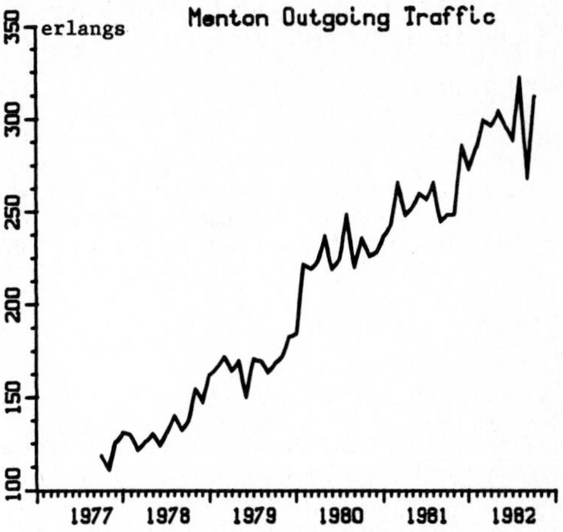

Fig.3 - Linear Trend Variations -

- the third class is defined when great but not seasonal variations exist. This last class cannot be modelled in view of forecasting and will not be considered (see Fig. 4).

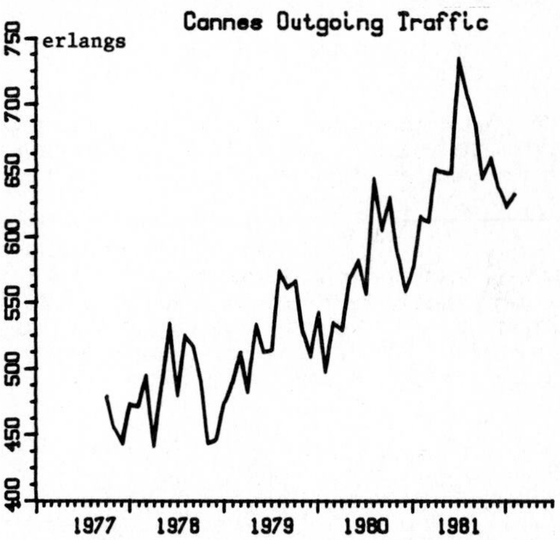

Fig.4 - Random Variations -

In this paper we shall first investigate mathematical models to represent seasonal and linear trend traffics. Next, we derive the Kalman filter and we discuss this approach. Then we propose an adaptive algorithm based on Kalman filtering. Finally, we present the behaviour and performance evaluation of the adaptive approach on traffic data of the french Riviera. A comparison is made with the Holt-Winters method that shows the adequacy of the adaptive traffic forecasting procedure.

2. MODELS OF TRAFFIC VARIATIONS

Two telephone traffic models are derived from Monthly Representative Values (MRVs) according to the existence or not of seasonal traffic variations. In the last case, only linear growth variations are considered. This part is mainly devoted to the description of the deterministic characteristics of the MRV series. Stochastic components are then added in the model to take into account the uncertainty of the modelling as well as the measurement errors.

2.1. Seasonal Traffic Model

Following the approach of Holt-Winters [2], we consider that the traffic is composed of two main components : a deseasonalized mean level, which is assumed to have a linear trend variation and a seasonal factor corresponding to the influence of the considered period (i.e. a month). Due to the periodicity of the time-series, the model is written as follows :

$$\begin{aligned} m_k &= m_{k-1} + r_{k-1} \\ r_k &= r_{k-1} \\ f_k &= f_{k-12} \\ y_k &= m_k + f_k \end{aligned} \qquad (2.1)$$

where y_k, m_k, r_k and f_k are respectively the traffic observation, the deseasonalized mean level, the trend term and the seasonal factor at time k.
However, as m_k is a mean variable, it implies that $\sum_{i=1}^{12} f_i = 0$. Thus introducing this constraint in the dynamical model (2.1), we get a redundant information allowing an order reduction of the model.
The new model based on the former one, can be represented into a state-space form where the 13 state variables are defined as :

$$x_k^T = [m_k \ r_k \ s_{1k} \ s_{2k} \ \cdots \ \cdots \ s_{11k}]^T \qquad (2.2)$$

where x^T stands for x-transpose.
In this model the terms s_{ik} are defined as the sum of the i first seasonal factor f_i at time k. Then s_{ik} should be understood as the cumulative seasonal factors during a year.

$$s_{ik} = \sum_{j=1}^{i} f_{jk} \qquad i = 1 \text{ to } 11$$

Of course, over one full year this cumulative factor is null. Then the state-form model is written as follows :

$$x_k = Fx_{k-1} \qquad (2.3a)$$

$$y_k = Hx_k \qquad (2.3b)$$

where the transition and the observation matrice are given by :

$$F = \begin{bmatrix} \begin{bmatrix}1 & 1\\ 0 & 1\end{bmatrix} & 0 \\ 0 & \begin{bmatrix} -1 & 1 & & & 0 \\ -1 & 0 & 1 & & \\ \vdots & & \ddots & \ddots & \\ -1 & & 0 & 0 & 1 \\ -1 & & & & 0 \end{bmatrix} \end{bmatrix}, \quad H^T = \begin{bmatrix} 1 \\ 0 \\ 1 \\ 0 \\ \vdots \\ \vdots \\ 0 \end{bmatrix} \qquad (2.4)$$

The model (2.3) is an ideal description of the traffic behavior when all the variables are well identified and when all the observations are perfect. In fact uncertainty in the modelling as well as the measurements errors make this model unrealistic. Stochastic components are then added in the model (2.3) to take into account the random phenomena. One component is included in the state equation (2.3a) to represent the modelling errors and another one is added in the observation equation (2.3b) to take into account the measurement errors. The full seasonal model is finally :

$$\begin{aligned} x_k &= Fx_{k-1} + w_{k-1} \\ y_k &= Hx_k + v_k \end{aligned} \qquad (2.5)$$

The noises w_k and v_k are generally gaussian white centered sequences.

2.2. Linear Trend Traffic Model

The model of linear trend traffic is easily deduced from the former model (2.5) by considering that no seasonal component exists. It only consists in the traffic load including a growth factor. The second order dynamical model is therefore :

$$\begin{aligned} x_k &= F'x_{k-1} + w_{k-1} \\ y_k &= H'x_k + v_k \end{aligned} \qquad (2.6)$$

where the system matrices F' and H' are the following :

$$F' = \begin{bmatrix}1 & 1\\ 0 & 1\end{bmatrix}, \qquad H^T = \begin{bmatrix}1 & 0\end{bmatrix} \qquad (2.7)$$

In fact this model is non-satisfactory since the trend term follows the white noise variations; as the forecasts are mainly dependent of the trend factor, the projection several steps ahead of the estimation is only an amplification of the noise and leads to uncorrect forecasts.

To avoid this problem which is illustrated in part 5., we propose to describe a linear trend traffic with a new first order model.
This model is obtained by considering that the noise disturbing the state equation has a non-zero mean which represents a smoothed trend factor. The first order model is simply :

$$\begin{aligned} x_k &= x_{k-1} + w_{k-1} \\ y_k &= x_k + v_k \end{aligned} \qquad (2.8)$$

where w_k has a non-zero mean value. The forecasting method that we propose automatically identifies this mean value and therefore leads to good results.

We have presented two models which describe seasonal and linear trend traffics respectively. These two models are written in the same state-space form and a unified forecasting method can therefore be derived using Kalman filtering.

3. KALMAN FILTER.

Kalman filter has been developped in the early 60's [6] to determine state estimation of linear dynamical systems. It is much used in control theory to design real-time control of stochastic processes (missile and rocket guidance, industrial processes, ...). Due to the markovian properties of the considered model, Kalman filter provides a simple and recursive solution that gives the optimal linear estimation of the state vector. In this approach it is assumed that both the model and the observation are disturbed with gaussian white noise. The filter is designed in a two-step way :

- a prediction step where **a priori** estimation determines the optimal one-step-ahead prediction of the former estimate.
- a correction step where the prediction is updated according to a new observation resulting then in the optimal **a posteriori** estimation of the state.

3.1. Formulation

Let us consider a discrete linear dynamical system described as :

$$x_{k+1} = Fx_k + w_k \tag{3.1a}$$

$$y_k = Hx_k + v_k \tag{3.1b}$$

where x_k, y_k, w_k and v_k are respectively the state, the observation and two white sequences of dimension n, p, n and p.

The moments of the variables are defined as :

$$E[x_0] = \bar{x}_0, \qquad E[(x_0 - \bar{x}_0)(x_0 - \bar{x}_0)^T] = P_0$$

$$E[w_k] = q_k, \qquad E[(w_k - q_k)(w_l - q_l)^T] = Q_k \delta_{kl}$$

$$E[v_k] = r_k, \qquad E[(v_k - r_k)(v_l - r_l)^T] = R_k \delta_{kl}$$

where δ_{kl} is the Kronecker delta. Generally it is assumed that w_k and v_k are independent of the past and of the state x_k as well. We also assume that the two noises have no correlation. Then the criterion to be minimized is :

$$J = \min \{\frac{1}{k_f} \sum_{k=0}^{k_f} (x_k - \bar{x}_k)(x_k - \bar{x}_k)^T\} \tag{3.2}$$

where $\bar{x}_k$ is the estimate of x_k and k_f the number of observations.

Then the two-step filter is defined as follows :

*** prediction stage**

$$x^*_{k+1} = F\bar{x}_k + q_k \tag{3.3}$$

where x^* and $\bar{x}$ are the one-step-ahead prediction and the estimate of the state x at step k.
Let P^*_k be the covariance matrix of the prediction error $\tilde{x}^*_k = x_k - x^*_k$ and $\bar{P}_k$ be the covariance matrix of the estimation error $\tilde{x}_k = x_k - \bar{x}_k$ then the following result holds :

$$P^*_{k+1} = F\,\bar{P}_k F^T + Q_k \tag{3.4}$$

*** correction stage**

$$\bar{x}_{k+1} = x^*_{k+1} + K_{k+1} \nu_{k+1} \tag{3.5}$$

where ν is the **innovation** term defined as :

$$\nu_{k+1} = y_{k+1} - Hx^*_{k+1} - r_{k+1}$$

The gain of the filter is :

$$K_{k+1} = P^*_{k+1} H^T [HP^*_{k+1} H^T + R_{k+1}]^{-1} \tag{3.6}$$

Finally the covariance matrix of the estimation error is updated with the equation :

$$\bar{P}_{k+1} = P^*_{k+1} - P^*_{k+1} H^T [HP^*_{k+1} H^T + R_{k+1}]^{-1} HP^*_{k+1} \tag{3.7}$$

Kalman filter exhibits very nice properties such as Simplicity, Optimality and Sequentiality. The filter first determines the linear solution of minimum variance. Second, due to the markovian modelling of the system, step k+1 depends only on step k, defining then a sequential algorithm leading to a very simple implementation.
In our problem the model is moreover invariant (matrices F and H are constant); this allows to calculate and implement the constant gain Kalman filter obtained by determining the steady state solution of the Riccati equation :

$$\bar{P} = F\bar{P}F^T + Q - F\bar{P}H^T [H\bar{P}H^T + R]^{-1} H\bar{P}F^T \tag{3.10}$$

However the determination of the optimal filter (3.3)-(3.7) requires that some preliminary conditions are satisfied.
The solution of Kalman filter exists if :

- The system is detectable (all the unstable modes of the system are observable). It means that the state variables can be reconstructed from the measurement.
- The covariance matrix of the measurement noise is positive definite.
- The covariance matrix of the model noise is semi-positive definite.

3.2. Kalman Filtering limitations

Kalman filtering assumes that the moments of the noises are known, which is often untrue. As the noises are usually centered, only variances are considered. Guessed values are to be fixed according to the physical knowledge of the system.
This approach becomes inaccurate as soon as there is an important change in the time series, which can often occur in telephone traffic (due to creation or deletion of trunk groups, changes in the subscriber behaviour). As in the model, the deterministic component is fixed, only the stochastic component can reflect that change. Therefore modifications of the statistical characteristics are expected.

It is then suggested in the paper to make adaptive the Kalman filter. It is realized by introducing in the derivation of the filter an algorithm which identifies the means and variances of the process noises.

4. ADAPTIVE KALMAN FILTER

The principle of the method is to identify the means and variances of the noises. The parameter estimation problem is converted to an optimization problem which is solved using the maximum a posteriori approach. The problem formulation results in a two point boundary value problem solved using the invariant imbedding technique. However in the interests of simplicity of implementation and of speed of computation a sequential near-optimal solution is derived which updates the statistical parameters step by step.

4.1. Problem formulation

We solve the identification problem using Sage-Husa approach [7]. In this approach it is assumed that mean and variance of the noises are unknown but constant. The principle of the algorithm is to maximize the density function according to the unknown parameters, i.e. :

$$J = \max_{r,R,q,Q} P\{X_{k_f}, r, R, q, Q \mid Y_{k_f}\} \tag{4.1}$$

where X_{k_f} and Y_{k_f} are the state and observation sequences over the interval $[0, k_f]$.

Using Bayes' theorem and considering that the noise sequences are gaussian with constant density function, the problem is converted to the minimization problem [8] :

$$J = \tfrac{1}{2}\| x_0 - \bar{x}_0 \|_P^2 + \tfrac{1}{2}\Sigma \| y_k - Hx_k - r_k \|_R^2 + \tfrac{1}{2}\Sigma \| x_k - Fx_{k-1} - q \|_Q^2 \tag{4.2}$$

The parameters r, R, q and Q are then obtained using the maximum likelihood estimation. For example the mean value of the observation noise is obtained as follows :

$$\bar{r}_{k-1} = \frac{1}{k-1} \sum_{j=1}^{k-1} \{y_j - H\bar{x}_{j/k}\} \tag{4.3}$$

which is approximated with :

$$\bar{r}_{k-1} = \frac{1}{k-1} \sum_{j=1}^{k-1} \{y_j - Hx_j^*\} = r_k^* \tag{4.4}$$

Finally the recursive equation can be written :

$$r_k^* = \frac{1}{k-1}[(k-2)r_{k-1}^* + y_{k-1} - Hx_{k-1}^*] \tag{4.5}$$

The other parameters are obtained in the same way and can be introduced in the former Kalman filter resulting in the adaptive Kalman filter.

4.2. Adaptive Kalman filter

The two steps of the filter are now :

Prediction step

$$x_{k+1}^* = F\bar{x}_k + \bar{q}_k \tag{4.6a}$$

$$P_{k+1}^* = F\bar{P}_k F^T + \bar{Q}_k \tag{4.6b}$$

$$r_{k+1}^* = \frac{1}{k}[(k-1)r_k^* + y_k - Hx_k^*] \tag{4.6c}$$

$$R_{k+1}^* = \frac{1}{k}[(k-1)R_k^* + \bar{\nu}_k \bar{\nu}_k^T - H P_k^* H^T] \tag{4.6d}$$

Estimation step

$$\bar{\nu}_{k+1} = y_{k+1} - Hx_{k+1}^* - r_{k+1}^* \tag{4.7a}$$

$$\bar{x}_{k+1} = x_{k+1}^* + K_{k+1}\bar{\nu}_{k+1} \tag{4.7b}$$

$$K_{k+1} = P_{k+1}^* H^T [HP_{k+1}^* H^T + R_{k+1}^*]^{-1} \tag{4.7c}$$

$$\bar{P}_{k+1} = P_{k+1}^* - K_{k+1} H P_{k+1}^* \tag{4.7d}$$

$$\bar{q}_{k+1} = \frac{1}{k+1}[k\bar{q}_k + \bar{x}_{k+1} - F\bar{x}_k] \tag{4.7e}$$

$$\bar{Q}_{k+1} = \frac{1}{k+1}[k\bar{Q}_k + K_{k+1}\bar{\nu}_{k+1}\bar{\nu}_{k+1}^T K_{k+1}^T + \bar{P}_{k+1} - F\bar{P}_k F^T] \tag{4.7f}$$

The equations (4.6)-(4.7) describe the adaptive estimation filter; we need now to extend it to forecasting.

4.3 Adaptive forecasting algorithm

The adaptive filter that we derived determines the near-optimal estimation and one-step-ahead prediction of the state-vector. Mathematically it reads :

$$x_k^* = \mathrm{Opt}_{lin}[x_k \mid y_{k-1}] \text{ and } \bar{x}_k = \mathrm{OPT}_{lin}[x_k \mid y_k]$$

Forecasting can therefore be implemented using equation (4.6a). At each step forecasts are updated according to the equations (4.6)-(4.7). For a forecasting n steps ahead, the prediction stage (4.6) is simply added with the projection equation :

$$x_{k+n}^* = F^n \bar{x}_k + n\bar{q}_k \tag{4.8}$$

5. PERFORMANCE EVALUATION OF THE ADAPTIVE METHOD

To analyze the behaviour of the procedure, we consider data from the french Riviera subnetwork. Three time-series were extracted from available traffic measurements corresponding to the long distance traffic outgoing of Menton, Nice and Ste-Maxime.
For each series, the six-month-ahead traffic forecast compared to the measured traffic is drawn. These curves are completed with relevant statistical results, which summarize the performance evaluation. The mean errors and the standard deviation of the errors are given on Table 1 for the forecasts from one month up to one year. These two parameters are sufficient for analysis as the models are assumed to be gaussian. The mean error determines the precision of the procedure and the standard deviation of the error provides the confidence interval of the forecasts. These two parameters are compared to those obtained when applying the Holt-Winters method [2].
All the results are obtained using a software tool designed at CNET to evaluate forecasting methods implementing Kalman filtering.

5.1 Menton case

When analyzing Menton data, it was estimated that the traffic is linear. Then the first-order model was primarily used which gave good forecasts (see Fig. 5). To bring out the inadequacy of the second order model, the six-month-ahead forecasts are displayed on Fig. 6 for this model. It is illustrated on Fig. 7 where the trend factors are drawn for first and second order traffic models. It can be shown the mean value of the white noise corresponds effectively to the smoothed value of the trend term. For the second order model the growth factor is sharply varying that points out the bad results with this model. Table 1 enhances these conclusions.

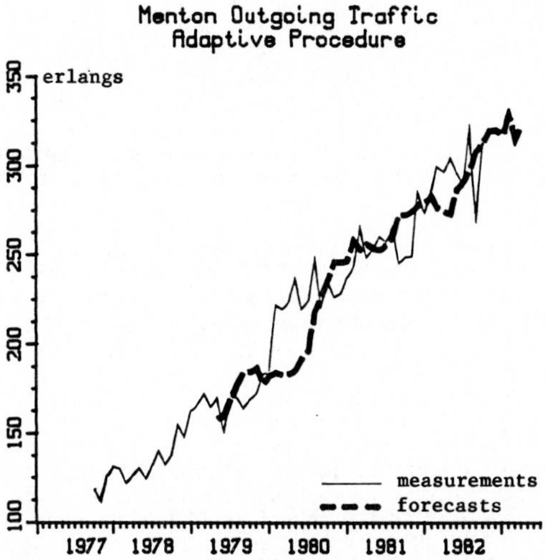

Fig.5 - Six-Month-Ahead Forecasts -

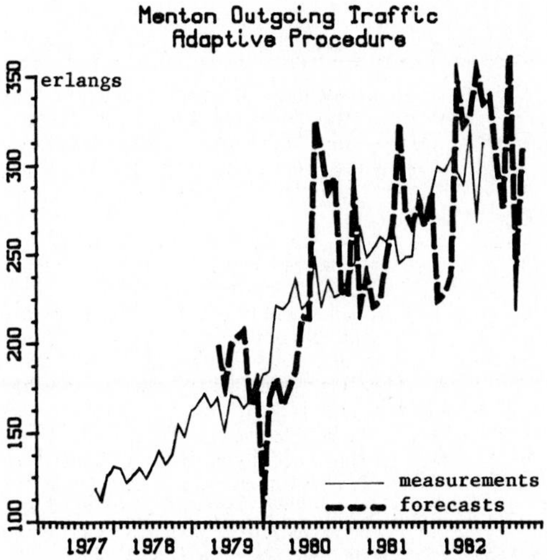

Fig.6 - Six-Month-Ahead Forecasts -

To check the assumption that no seasonal variation exists in Menton, we implemented the 13th-order seasonal model. Fig. 8 shows that the seasonal model gives good results although the reaction to the jump of December 1979 is slower than with the first-order model. However, in examining the results in Table 1, we can see that the seasonal model is more accurate and we can deduce Menton traffic is weakly seasonal.

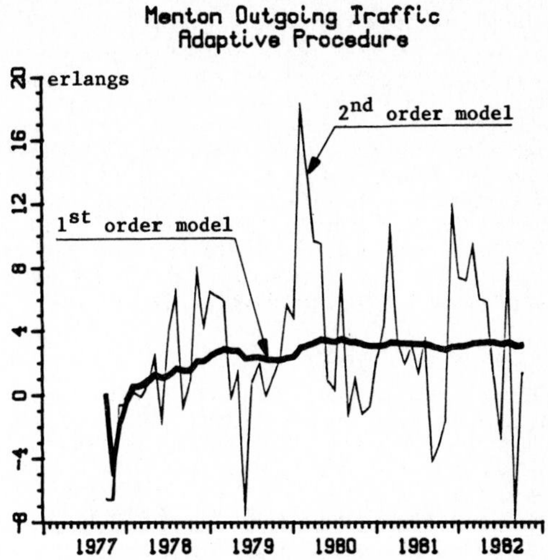

Fig.7 - Growth Factor -

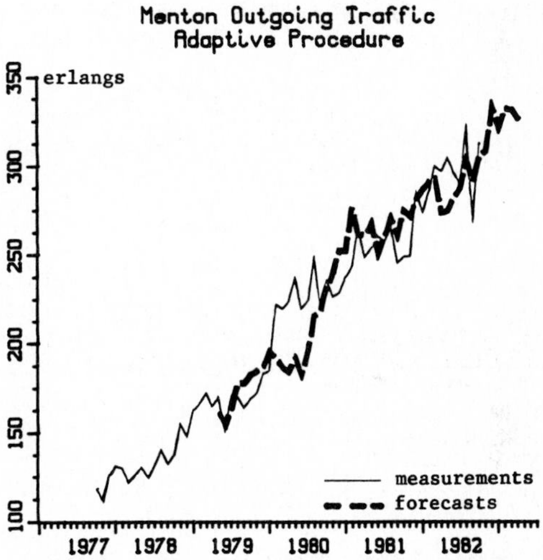

Fig.8 - Six-Month-Ahead Forecasts -

5.2 Nice case

Nice data show that a strong seasonality exists on August where telephone traffic is much lower than the other months. The seasonal forecasting model was then used. The six-month-ahead prediction is displayed in Fig. 9 and shows the good behaviour of the method. Table 1 gives the precision of the procedure up to one year (a mean prediction error from 2% to 4%). Furthermore it can be noticed that results are better than those obtained with the Holt-Winters method.

5.3 Ste-Maxime case

From graphical analysis, it is obvious that the Ste-Maxime traffic is strongly seasonal. Due to the importance of this resort centre, traffic is very high in summer and low in winter. The seasonal model was accordingly implemented. Figure 10 shows the accuracy of the forecasts. Table 1 emphazises the good performances of the forecasting method. The comparison with the Holt-Winters procedure shows the results are similar.

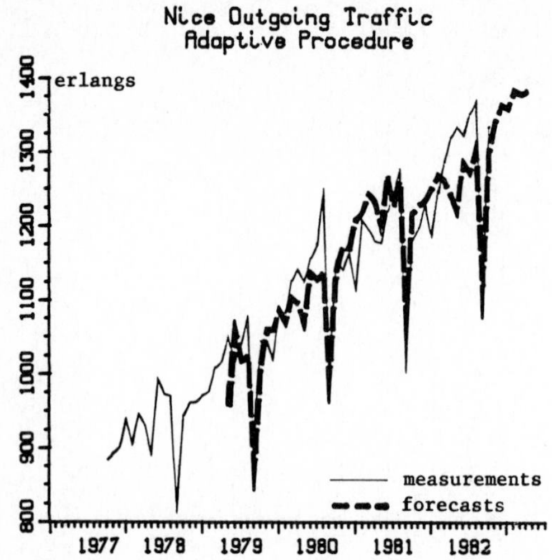

Fig.9 - Six-Month-Ahead Forecasts -

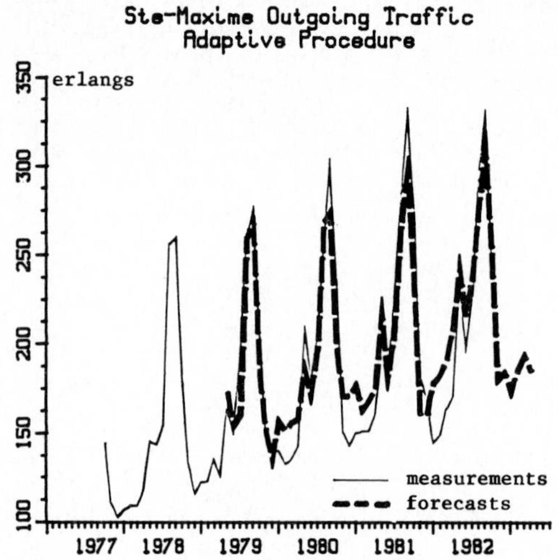

Fig.10 - Six-Month-Ahead Forecasts -

TOWN	METHOD (MONTH AHEAD FORECAST)	1 month	2 month	3 month	4 month	5 month	6 month	7 month	8 month	9 month	10 month	11 month	12 month
Menton	ADAPTIVE 1st order	4.9 13.1	5.4 14.3	6.3 16.3	7.1 18.2	7.8 19.5	8.4 20.8	8.9 21.8	9.3 22.7	9.5 23.6	9.7 24.2	10.1 25.8	10.9 27.6
	ADAPTIVE 2nd order	5.6 15.9	6.8 17.4	8.6 22.7	11.1 29.3	13.2 33.9	15.2 40.6	17.8 46.4	19.2 51.0	21.3 56.3	22.4 59.3	23.3 62.6	23.2 63.7
	ADAPTIVE 13th order	4.3 13.0	4.5 13.6	5.1 15.3	5.7 16.7	6.1 17.6	6.5 18.3	6.8 18.8	6.9 18.7	6.9 19.0	6.8 18.9	7.1 19.2	7.4 19.6
	Holt-Winters method	4.5 13.9	4.8 14.8	5.1 15.7	5.4 16.5	5.8 17.3	6.0 17.9	6.3 18.5	6.6 18.9	6.9 19.7	7.4 20.4	8.0 21.4	8.5 22.6
Nice	adaptive method	2.1 34.3	2.1 35.7	2.1 37.6	2.4 40.5	2.4 41.3	2.8 42.9	3.0 45.0	2.9 44.4	3.0 44.0	3.1 44.9	3.2 46.3	3.3 47.7
	Holt-Winters method	2.2 36.1	2.2 36.9	2.2 38.4	2.6 42.3	2.5 43.2	2.8 45.1	3.0 48.5	3.1 48.3	3.1 48.2	3.3 49.8	3.5 51.5	3.5 51.3
Ste-Maxime	adaptive method	6.7 16.6	7.2 17.5	8.3 18.7	8.7 19.5	9.0 19.9	9.5 20.0	9.1 19.5	8.4 18.5	8.2 18.1	8.3 18.7	8.1 19.1	8.5 19.1
	Holt-Winters method	6.8 15.7	7.2 16.2	7.5 16.7	7.9 17.0	8.1 17.1	8.3 17.4	8.3 17.5	8.5 17.8	8.6 18.1	8.9 18.7	9.5 19.3	9.8 19.8

a = mean error
b = standard error

a
b

Table 1. - Mean error (in %) and standard error (in erlang) of the forecasts -

It must be noticed that the smoothing parameters of the Holt-Winters method were fitted for the full observation period. Results are then optimal and the comparison proves the good behaviour of the adaptive method. Further results of this application concern the identification and improvement of the variances and can be found in [9].

6. CONCLUSION

In this paper we were concerned with a new adaptive traffic forecasting procedure. Contrarily to former methods, the model parameters are not set up using a smoothing algorithm, but they are improved at each step. Changes that appear in the observation series are then taken into account. The results we gave for the french Riviera subnetwork show the adequacy of the method.

REFERENCES

[1] Box (G.E.P.), Jenkins (G.M.) : "Time series analysis, forecasting and control. Holden-Day (1976)

[2] Passeron (A.), Eteve (E.) : "A forecasting model to set-up multi-season traffic matrices" Int. Teletraffic Congress, Montreal ITC'10, (1983).

[3] Winters (P.R.) : "Forecasting sales by exponentially weighted moving averages". Management Sci, Vol 6 (1960).

[4] Pack (C.D.), Whitaker (B.A.) : "Kalman filter models for network forecasting" Bell System Tech. Journal Vol. 61 (1982)

[5] Szelag (G.R.) : "A short-term forecasting algorithm for trunk demand servicing" Bell System Tech. Journal Vol. 61 (1982)

[6] Kalman (R.E.) : "A new approach to linear filtering and prediction problems" J. Basic Eng. vol. 82 (1960)

[7] Sage (A.P.), Husa (G.W.) : "Adaptive filtering with unknown prior statistics" 8th IEEE Symposium on adaptive process (1969)

[8] Sage (A.P.), Melsa (J.L.) :"System identification" Academic Press, New York (1971)

[9] Chemouil (P.) : "FAKIR : une méthode de prévision adaptative du trafic téléphonique" Tech. report CNET NT/PAA/ATR/SST/1467 (1984)

TELETRAFFIC ISSUES in an Advanced Information Society
ITC-11
Minoru Akiyama (Editor)
Elsevier Science Publishers B.V. (North-Holland)

SELECTION AND FORECASTING OF TRAFFIC MATRICES IN ITFS - A TRUNK FORECASTING SYSTEM FOR A NON-HIERARCHICAL NETWORK

W.H. SWAIN, J.A. POST, and E.K. FLINDALL

BELL CANADA
OTTAWA CANADA

ABSTRACT

This paper discusses features of the Intertoll Trunk Forecasting System (ITFS) currently under development by Bell Canada Engineering Economics for Telecom Canada. The system forecasts trunk requirements for a mixed Fixed Hierarchically Routed/High Performance (dynamically) Routed network. In particular, the paper describes the construction and forecasting of traffic demand matrices for multi-hour sizing. Although the characteristics of the Telecom Canada network led to the particular definition of the matrices used in ITFS, the method of construction easily generalizes to any network or environment. The forecasting method re-configures historical exchange code-to-exchange code demand to account for changes in network architecture. These demands are then projected using automatic techniques. The motivation for the choice of these techniques is discussed.

1. INTRODUCTION

In 1981 Telecom Canada, an association of the 9 major telephone companies across Canada, and Telesat Canada, the nation's satelite telecommunications carrier, set up a study to determine whether the development of a new computer-based forecasting system was feasible and required. Their existing system, then over 10 years old, could not accomodate all new services then offered without excessive manual intervention. The decision to develop a new software based system called the Intertoll Trunk Forecasting System (ITFS) was made in 1982 and a phase of research into required forecasting and sizing methodology and specification development which was to last for 2.5 years was started. Bell Canada Corporate Engineering Economics, an organization with experience in the development of Telecommunications Network Teletraffic models and who had developed previous Teletraffic models for Telecom Canada, was chosen to research potential methodology, assist the specifiers of the system, and to supply the system if it was decided to go ahead with development after completion of the specification.

This paper has been written to provide a brief illustration of the implementation of some of the techniques and methodologies discussed at the present and previous ITC's and at other forums [1,2,3,4,5]. The next section will briefly describe the modules of ITFS, list some of its important features and give the rationale for selecting these features. Section 3 will discuss in detail the selection of a set of of optimal traffic demand matrices to allow ITFS to size the network using multi-hour techniques and take advantage of non-coincident busy hours in the network. We will shed some light on the selection of an optimal set of demand matrices. Section 4 of the paper will describe a system of automated mechanized projection techniques used to provide seven year forecasts of trends in the network.

The development of systems the size and complexity of ITFS requires more than just one organixation with a few people working in isolation. We would like to acknowledge the contribution of the producers of the ITFS specification, Telecom Canada - Fundamental Planning (Projects) who conceptualized and defined much of the methodology described in this paper. Other key participants include Bell Northern Research who developed the algorithms and methodology for the sizing of the Non Hierarchical High Performance Routing system and the Bell Canada Corporate Systems Organization who have participated in the logical and physical database design of ITFS. Without their help and analysis this work could not have been carried out. Last but by no means least, we would like to acknowledge the contribution of our colleagues on the Bell Canada Engineering Economics ITFS Development Team.

2.0 OVERVIEW AND FEATURES OF ITFS

The Intertoll Trunk Forecasting System (ITFS) is logically partitioned into 5 high level business functions which are shown within the circles in Figure 2.1. It interfaces with outside systems, data bases and forecasters as characterized by the squares in Figure 2.1. These outside sources provide the information required to operate the system e.g existing service demand, structure and topology of the network to be forecasted, etc. The system is also capable of being of overridden at a number of points by forecasters and operators who may impose constraints on operation, fine-tune forecasted rates of growth based on economic considerations, future plans, etc. In addition, there are manual/mechanized interfaces to allow information beyond that provided in the standard reports of the system to be accessed by forecasters and/or operators.

As shown in Figure 2.1, the 5 high level business functions are, 1 - Network Architecture, 2 - Process PTP Demand, 3 - Project PTP Demand, 4 - Identify New Fixed Hierarchical Routing Trunk Groups, 5 - Dimension Network.

The Network Architecture function allows the

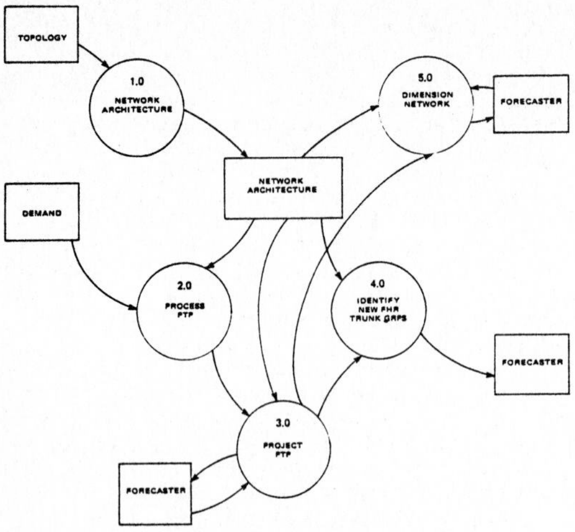

FIG. 2.I ITFS BUSINESS FUNCTION

user to input and validate data on the topology and correlations of the network to be sized.

The Process PTP Demand function combines the network architecture description and a point-to-point (PTP) demand data base (derived from network billing tapes) to produce the traffic matrices to be forecasted.

The Project PTP Demand function takes the 5 years of demand history derived in the second business function and projects it for 7 years.

The Identify New Fixed Hierarchical Groups function, identifies new pairs of points between which it has become economic to establish new trunk groups, using the forecast demand for a given year. (ECCS methodology is used.)

Dimension Network sizes the trunk groups in each future year using the forecast traffic matrices.

Features

A considerable amount of time was spent in the early specification phase on the selection of operational features to be incorporated into the system from a long list of potential features which could be implemented. Those features which were selected were based on future technological developments in the network, opportunities for increasing network efficiency, savings in forecasting cost, the need to make the system user-friendly and statements of functionality requirements by the forecasters. Some features thus introduced include the use of point-to-point demand, the ability to forecast a mixed non-hierarchical/hierarchical network, incorporation of online input and validation, multi-hour sizing techniques, mechanized creation of forecast entities and automated forecasting of demand.

High Performance Routing (HPR)

The decision to include non-hierarchical routing capability was taken as a result of plans by Telecom Canada to incorporate a High Performance Routing (HPR) into the Canadian network. This system which is a variant of dynamic routing developed by BNR is expected to produce trunk savings and make the network more resistant to equipment failure and unexpected overloads. The HPR system results in routing of calls sufficiently different to a fixed hierarchical network so that both needed to be reflected in the forecasting system. The developers of HPR, BNR were included in the ITFS system development to define the routing and sizing algorithms required. BNR's work is discussed in detail within another paper to be presented at this conference [1].

Point-to-Point Demand

Both point to point demand measured by sampling the network billing data of the Canadian network and measured trunk demand (peg-count, usage and overflow) are available to the system forecaster One of the early decisions made in the specification development was to use exclusively PTP demand in ITFS. The main reason for making this choice were the plans to incorporate HPR in the network. The HPR routing of calls varies based on a calculation of the optimum network routing given availability of facilities on the network at a given time, and is affected by abnormal demands, equipment failure etc. As a result usage measurements of trunk group demand on HPR trunk groups cannot be expected to be an accurate base for the forecasting of future demand. The exclusive use of PTP demand within ITFS expected to produce more stable, more efficient forecasts.

Online Entry and Validation

One of the most difficult and time consuming tasks associated with the existing forecasting system is the entry of the network architecture (topology and routing information) into the system. This information must be updated once each forecast period to reflect changing plans of all 9 operating member companies. The process is usually left as late as possible in the forecast cycle to ensure that the latest information is included and is accomplished by inputting large flat files of update information The providers of the information are geographically and organizationally dispersed and information must be centrally validated before being entered in the database.

ITFS has been designed to update network information in an online mode allowing for remote entry with immediate transaction verification. If the transaction is consistent with the current network view, then the network architecture database is immediately updated and the new version is ready for the next transaction. New information may be added as it becomes available, ameliorating the last minute rush and later vintage information may be entered since the requirement for several cycles of final validation and correction is eliminated. ITFS will use a 'snapshot' of the database taken at a preannounced time for each cycle of the forecast.

Multiple Hour Sizing

The Canadian network spans 5 time zones with a four-and-a-half hour time-difference. There are also significant differences in the

seasonal geographic distribution of the traffic. The result of these seasonality differences together with local peaks occurring at the same clock-hour but not the same network hour, results in a large number of non-coincident demand peaks.
The selection of a single database capturing all of the peaks would result in the over-provisioning of trunks for the network. To overcome this, a multi-hour sizing algorithm is used in ITFS. For this algorithm, a number of demand databases are created representing different periods of the year and different times of the day. The objective of database selection is to select periods and time slots which include all peaks in the network in at least one set of demand but which separate non-coincident peaks which affect the same facilities. A minimum number of demand data sets to accomplish the above must be determined since each additional data set leads to additional operating costs. On the other hand, too few data sets will lead to the over-provisioning of the network due to false coincidence. The criteria for minimum selection will be covered in greater detail.

Entity Creation

The PTP data source for the ITFS is a sample of 20 million call-by-call-records per month. The processing of this number of data items for trend forecasting, and for sizing would be expensive. Accordingly, it is desirable to find a means of aggregating the data without losing the ability to accurately reflect changes in the network architecture over the forecasting period. In ITFS, we search for groups of customers whose demand enters the network together throughout the life of the forecast. These groups are called ENTITIES. Manual choice of entities is a difficult process for the forecaster and often results in the creation of extra entities which are redundant. We have incorporated a process to examine the architecture of the network over the 5 year period and automatically select those entities which are required. The call records are then aggregated into demand from entity-to-entity. Automating entity creation will save both forecaster time and processing time, the latter because fewer entities are defined.

Automated Demand Forecasting

Because of the large volume of items to be forecasted (entity-to-entity parcels for about 1000 entities), automatic methods must be used. A complete description of the forecasting system is given in Section 4.

3.0 CONSTRUCTION OF TRAFFIC MATRICES

This section discusses the method of constructing representative traffic matrices. The first sub-section will motivate the approaches taken to the problem; the second sub-section will describe how these approaches were applied in an empirical study of data from the Telecom Canada network. Generalizations of the results to other networks and other environments will be made.

3.1 Motivation

As indicated in Section 2, the intent of this research was to find a compromise between the processing cost of using a large number of matrices and the loss of information resulting from using a small number. The objective was to find a small number of matrices that captured the important features of the traffic behaviour, especially time-of-day differences or seasonal differences in the busy periods of sub-networks, without distortion. From an initial examination of acceptable processing cost, the maximum number of matrices was set to 12.

When many traffic matrices are collapsed in any fashion, there are two risks to be considered: the risk of missing a peak somewhere in the network; and the risk of lumping two non-coincident peaks together which affect the same trunks and/or switches. The former leads to undersizing trunk groups and can result in quality-of-service problems. The latter, FALSE COINCIDENCE, leads to oversizing trunks because it does not take advantage of idle capacity in one sub-network to serve the peak in another.

To minimize the risk of missing a peak the method of construction should examine all periods of the day and year when a peak may occur. No periods should be eliminated a priori, as in a Busy Season approach, and the load for any hour should be eligible to be selected and placed in a traffic matrix. Furthermore, loads to be placed in the matrices should be selected on the basis of peaks, not washed out by averaging.

To minimize risk of false coincidence the method of construction should keep non-coincident peaks in different matrices. If all hours and all days were kept separate, there would be no false coincidence. Fortunately, hours and days may be combined. uppose for example that there are three identifiable sub-networks A, B, and C and suppose, furthermore, that all three have their busy period in the morning. Sub-networks A and B peak in the summer months and C peaks in the winter months. Then a representative set of matrices is one matrix representing the morning period in the summer and one matrix representing the morning period in the winter. We need not include the afternoon and evening periods when no peaks occur. No finer resolution of the summer is required if A and B have coincident peaks within this period. Inputting a summer and a winter matrix to the dimensioner enables it to find the idle capacity in the C sub-network during the summer to handle A's peak and B's peak, and the idle capacity in the A and B sub-networks during the winter to handle C's peak.

In ITFS, we divide the year into PERIODS and the hours of the day into TIME SLOTS. One traffic matrix, called a LOADSET, represents each combination of period and time-slot. The values placed in the matrices should represent the peak load. In the diagram below, one such division of the year into twelve loadsets has been presented.

LOADSET DEFINITION

```
       Jan        Apr        Jul        Oct
0900   +---------+---------+---------+--------+
       I         I         I         I        I
1200   +---------+---------+---------+--------+
       I         I         I         I        I
1800   +---------+---------+---------+--------+
       I         I         I         I        I
0800   +---------+---------+---------+--------+
```

By selecting peak loads from all hours of the entire year, this method ensures that all peaks are captured. False coincidence still exists, for example, if one sub-network peaks at 10:00 in the same period. In this case, the sizer will create idle capacity to serve what it assumes to be coincident peaks. The challenge lies in defining a set of loadsets capturing all peaks and minimizing false coincidence.

3.1.1 Selection of Peak Loads

Consider matrix entry or traffic parcel A-Z (the load from switch A to switch Z). The load when the parcel itself is busiest is not necessarily the best estimate of peak load especially when there is not a direct trunk group between the two switches. In the diagram below, the route from A to Z is shown with attention focussed on trunk group P-Q.

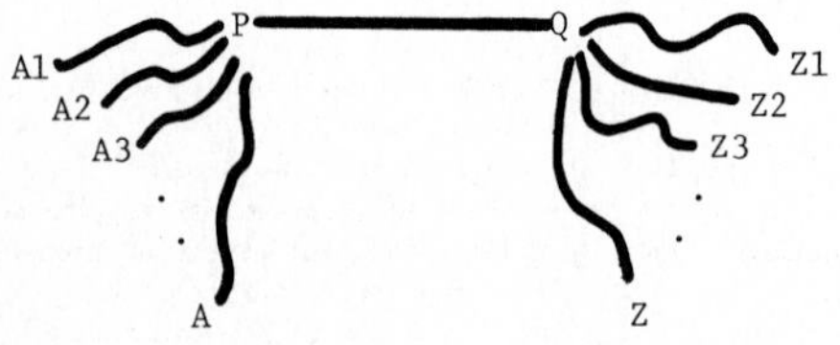

Trunk group P-Q is also part of the route from A1, A2, ... to Z1, Z2, ... If loadsets were constructed using the peak load of each parcel then trunk group P-Q could be oversized if the peaks of parcels A-Z, A1-Z1, A2-Z2, ... were non-coincident.

Trunk group P-Q busy hour is also not useful since many trunks are impacted by the A-Z parcel and these groups may all have different busy hours.

The load for the A-Z parcel could be taken when switch A is busiest. An additional advantage of this is that the HPR sizer is particularly concerned with the load when the 'bundle' of all trunks at A is busy [2]. Thus, picking A-Z, A-Z1, ... when A is busy is useful for the sizer. However, by the same argument, A-Z could be selected when Z is busy and this might not be at the same time.

Since neither of these alternatives have any obvious superiority, the solution was to build one matrix using A-Z loads when A is busy (called the Control-A matrix) and one matrix with A-Z loads when Z is busy (called the Control-Z matrix). These two matrices were then combined using the Kruithof method [6].

3.1.2 Selecting Loadset Definition

To pick the best loadset definition, metrics assessing goodness had to be selected. After examining some statistics from both a theoretical and an empirical basis, a simple one-dimensional statistic was selected which measured the distance of a particular definition from the worst case.

The worst case in terms of false coincidence occurs when only one matrix is used to represent the whole year. All peaks for all sub-networks appear in the same matrix, no idle capacity is observed, and trunks are oversized. This matrix was termed the PEAK-OF-PEAKS matrix and its volume ($VP\text{-}P = V(L) = \Sigma L(i,j)$ i.e. the total network load represented) is the worst case load.

In a Multi-Hour context, it is reasonable to assume that the cost of a network varies directly with the total network load of the largest (in terms of volume) matrix input to the sizer since the network created by a Multi-Hour sizer can be no smaller than the networks created by sizing each of the input matrices in a single hour fashion. This suggests using the ratio of the volume of the largest matrix to the peak-of-peak load ($R = \max(V(Lk))/VP\text{-}P$) as a measure of the goodness of a loadset definition. A value near 1.0 implies a definition with essentially the same amount of false coincidence present in the peak-of-peaks matrix. A value much less than 1.0 implies less false coincidence and lower total network cost. With this metric, the empirical task is to find a partition of the year into periods and time-slots such that R is minimized.

A lower bound for R, RLOW, is computed by constructing the largest (in terms of V(L)) matrix with zero false coincidence. To obtain zero false coincidence, the loadset definition must partition the year into 24 hours and all days. As a consequence, the largest matrix is that representing the overall network busy hour.

3.2 Results of the Empirical Study

A year of point-to-point data was obtained for all hours averaged by week. The process of finding the definition with the least value of the volume ratio (R) was iterative, the results of one definition suggesting the definition for the next iteration. All matrices were examined (not just the largest one) to see which should be split (resolution of periods or time-slots increased) or which could be combined (resolution decreased).

While searching, some useful groundrules for constructing good loadset definitions were found. Chief among these was to keep the overall network peak and the overall second-peak (and third-peak) separate. (The second-peak is the second highest load; the third-peak is the third highest load, etc.) These overall second-peaks are created by individual sub-networks peaking. By keeping the peak and second-peak in separate matrices, the chance of lumping together sub-neto-rks with non-coincident peaks and so creating false coincidence is minimized. This may best be done by examining the profile of network load over time-of-day and period of the year. Fig. 3.1 shows the pattern of peak-month usage for the Telecom Canada network by hour of the day. Fig. 3.2 below shows the pattern of peak-hour usage by month of the year.

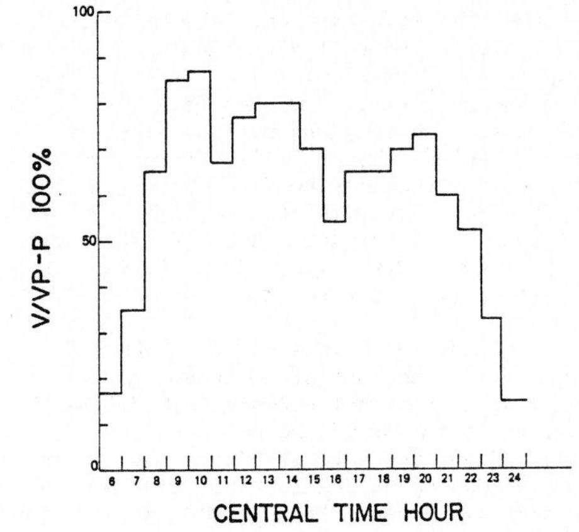

FIG. 3.1 - PEAK BY HOUR (ALL MONTHS)

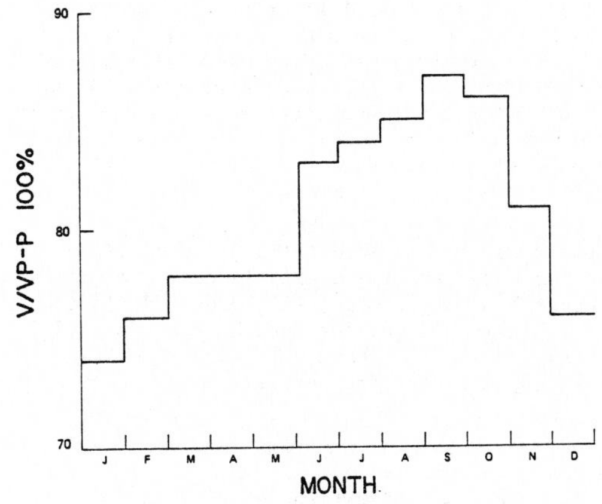

FIG. 3.2 - PEAK BY MONTH (ALL HOURS)

The network as a whole peaks in September. The spring of the year is the only recognizable low period. The profile of usage over hours of the day shows the traditional influence of business traffic on weekday loads - busy during the morning and early afternoon, a pronounced 'lunch hour' and a slight residential peak in the early evening.

The first-cut partitioning was based on an examination of these figures. The ratio of matrix volume to peak-of-peak load for each matrix is given in Fig. 3.3 below.

Fig. 3.3 - RATIO VALUES

	JAN-JUN	JUL-AUG	SEP-OCT	NOV-DEC
2400-1059CT	87.71%	91.39	93.95	86.72
1100-1559CT	83.02	85.90	90.28	85.00
1600-2359CT	74.60	76.00	79.78	78.78

The maximum value of the ratio was 94.0%. For comparison, the value of RLOW is 86.9% based on the network busy-hour of September 1000CT (Central Time).

The next definition was created by splitting the morning period in an attempt to reduce the false coincidence caused by lumping together all morning busy periods. Creating extra time-slots implied that fewer period definitions could be used in order to keep the total number of loadsets to 12. The ratios for the new definition were calculated and examined, new definitions suggested themselves and so on.

The selected loadset definition is given in Fig. 3.4 below.

Fig. 3.4 - Final Loadset Definition

	JAN-AUG	SEP-DEC
2400-0859CT	74.11%	72.17
0900-0959CT	85.68	85.28
1000-1059CT	86.85	87.80
1100-1259CT	79.15	80.05
1300-1659CT	84.50	87.79
1700-2359CT	78.05	79.99

The maximum value is 87.8% which is quite close to the value of RLOW, 86.9%. The impact of using Multi-Hour methods should give much less cost than using the peak-of-peak matrix and just slightly more cost than using only the network busy hour. However, because of their construction, sizing the loadsets will not risk undersizing trunks and compromising service by missing peaks, as would sizing the network using the network busy hour only.

3.3 Conclusions

The final loadset definition has finer resolution of the hours of the day than it does of the months or seasons of the year. This is in line with conventional wisdom about the peakedness of the business day but against common assumptions of the importance of seasonal effects. A definition with 6 time slots but no period split is almost as good as the final 12 loadset definition (maximum ratio = 88.4%). The 6 loadset definition is now being used for ITFS testing because it saves processing time while giving comparable results.

The simple one-dimensional metric used here was extremely valuable in the search for a good loadset definition. It was easy to relate to the goals of the empirical study, was easy to compute, and easy to assess - there was no information overload. The statistic could have been programmed into a search routine that would compute the best partition. This was not done, in our study, since we wanted to observe the results of each iteration. Furthermore, total network cost is not the sole criterion for assessing a definition. This metric (and any other one-dimensional metrics) should be supplemented with information on off-peak volumes, sub-network information etc.

It is important to look at sub-networks, especially when the overall network is dominated by one or two large markets (e.g. cities, states companies). The definition which is best for the overall network may be seriously deficient for sensitive sub-networks, creating false-coincidence and oversizing trunk groups.

4.0 FORECASTING

This section will discuss two different aspects of the forecasting problem: the way ITFS deals with an evolving network architecture, and the selection of an automatic forecasting technique.

4.1 Network Evolution

Trunk forecasting systems cannot ignore network evolution. One of the purposes of these systems is to calculate changes in trunk requirements as the network evolves.

A number of methods to handle evolution were examined. These included: forecasting at the exchange code-to-exchange code level, forecasting at switch-to-switch level and then adjusting for evolution, and re-configuring history to reflect the base architecture before forecasting.

Exchange code-to-exchange code level forecasting was eliminated because of the volume of items to be forecast, and the inaccuracy of forecasts at this level. In the Telecom Canada network there are approximately 5000 exchange codes implying the need to forecast 25 million items. The main source of point-to-point data for ITFS is a sampled base of billed calls. Studies of data from the Telecom Canada network showed that the estimation error for traffic parcels based on this data base varies inversely with the volume of the parcel. When the load for a point-to-point combination is large, the estimated load is quite accurate. When the actual load is small, as in the case of exchange code-to-exchange code level parcels, the variation is considerable. Moreland [7] studied the AT&T sample base from a theoretical viewpoint and found the same behaviour. As exchange code-to-exchange code parcels are aggregated to form entity-to-entity or switch-to-switch parcels, the errors at the low level tend to cancel out, producing more accurate estimates of load. Forecasting at as high a level of aggregation as possible reduces processing cost and improves accuracy.

Forecasting at switch-to-switch level is more accurate because of aggregation but the necessity to adjust base loads for changes in architecture introduces new sources of inaccuracy and complexity. Most adjustment methods are based on the assumption that the two switches involved, the switch-to-switch load, and the parcels being moved are all growing at the same rate. This is assumed either implicitly, by considering statistics like the ratio of parcel demand to total demand, or explicitly, by forecasting the parcel using either the switch-to-switch or end-switch growth rate. With these methods, the forecast of what is subtracted from one switch-to-switch parcel may not equal what is added to another, especially if many parcels and many switch-to-switch combinations are involved. This method may not CLOSE i.e. the forecast total network load, after adjusting for re-arrangements, may not equal the forecast total network load before adjustment. The method itself creates or loses load without regard to actual growth or decline.

Even if a method closes, the estimates of adjustment are typically based on the parcels affected by re-arrangement. These are likely to be highly disaggregated traffic parcels and, as discussed above under exchange code-to-exchange code level forecasting, are likely to be inaccurate.

The third method, which was adopted for ITFS, is to re-configure the demand history to reflect the network architecture in the base year of the forecast. The data is held at exchange code-to-exchange code level. Given the association of exchange codes to entities for the forecast run (recall the definition of entities in Section 2) a new entity-to-entity level history is created. Five years of exchange code-to-exchange code level data is re-configured to form five years of entity-to-entity level data each run. These entity-to-entity parcels are then forecasted and combined to form the switch-to-switch demand for the association of entities to switches of each forecast year. The re-configuration of the exchange code-to-exchange code history to entity-to-entity maintains a consistent data definition for demand forecasting; the aggregation of entity-to-entity forecasts to switch-to-switch forecasts based of future architecture ensures accurate tracking of future network evolution. This method guarantees that the re-arranged estimates close since different switch arrangements reflect only differing associations of entities to switches and that the data are aggregated above the statistically unreliable level of exchange code-to-exchange code.

This method requires large amounts of storage and processing but recent advances in hardware and software which lower processing cost and increase speed allow such methods to be considered as viable alternatives to approximations.

4.2 Automatic Forecasting Techniques

There are many techniques for automatic or 'hands off' forecasting in the statistical literature which are either simple or complex. (See Makridakis and Hibbon [8] for a recent bibliography.) Complex techniques have theoretical appeal, however, there are two problems with most of these methods: they have not been shown to forecast accurately in practice and their sophistication leads to a lack-of-confidence in their users.

The forecast accuracy of competing techniques was investigated by Makridakis et al [9]. The data being forecasted were mostly macro-economic series mixed with some conpany-level revenue and sales series. The results showed that simple techniques like linear regression and exponential smoothing out-performed their more complicated rivals.

When the forecaster lacks confidence in an automatic method he/she usually responds by overriding. The mechanical forecast is felt to be deficient and it must be corrected with an override. This effect has been ovserved in inventory forecasting systems (see [10]).

This almost constant overriding makes an automatic system redundant and degrades the forecast accuracy of the whole system. Other empirical studies have shown that 'seat-of-the-pants' forecasts are not as accurate as those produced by statistical methods.

The methods chosen for ITFS were simple time-based linear regression and exponential regression. The linear model assumes that point-to-point load grows by a constant number of erlangs (CCS, minutes etc.) every year. The exponential model assumes that point-to-point load grows at a constant rate every year (e.g. 10%). Both models are fit to the history and their predictive performance over the historical period are compared. The model with the best performance is selected for forecasting. (A simple one-pass check for outliers is performed in order to eliminate gross data errors or distortions in the series.)

Just because simple techniques work well on the majority of data series does not mean they work well in every case. The forecasting module includes override capability for the forecasters. This enables them to reconcile the forecast to economic and demographic trends, and market stimulations. It also enables them to override the effects of unusual occurrences, such as strikes in the postal system, disasters, and unusual weather conditions which result in one-time increases of demand for telephone service.

Tracking of growth rates produced by the automated system against actuals measured in succeeding years is also provided. This will provide information to fine-tune the projection algorithms and to keep track of the accuracy of overrides provided by the forecasters. We expect that the automatic forecasting feature will provide manpower savings, free the forecasters to spend additional time on the examination of economic and demographic effects, and provide a more accurate overall forecast.

5.0 SUMMARY

In this paper we have illustrated the tangible application in ITFS of some of the methodologies and techniques for forecasting and dimensioning presented in the recent literature. New methodologies have been presented for the construction of traffic demand matrices for multi-hour engineering. The concept of re-configuration of historical demands for changes in network architecture has also been introduced.

REFERENCES

[1] R. Huberman, S. Hurtubise, A. LeNir, T. Drwiega, "Multi-Hour Dimensioning for a Dynamically Routed Network", Eleventh International Teletraffic Congress, Kyoto, Japan, 1985.

[2] W.A. Cameron, J. Regnier, P. Galloy, A.M. Savoie, "Dynamic Routing for Intercity Networks", Tenth International Teletraffic Congress, Montreal, Canada, 1983.

[3] W.A. Cameron, P. Galloy, W.J. Graham, "Report on the Toronto Advanced Routing Concept Trial", Telecommunications Networks Planning Conference, Paris, 1980.

[4] M.E. Lavigne, J.R. Barry, "Administrative Concepts in an Advanced Routing Network", Ninth International Teletraffic Congress, Torremolinos, Spain, 1979.

[5] E. Szybicki, M.E. Lavigne, "The Introduction of an Advanced Routing System into Local Digital Networks and its Impact on the Networks' Economy, Reliability, and Grade of Service", International Switching Symposium Paris, 1979.

[6] R.S.Krupp, "Properties of Kruithof's Projection Method", Bell Syst. Tech. J., vol. 58, no. 2, pp 517-538, 1979.

[7] J.P. Moreland, "Estimation of Point-to-Point Telephone Traffic", Bell Syst. Tech. J., vol. 57, no. 8, pp 2847-2863, 1978.

[8] S. Makridakis, M. Hibon, "Accuracy of Forecasting: An Emperical Investigation", J. of Royal Statistical Society, vol. 142, part 2, pp 97-145, 1979.

[9] S. Makridakis et al, "The Accuracy of Extrapolation (Time Series) Methods: Results of a Forecasting Competition", J. of Forecasting, vol. 1, no. 2, pp 111-152, 1982.

[10] B.T. Smith, "Focus Forecasting - Computer Techniques for Inventory Control", CPI Publishing, Boston, 1978.

TELETRAFFIC ISSUES in an Advanced Information Society
ITC-11
Minoru Akiyama (Editor)
Elsevier Science Publishers B.V. (North-Holland)

TRAFFIC FORECASTING MODELS BASED ON TOP DOWN AND BOTTOM UP PROCEDURES

Kjell STORDAHL and Lars HOLDEN

Norwegian Telecommunications Administration, Oslo, Norway

Norwegian Computing Center, Oslo, Norway

ABSTRACT

The object of the present paper is to describe methods for making forecasts of the national traffic between areas in a country or of international traffic between countries.

The performance of two different methods: Kruithof's method and a new weighted least squares method for splitting the aggregated traffic is analysed. Simulations show that if the structure of the traffic matrix to some extent changes, then the weighted least squares method should be preferred.

A number of forecasting models is examined on aggregated level. The "Airline model" turned out to be the best common model for all time series. For short term forecasting introduction of explanatory variables did not improve the performance.

1. INTRODUCTION

New telecommunication exchanges are able to register not only the traffic through the exchange, but also the volume of traffic to the different destinations. These data form the basis for both short term and longer term traffic forecasts for dimensioning and network planning.

The forecasts can be regarded as a matrix, where each element is the forecast of the traffic from one exchange to another. Usually it is possible to forecast the total incoming and outgoing traffic by some direct method as well, and with a better precision than for the individual point-to-point forecasts. In practical use of the forecasts it is necessary that the forecasts of the matrix elements are consistent with the forecasts of the total incoming and outgoing traffic. It is therefore necessary to adjust the forecasts of the elements and/or the forecasts of the totals so that the row and column sums of the forecast matrix equal the forecasts for the totals.

One way to construct and adjust these forecasts is the well known Kruithof's method [1]. In this paper we propose to use a weighted least squares (WLSQ) method, as an alternative. The advantage of the WLSQ is its ability to use more information than the Kruithof's method. It will be possible to take special characteristics of each element into consideration. To be able to perform the WLSQ, separate forecasts for each element in the matrix must be available. We discuss some methods for constructing these forecasts.

We are comparing the two methods, to see when one method is superior to the other. The answer depends on the amount of information on each element.

2. THE KRUITHOF'S METHOD

The Kruithof's method is based on forecasts for the total volume of incoming and outgoing traffic in each exchange. Some methods for constructing such forecasts are presented in a separate paragraph. These forecasts are then used to adjust the last known traffic matrix so that the row/column sums are correct. Kruithof's method is defined in the following way:

Let $\{A_{i,j,0}\}$ be the last known traffic matrix and $B_{i,\cdot,t}$, $B_{\cdot,j,t}$ be forecasts for the row/column sums of the matrix one timestep ahead. The Kruithof's estimates are then defined as the matrix $\{B_{i,j,t}\}$ satisfying

$$B_{i,j,t} = E_{i,t} F_{j,t} A_{i,j,0},$$

$$B_{i,\cdot,t} = \sum_j B_{i,j,t} \text{ and } B_{\cdot,j,t} = \sum_i B_{i,j,t} \quad (2.1)$$

for some arrays E and F. The Kruithof's method is uniquely defined even though the arrays E and F are not. The method of calculating the Kruithof's matrix has been improved [2].

The Kruithof's method is not able to treat elements in the matrix individually. The method is therefore not able to use additional information on separate elements.

Occasionally, some elements are not known in the previous traffic matrix. It is then necessary to insert estimates for the missing elements. But this estimates should not be given the same weight as the known values for the other elements.

Another and more serious disadvantage is that prior knowledge may lead one to expect a larger increase in some elements than in others. This is not possible to handle in Kruithof's method.

Kruithof's method is only using one (usually the last) known traffic matrix. When many previous traffic matrices are known not only one should be used. This is very important if seasonal variation is present.

Kruithof's method uses the previous traffic matrix as a forecast for the next traffic matrix. It will often be possible to construct better forecasts for each element. At least one will have more confidence in some forecasts than the others. If the uncertainty varies, this information should be taken into consideration.

3. WEIGHTED LEAST SQUARES METHOD

In this paper we propose to use weighted least squares to construct consistent traffic matrix forecasts.
Let $C_{i,j,t}$ be a forecast for the traffic from exchange i to exchange j, and let $C_{i,.,t}$ and $C_{.,j,t}$ be forecasts for the total outgoing traffic from exchange i and incoming traffic to exchange j, respectively. $C_{i,.,t}$ and $C_{.,j,t}$ are equivalent to $B_{i,.,t}$ and $B_{.,j,t}$ in the last paragraph. We describe some methods to construct these forecasts in paragraph 5 and 6.

We define the weighted sum of squares Q(D,C) as

$$Q(D,C) = \sum_{i,j} a_{i,j}(C_{i,j,t} - D_{i,j,t})^2$$

$$+ \sum_i b_i(C_{i,.,t} - D_{i,.,t})^2$$

$$+ \sum_j c_j(C_{.,j,t} - D_{.,j,t})^2 \qquad (3.1)$$

We then define the weighted least squares forecast as $D = \{D_{i,j,t}\}$ which minimize Q(D,C) with respect to D under the constraints

$$D_{i,.,t} = \sum_j D_{i,j,t} \quad \text{and}$$

$$D_{.,j,t} = \sum_i D_{i,j,t} \qquad (3.2)$$

A natural choice of the weightes $a_{i,j}$, b_i and c_j is the inverse of the variances of the estimators $D_{i,j,t}$, $D_{i,.,t}$ and $D_{.,j,t}$. The solution of equation (3.1) under the constraints (3.2) is found using Lagranges multiplicator method.

The advantage of WLSQ is that all available information is used. Some of the information is contained in the forecasts and the knowledge about the forecast precision is accounted for in the weights.

One disadvantage of WLSQ is that if the forecasts for the row/column sum are very far from consistent with the forecasts for each element, and some elements are very small compared to others, the method may give negative forecasts for some elements. We do not believe this to be any practical problem.

4. COMPARISON OF THE TWO METHODS

4.1 Model assumptions

It is difficult to compare the two methods for forecasting traffic matrices. The result will depend heavily on the progress of the "true" matrix from one timepoint to the next, and on how precisely it is possible to forecast each element in the matrix and the row/column sums. The objective of this paper has been to give a rule of thumb as to when one method is superior to the other. The evaluation is done by using numerical simulation.

We have to notice an important difference between the two methods. In the Kruithof's method the row/column sums remain unaltered during the adjustment procedure. In the weighted least squares method these sums are changed in the same way as the other forecasts according to the uncertainty in each forecast.

In Table 4.1 a typical traffic matrix (5x5) for parts of the Norwegian telephone network is given as a start matrix at time t=0. Other simulation runs have shown that the conclusions do not depend significantly on the start traffic matrix nor on the size of the matrix.

Table 4.1 The traffic matrix used in the simulations

1.2	1.5	3.2	4.2	1.3
5.8	2.1	10.3	8.5	2.0
.7	4.8	2.3	5.2	1.0
3.1	2.0	5.1	8.7	2.0
1.9	5.6	6.1	7.2	3.1

In the simulations, we generate the "true" traffic matrix one step ahead from a known traffic matrix. Each element is generated by the same stocastic formula.

Let $A_0 = \{A_{i,j,0}\}$. The traffic matrix one step ahead, at time t, is denoted $A_t = \{A_{i,j,t}\}$. In the simulations A_t is generated by the formula

$$A_{i,j,t} = \alpha_{i,j} + \gamma_{i,j} A_{i,j,0}, \quad \text{where}$$

$$\gamma_{i,j} = e_i f_j + \omega_{i,j} A_{i,j,0} \qquad (4.1)$$

Here $\alpha_{i,j} \sim N(0,\sigma_{\alpha_{i,j}})$ and $\omega_{i,j} \sim N(0,\sigma_{\omega_{i,j}})$, and e and f are fixed arrays. The notation N(.,.) means that the variable is normally distributed with the given mean and standard deviation.

We then need some forecasts for A_t. Because we want to separate the influence of various forecasting methods from the test of the two adjustment methods, we assume that we have available unbiased forecasts $C_{i,j,t}$, $C_{i,.,t} = B_{i,.,t}$ and $C_{.,j,t} = B_{.,j,t}$ for $A_{i,j,t}$, $A_{i\,.,t}$ and $A_{.,j,t}$ respectively.

More precisely, we assume

$$C_{i,j,t} - A_{i,j,t} \sim N(0,(A_{i,j,0}(1-\delta)+\delta)\sigma_{i,j})$$

$$C_{i,.,t} - A_{i,.,t} \sim N(0,A_{i,.,0}\sigma_{i,.}) \text{ and}$$

$$C_{.,j,t} - A_{.,j,t} \sim N(0,A_{.,j,0}\sigma_{.,j}) \qquad (4.2)$$

where δ is equal to 1 or 0 depending on σ_α and σ_ω

In each simulation run no. n, n=1,2,...,N, the "true" traffic matrices

$A_t^n = \{A_{i,j,t}^n\}$ are generated, while the

Kruithof's estimates $B_t^n = \{B_{i,j,t}^n\}$ and the

WLSQ estimates $D_t^n = \{D_{i,j,t}^n\}$ are found.

The normes used in the tests are

$$KR = \frac{1}{N}\sum_{n=1}^{N}\sqrt{(\frac{1}{M^2}\sum_{i,j}(A_{i,j,t}^n - B_{i,j,t}^n)^2)} \qquad (4.3)$$

and

$$LS = \frac{1}{N}\sum_{n=1}^{N}\sqrt{(\frac{1}{M^2}\sum_{i,j}(A_{i,j,t}^n - B_{i,j,t}^n)^2)} \qquad (4.4)$$

where M is the number of exchanges in the network. Other norms for the forecasting precision have also been examined, without significantly different results.

In all these tests, we have done at least 100 simulations.

4.2 Simulation results based on exact forecasts for the row and column sums

In the first simulations, test one, we assume that $\sigma_{i,.} = \sigma_{.,j} = 0$ for all i and j, so that the marginals are forecasted exactly. The reason for this is the different handling of row/column sums in the two methods. The Kruithof's method does not change the forecast for the row/column sums, while the WLSQ method does. In this way we are able to separate the consequences of the different handling of these forecasts from the rest of the method. When the marginals are assumed to be forecasted exactly, it is natural to change the Q(D,C) in the least squares method to

$$Q(D,C) = \sum_{i,j} a_{i,j}(C_{i,j,t} - D_{i,j,t})^2$$

Then the parameters are chosen in the following way: $\sigma_{\omega_{i,j}} = 0$, $e_i = f_j = 1$

for all i and j, and $\delta = 0$. The rest of the parameters are given in Table 4.2.

Table 4.2 Test 1. Forecasting precision of Kruithof's method and WLSQ as a function of σ_ω and $\sigma_{i,j}$

σ_ω	.04	.08	.12	.16	.20
KR	.12	.22	.32	.42	.52

$\sigma_{i,j}$	.02	.04	.06	.08	.10	.12	.14	.16
LS	.083	.14	.20	.25	.32	.37	.44	.50

The forecasts for each element in the traffic matrix have the same standard deviation. $\sigma_{i,j} < \sigma_{\omega_{i,j}}$ means that we are able to construct a forecast for element i,j which is superior to the value from the previous matrix. The smaller $\sigma_{i,j}$ is compared to $<\sigma_{\omega_{i,j}}$ means that the better is WLSQ compared to Kruithof's method. When they are approximately of the same size, Kruithof's method is a little better than WLSQ.

The only change in the parameters from test one to test two is that e_i and f_j are no longer all 1, but is given as in Table 4.3.

Table 4.3 The e and f arrays in test 2

i/j	1	2	3	4	5
e_i	1.02	1.04	1.03	1.05	1.03
f_j	1.04	1.07	.99	1.01	1.06

When e_i and f_j are not all 1, one should expect Kruithof's method to perform almost exactly as in the previous test. This is because the Kruithof's method is exact for all e and f arrays, when $\sigma_\omega = 0$. The results in this test confirm this assumption. In this test we got the same results for WLSQ as in test one. The reason for this is that the estimates are the same in both tests.

Table 4.4 Results from test 2

σ_ω	.04	.08	.12	.16	.20
KR	.12	.22	.32	.42	.52

In the third simulation (see Table 4.5) we have chosen varying precision in the forecasts for each element. Thus $\sigma_{i,j}$ is chosen at random uniformly in the interval $(\sigma-.02,\sigma+.02)$.

Table 4.5 Results from test 3

σ	.02	.04	.06	.08	.10	.12	.14	.16
LS	.081	.14	.19	.25	.31	.37	.44	.50

The results from the WLSQ method are almost the same as in test 1. This means that when $\sigma_{i,j}$ is varying uniformly in an interval, the results are almost the same as when $\sigma_{i,j}$ is constant and equal to the midpoint in the interval.

In simulation four (see Table 4.6) we let $\alpha_{i,j}$ is vary. Because $\alpha_{i,j}$ is not multiplied with $A_{i,j,0}$, we use $\delta=1$.

Then it is possible to compare $\sigma_{\alpha_{i,j}}$ and $\sigma_{i,j}$. In this test we let $\sigma_{\omega_{i,j}}=0$. The rest of the parameters are defined in Table 4.6.

Table 4.6 Results from test 4

σ_α	.04	.08	.12	.16	.20
KR	.054	.078	.10	.13	.16

$\sigma_{i,j}$	.02	.04	.06	.08	.10	.12	.14	.16
LS	.043	.052	.063	.074	.087	.098	.11	.12

In this test WLSQ is better when $\sigma_{i,j} < \sigma_{\alpha_{i,j}}$ or the two variables are approximately of the same size. This is a very interesting result. In test 1-3 the stocastic parameter ω was multiplied with $A_{i,j,0}$. This is near Kruithof's formula (2.1). In test 1-3 Kruithof's method was a little better, when we had the same uncertainty in the forecasts compared to the precision of A_0 when used as an estimate for A_t. In this test the stocastic variable α is added to $A_{i,j,0}$. This is why WLSQ is better in this test when the uncertainty is the same in the forecasts used by WLSQ as when A_0 is used as a forecast for A_t.

4.3 Simulation results when the row and column sums are stocastic

We finally perform exactly the same tests as before, with the exception that the row and column sums now are stocastic with $\sigma_{\cdot,j} = \sigma_{i,\cdot} = \cdot$. This is more realistic than the previous test, as there are no reason to believe that the total future traffic will be known exactly.

Table 4.7 Test 5. Forecasting precision of Kruithof's method and WLSQ as a function of σ_ω and $\sigma_{i,j}$.

σ_ω	.04	.08	.12	.16	.20
KR	.25	.30	.39	.47	.56

$\sigma_{i,j}$	.02	.04	.06	.08	.10	.12	.14	.16
LS	.14	.22	.28	.35	.39	.45	.50	.56

The error is naturally a little larger in this test than in test 1, where the exact row and column sums were known. But the difference is decreasing with increasing σ_ω and $\sigma_{i,j}$.

In this test, as in test 1, WLSQ is the better when $\sigma_{i,j} < \sigma_\omega$.

Table 4.8 Results from test 6

σ_ω	.04	.08	.12	.16	.20
KR	.24	.30	.37	.45	.54

In test 6, the e and f arrays are not equal to 1. Just as in test 1 and 2, we see that the Kruithof's method is exactly as good as when the e and f arrays are equal to 1.

Table 4.9 Results from test 7

σ	.02	.04	.06	.08	.10	.12	.14	.16
LS	.17	.23	.28	.35	.40	.45	.50	.58

In test 7, $\sigma_{i,j}$ is chosen at random from a uniform distribution on the interval $(\sigma-.02,\sigma+.02)$. Again, the results are very close to the results from test 5. This means that when the standard deviation is varying uniformly in an interval, then WLSQ is approximately as good as when the standard deviation is fixed in the middle of the interval.

Table 4.10 Result from test 8

σ_α	.04	.08	.12	.16	.20
KR	.22	.22	.23	.25	.26

$\sigma_{i,j}$	.02	.04	.06	.08	.10	.12	.14	.16
LS	.12	.18	.20	.23	.25	.25	.26	.28

In test 8 the stocastic element is added to each element in A_0. When σ_α and $\sigma_{i,j}$ are small then the dominating error is in the row and column sums. The Kruithof's method, which does not take the uncertainty in each forecast into consideration, is not able to use this information. The error is almost as large for $\sigma_\alpha = .04$ as for .12.

In addition, we notice that WLSQ is better when it is possible to construct a better forecast for each element than the corresponding element in the previous matrix.

4.4 Conclusion

In this paper we have presented a weighted least squares method (WLSQ) for constructing consistent forecasts for traffic matrices. We have compared it with Kruithof's method, using simulation techniques. The simulations show that if we are able to give forecasts for each element which are better forecasts than the value in the last known matrix, then we should use weighted least squares. If we are unable to construct forecasts which are better than the values in the previous traffic matrix, then we should use the Kruithof's method when we expect the change to be close to the Kruithof's formula (4.5), else WLSQ should be used.

5. TRAFFIC FORECASTING MODEL BASED ON TIME SERIES ANALYSIS

5.1 Different methods

A number of different methods can be used to forecast the traffic on aggregated and local levels. The most relevant methods are:

- Time series models
- Kalmanfilter models
- Regression models
- Econometric models

The Kalmanfilter models are used by administrations in USA and are described in a number of papers, among others in [3], [4] and [5]. In this paper only the time series models will be examined.

Identification and estimation of the univaiate models resulted in different model structures in some cases. This is a drawback in practical situations. Therefore analysis were carried out to find the best common model for all time series.

The paper presents a comparison between the best univariate model, the best common univariate model and models based on number of main stations (subscribers) as explanatory variables.

5.2 The best univariate model

In the Norwegian telephone network traffic measurements are carried out four times a year. Quarterly observations of incoming and outgoing traffic from six groups exchanges for the last eight years have been studied. By using time series analysis the best univariate forecasting models were developed.

Let Y_t be the traffic at time t, B the backward shift operator, $\{\theta_i\}$ and $\{\psi_i\}$ parameters and a_t white noise at time t. Then the best univariate models are expressed by:

Outgoing traffic

$$\text{Sandnes} \quad (1-b)(1-\psi_1 B)Y_t = \theta_0+(1-\theta_1 B)(1-B^4)a_t$$
$$\text{Narvik} \quad (1-B)Y_t = \theta_0+(1-\theta_1 B)a_t$$
$$\text{Mandal} \quad (1-B)(1-B^4)Y_t = (1-\theta_1 B)(1-\theta_4 B^4)a_t$$
$$\text{Sandefjord} \quad (1-B)(1-\psi_1 B)Y_t = (1-\theta_1 B)a_t$$
$$\text{Tvedestrand} \quad (1-B)(1-\psi_1 B)Y_t = (1-\theta_4 B^4)a_t$$
$$\text{Kragerø} \quad (1-B)(1-B^4)Y_t = (1-\theta_1 B)(1-\theta_4 B^4)a_t \qquad (5.1)$$

Incoming traffic

$$\text{Sandnes} \quad (1-\psi_1 B)(1-B)(1-B^4)Y_t = a_t$$
$$\text{Narvik} \quad (1-B)Y_t = \theta_0+(1-\theta_1 B)(1-\theta_4 B^4)a_t$$
$$\text{Mandal} \quad (1-B^4)Y_t = \theta_0+(1-\theta_1 B)(1-\theta_4 B^4)a_t$$
$$\text{Sandefjord} \quad (1-B^4)Y_t = (1-\theta_1 B)(1-\theta_4 B^4)a_t$$
$$\text{Tvedestrand} \quad (1-B)(1\ B^4)Y_t = (1-\theta_1 B)(1-\theta_4 B^4)a_t$$
$$\text{Kragerø} \quad (1-B)(1-B^4)Y_t = (1-\theta_1 B)(1-\theta_4 B^4)a_t \qquad (5.2)$$

Only in a few of the models the log transform of the data turned out to give a better fitting.

5.3 The best common univariate model

The data base for network planning consists of a large number of time series. It would be extremely time-consuming to analyse every time series in order to find the best model. For such analysis it is also necessary to use special experts in this field. The most relevant solution in a practical situation will be to identify the best common model for the time series and then estimate the parameters in the models separately.

Analysis have shown that the so-called "Airline model" [6]:

$$(1-B)(1\ B^4)Y_t=(1-\theta_1 B)(1-\theta_4 B^4)a_t \qquad (5.3)$$

is the best common model for outgoing traffic. The results are similar to forecasting models developed in [7]. Table 5.1 contains the estimated parameter values and the Portmanteau test for the different time series. The Portmanteau test Q is based on estimation of 24 autocorrelation coefficients and has 22 degrees of freedom.

Table 5.1 Outgoing traffic. Estimated parameter values and the Portmanteau test

Exchange	$\hat{\theta}_1$	$\hat{\theta}_4$	Q_{22}
Sandnes	0.72	1.03x	37.6
Narvik	0.76	0.61	12.3
Mandal	0.55	0.76	20.3
Sandefjord	0.43	0.73	16.9
Tvedestrand	0.69	0.91	16.2
Kragerø	0.32	0.46	19.3

x Insignificant value.

As for outgoing traffic the best common model for incoming traffic is the Airline mode given by equation (5.3). The estimated parameters and the Portmanteau test are given in Table 5.2.

Table 5.2 Incoming traffic. Estimated parameter values and the Postmanteau test

Exchange	θ_1	θ_4	Q_{22}
Sandnes	-0.45	0.67	12.7
Narvik	0.41	0.55	20.8
Mandal	0.43	0.81	35.6
Sandefjord	0.55	0.86	19.8
Tvedestrand	0.61	0.63	15.7
Kragerø	0.79	0.17	11.3

5.4 Traffic models based on number of main stations

Explanatory variables like telephone tariffs, consumers price index, number of main stations may improve the forecasting models. In Norway the number of main stations within an exchange area has been the most significant variable during the last years. The telephone tariffs and the consumers price index have not affected the traffic in the same way because their relative increase is about the same.

On aggregated level the number of main stations is included as an explanatory variable. The first step has then been to make one model for the traffic per main station and then another for the number of main stations within the exchange area. The last model is used as an input to the first one.

Analysis have shown that the best univariate models are expressed by:

Outgoing traffic per main station

Sandnes	$Y_t-\mu = (1-\theta_4B^4)a_t$
Narvik	$Y_t-\mu = (1-\theta_1B-\theta_2B^2-\theta_3B^3)a_t$
Mandal	$(1-\psi_1B)(1-B^4)Y_t = (1-\theta_1B)(1-\theta_4B^4)a_t$
Sandefjord	$(1-B)(1-B^4)Y_t = (1-\theta_1B)(1-\theta_4B^4)a_t$
Tvedestrand	$(1-B^4)Y_t = (1-\theta_1B)(1-\theta_4B^4)a_t$
Kragerø	$1-\psi_1B)(1-B^4)Y_t = (1-\theta_4B^4)a_t$

(5.4)

Incoming traffic per main station

Sandnes	$(1-B)(1-B^4)Y_t = (1-\theta_1B)(1-\theta_4B^4)a_t$
Narvik	$(1-B)(1-B^4)Y_t = (1-\theta_1B)(1-\theta_4B^4)a_t$
Mandal	$(1-B^4)Y_t = (1-\theta_1B)(1-\theta_4B^4)a_t$
Sandefjord	$(1-B^4)Y_t = (1-\theta_1B)(1-\theta_4B^4)a_t$
Tvedestrand	$(1-B)(1-B^4)Y_t = (1-\theta_1B)(1-\theta_4B^4)a_t$
Kragerø	$(1-B)(1-B^4)Y_t = (1-\theta_1B)(1-\theta_4B^4)a_t$

(5.5)

Also for outgoing and incoming traffic per main station analysis have shown that the best common model is the Airline model given in (5.3).

The development of the number of main stations in an exchange area is well fitted by the Airline model. However, in some cases there are only annual observations available. The missing intermediate observations are then estimated by an interpolation procedure which obviously may affect the modelling. The best common model in such situations has turned out to be:

$$(1-B)^2\chi_t = (1-\theta_4B^4)a_t \qquad (5.6)$$

where χ_t is the number of main stations at time t.

6. COMPARISON OF THE VARIOUS FORECASTING METHODS

6.1 Methods for comparison

For evaluation of the different forecasting methods it is necessary to introduce a comparison measure. The root mean square error (RMSE) compares the forecasts with the true observations.

In all time series the last m=4 observations are removed before the modelling procedure. Then the forecasting model is used to forecast the traffic m steps ahead in order to compare the forecasts with the true observations. The root mean square error is given by:

$$\text{RMSE} = \sqrt{\frac{1}{m}\sum_{\ell=1}^{m}(Y_{n+\ell}-\hat{Y}_n(\ell))^2} \qquad (6.1)$$

where n is the actual time before forecasting, $Y_{n+\ell}$ the true value at time $n+\ell$, and $\hat{Y}_n(\ell)$ the forecast ℓ steps ahead.

6.2 Forecasting models for outgoing traffic

The root mean square error is calculated for:

- the best univariate models (BU), given by equation (5.1)
- the best common univariate model (BC), given by equation (5.3) and Table 5.1
- the best univariate model per main station given by equation (5.4), combined with the best common model for number of main stations (BUM)
- the best common univariate model per main station given by equation (5.3) and Table 5.3, combined with the best common model for number of main stations (BCM).

The results are presented in Table 6.1.

Table 6.1 Root mean square error for different forecasting models for outgoing traffic in Erlang

Model \ Exchange	Sand-nes	Narvik	Mandal	Sande-fjord	Tvede-strand	Kragerø	Mean
BU	3.9	3.0	1.1	3.5	2.8	9.8	4.0
BC	13.4	3.1	1.1	3.5	2.6	9.8	5.6
BUM	7.8	3.7	1.8	8.0	5.3	7.1	5.6
BCM	6.4	2.3	1.4	8.0	3.5	8.9	5.0

6.3 Forecasting models for incoming traffic

The root mean square error is calculated for:

- the best univariate models (BU), given by equation (5.2)
- the best common univariate model (BC), given by equation (5.3)
- the best univariate model per main station given by equation (5.5), combined with the best common model for number of main stations (BUM)
- the best common univariate model per main station given by equation (5.3), combined with the best common model for number of main stations (BCM).

The results are presented in Table 6.2.

Table 6.2 Root mean square error for different forecasting models for incoming traffic in Erlang

Model \ Exchange	Sand-nes	Narvik	Mandal	Sande-fjord	Tvede-strand	Kragerø	Mean
BU	37.3	0.3	0.9	6.6	1.4	4.5	8.5
BC	33.0	1.3	0.8	2.6	1.4	4.5	7.3
BUM	35.2	1.6	1.3	4.7	1.8	4.6	8.2
BCM	35.6	1.6	1.1	4.1	1.8	4.6	8.1

6.4 Evaluation of the forecasting models

Comparison between the best univariate models and the best common univariate model for both outgoing and incoming traffic shows that there is a small difference in the behaviour of the models. Both types of models perform very well except for Sandnes group exchange, where the traffic increased unexpectedly for the last four observations. Since there is only a small difference in the performance between the models, the conclusion is that the best common model is used for forecasting.

The root mean square error is based on four observations representing a time period of one year. Hence, the conclusions are valid for short term forecasting.

The results in Tables 6.1 and 6.2 show that introduction of the number of main stations as an explanatory variable does not improve the performance of the forecasting models to any great extent. This means that the best common univariate model, the Airline model, given by equation (5.3), is preferred for short term forecasting of outgoing and incoming traffic.

The outgoing and incoming traffic represents the column sum and the row sum respectively, in the traffic matrix. Concerning the traffic elements within the matrix, it is not recommended to have more complicated models for short term forecasting. The traffic elements will among other things depend on the geographical distance between the exchanges and the number of main stations in the exchange areas. But the fact that models without explanatory variable are preferred on higher levels indicates that models of no more than the same complexity should be used on the lower levels for short term forecasting.

For long term forecasting additional studies have to be carried out in order to examine the effect of using explanatory variables.

7. ACKNOWLEDGEMENT

The authors thank Eivind Damsleth for helpful discussions and exchange of view in preparing this paper.

REFERENCES

[1] J. Kruithof, "Telefoonverkeers-rekening". De Ingenieur 52, No 8, 1937.

[2] R.S. Krupp, "Properties of Kruithof's projection method". The Bell System Technical Journal, Vol. 58, No 2, 1979.

[3] A.J. David, C.D. Pack, "Time projection algorithm: A new improved forecasting procedure". ITC 9, Torremolinos, 1979.

[4] C.D. Pack, B.A. Whitaker, "Kalmanfilter models for network forecasting". The Bell System Technical Journal, Vol. 61, No 1, 1982.

[5] J.P. Moreland, "A robust sequential projection algorithm for traffic load forecasting". The Bell System Journal, Vol. 61, No 1, 1982.

[6] G.E.P. Box, G.M. Jenkins, "Time Series Analysis. Forecasting and Control". Holden Day, 1970.

[7] A. Passeron, E. Eteve, "A fore-casting model to set up multi season traffic matrices". ITC 10, Montreal 1983.

TELETRAFFIC ISSUES in an Advanced Information Society
ITC-11
Minoru Akiyama (Editor)
Elsevier Science Publishers B.V. (North-Holland)

BAYESIAN FORECASTING WITH MULTIPLE STATE SPACE MODEL

Takeo ABE and Hiroshi SAITO

Musashino Electrical Communication Laboratory, NTT,
Tokyo, Japan

ABSTRACT

In this paper, we propose a new forecasting method which we call Bayesian forecasting with multiple state space model. This method increases the robustness of forecasting, that is, it has the ability to adapt to the various situations. A multiple state space model is composed of several state space models called sub-models. Its forecasting value is given by the weighted summation of each sub-model's forecasting value, which is calculated individually by means of a Kalman filter. The weight of each sub-model takes a value proportional to Bayesian posterior probability, which is calculated from the likelihood. A good fitting sub-model's posterior probability increases as the number of observations increases. Prior to forecasting, the initial state and noise variances of each sub-model are needed in order to use the Kalman filter. In this paper these parameters are estimated by numerical maximization of the likelihood concerning these parameters. Examples of how this method may be applied to monthly telephone revenue data and trunk group load data are given, demonstrating the ability of the method for one-step ahead forecasting. The transitions of posterior probabilities of sub-models are also shown.

1. INTRODUCTION

This paper presents a forecasting method using multiple state space models. Many papers on traffic forecasting with state space models have been published [1]-[4]. A state space model has the advantage of being able to give a sequential projection with a Kalman filter, i.e. forecasting is possible even when only a small amount of data is available. The main difficulty of using a state space model in traffic forecasting is the building of the models themselves. In general, the structure of the social phenomena which moves dynamically (e.g. traffic trend), is unclear and likely to change and it is difficult to grasp it precisely. And it is not appropriate to apply a one state space model to forecast at all times. The varying trends can be represented more precisely when various models are constructed and the best of them is selected for forecast as the data is obtained. For example, random walk models are insensitive and tough for exceptional values, and therefore the robustness of a multiple state space model can be expected to increase when a random walk model is added to it as sub-model.

Forecasting with multiple state space models as a whole is based on this line of reasoning. In this method, several state space models for forecasting are first prepared as sub-models (trend model and random walk model are considered sub-models). Next, the observation time series is divided into appropriate time segments, which are the set of observation time series, and the likelihood of each sub-model for each time segment is calculated. If the prior probability for each sub-model is initially given, for example an ignorant prior (the prior probabilities of all the sub-model are equal), the posterior probability can be calculated from the prior probability and the likelihood of each sub-model according to the principles of Bayesian statistics [5]. When the posterior probability of each sub-model is obtained, the forecasting value of a multiple model can be obtained by a weighted summation of the forecasting value of each sub-model. The weights of the sub-models are obtained from their posterior probabilities. There are several advantages to the forecasting method proposed in this paper, e.g.,

(i) We can dynamically follow the variations in the target by calculating each sub-model's likelihood for each segment of the observation time series.

(ii) This method is robust to exceptional values when an insensitive sub-model is included.

(iii) This method provides accurate forecasting if at least one of the sub-models describes the structure of the forecasting target well.

Results obtained when applying this forecasting method to telephone revenue data which includes tariff changes and trunk group load data which is likely to change will be provided in a later section.

2. FORECASTING METHOD

Let us consider a multiple state space model composed of n kinds of state space models as sub-model.

2.1 State Space Model

M_j is the j-th state space model (j=1,2,....,n) and is described by a transition equation

$$x_j(t+1) = F_j x_j(t) + G_j u_j(t) \quad (1)$$

where,

$x_j(t)$ is a vector of the state at time t.

F_j is a system matrix.

G_j is a driving matrix.

$u_j(t)$ is system noise at time t.

y(t) denotes an observation value at time t (t=1,2,.....) and y(t) is assumed to be represented by an observation equation

$$y(t) = H_j' x_j(t) + v_j(t) \quad (2)$$

where,

H_j is an observation matrix.

$v_j(t)$ is observation noise at time t.

F_j, G_j and H_j are assumed to be known matrices. H_j' is a transpose matrix of H_j. $u_j(t)$ and $v_j(t)$ are assumed to be mutually independent Gaussian white noise distributed with mean zero and variance Q_j and R_j, respectively.

Next the observation data series $Y(1,\cdot)$ = [y(t) ; t=1,2,....] is divided into several time segments.

$$Y(t_{i-1}+1, t_i) = [y(t) \; ; t=t_{i-1}+1, t_{i-1}+2, .., t_i]$$

where $t_0=0$.

In Ref.[7] the conditional likelihood of model M_j for the i-th segment $Y(t_{i-1}+1, t_i)$ is defined as

$$L_{ij} = L_{ij}(Y(t_{i-1}+1, t_i) \mid Y(1, t_{i-1}), \theta_j, x_j(0))$$

$$= \prod_{t=t_{i-1}+1}^{t_i} f_j(y(t) \mid Y(1,t-1)), \quad (3)$$

where $f_j(y(t) \mid Y(1,t-1))$ is the conditional probability density function of y(t) given the past history Y(1,t-1), θ_j and $x_j(0)$.

$\theta_j = (Q_j, R_j)$ is a vector composed of the system noise and observation noise variances of model M_j. $x_j(0)$ is the initial state of model M_j. L_{ij} is considered the local likelihood of model M_j for the i-th segment.

Assuming that $x_j(0)$, $u_j(t)$ and $v_j(t)$ are Gaussian, the probability density function f_j given in the above expression is also Gaussian and Eq.(3) takes the form

$$L_{ij}(Y(t_{i-1}+1 , t_i) \mid Y(1,t_{i-1}), \theta_j, x_j(0))$$

$$= \prod_{t=t_{i-1}+1}^{t_i} (2\pi V_j(t,t-1))^{-1/2} \exp(- \frac{e_j(t)^2}{2V_j(t,t-1)}) \quad (4)$$

where, $e_j(t)$ is defined by

$$e_j(t) = y(t) - \hat{y}_j(t , t-1) \quad (5)$$

$e_j(t)$ is a one-step ahead forecasting error when adopting the model M_j and is called innovation process [9]. $e_j(t)$ is considered the part of the observation y(t) containing new information not carried y(t-1), y(t-2),.... .

$\hat{y}_j(t,t-1)$ and $V_j(t,t-1)$ are, respectively, the conditional mean and its error variance of y(t) given y(1), y(2),...., y(t-1), θ_j and $x_j(0)$. In the following section, we will show that $\hat{y}_j(t,t-1)$ and $V_j(t,t-1)$ are easily obtained by using Kalman filter recursive formula given θ_j and $x_j(0)$. Originally θ_j and $x_j(0)$ are unknown parameters and are estimated from numerical maximization of the conditional likelihood [7],[8] of the training data.

2.2 Recursive Filter [9]

As shown in the previous section, the likelihood of innovation process $e_j(t)$ of model M_j is defined by Eq.(4). From our model in Eqs. (1) and (2), $\hat{y}_j(t,t-1)$ and $V_j(t,t-1)$ are obtained as

$$\hat{y}_j(t,t-1) = H'_j \hat{x}_j(t,t-1) \quad (6)$$

$$V_j(t,t-1) = H'_j P_j(t,t-1) H_j + R_j \quad (7)$$

where $\hat{x}_j(t,t-1)$ and $P_j(t,t-1)$ are, respectively, the conditional mean and covariance of the state vector x(t) given the observations up to time t-1.

Given information on the initial state $x_j(0)$ and an estimate $\hat{\theta}_j$ of the noise parameters θ_j, the conditional mean and covariance of the state $x_j(t)$ of model M_j are obtained by a Kalman filter algorithm, as follows.

$$\hat{x}_j(t,t-1) = F_j \hat{x}_j(t-1,t-1) \quad (8)$$

$$\hat{x}_j(t,t) = K_j(t) (y(t) - H'_j \hat{x}_j(t,t-1)) + \hat{x}_j(t,t-1) \quad (9)$$

$$K_j(t) = P_j(t,t-1) H_j \cdot (H'_j P_j(t,t-1) H_j + R_j)^{-1} \quad (10)$$

$$P_j(t,t) = (I - K_j(t) H'_j) P_j(t,t-1) \quad (11)$$

$$P_j(t,t-1) = F_j P_j(t-1,t-1) F'_j + G_j Q_j G'_j \quad (12)$$

$\hat{x}_j(t,t)$ is the filtering of state x(t) and $P_j(t,t)$ is its error variance of model M_j given observations y(1),y(2),....,y(t). I is an elementary matrix.

2.3 Parameter Estimation

Using the training data y(1),y(2),....,y(n), we estimate noise parameters θ_j and initial state $x_j(o)$. To use the observation data efficiently in parameter estimation, we propose the use of the backward Kalman filtering technique. The procedure of this technique is as follows.

(1) We assume $\theta_j = \theta_j^{(o)}$, $\hat{x}_j(n,n)=0$ and its error covariance $P_j(n,n)$ = kxI, where k is large enough to weaken the effect of initial state $\hat{x}_j(n,n)$ which is assumed to be an arbitrary value.

(2) Using Kalman filter for the data y(n),y(n-1),....,y(1), we estimate $\hat{x}_j(1,1)$.

(3) Considering $\hat{x}_j(1,1)$ as the initial state, we calculate the likelihood $L_j^{(o)}$ by Eq.(4) using a Kalman filter for the data y(1), y(2), ..., y(n) and $\theta_j^{(o)}$.

Given $\theta_j^{(o)}$, we can calculate the likelihood $L_j^{(o)}$. We determine the optimal noise parameter θ_j by numerical maximization of the likelihood of the noise parameter. We use the Davidon method [6] for non-linear maximization. Because the stationary Kalman gain K_j depends only on the ratio of system noise variance and observation noise variance [7],[8], we normalize the observation noise variance R_j into 1, regard the system noise variance Q_j as the noise ratio Q_j/R_j, and maximize the likelihood of the system noise variance Q_j only.

2.4 Bayesian Forecasting

After Kalman filtering for the i-th segment, the conditional likelihoods L_{ij} of sub-model M_j (j=1,2,....,n) is calculated as shown in the previous sections. On the basis of these likelihoods, each sub-model's weight $w_{i+1,j}$ (j=1,2,..,n) for the (i+1)-th segment is calculated.

In Bayesian statistics, posterior probability is proportional to the product of the prior probability and the likelihood [5].

$$\text{posterior probability} \propto (\text{prior probability}) \times (\text{likelihood}) \quad (13)$$

In this Bayesian forecasting $w_{i+1,j}$ is considered the prior probability of the (i+1)-th segment and the posterior probability of the i-th segment in sub-model M_j. The proportional of the posterior probability can be obtained from Eq.(13). As the summation of all the posterior probabilities is one, the sub-model's weight is calculated by normalizing the product of prior probability and the likelihood of all sub-models in the multiple state space model, as follows.

$$w_{i+1,j} = w_{i,j} \times L_{i,j} / \left(\sum_{j=1}^{n} w_{i,j} \times L_{i,j} \right)$$

$$(j=1,\ldots.,n) \quad (14)$$

The prior probability for the first segment is defined as

$$w_{1,j} = 1/n \quad (j=1,\ldots.,n) \quad (15)$$

This prior probability is called ignorant prior in Bayesian statistics. In a multiple state space model, each sub-model's weight is equal when there are no observations. With these weights we calculate the Bayesian forecasting value $\hat{y}(t,t-1)$ ($t \in$ (i+1)-th segment). $\hat{y}(t,t-1)$ is calculated in the weighted summation of each sub-model's one-step ahead forecasting $\hat{y}_j(t,t-1)$ in Eq.(6).

$$\hat{y}(t,t-1) = \sum_{j=1}^{n} w_{i+1,j} \times \hat{y}_j(t,t-1) \quad (16)$$

$$(t = t_i+1, t_i+2, \ldots, t_{i+1})$$

The procedures of calculating posterior probability and Bayesian forecasting are shown Fig. 1 and 2.

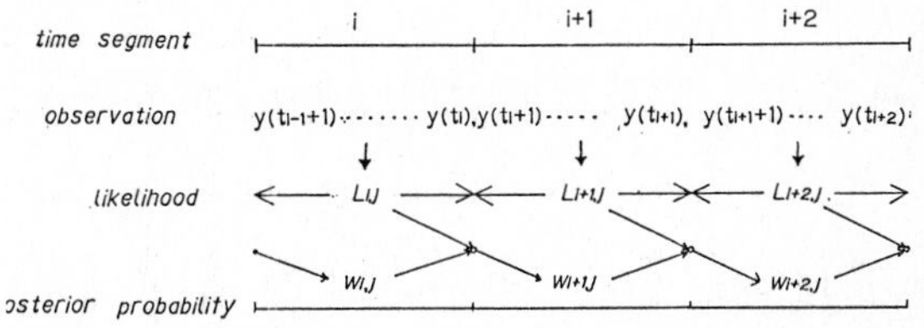

Fig.1 Posterior Probability of j-th model

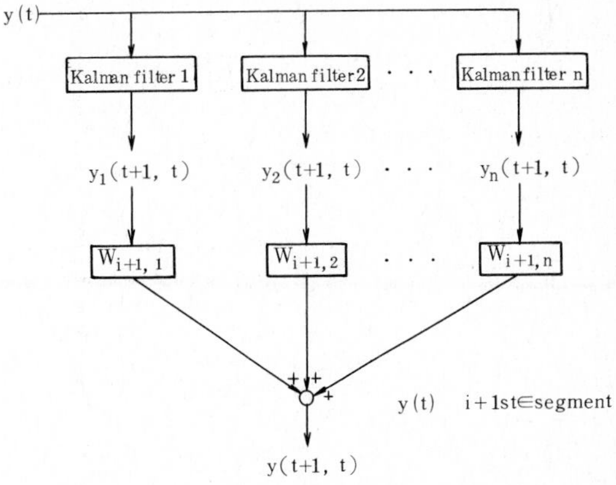

Fig. 2 Bayesian forecasting algorithm

3. NUMERICAL EXAMPLES

Let us apply the method described above to the forecasting of monthly telephone revenues and trunk group loads. In each example, we consider two different multiple state space models.

3.1 Monthly Telephone Revenue Forecasting

Monthly telephone revenue data is shown in Fig.3. The total number of months is 60 and a tariff change is imposed at the 21st month. The effect of a tariff change continues from the 21st to the 23rd month. In an earlier work [8] we analyzed the effect of a tariff change using a state space model and estimated it under the condition that it was exactly known when the effect appeared. In Sect.3.1.1 we show that the multiple state space model is able to adapt to sudden changes e.g., a tariff change, even if we do not know the time of change a priori. In Sect.3.1.2 parameter estimation of trend model is shown using a multiple state space model.

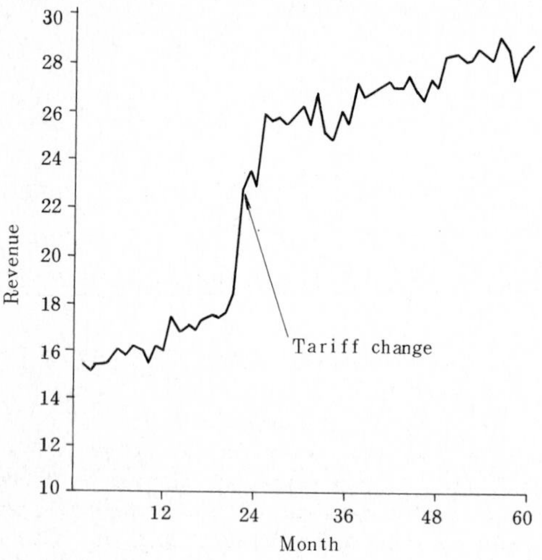

Fig.3 Monthly revenue data

3.1.1 Multiple Model with Trend Model and Random Walk Model

Let us consider a multiple model composed of a trend model and a random walk model. In a trend model the mean increment is assumed to be constant, that is, $E[X(t+1) - X(t)] = E[X(t) - X(t-1)]$. E is the operator to take the expectation. While in a random walk model the mean at time n+1 is assumed to be the value at time n, that is, $E[X(t+1)] = E[X(t)]$.

System matrices, driving matrices and observation matrices of each sub-model are as follows.

M_1 : trend model

$$F_1 = \begin{pmatrix} 2 & -1 \\ 1 & 0 \end{pmatrix} \quad G_1 = \begin{pmatrix} 1 \\ 0 \end{pmatrix} \quad H_1 = \begin{pmatrix} 1 \\ 0 \end{pmatrix}$$

$$Q_1 = 0.05$$

$$R_1 = 1.0$$

$$x_1(t) = (X_1(t), X_1(t-1))'$$

M_2 : random walk model

$$F_2 = \begin{pmatrix} 1 & 0 \\ 1 & 0 \end{pmatrix} \quad G_2 = \begin{pmatrix} 1 \\ 0 \end{pmatrix} \quad H_2 = \begin{pmatrix} 1 \\ 0 \end{pmatrix}$$

$$Q_2 = 0.25$$

$$R_2 = 1.0$$

$$x_2(t) = (X_2(t), X_2(t-1))'$$

M : Multiple state model (M_1, M_2)

Variances of system noise Q_1 and Q_2 are estimated by numerical maximization. The training data to estimate them is the initial 12 data of the observation series. The transitions of the posterior probability of M_1 and M_2 are shown in Fig.4. We find the posterior probability of the random walk model is relatively large near the tariff change. In general the forecasting by trend model tends to overshoot after a sudden change. In this case the forecasting accuracy of the random walk model is relatively high. Figure 4 shows that Bayesian multiple model M approaches trend model M_1, as the number of observations increases.

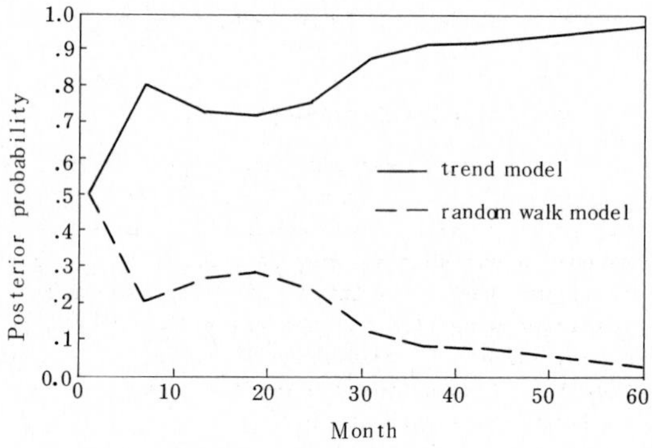

Fig.4 Posterior probability of multiple model

Table 1 shows one-step ahead forecasting accuracy. MS is the mean square forecasting error rate and MA is the mean absolute forecasting error rate defined as follows.

$$MS = \frac{1}{60} \sum_{i=1}^{60} [(y(i) - y(i,i-1))/y(i)]^2$$

$$MA = \frac{1}{60} \sum_{i=1}^{60} \left| (y(i) - y(i,i-1))/y(i) \right|$$

From Table 1 we know that Bayesian multiple model M is a better forecasting model than single state model M_1 or M_2.

Table 1 Forecasting accuracy.

	M_1	M_2	M
MS	0.017	0.021	0.015
MA	0.028	0.027	0.025

3.1.2 Multiple Model with Trend Models of Different Noise Variances

Let us consider a multiple state space model composed of 5 trend models of different system noises. We adopt several arbitrary values as the system noise candidates and apply a multiple state space model to parameter estimation. For the case when the unknown parameter is varies with time, this estimation is useful. System matrices, driving matrices, observation matrices and noise variances of sub-models are as follows.

M_j : the j-th trend model (j=1,2,..,5)

$$F_j = \begin{pmatrix} 2 & -1 \\ 1 & 0 \end{pmatrix} \quad G_j = \begin{pmatrix} 1 \\ 0 \end{pmatrix} \quad H_j = \begin{pmatrix} 1 \\ 0 \end{pmatrix}$$

$R_j = 1.0$

$Q_1 = 0, \quad Q_2 = 0.1 \quad ,Q_3 = 0.5, \quad Q_4 = 1.0,$

$Q_5 = 10$

$x_j(t) = (X_j(t), X_j(t-1))'$

Using this multiple state space model, we can obtain the sub-optimal system noise variance of trend model. The selected value by this method is the best as long as considered. Figure 5 shows the transition of each sub-model's posterior probability. The posterior probability of sub-model M_2 gradually increases. We know from this figure that the optimal system noise variance comes close to Q_2.

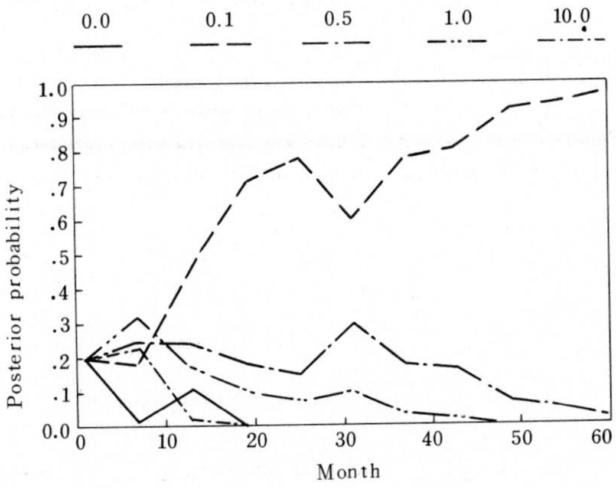

Fig. 5 Posterior probabilities of trend model with different noises

3.2 Monthly Trunk Group Load Forecasting

Monthly trunk group load data is shown in Fig.6. The total number of months is 48. In Sect.3.2.1 the multiple state space model is applied to this data and the forecasting ability is shown. In Sect.3.2.2 parameter estimation of a random walk model is shown using a multiple state space model.

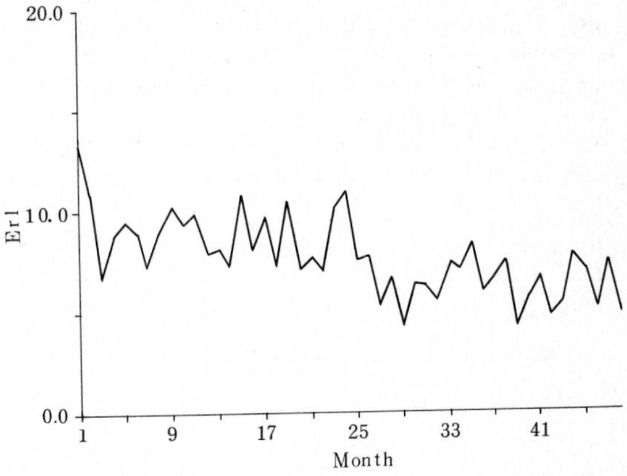

Fig.6 Trunk group load data

3.2.1 Multiple Model with Trend Model and Random Walk Model

Let us forecast trunk group load. We apply the same multiple state space model as in Sect.3.1.1. System matrices, driving matrices and observation matrices of each sub-model are the same as in sect 3.1.1 except a system noise. In this forecasting, we use $Q_1 = 0.1$, $Q_2 = 0.1$ as the noise variances. The transition of the posterior probability of M_1 and M_2 is shown in Fig.7. It shows that Bayesian multiple model M approaches random walk model M_2, as the number of observations increases.

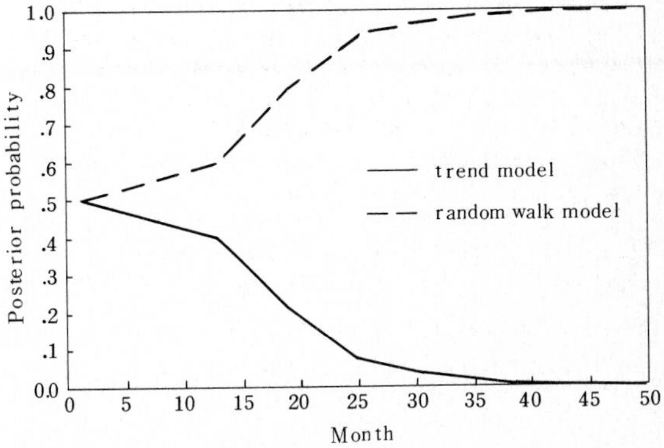

Fig. 7 Posterior probability of multiple model

Table 2 shows the one-step ahead forecasting accuracy.

Table 2 Forecasting accuracy.

	M_1	M_2	M
MS	0.068	0.059	0.057
MA	0.214	0.185	0.185

From these results we know that a random walk model has a good forecasting ability for changeable data compared with a trend model.

3.2.2 Multiple Model with Random Walk Models of Different Noise Variances

Let us consider a multiple state space model composed of 5 random walk models of different system noises. System matrices, driving matrices, observation matrices and noise variances of sub-models are as follows.

M_j : the j-th random walk model (j=1,2,..,5)

$$F_j = \begin{pmatrix} 1 & 0 \\ 1 & 0 \end{pmatrix} \quad G_j = \begin{pmatrix} 1 \\ 0 \end{pmatrix} \quad H_j = \begin{pmatrix} 1 \\ 0 \end{pmatrix}$$

$R_j = 1.0$

$Q_1 = 0,\ Q_2 = 0.1\ ,Q_3 = 0.5,\ Q_4 = 1.0,$

$Q_5 = 10$

$x_j(t) = (\ X_j(t), X_j(t-1))'$

Figure 8 shows the transition of each sub-model's posterior probability. The most suitable sub-model changed from M_1 to M_2 at the 31th month.

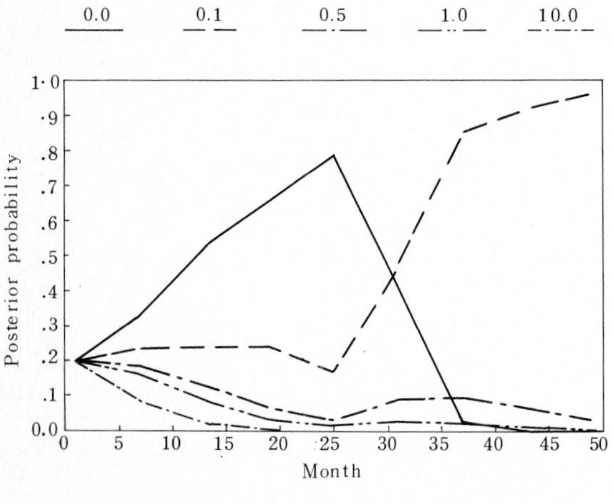

Fig. 8 Posterior probabilities of random walk model with different noise variances

This change can be explained as follows. As you can see in Fig.6, a level is about 8.0 for the first half (before the 31th month), and about 6.5 for the latter half. Though the levels are different for both periods, we can regard that it remains constant within each period. This means that the system noise is very small for all period. On the other hand, it is known that the observation noise variance is proportional to trunk group load [10]. Thus, the observation noise variance is greater for the first half than for the latter half. This turns out that the system noise variance becomes greater relatively for the latter half, that is, the most suitable sub-model changes from M_1 to M_2. The level change might occur as a result of trunk group replacement.

4. CONCLUSIONS

Bayesian forecasting with multiple state space models has been proposed. Using the likelihoods of state space model, Bayesian posterior probabilities were obtained. The structure of the social phenomenon like a traffic trend is unclear and likely to change. To conquer them in model selecting, several state space models were prepared as candidates to represent the true behavior. Using this Bayesian forecasting method, the posterior probability of the best fitting model increases automatically, as the number of observations increases. In the numerical example, it is demonstrated that forecasting with multiple state space models is robust and more accurate than that with single state space models. This method is applicable to situations which include exceptional data when an insensitive model is added to the multiple model. A problem remaining to be studied is how to develop more versatile sub-models.

5. ACKNOWLEDGEMENT

The authors would like to express their appreciation to Mr. Kunio Kodaira, Chief of Teletraffic Section in the Musashino Electrical Communication Laboratory, NTT, and to Dr Jun Matsuda, Staff Engineer of the same section, for their many helpful comments and encouragement.

REFERENCES

[1] A.J.David and C.D.Pack, "The Sequential Projection Algorithm : A new and improved traffic forecasting procedure," 9-th Int Teletraffic Congress, Malaga, Spain, 1979.

[2] J.P. Moreland, "A Robust Sequential Projection Algorithm for Traffic Load Forecasting," B.S.T.J., vol.61, no.1, pp.15-36, 1982.

[3] A. Ionescu-Graff, "A Sequential Projection Algorithm for Special Services Demand," B.S.T.J., vol.61, no.1, pp.37-66, 1982.

[4] C.R. Szelag, "A Short-Term Forecasting Algorithm for Trunk Demand Servicing," B.S.T.J., vol.61, no.1, pp.67-96, 1982.
[5] G. Box and G. Tiao, "Bayesian Inference in Statistical Analysis," Addison-Wesley publishing company, 1973.
[6] H. Akaike, G. Kitagawa, T. Ozaki, " TIMSAC-78," Computer Science Monograph, Stat. Math. Inst. in Japan, 1979.
[7] G. Kitagawa, "A nonstationary time series model and its fitting by a recursive filter," J. Time series Analysis, vol.2, no.2, 1982.
[8] T. Abe and T. Ueda, "Telephone Revenue Forecasting Using State Space Model" IECE-A, 1985.(to appear)
[9] B. Anderson and J. Moore, "Optimal Filtering," Prentice-Hall inc, 1979.
[10]V. E. Beneš," Mathematical Theory of Connecting Networks and Telephone Traffic," Academic press, 1965.

TELETRAFFIC ISSUES in an Advanced Information Society
ITC-11
Minoru Akiyama (Editor)
Elsevier Science Publishers B.V. (North-Holland)

TRAFFIC FORECASTING WITH A STATE-SPACE MODEL

TOMÉ, F. M. **CUNHA, J. A.**

T.L.P. — Telefones de Lisboa e Porto
Lisboa, Portugal

ABSTRACT

The improvement of traffic forecasting methods in TLP has been a reality since some years ago. As a matter of fact, network planning requires a greater performance in prediction traffic algorithms. In order to achieve this purpose we have been studying the use of Kalman filters.

The scope of the paper is to analyse the way we try to overcome some usual difficulties related to the implementation of these models, namely the problem of the knowledge of Q and R matrices, and to show an application of Kalman filters to traffic forecasting in TLP network.

1. INTRODUCTION

Our dynamic process was modeled using empirical data and the knowledge of telephone traffic behaviour in the routes we have been studying.

We assume to have a discrete linear system described by the following state-space equations concerning process and measurement models:

$$\underline{x}_{k+1} = \Phi_k \underline{x}_k + \Lambda_k \underline{u}_k + \underline{w}_k \qquad (1)$$

$$\underline{z}_k = H_k \underline{x}_k + \underline{v}_k \qquad (2)$$

where:

$\underline{x}_k$ - state vector

$\underline{u}_k$ - deterministic input vector

$\underline{w}_k$ - state error vector

Φ_k - transition matrix

Λ_k - input matrix

$\underline{z}_k$ - measurement vector

H_k - observation matrix

$\underline{v}_k$ - measurement noise

We supose that $\underline{w}_k$ and $\underline{v}_k$ are independent zero mean gaussian white processes and

$$E\left[\underline{w}_k , \underline{w}_l^T\right] = Q_k \cdot \delta_{kl}$$

$$E\left[\underline{v}_k , \underline{v}_l^T\right] = R_k \cdot \delta_{kl}$$

$$E\left[\underline{w}_k , \underline{v}_l^T\right] = 0 , \ \forall_{k,l}$$

It is possible to derive estimates for the process using a Kalman filter based upon the described model. The computation of optimal estimates, which yields to the minimum covariance of the error, can be obtained by the well known iterative procedure defined by the equations:

extrapolation
$$\hat{\underline{x}}_k(-) = \Phi_{k-1}\hat{\underline{x}}_{k-1}(+) + \Lambda_{k-1}\underline{u}_{k-1}$$

$$P_k(-) = \Phi_{k-1}P_{k-1}(+)\,\Phi_{k-1}^T + Q_{k-1}$$

update
$$\hat{\underline{x}}_k(+) = \hat{\underline{x}}_k(-) + K_k\underline{\nu}_k$$

$$P_k(+) = \left[\,I - K_k H_k\,\right] P_k(-)$$

where:

$K_k = P_k(-)H_k^T\left[\,H_k P_k(-) H_k^T + R_k\,\right]^{-1}$ is the Kalman gain

$\underline{\nu}_k = \underline{z}_k - H_k\hat{\underline{x}}_k(-)$ is the innovation sequence

P_k is the error covariance matrix.

In the sequel we will consider Φ_k , H_k and Λ_k as time invariants.

These equations assume the knowledge of all matrices involved. Although, in practice some of them are usually unknown and in fact in our model remains unsolved the problem of the identification of Q and R matrices. To identify their elements we began to study the behaviour of the model concerning Q and R variations, using a simulation procedure described in the next section.

2. SIMULATION

We are going to analyse the simulation of a traffic flow using a two-dimensional model, without deterministic events, whose state variables are the traffic value x_k and its growth rate $\dot{x}_k$.

If we consider the model

$$x_{k+1} = x_k + \dot{x}_k + w_{1k}$$

$$\dot{x}_{k+1} = \dot{x}_k + w_{2k}$$

and the measurement model

$$z_k = x_k + v_k$$

the transition and the observation matrices will be

$$\Phi = \begin{bmatrix} 1 & 1 \\ 0 & 1 \end{bmatrix} \qquad \text{and} \qquad H = \begin{bmatrix} 1 & 0 \end{bmatrix}$$

Let us consider a simulation procedure including a random process generator (RPG) which provides the errors simulated values, w_{1k}^s, w_{2k}^s, v_k^s, as well as the starting traffic value $\underline{x}^s_{in}$. If we assume Q and R time invariants it can be illustrated as follows:

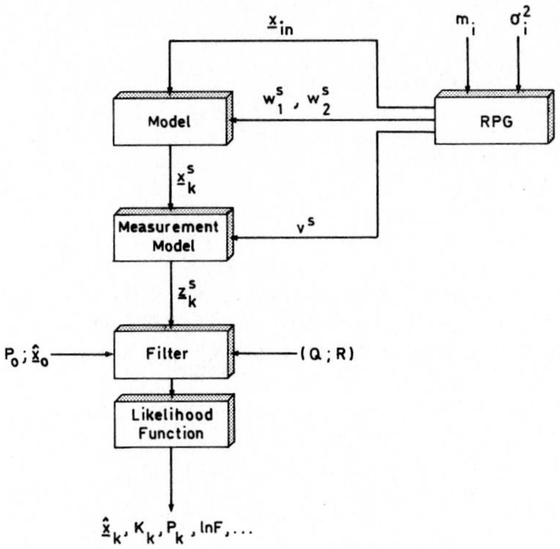

Fig. 1 - Simulation Procedure

where $\gamma_i = (q^i_{11}, q^i_{22}, r^i_{11})$, being q^i_{11}, q^i_{22}, and r^i_{11} the diagonal elements of Q and R.

The filtering initial conditions are:

$\hat{\underline{x}}_o = E\left[\underline{x}_o\right]$ initial estimate of the state vector $\underline{x}_o$

$P_o = E\left[(\underline{x}_o - \hat{\underline{x}}_o)(\underline{x}_o - \hat{\underline{x}}_o)^T\right]$ a priori covariance matrix of $\underline{x}_o$

The measurements simulated sequence $\underline{z}^s_k$ is filtered for several values of Q and R elements. To test the behaviour of the model consider the evaluation of the likelihood function F for each γ_i belonging to a discrete set $\gamma_1, \gamma_2, \ldots, \gamma_n$.

The likelihood function of γ, for N observations $\underline{z}_k$, is given by $F = F(\underline{z}_1, \underline{z}_2, \ldots, \underline{z}_N, \gamma) =$

$$= \prod_{k=1}^{N} p\,(\underline{z}_k \mid Z_{k-1}, \gamma) \qquad Z_k = \{\underline{z}_1, \underline{z}_2, \ldots, \underline{z}_N\}$$

If $\underline{z}_k$ is a d-vector and assuming that $p\,(\underline{z}_k \mid Z_{k-1}, \gamma)$ is gaussian with mean $\hat{\underline{z}}_{k|\gamma}$ and covariance $\sigma^2_{k|\gamma}$, for N independent $\underline{z}_k$ we can write

$$F = \left[\frac{1}{(2\pi)^{d/2}}\right]^N \frac{1}{\prod_{k=1}^{N} \left|\sigma^2_{k|\gamma}\right|^{1/2}} \cdot \exp\{-\frac{1}{2} \cdot$$

$$\cdot \sum_{k=1}^{N} (\underline{z}_{k|\gamma} - \hat{\underline{z}}_{k|\gamma})^T (\sigma^2_{k|1})^{-1} (\underline{z}_{k|1} - \hat{\underline{z}}_{k|1})\}$$

On the other side it is known from the Kalman filtering theory that the innovation sequence $\underline{\nu}_k = \underline{z}_k - H\hat{\underline{x}}_k$ is a zero mean gaussian white noise (optimal filter) and according to recursive equations of the filter we can write that

$$\hat{\underline{z}}_k = H\hat{\underline{x}}_k$$

$$\sigma^2_k = E\left[\underline{\nu}_k, \underline{\nu}^T_j\right] \qquad k = j$$

$$= HP_kH^T + R$$

and compute lnF as

$$\ln F = C - \frac{1}{2}\sum_{k=1}^{N}\left[\ln\left|HP_kH^T + R\right| + (\underline{z}_k - H\hat{\underline{x}}_k)^T(HP_kH^T + R)^{-1}(\underline{z}_k - H\hat{\underline{x}}_k)\right]$$

It is now possible to evaluate lnF after filtering the simulated sequence $\underline{z}^s_k$ for each γ_i. In Fig. 2 we represent the behaviour of lnF regarding Q and R variations using the results of one simulation procedure.

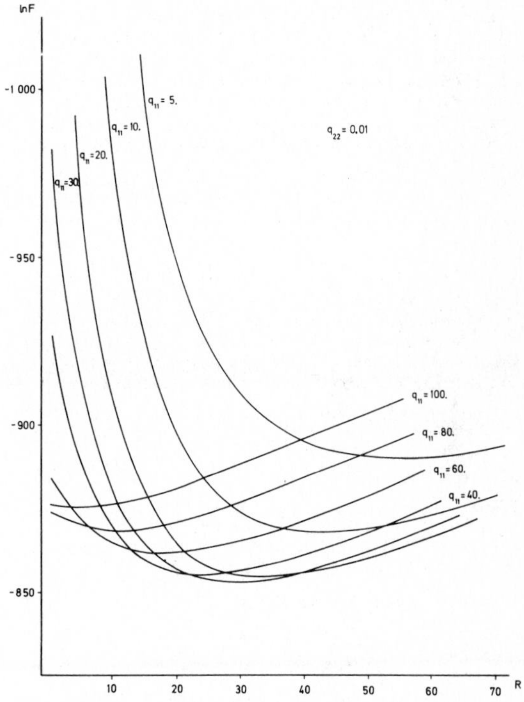

Fig. 2 - Graphical illustration of lnF

The representation of lnF when q_{22} varies is similar to the curves indicated above, however a smaller sensitivity was noticed.

This simulation process was accomplished using a Fortran 77 program - ALUMIS - whose flowchart can be seen in the Appendix.

According to this analysis we decided to implement the estimation process of Q and R that we are going to describe.

3. EVALUATION OF Q AND R

As we have seen, the choice of γ_i, i.e. the

joint estimation of Q and R elements, can be accomplished using a maximum likelihood process. So, we are going to use a batch of n Kalman filters each one tunned for γ_i belonging to a discrete set $\gamma_1, \gamma_2, \ldots, \gamma_n$ to process the measurement sequence $\underline{z}_k$, then we evaluate the likelihood function for each case and we choose γ_f that yields to the greatest value of lnF.

In Fig. 3 we show how we can reach the estimates of Q and R.

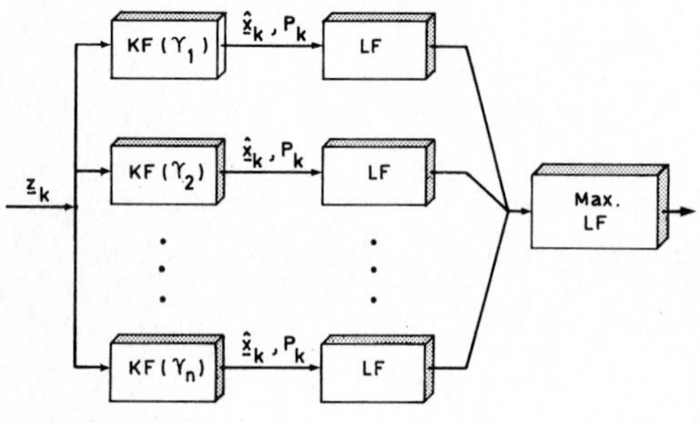

Fig. 3 - Block diagram of Q and R evaluation

To set up this process we developed a Fortran 77 program - LUO - whith about 500 instructions which allow us to choose Q and R, to filter the measurement data and to obtain forecasts for n periods. From its outputs, Kalman gain, r.m.s. error, innovation sequence, etc., we can analyse the performance of the filter.

With LUO we have the possibility of considering deterministic inputs and also, knowing R, to determine Q with the process described above as we show in the Appendix.

4. EXAMPLE

We will consider the application of a Kalman filter model to the measurement data of the traffic flow carried by one of a tandem SPC exchange outgoing routes. We have at our disposal 157 weekly measurements.

Analysing the measurement sequence we thought reasonable to consider a model with a linear trend and the superposition of a seasonal behaviour on summer weeks. In fact in

$$\underline{x}_{k+1} = \Phi \underline{x}_k + \Lambda \underline{u}_k + \underline{w}_k$$

we consider the system also excited by a deterministic time varying input $\underline{u}_k$ due to a disturbance occuring on summer, assuming that the uncertainty in $\underline{u}_k$ is zero. So, we will use as state variables x_k (traffic value) and $\dot{x}_k$ (growth rate), the transition, input and observation matrices being time invariants and having the form:

$$\Phi = \begin{bmatrix} 1 & 1 \\ 0 & 1 \end{bmatrix} \qquad \Lambda = \begin{bmatrix} 0 & 0 \\ 0 & 1 \end{bmatrix} \qquad H = \begin{bmatrix} 1 & 0 \end{bmatrix}.$$

We are now able to verify that our second order system is observable and controllable. In fact

$$\text{rank} \left[H^T \;\vdots\; \Phi^T H^T \right] = 2$$

$$\text{rank} \left[\Lambda \;\vdots\; \Phi \Lambda \right] = 2 \,.$$

Q and R matrices elements were reached using the method arising from the simulation process. The variation limits and the increase, according to the available data, were chosen and those values are:

$$q_{11} \in [\, 0.00 \; ; \; 200.00 \,] \qquad \text{inc} = 10.00$$

$$q_{22} \in [\, 0.00 \; ; \; 0.20 \,] \qquad \text{inc} = 0.01$$

$$r_{11} \in [\, 0.00 \; ; \; 200.00 \,] \qquad \text{inc} = 10.00 \,.$$

Concerning the initial values $\hat{\underline{x}}_o$ and P_o they were estimated with a linear regression model using the monthly means serie. Hence

$$\hat{\underline{x}}_o = \begin{bmatrix} 106.0 \\ 0.9 \end{bmatrix}$$

and

$$P_o = \begin{bmatrix} 250.00 & 0.00 \\ 0.00 & 0.64 \end{bmatrix} .$$

In Fig. 4 we plot the measured and the predicted values $\underline{z}_k$ and $\hat{\underline{x}}_{1k}$. We indicate also forecasts for 3 periods.

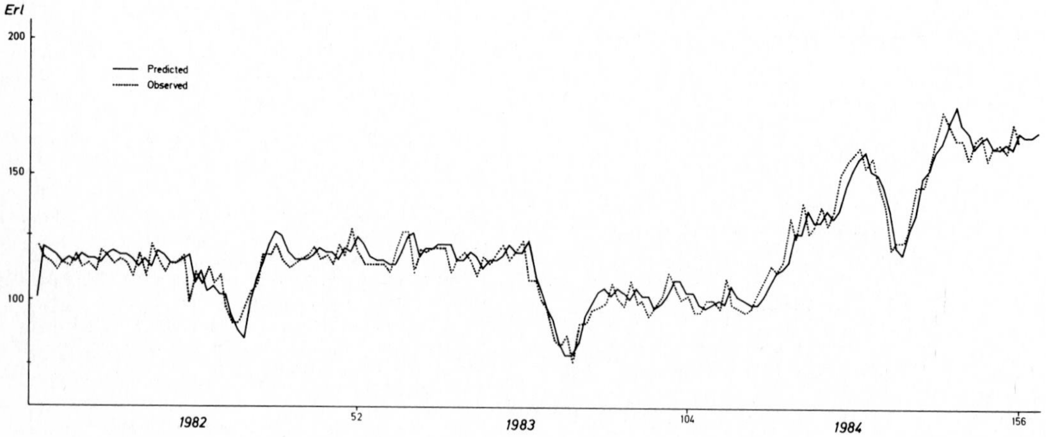

Fig. 4 - Measured and predicted values

It becomes clear the existence of the seasonal behaviour on summer and it is possible to notice also an increase in traffic load level, begining in April 1984, due to a change in traffic routing.

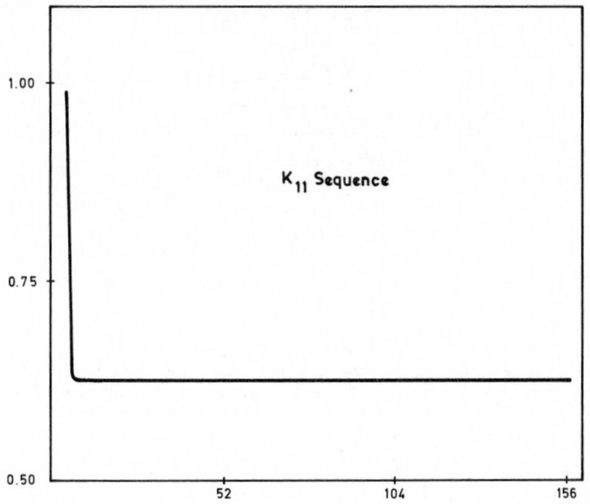

Fig. 5 - Kalman gain

As we know from Kalman filtering theory that the optimality of the filter is related to the gain and the r.m.s. error, take a look at the plots of the Kalman gain and the trace of P_k in Fig. 5 and 6. It is obvious that the system reaches the steady state very quickly.

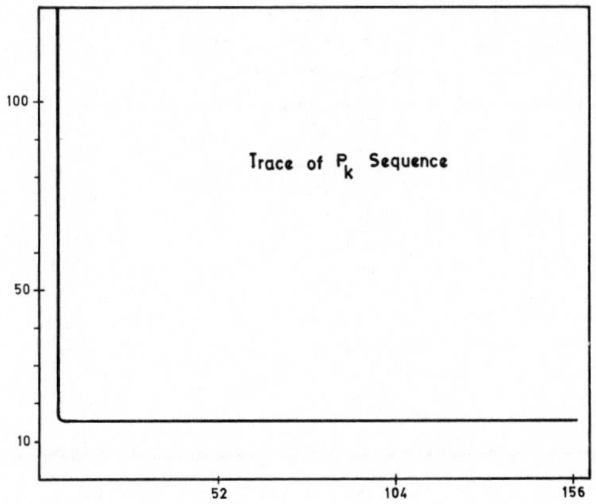

Fig. 6 - Trace of P_k

On the other hand, we can test the optimality of the filter keeping in mind that its necessary and sufficient condition is the whiteness of the innovation sequence $\underline{\nu}_k = \underline{z}_k - H\hat{\underline{x}}_k(-)$.

To do so let us estimate, for our example, the autocorrelation function of the innovation and submit it to a statistical test. We will assume that the most satisfactory estimate of the kth lag autocorrelation is

$$r_k = \frac{c_k}{c_o}$$

where

$$c_k = \frac{1}{N} \sum_{i=1}^{N-k} (\nu_i - \bar{\nu})(\nu_{i+k} - \bar{\nu})$$

In this case, N = 157, k = 0,1,...,40.

Considering r_k asymptotically normal, and a level of significance of 5%, the confidence limits for r_k are ± 0.16 which yields us to the acceptance of the whiteness of the innovation.

5. CONCLUSIONS

We have been discussing a possibility to overcome the problem of determining the unknown elements of Q and R matrices, and we show in the example an application of our process. There is, however, some situations for which we know from the analysis of the measurement methods the covariance structure of $\underline{v}_k$. In such cases R is not time invariant and knowing it we only have to determine Q estimates using the process described above.

The theory of Kalman filters only recently have been applied on traffic forecasting in spite of some difficulties related to its implementation, namely the ones we refer in this paper.

The solution we propose is easily reached yielding to a better accuracy of the forecasts. Besides its computer implementation is straightforward.

6. ACKNOWLEDGMENT

The authors whish to express their gratitude to Dr. J. Leitão of the Technical University of Lisbon for his suggestions and encouragement to this work.

7. REFERENCES

|1| B. O. Anderson and J. B. Moore, "Optimal Filtering", Prentice-Hall Inc., New Jersey, 1979.

|2| G. P. Box and G. M. Jenkins, "Time Series Analysis, Forecasting and Control", Holden--Day, San Francisco, 1976.

|3| R. G. Brown, "Introduction To Random Signal Analysis and Kalman Filtering", John Wiley & Sons, New York, 1983.

|4| A. Gelb, et al. "Applied Optimal Estimation", M.I.T. Press, Massachusetts, 1974.

|5| A. Ionesco-Graff, "A sequencial projection algorithm for special-services demand", Bell Syst. Tech. J., vol. 61, nº 1, pp. 39-66, January 1982.

|6| R. E. Kalman, "A new approach to linear filtering and prediction problems", Trans. Asme, J. Basic Eng., vol. 82, pp. 35-45, March 1960.

|7| R. K. Mehra, "On the identification of variances and adaptive Kalman filtering", IEEE Trans. Automat. Contr., vol. AC-15, pp. 175-184, April 1970

|8| R. K. Mehra, "Approaches to adaptive filtering", IEEE Trans Automat. Contr., pp. 693-698, October 1972.

|9| R. K. Mehra, "Kalman filters and their applications to forecasting", TIMS, vol. 12 (1979), pp. 75-94, North-Holland Publishing Company.

|10| J. P. Moreland, "A robust sequencial projection algorithm for traffic load forecasting", Bell Syst. Tech. J., vol. 61, nº 1, pp. 15-38, January 1982.

|11| C. Pack and B. Whitaker, "Kalman filters models for network forecasting", Bell Syst. Tech. J., vol. 61, nº 1, pp. 1-14, January 1982.

APPENDIX

Flowchart of ALUMIS

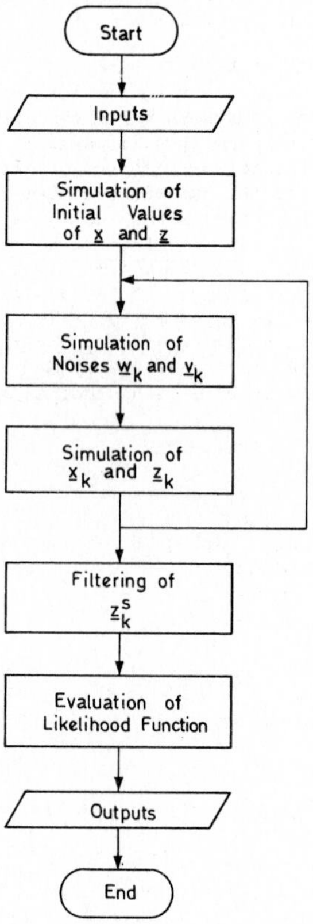

Flowchart of LUO

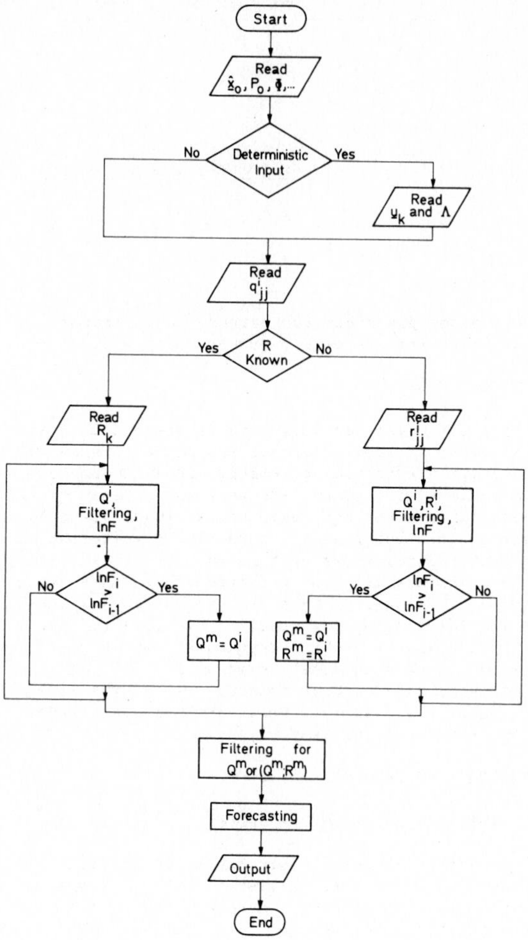

TELETRAFFIC ISSUES in an Advanced Information Society
ITC-11
Minoru Akiyama (Editor)
Elsevier Science Publishers B.V. (North-Holland)

PRIVATE NETWORK TRAFFIC FORECASTING MODELS

C.W. CHAN

ALBERTA GOVERNMENT TELEPHONES (AGT), CANADA

ABSTRACT

This paper describes two methods for forecasting the trunk traffic of user-owner facilities.

INTRODUCTION

The interconnection policy in Canada has a great impact to telecommunication industry. The advanced technologies in information sciences are also increasing the demand for telecommunications. A good private network could increase the effectiveness and image of a corporation. Therefore, private network planning is more important than ever. Forecasting of private networks has thus become more complex than it used to be. Inaccurate forecasting can result in misdirected capital investment, service degradation and the associated expense and lost revenue.
Two methods which have been used in other applications are described here for traffic forecasting for the first time.

FORECASTING METHODS

1. THE PARTIAL ADJUSTMENT MODEL

This model is used to suit the volatile economic situation. It uses different set of exogenous variables for different companies depending on the industry they are in. The basic variables are GNP, industry index and consumer index etc.. The endogenous variables is traffic volume.
The following equations have been used to derive a dynamic relationship:

$$Y^* = A + BX_t + E_t \qquad (1)$$

$$Y_t - Y_{t-1} = K(Y^* - Y_{t-1}), \quad 0<K<1 \qquad (2)$$

On combining (1) and (2), the equation can be written as,

$$Y_t = KA + KBX_t + K(1-K)Y_{t-1} + KE_t \qquad (3)$$

The application of logarithmic adjustment to (3),

$$\log Y_t = K(A + B(\log X_t)) + (1-K)\log Y_{t-1} + KE_t \qquad (4)$$

where,

- Y^* the desired traffic
- X_t the exogenous variables
- Y_t the endogenous variable, current traffic
- Y_{t-1} lagged traffic volume
- K the adjustment coefficient
- A, B coefficients
- E_t disturbances term

Three-stage least squares estimation method has been used to estimate parameters. The results show that independent variables are somewhat correlated with the exogenous variables. The predicting power is reasonably well.

2. THE BLACK BOX MODEL

The black box technique has been investigated. This technique is being used in other area of engineering practices, but it has not been used in forecasting. It assumes that the trend will continue in the future(short term). By using the known statistics such as GNP, consumers index and price index etc. and the traffic data, the value of an unknown black box can be determined. Once the black box is known, the estimated future traffic can be obtained. The results vary depending on data. Generally speaking, the results can be improved if the right statistic data being used.
The method is as following:

- A_{i1} GNP
- A_{i2} consumer index
- A_{i3} price index
- A_{ij} other indices
- F_i traffic volume
- X_i element in black box

$$A = \begin{vmatrix} a_{11} \dots\dots a_{1n} \\ : \\ a_{n1} \dots\dots a_{nn} \end{vmatrix}, \quad X = \begin{vmatrix} x_1 \\ : \\ x_n \end{vmatrix}, \quad F = \begin{vmatrix} f_1 \\ : \\ f_n \end{vmatrix}$$

$$A = X * F \qquad (5)$$

Thus, from the known statistics, X can be obtained,

$$X = A^{-1} * F \qquad (6)$$

If we know a_{ij} of the future, then we will know F

$$F = A * X \qquad (7)$$

CONCLUSION

These two methods are very easy to use and have good predictive power. Due to the changing aspects of a company's business and global economic condition, both methods are for short term forecasting. These methods with limited historical data produce reasonably accurate results. If more historical data is available, these methods should give a much better traffic prediction.

TELETRAFFIC ISSUES in an Advanced Information Society
ITC-11
Minoru Akiyama (Editor)
Elsevier Science Publishers B.V. (North-Holland)

AN ANALYSIS OF THE SENSITIVITY OF KRUITHOF BIPROPORTIONAL MATRIX SOLUTIONS TO TRAFFIC DATA PERTURBATIONS

Dallas R. WINGO

Bell Communications Research Inc.
Red Bank, New Jersey 07701, U.S.A.

The well-known Kruithof biproportional model is frequently used by traffic network planners to derive forecasts of point-to-point traffic demands (flows) between exchanges in a closed network. The model assumes that the $m \times n$ forecast matrix $X \equiv [x_{ij}]$ is related to the given $m \times n$ non-negative (base) matrix $A \equiv [a_{ij}]$ according to the functional form

$$x_{ij} = P_i a_{ij} Q_j \, . \tag{1}$$

The elements of X are required to satisfy the constraints

$$\sum_i x_{ij} = C_j \quad (j=1,\ldots,n) \tag{2}$$

$$\sum_j x_{ij} = L_i \quad (i=1,\ldots,m) \tag{3}$$

$$x_{ij} \geqq 0 \quad (\forall\ i,j) \tag{4}$$

$$a_{ij} = 0 \text{ implies } x_{ij} = 0 \quad (\forall\ i,j). \tag{5}$$

The strictly positive quantities L_i and C_j are, respectively, (known) estimates of future total originating and terminating traffic demands for each exchange. Condition (5) requires that zero elements in the base matrix A remain zero in the forecast matrix X. The L_i and C_j are also assumed to satisfy $\sum_i L_i = \sum_j C_j$.

A forecast matrix X can also be derived by minimizing with respect to x_{ij} the strictly convex functional

$$S = \sum_{i,j} x_{ij} \ell og(x_{ij}/a_{ij}) \tag{6}$$

subject to the constraints (2)-(4). The (unique) solution to this (convex) nonlinear program gives a "maximum entropy" solution [1]. Such a solution is said to give the "most probable" distribution X of point-to-point traffic demands for the given planning scenario, i.e., for given L_i, C_j and A.

A close relationship exists between the maximum entropy and Kruithof biproportional matrix adjustment problems: their optimal solution sets are identical and the iterative adjustment procedure [4]

$$P_i^{(k+1)} = L_i (\sum_j Q_j^{(k)} a_{ij})^{-1} \tag{7}$$

$$Q_j^{(k+1)} = C_j (\sum_i P_i^{(k+1)} a_{ij})^{-1} \tag{8}$$

($k \geqq 0$ and $Q_j^{(0)} = 1$) used to obtain P_i and Q_j in (1) can be easily derived from the maximum entropy model by affixing the constraints (2)-(3) to the objective function (6) by means of Lagrange multipliers. The necessary conditions for a stationary point of the resulting Lagrangian function lead, after a simple change of variable, to (7)-(8).

This paper exploits the close relationship between the Kruithof and maximum entropy models to quantify the effect on X of changes in the problem data L_i, C_j and A. This quantitative information is obtained by regarding the given data as "problem parameters" and carrying out a sensitivity analysis on the optimal solution set of the strictly convex separable nonlinear program which results when (6) is minimized subject to (2)-(4). By "sensitivity analysis" is meant an analysis of the effect on the optimal solution and optimal objective function value of small perturbations in the problem parameters [3]. Since we are interested in studying the effects on the elements of X of perturbations in some or all of the elements of L_i, C_j and A, our results pertain only to the prediction of changes in the optimal solution, given corresponding changes in the problem parameters.

REFERENCES

[1] D. Bear (1976). Principles of Telecommunication-Traffic Engineering. London: Peter Peregrinus.

[2] J. H. Bigelow and N. Z. Shapiro (1974). Implicit function theorems for mathematical programming and for systems of inequalities. Mathematical Programming 6, 141-156.

[3] A. V. Fiacco (1983). Introduction to Sensitivity and Stability Analysis in Nonlinear Programming. New York: Academic Press.

[4] S. M. Macgill (1977). Theoretical properties of biproportional matrix adjustments. Environment and Planning-A 9, 687-701.

TELETRAFFIC ISSUES in an Advanced Information Society
ITC-11
Minoru Akiyama (Editor)
Elsevier Science Publishers B.V. (North-Holland)

OSCAR, TRAFFIC MEASUREMENT, MAINTENANCE AND MANAGEMENT OF THE PORTUGUESE TELEPHONE NETWORK

Paulo NORDESTE, Lourenço MOURA

CTT, Centro de Estudos de Telecomunicações, PORTUGAL

INTRODUCTION

The transition from an analog network to a digital network is a commom problem for all Administrations.
The key word is "integration" not only in equipments and networks, but also in methods.
The transition strategy depends on multiple factors but one of the most important constraints is the large investment related with the existing network. One expects to have analog exchanges in operation during the next 20 years.
Taking these factors into account, the Portuguese Administration decided to develop a project, called "OSCAR", for the traffic measurement, maintenance and management of the telephone network.
The main aim is that the project must be commom, as far as possible, to the present analog network and to the future digital network.
Another aspect is that the development and installation of auxiliary equipments, using new technologies, contribute to the progressive training of personel both in hardware and software techniques.
This paper describes the main features of "OSCAR" project.

OSCAR NETWORK

The OSCAR network is a data network with four levels: one level for control, processing and storing the information (level 1); two levels for the utilization of the information (levels 2, 3); one level for collecting the data (level 4).
At the bottom of the network, level 4, one finds the TO3's (Terminal OSCAR 3rd hierarchie) which are located at the analog exchanges.
The most important TO3 is the so called MTGC, a traffic measurement equipment with a capacity from 256 to 8000 points. Beside the measurement of the traffic, the MTGC is able to collect information from other TO3's and send it to the different levels of the network.
At the top, level 1, there are minicomputers (Centres OSCAR) where the information is processed and stored.
The Centres OSCAR (CO) will be connected to the Portuguese packet switching network, TELEPAC.
The interconnection between the different levels of the network is supported by data links of 1200 bit/s (4 wire).
The levels 2 and 3 are coincident with the traffic engineering and maintenance operational centres, which can control the information related to the exchanges, routes and local networks under its supervision. These levels are known as TO1 and TO2, terminals of 1st and 2nd hierarchie.
The first CO cut over during 1984 at Picoas, the most important switching center of the trunk network.

SUBSYSTEMS OF OSCAR NETWORK (TO3)

- MTGC - Traffic measurement equipment of great capacity (256-8000)
- MTPA 250 - Traffic measurement equipment of small capacity (up to 250)
- DETA - Equipment for detection and transference of alarms.
- EVE - Equipment for supervision and testing of analog exchanges (registers, junctors).
- EAR - Equipment for supervision of strowger exchanges.
- TESLA - Equipment for supervision and testing of subscriber lines.
- EJ - Equipment for testing of transmission lines.

All subsystems have been designed and developed at the Portuguese Research Center (CET).
The O/M software of the future digital exchanges will be considered as a TO3.

FUNCTIONS AND METHODS

The functions performed by the OSCAR network, are mainly based on the collection of traffic data.
The measurement programs are prepared at the TO1's according to the TCBH method or the PSBH method.
After programming the MTGC the measurements take place and the data is stored on magnetic tape. The tape is read from the CO and the data is processed in order to obtain the traffic carried on the routes and the related busy hours. The results can be transferred automatically to the network planning systems used in the Administration.
Special traffic measurement programs, for instance dispersion analysis and measurements on routes with peack traffic, are now being studied.
For Maintenance purposes ICUP values are measured on each internal and external route as well as on the commom equipment of the exchanges. This action takes place during the morning and the afternoon producing two daily reports. An analysis algorithm allows to produce a list of the equipments with higher probability of being faulty. The algorithm takes into account all the information concerned with each type of equipment, collected via other subsystems of OSCAR network (alarms, routines). Each ICUP report is used to built up and update a distribution function in terms of "mean holding time" and "peg count" for the behaviour of the different types of equipment supervised.
At the same time, this report also gives the total traffic carried by each group or route, which allows the calculation of the loss.
The comparison with the forecasted loss for the route, is used for the management of the network.

TELETRAFFIC ISSUES in an Advanced Information Society
ITC-11
Minoru Akiyama (Editor)
Elsevier Science Publishers B.V. (North-Holland)

TOPOLOGIES AND ROUTING FOR 2-REGULAR RING NETWORKS

F. K. HWANG

AT&T Bell Laboratories
Murray Hill, New Jersey 07974

ABSTRACT

A ring network is a directed graph where each node (which can be a station, a computer, a processor, a memory unit, an interface...) has one inlink and one outlink and the links connect the nodes into a cycle. Due to its simple structure, easy implementation and expandability, the ring network is one of the most popular topologies for local computer networks. However, a ring network is also known to be unreliable and to have long delay. To be more specific, define the diametr of a ring network as the maximum distance over all pairs of nodes. Then a large diameter signifies the existence of pairs of nodes which can be connected to each other only through many other intermediate nodes and hence have long delay. A ring network of size n has diameter n-1 which is longest possible. In addition, a common measure of reliability is the connectivity of a network, which is defined to be the largest k such that every node has a directed path to every other node after the removal of any k nodes. The connectivity of a ring network is one, which means that the failure of one node can disrupt the whole network.

One way to shorten the diameter and increase the connectivity is to add links to a ring network. Two general principles are usually followed in adding links. The first is to add as few links as possible, since links can be costly and more links per node increase the control cost at the node. The second is to add links evenly to the nodes so that we can use standardized nodes to build the network. The simplest addition under these two principles is to add one inlink and one outlink to each node. We call such a network a 2-regular ring network. Note that the existence of a second outlink at a node makes the routing algorithm no longer trivial.

There are two general classes of 2-regular ring networks depending on whether the network is node-symmetric or not. The class of node-symmetric networks, also called double-loop networks, can be characterized by two parameters (s_1,s_2) such that the 2n links are $i \to i+s_1$, $i \to i+s_2$ (mod n), i = 0,1,...,n-1. Since the network must be a ring, any (s_1,s_2) network is equivalent to a (1,s) network for some $2 \le s \le n-1$. The (1,n-1) network was proposed by Liu[1] and is called the distributed double loop computer network.. The (1,n-2) network was proposed by Grnarov, Kleinrock and Gerla[2] and is called the daisy-chain loop. Wang and Coppersmith,[3] and Raghavandra and Gerla,[4] studied the problem of selecting s to minimize diameter (RGA also called a (1,s) network a forward loop backward hop network). Du, Hsu and Hwang studied a 2-regular ring network, called a doubly linked ring network, which is not node-symmetric.

It is easily shown that a (1,s) network has connectivity two. Furthermore, since a (1,s) network is node-symmetric, the routing algorithms stored at each node are mathematically equivalent. Therefore we need only discuss the routing algorithm at node 0. One algorithm which assures shortest-path routing requires the storage of n/2 numbers such that if the destination is one of those numbers, the link $0 \to 1$ is used; otherwise, the link $0 \to s$ is used. A simplier self-routing algorithm uses the link $0 \to 1$ if the destination lies between [1,s-1]; otherwise the link $0 \to s$ is used. This self-routing algorithm guarantees that the path length never exceeds s+n/s.

The problem of selecting s to minimize the diameter is a difficult mathematical problem. Wang and Coppersmith proposed the selection of $s=n^{1/2}$ which yields a diameter of value $2n^{1/2}$. They also gave $(3n)^{1/2}-2$ as a lower bound for the diameter. Recently Hwang and Xu showed that essentially for every n, the diameter $(3n)^{1/2}+2n^{1/4}$ can be achieved.

The doubly linked ring network proposed by Du, Hsu and Hwang also has connectivity two but has a much shorter diameter $\lceil \log_2 n \rceil$. We give a self-routing algorithm which assures that the path is not longer than $\lceil \log_2 n \rceil$ and a variation of it which assures shortest-path routing.

REFERENCES

1. M. T. Liu, "Distributed loop computer networks, " Advances in Computers, Vol. 17, Academic Press, pp. 163-211, 1978.

2. A. Grnarov, L. Kleinrock and M. Gerla, "A highly reliable distributed loop network architecture," Proc. 1980 Intern. Symp. Fault Tolerant Comput., Kyoto, pp. 1980, 319-324.

3. C. K. Wong, and D. Coppersmith, "A combinatorial problem related to multimodule memory organizations," J. Assoc. for Comput. Mach. Vol. 21, pp. 392-402, 1974.

4. C. S. Raghavendra and M. Gerla, "Optimal loop topologies for distributed systems," Proc. 7th Data Commun. Symp., Mexico City, pp. 218-223, 1981.

TELETRAFFIC ISSUES in an Advanced Information Society
ITC-11
Minoru Akiyama (Editor)
Elsevier Science Publishers B.V. (North-Holland)

A METHOD FOR TRAFFICABILITY/RELIABILITY ANALYSIS : APPLICATION TO TROPICO-R SWITCHING SYSTEM

Jorge MOREIRA DE SOUZA, Marta R. BASTOS MARTINI, Antonio C.LAVELHA

C.Pq.D. - TELEBRÁS, CAMPINAS - BRASIL

ABSTRACT

A method is presented to evaluate the point-to-point loss probability is switching systems, considering the presence of failures.

INTRODUCTION

The design of a switching network must follow specified requirements imposed on point-to-point loss probability.
The point-to-point loss probability is evaluated considering that the network is fully operational. This analysis gives no insight in the way the network perfoms in presence of failures.
To consider the effects of failures (reliability analysis) we must consider the states where the network is still operational but failures increase the poin-to-point loss probability. For each state we can evaluate the corresponding point-to point loss probability (trafficability analysis) reflecting the new way the network perfoms.
To carry out the trafficability/reliability analysis we define the possible states of the network characterizing the different degradation states experimented after successive failures, establish the network's Markov reliability model by defining the transition rates between the defined states and evaluate, for each state, the corresponding point-to-point loss probability.

DESCRIPTION OF THE SWITCHING NETWORK

The TROPICO-R has a 3-stage switching network. The first and the last stages are time switches which multiplex 30 time-slots and in the central stage there are 3 non-blocking time-space switches wich multiplex 16 PCM links.
The connection graph representation of the switching network is given in the figure, where the stages are represented by nodes and the interconnecting channels by directed branches.

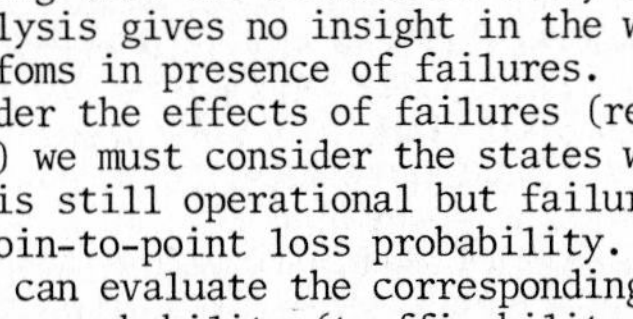

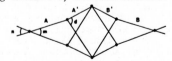

FIGURE 1 - CONNECTION GRAPH

TRAFFICABILITY/RELIABILITY/ANALYSIS

The blocking probability when the system is fully operational, under the hypothesis of homogeneous traffic distribution between the d central stages is :

$$B(d)=\sum_{i=0}^{m} P_A(i)\sum_{K=0}^{m} P_B(k)\left[b_A(i)+(1-b_A(i).b_B(k)\right]^d$$

where $b_A(i)$ is the the probability that i central stages are blocked at side A, and $P_A(i)$ is the probability that i links A are blocked. $P_B(i)$ and $b_B(i)$ have identical definitions corresponding to the B side.
The Markov reliability model is shown in figure-2. If a time (T) switch fails the system reaches an inoperable state (3,4,5). Occurring a failure in a time-space switch the system still functions but in degratation state (1,2/d=3). The failure rate of T and T stages are λ_1 and λ_2 respectively. The repair rate is u when the system is in operation. We suppose that in an inoperable state all failures are repaired simultaneously.

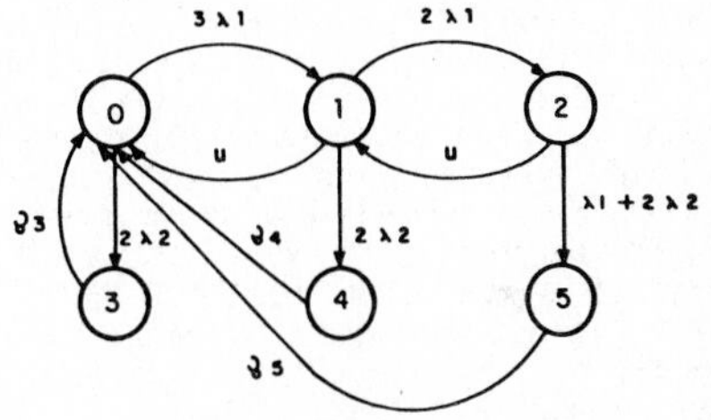

FIGURE 2 - TRANSITION DIAGRAM

To give quantitative results we use the mean blocking probability defined as : $\bar{B} = \sum_j P(j) . \emptyset (j)$, where P(j) is the Markov model steady-state probability that the system is in state j and $\emptyset$ (j) is the network blocking probability given that it is in state j. Figure-3 shows the mean blocking probability as a function of λ_1/u, considering $\lambda_1 = \lambda_2$, $u = \vartheta_3 = \vartheta_4 = \vartheta_5$

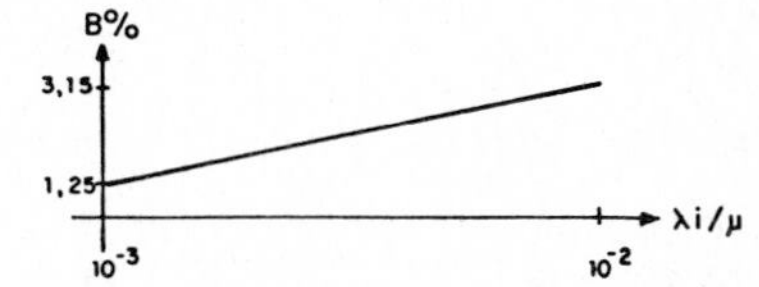

FIGURE 3 - BLOCKING PROBABILITY

The mean-loss probability increases as λ_1/u decreases. As a consequence the traffic capability of the switch must be reduced if the 1% loss probability is to be satisfied.

TELETRAFFIC ISSUES in an Advanced Information Society
ITC-11
Minoru Akiyama (Editor)
Elsevier Science Publishers B.V. (North-Holland)

HIGH RELIABLE NETWORKS

V.A.GARMASH, L.A.SHOR, A.I.ABRAMOVICH

Institute for Problems of Information Transmission
USSR Academy of Sciences Moscow, USSR

ABSTRACT

The paper considers methods to detect single and binary short duration failures in networks and to improve the reliability in the class of networks Σ when switching elements refuse.Thepaper is based on and develops further the authors' report on the 10th International Teletraffic Congress (ITC-10) in Montreal.

Two methods to improve the reliability of one-switching network is considered. The first method is to apply a built-in device to test and to self-regenerate rearrangeable networks. The constructing of such arrangement we shall illustrate by an example of threestage one-switching network $\gamma(n,r)$, where r is the number of quadratic nxn switches in the first and the third stages. Ennumerate inputs and outputs in i-th switches of stages 1 and 3 by numbers from n(i-1)+1 to ni and imputs and outputs of j-th switches of stage 2 by mumbers from r(j-1)+1 to rj. To the pair of numbers (i,j) where $1 \le i \le nr$, $1 \le j \le 2n+r$ corresponds a switching element (SE) from $\gamma(n,r)$, determined according to the following rule;

- if $1 \le j \le n$, the SE connects in the first stage input number i with output number $\lfloor (i-1)/n \rfloor n+j$;
- if $n+1 \le j \le n+r$, the SE connects in the second stage input number i with output number $\lfloor (i-1)/r \rfloor r+j-n$;
- if $n+r+1 \le j \le 2n+r$, then SE connects in third stage input number i with output number $\lfloor (i-1)/n \rfloor n+j-n-r$.

Let on $\gamma(n,r)$ be given a realization of mapping of a set of inputs to a set of outputs, that should be written as a matrix of nr x (2n+r) size, in which element a_{ij} is equal to one, if SE (i,j) is blocked, and to 0 otherwise.

Determine fro all i, $i=\overline{1,nr}$, $j=\overline{1,2n+r}$, $k=\overline{1,nr}$

$$b_i=\sum_{\alpha=1}^{2n+2} a_{i\alpha},\quad c_k=\sum_{\alpha=1}^{2n+2} a_{[k+\alpha-1]\alpha},\quad d_j=\sum_{\alpha=1}^{n^2} a_{\alpha j} \quad (*)$$

whwre all the sums are taken with respect to module 2. Denote by B,C,D the control sums, i.e. sets of numbers $\{b_i\}$, $\{c_k\}$, $\{d_j\}$ found according to (*) to realize mapping without mistake.

Theorem. If during the mapping realization on $\gamma(n,r)$ short duration failures in one or two SE take place, then, knowing sets of indexes I,J,K for which control sums B,C and D were changed, it can be determined simply, in which SE were short duration failures.

The other method is to introduce auxiliary SE into the reconstructed network that makes possible loss probability decrease.

Optimization is also taken into account, that consists in the following. Among all the networks with given level of reliability one should be constructed such thet has less number of auxiliary SE. In the report the problem of optimal synthesis is resolved in class of networks Σ in which reliable characteristics and the number of SE is determined by finite (and not very large) number of integer parameters.

The networks of class Σ are built as follows. Let R be a one-switching network with N inputs and N outputs. Denote by Σ_2 the network in which each SE of network R is replaced by a hammock with parameters r_2, l_2. We call hammock of length l and width r a network of rl SE that has one input and one output connected by chains of length l and to break all the paths of whic it is necessary to disconnect no less that r SE.

Networks, consisting of N unconnected hammocks each we denote by Σ_1 and Σ_3. All the parameters of Σ_1 and Σ_3 are the same. Network S with N inputs and N outputs is obtained by consequent connection of networks $\Sigma_1, \Sigma_2, \Sigma_3$. Denote by M a class of networks, obtained of (1,2N) polars by replacing all their SE by hammocks. Hammocks 1,2,...,N have the same parameters r_4 and l_4 and hammocks N+1,...,2N - the same parameters r_5, l_5. Denote by Σ_4 a circuit consisting of m networks $M_1, M_2, \ldots, M_m$ of class M with inputs and outputs switching in parallels. Denote by Σ a circuit obtained by parallel connection of networks S and Σ_4.Changing integer parameters r_i, $i=\overline{1,5}$ and integer m we obtain the class of networks Σ .

Synthesis algorirtm of the networks optimal on auxiliary SE number from class Σ is given, i.e. solving algorithm of integer non-linear optimization task. The algorithm consists of subsequent elimination of such values of psrsmeters r_i, $i=\overline{1,5}$ and m that do not solve the problen due to fault probability limitation.

TELETRAFFIC ISSUES in an Advanced Information Society
ITC-11
Minoru Akiyama (Editor)
Elsevier Science Publishers B.V. (North-Holland)

NETWORK FLOW DEPENDING ON AVAILABILITY OF NETWORK ELEMENTS

K. FISCHER and W. RÜGER

UNIVERSITY OF TRANSPORT AND COMMUNICATIONS "FRIEDRICH LIST"
DRESDEN, GDR

Efficiency as the most important system feature also depends on reliability and maintainability, resp., and unfortunately one must say, to a rather high degree. Therefore up to now efficiency and availability of system are calculated separately. But as a result of this classical way we obtain two different statements without a connection between them. That means that this manner is not sufficient because the influence of availability of system elements cannot be computed and so it is also impossible to estimate the economical importance of availability of system elements with regard to system efficiency. But only if we succeed in integrating of availability features into the system efficiency we are able to derive concrete requirements relative to reliability and maintainability of system elements. Therefore our research work aimed at such solutions where the availability of system elements can be integrated into the calculation of system efficiency. For solving this problem a new reliability concept for performance systems was developed and applied to communication systems. In our opinion this general way outlined here must also be used for network considerations. In this case the network flow between a source and a terminal can be used as the most important network feature. But it also depends on reliability and maintainability parameters of network elements (e.g. branches and nodes). The maximal network flow between two points is only given if all network elements are working free of failure. But in reality often some of the network elements are out of order. Therefore it is necessary to compute the instantaneous (deterministical consideration) and the expectation value (stochastical consideration) of the different network states.

To solve the given task for networks the following steps have to be realized:

- to compute the reliability that the maximal flow x is possible between source and terminal
- to compute the maximal flow x of each elementary state of the network which occurs as the result of disturbed network elements
- to compute the expectation value of network flow between two points by using the results from the calculations mentioned above. This can be realized by combining these two results based on the equation of the total probability.

The practical importance of this result achieved is very high because the calculation of the reliability that two points of the network are connected contains no information if it is possible to translate a determined quantity of data in a fixed period. Based on the new consideration and methods developed the dimension of networks can be realized on a more realistic foundation. The main problem consists in computing the probabilities of all elementary states with the network flow x. For solving this task a new reliability branching algorithm was developed (Based on pathsets or cutsets).

The main ideas of this algorithm are the following:

- all sets of network states are arranged by using two special rules
- the sets are generated one after the other by taking into account the rules of arrangement
- the sets are tested whether the network flow demanded is possible or not. The last one is marked as system failure
- if the set tested is not equal to a system failure the next set is generated
- if the set tested represents a system failure then a tree-node of the branching tree is fixed
- after finding out a set representing a system failure all uppersets of this set do not need to be generated, which can be shown by using the monotony of networks
- there exist some stopping rules for finishing the algorithm.

The developed algorithm is very fast and needs only a small capacity of memory which can be seen by the written program.

Summarizing a method was developed which permits the calculation of a network flow also depending on availability features. It is possible to go on developing these results to a generl Performance-Reliability-Model applicable to further system structures.

TELETRAFFIC ISSUES in an Advanced Information Society
ITC-11
Minoru Akiyama (Editor)
Elsevier Science Publishers B.V. (North-Holland)

TIME CONGESTION AND THE CONTINUED FRACTION EXPANSION OF THE ERLANG FORMULA

Per LINDBERG

Swedish Telecommunications Administration
Farsta, Sweden

This contribution gives formulas for the computation of the time congestion B_T of the secondary group in the simple overflow system shown in fig.1. By use of them B_T is computed with an arbitrary precision for all positive real j and all real i. This has a large practical application due to Wilkinson's Equivalent Random Theory.

$$A \rightarrow o\!-\!\!-i-\!\!-\!o \rightarrow m,v \rightarrow o\!-\!\!-j-\!\!-\!o \rightarrow \tilde{m}$$

Fig.1

Define λ_{ij} by $\lambda_{ij} = \tilde{m}/B_T$, where $\tilde{m} = A\ E_{i+j}(A)$.

Thus $\lambda_{0j} = A$ and $\lambda_{i0} = A\ E_i(A)$.

One recurrence formula (see [2]) for λ_{ij} is

$$\lambda_{ij} = A - i - j - 1 + A(j+1)/\lambda_{i\ j+1} \qquad (1)$$

From the state equations it is simple to prove the following generalization of the Erlang recurrence.

$$\lambda_{ij} = A/(1+i/(j+\lambda_{i-1\ j})) \qquad (2)$$

Starting with the initial value for i = 0, we can by (2) compute λ_{ij} for all positive integer i. That instead start from j = 0 and use (1) backwards is not as suitable since for $i+j+1 < A$ each step would increase the relative error. If (1) and (2) are combined we obtain the recursive step of the wellknown continued fraction expansion of Erlang's formula.

$$\lambda_{ij} = A - i/(1+(j+1)/\lambda_{i-1\ j+1}) \qquad (3)$$

When $i < A$ each step of (3) will decrease the relative error. Therefore, by using an approximate initial value and a number of steps large enough, we can compute λ_{ij} with an arbitrary precision for all $i < A$ and all j. By afterwards applying (2), the result is extended to all i. As initial value we use Frederick's approximation [2] with a slight modification.

$$\lambda_{i-K\ j+K} \approx \tfrac{1}{2}\left(A-i-j-\tfrac{1}{2} + \sqrt{(A-i-j-\tfrac{1}{2})^2 + 4A(j+K+\tfrac{1}{2})}\right)$$

In order to have a simple algorithm we must know in advance the required number of steps K. The following formulas for this have been proved analytically to give a relative error less than exp(-2r), i.e. 0.87r significant digits [3].

1. for $r \leqslant i \leqslant A$

$$K = ((A-i)^{1.5} + 1.5r\sqrt{A})^{2/3} - A + i \qquad (4)$$

2. for $0 \leqslant i \leqslant r$

$$K = r + (r-i)^2/(4A) \qquad (5)$$

3. for $i \leqslant 0$

$$K = r\sqrt{(A-i)/A} + r^2/(4A) \qquad (6)$$

With these formulas the continued fraction provides an efficient way of calculating the time congestion except when A is close to 0. For $A < 0.1$ the continued fraction gives only a limited precision in a reasonable computing time.

For the calculation of the Erlang formula. i.e. the special case j = 0, the formulas can be sharpened. Thus (4) can be substituted by (4') and (6) by (6'). These are based on the previous formulas and tested numerically.

$$K = r/2 + \frac{r^2 A}{(A-i)^2 + r\sqrt[3]{4rA^2/9}} \qquad (4')$$

$$K = r + \max(0, r+i/3)^2/(4A) \qquad (6')$$

A heuristic formula for the number of steps for a fixed precision was published by Farmer and Kaufman [1]. One for negative i was given by J Oppelstrup. An important feature of (4') is that it is uniformly bounded for $i \leqslant A - c\sqrt{A}$ if c is a positive constant. In the complementary area the Erlang formula can for a large A be calculated with a fixed precision by the normal distribution approximation recommended in [1]. Combining these two methods the Erlang formula is computed with at least six significant digits within a bounded CPU-time for arbitrarily large A and i.

REFERENCES

[1] R.F. Farmer, I. Kaufman, "On the numerical evaluation of some basic traffic formulæ," Networks 8:2 (1978), pp 153-186.

[2] A.A. Fredericks, "Approximating parcel blocking via state dependent birth rates," Proc. ITC 10, Montreal 1983.

[3] P. Lindberg, "Simple computation methods for Erlang's formula and Wilkinson's ERT," Report Nst85 008, Televerket Sweden 1985.

TELETRAFFIC ISSUES in an Advanced Information Society
ITC-11
Minoru Akiyama (Editor)
Elsevier Science Publishers B.V. (North-Holland)

A Method for Determining End-to-End Congestions in Non-Hierarchical Telecommunication Networks.

Andrew COYLE

The University of Adelaide. South Australia.

With the introduction of non-hierarchical telecommunication networks new methods of analysing these networks must be discovered. An approximate method for calculating the traffics lost from these networks has been devised. This method is based around an iterative procedure in which the calculated results for the traffics lost approach certain values, these values are approximately equal to the exact results.

Given any telephone network, the routing rules associated with this network, the number of circuits on each link in the network and the arrival traffic offered to this network we want to find out how much of the offered traffic is lost. We do this by looking at each offered stream in turn calculating how much traffic is lost from this particular stream's first choice route and then its second choice route and so on. These results depend not only on the chain,(a series of links that the call may use), that is being looked at but the other arrival streams which use the links in this chain. Initially it is assumed that no traffic is offered to this network, then we offer the first stream to the network and the amount of traffic lost from this stream is calculated. Since no other traffic streams are taken into consideration the first approximation for the traffic lost from this first stream will be too small. The second stream is now investigated. For this stream the results obtained for the first stream will be taken into consideration. The accuracy of the initial result for the traffic lost from the second stream should be better than if the first stream's results had not been taken into account. We repeat this process for all the arrival streams. Once this first iteration has been completed we begin the second iteration. We return to the first stream and repeat the above process. This result should be much more accurate than our initial approximation as now all the other streams can be taken into consideration. A more accurate result for this first stream should lead to more accurate results for all the other streams which in turn will lead to an even more accurate result for the first stream and so on. When the results for two succeeding iterations are sufficiently close to one another the procedure stops, the final results have been determined.

How accurate these results are depends mainly on how accurate the values of the lost traffics can be calculated. For networks in which the chains are more than one link long and in which other traffic streams are also offered to these chains this is a very difficult problem. An attempted solution has been to use the traffics carried on the chains in the network not associated with the arrival stream which we are looking at instead of the streams offered to these chains. We offer the arrival stream, for which the lost traffic is being calculated, to a link for which all the other traffics carried on this link are known. It is difficult to talk about traffic offered to a link in a chain as this depends on the other links in this chain and the other arrival streams in the network and so on. The traffic carried on any link in a chain from a particular arrival stream is the traffic carried on this chain from this arrival stream. It is not necessary when looking at a particular link in a chain to have to also look at the other links in the other chains which use this link. This method gives more accurate results than one which uses offered traffics from the individual streams that use this link but creates many problems which up until now few people have investigated.

This paper reports work performed by the author at the University of Adelaide under contract to Telecom Australia. The author wishes to acknowledge the support of both organisations and their kind permission to publish this work.

TELETRAFFIC ISSUES in an Advanced Information Society
ITC-11
Minoru Akiyama (Editor)
Elsevier Science Publishers B.V. (North-Holland)

APPROXIMATION TECHNIQUES FOR QUEUEING SYSTEMS WITH FINITE WAITING ROOM

B. Sanders, W.H. Haemers, R. Wilcke
Dr. Neher Laboratories, Leidschendam, The Netherlands

The paper presents a short overview of two approximation techniques that may be appropriate to estimate performance measures of queueing systems with non-Poisson offered traffic and finite queue. The techniques evolved from similar techniques originally designed for pure loss systems [1,3], recently generalized to systems with infinite queue [4].

Consider the queueing system GI/M/N/N+Q, in which renewal traffic is offered to N exponential servers and a finite number Q of waiting places. Customers that arrive when the total system capacity N+Q is fully occupied are lost from the system and do not return. We report a study on approximation techniques to estimate the performance measures of the GI/M/N/N+Q-queueing system, viz., the blocking probability B, the waiting probability W and the mean waiting time T, based on non-lost arrivals. In the approximation models offered traffic is supposed to be sufficiently described by its mean M and its peakedness Z (Z=V/M, where V represents the variance). The approximations are such that the performance measures of the original GI/M/N/N+Q-system are expressed in terms of the corresponding measures B', W' and T' of an adjoint system with Poisson input, viz., the Markovian queue M/M/N'/N'+Q' (quantities in the adjoint system are indicated with ' in the sequel). The adjoint system provides us with exact formulas [2] that can be advantageously expressed in terms of Erlang loss functions. Fast algorithms are available to compute the Erlang formula for integral as well as nonintegral values of N'. In order to evaluate the approximations, exact analysis of GI/M/N/N+Q [5], in general complicated, has been performed for hyperexponentially and gamma distributed interarrival times in case of peaked (Z>1) and smooth (Z<1) offered traffic, respectively.

The perfomance measures B', W' and T' of the adjoint system, with offered Poisson traffic M', are:

(1) $B'=b(M',N',Q')$, $W'=w(M',N',Q')$, $T'=t(M',N',Q')$,

where the functions b, w and t are known [2].

First, we introduce an approximation technique that is a new application of the decomposition method as explained in [3]. The approximation formulas are ('decom' in the example):

(2) $B \simeq Z.B' = Z.b(M',N',Q')$,

(3) $W \simeq \alpha.Z.W' = \alpha.Z.w(M',N',Q')$,

(4) $T \simeq Z.T' = Z.t(M',N',Q')$, where

(5) $M'=V$, $N'=N-M+V$, $Q'=Q$, $\alpha=N/N'$.

Formulas (2) and (4) follow from the rather simple decomposition argument [3,4]. The introduction of α in (3), however, needs a more complicated argument. Approximation (3) with $Q'=\infty$ and $\alpha=1$, for the class of pure waiting systems $GI/M/N/\infty$, has been envisaged in [4]. Also in this case $\alpha=N/N'$ gives a better estimate to the waiting probability.

Second, in the Fredericks-Hayward approximation to pure loss system [1], the notion 'equivalent congestion' can be generalized to 'equivalent delay'. The transformation formulas are ('EDM' in the example):

(6) $B \simeq B' = b(M',N',Q')$,

(7) $W \simeq W' = w(M',N',Q')$,

(8) $T \simeq T' = t(M',N',Q')$, where

(9) $M'= M/Z$, $N' = N/Z$, $Q' = Q/Z$.

For the class of pure waiting systems $GI/M/N/\infty$, approximation (7) with $Q'=\infty$ has also been introduced in [4].

Third, we take into consideration a Hayward-variation of (6), recently published by Whitt [6], where:

(10) $M' = M/Z^2$, $N' = N/Z^2$, $Q' = Q/Z$.

We present several numerical results of these approximations and exact analysis for Z<1. It can be seen that the approximations perform well (and often extremely well) in a large range of the server occupancy ρ=M/N (low and heavy traffic).

REFERENCES.

[1] Fredericks, A.A., "Congestion in Blocking Systems - A Simple Approximation Technique". BSTJ, Vol.59 (1980), No.6, pp.805-827.

[2] Syski, R., "Markovian Queues", ch.7 in: W.L. Smith and W.E. Wilkinson (eds.), "Proceedings of the Symposium on Congestion Theory", The University of North Carolina Press, Chapel Hill, 1964.

[3] Sanders, B., Haemers, W.H., Wilcke, R., "Simple approximation techniques for congestion functions for smooth and peaked traffic", ITC 10, Montreal 1983, doc.4.4b1/1-7.

[4] Sanders, B., Haemers, W.H., Wilcke, R., "A contribution to the techniques of traffic enigeering in communications networks with waiting facilities", ICC '84, Amsterdam, Conference Records Vol.1, pp.56-60.

[5] Takács, L., "On a combined waiting time and loss problem concerning telephone traffic", Ann. Univ. Sci. Budapest Eötvös, Sect. Math., Vol.1. (1958), pp.73-82.

[6] Whitt, W., "Heavy-Traffic Approximations for Service Systems With Blocking", AT&T BLTJ, Vol. 63 (1984), No.5, pp.689-708.

Example: N=25 Q=5 Z= .8

ρ	blocking probability B				waiting probability W			mean waiting time T		
	exact	decom	EDM	Whitt	exact	decom	EDM	exact	decom	EDM
0.6	.0001	.0001	.0001	.0000	.0047	.0057	.0053	.0003	.0004	.0004
0.7	.0012	.0014	.0013	.0007	.0334	.0359	.0359	.0026	.0032	.0030
0.8	.0081	.0086	.0085	.0064	.1236	.1252	.1284	.0110	.0125	.0122
0.9	.0301	.0304	.0306	.0268	.2788	.2781	.2848	.0289	.0317	.0310
1.0	.0715	.0708	.0717	.0677	.4418	.4427	.4483	.0530	.0583	.0560
1.1	.1258	.1245	.1259	.1228	.5581	.5641	.5649	.0775	.0866	.0812
1.2	.1838	.1824	.1837	.1818	.6181	.6284	.6247	.0984	.1126	.1026

TELETRAFFIC ISSUES in an Advanced Information Society
ITC-11
Minoru Akiyama (Editor)
Elsevier Science Publishers B.V. (North-Holland)

COMPARISON OF TRAFFIC CAPACITY CALCULATION METHODS FOR THE SWITCHING NETWORK OF SMALL DIGITAL PABX'S

József A. SCHULLER and Katalin SZENTIRMAI

BHG Telecommunication Works
Budapest, Hungary

1. INTRODUCTION

The design phase traffic investigations related to the switching network of the DIPEX system revealed that the traffic capacity calculated by the Engset-formula gives almost the same and safe side result as the calculation based on a call repetition model including the effect of the busy called party. The DIPEX family of small SPC digital PABX's was developed by BHG Telecommunication Works in Hungary [1].

2. SYSTEM ARCHITECTURE

In the switching network connections are established via 32-channel PCM busses. Inlets are arranged in groups each group being connected to one of the busses. Inlets represent different type traffic sources, on each bus 30 time slots are available for normal connections.
For connecting two inlets belonging to the same group two free time slots are necessary on the bus of the considered group. For connecting inlets of different groups a single free time slot is required on each bus involved. These slots must be coincident, i.e. the switching network forms a space switch.
From the individual inlet group/bus point of view "internal" and "external" traffics can be defined.

3. RESULTS

The switching network consisting of one PCM bus represents a simple full availability group with finite number of traffic sources and internal traffic. For this case three mathematical models were investigated: A. the ordinary Engset-model related to internal traffic, B. the modification of this model by taking the effect of the engaged called party into account just as in [2], and C. this latter one completed by allowing the presence of repeated calls. Here the approximate calculation method of call repetition of [3] was used. Inlets of the group were supposed to have uniform traffic characteristics. The results of Fig. 1 show that at the given level of the switching network blocking probability the simpler model can be well applied as a good approximation which remains on the safe side. In the case of the C. model it has also been stated that the subscriber's perseverance has practically no effect on the carried traffic.

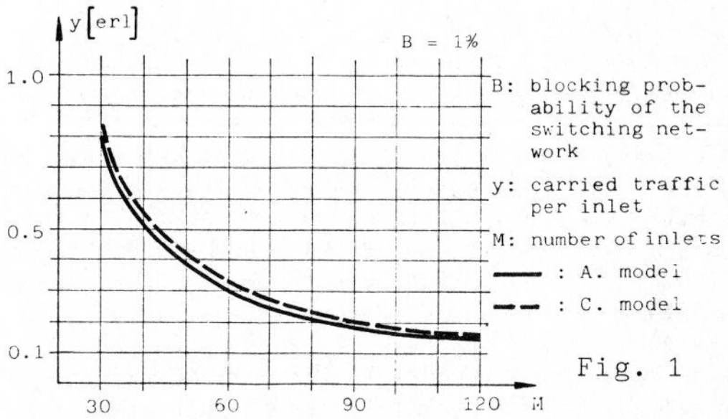

Fig. 1

For the investigation of switching networks consisting of several busses a general simulation program written in SIMULA language has been developed. An analytical model with state equations proved to be namely too much complex for practical applications. Based on the observation described above the simulation model could suppose simple Engset-type traffic process without called party busy phenomenon and call repetition.

CONCLUSION

The results show that at least in the range of small blocking probabilities of the switching network the more complex model being nearer to reality gives almost the same traffic capacity values as those gained by the well-known Engset-formula. This result of general interest can probably be explained by the stability of carried traffic. Keeping in mind that the Engset-formula is invariant to the distribution of the holding time, this observation could find some applications also in the case of mixed voice, non-voice traffic, i.e. in the estimation of the traffic capacity of small switching systems used in ISDN environment. /The input process must of course remain Poissonian./

REFERENCES

[1] I. L. Horváth, "Small capacity digital private automatic branch exchanges developed in Hungary," Proc. 11th Int. Switching Symposium, Florence, pp. SESSION 43 C PAPER 4/1-6, 1984.

[2] D. Bazlen, "Multi-stage switching systems with both-way connections for internal and external traffic," 18th Report on studies in congestion theory, ISDT, Univ. of Stuttgart Press, Stuttgart, 1973.

[3] G. Honi, G. Gosztony, "Some practical problems of the traffic engineering of overloaded telephone networks," Proc. 8th Int. Ieletraffic Congress, Melbourne, pp. Paper 141/1-8, 1976.

TELETRAFFIC ISSUES in an Advanced Information Society
ITC-11
Minoru Akiyama (Editor)
Elsevier Science Publishers B.V. (North-Holland)

THE OPTIMIZATION OF INTEGRATED DIGITAL NETWORKS WITH VOICE AND DATA

Tatsuki WATANABE and Masayuki MATSUMOTO

Department of Electrical Engineering, Toyo University
Kawagoe-Shi, Japan

1. INTRODUCTION The heuristic procedure to generate the optimized or near optimized structures of integrated voice/data networks as well as the comparison of two arrangements for voice, i.e., packet switching and line switching is presented. The load curves used to estimate the trunk capacity are shown that give greater capacities to the packet-switching for voice. The heuristic procedure used here is based on the similar principle to that developed by S.Lin, et. al. for the travelling-salesman problem [2].

2. PROCEDURE As shown in Fig.1, two kinds of terminals, telephone and data, are considered in the integrated voice/data exchange. Telephone calls are switched to one of VOCODER's connected to the CCU which generates voice-packets consisting of 240 bits each at every 90 ms during the talkspurts for each telephone call. Data terminals are assumed to be sources of data packets having 1024 bits each generated randomly from infinite number of sources. These voice/data packets are transmitted to the respective digital trunks together with the tandem-switched packets. The following two kinds of integration arrangements have been considered.

(A1)Packet switching with packetized voice and data: The required trunk speeds are dimensioned using the load curves produced on the basis of single-server model of pre-emptive-repeated-identical-priority queues [1], incorporating the traffic simulation results (Fig.2).

(A2) Hybrid witching where telephone calls are circuit-switched while data are packet-switched on the fixed boundary basis: The required transmission capacities are dimensioned using Fig.2 in which the load curves are produced using Erlang B formula (B=0.001) for voice taking the bit-rate of 2.667 kb/sec as one voice channel.
In either arrangement, the data packet queueing delays are calculated as less than 50 ms.

The optimization problem here is to find the optimized or near-optimized network having the minimum total cost estimated by the given cost functions when the traffic matrices required for voice and data, locations of switching nodes, etc. are given. The fully connected mesh network is taken as the starting solution. The algorithm is to find the optimized solution through repetition of a procedure such as to re-route the traffic carried through the edge (direct trunk) between the randomly selected pair of nodes to a selected tandem switching node (exchange) as shown in Fig.3.

3. RESULTS As the example, the results in the case of six different exchanges located uniformly on a circle whose radius is 200 km long are shown in Fig. 4. The following tendencies have been observed from the trials.

If no other constraint such as protective measures is incorporated, networks that are close to those of full tandem tree or star type tend to be generated as the optimum. The degree of cost reduction is usually greater for A1 arrangement than A2 and the absolute cost value for A2 is greater than A1. If two kinds of traffics are out of balance, by-path trunks dedicated to one of two kinds of traffics tend to appear especially in A2 arrangement.

REFERENCES
[1] D. P. Gaver, "A waiting line with interrupted service, including priorities", J. Roy. Statist. Soc. , ser. B, 24, pp. 73-90, 1962.
[2] S. Lin, et. al. , "Computer solutions of the traveling salesman problem", Bell Syst. Tech. J. , vol. 42, no. 12, pp. 2245-2269, 1965.

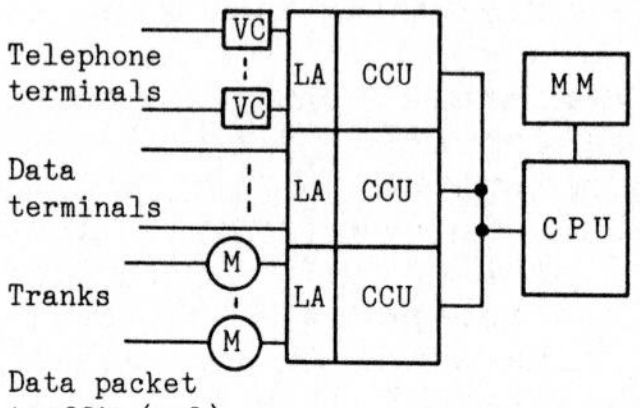

Fig.1 Integrated voice/data exchange

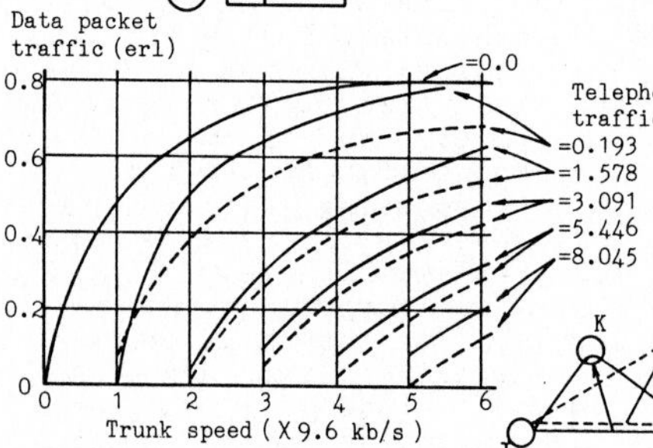

Fig.2 Load curves —— for A1 ----- for A2

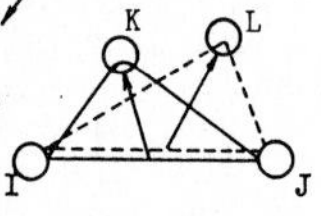

Fig.3 Rerouting

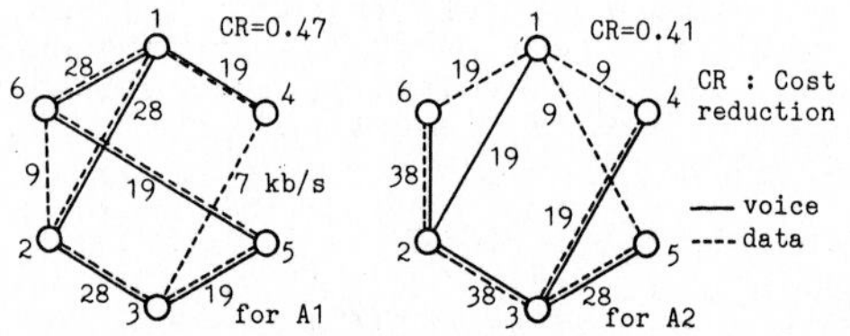

Fig.4 Optimized networks
(ratio of data to total packet traff. : 0.9)

TELETRAFFIC ISSUES in an Advanced Information Society
ITC-11
Minoru Akiyama (Editor)
Elsevier Science Publishers B.V. (North-Holland)

TELETRAFFIC ASPECTS OF VOICE/DATA INTEGRATION

Jozef LUBACZ & Michal JAROCINSKI

Technical University of Warsaw, Inst. of Telecommunications
ul.Nowowiejska 15/19, 00-665 Warsaw, Poland

The following remarks are a short supplement to the considerations presented in [1]. The scepticism presented therein concerning the practical utility of voice/data integration concepts for the core network (eg. hybrid switching, packet switching, burst switching intensively discussed in literature) is based mainly on traffic arguments - economical terms are treated fragmentarily. Here we enhance the latter viewpoint and combine it with traffic arguments.

As discussed in [1] the fraction of voice information throughput in the total information throughput in future integrated networks (both public and private) is not likely to fall below v/d = 95% (even if a great increase of data traffic and low voice digitalization rates are foreseen). We believe that smaller v/d values may be expected only in special eg. military networks which we do not consider here.

From this v/d evaluation it follows that any switching concepts proposed for voice/data integration must, in the first place, prove that it is well suited (justified from traffic and economical viewpoints) for telephony. Clearly, advanced voice processing and switching (APS) (eg. speech interpolation and burst or packet switching) will result in some improvement in transmission capacity utilisation (i.e. in decrease of transmission costs) but in the same time may result in processing and switching costs. We thus arrive at a cost optimisation problem which we treat below in a simplified but comprehensive way.

Let C = L+S+T be the total cost of a classically circuit switched (CCS) telephone network; L denotes the cost of the access layer, S and T denote switching and transmission costs respectively. Similarly, let C' = L'+S'+T' denote the cost of this network when realised with some APS; for convenience we include all voice processing costs (eg. talk spurts detection) in L'. The cost ratio

$$C'/C = l\cdot(L'/L) + s\cdot(S'/S) + t\cdot(T'/T)$$

where: $l=L/C$, $s=S/C$, $t=T/C$.

We are looking for conditions on which $C'/C \leqslant 1$.

We assume usual cost proportions: $l=s=t=.33$ for the network as a whole and $l=.45$, $s=.4$, $t=.15$ for its urban part. The speech activity factor equals approximately .4, thus we can expect at most twice better transmission capacity utilization as the result of even most sophisticated APS (with respect to CCS). Since transmission costs are approximately proportional to capacity utilization we assume $T'/T = .5$.

Taking account of the above said one obtains the results depicted in the figure. The shaded area corresponds to $C'/C \leqslant 1$. As can be seen there is not much hope in obtaining cost savings as a result of APS (compare "clouds" on figure indicating foreseable areas for packet and burst switching). Consequently, CCS proves to be the most resonable solution.

Now, since v/d .95 then clearly CCS remains also the most resonable solution for voice processing and switching in integrated networks. This, together with the results from [1], implies that voice/data integration based on eg. hybrid switching or packet switching can not be justified by traffic and/or economical arguments. Concluding, we believe that the future lies in "coexistence" rather than "integration" of switching in core network.

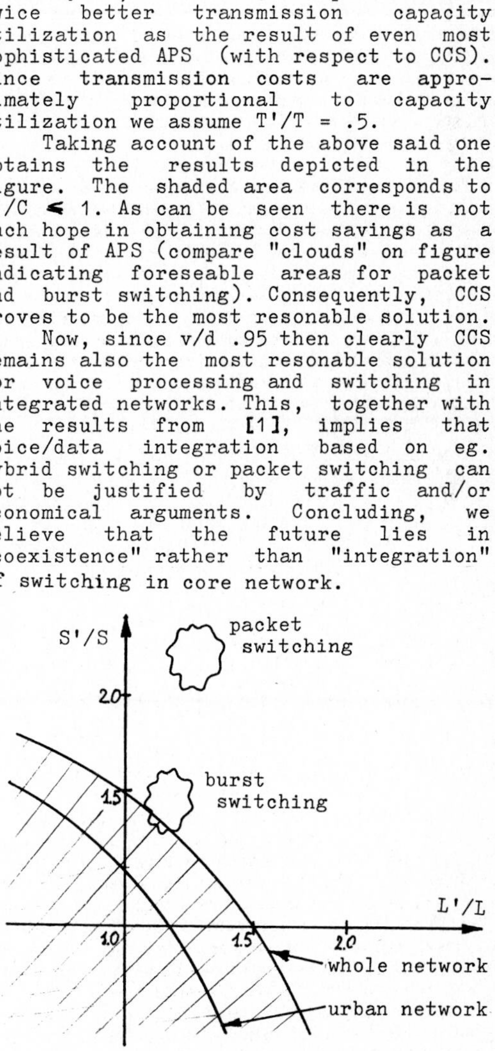

[1] S.Debaille, J.Lubacz: Delay and throughput comparisons of switching techniques in integrated networks. Proc. of the ICCC'84, Sydney 1984

TELETRAFFIC ISSUES in an Advanced Information Society
ITC-11
Minoru Akiyama (Editor)
Elsevier Science Publishers B.V. (North-Holland)

ON THE APPLICABILITY OF ERLANG'S B MODEL IN ISDN

Åke ARVIDSSON

Lund Institute of Technology, Dept. of Communication Systems,
Box 118, S-221 00 Lund, Sweden.

1. INTRODUCTION.

In today's rapid development of telecommunications several important trends can be observed. Among these are increasing trunk capacities, network digitalization and the merging of several kinds of traffic such as interactive data, file transfer, video etc. onto the same network. Further, as the development of the VLSI technique makes intelligence cheaper relative to transmission costs, more sophisticated network routing principles are becoming attractive. These facts implies that the traditionally employed trunk dimensioning formulas, such as the Erlang B-model (EB-model), cannot a priori be adopted, but an investigation of its validity under these new circumstances is called for. In this paper, two aspects of trunk dimensioning in modern networks are considdered. Firstly, we investigate the impact of network operation in discrete time and, secondly, we study the impact of so called service protection within networks.

2. DISCRETE TIME.

On a digital link, time is divided into small segments called frames, each of which consits of a number of even smaller segments named slots. The duration of a slot is selected so that one PCM-sample from a call can be transmitted during each slot. Each call carried by the trunk is then assigned a certain slot in every frame during which the information is transmitted. Call attempts are stored up in a queue during the frame. At the end of a frame, all disconnections detected during the frame are handled after which the idle slots are assigned to the waiting calls. If the queue is not empty when no idle slots are left, the rest of the queueing calls are rejected and the queue is cleared. Hence, arrivals and departures are handeled only at the discrete time instants marked by the transition between two frames.

In order to obtain the loss and utilization for such a link, the system is well described by a discrete time Markov chain. Assuming the number of call attempts per frame to be Poissonian and the holding times to be exponentially distributed, the transition probabilities for this chain are readily obtained, after which the linear equations to solve for the stationary state probabilities are easily defined. However, since the number of slots per frame can be expected to be in the order of several hundreds or more, numerical difficulties arise when solving the equations. In [1] a detailed investigation on the performance of several numerical algorithms to solve such a Markov chain is found. The main conclusion is that the methods that generate the most accurate solutions are the method of inverse iteration and the method of repeated matrix squaring with normaliza- tions, the latter developed in [1].

Due to the numerical difficulties, an alternative, continous time model is developed which approximates the beha- viour of the original discrete time system better than the EB-model does. As an example, a relative error of 90% in the loss value obtained from the EB-model decreases to about 5% with the new model. For more details, refer to [1].

3. SERVICE PROTECTION.

For example in a network with advanced alternative routing possibilities it is desireable to protect certain classes of calls from calls of other classes. This facility is refered to as service protection (SP) and has been proven both powerful and necessary in, for example, non-hierarchical networks. When a SP-device is active, calls of certain classes are rejected while those of other classes are accepted. However, during periods of heavy load, it is likely that these devices will be active for quite long periods of time and thus there is a risk of very long periods of blocking for the calls rejected by the SP-mechanism.

It has then been proposed to let the SP work in a somewhat modified manner: Unprotected calls are being rejected at random with a probability P, which might depend on the number of occupied circuits in the protected link. The advantage with this arrangement is that the total blocking of an unprotected traffic is less probable and will, when it occurs, be of shorter duration.

In [2], we study the distribution of blocking periods on links with a SP of the latter kind, which includes the former, primitive SP as as special case. It is shown by analytical means that the distribution of a blocking period is hyper-exponential and formulas for calculating the parameters of the distribution are given. Non-Poissonian traffics often occur as input processes to a link, but the results of the study also cover this case. The resulting, overflowing blocking process from a link with SP might occur as the input process to an alternative choice link. Hence it is important to characterize such a traffic, e.g. in terms of mean and variance. The results of a study of this subject are also presented in [2].

REFERENCES:

[1] Åke Arvidsson: "Numerical Solution Methods of Discrete Time Markov Chains" (1984), Lund Inst. of Tech., Dept. of Comm. Sys., Lund, Sweden.
[2] Åke Arvidsson: "On the Impacts of Service Protection Facilities" (1985), as above.

TELETRAFFIC ISSUES in an Advanced Information Society
ITC-11
Minoru Akiyama (Editor)
Elsevier Science Publishers B.V. (North-Holland)

ISDN----MODEL AND DEVELOPMENT

Wei Yuan-rui, Qu Yan-zhen, Shen De

Chinese Academia Sinica, PO Box 3908, Beijing, China

ABSTRACT

A new RM for ISDN which is closer to the current distributed system is proposed. A new concept: Soft--bus is introduced for interfacing various function subsystems. Programming language and User commands must be expanded or developed to fit in with the new application requirement in ISDN.

I. Introduction

With the rapid development of the computer system and communication system, ISDN has become an attractive project in the communication departments all over the world. In such a situation, the more reasonable reference model is considered to improve the system design and applications.

II. Reference Model of ISDN

A new Reference Model(RM) for Integrated Services Data Network(ISDN) is shown in the following figure. The new RM is based on OSI RM but is closer to the current distributed computer system.

Layered structure is adopted. The function of each station can be divided into two levels--- high level and low level. The high level is oriented to user's application and the low level is oriented to various function subsystems for graphic, voice, storage, as well as transmission service.

The high level function is composed of two layers: application layer and presentation layer. Application layer deals with distributed database, distributed calculation, and distributed real-time control procedure. Presentation layer includes virtual character terminal service, virtual voice terminal service, virtual graphic terminal service, and virtual text transfer service.

The low level function is composed of several function subsystems. Each of them adopts layered structure according to the different special function protocols.

The interface between high level and low level is what we called "Soft-bus" which provides a set of standard interface commands in order to enhance the flexibility of expanding various function subsystems.

III. Development Direction of ISDN

Two interfaces must be treated carefully. One is the interface to the system. The requirement to this interface is good compatibility both in hardware and software. The other is the interface to the users. The requirement to this interface is friendly convenience in operating various system functions.

The standardization of the interface among all the function subsystems is the key problem. A reasonable abstraction for the operating system makes a good base for this aspect.

A new kind of programming language including voice and graphic processing functions should be developed. An alternative is to combine the voice and graphic functions with a mature programming language. User commands also must be expanded according to the characteristics of ISDN. In future, some commands will be the picture commands or oral commands ,i.e. , the commands will be input directly by graphic system or voice system.

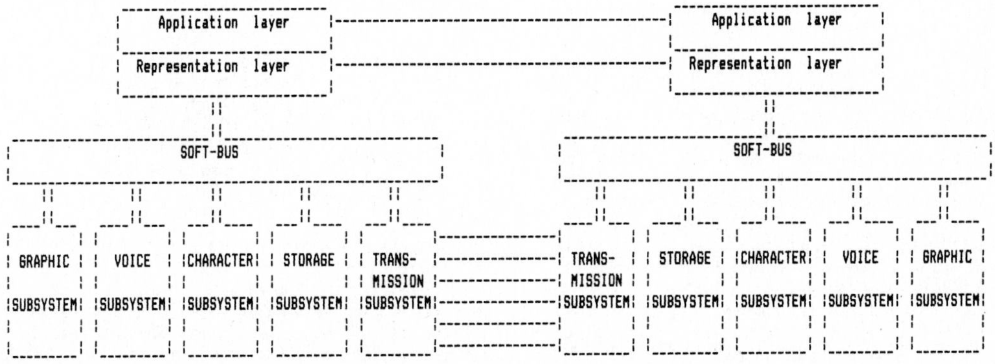

Figure Reference Model of ISDN

TELETRAFFIC ISSUES in an Advanced Information Society
ITC-11
Minoru Akiyama (Editor)
Elsevier Science Publishers B.V. (North-Holland)

ON A TWO PHASE MODEL FOR BURSTY TRAFFIC AND ITS APPLICATION TO TDMA NETWORKS WITH SERIAL NODES

Stephan Jobke

Institute for Electronic Systems and Switching
University of Dortmund Federal Republic of Germany

Abstract

In this paper, the performance of nodes in particular TDMA networks is investigated. For the calculation of the loss probabilities of nodes with multiple random inputs and of nodes with two bursty inputs (generated by two preceding nodes), exact methods are presented. Furthermore, an approximation method for nodes with two bursty inputs is derived which is well suited for practical application in dimensioning whole networks. Nodes with random input and nodes with bursty input, as well as networks with several nodes have also been investigated by means of traffic simulations. The approximation values are in good agreement with exact calculations and simulation results.

A queueing problem in a special TDMA network is investigated. In the considered system all subscribers are connected to a network of time division multiplex (TDM) highways. This system represents a decentralized switching system with a branched network containing no closed loops (1). The branches of the network are connected by means of nodes. As each subscriber may be connected to the network at an arbitrary point, all blocks arriving at a node must be transmitted through the whole network. If more than one information block arrives in a node at the same time slot, only one block can be transmitted to the outgoing highway whereas the other(s) must be stored. For this purpose a finite number of waiting places is provided. When all waiting places are occupied a store overflow may occur and information blocks get lost.
Formulae for the calculation of the probability that an arbitrary block is lost when arriving at a node are known for the special case of a network in which at most three pairs of highways are connected at each node and for the case that the blocks are distributed at random within the frames (2). In the present paper a method has been derived for the exact calculation of the loss probability of a general node with an arbitrary number of incoming and outgoing trunks (in case of very large frames).
The random distribution of blocks is only fulfilled in nodes which are situated in marginal districts of a TDMA network where the arriving information blocks have not yet passed any further nodes (and, of course in networks with only one node). If the blocks arriving at a node already have passed further nodes before, the outgoing traffic contains longer series of occupied time slots (bursty traffic). This results in an increased loss probability in nodes with bursty incoming traffic. A node with two bursty inputs (generated by two preceding nodes) has been investigated by means of exact calculations with the aid of the succesive overrelaxation (SOR) method.
Furthermore, an approximation formula for the calculation of the loss probability of nodes with bursty incoming traffic is presented. The incoming traffic is modeled by means of a two phase model. In the bursty incoming traffic two phases are distinguished: one phase with clustered arrivals (bursts) and another phase with arrivals distributed at random within the frame.
The transition probabilities between certain states of the phases are calculated from the output process of the preceding nodes. An approximation formula for the loss probability is derived, which is based on the two phase model for bursty traffic.
From the results it can be seen that the loss probability of a node with bursty incoming traffic is much higher than in a node in which the arriving blocks are distributed at random within the frames. By means of the new approximation method it is now possible to calculate the loss probabilities in all nodes of networks of the type considered here. In this calculations the properties of bursty traffic can be taken into account.

References

(1) Schenkel, K.D.
Entwurf eines integrierten digitalen Nachrichtensystems mit Vielfachzugriff für ein beliebig verzweigtes Breitbandnetz
AEÜ, Vol. 27 (1973), No. 4, pp. 168-176
(2) Schehrer, R.
On a Queuing Problem in TDMA Networks
AEÜ, Vol. 29 (1975), No. 2, pp. 62-68
(3) Jobke, S.
On Delay Loss Systems in Nodes of TDMA Networks with Bursty Traffic (to be published)
(4) Jobke, S.
Über Warte-Verlust-Systeme in Zeitmultiplexnetzen mit Vielfachzugriff und Burst-Verkehr
Ph.D.Thesis
University of Dortmund 1985

TELETRAFFIC ISSUES in an Advanced Information Society
ITC-11
Minoru Akiyama (Editor)
Elsevier Science Publishers B.V. (North-Holland)

Performance Analysis of a Virtual Circuit Switch Based LAN

A. A. Fredericks

AT&T - Bell Laboratories
Holmdel, New Jersey 07733

In this paper we analyze the performance of a Local Area Network based on a virtual circuit switch. The performance models that are developed have much broader applicability and provide insight into several important problems associated with the performance of computer/communications systems.

For the purpose of describing some of the key performance issues addressed, we may view the switch as consisting of N input and N output ports, each with an available bandwidth B (bits per second). For any point-to-point demand, a given fraction of the available bandwidth can be assigned at both the output and input ports, thus providing a "virtual" circuit communication facility with given bandwidth between the two points.

In order to establish a virtual circuit with a given dedicated bandwidth, say between port i (output) and port j (input) it is necessary that both of these ports have the needed available bandwidth. If we assume for the moment that all requests are for the same bandwidth, b, then we can view each port as having a total of nc = B/b channels available. A point-to-point connection thus requires an available channel at each port.

Perhaps the simplest strategy for managing system capacity is to reject any request for which the needed bandwidth on both ports is not available. (Rejected requests would result in subsequent retries.) This strategy tends to waste capacity since the needed bandwidth might free up and remain idle "waiting" for a retry. Queuing for resources tends to increase their efficient use. Two such strategies we consider are:

1. When a request for a circuit occurs, request a channel at each port and if none is available at either port, a request for a channel is placed in a queue at each of the ports. When the first of these requests is satisfiable, the free channel at that port is held until a channel becomes free at the other port allowing a connection to be established. While this tends to minimize the delay for setup (once the request is made) it does result in wasted channel capacity.

2. When a request for a circuit occurs which cannot be satisfied, wait until the first time that a channel is available on both ports. This strategy tends to eliminate wasted channel capacity, but a given request could be delayed indefinitely, particularly at high loads.

We formulate models for analyzing the performances of these as well as other strategies. An analytic approximation technique is developed which, when applied to these models, is shown to produce quite accurate performance predictions (based on comparisons with simulation). These models provide insight into the basic trade-offs between maximizing overall resource usage and providing fairness of treatment to individual customers.

The above "simultaneous resource possession" problem arises quite often in a variety of computer-communications systems. For example, it is closely related to the problem of obtaining available trunks in a multilink telephone call. Similar problems also arise in connection with shared peripherals in computer systems.

TELETRAFFIC ISSUES in an Advanced Information Society
ITC-11
Minoru Akiyama (Editor)
Elsevier Science Publishers B.V. (North-Holland)

AN ANALYSIS OF INFORMATION THROUGHPUT IN UNSLOTTED ALOHA WITH PARTIAL ACK OF RANDOM LENGTH PACKETS

Józef WOŹNIAK

Department of Electronics, Technical University of Gdańsk
Gdańsk, Poland

ABSTRACT

In the paper an unslotted ALOHA protocol with partial positive acknowledgment of packets and geometrically distributed packet lengths is discussed. The analysis is carried out with respect to the information throughput as a measure of performance quality. The results are compared with the pure ALOHA protocol.

1. GENERAL ASSUMPTIONS

Users of the ALOHA system transmit packets at random and hence the packets sent from different terminals may mutualy interfer and result in transmission errors. In the pure ALOHA protocol [1] all overlapped packets have to be wholly retransmitted. Most analysis is carried out basing on the assumption that the packet lengths are fixed. This assumption is due to the fact that in such a case the channel throughput reaches its maximum [2]. In this paper we are mainly interested in the improvement of the channel utilization for the case when packet lengths are geometrically distributed [3,4] and we allow a partial positive acknowledgment of the first of the interfering packets.

We consider three cases:

A) There is a possibility of partial ACK; we assume a geometrical distribution of information parts in packets.

B) The pure ALOHA protocol is used; information parts in packets are geometrically distributed.

C) The pure ALOHA protocol is used; packets have fixed lengths.

According to the principle of packet (message) switching each message I generated by a user is transmitted with an appropriate header H forming together a packet of length L=H+I. In cases A and B the lengths of I are geometrically distributed

$$P\left(I=kL_1\right) = q^{k-1}\left(1-q\right)\ ;\ k=1,2,\ldots;q \in \langle 0,1)$$

where L_1 denotes the length of one segmment (we assume $L_1=1$). In case C the length I is equal to the average length $\bar{I}$ obtain for case A or B. In all cases we make the assumption that message arrivals are governed by a Poisson process with arrival rate λ.

2. RESULTS

From Figures 1 and 2 one can see that the information throughput depends very strongly on the operational scheme as well as on the distribution of the packet lengths. Protocol A results in an

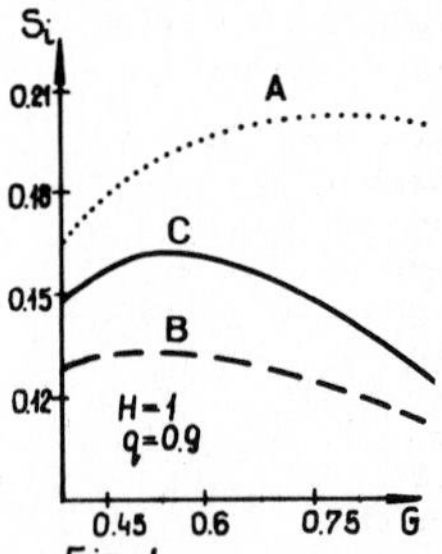

Fig. 1

The information throughput vs. the offered traffic

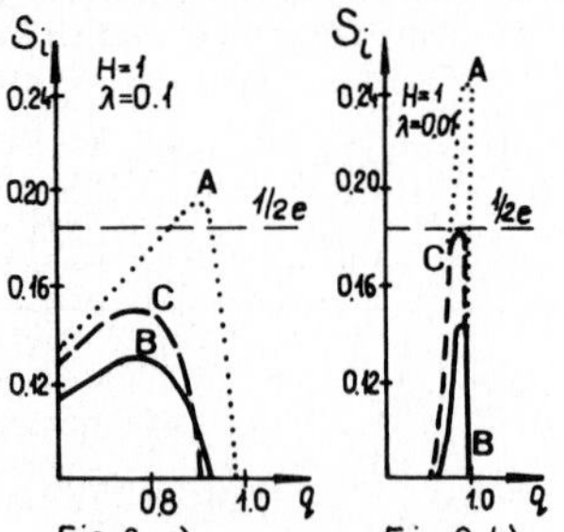

Fig. 2 a) Fig. 2 b)

The information throughput vs. q; where q the parameter of the geometrical distribution

increased value of the information throughput (or equivalently the channel throughput) in about 30% (for q→1). This means that such a modified protocol is preferable in the case of very long; at the mean, random length packets.

REFERENCES

[1] N. Abramson, The ALOHA System-Another Alternative for Computer Communications", Proc. AFIPS Fall Joint Comp. Conf., Vol. 37, pp. 281-285, 1970.

[2] M.J. Ferguson, "A Study of Unslotted ALOHA with Arbitrary Message Lengths", DATA COMM., Quebec, pp. 5.20-5.25, 1975.

[3] J. Woźniak, "Variations of the Channel Throughput for Geometrically Distributed Packets Lengths in a System with Random Access", Arch. Autom. Telemech., vol. 29, pp. 377-388, 1984.

[4] J. Woźniak, K. Pawlikowski, "Influence of Message Length on Information Throughput in Random Access Channel", to be published by Arch. Autom. Telemech. 1985

TELETRAFFIC ISSUES in an Advanced Information Society
ITC-11
Minoru Akiyama (Editor)
Elsevier Science Publishers B.V. (North-Holland)

COMPARISON OF COMMUNICATION SERVICES WITH CONNECTION-ORIENTED AND CONNECTIONLESS DATA TRANSMISSION

Ottmar GIHR and Paul J. KUEHN

Institute of Communications Switching and Data Technics
University of Stuttgart, Fed. Rep. of Germany

ABSTRACT

During the recent years, many advances have been made in the standardization of protocols and interfaces of communications systems. Within the communications-oriented levels 1-4 of the Basic Reference Model, level-specific connections with their functions and procedures for establishment and release are of central interest. Connections play a similar role within the new standards of the ISDN basic access (D-channel protocol).

Within wide area networks (WAN), specifically however within local area networks (LAN), modifications of the fully connection-oriented data transmission have been discussed. The reasons behind this discussion originate from the repeated application of basic mechanisms within level 2 (data link), level 3 (network) and level 4 (transport) such as sequence control, error recovery, and data flow control. *Especially in* the case of LAN with a limited spatial network extension, low bit error probability and completely distributed control, ECMA, ISO, and IEEE provide for an alternative of connectionless data transmission.

Connection -oriented and connectionless data transmissions differ in various aspects as connection establishment and release, negotiation on service parameters, data unit relationship, acknowledgement, and addressing. Such differences influence also the performance figures of the communication services. Our study is devoted to the modelling and analysis of both types of communication services to obtain quantitative criteria additional to the qualitative criteria for the selection of services for particular applications of network types, e.g., Local Area Networks (LAN).

1. MODELLING

Our modelling approach follows the layered protocol architecture, where a model for a lower layer is aggregated and nested into the higher layer model as a submodel. In LAN's, we particular distinguish the following levels:

MAC-Submodel:
Layers 1 (Physical Media) and 2 (Media Access). This model is a multi-queue, single-server model with a specific interqueue discipline according to the Media Access Control Mechanism (as Token-Passing or CSMA/CD).

L-Connection-Submodel:
A full-duplex link-level connection with flow control, acknowledgement signalling, and MAC-submodel is modelled by a priority queuing network with two flow-controlled open chains according to Fig. 1. For each direction, a LLC-Admission mechanism and a LLC-Priority Processor model is included. The MAC-Submodel is inserted only by its terminal behavior (delays).

LLC-Submodel:
The Network level uses the LLC-services which are modelled by connection-oriented or connectionless data unit transmissions.

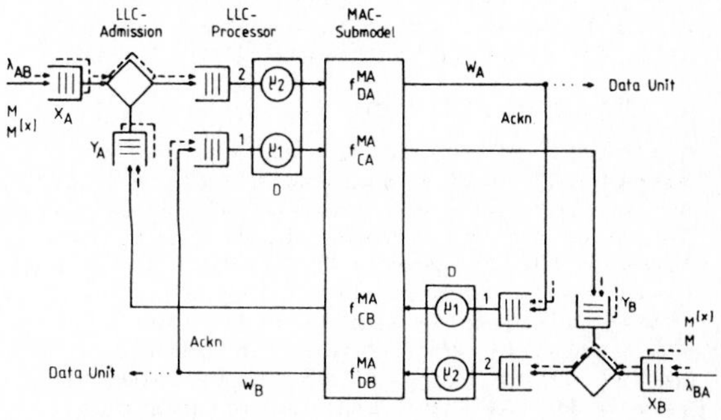

Fig.1. L-connection submodel for Local Area Networks

2. ANALYSIS

The analysis of the model follows the principles of Hierarchical Decomposition and Aggregation:

- Analysis of the MAC-Submodel for the total traffic flows from each station (control frames and data frames).
- Analysis of one FDX-LLC-connection through
  - decomposition of the two chains into two models, each for one direction only
  - aggregation of serial delays into a composite service station.

3. RESULTS

By this procedure we find typical results as end-to-end delays dependent on the LAN-parameters and window sizes. Such results can be applied for:

- selection of optimal window sizes
- delay estimations for file transfer- and inquiry/response-type traffics
- selection of the appropriate service type (connectionless/oriented).

The detailed description of the modelling method, analysis procedure, and results is included in a forthcoming paper.